INSTRUCTOR'S MANUAL FOR

CALCULUS
THIRD EDITION

By Berkey/Blanchard

Volume One for Chapters 1-11
by

Judy Coomes (William Paterson College)

Tohien Hoang

Dennis Kletzing (Stetson University)

Gloria Langer (University of Colorado)

David Wagner (University of Houston– University Park)

SAUNDERS COLLEGE PUBLISHING

Harcourt Brace Jovanovich College Publishers

Fort Worth Philadelphia San Diego New York Orlando Austin San Antonio
Toronto Montreal London Sydney Tokyo

Berkey/ Coomes, Hoang, Kletzing, Langer, Wagner: Instructor's Manual, Volume I
to accompany CALCULUS, 3/E

ISBN 0-03-051023-63

234 021 987654321

Preface

The Instructor's Manuals for Berkey/Blanchard's <u>Calculus</u>, Third Edition, contain detailed solutions to all the problems in the text, including the graphing calculator and computer problems. They are available in two volumes (for Chapters 1–11 and 12–21). We trust they will be a valuable aid as you undertake the challenging task of teaching calculus.

This volume has been the combined work of several people. We would like to thank:

* Lori Bosch, who did an thorough cross–check of all solutions to ensure accuracy;
* James Angelos (Central Michigan University), who provided the solutions to the graphing calculator problems; and
* Nancy Moore, who did a careful and conscientious job typing and proofreading.

Great care has been taken to avoid errors; however, we are only human. If you find an error or have a suggestion for improving the presentation of a problem, please send your comments to: Mathematics Editor, Saunders College Publishing, The Public Ledger Building, 620 Chestnut Street, Suite 560, Philadelphia, PA 19106.

Judy Coomes
William Paterson College

Tohein Hoang

Dennis Kletzing
Stetson University

Gloria Langer
University of Colorado

David Wagner
University of Houston – University Park

Guide to the Supplements

The following supplements for the instructor and student accompany Berkey/Blanchard's Calculus, Third Edition.

Instructor's Manuals

The Instructor's Manuals are availalbe in two volumes (Chapters 1–11 and 12–21) and are free to adopters of the text. They contain complete detailed solutions to all the problems in the text, including the graphing calculator and computer problems. Answers to most odd-numbered problems are also included at the end of the text.

Student Solutions Manual

The Student Solutions Manual is available in one volume for sale to students. It contains complete detailed solutions to most of the odd-numbered problems in the text.

Test Bank
Computerized Test Bank

The Test Bank offers instructors approximately 1300 open-ended test questions corresponding in level and difficulty with the examples and exercises in the text. Each chapter contains four test forms of approximately 15 questions each. The three final exams are each divided in two parts (for Chapters 1–11 and 12–21). Answers for each test are also provided.

The ExaMaster™ Computerized Test Bank is available in IBM^R and Macintosh^R formats. This software allows you to create tests using fewer keystrokes. Easy-to-follow screen prompts guide you step by step through test construction and allow you to select, edit and add questions and graphs, link related questions or select questions randomly, and print up to 99 versions of the same test and answer sheets. Included with each version of the software is ExamRecord™, our gradebook program. It allows you to record, curve, graph, and print out grades.

Overhead Transparencies
Transparency Masters

The transparencies and transparency masters are free to adopters and are designed to save you time and energy as you prepare for lecture. The 25 four-color overhead transparencies illustrate figures from the text that are important enough to be discussed in class but difficult to draw accurately and effectively. In addition, there are 100 transparency masters of other important and useful figures throughout the text.

Technology Supplements

For instructors interested in using graphing calculators or computers in the course, we offer several supplements for sale to students to complement the technology exercises and appendixes in the text.

Calculus and Mathematica

Calculus and Mathematica, by John Emert and Roger Nelson (both of Ball State University), consists of 30 projects to supplement the traditional calculus sequence. Introductions to Mathematica commands are included, accompanied immediately by examples. The experiments use Mathematica to lead a process of discovery, conjecture, and verification. They cover a wide variety of topics, from the expected Newton's Method and Riemann Sum explorations to investigations of symmetic derivatives, chaotic sequences, and global/local behavior.

Calculus and Derive

Calculus and Derive, by David Olwell and Pat Driscoll (both of the United States Military Academy), contains almost 40 exercises and projects designed specifically to illustrate symbolically and graphically the essential concepts of calculus. Each exercise has been structured so that the average student can complete it in 45 minutes in the Derive environment. The projects, which are designed to allow the student to explore the topic in greater depth, will take the average student about four hours to complete.

Supplements for Graphing Calculators

Calculator Enhancement for Single Variable Calculus, by James Nicholson and J.W. Kenelly, consultant, and Calculator Enhancement for Multivariable Calculus, by J.A. Reneke and D.R. LaTorre, consultant (all of Clemson University) are written for the HP–28S and HP–48S graphing calculators. These supplements provide procedures, calculator programs, examples, and exercises designed to remove the computational burden normally associated with these courses. Each supplement helps students appreciate the geometrical and graphical aspects of calculus, master its theory and methods, and explore new topics and applications.

CalcAide Software

CalcAide, by Elizabeth Chang (Hood College), is available for IBM^R systems with CGA capabilities and is free to instructors. This software provides 17 program modules and is intended for use in the classroom, in a mathematics laboratory, or on students' own computers. CalcAide can draw the graphs of most functions and provides numeric and graphic illustrations of many concepts of calculus. The flexible menu system allows the user to graph a function on any portion of the coordinate plane, change some features of the graph or function while leaving others unchanged, examine value, slope or other features at a specific point of the graph, and many other options.

If you have any questions about these ancillaries, please contact your local Saunders College Publishing sales representative.

Contents

(Solutions for Chapters 12–21 are available in Volume II of the Instructor's Manual.)

Chapter 1

Review of Precalculus Concepts

1.1　The Real Number System

1. Rational. The ratio of two integers is rational, by definition. See text, p. **3**.

2. Rational. In $3.1414\overline{14}\ldots$ the digit pattern 14 repeats. See text, p. **3**, and problem # 72.

3. Irrational. Although there is a pattern to the digits, there is no fixed repeating pattern of digits.

4. Irrational. The sum of rational number and an irrational number is always irrational. Proof: If $a = \frac{p_1}{q_1}$, with p_1 and q_1 integers, and b is irrational, and $c = a + b$, then if c were rational, say, $c = \frac{p_2}{q_2}$, $(p_2,\ q_2$ integers) then $b = c - a = \frac{p_2 q_1 - p_1 q_2}{q_1 q_2}$. Hence c would have to be rational. Since we assumed that c is irrational, b must then be irrational.

5. Rational. The digit pattern 163 repeats.

6. Rational, The ratio of two integers is rational, by definition.

7. $\sqrt{256} = 16$ which is an integer, and hence rational.

8. $\sqrt{2}$ is irrational. See problem # 65.

9. Irrational. The sum of rational number and an irrational number is always irrational. See problem # 4.

10. $[-2, 5] \cap (-1, 6) = (-1, 5]$

11. $[-2, 5] \cup (-1, 6) = [-2, 6)$

12. $[-2, 5] \cup (-\infty, 0) = (-\infty, 5]$

13. $[-2, 5] \cap (-\infty, 0) = [-2, 0)$

14. $(-1, 6) \cup (-\infty, 0) = (-\infty, 6)$

15. $(-1, 6) \cap (-\infty, 0) = (-1, 0)$

16. If $A = [2, 4)$ and $A \cup B = [2, 6]$, $A \cap B = (3, 4)$, then

$$
\begin{aligned}
B &= (A \cup B) \text{ minus } (A \text{ minus } A \cap B) \\
&= [2, 6] \text{ minus } [2, 3] \\
&= (3, 6].
\end{aligned}
$$

1

17. a True. Since the circumference $C = 2\pi r$, where r is the radius, C will be irrational whenever r is rational.

    b True. If A is the area, then $A = \pi r^2$. If r is rational, then so is r^2. Hence A must be irrational in this case.

    c True. $C = 2\pi r \Leftrightarrow r = \frac{C}{2\pi}$ Hence $A = \pi r^2 = \pi \times (\frac{C}{2\pi})^2 = \frac{C^2}{4\pi}$. If C is rational, then so is $\frac{C^2}{4}$. Hence A must be irrational in this case.

18. $2x + 3 < 9 \Leftrightarrow 2x < 6 \Leftrightarrow x < 3$

19. $x + 4 \leq 3x \Leftrightarrow 4 \leq 2x \Leftrightarrow x \geq 2$

20. $2x + 7 > 4x - 5 \Leftrightarrow 2x + 12 > 4x \Leftrightarrow 12 > 2x \Leftrightarrow x < 6$

21. $6x - 6 \leq 8x + 8 \Leftrightarrow 6x \leq 8x + 14 \Leftrightarrow -2x \leq 14 \Leftrightarrow x \geq -7$

22. $-4 \leq 2(x + 2) < 10 \Leftrightarrow -2 \leq x + 2 < 5 \Leftrightarrow -4 \leq x < 3$

23. The quantity $(x - 3)(x + 1)$ changes sign only where $(x - 3)(x + 1) = 0$, that is, at $x = 3$ or $x = -1$. If $x < -1$ then $x - 3 < 0$ and $x + 1 < 0$, hence the product of these factors is positive. If $-1 < x < 3$ then the factor $x + 1$ is positive while the other factor is negative. Hence in this case the product is negative. Similarly, when $x > 3$ both factors are positive, so the product is positive.

Hence the solution to $(x - 3)(x + 1) < 0$ is $-1 < x < 3$.

24. The quantity $(x + 7)(2x - 4)$ changes sign only where $(x + 7)(2x - 4) = 0$, that is, at $x = -7$ or $x = 2$. If $x < -7$ then $x + 7 < 0$ and $2x - 4 < 0$, hence the product of these two factors is positive. If $-7 < x < 2$ then the factor $x + 7$ is positive and the factor $2x - 4$ is negative. Hence in this case the product is negative. When $x > 2$ both factors are positive, so the product is positive.

Thus the solution to $(x + 7)(2x - 4) > 0$ is $\{x \mid x < -7 \text{ or } x > 2\}$.

25. $x(x + 6) > -8 \Leftrightarrow x^2 + 6x + 8 > 0 \Leftrightarrow (x + 4)(x + 2) > 0$ The quantity $(x + 4)(x + 2)$ changes sign only where $(x + 4)(x + 2) = 0$, that is, at $x = -4$ or $x = -2$. If $x < -4$ then $x + 4 < 0$ and $x + 2 < 0$, hence the product of these two factors is positive. If $-4 < x < -2$, then $x + 4$ is positive and $x + 2$ is negative. Hence in this case the product is negative. If $x > -2$, then both factors are positive and the product is positive.

Hence the solution to $x(x + 6) > -8$ is $\{x \mid x < -4 \text{ or } x > -2\}$.

26. Factor $x^2 + x$ as $x(x + 1)$. Both factors have the same sign if $x < -1$ or $x > 0$. They have opposite signs if $-1 < x < 0$. Hence $x^2 + x < 0$ if and only if $-1 < x < 0$.

27. Factor $x^3 + x^2 - 2x$ as $x(x^2 + x - 2) = x(x + 2)(x - 1)$. This quantity is positive if it is not zero and the number of negative factors is even. This is true if $-2 < x < 0$ or $x > 1$.

28. $x^2 + x + 7 > 19 \Leftrightarrow x^2 + x - 12 > 0$. Factor $x^2 + x - 12$ as $(x + 4)(x - 3)$. Both factors have the same sign if $x < -4$ or $x > 3$. Hence $x^2 + x + 7 > 19 \Leftrightarrow (x < -4 \text{ or } x > 3)$.

29. Factor $x^4 - 9x^2$ as $x^2(x - 3)(x + 3)$. Since $x^2 \geq 0$, $x^2(x - 3)(x + 3)$ is negative when the other two factors, $(x - 3)$ and $(x + 3)$, have opposite signs, and $x \neq 0$. Hence $x^4 - 9x^2 < 0 \Leftrightarrow -3 < x < 0$ or $0 < x < 3$, that is, $0 < |x| < 3$.

30. $|x - 7| = 2$ when either $x - 7 = 2$ or $x - 7 = -2$. Solving these two equations gives us the solutions $x = 9$ and $x = 5$.

31. $|5x - 2| = 0 \Leftrightarrow 5x - 2 = 0 \Leftrightarrow x = \frac{2}{5}$.

32. $x + |x| = 0 \Leftrightarrow |x| = -x \Leftrightarrow x \leq 0$

33. $2|3x - 1| = 22 \Leftrightarrow |3x - 1| = 11.$ The last equality holds when $3x - 1 = 11$ or $3x - 1 = -11.$ Solving these two equations gives the solutions $x = 4$ and $x = -\frac{10}{3}.$

34. $x + |-3| = 7 \Leftrightarrow x + 3 = 7 \Leftrightarrow x = 4$

35. $(|x| + 6)^2 = 49 \Leftrightarrow |x| + 6 = 7$ or $|x| + 6 = -7.$ The second equation has no solutions. The first equation is equivalent to $|x| = 1.$ Hence the solutions are $x = 1$ and $x = -1.$

36. First note that $|\pi - 6| = -(\pi - 6).$
$$x - 6 = 2x - |\pi - 6| \Leftrightarrow x - 6 = 2x + (\pi - 6) \Leftrightarrow x = -\pi$$

37. $x + 3|x| = 8 \Leftrightarrow |3x| = 8 - x.$ The last equation holds when $3x = 8 - x$ or $-3x = 8 - x.$ Solving these two equations gives the solutions $x = 2$ and $x = -4.$

38. The quantity between the absolute value signs is positive for all values of x. Therefore removing the absolute value sign does not alter the solution set. Hence,
$$|(x + 2)^2 + 3| = 12 \Leftrightarrow (x + 2)^2 + 3 = 12 \Leftrightarrow (x + 2)^2 = 9 \Leftrightarrow (x + 2) = \pm 3 \Leftrightarrow x = -2 \pm 3.$$
Hence, the solutions are $x = 1$ and $x = -5.$

39. $|x - 4| = |x - 7|$ when either $x - 4 = x - 7$ or $x - 4 = -(x - 7).$
$x - 4 = x - 7$ has no solution.
$x - 4 = -(x - 7) \Leftrightarrow 2x = 11 \Leftrightarrow x = \frac{11}{2}.$

40. True, since (a-b) = -(b-a) for all real numbers a and b.

41. $|x - 3| \leq 2 \Leftrightarrow -2 \leq (x - 3) \leq 2 \Leftrightarrow 1 \leq x \leq 5$

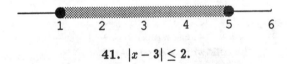

41. $|x - 3| \leq 2.$

42. $|x + 5| < 4 \Leftrightarrow -4 < x + 5 < 4 \Leftrightarrow -9 < x < -1$

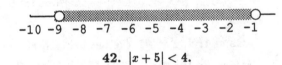

42. $|x + 5| < 4.$

43. $|x + 2| > 1$ when $x + 2 < -1$ or $x + 2 > 1.$ Solving these two inequalities gives the solution set $\{x \mid x < -3 \text{ or } x > -1\}.$

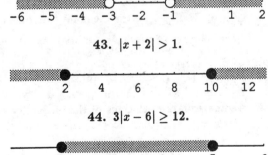

43. $|x + 2| > 1.$

44. $3|x - 6| \geq 12 \Leftrightarrow |x - 6| \geq 4.$ This inequality is true when $x - 6 \leq -4$ or $x - 6 \geq 4.$ Solving the two inequalities gives the solution set $\{x \mid x \leq 2 \text{ or } x \geq 10\}.$

44. $3|x - 6| \geq 12.$

45. $|2x - 7| \leq 3 \Leftrightarrow -3 \leq 2x - 7 \leq 3 \Leftrightarrow 4 \leq 2x \leq 10 \Leftrightarrow 2 \leq x \leq 5$

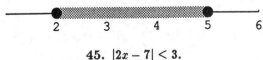

45. $|2x - 7| \leq 3.$

46. There is no solution. If $x < 0,$ then $x < |x|.$ If $x \geq 0,$ then $x = |x|.$

47. $|x - 1| = x - 1 \Leftrightarrow x - 1 \geq 0.$

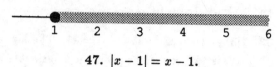

47. $|x - 1| = x - 1.$

48. The inequality $|x + 4| - |x - 1| < 4$ holds if and only if:

$$|x + 4| < |x - 1| + 4$$
$$\Leftrightarrow \quad (x + 4)^2 < (x - 1)^2 + 8|x - 1| + 16$$
$$\Leftrightarrow \quad x^2 + 8x + 16 < x^2 - 2x + 1 + 8|x - 1| + 16$$
$$\Leftrightarrow \quad 10x - 1 < 8|x - 1|$$

Case 1: If $x \geq 1$, then the last inequality becomes:

$$10x - 1 < 8(x - 1)$$
$$\Leftrightarrow \quad 2x < -7.$$

Thus there are no solutions in $x \geq 1$.

Case 2: If $x < 1$,

$$10x - 1 < 8|x - 1|$$
$$\Leftrightarrow \quad 10x - 1 < -8(x - 1)$$
$$\Leftrightarrow \quad 18x < 9$$
$$\Leftrightarrow \quad x < \frac{1}{2}$$

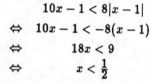

48. $|x + 4| - |x - 1| < 4.$

49. This inequality is true when $(8 - 3x) \leq -5$ or $8 - 3x \geq 5$. Solving these two inequalities give the solution $x \geq \frac{13}{3}$ or $x \leq 1$.

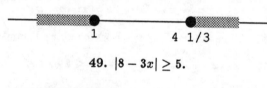

49. $|8 - 3x| \geq 5.$

50. To solve the inequality $|x + 2| + |x - 5| \geq 10$, consider the following three cases:

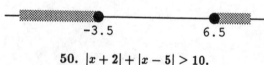

50. $|x + 2| + |x - 5| \geq 10.$

Case 1: If $x < -2$, the inequality becomes $-(x + 2) - (x - 5) \geq 10 \Leftrightarrow x \leq -\frac{7}{2}$

Case 2: If $-2 \leq x \leq 5$, the inequality becomes $(x + 2) - (x - 5) \geq 10 \Leftrightarrow 7 \geq 10$. Hence, there is no solution in this case.

Case 3: If $x > 5$, the inequality becomes $(x + 2) + (x - 5) \geq 10 \Leftrightarrow x \geq \frac{13}{2}$

Hence, the solution set is given by $x \leq -\frac{7}{2}$ or $x \geq \frac{13}{2}$.

51. $|x + 3| + |x - 2| < 7 \Leftrightarrow |x + 3| < 7 - |x - 2|$. Since the right hand side of the inequality is always positive, we can square both sides of the inequality to obtain:

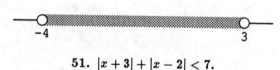

51. $|x + 3| + |x - 2| < 7.$

$$(x + 3)^2 < 49 - 14|x - 2| + (x - 2)^2$$
$$\Leftrightarrow \quad 14|x - 2| < 44 - 10x$$
$$\Leftrightarrow \quad 10x - 44 < 14(x - 2) < 44 - 10x$$
$$\Leftrightarrow \quad -4 < x < 3.$$

52. The inequality describes all the numbers x that lie more than 2 units away from 5. Hence $x < 3$ or $x > 7$.

53. The distance between x and 1 is the same as the distance between x and -3. Hence $x = -1$.

54. The distance between x and 1 is twice the distance between x and 3. There are two values of x which satisfy this geometric condition, $x = \frac{7}{3}$, which lies between 1 and 3, and $x = 5$, which lies to the right of 5.

55. The inequality describes all the numbers x that lie more than 2 and less than 5 units from 4. Hence, $-1 < x < 2$ or $6 < x < 9$.

56. $|x + 5| > 4$

57. $|x + 2| = 2|x - 12|$

58.

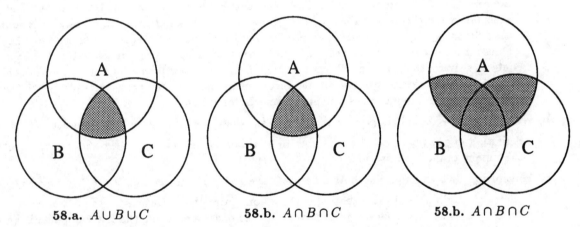

58.a. $A \cup B \cup C$ 58.b. $A \cap B \cap C$ 58.b. $A \cap B \cap C$

59. Exactly one of the following holds: $y - x$ is positive, $y - x$ is negative, or $x = y$.

    $y - x$ is positive if and only if $x < y$.

    $y - x$ is negative if and only if $x - y$ is positive if and only if $x > y$.

    Hence, exactly one of the following holds: $x = y$, $x < y$ or $x > y$.

60. We assume that $x < y$ and that $y < z$. We must prove that $x < z$.

    $x < y \Rightarrow y - x > 0$ by definition.

    $y < z \Rightarrow z - y > 0$ by definition.

    Hence, $z - x = (z - y) + (y - x) > 0$. $x < z$ by definition.

61. False. If $a = -1$ and $b = -2$, then $a^2 < b^2$, but $b < a$.

62. $a^2 < b^2 \Rightarrow (b - a)(b + a) = b^2 - a^2 > 0$. The latter inequality is true only if the factors, $(b - a)$ and $(b + a)$ have the same sign.

    If $b - a < 0$ and $b + a < 0$, then $2b = (b - a) + (b + a) < 0$, contradicting the condition that $b > 0$.

    Hence we must have $b - a > 0$ and $b + a > 0$. Combining the two inequalities gives us $-b < a < b$.

63. Let $a = x + 6$ and let $b = 7 > 0$. Then $(x + 6)^2 < 49$ is the same as $a^2 < b^2$. By exercise 62, $-7 < x + 6 < 7$. Solving for x gives us $-13 < x < 1$.

64. (iii) If $x < y$, then $y - x$ is positive by definition.

$(y+c) - (x+c) = y - x$ is also positive. Using the definition of inequality, we have $(x+c) < (y+c)$.

(iv) If $x < y$, then $y - x$ is positive by definition. Hence, $cy - cx = c(y - x)$ is also positive since $c > 0$. By definition, $cx < cy$.

(v) If $x < y$, then $y - x$ is positive. Since $c < 0$, $-c = 0 - c$ is positive. Hence, $cx - cy = -c(y - x)$ is positive and $cx > cy$.

65. a. Actually 'proof by contradiction' is more properly called 'proof by contrapositive'. The contrapositive of the implication $p \Rightarrow q$ is the implication $(\neg q) \Rightarrow (\neg p)$. Here $\neg p$ is the symbol for the negation of p. A logical implication is always logically equivalent to its contrapositive. Thus to prove $p \Rightarrow q$ by 'contradiction', we assume $\neg q$ and prove that this implies $\neg p$. Here p represents the hypotheses of the theorem, and q represents the conclusion. Constructing 'proofs by contradiction' in this manner, using careful reasoning, helps avoid mistakes where a 'contradiction' simply results from poor reasoning or a mathematical error. For the theorem at hand the hypothesis is the (true) statement that 2 is a prime number. The conclusion is the statement that $\sqrt{2}$ is not rational. The statement of the theorem is then "*If 2 is prime, then $\sqrt{2}$ is not rational.*" However it sounds a little silly to say "*If 2 is prime,*" because 2 is always prime. The theorem sounds much better if we pose it as: "*If x is prime, then $\sqrt{x}$ is not rational.*" This is also a much better theorem. The contrapositive of this, is then: "*If $\sqrt{x}$ is rational, then x is not prime.*" In order to prove this, we assume some important results from the theory of the natural numbers, particularly the theorem that every natural number has a unique prime factorization.

b. Alternatively, we could simply say that $\sqrt{x} = \frac{p}{q}$, where p and q are natural numbers. This is the definition of what it means for a number to be rational. Let $p = p_1^{j_1} \cdots p_n^{j_n}$ and $q = q_1^{k_1} \cdots q_m^{k_m}$ be the unique prime factorizations of p and q, respectively.

c. Multiplying by q and squaring both sides, we have $xq^2 = p^2$. Hence $xq_1^{2k_1} \cdots q_m^{2k_m} = p_1^{2j_1} \cdots p_n^{2j_n}$.

d.,e. Since the right side of this equation is the unique prime factorization for the left hand side, and since this prime factorization has an even number of factors, while the left hand side is an even number of prime factors, times x, it must be that x has an even number of prime factors.

f. This implies that x is not prime. Thus we have proved that *If the square root of x is rational, then x is not prime*, or, equivalently, *If x is prime, then the square root of x is not rational.* The exercise is the application of this statement to the case $x = 2$.

66. $\sqrt{x^2} = \sqrt{|x^2|} = \sqrt{|x|^2} = |x|$

The first equality follows from Definition 2 and the second equality follows from Theorem 3.

67. Adding the two inequalities $-|x| \leq x \leq |x|$ and $-|y| \leq y \leq |y|$, we obtain $-(|x| + |y|) \leq x + y \leq |x| + |y|$. Then Theorem 2 gives the result: $|x + y| \leq |x| + |y|$.

68. False. Let $a = -1$ and $b = 1$. Then $a < b$ and $\frac{1}{a} < \frac{1}{b}$.

69. If $a < b$ and $ab > 0$, then $\frac{1}{a} > \frac{1}{b}$

Proof: If $a < b$ and $ab > 0$, then both $(b - a)$ and ab are positive. Hence, $\frac{1}{a} - \frac{1}{b} = \frac{1}{ab}(b - a)$ is positive. By definition, $\frac{1}{a} - \frac{1}{b} > 0$

70. $|x| = |y + (x - y)| \leq |y| + |x - y|$. Subtract $|y|$ from both sides of the inequality to obtain the result $|x| - |y| \leq |x - y|$.

71. Yes, simply apply the theorem in the above solution for #65 to the case $x = 3$.

72a. Let $x = 1.341341\overline{341}$. Then $1000x = 1341.341\overline{341}$. So, $999x = (1000x - x) = 1340$. Thus, $x = \frac{1340}{999}$.

72b. $x = 0.\overline{a_1 a_2 ... a_n}$. Then $10^n x = a_1 a_2 ... a_n . \overline{a_1 a_2 ... a_n}$. So,

$(10^n - 1)x = a_1 a_2 ... a_n - 1$. Thus $x = \frac{a_1 a_2 ... a_n - 1}{10^n - 1}$

73. We will first prove: $|x| < a \Leftrightarrow -a < x < a$

    Case 1: $x \geq 0$. In this case $|x| = x$, so $|x| < a \Leftrightarrow x < a$.

    Case 2: $x < 0$. In this case $|x| = -x$, so $|x| < a \Leftrightarrow -x < a \Leftrightarrow -a < x$

    Hence, $|x| < a \Leftrightarrow -a < x < a$.

    Now, we will prove that the inequality $|x| > a$ holds if either $x > a$ or $x < -a$.

    If $x \geq 0$, then $|x| > a \Leftrightarrow x > a$.

    If $x < 0$, then $|x| > a \Leftrightarrow -x > a \Leftrightarrow x < -a$.

1.2   The Coordinate Plane, Distance, and Circles

1a. $\sqrt{(0-2)^2 + [2-(-1)]^2} = \sqrt{2^2 + 3^2} = \sqrt{13}$

1b. $\sqrt{(1-3)^2 + (3-1)^2} = \sqrt{2^2 + 2^2} = \sqrt{8} = 2\sqrt{2}$

1c. $\sqrt{(1-0)^2 + (-9-2)^2} = \sqrt{1^2 + 11^2} = \sqrt{122}$

1d. $\sqrt{(6-1)^2 + [6-(-3)]^2} = \sqrt{5^2 + 9^2} = \sqrt{106}$

1e. $\sqrt{(-1-1)^2 + (-1-1)^2} = \sqrt{2^2 + 2^2} = \sqrt{8} = 2\sqrt{2}$

1f. $\sqrt{[1-(-2)]^2 + [2-(-2)]^2} = \sqrt{3^2 + 4^2} = \sqrt{25} = 5$

2. Let $(5, y)$ be a point whose distance from the point $(1, 3)$ is 5. Then

$$\sqrt{(1-5)^2 + (3-y)^2} = 5$$
$$\Leftrightarrow \quad 4^2 + (3-y)^2 = 25$$
$$\Leftrightarrow \quad (3-y)^2 = 9$$
$$\Leftrightarrow \quad (3-y) = \pm 3$$
$$\Leftrightarrow \quad y = 0 \text{ or } y = 6$$

Hence the two points are $(5, 0)$ and $(5, 6)$.

3. Let (x, y) be a point which lies a distance of 4 units from both $(-2, 3)$ and $(2, 3)$. Then x and y must satisfy the two equations:

$$\sqrt{[x-(-2)]^2 + (y-3)^2} = 4$$
$$\sqrt{[(x-2)^2 + (y-3)^2} = 4$$

Squaring both sides of both equations gives us the equations:

$$(x+2)^2 + (y-3)^2 = 16 \qquad\qquad (1.1)$$
$$(x-2)^2 + (y-3)^2 = 16 \qquad\qquad (1.2)$$

Note that $(x+2)^2 = 16 - (y-3)^2 = (x-2)^2$. Hence, $(x+2) = \pm(x-2)$. Solving for x, we have $x = 0$. As equations 1.1 and 1.2 are the same when $x = 0$, we need only solve the following equation:

$$2^2 + (y-3)^2 = 16 \Leftrightarrow (y-3)^2 = 12 \Leftrightarrow y = 3 \pm 2\sqrt{3}.$$

Hence, the two points are $(0, 3 - 2\sqrt{3})$ and $(0, 3 + 2\sqrt{3})$.

4. **True. This follows from the definition of a circle.**

5. **A triangle is isosceles when any two sides have the same length.**

   Case 1: The distance from $(a, 0)$ to $(-1, 0)$ is the same as the distance from $(a, 0)$ to $(2, 3)$:

   $$\sqrt{(-1-a)^2 + 0^2} = \sqrt{(2-a)^2 + 3^2}$$
   $$\Leftrightarrow \qquad (a+1)^2 = (a-2)^2 + 9$$
   $$\Leftrightarrow \qquad a^2 + 2a + 1 = a^2 - 4a + 13$$
   $$\Leftrightarrow \qquad 6a = 12$$
   $$\Leftrightarrow \qquad a = 2$$

   Case 2: The distance from $(a, 0)$ to $(-1, 0)$ is the same as the distance from $(-1, 0)$ to $(2, 3)$:

   $$\sqrt{(-1-a)^2 + 0^2} = \sqrt{[2-(-1)]^2 + (3-0)^2}$$
   $$\Leftrightarrow \qquad (a+1)^2 = 18$$
   $$\Leftrightarrow \qquad a = -1 \pm 3\sqrt{2}$$

   Case 3: The distance from $(a, 0)$ to $(2, 3)$ is the same as the distance from $(-1, 0)$ to $(2, 3)$:

   $$\sqrt{(2-a)^2 + 3^2} = \sqrt{18}$$
   $$\Leftrightarrow \qquad (a-2)^2 + 9 = 18$$
   $$\Leftrightarrow \qquad a = 2 \pm 3$$

   Note that $a = -1$ is not a solution, for then $(a, 0) = (-1, 0)$. From the three cases we see that there are four solutions for a: $2, -1 \pm 3\sqrt{2}, 5$.

6. **A triangle is equilateral if all three sides have the same length:**

   $$\sqrt{(2-0)^2 + (0-0)^2} = \sqrt{(1-2)^2 + (b-0)^2} = \sqrt{(1-0)^2 + (b-0)^2}$$
   $$\Leftrightarrow \qquad 4 = 1 + b^2 = 1 + b^2$$
   $$\Leftrightarrow \qquad b = \pm\sqrt{3}$$

7. **Let** (x, y) **be any point lying equidistant from the points** $(-1, -1)$ **and** $(3, 1)$. **This means that:**

   $$\sqrt{(x+1)^2 + (y+1)^2} = \sqrt{(x-3)^2 + (y-1)^2}$$
   $$\Leftrightarrow \quad x^2 + 2x + 1 + y^2 + 2y + 1 = x^2 - 6x + 9 + y^2 - 2y + 1$$
   $$\Leftrightarrow \qquad 4y = -8x + 8$$
   $$\Leftrightarrow \qquad y = -2x + 2$$

8. **Let** (x, y) **be any point lying equidistant from the points** $(1, 2)$ **and** $(5, -1)$. **This means that**

   $$\sqrt{(x-1)^2 + (y-2)^2} = \sqrt{(x-5)^2 + (y+1)^2}$$
   $$\Leftrightarrow \quad x^2 - 2x + 1 + y^2 - 4y + 4 = x^2 - 10x + 25 + y^2 + 2y + 1$$
   $$\Leftrightarrow \qquad 8x - 6y = 21$$

9. The distance from (x_1, y_1) to $(\frac{x_1 + x_2}{2}, \frac{y_1 + y_2}{2})$ is $\sqrt{(x_1 - x_2)^2 + (y_1 - y_2)^2}/2$. The distance from $(\frac{x_1 + x_2}{2}, \frac{y_1 + y_2}{2})$ to (x_2, y_2) is the same.

10.a. $(\frac{-1 + 6}{2}, \frac{2 + 4}{2}) = (2.5, 3)$.

10.b. $(\frac{1 + (-7)}{2}, \frac{1 + (-3)}{2}) = (-3, -1)$.

11. The distance from (x, y) to $(-3, 1)$ is $\sqrt{(x + 3)^2 + (y - 1)^2}$. Thus the set of all points (x, y) that lie more that $\sqrt{2}$ units from $(-3, 1)$ is described by the inequality:

$$\sqrt{(x + 3)^2 + (y - 1)^2} > \sqrt{2} \qquad (1.3)$$
$$\text{or} \qquad (x + 3)^2 + (y - 1)^2 > 2 \qquad (1.4)$$

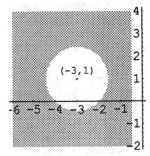

11. $(x + 3)^2 + (y - 1)^2 > 2$.

12. This set is described by two inequalities. A point lies at least 3 points to the right of the y-axis, if and only if $x \geq 3$. A point is at least 3 units from the x-axis, if and only if $|y| \geq 3$. To describe this set with a *single* inequality is quite a challenge!

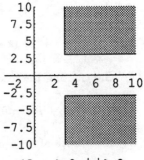

12. $x \geq 3$, $|y| \geq 3$.

13. Again, this set is most simply described by two inequalities. The distance from (x, y) to the x-axis is $|x|$. We need both of these quantities to be less than 2. This is expressed by the inequalities $|x| \leq 2$, $|y| \leq 2$.

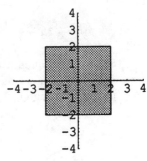

13. $|x| \leq 2$, $|y| \leq 2$

14. Since the distance from (x, y) to $(1, -2)$ is $\sqrt{(x-1)^2 + (y+2)^2}$, the desired inequality is: $9 < (x-1)^2 + (y+2)^2 < 25$.

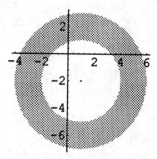

14. $9 < (x-1)^2 + (y+2)^2 < 25$.

15. The equation for a circle with center (h, k) and radius r is $(x-h)^2 + (y-k)^2 = r^2$. Here $h = k = 0$ and $r = 3$. So the equation for the circle is $x^2 + y^2 = 9$.

16. $(x+2)^2 + (y-4)^2 = 9$.

17. $(x+6)^2 + (y+4)^2 = 25$.

18. Since the center is given, we only need the radius to determine the equation. But the radius is the distance from the center to *any* point on the circle. Hence we may compute the radius by computing the distance from the center to the point given: $r^2 = (2-2)^2 + (-1-3)^2 = 16$. So the equation of the circle is $(x-2)^2 + (y-3)^2 = 16$.

19. There are two points whose distance to the points $(-1, -4)$ and $(4, 1)$, is 5. To solve for the coordinates (h, k) of these points, we write equations for the distance from the centers to the given points:

$$(h+1)^2 + (k+4)^2 = 25,$$

$$(h-4)^2 + (k-1)^2 = 25.$$

Subtracting the second equation from the first, we obtain:

$$10h + 10k = 0.$$

Thus both of the centers are on the line $k = -h$. Substituting this in the distance equations, we obtain

$$(h+1)^2 + (-h+4)^2 = 25.$$

or

$$h^2 - 3h - 4 = 0.$$

Solving this quadratic equation, we obtain $h = 4$ or $h = -1$. Since the centers lie on the line $k = -h$, the centers must have the coordinates $(4, -4)$, and $(-1, 1)$. Thus the equations of the two circles passing through $(-1, -4)$ and $(4, 1)$ are:

$$(x-4)^2 + (y+4)^2 = 25,$$

$$(x+1)^2 + (y-1)^2 = 25.$$

20. Complete squares:

$$
\begin{aligned}
x^2 - 2x + y^2 - 8 &= 0 \\
\Leftrightarrow \quad x^2 - 2x + 1 - 1 + y^2 - 8 &= 0 \\
\Leftrightarrow \quad (x - 1)^2 + y^2 &= 9.
\end{aligned}
$$

Thus the radius is 3 and the center is $(1, 0)$.

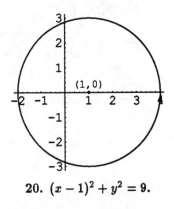

20. $(x - 1)^2 + y^2 = 9.$

21. Complete squares:

$$
\begin{aligned}
x^2 + y^2 + 4x + 2y - 11 &= 0 \\
\Leftrightarrow \quad x^2 + 4x + 4 - 4 + y^2 + 2y + 1 - 1 - 11 &= 0 \\
\Leftrightarrow \quad (x + 2)^2 + (y + 1)^2 &= 16.
\end{aligned}
$$

Thus the radius is 4 and the center is $(-2, -1)$.

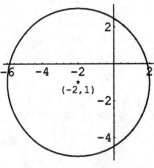

21. $(x + 2)^2 + (y + 1)^2 = 16.$

22. Complete squares:

$$
\begin{aligned}
x^2 + 14x + y^2 - 10y + 70 &= 0 \\
\Leftrightarrow \quad x^2 + 14x + 49 - 49 + y^2 - 10y + 25 - 25 + 70 &= 0 \\
\Leftrightarrow \quad (x + 7)^2 + (y - 5)^2 = 49 + 25 - 70 &= 4.
\end{aligned}
$$

Thus the radius is 2 and the center is $(-7, 5)$.

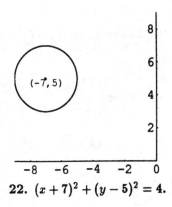

22. $(x + 7)^2 + (y - 5)^2 = 4.$

23. Complete squares:

$$x^2 + y^2 - 2x - 6y + 3 \qquad = 0$$
$$\Leftrightarrow \quad x^2 - 2x + 1 - 1 + y^2 - 6y + 9 - 9 + 3 \quad = 0$$
$$\Leftrightarrow \quad (x-1)^2 + (y-3)^2 = 1 + 9 - 3 \qquad = 7.$$

Thus the radius is $\sqrt{7}$ and the center is $(1, 3)$.

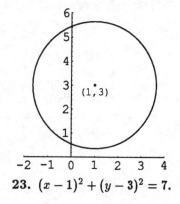

23. $(x-1)^2 + (y-3)^2 = 7.$

24. Complete squares:

$$x^2 - 2ax + y^2 + 4ay + 5a^2 - 1 \qquad = 0$$
$$\Leftrightarrow \quad x^2 - 2ax + a^2 + y^2 + 4ay + 4a^2 - 1 \quad = 0$$
$$\Leftrightarrow \qquad (x-a)^2 + (y+2a)^2 \qquad = 1.$$

Thus the radius is 1 and the center is $(a, -2a)$. The figure shows the special case $a = 1$.

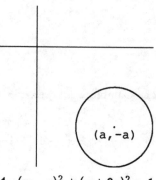

24. $(x-a)^2 + (y+2a)^2 = 1.$

25. Complete squares:

$$x^2 + y^2 - 2by + b^2 - a^2 \quad = 0$$
$$\Leftrightarrow \quad x^2 + (y-b)^2 = a^2.$$

Thus the radius is a and the center is $(0, b)$. The figure shows the case where $b^2 > a^2$ and $b > 0$.

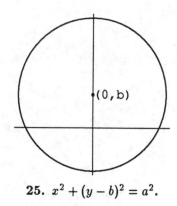

25. $x^2 + (y-b)^2 = a^2.$

26. Since the center is $(3, 2)$, the circle will be tangent to the y-axis at the point $(0, 2)$ and the radius is 3. The equation is $(x-3)^2 + (y-2)^2 = 9$.

27. We solve the first equation for y to get $y = x - 5$. Substituting for y in the equation for the circle gives:

$$x^2 - 8x + (x-5)^2 - 4(x-5) + 11 = 0$$
$$\Leftrightarrow \qquad x^2 - 11x + 28 = 0$$
$$\Leftrightarrow \qquad (x-7)(x-4) = 0$$
$$\Leftrightarrow \qquad x = 7 \text{ and } x = 4$$

Substituting these values for x in the equation $y = x - 5$ gives us the two points of intersection $(7, 2)$ and $(4, -1)$.

28. **False.** If the three points all lie on the same line, then they cannot determine a circle (unless one thinks of a line as a circle of infinite radius!) However any set of three distinct points which do not lie on a line will determine a circle.

29. Let (x, y) be any point whose distance from the point $(-2, 1)$ is twice the distance from the point $(4, -2)$. Then (x, y) satisfies the equation:

$$\sqrt{(x+2)^2 + (y-1)^2} = 2\sqrt{(x-4)^2 + (y+2)^2}$$
$$\Leftrightarrow \quad (x+2)^2 + (y-1)^2 = 4\left((x-4)^2 + (y+2)^2\right)$$
$$\Leftrightarrow \quad x^2 + 4x + 4 + y^2 - 2y + 1 = 4\left(x^2 - 8x + 16 + y^2 + 4y + 4\right)$$
$$\Leftrightarrow \quad 3x^2 - 36x + 3y^2 + 18y + 75 = 0$$
$$\Leftrightarrow \quad x^2 - 12x + y^2 + 6y + 25 = 0$$
$$\Leftrightarrow \quad (x-6)^2 - 36 + (y+3)^2 - 9 + 25 = 0$$
$$\Leftrightarrow \quad (x-6)^2 + (y+3)^2 = 20$$

This is the equation of the circle with center $(6, -3)$ and radius $2\sqrt{5}$.

30. Let L_1 be the set of all points which are equidistant from $(-2, 1)$ and $(1, 4)$. Let L_2 be the set of all points which are equidistant from $(1, 4)$ and $(4, 1)$. Then the equation for L_1 is given by:

$$\sqrt{(x+2)^2 + (y-1)^2} = \sqrt{(x-1)^2 + (y-4)^2}$$
$$\Leftrightarrow \quad x^2 + 4x + 4 + y^2 - 2y + 1 = x^2 - 2x + 1 + y^2 - 8y + 16$$
$$\Leftrightarrow \quad x + y = 2$$

The equation for L_2 is given by:

$$\sqrt{(x-1)^2 + (y-4)^2} = \sqrt{(x-4)^2 + (y-1)^2}$$
$$\Leftrightarrow \quad x^2 - 2x + 1 + y^2 - 8y + 16 = x^2 - 8x + 16 + y^2 - 2y + 1$$
$$\Leftrightarrow \quad x - y = 0$$

Solving the equations $x + y = 2$ and $x - y = 0$ gives us the center of the circle, $(1, 1)$. The radius is the distance from $(1, 1)$ to $(4, 1)$ which is 3. Hence the equation of the circle is:

$$(x-1)^2 + (y-1)^2 = 9.$$

31. All circles containing the two points $(1, 2)$ and $(-1, 2)$ have centers on the set of points, L_1, which are equidistant from the two points. The equation for L_1 is given by $x = 0$. Let $(0, k)$ be the center of any circle containing the point $(1, 2)$. The radius, r, is given by $r^2 = 1 + (k - 2)^2$. Hence, the family of circles containing the two given points is given by the equation:

$$x^2 + (y - k)^2 = 1 + (k - 2)^2,$$

where k is any real number.

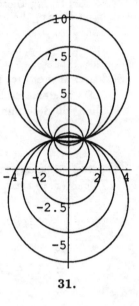

31.

32. (1) If $d > 0$, then the equation is that of a circle with center (h, k) and positive radius $\sqrt{d}$. Hence, the solution set has an infinite number of points.

(2) If $d = 0$, the only solution is the point (h, k).

(3) If $d < 0$, there are no solutions since $(x - h)^2 + (y - k)^2 \geq 0$ for all values of x and y.

33. Since $x^2 - x - 6 = (x - 3)(x + 2)$, and since the product of two real numbers is negative if and only if one number is positive and the other is negative, we have that $x^2 - x - 6 \leq 0 \Leftrightarrow -2 \leq x \leq 3$.

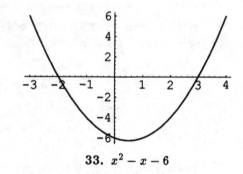

33. $x^2 - x - 6$

34. We have that $\frac{2x - 3}{x + 4} = 3 \Leftrightarrow 3(x + 4) - (2x - 3) = x + 15 = 0$. The only other place where the inequality can change direction is the point where the denominator $x + 4$ vanishes, that is, at $x = -4$. Using this information together with the graph, we find that $\frac{2x - 3}{x + 4} > 3 \Leftrightarrow -15 < x < -4$.

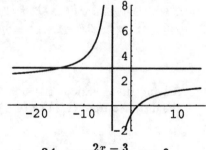

34. $y = \frac{2x - 3}{x + 4}$, $y = 3$.

35. $2x + 3 = -x^2 + 4x + 7 \Leftrightarrow x = 1 \pm \sqrt{5}$. These are the only points where the inequality $2x + 3 > -x^2 + 4x + 7$ can change direction. Using this fact together with the graphs of $2x + 3$ and $-x^2 + 4x + 7$, we see that $2x + 3 > -x^2 + 4x + 7$ if and only if $(x < 1 - \sqrt{5}$ or $x > 1 + \sqrt{5})$.

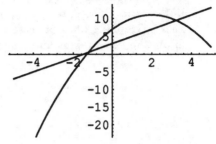

35. $y = 2x + 3$, $y = -x^2 + 4x + 7$

36. $\left| \dfrac{3 - 2x}{5x + 3} \right| = 1$ if and only if $x = 0$ or $x = -2$. The inequality $\left| \dfrac{3 - 2x}{5x + 3} \right| < 1$ can only change direction at these points or where the denominator $5x + 3$ vanishes, that is, at $x = -3/5$.

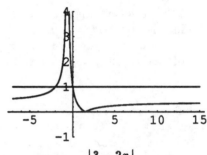

36. $y = \left| \dfrac{3 - 2x}{5x + 3} \right|$, $y = 1$.

37. Solving for y using the quadratic formula (or using the *Mathematica* "Solve" function), we have $y = \left(x \pm \sqrt{30 - 23 * x^2} \right)/6$. Graphing both of these solutions together we obtain the ellipse at right.

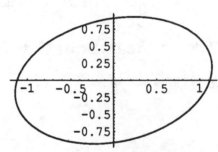

37. $y = \left(x \pm \sqrt{30 - 23 * x^2} \right)/6$

38. Solving for y using the quadratic formula (or using the *Mathematica* "Solve" function), we have $y = \left(x \pm \sqrt{-50 - 16 * x + 11 * x^2} \right)/4$. Graphing both of these solutions together we obtain the hyperbola at right.

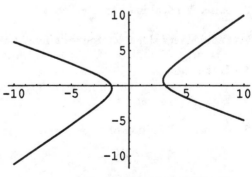

38. $y = \dfrac{x \pm \sqrt{-50 - 16 * x + 11 * x^2}}{4}$.

15

39. Solving for y using the quadratic formula (or using the *Mathematica* "Solve" function), we have $y = -1 - x \pm \sqrt{328}$. Graphing both of these solutions together we obtain the two straight lines (surprise!) at right.

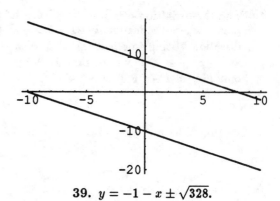

39. $y = -1 - x \pm \sqrt{328}$.

40. Solving for y using the quadratic formula (or using the *Mathematica* "Solve" function), we have $y = \left(-3 - 3*x \pm \sqrt{49 + 38*x - x^2}\right)/5$. Graphing both of these solutions together we obtain the ellipse at right

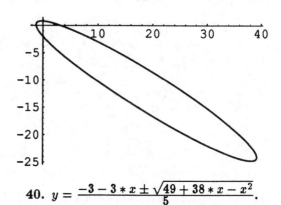

40. $y = \dfrac{-3 - 3*x \pm \sqrt{49 + 38*x - x^2}}{5}$.

1.3   Linear Equations

1. a. $m = \dfrac{2-1}{-1-3} = -\dfrac{1}{4}$

   b. $m = \dfrac{-1-(-2)}{-1-6} = -\dfrac{1}{7}$

   c. $m = \dfrac{a-b}{b-a} = -1$

   d. $m = \dfrac{a-1}{a-1} = 1$

2. False. Vertical lines have no slope.

3. False. Every slope determines a family of parallel lines.

4. True.

5. True.

6. From the definition of slope, we have the following equation:

$$3 = \frac{a-4}{-2-2}$$

Solving the equation gives $a = -8$.

16

7. From the definition of slope, we have the following equation:

$$2 = \frac{-5 - 1}{1 - b}$$

Solving the equation gives $b = 4$.

8. Use the slope-intercept form with $m = 4$ and $b = -2$ to obtain the equation $y = 4x - 2$.

9. Use the slope-intercept form with $m = -2$ and $b = 5$ to obtain the equation $y = -2x + 5$.

10. $y = -5$

11. We first compute the slope $m = \frac{12 - 6}{4 - (-1)} = \frac{6}{5}$. Now we use the point-slope form of the equation with $(x_1, y_1) = (-1, 6)$:

$$y - 6 = \frac{6}{5}(x + 1)$$
$$\text{or} \quad y = \frac{6}{5}x + \frac{36}{5}$$

12. Use the point-slope form of the equation:

$$y - 1 = 7(x - 4)$$
$$\text{or} \quad y = 7x - 27$$

13. $y - 3 = -3(x - 1)$ or $y = -3x + 6$.

14. Since the x-intercept is -3 and the y-intercept is 6, the line goes through the two points $(-3, 0)$ and $(0, 6)$. The slope is $m = \frac{6 - 0}{0 - (-3)} = 2$. Using the slope-intercept form of the equation gives $y = 2x + 6$.

15. $x = -3$

16. $y = 5$

17. $m = \frac{8 - 4}{-6 - (-2)} = -1$. We will use the point-slope form of the equation with the point $(-2, 4)$:

$$y - 4 = -1(x + 2)$$
$$\text{or} \quad y = -x + 2$$

18. $m = \frac{-4 - 2}{-1 - 0} = 6$. Using the point-slope form of the equation with the point $(0, 2)$, we obtain the equation:

$$y - 2 = 6(x - 0)$$
$$\text{or} \quad y = 6x + 2$$

19. To determine the slope, we put the equation of the line $2x - 6y + 5 = 0$ in slope-intercept form, $y = \frac{1}{3}x + \frac{5}{6}$. Using $\frac{1}{3}$ for the slope and $(1, 4)$ for the point, the equation of the line is:

$$y - 4 = \frac{1}{3}(x - 1)$$
$$\text{or} \quad y = \frac{1}{3}x + \frac{11}{3}$$

20. To determine the slope, we put the equation of the line $x - y = 2$ into slope-intercept form, $y = x - 2$. Using 1 for the slope and $(5, -2)$ for the point, the equation of the line is:

$$y + 2 = x - 5$$
$$\text{or} \quad y = x - 7$$

21. Putting the equation of the line $3x + y = 7$ in slope-intercept form, $y = -3x + 7$, we find that this line has slope -3. The negative reciprocal of -3 is $\frac{1}{3}$. Hence, the perpendicular line has slope $\frac{1}{3}$. Since it goes through the point $(1, 3)$, its equation is:

$$y - 3 = \tfrac{1}{3}(x - 1)$$
$$\text{or} \quad y = \tfrac{1}{3}x + \tfrac{8}{3}$$

22. The line through the points $(-2, 5)$ and $(-1, 9)$ has slope $\dfrac{9 - 5}{-1 - (-2)} = 4$. Hence, the perpendicular line has the negative reciprocal slope of $-\frac{1}{4}$. Since the perpendicular line goes through the point $(4, -1)$, its equation is:

$$y + 1 = -\tfrac{1}{4}(x - 4)$$
$$\text{or} \quad y = -\tfrac{1}{4}x$$

23. False. This is only true if $(0, 0)$ satisfies the linear equation.

24. False. A linear equation in x and y has the form $Ax + By + C = 0$. If B is 0, (vertical line), then y is not proportional to $x + a$ for any choice of a constant a. If A is 0, (horizontal line), then y is constant. If the constant value of y is not 0, then y is not proportional to $x + a$ for any choice of a. If the constant value of y is 0, then y is proportional to x in the sense that $y = 0x$. However, some might not consider this to be 'proportionality'. In any other case, the statement is true. Let x and y satisfy the linear equation $Ax + By + C = 0$. If A and B are not both 0, then $y = -\frac{A}{B}\left(x + \frac{C}{A}\right)$, so that y is proportional to $x + \frac{C}{A}$.

25. $y = -x + 7$. The slope is -1, the y-intercept is 7, and the x-intercept is 7.

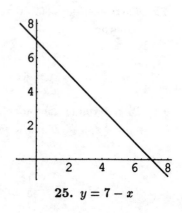

25. $y = 7 - x$

26. Letting $y = 0$ and solving the equation $3x = 6$ gives us the x-intercept of $x = 2$. Put the equation into the slope-intercept form of $y = \frac{3}{2}x - 3$, to get the slope of $\frac{3}{2}$ and y-intercept of -3.

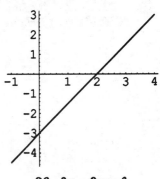

26. $3x - 2y = 6$

27. Let $y = 0$ and solve the equation $x + 3 = 0$ to obtain the x-intercept of -3. Put the equation into the slope-intercept form of $y = -x - 3$ to obtain the slope of -1 and the y-intercept of -3.

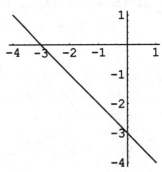

27. $x + y + 3 = 0$

28. Letting $y = 0$ and solving $2x = 10$ gives the x-intercept of $x = 5$. The slope-intercept form of the equation is $y = -\frac{2}{3}x + \frac{10}{3}$. The slope is $-\frac{2}{3}$ and the y-intercept is $\frac{10}{3}$.

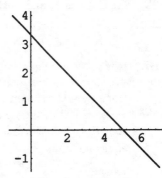

28. $2x = 10 - 3y$

29. There is no x-intercept. The slope is 0 and the y-intercept is 5.

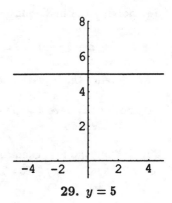

29. $y = 5$

30. $y = 2x + 9$. The slope is 2 and the y-intercept is 9. Setting $y = 0$ and solving for x gives us an x-intercept of $-\frac{9}{2}$.

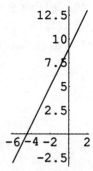

30. $y = 2x + 9$

31. There is neither a slope nor a y-intercept. The x-intercept is 4.

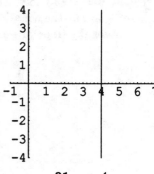

31. x=4

32. The slope is 1, and both intercepts are 0.

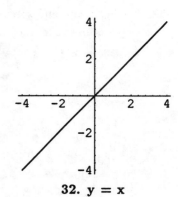

32. y = x

33. Solving the first equation for y, we have $y = 3x - 1$. Substitute this into the second equation:

$$x + 3x - 1 - 3 = 0$$

Solving this equation x gives us $x = 1$ and $y = 3x - 1 = 2$. The point of intersection is $(1, 2)$.

34. Both lines have a slope of $\frac{1}{2}$. Since the two lines are parallel and do not coincide, there are no points of intersection.

35. Solving the first equation for x gives us $x = 3y - 3$. Substituting this into the second equation gives us $2(3y - 3) - 3y + 6 = 0$. Solving for y, we have $y = 0$ and $x = 3y - 3 = -3$. The two lines intersect at $(-3, 0)$.

36. Solving the first equation for x gives us $x = 2y - 4$. Substitute this into the second equation to obtain $3(2y - 4) + 6y - 12 = 0$. Solving for y gives us $y = 2$ and $x = 0$. The point of intersection is $(0, 2)$.

37. Both lines have a slope of 2. Since the two lines are parallel and do not coincide, there are no points of intersection.

38. Solving the second equation for y gives us $y = 3x$. Substitute this into the first equation to get $3x - 6x + 2 = 0$. Solving for x we have $x = \frac{2}{3}$. Then $y = 3x = 3 * \frac{2}{3} = 2$. Thus the point of intersection is $\left(\frac{2}{3}, 2 \right)$.

39. If $b \neq 0$, then the equation $ax + by + c = 0$ is the same as $y = -\frac{a}{b}x - \frac{c}{b}$, and this is the slope-intercept form of the equation of a line. If $b = 0$, the equation becomes $x = -\frac{c}{a}$, and this is the equation of a vertical line.

40. One of the constants is unnecessary. If $b = 0$, we have two constants left. If $b \neq 0$, then the equation can be written in the form $y = -\frac{a}{b}x - \frac{c}{b}$ and we need only consider the two constants $-\frac{a}{b}$ and $-\frac{c}{b}$.

41. a. Let P, Q and R respectively be the points $(1, 3)$, $(-2, 0)$ and $(4, 6)$. The slope of the line PQ is $m_1 = \frac{0-3}{-2-1} = 1$. The slope of the line QR is $m_2 = \frac{6-0}{4+2} = 1$. Since PQ and QR have the same slope and a point in common, P, Q, and R must lie on the same line.

    b. Let P, Q and R be the points $(2, -7)$, $(-2, -3)$, and $(-1, -4)$. The slope of the line PQ is $\frac{-3+7}{-2-2} = -1$. The slope of the line QR is $\frac{-4+3}{-1+2} = -1$. Since PQ and QR have the same slope and a point in common, P, Q, and R must lie on the same line.

    c. Let P, Q and R be the points $(5, 15)$, $(0, 3)$ and $(-2, -7)$. The slope of the line PQ is $\frac{3-15}{0-5} = \frac{12}{5}$. The slope of the line QR is $\frac{-7-3}{-2-0} = 5$. Since the slopes of the two line segments are different, the three points do not lie on a common line.

42. Let P, Q and R be the three points $(1, 3)$, $(3, 5)$ and $(4, 0)$. The slope of PQ is $m_1 = \frac{5-3}{3-1} = 1$. The slope of PR is $m_2 = \frac{0-3}{4-1} = -1$. Since m_1 is the negative reciprocal of m_2, PQ and PR are perpendicular. Hence P, Q, and R form the vertices of a right triangle.

43. Let P, Q and R be the points $(-2, 2)$, $(4, 4)$ and $(0, a)$.

    a. A triangle has three vertices. The right angle could occur at any of these angles. The angle at $(0, a)$ could be either above or below the line segment PQ. Thus there are two possible values of a that will produce a right angle at this point. A line perpendicular to PQ through $(4, 4)$ will determine a third value of a, and a line perpendicular to PQ through $(-2, 2)$ will determine a fourth value of a. Thus there are 4 possible values of a.

    b. The slope of the line PQ is $m_1 = \frac{4-2}{4+2} = \frac{1}{3}$. The slope of the line QR is $m_2 = \frac{a-4}{0-4}$. The slope of the line PR is $m_3 = \frac{a-2}{0+2}$.

    - The lines PQ and QR are perpendicular when $m_2 = -\frac{1}{m_1}$, that is when $\frac{4-a}{4} = -3$, or $4 - a = -12$. Thus $a = 16$.
    - The line PR is perpendicular to PQ if $m_3 = -\frac{1}{m_1}$, or $\frac{a-2}{2} = -3$. This is equivalent to $a - 2 = -6$, or $a = -4$.
    - The line PR is perpendicular to QR if $m_3 = -\frac{1}{m_1}$, or $\frac{2}{a-2} = -3$. This is equivalent to $(a-2)(a-4) = 8$, or $a^2 - 6a = 0$. Thus $a = 0$ or $a = 6$.

44. Since corresponding parts of similar triangles are proportional, $\frac{\Delta y_1}{\Delta x_1} = \frac{\Delta y_2}{\Delta x_2}$.

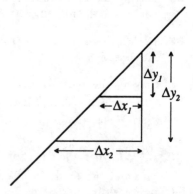

44. Slopes, similar triangles

45. Two values for (P, d) are $(20, 12)$ and $(60, 6)$. The slope of the line is $m = \frac{6-12}{60-20} = -\frac{3}{20}$. Using the point-slope form of the equation of a line, we have:

$$d - 12 = -\frac{3}{20}(P - 20)$$

or $\qquad d = -\frac{3}{20}P + 15$

When $d = 0$, we have $P = \frac{20}{3} \times 15 = 100$. Hence, the demand is 0 when the product is priced at $100.

46. a. $T = .05(i - 5000)$

b. When $T = 800$, we have $i = \frac{800}{.05} + 5000 = 21,000$. The income is $21,000$.

47. Let F be the temperature in degrees Fahrenheit, and let C be the temperature in degrees Celsius. The freezing point of water is 0 degrees Celsius and 32 degrees Fahrenheit. The boiling point of water is 100 degrees Celsius and 212 degrees Fahrenheit. Hence two values for (C, F) are $(0, 32)$ and $(100, 212)$. The F-intercept is 32 and the slope is $m = \frac{212-32}{100-0} = \frac{9}{5}$. The linear equation is:

$$F = \frac{9}{5}C + 32$$

48. $y = 2x + 1$ is the slope-intercept form of the equation for l.

a. $m = 2$

b. The slope of the perpendicular from $(4, -2)$ to the line l is $-\frac{1}{m} = -\frac{1}{2}$.

c. $y + 2 = -\frac{1}{2}(x - 4)$ or $x + 2y = 0$.

d. Sustituting $y = 2x + 1$ into $x + 2y = 0$, we have $x + 2(2x + 1) = 0$, or $x = -\frac{2}{5}$. Also $y = 2(-\frac{2}{5}) + 1 = \frac{1}{5}$. Hence, the point of intersection is $(-\frac{2}{5}, \frac{1}{5})$.

e. The distance between $(4, -2)$ and $(-\frac{2}{5}, \frac{1}{5})$ is:

$$d = \sqrt{\left(4 + \frac{2}{5}\right)^2 + \left(-2 - \frac{1}{5}\right)^2}$$

$$= \sqrt{\left(\frac{22}{5}\right)^2 + \left(\frac{11}{5}\right)^2}$$

22

$$= \sqrt{\frac{484}{25} + \frac{121}{25}} = \sqrt{\frac{605}{25}}$$
$$= \sqrt{\frac{121}{5}} = \frac{11\sqrt{5}}{5}$$

49. a. Two values for (t, T) are $(0, 22)$ and $(10, 42)$. The T-intercept is 22 and the slope is $m = \frac{42 - 22}{10 - 0} = 2$. Hence the linear equation is $T = 2t + 22$. Note here that $(20, 62)$ also satisfies this equation.

    b. When $t = 25$, $T = 2 \times 25 + 22 = 72°C$. When $t = 35$, $T = 2 \times 35 + 22 = 92°C$.

50. a. We know that when $t = 0$, $V = 40$. Hence $40 = k(0 + 273)$. Solving for k, we have $k = \frac{40}{273}$. The equation is:

$$V = \frac{40}{273}(t + 273) = \frac{40}{273}T$$

    b. When $T = 323$, $V = \frac{40}{273} \times 323 = \frac{12920}{273} = 47.326$ liters.

    c. When $t = 100$, $V = \frac{40}{273} \times (100 + 273) = \frac{14920}{273} = 54.652$ liters.

51. Since the rod is 100cm long at temperature $0°C$, we see that $l_0 = 100$. Substituting this into the equation $l = l_0(1 + at)$, gives us $l = 100(1 + at)$. Since the rod is 100.2 cm long at temperature $50°C$, we have $100(1 + 50a) = 100.2$. Thus, $a = .00004$.

52. By subtracting the second equation from the first, we have $2x - 2y = 0$, or $y = x$. Substitute $y = x$ in the first equation to get $2x^2 - 2x = 0$ or $x(x - 1) = 0$. The two points of intersection are $(0, 0)$ and $(1, 1)$. The equation for the line through these points is $y = x$.

53. We complete the squares for both equations to put them in the standard form for the equation of a circle:

$$(x - 1)^2 + (y - 2)^2 = 4$$
$$x^2 + (y + 1)^2 = 1$$

The two centers are $(1, 2)$ and $(0, -1)$. The y-intercept is -1, and the slope is $m = \frac{-1 - 2}{0 - 1} = 3$. Hence the equation of the line through the two centers is $y = 3x - 1$

54. If $m_1 \neq m_2$, then the two lines l_1 and l_2 are not parallel and must intersect. If the two lines intersect, then they are not parallel and hence $m_1 \neq m_2$.

55. a. $m = \frac{1.5 - 1.4}{2.25 - 1.96} = \frac{10}{29}$

    b. The equation for l is $y - 1.5 = .345(x - 2.25)$ or $y = .345x + .724$. When $x = 2$, $y = .345 + .724 = 1.414$. Hence the point is $(2, 1.414)$.

23

c. An estimate for $\sqrt{2}$ is 1.41.

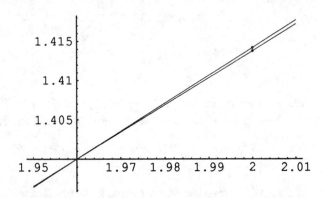

d. The secant line is *very* close to the curve

56. The data for $i(n)$, $r(n)$, and $b(n)$ certainly apear to fall on straight lines. Using the Mathematica least-squares fit function "Fit", we find that $i(n) = 37.9607 - 2.25396n$, $r(n) = 271.35 + 16.3134n$, and $b(n) = 4843.28 - 399.922n$.

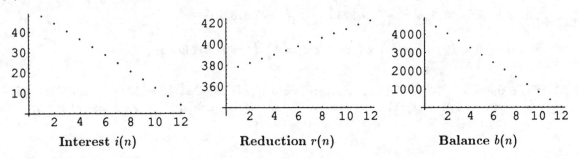

| Interest $i(n)$ | Reduction $r(n)$ | Balance $b(n)$ |

1.4 Functions

1.

$f(x)$	$f(-2)$	$f(0)$	$f(4)$	$f(5)$	$|f(3)|$	$f(f(0))$
$1 - 3x^2$	-11	1	-47	-74	26	-2
$\dfrac{1}{x+2}$	undefined	$\dfrac{1}{2}$	$\dfrac{1}{6}$	$\dfrac{1}{7}$	$\dfrac{1}{5}$	$\dfrac{2}{5}$
$\dfrac{(x-3)^2}{x^2+1}$	5	9	$\dfrac{1}{17}$	$\dfrac{4}{26}$	0	$\dfrac{36}{82}$
$\sqrt{x+4}$	$\sqrt{2}$	2	$2\sqrt{2}$	3	$\sqrt{7}$	$\sqrt{6}$
$\dfrac{1}{\sqrt{16-x^2}}$	$\dfrac{1}{2\sqrt{3}}$	$\dfrac{1}{4}$	undefined	undefined	$\dfrac{1}{\sqrt{7}}$	$\dfrac{4}{\sqrt{255}}$
$\begin{cases} 1-x, & x < -3 \\ x-1, & x > 1 \end{cases}$	undefined	undefined	3	4	2	undefined

2. $y = 2x - 7$ is a function of x.

3. $y = \frac{1}{x}$ is a function of x, $x \neq 0$.

4. y is not a function of x. For $x = 2$, y may be either 1 or -1.

5. y is not a function of x. For $x = 0$, y may either be $\sqrt{8}$ or $-\sqrt{8}$

6. The equation is equivalent to $y = 0$. Hence y is a function of x.

7. y is a function of x.

8. False. The value -2 is in the domain of g, but it is not in the domain of f.

9. The domain is the set of all real numbers $(-\infty, \infty)$.

10. $(3, \infty)$

11. We must have $x + 4 \geq 0$, or $x \geq -4$. Hence the domain is $[-4, \infty)$.

12. We must have $x(x - 1) \geq 0$. This occurs when $x \geq 1$ or $x \leq 0$. Hence the domain is $(-\infty, 0] \cup [1, \infty)$.

13. $[0, \infty)$

14. We must have $1 - |x| \neq 0$, that is x can neither be -1 nor 1. Hence, the domain is

$$(-\infty, -1) \cup (-1, 1) \cup (1, \infty).$$

15. We must have $16 - s^2 \geq 0$. Hence the domain is $[-4, 4]$.

16. $(t + 7)(t - 5) = t^2 + 2t - 35 \neq 0$. This condition is satisfied when $t \neq -7$ and $t \neq 5$. The domain is

$$(-\infty, -7) \cup (-7, 5) \cup (5, \infty)$$

17. The domain is given to be $(-\infty, 0) \cup (0, \infty)$.

18. $1 - s^2 \geq 0$ or $-1 \leq s \leq 1$. Hence, the domain is $[-1, 1]$.

19. Both conditions $x - 2 \neq 0$ and $x(x + 2) = x^2 + 2x \geq 0$ must be satisfied. The second condition is satisfied when both x and $x + 2$ have the same sign, that is, when $x \leq -2$ and when $x \geq 0$. The domain is

$$(-\infty, -2] \cup [0, 2) \cup (2, \infty).$$

20. $6 - |x + 2| \geq 0 \Leftrightarrow -6 \leq x + 2 \leq 6 \Leftrightarrow -8 \leq x \leq 4$. The domain is $[-8, 4]$.

21. $x - 2 \geq 0$. The domain is $[2, \infty)$.

22. $(-\infty, \infty)$

23. Since $s^2 + 1 \geq 0$ for all real numbers s, the domain is $(-\infty, \infty)$.

24. $(t - 2)(t - 1) = t^2 - 3t + 2 \geq 0$ is satisfied when both factors, $(t - 2)$ and $(t - 1)$, have the same sign. This occurs when $t \leq 1$ or $t \geq 2$. The domain is $(-\infty, 1] \cup [2, \infty)$.

25. $\{x : x \neq -1\}$. Since the exponent is negative, we $g(x)$ is not defined when $x \neq -1$.

26. The condition $(t - 1)(t + 1) = t^2 - 1 \geq 0$ must be satisfied. This happens when both factors have the same sign, that is, when $t \leq -1$ or $t \geq 1$. In addition, since the exponent is negative, $h(t)$ is not defined for $t = \pm 1$. The domain is $(-\infty, -1] \cup [1, \infty)$.

27. False. If $h(x) = x^2 + 1$, then $f(x)$ is defined for all real numbers x.

28. False. If a *vertical* line $x = a$ intersects the graph of the equation $y = f(x)$, then y has more than one value when $x = a$. Hence $y = f(x)$ cannot be determined as a function of x. Note that $y = c$ is a function that has the value c for all values of x, and the horizontal line $y = c$ intersects the graph of the function in more than one point.

29. We solve the equation, $3x^2 - 2y - 8 = 0$, for y: $y = \frac{3}{2}x^2 - 4$.

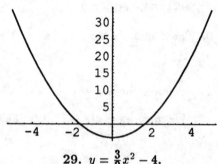

29. $y = \frac{3}{2}x^2 - 4$.

30. We solve the equation, $3x^2 - y - 7 = 0$ for y: $y = 3x^2 - 7$.

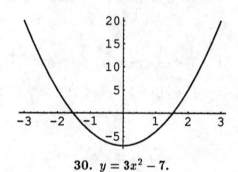

30. $y = 3x^2 - 7$.

31. We first solve the equation $y - 3x^2 + x - \frac{1}{2} = 0$ for y:

$$y = 3x^2 - x + \frac{1}{2}$$
$$= 3\left(x^2 - \frac{1}{3}x + \frac{1}{36}\right) - \frac{1}{12} + \frac{1}{2}$$
$$= 3\left(x - \frac{1}{6}\right)^2 + \frac{5}{12}$$

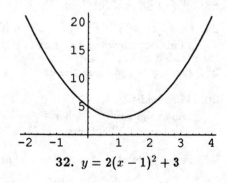

31. $y = 3\left(x - \frac{1}{6}\right)^2 + \frac{5}{12}$

32. We first solve the equation, $y - 2x^2 + 4x - 5 = 0$ for y:

$$y = 2x^2 - 4x + 5$$
$$= 2(x^2 - 2x + 1) - 2 + 5$$
$$= 2(x - 1)^2 + 3$$

32. $y = 2(x - 1)^2 + 3$.

33. Solve the equation, $4x^2 + y - 24x + 34 = 0$, for y:

$$y = -4x^2 + 24x - 34$$
$$= -4(x^2 - 6x + 9) + 36 - 34$$
$$= -4(x - 3)^2 + 2$$

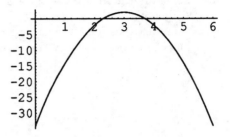

33. $y = -4(x - 3)^2 + 2$

34. Solve the equation, $y - \frac{1}{2}x^2 - \frac{1}{2}x - \frac{11}{24} = 0$ for y:

$$y = \frac{1}{2}x^2 + \frac{1}{2}x + \frac{11}{24}$$
$$= \frac{1}{2}\left(x^2 + x + \frac{1}{4}\right) - \frac{1}{8} + \frac{11}{24}$$
$$= \frac{1}{2}\left(x + \frac{1}{2}\right)^2 + \frac{1}{3}$$

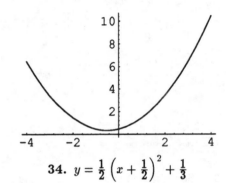

34. $y = \frac{1}{2}\left(x + \frac{1}{2}\right)^2 + \frac{1}{3}$

35. Solve the equation, $y^2 + x - 2 = 0$, for x: $x = -y^2 + 2$.

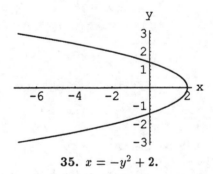

35. $x = -y^2 + 2$.

36. Solve the equation, $3y^2 - x - 4 = 0$ for x: $x = 3y^2 - 4$.

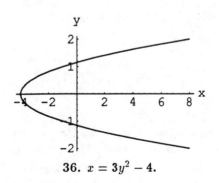

36. $x = 3y^2 - 4$.

37. Solve the equation, $3x + 2y^2 - 3 = 0$ for x: $x = -\frac{2}{3}y^2 + 1.$

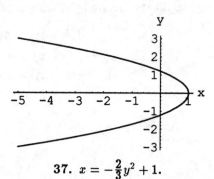

37. $x = -\frac{2}{3}y^2 + 1.$

38. Solve the equation, $x - 3y^2 + 6y - 4 = 0$, for x:

$$x = 3y^2 - 6y + 4 = 3(y^2 - 2y + 1) - 3 + 4$$
$$= 3(y - 1)^2 + 1$$

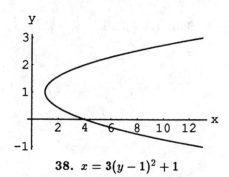

38. $x = 3(y - 1)^2 + 1$

39. Solve the equation, $x - 2y^2 + 4y - 2 = 0$: for x:

$$x = 2y^2 - 4y + 2$$
$$= 2(y^2 - 2y + 1) - 2 + 2$$
$$= 2(y - 1)^2$$

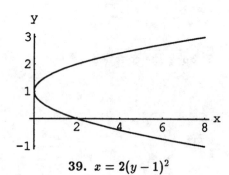

39. $x = 2(y - 1)^2$

40. Solve the equation $4y^2 - x - 24y + 34 = 0$ for x:

$$x = 4y^2 - 24y + 34 = 4(y^2 - 6y + 9) - 36 + 34$$
$$= 4(y - 3)^2 - 2$$

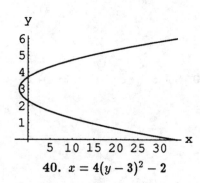

40. $x = 4(y - 3)^2 - 2$

41. Solve $x - y - 2 = 0$ for y to obtain $y = x - 2$. Now substitute for y in the equation $y = 4 - x^2$ and solve for x:

$$x - 2 = 4 - x^2 \Leftrightarrow x^2 + x - 6 = 0 \Leftrightarrow (x + 3)(x - 2) = 0 \Leftrightarrow x = -3, 2$$

When $x = -3$, $y = -3 - 2 = -5$. When $x = 2$, $y = 2 - 2 = 0$. Hence, the points of intersection are $(-3, -5)$ and $(2, 0)$.

42. We see that $6 - x^2 = y = x^2 - 2$. We solve this equation for x.

$$6 - x^2 = x^2 - 2 \Leftrightarrow 2x^2 - 8 = 0 \leftrightarrow 2(x - 2)(x + 2) = 0 \Leftrightarrow x = 2, -2$$

When $x = 2$ or $x = -2$, $y = x^2 - 2 = 4 - 2 = 2$. Hence the points of intersection are $(-2, 2)$ and $(2, 2)$.

43.

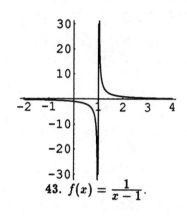

43. $f(x) = \dfrac{1}{x - 1}$.

44.

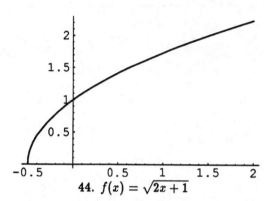

44. $f(x) = \sqrt{2x + 1}$

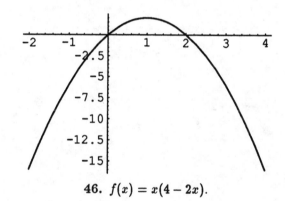

45. $f(x) = \begin{cases} 7 - x, & x \le 2 \\ 2x + 1, & x > 2 \end{cases}$

46. $f(x) = x(4 - 2x)$.

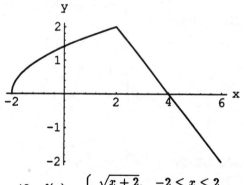

47. $f(x) = \dfrac{x^2 - 1}{x - 1} = $ (Surprise!) $x + 1$.

48. $f(x) = \begin{cases} \sqrt{x + 2}, & -2 \le x \le 2 \\ 4 - x, & x > 2 \end{cases}$

29

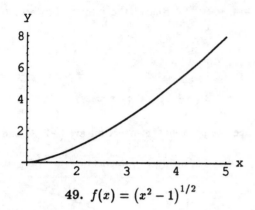

49. $f(x) = (x^2 - 1)^{1/2}$

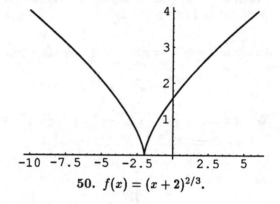

50. $f(x) = (x+2)^{2/3}$.

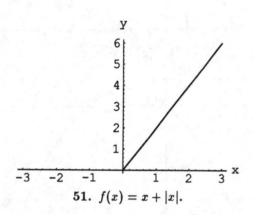

51. $f(x) = x + |x|$.

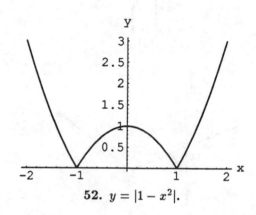

52. $y = |1 - x^2|$.

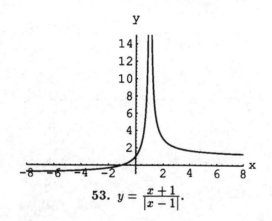

53. $y = \dfrac{x+1}{|x-1|}$.

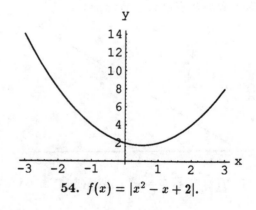

54. $f(x) = |x^2 - x + 2|$.

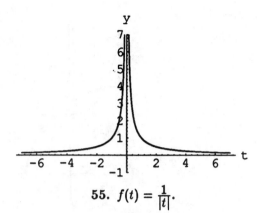

55. $f(t) = \frac{1}{|t|}$.

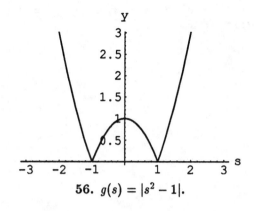

56. $g(s) = |s^2 - 1|$.

57. a. Since $f(x) = x^2$, $f(-x) = (-x)^2 = x^2 = f(x)$ for all real numbers x. Hence, $f(x)$ is even.

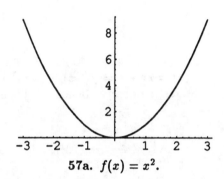

57a. $f(x) = x^2$.

b. Since $f(x) = 2 - x^2$, $f(-x) = 2 - (-x)^2 = 2 - x^2 = f(x)$ for all real numbers x. Hence, $f(x)$ is an even function.

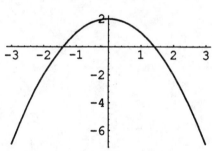

57b. $f(x) = 2 - x^2$.

c. $f(x) = x^3 \Rightarrow f(-x) = (-x)^3 = -x^3 = -f(x)$ for all real numbers x. Hence $f(x)$ is an odd function.

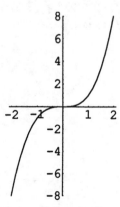

57c. $f(x) = x^3$.

d. $f(x) = 1 - x^3 \Rightarrow f(-x) = 1 - (-x)^3 = 1 + x^3$. Hence, $f(x)$ is neither odd nor even.

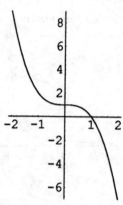

57d. $f(x) = 1 - x^3$.

e. $f(x) = x^3 + x \Rightarrow f(-x) = (-x)^3 + (-x) = -(x^3 + x) = -f(x)$ for all real number x. Hence, $f(x)$ is an odd function.

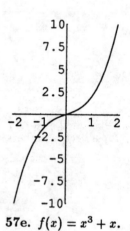

57e. $f(x) = x^3 + x$.

f. $f(x) = 2x^4 + x^2 \Rightarrow f(-x) = 2(-x)^4 + (-x)^2 = 2x^4 + x^2$. Hence, $f(x)$ is even.

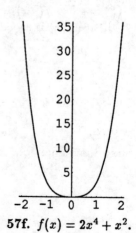

57f. $f(x) = 2x^4 + x^2$.

g. $f(x) = |x| + 2 \Rightarrow f(-x) = |-x| + 2 = |x| + 2 = f(x)$
for all real numbers x. Hence $f(x)$ is an even function.

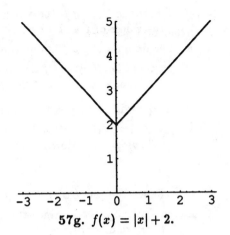

57g. $f(x) = |x| + 2$.

h. $f(x) = x^2 + x \Rightarrow f(-x) = (-x)^2 + (-x) = x^2 - x$. Hence,
$f(x)$ is neither odd nor even.

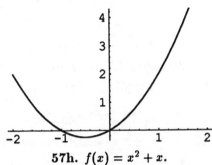

57h. $f(x) = x^2 + x$.

58. The graph of an even function is symmetric about the y-axis. That is, for every point (x, y) on the graph, the point $(-x, y)$ is also on the graph. The graph of an odd function is symmetric about the origin. This means that for every point (x, y) on the graph, the point $(-x, -y)$ is also on the graph.

In Exercises 59-66, we will use the functions:

$$f(x) = 3x + 1 \qquad g(x) = x^3 \qquad h(x) = \sqrt{x}$$

59. $f \circ g(x) = f(x^3) = 3x^3 + 1$

60. $f \circ h(x) = f(\sqrt{x}) = 3\sqrt{x} + 1$

61. $g \circ f(x) = g(3x + 1) = (3x + 1)^3$

62. $h \circ g(x) = h(x^3) = \sqrt{x^3} = x^{\frac{3}{2}}$

63. $h(f(x)) = h(3x + 1) = \sqrt{3x + 1}$

64. $h(f(g(x))) = h(f(x^3)) = h(3x^3 + 1) = \sqrt{3x^3 + 1}$

65. $f(g(h(x))) = f(g(\sqrt{x})) = f((\sqrt{x})^3) = f(x^{\frac{3}{2}}) = 3x^{\frac{3}{2}} + 1$

66. $g(h(f(x))) = g(h(3x + 1)) = g(\sqrt{3x + 1}) = (\sqrt{3x + 1})^3 = (3x + 1)^{\frac{3}{2}}$

67. **a.** Since the domain of $f(x) = x^2$ is $\{x | -1 \le x \le 3\}$ the domain of $f \circ g$ is restricted by

$$-1 \le g(x) \le 3 \Leftrightarrow -1 \le (2x + 6) \le 3 \Leftrightarrow -\frac{7}{2} \le x \le -\frac{3}{2}$$

Hence the domain of $f \circ g$ is $\{x | -\frac{7}{2} \le x \le -\frac{3}{2}\}$

b. The range of $f \circ g$ is the same as the range of f. Since $f(x) = x^2 \geq 0$ and $-1 \leq x \leq 0$ or $0 \leq x \leq 3$ the range of f and $f \circ g$ is $[0, 9]$.

c. The domain of $g \circ f$ is the same as the domain for f which is $\{x| -1 \leq x \leq 3\}$.

d. $g \circ f(x) = g(f(x)) = 2f(x) + 6$ From part b. we know that $0 \leq f(x) \leq 9$. Hence $6 \leq 2f(x) + 6 \leq 24$. The range of $g \circ f$ is $[6, 24]$.

68. a. The domain of $f \circ g$ is the same as the domain of g which is $\{x| -5 \leq x \leq 4\}$.

b. $f \circ g(x) = f(2x + 6) = (2x + 6)^2$. From part a we know the domain of $f \circ g$ which gives us the following restrictions on x:

$$-5 \leq x \leq 4 \Leftrightarrow -4 \leq 2x + 6 \leq 14 \Rightarrow 0 \leq (2x + 6)^2 \leq 196$$

Hence the range of $f \circ g$ is $[0, 196]$.

c. $g \circ f(x) = g(f(x))$. Since the domain of g is restricted by $-5 \leq x \leq 4$, we must have the folloing restrictions on $f(x)$:

$$-5 \leq f(x) \leq 4 \Leftrightarrow -5 \leq x^2 \leq 4 \Leftrightarrow 0 \leq x^2 \leq 4 \Leftrightarrow -2 \leq x \leq 2$$

Hence the domain of $g \circ f$ is $[-2, 2]$.

d. $g \circ f(x) = g(x^2) = 2x^2 + 6$. Using the domain we derived in part a, we see that:

$$-2 \leq x \leq 2 \Leftrightarrow 0 \leq x^2 \leq 4 \Leftrightarrow 6 \leq 2x^2 + 6 \leq 14$$

Hence the range of $g \circ f$ is $[6, 14]$.

69. $f(x) = \dfrac{1}{3 - x^2}$ and $g(x) = \sqrt{x^2 - 1}$. To find the domain of $f \circ g(x)$, we must consider the restrictions

$$x^2 - 1 \geq 0 \qquad\qquad 3 - (g(x))^2 \neq 0$$

From the first inequality we see that:

$$x^2 - 1 \geq 0 \Leftrightarrow x^2 \geq 1 \Leftrightarrow |x| \geq 1.$$

The second inequality gives us:

$$3 - (g(x))^2 \neq 0 \Leftrightarrow (g(x))^2 \neq 3 \Leftrightarrow x^2 - 1 \neq 3 \Leftrightarrow x^2 \neq 4 \Leftrightarrow x \neq \pm 2$$

Hence the domain of $f \circ g$ is

$$(-\infty, -2) \cup (-2, -1] \cup [1, 2) \cup (2, \infty)$$

To find the domain of $g \circ f(x)$, we must consider the restrictions:

$$3 - x^2 \neq 0 \qquad\qquad (f(x))^2 - 1 \geq 0$$

From the first inequality we see that:

$$3 - x^2 \neq 0 \Leftrightarrow x^2 \neq 3 \Leftrightarrow |x| \neq \sqrt{3}$$

From the second inequality, we see that:

$$(f(x))^2 \geq 1 \Leftrightarrow \left(\frac{1}{3 - x^2}\right)^2 \geq 1 \Leftrightarrow (3 - x^2)^2 \leq 1 \Leftrightarrow -1 \leq 3 - x^2 \leq 1$$

$$\Leftrightarrow -4 \leq -x^2 \leq -2 \Leftrightarrow 2 \leq x^2 \leq 4 \Leftrightarrow \sqrt{2} \leq |x| \leq 2$$

Hence the domain of $g \circ f$ is

$$[-2, -\sqrt{3}) \cup (-\sqrt{3}, -\sqrt{2}] \cup [\sqrt{2}, \sqrt{3}) \cup (\sqrt{3}, 2]$$

70. $f(x) = \frac{x}{x+1}$ and $g(x) = \frac{1}{x-1}$. To find the domain of $f \circ g(x)$, we must consider the restrictions

$$x - 1 \neq 0 \qquad\qquad g(x) + 1 \neq 0$$

From the first inequality we see that $x \neq 1$. We solve the second inequality:

$$g(x) + 1 \neq 0 \Leftrightarrow \frac{1}{x-1} + 1 \neq 0 \Leftrightarrow \frac{x}{x-1} \neq 0 \Leftrightarrow x \neq 0$$

Hence the domain of $f \circ g$ is

$$(-\infty, o) \cup (0, 1) \cup (1, \infty)$$

To find the domain of $g \circ f$, we must consider the restrictions:

$$x + 1 \neq 0 \qquad\qquad f(x) - 1 \neq 0$$

From the first inequality, we see that $x \neq -1$. We solve the second inequality:

$$f(x) - 1 \neq 0 \Leftrightarrow \frac{x}{x+1} - 1 \neq 0 \Leftrightarrow \frac{-1}{x+1} \neq 0$$

Since the last inequality holds for all real values of x, the domain for $g \circ f$ is:

$$(-\infty, -1) \cup (-1, \infty)$$

71. $f(x) = x(x^2+1)^{-\frac{1}{2}} - \frac{2\sqrt{x^2+1}}{x^3} + \frac{3}{x^3\sqrt{x^2+1}}$ Note that x^2+1 is always positive, so it will never cause any problems. The only restriction is that $x \neq 0$. Hence, the domain is $(-\infty, 0) \cup (0, +infty)$. Since $x \neq 0$, the range of f contains the number 0 if and only if the range of $x^3\sqrt{x^2+1}f$ contains 0, that is if and only if:

$$x^4 - 2(x^2+1) + 3 = 0 \Leftrightarrow x^4 - 2x^2 + 1 = 0 \Leftrightarrow (x^2-1)^2 = 0 \Leftrightarrow x = \pm 1$$

$$f(1) = f(-1) = \frac{1}{\sqrt{2}} - 2\sqrt{2} + \frac{3}{\sqrt{2}} = \frac{\sqrt{2}}{2} - 2\sqrt{2} + \frac{3\sqrt{2}}{2} = 0$$

Hence, the range of f does contain 0.

72. $f(x) = \frac{(x+4)^{\frac{2}{3}}}{\sqrt[5]{x+2}} + \frac{(x+2)^{\frac{4}{5}}}{\sqrt[3]{(x+4)}}$ Since any odd root function is defined on all real numbers, there are no restrictions from the numerators. We must be certain that the denominators are never 0. Hence $x \neq -2$ and $x \neq -4$, and the domain of f is

$$(-\infty, -4) \cup (-4, -2) \cup (-2, \infty)$$

We will now try to solve the equation $f(x) = 0$. This is equivalent to solving

$$\sqrt[5]{x+2}\sqrt[3]{x+4}f(x) = 0 \Leftrightarrow (x+2)^{\frac{1}{3}}(x+4)^{\frac{4}{5}}f(x) = 0 \Leftrightarrow (x+4) + (x+2) = 0 \Leftrightarrow x = -3$$

Since $f(-3) = 0$, 0 is in the range of f.

73. True. Every polynomial is the quotient of the polynomial and the constant polynomial $g(x) = 1$.

74. Since the area is $36cm^2$, $lw = 36$. Hence,

$$w(l) = \frac{36}{l}.$$

Since a width is always non-negative, and the denominator cannot be zero, the domain is $(-\infty, 0) \cup (0, \infty)$

75. Since the truck depreciates 30% per year, a truck costing $20,000 has a value $V(t) = \$20,000(.7)^t$ after t years. The cost of the insurance policy is 8% of V. Hence,

$$C(t) = .08(\$20,000)(.7)^t$$

76. Let V, r_0, h be the volume, radius and height. Then

$$V(h) = \pi r_0^2 h$$

77. a. $p - 30$ is the increase in the daily rental fee over $30. This causes $20(p - 30)$ fewer jackhammers to be rented per year. The number of jackhammers rented per year is therefore $500 - 20(p - 30)$. Hence the yearly revenue is $R = p[500 - 20(p - 30)]$.

b. Since $p \geq 30$, $R = 0$ when:

$$500 - 20(p - 30) = 0 \Leftrightarrow p - 30 = 25 \Leftrightarrow p = 55.$$

The revenues will be 0 when the price reaches $55.00.

78. Note that if $f(g(x_1)) = f(g(x_2))$, then either $x_1 = x_2$, or $g(x_1) = g(x_2)$, or $g(x_1) = y_1$, $g(x_2) = y_2$, with $y_1 \neq y_2$ and $f(y_1) = f(y_2)$. Thus for $f(g(x))$ to be constant, it must be that for any pair of number x_1 and x_2 in the domain, that one of these three conditions must hold. The simplest way to construct such a function with f and g non-constant is to make f constant on the range of g. For example, if

$$f(x) = \begin{cases} 0 & x \leq 0 \\ x & x > 0 \end{cases}, \qquad g(x) = \begin{cases} x & x \leq 0 \\ 0 & x > 0 \end{cases}$$

Then $f(g(x)) = 0$, even though neither f nor g is constant.

79. With $c(n) = \sqrt{1 + 0.2n}$, and $n(t) = 10,000 + 50t^{\frac{2}{3}}$, we compute

$$c \circ n(t) \quad = \quad \sqrt{1 + 0.2n(t)} \quad = \quad \sqrt{1 + 0.2(10,000 + 50t^{\frac{2}{3}})} \quad = \quad \sqrt{2001 + 10t^{\frac{2}{3}}}.$$

80. With $r(c) = 10 + 0.2c + 0.01c^2$ and $f(r) = \sqrt{r + 5}$, we compute:

$$f(r(c)) \quad = \quad \sqrt{10 + 0.2c + 0.01c^2 + 5} \quad = \quad \sqrt{15 + 0.2c + 0.01c^2}$$

81. Let V be the volume and l be the length of one of the edges of a cube. Then

$$V = f(l) = l^3$$

82. $l = \sqrt[3]{V}$

83. $\gamma = a(Z - b)^2 \qquad Z \geq 0$

This is the right half of a parabola with a vertex at $(b, 0)$ which opens upward.

84. Let R be the air resistance and let S be the speed. Then $R = aS^2$ where a is a constant and $S \geq 0$. This is the right half of a parabola with a vertex at $(0, 0)$ which opens upward.

85.

$$f(x) = Ax^2 + Bx + C$$
$$= A[x^2 + \frac{B}{A}x + (\frac{B}{2A})^2] - A(\frac{B}{2A})^2 + C$$
$$= A[x - (-\frac{B}{2A})]^2 + C - \frac{B^2}{4A}$$
$$= A(x - D)^2 + E,$$

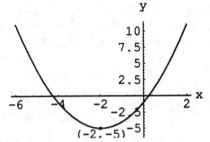

where $D = -\frac{B}{2A}$ and $E = C - \frac{B^2}{4A}$. The figure at right show the case where $A = 1$, $B = 4$, and $C = -1$. In this case $D = -2$ and $E = -5$. Thus the vertex of the parabola is at $(-2, -5)$.

85. $f(x) = x^2 + 4x - 1 = (x + 2)^2 - 5$

86. Note that in exercise 85, (D, E) is the vertex of a parabola. If $a > 0$, then f is a parabola opening upward from the vertex whose y-coordinate is $E = c - \frac{b^2}{4a}$. Hence, the range is $(c - \frac{b^2}{4a}, \infty)$. If $a < 0$, the parabola opens downward and the range ia $(-\infty, c - \frac{b^2}{4a})$.

87. a. $g(x) = \sqrt{x + 4}$. The range is $[0, \infty)$.

  b. $f(t) = 1 + t^2$, $t \geq 0$. The range is $[1, \infty)$.

  c. $f(x) = \frac{1}{|x|}$ for $x \neq 0$. The range is $(0, \infty)$.

  d. $f(x) = \sqrt{6 - |x + 2|}$. Since the quantity within the square root sign must always be non-negative, we have $0 \leq 6 - |x + 2| \leq 6$. Hence, the range of f is $[0, \sqrt{6}]$.

  e. $h(s) = (s^2 + 1)^{\frac{3}{2}}$. $s^2 + 1 \geq 1 \Leftrightarrow h(s) \geq 1$. Hence the range is $[1, \infty)$.

  f. $g(x) = (x + 1)^{-\frac{2}{3}}$. Since the exponent is negative, we must have $x + 1 \neq 0$. Since $(x + 1)^2 > 0$, we also have $g(x) = (x + 1)^{-\frac{2}{3}} > 0$. Hence the range is $(0, \infty)$.

  g. $f(x) = \sqrt{2x + 1}$. $2x + 1 \geq 0 \Leftrightarrow f(x) \geq 0$. Hence the range is $[0, \infty)$.

  h. $f(x) = (x + 2)^{\frac{2}{3}} = [(x + 2)^{\frac{1}{3}}]^2 \geq 0$. Hence, the range is $[0, \infty)$.

88. $f(x) = ax^2 + bx + c = a[x^2 + \frac{b}{a}x + (\frac{b}{2a})^2] - a(\frac{b}{2a})^2 + c = a(x + \frac{b}{2a})^2 + \frac{4ac - b^2}{4a}$. If $f(x) = 0$, then

$$\left(x + \frac{b}{2a}\right)^2 = -\frac{4ac - b^2}{4a^2} = \frac{b^2 - 4ac}{4a^2}$$

Taking the square root of both sides gives us

$$x + \frac{b}{2a} = \pm \frac{\sqrt{b^2 - 4ac}}{2a}$$

Hence, $x = \frac{-b \pm \sqrt{b^2 - 4ac}}{2a}$.

89. $y = a|b(x+c)| + d$. Since the graph opens downward, a must be negative; thus y attains its maximum value when $(x+c) = 0$. Since the highest point on the graph occurs at $(-2, 2)$, we must have $c = 2$ and $d = 2$. Thus $y = a|b||x + 2| + 2$. Since $(2, 0)$ is on the graph, we have $0 = 4a|b| + 2$. Hence $a|b| = -\frac{1}{2}$, and

$$y = -\frac{1}{2}|x + 2| + 2 = -\frac{1}{2}f(x + 2) + 2$$

90. $y = a(b(x+c))^2 + d$. The graph shows a parabola opening downwards with vertex at $(-2, 3)$. Hence, $c = 2$ and $d = 3$, and so $y = ab^2(x+2)^2 + 3$. The graph appears to go through the point $(-1, 0)$. Hence $0 = ab^2(-1 + 2)^2 + 3 \Leftrightarrow ab^2 = -2$. Thus

$$y = -2(x + 2)^2 + 3 = -2f(x_2) + 3.$$

91. $y = \frac{a}{b(x+c)} + d$. First note that 1 is not in the domain of the graph, so that $c = -1$. The range does not contain 0, so that $d = 0$. The funtion is $y = \frac{a}{b(x-1)}$. We see from the graph that $(0, -3)$ is on the graph of the function. Therefore, $-3 = \frac{a}{b(0-1)} \Leftrightarrow \frac{a}{b} = 3$. Hence the function is

$$y = \frac{3}{x - 1}.$$

92. $y = a\sin(b(x+c)) + d$. It appears from the graph that the largest value of y is 2.5 and the smallest value of y is -1.5. Since .5 is half way between -1.5 and 2.5, $d = .5$. The period of the graph is π, so that $b = 2$. Since the amplitude of the graph is 4, we must have $a = 2$. Substituting $a = 2$, $b = 2$, and $d = .5$, the function becomes $y = 2\sin(2(x+c)) + .5$. It is difficult to determine c from the graph. However we can estimate c as follows. Since the graph passes through $y = .5$ between $x = \frac{\pi}{8}$ and $x = \frac{\pi}{4}$, we must have that $\frac{\pi}{8} < c < \frac{\pi}{4}$. We might guess that $c = \frac{\pi}{6}$, but that would be a rather presumptious guess.

93. a. $f(x) = 2x + 1 \qquad g(x) = \sqrt{x}$
$(f + g)(x) = 2x + \sqrt{x} + 1$. The domain is $[0, \infty)$. Since the function is increasing on its domain, the minimum value occurs at $x = 0$. Hence the range is $[1, \infty)$.

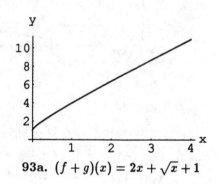

93a. $(f + g)(x) = 2x + \sqrt{x} + 1$

b. $(f - g)(x) = 2x - \sqrt{x} + 1 = 2\left(\sqrt{x} - \frac{1}{4}\right) + \frac{7}{8}$. The domain is $[0, \infty)$. This function attains a minimum value in its domain, at $x = \frac{1}{4}$. The range is $[\frac{7}{8}, \infty)$.

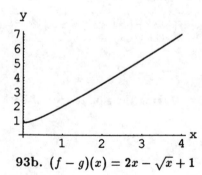

93b. $(f - g)(x) = 2x - \sqrt{x} + 1$

c. $(fg)(x) = (2x+1)\sqrt{x}$. The domain is $[0, \infty)$. The function takes on the value 0 at $x = 0$. At all other values of x the function is positive. It is continuous and increases without bound as $x \to \infty$. Hence the range is $[0, \infty)$.

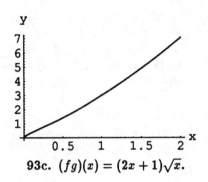

93c. $(fg)(x) = (2x+1)\sqrt{x}$.

d. $(\frac{f}{g})(x) = 2\sqrt{x} + \frac{1}{\sqrt{x}}$. The domain is $(0, \infty)$ and the range is $[2\sqrt{2}, \infty)$.

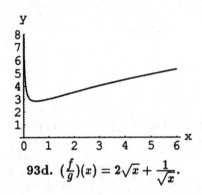

93d. $(\frac{f}{g})(x) = 2\sqrt{x} + \frac{1}{\sqrt{x}}$.

e. $(\frac{g}{f})(x) = \frac{\sqrt{x}}{2x+1}$. The domain is $[0, \infty)$ and the range is $[\frac{\sqrt{2}}{4}, \infty)$

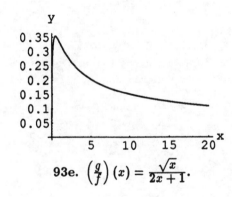

93e. $\left(\frac{g}{f}\right)(x) = \frac{\sqrt{x}}{2x+1}$.

f. $(f \circ g)(x) = f(\sqrt{x}) = 2\sqrt{x} + 1$. The domain is $[0, \infty)$ and the range is $[0, \infty)$.

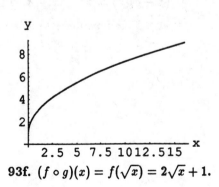

93f. $(f \circ g)(x) = f(\sqrt{x}) = 2\sqrt{x} + 1$.

g. $(g \circ f)(x) = g(2x+1) = \sqrt{x+1}$. The domain is $[-1,\infty)$ and the range is $[0,\infty)$.

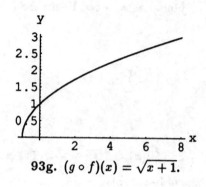

93g. $(g \circ f)(x) = \sqrt{x+1}$.

94. $f(x) = x^2 - 4$ $\qquad$ $g(x) = |x|$

a. $(f+g)(x) = x^2 + |x| - 4$. The domain is $(-\infty, \infty)$ and the range is $[-4, \infty)$.

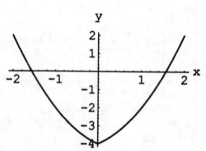

94a. $(f+g)(x) = x^2 + |x| - 4$.

b. $(f-g)(x) = x^2 - |x| - 4$. The domain is $(-\infty, \infty)$. The range is $[-\frac{17}{4}, \infty)$.

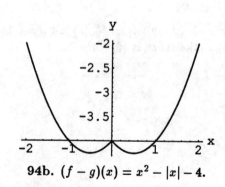

94b. $(f-g)(x) = x^2 - |x| - 4$.

40

c. $(fg)(x) = |x|(x^2 - 4)$. The domain is $(-\infty, \infty)$. The range is $[-\frac{16\sqrt{3}}{9}, \infty)$.

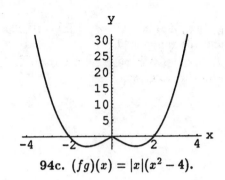

94c. $(fg)(x) = |x|(x^2 - 4)$.

d. $\left(\frac{f}{g}\right)(x) = \frac{x^2 - 4}{|x|}$. The domain is $(-\infty, 0) \cup (0, \infty)$. The range is $(-\infty, \infty)$

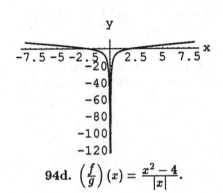

94d. $\left(\frac{f}{g}\right)(x) = \frac{x^2 - 4}{|x|}$.

e. $\left(\frac{g}{f}\right)(x) = \frac{|x|}{x^2 - 4}$. The domain is $(-\infty, -2) \cup (-2, 2) \cup (2, \infty)$ and the range is $(-\infty, \infty)$.

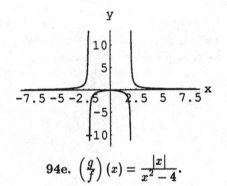

94e. $\left(\frac{g}{f}\right)(x) = \frac{|x|}{x^2 - 4}$.

f. $(f \circ g)(x) = f(|x|) = x^2 - 4$. The domain is $(-\infty, \infty)$ and the range is $[-4, \infty)$.

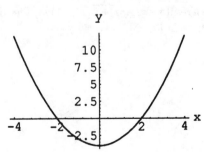

94f. $(f \circ g)(x) = f(|x|) = x^2 - 4$.

41

g. $(g \circ f)(x) = g(x^2 - 4) = |x^2 - 4|$. The domain is $(-\infty, \infty)$ and the range is $[0.\infty)$.

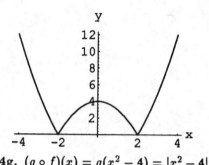

94g. $(g \circ f)(x) = g(x^2 - 4) = |x^2 - 4|$.

95. $f(x) = \sqrt{x+1}$, $g(x) = x^2 - 2x + 1$.

a. $(f+g)(x) = x^2 - 2x + \sqrt{x+1} + 1$. The domain is $[-1, \infty)$. The range includes at least the set $[\sqrt{2}, \infty)$. However the minimum value of this function is difficult to determine.

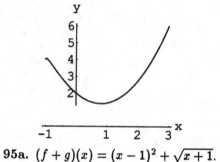

95a. $(f+g)(x) = (x-1)^2 + \sqrt{x+1}$.

b. $(f-g)(x) = \sqrt{x+1} - x^2 + 2x - 1$. The domain is $[-1, \infty)$. The range includes at least the set $(-\infty, \sqrt{2}]$. However the maximum value of this function is difficult to determine.

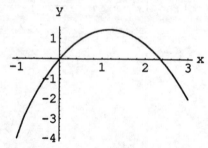

95b. $(f-g)(x) = \sqrt{x+1} - (x-1)^2$.

c. $(fg)(x) = \sqrt{x+1}(x-1)^2$. The domain is $[-1, \infty)$. The range is $[0, \infty)$.

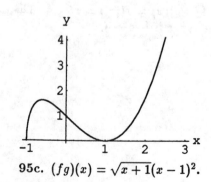

95c. $(fg)(x) = \sqrt{x+1}(x-1)^2$.

42

d. $(f/g)(x) = \dfrac{\sqrt{x+1}}{(x-1)^2}$. The domain is $\{x : x \neq 1$ and $-1 \leq x < \infty\}$. The range is $[0, \infty)$.

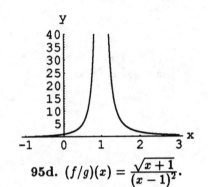

95d. $(f/g)(x) = \dfrac{\sqrt{x+1}}{(x-1)^2}$.

e. $(g/f)(x) = \dfrac{(x-1)^2}{\sqrt{x+1}}$. The domain is $(-1, \infty)$ and the range is $[0, \infty)$.

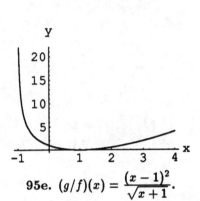

95e. $(g/f)(x) = \dfrac{(x-1)^2}{\sqrt{x+1}}$.

f. $(f \circ g)(x) = \sqrt{x^2 - 2x + 2}$. The domain is $(-\infty, \infty)$ and the range is $[1, \infty)$.

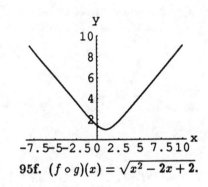

95f. $(f \circ g)(x) = \sqrt{x^2 - 2x + 2}$.

g. $(g \circ f)(x) = x + 2 - 2\sqrt{x+1}$. The domain is $[-1, \infty)$, and the range is $[0, \infty)$.

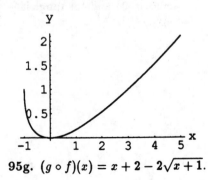

95g. $(g \circ f)(x) = x + 2 - 2\sqrt{x+1}$.

96. $f(x) = 4x^2 - x$, $g(x) = \frac{x+1}{x}$.

a. $(f+g)(x) = 4x^2 - x + \frac{x+1}{x}$. The domain is $\{x : x \neq 0\}$. The range is $(-\infty, \infty)$.

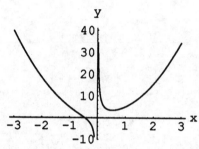

96a. $(f+g)(x) = 4x^2 - x + \frac{x+1}{x}$.

b. $(f-g)(x) = 4x^2 - x - \frac{x+1}{x}$. The domain is $\{x : x \neq 0\}$, and the range is $(-\infty, \infty)$.

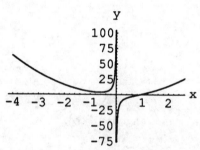

96b. $(f-g)(x) = 4x^2 - x - \frac{x+1}{x}$

c. $(fg)(x) = (4x-1)x\frac{x+1}{x} = 4x^2 + 3x - 1$. The domain is $\{x : x \neq 0\}$ and the range is $[-\frac{25}{16}, \infty)$.

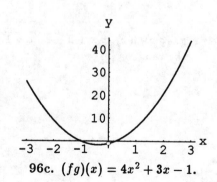

96c. $(fg)(x) = 4x^2 + 3x - 1$.

d. $(f/g)(x) = \dfrac{4x^3 - x^2}{x+1}$. The domain is $\{x : x \neq -1 \text{ and } x \neq 0\}$. The range is $\{y : y \neq 0\}$.

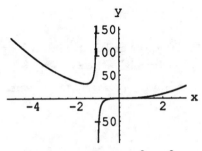

96d. $(f/g)(x) = \dfrac{4x^3 - x^2}{x+1}$.

e. $(g/f)(x) = \dfrac{x+1}{4x^3 - x^2}$. The domain is $\left\{x : x \neq 0 \text{ and } x \neq \frac{1}{4}\right\}$. The range is $\{y : y \neq 0\}$.

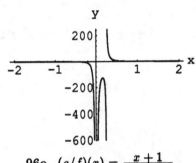

96e. $(g/f)(x) = \dfrac{x+1}{4x^3 - x^2}$.

f. $(f \circ g)(x) = 4\left(\dfrac{x+1}{x}\right)^2 - \dfrac{x+1}{x} = \dfrac{4x^2 + 8x + 4 - x^2 - x}{x^2} = \dfrac{3x^2 + 7x + 4}{x^2}$. The domain is $\{x : x \neq 0\}$, and the range is $[-\frac{1}{16}, \infty)$.

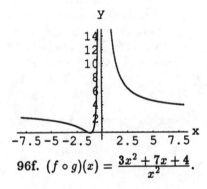

96f. $(f \circ g)(x) = \dfrac{3x^2 + 7x + 4}{x^2}$.

g. $(g \circ f)(x) = \dfrac{4x^2 - x + 1}{4x^2 - x} = 1 + \dfrac{1}{4x^2 - x}$. The domain is $\left\{x : x \neq \frac{1}{4} \text{ and } x \neq 0\right\}$, and the range is $(0, \infty) \cup (-\infty, -16]$.

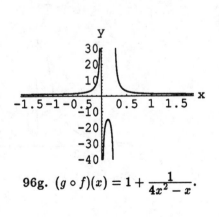

96g. $(g \circ f)(x) = 1 + \dfrac{1}{4x^2 - x}$.

97. From the graph at right, it is apparent that the interpolating polynomial $p(x) = -\frac{1}{520}x^6 + \frac{9}{130}x^4 - \frac{59}{104}x^2 + 1$ does not faithfully represent the rational function $f(x) = \frac{1}{1+x^2}$.

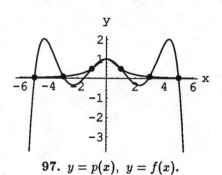

97. $y = p(x), \; y = f(x)$.

98. b. The exact formula for A in terms of a is $A = \pi * a^{3/2}$.

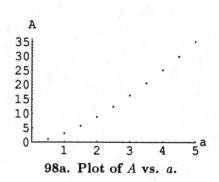

98a. Plot of A vs. a.

c. There is no exact formula for P as a function of a in terms of elementary functions. However a fit of the given data in terms of the functions $\sqrt{x}$, x, $x^{3/2}$, yields $P \approx 3.31375\sqrt{x} + 2.76979x + 0.207718x^{3/2}$.

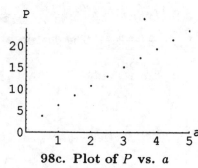

98c. Plot of P vs. a

99.

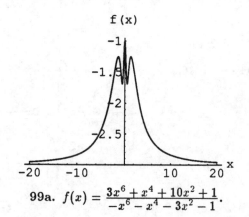

99a. $f(x) = \dfrac{3x^6 + x^4 + 10x^2 + 1}{-x^6 - x^4 - 3x^2 - 1}$.

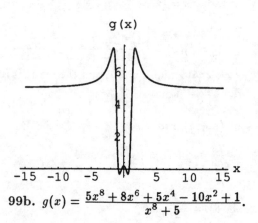

99b. $g(x) = \dfrac{5x^8 + 8x^6 + 5x^4 - 10x^2 + 1}{x^8 + 5}$.

100. a. The functions $f(x) = \sqrt{x+1}$ and $g(x) = \dfrac{x^2 + 8x + 8}{4x + 8}$ are plotted at right. They are so close to each other that one can hardly tell that two curves are plotted, rather than just one.

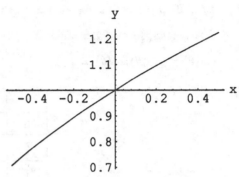

100a. A plot of f and g on $\left(-\frac{1}{2}, \frac{1}{2}\right)$.

b. A graph of $h(x) = |f(x) - g(x)|$ on the interval $\left(-\frac{1}{2}, \frac{1}{2}\right)$ makes it apparent that the maximum value occurs either at $x = \frac{1}{2}$ or at $x = -\frac{1}{2}$. Comparing the values of $h(x)$ at these points, we find that the maximum value is $h(-.5) \approx 0.00122$.

c. Note that this procedure cannot be used for numbers which lie between $\frac{3}{2}$ and 2. For such numbers x there is no integer n and there is no real number u such that $x = 2^{2n}u = 4^n u$. Similary this procedure will not work for numbers y of the form $4^n x$, where $\frac{3}{2} < x < 2$.

1.5   Trigonometric Functions

1. a. 30 degrees $= (30 \text{ degrees}) * \dfrac{2\pi \text{ radians}}{360 \text{ degrees}} = \frac{\pi}{6}$ radians.

  b. 75 degrees $= (75 degrees) * \dfrac{2\pi \text{ radians}}{360 \text{ degrees}} = \frac{5\pi}{12}$ radians.

  c. -15 degrees $= (-15 \text{ degrees}) * \dfrac{2\pi \text{ radians}}{360 \text{ degrees}} = -\frac{\pi}{12}$ radians.

  d. 315 degrees $= (315 \text{ degrees}) * \dfrac{2\pi \text{ radians}}{360 \text{ degrees}} = \frac{7\pi}{4}$ radians.

  e. $x + 30$ degrees $= (x + 30 \text{ degrees}) * \dfrac{2\pi \text{ radians}}{360 \text{ degrees}} = \left(\frac{x}{180} + \frac{1}{6}\right)\pi$ radians.

  f. 2910 degrees $= (2910 \text{ degrees}) * \dfrac{2\pi \text{ radians}}{360 \text{ degrees}} = 16\frac{1}{6}\pi$ radians.

2. a. π radians $= (\pi \text{ radians}) * \dfrac{180 \text{ degrees}}{\pi \text{ radians}} = 180$ degrees.

  b. 1 radian $= (1 \text{ radian}) * \dfrac{180 \text{ degrees}}{\pi \text{ radians}} = \frac{180}{\pi}$ degrees.

  c. $\frac{9\pi}{2}$ radians $= \left(\frac{9\pi}{2} \text{ radians}\right) * \dfrac{180 \text{ degrees}}{\pi \text{ radians}} = 810$ degrees.

  d. $-\frac{21\pi}{6}$ radians $= \left(-\frac{21\pi}{6} \text{ radians}\right) * \dfrac{180 \text{ degrees}}{\pi \text{ radians}} = -630$ degrees.

  e. $(a + \pi)$ radians $= ((a + \pi) \text{ radians}) * \dfrac{180 \text{ degrees}}{\pi \text{ radians}} = \left(\frac{a}{\pi} + 1\right) 180$ degrees.

  f. $\frac{7\pi}{16}$ radians $= \left(\frac{7\pi}{16} \text{ radians}\right) * \dfrac{180 \text{ degrees}}{\pi \text{ radians}} = \frac{315}{4}$ degrees.

3. If $\sin(\theta) = 0$ and $\cos(\theta) = 1$, then θ must be $\pi/2$. Then $\tan(\theta) = 0$ and $\sec(\theta) = 1$. Similarly, if $\sin(\theta) = \frac{\sqrt{2}}{2}$ and $\tan(\theta) = 1$, then $\cos(\theta) = \sin(\theta) = \frac{\sqrt{2}}{2}$ and $\sec(\theta) = \sqrt{2}$. In addition, θ must be

$\pi/4$. If $\sin(\theta) = \frac{3}{5}$ and $\sec(\theta) = \frac{5}{4}$, then $\cos(\theta) = \frac{1}{\sec(\theta)} = \frac{4}{5}$, and $\tan(\theta) = \frac{\sin(\theta)}{\cos(\theta)} = \frac{3}{4}$. In this case θ is $\arctan(\frac{3}{4}) \approx .6435$ radians. If $\tan(\theta) = \frac{5}{12}$ and $\sec(\theta) = \frac{13}{12}$, then $\cos(\theta) = \frac{1}{\sec(\theta)} = \frac{12}{13}$, and $\sin(\theta) = \tan(\theta)\cos(\theta) = \frac{5}{12}$.

θ	$\sin(\theta)$	$\cos(\theta)$	$\tan(\theta)$	$\sec(\theta)$
0	0	1	0	1
$\frac{\pi}{4}$	$\frac{\sqrt{2}}{2}$	$\frac{\sqrt{2}}{2}$	1	$\sqrt{2}$
???	$\frac{3}{5}$	$\frac{4}{5}$	$\frac{3}{4}$	$\frac{5}{4}$
????	$\frac{5}{13}$	$\frac{12}{13}$	$\frac{5}{12}$	$\frac{13}{12}$

1.5.3

4.

θ $0 \le \theta < 2\pi$	$\sin(\theta)$	$\cos(\theta)$	$\tan(\theta)$	$\sec(\theta)$
$\frac{3\pi}{2}$	0	-1	0	-1
$\frac{\pi}{3}$	$\frac{\sqrt{3}}{2}$	$\frac{1}{2}$	$\sqrt{3}$	2
$\frac{3\pi}{4}$	$\frac{\sqrt{2}}{2}$	$-\frac{\sqrt{2}}{2}$	-1	$-\sqrt{2}$
$\frac{3\pi}{2}$	-1	0	undefined	undefined
$-\frac{\pi}{6}$	$\frac{1}{2}$	$-\frac{\sqrt{3}}{2}$	$\frac{1}{\sqrt{3}}$	$-\frac{2}{\sqrt{3}}$
$\frac{7\pi}{4}$	$-\frac{\sqrt{2}}{2}$	$\frac{\sqrt{2}}{2}$	-1	$\sqrt{2}$

5. a. $\tan(\pi/6) = \frac{\sin(\pi/6)}{\cos(\pi/6)} = \frac{1/2}{\sqrt{3}/2} = \frac{1}{\sqrt{3}}$. $\cot(\pi/6) = \frac{1}{\tan(\pi/6)} = \sqrt{3}$. $\sec(\pi/6) = \frac{1}{\cos(\pi/6)} = \frac{2\sqrt{3}}{3}$. $\csc(\pi/6) = \frac{1}{\sin(\pi/6)} = 2$.

b. $\tan(5\pi/2) = \tan(\pi/2)$ and is undefined. $\sec(5\pi/2) = \frac{1}{\cos(\pi/2)}$ and is undefined. $\cot(5\pi/2) = \frac{\cos(\pi/2)}{\sin(\pi/2)} = 0$. $\csc(5\pi/2) = \frac{1}{\sin(\pi/2)} = 1$.

c. Since $-\frac{13\pi}{3} = -(4\frac{1}{3})\pi$, $\sin(x) = \sin(-\pi/3) = -\frac{\sqrt{3}}{2}$, and $\cos(x) = \cos(-\pi/3) = \frac{1}{2}$. Thus $\tan(x) = -\sqrt{3}$, $\cot(x) = \frac{1}{\tan(x)} = -\frac{\sqrt{3}}{3}$, $\sec(x) = 2$ and $\csc(x) = -\frac{2}{\sqrt{3}}$.

d. $\sin(9\pi) = \sin(\pi) = 0$, and $\cos(9\pi) = \cos(\pi) = -1$. Then $\tan(9\pi) = \tan(\pi) = 0$, $\cot(9\pi) = \frac{1}{\tan(9\pi)}$ is undefined, $\sec(9\pi) = \sec(\pi) = -1$, and $\csc(9\pi)$ is undefined.

e. $\sin(-9\pi/4) = \sin(-\pi/4) = -\sin(\pi/4) = -\frac{\sqrt{2}}{2}$. $\cos(-9\pi/4) = \cos(-\pi/4) = \cos(\pi/4) = -\frac{\sqrt{2}}{2}$. Thus $\tan(-9\pi/4) = -1$, $\cot(-9\pi/4) = \frac{1}{\tan(-9\pi/4)} = -1$, $\sec(-9\pi/4) = \sqrt{2}$, and $\csc(-9\pi/4) = -\sqrt{2}$.

f. $\sin(7\pi/6) = \sin(\pi - 7\pi/6) = \sin(-\pi/6) = -\sin(\pi/6) = -\frac{1}{2}$. Also $\cos(7\pi/6) = \cos(7\pi/6 - 2\pi) = \cos(-5\pi/6) = \cos(5\pi/6) = -\frac{\sqrt{3}}{2}$. Thus $\tan(7\pi/6) = -\frac{1}{\sqrt{3}}$, $\cot(7\pi/6) = \frac{1}{\tan(7\pi/6)} = -\sqrt{3}$, $\sec(7\pi/6) = \frac{2}{\sqrt{3}}$, and $\csc(7\pi/6) = -2$.

6. In $0 \le x < 2\pi$,

    a. $\sin(x) = 0$ if and only if $x = 0$ or $x = \pi$.

    b. $\cos(x) = \sqrt{2}/2$ if and only if $x = \pi/4$, or $x = 7\pi/4$.

    c. $\csc(x) = \sqrt{2}$ if and only if $x = \pi/4$, or $x = 3\pi/4$.

    d. $\cos(\pi/2 + 2x) = 1$ if and only if $\pi/2 + 2x = 2n\pi$. This is true if and only if $x = n\pi - \pi/4$. The corresponding values in $0 \le x < 2\pi$ are $x = 3\pi/4, 7\pi/4$.

    e. $2\cos^2(x) - \cos(x) - 1 = 0$ is a quadratic equation in $y = \cos(x)$. This equation factors as $(2y + 1)(y - 1) = 0$. The solutions are $\cos(x) = y = 1$ or $-1/2$. The corresponding values of x in $0 \le x < 2\pi$ are $x = 0$, $x = 2\pi/3$, and $x = 4\pi/3$.

    f. $\cos(2x) + \cos(x) + 1 = 0$ is equivalent to the equation $2\cos^2(x) - 1 + \cos(x) + 1 = 0$, using the double angle formula. This equation factors as $\cos(x)(2\cos(x) + 1) = 0$. This implies $\cos(x) = 0$ or $\cos(x) = -1/2$. Thus $x = \pi/2, 3\pi/2, 2\pi/3$, or $4\pi/3$.

    g. $|\cos(x)| = \sqrt{3}/2$ if $\cos(x) = \pm\sqrt{3}/2$. This is true if and only if $x = \pi/6, 5\pi/6, 7\pi/6$, or $11\pi/6$.

7. In $-2\pi \le x < 2\pi$,

    a. $\cos(x) = 0$ if and only if $-3\pi/2, -\pi/2, \pi/2, 3\pi/2$.

    b. $\sin(x) = \cos(x)$ if and only if $x = -7\pi/4, -3\pi/4, \pi/4$, or $5\pi/4$.

    c. $\tan(x) = -1$ if and only if $x = -5\pi/4, -\pi/4, 3\pi/4$, or $7\pi/4$.

    d. $\sin(3x) = 0$ if and only if $3x = n\pi$ for some integer n. Then $x = n\pi/3$. The corresponding values in $-2\pi \le x < 2\pi$ are $x = -2\pi, -5\pi/3, -4\pi/3, -\pi, -2\pi/3, -\pi/3, 0, \pi/3, 2\pi/3, \pi, 4\pi/3, 5\pi/3, 2\pi$.

    e. $\sin(2x) = \cos(x)$ is equivalent to the equation $2\sin(x)\cos(x) = \cos(x)$. This equation factors as $\cos(x)(2\sin(x) - 1) = 0$. The solutions are $\cos(x) = 0$, $\rightarrow x = -3\pi/2, -\pi/2, \pi/2, 3\pi/2$, and $\sin(x) = 1/2$, $\rightarrow x = -11\pi/6, -7\pi/6, \pi/6, 5\pi/6$.

    f. $3\cos^2(x) - \sin^2(x) = 4\cos^2(x) - 1 = 0$ if and only if $\cos(x) = \pm 1/2$. The solutions are $x = -5\pi/3, -4\pi/3, -2\pi/3, -\pi/3, \pi/3, 2\pi/3, 4\pi/3, 5\pi/3$.

    g. $\cos^2(x) + \sin(x) = 0 \Leftrightarrow 1 - \sin^2(x) + \sin(x) = 0$. This is a quadratic equation in $y = \sin(x)$. Using the quadratic formula we find that the solutions of this equation are $y = 1/2 \pm \sqrt{5}/2$. The value $y = 1/2 + \sqrt{5}/2$ cannot be a value of $\sin(x)$ since it is greater than 1. The other value is ≈ -0.618034. Computing $\arcsin(-0.618034)$ we obtain $-.666239$ radians. Other solutions in $-2\pi < x < 2\pi$ are $x = -.666239 + 2\pi$ radians, $x = \pi - (-.666239)$ radians, and $x = .666239 - \pi$ radians.

8. $\sin(t) = \cos(t) \Leftrightarrow \tan(t) = 1 \Rightarrow \tan(t) > 0$. In $0 \le t < 2\pi$ the solutions are $t = \pi/4, 5\pi/4$.

9. $\sin(t) = \sqrt{3}\cos(t) \Leftrightarrow \tan(t) = \sqrt{3}$. In $0 \le t < 2\pi$, $\cos(t) < 0 \Leftrightarrow \pi/2 < t < 3\pi/2$. The solution to these two conditions is $t = 4\pi/3$.

10. $\sec(t) = 2 \Leftrightarrow \cos(t) = 1/2$. In $0 \le t < 2\pi$, $\cot(t) < 0 \Leftrightarrow \pi/2 < t < \pi/$ or $3\pi/2 < t < 2\pi$. Of these two intervals, $\cos(t) > 0$ only in $3\pi/2 < t < 3\pi$. The solution to these two conditions is $t = 5\pi/3$.

11. In $0 \leq t < 2\pi$, $\sin(t) > 0 \Leftrightarrow 0 < t < \pi$, and $\tan(t) > 0 \Leftrightarrow 0 < t < \pi/2$, or $\pi < t < 3\pi/2$. The intersection of these two solutions sets is $0 < t < \pi/2$.

12. $\sin(t) \geq \tan(t) \Leftrightarrow \dfrac{\sin(t)(\cos(t) - 1)}{\cos(t)} \geq 0$. Since for all t, $\cos(t) \leq 1$, and we assume $\sin(t) \geq 0$, this implies $\cos(t) \leq 0$. The combination of these last two inequalities implies $\pi/2 \leq t \leq \pi$.

13. Since $\sec(t) = 1/\cos(t)$, these two inequalities are mutually exclusive. Note that if $\cos(t) = 0$, then $\sec(t)$ is undefined.

14. If $\cos(t) < 0$ and $\tan(t) < 0$, then $\sin(t) > 0$. In the interval $0 \leq t < 2\pi$, $\sin(t) > 0$ and $\cos(t) < 0$ if and only if $\pi/2 < t < \pi$.

15. $\sin(5\pi/12) = \sin(2\pi/3 - \pi/4)$

$$
\begin{aligned}
&= \sin(2\pi/3)\cos(\pi/4) - \sin(\pi/4)\cos(2\pi/3) \\
&= \frac{\sqrt{3}}{2}\frac{\sqrt{2}}{2} - \frac{\sqrt{2}}{2}\frac{-1}{2} \\
&= \frac{\sqrt{6}}{4} + \frac{\sqrt{2}}{4} \\
&= \frac{\sqrt{6} + \sqrt{2}}{4}.
\end{aligned}
$$

16. $\tan(5\pi/12) = \dfrac{\sin(5\pi/12)}{\cos(5\pi/12)}$. We compute:

$$
\begin{aligned}
\cos(5\pi/12) &= \cos(2\pi/3 - \pi/4) \\
&= \cos(2\pi/3)\cos(\pi/4) + \sin(2\pi/3)\sin(\pi/4) \\
&= \frac{-1}{2}\frac{\sqrt{2}}{2} + \frac{\sqrt{3}}{2}\frac{\sqrt{2}}{2} \\
&= \frac{\sqrt{6} - \sqrt{2}}{4}
\end{aligned}
$$

Since $\sin(5\pi/12) = \dfrac{\sqrt{6} + \sqrt{2}}{4}$, (#16) we find that $\tan(5\pi/12) = \dfrac{\sqrt{6} + \sqrt{2}}{\sqrt{6} - \sqrt{2}}$

17. Using the half-angle formula, we obtain $\cos(\pi/12) = \cos\left(\dfrac{\pi/6}{2}\right) = \sqrt{\dfrac{1 + \cos(\pi/6)}{2}}$, Evaluating this expression, we obtain $\sqrt{\dfrac{2 + \sqrt{3}}{4}}$. Alternatively, using methods like those used above for #15, we obtain $\cos(\pi/12) = \dfrac{\sqrt{6} + \sqrt{2}}{4}$. Can you reconcile these two answers?

18. $\tan(13\pi/12) = \tan(3\pi/4 + \pi/3)$

$$
\begin{aligned}
&= \frac{\tan(3\pi/4) + \tan(\pi/3)}{1 - \tan(3\pi/4)\tan(\pi/3)} \\
&= \frac{-1 + \sqrt{3}}{1 + \sqrt{3}}
\end{aligned}
$$

19. $\sec(11\pi/12) = \dfrac{1}{\cos(11\pi/12)} = -\dfrac{1}{\cos(\pi/12)} = -\dfrac{4}{\sqrt{6} + \sqrt{2}}$. See #17.

20. $\csc(\pi/12) = \dfrac{1}{\sin(\pi/12)}.$ $\sin(\pi/12) = \sin(\pi/3 - \pi/4)$

$$= \sin(\pi/3)\cos(\pi/4) - \sin(\pi/4)\cos(\pi/3)$$

$$= \frac{\sqrt{3}}{2}\frac{\sqrt{2}}{2} - \frac{\sqrt{2}}{2}\frac{1}{2}$$

$$= \frac{\sqrt{6} - \sqrt{2}}{4}$$

So $\csc(\pi/12) = \dfrac{4}{\sqrt{6} - \sqrt{2}}.$

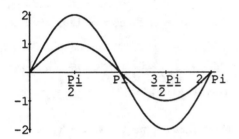

21. $\sin(x),\ 2\sin(x).$

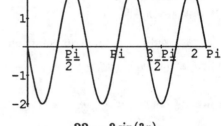

22. $-2\sin(3x).$

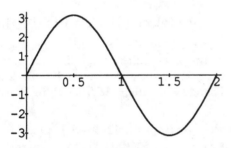

23. $\pi\sin(\pi x).$

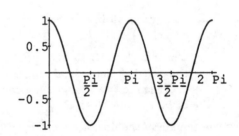

24. $\sin(2x + \pi/2).$

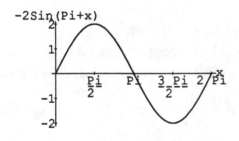

25. $-2\sin(\pi + x).$

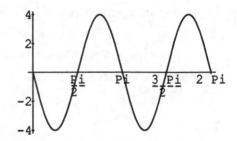

26. $4\sin(2x + \pi).$

27. The functions $\sin(x)$ and $\cos(x)$ are defined for any real number x. Both functions have range $-1 \le x \le 1$. The functions $\tan(x)$ and $\sec(x)$ are defined whenever $\cos(x) \ne 0$, that is, for $x \ne \pi/2 + n\pi$, $n = 0, \pm1, \pm2, \ldots$. The functions $\cot(x)$ and $\csc(x)$ are defined whenever $\sin(x) \ne 0$, that is, for $x \ne n\pi$, $n = 0, \pm1, \pm2, \ldots$ The functions $\tan(x)$ and $\cot(x)$ both have range $(-\infty, \infty)$. The functions $\sec(x)$ and $\csc(x)$ both have ranges $(-\infty, -1] \cup [1, \infty)$.

28. Since $\sin(-x) = -\sin(x)$, sin is an odd function.

29. Since $\cos(-x) = \cos(x)$, cos is an even function.

30. Since $\tan(-x) = \dfrac{\sin(-x)}{\cos(-x)} = \dfrac{-\sin(x)}{\cos(x)} = -\tan(x)$, tan is an odd function.

31. Here $y = f(x) = \dfrac{\sin(x)}{x}$, and $f(-x) = \dfrac{\sin(-x)}{-x} = \dfrac{\sin(x)}{x} = f(x)$. Therefore f is an even function.

32. If $y = f(x) = 3\sec(x)$, then $f(-x) = 3\sec(-x) = \dfrac{3}{\cos(-x)} = 3\sec(x) = f(x)$. Thus f is an even function.

33. Theorem: The product of two odd functions is always even (verify this for yourself!). Thus $f(x) = \sin^2(x)$ is an even function.

34. If $x° = y$ *(radians)*, and $x = y$, then we also have that $y = \dfrac{\pi x}{180}$. Since $\pi \neq 180$, we must have in this case that $x = y = 0$.

35. Let (x_1, y_1) and (x_2, y_2) be two points on this line. Then $m = \dfrac{y_2 - y_1}{x_2 - x_1}$. If we form a right triangle with vertices at (x_1, y_1), (x_2, y_2), and (x_2, y_1), then we compute that $\tan(\theta) = \dfrac{y_2 - y_1}{x_2 - x_1} = m$.

36. The area of the sector is $\dfrac{7\pi/12}{2\pi}$ of the area of the whole circle. Thus the sector has area $\dfrac{7\pi}{6}$.

37. To prove: $\cos(\theta + \phi) = \cos(\theta)\cos(\phi) - \sin(\theta)\sin(\phi)$, for θ and ϕ between 0 and $\pi/2$.

    a. Refer to figure 5.13 in the text. $P_1 = (1, 0)$, $P_2 = (\cos(\phi), \sin(\phi))$, $P_3 = (\cos(\phi + \theta), \sin(\phi + \theta))$, $P_4 = (\cos(-\theta), \sin(-\theta))$.

    b. Since the triangles OP_1P_3 and OP_4P_2 both have two sides which are radii of the unit circle, they are congruent if the central angles $\angle P_1OP_3$ and $\angle P_4OP_2$ are the same. But both angles equal $\theta + \phi$. Thus the two triangles are congruent. Therefore the distance from P_3 to P_1 is the same as the distance from P_2 to P_4.

    c. The square of the distance from $P_1 = (1, 0)$ to $P_3 = (\cos(\phi + \theta), \sin(\phi + \theta))$ is $(\cos(\phi + \theta) - 1)^2 + (\sin(\phi + \theta))^2$. The square of the distance from $P_4 = (\cos(-\theta), \sin(-\theta))$ to $P_2 = (\cos(\phi), \sin(\phi))$ is $(\cos(\theta) - \cos(\phi))^2 + (\sin(\phi) - \sin(-\theta))^2$. By part b, these two distances must be equal.

    d. Expanding the squares, we obtain
$$\cos^2(\theta + \phi) - 2\cos(\theta + \phi) + 1 + \sin^2(\theta + \phi) =$$
$$\cos^2(\phi) - 2\cos(\phi)\cos(\theta) + \cos^2(\theta) + \sin^2(\phi) + 2\sin(\phi)\sin(\theta) + \sin^2(\theta).$$
Successively applying the identity $\sin^2(\alpha) + \cos^2(\alpha) = 1$ with $\alpha = \theta + \phi$, ϕ, and θ, we obtain:
$$2 - 2\cos(\theta + \phi) = 2 - 2\cos(\theta)\cos(\phi) + 2\sin(\theta)\sin(\phi)$$

    e. Thus
$$\cos(\theta + \phi) = \cos(\theta)\cos(\phi) - \sin(\theta)\sin(\phi)$$

38. The area of a triangle is $\dfrac{BH}{2}$, where H is the height, and B is the length of the base, of the triangle. If h is the length of the hypotenuse, and θ measures one acute angle, then the base B is $h\cos(\theta)$ and the height H is $h\sin(\theta)$. Thus the area is $\dfrac{h\sin(\theta)\cos(\theta)}{2} = \dfrac{hx\sin(\theta)}{2}$.

39. By definition, $\tan(x + T) = \dfrac{\sin(x + T)}{\cos(x + T)}$. This, in turn, equals $\dfrac{\sin(x)\cos(T) + \sin(T)\cos(x)}{\cos(x)\cos(T) - \sin(x)\sin(T)}$. Dividing both numerator and denominator by $\cos(x)\cos(T)$, this becomes $\dfrac{\tan(x) + \tan(T)}{1 - \tan(x)\tan(T)}$ (another formula to remember!). Thus $\tan(x + T) = \tan(x)$, if

$$\frac{\tan(x) + \tan(T)}{1 - \tan(x)\tan(T)} = \tan(x)$$

$$\frac{\tan(x) + \tan(T) - \tan(x) + \tan(x)^2\tan(T)}{1 - \tan(x)\tan(T)} = 0$$

$$\tan(T)(1 + \tan(x)^2) = \tan(T)\sec(x)^2 = 0$$

Thus we conclude that $\tan(x+T) = \tan(x)$, if and only if $\tan(T) = 0$. The solution set of this equation is $\{T|\ T = n\pi,\ n = 0,\ \pm1,\ \pm2,\ \pm3, \cdots\}$. The minimum positive such value of T is π. Thus tan has period π.

Since $\cot(x) = \dfrac{1}{\tan(x)}$, the period of cot is the same as the period of tan.

40. False! $\sin(x + y) = \sin(x)\cos(y) + \sin(y)\cos(x)$.

41. The pendulum travels a distance equal to the change in the angle, measured in radians, times the radius. The total change in the angle over one cycle is $2\pi/3$. Thus the distance travelled is $4\pi/3m$.

42. The distance from B to C is $(50m) \times \tan(\pi/6) = \dfrac{50m}{\sqrt{3}}$.

43. The distance from B to C is $(50m) \times \tan(\pi/3) = 50m\sqrt{3}$.

44. Since the altitude is the length of the side of a right triangle, and this side is the side opposite an angle measuring $\pi/6$ radians, and the horizontal distance is 300 meters, and this is the length of the adjacent side, the altitude A equals $\tan(\pi/6) * (300 \text{ meters}) = \dfrac{300 \text{ meters}}{\sqrt{3}}$

45. Since the altitude is the length of the side of a right triangle, and this side is the side opposite an angle measuring $\pi/4$ radians, and the horizontal distance is 300 meters, and this is the length of the adjacent side, the altitude A equals $\tan(\pi/4) * (300 \text{ meters}) = 300 \text{ meters}$

46.
   - In Figure 5.17, $\angle \bar{C} = 180° - (45° + 65°) = 70°$. By the Law of Sines, $\dfrac{c}{\sin(C)} = \dfrac{10m}{\sin(45°)} = 5\sqrt{2}m$. Thus $c = 5\sqrt{2}\sin(70°)m. \approx 6.64463m$.

   - In Figure 5.18, $a^2 = (6m)^2 + (10m)^2 - 2(6m)(10m)\cos(40°)$ by the Law of Cosines. Thus $a \approx \sqrt{136 - 120 * .76604} = 6.638875$.

47. The ladder would form the hypotenuse of a $30° - 60° - 90°$ triangle, with a 3 meter side adjacent to the $30°$ ($\pi/6$) angle. Thus the length of the ladder must be $(3m)/\cos(\pi/6) = (3m)2/\sqrt{3} = 2\sqrt{3}m$.

48. Let h be the height of the building. Then h is the length of the side opposite an angle of $\alpha = \pi/4$ and an angle $\beta = \pi/6$ for two different right triangles. The distance d is the difference between the lengths of the sides adjacent to these angles. Thus $h * (\cot(\pi/6) - \cot(\pi/4)) = 100m$. Since $\cot(\pi/6) = \sqrt{3}$ and $\cot(\pi/4) = 1$, we have that $h * (\sqrt{3} - 1) = 100m$. So $h = \dfrac{100m}{\sqrt{3} - 1}$.

49. Examining the graph at right, we see that the solution lies somewhere between 0.7 and 0.8. An approximate solution, accurate to 6 digits, is $x = 0.739085$.

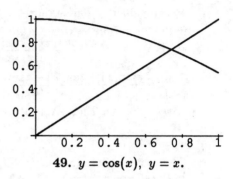

49. $y = \cos(x)$, $y = x$.

50. Conjecture: As $n \to \infty$, $\cos^n(x) \to x_0$, where $\cos(x_0) = x_0 \approx 0.739085$.

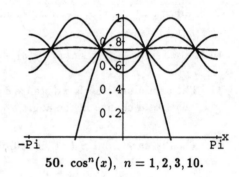

50. $\cos^n(x)$, $n = 1, 2, 3, 10$.

51. Since $y \to \infty$ as $x \to \infty$ on the line $y = x$, the points of intersection of this line with the graph of $y = \tan(x)$ will approach the vertical asymptotes of this graph. These asymptotes occur at $x = \dfrac{(2n + 1)\pi}{2}$. Thus the solutions tend to these values of x as $x \to \infty$.

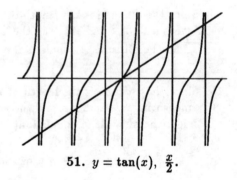

51. $y = \tan(x)$, $\frac{x}{2}$.

52a. These sums are converging to a 'sawtooth' function which is the 2π periodic extension of the function defined on $(0, 2\pi)$ as $f(x) = |x - \pi|$.

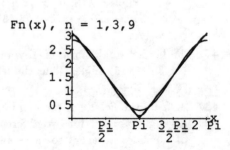

52a. $F_n(x), n = 1, 3, 9$.

52b. These sums are converging to a 'square wave' function which is the 2π periodic extension of the function defined on $[0, 2\pi]$ as $g(x) = 1$ if $0 \leq x \leq \pi$, and $g(x) = -1$ if $\pi \leq x \leq 2\pi$. Notice the high degree of oscillation in the graph for $n = 9$. This is due to the difficulty of approximating a jump discontinuity with smooth functions. This phenomenon is known as the *Gibb's phenomenon*.

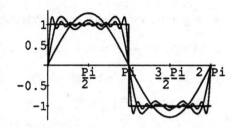

52b. $G_n(x)$ $n = 1,\ 3,\ 9,\ \infty$.

52c. These sums converge to another 'sawtooth' function like the one in 52a, but with a smaller *amplitude*. Also notice that the average of this function over $[0, 2\pi]$ os 0, while the function in 52a has an average value of $\pi/2$.

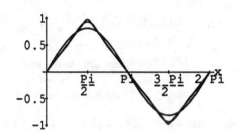

52c. $H_n(x)$, $n = 1,\ 3,\ 9$.

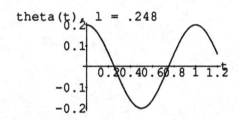

53.a $l = .248$, $T = 1s$.

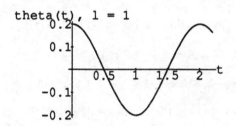

53.b $l = 1$, $T = 2s$.

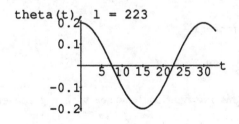

53.c $l = 223$, $T = 30s$

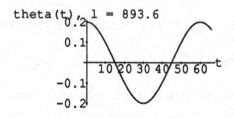

53.d $l = 893.6$, $T = 1$ min. .

Review Exercises–Chapter 1

1. If $A = [2, 7)$ and $B = (3, 9)$, then $A \cup B = [2, 9)$, and $A \cap B = (3, 7)$.

2. $\{1\}$, $\{7\}$, $\{19\}$, $\{\pi\}$, $\{1, 7\}$, $\{1, 19\}$, $\{1, \pi\}$, $\{7, 19\}$, $\{7, \pi\}$, $\{19, \pi\}$, $\{1, 7, 19\}$, $\{1, 7, \pi\}$, $\{1, 19, \pi\}$, $\{7, 19, \pi\}$, $\{1, 7, 19, \pi\}$.

3. $2x - 7 \geq 9 \Leftrightarrow 2x \geq 16 \Leftrightarrow x \geq 8$.

4. $6 \leq x^2 - 3 \leq 22 \Leftrightarrow 9 \leq x^2 \leq 25$. Hence the solution set is $[-5, -3] \cup [3, 5]$.

5. $A = \{x | x \geq 4\}, B = \{x | -3 \leq x \leq 3\}, C = \{-4, -3, -2, -1, 0, 1, 2, 3, 4\}$

    a. $A \cap B = [-3, -2] \cup [2, 3]$

    b. $A \cup B = (-\infty, \infty)$

    c. $A \cap C = \{-4, -3, -2, 2, 3, 4\}$

    d. $B \cap C = \{-3, -2, -1, 0, 1, 2, 3\}$

    e. $A \cup (B \cap C) = (-\infty, -2] \cup \{-1, 0, 1\} \cup [2, \infty)$

    f. $A \cap (B \cup C) = [-3, -2] \cup [2, 3] \cup \{-4, 4\}$

6. $4 \leq (x + 2)^2 \leq 36 \Leftrightarrow 2 \leq |x + 2| \leq 6 \Leftrightarrow 2 \leq x + 2 \leq 6$ or $2 \leq -x - 2 \leq 6$.
The first inequality gives us $0 \leq x \leq 4$. We evaluate the second inequality:
$2 \leq -x - 2 \leq 6 \Leftrightarrow 4 \leq -x \leq 8 \Leftrightarrow -8 \leq x \leq -4$
Hence the solution set is $[-8, -4] \cup [0, 4]$

7. $|x - 7| \leq 12 \Leftrightarrow -12 \leq (x - 7) \leq 12 \Leftrightarrow -5 \leq x \leq 19$.

8. $2 \leq |x^2 - 2| \leq 7 \Leftrightarrow 2 \leq x^2 - 2 \leq 7$ or $2 \leq 2 - x^2 \leq 7$.
$2 \leq x^2 - 2 \leq 7 \Leftrightarrow 4 \leq x^2 \leq 9 \Leftrightarrow -3 \leq x \leq -2$ or $2 \leq x \leq 3$.
$2 \leq 2 - x^2 \leq 7 \Leftrightarrow 0 \leq -x^2 \leq 5 \Leftrightarrow x = 0$.
The solution set is $[-3, -2] \cup \{0\} \cup [2, 3]$.

9. $A \cup B = A$ and $A \cap B = B$

10. $|x - 3| \leq 5 \Leftrightarrow -5 \leq x - 3 \leq 5 \Leftrightarrow -2 \leq x \leq 8$.

11. $-3 \leq |2x + 1| \leq 11 \Leftrightarrow 0 \leq |2x + 1| \leq 11 \Leftrightarrow -11 \leq 2x + 1 \leq 11 \Leftrightarrow -6 \leq x \leq 5$

12. $|x + 2| \geq 6 \Leftrightarrow x + 2 \geq 6$ or $x + 2 \leq -6$. The solution set is $(-\infty, -8) \cup (4, \infty)$.

13. $|2x^2 + 1| \leq 9 \Leftrightarrow 2x^2 + 1 \leq 9 \Leftrightarrow x^2 \leq 4 \Leftrightarrow -2 \leq x \leq 2$.

14. $|9 - x^2| \geq 0$. This inequality is true for all real numbers, x.

15. $\cos x < \frac{\sqrt{3}}{2}$, $0 \leq x \leq 2\pi \Leftrightarrow \frac{\pi}{6} < x < 2\pi - \frac{\pi}{6} = \frac{11\pi}{6}$.

16. $|\sin(x)| = \frac{1}{2}$ if $x = \frac{\pi}{6}, \frac{5\pi}{6}, \frac{7\pi}{6}, \frac{11\pi}{6}$. Using this information together with a graph of $\sin(x)$ for $0 \leq x \leq 2\pi$, we find that $|\sin(x)| > \frac{1}{2}$, $0 < x < 2\pi \Leftrightarrow \frac{\pi}{6} < x < \frac{5\pi}{6}$, or $\frac{7\pi}{6} < x < \frac{11\pi}{6}$.

17. $\left(\sin(x) - \frac{1}{2}\right)\left(\cos(x) + \frac{1}{2}\right) = 0 \Leftrightarrow \sin(x) = \frac{1}{2}$ or $\cos(x) = -\frac{1}{2}$. This occurs at $x = \frac{\pi}{6}, \frac{2\pi}{3}, \frac{5\pi}{6}$ and at $x = \frac{4\pi}{3}$. These are the only points where the quantity $\left(\sin(x) - \frac{1}{2}\right)\left(\cos(x) + \frac{1}{2}\right)$ can change sign. Evaluating $\left(\sin(x) - \frac{1}{2}\right)\left(\cos(x) + \frac{1}{2}\right)$ at points in the intervals $(0, \pi/6)$, $(\pi/6, 2\pi/3)$, $(2\pi/3, 5\pi/6)$, $(5\pi/6, 4\pi/3)$, and $(4\pi/3, 2\pi)$, we find that this quantity is positive if and only if $\frac{\pi}{6} < x < \frac{2\pi}{3}$, or $\frac{5\pi}{6} < x < \frac{4\pi}{3}$.

18. $\cos(2x) + \sin(x) = 1 - 2\sin^2(x) + \sin(x)$. This is a quadratic expression in $\sin(x)$. This expression is 0 if and only if $\sin(x) = \frac{1}{4} \pm \frac{3}{4}$. $\sin(x) = 1 \Leftrightarrow x = \frac{\pi}{2}, \frac{3\pi}{2}$. $\sin(x) = -\frac{1}{2} \Leftrightarrow x = \frac{7\pi}{6}$, or $\frac{11\pi}{6}$. These are the only points in $0 \le x \le 2\pi$ where $\cos(2x) + \sin(x)$ can change sign. Evaluating $1 - 2\sin^2(x) + \sin(x)$ at point selected from the intervals $(0, \pi/2)$, $(\pi/2, 7\pi/6)$, $(7\pi/6, 3\pi/2)$, $(3\pi/2, 11\pi/6)$, and $(11\pi/6, 2\pi)$, we find that the solution to the inequality is $0 < x < \frac{\pi}{2}$, or $\frac{7\pi}{6} < x < \frac{3\pi}{2}$, or $\frac{11\pi}{6} < x < 2\pi$.

19. $|x - 2| = |x + 2| \Leftrightarrow x - 2 = x + 2$ or $x - 2 = -x - 2$. The first equation has no solutions. Hence $x = 0$ (from the second equation).

20. a. Let P, Q and R be the points $(0, 0), (-3, 6) and (2, 4)$. The slope of PQ is $m_1 = \frac{6-0}{-3-0} = -2$. The slope of PR is $\frac{4-0}{2-0} = 2$. Since the two slopes are different, P, Q and R do not lie on a common line.

   b. Let P, Q, and R be the points $(-4, 2)$, $(-1, 5)$, and $(6, 10)$. The slope of PQ is $m_1 = \frac{5-2}{-1+4} = 1$. The slope of PR is $m_2 = \frac{10-2}{6+4} = \frac{8}{10}$. Since the two slopes are not the same, P, Q, and R do not lie on a common line.

   c. Let P, Q and R be the points $(-3, 2)$, $(-1, 1)$ and $(2, -3)$. The slope of PQ is $m_1 = \frac{1-2}{-1+3} = -\frac{1}{2}$. The slope of PR is $m_2 = \frac{-3-2}{2+3} = -1$. Since the two slopes are not the same, P, Q and R do not lie on a common line.

   d. Let P, Q and R be the points $(1, 3)$, $(2, 4)$ and $(-2, 0)$. The slope of PQ is $m_1 = \frac{4-3}{2-1} = 1$. The slope of QR is $m_2 = \frac{0-4}{-2-2} = 1$. Since the two slopes are the same, P, Q and R do lie on a common line.

21. Consider the line connecting $(0, 0)$ and $(5, 0)$ to be the base, b, of the triangle. Then $b = 5$. Since the third vertex is at $(2, 3)$, the height of the triangle is $h = 3$. Hence the area is

$$A = \frac{1}{2}bh = \frac{15}{2}.$$

22. Let P, Q and R be the points $(0, 0)$, $(2, 4)$ and $(a, 0)$.

   a. Note that PR is a horizontal line. If there is a right angle at the vertex $(a, 0)$, then QR must be a vertical line. Hence $a = 2$.

   b. If the right angle is at $(2, 4)$, then the slopes m_1 of PQ and m_2 of QR must be negative reciprocals of each other.
   $m_1 = \frac{4-0}{2-0} = 2.$ $\qquad m_2 = \frac{0-4}{a-2} = -\frac{4}{a-2}$
   Since $m_1 = -\frac{1}{m_2}$, we have $2 = \frac{a-2}{4} \Leftrightarrow a = 10$

23. Since the diameter is 10, the radius is $r = 5$. Since the center is at $(2, 4)$, the equation of the circle is $(x - 2)^2 + (y - 4)^2 = 25$.

24. Center: $(0, 0)$ $\qquad\qquad$ Radius: 7

25. $x^2 - 4x + y^2 = 0 \Leftrightarrow (x - 2)^2 + y^2 = 4$. Hence the center is at $(2, 0)$ and the radius is 2.

26. $x^2 - 2x + y^2 + 6y = -9 \Leftrightarrow (x^2 - 2x + 1) + (y^2 + 6y + 9) = 10 - 9 \Leftrightarrow (x - 1)^2 + (y + 3)^2 = 1$. Hence the center is at $(1, -3)$ and the radius is 1.

27. $x^2 - 2x + y^2 + 2y = 14 \Leftrightarrow (x^2 - 2x + 1) + (y^2 + 2y + 1) = 14 + 2 \Leftrightarrow (x - 1)^2 + (y + 1)^2 = 16$. Hence the center is $(1, -1)$ and the radius is 4.

28. $y = -x + 1$. The slope is -1, the y-intercept is 1 and the x-intercept is 1.

29. $x - 3y + 4 = 0 \Leftrightarrow y = \frac{1}{3}x + \frac{4}{3}$. Note that when $y = 0$, $x = -4$. Hence the slope is $\frac{1}{3}$, the y-intercept is $\frac{4}{3}$ and the x-intercept is -4.

30. $y = 4$. The line is horizontal and has no x-intercept. The slope is 0 and the y-intercept is 4.

31. $7x - 7y + 21 = 0 \Leftrightarrow y = x + 3$. The slope is 1, the y-intercept is 3 and the x-intercept is -3.

32. $x - y = 5 \Leftrightarrow y = x - 5$. The slope is 1, the y-intercept is -5 and the x-intercept is 5.

33. $3x = 6 \Leftrightarrow x = 2$. The line is vertical and has neither a slope nor a y-intercept. The x-intercept is 2.

34. $(-2, 1), (3, 3)$. The slope of the line is $\frac{3-1}{3+2} = \frac{2}{5}$. Using the point-slope form of the equation of the line with the point $(3,3)$, we have:
$$y - 3 = \frac{2}{5}(x - 3) \Leftrightarrow y = \frac{2}{5}x + \frac{9}{5}.$$

35. $2x - 4y = 14 \Leftrightarrow y = \frac{1}{2}x - \frac{7}{2}$. The slope of the parallel line is $\frac{1}{2}$. Since the line passes through the point $(2, -4)$, the equation is given by:
$$y + 4 = \frac{1}{2}(x - 2) \Leftrightarrow y = \frac{1}{2}x - 5.$$

36. $y - ax = 1 \Leftrightarrow y = ax + 1$.
$3y = 6x + 12 \Leftrightarrow y = 2x + 4$.
Since the two lines are parallel, the slopes must be equal. Hence, $a = 2$.

37. $3x - 6y = 8 \Leftrightarrow y = \frac{1}{2}x - \frac{4}{3}$. The slope of this line is $\frac{1}{2}$. The slope of the perpendicular line is the negative reciprocal of $\frac{1}{2}$, which is -2. Since the line passes through the origin, its equation is $y = -2x$.

38. The slope of the line through the points $(1, 3)$ and $(6, -4)$ is $\frac{-4 - 3}{6 - 1} = -\frac{7}{5}$. Hence the equation of the parallel line passing through $(-1, 3)$ is:
$$y - 3 = -\frac{7}{5}(x + 1) \Leftrightarrow y = -\frac{7}{5}x + \frac{8}{5}.$$

39. The line $ax + by + c = 0$ crosses the x-axis when $y = 0$:
$ax + 0y + c = 0 \Leftrightarrow x = -\frac{c}{a}$. Hence, the line crosses the x-axis at the point $(-\frac{c}{a}, 0)$.

40. Substituting $(2, 2)$ and $(4, -5)$ into the equation $y = ax + b$ gives us the equations $2 = 2a + b$ and $-5 = 4a + b$. Subtract the second equation from the first to obtain $7 = -2a$ or $a = -\frac{7}{2}$. Substituting this value of a in the first equation gives us $2 = -7 + b$ or $b = 9$.

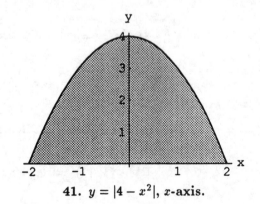

41. $y = |4 - x^2|$, x-axis.

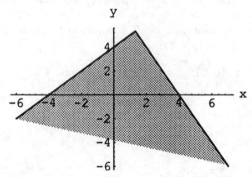

42. $R = \{(x, y) | y \le x + 4$, and $y \le -2x + 8\}$

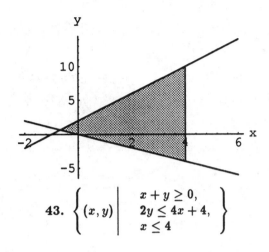

43. $\left\{ (x,y) \,\middle|\, \begin{array}{l} x+y \geq 0, \\ 2y \leq 4x+4, \\ x \leq 4 \end{array} \right\}$

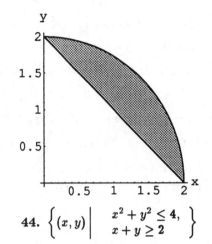

44. $\left\{ (x,y) \,\middle|\, \begin{array}{l} x^2+y^2 \leq 4, \\ x+y \geq 2 \end{array} \right\}$

45.

θ	$\sin(\theta)$	$\cos(\theta)$	$\tan(\theta)$	$\cot(\theta)$	$\sec(\theta)$	$\csc(\theta)$
$\frac{\pi}{3}$	$\frac{\sqrt{3}}{2}$	$\frac{1}{2}$	$\sqrt{3}$	$\frac{1}{\sqrt{3}}$	2	$\frac{2}{\sqrt{3}}$
$\frac{-3\pi}{4}$	$-\frac{\sqrt{2}}{2}$	$-\frac{\sqrt{2}}{2}$	1	1	$-\sqrt{2}$	$-\sqrt{2}$
$\frac{7\pi}{3}$	$\frac{\sqrt{3}}{2}$	$\frac{1}{2}$	$\sqrt{3}$	$\frac{1}{\sqrt{3}}$	2	$\frac{2}{\sqrt{3}}$
$\frac{9\pi}{4}$	$\frac{\sqrt{2}}{2}$	$\frac{\sqrt{2}}{2}$	1	1	$\sqrt{2}$	$\sqrt{2}$
$\frac{\pi}{12}$	$\frac{\sqrt{6}-\sqrt{2}}{4}$	$\frac{\sqrt{6}+\sqrt{2}}{4}$	$2-\sqrt{3}$	$2+\sqrt{3}$	$\sqrt{6}-\sqrt{2}$	$\sqrt{6}+\sqrt{2}$
$\frac{5\pi}{12}$	$\frac{\sqrt{6}+\sqrt{2}}{4}$	$\frac{\sqrt{6}-\sqrt{2}}{4}$	$2+\sqrt{3}$	$2-\sqrt{3}$	$\sqrt{6}+\sqrt{2}$	$\sqrt{6}-\sqrt{2}$

46. Let $x = 32.6161\overline{61}$. Then $100x = 3261.6161\overline{61}$.
$99x = (100-1)x = 3261.6161\overline{61} - 32.6161\overline{61} = 3229$.
Hence $32.6161\overline{61} = \frac{3229}{99}$.

47. Let $x = -6.214\overline{214}$. Then $1000x = -6214.214\overline{214}$.
$999x = (1000-1)x = -6214.214\overline{214} + 6.214\overline{214} = -6208$.
Hence $-6.214\overline{214} = -\frac{6208}{999}$

48.a.

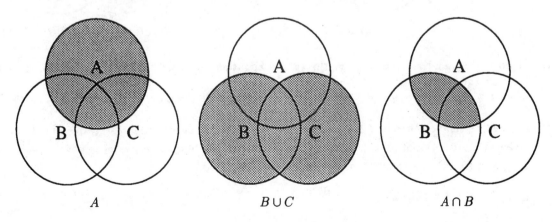

A $\qquad\qquad$ $B \cup C$ $\qquad\qquad$ $A \cap B$

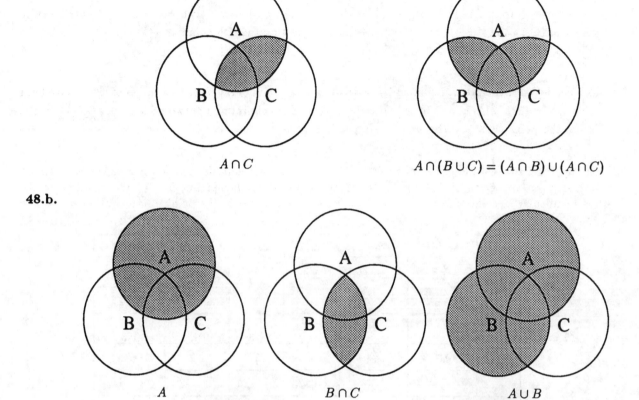

$A \cap C$

$A \cap (B \cup C) = (A \cap B) \cup (A \cap C)$

48.b.

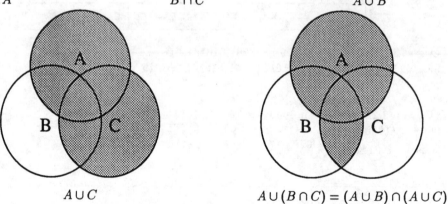

A

$B \cap C$

$A \cup B$

$A \cup C$

$A \cup (B \cap C) = (A \cup B) \cap (A \cup C)$

49. $x + 3y - 6 = 0$ and $x - y = 2$. solve the second equation for x, we have $x = y + 2$. substitute this into the first equation:

$y + 2 + 3y - 6 = 0 \leftrightarrow 4y = 4 \leftrightarrow y = 1$. also $x = y + 2 = 3$. the point of intersection is $(3, 1)$.

50. let c be the temperature in celsius and t be the time in minutes. then two values for (t, c) has been given to be $(0, 10)$ and $(3, 25)$. the slope of the line connecting these two points is $\frac{25 - 10}{3 - 0} = 5$. using the point $(0, 10)$, the equation of the line is $c = 5t + 10$.

a. $c(5) = 5 \times 5 + 10 = 35°c$.

b. $c(30) = 5 \times 30 + 10 = 160°c$. we know this cannot be true, because the boiling point of water is $100°c$.

51. let k be the temperature in kelvin and c be the temperature in cesius. two values for (k, c) are given to be $(373, 100)$ and $(273, 0)$. the slope of the line connecting the two points is $\frac{100 - 0}{373 - 273} = 1$. using the point $(273, 0)$, the equation of the line is:
$c = k - 273$.

52. let k be the temperature in kelvin and f be the temperature in fahrenheit. two values of (k, f) have been given to be $(373, 212)$ and $(273, 32)$. the slope of the line connecting the two points is $\frac{212 - 32}{373 - 273} = 1.8$. using the point $(273, 32)$, the equation of the line is:
$f - 32 = 1.8(k - 273) \leftrightarrow f = 1.8k - 459.4$.

53. first we find the equation of the line through $(4, 1)$ which is perpendicular to $x + y = 2$, that is $y = -x + 2$. the slope of the given line is -1, so the slope of the perpendicular line is $+1$. the equation of the perpendicular line through $(4, 1)$ is $y - 1 = (x - 4)$ or $y = x - 3$. the point of intersection of the two lines is the desired point. $-x + 2 = y = x - 3$. hence, $x = \frac{5}{2}$ and $y = -\frac{1}{2}$. the point on the line $x + y = 2$ nearest to $(4, 1)$ is $(\frac{5}{2}, -\frac{1}{2})$.

54. a. $p_s = \frac{n_s}{n_t} p_p$.

   b. we are given that $\frac{n_s}{n_t}$ is .85. hence, $p_s = .85 p_p$. when $p_p = 0.25$, we have
   $p_s = .85 \times 0.25 = 0.2125$.

55. no, the equation is that of a circle. the value of y is not uniquely determined for every value of x.

56. yes. $x^2 + y^3 = 3 \leftrightarrow y = \sqrt[3]{x^2 - 3}$.

57. yes. y is a constant function of x.

58. no. for each value of x, there are an infinite number of values of y since $x = sin y$ is periodic.

59. we must have $x^2 - 1 \geq 0$. the domain is $(-\infty, 1] \cup [1, \infty)$.

60. we must have $x + 2 \neq 0$. the domain is $(-\infty, -2) \cup (-2, \infty)$

61. since the domain of the sin function is the set of real numbers, there are no restrictions on the domain. the domain is the set of all real numbers x.

62. we must have $1 + \cos x \neq 0$ or $\cos x \neq \pi + 2\pi k$ for any integer k. hence the domain is $\{x | x \neq \pi + 2\pi k, \ k \ any \ integer\}$.

63. we must have $1 - \sin^2 x > 0$. this inequality holds for all real numbers $x \neq \frac{\pi}{2} + \pi k$ for some integer k. hence the domain is $\{x | x \neq \frac{\pi}{2} + \pi \ k, \ k \ any \ integer\}$.

64. we must have $x(x + 2) > 0$. this occurs when both factors x and $(x + 2)$ have the same sign. both factors are negative when $x < -2$ and both factors are positive when $x > 0$. hence the domain is $(-\infty, -2) \cup (0, \infty)$.

65. since $f(x) = x^3 \sin x$, $f(-x) = (-x)^3 \sin(-x) = (-x^3)(-\sin x) = x^3 \sin x = f(x)$. hence f is an even funtion.

66. $f(x) = \frac{1 - \cos x}{x^2}$.
   $f(-x) = \frac{1 - \cos(-x)}{(-x)^2} = \frac{1 - \cos x}{x^2} = f(x)$. hence f is an even function.

67. $f(x) = (x-1)(x+1)$

$f(-x) = (-x-1)(-x+1) = (x+1)(x-1) = f(x)$. hence f is an even function.

68. $f(x) = \frac{x \tan x}{1 + x^3}$.

$f(-x) = \frac{-x \tan(-x)}{1 + (-x)^3} = \frac{x \tan x}{1 - x^3}$. hence, f is neither odd nor even.

69. Completing squares:

$$y = 2x^2 - 2x + \frac{7}{2}$$
$$= 2(x^2 - x + \frac{1}{4}) - \frac{1}{2} + \frac{7}{2}$$
$$= 2(x - \frac{1}{2})^2 + 3.$$

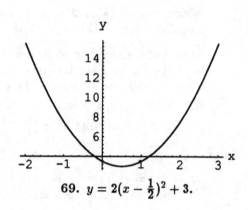

69. $y = 2(x - \frac{1}{2})^2 + 3.$

70. Completing squares:

$$y^2 - 2x + 4y + 6 = 0$$
$$\Leftrightarrow y^2 + 4y + 4 = 2x - 2$$
$$\Leftrightarrow (y + 2)^2 = 2(x - 1)$$
$$\Leftrightarrow x = \frac{1}{2}(y + 2)^2 + 1.$$

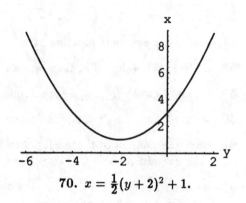

70. $x = \frac{1}{2}(y + 2)^2 + 1.$

71. Completing squares:

$$y^2 + 2x - y + \frac{3}{4} = 0$$
$$\Leftrightarrow y^2 - y + \frac{1}{4} = -2x - \frac{1}{2}$$
$$\Leftrightarrow (y - \frac{1}{2})^2 = -2(x + \frac{1}{4})$$
$$\Leftrightarrow x = -\frac{1}{2}(y - \frac{1}{2})^2 - \frac{1}{4}.$$

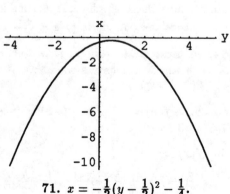

71. $x = -\frac{1}{2}(y - \frac{1}{2})^2 - \frac{1}{4}.$

72. Completing squares:

$$y = \tfrac{1}{6}x^2 + x + \tfrac{5}{2}$$
$$= \tfrac{1}{6}(x^2 + 6x + 9) - \tfrac{3}{2} + \tfrac{5}{2}$$
$$= \tfrac{1}{6}(x + 3)^2 + 1.$$

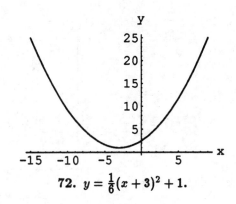

72. $y = \tfrac{1}{6}(x + 3)^2 + 1.$

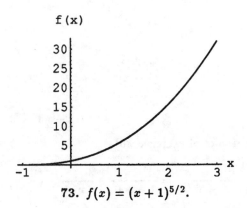

73. $f(x) = (x + 1)^{5/2}.$

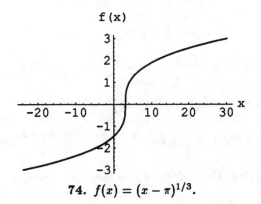

74. $f(x) = (x - \pi)^{1/3}.$

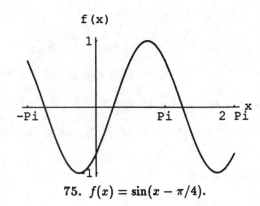

75. $f(x) = \sin(x - \pi/4).$

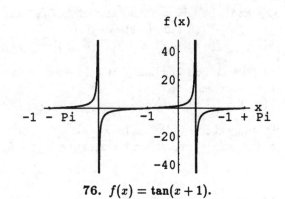

76. $f(x) = \tan(x + 1).$

77. a. if $\cos(\theta) < 0$ then $\pi < \theta < 2\pi$. then $\sin(\theta) = \sqrt{3}/2 \to \theta = \tfrac{4\pi}{3}, \tfrac{5\pi}{3}.$

b. if $\sec(\theta) = \sqrt{2}$ then $\theta = \tfrac{\pi}{4}, \tfrac{7\pi}{4}.$ but $\tan(\pi/4) = 1$, while $\tan(7\pi 4) = -1.$ hence $\theta = \tfrac{7\pi}{4}.$

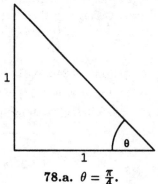

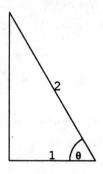

78.a. $\theta = \frac{\pi}{4}$.   **78.b,c.** $\theta = \frac{\pi}{3}$.

In exercises 79–80, we use $f(x) = x^3 - x$, $g(x) = \frac{1}{1-x}$, and $h(x) = \sin x$.

79. $g(h(x)) = g(\sin x) = \frac{1}{1 - \sin x}$.

80. $f(g(x)) = f(\frac{1}{1-x}) = (\frac{1}{1-x})^3 - \frac{1}{1-x} = \frac{1}{(1-x)^3}[1 - (1-x)^2] = \frac{x^2 - 2x}{(1-x)^3}$

81. $h(g(x)) = h(\frac{1}{1-x}) = \sin(\frac{1}{1-x})$

82. $f(g(h(x))) = f(g(\sin x)) = \frac{(\sin x)^2 - 2\sin x}{(1 - \sin x)^3}$ (use the results of exercis'e 80).

83. $g(f(h(x))) = g(f(\sin x)) = g(\sin^3 x - \sin x) = \frac{1}{1 - (\sin^3 x - \sin x)} = \frac{1}{1 - \sin^3 x + \sin x}$

84. $g(h(g(x))) = g(h(\frac{1}{1-x})) = g(\sin(\frac{1}{1-x})) = \frac{1}{1 - \sin(\frac{1}{1-x})}$

85. let f and g be odd functions and let $h = fg$. since f and g are odd, we know that
$f(-x) = -f(x)$ and $g(-x) = -g(x)$.
$h(-x) = f(-x)g(-x) = [-f(x)][-g(x)] = f(x)g(x) = h(x)$. Hence, h is an even function.

86. Let f be an odd function. Let g be an even function. Then $f(-x) = -f(x)$ and $g(-x) = g(x)$. Let $h = fg$. Then
$h(-x) = f(-x)g(-x) = -f(x)g(x) = -h(x)$. Hence, h is an odd function.

87. The tangent line to the circle $(x - 6)^2 + (y - 4)^2 = 25$ goes through the point $(3, 8)$. Let m be the slope of the tangent line. Then the equation is $(y - 8) = m(x - 3)$ or

$$y = m(x - 3) + 8.$$

We substitute this value of y into the equation of the circle and solve for x:

$$(x - 6)^2 + [m(x - 3) + 4]^2 = 25$$
$$\Leftrightarrow \quad x^2 - 12x + 36 + m^2 x^2 - 6m^2 x + 9m^2 + 8mx - 24m + 16 = 25$$
$$\Leftrightarrow \quad (1 + m^2)x^2 + (-12 - 6m^2 + 8m)x + 9m^2 - 24m + 27 = 0$$

We recall that the solution to the quadratic equation $ax^2 + bx + c = 0$ is $x = \frac{-b \pm \sqrt{b^2 - 4ac}}{2a}$. The number of solutions for x depends on the value of $D = b^2 - 4ac$. If $D > 0$ there are 2 solutions, if

$D = 0$, there is 1 solution, and if $D < 0$ there are no solutions. In our case, we are looking for a tangent line to a circle. There must be only one point of intersection and therefore we must have $D = 0$. In our case, $a = (1 + m^2)$, $b = -12 - 6m^2 + 8m$, and $c = 9m^2 - 24m + 27$. We now solve the equation $D = 0$:

$$(-12 - 6m^2 + 8m)^2 - 4(1 + m^2)(9m^2 - 24m + 27) = 0$$
$$\Leftrightarrow \qquad (-6 - 3m^2 + 4m)^2 - (1 + m^2)(9m^2 - 24m + 27) = 0$$
$$\Leftrightarrow \quad 16m^2 - 48m - 24m^3 + 36 + 36m^2 + 9m^4 - 9m^4 + 24m^3 - 36m^2 + 24m - 27 = 0$$
$$\Leftrightarrow \qquad\qquad 16m^2 - 24m + 9 = 0$$
$$\Leftrightarrow \qquad\qquad (4m - 3) = 0$$
$$\Leftrightarrow \qquad\qquad m = \tfrac{3}{4}$$

Hence the equation for the tangent line is

$$y = \frac{3}{4}(x - 3) + 8$$

88.

$$y \; = \; ax^2 + bx + c \tag{1.5}$$
$$= \; a\left(x^2 + \frac{b}{a}x + \frac{b^2}{4a^2}\right) - \frac{b^2}{4a} + c \tag{1.6}$$
$$= \; a\left(x + \frac{b}{2a}\right)^2 + c - \frac{b^2}{4a} \tag{1.7}$$

Since the y-intercept is $y = -6$. Using equation (1.1), we see that $c = -6$. Since the parabola has vertex at $x = 2$, we see that $-\frac{b}{2a} = 2$ in equation (1.2). Hence $4a + b = 0$. Since the x-intercept is $x = 3$, we substitute this, $y = 0$, and $c = -6$ into equation (1.1) to get $0 = 9a + 3b - 6$ or $3a + b = 2$. To find a and b we solve the equations:

$$4a + b = 0$$
$$3a + b = 2$$

Subtract the second equation from the first to get $a = -2$. Then $b = 8$ and $c = -6$.

89. $f(x) = 4 - x^2$, $-\infty < x < \infty$ and $g(x) = \sin x$, $0 \le x \le 2\pi$. Since the domain of f is unrestricted, the domain of $f \circ g$ is the same as the domain of g. Hence the domain of $f \circ g$ is $[0, 2\pi]$. Now, $f \circ g(x) = f(\sin x) = 4 - \sin^2 x$.

$$-1 \le \sin x \le 1 \;\Rightarrow\; 0 \le \sin^2 x \le 1 \;\Rightarrow\; 3 \le 4 - \sin^2 x \le 4.$$

Hence the range of $f \circ g$ is $[3, 4]$.

90. $f(x) = \sqrt{x - \tfrac{1}{2}}$ and $g(x) = \sin x$, $0 \le x \le 2\pi$.

$$f \circ g(x) = f(\sin x) = \sqrt{\sin x - \frac{1}{2}}$$

The square root function restricts the domain of $f \circ g$.

$$\sin x - \frac{1}{2} \ge 0 \;\Rightarrow\; \sin x \ge \frac{1}{2} \;\Rightarrow\; \frac{\pi}{6} \le x \le \frac{5\pi}{6}$$

Hence the domain is $[\frac{\pi}{6}, \frac{5\pi}{6}]$. Now we find the range:

$$\frac{1}{2} \le \sin x \le 1 \;\Rightarrow\; 0 \le \sin x - \frac{1}{2} \le \frac{1}{2} \;\Rightarrow\; 0 \le \sqrt{\sin x - \frac{1}{2}} \le \frac{\sqrt{2}}{2}$$

Hence the range is $[0, \frac{\sqrt{2}}{2}]$

91. Let P, Q and R be the points $(4,3)$, $(9,13)$ and $(20,15)$. We will consider the base, b, of the triangle to be the line segment, PR. We will compute the height, h, of the triangle by finding the line through Q which is perpendicular to PR. The slope of the line PR is $m = \frac{15-3}{20-4} = \frac{3}{4}$. The equation of the line PR is $y - 3 = \frac{3}{4}(x-4)$ or

$$y = \frac{3}{4}x$$

The slope of the perpendicular line is $-\frac{1}{m} = -\frac{4}{3}$. This line goes through $Q = (9,13)$ and its equation is $y - 13 = -\frac{4}{3}(x - 9)$ or

$$y = -\frac{4}{3}x + 25$$

To find the intersection of the two lines, sustitute $y = \frac{3}{4}x$ into the last equation:

$$\frac{3}{4}x = -\frac{4}{3}x + 25 \;\Leftrightarrow\; 9x = -16x + 12 \times 25 \;\Leftrightarrow\; x = 12$$

$y = \frac{3}{4}x = 9$. The height of the triangle is the distance from $(12,9)$ to $Q = (9,13)$. Hence, $h = \sqrt{(13-9)^2 + (9-12)^2} = \sqrt{25} = 5$. The base of the triangle is the distance from $P = (4,3)$ to $R = (20,15)$ which is $b = \sqrt{(15-3)^2 + (20-4)^2} = \sqrt{400} = 20$. Hence the area of the triangle is

$$\frac{1}{2}bh = \frac{1}{2} \times 20 \times 5 = 50$$

92. a.

$$p = \begin{cases} 500 & \text{if } x \le 5 \\ 500 - 10(x-5) & \text{if } 5 < x \le 25 \\ 300 & \text{if } x > 25 \end{cases}$$

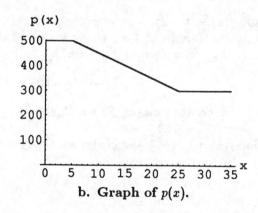

b. Graph of $p(x)$.

93.
$$R = \begin{cases} 500x & \text{if } x \le 5 \\ (500 - 10(x-5))x & \text{if } 5 < x \le 25 \\ 300x & \text{if } x > 25 \end{cases}$$

Readiness Test

1. Completing squares,

$$x^2 - 2x + y^2 + 3y = 4$$
$$\Leftrightarrow (x-1)^2 - 1 + \left(y + \tfrac{3}{2}\right)^2 - \tfrac{9}{4} = 4$$
$$\Leftrightarrow (x-1)^2 + \left(y + \tfrac{3}{2}\right)^2 = \tfrac{29}{4}$$

Thus the center is $(1, -3/2)$ and the radius is $\sqrt{29}/2$.

2. Multiplying the first equation by 2 and subtracting from the second equation, we find that $x = 0$. Consequently $y = 7$. The point of intersection is $(0, 7)$.

3. The slope of the given line is $-\frac{3}{2}$. The point–slope form of the parallel line through $(-1, 2)$ is

$$y - 2 = -\frac{3}{2}(x + 1).$$

The slope–intercept form of this equation is

$$y = -\frac{3}{2}x + \frac{1}{2}.$$

4. The side adjacent to θ has length $\frac{5\sqrt{3}}{2}$. The side opposite to θ has length $\frac{5}{2}$

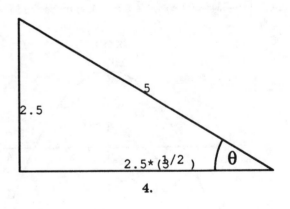

4.

5. $|x + 3| > 3 \Leftrightarrow -3 > x + 3$ or $x + 3 > 3$. This is equivalent to $x < -6$ or $x > 0$. The solution set is $(-\infty, -6) \cup (0, \infty)$.

6. The slope of the line $x - 2y = 4$ is $\frac{1}{2}$. A perpendicular line will have slope -2. Note that the second equation given in the problem is the same as the first equation. The y–intercept of this line is $y = -2$. Using the slope–intercept form, we find that the equation of the perpendicular line with this y–intercept is $y = -2x - 2$.

7. Completing squares, $2x^2 - 4x + 3 = 2(x-1)^2 + 1$. Thus the vertex of the parabola is at $(1, 1)$.

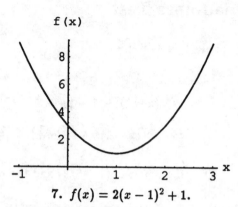

f (x)

7. $f(x) = 2(x-1)^2 + 1$.

8. For $g(x) = \sqrt{x^2 - 2x}$ to be defined, we must have $x^2 - 2x \geq 0$. This is true if and only if $(x-1)^2 \geq 1$. Thus the domain is $(-\infty, 0] \cup [2, \infty)$.

9. $x^2 - 2x - 3 > 0 \Leftrightarrow (x-1)^2 > 4 \Leftrightarrow |x-1| > 2$. Thus $\{x \mid x^2 - 2x - 3 > 0\} = (-infty, -1) \cup (3, \infty)$.

10. Given $f(x) = \sqrt{x}$ and $g(x) = x^2 + 1$, we first note that the domain of f is $[0, \infty)$ and the domain of g is $(-\infty, \infty)$.

   a. The domain of $2f/g$ is $[0, \infty)$ since $g(x)$ never vanishes. $(2f/g)(x) = \dfrac{2\sqrt{x}}{x^2 + 1}$.

   b. The domain of g/f is $(0, \infty)$ since $f(0) = 0$. $(g/f)(x) = x^{3/2} + x^{-1/2}$.

   c. The domain of fg is $[0, \infty)$. $(fg)(x) = x^{5/2} + \sqrt{x}$.

   d. The domain of $f \circ g$ is $(-\infty, \infty)$ since $g(x)$ is always positive. $(f \circ g)(x) = \sqrt{x^2 + 1}$.

   e. The domain of $g \circ f$ is equal to the domain of f, namely, $[0, \infty)$, since g is defined for all x. $(g \circ f)(x) = x + 1$.

11. $(|x| - 3)^2 = 9 \Leftrightarrow |x| - 3 = \pm 3$. This is true if and only if $|x| = 3 \pm 3 \Leftrightarrow x = \pm 3 \pm 3 = -6,\ 0,\ 6$.

12.

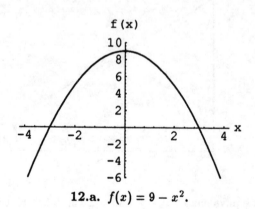

12.a. $f(x) = 9 - x^2$.

12.b. $g(x) = |9 - x^2|$.

13. $\sin(\theta) < 0 \Leftrightarrow (2n-1)\pi < \theta < 2n\pi$ for some integer n. For such values of θ, $\cos(\theta) = \dfrac{\sqrt{3}}{2} \Leftrightarrow \theta = 2n\pi - \dfrac{\pi}{6}$.

14. $\cos(2\theta) = 1/4 \Leftrightarrow \cos(\theta) = \pm\sqrt{\left(1 + \frac{1}{4}\right)\frac{1}{8}}$, by the half angle formula. Since θ is an angle in the third quadrant, $\cos(\theta) < 0$. Thus $\cos(\theta) = \sqrt{\left(1 + \frac{1}{4}\right)\frac{1}{8}}$.

15. The simplest way to do this is to let the opposite side have length 2, and let the hypotenuse have length 5. Then the adjacent side has length $\sqrt{21}$.

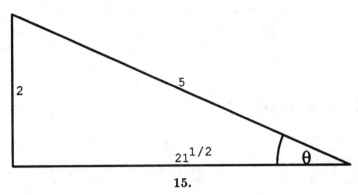

15.

16. $\cos(x) > 1/2 \Leftrightarrow 0 < x < \pi/3$ or $5\pi/3 < x < 2\pi$.

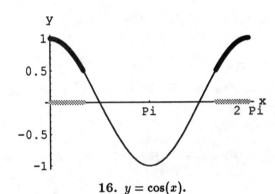

16. $y = \cos(x)$.

17. The domain of $f(x) = \frac{1}{1+\sin(x)}$ is $\{x \mid \sin(x) \neq -1\}$. This set is $\left\{x \mid x \neq \frac{3\pi}{2} + 2n\pi\right\}$.

18. The domain of the function $f(x) = (x+2)^{3/2}$ is $\{x \mid x \geq -2\}$.

19. The desired set is the perpendicular bisector of the line segment between $(-2, 1)$ and $(0, 0)$. This line segment has slope $-\frac{1}{2}$ and midpoint $(-1, 1/2)$. Thus the desired set is $\{(x, y) \mid y - 1/2 = 2(x+1)\}$.

20. $0 \leq x^2 - 4x < 5 \Leftrightarrow 4 \leq (x-2)^2 < 9$. This is equivalent to $2 \leq |x-2| < 3$. This describes the set $(-1, 0] \cup [4, 5)$.

21. a. $x^4 - 1$ is a polynomial of degree 4.

    b. $\frac{x}{x^2+1}$ is a rational function which is not a polynomial.

22. If the temperature rises linearly, at a rate of $1.2°F$ every forty years, then in twenty years the temperature will rise $.6°F$. Thus the temperature in the year 2000 will be $59.8°F$.

23. The domain of $f(x) = (x-2)^{-2/3}$ is $\{x \mid x \neq 2\}$. $f(2)$ is not defined due to the negative exponent.

24. The range of the function $f(x) = \sec(x^2)$ is $\{y \mid |y| \geq 1\}$.

25. $Y(x) = 400 - 2x$ pounds.

Chapter 2

Limits of Functions

2.1 Tangents, Areas, and Limits

1. Here $x_0 = 2$. Hence the slope of the tangent line is
$$m = \lim_{h \to 0} \frac{f(2+h) - f(2)}{h} = \lim_{h \to 0} \frac{[3(2+h) - 2] - 4}{h} = \lim_{h \to 0} 3 = 3.$$

2. Here $x_0 = 1$. Hence the slope of the tangent line is
$$m = \lim_{h \to 0} \frac{f(1+h) - f(1)}{h} = \lim_{h \to 0} \frac{[7 - 3(1+h)] - 4}{h} = \lim_{h \to 0} (-3) = -3.$$

3. Here $x_0 = 3$. Hence the slope of the tangent line is
$$m = \lim_{h \to 0} \frac{f(3+h) - f(3)}{h} = \lim_{h \to 0} \frac{2(3+h)^2 - 18}{h} = \lim_{h \to 0} \frac{12h + 2h^2}{h} = \lim_{h \to 0} (12 + 2h) = 12.$$

4. Here $x_0 = -1$. Hence the slope of the tangent line is
$$m = \lim_{h \to 0} \frac{f(-1+h) - f(-1)}{h} = \lim_{h \to 0} \frac{9(-1+h)^2 - 9}{h} = \lim_{h \to 0} \frac{-18h + 9h^2}{h} = \lim_{h \to 0} (-18h + 9h) = -18.$$

5. Here $x_0 = 1$. Hence the slope of the tangent line is
$$m = \lim_{h \to 0} \frac{f(1+h) - f(1)}{h} = \lim_{h \to 0} \frac{[2(1+h)^2 + 3] - 5}{h} = \lim_{h \to 0} \frac{4h + 2h^2}{h} = \lim_{h \to 0} (4 + 2h) = 4.$$

6. Here $x_0 = -2$. Hence the slope of the tangent line is
$$m = \lim_{h \to 0} \frac{f(-2+h) - f(-2)}{h} = \lim_{h \to 0} \frac{[5 - (-2+h)^2] - 1}{h} = \lim_{h \to 0} \frac{4h - h^2}{h} = \lim_{h \to 0} (4 - h) = 4.$$

7. Here $x_0 = -2$. Hence the slope of the tangent line is
$$m = \lim_{h \to 0} \frac{f(-2+h) - f(-2)}{h} = \lim_{h \to 0} \frac{[3(-2+h)^2 + 4(-2+h) + 2] - 6}{h}$$
$$= \lim_{h \to 0} \frac{-8h + 3h^2}{h} = \lim_{h \to 0} (-8 + 3h) = -8.$$

70

8. Here $x_0 = 2$. Hence the slope of the tangent line is

$$m = \lim_{h \to 0} \frac{f(2+h) - f(2)}{h} = \lim_{h \to 0} \frac{[(2+h)^2 - 6(2+h) + 3] - [-5]}{h}$$

$$= \lim_{h \to 0} \frac{-2h + h^2}{h} = \lim_{h \to 0} (-2 + h) = -2.$$

9. Here $x_0 = 2$. Hence the slope of the tangent line is

$$m = \lim_{h \to 0} \frac{f(2+h) - f(2)}{h} = \lim_{h \to 0} \frac{[(2+h)^3 + 3] - 11}{h}$$

$$= \lim_{h \to 0} \frac{12h + 6h^2 + h^3}{h} = \lim_{h \to 0} (12 + 6h + h^2) = 12.$$

10. Here $x_0 = 1$. Hence the slope of the tangent line is

$$m = \lim_{h \to 0} \frac{f(1+h) - f(1)}{h} = \lim_{h \to 0} \frac{[(1+h)^3 - 2(1+h) + 6] - 5}{h}$$

$$= \lim_{h \to 0} \frac{h + 3h^2 + h^3}{h} = \lim_{h \to 0} (1 + 3h + h^2) = 1.$$

11. Here $x_0 = -2$. Hence the slope of the tangent line is

$$m = \lim_{h \to 0} \frac{f(-2+h) - f(-2)}{h} = \lim_{h \to 0} \frac{(-2+h)^4 - 16}{h}$$

$$= \lim_{h \to 0} \frac{-32h + 24h^2 - 8h^3 + h^4}{h} = \lim_{h \to 0} (-32 + 24h - 8h^2 = h^3) = -32.$$

12. Here $x_0 = 1$. Hence the slope of the tangent line is

$$m = \lim_{h \to 0} \frac{f(1+h) - f(1)}{h} = \lim_{h \to 0} \frac{[a(1+h)^2 + b(1+h) + c] - [a+b+c]}{h}$$

$$= \lim_{h \to 0} \frac{2ah + ah^2 + bh}{h} = \lim_{h \to 0} (2a + ah + b) = 2a + b.$$

13. Here $x_0 = 1$. Hence the slope of the tangent line is

$$m = \lim_{h \to 0} \frac{f(1+h) - f(1)}{h} = \lim_{h \to 0} \frac{[a(1+h)^3 + b(1+h)^2 + c(1+h) + d] - [a+b+c+d]}{h}$$

$$= \lim_{h \to 0} \frac{3ah + 3ah^2 + ah^3 + 2bh + bh^2 + ch}{h} = \lim_{h \to 0} (3a + 3ah + ah^2 + 2b + bh + c) = 3a + 2b + c.$$

14. Here $x_0 = 1$. Hence the slope of the tangent line is

$$m = \lim_{h \to 0} \frac{f(1+h) - f(1)}{h} = \lim_{h \to 0} \frac{\frac{1}{1+h} - 1}{h} = \lim_{h \to 0} \frac{1 - (1+h)}{h(1+h)} = \lim_{h \to 0} \frac{-h}{h(1+h)} = \lim_{h \to 0} \frac{-1}{1+h} = -1.$$

15. Here $x_0 = -2$. Hence the slope of the tangent line is

$$m = \lim_{h \to 0} \frac{f(-2+h) - f(-2)}{h} = \lim_{h \to 0} \frac{\frac{1}{h+1} - 1}{h} = \lim_{h \to 0} \frac{1-(h+1)}{h(h+1)} = \lim_{h \to 0} \frac{-h}{h(h+1)} = \lim_{h \to 0} \frac{-1}{h+1} = -1.$$

16. Here $x_0 = 2$. Hence the slope of the tangent line is

$$m = \lim_{h \to 0} \frac{f(2+h) - f(2)}{h} = \lim_{h \to 0} \frac{\frac{3}{2(2+h)-1} - 1}{h} = \lim_{h \to 0} \frac{\frac{3}{2h+3} - 1}{h}$$

$$= \lim_{h \to 0} \frac{3-(2h+3)}{h(2h+3)} = \lim_{h \to 0} \frac{-2}{2h+3} = -\frac{2}{3}.$$

17. Here $x_0 = 2$. Hence the slope of the tangent line is

$$m = \lim_{h \to 0} \frac{f(2+h) - f(2)}{h} = \lim_{h \to 0} \frac{\frac{4}{(2+h)^2} - 1}{h} = \lim_{h \to 0} \frac{4-(2+h)^2}{h(2+h)^2}$$

$$= \lim_{h \to 0} \frac{-4h + h^2}{h(2+h)^2} = \lim_{h \to 0} \frac{-4+h}{(2+h)^2} = \frac{-4}{4} = -1.$$

18. Here $x_0 = 2$. Therefore the slope of the tangent line is

$$m = \lim_{h \to 0} \frac{f(2+h) - f(2)}{h} = \lim_{h \to 0} \frac{2(2+h)^2 - 8}{h} = \lim_{h \to 0} \frac{8h + 2h^2}{h} = \lim_{h \to 0} (8 + 2h) = 8.$$

Hence the equation of the tangent line at $(2,8)$ is $y - 8 = 8(x - 2)$, or $y = 8x - 8$.

19. Here $x_0 = -1$. Therefore the slope of the tangent line is

$$m = \lim_{h \to 0} \frac{f(-1+h) - f(-1)}{h} = \lim_{h \to 0} \frac{3(-1+h)^2 - 3}{h} = \lim_{h \to 0} \frac{-6h + 3h^2}{h} = \lim_{h \to 0} (-6 + 3h) = -6.$$

Hence the equation of the tangent line at $(-1, 3)$ is $y - 3 = -6(x + 1)$, or $y = -6x - 3$.

20. Here $x_0 = 1/2$. Therefore the slope of the tangent line is

$$m = \lim_{h \to 0} \frac{f(1/2+h) - f(1/2)}{h} = \lim_{h \to 0} \frac{\left[\left(\frac{1}{2}+h\right)^2 - 4\left(\frac{1}{2}+h\right)\right] - \left[\frac{i}{4} - 2\right]}{h}$$

$$= \lim_{h \to 0} \frac{-3h + h^2}{h} = \lim_{h \to 0} (-3 + h) = -3.$$

Hence the equation of the tangent line at $(1/2, -7/4)$ is $y - 1/2 = -3(x + 7/4)$, or $y = -3x - 19/4$.

21. Here $x_0 = 1$. Therefore the slope of the tangent line is

$$m = \lim_{h \to 0} \frac{f(1+h) - f(1)}{h} = \lim_{h \to 0} \frac{a(1+h)^2 - a}{h} = \lim_{h \to 0} \frac{2ah + ah^2}{h} = \lim_{h \to 0} (2a + ah) = 2a.$$

Hence the equation of the tangent line at $(1, a)$ is $y - a = 2a(x - 1)$, or $y = 2ax - a$.

22. Here $x_0 = -2$. Therefore the slope of the tangent line is

$$m = \lim_{h\to 0} \frac{f(-2+h) - f(-2)}{h} = \lim_{h\to 0} \frac{[5 - (-2+h)^2] - 1}{h} = \lim_{h\to 0} \frac{4h - h^2}{h} = \lim_{h\to 0}(4 - h) = 4.$$

Hence the equation of the tangent line at $(-2, 1)$ is $y - 1 = 4(x + 2)$, or $y = 4x + 9$.

23. Here $x_0 = 1$. Therefore the slope of the tangent line is

$$m = \lim_{h\to 0} \frac{f(1+h) - f(1)}{h} = \lim_{h\to 0} \frac{[2(1+h)^3 + (1+h)] - 3}{h}$$

$$= \lim_{h\to 0} \frac{7h + 6h^2 + 2h^3}{h} = \lim_{h\to 0}(7 + 6h + 2h^2) = 7.$$

Since $f(1) = 3$, the equation of the tangent line at $(1, 3)$ is $y - 3 = 7(x - 1)$, or $y = 7x - 4$.

24. Here $x_0 = 1$. Therefore the slope of the tangent line is

$$m = \lim_{h\to 0} \frac{f(1+h) - f(1)}{h} = \lim_{h\to 0} \frac{a(1+h)^3 - a}{h} = \lim_{h\to 0} \frac{3ah + 3ah^2 + h^3}{h} = \lim_{h\to 0}(3a + 3ah + h^2) = 3a.$$

Hence the equation of the tangent line at $(1, a)$ is $y - a = 3a(x - 1)$, or $y = 3ax - 2a$.

25. False. The tangent line is horizontal at a point x_0 if the slope $m = 0$. For example, if $f(x) = x^2$, then the slope of the tangent line at the point $(0, 0)$ is

$$m = \lim_{h\to 0} \frac{f(h) - f(0)}{h} = \lim_{h\to 0} \frac{h^2}{h} = \lim_{h\to 0} h = 0,$$

and hence the equation of the tangent line is $y = 0$, which is horizontal.

26. In general, we find that

$$\frac{f(x_0 + h) - f(x_0)}{h} = \frac{[(x_0 + h)^2 - 4(x_0 + h) + 3] - [x_0^2 - 4x_0 + 3]}{h} = \frac{2hx_0 + h^2 - 4h}{h} = 2x_0 + h - 4,$$

and hence the slope of the tangent line at $(x_0, f(x_0))$ is

$$m = \lim_{h\to 0}(2x_0 + h - 4) = 2x_0 - 4.$$

For example, if $x_0 = -2$, then $m = 2(-2) - 4 = -8$. Table 1.2, below on the left, summarizes these values for $x_0 = -2, -1, 0, 1, 2, 3$.

Table 1.2

x_0	$f(x_0)$	$\dfrac{f(x_0 + h) - f(x_0)}{h}$	Slope of tangent at $(x_0, f(x_0))$
-2	15	$-8 + h$	-8
-1	8	$-6 + h$	-6
0	3	$-4 + h$	-4
1	0	$-2 + h$	-2
2	-1	h	0
3	0	$2 + h$	2

Table 1.3

x_0	h	$\dfrac{f(x_0 + h) - f(x_0)}{h}$
2	0.1000	9.61
2	0.0100	9.0601
2	0.0010	9.006001
2	0.0001	9.00060001
2	-0.0001	8.99940001
2	-0.0010	8.994001
2	-0.0100	8.9401
2	-0.1000	8.41

Exercise 26

Exercise 27

27. (a) In general, we find that

$$\frac{f(2+h) - f(2)}{h} = \frac{[(2+h)^3 - 3(2+h)] - 2}{h} = \frac{9h + 6h^2 + h^3}{h} = 9 + 6h + h^2.$$

For example, if $h = 0.1000$,

$$\frac{f(2+h) - f(2)}{h} = 9 + 6(0.1) + (0.1)^2 = 9.61.$$

Table 1.3, above on the right, summarizes the slopes of the secant lines for the indicated values of h.

(b) The slope of the tangent line at $x_0 = 2$ is

$$m = \lim_{h \to 0} \frac{f(2+h) - f(2)}{h} = \lim_{h \to 0} (9 + 6h + h^2) = 9.$$

The values in Table 1.3 show that the slopes of the secant lines become closer to the limiting value as $h \to 0$.

28. (a) In general, we find that

$$\frac{f(1+h) - f(1)}{h} = \frac{3(1+h)^4 - 3}{h} = \frac{12h + 18h^2 + 12h^3 + 3h^4}{h} = 12 + 18h + 12h^2 + 3h^3.$$

For example, if $h = 0.1000$,

$$\frac{f(1+h) - f(1)}{h} = 12 + 18(0.1) + 12(0.1)^2 + 3(0.1)^3 = 13.923.$$

Table 1.3, below on the left, summarizes the slopes of the secant lines for the indicated values of h.

(b) The slope of the tangent line at $x_0 = 1$ is

$$m = \lim_{h \to 0} \frac{f(1+h) - f(1)}{h} = \lim_{h \to 0} (12 + 18h + 12h^2 + 3h^3) = 12.$$

The values in Table 1.3 show that the slopes of the secant lines become closer to this limiting value as $h \to 0$.

Table 1.3

x_0	h	$\dfrac{f(x_0 + h) - f(x_0)}{h}$
1	0.1000	13.923
1	0.0100	12.181203
1	0.0010	12.018012
1	0.0001	12.0018001
1	-0.0001	11.9982001
1	-0.0010	11.982012
1	-0.0100	11.821197
1	-0.1000	10.317

Exercise 28

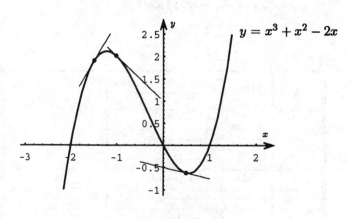

Exercise 29

74

29. (a) Estimating the slopes at $x = -3/2$, -1, and $1/2$, we obtain

$$m_{-3/2} = -7/(-4) = 7/4, \quad m_{-1} = -1, \quad m_{1/2} = 1/(-5) = -1/5.$$

(b) In general, the slope of the tangent line at $x = x_0$ is

$$m = \lim_{h \to 0} \frac{f(x_0 + h) - f(x_0)}{h} = \lim_{h \to 0} \frac{[(x_0 + h)^3 + (x_0 + h)^2 - 2(x_0 + h)] - [x_0^3 + x_0^2 - 2x_0]}{h}$$

$$= \lim_{h \to 0} \frac{3x_0^2 h + 3x_0 h^2 + h^3 + 2x_0 h + h^2 - 2h}{h} = \lim_{h \to 0} (3x_0^2 + 3x_0 h + h^2 + 2x_0 + h - 2) = 3x_0^2 + 2x_0 - 2.$$

Hence the slopes of the tangent lines at $x_0 = -3/2$, -1, and $1/2$ are:

$$\begin{aligned} x_0 = -3/2 &\implies m = 3(-3/2)^2 + 2(-3/2) - 2 = 7/4 \\ x_0 = -1 &\implies m = 3(-1)^2 + 2(-1) - 2 = -1 \\ x_0 = 1/2 &\implies m = 3(1/2)^2 + 2(1/2) - 2 = -1/4. \end{aligned}$$

30. (a) In general, we find that

$$\frac{f(1/2 + h) - f(1/2)}{h} = \frac{\dfrac{1}{1/2 + h} - 2}{h} = \frac{-2}{1/2 + h}.$$

For example, if $h = 0.1000$,

$$\frac{f(1/2 + h) - f(1/2)}{h} = \frac{-2}{1/2 + 0.1} = -3.3333.$$

Table 1.3 summarizes the slopes of the secant lines for the indicated values of h and shows that the slope of the tangent line at $x_0 = 1/2$ is approximately -4.

(b) The slope of the tangent line at $x_0 = 1/2$ is

$$m = \lim_{h \to 0} \frac{f(1/2 + h) - f(1/2)}{h} = \lim_{h \to 0} \frac{-2}{1/2 + h} = -4.$$

Table 1.3

x_0	h	$\dfrac{f(x_0 + h) - f(x_0)}{h}$
1/2	0.1000	-3.3333333
1/2	0.0100	-3.9215686
1/2	0.0010	-3.9920160
1/2	0.0001	-3.9992002
1/2	-0.0001	-4.0008002
1/2	-0.0010	-4.0080160
1/2	-0.0100	-4.0816327
1/2	-0.1000	-5

Exercise 30

Table 1.3

x_0	h	$\dfrac{f(x_0 + h) - f(x_0)}{h}$
0	0.1000	0.99833417
0	0.0100	0.99998333
0	0.0010	0.99999983
0	0.0001	1
0	-0.0001	1
0	-0.0010	0.99999983
0	-0.0100	0.99998333
0	-0.1000	0.99833417

Exercise 31

31. (a) In general, we find that

$$\frac{f(h) - f(0)}{h} = \frac{\sin h - \sin 0}{h} = \frac{\sin h}{h}.$$

For example, if $h = 0.1000$,

$$\frac{f(h) - f(0)}{h} = \frac{\sin 0.1}{0.1} = 0.99833417.$$

Table 1.3 summarizes the slopes of the secant lines for the indicated values of h and shows that the slope of the tangent line at $x_0 = 0$ is approximately 1.

(b) The slope of the tangent line at $x_0 = 0$ is

$$m = \lim_{h \to 0} \frac{f(h) - f(0)}{h} = \lim_{h \to 0} \frac{\sin h}{h}.$$

At present, we have no method for evauating this expression.

32. In general, we find that

$$\frac{f(\pi/4 + h) - f(\pi/4)}{h} = \frac{\sin(\pi/4 + h) - \sin(\pi/4)}{h}.$$

For example, if $h = 0.1000$,

$$\frac{f(\pi/4 + h) - f(\pi/4)}{h} = \frac{\sin(\pi/4 + 0.1) - \sin(\pi/4)}{h}$$

$$= 0.67060297.$$

The table to the right summarizes the slopes of the secant lines for the indicated values of h and shows that the slope of the tangent line at $x_0 = \pi/4$ is approximately 0.707.

x_0	h	$\dfrac{f(x_0 + h) - f(x_0)}{h}$
$\pi/4$	0.1000	0.67060297
$\pi/4$	0.0100	0.70355949
$\pi/4$	0.0010	0.70675311
$\pi/4$	0.0001	0.70707142
$\pi/4$	-0.0001	0.70714214
$\pi/4$	-0.0010	0.70746022
$\pi/4$	-0.0100	0.71063050
$\pi/4$	-0.1000	0.74125475

33. The slope of the tangent line at $x = a$ is

$$m = \lim_{h \to 0} \frac{f(a + h) - f(a)}{h} = \lim_{h \to 0} \frac{[(a + h)^2 + 6(a + h) + 1] - [a^2 + 6a + 1]}{h}$$

$$= \lim_{h \to 0} \frac{2ah + h^2 + 6h}{h} = \lim_{h \to 0} (2a + h + 6) = 2a + 6.$$

In particular, the slope of the tangent line is $0 \iff m = 2a + 6 = 0 \iff a = -3$.

34. In general, the slope of the tangent line at $x = a$ is

$$m = \frac{f(a + h) - f(a)}{h} = \lim_{h \to 0} \frac{[(a + h)^2 + 2(a + h) + 3] - [a^2 + 2a + 3]}{h}$$

$$= \lim_{h \to 0} \frac{2ah + h^2 + 2h}{h} = \lim_{h \to 0} (2a + h + 2) = 2a + 2.$$

Now, the tangent line is parallel to the line $y = 6x + 1 \iff m = 2a + 2 = 6 \iff a = 2$. Since $f(2) = 11$, the equation of the tangent line is $y - 11 = 6(x - 2)$, or $y = 6x - 1$.

76

35. In general, the slope of the tangent line at $x = a$ is

$$m = \frac{f(a+h) - f(a)}{h} = \lim_{h \to 0} \frac{[(a+h)^2 - 3(a+h) + 1] - [a^2 - 3a + 1]}{h}$$

$$= \lim_{h \to 0} \frac{2ah + h^2 - 3h}{h} = \lim_{h \to 0} (2a + h - 3) = 2a - 3.$$

Hence, the slope of the tangent line at the point $(a, f(a))$ is equal to the y-coordinate $\iff m = 2a - 3$ $= f(a) = a^2 - 3a + 1 \iff a^2 - 5a + 4 = (a - 4)(a - 1) = 0 \iff a = 1, 4$. Thus, there are two points at which the slope of the tangent line is equal to the y-coordinate: $(1, f(1)) = (1, -1)$ and $(4, f(4)) = (4, 5)$.

36. (a) The slope of the tangent line at $(x, f(x))$ is

$$m = \lim_{h \to 0} \frac{f(x+h) - f(x)}{h} = \lim_{h \to 0} \frac{[a(x+h)^2 + b(x+h) + c] - [ax^2 + bx + c]}{h}$$

$$= \lim_{h \to 0} \frac{2ahx + ah^2 + bh}{h} = \lim_{h \to 0} (2ax + ah + b) = 2ax + b.$$

(b) Since the x-coordinate of the vertex is $-b/2a$, the slope of the tangent line at the vertex is $2a(-b/2a) + b = 0$.

37. y-intercept $(0, 5) \implies f(0) = c = 5$; graph contains the point $(1, 2) \implies f(1) = a + b + c = 2$; slope of tangent line at $x = 2$ is $3 \implies 2a(2) + b = 3$. Hence $a + b = -3$, $4a + b = 3$ and therefore $a = 2$, $b = -5$, $c = 5$.

38. The slope of the tangent line at $(x_0, f(x_0))$ is

$$m_{x_0} = \lim_{h \to 0} \frac{f(x_0 + h) - f(x_0)}{h} = \lim_{h \to 0} \frac{(x_0 + h)^3 - x_0^3}{h}$$

$$= \lim_{h \to 0} \frac{3x_0^3 h + 3x_0 h^2 + h^3}{h} = \lim_{h \to 0} (3x_0^2 + 3x_0 h + h^2) = 3x_0^2.$$

Since $m_{-x_0} = 3(-x_0)^2 = 3x_0^2 = m_{x_0}$, the slope of the tangent line at $(x_0, f(x_0))$ is the same as at $(-x_0, f(-x_0))$. This result is in fact true for any odd function for which the limit exists; for if $f(x)$ is an odd function, then

$$m_{-x_0} = \lim_{h \to 0} \frac{f(-x_0 + h) - f(-x_0)}{h}$$

$$= \lim_{h' \to 0} \frac{f(-x_0 - h') - f(-x_0)}{-h'} \qquad \text{where } h' = -h$$

$$= \lim_{h' \to 0} \frac{-f(x_0 + h') + f(x_0)}{-h'} \qquad \text{since } f \text{ is odd}$$

$$= \lim_{h' \to 0} \frac{f(x_0 + h') - f(x_0)}{h'}$$

$$= m_{x_0}.$$

77

39. In general, we find that

$$\frac{f(0+h) - f(0)}{h} = \frac{|h|}{h} = \begin{cases} +1, & \text{if } h > 0 \\ -1, & \text{if } h < 0. \end{cases}$$

Table 1.3 summarizes the slopes of the secant lines for the indicated values of h and shows that the absolute value function does not have a tangent line at the point $(0,0)$.

Table 1.3

x_0	h	$\dfrac{f(x_0 + h) - f(x_0)}{h}$
0	0.1000	1
0	0.0100	1
0	0.0010	1
0	0.0001	1
0	−0.0001	−1
0	−0.0010	−1
0	−0.0100	−1
0	−0.1000	−1

Exercise 39

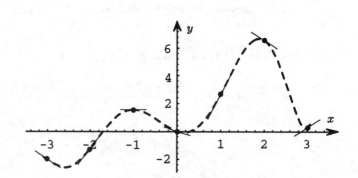

Exercise 40

40. Through each indicated point on the above graph a small line segment has been drawn having the required slope. The dotted curve is the graph of one possible function having these slopes.

41. (a) The table below summarizes the values of h, m_h^+, and m_h^- for $h = 1$, 0.1, 0.01, 0.001, and 0.0001.

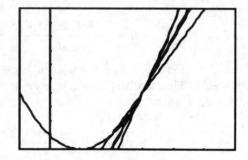

h	m_h^+	m_h^-
1	5	3
0.1	4.1	3.9
0.01	4.01	3.99
0.001	4.001	3.999
0.0001	4.0001	3.9999

(b) The graphs of $y = f(x)$, $y = m_h^+(x - 3) + 4$, and $y = m_h^-(x - 3) + 4$ with range $[-1, 6] \times [0, 10]$ and **Xscl** = **Yscl** = 1 are shown above on the right.

(c) Slope of tangent line at $a = 3$ is 4.

42. (a) The table at the top of the following page summarizes the values of h, m_h^+, and m_h^- for $h = 1$, 0.1, 0.01, 0.001, and 0.0001.

(b) The graphs of $y = f(x)$, $y = m_h^+(x - \pi/2) + 1$, and $y = m_h^-(x - \pi/2) + 1$ with range $[-\pi/2, \pi/2] \times [-1.5, 1.5]$ and **Xscl** = **Yscl** = 1 are shown at the top of the following page on the right.

(c) Slope of tangent line at $a = \pi/2$ is 0.

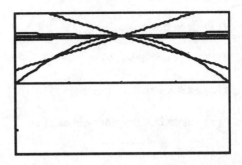

h	m_h^+	m_h^-
1	−0.459698	0.459698
0.1	−0.049958	0.049958
0.01	−0.004999	0.004999
0.001	−0.0005	0.0005
0.0001	−0.00005	0.00005

43. (a) The table below summarizes the values of h, m_h^+, and m_h^- for $h = 1, 0.1, 0.01, 0.001$, and 0.0001.

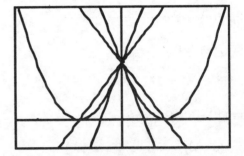

h	m_h^+	m_h^-
1	−1	1
0.1	−1.9	1.9
0.01	−1.99	1.99
0.001	−1.999	1.999
0.0001	−1.9999	1.9999

(b) The graphs of $y = f(x)$, $y = m_h^+ x + 1$, and $y = m_h^- x + 1$ with range $[-2.5, 2.5] \times [-1/2, 2]$ and **Xscl** = **Yscl** = 1 are shown above on the right.

(c) The tangent line at $a = 0$ does not exist.

44. The Instructor's Preliminary Edition and the first printing of the Student Edition used $-x^2 + 2x + 3$ for the function $f(x)$ when $x > 0$. Subsequent editions used $-x^2 + 2x + 3$ when $x > 0$. The solution given here uses $-x^2 + 2x + 1$ when $x > 0$.

(a) The table below summarizes the values of h, m_h^+, and m_h^- for $h = 1, 0.1, 0.01, 0.001$, and 0.0001.

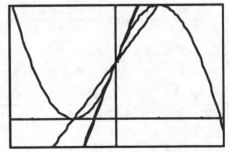

h	m_h^+	m_h^-
1	−1	1
0.1	1.9	1.9
0.01	1.99	1.99
0.001	1.999	1.999
0.0001	1.9999	1.9999

(b) The graphs of $y = f(x)$, $y = m_h^+ x + 1$, and $y = m_h^- x + 1$ with range $[-2.5, 2.5] \times [-1/2, 3]$ and **Xscl** = **Yscl** = 1 are shown above on the right.

(c) The slope of the tangent line at $a = 0$ is 2.

2.2 Limits of Functions

1. $\lim\limits_{x \to a} f(x)$ exists and equals $f(a)$.

2. $\lim_{x \to a} f(x)$ exists and equals $f(a)$.

3. $\lim_{x \to a} f(x)$ does not exist.

4. $\lim_{x \to a} f(x)$ does not exist.

5. $\lim_{x \to a} f(x)$ exists but does not equal $f(a)$.

6. $\lim_{x \to a} f(x)$ exists but does not equal $f(a)$.

7. $\lim_{x \to 0} f(x) = 2$

8. $\lim_{x \to 0} f(x) = 2$

9. $\lim_{x \to 0} f(x) = 1$

10. $\lim_{x \to 0} f(x) = 0$

11. $\lim_{x \to 2} (3 + 7x) = 17$

12. $\lim_{x \to 0} (x^3 - 7x + 5) = 5$

13. $\lim_{x \to 0} \dfrac{(3 + x)^2 - 9}{x} = \lim_{x \to 0} \dfrac{x(x + 6)}{x} = \lim_{x \to 0} (x + 6) = 6$

14. $\lim_{x \to 3} \dfrac{x^2 - 9}{x - 3} = \lim_{x \to 3} \dfrac{(x - 3)(x + 3)}{x - 3} = \lim_{x \to 3} (x + 3) = 6$

15. $\lim_{h \to 0} \dfrac{h^2 - 1}{h - 1} = \dfrac{-1}{-1} = 1$

16. $\lim_{x \to 0} \dfrac{(x + 3)^3 - 27}{x} = \lim_{x \to 0} \dfrac{x(x^2 + 9x + 27)}{x} = \lim_{x \to 0} (x^2 + 9x + 27) = 27$

17. $\lim_{x \to -2} \dfrac{x^2 - x - 6}{x + 2} = \lim_{x \to -2} \dfrac{(x - 3)(x + 2)}{x + 2} = \lim_{x \to -2} (x - 3) = -5$

18. $\lim_{h \to 0} \dfrac{(4 + h)^2 - 16}{h} = \lim_{h \to 0} \dfrac{h(h + 8)}{h} = \lim_{h \to 0} (h + 8) = 8$

19. $\lim_{x \to 0} \dfrac{1 - \cos^2 x}{\sin x \cos x} = \lim_{x \to 0} \dfrac{\sin^2 x}{\sin x \cos x} = \lim_{x \to 0} \dfrac{\sin x}{\cos x} = \dfrac{0}{1} = 0$

20. $\lim_{x \to 0} \dfrac{\tan x}{\sin x} = \lim_{x \to 0} \dfrac{1}{\cos x} = \dfrac{1}{1} = 1$

21. $\lim_{h \to 0} \dfrac{\sin 2h}{\sin h} = \lim_{h \to 0} \dfrac{2 \sin h \cos h}{\sin h} = \lim_{h \to 0} (2 \cos h) = 2$

22. $\lim_{\theta \to 2} \dfrac{\theta^4 - 2^4}{\theta - 2} = \lim_{\theta \to 2} \dfrac{(\theta - 2)(\theta + 2)(\theta^2 + 2^2)}{\theta - 2)} = \lim_{\theta \to 2} (\theta + 2)(\theta^2 + 2^2) = (4)(8) = 32$

23. $\lim_{x \to \pi/2} \sin 2x \csc x = \lim_{x \to \pi/2} \dfrac{\sin 2x}{\sin x} = \dfrac{0}{1} = 0$

24. $\lim\limits_{x\to\pi/2}\dfrac{\sin 2x}{\cos x} = \lim\limits_{x\to\pi/2}\dfrac{2\sin x\cos x}{\cos x} = \lim\limits_{x\to\pi/2}(2\sin x) = 2$

25. $\lim\limits_{x\to-1}\dfrac{x^2-2x-3}{x+1} = \lim\limits_{x\to-1}\dfrac{(x-3)(x+1)}{x+1} = \lim\limits_{x\to-1}(x-3) = -4$

26. $\lim\limits_{x\to4}\dfrac{x^2-2x-8}{x-4} = \lim\limits_{x\to4}\dfrac{(x-4)(x+2)}{x-4} = \lim\limits_{x\to4}(x+2) = 6$

27. $\lim\limits_{x\to\pi/2}\dfrac{\sec x\cos x}{x} = \lim\limits_{x\to\pi/2}\dfrac{1}{x} = \dfrac{2}{\pi}$

28. $\lim\limits_{x\to1}\dfrac{x^3+3x^2-x-3}{x^2-1} = \lim\limits_{x\to1}\dfrac{x^2(x+3)-(x+3)}{x^2-1} = \lim\limits_{x\to1}\dfrac{(x+3)(x^2-1)}{x^2-1} = \lim\limits_{x\to1}(x+3) = 4$

29. $\lim\limits_{x\to-1}\dfrac{x^3+3x^2-x-3}{x^2-1} = \lim\limits_{x\to-1}\dfrac{x^2(x+3)-(x+3)}{x^2-1} = \lim\limits_{x\to-1}\dfrac{(x+3)(x^2-1)}{x^2-1} = \lim\limits_{x\to-1}(x+3) = 2$

30. $\lim\limits_{x\to1}\dfrac{x^3-7x+6}{x^2+2x-3} = \lim\limits_{x\to1}\dfrac{(x-1)(x^2+x-6)}{(x+3)(x-1)} = \lim\limits_{x\to1}\dfrac{x^2+x-6}{x+3} = \dfrac{-4}{4} = -1$

31. $\lim\limits_{x\to-3}\dfrac{x^3-7x+6}{x^2+2x-3} = \lim\limits_{x\to-3}\dfrac{(x-1)(x+3)(x-2)}{(x+3)(x-1)} = \lim\limits_{x\to-3}(x-2) = -5$

32. $\lim\limits_{x\to1}\dfrac{x^{5/2}-x^{1/2}}{x^{3/2}-x^{1/2}} = \lim\limits_{x\to1}\dfrac{x^{1/2}(x^2-1)}{x^{1/2}(x-1)} = \lim\limits_{x\to1}(x+1) = 2$

33. $\lim\limits_{x\to2}\dfrac{x^{13/4}-2x^{9/4}}{x^{5/4}-2x^{1/4}} = \lim\limits_{x\to2}\dfrac{x^{9/4}(x-2)}{x^{1/4}(x-2)} = \lim\limits_{x\to2}x^2 = 4$

34. $\lim\limits_{x\to-2}\dfrac{x^{7/3}+x^{4/3}-2x^{1/3}}{x^{4/3}+2x^{1/3}} = \lim\limits_{x\to-2}\dfrac{x^{1/3}(x^2+x-2)}{x^{1/3}(x+2)} = \lim\limits_{x\to-2}\dfrac{(x+2)(x-1)}{x+2} = \lim\limits_{x\to-2}(x-1) = -3$

35. $\lim\limits_{h\to0}\dfrac{3-\sqrt{9+h}}{h} = \lim\limits_{h\to0}\dfrac{3-\sqrt{9+h}}{h}\cdot\dfrac{3+\sqrt{9+h}}{3+\sqrt{9+h}} = \lim\limits_{h\to0}\dfrac{-h}{h(3+\sqrt{9+h})} = \lim\limits_{h\to0}\dfrac{-1}{3+\sqrt{9+h}} = -\dfrac{1}{6}$

36. $\lim\limits_{x\to4}\dfrac{\sqrt{x}-2}{x-4} = \lim\limits_{x\to4}\dfrac{\sqrt{x}-2}{(\sqrt{x}-2)(\sqrt{x}+2)} = \lim\limits_{x\to4}\dfrac{1}{\sqrt{x}+2} = \dfrac{1}{4}$

37. $\lim\limits_{x\to1}\dfrac{(1/x)-1}{x-1} = \lim\limits_{x\to1}\dfrac{1-x}{x(x-1)} = \lim\limits_{x\to1}\dfrac{-1}{x} = -1$

38. $\lim\limits_{x\to-1}\dfrac{x^{-2}-1}{x+1} = \lim\limits_{x\to-1}\dfrac{1-x^2}{x^2(x+1)} = \lim\limits_{x\to-1}\dfrac{(1-x)(1+x)}{x^2(x+1)} = \lim\limits_{x\to-1}\dfrac{1-x}{x^2} = 2$

39. $\lim\limits_{x\to0}f(x) = 2$

40. $\lim\limits_{x\to0}f(x)$ does not exist since, if x is close to zero and positive, $f(x)$ is close to 2, but if x is close to zero and negative, $f(x)$ is close to 1.

41. $\lim\limits_{x\to0}f(x)$ does not exist since, if x is close to zero and positive, $f(x)$ is close to 8, but if x is close to zero and negative, $f(x)$ is close to 4.

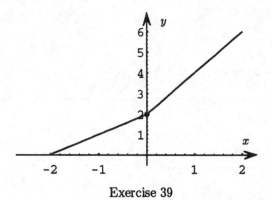

Exercise 39

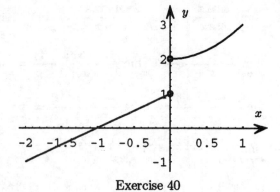

Exercise 40

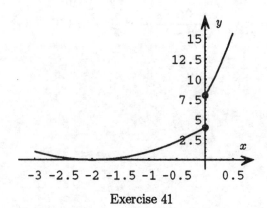

Exercise 41

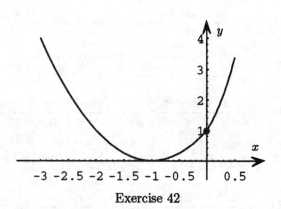

Exercise 42

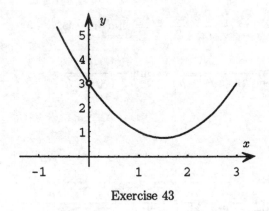

Exercise 43

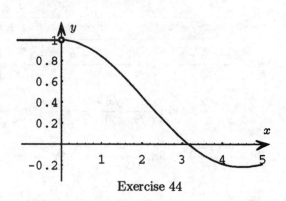

Exercise 44

42. $\lim_{x \to 0} f(x) = 1$

43. First observe that if $x \neq 0$, $\dfrac{(x-1)^3 + 1}{x} = \dfrac{x^3 - 3x^2 + 3x}{x} = x^2 - 3x + 3$. Hence $\lim_{x \to 0} f(x) = 3$.

44. $\lim_{x \to 0} f(x) = 1$

45. $\lim_{x \to 0} f(x) = 1$

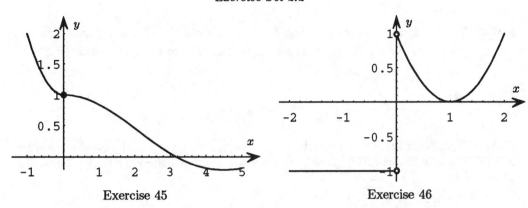

Exercise 45   Exercise 46

46. $\lim\limits_{x \to 0} f(x)$ does not exist since, if x is close to zero and positive, $f(x)$ is close to 1, but if x is close to zero and negative, $f(x)$ is close to -1.

47. $\lim\limits_{x \to 0} \dfrac{x^2}{1 - \cos x} = 2$; see the table of values below.

48. $\lim\limits_{x \to 0} \dfrac{\cos 2x - 1}{x^2} = -2$; see the table of values below.

x	$\dfrac{x^2}{1 - \cos x}$	$\dfrac{\cos 2x - 1}{x^2}$	$\dfrac{x - \sin x}{x^3}$	$\dfrac{x^2 - \sin x^2}{x^6}$	$\dfrac{1 - \cos x^2}{x^4}$	$\dfrac{x^6}{x^2 - \tan x^2}$
1.000	2.17534265	-1.4161468	0.15852902	0.15852902	0.45969769	-1.7940189
0.500	2.04219271	-1.8387908	0.16459569	0.16614661	0.49740125	-2.9249776
0.100	2.00166750	-1.9933422	0.16658335	0.16666583	0.49999583	-2.9998800
0.050	2.00041672	-1.9983339	0.16664583	0.16666661	0.49999974	-2.9999925
0.010	2.00001667	-1.9999333	0.16666583	0.16666666	0.50000000	-3.0000000
0.005	2.00000417	-1.9999833	0.16666646	0.16666659	0.50000004	-2.9999994
-0.005	2.00000417	-1.9999833	0.16666646	0.16666659	0.50000004	-2.9999994
-0.010	2.00001667	-1.9999333	0.16666583	0.16666666	0.50000000	-3.0000000
-0.050	2.00041672	-1.9983339	0.16664583	0.16666661	0.49999974	-2.9999926
-0.100	2.00166750	-1.9933422	0.16658335	0.16666583	0.49999583	-2.9998800
-0.500	2.04219271	-1.8387908	0.16459569	0.16614661	0.49740125	-2.9249776
-1.000	2.17534265	-1.4161468	0.15852902	0.15852902	0.45969769	-1.7940189

49. $\lim\limits_{x \to 0} \dfrac{x - \sin x}{x^3} = \dfrac{1}{6}$; see the table of values above.

50. $\lim\limits_{x \to 0} \dfrac{x^2 - \sin x^2}{x^6} = \dfrac{1}{6}$; see the table of values above.

51. $\lim\limits_{x \to 0} \dfrac{1 - \cos x^2}{x^4} = \dfrac{1}{2}$; see the table of values above.

52. $\lim\limits_{x \to 0} \dfrac{x^6}{x^2 - \tan x^2} = -3$; see the table of values above.

53. $c = \pm\sqrt{2}/2$; for, if x is close to zero and negative, $f(x)$ is close to $1/2$, while if x is close to zero and is positive, $f(x)$ is close to c^2. Hence, in order for $\lim_{x \to 0} f(x)$ to exist, $c^2 = 1/2$, or $c = \pm\sqrt{1/2} = \pm\sqrt{2}/2$.

54. (a) First observe that

$$\frac{(x+1)^2 - 2}{x} = \frac{x^2 + 2x - 1}{x} = x + 2 - \frac{1}{x}.$$

Hence, if x is close to zero and positive, $f(x)$ becomes larger without bound negatively, while if x is close to zero and negative, $f(x)$ becomes larger without bound positively. Thus, $\lim_{x \to 0} f(x)$ does not exist.

(b) $c = 1$; for, since

$$g(x) = \frac{(x+1)^2 - c}{x} = \frac{x^2 + 2x + 1 - c}{x} = x + 2 + \frac{1-c}{x},$$

the limit of $f(x)$ does not exist, as in part (a), unless the term $(1-c)/x$ is identically zero for all x, which means that $c = 1$.

55. False; the function whose graph appears in Exercise 10 has the property that $f(0)$ does not exist but $\lim_{x \to 0} f(x) = 0$.

56. (a) Yes; $\lim_{x \to 0} x\sin(1/x) = 0$ since $x\sin(1/x)$ lies between $-x$ and x and hence, if x is close to zero, either positive or negative, $x\sin(1/x)$ is close to $\pm x$ and hence close to zero.

(b) $\lim_{x \to 0} x^2\sin(1/x) = 0$; for, as in part (a), $x^2\sin(1/x)$ lies between $\pm x^2$ and hence, if x is close to zero, either positive or negative, $x^2\sin(1/x)$ is close to $\pm x^2$ and hence close to zero. Note that in Example 7, $\sin(1/x)$ oscillates between ± 1 and does not become arbitrarily close to zero; on the other hand, the presence of x and x^2 in the functions $x\sin(1/x)$ and $x^2\sin(1/x)$ forces these functions to lie between $\pm x$ and $\pm x^2$, respectively, and hence forces them to become arbitrarily close to zero as x becomes close to zero.

57. Graph with the range: $[-6, -2] \times [-15, 5]$, **Xscl** $= 1$, **Yscl** $= 1$. $\lim_{x \to -4} f(x) = -11$.

h	1	0.1	0.01	0.001	0.0001	0.00001
$f(a+h)$	-9	-10.8	-10.98	-10.998	-10.9998	-10.99998
$f(a-h)$	-13	-11.2	-11.02	-11.002	-11.0002	-11.00002

58. Graph with the range: $[-1, 1] \times [-3, 3]$, **Xscl** $= 1$, **Yscl** $= 1$. $\lim_{x \to 0} f(x) = 0$.

h	1	0.1	0.01	0.001	0.0001	0.00001
$f(a+h)$	2.524413	-0.163206	-0.015191	0.002481	-0.000092	.0000001
$f(a-h)$	2.524413	-0.163206	-0.015191	0.002481	-0.000092	.0000001

59. Graph with the range: $[-1.5, 1.5] \times [0, 1/2]$, **Xscl** $= 0.5$, **Yscl** $= 0.25$. $\lim_{x \to 0} f(x) = 1/3$.

h	1	0.1	0.01	0.001	0.0001	0.00001
$f(a+h)$	0.280490	0.332778	0.333328	0.333333	0.333333	0.333333
$f(a-h)$	0.280490	0.332778	0.333328	0.333333	0.333333	0.333333

60. Graph with the range: $[0, 4] \times [0, 6]$, **Xscl** $= 1$, **Yscl** $= 1$. $\lim_{x \to 2} f(x) = 5$.

h	1	0.1	0.01	0.001	0.0001	0.00001
$f(a+h)$	1.75	4	4.873786	4.987039	4.998700	4.999870
$f(a-h)$	3	4.62	4.9602	4.996002	4.999600	4.999960

61. 3/2

62. $2/\pi = 0.6366197724$

63. 16.4852814

64. $-\pi/2 = -1.570796327$

2.3 The Formal Definition of Limit

1. (a) $|(2x + 5) - 9| < \varepsilon \iff |2x - 4| < \varepsilon \iff |x - 4| < \varepsilon/2$. Thus, in general, set $\delta = \varepsilon/2$, although any smaller value will do.

   (b) If $\varepsilon = 2$, set $\delta = \varepsilon/2 = 1$
   If $\varepsilon = 0.4$, set $\delta = \varepsilon/2 = 0.2$
   If $\varepsilon = 0.05$, set $\delta = \varepsilon/2 = 0.025$

2. (a) $|(1 - 4x) - 13| < \varepsilon \iff |-4x - 12| < \varepsilon \iff |x + 3| < \varepsilon/4$. Thus, in general, set $\delta = \varepsilon/4$, although any smaller value will do.

   (b) If $\varepsilon = 2$, set $\delta = \varepsilon/4 = 0.5$
   If $\varepsilon = 0.4$, set $\delta = \varepsilon/4 = 0.1$
   If $\varepsilon = 0.1$, set $\delta = \varepsilon/4 = 0.025$

3. (a) $|(x^2 + 3) - 3| < \varepsilon \iff |x|^2 < \varepsilon \iff |x| < \sqrt{\varepsilon}$. Thus, in general, set $\delta = \sqrt{\varepsilon}$, although any smaller value will do.

   (b) If $\varepsilon = 2$, set $\delta = \sqrt{2}$
   If $\varepsilon = 1$, set $\delta = 1$
   If $\varepsilon = 0.3$, set $\delta = \sqrt{0.3}$

4. (a) First observe that

$$\left| \frac{1}{x - 2} + \frac{1}{4} \right| = \left| \frac{4 + x - 2}{4(x - 2)} \right| = \frac{1}{4} \left| \frac{x + 2}{x - 2} \right|.$$

Now, since $-3 < x < -1$, $-5 < x - 2 < -3$. Therefore $|x - 2| > 3$ and hence

$$\left| \frac{1}{x - 2} \right| < \frac{1}{3}.$$

Hence, if $|x + 2| < 12\varepsilon$,

$$\left| \frac{1}{x - 2} + \frac{1}{4} \right| = \frac{1}{4} |x + 2| \left| \frac{1}{x - 2} \right| \leq \frac{1}{4} (12\varepsilon) \left(\frac{1}{3} \right) = \varepsilon.$$

Thus, in general, set $\delta = \min\{1, 12\varepsilon\}$.

   (b) If $\varepsilon = 2$, set $\delta = 1$.
   If $\varepsilon = 0.4$, set $\delta = 1$.
   If $\varepsilon = 0.05$, set $\delta = 12\varepsilon = 0.6$.

5. First observe that if $x \neq 1$,
$$\frac{x^2 - 1}{x - 1} = \frac{(x - 1)(x + 1)}{x - 1} = x + 1.$$

Now, $|(x + 1) - 2| < \varepsilon \iff |x - 1| < \varepsilon$. Hence, for $x \neq 1$,

$$0 < |x - 1| < \varepsilon \implies \left| \frac{x^2 - 1}{x - 1} - 2 \right| = |x - 1| < \varepsilon.$$

Thus, in general, let $\delta = \varepsilon$, although any smaller value will do. In particular, if $\varepsilon = 2$, set $\delta = 2$; if $\varepsilon = 0.8$, set $\delta = 0.8$; and if $\varepsilon = 0.05$, set $\delta = 0.05$.

6. First observe that if we are given $\varepsilon > 0$ and x is sufficiently close to 4, say within one unit, then $\left| \sqrt{2x+1} + 3 \right| > \sqrt{7} + 3$ and hence

$$\left| \sqrt{2x+1} - 3 \right| = \left| \frac{(\sqrt{2x+1} - 3)(\sqrt{2x+1} + 3)}{\sqrt{2x+1} + 3} \right| = \frac{2|x-4|}{\left| \sqrt{2x+1} + 3 \right|} < \frac{2|x-4|}{\sqrt{7} + 3} < \varepsilon$$

$$\Longleftrightarrow \quad |x - 4| < \frac{\varepsilon}{2} \left(\sqrt{7} + 3 \right).$$

Thus, if we let δ be the smaller of the two numbers 1 and $(\varepsilon/2)\left(\sqrt{7} + 3 \right)$, it follows that

$$0 < |x - 4| < \delta \quad \Longrightarrow \quad \left| \sqrt{2x+1} - 3 \right| < \frac{2|x-4|}{\sqrt{7}+3} < \frac{2}{\sqrt{7}+3} \frac{\varepsilon}{2} \left(\sqrt{7} + 3 \right) = \varepsilon.$$

In particular, if $\varepsilon = 2$, set $\delta = 1$; if $\varepsilon = 0.4$, set $\delta = 1$; and if $\varepsilon = 0.1$, set $\delta = 0.28$.

7. Let $\varepsilon > 0$ be given. Then

$$|(x + 3) - 6| < \varepsilon \quad \Longleftrightarrow \quad |x - 3| < \varepsilon.$$

Thus, if we set $\delta = \varepsilon$, it follows that

$$0 < |x - 3| < \delta \quad \Longrightarrow \quad |(x + 3) - 6| < \varepsilon$$

and hence $\lim_{x \to 3} (x + 3) = 6$.

8. Let $\varepsilon > 0$ be given. Then

$$|(2x - 3) + 1| = |2x - 2| = 2|x - 1| < \varepsilon \quad \Longleftrightarrow \quad |x - 1| < \varepsilon/2.$$

Thus, if we set $\delta = \varepsilon/2$, it follows that

$$0 < |x - 1| < \delta \quad \Longrightarrow \quad |(2x - 3) + 1| < \varepsilon$$

and hence $\lim_{x \to 1} (2x - 3) = -1$.

9. Let $\varepsilon > 0$ be given. Then

$$|(7 - 3x) + 5| = |-3x + 12| = 3|x - 4| < \varepsilon \quad \Longleftrightarrow \quad |x - 4| < \varepsilon/3.$$

Thus, if we set $\delta = \varepsilon/3$, it follows that

$$0 < |x - 4| < \delta \quad \Longrightarrow \quad |(7 - 3x) + 5| < \varepsilon$$

and hence $\lim_{x \to 4} (7 - 3x) = -5$.

10. Let $\varepsilon > 0$ be given. Then

$$|(3x + 5) - 11| = |3x - 6| = 3|x - 2| < \varepsilon \quad \Longleftrightarrow \quad |x - 2| < \varepsilon/3.$$

Thus, if we set $\delta = \varepsilon/3$, it follows that

$$0 < |x - 2| < \delta \quad \Longrightarrow \quad |(3x + 5) - 11| < \varepsilon$$

and hence $\lim_{x \to 2} (3x + 5) = 11$.

11. Let ε be given. Then, if $x \neq 3$,

$$\left| \frac{x^2 - 9}{x - 3} - 6 \right| = \left| \frac{(x-3)(x+3)}{x-3} - 6 \right| = |x - 3|.$$

Thus, if we set $\delta = \varepsilon$, it follows that

$$0 < |x - 3| < \delta \quad \Longrightarrow \quad \left| \frac{x^2 - 9}{x - 3} - 6 \right| < \varepsilon$$

and hence $\lim\limits_{x \to 3} \dfrac{x^2 - 9}{x - 3} = 6$.

12. Let $\varepsilon > 0$ be given. Then, if $x \neq 2$,

$$\left| \frac{x^2 - 4}{x - 2} - 4 \right| = \left| \frac{(x-2)(x+2)}{x-2} - 4 \right| = |x - 2|.$$

Thus, if we set $\delta = \varepsilon$, it follows that

$$0 < |x - 2| < \delta \quad \Longrightarrow \quad \left| \frac{x^2 - 4}{x - 2} - 4 \right| < \varepsilon$$

and hence $\lim\limits_{x \to 2} \dfrac{x^2 - 4}{x - 2} = 4$.

13. Let $\varepsilon > 0$ be given. Then, if x is sufficiently close to 3, say within one unit, $|x + 3| < 7$ and hence

$$|x^2 - 9| = |(x - 3)(x + 3)| = |x - 3|\,|x + 3| < 7|x - 3| < \varepsilon \quad \Longleftrightarrow \quad |x - 3| < \varepsilon/7.$$

Thus, if we set δ equal to the smaller of the numbers 1 and $\varepsilon/7$, then

$$0 < |x - 3| < \delta \quad \Longrightarrow \quad |x^2 - 9| < 7(\varepsilon/7) = \varepsilon$$

and hence $\lim\limits_{x \to 3} x^2 = 9$.

14. Let $\varepsilon > 0$ be given. Then, if x is sufficiently close to 2, say within one unit, $|x + 2| < 5$ and hence

$$|3x^2 - 12| = 3|x - 2|\,|x + 2| < 3(5)|x - 2| < \varepsilon \quad \Longleftrightarrow \quad |x - 2| < \varepsilon/15.$$

Thus, if we set δ equal to the smaller of the numbers 1 and $\varepsilon/15$, it follows that

$$0 < |x - 2| < \delta \quad \Longrightarrow \quad |3x^2 - 12| < 15(\varepsilon/15) = \varepsilon$$

and hence $\lim\limits_{x \to 2} 3x^2 = 12$.

15. Let $\varepsilon > 0$ be given. Then

$$|(x^2 - 2x + 4) - 3| = |x^2 - 2x + 1| = |x - 1|^2 < \varepsilon \quad \Longleftrightarrow \quad |x - 1| < \sqrt{\varepsilon}.$$

Thus, if we set $\delta = \sqrt{\varepsilon}$, it follows that

$$0 < |x - 1| < \delta \quad \Longrightarrow \quad |(x^2 - 2x + 4) - 3| < \varepsilon$$

and hence $\lim\limits_{x \to 1} (x^2 - 2x + 4) = 3$.

16. Let $\varepsilon > 0$ be given. Then

$$\left|(x^2 - 6x + 13) - 4\right| = \left|x^2 - 6x + 9\right| = |x - 3|^2 < \varepsilon \quad \Longleftrightarrow \quad |x - 3| < \sqrt{\varepsilon}.$$

Thus, if we set $\delta = \sqrt{\varepsilon}$, it follows that

$$0 < |x - 3| < \delta \quad \Longrightarrow \quad \left|(x^2 - 6x + 13) - 4\right| < \varepsilon$$

and hence $\lim\limits_{x \to 3}(x^2 - 6x + 13) = 4$.

17. Let $\varepsilon > 0$ be given. Then, if x is sufficiently close to 4, say within one unit, $|\sqrt{x} + 2| > \sqrt{3} + 2$ and hence

$$\left|\sqrt{x} - 2\right| = \left|\frac{(\sqrt{x} - 2)(\sqrt{x} + 2)}{\sqrt{x} + 2}\right| = \frac{|x - 4|}{|\sqrt{x} + 2|} < \frac{|x - 4|}{\sqrt{3} + 2} < \varepsilon \quad \Longleftrightarrow \quad |x - 4| < \varepsilon(\sqrt{3} + 2).$$

Thus, if we set δ equal to the smaller of the numbers 1 and $\varepsilon\left(\sqrt{3} + 2\right)$, it follows that

$$0 < |x - 4| < \delta \quad \Longrightarrow \quad \left|\sqrt{x} - 2\right| < \frac{|x - 4|}{\sqrt{3} + 2} < \frac{\varepsilon\left(\sqrt{3} + 2\right)}{\sqrt{3} + 2} = \varepsilon$$

and hence $\lim\limits_{x \to 4} \sqrt{x} = 2$.

18. Let $\varepsilon > 0$ be given and let $\delta = \varepsilon$. Then

$$0 < |x - 3| < \delta \quad \Longrightarrow \quad \left||x - 3| - 0\right| < \varepsilon$$

and hence $\lim\limits_{x \to 3} |x - 3| = 0$.

19. Let $\varepsilon > 0$ be given. Then, if x is sufficiently close to 2, say within one unit, $|x| < 3$ and hence

$$\left|(x^2 - 2x + 2) - 2\right| = |x|\,|x - 2| < 3|x - 2| < \varepsilon \quad \Longleftrightarrow \quad |x - 2| < \varepsilon/3.$$

Thus, if we let δ be the smaller of the numbers 1 and $\varepsilon/3$, it follows that

$$0 < |x - 2| < \delta \quad \Longrightarrow \quad \left|(x^2 - 2x + 2) - 2\right| = |x|\,|x - 2| < 3(\varepsilon/3) = \varepsilon$$

and hence $\lim\limits_{x \to 2}(x^2 - 2x + 2) = 2$.

20. Let $\varepsilon > 0$. Then, if x is sufficiently close to 1, say within one unit, $|x - 3| < 3$ and hence

$$\left|(x^2 - 4x + 1) + 2\right| = |(x - 1)(x - 3)| < 3|x - 1| < \varepsilon \quad \Longleftrightarrow \quad |x - 1| < \varepsilon/3.$$

Thus, if we let δ be the smaller of the numbers 1 and $\varepsilon/3$, it follows that

$$0 < |x - 1| < \delta \quad \Longrightarrow \quad \left|(x^2 - 4x + 1) + 2\right| = |x - 1|\,|x - 3| < (3)(\varepsilon/3) = \varepsilon$$

and hence $\lim\limits_{x \to 1}(x^2 - 4x + 1) = -2$.

21. Let $\varepsilon > 0$ and set $\delta = \varepsilon/2$. Using the inequality $|\sin x| \leq |x|$ from Example 5, it follows that

$$0 < |x| < \delta \quad \Longrightarrow \quad |\sin 2x| \leq |2x| = 2|x| < 2(\varepsilon/2) = \varepsilon$$

and hence $\lim\limits_{x \to 0} \sin 2x = 0$.

22. Let $\varepsilon > 0$ be given and set $\delta = \varepsilon$. Then

$$0 < |x| < \delta \quad \Longrightarrow \quad \big||x| - 0\big| < \varepsilon$$

and hence $\lim\limits_{x \to 0} |x| = 0$.

23. Let $\varepsilon > 0$ be given and set $\delta = \sqrt{\varepsilon}$. Then, since $|\sin x| \le |x|$,

$$0 < |x| < \delta \quad \Longrightarrow \quad |x \sin x - 0| = |x||\sin x| < |x|^2 < \left(\sqrt{\varepsilon}\right)^2 = \varepsilon$$

and hence $\lim\limits_{x \to 0} x \sin x = 0$.

24. (a) Since $-1 \le \sin x \le 1$ for all x, $|\sin x| \le 1$ and hence

$$|f(x)| = |x \sin(1/x)| = |x||\sin(1/x)| \le |x| \quad \text{for all } x \ne 0.$$

  (b) Let $\varepsilon > 0$ be given and set $\delta = \varepsilon$. Then

$$0 < |x| < \delta \quad \Longrightarrow \quad |x \sin(1/x) - 0| \le |x| < \varepsilon$$

   and hence $\lim\limits_{x \to 0} x \sin(1/x) = 0$.

25. Let $\varepsilon > 0$ be given and let $\delta = \varepsilon/3$. Then

$$
\begin{aligned}
-\delta < x < 0 \quad &\Longrightarrow \quad |f(x) - 2| = 2|x| < 2\delta < 2(\varepsilon/2) < \varepsilon \\
0 < x < \delta \quad &\Longrightarrow \quad |f(x) - 2| = 3|x| < 3\delta = \varepsilon.
\end{aligned}
$$

Therefore $0 < |x| < \delta \Longrightarrow |f(x) - 2| < \varepsilon$ and hence $\lim\limits_{x \to 0} f(x) = 2$.

Remark: In this case there are two choices for δ, namely $\varepsilon/2$ and $\varepsilon/3$, depending on whether x is positive or negative. We choose the smaller of the two values, $\varepsilon/3$, to insure that it works for both cases.

26. Let $\varepsilon > 0$ be given and let δ be the smaller of the numbers $\varepsilon/2$ and $\sqrt{\varepsilon}$. Then

$$
\begin{aligned}
-\delta < x < 0 \quad &\Longrightarrow \quad |f(x) + 1| = 2|x| < 2\delta < 2(\varepsilon/2) = \varepsilon \\
0 < x < \delta \quad &\Longrightarrow \quad |f(x) + 1| = |x|^2 < \delta^2 < \left(\sqrt{\varepsilon}\right)^2 = \varepsilon.
\end{aligned}
$$

Therefore $0 < |x| < \delta \Longrightarrow |f(x) + 1| < \varepsilon$ and hence $\lim\limits_{x \to 0} f(x) = -1$.

27. $\lim_{x \to 1/2}(2x^2 - x - 1) = 1$.

ε	0.5	0.25	0.1	0.05
δ	0.25	0.125	0.0625	0.3125

28. $\lim_{x \to 2/3}(-3x^2 + 4x + 1) = 7/3$.

ε	0.5	0.25	0.1	0.05
δ	0.25	0.25	0.125	0.125

29. $\lim_{x \to 1/10}(1/x) = 10$.

ε	0.5	0.25	0.1	0.05
δ	.003906	0.001953	0.000977	0.000488

30. $\lim_{x\to 1/10} f(x) = 2$.

ε	0.5	0.25	0.1	0.05
δ	0.125	0.0625	0.03125	0.015625

31. Yes, it is possible to find a δ that works for all values of a. At $a = 0$ a value of $\delta = 0.0078$ works. If you now draw the graph of $f(x) = \sin x$ using the range $[a - \delta, a + \delta] \times [f(a) - 0.01, f(a) + 0.01]$ with a any real number, you will see that $f(x)$ stays between the lines $y = f(x) \pm 0.01$. It is not possible to do this with $f(x) = 1/x$. For example, if $a = 1$ then $\delta = 0.0078$ will work, but this value of δ will not work at $a = 0.5$. To see this, draw the graph of $f(x) = 1/x$ on the range $[0.4922, 0.5078] \times [1.99, 2.01]$ and observe that the graph of $f(x)$ is not contained in this graphing window.

2.4 Properties of Limits

1. $\lim_{x\to 3}(3x - 7) = 3 \lim_{x\to 3} x - \lim_{x\to 3} 7 = 3(3) - 7 = 2$

2. $\lim_{x\to 2}(x^3 - 4x + 1) = \lim_{x\to 2} x^3 - 4 \lim_{x\to 2} x + \lim_{x\to 2} 1 = 2^3 - 4(2) + 1 = 1$

3. $\lim_{x\to 2} \dfrac{x^2 + 3x}{x - 3} = \dfrac{\lim_{x\to 2}(x^2 + 3x)}{\lim_{x\to 2}(x - 3)} = \dfrac{\lim_{x\to 2} x^2 + 3 \lim_{x\to 2} x}{\lim_{x\to 2} x - \lim_{x\to 2} 3} = \dfrac{2^2 + 3(2)}{2 - 3} = -10$

4. $\lim_{x\to 5} \dfrac{x^2 - 5x}{x - 5} = \lim_{x\to 5} \dfrac{x(x - 5)}{x - 5} = \lim_{x\to 5} x = 5$

5. $\lim_{x\to 4} \sqrt{x}\,(1 - x^2) = \left(\lim_{x\to 4} \sqrt{x}\right)\left(\lim_{x\to 4}(1 - x^2)\right) = \sqrt{4}\,(1 - 4^2) = -30$

6. $\lim_{x\to 2}\left(4\sqrt{x} - 5x^2\right) = 4 \lim_{x\to 2} \sqrt{x} - 5 \lim_{x\to 2} x^2 = 4\sqrt{2} - 5\left(2^2\right) = 4\sqrt{2} - 20$

7. $\lim_{x\to 1}\left(3x^9 - \sqrt[3]{x}\right) = 3 \lim_{x\to 1} x^9 - \lim_{x\to 1} \sqrt[3]{x} = 3(1)^9 - \sqrt[3]{1} = 2$

8. $\lim_{x\to 2} \dfrac{x^3 - 6x + 5}{x^2 + 2x + 2} = \dfrac{\lim_{x\to 2} x^3 - 6 \lim_{x\to 2} x + \lim_{x\to 2} 5}{\lim_{x\to 2} x^2 + 2 \lim_{x\to 2} x + \lim_{x\to 2} 2} = \dfrac{2^3 - 6(2) + 5}{2^2 + 2(2) + 2} = \dfrac{1}{10}$

9. $\lim_{x\to 3} \dfrac{x^2 - 2x - 3}{x - 3} = \lim_{x\to 3} \dfrac{(x - 3)(x + 1)}{x - 3} = \lim_{x\to 3}(x + 1) = 4$

10. $\lim_{x\to 3} \dfrac{x^2 - 10x + 21}{x - 3} = \lim_{x\to 3} \dfrac{(x - 3)(x - 7)}{x - 3} = \lim_{x\to 3}(x - 7) = -4$

11. $\lim_{x\to 4} \dfrac{x^{3/2} + 2\sqrt{x}}{x^{5/2} + \sqrt{x}} = \dfrac{\lim_{x\to 4} x^{3/2} + 2 \lim_{x\to 4} \sqrt{x}}{\lim_{x\to 4} x^{5/2} + \lim_{x\to 4} \sqrt{x}} = \dfrac{4^{3/2} + 2\sqrt{4}}{4^{5/2} + \sqrt{4}} = \dfrac{6}{17}$

12. $\lim_{x\to 0} \dfrac{\sin x}{\sin 2x} = \lim_{x\to 0} \dfrac{\sin x}{2 \sin x \cos x} = \dfrac{1}{2} \dfrac{1}{\lim_{x\to 0} \cos x} = \dfrac{1}{2}$

13. $\lim_{x\to 0} \dfrac{\tan x}{\sin 2x} = \lim_{x\to 0} \dfrac{\sin x}{\cos x} \dfrac{1}{2 \sin x \cos x} = \dfrac{1}{2} \lim_{x\to 0} \dfrac{1}{\cos^2 x} = \dfrac{1}{2}$

14. $\lim_{x\to 1} \dfrac{x^3 + 2x^2 + 2x - 5}{x^2 - 1} = \lim_{x\to 1} \dfrac{(x - 1)(x^2 + 3x + 5)}{(x - 1)(x + 1)} = \lim_{x\to 1} \dfrac{x^2 + 3x + 5}{x + 1} = \dfrac{1^2 + 3(1) + 5}{1 + 1} = \dfrac{9}{2}$

15. $\lim\limits_{x \to -8} \dfrac{x^{2/3} - x}{x^{5/3}} = \dfrac{(-8)^{2/3} - (-8)}{(-8)^{5/3}} = -\dfrac{3}{8}$

16. $\lim\limits_{x \to 4} \dfrac{x^{3/2} - x^{5/2}}{4 + \sqrt{x}} = \dfrac{4^{3/2} - 4^{5/2}}{4 + \sqrt{4}} = -4$

17. $\lim\limits_{x \to 4} \dfrac{x - 4}{\sqrt{x} - 2} = \lim\limits_{x \to 4} \dfrac{(\sqrt{x} - 2)(\sqrt{x} + 2)}{\sqrt{x} - 2} = \lim\limits_{x \to 4} (\sqrt{x} + 2) = \sqrt{4} + 2 = 4$

18. $\lim\limits_{x \to -3} \dfrac{x^2 - 6x - 27}{x^2 + 3x} = \lim\limits_{x \to -3} \dfrac{(x + 3)(x - 9)}{x(x + 3)} = \lim\limits_{x \to -3} \dfrac{x - 9}{x} = \dfrac{-3 - 9}{-3} = 4$

19. $\lim\limits_{x \to -1} \left(x^{7/3} - 2x^{2/3} \right)^2 = \left[(-1)^{7/3} - 2(-1)^{2/3} \right]^2 = 9$

20. $\lim\limits_{x \to -1} \dfrac{x^3 + x^2 - x - 1}{x^3 + 2x^2 + 2x + 1} = \lim\limits_{x \to -1} \dfrac{(x + 1)(x^2 - 1)}{(x + 1)(x^2 + x + 1)} = \lim\limits_{x \to -1} \dfrac{x^2 - 1}{x^2 + x + 1} = \dfrac{(-1)^2 - 1}{(-1)^2 + (-1) + 1} = 0$

21. $\lim\limits_{x \to a} 3 \cdot f(x) \cdot g(x) = \left(\lim\limits_{x \to a} 3 \right) \left(\lim\limits_{x \to a} f(x) \right) \left(\lim\limits_{x \to a} g(x) \right) = 3(2)(-3) = -18$

22. $\lim\limits_{x \to a} \dfrac{f(x) - g(x)}{[h(x)]^2} = \dfrac{\lim\limits_{x \to a} f(x) - \lim\limits_{x \to a} g(x)}{\left[\lim\limits_{x \to a} h(x) \right]^2} = \dfrac{2 - (-3)}{5^2} = \dfrac{1}{5}$

23. $\lim\limits_{x \to a} \dfrac{6f(x) - 4[g(x)]^2}{g(x) - 4f(x)} = \dfrac{6 \lim\limits_{x \to a} f(x) - 4 \left[\lim\limits_{x \to a} g(x) \right]^2}{\lim\limits_{x \to a} g(x) - 4 \lim\limits_{x \to a} f(x)} = \dfrac{6(2) - 4(-3)^2}{(-3) - 4(2)} = \dfrac{24}{11}$

24. $\lim\limits_{x \to a} [f(x) + g(x)]^3 = \left[\lim\limits_{x \to a} f(x) + \lim\limits_{x \to a} g(x) \right]^3 = [2 + (-3)]^3 = -1$

25. $\lim\limits_{x \to 0} \dfrac{\sin x}{2x} = \dfrac{1}{2} \lim\limits_{x \to 0} \dfrac{\sin x}{x} = \dfrac{1}{2} (1) = \dfrac{1}{2}$

26. $\lim\limits_{x \to 0} \dfrac{3 \sin x}{x} = 3 \lim\limits_{x \to 0} \dfrac{\sin x}{x} = 3(1) = 3$

27. $\lim\limits_{x \to 0} \dfrac{\tan x}{4x} = \dfrac{1}{4} \lim\limits_{x \to 0} \left(\dfrac{\sin x}{x} \dfrac{1}{\cos x} \right) = \dfrac{1}{4} (1)(1) = \dfrac{1}{4}$

28. $\lim\limits_{x \to 0} \dfrac{\sec x \tan x}{x} = \lim\limits_{x \to 0} \dfrac{\sin x}{x} \dfrac{1}{\cos^2 x} = (1)(1) = 1$

29. $\lim\limits_{x \to 0} x^2 \cot x = \lim\limits_{x \to 0} \left(\dfrac{x}{\sin x} \right) (x \cos x) = (1)(0)(1) = 0$

30. $\lim\limits_{x \to 0} \dfrac{\sin^2 x}{x} = \lim\limits_{x \to 0} \left(\dfrac{\sin x}{x} \right)^2 x = (1)(0) = 0$

31. $\lim\limits_{x \to 0} \dfrac{\sin x}{\sqrt[5]{x}} = \lim\limits_{x \to 0} \left(\dfrac{\sin x}{x} \right) x^{4/5} = (1)(0) = 0$

32. $\lim\limits_{x \to 0} \dfrac{\sin^2 x}{x^{4/3}} = \lim\limits_{x \to 0} \left(\dfrac{\sin x}{x} \right)^2 x^{2/3} = (1)(0) = 0$

33. $\lim\limits_{x \to 0} f(x) = 1$ since $1 = \lim\limits_{x \to 0} (1 - x^2) \le \lim\limits_{x \to 0} f(x) \le \lim\limits_{x \to 0} (1 + x^2) = 1$

34. $\lim\limits_{x \to 3} f(x) = 9$ since $9 = \lim\limits_{x \to 3} (6x - x^2) \le \lim\limits_{x \to 3} f(x) \le \lim\limits_{x \to 3} (x^2 - 6x + 18) = 9$

35. $\lim\limits_{x \to 0} f(x) = 1$ since $1 = \lim\limits_{x \to 0} (1 - x^4) \le \lim\limits_{x \to 0} f(x) \le \lim\limits_{x \to 0} \sec x = \lim\limits_{x \to 0} \dfrac{1}{\cos x} = 1$

36. $\lim\limits_{x \to a} [f(x) + g(x) + h(x)] = \lim\limits_{x \to a} [f(x) + g(x)] + \lim\limits_{x \to a} h(x) = \lim\limits_{x \to a} f(x) + \lim\limits_{x \to a} g(x) + \lim\limits_{x \to a} h(x)$

37. $\lim\limits_{x \to a} [f(x)g(x)h(x)] = \left(\lim\limits_{x \to a} [f(x)g(x)] \right) \left(\lim\limits_{x \to a} h(x) \right) = \left(\lim\limits_{x \to a} f(x) \right) \left(\lim\limits_{x \to a} g(x) \right) \left(\lim\limits_{x \to a} h(x) \right)$

38. (a) If $n = 1$, $\lim\limits_{x \to a} x^n = \lim\limits_{x \to a} x = a = a^n$.

    (b) Let $m > 1$ and assume that $\lim\limits_{x \to a} x^m = a^m$. Then

$$
\begin{aligned}
\lim\limits_{x \to a} x^{m+1} &= \lim\limits_{x \to a} (x^m x) \\
&= \left(\lim\limits_{x \to a} x^m \right) \left(\lim\limits_{x \to a} x \right) && \text{since the limit of a product is the product of the limits} \\
&= (a^m)(a) && \text{by the induction hypothesis} \\
&= a^{m+1}.
\end{aligned}
$$

    (c) It follows by mathematical induction that $\lim\limits_{x \to a} x^n = a^n$ for all positive integers n.

39. (a) If $n = 1$, $\lim\limits_{x \to a} [f(x)]^n = \lim\limits_{x \to a} f(x) = L = L^n$.

    (b) Let $m > 1$ and assume that $\lim\limits_{x \to a} [f(x)]^m = L^m$. Then

$$
\begin{aligned}
\lim\limits_{x \to a} [f(x)]^{m+1} &= \lim\limits_{x \to a} [f(x)]^m f(x) \\
&= \left(\lim\limits_{x \to a} [f(x)]^m \right) \left(\lim\limits_{x \to a} f(x) \right) && \text{since the limit of a product} \\
& && \text{is the product of the limits} \\
&= (L^m)(L) && \text{by the induction hypothesis} \\
&= L^{m+1}.
\end{aligned}
$$

    (c) It follows by mathematical induction that $\lim\limits_{x \to a} [f(x)]^n = L^n$ for all positive integers n.

40. If $n = 1$, then $\lim\limits_{x \to c} Q(x) = \lim\limits_{x \to c} (a_1 x + a_0) = a_1 \left(\lim\limits_{x \to c} x \right) + \lim\limits_{x \to c} a_0 = a_1 c + a_0$. Now, let $m > 1$ and assume that

$$\lim\limits_{x \to c} (a_m x^m + a_{m-1} x^{m-1} + \cdots + a_1 x + a_0) = a_m c^m + \cdots + a_1 c + a_0.$$

92

Then

$$\lim_{x \to c} \left(a_{m+1} x^{m+1} + \cdots + a_1 x + a_0 \right) = \lim_{x \to c} \left[a_{m+1} x^{m+1} + (a_m x^m + \cdots + a_1 x + a_0) \right]$$

$$= a_{m+1} \left(\lim_{x \to c} x^{m+1} \right) + \lim_{x \to c} (a_m x^m + \cdots + a_1 x + a_0)$$
since the limit of a sum is the sum of the limits

$$= a_{m+1} c^{m+1} + (a_m c^m + \cdots + a_1 c + a_0)$$
by the induction hypothesis and Exercise 38

$$= a_{m+1} c^{m+1} + a_m c^m + \cdots + a_1 c + a_0.$$

It now follows by mathematical induction that

$$\lim_{x \to c} Q(x) = a_n c^n + a_{n-1} c^{n-1} + \cdots + a_1 c + a_0$$

for all positive integers n.

41. Let $f(x) = x^2$, $a = 1$, $b = -1$. Then $\lim_{x \to 1} x^2 = \lim_{x \to -1} x^2 = 1$.

42. (a) $h(x) = f(x) + g(x) = \dfrac{|x|}{x} + \left(-\dfrac{|x|}{x} \right) = 0$ for $x \neq 0$.

   (b) $\lim_{x \to 0} h(x) = 0$ since, if x is close to zero and positive, $h(x)$ is zero, while if x is close to zero and negative, $h(x)$ is also zero. But $\lim_{x \to 0} f(x)$ does not exist since, if x is close to zero and positive, $f(x)$ is equal to 1, while if x is close to zero and negative, $f(x)$ is equal to -1. Similarly, $\lim_{x \to 0} g(x)$ does not exist since it is equal to -1 for $x > 0$ and 1 for $x < 0$.

43. Let $f(x) = |x|/x$ and $g(x) = -|x|/x$, $x \neq 0$. Then $f(x)g(x) = -1$ for all $x \neq 0$ and hence $\lim_{x \to 0} f(x)g(x) = -1$. But neither $\lim_{x \to 0} f(x)$ nor $\lim_{x \to 0} g(x)$ exist.

44. Yes; if $f(x) = |x|/x$, $x \neq 0$, and $c = 0$, then $\lim_{x \to 0} cf(x) = \lim_{x \to 0} 0 = 0$, but $\lim_{x \to 0} f(x)$ does not exist. However, if $c \neq 0$, then it is not possible to find such a function; for, if $\lim_{x \to 0} cf(x)$ exists and $c \neq 0$, then

$$\lim_{x \to 0} f(x) = \lim_{x \to 0} \frac{1}{c} \left(cf(x) \right)$$

exists.

45. $\lim_{x \to a} (x^2 - 2x - 5) = a^2 - 2a - 5 = 10 \implies a^2 - 2a - 15 = (a - 5)(a + 3) = 0 \implies a = -3, 5$.

46. Let c be any constant and let $\varepsilon > 0$. Set $\delta = \varepsilon$. Then

$$0 < |x - a| < \delta \implies |c - c| = 0 < \varepsilon$$

and hence $\lim_{x \to a} c = c$.

47. Let a be any number and let $\varepsilon > 0$. Set $\delta = \varepsilon$. Then

$$0 < |x - a| < \delta \implies |x - a| < \varepsilon$$

and hence $\lim_{x \to a} x = a$.

48. (b) Since $1 - \cos^2 x = \sin^2 x$, it follows from part (a) that

$$\frac{1 - \cos x}{x} = \frac{1 - \cos^2 x}{x(1 + \cos x)} = \frac{\sin^2 x}{x(1 + \cos x)}.$$

(c) It follows from part (b) that

$$\lim_{x \to 0} \frac{1 - \cos x}{x} = \left(\lim_{x \to 0} \frac{\sin x}{x}\right)\left(\lim_{x \to 0} \frac{\sin x}{1 + \cos x}\right) = 1 \cdot \frac{0}{1 + 1} = 0.$$

49. $\displaystyle\lim_{x \to 0} \frac{1 - \cos x}{x^{1/3}} = \lim_{x \to 0} \left(\frac{1 - \cos x}{x} \cdot x^{2/3}\right) = \left(\lim_{x \to 0} \frac{1 - \cos x}{x}\right)\left(\lim_{x \to 0} x^{2/3}\right) = 0 \cdot 0 = 0.$

50. $\displaystyle\lim_{x \to 0} \frac{\cos^2 x - 2\cos x + 1}{x^{5/3}} = \lim_{x \to 0} \left[\left(\frac{1 - \cos x}{x}\right)^2 \cdot x^{1/3}\right] = \left(\lim_{x \to 0} \frac{1 - \cos x}{x}\right)^2 \left(\lim_{x \to 0} x^{1/3}\right) = 0 \cdot 0 = 0.$

51. Let $-\pi/2 < x < 0$. Then $0 < -x < \pi/2$ and hence, by inequality (8),

$$\frac{1}{\cos(-x)} \ge \frac{\sin(-x)}{-x} \ge \cos(-x).$$

But $\sin(-x) = -\sin x$ and $\cos(-x) = \cos x$. Therefore

$$\frac{1}{\cos x} \ge \frac{-\sin x}{-x} = \frac{\sin x}{x} \ge \cos x.$$

Hence inequality (8) is true for $-\pi/2 < x < 0$.

52. Let $n = -s$, $s > 0$, and $m > 0$ be integers.

(i) If m is even and $0 < a < \infty$, it follows from Theorem 3(i) that

$$\lim_{x \to a} x^{n/m} = \lim_{x \to a} x^{-s/m} = \lim_{x \to a} \frac{1}{x^{s/m}} = \frac{\displaystyle\lim_{x \to a} 1}{\displaystyle\lim_{x \to a} x^{s/m}} = \frac{1}{a^{s/m}} = a^{-s/m} = a^{n/m}.$$

(ii) If m is odd and $-\infty < a < 0$, it follows from Theorem 3(ii) that

$$\lim_{x \to a} x^{n/m} = \lim_{x \to a} x^{-s/m} = \lim_{x \to a} \frac{1}{x^{s/m}} = \frac{\displaystyle\lim_{x \to a} 1}{\displaystyle\lim_{x \to a} x^{s/m}} = \frac{1}{a^{s/m}} = a^{-s/m} = a^{n/m}.$$

53. The graph of these functions with the range $[-1, 1] \times [0, 2]$ is shown below. The graphs show that $\lim_{x \to 0}(\sin x/x) = 1$.

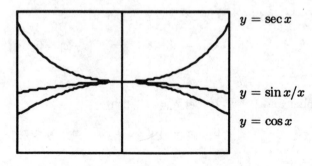

$y = \sec x$

$y = \sin x/x$

$y = \cos x$

2.5   One-Sided Limits

1. (a) $\lim\limits_{x \to 1^-} f(x) = 2$

   (b) $\lim\limits_{x \to 1} f(x) = 2$

   (c) $\lim\limits_{x \to 0} f(x) = 0$

2. (a) $\lim\limits_{x \to 1^-} f(x) = -1$

   (b) $\lim\limits_{x \to 1^+} f(x) = 0$

   (c) $\lim\limits_{x \to 1} f(x)$ does not exist.

3. (a) $\lim\limits_{x \to 2} f(x) = 0$

   (b) $\lim\limits_{x \to 2^-} f(x) = 2$

   (c) $\lim\limits_{x \to 2^+} f(x) = -1$

4. (a) All numbers except $a = 1$.

   (b) All numbers except $a = 1$.

   (c) All numbers.

   (d) All numbers except $a = 1$.

5. (a) All numbers except $a = 1$.

   (b) All numbers.

   (c) All numbers except $a = 1$.

   (d) All numbers except $a = 1$.

6. (a) All numbers.

   (b) All numbers except $a = 2$.

   (c) All numbers except $a = 2$.

   (d) All numbers except $a = 2$.

7. $\lim\limits_{x \to 2^+} \sqrt{x - 2} = 0$

8. $\lim\limits_{x \to 3^-} \dfrac{|x - 3|}{x - 3} = -1$

9. $\lim\limits_{x \to 0^+} (x^{5/2} - 5x^{3/2}) = 0$

10. $\lim\limits_{x \to 1^-} \sqrt{1 - x} = 0$

11. $\lim\limits_{x \to 2^-} \dfrac{x^2 - 4x + 4}{x - 2} = \lim\limits_{x \to 2^-} \dfrac{(x - 2)^2}{x - 2} = \lim\limits_{x \to 2^-} (x - 2) = 0$

12. $\lim\limits_{x \to 0^+} \dfrac{x - 1}{\sqrt{x} - 1} = 1$

13. $\lim\limits_{x \to 0^+} \dfrac{\sqrt{x} + x^{4/3}}{6 - x^{3/4}} = 0$

14. $\lim\limits_{x \to 3^+} [\![1 + x]\!] = 4$

15. $\lim\limits_{x \to 3^-} [\![1 + x]\!] = 3$

16. $\lim\limits_{x \to 0^+} \dfrac{3\sqrt{x} + 4x^2 - 6}{\sqrt{x} + \sqrt{x + 4}} = -3$

17. $\lim\limits_{x \to 3^-} x[\![x]\!] = 6$

18. $\lim\limits_{x \to 0^-} [\![x^2 - 6x + 5]\!] = \lim\limits_{x \to 0^-} [\![(x - 5)(x - 1)]\!] = 5$

19. $\lim\limits_{x \to 4^+} \dfrac{\sqrt{x - 4}}{x + 2} = 0$

20. $\lim\limits_{x \to 2^-} [\![x^2 - 6x + 5]\!] = \lim\limits_{x \to 2^-} [\![(x - 5)(x - 1)]\!] = -3$

21. $\lim\limits_{x \to 3^+} \dfrac{[\![x - 7]\!]}{[\![x + 4]\!]} = -\dfrac{4}{7}$

22. $\lim\limits_{x \to 0^+} \sin \dfrac{1}{x}$ does not exist

23. $\lim\limits_{x \to 0^+} [\![2 - x^2]\!] = 1$

24. $\lim\limits_{x \to 2^-} [\![1 + 4x - x^2]\!] = 4$

25. $\lim\limits_{x \to 0} [\![2 - x^2]\!] = 1$

26. (a) $\lim\limits_{x \to -1^-} f(x) = \lim\limits_{x \to -1^-} (x + 2) = 1$

    (b) $\lim\limits_{x \to -1^+} f(x) = \lim\limits_{x \to -1^+} (-x) = 1$

    (c) Yes; $\lim\limits_{x \to -1} f(x) = 1$

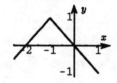

27. (a) $\lim\limits_{x \to 0^-} f(x) = \lim\limits_{x \to 0^-} \cos x = 1$

    (b) $\lim\limits_{x \to 0^+} f(x) = \lim\limits_{x \to 0^+} (1 - x) = 1$

    (c) Yes; $\lim\limits_{x \to 0} f(x) = 1$

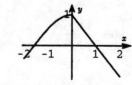

28. (a) $\lim\limits_{x \to 0^-} f(x) = \lim\limits_{x \to 0^-} [\![x + 2]\!] = 1$

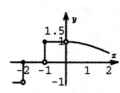

   (b) $\lim\limits_{x \to 0^+} f(x) = \lim\limits_{x \to 0^+} \dfrac{\sin x}{x} = 1$

   (c) Yes; $\lim\limits_{x \to 0} f(x) = 1$

29. (a) $\lim\limits_{x \to 0^-} f(x) = \lim\limits_{x \to 0^-} [\![x + 3]\!] = 2$

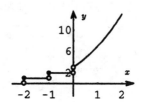

   (b) $\lim\limits_{x \to 0^+} f(x) = \lim\limits_{x \to 0^+} \dfrac{(x + 1)^3 - 1}{x}$
   $= \lim\limits_{x \to 0^+} (x^2 + 3x + 3) = 3$

   (c) No; the one-sided limits are not equal.

30. (a) $\lim\limits_{x \to 0^-} f(x) = \lim\limits_{x \to 0^-} \dfrac{\sin x}{-x} = -1$

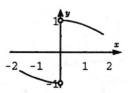

   (b) $\lim\limits_{x \to 0^+} f(x) = \lim\limits_{x \to 0^+} \dfrac{\sin x}{x} = 1$

   (c) No; the one-sided limits are not equal.

31. Since
$$\lim\limits_{x \to 3^-} f(x) = \lim\limits_{x \to 3^-} (x + 2) = 5 \quad \text{and} \quad \lim\limits_{x \to 3^+} f(x) = \lim\limits_{x \to 3^+} (2x - 1) = 5$$
are equal, $\lim\limits_{x \to 3} f(x)$ exists and is equal to 5.

32. Since
$$\lim\limits_{x \to 1^-} f(x) = \lim\limits_{x \to 1^-} x^2 = 1 \quad \text{and} \quad \lim\limits_{x \to 1^+} f(x) = \lim\limits_{x \to 1^+} x^5 = 1$$
are equal, $\lim\limits_{x \to 1} f(x)$ exists and is equal to 1.

33. (a) The graph is shown to the right.

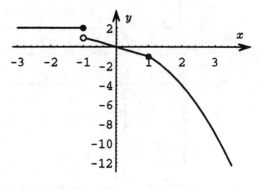

   (b) No; $\lim\limits_{x \to -1^-} f(x) = \lim\limits_{x \to -1^-} 2 = 2$ and $\lim\limits_{x \to -1^+} f(x)$
   $= \lim\limits_{x \to -1^+} (-x) = 1$. Therefore $\lim\limits_{x \to -1} f(x)$ does not
   exist since the one-sided limits are not equal.

   (c) Yes; $\lim\limits_{x \to 1^-} f(x) = \lim\limits_{x \to 1^-} (-x) = -1$ and $\lim\limits_{x \to 1^+} f(x)$
   $= \lim\limits_{x \to 1^+} (-x^2) = -1$. Therefore $\lim\limits_{x \to 1} f(x)$ exists and
   is equal to -1.

   (d) See (b) and (c).

34. The two one-sided limits must be equal:
$$\lim\limits_{x \to -1^-} f(x) = \lim\limits_{x \to -1^-} (x + 2) = 1 = \lim\limits_{x \to -1^+} f(x) = \lim\limits_{x \to -1^+} cx^2 = c.$$

Hence $c = 1$.

35. The two one-sided limits must be equal:

$$\lim_{x \to -2^-} f(x) = \lim_{x \to -2^-} (3 - x^2) = -1 = \lim_{x \to -2^+} f(x) = \lim_{x \to -2^+} (ax + b) = -2a + b$$

and

$$\lim_{x \to 2^-} f(x) = \lim_{x \to 2^-} (ax + b) = 2a + b = \lim_{x \to 2^+} f(x) = \lim_{x \to 2^+} (x^2/2) = 2.$$

Hence $-2a + b = -1$ and $2a + b = 2 \implies a = 3/4, b = 1/2$.

36. (b) Let $\varepsilon > 0$ be given. Since $\lim_{x \to a^+} f(x) = L$, there is a number δ_1 such that

$$a < x < a + \delta_1 \implies |f(x) - L| < \varepsilon.$$

(c) Let $\varepsilon > 0$ be given. Since $\lim_{x \to a^-} f(x) = L$, there is a number δ_2 such that

$$a - \delta_2 < x < a \implies |f(x) - L| < \varepsilon.$$

(d) Let $\varepsilon > 0$ be given and let δ be the smaller of the numbers δ_1, δ_2. Suppose $0 < |x - a| < \delta$. If $a < x < a + \delta$, then $a < x < a + \delta_1$ and hence $|f(x) - L| < \varepsilon$ by part (b). Otherwise $a - \delta < x < a$, in which case $a - \delta_2 < x < a$ and hence $|f(x) - L| < \varepsilon$ by part (c). Therefore

$$0 < |x - a| < \delta \implies |f(x) - L| < \varepsilon$$

and hence $\lim_{x \to a} f(x) = L$.

37. The graph below on the left is that of $f(x) = \sin x$ and $g(x) = (1 - \cos x)/x$ with the range $[0, \pi/2] \times [-1/2, 1]$. It shows that $0 < g(x) < \sin x$. Therefore $\lim_{x \to 0^+} g(x) = 0$. The graph on the right is that of the same functions with the range $[-\pi/2, 0] \times [-1, 1/2]$. This graph shows that $\sin x < g(x) < 0$ and hence $\lim_{x \to 0^-} g(x) = 0$.

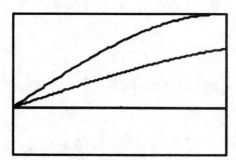
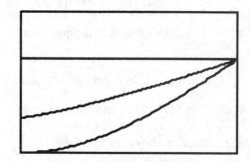

38. The graph is shown below on the left with range $[-2, 0] \times [0, 3.5]$ and indicates that $\lim_{x \to -1^+} f(x) = 0$ and $\lim_{x \to -1^-} f(x) = \pi$.

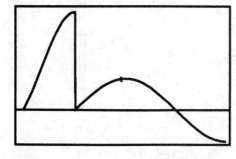
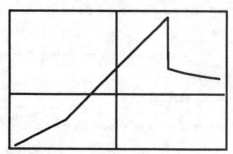

39. The graph is shown above on the right with range $[-3, 3] \times [-3, 9]$ and indicates that $\lim_{x \to -2^-} f(x) = \lim_{x \to -2^+} f(x) = 2$ and $\lim_{x \to 2^-} f(x) = 6$, $\lim_{x \to 2^+} f(x) = 2$.

2.6   Continuity

1. (a) Discontinuous at $x = 1, 2, 4, 5, 6$.

    (b) The table below summarizes continuity or discontinuity on each interval.

$(0, 1)$	$(1, 2)$	$(2, 3)$	$(3, 4)$	$(4, 5)$	$(5, 6)$	$(6, 7)$
cont	cont	cont	cont	cont	cont	cont
$[0, 1)$	$[1, 2)$	$[2, 3)$	$[3, 4)$	$[4, 5)$	$[5, 6)$	$[6, 7)$
cont	cont	cont	cont	disc	disc	disc
$(0, 1]$	$(1, 2]$	$(2, 3]$	$(3, 4]$	$(4, 5]$	$(5, 6]$	$(6, 7]$
disc	disc	cont	cont	disc	disc	cont
$[0, 1]$	$[1, 2]$	$[2, 3]$	$[3, 4]$	$[4, 5]$	$[5, 6]$	$[6, 7]$
disc	disc	cont	cont	disc	disc	disc

2. $(-\infty, 1), (1, +\infty)$

3. $\left((k - 1/2)\pi, (k + 1/2)\pi\right)$, k an integer

4. $(-\infty, -2), (-2, 2), (2, +\infty)$

5. $\left(k\pi, (k + 1)\pi\right)$, k an integer

6. $\left((k - 1/2)\pi, (k + 1/2)\pi\right)$, k an integer

7. $(-\infty, -1), (-1, 2), (2, +\infty)$

8. $(-\infty, -3), (-3, 0), (0, 1), (1, +\infty)$

9. $(-\infty, 0), (0, +\infty)$

10. $(\infty, +\infty)$

11. $(-\infty, 2], (2, +\infty)$

12. $(-\infty, +\infty)$

13. $(-\infty, -1], (-1, +\infty)$

14. $(-\infty, +\infty)$

15. $\left(k\pi, (k + 1)\pi\right)$, k an integer

16. $[-7, +\infty)$

17. $(-7, +\infty)$

18. $[-2, 0], [2, +\infty)$

19. $(-\infty, +\infty)$

20. $(0, +\infty)$

21. $\left((k - 1/2)\pi, (k + 1/2)\pi\right)$, k an integer

22. $\left[k\pi, (k + 1/2)\pi\right)$, k an integer

23. $\ldots, \; (-3\pi/2, -\pi/2), \; (-\pi/2, -1), \; (-1, \pi/2), \; (\pi/2, 2), \; (2, 3\pi/2), \; (3\pi/2, 5\pi/2), \ldots$

24. $(-\infty, +\infty)$

25. Since
$$\frac{x^2 - 1}{x - 1} = \frac{(x-1)(x+1)}{x-1} = x + 1 \quad \text{for } x \neq 1,$$
set
$$f(1) = \lim_{x \to 1} \frac{x^2 - 1}{x - 1} = \lim_{x \to 1}(x + 1) = 2$$
to remove the discontinuity at $a = 1$.

26. Since $\lim\limits_{x \to 1^-} f(x) = \lim\limits_{x \to 1^-}(x^2 + 1) = 2$ and $\lim\limits_{x \to 1^+} f(x) = \lim\limits_{x \to 1^+} \sqrt{3 + x} = 2$, set
$$f(1) = \lim_{x \to 1} f(x) = 2$$
to remove the discontinuity at $a = 1$.

27. Since
$$\frac{\cos^2 x - 1}{\sin x} = \frac{-\sin^2 x}{\sin x} = -\sin x \quad \text{for } \sin x \neq 0,$$
set
$$f(0) = \lim_{x \to 0} \frac{\cos^2 x - 1}{\sin x} = \lim_{x \to 0}(-\sin x) = 0$$
to remove the discontinuity at $a = 0$.

28. Since
$$\frac{x^2 + x - 2}{x^3 - x^2 - 6x} = \frac{(x+2)(x-1)}{x(x-3)(x+2)} = \frac{x-1}{x(x-3)} \quad \text{for } x \neq -2,$$
set
$$f(-2) = \lim_{x \to -2} \frac{x^2 + x - 2}{x^3 - x^2 - 6x} = \lim_{x \to -2} \frac{x-1}{x(x-3)} = -\frac{3}{10}$$
to remove the discontinuity at $a = -2$.

29. $\lim\limits_{x \to 2^-} f(x) = \lim\limits_{x \to 2^-} x^k = 2^k$ and $\lim\limits_{x \to 2^+} f(x) = \lim\limits_{x \to 2^+}(10 - x) = 8 \Longrightarrow 2^k = 8 \Longrightarrow k = 3$.

30. $\lim\limits_{x \to 1^-} f(x) = \lim\limits_{x \to 1^-} \left(1/\sqrt{kx^2 + k}\right) = 1/\sqrt{2k}$ and $\lim\limits_{x \to 1^+} f(x) = \lim\limits_{x \to 1^+} k = k \Longrightarrow 1/\sqrt{2k} = k \Longrightarrow 1 = 2k^3$
$\Longrightarrow k = \sqrt[3]{1/2}$.

31. $\lim\limits_{x \to 2^-} f(x) = \lim\limits_{x \to 2^-}(x-k)(x+k) = (2-k)(2+k)$ and $\lim\limits_{x \to 2^+} f(x) = \lim\limits_{x \to 2^+}(kx + 5) = 2k + 5 \Longrightarrow 2k + 5$
$= (2-k)(2+k) = 4 - k^2 \Longrightarrow k^2 + 2k + 1 = (k+1)^2 = 0 \Longrightarrow k = -1$.

32. $\lim\limits_{x \to 0} \sin(\pi - x/2) = \sin \lim\limits_{x \to 0}(\pi - x/2) = \sin \pi = 0$

33. $\lim\limits_{x \to 8} \left(1 + \sqrt[3]{x}\right)^5 = \left(\lim\limits_{x \to 8}\left(1 + \sqrt[3]{x}\right)\right)^5 = 3^5 = 243$

34. $\lim\limits_{x \to 0} \sqrt{\dfrac{1-x}{1+x}} = \sqrt{\lim\limits_{x \to 0} \dfrac{1-x}{1+x}} = \sqrt{1} = 1$

35. $\lim\limits_{x\to\pi/2} \cos(\pi + x) = \cos\left(\lim\limits_{x\to\pi/2}(\pi + x)\right) = \cos(3\pi/2) = 0$

36. $\lim\limits_{x\to 0}\left(\dfrac{3x + \sin x}{x}\right)^3 = \left(\lim\limits_{x\to 0}\dfrac{3x + \sin x}{x}\right)^3 = \left(\lim\limits_{x\to 0}\left(3 + \dfrac{\sin x}{x}\right)\right)^3 = (3+1)^3 = 64$

37. $\lim\limits_{x\to 1}\left(3x^9 - \sqrt[3]{x}\right)^6 = \left(\lim\limits_{x\to 1}\left(3x^9 - 3\sqrt[3]{x}\right)\right)^6 = (3-1)^6 = 64$

38. $\lim\limits_{x\to 2}\dfrac{\sqrt{4-x}}{\sqrt{x+2}} = \sqrt{\lim\limits_{x\to 2}\dfrac{4-x}{x+2}} = \sqrt{1/2}$

39. $\lim\limits_{x\to 0}\cos\pi(x + |x|) = \cos\left(\lim\limits_{x\to 0}\pi(x + |x|)\right) = \cos 0 = 1$

40. Self-explanatory

41. Self-explanatory

42. $\lim\limits_{x\to 0}\dfrac{\sin 6x}{x} = 6$

43. $\lim\limits_{x\to 0}\dfrac{2x}{\sin x} = 2\Big/\lim\limits_{x\to 0}(\sin x/x) = 2$

44. $\lim\limits_{x\to 0}\dfrac{\tan 2x}{x} = \lim\limits_{x\to 0}\dfrac{\sin 2x}{x}\dfrac{1}{\cos 2x} = (2)(1) = 2$

45. $\lim\limits_{x\to 0}\dfrac{\sin x}{\sin 3x} = \lim\limits_{x\to 0}\left(\dfrac{\sin x}{x}\Big/\dfrac{\sin 3x}{x}\right) = 1/3$

46. $\lim\limits_{x\to 0}\dfrac{\sin ax}{\sin bx} = \lim\limits_{x\to 0}\left(\dfrac{\sin ax}{x}\Big/\dfrac{\sin bx}{x}\right) = a/b$

47. $\lim\limits_{x\to 0} x\csc 3x = \lim\limits_{x\to 0}\dfrac{x}{\sin 3x} = 1\Big/\lim\limits_{x\to 0}(\sin 3x/x) = 1/3$

48. $f(x) = x^2 - 4x - 5 = (x-5)(x+1) = 0 \iff x = -1, 5$. The table below indicates the sign changes of $f(x)$.

Interval	$(-\infty, -1)$	$(-1, 5)$	$(5, +\infty)$
Sign f	+	−	+

$\implies$ $f > 0$ on $(-\infty, -1)$, $(5, +\infty)$
$f < 0$ on $(-1, 5)$

49. $f(x) = 9 - x^2 = 0 \iff x = -3, 3$. The table below indicates the sign changes for $f(x)$.

Interval	$(-\infty, -3)$	$(-3, 3)$	$(3, +\infty)$
Sign f	−	+	−

$\implies$ $f > 0$ on $(-3, 3)$
$f < 0$ on $(-\infty, -3)$, $(3, +\infty)$

50. $f(x) = x^3 - 3x - 2 = (x+1)(x^2 - x - 2) = (x+1)(x-2)(x+1) = 0 \iff x = -1, 2$. The table below indicates the sign changes of $f(x)$.

Interval	$(-\infty, -1)$	$(-1, 2)$	$(2, +\infty)$
Sign f	−	−	+

$\implies$ $f > 0$ on $(2, +\infty)$
$f < 0$ on $(-\infty, -1)$, $(-1, 2)$

51. $f(x) = x^3 + 2x^2 - x - 2 = x^2(x+2) - (x+2) = (x+2)(x-1)(x+1) = 0 \Longleftrightarrow x = -2, -1, 1$. The table below indicates the sign changes of $f(x)$.

Interval	$(-\infty, -2)$	$(-2, -1)$	$(-1, 1)$	$(1, +\infty)$
Sign f	$-$	$+$	$-$	$+$

$\Longrightarrow$ $f > 0$ on $(-2, -1)$, $(1, +\infty)$
$f < 0$ on $(-\infty, -2)$, $(-1, 1)$

52. $f(x) = \sin \pi x = 0 \Longleftrightarrow x$ is an integer. The table below indicates the sign changes of $f(x)$.

Interval	$\cdots$	$(-2, -1)$	$(-1, 0)$	$(0, 1)$	$(1, 2)$	$\cdots$
Sign f	$\cdots$	$+$	$-$	$+$	$-$	$\cdots$

$\Longrightarrow$ $f > 0$ on the intervals $(2k, 2k+1)$, k an integer
$f < 0$ on the intervals $(2k-1, 2k)$, k an integer

53. $f(x) = \cos(x + \pi) = 0 \Longleftrightarrow x + \pi = (k + 1/2)\pi \Longleftrightarrow x = (k - 1/2)\pi$, k an integer. The table below indicates the sign changes of $f(x)$.

Interval	$\cdots$	$(-3\pi/2, -\pi/2)$	$(-\pi/2, \pi/2)$	$(\pi/2, 3\pi/2)$	$\cdots$
Sign f	$\cdots$	$+$	$-$	$+$	$\cdots$

$\Longrightarrow$ $f > 0$ on the intervals $((2k+1/2)\pi, (2k+3/2)\pi)$, k an integer
$f < 0$ on the intervals $((2k-1/2)\pi, (2k+1/2)\pi)$, k an integer

54. Let $f(x) = (x-3)(x+1)$. Then $f(x) = 0 \Longleftrightarrow x = -1, 3$. The table below indicates the sign changes of $f(x)$.

Interval	$(-\infty, -1)$	$(-1, 3)$	$(3, +\infty)$
Sign f	$+$	$-$	$+$

$\Longrightarrow$ $(x-3)(x+1) < 0 \Longleftrightarrow -1 < x < 3$

55. Let $f(x) = x(x+6) + 8$. Then $x(x+6) > -8 \Longleftrightarrow f > 0$. Now, $f(x) = x^2 + 6x + 8 = (x+4)(x+2) = 0 \Longleftrightarrow x = -4, -2$. The table below indicates the sign changes of $f(x)$.

Interval	$(-\infty, -4)$	$(-4, -2)$	$(-2, +\infty)$
Sign f	$+$	$-$	$+$

Hence $f > 0$ on $(-\infty, -4)$ or $(-2, +\infty)$ and therefore $x(x+6) > -8 \Longleftrightarrow x < -4$ or $x > -2$.

56. $f(x) = x^2 + x = x(x+1) = 0 \Longleftrightarrow x = 0, -1$.

Interval	$(-\infty, -1)$	$(-1, 0)$	$(0, +\infty)$
Sign f	$+$	$-$	$+$

Hence $x^2 + x < 0 \Longleftrightarrow -1 < x < 0$.

57. $f(x) = x^3 + x^2 - 2x = x(x^2 + x - 2) = x(x+2)(x-1) = 0 \Longleftrightarrow x = -2, 0, 1$.

Interval	$(-\infty, -2)$	$(-2, 0)$	$(0, 1)$	$(1, +\infty)$
Sign f	$-$	$+$	$-$	$+$

Hence $x^3 + x^2 - 2x > 0 \iff -2 < x < 0$ or $x > 1$.

58. Let $f(x) = (x^2 + x + 7) - 19$. Then $x^2 + x + 7 > 19 \iff f(x) > 0$. Now, $f(x) = x^2 + x - 12 = (x + 4)(x - 3) = 0 \iff x = -4, 3$.

Interval	$(-\infty, -4)$	$(-4, 3)$	$(3, +\infty)$
Sign f	$+$	$-$	$+$

Hence $f > 0$ on $(-\infty, -4)$ or $(3, +\infty)$ and therefore $x^2 + x + 7 > 19 \iff x < -4$ or $x > 3$.

59. $f(x) = x^4 - 9x^2 = x^2(x^2 - 9) = 0 \iff x = -3, 3$.

Interval	$(-\infty, -3)$	$(-3, 0)$	$(0, 3)$	$(3, +\infty)$
Sign f	$+$	$-$	$-$	$+$

Hence $x^4 - 9x^2 > 0 \iff x < -3$ or $x > 3$.

60. Let $f(x) = x \sec x$, where $-2\pi < x < 2\pi$. Then $f(x) = 0 \iff x = 0$. However, $f(x)$ is discontinuous at $x = -3\pi/2, -\pi/2, \pi/2, 3\pi/2$.

Interval	$(-2\pi, -3\pi/2)$	$(-3\pi/2, -\pi/2)$	$(\pi/2, 0)$	$(0, \pi/2)$	$(\pi/2, 3\pi/2)$	$(3\pi/2, 2\pi)$
Sign f	$-$	$+$	$-$	$+$	$-$	$+$

Hence $x \sec x > 0 \iff -3\pi/2 < x < -\pi/2$ or $0 < x < \pi/2$ or $3\pi/2 < x < 2\pi$.

61. Let $f(x) = \sin x \cos x$, where $-2\pi < x < 2\pi$. Then $f(x) = 0 \iff x = -3\pi/2, -\pi, -\pi/2, 0, \pi/2, \pi, 3\pi/2$.

Interval	$(-2\pi, -3\pi/2)$	$(-3\pi/2, -\pi)$	$(-\pi, -\pi/2)$	$(-\pi/2, 0)$	$(0, \pi/2)$	$(\pi/2, \pi)$	$(\pi, 3\pi/2)$	$(3\pi/2, 2\pi)$
Sign f	$+$	$-$	$+$	$-$	$+$	$-$	$+$	$-$

Hence $\sin x \cos x > 0 \iff -2\pi < x < -3\pi/2$ or $-\pi < x < -\pi/2$ or $0 < x < \pi/2$ or $\pi < x < 3\pi/2$.

62. There is no number x such that $f(x) = 0$. This does not contradict the Intermediate Value Theorem since $f(x)$ is not continuous on $[0, 3]$.

63. $f(1) = 1$, $f(3/2) = 9/4 = 2.25 \implies f(c) = c^2 = 2$ for some $c \in [1, 3/2]$. Hence $c = \sqrt{2}$ is between 1 and 3/2. Similarly, $f(3/2) = 2.25$ and $f(2) = 4 \implies f(d) = d^2 = 3$ for some $d \in [3/2, 2] \implies d = \sqrt{3}$ is between 3/2 and 2.

64. (a) r is discontinuous at any number of the form $0.ab5$, where a and b are arbitrary digits.

    (b) $r(x) = (1/100)(\llbracket 100x \rrbracket + \llbracket 2(100x - \llbracket 100x \rrbracket) \rrbracket)$. We obtain this formula as follows. First, $100x - \llbracket 100x \rrbracket$ gives the decimal part of the number x beginning with the third decimal position. To determine if this is < 5 or > 5, double it and take the integer part; if $\llbracket 2(100x - \llbracket 100x \rrbracket) \rrbracket = 0$, then the third decimal position is < 5, otherwise it is > 5. In either case, add this "round off" term to the base number $\llbracket 100x \rrbracket$ to obtain the "rounded up" number, then divide by 100 to obtain the original number "rounded up."

65. Let $f(x) = x + \sin x - 4$. Then $f(\pi) = \pi - 4 < 0$ and $f(2\pi) = 2\pi - 4 > 0$. Hence f has a zero between π and 2π; *i.e.*, the equation $x + \sin x = 4$ has at least one solution in the interval $[\pi, 2\pi]$.

66. Let $f(x) = 1$ for all x and let

$$g(x) = \begin{cases} 1, & x \neq 0 \\ 0, & x = 0. \end{cases}$$

Then $(f \circ g)(x) = f\big(g(x)\big) = 1$ for all x. Since

$$\lim_{x \to 0}(f \circ g)(x) = \lim_{x \to 0} 1 = 1 = f\big(g(1)\big),$$

$f \circ g$ is continuous at $x = 1$. But $\lim_{x \to 0} g(x) = 1 \neq g(0)$. This does not contradict Theorem 11, however, since Theorem 11 shows that the continuity of f and g are only sufficient conditions for $f \circ g$ to be continuous.

67. Let $x = a$ be any number. Then $\lim_{x \to a} f(x) = \lim_{x \to a} c = c = f(a)$. Hence the constant function $f(x) = c$ is continuous at all x.

68. Let $x = a$ be any number. Then $\lim_{x \to a} f(x) = \lim_{x \to a} x = a = f(a)$. Hence the function $f(x) = x$ is continuous at all x.

69. (ii) Since $(cf)(a) = cf(a)$, cf is defined at $x = a$. Moreover,

$$
\begin{aligned}
\lim_{x \to a}(cf)(x) &= \lim_{x \to a} cf(x) & \text{definition of } cf \\
&= c \lim_{x \to a} f(x) & \text{Theorem 1(ii)} \\
&= cf(a) & \text{since } f \text{ is continuous at } x = a \\
&= (cf)(a) & \text{definition of } f.
\end{aligned}
$$

Hence cf is continuous at $x = a$.

(iii) Since $(fg)(a) = f(a)g(a)$, fg is defined at $x = a$. Moreover,

$$
\begin{aligned}
\lim_{x \to a}(fg)(x) &= \lim_{x \to a} f(x)g(x) & \text{definition of } fg \\
&= \left(\lim_{x \to a} f(x)\right)\left(\lim_{x \to a} g(x)\right) & \text{Theorem 1(iii)} \\
&= f(a)g(a) & \text{since } f, g \text{ are continuous at } x = a \\
&= (fg)(a) & \text{definition of } fg.
\end{aligned}
$$

Hence fg is continuous at $x = a$.

(iv) Since $(f/g)(a) = f(a)/g(a)$ and $g(a) \neq 0$, f/g is defined at $x = a$. Moreover,

$$
\begin{aligned}
\lim_{x \to a}(f/g)(x) &= \lim_{x \to a} f(x)/g(x) & \text{definition of } f/g \\
&= \frac{\lim_{x \to a} f(x)}{\lim_{x \to a} g(x)} & \text{Theorem 1(iv)} \\
&= f(a)/g(a) & \text{since } f, g \text{ are continuous at } x = a \text{ and } g(a) \neq 0 \\
&= (f/g)(a) & \text{definition of } f/g.
\end{aligned}
$$

Hence f/g is continuous at $x = a$.

70. (i) Let n be a positive integer and let $x = a$ be a number. Then $\lim\limits_{x \to a} f(x) = \lim\limits_{x \to a} x^n = a^n$ by Theorem 2(i). Hence the function $f(x) = x^n$ is continuous at all numbers x.

   (ii) Let n be a positive integer and let $x = a$ be a number at which g is continuous. Then $\lim\limits_{x \to a} g(x) = g(a)$ and hence, by Theorem 2(ii),

$$\lim_{x \to a} f(x) = \lim_{x \to a} g(x)^n = \left(\lim_{x \to a} g(x) \right)^n = g(a)^n = f(a).$$

   Therefore the function $f(x) = g(x)^n$ is continuous at $x = a$.

71. Let $\varepsilon > 0$ be a given number and set $\delta = \varepsilon^{m/n}$. Then

$$0 < x < \delta \quad \Longrightarrow \quad x^{n/m} < \delta^{n/m} = \varepsilon \quad \Longrightarrow \quad |x^{n/m} - 0| < \varepsilon.$$

   Hence $\lim\limits_{x \to 0^+} x^{n/m} = 0$.

72. (a) Graph with the range $[-0.4, 4.35] \times [-1.6, 1.6]$. The roots approach $x = 1$ and $x = 3$.

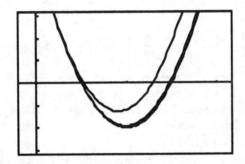

   (b) Set the range to be $[0.95, 1.05] \times [-0.1, 0.1]$ and graph $f(x) = ax^2 - 4x + 3$ for $a = 1 \pm h$; $0 \le h \le 0.09$. You will see that $f(x) = 0$ in $[0.95, 1.05]$. Therefore $0.91 \le a \le 1.09$. Now set the range to be $[2.95, 3.05] \times [-0.1, 0.1]$ and graph $f(x) = ax^2 - 4x + 3$ for $a = 1 \pm h$; $0 \le h \le 0.01$. You will see the $f(x) = 0$ in $[2.95, 3.05]$ and hence $0.99 \le a \le 1.01$. Therefore $f(x)$ has one root in $[0.95, 1.05]$ and the other in $[2.95, 3.05]$.

   (c) For $a = 0.1$ and range $[0, 40] \times [-40, 10]$, the graph of $f(x) = ax^2 - 4x + 3$ is shown below on the left; for $a = 0.001$ and range $[0, 4000] \times [-4000, 100]$, the graph of $f(x) = ax^2 - 4x + 3$ is below on the right.

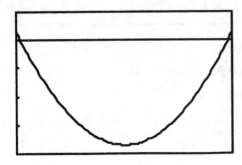 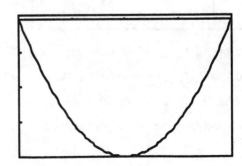

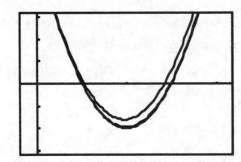

One root approaches 3/4 while the other approaches $+\infty$. To see this note that, by the quadratic formula,

$$x = \frac{4 \pm \sqrt{16 - 4(3)a}}{2a} = \frac{4 \pm \sqrt{16 - 12a}}{2x} = \frac{1}{a}(2 \pm \sqrt{4 - 3a}).$$

Therefore

$$\lim_{a \to 0^+} \frac{2 + \sqrt{4 - 3a}}{a} = \lim_{a \to 0^+} \frac{4 - (4 - 3a)}{a(2 - \sqrt{4 - 3a})} = \lim_{a \to 0^+} \frac{3}{2 - \sqrt{4 - 3a}} = +\infty$$

and

$$\lim_{a \to 0^+} \frac{2 - \sqrt{4 - 3a}}{a} = \lim_{a \to 0^+} \frac{4 - (4 - 3a)}{a(2 + \sqrt{4 - 3a})} = \lim_{a \to 0^+} \frac{3}{2 + \sqrt{4 - 3a}} = 3/4.$$

(d) The graphs of $f(x) = x^2 + bx + 3$, for $b = -4 \pm h$ and $h = 0.1, 0.01$ with range $[-0.4, 4.35] \times [-1.6, 1.55]$ are shown at the top of the page. In this case the roots approach $x = 1$ and $x = 3$. If $-4.032 \le b \le -3.968$, the roots lie in the intervals $[0.95, 1.05]$ and $[2.95, 3.05]$. Note that as b approaches zero the function becomes $g(x) = x^2 + 3$, which has no real roots. In fact, if $-2\sqrt{3} < b \le 0$, $g(x)$ will have no real roots.

(e) The graph of $p(x)$ and $q(x) = x^5 - 15.01x^4 + 85x^3 - 225x^2 + 274x - 120$ with range $[0, 6] \times [-4, 4]$ is shown below on the left. The graph of $p(x)$ and $q(x) = x^5 - 14.99x^4 + 85x^3 - 225x^2 + 274x - 120$ with the same range is on the right.

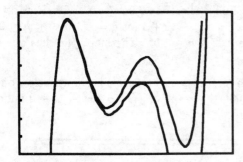 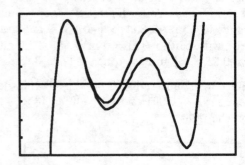

73. $c_{14} = 2.28680$, $f(c_{14}) = -0.00033$, $(b_{14} - a_{14})/2 = 0.00006$.

74. $c_{14} = 0.73907$, $f(c_{14}) = -0.000027$, $(b_{14} - a_{14})/2 = 0.00006$

75. $c_{14} = 2.53516$, $f(c_{14}) = 0.00009$, $(b_{14} - a_{14})/2 = 0.00006$

76. The smallest positive root is in the interval $[4, 5]$. $c_{14} = 4.27484$, $f(c_{14}) = 0.0003$, $(b_{14} - a_{14})/2 = 0.00006$

Review Exercises – Chapter 2

1. $\lim\limits_{x\to 2}(x^2 - x + 2) = 4$

2. $\lim\limits_{x\to 2}\dfrac{2x-1}{x+6} = \dfrac{3}{8}$

3. $\lim\limits_{x\to 3}(x^4 - x - 1) = 77$

4. $\lim\limits_{x\to 2}(3x - 2)^{3/2} = 4^{3/2} = 8$

5. $\lim\limits_{x\to 3}(4 - 4x)^{1/3} = (-8)^{1/3} = -2$

6. $\lim\limits_{x\to -5}\dfrac{x^2 - 25}{x+5} = \lim\limits_{x\to -5}\dfrac{(x-5)(x+5)}{x+5} = \lim\limits_{x\to -5}(x-5) = -10$

7. $\lim\limits_{x\to 3/2}\dfrac{4x^2 - 9}{2x - 3} = \lim\limits_{x\to 3/2}\dfrac{(2x-3)(2x+3)}{2x-3} = \lim\limits_{x\to 3/2}(2x+3) = 6$

8. $\lim\limits_{x\to 3}\dfrac{x-3}{x^2 - 9} = \lim\limits_{x\to 3}\dfrac{x-3}{(x-3)(x+3)} = \lim\limits_{x\to 3}\dfrac{1}{x+3} = \dfrac{1}{6}$

9. $\lim\limits_{x\to -2}\dfrac{x^2 + x - 2}{x+2} = \lim\limits_{x\to -2}\dfrac{(x+2)(x-1)}{x+2} = \lim\limits_{x\to -2}(x-1) = -3$

10. $\lim\limits_{x\to 1}\dfrac{x^2 + 6x - 7}{x-1} = \lim\limits_{x\to 1}\dfrac{(x+7)(x-1)}{x-1} = \lim\limits_{x\to 1}(x+7) = 8$

11. $\lim\limits_{x\to 1}\dfrac{x^2 + 2x - 3}{x^2 + x - 2} = \lim\limits_{x\to 1}\dfrac{(x+3)(x-1)}{(x+2)(x-1)} = \lim\limits_{x\to 1}\dfrac{x+3}{x+2} = \dfrac{4}{3}$

12. $\lim\limits_{x\to -2}\dfrac{x^2 + 2x}{x^2 + x - 2} = \lim\limits_{x\to -2}\dfrac{x(x+2)}{(x+2)(x-1)} = \lim\limits_{x\to -2}\dfrac{x}{x-1} = \dfrac{2}{3}$

13. $\lim\limits_{x\to 1^-}\sqrt{\dfrac{1 - x^2}{1 + x}} = 0$

14. $\lim\limits_{x\to 8}\sqrt{\dfrac{x - 7}{x + 2}} = \sqrt{1/10}$

15. $\lim\limits_{x\to 7^+}\sqrt{\dfrac{x - 7}{x + 2}} = 0$

16. $\lim\limits_{x\to 0^-}\sqrt{\dfrac{x}{x - 1}} = 0$

17. $\lim\limits_{x\to 0}\dfrac{\tan x}{\sin x} = \lim\limits_{x\to 0}\dfrac{1}{\cos x} = 1$

18. $\lim\limits_{x\to 0}\dfrac{3x}{\sin 2x} = \lim\limits_{x\to 0}\dfrac{3}{(\sin 2x)/x} = 3/2$

19. $\lim\limits_{x\to 0}\dfrac{(2+x)^2 - 4}{x} = \lim\limits_{x\to 0}\dfrac{x^2 + 4x}{x} = \lim\limits_{x\to 0}(x+4) = 4$

20. $\displaystyle\lim_{x\to0}\frac{\sqrt{4+x}-2}{x}=\lim_{x\to0}\frac{\sqrt{4+x}-2}{x}\cdot\frac{\sqrt{4+x}+2}{\sqrt{4+x}+2}=\lim_{x\to0}\frac{1}{\sqrt{4+x}+2}=1/4$

21. $\displaystyle\lim_{x\to0}\frac{|x|}{2x}$ does not exist

22. $\displaystyle\lim_{x\to2}\frac{3x^2+x+1}{1-x^3}=-15/7$

23. $\displaystyle\lim_{x\to0}3x\csc 4x=\lim_{x\to0}\frac{3x}{\sin 4x}=\lim_{x\to0}\frac{3}{(\sin 4x)/x}=3/4$

24. $\displaystyle\lim_{x\to0}\frac{\sqrt{4x^2+1}}{x^2+2}=1/2$

25. $\displaystyle\lim_{x\to1}\frac{x^3-1}{x-1}=\lim_{x\to1}\frac{(x-1)(x^2+x+1)}{x-1}=\lim_{x\to1}(x^2+x+1)=3$

26. $\displaystyle\lim_{x\to0}\frac{3x+5x^2}{x}=\lim_{x\to0}(3+5x)=3$

27. $\displaystyle\lim_{x\to4^-}\sqrt{8-2x}=0$

28. $\displaystyle\lim_{x\to0^+}\frac{1-x}{\sqrt{x}-1}=-1$

29. $\displaystyle\lim_{x\to1}\frac{|x-1|}{x-1}$ does not exist

30. $\displaystyle\lim_{x\to2}\left(1-[\![2x]\!]\right)$ does not exist

31. $\displaystyle\lim_{x\to3}\left(2[\![x/2]\!]+3\right)=5$

32. $\displaystyle\lim_{x\to0^+}\sin(1/x)$ does not exist

33. $\displaystyle\lim_{x\to1^+}\frac{|x-1|}{x-1}=1$

34. $\displaystyle\lim_{x\to1^-}\frac{|x-1|}{x-1}=-1$

35. $\displaystyle\lim_{x\to0^+}\frac{2+\sqrt{x}}{2-\sqrt{x}}=1$

36. $\displaystyle\lim_{x\to1^-}\sqrt{1-[\![x]\!]}=1$

37. $\displaystyle\lim_{x\to1^+}\sqrt{1-[\![x]\!]}=0$

38. $\displaystyle\lim_{x\to0^+}\frac{x-[\![x]\!]}{x}=1$

39. $\displaystyle\lim_{x\to0}\frac{\sqrt[3]{x+1}-1}{x}=\lim_{x\to0}\frac{\sqrt[3]{x+1}-1}{x}\cdot\frac{\sqrt[3]{(x+1)^2}+\sqrt[3]{x+1}+1}{\sqrt[3]{(x+1)^2}+\sqrt[3]{x+1}+1}=\lim_{x\to0}\frac{1}{\sqrt[3]{(x+1)^2}+\sqrt[3]{x+1}+1}=\frac{1}{3}$

40. The slope of the tangent line at $x_0 = 3$ is

$$m = \lim_{h \to 0} \frac{f(3+h) - f(3)}{h} = \lim_{h \to 0} \frac{[4 - 2(3+h)] - (-2)}{h} = \lim_{h \to 0}(-2) = -2.$$

Therefore the equation of the tangent line at $(3, -2)$ is $y + 2 = -2(x - 3)$, or $y = -2x + 4$.

41. The slope of the tangent line at $x_0 = 1$ is

$$m = \lim_{h \to 0} \frac{f(1+h) - f(1)}{h} = \lim_{h \to 0} \frac{[3(1+h)^2 + 4] - 7}{h} = \lim_{h \to 0}(6 + 3h) = 6.$$

Therefore the equation of the tangent line at $(1, 7)$ is $y - 7 = 6(x - 1)$, or $y = 6x + 1$.

42. The slope of the tangent line at $x_0 = 2$ is

$$m = \lim_{h \to 0} \frac{f(2+h) - f(2)}{h} = \lim_{h \to 0} \frac{[9 - (2+h)^2] - 5}{h} = \lim_{h \to 0}(-4 - h) = -4.$$

Therefore the equation of the tangent line at $(2, 5)$ is $y - 5 = -4(x - 2)$, or $y = -4x + 13$.

43. The slope of the tangent line at $x_0 = 2$ is

$$m = \lim_{h \to 0} \frac{f(2+h) - f(2)}{h} = \lim_{h \to 0} \frac{\dfrac{1}{2+h} - \dfrac{1}{2}}{h} = \lim_{h \to 0} \frac{2 - (2+h)}{2h(2+h)} = \lim_{h \to 0} \frac{-1}{2(2+h)} = -\frac{1}{4}.$$

Therefore the equation of the tangent line at $(2, 1/2)$ is $y - 1/2 = (-1/4)(x - 2)$.

44. The slope of the tangent line at $x_0 = 1$ is

$$m = \lim_{h \to 0} \frac{f(1+h) - f(1)}{h} = \lim_{h \to 0} \frac{\dfrac{3}{3+h} - 1}{h} = \lim_{h \to 0} \frac{-1}{3+h} = -\frac{1}{3}.$$

Therefore the equationof the tangent line at $(1, 1)$ is $y - 1 = (-1/3)(x - 1)$.

45. The slope of the tangent line at $x_0 = 1$ is

$$m = \lim_{h \to 0} \frac{f(1+h) - f(1)}{h} = \lim_{h \to 0} \frac{(2+h)^4 - 16}{h} = \lim_{h \to 0}(32 + 24h + 8h^2 + h^3) = 32.$$

Therefore the equation of the tangent line at $(1, 16)$ is $y - 16 = 32(x - 1)$.

46. $(-\infty, 2)$, $(2, +\infty)$ since f is continuous at all $x \neq 2$.

47. $(-\infty, -2)$, $(-2, 3)$, $(3, +\infty)$ since f is continuous at all x for which $x^2 - x - 6 = (x - 3)(x + 2) \neq 0$.

48. $(-\infty, -2)$, $(-2, +\infty)$ since f is continuous at all $x \neq -2$.

49. $[0, \pi/2)$, $(\pi/2, 3\pi/2)$, $(3\pi/2, 2\pi]$ since y is continuous at all x in the interval $[0, 2\pi]$ for which $\cos x \neq 0$, namely $x \neq \pi/2$ and $x \neq 3\pi/2$.

50. $(-\infty, 0)$, $(0, +\infty)$ since f is continuous at all $x \neq 0$.

51. $(-\infty, -1]$, $[4, +\infty)$ since f is continuous at all x for which $x^2 - 3x - 4 = (x - 4)(x + 1) \geq 0$.

52. $(-\infty, +\infty)$ since $\lim\limits_{x \to 2^-} f(x) = 2 - 2^2 = -2 = \lim\limits_{x \to 2^+} f(x)$.

53. $(-\infty, \pi/4]$, $(\pi/4, +\infty)$ since $\lim\limits_{x \to \pi/4^-} y = \sqrt{2}/2 = y(\pi/4)$ but $\lim\limits_{x \to \pi/4^+} y = 1 - \sqrt{2}/2 \neq y(\pi/4)$.

54. $[0, 4]$, $(4, \infty)$ since $\lim\limits_{x \to 4^-} f(x) = \sqrt{4} = 2 = f(4)$, but $\lim\limits_{x \to 4^+} f(x) = 3 \neq f(4)$.

55. $(-\infty, -2)$, $[-2, +\infty)$ since $\lim\limits_{x \to -2^-} f(x) = 1 - (-2) = 3 \neq f(-2)$ but $\lim\limits_{x \to -2^+} f(x) = -(-2) = 2 = f(-2)$, while $\lim\limits_{x \to 1^-} f(x) = -1 = \lim\limits_{x \to 1^+} f(x) = f(1)$.

56. $(-\infty, 1)$, $(1, +\infty)$ since $\lim\limits_{x \to 1} f(x)$ does not exist

57. $(-\infty, +\infty)$ since $\lim\limits_{x \to -5} f(x) = \lim\limits_{x \to -5} \dfrac{(x-2)(x+5)}{x+5} = \lim\limits_{x \to -5} (x-2) = -7 = f(-5)$.

58. Let $f(x) = x^3 + x - 3$. Then $f(1) = -1$ and $f(2) = 7$. Since $f(x)$ is continuous, it follows from the Intermediate Value Theorem that there is a number x_0, between 1 and 2, such that $x_0^3 + x_0 - 3 = 0$ or, equivalently, $x_0^3 + x_0 = 3$. Moreover, since $f(3/2) = 15/8 > 0$, x_0 lies between 1 and 3/2 and hence is closer to 1 than 2.

59. (a) $\lim\limits_{x \to -1^-} f(x) = \lim\limits_{x \to -1^-} 1 = 1$

    (b) $\lim\limits_{x \to -1^+} f(x) = \lim\limits_{x \to -1^+} \sqrt{x+2} = \sqrt{1} = 1$

    (c) $\lim\limits_{x \to -1} f(x) = 1$

    (d) $\lim\limits_{x \to 2^-} f(x) = \lim\limits_{x \to 2^-} \sqrt{x+2} = \sqrt{4} = 2$

    (e) $\lim\limits_{x \to 2^+} f(x) = \lim\limits_{x \to 2^+} \cos \pi x = \cos 2\pi = 1$

    (f) $\lim\limits_{x \to 2} f(x)$ does not exist since the one-sided limits are not equal

    (g) Yes, since $\lim\limits_{x \to -1} f(x) = 1 = f(-1)$.

    (h) No, since $\lim\limits_{x \to 2} f(x)$ does not exist.

60. (a) $\lim\limits_{x \to 0^-} g(x) = \lim\limits_{x \to 0^-} \tan x = 0$

    (b) $\lim\limits_{x \to 0^+} g(x) = \lim\limits_{x \to 0^+} \sin x = 0$

    (c) $\lim\limits_{x \to 0} g(x) = 0$

    (d) $\lim\limits_{x \to \pi/2^-} g(x) = \lim\limits_{x \to \pi/2^-} \sin x = 1$

    (e) $\lim\limits_{x \to \pi/2^+} g(x) = \lim\limits_{x \to \pi/2^+} 2x/\pi = 1$

    (f) $\lim\limits_{x \to \pi/2} g(x) = 1$

    (g) No, since $g(0)$ does not exist.

    (h) Yes, since $\lim\limits_{x \to \pi/2} g(x) = 1 = g(\pi/2)$

61. $\lim\limits_{x \to 1^-} f(x) = \lim\limits_{x \to 1^-} (4 - x + x^2) = 4 = \lim\limits_{x \to 1^+} f(x) = \lim\limits_{x \to 1^+} (9 - ax^2) = 9 - a \implies a = 5$

62. $\lim\limits_{x\to2}(1+4x-x^2)=5 \leq \lim\limits_{x\to2}f(x) \leq \lim\limits_{x\to2}(x^2-4x+9)=5 \implies \lim\limits_{x\to2}f(x)=5$

63. $\lim\limits_{x\to0}(1-|x|)=1 \leq \lim\limits_{x\to0}f(x) \leq \lim\limits_{x\to0}(\sec x)=1 \implies \lim\limits_{x\to0}f(x)=1$

64. Let $\varepsilon > 0$ be given. Then

$$|(2x+3)-9| = 2|x-3| < \varepsilon \iff |x-3| < \varepsilon/2.$$

Thus, if we set $\delta = \varepsilon/2$, it follows that

$$0 < |x-3| < \delta \implies |(2x+3)-9| < \varepsilon$$

and hence $\lim\limits_{x\to3}(2x+3)=9$.

65. Let $\varepsilon > 0$ be given. Then

$$|(4x-3)-1| = 4|x-1| < \varepsilon \iff |x-1| < \varepsilon/4.$$

Thus, if we set $\delta = \varepsilon/4$, it follows that

$$0 < |x-1| < \delta \implies |(4x-3)-1| < \varepsilon$$

and hence $\lim\limits_{x\to1}(4x-3)=1$.

66. Let $\varepsilon > 0$ be given. Then, if x is sufficiently close to 2, say within one unit, $|x+2| < 5$ and hence

$$\left|(x^2+3)-7\right| = |(x-2)(x+2)| < 5|x-2| < \varepsilon \iff |x-2| < \varepsilon/5.$$

Thus, if we let δ be the smaller of 1 and $\varepsilon/5$, it follows that

$$0 < |x-2| < \varepsilon/5 \implies \left|(x^2+3)-7\right| < \varepsilon$$

and hence $\lim\limits_{x\to2}\left(x^2+3\right)=7$.

67. Let $\varepsilon > 0$ be given and let x be sufficiently close to 4, say within one unit. Then $1/|x| < 1/3$ and hence

$$\left|\frac{1}{x}-\frac{1}{4}\right| = \frac{|x-4|}{4|x|} < \frac{|x-4|}{12} < \varepsilon \iff |x-4| < 12\varepsilon.$$

Thus, if we let δ be the smaller of 1 and 12ε, it follows that

$$0 < |x-4| < \delta \implies \left|\frac{1}{x}-\frac{1}{4}\right| < \varepsilon$$

and hence $\lim\limits_{x\to4}(1/x)=1/4$.

68. Let $\varepsilon > 0$ be given. Then, for $x > 2$,

$$\left|\sqrt{x-2}\right| < \varepsilon \iff x-2 < \varepsilon^2.$$

Thus, if we let $\delta = \varepsilon^2$, then

$$0 < x-2 < \delta \implies \left|\sqrt{x-2}\right| < \varepsilon$$

and hence $\lim\limits_{x\to2^+}\sqrt{x-2}=0$.

69. Let $\varepsilon > 0$ be given and let $\delta = \varepsilon$. Then

$$0 < |x + 3| < \delta \quad \Longrightarrow \quad \big||x + 3|\big| = |x + 3| < \varepsilon$$

and therefore $\lim\limits_{x \to -3} |x + 3| = 0$.

70. Let $\varepsilon > 0$ be given. Then, if x is sufficiently close to 1, say within one unit, $|x + 3| < 5$ and hence

$$\big|(x^2 + 2x) - 3\big| = |x + 3|\,|x - 1| < 5|x - 1| < \varepsilon \quad \Longleftrightarrow \quad |x - 1| < \varepsilon/5.$$

Thus, if we let δ be the smaller of 1 and $\varepsilon/5$, it follows that

$$0 < |x - 1| < \delta \quad \Longrightarrow \quad \big|(x^2 + 2x) - 3\big| < \varepsilon$$

and therefore $\lim\limits_{x \to 1} (x^2 + 2x) = 3$.

71. Let $\varepsilon > 0$ be given. Then, if x is sufficiently close to -2, say within one unit, $|x - 1| < 2$ and hence

$$\big|(x^2 + x + 1) - 3\big| = |x + 2|\,|x - 1| < 2|x + 2| < \varepsilon \quad \Longleftrightarrow \quad |x + 2| < \varepsilon/2.$$

Thus, if we let δ be the smaller of 1 and $\varepsilon/2$, it follows that

$$0 < |x + 2| < \delta \quad \Longrightarrow \quad \big|(x^2 + x + 1) - 3\big| < \varepsilon$$

and therefore $\lim\limits_{x \to -2} (x^2 + x + 1) = 3$.

72. $\lim\limits_{x \to 3^-} f(x) = \lim\limits_{x \to 3^-} (2ax + b) = 6a + b$ and $\lim\limits_{x \to 3^+} f(x) = \lim\limits_{x \to 3^+} (ax + 3b) = 3a + 3b$. Hence

$$\lim\limits_{x \to 3} f(x) = 10 \quad \Longleftrightarrow \quad 6a + b = 3a + 3b = 10.$$

Solving these equations, we find that $a = 4/3$ and $b = 2$.

73. $\lim\limits_{x \to a} (x^3 - 4x^2 + 3x - 2) = a^3 - 4a^2 + 3a - 2 = -10 \Longrightarrow a^3 - 4a^2 + 3a + 8 = (a + 1)(a^2 - 5a + 8) = 0$
$\Longrightarrow a = -1$ since the equation $a^2 - 5a + 8$ has no real solutions.

74. $\lim\limits_{x \to 1^-} f(x) = \lim\limits_{x \to 1^-} (a^2x^2 + bx - 12) = a^2 + b - 12$ and $\lim\limits_{x \to 1^+} f(x) = \lim\limits_{x \to 1^+} (ax + b) = a + b$. Hence

$$\lim\limits_{x \to 1} f(x) = 2 \quad \Longleftrightarrow \quad a^2 + b - 12 = a + b = 2.$$

Therefore $b = 2 - a$ and hence

$$a^2 + b - 12 = a^2 + (2 - a) - 12 = 2 \Longrightarrow a^2 - a - 12 = (a - 4)(a + 3) = 0 \Longrightarrow a = -3, 4; b = 5, -2.$$

Thus, $a = -3$ and $b = 5$, or $a = 4$ and $b = -2$.

75. False. Let $f(x) = -2$, $g(x) = 2$, and $h(x) = x/|x|$ for $x \neq 0$, $h(0) = 0$. Then f and g are continuous at $x = 0$ and $f(x) \leq h(x) \leq g(x)$ for all x, but h is not continuous at $x = 0$.

76. Suppose that $L \neq M$ and let $\varepsilon > 0$ be a number such that $\varepsilon < |L - M|$. Since $\lim\limits_{x \to a} f(x) = L$ and $\lim\limits_{x \to a} f(x) = M$, there are numbers δ_1, δ_2 such that

$$0 < |x - a| < \delta_1 \quad \Longrightarrow \quad |f(x) - L| < \varepsilon/2$$
$$0 < |x - a| < \delta_2 \quad \Longrightarrow \quad |f(x) - M| < \varepsilon/2.$$

Let δ be the smaller of δ_1 and δ_2. Then, if x is any number for which $0 < |x - a| < \delta$, we have that

$$|L - M| = \big|(f(x) - M) - (f(x) - L)\big| \leq |f(x) - M| + |f(x) - L| < \varepsilon/2 + \varepsilon/2 = \varepsilon.$$

Since this contradicts the choice of ε, it follows that no such ε exists. Hence $L = M$.

77. Suppose that $\lim\limits_{x\to a} f(x) = M < L$ and let $\varepsilon > 0$ be a number such that $\varepsilon < L - M$. Since $\lim\limits_{x\to a} f(x) = M$ and $\lim\limits_{x\to a} g(x) = L$, there are numbers δ_1, δ_2 such that

$$0 < |x - a| < \delta_1 \quad \Longrightarrow \quad |f(x) - M| < \varepsilon/2$$
$$0 < |x - a| < \delta_2 \quad \Longrightarrow \quad |g(x) - L| < \varepsilon/2.$$

Let δ be the smaller of δ_1 and δ_2. Then, if x is any number for which $0 < |x - a| < \delta$, we have that

$$f(x) < M + \varepsilon/2 < M + (1/2)(L - M) = L - (1/2)(L - M) < L - \varepsilon/2 < g(x),$$

which contradicts the fact that $g(x) \leq f(x)$. Hence $\lim\limits_{x\to a} f(x) \geq L$.

78. g must be discontinuous at $x = a$; for if g is continuous at $x = a$, then $f + g$ is continuous at $x = a$.

Chapter 3

The Derivative

3.1 The Derivative as a Function

1. a $\leftrightarrow$ iv, b $\leftrightarrow$ v, c $\leftrightarrow$ iii, d $\leftrightarrow$ i, e $\leftrightarrow$ ii.

2. a $\leftrightarrow$ vi, b $\leftrightarrow$ ii, c $\leftrightarrow$ viii, d $\leftrightarrow$ v.

3. $f(x) = 6x + 5 \Longrightarrow$

$$f'(x) = \lim_{h \to 0} \frac{f(x+h) - f(x)}{h} = \lim_{h \to 0} \frac{[6(x+h) + 5] - [6x + 5]}{h} = \lim_{h \to 0} 6 = 6$$

4. $f(x) = x^2 - 7 \Longrightarrow$

$$f'(x) = \lim_{h \to 0} \frac{f(x+h) - f(x)}{h} = \lim_{h \to 0} \frac{[(x+h)^2 - 7] - [x^2 - 7]}{h} = \lim_{h \to 0} (2x + h) = 2x$$

5. $f(x) = 2x^3 + 3 \Longrightarrow$

$$f'(x) = \lim_{h \to 0} \frac{f(x+h) - f(x)}{h} = \lim_{h \to 0} \frac{[2(x+h)^3 + 3] - [2x^3 + 3]}{h} = \lim_{h \to 0} (6x^2 + 6xh + 2h^2) = 6x^2$$

6. $f(x) = 1 - x^3 \Longrightarrow$

$$f'(x) = \lim_{h \to 0} \frac{f(x+h) - f(x)}{h} = \lim_{h \to 0} \frac{[1 - (x+h)^3] - [1 - x^3]}{h} = \lim_{h \to 0} (-3x^2 - 3xh - h^2) = -3x^2$$

7. $f(x) = ax^3 + bx^2 + cx + d \Longrightarrow$

$$f'(x) = \lim_{h \to 0} \frac{f(x+h) - f(x)}{h} = \lim_{h \to 0} \frac{[a(x+h)^3 + b(x+h)^2 + c(x+h) + d] - [ax^3 + bx^2 + cx + d]}{h}$$

$$= \lim_{h \to 0} (3ax^2 + 3axh + ah^2 + 2bx + bh + c) = 3ax^2 + 2bx + c$$

8. $f(x) = \dfrac{1}{x - 1} \Longrightarrow$

$$f'(x) = \lim_{h \to 0} \frac{f(x+h) - f(x)}{h} = \lim_{h \to 0} \frac{\dfrac{1}{x+h-1} - \dfrac{1}{x-1}}{h} = \lim_{h \to 0} \frac{-1}{(x-1)(x+h-1)} = \frac{-1}{(x-1)^2}$$

9. $f(x) = \dfrac{1}{2x+3} \implies$

$$f'(x) = \lim_{h\to 0} \frac{f(x+h)-f(x)}{h} = \lim_{h\to 0} \frac{\dfrac{1}{2(x+h)+3} - \dfrac{1}{2x+3}}{h}$$

$$= \lim_{h\to 0} \frac{-2}{(2x+3)(2x+2h+3)} = \frac{-2}{(2x+3)^2}$$

10. $f(x) = \dfrac{1}{x^2-9} \implies$

$$f'(x) = \lim_{h\to 0} \frac{f(x+h)-f(x)}{h} = \lim_{h\to 0} \frac{\dfrac{1}{(x+h)^2-9} - \dfrac{1}{x^2-9}}{h}$$

$$= \lim_{h\to 0} \frac{-2x-h}{(x^2-9)(x^2+2xh+h^2-9)} = \frac{-2x}{(x^2-9)^2}$$

11. $f(x) = \sqrt{x+1} \implies$

$$f'(x) = \lim_{h\to 0} \frac{f(x+h)-f(x)}{h} = \lim_{h\to 0} \frac{\sqrt{x+h+1}-\sqrt{x+1}}{h}$$

$$= \lim_{h\to 0} \frac{\sqrt{x+h+1}-\sqrt{x+1}}{h} \cdot \frac{\sqrt{x+h+1}+\sqrt{x+1}}{\sqrt{x+h+1}+\sqrt{x+1}} = \lim_{h\to 0} \frac{1}{\sqrt{x+h+1}+\sqrt{x+1}} = \frac{1}{2\sqrt{x+1}}$$

12. $f(x) = \sqrt{2x+3} \implies$

$$f'(x) = \lim_{h\to 0} \frac{f(x+h)-f(x)}{h} = \lim_{h\to 0} \frac{\sqrt{2(x+h)+3}-\sqrt{2x+3}}{h}$$

$$= \lim_{h\to 0} \frac{\sqrt{2(x+h)+3}-\sqrt{2x+3}}{h} \cdot \frac{\sqrt{2(x+h)+3}+\sqrt{2x+3}}{\sqrt{2(x+h)+3}+\sqrt{2x+3}}$$

$$= \lim_{h\to 0} \frac{2}{\sqrt{2(x+h)+3}+\sqrt{2x+3}} = \frac{1}{\sqrt{2x+3}}$$

13. $f(x) = \dfrac{1}{\sqrt{x+1}} \implies$

$$f'(x) = \lim_{h\to 0} \frac{f(x+h)-f(x)}{h} = \lim_{h\to 0} \frac{\dfrac{1}{\sqrt{x+h+1}} - \dfrac{1}{\sqrt{x+1}}}{h}$$

$$= \lim_{h\to 0} \frac{\sqrt{x+1}-\sqrt{x+h+1}}{h\sqrt{x+1}\sqrt{x+h+1}} \cdot \frac{\sqrt{x+1}+\sqrt{x+h+1}}{\sqrt{x+1}+\sqrt{x+h+1}}$$

$$= \lim_{h\to 0} \frac{-1}{\sqrt{x+1}\sqrt{x+h+1}\left(\sqrt{x+1}+\sqrt{x+h+1}\right)} = \frac{-1}{2(x+1)\sqrt{x+1}}$$

14. $f(x) = \dfrac{1}{ax+b} \implies$

$$f'(x) = \lim_{h \to 0} \frac{f(x+h) - f(x)}{h} = \lim_{h \to 0} \frac{\dfrac{1}{a(x+h)+b} - \dfrac{1}{ax+b}}{h} = \lim_{h \to 0} \frac{-a}{(ax+b)(ax+ah+b)} = \frac{-a}{(ax+b)^2}$$

15. $f(x) = 1/x^2 \implies$

$$f'(x) = \lim_{h \to 0} \frac{f(x+h) - f(x)}{h} = \lim_{h \to 0} \frac{\dfrac{1}{(x+h)^2} - \dfrac{1}{x^2}}{h} = \lim_{h \to 0} \frac{-2x-h}{x^2(x+h)^2} = -\frac{2}{x^3}$$

16. $f(x) = x^4 \implies$

$$f'(x) = \lim_{h \to 0} \frac{f(x+h) - f(x)}{h} = \lim_{h \to 0} \frac{(x+h)^4 - x^4}{h} = \lim_{h \to 0} \left(4x^3 + 6x^2h + 4xh^2 + h^3\right) = 4x^3$$

17. $f(x) = (x+3)^3 \implies$

$$f'(x) = \lim_{h \to 0} \frac{f(x+h) - f(x)}{h} = \lim_{h \to 0} \frac{(x+h+3)^3 - (x+3)^3}{h}$$

$$= \lim_{h \to 0} \frac{(x+3)^3 + 3(x+3)^2h + 3(x+3)h^2 + h^3 - (x+3)^3}{h}$$

$$= \lim_{h \to 0} \left[3(x+3)^2 + 3(x+3)h + h^2\right] = 3(x+3)^2$$

18. $f(x) = (x-1)^2 \implies$

$$f'(x) = \lim_{h \to 0} \frac{f(x+h) - f(x)}{h} = \lim_{h \to 0} \frac{(x+h-1)^2 - (x-1)^2}{h}$$

$$= \lim_{h \to 0} \frac{(x-1)^2 + 2h(x-1) + h^2 - (x-1)^2}{h} = \lim_{h \to 0} [2(x-1) + h] = 2(x-1)$$

19. $f(x) = \dfrac{1}{\sqrt{x+5}} \implies$

$$f'(x) = \lim_{h \to 0} \frac{f(x+h) - f(x)}{h} = \lim_{h \to 0} \frac{\dfrac{1}{\sqrt{x+h+5}} - \dfrac{1}{\sqrt{x+5}}}{h}$$

$$= \lim_{h \to 0} \frac{\sqrt{x+5} - \sqrt{x+h+5}}{h\sqrt{x+5}\sqrt{x+h+5}} \cdot \frac{\sqrt{x+5} + \sqrt{x+h+5}}{\sqrt{x+5} + \sqrt{x+h+5}}$$

$$\lim_{h \to 0} \frac{-1}{\sqrt{x+5}\sqrt{x+h+5}\left(\sqrt{x+5} + \sqrt{x+h+5}\right)} = \frac{-1}{2(x+5)\sqrt{x+5}}$$

20. $f(x) = \dfrac{1}{(x+2)^2} \implies$

$$f'(x) = \lim_{h \to 0} \frac{f(x+h) - f(x)}{h} = \lim_{h \to 0} \frac{\dfrac{1}{(x+h+2)^2} - \dfrac{1}{(x+2)^2}}{h}$$

$$= \lim_{h \to 0} \frac{(x+2)^2 - (x+h+2)^2}{h(x+2)^2(x+h+2)^2} = \lim_{h \to 0} \frac{-(2x+h+4)}{(x+2)^2(x+h+2)^2} = \frac{-2}{(x+2)^3}$$

116

21. $f(x) = \dfrac{3}{(x-1)^2} \implies$

$$f'(x) = \lim_{h \to 0} \frac{f(x+h) - f(x)}{h} = \lim_{h \to 0} \frac{\dfrac{3}{(x+h-1)^2} - \dfrac{3}{(x-1)^2}}{h}$$

$$= \lim_{h \to 0} \frac{3(x-1)^2 - 3(x+h-1)^2}{h(x-1)^2(x+h-1)^2} = \lim_{h \to 0} \frac{-3(2x+h-2)}{(x-1)^2(x+h-1)^2} = \frac{-6}{(x-1)^3}$$

22. $f(x) = \dfrac{-2}{\sqrt{x+1}} \implies$

$$f'(x) = \lim_{h \to 0} \frac{f(x+h) - f(x)}{h} = \lim_{h \to 0} \frac{\dfrac{-2}{\sqrt{x+h+1}} + \dfrac{2}{\sqrt{x+1}}}{h}$$

$$= -2 \lim_{h \to 0} \frac{\sqrt{x+1} - \sqrt{x+h+1}}{h\sqrt{x+1}\sqrt{x+h+1}} \cdot \frac{\sqrt{x+1} + \sqrt{x+h+1}}{\sqrt{x+1} + \sqrt{x+h+1}}$$

$$= -2 \lim_{h \to 0} \frac{-h}{h\sqrt{x+1}\sqrt{x+h+1}\left(\sqrt{x+1} + \sqrt{x+h+1}\right)} = \frac{2}{(x+1)\left(2\sqrt{x+1}\right)} = \frac{1}{(x+1)\sqrt{x+1}}$$

23. By Exercise 9, $y' = -2/(2x+3)^2$. Hence the slope of the tangent line at $(0, 1/3)$ is $y'(0) = -2/9$ and the equation is $y - 1/3 = (-2/9)x$.

24. By Exercise 11, $f'(x) = (1/2)(x+1)^{-1/2}$. Hence the slope of the tangent line at $(3, 2)$ is $f'(3) = 1/4$ and the equation is $y - 2 = (1/4)(x - 3)$.

25. By Exercise 4, $f'(x) = 2x$. Hence the slope of the tangent line at $P = (3, 2)$ is $f'(3) = 6$. Therefore the slope of the normal is $-1/6$ and the equation is $y - 2 = (-1/6)(x - 3)$.

26. By Example 2 in the text, $f'(x) = 2x$. Hence the slopes of the tangent lines at $x = x_1$ and $x = x_2$ are $2x_1$ and $2x_2$, and these lines are perpendicular if and only if $(2x_1)(2x_2) = -1$. Now, the equations of the tangent lines are $y - x_1^2 = (2x_1)(x - x_1)$ and $y - x_2^2 = (2x_2)(x - x_2)$. The y-intercepts of these lines are $-x_1^2$ and $-x_2^2$. Since they have the same y-intercept, $-x_1^2 = -x_2^2$ and hence $x_2 = \pm x_1$. Therefore

$$(2x_1)(2x_2) = -1 \implies 4x_1(-x_1) = -1 \implies x_1^2 = 1/4 \implies x_1 = \pm 1/2.$$

Hence the equations of the lines are $y - 1/4 = x - 1/2$ and $y - 1/4 = -(x - 1/2)$.

27. Here $f'(x) = 1' - \left(x^2\right)' = -2x$. Hence the slopes of the tangent lines at $x = x_1$ and $x = x_2$ are $-2x_1$ and $-2x_2$. Now, if the triangle is equilateral, then each of the angles is $60°$ and hence

$$\text{slope of tangent at } x_1 = f'(x_1) = \tan 60° \implies -2x_1 = \sqrt{3} \implies x_1 = -\sqrt{3}/2.$$

Therefore $x_2 = \sqrt{3}/2$ since x_1 and x_2 are symmetric about the y-axis.

28. $y = x^2 + 4x + 4 \implies y' = (x^2)' + (4x + 4)' = 2x + 4 = 4 \implies x = 0$. Hence the point is $(0, 4)$.

29. $y = 2 - ax^2 \implies y' = 2' - a(x^2)' = -2ax$. Hence the slope of the tangent line is 6 at $x = -1$ when $-2a(-1) = 6 \implies a = 3$.

30. $f(x) = ax^2 + bx + 3 \implies f'(x) = a(x^2)' + (bx + 3)' = 2ax + b$. Hence, if the slope of the tangent line at the point $(1, 5)$ is 1, it follows that $2a + b = 1$. Moreover, $f(1) = 5 \implies a + b + 3 = 5$. Therefore $b = 2 - a$ and hence $2a + b = 2a + (2 - a) = 1 \implies a = -1, b = 3$.

31. $y = ax^2 + b \implies y' = a(x^2)' + (b)' = 2ax$. Now, if $y = 4x$ is the tangent line at $(1, 4)$, then the slope is 4 and hence $y'(1) = 2a = 4 \implies a = 2$. Since $y(1) = a + b = 4$, $b = 4 - a = 2$.

32. Let $m_h = \big(f(x + h) - f(x)\big)/h$ stand for the difference quotient. The values of m_h for $f(x) = \sin x$ and $h = \pm 0.5, \pm 0.1, \pm 0.05,$ and ± 0.01 are listed below in the table, together with the estimate of the slope.

h	0	$\pi/4$	$\pi/2$	$3\pi/4$	π	$5\pi/4$	$3\pi/2$	$7\pi/4$	2π
0.5000	0.9588	0.5049	−0.2448	−0.8511	−0.9589	−0.5049	0.2448	0.8511	0.9589
0.1000	0.9983	0.6706	−0.0500	−0.7412	−0.9983	−0.6706	0.0500	0.7413	0.9983
0.0500	0.9996	0.6891	−0.0250	−0.7245	−0.9996	−0.6891	0.0250	0.7245	0.9996
0.0100	1.0000	0.7036	−0.0050	−0.7106	−1.0000	−0.7036	0.0050	0.7106	1.0000
−0.0100	1.0000	0.7106	0.0050	−0.7036	−1.0000	−0.7106	−0.0050	0.7036	1.0000
−0.0500	0.9996	0.7245	0.0251	−0.6891	−0.9996	−0.7245	−0.0250	0.6891	0.9996
−0.1000	0.9983	0.7412	0.0500	−0.6706	−0.9983	−0.7413	−0.0500	0.6706	0.9983
−0.5000	0.9588	0.8511	0.2448	−0.5049	−0.9589	−0.8511	−0.2448	0.5049	0.9589
Estimate	1	0.707	0	−0.707	−1	−0.707	0	0.707	1

The last row of the table lists the estimate of the slope for each value of x. Below is the graph of the function $f(x) = \sin x$ with the tangents lines. The graph on the right sketches the slope function.

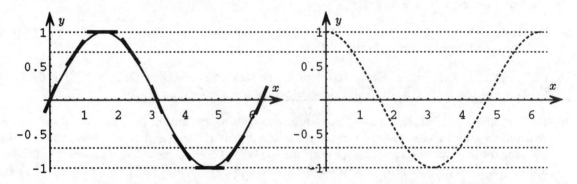

33. Let $m_h = \big(f(x + h) - f(x)\big)/h$ stand for the difference quotient. The values of m_h for $f(x) = \cos x$ and $h = \pm 0.5, \pm 0.1, \pm 0.05,$ and ± 0.01 are listed in the table below, together with the estimate of the slope.

h	0	$\pi/4$	$\pi/2$	$3\pi/4$	π	$5\pi/4$	$3\pi/2$	$7\pi/4$	2π
0.5000	−0.2448	−0.8511	−0.9588	−0.5049	0.2448	0.8511	0.9589	0.5049	−0.2448
0.1000	−0.0500	−0.7413	−0.9983	−0.6706	0.0500	0.7413	0.9983	0.6706	−0.0500
0.0500	−0.0250	−0.7245	−0.9996	−0.6891	0.0250	0.7245	0.9996	0.6891	−0.0250
0.0100	−0.0050	−0.7106	−1.0000	−0.7036	0.0050	0.7106	1.0000	0.7036	−0.0050
−0.0100	0.0050	−0.7036	−1.0000	−0.7106	−0.0050	0.7036	1.0000	0.7106	0.0050
−0.0500	0.0250	−0.6891	−0.9996	−0.7245	−0.0250	0.6891	0.9996	0.7245	0.0250
−0.1000	0.0500	−0.6706	−0.9983	−0.7412	−0.0500	0.6706	0.9983	0.7413	0.0500
−0.5000	0.2448	−0.5049	−0.9588	−0.8511	−0.2448	0.5049	0.9589	0.8511	0.2448
Estimate	0	−0.707	−1	−0.707	0	0.707	1	0.707	0

The last row of the table gives the estimate for the slope of the tangent line. It appears that the slope is equal to $-\sin x$ for each x.

34. $f(x) = x^3 + 2x^2 - 4x + 10 \implies$

$$f'(x) = \lim_{h \to 0} \frac{f(x+h) - f(x)}{h} = \lim_{h \to 0} \frac{\left[(x+h)^3 + 2(x+h)^2 - 4(x+h) + 10\right] - \left[x^3 + 2x^2 - 4x + 10\right]}{h}$$

$$= \lim_{h \to 0} \left(3x^2 + 3xh + h^2 + 4x + 2h - 4\right) = 3x^2 + 4x - 4 = (3x-2)(x+2).$$

Therefore $f' = 0 \iff x = 2/3, -2$, and hence f' can change sign only when $x = 2/3$ or $x = -2$. The table below summarizes the sign changes for f'

Sign f'	+	−	+
Interval	$(-\infty, -2)$	$(-2, 2/3)$	$(2/3, +\infty)$

Hence the slope of the tangent line is positive when $x < -2$ or $x > 2/3$.

35. $f(x) = x^2 - 6x + 2 \implies f'(x) = (x^2)' + (-6x+2)' = 2x - 6$. Hence $f' < 0 \iff x < 3$.

36. (a) Yes, $f'(3) = 6$ since

$$\lim_{h \to 0^+} \frac{f(3+h) - f(3)}{h} = \lim_{h \to 0^+} \frac{\left[(3+h)^2 + 1\right] - 10}{h} = \lim_{h \to 0^+} (6+h) = 6$$

and

$$\lim_{h \to 0^-} \frac{f(3+h) - f(3)}{h} = \lim_{h \to 0^-} \frac{\left[6(3+h) - 8\right] - 10}{h} = \lim_{h \to 0^-} 6 = 6.$$

(b) Yes, since f is differentiable at $x = 3$ (Theorem 2).

37. (a) No, $f'(1)$ does not exist since

$$\lim_{h \to 0^+} \frac{f(1+h) - f(1)}{h} = \lim_{h \to 0^+} \frac{\left[(1+h)^2 - (1+h)\right] - 0}{h} = \lim_{h \to 0^+} (1+h) = 1$$

but

$$\lim_{h \to 0^-} \frac{f(1+h) - f(1)}{h} = \lim_{h \to 0^-} \frac{\left[2(1+h) - 2\right] - 0}{h} = \lim_{h \to 0^-} 2 = 2.$$

(b) Yes, $\lim_{x \to 1} f(x) = 0 = f(1)$ since

$$\lim_{x \to 1^+} f(x) = \lim_{x \to 1^+} \left(x^2 - x\right) = 0$$

and

$$\lim_{x \to 1^-} f(x) = \lim_{x \to 1^-} (2x - 2) = 0.$$

38. (a) No, $f'(0)$ does not exist since

$$\lim_{h \to 0^+} \frac{f(0+h) - f(0)}{h} = \lim_{h \to 0^+} \frac{(2h^2 - 1) - (-1)}{h} = \lim_{h \to 0^+} (2h) = 0$$

but

$$\lim_{h \to 0^-} \frac{f(0+h) - f(0)}{h} = \lim_{h \to 0^-} \frac{(h - 1) - (-1)}{h} = \lim_{h \to 0^-} 1 = 1.$$

(b) Yes, $\lim_{x \to 0} f(x) = -1 = f(0)$ since

$$\lim_{x \to 0^+} f(x) = \lim_{x \to 0^+} (2x^2 - 1) = -1$$

and

$$\lim_{x \to 0^-} f(x) = \lim_{x \to 0^-} (x - 1) = -1.$$

39. $f'(0) = \lim_{h \to 0} \dfrac{\sin(h + 0) - \sin 0}{h} = \lim_{h \to 0} \dfrac{\sin h}{h} = 1.$

40. (a) The graph of $f(x) = \sqrt{|x|}$ is shown below on the left.

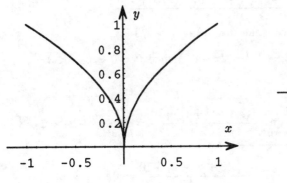

$$y = f(x) = \sqrt{|x|}$$

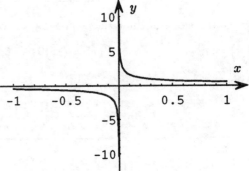

$$y = f'(x) = |x|/\left(2x\sqrt{|x|}\right)$$

(b) f is differentiable at all $x \neq 0$; for, if $x > 0$, then $x + h > 0$ for h sufficiently close to zero and hence

$$f'(x) = \lim_{h \to 0} \frac{\sqrt{x + h} - \sqrt{x}}{h} = \lim_{h \to 0} \frac{1}{\sqrt{x + h} + \sqrt{x}} = \frac{1}{2\sqrt{x}},$$

while if $x < 0$, $x + h < 0$ for h sufficiently close to zero and hence

$$f'(x) = \lim_{h \to 0} \frac{\sqrt{-(x + h)} - \sqrt{-x}}{h} = \lim_{h \to 0} \frac{-1}{\sqrt{-(x + h)} + \sqrt{-x}} = \frac{-1}{2\sqrt{-x}}.$$

(c) Since $f'(x)$ is equal to $1/(2\sqrt{x})$ for $x > 0$ and $-1/(2\sqrt{-x})$ for $x < 0$, it follows that

$$f'(x) = \frac{1}{2\sqrt{|x|}} \frac{|x|}{x} = \begin{cases} 1/(2\sqrt{x}), & x > 0 \\ -1/(2\sqrt{-x}), & x < 0. \end{cases}$$

41. (a) False; $f(a)$ must also exist and equal the limit. For example, if $f(x) = (x^2 - 1)/(x - 1)$, then $\lim_{x \to 1} f(x) = 2$, but f is not continuous at $x = 1$ since $f(1)$ does not exist.

(b) True; continuity means $\lim_{x \to a} f(x)$ exists and is equal to $f(a)$.

(c) False; if $f(x) = |x|$, then $\lim_{x \to 0} f(x) = 0$ but f is not differentiable at $x = 0$ (Example 5 in the text).

(d) True; if f is differentiable at $x = a$, then, by Theorem 2, f is continuous at $x = a$ and hence $\lim\limits_{x \to a} f(x) = f(a)$.

(e) False; $f(x) = |x|$ is continuous but not differentiable at $x = 0$.

(f) True; this is Theorem 2.

42. It follows from the given property of $f(x)$ that $f(x + h) = f(x) + 2xh + 3h + h^2$. Therefore

$$f'(x) = \lim_{h \to 0} \frac{f(x + h) - f(x)}{h} = \lim_{h \to 0} \frac{f(x) + 2xh + 3h + h^2 - f(x)}{h} = \lim_{h \to 0} (2x + 3 + h) = 2x + 3.$$

43. $m_h(x) = [2(x + h) + 1 - (2x + 1)]/h = 2h/h = 2$. Range: $[-4.7, 4.7] \times [-3.1, 3.1]$. This graph is the true picture of the derivative for this function.

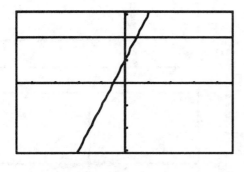

Exercise 43

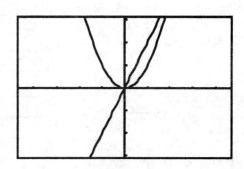

Exercise 44

For exercises 44-46, the graph of $m_h(x)$ is an approximation of the true graph of $f'(x)$, but as h gets smaller and smaller, the approximation becomes better.

44. $m_h(x) = ((x + h)^2 - x^2)/h = 2x + h$. Range: $[-4.7, 4.7] \times [-3.1, 3.1]$ with $h = 0.01$.

45. $m_h(x) = (\sqrt{x + h} - \sqrt{x})/h$. Range: $[-2.8, 6.7] \times [-2.2, 4.1]$ with $h = 0.01$.

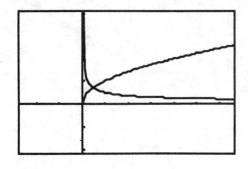

Exercise 45

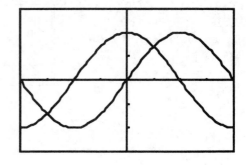

Exercise 46

46. $m_h(x) = (\sin(x + h) - \sin x)/h$. Range: $[-\pi, \pi] \times [-1.5, 1.5]$ with $h = 0.01$.

47. $f(x) = x^2 - 2x + 1$, $a = 1$. The table to the right summarizes the values of m_h for $h = 2, -1.5, -1, 0.1, -0.1$, and 0.001. The graphs are shown for $h = 2, -1$ with range $[-3.8, 5.7] \times [-3.2, 3.1]$. $f(x)$ is differentiable at $a = 1$.

h	m_h
2	2
−1.5	−1.5
−1	−1
−0.1	0.1
0.001	0.001

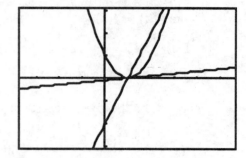

48. $f(x) = \sqrt{x}$, $a = 0$. The table to the right summarizes the values of m_h for $h = 2, 1.5, 1, 0.1, 0.01$, and 0.001. The graphs are shown for $h = 1$, 0.01 with range $[-1.4, 3.35] \times [-0.6, 2.55]$. $f(x)$ is differentiable at $a = 0$.

h	m_h
2	$1/\sqrt{2}$
1.5	0.8165
1	1
0.1	3.1623
0.01	10
0.001	31.6228

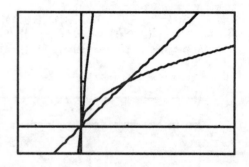

49. $f(x) = (\sin x)/x$, $a = 0$. The table to the right summarizes the values of m_h for $h = -2, -1.5, 1, -0.1, 0.01$, and 0.001. The graphs are shown for $h = 1$, 0.001 with range $[-2.35, 2.35] \times [-0.55, 2.55]$. $f(x)$ is differentiable at $a = 0$.

h	m_h
−2	0.2727
−1.5	0.2233
1	−0.1585
−0.1	0.0167
0.01	−0.0017
0.001	−0.0002

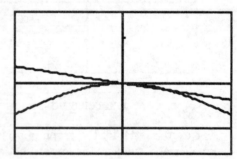

50. $f(x) = \begin{cases} 2x - 1, & \text{if } x \le 1 \\ x^2, & \text{if } x > 1 \end{cases}$, $a = 1$. The table to the right summarizes the values of m_h for $h = 2, 1.5, 1, 0.1, 0.01$, and 0.001. If $h < 0$, $m_h = 2$. The graphs are shown for $h = 1, 0.1$ with range $[0, 3] \times [-1, 9]$. $f(x)$ is differentiable at $a = 1$.

h	m_h
2	4
1.5	3.5
1	3
0.1	2.1
0.01	2.01
0.001	2.001

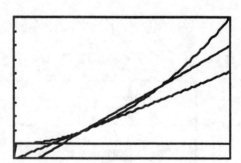

51. $f'(x) = 1$

52. $f'(x) = 1$

53. $f'(x) = 2$

54. $f'(x) = 2x$

55. $f'(x) = 2x^2 + 2$

56. $f'(x) = 3x^2$

57. $f'(x) = 3x^2 - 10x$

58. $f'(x) = -1/x^2$

59. $f'(x) = -1/(x+1)^2$

60. $f'(x) = \cos x$

61. $f'(x) = -2\sin 2x$

3.2 Rules for Calculating Derivatives

1. $f(x) = 8x^3 - x^2 \implies f'(x) = 24x^2 - 2x$

2. $f(x) = x - x^5 \implies f'(x) = 1 - 5x^4$

3. $f(x) = ax^3 + bx \implies f'(x) = 3ax^2 + b$

4. $f(x) = a^5 + 3a^2x^2 + x^3 \implies f'(x) = 6a^2x + 3x^2$

5. $f(x) = (x^2 + 5)/2 = (1/2)x^2 + 5/2 \implies f'(x) = x$

6. $f(x) = (2/3)x^3 + (1/2)x^2 + x \implies f'(x) = 2x^2 + x + 1$

7. $f(x) = (x-1)(x+2) = x^2 + x - 2 \implies f'(x) = 2x + 1$

8. $f(x) = (x^2 - 1)(2 - x) \implies f'(x) = (x^2 - 1)(-1) + (2x)(2 - x) = -3x^2 + 4x + 1$

9. $f(x) = (3x^2 - 8x)(x^2 + 2) \implies f'(x) = (3x^2 - 8x)(2x) + (6x - 8)(x^2 + 2) = 12x^3 - 24x^2 + 12x - 16$

10. $f(x) = (x^2 + x + 1)(x + 1) \implies f'(x) = (x^2 + x + 1)(1) + (2x + 1)(x + 1) = 3x^2 + 4x + 2$

11. $f(x) = (x^3 - x)^2 = x^6 - 2x^4 + x^2 \implies f'(x) = 6x^5 - 8x^3 + 2x$

12. $f(x) = \left(x^2 - \dfrac{3}{x^2}\right)^2 = x^4 - 6 + 9x^{-4} \implies f'(x) = 4x^3 - 36x^{-5}$

13. $f(x) = 5x^2 + 2x + \dfrac{3}{x} - \dfrac{4}{x^2} = 5x^2 + 2x + 3x^{-1} - 4x^{-2} \implies f'(x) = 10x + 2 - 3x^{-2} + 8x^{-3}$

14. $f(x) = \dfrac{x+2}{x-2} \implies f'(x) = \dfrac{(x-2)(1) - (x+2)(1)}{(x-2)^2} = \dfrac{-4}{(x-2)^2}$

15. $f(x) = \dfrac{6}{3-x} \implies f'(x) = \dfrac{(3-x)(0) - 6(-1)}{(3-x)^2} = \dfrac{6}{(3-x)^2}$

16. $f(x) = \dfrac{(8x+2)(x+1)}{x-3} = \dfrac{8x^2 + 10x + 2}{x-3} \implies$

$$f'(x) = \dfrac{(x-3)(16x+10) - (8x^2 + 10x + 2)(1)}{(x-3)^2} = \dfrac{8(x^2 - 6x - 4)}{(x-3)^2}$$

17. $f(x) = \dfrac{x^4 + 4x + 4}{1 - x^3} \implies f'(x) = \dfrac{(1-x^3)(4x^3 + 4) - (x^4 + 4x + 4)(-3x^2)}{(1-x^3)^2} = \dfrac{-x^6 + 12x^3 + 12x^2 + 4}{(1-x^3)^2}$

18. $f(x) = \dfrac{(2x+1)(3x+2)}{(x+1)(x-1)} = \dfrac{6x^2+7x+2}{x^2-1} \Longrightarrow$

$$f'(x) = \frac{(x^2-1)(12x+7)-(6x^2+7x+2)(2x)}{(x^2-1)^2} = \frac{-7x^2-16x-7}{(x^2-1)^2}$$

19. $f(x) = 5x^{-3} - 2x^{-5} \Longrightarrow f'(x) = -15x^{-4} + 10x^{-6}$

20. $f(x) = \dfrac{x^2-4}{x+3} \Longrightarrow f'(x) = \dfrac{(x+3)(2x)-(x^2-4)(1)}{(x+3)^2} = \dfrac{x^2+6x+4}{(x+3)^2}$

21. $f(x) = (x-2)\left(x+\dfrac{1}{x}\right) = x^2 - 2x + 1 - 2x^{-1} \Longrightarrow f'(x) = 2x - 2 + 2x^{-2}$

22. $f(x) = \dfrac{1}{(x-3)^2} = \dfrac{1}{x^2-6x+9} \Longrightarrow$

$$f'(x) = \frac{(x^2-6x+9)(0)-(1)(2x-6)}{(x^2-6x+9)^2} = \frac{-2(x-3)}{(x-3)^4} = \frac{-2}{(x-3)^3}$$

23. $f(x) = \left(1+\dfrac{3}{x}\right)^2 = 1 + 6x^{-1} + 9x^{-2} \Longrightarrow f'(x) = -6x^{-2} - 18x^{-3}$

24. $g(t) = \dfrac{t}{t^2+t+1} \Longrightarrow g'(t) = \dfrac{(t^2+t+1)(1)-t(2t+1)}{(t^2+t+1)^2} = \dfrac{-t^2+1}{(t^2+t+1)^2}$

25. $f(x) = \left(\dfrac{x+1}{x-1}\right)^2 = \dfrac{x^2+2x+1}{x^2-2x+1} \Longrightarrow$

$$f'(x) = \frac{(x^2-2x+1)(2x+2)-(x^2+2x+1)(2x-2)}{(x^2-2x+1)^2} = \frac{-4x^2+4}{(x-1)^4} = \frac{-4(x+1)}{(x-1)^3}$$

26. $f(t) = \dfrac{t-4+t^2}{t^3+3t^2+3} \Longrightarrow$

$$f'(t) = \frac{(t^3+3t^2+3)(1+2t)-(t-4+t^2)(3t^2+6t)}{(t^3+3t^2+3)^2} = \frac{-t^4-2t^3+9t^2+30t+3}{(t^3+3t^2+3)^2}$$

27. $f(x) = \left(x^5+x^{-2}\right)\left(x^3-x^{-7}\right) = x^8 + x - x^{-2} - x^{-9} \Longrightarrow f'(x) = 8x^7 + 1 + 2x^{-3} + 9x^{-10}$

28. $f(x) = \dfrac{ax+b}{cx^2+d} \Longrightarrow f'(x) = \dfrac{(cx^2+d)(a)-(ax+b)(2cx)}{(cx^2+d)^2} = \dfrac{-acx^2-2bcx+ad}{(cx^2+d)^2}$

29. $f(x) = \dfrac{x^{-3}-x^4}{x^5} = x^{-8} - x^{-1} \Longrightarrow f'(x) = -8x^{-9} + x^{-2}$

30. By the Product Rule: $f'(x) = x(1) + (x+1)(1) = 2x + 1$
Expanding first: $f(x) = x(x+1) = x^2 + x \Longrightarrow f'(x) = 2x + 1$

31. By the Product Rule: $f'(x) = (x^2+2)(2x) + (x^2-2)(2x) = 4x^3$
Expanding first: $f(x) = (x^2+2)(x^2-2) = x^4 - 4 \Longrightarrow f'(x) = 4x^3$

32. By the Product Rule: $f(x) = \left(\dfrac{1}{x} + 1\right)\left(3 - \dfrac{2}{x^2}\right) = (x^{-1} + 1)(3 - 2x^{-2}) \implies$

$$f'(x) = (x^{-1} + 1)(4x^{-3}) + (3 - 2x^{-2})(-x^{-2}) = 6x^{-4} + 4x^{-3} - 3x^{-2}$$

Expanding first: $f(x) = \dfrac{3}{x} + 3 - \dfrac{2}{x^3} - \dfrac{2}{x^2} = 3x^{-1} + 3 - 2x^{-3} - 2x^{-2} \implies f'(x) = -3x^{-2} + 6x^{-4} + 4x^{-3}$

33. By the Product Rule: $f'(x) = (x^3 + 7)(12x^3 + 1) + (3x^4 + x + 9)(3x^2) = 21x^6 + 88x^3 + 27x^2 + 7$
    Expanding first: $f(x) = (x^3 + 7)(3x^4 + x + 9) = 3x^7 + 22x^4 + 9x^3 + 7x + 63 \implies f'(x) = 21x^6 + 88x^3 + 27x^2 + 7$

34. By the Product Rule: $f(x) = (1 - x^2)(1 + x^2) \implies f'(x) = (1 - x^2)(2x) + (1 + x^2)(-2x) = -4x^3$
    Expanding first: $f(x) = (1 - x^2)(1 + x^2) = 1 - x^4 \implies f'(x) = -4x^3$

35. By the Product Rule: $f(x) = \left(\dfrac{1}{x+1}\right)\left(\dfrac{2}{x+2}\right) \implies$

$$f'(x) = \frac{1}{x+1}\frac{(x+2)(0) - 2(1)}{(x+2)^2} + \frac{2}{x+2}\frac{(x+1)(0) - (1)(1)}{(x+1)^2} = \frac{-4x - 6}{(x+1)^2(x+2)^2}$$

Expanding first: $f(x) = \dfrac{2}{x^2 + 3x + 2} \implies f'(x) = \dfrac{(x^2 + 3x + 2)(0) - 2(2x + 3)}{(x^2 + 3x + 2)^2} = \dfrac{-4x - 6}{(x^2 + 3x + 2)^2}$

36. By the Product Rule: $f(s) = \left(s - \dfrac{3}{s^2}\right)\left(s + \dfrac{5}{s^2}\right) = (s - 3s^{-2})(s + 5s^{-2}) \implies$

$$f'(s) = (s - 3s^{-2})(1 - 10s^{-3}) + (s + 5s^{-2})(1 + 6s^{-3}) = 2s - 2s^{-2} + 60s^{-5}$$

Epanding first: $f(s) = (s - 3s^{-2})(s + 5s^{-2}) = s^2 + 2s^{-1} - 15s^{-4} \implies f'(s) = 2s - 2s^{-2} + 60s^{-5}$

37. By the Product Rule: $f'(u) = (u^2 + u + 1)(2u - 1) + (u^2 - u - 1)(2u + 1) = 4u^3 - 2u - 2$
    Expanding first: $f(u) = (u^2 + u + 1)(u^2 - u - 1) = u^4 - u^2 - 2u - 1 \implies f'(u) = 4u^3 - 2u - 2$

38. By the Product Rule: $f'(x) = (x^{-2} + x^3)(2x + 4x^{-5}) + (x^2 - x^{-4})(-2x^{-3} + 3x^2) = 6x^{-7} + 5x^4 + x^{-2}$
    Expanding first: $f(x) = 1 + x^5 - x^{-6} - x^{-1} \implies f'(x) = 5x^4 + 6x^{-7} + x^{-2}$

39. Using the Product Rule, we obtain

$$(fgh)'(x) = \left[(fg)h\right]'(x) = \left[(fg)'h + (fg)h'\right](x)$$

$$= \left[(f'g + fg')h + fgh'\right](x) = f'(x)g(x)h(x) + f(x)g'(x)h(x) + f(x)g(x)h'(x).$$

40. $f(x) = x(x + 1)(x + 2) \implies f'(x) = (1)(x + 1)(x + 2) + x(1)(x + 2) + x(x + 1)(1) = 3x^2 + 6x + 2$

41. $f(s) = (2s - 1)(s - 3)(s^2 + 4) \implies$

$$f'(s) = (2)(s - 3)(s^2 + 4) + (2s - 1)(1)(s^2 + 4) + (2s - 1)(s - 3)(2s) = 8s^3 - 21s^2 + 22s - 28$$

42. $f(t) = (t^2 - 7)(3t^5 + t)(t^3 - 9) \implies$

$$f'(t) = (2t)(3t^5 + t)(t^3 - 9) + (t^2 - 7)(15t^4)(t^3 - 9) + (t^2 - 7)(3t^5 + t)(3t^2)$$

$$= 30t^9 - 168t^7 - 189t^6 + 6t^5 + 945t^4 - 28t^3 - 27t^2 + 63$$

43. $f(x) = \left(\dfrac{1}{x}\right)\left(\dfrac{1}{x+1}\right)\left(\dfrac{1}{x+2}\right) \implies$

$$f'(x) = \left(\dfrac{-1}{x^2}\right)\left(\dfrac{1}{x+1}\right)\left(\dfrac{1}{x+2}\right) + \left(\dfrac{1}{x}\right)\left(\dfrac{-1}{(x+1)^2}\right)\left(\dfrac{1}{x+2}\right) + \left(\dfrac{1}{x}\right)\left(\dfrac{1}{x+1}\right)\left(\dfrac{-1}{(x+2)^2}\right)$$

$$= \dfrac{-3x^2 - 6x - 2}{x^2(x+1)^2(x+2)^2}$$

44. $f(u) = (u^2 - 4)^3 = (u^2 - 4)(u^2 - 4)(u^2 - 4) \implies$

$$f'(u) = (2u)(u^2 - 4)(u^2 - 4) + (u^2 - 4)(2u)(u^2 - 4) + (u^2 - 4)(u^2 - 4)(2u) = 6u(u^2 - 4)^2$$

45. $f(x) = (2x^3 - 6x + 9)^3 = (2x^3 - 6x + 9)(2x^3 - 6x + 9)(2x^3 - 6x + 9) \implies$

$$f'(x) = (6x^2 - 6)(2x^3 - 6x + 9)^2 + (6x^2 - 6)(2x^3 - 6x + 9)^2 + (6x^2 - 6)(2x^3 - 6x + 9)^2$$

$$= 3(6x^2 - 6)(2x^3 - 6x + 9)^2$$

46. (a) $(f^2)'(x) = (ff')(x) = f'(x)f(x) + f(x)f'(x) = 2f(x)f'(x)$

    (b) $(f^3)'(x) = (ff^2)'(x) = f'(x)f^2(x) + f(x)(f^2)'(x) = f'(x)f^2(x) + f(x)[2f(x)f'(x)] = 3f^2(x)f'(x)$

    (c) $(f^4)'(x) = (ff^3)'(x) = f'(x)f^3(x) + f(x)(f^3)'(x) = f'(x)f^3(x) + f(x)[3f^2(x)f'(x)] = 4f'(x)f^3(x)$

    (d) $(f^n)'(x) = nf'(x)f^{n-1}(x)$. If $n = 1$, this formula is clearly true. Now let $n > 1$ and assume as induction hypothesis that $(f^n)'(x) = nf'(x)f^{n-1}(x)$. Then

$$
\begin{aligned}
(f^{n+1})'(x) &= (ff^n)'(x) \\
&= f'(x)f^n(x) + f(x)(f^n)'(x) \\
&= f'(x)f^n(x) + f(x)\left[nf^{n-1}(x)f'(x)\right] \quad \text{by the induction hypothesis} \\
&= (n+1)f'(x)f^n(x).
\end{aligned}
$$

Therefore $(f^n)'(x) = nf'(x)f^{n-1}(x)$ for all positive integers n.

47. Using the derivative for $f(x)$ found in Exercise 11, $f'(2) = 2(11)(6) = 132$.

48. Using the derivative for $f(x)$ found in Exercise 20, $f'(-1) = (1 - 6 + 4)/4 = -1/4$.

49. $f(x) = \left(\dfrac{x-1}{x}\right)^3 = (1 - x^{-1})^3 \implies f'(x) = 3(1 - x^{-1})^2(x^{-2})$

50. $f(s) = \dfrac{1 - s}{(1 + s)^2} \implies$

$$f'(s) = \dfrac{(1+s)^2(-1) - (1-s)[2(1+s)(1)]}{(1+s)^4} = \dfrac{-3 - 2s + s^2}{(1+s)^4} = \dfrac{(s+1)(s-3)}{(s+1)^4} = \dfrac{s-3}{(s+1)^3}$$

51. $f(x) = \dfrac{1}{(x-6)^3} \implies f'(x) = \dfrac{(x-6)^3(0) - (1)\left[3(x-6)^2(1)\right]}{(x-6)^6} = \dfrac{-3}{(x-6)^4}$

52. $f(u) = (u^2 + 4)^3 \implies f'(u) = 3(u^2 + 4)^2(2u) = 6u(u^2 + 4)^2$

53. $f(x) = (x+2)(x-1)^2 \implies f'(x) = (1)(x-1)^2 + (x+2)[2(x-1)(1)] = 3x^2 - 3 = 3(x-1)(x+1)$

54. $f(x) = 3x^3 - 7 \implies f'(x) = 9x^2 \implies f'(1) = 9$. Hence the equation of the tangent line at $(1, -4)$ is $y + 4 = 9(x - 1)$.

55. $f(x) = \dfrac{x-1}{x+1} \implies f'(x) = \dfrac{(x+1)(1) - (x-1)(1)}{(x+1)^2} = \dfrac{2}{(x+1)^2} \implies f'(1) = 1/2$. Hence the equation of the tangent line at $(1, 0)$ is $y = (1/2)(x - 1)$.

56. $f(x) = (x^2 + x)(1 - x^3) = -x^5 - x^4 + x^2 + x \implies f'(x) = -5x^4 - 4x^3 + 2x + 1 \implies f'(2) = -107$. Hence the equation of the tangent line at $(2, -42)$ is $y + 42 = -107(x - 2)$.

57. $f(x) = \left(1 - \dfrac{1}{x}\right)^2 = (1 - x^{-1})^2 \implies f'(x) = 2(1 - x^{-1})(x^{-2}) \implies f'(1) = 0$. Hence the equation of the tangent line at $(1, 0)$ is $y = 0$.

58. $f(x) = \dfrac{1}{x^3 + x} \implies f'(x) = \dfrac{(x^3 + x)(0) - (1)(3x^2 + 1)}{(x^3 + x)^2} = -\dfrac{3x^2 + 1}{(x^3 + x)^2} \implies f'(2) = -13/100$. Hence the equation of the tangent line at $(2, 1/10)$ is $y - 1/10 = (-13/100)(x - 2)$.

59. $f(x) = 2x^3 + 1 \implies f'(x) = 6x^2 \implies f'(1) = 6$. Hence the slope of the normal line to the graph at $(1, 3)$ is $-1/6$ and the equation is $y - 3 = (-1/6)(x - 1)$.

60. $f(x) = (x - 2)/(x + 1) \implies f'(x) = [(x + 1)(1) - (x - 2)(1)]/(x + 1)^2 = 3/(x + 1)^2 \implies f'(2) = 1/3$. Hence the slope of the normal line to the graph at $(2, 0)$ is -3 and the equation is $y = -3(x - 2)$.

61. $f(x) = 2/x^2 = 2x^{-2} \implies f'(x) = -4x^{-3} \implies f'(2) = -1/2$. Hence the slope of the normal line to the graph at $(2, 1/2)$ is 2 and the equation is $y - 1/2 = 2(x - 2)$.

62. $f(x) = \sqrt{x}(x - 1) \implies f'(x) = \sqrt{x}(1) + [1/(2\sqrt{x})](x - 2) \implies f'(3) = \sqrt{3} + 1/\sqrt{3} = 4\sqrt{3}/3$. Hence the slope of the normal line to the graph at $(3, 2\sqrt{3})$ is $-\sqrt{3}/4$ and the equation is $y - 2\sqrt{3} = (-\sqrt{3}/4)(x - 3)$.

63. $f(x) = \dfrac{3x}{2x - 4} \implies$

$$f'(x) = \frac{(2x - 4)(3) - (3x)(2)}{(2x - 4)^2} = \frac{-12}{(2x - 4)^2} = -3$$

$$\implies \quad (2x - 4)^2 = 4 \quad \implies \quad 2x - 4 = \pm 2 \quad \implies \quad x = 3, 1.$$

Hence there are two points where the slope of the tangent line is -3: $(3, 9/2)$ and $(1, -3/2)$.

64. $y = 1/(ax + 2) \implies y' = [(ax + 2)(0) - (1)(a)]/(ax + 2)^2 = -a/(ax + 2)^2$. Now, the slope of the line $4y + 3x - 2 = 0$ is $-3/4$. Hence $y'(0) = -a/4 = -3/4 \implies a = 3$.

65. $y = b/x^2 = bx^{-2} \implies y' = -2bx^{-3} \implies y'(-2) = b/4$. Since the slope of the line $4y - bx - 21 = 0$ is also $b/4$, this line is parallel to the tangent line at $x = -2$ for all values of b. If $x = -2$, $y = b/(-2)^2 = b/4$ and hence this line touches the graph when $4(b/4) - b(-2) - 21 = 0 \implies b = 7$.

66. Since $f^3(x) = f(x)f(x)f(x) = x$, it follows from Exercise 46 that $3f^2(x)f'(x) = 1$. Hence $f'(x) = 1/(3f^2(x)) = (1/3)x^{-2/3}$.

67. If $f(x) = x^{1/n}$, then $f^n(x) = x$ and hence $nf^{n-1}(x)f'(x) = 1$. Therefore

$$f'(x) = \frac{1}{nf^{n-1}(x)} = \frac{1}{nx^{(n-1)/n}} = (1/n)x^{(1/n)-1}.$$

68. If $f(x) = x^{n/m}$, then $f^m(x) = x^n$ and hence $mf^{m-1}(x)f'(x) = nx^{n-1}$. Therefore

$$f'(x) = \frac{nx^{n-1}}{mf^{m-1}(x)} = \frac{nx^{n-1}}{mx^{n(m-1)/m}} = (n/m)x^{(n/m)-1}.$$

69. The slope of the secant line through $(1,1)$ and $(2,8)$ is 7. Hence, $f'(x) = 3x^2 = 7 \implies x = \pm\sqrt{7/3}$. The graph of $f(x)$, the secant line and tangent line are shown below.

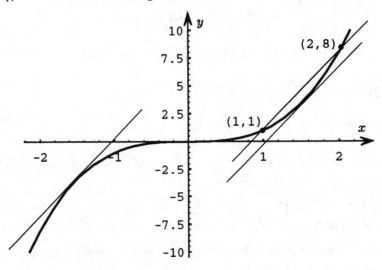

70. Let $f(x) = x^{100}$. Then $f'(x) = 100x^{99}$ and hence

$$\lim_{h \to 0} \frac{(3+h)^{100} - 3^{100}}{h} = f'(3) = 100(3)^{99}.$$

71. (a) By definition, the difference quotient of $1/g$ at x_0 is

$$\frac{\dfrac{1}{g(x_0+h)} - \dfrac{1}{g(x_0)}}{h} = \frac{g(x_0) - g(x_0+h)}{hg(x_0)g(x_0+h)}.$$

   (b) By definition,

$$\left(\frac{1}{g}\right)'(x) = \lim_{h \to 0} \frac{\dfrac{1}{g(x+h)} - \dfrac{1}{g(x)}}{h} = \lim_{h \to 0} \frac{g(x) - g(x+h)}{hg(x)g(x+h)}$$

$$= \lim_{h \to 0} \frac{-[g(x+h) - g(x)]}{h} \frac{1}{g(x)} \frac{1}{g(x+h)} = -g'(x)\frac{1}{g(x)}\frac{1}{g(x)} = -\frac{g'(x)}{[g(x)]^2}.$$

   (c) Using the Product Rule, we obtain

$$\left(\frac{f}{g}\right)'(x) = \left[f\left(\frac{1}{g}\right)\right]'(x) = f'(x)\left(\frac{1}{g(x)}\right) + f(x)\left(\frac{1}{g}\right)'(x)$$

$$= \frac{f'(x)}{g(x)} + f(x)\frac{-g'(x)}{[g(x)]^2} = \frac{f'(x)}{g(x)} - \frac{f(x)g'(x)}{[g(x)]^2} = \frac{f'(x)g(x) - f(x)g'(x)}{[g(x)]^2}.$$

72. (a) If $n = 1$, then, for any x,

$$f'(x) = \lim_{h \to 0} \frac{f(x+h) - f(x)}{h} = \lim_{h \to 0} \frac{(x+h) - x}{h} = \lim_{h \to 0} 1 = 1 = 1 \cdot x^0.$$

(b) Let $n = k > 1$ and assume as induction hypothesis that $f'(x) = kx^{k-1}$.

(c) Let $f(x) = x^{k+1}$. Then $f(x) = x(x^k)$ and hence $f(x)$ is differentiable since both x and x^k are differentiable.

(d) By the Product Rule, $f'(x) = (1)(x^k) + (x)(kx^{k-1}) = x^k + kx^k = (k+1)x^k$.

(e) By mathematical induction, it follows that if $f(x) = x^n$, then $f'(x) = nx^{n-1}$ for all integers $n \geq 1$.

3.3   The Derivative as a Rate of Change

1. Average rate of change from $t = 2$ to $t = 3$ is

$$\frac{s(3) - s(2)}{3 - 2} = \frac{16 - 9}{1} = 7.$$

Since $s'(t) = 2(1 + t)$, the instantaneous rate of change at $t_0 = 5/2$ is $2(1 + 5/2) = 7$.

2. Average rate of change from $t = 2$ to $t = 2.5$ is

$$\frac{s(2.5) - s(2)}{2.5 - 2} = \frac{1/(-1.5) - 1/(-1)}{0.5} = 2/3.$$

Since $s'(t) = [(1-t)(0) - (1)(-1)]/(1-t)^2 = 1/(1-t)^2$, the instantaneous rate of change at $t_0 = 2$ is $1/(1-2)^2 = 1$.

3. Average rate of change from $t = 0$ to $t = 1$ is

$$\frac{f(1) - f(0)}{1 - 0} = \frac{0 - 0}{1} = 0.$$

Since $f'(t) = [1/(2\sqrt{t})](1 - t^3) + \sqrt{t}(-3t^2)$, the instantaneous rate of change at $t_0 = 1$ is $f'(1) = 0 + (-3) = -3$.

4. Average rate of change from $t = 1$ to $t = 2$ is

$$\frac{f(2) - f(1)}{2 - 1} = \frac{(2^{-2} + 2)^3 - 27}{1} = -15.6094.$$

Since $f'(x) = 3(t^{-2} + 2)^2(-2t^{-3})$ (see Exercise 46(b), Section 3.2), the instantaneous rate of change at $t_0 = 3/2$ is $f'(3/2) = 3(4/9 + 2)^2(-2)(8/27) = -10.6228$.

5. (a) $v(t) = s'(t) = 3$

   (b) $v(t)$ is never zero

   (c) $[0, +\infty)$ since $v(t) > 0$ for all $t \geq 0$.

6. (a) $v(t) = s'(t) = 2t - 6 = 2(t - 3)$

   (b) $v(t) = 0 \iff t = 3$

   (c) $(3, +\infty)$ since $v(t) > 0 \iff t > 3$.

7. (a) $v(t) = s'(t) = [(1+t)(0) - (1)(1)]/(1+t)^2 = -1/(1+t)^2$

   (b) $v(t)$ is never zero

   (c) $v(t)$ is never positive

8. (a) $v(t) = s'(t) = 2t - 3 = 2(t - 3/2)$

   (b) $v(t) = 0 \iff t = 3/2$

   (c) $(3/2, +\infty)$ since $v(t) > 0 \iff t > 3/2$.

9. (a) $v(t) = s'(t) = 3t^2 - 18t + 24 = 3(t - 4)(t - 2)$

   (b) $v(t) = 0 \iff t = 2, 4$

   (c) $[0, 2), (4, +\infty)$ since, for positive t, $v(t) > 0 \iff 0 < t < 2$ or $t > 4$.

10. (a) $s(t) = t(t - 1)(t + 2) = t^3 + t^2 - 2t \implies v(t) = s'(t) = 3t^2 + 2t - 2$

    (b) For $t > 0$, $v(t) = 0 \iff 3t^2 + 2t - 2 = 0 \iff t = (-2 + \sqrt{28})/6 = (-1 + \sqrt{7})/3$.

    (c) $((-1 + \sqrt{7})/3, +\infty)$ since, for positive t, $v(t) > 0 \iff t > (-1 + \sqrt{7})/3$.

11. (a) $v(t) = s'(t) = 3t^2 - 12t + 9 = 3(t - 3)(t - 1)$

    (b) $v(t) = 0 \iff t = 1, 3$

    (c) $[0, 1), (3, +\infty)$ since, for positive t, $v(t) > 0 \iff 0 < t < 1$ or $t > 3$.

12. (a) $v(t) = s'(t) = 4t^3 - 4 = 4(t - 1)(t^2 + t + 1)$

    (b) $v(t) = 0 \iff t = 1$

    (c) $(1, +\infty)$ since $v(t) > 0 \iff t > 1$.

13. $s(t) = \dfrac{t^2 + 2}{t + 1} \implies v(t) = s'(t) = \dfrac{(t + 1)(2t) - (t^2 + 2)(1)}{(t + 1)^2} = \dfrac{t^2 + 2t - 2}{(t + 1)^2}$. Hence $v(3) = (9 + 6 - 2)/4^2$
    $= 13/16$.

14. $s(t) = 6 + 5t - t^2 \implies v(t) = s'(t) = 5 - 2t$. Hence $v(t) = 0 \iff$
    $t = 2.5$ seconds. The graph to the right illustrates the distance
    of the particle from the origin from $t = 0$ to $t = 6$. Initially
    it is $s(0) = 6$ units from the origin. After 6 seconds, it is at
    $s(6) = 0$; that is, it is at the origin. The maximum distance to
    the right of the origin occurs when $v(t) = 5 - 2t = 0 \implies t =$
    $2.5 \implies s(2.5) = 6 + 5(2.5) - (2.5)^2 = 12.25$ units. Thus, its
    maximum distance from the origin is 12.25 units when $t = 2.5$
    seconds.

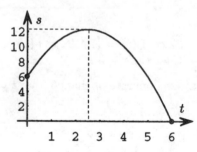

15. Let $V(t)$ stand for the volume of the snowball at time t. Then $V(t) = (4/3)\pi r^3 = (4/3)\pi(4 - t^2)^3$.

    (a) The average rate of change of the volume from $t = 0$ to $t = 1$ is

$$\frac{V(1) - V(0)}{1 - 0} = (4/3)\pi(27) - (4/3)\pi(64) = (4/3)\pi(-37) \approx -154.99 \text{ in}^3/\text{min}.$$

    The average rate of change of the volume from $t = 1$ to $t = 2$ is

$$\frac{V(2) - V(1)}{2 - 1} = (4/3)\pi(0) - (4/3)\pi(27) = -36\pi \approx -113.1 \text{ in}^3/\text{min}.$$

(b) $V'(t) = (4/3)\pi \left[3(4-t^2)^2(-2t)\right] = -8\pi t(4-t^2)^2$ (see Exercise 46(b), Section 3.2). Hence the instantaneous rate of change of the volume at $t = 1$ is $V'(1) = -8\pi(1)(9) = -72\pi \approx -226.2$ in^3/min.

16. Let $V(t)$ stand for the volume of the jar at time t. Then $V(t) = \pi r^2 h = \pi(3)^2\sqrt{t} = 9\pi\sqrt{t}$.

    (a) The average rate of change of the volume from $t = 0$ to $t = 1$ is

    $$\frac{V(1) - V(0)}{1 - 0} = 9\pi(1) - 9\pi(0) \approx 28.27 \text{ in}^3/\text{min}.$$

    (b) $V'(t) = 9\pi/(2\sqrt{t}) = 4.5\pi/\sqrt{t}$. Hence the instantaneous rate of change of the volume at $t = 1$ is $4.5\pi/1 \approx 14.14$ in^3/min.

17. (a) $v(t) = s'(t) = 6 - 2t \Longrightarrow v_0 = v(0) = 6$.

    (b) $v(t) = 0 \Longleftrightarrow 6 - 2t = 0 \Longleftrightarrow t = 3$. Since $v(t) > 0$ for $t < 3$ and $v(t) < 0$ for $t > 3$, velocity changes sign at $t = 3$ and hence the particle changes direction at $t = 3$.

    (c) The particle is at the origin when $s(t) = 0 \Longrightarrow t(6 - t) = 0 \Longrightarrow t = 0, 6$. Thus, the second time it is at the origin occurs when $t = 6$ and its velocity is $v(6) = 6 - 2(6) = -6$.

18. (a) $v(t) = s'(t) = v_0 - 9.8t = 98 - 9.8t$.

    (b) $v(t) = 0 \Longleftrightarrow 98 - 9.8t = 0 \Longleftrightarrow t = 10$. Since $v(t) > 0$ for $t < 10$ and $v(t) < 0$ for $t > 10$, velocity changes sign at $t = 10$ and hence maximum height occurs when $t = 10$.

    (c) Maximum height $= s(10) = 98(10) - 4.9(10)^2 = 490$ meters.

    (d) $s(t) = 0 \Longrightarrow t(98 - 4.9t) = 0 \Longrightarrow t = 0, 20$. Hence impact occurs when $t = 20$. Impact velocity $= v(20) = 98 - 9.8(20) = -98$ m/s.

19. Let $s(t)$ stand for the position of the object after t seconds. Then $s(t) = -4.9t^2$ (see Example 2 in the text).

    (a) $v(t) = s'(t) = -9.8t = -49 \Longrightarrow t = 5$. Hence the object fell for 5 seconds.

    (b) $s(5) = -4.9(5)^2 = -122.5$. Hence the object was dropped from 122.5 meters.

    (c) Average velocity from $t = 0$ to $t = 5$ is

    $$\frac{s(5) - s(0)}{5 - 0} = \frac{-122.5 - 0}{5} = -24.5 \text{ m/s}.$$

20. (a) $v(t) = s'(t) = v_0 - 9.8t \Longrightarrow v(4) = v_0 - 9.8(4) = v_0 - 39.2$ m/s.

    (b) Speed after 4 seconds $= |v(4)| = |v_0 - 39.2|$.

21. (a) $s(t) = 40 + 400t - 4.9t^2$

    (b) $v(t) = s'(t) = 400 - 9.8t = 0 \Longrightarrow t = 40.82 \Longrightarrow$ Maximum height $= s(40.82) = 40 + 400(40.82) - 4.9(40.82)^2 = 8203.3$ meters.

    (c) For t positive,

    $$s(t) = 40 + 400t - 4.9t^2 = 0 \quad \Longrightarrow \quad t = \frac{-400 - \sqrt{(400)^2 - 4(40)(-4.9)}}{2(-4.9)} = 81.73.$$

    Thus, the time to impact is 81.73 seconds. Impact speed $= |v(81.73)| = |400 - 9.8(81.73)| = 400.95$ m/s.

22. (a) $s(t) = -4.9t^2$. The tail section strikes the ground when $s(t) = -10000 \Longrightarrow -4.9t^2 = -10000 \Longrightarrow$ $t = 45.18$ seconds.

    (b) $v(t) = s'(t) = -9.8t \Longrightarrow$ Impact velocity $= v(45.18) = -9.8(45.18) = -442.76$ m/sec.

23. (a) The graph of $s(t)$ is shown to the right.

    (b) $v(t) = s'(t) = 800 - 32t$.

    (c) The object is rising $\Longleftrightarrow v(t) > 0 \Longleftrightarrow 800 - 32t > 0$ $\Longleftrightarrow 0 < t < 25$.

    (d) The object is falling $\Longleftrightarrow v(t) < 0 \Longleftrightarrow 800 - 32t < 0$ $\Longleftrightarrow 25 < t < 50$.

    (e) The object is at rest $\Longleftrightarrow v(t) = 0 \Longleftrightarrow 800 - 32t = 0$ $\Longleftrightarrow t = 25$.

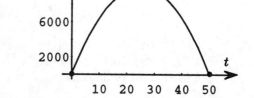

    (f) Maximum height $= s(25) = 800(25) - 16(25)^2 = 10,000$ feet.

    (g) $s(t) = 9600 \Longrightarrow 800t - 16t^2 = 9600 \Longrightarrow t^2 - 50t + 600 = (t - 20)(t - 30) = 0 \Longrightarrow t = 20, 30$. Thus, it takes 20 seconds to reach a height of 9600 feet.

    (h) For positive t, $s(t) = 0 \Longrightarrow 800t - 16t^2 = t(800 - 16t) = 0 \Longrightarrow t = 50$. Hence impact occurs at 50 seconds.

24. For more accuracy, evaluate both $p(t)$ and $p'(t)$ in nested form:

$$p(t) = (((((-0.0614t + 1.1502)t - 7.0039)t + 15.3990)t + 6.4932)t + 75.995$$

$$p'(t) = ((((-0.307t + 4.6008)t - 21.0117)t + 30.798)t + 6.4932.$$

    (a) The graph of $p'(t)$ with the range $[0, 10] \times [-200, 245]$ is shown below on the left.

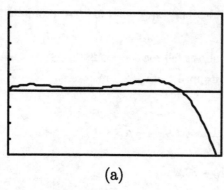

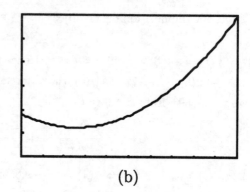

(a)   (b)

    (b) Between 1960 and 1970.

    (c) For $t = 5.5$ (1955), $p'(5.5) = 24.818$, or 24.8 million people per decade. The average rate $= (p(6) - p(5))/(6 - 5) = 24.740$ million people per decade.

    (d) The average rate $= (p(10) - p(8))/2 = -93.908$, i.e., losing about 93.9 million people per decade. The instantaneous rate in 1990 is $p'(9) = -78.5163$. This model predicts a decreasing population for the years 1980–2000. As you can tell from the census taken in 1990 this is not the case.

    (e) Evaluate $q(t)$ and $q'(t)$ in the nested forms

$$q(t) = ((0.297t - 2.0101)t + 17.6901)t + 75.995,$$

$$q'(t) = (0.891t - 4.0202) + 17.6901.$$

132

The graph of $q'(t)$ with the range $[0, 10] \times [0, 60]$ is shown above on the right. Fastest growth is between 1970–1980. For $t = 5.5$ (1955), $q'(5.5) = 22.532$ million people per decade. For the years 1950–1960 the average rate $= (q(6) - q(5))/(6 - 5) = 22.606$ million people per decade. For the years 1980-2000 the average rate $= (q(10) - q(8))/(10 - 8) = 53.976$ million people per decade. For $t = 9$ (1990), $q'(9) = 53.679$ million people per decade.

(f) The polynomial $q(t)$ is a better model for the U.S. population since the population will probably continue to increase through out the 1990's.

3.4 Derivatives of the Trigonometric Functions

1. $f(x) = 4\cos x \Longrightarrow f'(x) = -4\sin x$

2. $f(x) = x\sin x \Longrightarrow f'(x) = (1)\sin x + x(\cos x)$

3. $f(x) = x^3 \tan x \Longrightarrow f'(x) = (3x^2)\tan x + x^3(\sec^2 x)$

4. $f(x) = \sin x \cos x \Longrightarrow f'(x) = (\cos x)\cos x + (\sin x)(-\sin x) = \cos^2 x - \sin^2 x$

5. $f(x) = (x^3 - 2)\cot x \Longrightarrow f'(x) = (3x^2)\cot x + (x^3 - 2)(-\csc^2 x)$

6. $f(x) = \cot x \csc x \Longrightarrow f'(x) = (-\csc^2 x)\csc x + (\cot x)(-\csc x \cot x) = -\csc^3 x - \csc x \cot^2 x$

7. $f(x) = \sin x \sec x = \tan x \Longrightarrow f'(x) = \sec^2 x$

8. $f(x) = \sin^2 x = (\sin x)(\sin x) \Longrightarrow f'(x) = (\cos x)(\sin x) + (\sin x)(\cos x) = 2\sin x \cos x$

9. $f(x) = x\cos x - x\sin x = x(\cos x - \sin x) \Longrightarrow f'(x) = (1)(\cos x - \sin x) + x(-\sin x - \cos x)$

10. $f(x) = (\cos x)(x - \cot x) \Longrightarrow f'(x) = (-\sin x)(x - \cot x) + (\cos x)(1 + \csc^2 x)$

11. $f(x) = \sec x \tan x \Longrightarrow f'(x) = (\sec x \tan x)\tan x + (\sec x)(\sec^2 x) = \sec x(\tan^2 x + \sec^2 x)$

12. $f(x) = \csc^2 x \cot x \Longrightarrow$

$$f'(x) = [2\csc x(-\csc x \cot x)]\cot x + (\csc^2 x)(-\csc^2 x) = -(\csc^2 x)(2\cot^2 x + \csc^2 x)$$

13. $f(x) = \dfrac{x}{2 + \sin x} \Longrightarrow f'(x) = \dfrac{(2 + \sin x)(1) - x(\cos x)}{(2 + \sin x)^2}$

14. $f(x) = \dfrac{\tan x}{1 + x^2} \Longrightarrow f'(x) = \dfrac{(1 + x^2)(\sec^2 x) - (\tan x)(2x)}{(1 + x^2)^2}$

15. $f(x) = \dfrac{\sin x - \cos x}{1 + \tan x} \Longrightarrow f'(x) = \dfrac{(1 + \tan x)(\cos x + \sin x) - (\sin x - \cos x)(\sec^2 x)}{(1 + \tan x)^2}$

16. $f(x) = \dfrac{1 - \sin x}{1 + \sin x} \Longrightarrow f'(x) = \dfrac{(1 + \sin x)(-\cos x) - (1 - \sin x)(\cos x)}{(1 + \sin x)^2} = \dfrac{-2\cos x}{(1 + \sin x)^2}$

17. $f(x) = \dfrac{x^2 + 4\cot x}{x + \tan x} \Longrightarrow f'(x) = \dfrac{(x + \tan x)(2x - 4\csc^2 x) - (x^2 + 4\cot x)(1 + \sec^2 x)}{(x + \tan x)^2}$

18. $f(x) = \dfrac{x^2 + 4}{2 + \sec x} \Longrightarrow f'(x) = \dfrac{(2 + \sec x)(2x) - (x^2 + 4)(\sec x \tan x)}{(2 + \sec x)^2}$

19. $f(x) = \dfrac{3\csc x}{4x^2 - 5\tan x} \implies f'(x) = \dfrac{(4x^2 - 5\tan x)(-3\csc x \cot x) - (3\csc x)(8x - 5\sec^2 x)}{(4x^2 - 5\tan x)^2}$

20. $f(x) = x\csc x - \dfrac{x}{\cot x} = x(\csc x - \tan x) \implies f'(x) = (1)(\csc x - \tan x) + x(-\csc x \cot x - \sec^2 x)$

21. $y = x\sin x \implies y' = \sin x + x\cos x \implies y'(\pi) = \sin\pi + \pi\sin\pi = -\pi$. Hence the slope of the tangent line at $(\pi, 0)$ is $-\pi$ and the equation is $y = -\pi(x - \pi)$.

22. $f(x) = \csc x \cot x \implies f'(x) = (-\csc x \cot x)\cot x + \csc x(-\csc^2 x) = -(\csc x)(\cot^2 x + \csc^2 x) \implies$
    $f'(\pi/4) = -\big(\csc(\pi/4)\big)\big(\cot^2(\pi/4) + \csc^2(\pi/4)\big) = -3\sqrt{2}$. Hence the slope of the tangent line at $(\pi/4, \sqrt{2})$ is $-3\sqrt{2}$ and the equation is $y - \sqrt{2} = -3\sqrt{2}\,(x - \pi/4)$.

23. The tangent line is horizontal $\iff y' = \sec x \tan x = 0 \iff \tan x = 0 \iff x = 0,\ \pi,\ 2\pi,\ 3\pi,\ 4\pi$ in the interval $[0, 4\pi]$.

24. $f(x) = \sin 2x = (2\sin x)(\cos x) \implies f'(x) = 2\cos x)(\cos x) + (2\sin x)(-\sin x) = 2(\cos^2 x - \sin^2 x) = 2\cos 2x$

25. The piston changes direction when its velocity changes sign. Now, $s(t) = 5 - 4\sin t \implies v(t) = s'(t) = -4\cos t = 0 \implies t = \pi/2,\ 3\pi/2,\ 5\pi/2,\ \dots$.

Sign v	$-$	$+$	$-$
Interval	$[0, \pi/2]$	$[\pi/2, 3\pi/2]$	$[3\pi/2, 5\pi/2]$
Direction of piston	up	down	up

Hence the piston changes direction when $t = \pi/2,\ 3\pi/2,\ 5\pi/2,\ \dots$.

26. $\dfrac{d}{dx}(\tan^2 x) = \dfrac{d}{dx}(\tan x)(\tan x) = (\sec^2 x)(\tan x) + (\tan x)(\sec^2 x) = 2\sec^2 x \tan x$

$\dfrac{d}{dx}(\sec^2 x) = \dfrac{d}{dx}(\sec x)(\sec x) = (\sec x \tan x)(\sec x) + (\sec x)(\sec x \tan x) = 2\sec^2 x \tan x = \dfrac{d}{dx}(\tan^2 x)$

27. (a) The tangent lines are parallel when the derivatives are equal:

$$\frac{d}{dx}(\sin x) = \frac{d}{dx}(\cos x) \implies \cos x = -\sin x \implies \tan x = -1 \implies x = 3\pi/4, 7\pi/4.$$

Note that if $\cos x = -\sin x$, then $\cos x \neq 0$ since, if $\cos x = 0$, $\sin x = \pm 1$, and hence we may divide by $\cos x$.

(b) The tangent lines are perpendicular when the product of the derivatives is equal to -1:

$$\left[\frac{d}{dx}(\sin x)\right]\left[\frac{d}{dx}(\cos x)\right] = -1 \implies (\cos x)(-\sin x) = -1 \implies (1/2)\sin 2x = 1 \implies \sin 2x = 2.$$

There are no such numbers x since $|\sin x| \leq 1$ for all x. Hence there are no points where the tangent lines are perpendicular.

28. (a) $f'(x) = -\sin x = 0 \implies x = 0,\ \pi,\ 2\pi$

(b) $f'(x) = \sec x \tan x = 0 \implies x\sin x = 0 \implies x = 0,\ \pi,\ 2\pi$

(c) $f'(x) = -\csc x \cot x = 0 \implies \cos x = 0 \implies x = \pi/2,\ 3\pi/2$

29. $\dfrac{d}{dx}\cot x = \dfrac{d}{dx}\left(\dfrac{\cos x}{\sin x}\right) = \dfrac{(\sin x)(-\sin x) - (\cos x)(\cos x)}{\sin^2 x} = \dfrac{-1}{\sin^2 x} = -\csc^2 x$

$\dfrac{d}{dx}\sec x = \dfrac{d}{dx}\left(\dfrac{1}{\cos x}\right) = \dfrac{(\cos x)(0) - (1)(-\sin x)}{\cos^2 x} = \dfrac{1}{\cos x}\dfrac{\sin x}{\cos x} = \sec x \tan x$

$\dfrac{d}{dx}\csc x = \dfrac{d}{dx}\left(\dfrac{1}{\sin x}\right) = \dfrac{(\sin x)(0) - (1)(\cos x)}{\sin^2 x} = -\dfrac{1}{\sin x}\dfrac{\cos x}{\sin x} = -\csc x \cot x$

30. $\cot x = \cos x/\sin x$ is differentiable at those x for which $\sin x \neq 0$, namely all $x \neq 0, \pi, 2\pi$.
$\sec x = 1/\cos x$ is differentiable at those x for which $\cos x \neq 0$, namely all $x \neq \pi/2, 3\pi/2$.
$\csc x = 1/\sin x$ is differentiable at those x for which $\sin x \neq 0$, namely all $x \neq 0, \pi, 2\pi$.

31. (a) $v(t) = s'(t) = -10 \sin t$

(b) The mass changes direction when $v(t)$ changes sign, as indicated in the table below.

Sign v	$-$	$+$
Interval	$[0, \pi]$	$[\pi, 2\pi]$
Direction of motion	left	right

Hence the mass changes its direction of motion at $t = \pi, 2\pi, \ldots$.

3.5 The Chain Rule

1. $y = f(u) = u^3 + 1 \implies dy/du = 3u^2$; $u = g(x) = 1 - x^2 \implies du/dx = -2x$. Hence $y = f(g(x)) = (1 - x^2)^3 + 1$ and

$$\frac{dy}{dx} = \frac{dy}{du}\frac{du}{dx} = (3u^2)(-2x) = 3(1 - x^2)^2(-2x) = -6x(1 - x^2)^2.$$

2. $y = f(u) = 1/(1 + u) \implies dy/du = \left[(1 + u)(0) - (1)(1)\right]/(1 + u)^2 = -1/(1 + u)^2$; $u = g(x) = 3x^2 - 7$ $\implies du/dx = 6x$. Hence $y = f(g(x)) = 1/(1 + 3x^2 - 7) = 1/(3x^2 - 6)$ and

$$\frac{dy}{dx} = \frac{dy}{du}\frac{du}{dx} = \frac{-1}{(1 + u)^2}(6x) = \frac{-6x}{(3x^2 - 6)^2}.$$

3. $y = f(u) = u(1 - u^2) = u - u^3 \implies dy/du = 1 - 3u^2$; $u = g(x) = 1/x \implies du/dx = -1/x^2$. Hence $y = f(g(x)) = 1/x - 1/x^3$ and

$$\frac{dy}{dx} = \frac{dy}{du}\frac{dy}{du} = (1 - 3u^2)\left(-\frac{1}{x^2}\right) = \left(1 - \frac{3}{x^2}\right)\left(-\frac{1}{x^2}\right).$$

4. $y = f(u) = (u + 1)/(u - 1) \implies dy/du = \left[(u - 1)(1) - (u + 1)(1)\right]/(u - 1)^2 = -2/(u - 1)^2$; $u = g(x) = \sin x \implies du/dx = \cos x$. Hence $y = f(g(x)) = (\sin x + 1)/(\sin x - 1)$ and

$$\frac{dy}{dx} = \frac{dy}{du}\frac{du}{dx} = \frac{-2}{(u - 1)^2}(\cos x) = \frac{-2\cos x}{(\sin x - 1)^2}.$$

5. $y = f(u) = \tan u \implies dy/du = \sec^2 u$; $u = g(x) = 1 + x^4 \implies du/dx = 4x^3$. Hence $y = f(g(x)) = \tan(1 + x^4)$ and

$$\frac{dy}{dx} = \frac{dy}{du}\frac{du}{dx} = \left(\sec^2 u\right)\left(4x^3\right) = 4x^3 \sec^2\left(1 + x^4\right).$$

6. $y = f(u) = 3u^2 + 5 \implies dy/du = 6u$; $u = g(x) = \cos x \implies du/dx = -\sin x$. Hence $y = f\big(g(x)\big) = 3\cos^2 x + 5$ and

$$\frac{dy}{dx} = \frac{dy}{du}\frac{du}{dx} = (6u)(-\sin x) = -6\cos x \sin x.$$

7. $y = f(u) = 1/(1+u) \implies dy/du = -1/(1+u)^2$; $u = g(x) = \sin^2 x \implies du/dx = 2\sin x \cos x$ (see Exercise 8, Section 3.4). Hence $y = f\big(g(x)\big) = 1/(1+\sin^2 x)$ and

$$\frac{dy}{dx} = \frac{dy}{du}\frac{du}{dx} = \frac{-1}{(1+u)^2}(2\sin x \cos x) = \frac{-2\sin x \cos x}{(1+\sin^2 x)^2}.$$

8. $y = f(u) = u^2 - 1 \implies dy/du = 2u$; $u = g(x) = 1 + \sec x \implies du/dx = \sec x \tan x$. Hence $y = f\big(g(x)\big) = (1+\sec x)^2 - 1$ and

$$\frac{dy}{dx} = \frac{dy}{du}\frac{du}{dx} = (2u)(\sec x \tan x) = 2(1+\sec x)(\sec x \tan x).$$

9. $y = f(u) = \cos u \implies dy/du = -\sin u$; $u = g(x) = \sin x \implies du/dx = \cos x$. Hence $y = f\big(g(x)\big) = \cos(\sin x)$ and

$$\frac{dy}{dx} = \frac{dy}{du}\frac{du}{dx} = (-\sin u)(\cos x) = -\sin(\sin x)\cos x.$$

10. $y = f(u) = u^3 - 3u + 1/u \implies dy/du = 3u^2 - 3 - 1/u^2$; $u = g(x) = \tan x \implies du/dx = \sec^2 x$. Hence $y = f\big(g(x)\big) = \tan^3 x - 3\tan x + 1/\tan x$ and

$$\frac{dy}{dx} = \frac{dy}{du}\frac{du}{dx} = \left(3u^2 - 3 - \frac{1}{u^2}\right)(\sec^2 x) = \left(3\tan^2 x - 3 - \frac{1}{\tan^2 x}\right)(\sec^2 x).$$

11. $y = f(u) = \sin u$. Hence

$$\frac{dy}{dx} = \frac{dy}{du}\frac{du}{dx} = (\cos u)(2x+1) = \big[\cos(x^2 + x)\big](2x+1).$$

12. $y = f(u) = \tan u$. Hence

$$\frac{dy}{dx} = \frac{dy}{du}\frac{du}{dx} = (\sec^2 u)\frac{(1-x)(1) - (1+x)(-1)}{(1-x)^2} = \left[\sec^2\left(\frac{1+x}{1-x}\right)\right]\frac{2}{(1-x)^2}.$$

13. $y = f(u) = (1+u)/(1-u)$. Hence

$$\frac{dy}{dx} = \frac{dy}{du}\frac{du}{dx} = \frac{(1-u)(1) - (1+u)(-1)}{(1-u)^2}(\cos x) = \frac{2\cos x}{(1-\sin x)^2}.$$

14. $y = f(u) = u/(1+u^2)$. Hence

$$\frac{dy}{dx} = \frac{dy}{du}\frac{du}{dx} = \frac{(1+u^2)(1) - u(2u)}{(1+u^2)^2}(\sec x \tan x) = \frac{1-u^2}{(1+u^2)^2}(\sec x \tan x) = \frac{1-\sec^2 x}{(1+\sec^2 x)^2}(\sec x \tan x).$$

15. $y = f(u) = u^4$. Hence

$$\frac{dy}{dx} = \frac{dy}{du}\frac{du}{dx} = (4u^3)\frac{(3x-7)(3) - (3x+7)(3)}{(3x-7)^2} = 4\left(\frac{3x+7}{3x-7}\right)^3\left[\frac{-42}{(3x-7)^2}\right] = \frac{-168(3x+7)^3}{(3x-7)^5}.$$

16. $y = f(u) = (1 + 3u)/(1 - 3u)$. Hence

$$\frac{dy}{dx} = \frac{dy}{du}\frac{du}{dx} = \frac{(1 - 3u)(3) - (1 + 3u)(-3)}{(1 - 3u)^2}(-\sin x) = \frac{-6\sin x}{(1 - 3\cos x)^2}.$$

17. $y = f(u) = [(u + 7)/(u - 7)]^4$. To find dy/du, let $v = (u + 7)/(u - 7)$. Then $y = v^4$ and hence

$$\frac{dy}{du} = \frac{dy}{dv}\frac{dv}{du} = (4v^3)\frac{(u - 7)(1) - (u + 7)(1)}{(u - 7)^2} = 4\left(\frac{u + 7}{u - 7}\right)^3\left[\frac{-14}{(u - 7)^2}\right].$$

Therefore

$$\frac{dy}{dx} = \frac{dy}{du}\frac{du}{dx} = 4\left(\frac{u + 7}{u - 7}\right)^3\left[\frac{-14}{(u - 7)^2}\right](3) = \frac{-168(3x + 7)^3}{(3x - 7)^5}.$$

18. $y = f(u) = (1 + u)/(1 - u)$. Hence

$$\frac{dy}{dx} = \frac{dy}{du}\frac{du}{dx} = \frac{(1 - u)(1) - (1 + u)(-1)}{(1 - u)^2}(-3\sin x) = \frac{2}{(1 - u)^2}(-3\sin x) = \frac{-6\sin x}{(1 - 3\cos x)^2}.$$

19. $y = (x^2 + 4)^3 \implies dy/dx = 3(x^2 + 4)^2(2x) = 6x(x^2 + 4)^2$

20. $y = (2 - 3x)^4 \implies dy/dx = 4(2 - 3x)^3(-3) = -12(2 - 3x)^3$

21. $y = (\cos x - x)^6 \implies dy/dx = 6(\cos x - x)^5(-\sin x - 1)$

22. $y = (x^2 + 8x + 6)^5 \implies dy/dx = 5(x^2 + 8x + 6)^4(2x + 8)$

23. $y = \sin(x^3 + 3x) \implies dy/dx = \left[\cos(x^3 + 3x)\right](3x^2 + 3)$

24. $y = \cos(\pi - x^2) \implies dy/dx = -\left[\sin(\pi - x^2)\right](-2x) = 2x\sin(\pi - x^2)$

25. $y = \tan(x^2 + x) \implies dy/dx = \left[\sec^2(x^2 + x)\right](2x + 1)$

26. $y = \sec(x^2 + 4) \implies dy/dx = \left[\sec(x^2 + 4)\tan(x^2 + 4)\right](2x)$

27. $y = x(x^4 - 5)^3 \implies dy/dx = x\left[3(x^4 - 5)^2(4x^3)\right] + (x^4 - 5)^3(1) = (x^4 - 5)^2(13x^4 - 5)$

28. $y = (x^2 - 7)^3(5 - x)^2 \implies$

$$\frac{dy}{dx} = (x^2 - 7)^3[2(5 - x)(-1)] + (5 - x)^2\left[3(x^2 - 7)^2(2x)\right]$$

$$= (x^2 - 7)^2(5 - x)\left[-2(x^2 - 7) + 6x(5 - x)\right] = (x^2 - 7)^2(5 - x)(-8x^2 + 30x + 14)$$

29. $y = \dfrac{1}{(x^2 - 9)^3} = (x^2 - 9)^{-3} \implies dy/dx = -3(x^2 - 9)^{-4}(2x) = \dfrac{-6x}{(x^2 - 9)^4}$

30. $y = \dfrac{x + 3}{(x^2 - 6x + 2)^2} \implies$

$$\frac{dy}{dx} = \frac{(x^2 - 6x + 2)^2(1) - (x + 3)\left[2(x^2 - 6x + 2)(2x - 6)\right]}{(x^2 - 6x + 2)^4}$$

$$= \frac{(x^2 - 6x + 2)[(x^2 - 6x + 2) - 2(x + 3)(2x - 6)]}{(x^2 - 6x + 2)^4} = \frac{-3x^2 - 6x + 38}{(x^2 - 6x + 2)^3}$$

31. $y = (3\tan x - 2)^4 \implies dy/dx = 4(3\tan x - 2)^3(3\sec^2 x)$

32. $y = (x^6 - x^2 + 2)^{-4} \implies dy/dx = -4(x^6 - x^2 + 2)^{-5}(6x^5 - 2x)$

33. $y = \left(\dfrac{x - 3}{x + 3}\right)^4 \implies$

$$\frac{dy}{dx} = 4\left(\frac{x-3}{x+3}\right)^3 \frac{(x+3)(1) - (x-3)(1)}{(x+3)^2} = 4\left(\frac{x-3}{x+3}\right)^3 \frac{6}{(x+3)^2} = \frac{24(x-3)^3}{(x+3)^5}$$

34. $y = \left(\dfrac{1 + \sin x}{1 - \sin x}\right)^{-6} \implies$

$$\frac{dy}{dx} = -6\left(\frac{1+\sin x}{1-\sin x}\right)^{-7} \frac{(1-\sin x)(\cos x) - (1+\sin x)(-\cos x)}{(1-\sin x)^2} = \left(\frac{1+\sin x}{1-\sin x}\right)^{-7} \frac{-12\cos x}{(1-\sin x)^2}$$

35. $y = x\cos(1 - x^2) \implies dy/dx = x\left[(-\sin(1-x^2))(-2x)\right] + (1)\cos(1-x^2) = 2x^2\sin(1-x^2) + \cos(1-x^2)$

36. $y = (x\cos x - \sin x)^5 \implies$

$$\frac{dy}{dx} = 5(x\cos x - \sin x)^4[x(-\sin x) + (1)\cos x - \cos x] = 5(x\cos x - \sin x)^4(-x\sin x)$$

37. $y = \dfrac{\tan^2 x + 1}{1 - x} \implies$

$$\frac{dy}{dx} = \frac{(1-x)\left[2(\tan x)(\sec^2 x)\right] - (\tan^2 x + 1)(-1)}{(1-x)^2} = \frac{2(1-x)\tan x \sec^2 x + \tan^2 x + 1}{(1-x)^2}$$

38. $y = \dfrac{x + 2}{1 + \sec(1 + x^2)} \implies$

$$\frac{dy}{dx} = \frac{\left[1 + \sec(1+x^2)\right](1) - (x+2)\left[\sec(1+x^2)\tan(1+x^2)\right](2x)}{\left[1 + \sec(1+x^2)\right]^2}$$

39. $y = \dfrac{\tan^2(\pi/4) + 1}{\sec(0)} \implies dy/dx = 0$ since y is constant.

40. $y = \dfrac{\sin^2(\pi/6) + 1}{\sin^2(\pi/6) - 1} \implies dy/dx = 0$ since y is constant.

41. $y = \dfrac{x}{(x^2 + x + 1)^6} \implies$

$$\frac{dy}{dx} = \frac{(x^2+x+1)^6(1) - x\left[6(x^2+x+1)^5(2x+1)\right]}{(x^2+x+1)^{12}} = \frac{-11x^2 - 5x + 1}{(x^2+x+1)^7}$$

42. $y = \dfrac{2 + \csc(1 + x^2)}{1 - \csc(1 + x^2)} \implies \dfrac{dy}{dx} =$

$$\frac{\left[1 - \csc(1+x^2)\right]\left[-\csc(1+x^2)\cot(1+x^2)\right](2x) - \left[2 + \csc(1+x^2)\right]\left[\csc(1+x^2)\cot(1+x^2)\right](2x)}{\left[1 - \csc(1+x^2)\right]^2}$$

$$= \frac{2x\csc(1+x^2)\cot(1+x^2)\left[-1 + \csc(1+x^2) - 2 - \csc(1+x^2)\right]}{\left[1 - \csc(1+x^2)\right]^2} = \frac{-6x\csc(1+x^2)\cot(1+x^2)}{\left[1 - \csc(1+x^2)\right]^2}$$

43. $y = \left(\dfrac{ax+b}{cx+d}\right)^4 \implies dy/dx = 4\left(\dfrac{ax+b}{cx+d}\right)^3 \dfrac{(cx+d)(a)-(ax+b)(c)}{(cx+d)^2} = \dfrac{4(ax+b)^3(ad-bc)}{(cx+d)^5}$

44. $y = \left(\dfrac{a-bx}{c-dx}\right)^5 \implies dy/dx = 5\left(\dfrac{a-bx}{c-dx}\right)^4 \dfrac{(c-dx)(-b)-(a-bx)(-d)}{(c-dx)^2} = 5\left(\dfrac{a-bx}{c-dx}\right)^4 \dfrac{ad-bc}{(c-dx)^2}$

45. $y = \dfrac{1}{1+\cos^3 x} = (1+\cos^3 x)^{-1} \implies dy/dx = -(1+\cos^3 x)^{-2}3(\cos^2 x)(-\sin x) = \dfrac{3\cos^2 x \sin x}{(1+\cos^3 x)^2}$

46. $y = \dfrac{x-\sin \pi x}{4+\cos \pi x} \implies$

$$\dfrac{dy}{dx} = \dfrac{(4+\cos \pi x)(1-\pi \cos \pi x)-(x-\sin \pi x)(-\pi \sin \pi x)}{(4+\cos \pi x)^2}$$

$$= \dfrac{4+(1-4\pi)\cos \pi x - \pi(\cos^2 \pi x + \sin^2 \pi x)+\pi x \sin \pi x}{(4+\cos \pi x)^2} = \dfrac{4-\pi+(1-4\pi)\cos \pi x + \pi x \sin \pi x}{(4+\cos \pi x)^2}$$

47. $y = \dfrac{1}{(1+x^4)^3} = (1+x^4)^{-3} \implies dy/dx = -3(1+x^4)^{-4}(4x^3) = \dfrac{-12x^3}{(1+x^4)^4}$

48. $y = x\cot\left(\dfrac{1+x}{1-x}\right) \implies$

$$\dfrac{dy}{dx} = x\left\{\left[-\csc^2\left(\dfrac{1+x}{1-x}\right)\right]\dfrac{(1-x)(1)-(1+x)(-1)}{(1-x)^2}\right\}+\cot\left(\dfrac{1+x}{1-x}\right)$$

$$= \dfrac{-2x}{(1-x)^2}\csc^2\left(\dfrac{1+x}{1-x}\right)+\cot\left(\dfrac{1+x}{1-x}\right)$$

49. $y = \tan(6x)-6\tan x \implies dy/dx = \left[\sec^2(6x)\right](6)-6\sec^2 x = 6\sec^2(6x)-6\sec^2 x$

50. $y = f\Big(g\big(h(x)\big)\Big) \implies dy/dx = f'\Big(g\big(h(x)\big)\Big)\left[g\big(h(x)\big)\right]'= f'\Big(g\big(h(x)\big)\Big)g'\big(h(x)\big)h'(x)$

51. $y = \sec^3 4x \implies dy/dx = 3(\sec^2 4x)(\sec 4x \tan 4x)(4) = 12\sec^3 4x \tan 4x$

52. $y = \sin^3(2x-\pi/2) \implies dy/dx = 3[\sin^2(2x-\pi/2)][\cos(2x-\pi/2)](2) = 6\sin^2(2x-\pi/2)\cos(2x-\pi/2)$

53. $y = \tan^2(\pi-x^2) \implies dy/dx = 2[\tan(\pi-x^2)][\sec^2(\pi-x^2)](-2x) = -4x\tan(\pi-x^2)\sec^2(\pi-x^2)$

54. $y = \dfrac{x-\sin(\pi x)}{\tan^2(\pi x)} \implies$

$$\dfrac{dy}{dx} = \dfrac{\tan^2(\pi x)[1-\pi\cos(\pi x)]-[x-\sin(\pi x)][2\tan(\pi x)\sec^2(\pi x)](\pi)}{\tan^4(\pi x)}$$

$$= \dfrac{\tan(\pi x)[1-\pi\cos(\pi x)]-2\pi[x-\sin(\pi x)]\sec^2(\pi x)}{\tan^3(\pi x)}$$

55. $y = (\sin \pi x - \cos \pi x)^4 \implies dy/dx = 4(\sin \pi x - \cos \pi x)^3[\pi \cos \pi x + \pi \sin \pi x]$

56. $y = \sin^4(x^3+\pi x) \implies dy/dx = 4[\sin^3(x^3+\pi x)][\cos(x^3+\pi x)](3x^2+\pi)$

139

57. $y = \left[1 + (x^2 - 3)^4\right]^6 \implies dy/dx = 6\left[1 + (x^2 - 3)^4\right]^5 \left[4(x^2 - 3)^3(2x)\right] = 48x(x^2 - 3)^3\left[1 + (x^2 - 3)^4\right]^5$

58. $y = \dfrac{1}{a + (bx + c)^4} = [a + (bx + c)^4]^{-1} \implies$

$$\frac{dy}{dx} = -[a + (bx + c)^4]^{-2}\left[4(bx + c)^3(b)\right] = \frac{-4b(bx + c)^3}{[a + (bx + c)^4]^2}$$

59. $y = \cos^2\left(\dfrac{1+x}{1-x}\right) \implies$

$$\frac{dy}{dx} = 2\left[\cos\left(\frac{1+x}{1-x}\right)\right]\left[-\sin\left(\frac{1+x}{1-x}\right)\right]\left[\frac{(1-x)(1) - (1+x)(-1)}{(1-x)^2}\right]$$

$$= \frac{-4}{(1-x)^2}\cos\left(\frac{1+x}{1-x}\right)\sin\left(\frac{1+x}{1-x}\right)$$

60. $y = (\tan \pi x - \cos \pi x)^6 \implies$

$$\frac{dy}{dx} = 6(\tan \pi x - \cos \pi x)^5\left[(\sec^2 \pi x)(\pi) - (-\sin \pi x)(\pi)\right] = 6\pi(\tan \pi x - \cos \pi x)^5\left(\sec^2 \pi x + \sin \pi x\right)$$

61. $y = 1 + \left[1 + (1 + x^2)^2\right]^2 \implies$

$$\frac{dy}{dx} = 2\left[1 + (1 + x^2)^2\right]\left[2(1 + x^2)\right](2x) = 8x(1 + x^2)\left[1 + (1 + x^2)^2\right]$$

62. $y = \cos\left(\cos(x^2 + 1)\right) \implies$

$$\frac{dy}{dx} = \left[-\sin\left(\cos(x^2 + 1)\right)\right]\left[-\sin(x^2 + 1)\right](2x) = 2x\sin\left(\cos(x^2 + 1)\right)\sin\left(x^2 + 1\right)$$

63. $y = \tan\sqrt{1 + x^2} \implies$

$$\frac{dy}{dx} = \left[\sec^2\sqrt{1 + x^2}\right]\left[\frac{1}{2\sqrt{1 + x^2}}\right](2x) = \frac{x}{\sqrt{1 + x^2}}\sec^2\sqrt{1 + x^2}$$

64. $y = (1 - x^2)^3 \implies y' = 3(1 - x^2)^2(-2x) \implies y'(1) = 0$. Hence the slope of the tangent line at $(1, 0)$ is 0 and its equation is $y = 0$.

65. $y = \left(\dfrac{x}{x+1}\right)^4 \implies$

$$y' = 4\left(\frac{x}{x+1}\right)^3\frac{(x+1)(1) - x(1)}{(x+1)^2} = 4\left(\frac{x}{x+1}\right)^3\frac{1}{(x+1)^2} \implies y'(0) = 0.$$

Hence the slope of the tangent line at $(0, 0)$ is 0 and its equation is $y = 0$.

66. $y = \sin(\pi + x^3) \implies y' = \left[\cos(\pi + x^3)\right](3x^2) \implies y'(0) = 0$. Hence the slope of the tangent line at $(0, 0)$ is 0 and its equation is $y = 0$.

67. $y = x\cos(\pi + x^2) \implies y' = x\left[-\sin(\pi + x^2)\right](2x) + \cos(\pi + x^2) \implies y'(0) = \cos\pi = -1$. Hence the slope of the tangent line at $(0, 0)$ is -1 and its equation is $y = -x$.

68. $y = \left(\dfrac{3x^2 + 1}{x + 3}\right)^2 \implies$

$$y' = 2\left(\frac{3x^2 + 1}{x + 3}\right)\frac{(x + 3)(6x) - (3x^2 + 1)(1)}{(x + 3)^2} = 2\left(\frac{3x^2 + 1}{x + 3}\right)\frac{3x^2 + 18x - 1}{(x + 3)^2} \implies y'(-1) = -16.$$

Hence the slope of the tangent line at $(-1, 4)$ is -16 and its equation is $y - 4 = -16(x + 1)$.

69. $y = \tan\left(\dfrac{\pi/4 - x}{1 + x}\right) \implies$

$$y' = \left[\sec^2\left(\frac{\pi/4 - x}{1 + x}\right)\right]\frac{(1 + x)(-1) - (\pi/4 - x)(1)}{(1 + x)^2} = \sec^2\left(\frac{\pi/4 - x}{1 + x}\right)\frac{-1 - \pi/4}{(1 + x)^2}$$

$$\implies y'(0) = \left[\sec^2(\pi/4)\right](-1 - \pi/4) = -2(1 + \pi/4).$$

Hence the slope of the tangent line at $(0, 1)$ is $-2(1 + \pi/4)$ and its equation is $y - 1 = -2(1 + \pi/4)x$.

70. $h'(2) = f'(g(2))g'(2) = f'(3)g'(2) = (-2)(7) = -14$

71.   (a) $(f \circ g)'(1) = f'(g(1))g'(1) = f'(2)g'(1) = (5)(2) = 10$

     (b) $(g \circ f)'(1) = g'(f(1))f'(1) = g'(1)f'(1) = (2)(-2) = -4$

72. $h(t) = [g^2(t) + 1]^3 \implies h'(t) = 3[g^2(t) + 1]^2[2g(t)g'(t)] \implies$

$$h'(3) = 3[g^2(3) + 1]^2[2g(3)g'(3)] = 3[(-2)^2 + 1]^2[2(-2)(-1)] = 300$$

73. $h(t) = \sin g(t) \implies h'(t) = [\cos g(t)]g'(t) \implies$

$$h'(2) = [\cos g(2)]g'(2) = [\cos(3\pi/4)]\sqrt{2} = (-\sqrt{2}/2)\sqrt{2} = -1$$

74. $h(x) = \tan^2 g(x) \implies h'(x) = 2[\tan g(x)][\sec^2 g(x)]g'(x) \implies$

$$h'(\pi/4) = 2[\tan g(\pi/4)][\sec^2 g(\pi/4)]g'(\pi/4) = 2[\tan(\pi/6)][\sec^2(\pi/6)](\pi) = 2(\sqrt{3}/3)(4/3)\pi = 8\pi\sqrt{3}/9$$

75. $f(x) = \sin x^2 \implies f'(x) = (\cos x^2)(2x) \implies$

$$f'(\sqrt{\pi}/2) = \left[\cos(\sqrt{\pi}/2)^2\right](2\sqrt{\pi}/2) = [\cos(\pi/4)]\sqrt{\pi} = (\sqrt{2}/2)\sqrt{\pi} = \sqrt{\pi/2}.$$

Hence the slope of the tangent line at $(\sqrt{\pi}/2, \sqrt{2}/2)$ is $\sqrt{\pi/2}$ and hence the equation is $y - \sqrt{2}/2 = \sqrt{\pi/2}\left(x - \sqrt{\pi}/2\right)$.

76. $f(x) = \csc(\pi x) \implies f'(x) = -[\csc(\pi x)\cot(\pi x)](\pi) \implies$

$$f'(1/2) = -[\csc(\pi/2)\cot(\pi/2)](\pi) = 0.$$

Hence the slope of the tangent line at $(1/2, 1)$ is 0 and its equation is $y - 1 = 0$.

77. $v(t) = s'(t) = 5[\cos(4t)](4) = 20\cos(4t)$ cm/sec.

78.   (a) $s(t) = [t/(t + 1)]^2 + 10\sin(\pi t) \implies$

$$v(t) = s'(t) = 2\left(\frac{t}{t + 1}\right)\frac{(t + 1)(1) - t(1)}{(t + 1)^2} + 10[\cos(\pi t)](\pi) = \frac{2t}{(1 + t)^3} + 10\pi\cos(\pi t).$$

(b) $v(0) = 0 + 10\pi \cos(0) = 10\pi$ m/sec.

79. Let $V(t)$ and $r(t)$ stand for the volume and radius of the balloon after t seconds, respectively. Then $V(t) = (4/3)\pi r(t)^3 \implies dV/dt = 4\pi r(t)^2 (dr/dt)$. From the given information, $dr/dt = 0.5$. Hence, when $r = 5$ inches,

$$\frac{dV}{dt}\bigg|_{r=5} = 4\pi(5)^2(0.5) = 50\pi.$$

Thus, when the radius of the balloon is 5 inches the volume is increasing at a rate of 50π in^3/sec.

80. Let $A(t)$ stand for the area enclosed by the wave of radius $r(t)$ after t seconds. Then $A(t) = \pi r(t)^2 \implies dA/dt = 2\pi r(t)(dr/dt)$. From the given information, $dr/dt = 5$. Since, after 3 seconds, $r = 15$, it follows that

$$\frac{dA}{dt}\bigg|_{r=15} = 2\pi(15)(5) = 150\pi.$$

Thus, after 3 seconds the area enclosed by the wave is increasing at the rate of 150π in.^{2}/s.

81. (a) $|x|^2 = |x^2| = x^2$ since $x^2 \geq 0 \implies |x| = \sqrt{x^2}$ since $|x| \geq 0$.

(b) $f(x) = \sqrt{x^2} = \left(x^2\right)^{1/2} \implies$

$$f'(x) = \frac{1}{2}\left(x^2\right)^{-1/2}(2x) = \frac{x}{\sqrt{x^2}} = \frac{x}{|x|}.$$

(c) Because $\sqrt{x^2} = g(h(x))$, where $g(x) = \sqrt{x}$ and $h(x) = x^2$, and $g'(h(0)) = g'(0)$ does not exist.

82. (a) $\dfrac{d}{dx}\big(\text{sindeg}(x)\big) = \dfrac{d}{dx}\sin(\pi x/180) = [\cos(\pi x/180)](\pi/180)$.

(b) $\text{cosdeg}(x) = \cos(\pi x/180)$

(c) $\dfrac{d}{dx}\big(\text{sindeg}(x)\big) = (\pi/180)\text{cosdeg}(x)$.

(d) To avoid the factor $\pi/180$.

83. (c) We have that

$$\begin{aligned}
(g^{n+1})'(x) &= (g \cdot g^n)'(x) \\
&= (g' \cdot g^n)(x) + g \cdot (g^n)'(x) && \text{by the Product Rule} \\
&= (g' \cdot g^n)(x) + g \cdot (ng^{n-1} \cdot g')(x) && \text{by the induction hypothesis} \\
&= g'(x)g^n(x) + ng^n(x)g'(x) \\
&= (n+1)g^n(x)g'(x).
\end{aligned}$$

84. (a) $t_1 = 0.5943603$, $t_2 = 1.8059974$, $t_3 = 3.0626344$, $t_4 = 4.3192715$.

(b) $t_1 = 1.1776793$, $t_2 = 2.4343158$.

(c) These times are the same as in part (b). The pendulum is the farthest from the center.

85. (a) The graph with range $[-1, 1] \times [-2, 2]$ is shown below on the left. Note that since $\lim\limits_{x \to 0} \dfrac{f(x)}{x} = 0$,

$$f'(0) = \lim_{x \to 0}\frac{f(x) - f(0)}{x - 0} = \lim_{x \to 0}\frac{f(x)}{x} = 0.$$

Therefore $f(x)$ is differentiable at $x = 0$.

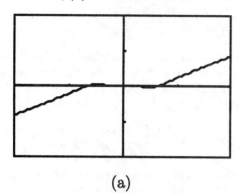

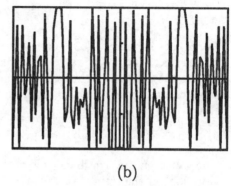

(a)   (b)

(b) $f'(x) = \begin{cases} 2x\sin(1/x) - \cos(1/x), & \text{if } x \neq 0 \\ 0, & \text{if } x = 0 \end{cases}$. The graph with the range $[-.01, .01] \times [-1, 1]$ is shown above on the right. It appears that $f'(x)$ is not continuous at $x = 0$.

(c) $f'(x) = \dfrac{2}{n\pi}\sin(n\pi) - \cos(n\pi) = (-1)^{n+1}$. Hence $f'(x)$ is alternatively 1 and -1 for an infinite number of points arbitrarily close to 0.

86. (a) The graphs are shown below. The range for each is $[0, 8\pi/\sqrt{32}] \times [-30, 30]$ with **Xscl**=1 and **Yscl**=5. This range is chosen because $2\pi/\sqrt{32}$ is the period of the function $\theta(t) = 5\sin\omega t$ when $l = 1$. Thus, the first graph depicts four periods.

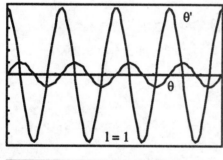

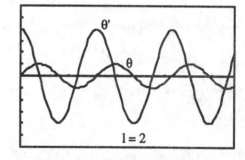

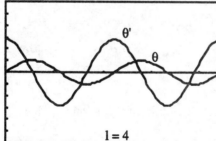

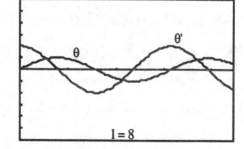

We see from these graphs that if $0 \leq t \leq 8\pi/\sqrt{32}$ then when $l = 1$ there are 4 periods or full swings completed. For that same time interval, when $l = 2$ there are approximately 2.75 full swings completed, when $l = 4$ there are approximately 2 full swings completed, and when $l = 8$ there are approximately 1.5 full swings completed. This indicates that the pendulum slows its rate of swing when its length, l, increases. This is also evident from the graphs of θ', since as l increases, the maximum of $|\theta'|$ decreases.

(b) By the Chain Rule,

$$\theta'(t) = 5\omega\cos\omega t = 5\sqrt{\frac{32}{l}}\cos\left(\sqrt{\frac{32}{l}}\,t\right).$$

Hence as l increases $\sqrt{32/l}$ decreases, i.e., the rate of change of the pendulum decreases. This means that it swings slower. To be more precise, let $\theta'_l(t)$ denote the derivative for l and let $\theta'_{2l}(t)$ denote the derivative when l is changed to $2l$. Since each of these functions involve the cosine function and $|\cos x| \le 1$ we have from the derivative of $\theta(t)$

$$|\theta'_l(t)| \le 5\sqrt{\frac{32}{l}} \quad \text{and} \quad |\theta'_{2l}(t)| \le 5\sqrt{\frac{32}{2l}} = \frac{20}{\sqrt{l}}.$$

The values on the right hand side of these expressions represent the extreme values of θ'_l and θ'_{2l}. If we look at the ratio of these two values we find that as l increases to $2l$ then

$$\frac{\max|\theta'_l(t)|}{\max|\theta'_{2l}(t)|} = \frac{5\sqrt{32/l}}{20/\sqrt{l}} = \sqrt{2}.$$

Thus as l increases to $2l$ the pendulum swings slower by $\sqrt{2}$.

3.6   Related Rates

1. $f(t) = 2[g(t)]^3 + 5 \implies f'(t) = 6[g(t)]^2 g'(t) \implies$

$$f'(1) = 6[g(1)]^2 g'(1) = 6(3)^2(-2) = -108.$$

2. $f(t) = \sqrt{2 + g(t)} = [2 + g(t)]^{1/2} \implies f'(t) = (1/2)[2 + g(t)]^{-1/2} g'(t) \implies$

$$f'(0) = (1/2)[2 + g(0)]^{-1/2} g(0) = (1/2)[2 + 3]^{-1/2}(4) = 2/\sqrt{5}.$$

3. $f(t) = \dfrac{1}{1 + g(t)} = [1 + g(t)]^{-1} \implies f'(t) = -[1 + g(t)]^{-2} g'(t) \implies$

$$f'(2) = -[1 + g(2)]^{-2} g'(2) = -[1 + 3]^{-2}(-2) = 1/8.$$

4. $[f(t)]^2 + [g(t)]^3 = 265 \implies 2f(t)f'(t) + 3[g(t)]^2 g'(t) = 0 \implies$

$$2f(1)f'(1) + 3[g(1)]^2 g'(1) = 0 \quad \implies \quad 2(7)f'(1) + 3(6)^2(-2) = 0 \quad \implies \quad f'(1) = 108/7.$$

5. $\sin(f(t)) = [g(t)]^2 \implies [\cos(f(t))]f'(t) = 2g(t)g'(t) \implies [\cos(f(0))]f'(0) = 2g(0)g'(0) \implies$

$$[\cos(\pi/6)]f'(0) = 2(1)(-2) \quad \implies \quad (\sqrt{3}/2)f'(0) = -4 \quad \implies \quad f'(0) = -8\sqrt{3}/3.$$

6. Let $V(t)$ stand for the volume of the balloon and $r(t)$ its radius after t seconds. Then

$$V(t) = \frac{4}{3}\pi[r(t)]^3 \quad \implies \quad \frac{dV}{dt} = 4\pi[r(t)]^2 \frac{dr}{dt}.$$

From the given information, $dr/dt = 3$. Hence, when $r = 10$,

$$\left.\frac{dV}{dt}\right|_{r=10} = 4\pi(10)^2(3) = 1200\pi.$$

Thus, when the radius is 10 cm, the volume is increasing at a rate of 1200π cm^3/s.

7. Let $D(x)$ stand for the length of the diagonal of a cube of edge length x. Then $D(x)^2 = x^2 + x^2 + x^2 = 3x^2 \implies D(x) = x\sqrt{3} \implies dD/dt = \sqrt{3}\,(dx/dt)$. From the given information, $dx/dt = 2$. Hence $dD/dt = 2\sqrt{3}$. Thus, the diagonal is increasing at the rate of $2\sqrt{3}$ cm/s.

8. Let $A(t)$ and $r(t)$ stand for the area and radius of the disturbed water, respectively. Then $A(t) = \pi r(t)^2 \implies dA/dt = 2\pi r(t)(dr/dt)$. From the given information, $dr/dt = 2$. Hence, when $r = 10$,

$$\frac{dA}{dt}\bigg|_{r=10} = 2\pi(10)(2) = 40\pi.$$

Thus, when the radius is 10 m, the area of disturbed water is increasing at a rate of 40π m^2/s.

9. Let $V(t)$ and $r(t)$ stand for the volume and radius of the snowball after t seconds, respectively. Then $V(t) = (4/3)\pi r(t)^3 \implies dV/dt = 4\pi r(t)^2(dr/dt)$. From the given information, $dr/dt = -1$. Hence, when $r = 6$,

$$\frac{dV}{dt}\bigg|_{r=6} = 4\pi(6)^2(-1) = -144\pi.$$

Thus, when the radius is 6 cm, the volume is decreasing at a rate of 144π cm^3/s.

10. $y = x^{5/2} \implies dy/dt = (5/2)x^{3/2}(dx/dt)$. From the given information, $dx/dt = 2$. Hence, at the point $(4, 32)$,

$$\frac{dy}{dt}\bigg|_{x=4} = \frac{5}{2}\left(4^{3/2}\right)(2) = 40.$$

Thus, at the point $(4, 32)$, the y-coordinate is increasing at a rate of 40 units/s.

11. Let $V(t)$, $h(t)$, and $r(t)$ stand for the volume, depth, and radius of water in the tank after t minutes, as shown in figures below.

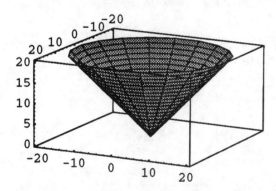 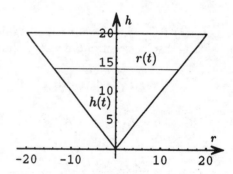

Then $V(t) = (1/3)\pi[r(t)]^2 h(t)$. To eliminate $r(t)$ from this equation, observe that, by similar triangles,

$$\frac{r(t)}{h(t)} = \frac{20}{20} \implies r(t) = h(t).$$

Hence

$$V(t) = \frac{1}{3}\pi[h(t)]^2 h(t) = \frac{\pi}{3}[h(t)]^3 \implies \frac{dV}{dt} = \pi[h(t)]^2 \frac{dh}{dt}.$$

From the given information, $dV/dt = 40$. Hence, when $h = 8$,

$$\frac{dV}{dt}\bigg|_{h=8} = \pi(8)^2 \frac{dh}{dt} = 40 \implies \frac{dh}{dt} = \frac{5}{8\pi} \approx 0.199.$$

Thus, when the water is 8 meters deep, it is rising at the rate of approximately 0.199 m/min.

12. Let $x(t)$ stand for the horizontal distance from the transmitter to the truck and $D(t)$ the direct line distance from the transmitter to truck, as illustrated in the figure below. Then

$$[x(t)]^2 + 3^2 = [D(t)]^2$$

$$\implies \quad 2x(t)\frac{dx}{dt} = 2D(t)\frac{dD}{dt}$$

$$\implies \quad \frac{dD}{dt} = \frac{x(t)}{D(t)}\frac{dx}{dt}.$$

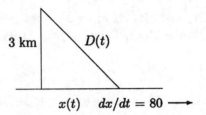

From the given information, $dx/dt = 80$. Hence, when $D = 5$, $x = \sqrt{5^2 - 3^2} = 4$ and therefore

$$\left.\frac{dD}{dt}\right|_{D=5} = \frac{4}{5}(80) = 64.$$

Thus, when the truck is 5 km from the transmitter its distance from the transmitter is **increasing at the rate of 64 km/h.**

13. Let $A(t)$ stand for the area of a rectangle of width $y(t)$ and length $2y(t)$ at time t. Then

$$A(t) = [2y(t)][y(t)] = 2[y(t)]^2 \quad \implies \quad \frac{dA}{dt} = 4y(t)\frac{dy}{dt} = 8 \quad \implies \quad \frac{dy}{dt} = \frac{2}{y(t)}.$$

Hence, when $y = 5$, $dy/dt = 2/5$ and therefore the rate of increase of the length $2y$ is

$$\frac{d}{dt}[2y(t)] = 2\left.\frac{dy}{dt}\right|_{y=5} = 2\left(\frac{2}{5}\right) = \frac{4}{5} \text{ cm/s.}$$

14. Let $x(t)$ stand for the horizontal displacement of the base of the ladder from the wall and $y(t)$ the vertical displacement of the top, as illustrated in the figure below. Then

$$[x(t)]^2 + [y(t)]^2 = 5^2$$

$$\implies \quad 2x(t)\frac{dx}{dt} + 2y(t)\frac{dy}{dt} = 0$$

$$\implies \quad \frac{dy}{dt} = -\frac{x(t)}{y(t)}\frac{dx}{dt}.$$

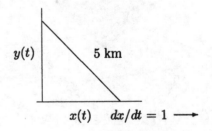

From the given information, $dx/dt = 1$. Hence, when $x = 3$, $y = \sqrt{5^2 - 3^2} = 4$ and therefore

$$\left.\frac{dy}{dt}\right|_{x=3} = \frac{3}{4}(1) = -\frac{3}{4}.$$

Thus, when the base of the ladder is 3 feet from the wall its top is falling at a rate of 3/4 m/s.

15. Let $x(t)$ be the horizontal distance from the shore to the boat and $s(t)$ the length of rope at time t. Then

$$[x(t)]^2 + 6^2 = [s(t)]^2 \quad \implies \quad 2x(t)\frac{dx}{dt} = 2s(t)\frac{ds}{dt} \quad \implies \quad \frac{dx}{dt} = \frac{s(t)}{x(t)}\frac{ds}{dt}.$$

From the given information, $ds/dt = -5$. Hence, when $x = 8$, $s = \sqrt{8^2 + 6^2} = 10$ and therefore

$$\frac{dx}{dt}\bigg|_{x=8} = \frac{10}{8}(-5) = -6.25.$$

Thus, the boat is approaching the shore at 6.25 m/min.

16. Since $A = 1$ and $B = 2$, it follows that

$$C^2 = 1^2 + 2^2 - 2(1)(2)\cos\theta = 5 - 4\cos\theta \implies 2C\frac{dC}{dt} = 4(\sin\theta)\frac{d\theta}{dt} \implies \frac{dC}{dt} = \frac{2\sin\theta}{C}\frac{d\theta}{dt}.$$

From the given information, $d\theta/dt = 0.2$. Hence, when $C = \sqrt{3}$, $3 = 5 - 4\cos\theta \implies \cos\theta = 1/2 \implies \sin\theta = \sqrt{3}/2$ and hence

$$\frac{dC}{dt}\bigg|_{C=\sqrt{3}} = \frac{2\left(\sqrt{3}/2\right)}{\sqrt{3}}(0.2) = 0.2.$$

Thus, when $C = \sqrt{3}$, side C is increasing at a rate of 0.2 units/min.

17. Let $x(t)$ be the distance from the lamp post to the woman and $s(t)$ the length of her shadow at time t, as illustrated in the figure below. Then, by similar triangle,

$$\frac{10}{1.6} = \frac{x(t) + s(t)}{s(t)}$$

$$\implies \quad 10s(t) = 1.6x(t) + 1.6s(t)$$

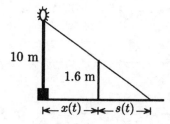

10 m

1.6 m

$\longleftarrow x(t) \longrightarrow\!\longleftarrow s(t) \longrightarrow$

$$\implies \quad s(t) = \frac{4}{21}x(t) \implies \frac{ds}{dt} = \frac{4}{21}\frac{dx}{dt}.$$

Observe that the formula for ds/dt does not involve x, only dx/dt, and hence the rate of change of the woman's shadow is independent of how far she is from the lamp post. From the given information, $dx/dt = 1.2$. Hence

$$\frac{ds}{dt} = \frac{4}{21}(1.2) \approx 0.23.$$

Thus, her shadow is increasing at a rate of approximately 0.23 m/s regardless of how far she is from the lamp post.

18. Let $l(t)$ stand for the length of the two equal sides of the triangle, $h(t)$ its height, and $A(t)$ the area of the triangle at time t. Then $h = \sqrt{l^2 - 81}$ and hence

$$A(t) = \frac{1}{2}(18)h = 9\sqrt{l^2 - 81}$$

$$\implies \quad \frac{dA}{dt} = 9\frac{1}{2}\left(l^2 - 81\right)^{-1/2}(2l)\frac{dl}{dt}$$

$$= \frac{9l}{\sqrt{l^2 - 81}}\frac{dl}{dt}.$$

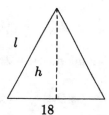

l

h

18

From the given information, $dl/dt = 2$. Hence, when $h = 12$, $l = \sqrt{9^2 + 12^2} = 15$ and therefore

$$\left.\frac{dA}{dt}\right|_{h=12} = \frac{9(15)}{12}(2) = 22.5.$$

Thus, when the height is 12 cm the area is increasing at the rate of 22.5 cm/s.

From the given information, $dl/dt = 2$. Hence, when $h = 12$, $l = \sqrt{9^2 + 12^2} = 15$ and therefore

$$\left.\frac{dA}{dt}\right|_{h=12} = \frac{9(15)}{12}(2) = 22.5.$$

Thus, when the height is 12 cm the area is increasing at the rate of 22.5 cm/s.

19. $N(t) = I(t) + S(t) \implies dN/dt = dI/dt + dS/dt$. From the given information, $dI/dt = 24$ and $dS/dt = -20$. Hence $dN/dt = 4$. Thus, the population is increasing at a rate of 4 persons per day.

20. Let $V(t)$, $D(t)$, and $h(t)$ stand for the volume, radius, and height of the conical sand pile at time t. From the given information, $D(t) = 3h(t)$. Hence

$$V(t) = \frac{1}{3}\pi \left[\frac{D(t)}{2}\right]^2 h(t) = \frac{1}{3}\pi \left[\frac{3h(t)}{2}\right]^2 h(t) = \frac{3\pi}{4}h(t)^3$$

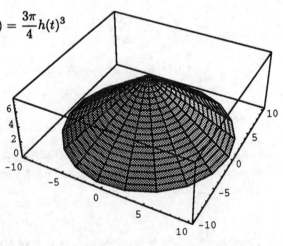

$$\implies \quad \frac{dV}{dt} = \frac{9\pi}{4}h(t)^2\frac{dh}{dt}$$

$$\implies \quad \frac{dh}{dt} = \frac{4}{9\pi h(t)^2}\frac{dV}{dt}.$$

Since $dV/dt = 180$, it follows that when $h = 6$,

$$\left.\frac{dD}{dt}\right|_{h=6} = 3\left.\frac{dh}{dt}\right|_{h=6} = 3\frac{4}{9\pi(6)^2}(180) = \frac{20}{3\pi} \approx 2.12.$$

Thus, when the sand pile is 6 m high the base diameter is increasing at the rate of approximately 2.12 m/min.

21. Let $x(t)$ be the distance east from Columbus of the first truck, $y(t)$ the distance north of the second truck, and $D(t)$ the distance between the trucks t hours after noon. Note that $y(t) = 0$ for $0 \leq t \leq 1$ and $60(t-1)$ for $t > 1$. Now,

$$D(t)^2 = x(t)^2 + y(t)^2 \quad \implies \quad 2D(t)\frac{dD}{dt} = 2x(t)\frac{dx}{dt} + 2y(t)\frac{dy}{dt}.$$

From the given information, $dx/dt = 40$ and $dy/dt = 60$. Hence

$$2D(t)\frac{dD}{dt} = 2x(t)(40) + 2y(t)(60) \quad \implies \quad \frac{dD}{dt} = \frac{40x(t) + 60y(t)}{D(t)}.$$

At 2:00 pm, $t = 2$, $x = 40(2) = 80$, $y = 60(1) = 60$, and $D = \sqrt{80^2 + 60^2}$. Hence

$$\left.\frac{dD}{dt}\right|_{t=2} = \frac{40(80) + 60(60)}{100} = 68.$$

Thus, at 2:00 pm the distance between the trucks is increasing at the rate of 68 km/h.

22. Let $V(t)$ and $h(t)$ stand for the volume and depth of water in the trough at time t, and let $2x(t)$ be the width of the water. Then, recalling that 4 m = 400 cm,

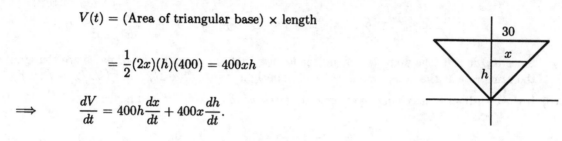

$$V(t) = (\text{Area of triangular base}) \times \text{length}$$

$$= \frac{1}{2}(2x)(h)(400) = 400xh$$

$$\implies \quad \frac{dV}{dt} = 400h\frac{dx}{dt} + 400x\frac{dh}{dt}.$$

Now, by similar triangles,

$$\frac{30}{x} = \frac{40}{h} \quad \implies \quad x = \frac{3}{4}h \quad \implies \quad \frac{dx}{dt} = \frac{3}{4}\frac{dh}{dt}.$$

Therefore

$$\frac{dV}{dt} = 400h\left(\frac{3}{4}\frac{dh}{dt}\right) + 400\left(\frac{3}{4}h\right)\frac{dh}{dt} = 600h\frac{dh}{dt} \implies \frac{dh}{dt} = \frac{1}{600h}\frac{dh}{dt}.$$

From the given information, $dV/dt = 9000$. Hence, when $h = 10$,

$$\left.\frac{dh}{dt}\right|_{h=10} = \frac{1}{600(10)}(9000) = 1.5.$$

Thus, when the water is 10 cm deep it is rising at the rate of 1.5 cm/s.

23. In this case, $x = h/\sqrt{3}$ and hence

$$V(t) = \frac{1}{2}(2x)(h)(400) = \frac{400h^2}{\sqrt{3}} \implies \frac{dV}{dt} = \frac{800h}{\sqrt{3}}\frac{dh}{dt} \implies \frac{dh}{dt} = \frac{\sqrt{3}}{800h}\frac{dV}{dt}.$$

Hence, when $h = 10$,

$$\left.\frac{dh}{dt}\right|_{h=10} = \frac{\sqrt{3}}{800(10)}(9000) = \frac{9}{8}\sqrt{3} \approx 1.95.$$

Thus, when the water is 10 cm deep it is rising at the rate of approximately 1.95 cm/s.

24. Let $s(t)$ stand for the height of the rocket at time t and $\theta(t)$ the angle of elevation from the observer's eye. Then

$$s(t) = 150\tan\theta(t) \implies \frac{ds}{dt} = 150\left[\sec^2\theta(t)\right]\frac{d\theta}{dt} \implies \frac{d\theta}{dt} = \frac{1}{150\sec^2\theta(t)}\frac{ds}{dt}.$$

From the information given, $ds/dt = 12$ when $s = 200$. Therefore $\tan\theta = 200/150 = 4/3 \implies \sec^2\theta = 1 + (4/3)^2 = 25/9$ and hence

$$\frac{d\theta}{dt} = \frac{1}{150(25/9)}(12) = \frac{18}{625} \approx 0.029.$$

149

Thus, when the rocket reaches a height of 200 m, the angle of elevation is increasing at the rate of approximately 0.029 rad/s, or 1.66°/s.

25. Let $(x(t), y(t))$ stand for the position of the point at time t. Then

$$x^2 + y^2 = 1 \quad \Longrightarrow \quad 2x\frac{dx}{dt} + 2y\frac{dy}{dt} = 0.$$

(a) If the coordinates are changing at the same rate, then $dx/dt = dy/dt$ and hence

$$2x\frac{dx}{dt} + 2y\frac{dx}{dt} = 2(x+y)\frac{dx}{dt} = 0 \quad \Longrightarrow \quad x+y = 0 \quad \text{or} \quad \frac{dx}{dt} = \frac{dy}{dt} = 0.$$

Since $dx/dt \neq 0$ (the point is moving), it follows that the x- and y-coordinates are changing at the same rate at the points where $y = -x$, namely at $(-\sqrt{2}/2, \sqrt{2}/2)$ and $(\sqrt{2}/2, -\sqrt{2}/2)$.

(b) If the coordinates are changing at opposite rates, $dx/dt = -dy/dt$ and hence

$$2x\frac{dx}{dt} - 2y\frac{dx}{dt} = 2(x-y)\frac{dx}{dt} = 0.$$

Thus, if the coordinates are changing at opposite rates, $x - y = 0$, or $x = y$, and hence is at $(\sqrt{2}/2, \sqrt{2}/2)$ and $(-\sqrt{2}/2, -\sqrt{2}/2)$.

26. $D = kV^2S^2 \Longrightarrow dD/dt = 2kVS^2(dV/dt)$. If the acceleration is 8 m/s^2 when the velocity is 30 m/s, then $dV/dt = 8$, $V = 30$ and hence

$$\left.\frac{dD}{dt}\right|_{V=30} = 2k(30)S^2(8) = 480kS^2.$$

Thus, the drag is increasing at the rate of $480kS^2$ units.

27. Let $s(\theta)$ stand for the length of the shadow when the sun is θ radians above the horizon. Then

$$s(\theta) = 30\cot\theta \quad \Longrightarrow \quad \frac{ds}{dt} = -30\csc^2\theta\frac{d\theta}{dt}.$$

From the given information, $d\theta/dt = -15(\pi/180)$. Hence, when $\theta = 30° = \pi/6$,

$$\left.\frac{ds}{dt}\right|_{\theta=\pi/6} = -30\csc^2(\pi/6)\left(\frac{-15\pi}{180}\right) = 10\pi.$$

Thus, the shadow length is increasing at the rate of 10π m/h.

28. $PV = kT \Longrightarrow (dP/dt)V + P(dV/dt) = k(dT/dt)$. Since volume is constant, $dV/dt = 0$ and therefore $(dP/dt)V = k(dT/dt)$. Hence, if $V = 1000$ and $dT/dt = 2$,

$$\left(\frac{dP}{dt}\right)(1000) = k(2) \quad \Longrightarrow \quad \frac{dP}{dt} = \frac{k}{500}.$$

Thus, the pressure is increasing at the rate of $k/500$ units/min.

29. Let $V(t)$, $r(t)$, and $h(t)$ be the volume, radius, and height of the coffee in the bottom container. Since coffee fills only the bottom portion of the cone,

$$V(t) \quad = \quad \text{Total volume of cone} - \text{Volume of unfilled portion of cone}$$

$$= \quad \frac{1}{3}\pi(4)^2(10) - \frac{1}{3}\pi r^2(10 - h)$$

$$= \quad \frac{1}{3}\pi\left[160 - r^2(10 - h)\right].$$

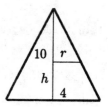

Now, $r/4 = (10 - h)/10 \implies r = (2/5)(10 - h)$. Hence

$$V(t) = \frac{1}{3}\pi\left[160 - \frac{4}{25}(10 - h)^3\right]$$

and therefore

$$\frac{dV}{dt} = -\frac{\pi}{3}\frac{4}{25}\left[3(10 - h)^2\left(-\frac{dh}{dt}\right)\right] = 4 \implies \frac{dh}{dt} = \frac{25}{\pi(10 - h)^2}.$$

Hence, when $h = 4$,

$$\left.\frac{dh}{dt}\right|_{h=4} = \frac{25}{\pi(36)} \approx 0.22.$$

Thus, when the depth of coffee in the bottom container is 4 cm, it is rising at the rate of approximately 0.22 cm/s.

30. $F = Gm_1m_2/r^2 = Gm_1m_2r^{-2} \implies dF/dt = Gm_1m_2(-2r^{-3})dr/dt$. If $dF/dt = \alpha$,

$$\alpha = -\frac{2Gm_1m_2}{r^3}\frac{dr}{dt} \implies \frac{dr}{dt} = -\frac{\alpha r^3}{2Gm_1m_2}.$$

Thus, the rate at which the distance between the bodies is changing is equal to $-\alpha r^3/2Gm_1m_2$.

31. Let $x(t)$ be the distance along the shore from the piling to the light beam and $\theta(t)$ the angle between the light beam and the piling on shore. Then

$$x = 200\tan\theta \implies \frac{dx}{dt} = 200\left(\sec^2\theta\right)\frac{d\theta}{dt}.$$

From the given information, $d\theta/dt = 2$ rev/min $= 2(2\pi)$ rad/min $= 4\pi$ rad/min. Hence, when $x = 200$, $\tan\theta = 1 \implies \sec^2\theta = 2$ and therefore

$$\frac{dx}{dt} = 200(2)(4\pi) = 1600\pi \approx 5026.5.$$

Thus, when the light beam is 200 m from the piling, it is moving at the rate of 1600π m/min, or equivalently, $80\pi/3$ m/s.

32. $E = 1 - \dfrac{T_c}{T_h} \implies T_h = \dfrac{T_c}{1 - E} \implies \dfrac{dT_h}{dt} = \dfrac{(1 - E)(dT_c/dt) - T_c(-dE/dt)}{(1 - E)^2}$

(a) If $dT_c/dt = 3$ and efficiency is not changing, $dE/dt = 0$ and therefore

$$\frac{dT_h}{dt} = \frac{(1 - E)(3) - T_c(0)}{(1 - E)^2} = \frac{3}{1 - E} \ \text{°K/min}.$$

(b) If efficiency is decreasing at 2%/h and T_c is fixed, then $dE/dt = -0.02E$, $dT_c/dt = 0$, and hence

$$\frac{dT_h}{dt} = \frac{(1-E)(0) - T_c(0.02E)}{(1-E)^2} = -\frac{0.02ET_c}{(1-E)^2} \ °\text{K/min}.$$

33. $(\sin\alpha)/(\sin\beta) = C_a \Longrightarrow \sin\beta = (1/C_a)\sin\alpha \Longrightarrow (\cos\beta)(d\beta/dt) = (1/C_a)(\cos\alpha)(d\alpha/dt)$. From the given information, $C_a = 1.33$, $d\alpha/dt = 0.2$, and hence

$$\cos\beta\frac{d\beta}{dt} = \frac{1}{1.33}(\cos\alpha)(0.2) = 0.15\cos\alpha \quad \Longrightarrow \quad \frac{d\beta}{dt} = 0.15\frac{\cos\alpha}{\cos\beta} \ \text{rad/s}.$$

34. Let $x(t)$ and $y(t)$ be the distances traveled by the ship and submarine, respectively, as illustrated below, and let $D(t)$ be the distance between them after t hours. Then

$$D(t)^2 = 1^2 + x(t)^2 + y(t)^2$$

$$\Longrightarrow \qquad 2D(t)\frac{dD}{dt} = 2x(t)\frac{dx}{dt} + 2y(t)\frac{dy}{dt}$$

$$\Longrightarrow \qquad \frac{dD}{dt} = \frac{x(t)}{D(t)}\frac{dx}{dt} + \frac{y(t)}{D(t)}\frac{dy}{dt}.$$

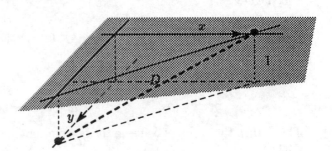

From the given information, $dx/dt = 30$, $dy/dt = 60$. When $t = 20$ min $= 1/3$ hour, $x = 30(1/3) = 10$, $y = 60(1/3) = 20$, and hence $D = \sqrt{1 + 10^2 + 20^2} = \sqrt{501}$. Therefore

$$\frac{dD}{dt}\bigg|_{t=1/3} = \frac{10}{\sqrt{501}}(30) + \frac{20}{\sqrt{501}}(60) = \frac{1500}{\sqrt{501}} \approx 67.02.$$

Thus, after 20 min, the distance between the ship and submarine is increasing at the rate of approximately 67.02 km/h.

35. (a) Here $L = 25$, $r = 0$, $dR/dt = 0.2$. Therefore

$$\nu = \frac{\alpha}{25}R^2 \quad \Longrightarrow \quad \frac{d\nu}{dt} = \frac{\alpha}{25}\left(2R\frac{dR}{dt}\right) = \frac{\alpha}{25}(2R)(0.2) = 0.016\alpha R.$$

When $R = 10$,

$$\frac{d\nu}{dt}\bigg|_{R=10} = 0.016\alpha(10) = 0.16\alpha.$$

Thus, when $R = 10$, the acceleration at the center is 0.16α cm/min^2.

(b) In this case $R = 10$, $r = 0$, $dL/dt = 0.5$. Therefore

$$\nu = \frac{\alpha}{L}(10^2) \quad \Longrightarrow \quad \frac{d\nu}{dL} = -\frac{100\alpha}{L^2}\frac{dL}{dt} = -\frac{100\alpha}{L^2}(0.5) = -\frac{50\alpha}{L^2}.$$

When $L = 25$,

$$\frac{d\nu}{dL}\bigg|_{L=25} = -\frac{50\alpha}{25^2} = -0.08\alpha.$$

Thus, when $L = 25$, the acceleration at the center is -0.08α cm/s^2.

3.7 Linear Approximation

1. (a) $f(x) = \sqrt{x} \implies f'(x) = 1/(2\sqrt{x})$. Hence, if $x_0 = 4$,

$$f(x_0) + f'(x_0)\Delta x = \sqrt{4} + \frac{1}{2\sqrt{4}}\Delta x = 2 + \frac{1}{4}\Delta x.$$

(b) Here $\Delta x = 0.1$. Hence

$$\sqrt{4.1} \approx 2 + \frac{1}{4}(0.1) = 2.025.$$

2. (a) $f(x) = \sqrt{x} \implies f'(x) = 1/(2\sqrt{x})$. Hence, if $x_0 = 9$,

$$f(x_0) + f'(x_0)\Delta x = \sqrt{9} + \frac{1}{2\sqrt{9}}\Delta x = 3 + \frac{1}{6}\Delta x.$$

(b) Here $\Delta x = 0.1$. Hence

$$\sqrt{9.1} \approx 3 + \frac{1}{6}(0.1) = 3.0167.$$

3. (a) $f(x) = 1/x \implies f'(x) = -1/x^2$. Hence, if $x_0 = 2$,

$$f(x_0) + f'(x_0)\Delta x = \frac{1}{2} - \frac{1}{2^2}\Delta x = \frac{1}{2} - \frac{1}{4}\Delta x.$$

(b) Since $13/30 = 1/(30/13)$, $\Delta x = 30/13 - 2 = 0.308$ and hence

$$\frac{13}{30} \approx \frac{1}{2} - \frac{1}{4}(0.308) = 0.423.$$

4. (a) $f(x) = 1/x^2 \implies f'(x) = -2/x^3$. Hence, if $x_0 = 3$,

$$f(x_0) + f'(x_0)\Delta x = \frac{1}{3^2} - \frac{2}{3^3}\Delta x = \frac{1}{9} - \frac{2}{27}\Delta x.$$

(b) Since $4/25 = 1/(5/2)^2$, $\Delta x = 5/2 - 3 = -0.5$ and hence

$$\frac{4}{25} \approx \frac{1}{9} - \frac{2}{27}(-0.5) = 0.148.$$

5. (a) $f(x) = x^3 \implies f'(x) = 3x^2$. Hence, if $x_0 = 1/2$,

$$f(x_0) + f'(x_0)\Delta x = \left(\frac{1}{2}\right)^3 + 3\left(\frac{1}{2}\right)^2 \Delta x = \frac{1}{8} + \frac{3}{4}\Delta x.$$

(b) Here $\Delta x = 0.48 - 0.5 = -0.02$. Hence

$$(0.48)^3 \approx \frac{1}{8} + \frac{3}{4}(-0.02) = 0.11.$$

6. (a) $f(x) = x^{10} \implies f'(x) = 10x^9$. Hence, if $x_0 = 1$,

$$f(x_0) + f'(x_0)\Delta x = 1^{10} + 10(1)^9 \Delta x = 1 + 10\Delta x.$$

(b) Here $\Delta x = 0.98 - 1 = -0.02$. Hence

$$(0.98)^{10} \approx 1 + 10(-0.02) = 0.8$$

7. (a) $f(x) = 2\sin x \implies f'(x) = 2\cos x$. Hence, if $x_0 = 0$,

$$f(x_0) + f'(x_0)\Delta x = 2\sin 0 + 2\cos 0(\Delta x) = 2\Delta x.$$

(b) Here $2° = 2(\pi/180) = 0.035$ radian and $\Delta x = 0.035$. Hence $2\sin 2° \approx 2(0.035) = 0.07$.

8. (a) $f(x) = \tan x \implies f'(x) = \sec^2 x$. Hence, if $x_0 = 0$,

$$f(x_0) + f'(x_0) = \tan 0 + \sec^2 0(\Delta x) = \Delta x.$$

(b) Here $\Delta x = 0.1$. Hence $\tan 0.1 \approx 0.1$.

9. (a) $f(x) = \cos x \implies f'(x) = -\sin x$. Hence, if $x_0 = \pi/6$,

$$f(x_0) + f'(x_0)\Delta x = \cos(\pi/6) - [\sin(\pi/6)]\Delta x = \frac{\sqrt{3}}{2} - \frac{1}{2}\Delta x.$$

(b) Here $31° = 31(\pi/180) = 0.541$ radian and $\Delta x = 0.541 - \pi/6 = 0.017$. Hence

$$\cos 31° \approx \frac{\sqrt{3}}{2} - \frac{1}{2}(0.017) \approx 0.858.$$

10. (a) $f(x) = 2\sec x \implies f'(x) = 2\sec x \tan x$. Hence, if $x_0 = \pi/4$,

$$f(x_0) + f'(x_0)\Delta x = 2\sec\frac{\pi}{4} + \left(2\sec\frac{\pi}{4}\tan\frac{\pi}{4}\right)\Delta x = 2\sqrt{2} + 2\sqrt{2}\,\Delta x.$$

(b) Here $43° = 43(\pi/180) = 0.75$ radian and $\Delta x = 0.75 - \pi/4 = -0.035$. Hence

$$2\sec 43° \approx 2\sqrt{2} + 2\sqrt{2}\,(-0.035) \approx 2.729.$$

11. Let $f(x) = \sqrt{x}$. Then $f'(x) = 1/(2\sqrt{x})$ and hence

$$f(x) \approx f(x_0) + f'(x_0)\Delta x = \sqrt{x_0} + \frac{1}{2\sqrt{x_0}}\,\Delta x.$$

Setting $x = 48$ and $x_0 = 49$, $\Delta x = -1$ and hence

$$\sqrt{48} \approx \sqrt{49} + \frac{1}{2\sqrt{49}}(-1) \approx 6.929.$$

12. Let $f(x) = \sqrt{x}$. Then $f'(x) = 1/(2\sqrt{x})$ and hence

$$f(x) \approx f(x_0) + f'(x_0)\Delta x = \sqrt{x_0} + \frac{1}{2\sqrt{x_0}}\,\Delta x.$$

Setting $x = 62$ and $x_0 = 64$, $\Delta x = -2$ and hence

$$\sqrt{62} \approx \sqrt{64} + \frac{1}{2\sqrt{64}}(-2) = 7.875.$$

13. Let $f(x) = 1/x$. Then $f'(x) = -1/x^2$ and hence

$$f(x) \approx f(x_0) + f'(x_0)\Delta x = \frac{1}{x_0} - \frac{1}{x_0^2}\, \Delta x.$$

Setting $x = 4.1$ and $x_0 = 4$, $\Delta x = 0.1$ and hence

$$\frac{1}{4.1} \approx \frac{1}{4} - \frac{1}{4^2}(0.1) = 0.244.$$

14. Let $f(x) = \sin x$. Then $f'(x) = \cos x$ and hence

$$f(x) \approx f(x_0) + f'(x_0)\Delta x = \sin x_0 + (\cos x_0)\Delta x.$$

Setting $x = 46°$ and $x_0 = 45°$, $\Delta x = 1° = \pi/180 = 0.017$ radian and hence

$$\sin 46° \approx \sin 45° + (\cos 45°)(0.017) \approx 0.719.$$

15. Let $f(x) = \cos x$. Then $f'(x) = -\sin x$ and hence

$$f(x) \approx f(x_0) + f'(x_0)\Delta x = \cos x_0 - (\sin x_0)\Delta x.$$

Setting $x = 59°$ and $x_0 = 60°$, $\Delta x = -1° = -\pi/180 = -0.017$ radian and hence

$$\cos 59° \approx \cos 60° - (\sin 60°)(-0.017) \approx 0.515.$$

16. Let $f(x) = x^6$. Then $f'(x) = 6x^5$ and hence

$$f(x) \approx f(x_0) + f'(x_0)\Delta x = x_0^6 + 6x_0^5\Delta x.$$

Setting $x = 2.03$ and $x_0 = 2$, $\Delta x = 0.03$ and hence

$$(2.03)^6 \approx 2^6 + 6(2)^5(0.03) = 69.76.$$

17. Let $f(x) = x^3$. Then $f'(x) = 3x^2$ and hence

$$f(x) \approx f(x_0) + f'(x_0)\Delta x = x_0^3 + 3x_0^2\Delta x.$$

Setting $x = 2.97$ and $x_0 = 3$, $\Delta x = -0.03$ and hence

$$(2.97)^3 \approx 3^3 + 3(3)^2(-0.03) = 26.19.$$

18. Let $f(x) = \cos x$. Then $f'(x) = -\sin x$ and hence

$$f(x) \approx f(x_0) + f'(x_0)\Delta x = \cos x_0 - (\sin x_0)\Delta x.$$

Setting $x = 91°$ and $x_0 = 90°$, $\Delta x = 1° = \pi/180 = 0.017$ radian and hence

$$\cos 91° \approx \cos 90° - (\sin 90°)(0.017) \approx -0.017.$$

19. The table below summarizes the calculator value, linear approximation, relative error, and percentage error for the approximations in Exercises 11, 13, 15, and 17.

		calculator value y	linear approx y_1	relative error $|(y - y_1)/y|$	% error
Ex. 11	$\sqrt{48}$	6.9282	6.929	0.000115	0.0115%
Ex. 13	$1/4.1$	0.2439	0.244	0.0004	0.04%
Ex. 15	$\cos 59°$	0.51504	0.515	0.000074	0.0074%
Ex. 17	$(2.97)^3$	26.198073	26.19	0.00031	0.031%

20. The table below summarizes the calculator value, linear approximation, relative error, and percentage error for the approximations in Exercises 12, 14, 16, and 18.

		calculator value y	linear approx y_1	relative error $|(y - y_1)/y|$	% error
Ex. 12	$\sqrt{62}$	7.87401	7.875	0.00013	0.013%
Ex. 14	$\sin 46°$	0.71934	0.719	0.00047	0.047%
Ex. 16	$(2.03)^6$	69.9804	69.76	0.00315	0.315%
Ex. 18	$\cos 91°$	-0.01745	-0.017	0.0259	2.59%

21. Relative error $= |(y - y_1)/y| = \left|\left(\sqrt{2}/2 - 0.6701\right)/\sqrt{2}/2\right| = 0.0523$, or 5.23%.

22. $f(9) \approx f(10) + f'(10)(9 - 10) = 6 + (-2)(-1) = 8$

23. $f(36) \approx f(39) + f'(39)(36 - 39) = 7 + 0.65(-3) = 5.05$

24.　(a) $V(r) = 3\pi r^2$

　(b) $V(4) = 3\pi(4)^2 = 48\pi$ L/min.

　(c) $\Delta V \approx dV = 6\pi r dr$. If $\Delta r = dr = 4.5 - 4 = 0.5$ and $r = 4$, $dV = 6\pi(4)(0.5) = 12\pi$. Thus, the volume increases by approximately 12π L.

25. Let $V(r)$ stand for the volume of the bearing of radius r and length $\ell = 5$. Then

$$V(r) = \pi r^2(5) = 5\pi r^2 \quad \Longrightarrow \quad \Delta V \approx dV = 10\pi r dr.$$

If $r = 2$ and $\Delta r = dr = 0.02$,

$$\Delta V \approx 10\pi(2)(0.02) = 0.4\pi \approx 1.26.$$

Thus, the volume varies by approximately 1.26 cm^3.

26. Let $A(s)$ be the area of the square. Then

$$A(s) = s^2 \quad \Longrightarrow \quad \Delta A \approx dA = 2sds.$$

If $s = 200$ and $\Delta s = ds = 2$, $\Delta A \approx 2(200)(2) = 800$. Thus, the area varies by approximately 800 ft^2.

27.　(a) $a(\theta) = -32\sin\theta \Longrightarrow a'(\theta) = -32\cos\theta \Longrightarrow a'(0) = -32 \Longrightarrow a(\theta) = a(0) + a'(0)(\theta - 0) = -32\theta$.

　(b) $a(\pi/6) \approx -32(\pi/6) \approx -16.76$.

(c) Let y = actual value = $a(\pi/6) = -16$ and y_1 = linear approximation = $a(0) + a'(0)(\pi/6 - 0) = -16.76$. Then the relative error in the linear approximation is equal to

$$\left| \frac{y - y_1}{y} \right| = \frac{0.76}{16} = 0.0475.$$

Remark: The relative error in approximating the acceleration $a(\pi/6)$, calculated above, should not be confused with the relative error in approximating the *change* in acceleration, as discussed in Example 4 of the text. The relative error in approximating the change in acceleration is equal to

$$\left| \frac{\Delta a - da}{\Delta a} \right| = \left| \frac{-16 - (-16.76)}{-16} \right| = 0.0475.$$

Although the two relative errors are the same in this case, they are conceptually different ideas and, in general, are not equal. See also Exercise 95 in the review exercises for Chapter 3.

28. (a) $s(t) = 10 + 5\cos t \implies s'(t) = -5\sin t$. Hence

$$s(t) \approx s(\pi/4) + s'(\pi/4)(t - \pi/4) = 10 + 5\sqrt{2}/2 - (5\sqrt{2}/2)(t - \pi/4).$$

(b) $s(\pi/2) \approx 10 + (5\sqrt{2}/2) - (5\sqrt{2}/2)(\pi/2 - \pi/4) \approx 10.7587.$

(c) Let y = actual value = $s(\pi/2) = 10$ and y_1 = linear approximation = $s(\pi/4) + s'(\pi/4)(\pi/2 - \pi/4)$ = 10.759. Then the relative error in the linear approximation is equal to

$$\left| \frac{y - y_1}{y} \right| = \frac{0.759}{10} = 0.0759.$$

Remark: The relative error in approximating $s(\pi/2)$ should not be confused with the relative error in approximating the *change* in s. See the remark following Exercise 27, this section.

29. $f(x) = \sqrt{x} \implies f'(x) = 1/(2\sqrt{x})$.

(a) If $x_0 = 49$, $\Delta x = 75 - 49 = 8$ and hence

$$\sqrt{57} = f(49 + 8) \approx f(49) + f'(49)(8) = \sqrt{49} + \frac{1}{2\sqrt{49}}(8) = 7.57.$$

(b) If $x_0 = 64$, $\Delta x = 75 - 64 = -7$ and hence

$$\sqrt{57} = f(64 + (-7)) \approx f(64) + f'(64)(-7) = \sqrt{64} + \frac{1}{2\sqrt{64}}(-4) = 7.56.$$

(c) The value in (a) is larger.

(d) The graph of $y = \sqrt{x}$ is shown below with the tangent lines at $x_0 = 49$ and $x_0 = 64$.

(e) $\sqrt{57}$ lies to the left (below) both of the approximations, as shown above on the number line on the right.

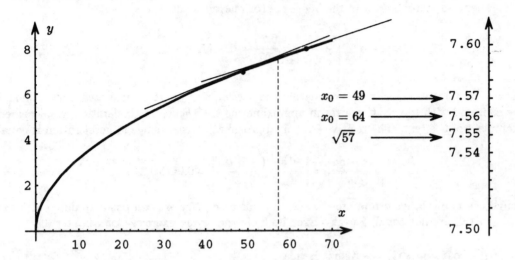

30. (a) $f'(x) = 3x^2/8 - 1 \Longleftrightarrow f'(-2) = 3(4)/8 - 1 = 1/2$. Therefore $t_1(x) = x/2 + 2$.

    (b) The graph with range $[-7, 3] \times [-2, 4]$ is shown below on the left.

    (c) Recall that $t_1(x)$ is the best linear approximation to $f(x)$ at the point $(-2, 1)$.

    (d) $h = 0.1$.

    (e) $f'(0) = -1 \Longleftrightarrow t_2(x) = -x$. The value of h for $t_2(x)$ is 0.425.

    (f) The value of h in part (e) is larger. Near $x = 0$ the graph of $f(x)$ is nearly a straight line.

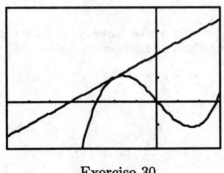

Exercise 30

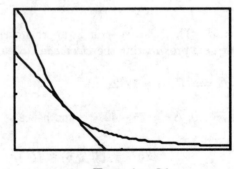

Exercise 31

31. (a) $f'(x) = -100x/(1 + 10x^2)^2 \Longleftrightarrow f'(1/2) = -200/49$. Therefore $t_1(x) = -200(x - 1/2)/49 + 10/7$.

    (b) The graph with range $[0, 2] \times [0, 5]$ is shown above on the right.

    (c) $t_1(x)$ is the best linear approximation to $f(x)$ at the point $(1/2, 10/7)$.

    (d) $h = 0.03$.

    (e) $f'(1) = -100/121 \Longleftrightarrow t_2(x) = -100(x - 1)/121 + 5/11$. The value of h for $t_2(x)$ is 0.09.

    (f) The function $f(x)$ is flatter near $x = 1$ than it is near $x = 1/2$.

3.8 Implicit Differentiation and Rational Power Functions

1. $f(x) = (x+2)^{4/3} \implies f'(x) = (4/3)(x+2)^{1/3}$

2. $f(x) = \sqrt[3]{x^2+1} = (x^2+1)^{1/3} \implies f'(x) = (1/3)(x^2+1)^{-2/3}(2x) = (2x/3)(x^2+1)^{-2/3}$

3. $f(x) = \sqrt{x} + \dfrac{1}{\sqrt{x}} = x^{1/2} + x^{-1/2} \implies f'(x) = (1/2)x^{-1/2} - (1/2)x^{-3/2} = \dfrac{x-1}{2x\sqrt{x}}$

4. $f(x) = \dfrac{\sqrt[3]{x+2}}{1+\sqrt{x}} = \dfrac{(x+2)^{1/3}}{1+x^{1/2}} \implies$

$$f'(x) = \frac{(1+x^{1/2})\left[(1/3)(x+2)^{-2/3}\right] - (x+2)^{1/3}(1/2)x^{-1/2}}{(1+x^{1/2})^2}$$

$$= \frac{(1/3)x^{1/2}(1+x^{1/2}) - (1/2)(x+2)}{x^{1/2}(1+x^{1/2})^2(x+2)^{2/3}} = \frac{(1/3)\sqrt{x} - (1/6)x - 1}{\sqrt{x}\,(1+\sqrt{x})^2\sqrt[3]{(x+2)^2}}$$

5. $f(x) = x^{2/3} + x^{-2/3} \implies f'(x) = (2/3)x^{-1/3} - (2/3)x^{-5/3}$

6. $f(x) = (3x^4 + 4x^3)^{-2} \implies f'(x) = -2(3x^4+4x^3)^{-3}(12x^3+12x^2) = -24x^2(x+1)(3x^4+4x^3)^{-3}$

7. $f(x) = (x^2-1)^{-2}(x^2+1)^2 \implies$

$$f'(x) = (x^2-1)^{-2}\left[2(x^2+1)(2x)\right] + (x^2+1)^2\left[-2(x^2-1)^{-3}(2x)\right]$$

$$= (x^2-1)^{-3}(x^2+1)\left[4x(x^2-1) - 4x(x^2+1)\right] = -8x(x^2-1)^{-3}(x^2+1)$$

8. $f(x) = (7-x^2)^{2/3} \implies f'(x) = (2/3)(7-x^2)^{-1/3}(-2x) = -(4x/3)(7-x^2)^{-1/3}$

9. $f(x) = \sqrt{1+\sqrt[3]{x}} = (1+x^{1/3})^{1/2} \implies$

$$f'(x) = (1/2)(1+x^{1/3})^{-1/2}\left[(1/3)x^{-2/3}\right] = (1/6)x^{-2/3}(1+x^{1/3})^{-1/2}$$

10. $f(x) = x^{5/2} + 3x^{2/3} + x^{-4/3} \implies f'(x) = (5/2)x^{3/2} + 2x^{-1/3} - (4/3)x^{-7/3}$

11. $f(x) = \dfrac{\sqrt[3]{x}}{\sqrt[3]{x}+x} = \dfrac{x^{1/3}}{x^{1/3}+x} \implies$

$$f'(x) = \frac{(x^{1/3}+x)(1/3)x^{-2/3} - x^{1/3}\left[(1/3)x^{-2/3}+1\right]}{(x^{1/3}+x)^2} = \frac{(-2/3)x^{1/3}}{(x^{1/3}+x)^2}$$

12. $f(x) = \sqrt{x^3}\,(x^{-2}+2x^{-1}+1)^4 = x^{3/2}(x^{-2}+2x^{-1}+1)^4 \implies$

$$f'(x) = x^{3/2}\left[4(x^{-2}+2x^{-1}+1)^3(-2x^{-3}-2x^{-2})\right] + (x^{-2}+2x^{-1}+1)^4\left[(3/2)x^{1/2}\right]$$

$$= (x^{-2}+2x^{-1}+1)^3x^{1/2}\left[4x(-2x^{-3}-2x^{-2}) + (3/2)(x^{-2}+2x^{-1}+1)\right]$$

$$= (1/2)x^{1/2}(x^{-2}+2x^{-1}+1)^3(-13x^{-2}-10x^{-1}+3)$$

$$= (1/2)x^{-15/2}(3x-13)(x+1)^7$$

13. $f(x) = \sqrt{x} + \sqrt[4]{x} + \sqrt[8]{x} = x^{1/2} + x^{1/4} + x^{1/8} \implies f'(x) = (1/2)x^{-1/2} + (1/4)x^{-3/4} + (1/8)x^{-7/8}$

14. $f(x) = \dfrac{x^{2/3}}{\sqrt[3]{x^2+1}} = x^{2/3}(x^2+1)^{-1/3} \Longrightarrow$

$$f'(x) = x^{2/3}\left[(-1/3)(x^2+1)^{-4/3}(2x)\right] + (x^2+1)^{-1/3}\left[(2/3)x^{-1/3}\right]$$

$$= (2/3)x^{-1/3}(x^2+1)^{-4/3}\left[-x^2+(x^2+1)\right] = (2/3)x^{-1/3}(x^2+1)^{-4/3}$$

15. $f(x) = \dfrac{(x^2+1)^{2/3}}{9-x^2)^{4/3}} = (1+x^2)^{2/3}(1-x^2)^{-4/3} \Longrightarrow$

$$f'(x) = (1+x^2)^{2/3}\left[(-4/3)(1-x^2)^{-7/3}(-2x)\right] + (1-x^2)^{-4/3}\left[(2/3)(1+x^2)^{-1/3}(2x)\right]$$

$$= (4/3)x(1+x^2)^{-1/3}(1-x^2)^{-7/3}\left[2(1+x^2)+(1-x^2)\right]$$

$$= (4/3)x(1+x^2)^{-1/3}(1-x^2)^{-7/3}(x^2+3)$$

16. $f(x) = \dfrac{(x-1)^{1/2}+(x+1)^{1/3}}{(x+2)^{1/4}} \Longrightarrow$

$$f'(x) = \dfrac{(x+2)^{1/4}\left[(1/2)(x-1)^{-1/2}+(1/3)(x+1)^{-2/3}\right] - \left[(x-1)^{1/2}+(x+1)^{1/3}\right](1/4)(x+2)^{-3/4}}{(x+2)^{1/2}}$$

17. $f(x) = \sqrt[3]{\sin x^2} = (\sin x^2)^{1/3} \Longrightarrow f'(x) = (1/3)(\sin x^2)^{-2/3}(\cos x^2)(2x)$

18. $f(x) = \sqrt{\dfrac{ax^2+b}{cx+d}} = \left(\dfrac{ax^2+b}{cx+d}\right)^{1/2} \Longrightarrow$

$$f'(x) = \dfrac{1}{2}\left(\dfrac{ax^2+b}{cx+d}\right)^{-1/2}\left[\dfrac{(cx+d)(2ax)-(ax^2+b)(c)}{(cx+d)^2}\right]$$

$$= \dfrac{1}{2}\left(\dfrac{ax^2+b}{cx+d}\right)^{-1/2}\dfrac{acx^2+2adx-bc}{(cx+d)^2}$$

19. $f(x) = \dfrac{x^{1/4}(x^{4/3}-x)}{(\sin^3 x - \sqrt{x})^{1/3}} = \dfrac{x^{19/12}-x^{5/4}}{(\sin^3 x - x^{1/2})^{1/3}} \Longrightarrow$

$$f'(x) = \dfrac{(\sin^3 x - x^{1/2})^{1/3}\left[(19/12)x^{7/12}-(5/4)x^{1/4}\right]}{(\sin^3 x - x^{1/2})^{2/3}}$$

$$-\dfrac{(x^{19/12}-x^{5/4})(1/3)(\sin^3 x - x^{1/2})^{-2/3}\left[3\sin^2 x \cos x - (1/2)x^{-1/2}\right]}{(\sin^3 x - x^{1/2})^{2/3}}$$

$$= \dfrac{(19/12)x^{7/12}-(5/4)x^{1/4}}{(\sin^3 x - x^{1/2})^{1/3}} - \dfrac{(x^{19/12}-x^{5/4})(1/3)\left[3\sin^2 x \cos x - (1/2)x^{-1/2}\right]}{(\sin^3 x - x^{1/2})^{4/3}}$$

20. $f(x) = \left[x^2 + x\tan(x^{3/2}-x^{1/2})\right]^{4/3} \Longrightarrow$

$$f'(x) = (4/3)\left[x^2 + x\tan(x^{3/2}-x^{1/2})\right]^{1/3} \cdot$$

$$\left\{2x + \tan(x^{3/2}-x^{1/2}) + x\left[\sec^2(x^{3/2}-x^{1/2})\right]\left[(3/2)x^{1/2}-(1/2)x^{-1/2}\right]\right\}$$

21. $f(x) = \sin^{5/2}(x^{2/3}) \implies$

$$f'(x) = (5/2)\sin^{3/2}(x^{2/3})\left[\cos(x^{2/3})\right](2/3)x^{-1/3} = (5/3)x^{-1/3}\sin^{3/2}(x^{2/3})\cos x^{2/3}$$

22. $f(x) = (x^{-2/3} - x^{3/2})^{3/5}(6 - x^{5/3})^{-2/5} \implies$

$$\begin{aligned} f'(x) = \ & (x^{-2/3} - x^{3/2})^{3/5}\left[(-2/5)(6 - x^{5/3})^{-7/5}(-5/3)x^{2/3}\right] \\ & + (6 - x^{5/3})^{-2/5}\left\{(3/5)(x^{-2/3} - x^{3/2})^{-2/5}\left[(-2/3)x^{-5/3} - (3/2)x^{1/2}\right]\right\} \end{aligned}$$

23. $f(x) = \sin\sqrt[3]{x} + \sqrt[3]{\sin x} = \sin(x^{1/3}) + (\sin x)^{1/3} \implies$

$$f'(x) = \left[\cos(x^{1/3})\right](1/3)x^{-2/3} + (1/3)(\sin x)^{-2/3}\cos x$$

24. $f(x) = \sqrt[3]{\sin^2 x} = (\sin x)^{2/3} \implies f'(x) = (2/3)(\sin x)^{-1/3}\cos x$

25. $f(x) = \sec\sqrt[4]{3x+1} = \sec(3x+1)^{1/4} \implies$

$$f'(x) = \left[\sec(3x+1)^{1/4}\tan(3x+1)^{1/4}\right](1/4)(3x+1)^{-3/4}(3)$$

$$= (3/4)(3x+1)^{-3/4}\sec(3x+1)^{1/4}\tan(3x+1)^{1/4}$$

26. $f(x) = \tan\sqrt[3]{x^2} = \tan x^{2/3} \implies f'(x) = (\sec^2 x^{2/3})(2/3)x^{-1/3}$

27. $x^2 + y^2 = 25 \implies 2x + 2y(dy/dx) = 0 \implies dy/dx = -x/y$

28. $x = \sin y \implies 1 = (\cos y)(dy/dx) \implies dy/dx = \sec y$

29. $x^{1/2} + y^{1/2} = 4 \implies (1/2)x^{-1/2} + (1/2)y^{-1/2}(dy/dx) = 0 \implies dy/dx = -\sqrt{y/x}$

30. $x^2 + 2xy + y^2 = 8 \implies 2x + (2x)(dy/dx) + (2)(y) + 2y(dy/dx) = 0 \implies (dy/dx)(2x + 2y) = -2x - 2y$
    $\implies dy/dx = -2(x+y)/2(x+y) = -1$. Note that $x + y \neq 0$; for if $x + y = 0$, then $x^2 + 2xy + y^2 = (x+y)^2 = 0$, not 8.

31. $x = \tan y \implies 1 = (\sec^2 y)(dy/dx) \implies dy/dx = \cos^2 y$

32. $x^2 y + xy^2 = 2 \implies$

$$x^2\frac{dy}{dx} + (2x)y + x(2y)\frac{dy}{dx} + y^2 = 0$$

$$\implies \frac{dy}{dx}(x^2 + 2xy) = -2xy - y^2$$

$$\implies \frac{dy}{dx} = -\frac{y(2x+y)}{x(x+2y)}$$

33. $x\sin y = y\cos x \implies$

$$x(\cos y)\frac{dy}{dx} + \sin y = y(-\sin x) + (\cos x)\frac{dy}{dx}$$

$$\implies \frac{dy}{dx}(x\cos y - \cos x) = -\sin y - y\sin x$$

$$\implies \frac{dy}{dx} = -\frac{y\sin x + \sin y}{x\cos y - \cos x}$$

161

34. $x = y(y-1) = y^2 - y \implies 1 = 2y(dy/dx) - dy/dx = (dy/dx)(2y-1) \implies dy/dx = 1/(2y-1)$

35. $(xy)^{1/2} = xy - x \implies$

$$(1/2)(xy)^{-1/2}\left[x\frac{dy}{dx} + y\right] = x\frac{dy}{dx} + y - 1$$

$$\implies \frac{dy}{dx}\left[(1/2)x(xy)^{-1/2} - x\right] = y - 1 - (1/2)y(xy)^{-1/2}$$

$$\implies \frac{dy}{dx} = \frac{y - 1 - (1/2)y(xy)^{-1/2}}{(1/2)x(xy)^{-1/2} - x} = \frac{2(y-1)\sqrt{xy} - y}{x(1 - 2\sqrt{xy})}$$

36. $y^2 = \sin^2 x - \cos^2 2x \implies$

$$2y\frac{dy}{dx} = 2\sin x \cos x - 2(\cos 2x)(-\sin 2x)(2)$$

$$\implies \frac{dy}{dx} = \frac{2\sin x \cos x + 4\sin 2x \cos 2x}{2y} = \frac{\sin 2x + 2\sin 4x}{2y}$$

37. $x^3 + x^2 y + xy^2 + y^3 = 15 \implies$

$$3x^2 + x^2\frac{dy}{dx} + 2xy + x\left(2y\frac{dy}{dx}\right) + y^2 + 3y^2\frac{dy}{dx} = 0$$

$$\implies (x^2 + 2xy + 3y^2)\frac{dy}{dx} = -3x^2 - 2xy - y^2$$

$$\implies \frac{dy}{dx} = -\frac{3x^2 + 2xy + y^2}{x^2 + 2xy + 3y^2}$$

38. $\cos(x+y) = y\sin x \implies$

$$-[\sin(x+y)]\left(1 + \frac{dy}{dx}\right) = y\cos x + (\sin x)\left(\frac{dy}{dx}\right)$$

$$\implies [-\sin(x+y) - \sin x]\left(\frac{dy}{dx}\right) = y\cos x + \sin(x+y)$$

$$\implies \frac{dy}{dx} = -\frac{y\cos x + \sin(x+y)}{\sin(x+y) + \sin x}$$

39. $\sqrt{x+y} = xy - x \implies$

$$(1/2)(x+y)^{-1/2}\left(1 + \frac{dy}{dx}\right) = x\frac{dy}{dx} + y - 1$$

$$\implies \frac{dy}{dx}\left[(1/2)(x+y)^{-1/2} - x\right] = y - 1 - (1/2)(x+y)^{-1/2}$$

$$\implies \frac{dy}{dx} = \frac{y - 1 - (1/2)(x+y)^{-1/2}}{(1/2)(x+y)^{-1/2} - x} = \frac{2(y-1)\sqrt{x+y} - 1}{1 - 2x\sqrt{x+y}}$$

40. $y^2 = \dfrac{x+1}{x^2+1} \implies$

$$2y\frac{dy}{dx} = \frac{(x^2+1)(1) - (x+1)(2x)}{(x^2+1)^2} = \frac{-x^2 - 2x + 1}{(x^2+1)^2}$$

$$\implies \frac{dy}{dx} = \frac{-x^2 - 2x + 1}{2y(x^2+1)^2}$$

41. $\cot y = 3x^2 + \cot(x+y) \Longrightarrow$

$$-(\csc^2 y)\frac{dy}{dx} = 6x - [\csc^2(x+y)]\left(1 + \frac{dy}{dx}\right)$$

$$\Longrightarrow \quad \frac{dy}{dx}\left[-\csc^2 y + \csc^2(x+y)\right] = 6x - \csc^2(x+y)$$

$$\Longrightarrow \quad \frac{dy}{dx} = \frac{6x - \csc^2(x+y)}{\csc^2(x+y) - \csc^2 y}$$

42. $x^2 y^2 = 4 \Longrightarrow x^2(2y)(dy/dx) + (2x)y^2 = 0 \Longrightarrow dy/dx = -2xy^2/2x^2 y = -y/x$

43. $\sin(xy) = 1/2 \Longrightarrow [\cos(xy)][x(dy/dx)+y] = 0 \Longrightarrow dy/dx = -y/x$. Note that $\cos(xy) \neq 0$; for if $\cos(xy) = 0$, then $\sin(xy) = \pm 1$, not $1/2$.

44. $\dfrac{1}{x} + \dfrac{1}{y} + \dfrac{1}{4} = 0 \Longrightarrow -x^{-2} - y^{-2}(dy/dx) = 0 \Longrightarrow dy/dx = -x^{-2}/y^{-2} = -y^2/x^2$

45. $y^4 = x^5 \Longrightarrow 4y^3(dy/dx) = 5x^4 \Longrightarrow dy/dx = 5x^4/4y^3 = 5x^5/4xy^3 = 5y^4/4xy^3 = 5y/4x$

46. $\sin(x+y) + \cos(x-y) = 1 \Longrightarrow$

$$[\cos(x+y)]\left(1 + \frac{dy}{dx}\right) - [\sin(x-y)]\left(1 - \frac{dy}{dx}\right) = 0$$

$$\Longrightarrow \quad \frac{dy}{dx}[\cos(x+y) + \sin(x-y)] = \sin(x-y) - \cos(x+y)$$

$$\Longrightarrow \quad \frac{dy}{dx} = \frac{\sin(x-y) - \cos(x+y)}{\sin(x-y) + \cos(x+y)}$$

47. $y^2 + 3 = x \sec y \Longrightarrow$

$$2y\frac{dy}{dx} = x(\sec y \tan y)\frac{dy}{dx} + \sec y$$

$$\Longrightarrow \quad \frac{dy}{dx}(2y - x\sec y \tan y) = \sec y$$

$$\Longrightarrow \quad \frac{dy}{dx} = \frac{\sec y}{2y - x\sec y \tan y}$$

48. $y^4 + 1 = \tan xy \Longrightarrow$

$$4y^3 \frac{dy}{dx} = (\sec^2 xy)\left(x\frac{dy}{dx} + y\right)$$

$$\Longrightarrow \quad \frac{dy}{dx}\left(4y^3 - x\sec^2 xy\right) = y\sec^2 xy$$

$$\Longrightarrow \quad \frac{dy}{dx} = \frac{y\sec^2 xy}{4y^3 - x\sec^2 xy}$$

49. $x = \sin 2xy \Longrightarrow 1 = (\cos 2xy)[2x(dy/dx) + 2y] \Longrightarrow dy/dx = (\sec 2xy - 2y)/2x$

50. $\cos^2 2y = y - x \Longrightarrow 2(\cos 2y)(-\sin 2y)(2dy/dx) = dy/dx - 1 \Longrightarrow dy/dx = 1/(1 + 4\sin 2y \cos 2y)$

163

51. $xy = 9 \Longrightarrow$

$$x\frac{dy}{dx} + y = 0 \quad \Longrightarrow \quad \frac{dy}{dx} = -\frac{y}{x} \quad \Longrightarrow \quad \frac{dy}{dx}\Big|_{(3,3)} = -1.$$

Hence the slope of the tangent line is -1 and the equation is $y - 3 = -(x - 3)$, or $y = -x + 6$.

52. $x^2 + y^2 = 4 \Longrightarrow$

$$2x + 2y\frac{dy}{dx} = 0 \quad \Longrightarrow \quad \frac{dy}{dx} = -\frac{x}{y} \quad \Longrightarrow \quad \frac{dy}{dx}\Big|_{(\sqrt{2},\sqrt{2})} = -1.$$

Hence the slope of the tangent line is -1 and the equation is $y - \sqrt{2} = -(x - \sqrt{2})$, or $y = -x + 2\sqrt{2}$.

53. $x^3 + y^3 = 16 \Longrightarrow$

$$3x^2 + 3y^2\frac{dy}{dx} = 0 \quad \Longrightarrow \quad \frac{dy}{dx} = -\frac{x^2}{y^2} \quad \Longrightarrow \quad \frac{dy}{dx}\Big|_{(2,2)} = -1.$$

Hence the slope of the tangent line is -1 and the equation is $y - 2 = -(x - 2)$, or $y = -x + 4$.

54. $x^2 y^2 = 16 \Longrightarrow$

$$x^2\left(2y\frac{dy}{dx}\right) + 2xy^2 = 0 \quad \Longrightarrow \quad \frac{dy}{dx} = -\frac{y}{x} \quad \Longrightarrow \quad \frac{dy}{dx}\Big|_{(-1,4)} = 4.$$

Hence the slope of the tangent line is 4 and the equation is $y - 4 = 4(x + 1)$, or $y = 4x + 8$.

55. $\dfrac{x+y}{x-y} = 4 \Longrightarrow x + y = 4(x - y) \Longrightarrow 3x = 5y \Longrightarrow dy/dx = 3/5$. Thus the slope of the tangent line is $3/5$ at all points. Hence the equation of the tangent line at $(5, 3)$ is $y - 3 = (3/5)(x - 5)$, or $y = 3x/5$.

56. $(y - x)^2 = x \Longrightarrow$

$$2(y - x)\left(\frac{dy}{dx} - 1\right) = 1 \quad \Longrightarrow \quad \frac{dy}{dx} = \frac{1}{2(y-x)} + 1 \quad \Longrightarrow \quad \frac{dy}{dx}\Big|_{(9,12)} = \frac{7}{6}.$$

Hence the slope of the tangent line is $7/6$ and the equation is $y - 12 = (7/6)(x - 9)$, or $y = (7/6)x + 3/2$.

57. $y^4 + 3x - x^2 \sin y = 3 \Longrightarrow$

$$4y^3\frac{dy}{dx} + 3 - x^2\cos y\frac{dy}{dx} - 2x\sin y = 0 \quad \Longrightarrow \quad \frac{dy}{dx} = \frac{2x\sin y - 3}{4y^3 - x^2\cos y} \quad \Longrightarrow \quad \frac{dy}{dx}\Big|_{(1,0)} = 3.$$

Hence the slope of the tangent line is 3 and the equation is $y = 3(x - 1)$.

58. $y = x^2 + \sin y \Longrightarrow$

$$\frac{dy}{dx} = 2x + \cos y\frac{dy}{dx} \quad \Longrightarrow \quad \frac{dy}{dx} = \frac{2x}{1 - \cos y} \quad \Longrightarrow \quad \frac{dy}{dx}\Big|_{(\sqrt{\pi},\pi)} = \sqrt{\pi}.$$

Hence the slope of the tangent line is $\sqrt{\pi}$ and the equation is $y - \pi = \sqrt{\pi}\,(x - \sqrt{\pi})$.

59. By definition, the equation of the tangent line at the point (x_0, y_0) has the form

$$y - y_0 = \left.\frac{dy}{dx}\right|_{(x_0,y_0)} (x - x_0).$$

It follows that if the tangent line is horizontal, then $y - y_0 = 0$ and hence $\left.\dfrac{dy}{dx}\right|_{(x_0,y_0)} = 0.$

60. In this case we may write the equation of the tangent line at the point (x_0, y_0) in the form

$$x - x_0 = \left.\frac{dx}{dy}\right|_{(x_0,y_0)} (y - y_0).$$

It follows that if the tangent line is vertical, then $x - x_0 = 0$ and hence $\left.\dfrac{dx}{dy}\right|_{(x_0,y_0)} = 0.$

61. $x^2 + y^2 = 2 \implies 2x + 2y(dy/dx) = 0 \implies dy/dx = -x/y.$ Hence

$$\frac{dy}{dx} = -\frac{x}{y} = 0 \implies x = 0 \implies y = \pm\sqrt{2}$$

$$\implies \quad \text{horizontal tangents: } y = \sqrt{2} \text{ at } (0, \sqrt{2}) \text{ and } y = -\sqrt{2} \text{ at } (0, -\sqrt{2});$$

$$\frac{dx}{dy} = -\frac{y}{x} = 0 \implies y = 0 \implies x = \pm\sqrt{2}$$

$$\implies \quad \text{vertical tangents: } x = \sqrt{2} \text{ at } (\sqrt{2}, 0) \text{ and } x = -\sqrt{2} \text{ at } (-\sqrt{2}, 0).$$

62. $x^2 + 4y^2 = 4 \implies 2x + 8y(dy/dx) = 0 \implies dy/dx = -x/4y.$ Hence

$$\frac{dy}{dx} = -\frac{x}{4y} = 0 \implies x = 0 \implies y = \pm 1$$

$$\implies \quad \text{horizontal tangents: } y = 1 \text{ at } (0, 1) \text{ and } y = -1 \text{ at } (0, -1);$$

$$\frac{dx}{dy} = -\frac{4y}{x} = 0 \implies y = 0 \implies x = \pm 2$$

$$\implies \quad \text{vertical tangents: } x = 2 \text{ at } (2, 0) \text{ and } x = -2 \text{ at } (-2, 0).$$

63. $x^2 - y^2 = 1 \implies 2x - 2y(dy/dx) = 0 \implies dy/dx = x/y.$ Now, if $dy/dx = 0$, then $x = 0$, in which case $-y^2 = 1$, which has no real solutions. Thus, there are no points where the tangent line is horizontal. If $dx/dy = y/x = 0$, then $y = 0$ and hence $x = \pm 1$. Thus, the graph has vertical tangents $x = \pm 1$ at $(\pm 1, 0)$.

64. $(x-1)^2 + (y+2)^2 = 9 \implies 2(x-1) + 2(y+2)(dy/dx) = 0 \implies dy/dx = -(x-1)/(y+2).$ Hence

$$\frac{dy}{dx} = -\frac{x-1}{y+2} = 0 \implies x = 1 \implies y = -2 \pm 3 = 1, -5$$

$$\implies \quad \text{horizontal tangents: } y = 1 \text{ at } (1, 1) \text{ and } y = -5 \text{ at } (1, -5);$$

$$\frac{dx}{dy} = -\frac{y+2}{x-1} = 0 \implies y = -2 \implies x = 1 \pm 3 = 4, -2$$

$$\implies \quad \text{vertical tangents: } x = 4 \text{ at } (4, -2) \text{ and } x = -2 \text{ at } (-2, -2).$$

65. $x^2 + y^2 + 2x + 4y = -4 \implies 2x + 2y(dy/dx) + 2 + 4(dy/dx) = 0 \implies dy/dx = -(2x+2)/(2y+4) = -(x+1)/(y+2)$. Hence

$$\frac{dy}{dx} = -\frac{x+1}{y+2} = 0 \implies x = -1 \implies y^2 + 4y + 3 = (y+3)(y+1) = 0 \implies y = -1, -3$$

$$\implies \text{horizontal tangents: } y = -1 \text{ at } (-1,-1) \text{ and } y = -3 \text{ at } (-1,-3)$$

and

$$\frac{dx}{dy} = -\frac{y+2}{x+1} = 0 \implies y = -2 \implies x^2 + 2x = x(x+2) = 0 \implies x = 0, -2$$

$$\implies \text{vertical tangents: } x = 0 \text{ at } (0,-2) \text{ and } x = -2 \text{ at } (-2,-2).$$

66. $(x+1)(y-1) = 1 \implies (x+1)(dy/dx) + (y-1) = 0 \implies dy/dx = -(y-1)/(x+1)$. Now, if $dy/dx = 0$, then $y = 1$, in which case $(x+1)(0) = 1$, which has no solutions. Thus, there are no points where the tangent line is horizontal. If $dx/dy = -(x+1)/(y-1) = 0$, then $x = -1$, in which case $(0)(y-1) = 1$, which has no solutions. Thus, the graph has no vertical or horizontal tangents.

67. $a = 1$. To obtain this value, we first find the intersection points of the two circles by solving their equations simultaneously. Subtracting the equations, we find that:

$$(x-a)^2 - (x+a)^2 = 0 \implies (x^2 - 2ax + a^2) - (x^2 + 2ax + a^2) = -4ax = 0 \implies x = 0.$$

Note that $a \neq 0$ since, if $a = 0$, there is only one circle. Now, if $x = 0$, then $a^2 + y^2 = 2 \implies y = \pm\sqrt{2 - a^2}$. Thus, the two circles intersect at the points $(0, \sqrt{2-a^2})$ and $(0, -\sqrt{2-a^2})$. Now,

$$(x-a)^2 + y^2 = 2 \implies 2(x-a) + 2y\frac{dy}{dx} = 0 \implies \frac{dy}{dx} = -\frac{x-a}{y} \implies \frac{dy}{dx}\bigg|_{(0,\sqrt{2-a^2})} = \frac{a}{\sqrt{2-a^2}}$$

and

$$(x+a)^2 + y^2 = 2 \implies 2(x+a) + 2y\frac{dy}{dx} = 0 \implies \frac{dy}{dx} = -\frac{x+a}{y} \implies \frac{dy}{dx}\bigg|_{(0,\sqrt{2-a^2})} = \frac{-a}{\sqrt{2-a^2}}.$$

Thus, the slopes of the tangent lines at $(0, \sqrt{2-a^2})$ are $a/\sqrt{2-a^2}$ and $-a/\sqrt{2-a^2}$. If the tangent lines are perpendicular, then

$$\frac{a}{\sqrt{2-a^2}} \cdot \frac{-a}{\sqrt{2-a^2}} = -1 \implies a^2 = 2 - a^2 \implies a = \pm 1.$$

At the point $(0, -\sqrt{2-a^2})$, the slopes of the tangent lines are the same numbers, except in reverse order:

$$\frac{a}{-\sqrt{2-a^2}} \quad \text{and} \quad \frac{a}{\sqrt{2-a^2}}.$$

Thus, if the tangent lines at the intersection points of the two circles are perpendicular, then $a = \pm 1$. But the two circles for $a = 1$ are the same as the two circles for $a = -1$. Hence $a = 1$.

68. $9x^2 + 16y^2 = 144 \implies 18x + 32y(dy/dx) = 0 \implies dy/dx = -9x/16y$. Hence the slope of the tangent line at $(2, 3\sqrt{3}/2)$ is

$$\frac{dy}{dx}\bigg|_{(2,3\sqrt{3}/2)} = \frac{-9(2)}{16(3\sqrt{3}/2)} = -\frac{\sqrt{3}}{4}.$$

Therefore the slope of the normal line is $4/\sqrt{3}$ and the equation is $y - 3\sqrt{3}/2 = (4/\sqrt{3})(x-2)$.

69. $x^2 + y^2 = a^2 \implies 2x + 2y(dy/dx) = 0 \implies dy/dx = -x/y$. Thus the slope of the tangent line at a typical point (x_0, y_0) on the curve is

$$\left.\frac{dy}{dx}\right|_{(x_0, y_0)} = -\frac{x_0}{y_0}.$$

Therefore the slope of the normal line is y_0/x_0 and the equation is $y - y_0 = (y_0/x_0)(x - x_0)$. Letting $x = 0$, we find that $y - y_0 = -y_0 \implies y = 0$. Hence the normal line passes through the origin $(0, 0)$.

70. $4 + x^2 + y^2 = 2xy \implies x^2 - 2xy + y^2 = -4 \implies (x - y)^2 = -4$. There are no real numbers that satisfy this equation since $a^2 \geq 0$ for all real a.

71. $5x^2 + 6xy + 5y^2 = 8 \implies$

$$10x + 6x\frac{dy}{dx} + 6y + 10y\frac{dy}{dx} = 0 \implies \frac{dy}{dx} = \frac{-10x - 6y}{6x + 10y} = \frac{-5x - 3y}{3x + 5y}.$$

Now, the slope of the line $x - y = 1$ is 1. Hence

$$\frac{dy}{dx} = \frac{-5x - 3y}{3x + 5y} = 1 \implies -5x - 3y = 3x + 5y \implies y = -x.$$

Thus, if (x, y) is a point on the curve at which the tangent line is parallel to $x - y = 1$, then $y = -x$. In this case,

$$5x^2 + 6xy + 5y^2 = 5x^2 + 6x(-x) + 5(-x)^2 = 4x^2 = 8 \implies x = \pm\sqrt{2} \implies y = \mp\sqrt{2}.$$

Hence there are two points on the curve at which the tangent line is parallel to $x - y = 1$, namely $(\sqrt{2}, -\sqrt{2})$ and $(-\sqrt{2}, \sqrt{2})$.

72. (a) $y^2 - x^2 = 1 \implies y = \pm\sqrt{1 + x^2}$. Thus, two functions defined implicitly by the equation $y^2 - x^2 = 1$ are

$$y = f_1(x) = \sqrt{1 + x^2} \quad \text{and} \quad y = f_2(x) = -\sqrt{1 + x^2}.$$

   (b) $f_1'(x) = (1/2)(1 + x^2)^{-1/2}(2x) = x/\sqrt{1 + x^2}$
   $f_2'(x) = -(1/2)(1 + x^2)^{-1/2}(2x) = -x/\sqrt{1 + x^2}$

   (c) $y^2 - x^2 = 1 \implies 2y(dy/dx) - 2x = 0 \implies dy/dx = x/y$

   (d) $y = f_1(x) = \sqrt{1 + x^2} \implies f_1'(x) = x/\sqrt{1 + x^2} = x/y = dy/dx$
   $y = f_2(x) = -\sqrt{1 + x^2} \implies f_2'(x) = -x/\sqrt{1 + x^2} = x/y = dy/dx$

   (e) $|dy/dx| = |x/y| = x/\sqrt{1 + x^2}$. If $x = 0$, $dy/dx = 0 < 1$. If $x \neq 0$, $|dy/dx| = 1/\sqrt{1 + (1/x)^2} < 1$. Hence $|dy/dx| < 1$ at all points on the hyperbola.

73. $xy^2 + x^2y = 4x^2 \implies x(y^2 + xy - 4x) = 0$. Thus, the graph of $xy^2 + x^2y = 4x^2$ consists of all points on the graph of $y^2 + xy = 4x$ together with those points at which $x = 0$, namely the y-axis. Now,

$$y^2 + xy - 4x = y^2 + x(y - 4) = 0 \implies x = \frac{y^2}{4 - y}.$$

Hence, as $y \to 0$, $x \to 0$. Moreover,

$$\frac{dx}{dy} = \frac{(4 - y)(2y) - y^2(-1)}{(4 - y)^2} = \frac{8y - y^2}{(4 - y)^2}.$$

Hence, as $y \to 0$, $dx/dy \to 0$. In other words, the tangent lines are approaching the vertical for x near zero.

74. (a) The graph of $x^2 + y^2 = 25$ is a circle of radius 5. Hence, by geometry, the tangent line at $(3, 4)$ is perpendicular to the radius through $(0, 0)$ and $(3, 4)$, whose slope is 4/3. Therefore the slope of the tangent line is $-3/4$.

(b) $y = \sqrt{25 - x^2} \Longrightarrow$

$$\frac{dy}{dx} = (1/2)(25 - x^2)^{-1/2}(-2x) = \frac{-x}{\sqrt{25 - x^2}} \quad \Longrightarrow \quad \frac{dy}{dx}\bigg|_{(3,4)} = -\frac{3}{4}.$$

(c) $x^2 + y^2 = 25 \Longrightarrow 2x + 2y(dy/dx) = 0 \Longrightarrow dy/dx = -x/y \Longrightarrow (dy/dx)|_{(3,4)} = -3/4$.

3.9   Higher Order Derivatives

1. $f(x) = 2x^3 - 6x \Longrightarrow f'(x) = 6x^2 - 6 \Longrightarrow f''(x) = 12x$

2. $f(x) = a - bx - cx^2 \Longrightarrow f'(x) = -b - 2cx \Longrightarrow f''(x) = -2c$

3. $f(x) = x^5 - 3x^{-2} \Longrightarrow f'(x) = 5x^4 + 6x^{-3} \Longrightarrow f''(x) = 20x^3 - 18x^{-4}$

4. $f(x) = \tan x \Longrightarrow f'(x) = \sec^2 x \Longrightarrow f''(x) = 2\sec x(\sec x \tan x) = 2\sec^2 x \tan x$

5. $f(x) = 1/(1 + x) \Longrightarrow (1 + x)^{-1} \Longrightarrow f'(x) = -(1 + x)^{-2} \Longrightarrow f''(x) = 2(1 + x)^{-3}$

6. $f(x) = \dfrac{ax + b}{cx + d} \Longrightarrow$

$$f'(x) = \frac{(cx + d)(a) - (ax + b)(c)}{(cx + d)^2} = \frac{ad - bc}{(cx + d)^2} = (ad - bc)(cx + d)^{-2}$$

$$\Longrightarrow \quad f''(x) = -2(ad - bc)(cx + d)^{-3}(c) = \frac{-2c(ad - bc)}{(cx + d)^3}$$

7. $f(x) = \sin^2 x \Longrightarrow f'(x) = 2\sin x \cos x \Longrightarrow f''(x) = (2\sin x)(-\sin x) + (2\cos x)\cos x = -2\sin^2 x + 2\cos^2 x$

8. $f(x) = \sin x \tan x \Longrightarrow$

$$\begin{aligned} f'(x) &= (\sin x)(\sec^2 x) + (\cos x)\tan x = \sin x \sec^2 x + \sin x \\ \Longrightarrow \quad f''(x) &= (\sin x)[2(\sec x)(\sec x \tan x)] + (\cos x)(\sec^2 x) + \cos x \\ &= 2\sin^2 x \sec^3 x + \sec x + \cos x \end{aligned}$$

9. $f(x) = 1/(1 + \cos x) = (1 + \cos x)^{-1} \Longrightarrow$

$$\begin{aligned} f'(x) &= -(1 + \cos x)^{-2}(-\sin x) = (\sin x)(1 + \cos x)^{-2} \\ \Longrightarrow \quad f''(x) &= (\sin x)\left[(-2)(1 + \cos x)^{-3}(-\sin x)\right] + (\cos x)(1 + \cos x)^{-2} \\ &= (1 + \cos x)^{-2}\left[2\sin^2 x(1 + \cos x)^{-1} + \cos x\right] \\ &= (1 + \cos x)^{-2}\left[2(1 - \cos^2 x)(1 + \cos x)^{-1} + \cos x\right] \\ &= (1 + \cos x)^{-2}(2 - \cos x) \end{aligned}$$

10. $f(x) = x\cos x \implies f'(x) = x(-\sin x) + \cos x = -x\sin x + \cos x \implies$

$$f''(x) = (-x)(\cos x) + (-1)(\sin x) - \sin x = -x\cos x - 2\sin x$$

11. $f(x) = \tan^2 x \implies f'(x) = 2(\tan x)\sec^2 x \implies$

$$f''(x) = (2\tan x)[2\sec x(\sec x \tan x)] + (2\sec^2 x)(\sec^2 x) = 2\sec^2 x(2\tan^2 x + \sec^2 x)$$

12. $f(x) = x^2\sin^2 x \implies f'(x) = x^2(2\sin x\cos x) + 2x\sin^2 x = x^2\sin 2x + 2x\sin^2 x \implies$

$$\begin{aligned} f''(x) &= x^2[(\cos 2x)(2)] + 2x\sin 2x + 2x(2\sin x\cos x) + 2\sin^2 x \\ &= 2x^2\cos 2x + 4x\sin 2x + 2\sin^2 x \end{aligned}$$

13. $f(x) = x - x\sec x \implies f'(x) = 1 - (x\sec x\tan x + \sec x) = 1 - (\sec x)(x\tan x + 1) \implies$

$$\begin{aligned} f''(x) &= -(\sec x)\left[x\sec^2 x + \tan x\right] - (\sec x\tan x)(x\tan x + 1) \\ &= -x\sec^3 x - 2\sec x\tan x - x\sec x\tan^2 x \end{aligned}$$

14. $f(x) = \dfrac{1 - \sin x}{\cos x} = \sec x - \tan x \implies f'(x) = \sec x\tan x - \sec^2 x \implies$

$$f''(x) = \sec x(\sec^2 x) + (\sec x\tan x)\tan x - 2(\sec x)(\sec x\tan x) = \sec x(\sec x - \tan x)^2$$

15. $f(x) = (x^3 + 1)^5 \implies f'(x) = 5(x^3 + 1)^4(3x^2) = 15x^2(x^3 + 1)^4 \implies$

$$f''(x) = (15x^2)\left[4(x^3 + 1)^3(3x^2)\right] + (30x)(x^3 + 1)^4 = 30x(x^3 + 1)^3(7x^3 + 1)$$

16. $f(x) = \sec^2 x + \tan^2 x = 1 + 2\tan^2 x \implies f'(x) = 4\tan x\sec^2 x \implies$

$$f''(x) = (4\tan x)[2\sec x(\sec x\tan x)] + (4\sec^2 x)\sec^2 x = 4\sec^2 x(2\tan^2 x + \sec^2 x)$$

17. $y = x^{-2} - 2x^{-4} \implies dy/dx = -2x^{-3} + 8x^{-5} \implies d^2y/dx^2 = 6x^{-4} - 40x^{-6}$

18. $y = x^2 + 1/x^2 = x^2 + x^{-2} \implies dy/dx = 2x - 2x^{-3} \implies d^2y/dx^2 = 2 + 6x^{-4}$

19. $dy/dx = \sec x\tan x \implies d^2y/dx^2 = (\sec x)(\sec^2 x) + (\sec x\tan x)\tan x = (\sec x)(\sec^2 x + \tan^2 x)$

20. $dy/dx = \dfrac{x}{x + \cot x} \implies$

$$\frac{d^2y}{dx^2} = \frac{(x + \cot x)(1) - x(1 - \csc^2 x)}{(x + \cot x)^2} = \frac{\cot x - x\csc^2 x}{(x + \cot x)^2}$$

21. $y = x^3\cos x \implies dy/dx = x^3(-\sin x) + (3x^2)\cos x = x^2(-x\sin x + 3\cos x) \implies$

$$\frac{d^2y}{dx^2} = x^2[(-x)\cos x + (-1)\sin x - 3\sin x] + (2x)(-x\sin x + 3\cos x) = -x^3\cos x - 6x^2\sin x + 6x\cos x$$

22. $dy/dx = x^4(1 - x^4) = x^4 - x^8 \implies d^2y/dx^2 = 4x^3 - 8x^7$

23. $y = 1/(1 - x^2) = (1 - x^2)^{-1} \implies dy/dx = -(1 - x^2)^{-2}(-2x) = 2x(1 - x^2)^{-2} \implies$

$$\frac{d^2y}{dx^2} = (2x)\left[-2(1 - x^2)^{-3}(-2x)\right] + (2)(1 - x^2)^{-2} = 2(1 - x^2)^{-3}(3x^2 + 1)$$

24. $y = x^m + x^n \implies dy/dx = mx^{m-1} + nx^{n-1} \implies d^2y/dx^2 = m(m-1)x^{m-2} + n(n-1)x^{n-2}$

25. $f(x) = \dfrac{x}{x+2} \implies f'(x) = \dfrac{(x+2)(1) - x(1)}{(x+2)^2} = 2(x+2)^{-2} \implies f''(x) = -4(x+2)^{-3}$

26. $f(x) = x^3 \sin x \implies f'(x) = x^3(\cos x) + (3x^2)\sin x = x^2(x\cos x + 3\sin x) \implies$

$$f''(x) = x^2 [x(-\sin x) + \cos x + 3\cos x] + (2x)(x\cos x + 3\sin x) = -x^3 \sin x + 6x^2 \cos x + 6x\sin x$$

27. $y = (x^2 - 1)^4 \implies dy/dx = 4(x^2-1)^3(2x) = 8x(x^2-1)^3 \implies$

$$\frac{d^2y}{dx^2} = (8x)\left[3(x^2-1)^2(2x)\right] + 8(x^2-1)^3 = 8(x^2-1)^2(7x^2-1)$$

$$\frac{d^3y}{dx^3} = 8(x^2-1)^2(14x) + \left[16(x^2-1)(2x)\right](7x^2-1) = 48(x^2-1)(7x^3-3x)$$

$$\frac{d^4y}{dx^4} = 48(x^2-1)(21x^2-3) + (96x)(7x^3-3x) = 48(35x^4 - 30x^2 + 3)$$

28. $y = x^3 - 3x^{-2} \implies dy/dx = 3x^2 + 6x^{-3} \implies d^2y/dx^2 = 6x - 18x^{-4} \implies d^3y/dx^4 = 6 + 72x^{-5}$

29. $y^2 = 4x \implies 2y(dy/dx) = 4 \implies dy/dx = 2y^{-1} \implies d^2y/dx^2 = -2y^{-2}(dy/dx) = -2y^{-2}(2y^{-1}) = -4y^{-3}$

30. $x^2 + y^2 = 1 \implies 2x + 2y(dy/dx) = 0 \implies dy/dx = -x/y \implies$

$$\frac{d^2y}{dx^2} = \frac{y(-1) - (-x)(dy/dx)}{y^2} = \frac{-y + x(-x/y)}{y^2} = -\frac{x^2+y^2}{y^3} = -\frac{1}{y^3}$$

31. $y^2 - xy = 4 \implies$

$$2y\frac{dy}{dx} - x\frac{dy}{dx} - y = 0 \implies \frac{dy}{dx} = \frac{y}{2y-x}$$

$$\implies \frac{d^2y}{dx^2} = \frac{(2y-x)(dy/dx) - y[2(dy/dx) - 1]}{(2y-x)^2} = \frac{-x[y/(2y-x)] + y}{(2y-x)^2} = \frac{2y(y-x)}{(2y-x)^3}$$

32. $\sqrt{x} + \sqrt{y} = 1 \implies x^{1/2} + y^{1/2} = 1 \implies$

$$(1/2)x^{-1/2} + (1/2)y^{-1/2}\frac{dy}{dx} = 0 \implies \frac{dy}{dx} = -\frac{x^{-1/2}}{y^{-1/2}} = -x^{-1/2}y^{1/2}$$

$$\implies \frac{d^2y}{dx^2} = (-x^{-1/2})\left[(1/2)y^{-1/2}\frac{dy}{dx}\right] + \left[(1/2)x^{-3/2}\right]y^{1/2}$$

$$= (-1/2)x^{-1/2}y^{-1/2}\left(-x^{-1/2}y^{1/2}\right) + (1/2)x^{-3/2}y^{1/2} = \frac{1}{2x} + \frac{\sqrt{y}}{2x\sqrt{x}} = \frac{\sqrt{x}+\sqrt{y}}{2x\sqrt{x}} = \frac{1}{2x\sqrt{x}}$$

33. Because the derivative of a polynomial is a polynomial. Note that if $p(x)$ has degree n, then $p^{(k)}(x) = 0$ for all $k > n$.

34. (a) $v(t) = s'(t) = 4t^3 - 16t = 4t(t^2 - 4)$

    (b) $v(t) = 0 \iff t = 0, -2, 2$. The table below summarizes the sign changes for $v(t)$.

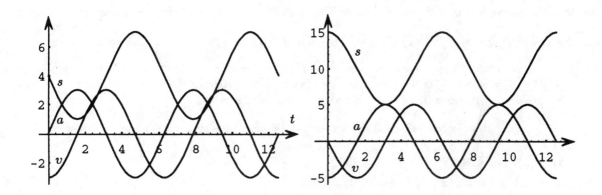

Exercise 36   Exercise 37

Sign $v(t)$	$-$	$+$	$-$	$+$
Interval	$(-\infty, -2)$	$(-2, 0)$	$(0, 2)$	$(2, +\infty)$

Velocity is positive when $-2 < t < 0$ or $t > 2$.

(c) $a(t) = v'(t) = 12t^2 - 16 = 4(3t^2 - 4)$

(d) $a(t) = 0 \Longleftrightarrow t = \pm\sqrt{4/3}$. The table below summarizes the sign changes for $a(t)$.

Sign $a(t)$	$+$	$-$	$+$
Interval	$(-\infty, -\sqrt{4/3}\,)$	$(-\sqrt{4/3}, \sqrt{4/3}\,)$	$(\sqrt{4/3}, +\infty)$

Acceleration is positive when $t < -\sqrt{4/3}$ or $t > \sqrt{4/3}$.

35. (a) $v(t) = s'(t) = 3t^2 - 12t - 30 = 3(t^2 - 4t - 10)$

(b) $v(t) = 0 \Longleftrightarrow t = (4 \pm \sqrt{56})/2 = 2 \pm \sqrt{14}$. The table below summarizes the sign changes for $v(t)$.

Sign $v(t)$	$+$	$-$	$+$
Interval	$(-\infty, 2 - \sqrt{14}\,)$	$(2 - \sqrt{14}, 2 + \sqrt{14}\,)$	$(2 + \sqrt{14}, +\infty)$

Velocity is negative when $2 - \sqrt{14} < t < 2 + \sqrt{14}$.

(c) $a(t) = v'(t) = 6t - 12 = 6(t - 2)$

(d) Acceleration is positive when $t > 2$.

36. $s(t) = 4 - 3\sin t \Longrightarrow v(t) = s'(t) = -3\cos t \Longrightarrow a(t) = v'(t) = 3\sin t$. The graphs, below on the left, show that when $v(t) > 0$, $s(t)$ is increasing, that is, the piston is travelling from the bottom of its stroke to the top, while when $a(t) > 0$, $v(t)$ is increasing, that is, the piston is travelling down to the bottom of its stroke then up to the middle position.

37. $s(t) = 10 + 5\cos t \Longrightarrow v(t) = s'(t) = -5\sin t \Longrightarrow a(t) = v'(t) = -5\cos t$. The graphs, below on the right, show that when $v(t) > 0$, $s(t)$ is increasing, while when $a(t) > 0$, $v(t)$ is increasing.

38. $f(x) = (x^2 + 1)^{50}$, when expanded, is a polynomial of degree 100. Therefore $f^{(200)}(x) = 0$.

171

39. For any integer $k \geq 0$, $f^{(4k)}(x) = -\sin x$. Therefore $f^{(161)}(x) = \left(f^{(160)}(x)\right)' = (-\sin x)' = -\cos x$.

40. For any integer $k \geq 0$, $f^{(4k)}(x) = \cos x$. Therefore $f^{(33)}(x) = \left(f^{(32)}(x)\right)' = (\cos x)' = -\sin x$.

41. $f(x) = x \cos x \implies f'(x) = -x \sin x + \cos x \implies f''(x) = -x \cos x - 2 \sin x \implies f'''(x) = x \sin x - 3 \sin s$ $\implies f^{(4)}(x) = x \cos x + 4 \sin x \implies f^{(8)}(x) = (x \cos x)^{(4)} + 4(\sin x)^{(4)} = x \cos x + 4 \sin x + 4 \sin x = x \cos x + 8 \sin x$. In general, $f^{(4k)}(x) = x \cos x + 4k \sin x$. Therefore

$$f^{(50)}(x) = (x \cos + 48 \sin x)'' = -x \cos x - 2 \sin x + 48(-\sin x) = -x \cos x - 50 \sin x.$$

42. (a) $f'(x) = 100(x+1)^{99} \implies f'(0) = 100$

(b) $f''(x) = 99 \cdot 100(x+1)^{98} \implies f''(0) = 9900$

(c) $a_1 = f'(0) = 100$

(d) $a_2 = f''(0)/2 = 4950$

(e) $f^{(50)}(x) = (100 \cdot 99 \cdots 51)x^{50} + \cdots + (50 \cdots 2 \cdot 1)a_{50} \implies f^{(50)}(0) = 50! \, a_{50}$. But $f(x) = (x+1)^{50}$ and hence $f^{(50)}(x) = (100 \cdot 99 \cdots 51)(x+1)^{50}$. Therefore $f^{(50)}(0) = 100 \cdot 99 \cdots 51$ and hence

$$50! a_{50} = 100 \cdot 99 \cdots 51 \implies a_{50} = \frac{100 \cdot 99 \cdots 51}{50!} = \frac{100!}{50! \, 50!}.$$

43. First recall that $d(|x|)/dx = |x|/x$. Hence, by the Chain Rule,

$$f'(x) = \frac{|x^2 - 1|}{x^2 - 1}(2x) = \begin{cases} 2x, & x < -1 \\ -2x, & -1 < x < 1 \\ 2x, & x > 1 \end{cases}$$

and

$$f''(x) = \frac{|x^2 - 1|}{x^2 - 1}(2) + (2x)\frac{(x^2 - 1)\left[|x^2 - 1|/(x^2 - 1)\right](2x) - |x^2 - 1|(2x)}{(x^2 - 1)^2}$$

$$= 2\frac{|x^2 - 1|}{x^2 - 1} = \begin{cases} 2, & x < -1 \\ -2, & -1 < x < 1 \\ 2, & x > 1. \end{cases}$$

The domain of $f(x)$ is all real numbers, while the domain of $f'(x)$ and $f''(x)$ is all $x \neq \pm 1$.

44. $y = (1+x)^{-1} \implies y' = (-1)^1(1+x)^{-2} \implies y'' = (-1)^2(2)(1+x)^{-3} \implies y''' = (-1)^3 1 \cdot 2 \cdot 3(1+x)^{-4}$. In general,

$$\frac{d^n y}{dx^n} = (-1)^n n! \, (1+x)^{-(n+1)}.$$

45. Let $f(x) = ax^2 + bx + c$. Then
$$f(2) = 4a + 2b + c = 2$$
$$f'(x) = 2ax + b \implies f'(2) = 4a + b = 4$$
$$f''(x) = 2a \implies f''(2) = 2a = 6.$$

Therefore $a = 3$, $b = -8$, $c = 6$. Hence $f(x) = 3x^2 - 8x + 6$.

46. If acceleration a is constant, then

$$a = \frac{v(5) - v(2)}{5 - 2} = \frac{15 - 6}{5 - 2} = 3\text{m/sec}.$$

47. The table below summarizes the first four derivatives of $\tan x$, $\cot x$, $\sec x$ and $\csc x$. Clearly, there is no apparent pattern to the derivatives.

$f(x)$	$f'(x)$	$f''(x)$	$f'''(x)$	$f^{(4)}(x)$
$\tan x$	$\sec^2 x$	$2\sec^2 x \tan x$	$2\sec^2 x(\sec^2 x + 2\tan^2 x)$	$8\sec^2 x \tan x(2\sec^2 x + \tan^2 x)$
$\cot x$	$-\csc^2 x$	$2\csc^2 x \cot x$	$-2\csc^2 x(\csc^2 x + 2\cot^2 x)$	$8\csc^2 x \cot x(\cot^2 x + 2\csc^2 x)$
$\sec x$	$\sec x \tan x$	$\sec x(\sec^2 x + \tan^2 x)$	$\sec x \tan x(5\sec^2 x + \tan^2 x)$	$\sec x(5\sec^4 x + 18\sec^2 x \tan^2 x + \tan^4 x)$
$\csc x$	$-\csc x \cot x$	$\csc x(\csc^2 x + \cot^2 x)$	$-\csc x \cot x(\cot^2 x + 5\csc^2 x)$	$\csc x(\cot^4 x + 18\cot^2 x \csc^2 x + 5\csc^4 x)$

48. (a) The graph of $q'(t)$ with range $[0, 10] \times [0, 60]$ is shown below on the left. The minimum of $q'(t)$ is 13.16 at $t = 2.26$. The year 1923.

   (b) The graph of $q''(t)$ with range $[0, 10] \times [-10, 60]$ is shown above on the right. $q''(t)$ is approximately zero at $t = 2.26$. $q'(2.26) = 13.16$.

   (c) The population growth rate is decreasing if $q'' < 0$ and the population growth rate is increasing if $q'' > 0$.

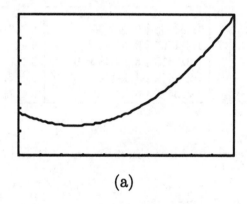

(a)

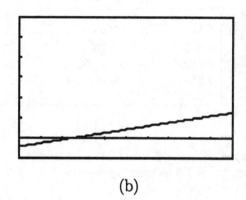

(b)

3.10   Newton's Method

1. $f(x) = x^2 - (7/2)x + 3/4 \implies f(0) = 3/4$ and $f(2) = -9/4$. Hence $f(x)$ has a zero between $a = 0$ and $b = 2$. Using $x_0 = 0.5$ as initial guess, set

$$x_{n+1} = x_n - \frac{f(x_n)}{f'(x_n)} = x_n - \frac{x_n^2 - (7/2)x_n + 3/4}{2x_n - 7/2}$$

for $n \geq 0$. The table below summarizes the successive iterations and shows that the root, to within six decimal places, is 0.229309.

n	x_n	$f(x_n)$	$f'(x_n)$
0	0.8	-1.41	-1.9
1	0.0578947	0.55072	-3.38421
2	0.220627	0.0264818	-3.05875
3	0.229285	0.0000749562	-3.04143
4	0.229309	6.07370×10^{-10}	-3.04138
5	0.229309		

n	x_n	$f(x_n)$	$f'(x_n)$
0	-1.5	0.25	2.
1	-1.625	-0.015625	2.25
2	-1.618056	-0.0000482253	2.23611
3	-1.618034	-4.65118×10^{-10}	2.23607
4	-1.618034		

Exercise 1   Exercise 2

2. $f(x) = 1 - x - x^2 \implies f(-2) = -1$ and $f(-1) = 1$. Hence $f(x)$ has a zero between $a = -2$ and $b = -1$. Using $x_0 = -1.5$ as initial guess, set

$$x_{n+1} = x_n - \frac{f(x_n)}{f'(x_n)} = x_n - \frac{1 - x_n - x_n^2}{-1 - 2x_n}$$

for $n \geq 0$. The table above summarizes the successive iterations and shows that the root, to within six decimal places, is -1.618034.

3. $f(x) = 1 - x - x^2 \implies f(0) = 1$ and $f(1) = -1$. Hence $f(x)$ has a zero between $a = 0$ and $b = 1$. Using $x_0 = 0.5$ as initial guess, set

$$x_{n+1} = x_n - \frac{f(x_n)}{f'(x_n)} = x_n - \frac{1 - x_n - x_n^2}{-1 - 2x_n}$$

for $n \geq 0$. The table below summarizes the successive iterations and shows that the root, to within six decimal places, is 0.618034.

n	x_n	$f(x_n)$	$f'(x_n)$
0	0.5	0.25	$-2.$
1	0.625	-0.015625	-2.25
2	0.6180556	-0.0000482253	-2.23611
3	0.618034	-4.65118×10^{-10}	-2.23607
4	0.618034		

n	x_n	$f(x_n)$	$f'(x_n)$
0	1.5	1.875	7.75
1	1.258065	0.249236	5.74818
2	1.214705	0.00701404	5.42653
3	1.213413	6.08598×10^{-6}	5.41711
4	1.213412	4.59468×10^{-12}	5.4171
5	1.213412	$0.$	5.4171

Exercise 3   Exercise 4

4. $f(x) = x^3 + x - 3 \implies f(1) = -1$ and $f(2) = 7$. Hence $f(x)$ has a zero between $a = 1$ and $b = 2$. Using $x_0 = 1.5$ as initial guess, set

$$x_{n+1} = x_n - \frac{f(x_n)}{f'(x_n)} = x_n - \frac{x_n^3 + x_n - 3}{3x_n^2 + 1}$$

for $n \geq 0$. The table above summarizes the successive iterations and shows that the root, to within six decimal places, is 1.213412.

5. $f(x) = \sqrt{x + 3} - x \implies f(1) = 1$ and $f(3) = \sqrt{6} - 3 < 0$. Hence $f(x)$ has a zero between $a = 1$ and $b = 3$. Using $x_0 = 1.5$ as initial guess, set

$$x_{n+1} = x_n - \frac{f(x_n)}{f'(x_n)} = x_n - \frac{\sqrt{x_n + 3} - x_n}{\left(1/2\sqrt{x_n + 3}\right) - 1}$$

for $n \geq 0$. The table below summarizes the successive iterations and shows that the root, to within six decimal places, is 2.302776.

n	x_n	$f(x_n)$	$f'(x_n)$
0	1.5	0.62132	-0.764298
1	2.31293	-0.00795038	-0.783078
2	2.302777	-1.05315×10^{-6}	-0.782871
3	2.302776	-1.85249×10^{-14}	-0.782871
4	2.302776		

n	x_n	$f(x_n)$	$f'(x_n)$
0	$-2.$	$-1.$	$8.$
1	-1.875	-0.0761719	6.79687
2	-1.863793	-0.000579467	6.69359
3	-1.863707	-3.44092×10^{-8}	6.69279
4	-1.863707		

Exercise 5   Exercise 6

6. $f(x) = x^3 + x^2 + 3 \implies f(-3) = -15$ and $f(-1) = 3$. Hence $f(x)$ has a zero between $a = -3$ and $b = -1$. Using $x_0 = -2$ as initial guess, set

$$x_{n+1} = x_n - \frac{f(x_n)}{f'(x_n)} = x_n - \frac{x_n^3 + x_n^2 + 3}{3x_n^2 + 2x_n}$$

for $n \geq 0$. The table above summarizes the successive iterations and shows that the root, to within six decimal places, is -1.863707.

7. $f(x) = x^4 - 5 \implies f(-2) = 11$ and $f(-1) = -4$. Hence $f(x)$ has a zero between $a = -2$ and $b = -1$. Using $x_0 = -1.5$ as initial guess, set

$$x_{n+1} = x_n - \frac{f(x_n)}{f'(x_n)} = x_n - \frac{x_n^4 - 5}{4x_n^3}$$

for $n \geq 0$. The table below summarizes the successive iterations and shows that the root, to within six decimal places, is -1.495349.

n	x_n	$f(x_n)$	$f'(x_n)$
0	-1.5	0.0625	-13.5
1	-1.49537	0.000288757	-13.3754
2	-1.495349	6.25312×10^{-9}	-13.3748
3	-1.495349		

n	x_n	$f(x_n)$	$f'(x_n)$
0	1.5	-0.0625	-13.5
1	1.49537	-0.000288757	-13.3754
2	1.495349	-6.25312×10^{-9}	-13.3748
3	1.495349		

Exercise 7   Exercise 8

8. $f(x) = 5 - x^4 \implies f(1) = 4$ and $f(2) = -11$. Hence $f(x)$ has a zero between $a = 1$ and $b = 2$. Using $x_0 = 1.5$ as initial guess, set

$$x_{n+1} = x_n - \frac{f(x_n)}{f'(x_n)} = x_n - \frac{5 - x_n^4}{-4x_n^3}$$

for $n \geq 0$. The table above summarizes the successive iterations and shows that the root, to within six decimal places, is 1.495349.

9. The two graphs intersect when $\sqrt{x + 3} = x$ or, equivalently, when $\sqrt{x + 3} - x = 0$. By Exercise 5, one of the solutions of this equation is 2.302776. Hence one of the intersection points is $(2.302776, 2.302776)$.

10. Let $f(x) = x^2 - 40$. Then $f(x)$ has a root between 6 and 7, namely $\sqrt{40}$. Using $x_0 = 6.5$ as initial guess, set

$$x_{n+1} = x_n - \frac{f(x_n)}{f'(x_n)} = x_n - \frac{x_n^2 - 40}{2x_n}$$

for $n \geq 0$. Then we find that

$$x_1 = 6.326923, \quad x_2 = 6.324556, \quad x_3 = 6.324555, \quad x_4 = 6.324555.$$

Hence, to six decimals, $\sqrt{40} \approx 6.324555$.

11. Let $f(x) = x^2 - 37$. Then $f(x)$ has a root between 6 and 7, namely $\sqrt{37}$. Using $x_0 = 6.5$ as initial guess, set

$$x_{n+1} = x_n - \frac{f(x_n)}{f'(x_n)} = x_n - \frac{x_n^2 - 37}{2x_n}$$

for $n \geq 0$. Then we find that

$$x_1 = 6.096154, \quad x_2 = 6.082777, \quad x_3 = 6.082763, \quad x_4 = 6.082763.$$

Hence, to six decimals, $\sqrt{37} \approx 6.082763$.

12. Let $f(x) = x^3 - 49$. Then $f(x)$ has a root between 3 and 4, namely $\sqrt[3]{49}$. Using $x_0 = 3.5$ as initial guess, set

$$x_{n+1} = x_n - \frac{f(x_n)}{f'(x_n)} = x_n - \frac{x_n^3 - 49}{3x_n^2}$$

for $n \geq 0$. Then we find that

$$x_1 = 3.666667, \quad x_2 = 3.65932, \quad x_3 = 3.659306, \quad x_4 = 3.659306.$$

Hence, to six decimals, $\sqrt[3]{49} \approx 3.659306$.

13. Let $f(x) = x^5 - 18$. Then $f(x)$ has a root between 1 and 2, namely $\sqrt[5]{18}$. Using $x_0 = 1.5$ as initial guess, set

$$x_{n+1} = x_n - \frac{f(x_n)}{f'(x_n)} = x_n - \frac{x_n^5 - 18}{5x_n^4}$$

for $n \geq 0$. Then we find that

$$x_1 = 1.911111, \quad x_2 = 1.798761, \quad x_3 = 1.78289, \quad x_4 = 1.782603, \quad x_5 = 1.782602, \quad x_6 = 1.782602.$$

Hence, to six decimals, $\sqrt[5]{18} \approx 1.782602$.

14. $x_n = x_1$ for all n since, if x_1 is a zero of $f(x)$, then $x_2 = x_1 - f(x_1)/f'(x_1) = x_1$.

15. The graph of $f(x) = 4x^3 - 3x^2 - 11x + 7$, shown below, indicates that the function has three zeros; one between -1 and -2, another between 0 and 1, and another between 1 and 2. Choose $x_0 = 1.5$, 0.5, and -1.5 as initial guesses and, for $n \geq 0$, set

$$x_{n+1} = x_n - \frac{f(x_n)}{f'(x_n)} = x_n - \frac{4x_n^3 - 3x_n^2 - 11x_n + 7}{12x_n^2 - 6x_n - 11}.$$

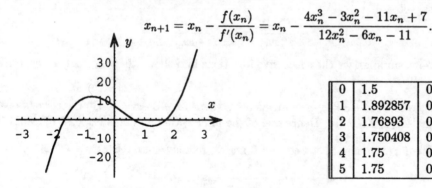

0	1.5	0.5	−1.5
1	1.892857	0.6136364	−1.63
2	1.76893	0.6180257	−1.618139
3	1.750408	0.618034	−1.618034
4	1.75	0.618034	−1.618034
5	1.75	0.618034	−1.618034

The table on the right summarizes the successive iterations using the three initial guesses and shows that, to six decimal places, the roots are 1.75, 0.618034, and −1.618034.

16. The graph of $f(x) = 3x^3 - 2x^2 - 9x + 4$, shown below, indicates that the function has three zeros; one between -1 and -2, another between 0 and 1, and another between 1 and 2. Choose $x_0 = 1.5$, 0.5, and -1.5 as initial guesses and, for $n \geq 0$, set

$$x_{n+1} = x_n - \frac{f(x_n)}{f'(x_n)} = x_n - \frac{3x^3 - 2x^2 - 9x + 4}{9x^2 - 4x - 9}.$$

The table on the right summarizes the successive iterations using the three initial guesses and shows that, to five decimal places, the roots are −1.64676, 0.429859, and 1.88357.

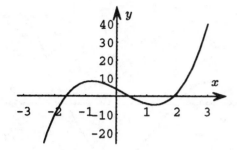

0	−1.5	0.5	1.5
1	−1.66667	0.428571	2.2381
2	−1.64706	0.429858	1.9627
3	−1.64676	0.429859	1.88899
4	−1.64676	0.429859	1.8836
5	−1.64676	0.429859	1.88357
6	−1.64676	0.429859	1.88357

17. The graph of $f(x) = x^4 + 4x^3 + 3x^2 - 2x - 1$, shown below, indicates that the function has four zeros; one between −3 and −2, another between −2 and −1, another between −1 and 0, and one between 0 and 1. Choose $x_0 = -2.5$, −1.5, −0.5, and 0.5 as initial guesses and, for $n \geq 0$, set

$$x_{n+1} = x_n - \frac{f(x_n)}{f'(x_n)} = x_n - \frac{x^4 + 4x^3 + 3x^2 - 2x - 1}{4x^3 + 12x^2 + 6x - 2}.$$

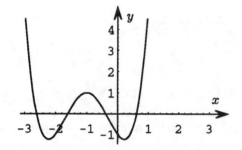

0	−2.5	−1.5	−0.5	0.5
1	−2.65278	−1.625	−0.375	0.652778
2	−2.61999	−1.61805	−0.381954	0.619985
3	−2.61804	−1.61803	−0.381966	0.618041
4	−2.61803	−1.61803	−0.381966	0.618034
5	−2.61803	−1.61803	−0.381966	0.618034

The table on the right summarizes the successive iterations using the three initial guess and shows that, to five decimal places, the roots are −2.61803, −1.61803, −0.38197, and 0.61803.

18. The graph of $f(x) = 6x^4 - 2x^3 - 12x^2 + 6x + 1$, shown below, indicates that the function has four zeros; one between −2 and −1, another between −1 and 0, another between 0 and 1, and another between 1 and 2. Choose $x_0 = -1.5$, −0.5, 0.5, and 1.5 as initial guesses and, for $n \geq 0$, set

$$x_{n+1} = x_n - \frac{f(x_n)}{f'(x_n)} = x_n - \frac{6x^4 - 2x^3 - 12x^2 + 6x + 1}{24x^3 - 6x^2 - 24x + 6}.$$

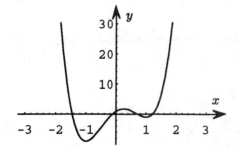

0	−1.5	−0.5	0.5	1.5
1	−1.45952	−0.175926	0.75	1.32333
2	−1.45682	−0.134504	0.712798	1.23595
3	−1.45681	−0.132596	0.713599	1.21114
4	−1.45681	−0.132592	0.713599	1.20915
5	−1.45681	−0.132592	0.713599	1.20914
6	−1.45681	−0.132592	0.713599	1.20914

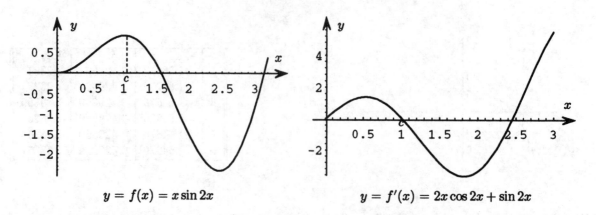

$$y = f(x) = x \sin 2x \qquad\qquad y = f'(x) = 2x \cos 2x + \sin 2x$$

The table on the right summarizes the successive iterations using the three initial guesses and shows that, to five decimal places, the roots are -1.45681, -0.13259, 0.71360, and 1.20914.

19. $f(x) = x \sin 2x \implies f'(x) = 2x \cos 2x + \sin 2x$. The graph of $y = f'(x)$ is shown at the top of the next page on the right. The smallest positive number c such that $f'(c) = 0$ is approximately 1. Using $x_0 = 1$ and setting

$$x_{n+1} = x_n - \frac{f'(x_n)}{f''(x_n)} = x_n - \frac{2x_n + \sin 2x_n}{-4x \sin 2x + 4 \cos 2x},$$

for $n \geq 0$, we find that

$$x_0 = 1, \quad x_1 = 1.01452, \quad x_2 = 1.01438, \quad x_3 = 1.01438.$$

Hence $c \approx 1.01438$. Geometrically, the graph of $y = f(x) = x \sin 2x$ has a horizontal tangent at c.

20. Let $f(x) = x^2 - 2$. Using Newton's Method, with $x_1 = 1$ and

$$x_{n+1} = x_n - \frac{f(x_n)}{f'(x_n)} = x_n - \frac{x_n^2 - 2}{2x_n},$$

for $n \geq 0$, we find that

$$
\begin{aligned}
x_1 &= 1 & &= 1.0000000 \\
x_2 &= 3/2 & &= 1.5000000 \\
x_3 &= 17/12 & &= 1.\underline{41}66666666667 \\
x_4 &= 577/408 & &= 1.\underline{41421}56862745 \\
x_5 &= 665857/470832 & &= 1.\underline{4142135623}747.
\end{aligned}
$$

Now, to thirteen decimal places, $\sqrt{2} = 1.4142135623731$. Thus, x_3 is twice as accurate as x_2, x_4 twice as accurate as x_3, and x_5 more than twice as accurate as x_4.

21. Let $f(x) = x^4$. Using $x_1 = 0.01$ and setting

$$x_{n+1} = x_n - \frac{f(x_n)}{f'(x_n)} = x_n - \frac{x_n^4}{4x_n^3} = 0.75x_n,$$

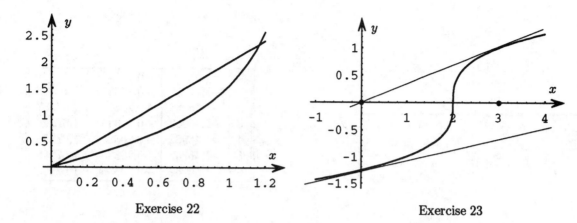

Exercise 22   Exercise 23

we find that

$$
\begin{aligned}
x_1 &= 0.01 && \text{1 decimal place accuracy}\\
x_2 &= 0.0075 && \text{1 decimal place accuracy}\\
x_3 &= 0.005625 && \text{1 decimal place accuracy}\\
x_4 &= 0.00421875 && \text{2 decimal place accuracy}\\
x_5 &= 0.00316406 && \text{2 decimal place accuracy}\\
x_6 &= 0.00237305 && \text{2 decimal place accuracy}\\
x_7 &= 0.00177979 && \text{2 decimal place accuracy}\\
x_8 &= 0.00133484 && \text{2 decimal place accuracy}\\
x_9 &= 0.00100113 && \text{2 decimal place accuracy}\\
x_{10} &= 0.000750847 && \text{2 decimal place accuracy.}
\end{aligned}
$$

If convergence had been quadratic, one would expect x_{10} to be accurate to 2^9, or 512 decimal places.

22. Let $h(x) = 2x - \tan x$. The graph on the following page shows that $h(x)$ has a zero between 1 and 1.2. Using $x_0 = 1.1$ and setting

$$
x_{n+1} = x_n - \frac{f(x_n)}{f'(x_n)} = x_n - \frac{2x - \tan x}{2 - \sec^2 x},
$$

we find that

$$
x_0 = 1.1, \quad x_1 = 1.18224, \quad x_2 = 1.16648, \quad x_3 = 1.16556, \quad x_4 = 1.16556.
$$

Thus, the root of $h(x)$ between 0 and $\pi/2$ is 1.16556, to five decimal places, and hence the graphs of $f(x) = 2x$ and $g(x) = \tan x$ intersect, between 0 and $\pi/2$, at 1.16556.

23. Here $x_1 = 3$ and

$$
x_{n+1} = x_n - \frac{(x-2)^{1/3}}{(1/3)(x-2)^{-2/3}} = x_n - 3(x_n - 2) = 6 - 2x_n.
$$

Hence

$$
x_1 = 3, \quad x_2 = 0, \quad x_3 = 6, \quad x_4 = -6, \quad x_5 = 18, \quad x_6 = -30, \quad x_7 = 66.
$$

The iterates continue to diverge as n increases. The graph of $y = f(x)$ on the following page shows that the tangent lines are becoming flatter as x increases, thus forcing the x-intercept to increase without bound rather than converge.

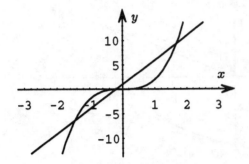

0	−1.5	−0.5	1.5
1	−1.47059	−0.142857	1.70588
2	−1.46962	−0.20263	1.67387
3	−1.46962	−0.203364	1.67298
4	−1.46962	−0.203364	1.67298
5	−1.46962	−0.203364	1.67298
6	−1.46962	−0.203364	1.67298

24. The graphs of $y = f(x)$ and $y = g(x)$ are shown below. Clearly, there are intersection points between −1 and −2, 0 and −1, and 1 and 2. To find them, let $h(x) = f(x) - g(x) = 2x^3 - 5x - 1$. Using −1.5, −0.5 and 1.5 as initial guesses, and setting

$$x_{n+1} = x_n - \frac{f(x_n)}{f'(x_n)} = x_n - \frac{2x^3 - 5x - 1}{6x^2 - 5},$$

the table below on the right summarizes the successive iterations and shows that the roots, to within five decimal places, are −1.46962, −0.203364 and 1.67298.

25. (a) $x_1 = 0$, $x_2 = 1.258229874$, $x_3 = 2.768911636$, $x_4 = 4.235147453$.

(b) The roots are near the asymptotes of $t = \tan 2x$. These asymptotes are $x = \pi/4 + k\pi/2$, $k = 0, 1, 2, \ldots$.

Review Exercises – Chapter 3

1. $f(x) = \dfrac{x}{3x - 7} \implies f'(x) = \dfrac{(3x - 7)(1) - x(3)}{(3x - 7)^2} = \dfrac{-7}{(3x - 7)^2}$

2. $y = \dfrac{\sqrt{x}}{1 + \sqrt{x}} = \dfrac{x^{1/2}}{1 + x^{1/2}} \implies \dfrac{(1 + x^{1/2})(1/2)x^{-1/2} - x^{1/2}(1/2)x^{-1/2}}{(1 + x^{1/2})^2} = \dfrac{1}{2\sqrt{x}(1 + \sqrt{x})^2}$

3. $y = \dfrac{6x^3 - x^2 + x}{3x^5 + x^3} \implies$

$$y' = \frac{(3x^5 + x^3)(18x^2 - 2x + 1) - (6x^3 - x^2 + x)(15x^4 + 3x^2)}{(3x^5 + x^3)^2} = \frac{-36x^4 + 9x^3 - 12x^2 + x - 2}{x^3(3x^2 + 1)^2}$$

4. $f(x) = \dfrac{x^3 - 1}{x^3 + 1} \implies \dfrac{(x^3 + 1)(3x^2) - (x^3 - 1)(3x^2)}{(x^3 + 1)^2} = \dfrac{6x^2}{(x^3 + 1)^2}$

5. $f(t) = \sqrt{t} \sin t \implies f'(t) = \sqrt{t} \cos t + \dfrac{\sin t}{2\sqrt{t}}$

6. $y = (6x - 4)^{-3} \implies y' = -3(6x - 4)^{-4}(6) = -18(6x - 4)^{-4}$

7. $y = \dfrac{1}{x + x^2 + x^3} = (x + x^2 + x^3)^{-1} \implies y' = -(x + x^2 + x^3)^{-2}(1 + 2x + 3x^2)$

180

8. $f(s) = \left(s + \dfrac{1}{s}\right)^5 = (s + s^{-1})^5 \Longrightarrow f'(s) = 5(s + s^{-1})^4(1 - s^{-2})$

9. $g(t) = (t^{-2} - t^{-3})^{-1} \Longrightarrow g'(t) = -(t^{-2} - t^{-3})^{-2}(-2t^{-3} + 3t^{-4})$

10. $y = \left(\dfrac{4x^3 + 3}{x + 2}\right)^4 \Longrightarrow 4\left(\dfrac{4x^3 + 3}{x + 2}\right)^3 \left[\dfrac{(x + 2)(12x^2) - (4x^3 + 3)(1)}{(x + 2)^2}\right] = 4\left(\dfrac{4x^3 + 3}{x + 2}\right)^3 \dfrac{8x^3 + 24x^2 - 3}{(x + 2)^2}$

11. $y = x^2 \sin x^2 \Longrightarrow y' = x^2[(\cos x^2)(2x)] + 2x \sin x^2 = 2x(x^2 \cos x^2 + \sin x^2)$

12. $f(x) = \tan^2 x^2 \Longrightarrow f'(x) = 2\tan x^2[(\sec^2 x^2)(2x)] = 4x \tan x^2 \sec^2 x^2$

13. $y = 1/(x^4 + 4x^2)^3 = (x^4 + 4x^2)^{-3} \Longrightarrow y' = -3(x^4 + 4x^2)^{-4}(4x^3 + 8x) = \dfrac{-12(x^2 + 2)}{x^7(x^2 + 4)^4}$

14. $f(t) = \tan t \csc t = \sec t \Longrightarrow f'(t) = \sec t \tan t$

15. $f(x) = \dfrac{2x - 7}{\sqrt{x^2 + 1}} = \dfrac{(x^2 + 1)^{1/2}(2) - (2x - 7)\left[(1/2)(x^2 + 1)^{-1/2}(2x)\right]}{x^2 + 1} = \dfrac{7x + 2}{(x^2 + 1)^{3/2}}$

16. $g(t) = 4t^2\sqrt{t} + t^3\sqrt{t} = 4t^{5/2} + t^{7/2} \Longrightarrow g'(t) = 10t^{3/2} + (7/2)t^{5/2}$

17. $y = \left[(x^2 + 1)^3 - 7\right]^5 \Longrightarrow y' = 5\left[(x^2 + 1)^3 - 7\right]^4 \left[3(x^2 + 1)^2(2x)\right] = 30x(x^2 + 1)^2 \left[(x^2 + 1)^3 - 7\right]^4$

18. $y = \left[\sin(1 + x^2) + x\right]^3 \Longrightarrow y' = 3\left[\sin(1 + x^2) + x\right]^2 \left[\left[\cos(1 + x^2)\right](2x) + 1\right]$

19. $f(x) = \sqrt{\dfrac{3x - 9}{x^2 + 3}} \Longrightarrow$

$$f'(x) = \dfrac{1}{2}\left(\dfrac{3x - 9}{x^2 + 3}\right)^{-1/2} \dfrac{(x^2 + 3)(3) - (3x - 9)(2x)}{(x^2 + 3)^2} = \dfrac{1}{2}\left(\dfrac{3x - 9}{x^2 + 3}\right)^{-1/2} \dfrac{-3x^2 + 18x + 9}{(x^2 + 3)^2}$$

20. $y = \csc(\cot 3x) \Longrightarrow y' = -\csc(\cot 3x)\cot(\cot 3x)\left[-\csc^2 3x\right](3) = 3\csc(\cot 3x)\cot(\cot 3x)\csc^2 3x$

21. $f(s) = \tan^2 s \cot^5 s = \cot^3 s \Longrightarrow f'(s) = 3\cot^2 s(-\csc^2 s) = -3\cot^2 s \csc^2 s$

22. $y = \sqrt[3]{\sin x} = (\sin x)^{1/3} \Longrightarrow y' = (1/3)(\sin x)^{-2/3}\cos x$

23. $y = \dfrac{\sqrt{1 + \sin x}}{\cos x} = \dfrac{(1 + \sin x)^{1/2}}{\cos x} \Longrightarrow$

$$y' = \dfrac{(\cos x)\left[(1/2)(1 + \sin x)^{-1/2}(\cos x)\right] - (1 + \sin x)^{1/2}(-\sin x)}{\cos^2 x} = \dfrac{(1/2)\cos^2 x + \sin x + \sin^2 x}{(\cos^2 x)\sqrt{1 + \sin x}}$$

24. $f(x) = \dfrac{x + \sqrt{x}}{\sqrt{1 + x^3}} = \dfrac{x + x^{1/2}}{(1 + x^3)^{1/2}} \Longrightarrow$

$$f'(x) = \dfrac{(1 + x^3)^{1/2}\left[1 + (1/2)x^{-1/2}\right] - (x + x^{1/2})\left[(1/2)(1 + x^3)^{-1/2}(3x^2)\right]}{1 + x^3}$$

$$= \dfrac{(1 + x^3)(\sqrt{x} + 1/2) - (3/2)x^2(x\sqrt{x} + x)}{\sqrt{x}(1 + x^3)^{3/2}} = \dfrac{1 + 2\sqrt{x} - 2x^3 - x^3\sqrt{x}}{2\sqrt{x}(1 + x^3)^{3/2}}$$

25. $f(t) = t^{9/2} - 6t^{5/2} \implies f'(t) = (9/2)t^{7/2} - 15t^{3/2}$

26. $y = \dfrac{\sqrt[3]{x}}{x + \sqrt[3]{x}} = \dfrac{x^{1/3}}{x + x^{1/3}} \implies$

$$y' = \frac{(x + x^{1/3})\left[(1/3)x^{-2/3}\right] - x^{1/3}\left[1 + (1/3)x^{-2/3}\right]}{(x + x^{1/3})^2} = \frac{-2\sqrt[3]{x}}{3(x + \sqrt[3]{x})^2}$$

27. $h(t) = \cos^2(t^2 - 1) \implies h'(t) = 2\left[\cos(t^2 - 1)\right]\left[-\sin(t^2 - 1)\right](2t) = -4t\cos(t^2 - 1)\sin(t^2 - 1)$

28. $f(s) = \pi^2 \implies f'(s) = 0$

29. $y = \sqrt{x^3}\,(x^{-3} - 2x^{-1} + 2)^2 = x^{3/2}(x^{-3} - 2x^{-1} + 2)^2 \implies$

$$y' = x^{3/2}\left[2(x^{-3} - 2x^{-1} + 2)(-3x^{-4} + 2x^{-2})\right] + (3/2)x^{1/2}(x^{-3} - 2x^{-1} + 2)^2$$

$$= x^{1/2}(x^{-3} - 2x^{-1} + 2)\left[(-9/2)x^{-3} + x^{-1} + 3\right]$$

$$= (1/2)x^{-11/2}(12x^6 - 8x^5 - 4x^4 - 12x^3 + 20x^2 - 9)$$

30. $f(x) = x|x| \implies f'(x) = x(|x|/x) + |x| = 2|x|$

31. $f(x) = x\sin x \implies$

$$f'(x) = x\cos x + \sin x$$

$$f''(x) = x(-\sin x) + \cos x + \cos x = -x\sin x + 2\cos x$$

32. $f(x) = x/(x + 1) \implies$

$$f'(x) = \frac{(x+1)(1) - x(1)}{(x+1)^2} = \frac{1}{(x+1)^2}$$

$$f''(x) = -2(x+1)^{-3} = \frac{-2}{(x+1)^3}$$

33. $f(x) = \sec x \tan x \implies$

$$f'(x) = (\sec x)(\sec^2 x) + (\tan x)(\sec x \tan x) = (\sec x)(\sec^2 x + \tan^2 x)$$

$$f''(x) = (\sec x)\left[2(\sec x)(\sec x \tan x) + 2(\tan x)(\sec^2 x)\right] + (\sec x \tan x)(\sec^2 x + \tan^2 x)$$

$$= \sec x \tan x(5\sec^2 x + \tan^2 x)$$

34. $f(x) = 1/\sqrt{1 + x^2} = (1 + x^2)^{-1/2} \implies$

$$f'(x) = (-1/2)(1 + x^2)^{-1/2}(2x) = -x(1 + x^2)^{-3/2}$$

$$f''(x) = (-x)\left[(-3/2)(1 + x^2)^{-5/2}(2x)\right] + (1 + x^2)^{-3/2}(-1) = (2x^2 - 1)(1 + x^2)^{-5/2}$$

35. $s(t) = t^2 - 1 + t^{1/2} \implies v(t) = 2t + (1/2)t^{-1/2} \implies a(t) = 2 - (1/4)t^{-3/2}.$

36. $s(t) = \sqrt{t}\,\sin t \implies$

$$v(t) = t^{1/2}\cos t + (\sin t)(1/2)t^{-1/2}$$

$$a(t) = t^{1/2}(-\sin t) + (\cos t)(1/2)t^{-1/2} + (\sin t)(-1/4)t^{-3/2} + (1/2)t^{-1/2}(\cos t)$$

$$= -t^{1/2}\sin t + t^{-1/2}\cos t - (1/4)t^{-3/2}\sin t$$

37. $s(t) = \sqrt{2 + t^2} = (2 + t^2)^{1/2} \Longrightarrow$

$$v(t) = (1/2)(2 + t^2)^{-1/2}(2t) = t(2 + t^2)^{-1/2}$$

$$a(t) = t\left[(-1/2)(2 + t^2)^{-3/2}(2t)\right] + (2 + t^2)^{-1/2} = 2(2 + t^2)^{-3/2}$$

38. $s(t) = \dfrac{\sin t}{1 + \sqrt{t}} \Longrightarrow$

$$v(t) = \frac{(1 + \sqrt{t})\cos t - (\sin t)(1/2)t^{-1/2}}{(1 + \sqrt{t})^2} = \frac{\cos t}{1 + \sqrt{t}} - \frac{\sin t}{2\sqrt{t}(1 + \sqrt{t})^2}$$

$$a(t) = \frac{(1 + t^{1/2})(-\sin t) - (\cos t)(1/2)t^{-1/2}}{(1 + t^{1/2})^2}$$

$$-\frac{\left[2\sqrt{t}(1 + \sqrt{t})^2\right](\cos t) - (\sin t)\left[(2\sqrt{t})\,2(1 + t^{1/2})(1/2)t^{-1/2} + t^{-1/2}(1 + \sqrt{t})^2\right]}{4t(1 + t^{1/2})^4}$$

$$= -\frac{\sin t}{1 + \sqrt{t}} - \frac{\cos t}{\sqrt{t}(1 + \sqrt{t})^2} + \frac{\sin t}{2t(1 + t^{1/2})^3} + \frac{\sin t}{4t\sqrt{t}(1 + t^{1/2})^2}$$

39. $y = 6x - 2 \Longrightarrow$

$$\frac{dy}{dx} = \lim_{h \to 0} \frac{f(x + h) - f(x)}{h} = \lim_{h \to 0} \frac{[6(x + h) - 2] - [6x - 2]}{h} = \lim_{h \to 0} 6 = 6$$

40. $y = 1/(x + 1) \Longrightarrow$

$$\frac{dy}{dx} = \lim_{h \to 0} \frac{f(x + h) - f(x)}{h} = \lim_{h \to 0} \frac{1/(x + h + 1) - 1/(x + 1)}{h}$$

$$= \lim_{h \to 0} \frac{(x + 1) - (x + h + 1)}{h(x + 1)(x + h + 1)} = \lim_{h \to 0} \frac{-1}{(x + 1)(x + h + 1)} = \frac{-1}{(x + 1)^2}$$

41. $y = \sqrt{x + 1} \Longrightarrow$

$$\frac{dy}{dx} = \lim_{h \to 0} \frac{f(x + h) - f(x)}{h} = \lim_{h \to 0} \frac{\sqrt{x + h + 1} - \sqrt{x + 1}}{h}$$

$$= \lim_{h \to 0} \frac{\sqrt{x + h + 1} - \sqrt{x + 1}}{h} \cdot \frac{\sqrt{x + h + 1} + \sqrt{x + 1}}{\sqrt{x + h + 1} + \sqrt{x + 1}}$$

$$= \lim_{h \to 0} \frac{1}{\sqrt{x + h + 1} + \sqrt{x + 1}} = \frac{1}{2\sqrt{x + 1}}$$

42. $y = 1/\sqrt{2x + 1} \Longrightarrow$

$$\frac{dy}{dx} = \lim_{h \to 0} \frac{f(x + h) - f(x)}{h} = \lim_{h \to 0} \frac{1/\sqrt{2x + 2h + 1} - 1/\sqrt{2x + 1}}{h}$$

$$= \lim_{h \to 0} \frac{\sqrt{2x + 1} - \sqrt{2x + 2h + 1}}{h\sqrt{2x + 1}\sqrt{2x + 2h + 1}} \cdot \frac{\sqrt{2x + 1} + \sqrt{2x + 2h + 1}}{\sqrt{2x + 1} + \sqrt{2x + 2h + 1}}$$

$$= \lim_{h \to 0} \frac{-2}{\sqrt{2x + 1}\sqrt{2x + 2h + 1}\left(\sqrt{2x + 1} + \sqrt{2x + 2h + 1}\right)} = \frac{-1}{(2x + 1)\sqrt{2x + 1}}$$

43. $y = x^2 + x + 1 \implies$

$$\frac{dy}{dx} = \lim_{h \to 0} \frac{f(x+h) - f(x)}{h} = \lim_{h \to 0} \frac{[(x+h)^2 + (x+h) + 1] - [x^2 + x + 1]}{h}$$

$$= \lim_{h \to 0} \frac{2xh + h^2 + h}{h} = \lim_{h \to 0} (2x + h + 1) = 2x + 1$$

44. $y = x^3 \implies$

$$\frac{dy}{dx} = \lim_{h \to 0} \frac{f(x+h) - f(x)}{h} = \lim_{h \to 0} \frac{(x+h)^3 - x^3}{h}$$

$$= \lim_{h \to 0} \frac{3x^2 h + 3xh^2 + h^3}{h} = \lim_{h \to 0} (3x^2 + 3xh + h^2) = 3x^2$$

45. $6x^2 - xy - 4y^2 = 0 \implies$

$$12x - x\frac{dy}{dx} - y - 8y\frac{dy}{dx} = 0$$

$$\implies \frac{dy}{dx}(-x - 8y) = y - 12x$$

$$\implies \frac{dy}{dx} = \frac{12x - y}{x + 8y}$$

46. $6x^3 - 2y^2 = x \implies 18x^2 - 4y(dy/dx) = 1 \implies dy/dx = (18x^2 - 1)/4y$

47. $x = y + y^2 + y^3 \implies$

$$1 = \frac{dy}{dx} + 2y\frac{dy}{dx} + 3y^2\frac{dy}{dx} = \frac{dy}{dx}(1 + 2y + 3y^2) \implies \frac{dy}{dx} = \frac{1}{1 + 2y + 3y^2}$$

48. $\sin \sqrt{x} + \sin \sqrt{y} = 1 \implies$

$$(\cos \sqrt{x})(1/2)x^{-1/2} + (\cos \sqrt{y})(1/2)y^{-1/2}\frac{dy}{dx} = 0 \implies \frac{dy}{dx} = -\frac{(\cos \sqrt{x})x^{-1/2}}{(\cos \sqrt{y})y^{-1/2}} = -\frac{\sqrt{y}\,\cos \sqrt{x}}{\sqrt{x}\,\cos \sqrt{y}}$$

49. $xy = \cot xy \implies$

$$x\frac{dy}{dx} + y = (-\csc^2 xy)\left(x\frac{dy}{dx} + y\right) = -x(\csc^2 xy)\frac{dy}{dx} - y\csc^2 xy$$

$$\implies \frac{dy}{dx}(x + x\csc^2 xy) = -y - y\csc^2 xy$$

$$\implies \frac{dy}{dx} = \frac{-y - \csc^2 xy}{x + x\csc^2 xy} = \frac{-y(1 + \csc^2 xy)}{x(1 + \csc^2 xy)} = -\frac{y}{x}$$

50. $(2x^2 y^3)^{1/3} = 1 \implies 2x^2 y^3 = 1 \implies 2x^2 \left[(3y^2)dy/dx\right] + 4xy^3 = 0 \implies dy/dx = -4xy^3/6x^2 y^2 = -2y/3x$

51. $x \sin y = y \implies x \cos(dy/dx) + \sin y = dy/dx \implies dy/dx = -(\sin y)/(x \cos y - 1)$

52. $\sqrt{x} + \sqrt{y} = K \implies (1/2)x^{-1/2} + (1/2)y^{-1/2}(dy/dx) = 0 \implies dy/dx = -x^{-1/2}/y^{-1/2} = -\sqrt{y/x}$

53. $x^2 + y^2 = 4 \implies$

$$2x + 2y\frac{dy}{dx} = 0 \quad \implies \quad \frac{dy}{dx} = -\frac{x}{y}$$

$$\implies \quad \frac{d^2y}{dx^2} = -\frac{y - x(dy/dx)}{y^2} = -\frac{y - x(-x/y)}{y^2} = -\frac{y^2 + x^2}{y^3} = -\frac{4}{y^3}$$

54. $\sin xy = 1/2 \implies$

$$(\cos xy)\left(x\frac{dy}{dx} + y\right) = 0 \quad \implies \quad \frac{dy}{dx} = -\frac{y}{x}$$

$$\implies \quad \frac{d^2y}{dx^2} = -\frac{x(dy/dx) - y}{x^2} = -\frac{x(-y/x) - y}{x^2} = \frac{2y}{x^2}$$

55. $\dfrac{1}{x} + \dfrac{1}{y} = 4 \implies$

$$-\frac{1}{x^2} - \frac{1}{y^2}\frac{dy}{dx} = 0 \quad \implies \quad \frac{dy}{dx} = -\frac{y^2}{x^2} \quad \implies \quad \left.\frac{dy}{dx}\right|_{(1/2,1/2)} = -\frac{(1/2)^2}{(1/2)^2} = -1.$$

Therefore the slope of the tangent line at $(1/2, 1/2)$ is -1 and the equation is $y - 1/2 = -(x - 1/2)$, or $y = -x + 1$.

56. $y = \tan x - x \implies$

$$\frac{dy}{dx} = \sec^2 x - 1 \quad \implies \quad \left.\frac{dy}{dx}\right|_{(\pi/4, 1-\pi/4)} = \sec^2(\pi/4) - 1 = 1.$$

Therefore the slope of the tangent line at $(\pi/4, 1 - \pi/4)$ is 1 and the equation is $y - (1 - \pi/4) = 1(x - \pi/4)$, or $y = x + 1 - \pi/2$.

57. $f(x) = \dfrac{\cos x}{\sin x} = \cot x \implies f'(x) = -\csc^2 x \implies f'(\pi/4) = -\csc^2(\pi/4) = -2$. Therefore the slope of the tangent line at $(\pi/4, 1)$ is -2 and the equation is $y - 1 = -2(x - \pi/4)$, or $y = -2x + 1 + \pi/2$.

58. $\sqrt{x} + \sqrt{y} = 1 \implies$

$$\frac{dy}{dx} = -\sqrt{\frac{y}{x}} \quad \text{(see Exercise 52)} \quad \implies \quad \left.\frac{dy}{dx}\right|_{(1/4,1/4)} = -\sqrt{\frac{1/4}{1/4}} = -1.$$

Therefore the slope of the tangent line at $(1/4, 1/4)$ is -1 and the equation is $y - 1/4 = -1(x - 1/4)$, or $y = -x + 1/2$.

59. $f(x) = \sec x \tan x \implies$

$$f'(x) = (\sec x)\sec^2 x + (\tan x)(\sec x \tan x) \quad \implies \quad f'(\pi/4) = \sec^3(\pi/4) + \sec(\pi/4)\tan^2(\pi/4) = 3\sqrt{2}.$$

Therefore the slope of the tangent line is $3\sqrt{2}$ and the equation is $y - \sqrt{2} = 3\sqrt{2}(x - \pi/4)$.

60. $y = ax + b \implies dy = (dy/dx)dx = a\,dx$

61. $y = x\sin x \implies dy = (dy/dx)dx = (x\cos x + \sin x)dx$

62. $y = \dfrac{x+1}{x+2} \implies dy = (dy/dx)dx = \dfrac{(x+2)(1) - (x+1)(1)}{(x+2)^2}\,dx = \dfrac{1}{(x+2)^2}\,dx$

63. $y = \sqrt{1 + x^2} \implies dy = (dy/dx)dx = (1/2)(1 + x^2)^{-1/2}(2x)dx = x(1 + x^2)^{-1/2}dx$

64. $y = \dfrac{\sqrt{x + 1}}{\sqrt{x + 2}} = \left(\dfrac{x + 1}{x + 2}\right)^{1/2} \implies$

$$dy = \frac{dy}{dx}dx = \frac{1}{2}\left(\frac{x + 1}{x + 2}\right)^{-1/2}\frac{(x + 2)(1) - (x + 1)(1)}{(x + 2)^2}dx = \frac{1}{2}\left(\frac{x + 1}{x + 2}\right)^{-1/2}\frac{1}{(x + 2)^2}dx$$

65. $y = \cos\sqrt{x} \implies dy = (dy/dx)dx = (-\sin\sqrt{x})(1/2)x^{-1/2}dx$

66. Let $f(x) = \sqrt{x}$. Then $f'(x) = 1/(2\sqrt{x})$ and hence

$$f(x) \approx f(x_0) + f'(x_0)\Delta x = \sqrt{x_0} + \frac{1}{2\sqrt{x_0}}\Delta x.$$

Setting $x_0 = 64$ and $x = 65$, $\Delta x = 1$ and hence

$$\sqrt{65} \approx \sqrt{64} + \frac{1}{2\sqrt{64}}(1) = 8.0625.$$

67. Let $f(x) = x^{2/3}$. Then $f'(x) = (2/3)x^{-1/3}$ and hence

$$f(x) \approx f(x_0) + f'(x_0)\Delta x = x_0^{2/3} + (2/3)x_0^{-1/3}\Delta x.$$

Setiing $x_0 = 8$ and $x = 8.2$, $\Delta x = 0.2$ and hence

$$(8.2)^{2/3} \approx 8^{2/3} + (2/3)8^{-1/3}(0.2) = 4.0667.$$

68. Let $f(x) = \cos x$. Then $f'(x) = -\sin x$ and hence

$$f(x) \approx f(x_0) + f'(x_0)\Delta x = \cos x_0 - (\sin x_0)\Delta x.$$

Setting $x_0 = \pi/4$ and $x = 44° = 44(\pi/180) = 0.7679$ radian, $\Delta x = -1° = -\pi/180 = -0.0175$ radian and hence

$$\cos 44° = \cos(0.7679) \approx \cos(\pi/4) - [\sin(\pi/4)](-0.0175) = 0.7194.$$

69. Let $f(x) = x^3$. Then $f'(x) = 3x^2$. and hence

$$f(x) \approx f(x_0) + f'(x_0)\Delta x = x_0^3 + 3x_0^2\Delta x.$$

Setting $x_0 = 1$ and $x = 0.96$, $\Delta x = -0.04$ and hence

$$(0.96)^3 \approx 1^3 + 3(1)^2(-0.04) = 0.88.$$

70. No;

$$\lim_{h \to 0^+}\frac{f(2 + h) - f(2)}{h} = \lim_{h \to 0^+}\frac{|h|}{h} = \lim_{h \to 0^+}\frac{h}{h} = 1$$

but

$$\lim_{h \to 0^-}\frac{f(2 + h) - f(2)}{h} = \lim_{h \to 0^-}\frac{|h|}{h} = \lim_{h \to 0^-}\frac{-h}{h} = -1.$$

71. No;

$$\lim_{h \to 0^+} \frac{f(2+h) - f(2)}{h} = \lim_{h \to 0^+} \frac{\left|(2+h)^2 - 6(2+h) + 8\right|}{h} = \lim_{h \to 0^+} \frac{-h(h-2)}{h} = 2$$

but

$$\lim_{h \to 0^-} \frac{f(2+h) - f(2)}{h} = \lim_{h \to 0^-} \frac{\left|(2+h)^2 - 6(2+h) + 8\right|}{h} = \lim_{h \to 0^-} \frac{h(h-2)}{h} = -2.$$

72. No;

$$\lim_{h \to 0^+} \frac{f(2+h) - f(2)}{h} = \lim_{h \to 0^+} \frac{[6(2+h) - 12] - 0}{h} = \lim_{h \to 0^+} \frac{6h}{h} = 6$$

but

$$\lim_{h \to 0^-} \frac{f(2+h) - f(2)}{h} = \lim_{h \to 0^-} \frac{(2+h)^3 - 8}{h} = \lim_{h \to 0^-} \frac{h(12 + 6h + h^2)}{h} = 12.$$

73. No;

$$\lim_{h \to 0^+} \frac{f(2+h) - f(2)}{h} = \lim_{h \to 0^+} \frac{2(2+h) - 4}{h} = 2$$

but

$$\lim_{h \to 0^-} \frac{f(2+h) - f(2)}{h} = \lim_{h \to 0^-} \frac{4 - (2+h)^2}{h} = \lim_{h \to 0^-} (-4 - h) = -4.$$

74. If the position function $s(t)$ is a polynomial of degree 2, say $s(t) = at^2 + bt + c$, where $a \neq 0$, then $v(t) = 2at + b$ is a polynomial of degree one and hence has only one zero, namely $-b/2a$. Since v changes sign at $t = -b/2a$, the particle must change its direction of travel.

75. a $\leftrightarrow$ v, b $\leftrightarrow$ viii, c $\leftrightarrow$ i

76. Let $A(t)$ and $r(t)$ be the area and radius of the circle at time t. Then

$$A = \pi r^2 \quad \Longrightarrow \quad \frac{dA}{dt} = 2\pi r \frac{dr}{dt}.$$

From the given information, $dr/dt = 2$. Moreover, when $t = 5$, $r = 10$. Hence

$$\frac{dA}{dt}\bigg|_{t=5} = 2\pi(10)(2) = 40\pi.$$

Thus, after 5 seconds the area is increasing at the rate of 40π m^2/s.

77. Let $A(t)$ be the area of a hexagon whose side is $s(t)$ meters at time t. In the figure below, each central angle is 60° and hence each triangle is equilateral. Hence

$$A = 6 \text{ Area(triangle)}$$

$$= 6 \left[\left(\frac{1}{2} s \right) \left(\frac{\sqrt{3}}{2} s \right) \right]$$

$$= \frac{3s^2 \sqrt{3}}{2}$$

$$\Longrightarrow \quad \frac{dA}{dt} = 3s\sqrt{3} \frac{ds}{dt}.$$

From the given information, $ds/dt = 2$. Hence

$$\left.\frac{dA}{dt}\right|_{s=10} = 3(10)\sqrt{3}\,(2) = 60\sqrt{3}.$$

Thus, when the side is 10 cm, the area is increasing at the rate of $60\sqrt{3}$ cm^2/s.

78. Let $V(t)$ and $r(t)$ be the volume and radius of the sphere at time t. Then

$$V = \frac{4}{3}\pi r^3 \quad\Longrightarrow\quad \frac{dV}{dt} = 4\pi r^2 \frac{dr}{dt}.$$

From the given information, $dV/dt = 500$ when $r = 25$. Hence

$$500 = 4\pi(25)^2\frac{dr}{dt} \quad\Longrightarrow\quad \frac{dr}{dt} = \frac{500}{4\pi(25)^2} \approx 0.064.$$

Thus, when the radius is 25 cm it is increasing at the rate of 0.064 cm/s.

79. Let $x(t)$ stand for the horizontal displacement of the base of the ladder from the wall and $y(t)$ the vertical displacement of the top, as illustrated in the figure below. Then

$$A = \frac{1}{2}xy = \frac{1}{2}x\sqrt{100 - x^2}$$

$$\Longrightarrow\quad \frac{dA}{dt} = \left[\left(\frac{x}{2}\right)\left(\frac{1}{2}\right)(100 - x^2)^{-1/2}(-2x) + \frac{1}{2}\sqrt{100 - x^2}\right]\frac{dx}{dt}$$

$$= \frac{50 - x^2}{\sqrt{100 - x^2}}\frac{dx}{dt}.$$

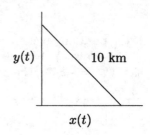

From the given information, $dx/dt = 1$. Hence, when $x = 6$,

$$\left.\frac{dA}{dt}\right|_{x=6} = \frac{50 - 6^2}{\sqrt{100 - 6^2}}(1) = 1.75.$$

Thus, when the base of the ladder is 6 feet from the wall, the area of the triangle is increasing at a rate of 1.75 m^2/s.

80. Let $r(t)$ and $h(t)$ be the radius and thickness of the disk at time t. Then

$$V = \pi r(t)^2 h(t) \quad\Longrightarrow\quad h(t) = (V/\pi)r(t)^{-2} \quad\Longrightarrow\quad \frac{dh}{dt} = (V/\pi)(-2)r(t)^{-3}\frac{dr}{dt}.$$

From the given information, $dr/dt = a/\sqrt{t}$. Hence, if $r(9) = 6a$,

$$\left.\frac{dh}{dt}\right|_{t=9} = (V/\pi)(-2)(6a)^{-3}\left(a/\sqrt{9}\right) = -\frac{V}{324\pi a^2}.$$

Thus, after 9 seconds, the thickness is decreasing at the rate of $V/(324\pi a^2)$ m^2/s.

81. Let $V(t)$ and $r(t)$ be the volume and radius of the snowball at time t. From the given information, $dV/dt = -k(4\pi r^2)$ for some constant $k > 0$, where $4\pi r^2$ is the surface area of the snowball. Since

$$V = \frac{4}{3}\pi r^3 \quad\Longrightarrow\quad \frac{dV}{dt} = 4\pi r^2\frac{dr}{dt} = -k(4\pi r^2) \quad\Longrightarrow\quad \frac{dr}{dt} = -k,$$

it follows that the radius of the snowball decreases at a constant rate.

82. Let $f(x) = x^{200}$. Then

$$f'(2) = \lim_{h \to 0} \frac{f(2+h) - f(2)}{h} = \lim_{h \to 0} \frac{(2+h)^{200} - 2^{200}}{h}.$$

But $f'(x) = 200x^{199} \Longrightarrow f'(2) = 200(2)^{199}$. Hence

$$\lim_{h \to 0} \frac{(2+h)^{200} - 2^{200}}{h} = 200(2)^{199}.$$

83. $y = \sin x^2 \Longrightarrow y' = (\cos x^2)(2x) \Longrightarrow y'(\sqrt{\pi/4}) = [\cos(\pi/4)](2\sqrt{\pi/4}) = \sqrt{\pi/2}$. Therefore the slope of the normal line at $(\sqrt{\pi/4}, \sqrt{2}/2)$ is $-\sqrt{2/\pi}$ and the equation is $y - \sqrt{2}/2 = -\sqrt{2/\pi}\,(x - \sqrt{\pi/4})$.

84. $y = 1/x^2 \Longrightarrow y' = -2x^{-3} \Longrightarrow y'(1) = -2$. Therefore the slope of the normal line at $(1,1)$ is $1/2$ and the equation is $y - 1 = (1/2)(x - 1)$.

85. $y = x^3 - 2x \Longrightarrow y' = 3x^2 - 2 \Longrightarrow y'(0) = -2$. Therefore the slope of the normal line at $(0,0)$ is $1/2$ and the equation is $y = (1/2)x$.

86. $y = \tan x \Longrightarrow y' = \sec^2 x \Longrightarrow y'(\pi) = 1$. Therefore the slope of the normal line at $(\pi, 0)$ is -1 and the equation is $y = -(x - \pi)$.

87. The average velocity during the time interval $0 \le t \le \pi$ is

$$\frac{s(\pi) - s(0)}{\pi - 0} = \frac{[-4\cos \pi] - [-4\cos 0]}{\pi - 0} = \frac{8}{\pi}.$$

Since $v(t) = s'(t) = 4\sin t$, the instantaneous velocity at $t = \pi/2$ is $v(\pi/2) = 4\sin(\pi/2) = 4$.

88. $f(x) = 1/x \Longrightarrow f'(x) = -1/x^2$. Hence, if $x_0 = 4$,

$$f(x_0) + f'(x_0)\Delta x = \frac{1}{x_0} + \left(-\frac{1}{x_0^2}\right)(x - x_0) = \frac{1}{4} - \frac{1}{16}(x - 4).$$

89. $f(x) = \sqrt{x} \Longrightarrow f'(x) = (1/2)x^{-1/2}$. Hence, if $x_0 = 8$,

$$f(x_0) + f'(x_0)\Delta x = \sqrt{x_0} + \frac{1}{2\sqrt{x_0}}(x - x_0) = \sqrt{8} + \frac{1}{2\sqrt{8}}(x - 8).$$

90. $f(x) = \sin 2x \Longrightarrow f'(x) = 2\cos 2x$. Hence, if $x_0 = 0$,

$$f(x_0) + f'(x_0)\Delta x = \sin 2x_0 + (2\cos 2x_0)(x - x_0) = \sin 0 + (2\cos 0)(x - 0) = 2x.$$

91. $f(x) = x^2 \Longrightarrow f'(x) = 2x$. Hence, if $x_0 = 1$,

$$f(x_0) + f'(x_0)\Delta x = x_0^2 + 2x_0(x - x_0) = 1 + 2(x - 1) = 2x - 1.$$

92. We may assume the circle has radius a, is centered at the origin, and hence has equation $x^2 + y^2 = a^2$. Let (x_0, y_0) be a typical point on the circle. Then

$$2x + 2y\frac{dy}{dx} = 0 \quad \Longrightarrow \quad \frac{dy}{dx} = -\frac{x}{y} \quad \Longrightarrow \quad \frac{dy}{dx}\bigg|_{(x_0,y_0)} = -\frac{x_0}{y_0}.$$

It follows that the slope of the normal line at (x_0, y_0) is y_0/x_0 and the equation of the normal is $y - y_0 = (y_0/x_0)(x - x_0)$. If $x = 0$, $y - y_0 = (y_0/x_0)(-x_0) = -y_0 \implies y = 0$. Hence the normal line passes through the origin, that is, the center of the circle. This is not necessarily true, however, for an ellipse. For example, consider the ellipse $2x^2 + y^2 = 4$. Here, $4x + 2y(dy/dx) = 0 \implies dy/dx = -2x/y$. Hence the slope of the normal at $(1, \sqrt{2})$ is $\sqrt{2}/2$ and the equation is $y - \sqrt{2} = (\sqrt{2}/2)(x - 1)$. This line does not pass through the center of the ellipse since, if $x = 0$, $y = \sqrt{2} - \sqrt{2}/2 = \sqrt{2}/2$, not 0.

93. $h(x) = f(g(x)) \implies h'(x) = f'(g(x))g'(x) \implies h'(3) = f'(g(3))g'(3) = f'(4)g'(3) = (5)(2) = 10$.

94. Let $A(\theta)$ be the area of an isosceles triangle whose base is 10 cm and whose equal base angles are θ radians. Then the height of the triangle is $5\tan\theta$ and hence

$$A = (1/2)(10)(5\tan\theta) = 25\tan\theta \implies dA = 25\sec^2\theta \, d\theta.$$

If $\theta_0 = \pi/4$ and $\Delta\theta = 0.05$ radian,

$$\Delta A = \text{change in area} \approx dA = 25\sec^2(\pi/4)(0.05) = 2.5.$$

Thus, the area of the triangle increases by approximately 2.5 cm^2.

95. Let $V(r)$ be the volume of a sphere of radius r. Then

$$V = (4/3)\pi r^3 \implies dV = 4\pi r^2 \, dr.$$

If $r_0 = 20$ and $r = 20.5$, $\Delta r = 0.5$ and hence

$$dV = \text{approximate change in volume} = 4\pi(20)^2(0.5) = 2513.27.$$

Thus, the volume increases by approximately 2513.27 cm^3. Since the actual change in volume is $\Delta V = V(20.5) - V(20) = (4/3)\pi(20.5^3 - 20^3) = 2576.63$, the relative error is

$$\left|\frac{\Delta V - dV}{\Delta V}\right| = \frac{2576.63 - 2513.27}{2576.63} = 0.0246$$

and the percentage error is 2.46 %

Remark: Note that the problem asks for the relative error in approximating the *change* in volume when r increases from 20 cm to 20.5 cm, not the relative error in approximating $V(20.5)$. The relative error in approximating $V(20.5)$ is

$$\left|\frac{y - y_1}{y}\right| = 0.0018,$$

where $y = $ actual volume $= V(20.5) = 36086.9512$ and $y_1 = $ linear approximation to $V(20.5) = 36023.5957$.

96. $(x + 2)^2 + (y - 3)^2 = 4 \implies 2(x + 2) + 2(y - 3)\dfrac{dy}{dx} = 0 \implies$

$$\frac{dy}{dx} = -\frac{x + 2}{y - 3} \implies \left.\frac{dy}{dx}\right|_{(-1, 3+\sqrt{3})} = -\frac{-1 + 2}{(3 + \sqrt{3}) - 3} = \frac{1}{\sqrt{3}} = \frac{\sqrt{3}}{3}.$$

Hence the slope of the tangent line at $(-1, 3 + \sqrt{3})$ is $\sqrt{3}/3$.

97. $y = 4x^3 - 4x + 4 \implies y' = 12x^2 - 4$. Since the slope of the line $y - 8x + 6 = 0$ is 8, the tangent line is parallel to this line $\iff 12x^2 - 4 = 8 \iff x^2 = 1 \iff x = \pm 1$. Hence the tangent line is parallel to $y - 8x + 6 = 0$ at two points: $(1, 4)$ and $(-1, 4)$.

98. $f(x) = \sin 2x \implies f'(x) = 2\cos 2x \implies f''(x) = -4\sin 2x$. Hence, for x in the interval $[-\pi, \pi]$, the slope of the tangent line to $y = f'(x)$ is 2 when

$$-4\sin 2x = 2 \implies \sin 2x = -1/2 \implies 2x = -5\pi/6, -\pi/6, 7\pi/6, 11\pi/6$$

$$\implies x = -5\pi/12, -\pi/12, 7\pi/12, 11\pi/12.$$

99. $f(x) = (8/5)x^{5/4}$

100. The graphs of $y = f(x) = x^2 - 4x + 4$ and $y = f'(x) = 2x - 4$ intersect when

$$x^2 - 4x + 4 = 2x - 4 \implies x^2 - 6x + 8 = (x-4)(x-2) = 0 \implies x = 2, 4; \; y = 0, 4.$$

Thus, the graphs intersect at the points $(2, 0)$ and $(4, 4)$. Since $f'(2) = 0$ and $f'(4) = 4$, the equations of the tangent lines to the graph of $y = x^2 - 4x + 4$ at these points are $y = 0$ and $y - 4 = 4(x - 4)$. Letting $y = 0$, it follows that the tangent lines intersect when $-4 = 4(x - 4) \implies x = 3, \; y = 0$.

101. (a) $v(t) = s'(t) = t^3 - 9t^2 + 24t - 20$

   (b) $a(t) = v'(t) = 3t^2 - 18t + 24$

   (c) To find sign changes for $v(t)$, we first solve $v(t) = 0$. Observe that $t = 2$ is a root of $v = 0$. Hence

$$v(t) = (t-2)(t^2 - 7t + 10) = (t-2)(t-5)(t-2) = 0 \implies t = 2, 5.$$

Sign v	$-$	$-$	$+$
Interval	$(-\infty, 2)$	$(2, 5)$	$(5, +\infty)$

   Hence the particle is moving to the right when $t > 5$.

   (d) From the table in part (c), the particle is moving to the left when $t < 2$ and when $2 < t < 5$.

   (e) By part (c), $v(t) = 0 \iff t = 2, 5$.

   (f) Only when $t = 5$.

   (g) $a(t) = 3t^2 - 18t + 24 = 3(t-4)(t-2) = 0 \iff t = 2, 4$.

102. $y = A\sin x + B\tan x \implies y' = A\cos x + B\sec^2 x$. Hence

$$y'(0) = A + B = m_1 = 4$$
$$y'(\pi/4) = A\left(\sqrt{2}/2\right) + B(2) = m_2 = 4 + \sqrt{2}.$$

Setting $B = 4 - A$ in the second equation, we find that

$$A\left(\sqrt{2}/2\right) + 2(4 - A) = 4 + \sqrt{2}$$

$$\implies A\left(\sqrt{2}/2 - 2\right) = -4 + \sqrt{2}$$

$$\implies A = \frac{-4 + \sqrt{2}}{-2 + \sqrt{2}/2} = \frac{-8 + 2\sqrt{2}}{-4 + \sqrt{2}} \cdot \frac{-4 - \sqrt{2}}{-4 - \sqrt{2}} = 2$$

$$\implies B = 4 - A = 2.$$

Hence $y = 2\sin x + 2\tan x$.

103. $s(t) = t^3 + 3t + 3 \implies v(t) = 3t^2 + 3 \implies a(t) = 6t$. Hence

$$|v(t)| = |a(t)| \implies 3t^2 + 3 = \pm 6t \implies 3(t^2 \mp 2t + 1) = 3(t \mp 1)^2 = 0 \implies t = \pm 1.$$

Thus, velocity and acceleration have the same magnitude when $t = \pm 1$.

104. $PV^{1.4} = C \implies P = CV^{-1.4} \implies dP = -1.4CV^{-2.4}dV$. If V changes by 1%, $\Delta V = \pm 0.01V$ and hence

$$dP = -1.4CV^{-2.4}\Delta V = -1.4CV^{-2.4}(\pm 0.01V) = \mp 0.014CV^{-1.4} = \mp 0.014P.$$

Thus, if V increases (decreases) by 1%, P decreases (increases) by 1.4%.

105. Let $f(x)$ be an odd function. Then

$$f'(-x) = \lim_{h \to 0} \frac{f(-x+h) - f(-x)}{h}.$$

Replacing h by $-h$, and noting that $h \to 0 \iff -h \to 0$,

$$
\begin{aligned}
f'(-x) &= \lim_{h \to 0} \frac{f(-x-h) - f(-x)}{-h} \\
&= \lim_{h \to 0} \frac{-f(x+h) + f(x)}{-h} \qquad \text{since } f \text{ is odd} \\
&= \lim_{h \to 0} \frac{f(x+h) - f(x)}{h} \\
&= f'(x).
\end{aligned}
$$

Hence $f'(x)$ is an even function.

106. No. For example, let $f(x) = g(x) = 1/x$. Then $y = f\big(g(x)\big) = 1/(1/x) = x$, which is differentiable at $x = 0$, but neither $f(x)$ nor $g(x)$ are defined or differentiable at $x = 0$.

Chapter 4

Applications of the Derivative

4.1 Extreme Values

1. a) Critical numbers: x_1, x_2.
   b) Numbers at which f has a relative minimum on $[a, b]$: a, b.
   c) Numbers at which f has an absolute maximum on $[a, b]$: x_2.
   d) Numbers at which f has an absolute extremum on $[a, b]$: a, x_2.

2. a) x_1, x_2, x_3
   b) x_2, x_3, a, b
   c) none
   d) a

3. a) x_1, x_2, x_3
   b) a, b
   c) x_3
   d) x_3, b

4. a) x_1, x_2, x_3

   b) x_1, x_3 (function constant on an open interval around x_3), b

   c) a, x_2

   d) a, x_2, b

5. $f(x) = x^2 - 4x + 3 \quad [0, 4] \quad c = 2$
   $f'(x) = 2x - 4$

   a) $f'(c) = f'(2) = 2(2) - 4 = 0$

   b)

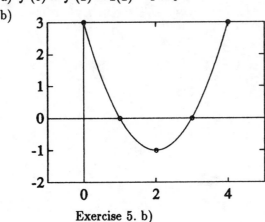

   Exercise 5. b)

   c) $f(c) = f(2) = -1$ is the absolute minimum of f on $[0, 4]$.

193

6) $f(x) = \sin\left(x + \dfrac{\pi}{4}\right)$ $[0, 2\pi]$ $c = \dfrac{\pi}{4}$

$f'(x) = \cos\left(x + \dfrac{\pi}{4}\right)$

a) $f'\left(\dfrac{\pi}{4}\right) = \cos\left(\dfrac{\pi}{4} + \dfrac{\pi}{4}\right) = \cos\left(\dfrac{\pi}{2}\right) = 0$

b)

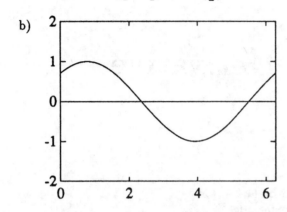

Exercise 6. b)

c) $f\left(\dfrac{\pi}{4}\right) = \sin\left(\dfrac{\pi}{4} + \dfrac{\pi}{4}\right) = \sin\dfrac{\pi}{2} = 1$ is the absolute maximum of f on $[0, 2\pi]$.

7. $f(x) = x^2 - 7$ $[-3, 4]$ $c = 0$

$f'(x) = 2x$

a) $f'(0) = (2)(0) = 0$

b)

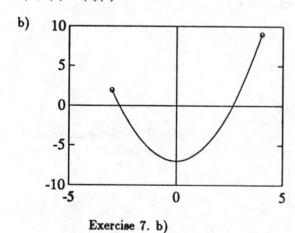

Exercise 7. b)

c) $f(0) = -7$ is the absolute minimum of f on $[-3, 4]$.

8. $f(x) = 6 - (x - 2)^2$ $[0, 5]$ $c = 2$

$f'(x) = -2(x - 2)$

a) $f'(2) = -2(2 - 2) = 0$

b)

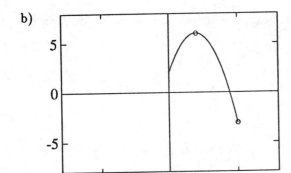

Exercise 8. b)

c) $f(2) = 6$ is the absolute maximum of f on $[0,5]$.

9. $f(x) = \cos(\pi x) \qquad [0,2] \qquad c = 1$
    $f'(x) = -\pi \sin(\pi x)$

    a) $f'(1) = -\pi \sin(\pi) = 0$

    b)

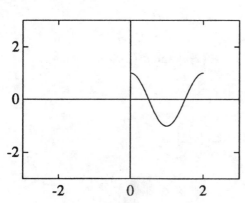

Exercise 9. b)

c) $f(1) = -1$ is the absolute minimum of f on $[0,2]$.

10. $f(x) = 4 + 2x^2 \qquad [-2,2] \qquad c = 0$
    $f'(x) = 4x$

    a) $f'(0) = (4)(0) = 0$

    b)

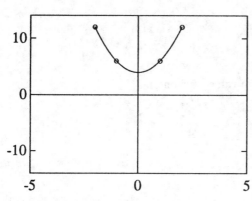

Exercise 10. b)

c) $f(0) = 4$ is the absolute minimum of f on $[-2,2]$.

In Exercises 11–28, find all critical numbers and the maximum and minimum values for f on the given interval.

11. $f(x) = x^2(x-4), \quad x \in [-1, 3]$
    $f'(x) = 2x(x-4) + x^2 = 3x^2 - 8x$
    Let $f'(x) = 0$, we have $3x^2 - 8x = 0$
    $\Rightarrow x(3x - 8) = 0$
    Thus, the critical numbers are 0 and $\frac{8}{3}$.
    $f(-1) = -5$
    $f(0) = 0$
    $f\left(\frac{8}{3}\right) = -\frac{256}{27}$
    $f(3) = -9$
    Hence, $f(0) = 0$ is the maximum value of f on $[-1, 3]$

    and $f\left(\frac{8}{3}\right) = -\frac{256}{27}$ is the minimum value of f on $[-1, 3]$.

12. $f(x) = |x - 2|, \quad x \in [0, 5]$
    $f(x) = \begin{cases} 2 - x & \text{for } x \in [0, 2] \\ x - 2 & \text{for } x \in [2, 5] \end{cases}$
    Clearly, $f'(x) = -1$ for $x \in [0, 2)$ and $f'(x) = 1$ for $x \in (2, 5]$ and $f'(2)$ is undefined.
    Thus $x = 2$ is the only critical number of f.
    $f(0) = 2$
    $f(2) = 0$
    $f(5) = 3$
    Therefore $f(5) = 3$ is the maximum value of f on $[0, 5]$
    and $f(2) = 0$ is the minimum value of f on $[0, 5]$.

13. $f(x) = \frac{1}{x(x-4)}, \quad x \in [1, 3]$
    $f'(x) = \frac{-(x - 4 + x)}{x^2(x-4)^2} = \frac{-2x + 4}{x^2(x-4)^2}$
    Let $f'(x) = 0$, we have: $\frac{-2x + 4}{x^2(x-4)^2} = 0 \Rightarrow x = 2$. Thus $x = 2$ is the only critical number of f.
    $f(1) = -\frac{1}{3}$

    $f(2) = -\frac{1}{4}$

    $f(3) = -\frac{1}{3}$

    Therefore $f(1) = f(3) = -\frac{1}{3}$ is the minimum value of f on $[1, 3]$
    and $f(2) = -\frac{1}{4}$ is the maximum value of f on $[1, 3]$.

14. $f(x) = x(x^2 - 2), \quad x \in [-1, 2]$
    $f'(x) = x^2 - 2 + x(2x) = 3x^2 - 2$
    $3x^2 - 2 = 0 \Rightarrow x = \pm\frac{\sqrt{6}}{3}$
    Thus the critical numbers are $\pm\frac{\sqrt{6}}{3}$.

196

$$f(-1) = 1$$

$$f\left(-\frac{\sqrt{6}}{3}\right) = \frac{4\sqrt{6}}{9}$$

$$f\left(\frac{\sqrt{6}}{3}\right) = \frac{-4\sqrt{6}}{9}$$

$$f(2) = 4$$

Therefore $f(2) = 4$ is the maximum value of f on $[-1, 2]$

and $f\left(\frac{\sqrt{6}}{3}\right) = \frac{-4\sqrt{6}}{9}$ is the minimum value of f on $[-1, 2]$.

15. $f(x) = x^2(x-2), \quad x \in [-1, 2]$
   $f'(x) = 2x(x-2) + x^2 = 3x^2 - 4x$
   $3x^2 - 4x = 0 \Rightarrow x = 0, \frac{4}{3}$.
   Therefore the critical numbers are 0 and $\frac{4}{3}$.

   $f(-1) = -3$ (minimum value of f on $[-1, 2]$) $f(0) = 0$ (maximum value of f on $[-1, 2]$)
   $f\left(\frac{4}{3}\right) = -\frac{32}{27}$ $f(2) = 0$ (maximum value of f on $[-1, 2]$)

16. $f(x) = \sin\left(\frac{x}{2}\right), \quad x \in [0, 4\pi]$
   $f'(x) = \frac{1}{2}\cos\left(\frac{x}{2}\right); \qquad \frac{1}{2}\cos\left(\frac{x}{2}\right) = 0$

   $$\Rightarrow \cos\left(\frac{x}{2}\right) = 0$$
   $$\Rightarrow \frac{x}{2} = \frac{\pi}{2}, \frac{3\pi}{2}$$
   $$\Rightarrow x = \pi, 3\pi$$

   Therefore the critical numbers are π and 3π.

   $f(0) = 0$ $f(\pi) = 1$ (maximum value of f on $[0, 4\pi]$)
   $f(3\pi) = -1$ (minimum value of f on $[0, 4\pi]$) $f(4\pi) = 0$

17. $f(x) = 1 - \tan^2 x, \quad x \in \left[-\frac{\pi}{4}, \frac{\pi}{4}\right]$

   $f'(x) = -2\tan x \sec^2 x = -2\frac{\sin x}{\cos x}\frac{1}{\cos^2 x} = \frac{-2\sin x}{\cos^3 x}$
   $f'(x) = 0 \Rightarrow -2\sin x = 0$
   $$\Rightarrow \sin x = 0$$
   $$\Rightarrow x = 0$$
   $f'(x)$ is undefined implies $\cos^3 x = 0$
   $$\Rightarrow \cos x = 0.$$

   Since $\cos x \neq 0$ in $\left[-\frac{\pi}{4}, \frac{\pi}{4}\right]$, the only critical number is 0.

   $f\left(-\frac{\pi}{4}\right) = 0 \longleftarrow$ minimum value of f on $\left[-\frac{\pi}{4}, \frac{\pi}{4}\right]$

   $f(0) = 1 \longleftarrow$ maximum value of f on $\left[-\frac{\pi}{4}, \frac{\pi}{4}\right]$

   $f\left(\frac{\pi}{4}\right) = 0 \longleftarrow$ minimum value of f on $\left[-\frac{\pi}{4}, \frac{\pi}{4}\right]$

18. $f(x) = \sin x \cos x, \quad x \in \left[-\frac{\pi}{2}, \frac{\pi}{2}\right]$
   $f'(x) = \cos x \cos x + \sin x(-\sin x) = \cos^2 x - \sin^2 x$

197

$\cos^2 x - \sin^2 x = 0$ when $\cos^2 x = \sin^2 x$

$\Rightarrow \quad \cos x = \pm \sin x$

$\Rightarrow \quad x = -\frac{\pi}{4}, \frac{\pi}{4}$

[Alternatively, we have $f(x) = \sin x \cos x = \frac{1}{2}\sin 2x$. Therefore $f'(x) = \frac{1}{2}(2)\cos 2x = \cos 2x$.

$\cos 2x = 0$ when $2x = -\frac{\pi}{2}, \frac{\pi}{2}$

$\Rightarrow \quad x = -\frac{\pi}{4}, \frac{\pi}{4}]$.

Thus, the critical numbers are $-\frac{\pi}{4}$ and $\frac{\pi}{4}$.

$f\left(-\frac{\pi}{2}\right) = 0$

$f\left(-\frac{\pi}{4}\right) = -\frac{1}{2}$ ⟵ minimum value of f on $\left[-\frac{\pi}{2}, \frac{\pi}{2}\right]$

$f\left(\frac{\pi}{4}\right) = \frac{1}{2}$ ⟵ maximum value of f on $\left[-\frac{\pi}{2}, \frac{\pi}{2}\right]$

$f\left(\frac{\pi}{2}\right) = 0$

19. $f(x) = \sin x + \cos x, \quad x \in [0, \pi]$
$f'(x) = \cos x - \sin x$
$\cos x - \sin x = 0$ when $\cos x = \sin x$

$\Rightarrow \quad x = \frac{\pi}{4}$.

Therefore the only critical number is $\frac{\pi}{4}$.

$f(0) = 1$

$f\left(\frac{\pi}{4}\right) = \sqrt{2}$ ⟵ maximum value of f on $[0, \pi]$

$f(\pi) = -1$ ⟵ minimum value of f on $[0, \pi]$

20. $f(x) = x^{2/3} + 2, \quad x \in [-2, 1]$

$f'(x) = \frac{2}{3}x^{-1/3} = \frac{2}{3x^{1/3}}$

$f'(x)$ is undefined when $3x^{1/3} = 0 \Rightarrow x = 0$. Therefore the only critical number is 0.

$f(-2) = 2 + \sqrt[3]{4}$

$f(0) = 2$ ⟵ minimum value of f on $[-2, 1]$

$f(1) = 3$ ⟵ maximum value of f on $[-2, 1]$

21. $f(x) = x + \frac{1}{x}, \quad x \in \left[\frac{1}{2}, 2\right]$

$f'(x) = 1 - \frac{1}{x^2} = \frac{x^2 - 1}{x^2}$

$\frac{x^2 - 1}{x^2} = 0$ when $x^2 - 1 = 0 \Rightarrow (x+1)(x-1) = 0 \Rightarrow x = \pm 1$.

Therefore the only critical number in $\left[\frac{1}{2}, 2\right]$ is 1.

$f\left(\frac{1}{2}\right) = \frac{5}{2}$ ⟵ maximum value of f on $\left[\frac{1}{2}, 2\right]$

$f(1) = 2$ ⟵ minimum value of f on $\left[\frac{1}{2}, 2\right]$

$f(2) = \frac{5}{2}$ ⟵ maximum value of f on $\left[\frac{1}{2}, 2\right]$

22. $f(x) = \left(\dfrac{x+1}{x-1}\right)^2, \quad x \in [-3, 0]$

$f'(x) = 2\left(\dfrac{x+1}{x-1}\right)\left(\dfrac{(x-1)-(x+1)}{(x-1)^2}\right) = \dfrac{-4(x+1)}{(x-1)^3}$

$f'(x) = 0$ when $-4(x+1) = 0 \Rightarrow x = -1$. Therefore the only critical number is -1.

$f(-3) = \frac{1}{4}$

$f(-1) = 0 \quad \longleftarrow$ minimum value of f on $[-3, 0]$
$f(0) = 1 \quad \longleftarrow$ maximum value of f on $[-3, 0]$

23. $f(x) = \sec x, \quad x \in \left[-\frac{\pi}{4}, \frac{\pi}{4}\right]$

$f'(x) = \sec x \tan x = \dfrac{\sin x}{\cos^2 x}$

$f'(x) = 0$ when $\sin x = 0 \Rightarrow x = 0$ in $\left[-\frac{\pi}{4}, \frac{\pi}{4}\right]$.

$f'(x)$ is undefined when $\cos^2 x = 0 \Rightarrow \cos x = 0$.

Since $\cos x \neq 0$ in $\left[-\frac{\pi}{4}, \frac{\pi}{4}\right]$, the only critical number is 0.

$f\left(-\frac{\pi}{4}\right) = \sqrt{2} \quad \longleftarrow$ maximum value of f on $\left[-\frac{\pi}{4}, \frac{\pi}{4}\right]$

$f(0) = 1 \quad \longleftarrow$ minimum value of f on $\left[-\frac{\pi}{4}, \frac{\pi}{4}\right]$

$f\left(\frac{\pi}{4}\right) = \sqrt{2} \quad \longleftarrow$ maximum value of f on $\left[-\frac{\pi}{4}, \frac{\pi}{4}\right]$

24. $f(x) = \dfrac{x^2}{1+x}, \quad x \in \left[-\frac{1}{2}, 2\right]$

$f'(x) = \dfrac{2x(1+x) - x^2}{(1+x)^2} = \dfrac{x^2 + 2x}{(1+x)^2}$

$f'(x) = 0$ when $x^2 + 2x = 0 \Rightarrow x(x+2) = 0$
$\Rightarrow x = 0$ or $x = -2$.

Therefore the only critical number in $\left[-\frac{1}{2}, 2\right]$ is 0.

$f\left(-\frac{1}{2}\right) = \frac{1}{2}$

$f(0) = 0 \quad \longleftarrow$ minimum value of f on $\left[-\frac{1}{2}, 2\right]$

$f(2) = \frac{4}{3} \quad \longleftarrow$ maximum value of f on $\left[-\frac{1}{2}, 2\right]$

25. $f(x) = \dfrac{x^3}{2+x}, \quad x \in [-1, 1]$

$f'(x) = \dfrac{3x^2(2+x) - x^3}{(2+x)^2} = \dfrac{2x^3 + 6x^2}{(2+x)^2} = \dfrac{2x^2(x+3)}{(2+x)^2}$

$f'(x) = 0$ when $2x^2(x+3) = 0 \Rightarrow x = 0 \quad$ or $\quad x = -3$.

Therefore the only critical number in $[-1, 1]$ is 0.

$f(-1) = -1 \quad \longleftarrow$ minimum value of f on $[-1, 1]$
$f(0) = 0$
$f(1) = \frac{1}{3} \quad \longleftarrow$ maximum value of f on $[-1, 1]$

26. $f(x) = x \sin x, \quad x \in \left[-\frac{\pi}{2}, \frac{\pi}{2}\right]$

    $f'(x) = \sin x + x \cos x$

    Note that f' is positive for $x \in \left(0, \frac{\pi}{2}\right)$

    and f' is negative for $x \in \left(-\frac{\pi}{2}, 0\right)$.

    Thus $f'(x) = 0$ only when $x = 0$. Therefore the only critical number is 0.

    $f\left(-\frac{\pi}{2}\right) = \frac{\pi}{2}$ ⟵ maximum value of f on $\left[-\frac{\pi}{2}, \frac{\pi}{2}\right]$

    $f(0) = 0$ ⟵ minimum value of f on $\left[-\frac{\pi}{2}, \frac{\pi}{2}\right]$

    $f\left(\frac{\pi}{2}\right) = \frac{\pi}{2}$ ⟵ maximum value of f on $\left[-\frac{\pi}{2}, \frac{\pi}{2}\right]$

27. $f(x) = 8x^{1/3} - 2x^{4/3}, \quad x \in [-1, 8]$

    $f'(x) = \frac{8}{3x^{2/3}} - \frac{8x^{1/3}}{3} = \frac{8 - 8x}{3x^{2/3}}$

    $f'(x) = 0$ when $8 - 8x = 0 \Rightarrow x = 1$.

    $f'(x)$ is undefined when $3x^{2/3} = 0 \Rightarrow x = 0$. Therefore the critical numbers are $0, 1$.

    $f(-1) = -10$ $\qquad\qquad\qquad\qquad$ $f(0) = 0$

    $f(1) = 6$ (maximum value of f on $[-1, 8]$) $\qquad$ $f(8) = -16$ (minimum value of f on $[-1, 8]$)

28. $f(x) = \frac{x^{2/3}}{2 + \sqrt[3]{x}}, \quad x \in [-1, 8]$

    $f'(x) = \frac{\frac{2}{3}x^{-1/3}\left(2 + \sqrt[3]{x}\right) - \frac{1}{3}x^{-2/3}(x^{2/3})}{\left(2 + \sqrt[3]{x}\right)^2}$

    $f'(x) = \frac{\frac{4}{3x^{1/3}} + \frac{2}{3} - \frac{1}{3}}{\left(2 + \sqrt[3]{x}\right)^2} = \frac{4 + x^{1/3}}{3x^{1/3}\left(2 + \sqrt[3]{x}\right)^2}$

    $f'(x) = 0$ when $4 + x^{1/3} = 0 \Rightarrow x^{1/3} = -4$
    $\qquad\qquad\qquad\qquad\qquad\qquad\qquad \Rightarrow x = -64$.

    $f'(x)$ is undefined when $3x^{1/3} = 0 \Rightarrow x = 0$. Therefore the only critical number in $[-1, 8]$ is 0.

    $f(-1) = 1$ ⟵ maximum value of f on $[-1, 8]$

    $f(0) = 0$ ⟵ minimum value of f on $[-1, 8]$

    $f(8) = 1$ ⟵ maximum value of f on $[-1, 8]$

In Exercises 29–32, find the maximum and minimum values for f on the given interval.

29. $f(x) = \left(\sqrt{x} - x\right)^2, \quad x \in [0, 4]$

    $f'(x) = 2\left(\sqrt{x} - x\right)\left(\frac{1}{2\sqrt{x}} - 1\right) = 2\sqrt{x}\left(1 - \sqrt{x}\right)\left(\frac{1 - 2\sqrt{x}}{2\sqrt{x}}\right) = \left(1 - \sqrt{x}\right)\left(1 - 2\sqrt{x}\right)$

    $f'(x) = 0$ when $1 - \sqrt{x} = 0 \Rightarrow x = 1$ or when $1 - 2\sqrt{x} = 0 \Rightarrow x = \frac{1}{4}$.

    Therefore the critical numbers are 1 and $\frac{1}{4}$.

    $f(0) = 0$ (minimum value of f on $[0, 4]$) $\quad$ $f(1) = 0$ (minimum value of f on $[0, 4]$)

    $f\left(\frac{1}{4}\right) = \frac{1}{16}$ $\qquad\qquad\qquad\qquad\qquad\quad$ $f(4) = 4$ (maximum value of f on $[0, 4]$)

30. $f(x) = \sqrt{x}(1-x^2), \quad x \in [0,4]$

$f'(x) = \frac{1}{2\sqrt{x}}(1-x^2) + \sqrt{x}(-2x) = \frac{1-x^2-4x^2}{2\sqrt{x}} = \frac{1-5x^2}{2\sqrt{x}}$

$f'(x) = 0$ when $1-5x^2 = 0 \Rightarrow x = \pm\frac{\sqrt{5}}{5}$.

$f'(x)$ is undefined when $2\sqrt{x} = 0 \Rightarrow x = 0$. Therefore the critical number in $[0,4]$ is $\frac{\sqrt{5}}{5}$.

$f(0) = 0$

$f\left(\frac{\sqrt{5}}{5}\right) = \frac{4}{5^{5/4}} \longleftarrow$ maximum value of f on $[0,4]$

$f(4) = -30 \longleftarrow$ minimum value of f on $[0,4]$

31. $f(x) = \frac{\sqrt{x}}{1+x}, \quad x \in [0,4]$

$f'(x) = \frac{\frac{1}{2\sqrt{x}}(1+x) - \sqrt{x}}{(1+x)^2} = \frac{1-x}{2\sqrt{x}(1+x)^2}$

$f'(x) = 0$ when $x = 1$ and $f'(x)$ is undefined when $x = 0$.

$f(0) = 0 \longleftarrow$ minimum value of f on $[0,4]$

$f(1) = \frac{1}{2} \longleftarrow$ maximum value of f on $[0,4]$

$f(4) = \frac{2}{5}$

32. $f(x) = \frac{\sqrt{x}}{1+\sqrt[3]{x}}, \quad x \in [0,8]$

$f'(x) = \frac{\frac{1}{2x^{1/2}}\left(1+\sqrt[3]{x}\right) - x^{1/2}\left(\frac{1}{3x^{2/3}}\right)}{\left(1+\sqrt[3]{x}\right)^2}$

$f'(x) = \frac{\frac{1+x^{1/3}}{2x^{1/2}} - \frac{1}{3x^{1/6}}}{\left(1+\sqrt[3]{x}\right)^2} = \frac{3+3x^{1/3}-2x^{1/3}}{6x^{1/2}\left(1+\sqrt[3]{x}\right)^2} = \frac{3+x^{1/3}}{6x^{1/2}\left(1+\sqrt[3]{x}\right)^2}$

$f'(x) = 0$ when $3 + x^{1/3} = 0 \Rightarrow x = -27$.

$f'(x)$ is undefined when $x = 0$. Therefore there are no critical numbers in $[0,8]$.

$f(0) = 0 \longleftarrow$ minimum value of f on $[0,8]$

$f(8) = \frac{2\sqrt{2}}{3} \longleftarrow$ maximum value of f on $[0,8]$

33. a) $g'(x) = c\,f'(x)$ and $c \neq 0$ imply that $g'(x) = 0$ exactly when $f'(x) = 0$; so: the critical numbers of f and g are the same.

b) $h'(x) = f'(x+c)$ implies that x is a critical number for h if and only if $x+c$ is a critical number for f. So: the critical numbers for h are those for f shifted c units to the left.

c) $k'(x) = c\,f'(cx)$ and $c \neq 0$ imply that x is a critical number for k if and only if cx is a critical number for f. So: the critical numbers for k are those for f divided by c.

34. No: for example, on the interval $[-1,1]$ the function $f(x) = x$ has maximum value $+1$ at $x = +1$ and $g(x) = -x$ has maximum value $+1$ at $x = -1$, but $f(x) + g(x)$ is identically zero. The maximum value of $h = f + g$ on $[a,b]$ will equal $L + M$ if the maximum values of f and g are attained at the same point. The maximum value of $h = f + g$ is always less than or equal to $L + M$.

35. We are told that $f(3) = -7$ is the minimum value of $f(x) = x^2 + bx + c$ on $[0, 5]$. Since $x = 3$ is an interior point, $f'(3) = 0$. Now $f'(x) = 2x + b$ implies

$$6 + b = f'(3) = 0$$

and $f(x) = x^2 + bx + c$ implies

$$9 + 3b + c = f(3) = -7$$

So $b = -6$ and $c = -7 - 9 - 3(-6) = 2$.

36. Since the maximum value of $f(x) = ax^3 - bx$ is achieved at the interior point $x = 2$ of $[0, 4]$ it follows that $f'(2) = 0$. Now $f'(x) = 3ax^2 - b$ implies

$$12a - b = f'(2) = 0 \text{ and}$$
$$f(x) = ax^3 - bx \quad \text{implies}$$
$$8a - 2b = f(2) = 4$$

so $b = 12a$ and then $4 = 8a - 24a = -16a$. Thus $a = -\frac{1}{4}$ and $b = -3$.

37. a) $f(x) = \sin x - \cos x \Rightarrow f'(x) = \cos x + \sin x$. Now $f'(x) = 0 \Rightarrow \sin x = -\cos x \Rightarrow \tan x = -1 \Rightarrow x = \frac{3\pi}{4}$ or $x = \frac{7\pi}{4}$ in $[0, 2\pi]$. Evaluating f at these critical numbers: $f\left(\frac{3\pi}{4}\right) = \sqrt{2}$ and $f\left(\frac{7\pi}{4}\right) = -\sqrt{2}$. Evaluating f at the endpoints: $f(0) = f(2\pi) = -1$.
So: maximum value of f is $\sqrt{2}$, minimum value of f is $-\sqrt{2}$.

b) Notice (from trigonometry) that

$$f(x) = 2\sin\left(\frac{\pi}{2} - x\right) = 2\cos x$$
$$\text{so } f'(x) = -2\sin x.$$

Now $f'(x) = 0 \Rightarrow x \in \{0, \pi, 2\pi\}$ in $[0, 2\pi]$. Evaluating: $f(0) = f(2\pi) = 2$ and $f(\pi) = -2$.
So: maximum value of f is 2, minimum value of f is -2.

c) The double angle formula yields

$$f(x) = 2\sin x \cos x = \sin 2x$$

which has period π; so we need only look at $[0, \pi]$. Here,

$$f'(x) = 2\cos 2x$$
$$\text{so } f'(x) = 0 \Rightarrow x \in \left\{\frac{\pi}{4}, \frac{3\pi}{4}\right\} \text{ in } [0, \pi].$$

Evaluating: $f(0) = f(\pi) = 0$, $f\left(\frac{\pi}{4}\right) = 1$ and $f\left(\frac{3\pi}{4}\right) = -1$.
So: maximum value of f is 1, minimum value of f is -1.

38. a) $x^2 - 5x + 4 = (x - 4)(x - 1) = 0$ at $x = 1, 4$.

$$
\begin{array}{ccc}
+\ +\ + & -\ -\ - & +\ +\ + \\
\hline
& & \\
1 & 4 & (x-4)(x-1)
\end{array}
$$

Thus $f(x) = |x^2 - 5x + 4| = \begin{cases} x^2 - 5x + 4 & \text{if } x \leq 1 \\ -x^2 + 5x - 4 & \text{if } 1 < x < 4 \\ x^2 - 5x + 4 & \text{if } 4 \leq x. \end{cases}$

202

b) In each of the open intervals $(-\infty, 1)$, $(1, 4)$ and $(4, \infty)$, the function $f(x)$ is given by polynomial formulae; its derivative on these intervals is computed as usual by means of **Theorem 1 and Theorem 4** of Chapter 3.

c) $\displaystyle \lim_{x \to 1^-} \frac{f(x) - f(1)}{x - 1} = \lim_{x \to 1^-} \frac{(x^2 - 5x + 4) - 0}{x - 1} = \lim_{x \to 1^-} (x - 4) = -3$

$\displaystyle \lim_{x \to 1^+} \frac{f(x) - f(1)}{x - 1} = \lim_{x \to 1^+} \frac{-x^2 + 5x - 4 - 0}{x - 1} = \lim_{x \to 1^+} -(x - 4) = 3$

Thus $f'(1)$ does not exist.

$\displaystyle \lim_{x \to 4^-} \frac{f(x) - f(4)}{x - 4} = \lim_{x \to 4^-} \frac{-x^2 + 5x - 4 - 0}{x - 4} = \lim_{x \to 4^-} -(x - 1) = -3.$

$\displaystyle \lim_{x \to 4^+} \frac{f(x) - f(4)}{x - 4} = \lim_{x \to 4^+} \frac{x^2 - 5x + 4 - 0}{x - 4} = 3$

Thus $f'(4)$ does not exist.

d) Find all critical numbers of $h(x)$ in (a, b) and all the roots of $h(x) = 0$ in (a, b). Then find the absolute values of $h(x)$ at those critical numbers, roots and also at the endpoints a and b. Compare these values: the largest is the maximum, and the smallest is the minimum.

39. $f(x) = |x - 3|$, $x \in [-2, 6]$
$x - 3 = 0 \Leftrightarrow x = 3$. Let $g(x) = x - 3$; $g'(x) = 1 \neq 0$. Thus $g(x)$ has no critical numbers.
$\begin{aligned} f(3) &= 0 &\longleftarrow \text{ minimum value} \\ f(-2) &= 5 &\longleftarrow \text{ maximum value} \\ f(6) &= 3 \end{aligned}$

40. $f(x) = |3x - 5|$, $x \in [0, 4]$
$3x - 5 = 0 \Leftrightarrow x = \frac{5}{3}$. Let $g(x) = 3x - 5$; $g'(x) = 3 \neq 0$. Thus $g(x)$ has no critical numbers.
$\begin{aligned} f\left(\tfrac{5}{3}\right) &= 0 &\longleftarrow \text{ minimum value} \\ f(0) &= 5 \\ f(4) &= 7 &\longleftarrow \text{ maximum value} \end{aligned}$

41. $f(x) = |x^2 - x - 6|$, $x \in [-3, 5]$
$x^2 - x - 6 = (x - 3)(x + 2) = 0 \Leftrightarrow x = 3, -2$. Let $g(x) = x^2 - x - 6$; $g'(x) = 2x - 1 = 0$ at $x = \frac{1}{2}$.
$\begin{aligned} f(3) &= 0 &\longleftarrow \text{ minimum value} \qquad & f(-3) &= 6 \\ f(-2) &= 0 &\longleftarrow \text{ minimum value} \qquad & f(5) &= 14 &\longleftarrow \text{ maximum value} \\ f\left(\tfrac{1}{2}\right) &= \tfrac{25}{4} \end{aligned}$

42. $f(x) = |\cos x|$, $x \in \left[\frac{\pi}{4}, \pi\right]$
$\cos x = 0 \Leftrightarrow x = \frac{\pi}{2}$. Let $g(x) = \cos x$; $g'(x) = -\sin x \neq 0$ for $x \in \left(\frac{\pi}{4}, \pi\right)$. Thus $g(x)$ has no critical numbers on $\left[\frac{\pi}{4}, \pi\right]$.
$\begin{aligned} f\left(\tfrac{\pi}{2}\right) &= 0 &\longleftarrow \text{ minimum value} \\ f\left(\tfrac{\pi}{4}\right) &= \tfrac{\sqrt{2}}{2} \\ f(\pi) &= 1 &\longleftarrow \text{ maximum value} \end{aligned}$

43. $f(x) = \dfrac{|x|}{4 + x}$; $x \in [-2, 4]$

Note that $4 + x > 0$ on $[-2, 4]$. Therefore $f(x)$ can be rewritten as $\left|\frac{x}{4+x}\right| \cdot \frac{x}{4+x} = 0 \Leftrightarrow x = 0$. Let $g(x) = \frac{x}{4+x}$, $g'(x) = \frac{4}{(4+x)^2} > 0$ for $x \in (-2, 4)$. Thus $g(x)$ has no critical numbers on $(-2, 4)$.

$$\begin{aligned} f(0) &= 0 &&\longleftarrow \text{ minimum value} \\ f(-2) &= 1 &&\longleftarrow \text{ maximum value} \\ f(4) &= \tfrac{1}{2} \end{aligned}$$

44. $f(x) = 2x^{73} + 7x^{15} + 4x$

$f'(x) = 146x^{72} + 105x^{14} + 4 > 0$ for all $x \in (-\infty, \infty)$. Thus $f(x)$ has no critical numbers in $(-\infty, \infty)$. Therefore f has no relative extrema on the interval $(-\infty, \infty)$.

45. Let $p_2(x) = ax^2 + bx + c$ and let $a \neq 0$. In fact, suppose first that $a > 0$. From $p_2'(x) = 2ax + b$ we see that p_2 has exactly one critical number: namely, $x = -\frac{b}{2a}$. If $x < -\frac{b}{2a}$ than $p_2(x) < 0$ whilst if $x > -\frac{b}{2a}$ then $p_2(x) > 0$; so p_2 has a relative minimum at $x = -\frac{b}{2a}$. This is actually an absolute minimum. The easiest way to see this is by completing the square:

$$\begin{aligned} p_2(x) \ + \ & a\left(x^2 + \frac{b}{a}x + \frac{c}{a}\right) \\ = \ & a\left\{\left(x + \frac{b}{2a}\right)^2 + \frac{c}{a} - \frac{b^2}{4a^2}\right\} \\ = \ & a\left(x + \frac{b}{2a}\right)^2 + \frac{4ac - b^2}{4a} \end{aligned}$$

which assumes a minimum value of $\dfrac{4ac - b^2}{4a}$ when $x = -\dfrac{b}{2a}$.

Similarly, if $a < 0$ then p_2 has an absolute maximum value of $\dfrac{4ac - b^2}{4a}$ at its unique critical number $x = -\dfrac{b}{2a}$.

46. The critical numbers of p_3 are the points x at which $p_3'(x) = 0$. Since p_3 is cubic, its derivative p_3' is quadratic and so has at most $N_{\max} = 2$ roots and at least $N_{\min} = 0$ roots.

Examples:

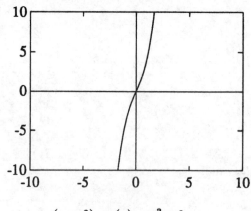

$(n = 0)$ $p_3(x) = x^3 + 3x$

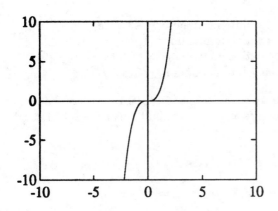

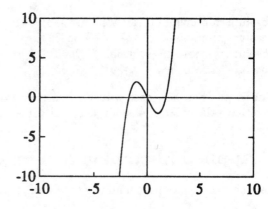

$(n = 1)$ $p_3(x) = x^3$ $\qquad\qquad$ $(n = 2)$ $p_3(x) = x^3 - 3x$

47. a) $p'^{+}(0) = \lim_{h \to 0^+} \dfrac{p(0 + h) - p(0)}{h} = \lim_{h \to 0^+} \dfrac{h^{3/2} - 0}{h}$

$\qquad\qquad = \lim_{h \to 0^+} h^{1/2} = 0.$

The function $p(x) = x^{3/2}$ can be said to have a critical number at 0 since its right-hand derivative is zero at 0.

b) $f'^{+}(a) = \lim_{h \to 0^+} \dfrac{f(a + h) - f(a)}{h}$

$\qquad\quad = \lim_{h \to 0^+} \dfrac{g(a + h) - g(a)}{h}$

$\qquad\quad = g'(a)$

since g is assumed differentiable at a and since $f(x) = g(x)$ if $x \in [a, b]$.

c) In the notation of part b) we let $f(x) = x$ and $g(x) = p(x) = |x|$. Since $f(x) = g(x)$ for $x \geq 0$ and since $f'(0)$ is surely 1, we deduce that $p'^{+}(0) = g'^{+}(0) = f'(0) = 1$.

d) The left-hand derivative $f'^{-}(b)$ of f at b is equal to

$$\lim_{h \to 0^-} \frac{f(b + h) - f(b)}{h}.$$

e) $\quad p(x) = x^4 - 2x^2$

$\Rightarrow \ p'(x) = 4x^3 - 4x$

$\qquad\qquad = 4x(x^2 - 1)$

$\Rightarrow \ p'(x) = 0$ when $x = 0, +1, -1$.

Thus, p has critical numbers at $0, +1, -1$. Strictly speaking, we deal with $x = -1$ using part b) and with $x = +1$ using the corresponding argument based on part c), since these are endpoints of $[-1, +1]$.

48. absolute minimum: $(-1, -227)$
relative minimum: $(3.06810, -133.07834)$
relative maximum: $(0.31778, 125.28998)$
absolute maximum: $(4, 227)$

49. absolute minimum: $(-3, -12 - \sqrt[3]{7})$
relative minimum: $(0.46987, -2.63767)$
relative minimum: $(1.22175, -3.15492)$
relative maximum: $(-1.25314, 2.77862)$
relative maximum: $(0.532, -2.10943)$
absolute maximum: $(3, 12 + \sqrt[3]{5})$

50. absolute minimum: $(2.11587, -1.96372)$
    absolute minimum: $(5.25746, -1.96372)$
    absolute maximum: $(0.90831, 2.37665)$
    absolute maximum: $(4.04991, 2.37665)$

51. absolute minimum: $(-3, -18)$
    relative minimum: $(1, -2)$
    relative maximum: $(-1, 2)$
    absolute maximum: $(3, 6)$

52. The times where the speed will be a relative maximum are where $s(t)$ is steepest, i.e., where the tangent lines have the greatest slope in absolute value. This occurs at $t = 1.5$, and at this time $s(1.5) = \frac{1}{8}$.

4.2   Applied Maximum-Minimum Problems

1. Let the two nonnegative numbers be x and y; thus $x + y = 20$ and so $y = 20 - x$.

   a) The product of the numbers is

   $$P \;=\; xy \;=\; x(20 - x)$$

   therefore
   $$\frac{dP}{dx} \;=\; 20 - 2x$$

   which is zero if $x = 10$, $P(10) = 100$, $P(0) = 0$, $P(20) = 0$. Thus the product is least when $x = 0$, $y = 20$.

   b) and c) From

   $$\begin{aligned} S \;&=\; x^2 + y^2 \;=\; x^2 + (20 - x)^2 \\ &=\; 2x^2 - 40x + 400 \end{aligned}$$

   we deduce that

   $$\frac{dS}{dx} \;=\; 4x - 40$$

   which is zero when $x = 10$. Now, $x = 10 \Rightarrow S = 200$, $x = 0$ (one endpoint) $\Rightarrow S = 400$ and $x = 20$ (other endpoint) $\Rightarrow S = 400$. So: the sum of the squares is

   a) least when the two numbers are both 10.

   b) greatest when one is 0 and the other 20.

2. If the numbers are x and y then $x + y = 36$, so $y = 36 - x$. The quantity to be maximized/minimized is

   $$\begin{aligned} f(x) \;&=\; x + y^2 \;=\; x + (36 - x)^2 \\ &=\; x^2 - 71x + 36^2. \end{aligned}$$

   Now
   $$f'(x) = 2x - 71$$

   which is zero when $x = \frac{71}{2}$ so that $f(x) = \frac{143}{4}$. Examining the endpoints of the relevant interval $[0, 36]$, $f(0) = 36^2$ and $f(36) = 36$. Thus: $f(x)$ is

   a) a maximum when $x = 0$ and $y = 36$,

   b) a minimum when $x = \frac{71}{2}$ and $y = \frac{1}{2}$.

3. Let the three sides of the fence have lengths (in meters) x, y and y, so that $x + 2y = 20$. The function to be optimized is the area enclosed, which is

$$A \; = \; xy \; = \; (20 - 2y)y$$

therefore
$$\frac{dA}{dy} \; = \; 20 - 4y$$

which is zero if $y = 5$. Also, A is zero at the endpoints of the relevant interval $[0, 10]$. Thus: the area enclosed is greatest when the rectangle is $10\,\mathrm{m}$ (measured paralled to the house) by $5\,\mathrm{m}$ (measured at right angles to the house).

4. If u denotes the number of uninfected individuals then $N - u$ is the number infected, so the rate of spread R is given by

$$R = ku(N - u)$$

for some constant of proportionality k. This function is to be optimized over $[0, N]$. Supposing u to be a continuous (rather than discrete) variable,

$$\frac{dR}{du} = k(N - 2u)$$

which is zero if $u = \frac{1}{2}N$. Also, $R = 0$ at the endpoints of $[0, N]$. Thus: the disease is spreading most rapidly (R is maximized) when $u = \frac{1}{2}N$, so that half the population is uninfected and half infected.

5. With the notation of the figure in the textbook, the total length of fence used is

$$120 \; = \; 6l + 3w$$
therefore
$$w \; = \; 40 - 2l.$$

Now, the total area of the pen is the function to optimize, namely

$$A \; = \; 3lw \; = \; 6l(20 - l)$$

therefore
$$\frac{dA}{dl} \; = \; 120 - 12l$$

which is zero if $l = 10$. Also, $A = 0$ at the endpoints $l = 0$ and $l = 20$ of $[0, 20]$. Thus: A is greatest when $l = 10$ and $w = 20$.

6. The slope of the graph of $y = x^3 - 9x^2 + 7x - 6$ at the point (x, y) is

$$m = \frac{dy}{dx} = 3x^2 - 18x + 7.$$

Now

$$\frac{dm}{dx} = 6x - 18$$

so that $\frac{dm}{dx} = 0$ if $x = 3$ and so $m = -20$. Examining the endpoints of $[1, 4]$, $x = 1 \Rightarrow m = -8$ and $x = 4 \Rightarrow m = -17$. Thus: slope is a minimum of -20 at $(3, -39)$ and a maximum of -8 at $(1, -7)$.

7. The box has height x and has base of length $24 - 2x$ and width $16 - 2x$; its volume is thus given by

$$\begin{aligned} V \; &= \; x(16 - 2x)(24 - 2x) \\ &= \; 4(x^3 - 20x^2 + 96x) \end{aligned}$$

where $0 \le x \le 8$ (see figure in textbook). Now

$$V'(x) = 4(3x^2 - 40x + 96)$$

which is zero when $x = \frac{40 \pm 8\sqrt{7}}{6} = \frac{20 \pm 4\sqrt{7}}{3}$. The critical number in $[0, 8]$ is $x = \frac{20 - 4\sqrt{7}}{3}$, at which $V > 0$; at the endpoints, $V = 0$. Consequently, the volume of the box is a maximum when the box has height $\frac{20 - 4\sqrt{7}}{3}$, width $\frac{8}{3}(1 + \sqrt{7})$ and length $\frac{8}{3}(4 + \sqrt{7})$.

8. Refer to the figure in the textbook. Let the rectangle have width x and height y. The rightmost small triangle then has width $8 - x$ and height y. Since this small triangle is similar to the large one, we see that $\frac{y}{8-x} = \frac{6}{8} = \frac{3}{4}$ so $y = \frac{3}{4}(8 - x)$. Clearly, the appropriate range for x is $0 \leq x \leq 8$. The rectangle has area

$$A \;=\; \text{(base)(height)} \;=\; xy \;=\; \frac{3}{4}x(8 - x) \;=\; 6x - \frac{3}{4}x^2$$

so

$$\frac{dA}{dx} \;=\; 6 - \frac{3}{2}x$$

which is zero when $x = 4$ (and $y = 3$) so that $A = 12 > 0$. At the endpoints ($x = 0$ and $x = 8$) we see that $A = 0$. Thus, area is maximized when the rectangle has width 4cm and height 3cm.

9. From the figure in the textbook, note that $l + 2h = 10$ and $2h + 2w = 16$, whilst the box has volume $V = wlh$. Choose one dimension in terms of which to express the other two: say h. Then $l = 10 - 2h$ and $w = 8 - h$, so $V = (8 - h)(10 - 2h)h = 2h(5 - h)(8 - h) = 2(h^3 - 13h^2 + 40h)$; the appropriate range of values for h is $0 \leq h \leq 5$. Now $\frac{dV}{dh} = 2(3h^2 - 26h + 40) = 2(h - 2)(3h - 20)$ so V has critical numbers $h = 2$ and $h = \frac{20}{3}$, of which only $h = 2$ lies in $[0, 5]$. At the endpoints, $V = 0$; so the volume of the box is a maximum of $72\,\text{cm}^3$ when $h = 2$, $l = 6$, $w = 6$.

10. Let the window have width $2r$ and overall height $h + r$ (in metres). Its perimeter is then $2r + 2h + \pi r = 10$, so $h = S - \left(\frac{\pi}{2} + 1\right)r$. Its area is

$$
\begin{aligned}
A \;&=\; 2rh + \frac{1}{2}\pi r^2 \\
&=\; 2r\left(5 - \left(\frac{\pi}{2} + 1\right)r\right) + \frac{1}{2}\pi r^2 \\
&=\; 10r - \left(2 + \frac{\pi}{2}\right)r^2,
\end{aligned}
$$

so

$$\frac{dA}{dr} \;=\; 10 - (4 + \pi)r$$

which is zero when $r = \frac{10}{4 + \pi}$. The appropriate range of values for r is from $r = 0$ (window has no width) to $r = \frac{10}{2 + \pi}$ (window is semicircular). Now

$$
\begin{aligned}
r \;&=\; 0 \;\Rightarrow\; A = 0 \\
r \;&=\; \frac{10}{4 + \pi} \;\Rightarrow\; A = \frac{50}{4 + \pi} \\
r \;&=\; \frac{10}{2 + \pi} \;\Rightarrow\; A = \frac{50\pi}{(2 + \pi)^2}.
\end{aligned}
$$

Of these values, $\frac{50}{4 + \pi}$ is the greatest; for example,

$$\frac{\frac{50}{4 + \pi}}{\frac{50\pi}{(2 + \pi)^2}} \;=\; \frac{(2 + \pi)^2}{(4 + \pi)\pi} \;=\; \frac{\pi^2 + 4\pi + 4}{\pi^2 + 4\pi} \;>\; 1.$$

Thus the area of the window is maximized when the semicircular part has radius $r = \frac{10}{4+\pi}$ and the rectangular part has height $h = 5 - \left(\frac{\pi+2}{2}\right)\left(\frac{10}{\pi+4}\right) = \frac{10}{4+\pi}$.

11. Bisect the triangle vertically as in the accompanying figure. Let the whole rectangle have width $2x$ and height y (in centimeters). Since the large triangle and the lower right triangle are similar, we have

$$\frac{y}{3-x} = \frac{4}{3}$$

therefore

$$y = 4 - \frac{4}{3}x.$$

The area A of the whole rectangle is the function to optimize, namely

$$A = 2xy = 2x\left(4 - \frac{4}{3}x\right)$$

so

$$\frac{dA}{dx} = 8 - \frac{16}{3}x$$

which is zero when $x = \frac{3}{2}$. Also, $A = 0$ at the endpoints of the relevant interval $[0,3]$ for x. Thus the rectangle's area is greatest when its width is 3cm and its height 2 cm.

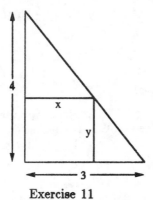

Exercise 11

12. Let the cone have base radius r and height h, so that (Pythagoras) $r^2 + h^2 = 10^2$. The volume of the cone is the function to be optimized, namely

$$V = \frac{1}{3}\pi r^2 h = \frac{1}{3}\pi h(100 - h^2)$$

so

$$\frac{dV}{dh} = \frac{100\pi}{3} - \pi h^2$$

which is 0 when $h = \frac{10}{\sqrt{3}}$. At the endpoints of the relevant interval $[0,10]$ for h, the volume V is zero. The volume of the cone is therefore greatest when the sides of the triangle are $\frac{10}{\sqrt{3}}$ cm (about which the triangle is rotated) and $10\sqrt{\frac{2}{3}}$ cm.

13. Let the rectangle have width $2x$ and height y so that $y = 4 - x^2$ as in the figure. Its area A is then the function to optimize, given by

$$A = 2xy = 2x(4 - x^2)$$

therefore

$$\frac{dA}{dx} = 8 - 6x^2$$

which is 0 if $x = \frac{2}{\sqrt{3}}$. Also, $A = 0$ at the endpoints of the relevant interval $0 \le x \le 2$. Thus the area of the rectangle is greatest when its width is $\frac{4}{\sqrt{3}}$ units and its height $\frac{8}{3}$ units.

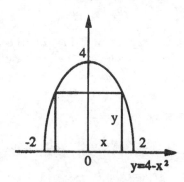

Exercise 13

14. Drop a perpendicular of height h units as in the figure. By trigonometry, $h = a\sin\theta$ so that the area of the triangle (the function to optimize) is

$$A \;=\; \frac{1}{2}\,bh \;=\; \frac{1}{2}\,ab\sin\theta$$

therefore

$$\frac{dA}{d\theta} \;=\; \frac{1}{2}\,ab\cos\theta$$

which is zero when $\cos\theta = 0$ therefore $\theta = \frac{\pi}{2}$. At the endpoints of the appropriate interval $0 \le \theta \le \pi$, the area is zero. Thus the area of the triangle is greatest when $\theta = \frac{\pi}{2}$.

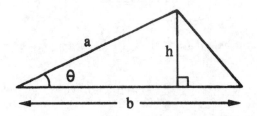

Exercise 14

15. The fixed perimeter of the sector is given by $P = 2r + r\theta$ and its area (the function to optimize) by

$$A = \frac{1}{2}r^2\theta = \frac{1}{2}r(P - 2r).$$

Now

$$\frac{dA}{dr} = \frac{1}{2}P - 2r$$

which is zero when $r = \frac{1}{4}P$. At the endpoints of the relevant interval $\left[0, \frac{P}{2}\right]$ for r, the area is zero. Thus the sectorial area is maximized when $r = \frac{1}{4}P$ and $\theta = \frac{P}{r} - 2 = 2$.

16. Let the legs of the right triangle have lengths x and y units, so (Pythagoras) $x^2 + y^2 = 2^2$. The area of the triangle is the function to optimize, given by

$$A = \frac{1}{2}xy = \frac{1}{2}x\sqrt{4-x^2}$$

therefore

$$\frac{dA}{dx} = \frac{1}{2}\sqrt{4-x^2} - \frac{\frac{1}{2}x^2}{\sqrt{4-x^2}}$$
$$= \frac{2-x^2}{\sqrt{4-x^2}}$$

which is zero when $x = \sqrt{2}$. Also, $A = 0$ at the endpoints of the relevant x interval $[0, 2]$. Thus the triangle has greatest area when its legs have equal lengths of $\sqrt{2}$ units.

17. Let T be the number of trees per acre and Y the yield per tree; we are told that $\frac{dY}{dT} = -15$ and that $Y = 495$ when $T = 25$, so

$$Y - 495 = (-15)(T - 25)$$

therefore

$$Y = 870 - 15T.$$

If A (the function to optimize) is the yield per acre then

$$A = TY = T(870 - 15T)$$

therefore

$$\frac{dA}{dT} = 870 - 30T$$

which is zero when $T = 29$. The relevant interval is $0 \leq T \leq 58$, at the endpoints of which A is zero. Thus, yield per acre will be maximized by planting four more trees per acre.

18. Note that the distance from the center of the circle to one of the vertices lying on the semicircle is 8 cm. If we let the rectangle have width $2x$ and height y (in centimeters) then it follows (Pythagoras) that $x^2 + y^2 = 8^2$. The area of the rectangle is the function to optimize, given by

$$A = 2xy = 2x\sqrt{64 - x^2}$$
$$\frac{dA}{dx} = \frac{4(32 - x^2)}{\sqrt{64 - x^2}}.$$

therefore

This is zero when $x = 4\sqrt{2}$. At the endpoints of the relevant interval $[0, 8]$, the area is zero. Consequently, the rectangle of largest area has width $8\sqrt{2}$ cm and height $4\sqrt{2}$ cm.

19. Let the northeast vertex of the rectangle have coordinates (x, y) when the circle is centered at the origin, so that $x^2 + y^2 = 16$. The area of the rectangle is the function to maximize, namely

$$A = 4xy = 4x\sqrt{16 - x^2}$$

therefore

$$\frac{DA}{dx} = \frac{8(8 - x^2)}{\sqrt{16 - x^2}}.$$

If $x = 2\sqrt{2}$ then $\frac{dA}{dx} = 0$; at the endpoints of the relevant interval $0 \leq x \leq 4$, $A = 0$. Thus the area of the rectangle is greatest when the rectangle is a square of side $4\sqrt{2}$ and area 32 square units.

20. Let the beam have width w and depth d, so that $w^2 + d^2 = 20^2$. The strength of the beam (the function to maximize) is given by

$$S = kwd^2 = kw(400 - w^2)$$

for some constant of proportionality k, so

$$\frac{dS}{dw} = k(400 - 3w^2)$$

which is zero when $w = \frac{20}{\sqrt{3}}$. At the endpoints of the relevant w-interval $[0, 20]$, strength is zero. Thus: the beam is strongest when its width is $\frac{20}{\sqrt{3}}$ cm and its depth $20\sqrt{\frac{2}{3}}$ cm.

21. Let the inscribed cylinder have base radius r and height h (in centimeters). The accompanying figure shows a vertical section of half the cone through its axis. The large triangle is similar to that on the lower right, so

$$\frac{h}{3-r} = \frac{5}{3} \quad \text{therefore } h = 5 - \frac{5}{3}r.$$

The cylinder has volume (the function to maximize) given by

$$\begin{aligned} V &= \frac{1}{3}\pi r^2 h = \frac{1}{3}\pi r^2\left(5 - \frac{5}{3}r\right) \\ &= \frac{5}{3}\pi\left(r^2 - \frac{1}{3}r^3\right) \end{aligned}$$

therefore

$$\frac{dV}{dr} = \frac{5}{3}\pi(2r - r^2)$$

which is zero when $r = 0$ and $r = 2$. Also, $V = 0$ at the endpoints of the relevant interval $0 \le r \le 3$. Thus the volume is maximized when the cylinder has base radius 2 cm and height $\frac{5}{3}$ cm.

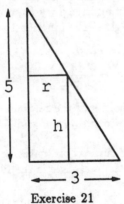

5

r

h

3

Exercise 21

22. Let the semivertex angle of the cone be θ, let its base radius be r and let its height be h. Its slant length is then $20\cos\theta$, so that $h = 20\cos^2\theta$ and $r = 20\cos\theta\sin\theta$, and its volume is then

$$V = \frac{1}{3}\pi r^2 h = \frac{20^3}{3}\pi\cos^4\theta\sin^2\theta.$$

Now to maximize V over the interval $0 \le \theta \le \frac{\pi}{2}$, consider first

$$\begin{aligned} \frac{dV}{d\theta} &= \frac{20^3}{3}\pi[-4\cos^3\theta\sin^3\theta + 2\cos^5\theta\sin\theta] \\ &= 2\frac{20^3}{3}\pi\cos^3\theta\sin\theta(\cos^2\theta - 2\sin^2\theta) \end{aligned}$$

which is zero when $\cos\theta = 0$, $\sin\theta = 0$ or $\cos^2\theta = 2\sin^2\theta$. Here, $\cos\theta = 0 \Leftrightarrow \theta = \frac{\pi}{2}$ and $\sin\theta = 0 \Leftrightarrow \theta = 0$ (corresponding to the physical limits: if $\theta = \frac{\pi}{2}$ then the cone becomes a point; if $\theta = 0$ then the cone is a rod) and $V = 0$ in both cases. If $\cos^2\theta = 2\sin^2\theta$ then in fact $\cos^2\theta = \frac{2}{3}$ and $\sin^2\theta = \frac{1}{3}$ (since $\cos^2\theta + \sin^2\theta = 1$) so that $V = 4\pi\frac{20^3}{3^5}$, which is a maximum. Thus: the volume of the cone is greatest when its height is $h = 20\cos^2\theta = \frac{40}{3}$ and its base radius is $r = 20\sin\theta\cos\theta = \frac{20\sqrt{2}}{3}$ (in centimeters).

23. With the given data, the growth rate is

$$\frac{dy}{dt} = \alpha + \beta\gamma\pi\cos(\gamma\pi t)$$

$$= 3 + \frac{2}{3}\pi\cos\left(\frac{1}{3}\pi t\right)$$

therefore

$$\frac{d}{dt}\left(\frac{dy}{dt}\right) = -\frac{2\pi^2}{9}\sin\left(\frac{1}{3}\pi t\right)$$

which (for $0 \le t \le 9$) is zero when $t = 0, 3, 6, 9$.

t	0	3	6	9
$\frac{dy}{dt}$	$3 + \frac{2}{3}\pi$	$3 - \frac{2}{3}\pi$	$3 + \frac{2}{3}\pi$	$3 - \frac{2}{3}\pi$

From the above table, minimum growth occurs at 3 months and 9 months whilst maximum growth occurs at birth and 6 months.

24. $f(x) = x - \sin x \Rightarrow f'(x) = 1 - \cos x$, which is zero when $x = 0$ in $[0, 1]$. Since $f(0) = 0$ and $f(1) = 1 - \sin 1 > 0$, it follows that $f(0) = 0$ is the minimum value of f on $[0, 1]$ so that if $x \in [0, 1]$ then $f(x) \ge 0$, that is, $x \ge \sin x$. If $x > 1$ then of course $x > \sin x$ since $|\sin x| \le 1$. Thus: $x \ge 0 \Rightarrow x \ge \sin x$, as required.

25. Let the northeast vertex of the rectangle be at (x, y), so that $x^2 + 4y^2 = 4$ and the area of the rectangle is the function to optimize, given by

$$A = 4xy = 8y\sqrt{1 - y^2}.$$

Now

$$\frac{dA}{dy} = 8\frac{(1 - 2y^2)}{\sqrt{1 - y^2}}$$

which is zero if $y = \frac{1}{\sqrt{2}}$. Further, $A = 0$ at the endpoints of the appropriate interval $0 \le y \le 1$. Thus: the rectangle of maximum area has height $2y = \sqrt{2}$ units and width $2x = 2\sqrt{2}$ units.

26. If S (the function to minimize) denotes the square of the distance from $(1, 0)$ to (x, y) on the ellipse then

$$S = (x - 1)^2 + y^2$$

$$= x^2 - 2x + 1 + 1 - \frac{x^2}{4}$$

$$= \frac{3}{4}x^2 - 2x + 2$$

so

$$\frac{dS}{dx} = \frac{3}{2}x - 2$$

which is zero when $x = \frac{4}{3}$. Now, since x is restricted to $[-2, 2]$ we examine S when $x = \pm 2$ (endpoints) and $x = \frac{4}{3}$ (critical). If $x = \frac{4}{3}$ then $S = \frac{2}{3}$, if $x = -2$ then $S = 9$ and if $x = 2$ then $S = 1$; so the points sought for are

$$\left(\frac{4}{3}, \pm\sqrt{1 - \frac{\left(\frac{4}{3}\right)^2}{4}} \right) = \left(\frac{4}{3}, \pm\frac{\sqrt{5}}{3} \right).$$

See also **Remark 3** in Section 4.4.

27. Let the circle have radius r and the square have side s (in centimeters). The respective perimeters are then $2\pi r$ (circle) and $4s$ (square) and we are given that $2\pi r + 4s = 50$. Now, the area enclosed by the circle and square together is to be optimized, given by

$$\begin{aligned} A &= \pi r^2 + s^2 \\ &= \pi r^2 + \left(\frac{25}{2} - \frac{\pi r}{2} \right)^2 \\ &= \left(\pi + \frac{\pi^2}{4} \right) r^2 - \frac{25\pi}{2} r + \frac{625}{4} \end{aligned}$$

so

$$\frac{dA}{dr} = \left(2\pi + \frac{\pi^2}{2} \right) r - \frac{25\pi}{2}$$

which is zero when $r = \frac{25}{4+\pi}$ and then $A = \frac{625}{4+\pi}$. The physical limits to the problem are $r = 0$ (no circle: $A = \frac{625}{4}$) and $r = \frac{25}{\pi}$ (no square: $A = \frac{625}{\pi}$). Inspecting these three values for A, the figures will enclose

a) the greatest area $\left(\frac{625}{\pi} \text{ cm}^2 \right)$ when $r = \frac{25}{\pi}$ (do not cut the wire, form a circle) and

b) the least $\left(\frac{625}{4+\pi} \text{ cm}^2 \right)$ when $r = \frac{25}{4+\pi}$ (cut the wire at $\frac{50\pi}{4+\pi}$ cm).

28. Let the parcel have length l, width w and height h, so that its girth g is $2w + 2h$ and so that $l + 2w + 2h = 100$. If $g = 2w + 2h$ is fixed, then the parcel has cross-sectional area

$$A = wh = w\left(\frac{g}{2} - w \right)$$

therefore
$$\frac{dA}{dw} = \frac{g}{2} - 2w$$

which is zero (and, checking endpoints $w = 0$ and $w = \frac{g}{2}$, maximizes A) when $w = \frac{g}{4}$ so that $w = h$. Next, taking $w = h$ we see that $l + 4w = 100$ and that the parcel has volume

$$V = lwh = (100 - 4w)w^2$$

therefore
$$\frac{dV}{dw} = 200w - 12w^2$$

which is zero when $w = 0$ and $w = \frac{200}{12} = \frac{50}{3}$. The size of w to consider is over the interval $[0, 25]$ ($w = 0$ corresponds to a "long,thin" parcel; $w = 25$ to a flat parcel). Now $w = \frac{50}{3} \Rightarrow V = \frac{250{,}000}{27}$, whilst $V = 0$ when $w = 0$ and when $w = 25$. Thus the parcel has maximum volume when its width and height are $\frac{50}{3}$ inches and its length is $\frac{100}{3}$ inches.

29. With the notation of the figure in the textbook, the route shown covers a distance $300 - x$ on land and a distance $\sqrt{100^2 + x^2}$ by sea, and so takes a time

$$T = \frac{300 - x}{5} + \frac{\sqrt{100^2 + x^2}}{3},$$

which is the function to minimize. Now

$$\frac{dT}{dx} = -\frac{1}{5} + \frac{x}{3\sqrt{100^2 + x^2}}$$

which is zero when

$$3\sqrt{100^2 + x^2} = 5x$$

therefore

$$9(100^2 + x^2) = 25x^2$$

therefore

$$9 \times 100^2 = 16x^2$$

therefore

$$x = 75.$$

If $x = 75$ then $T = \frac{260}{3}$, if $x = 0$ then $T = \frac{280}{3}$ and if $x = 300$ then $T = \frac{100\sqrt{10}}{3}$. Thus: in order to minimize time, the swimmer should aim to reach the shore $300 - 75 = 225$ meters from the person in distress.

30. Revenue per month is the function to maximize, given by

$$R = px = p(2000 - 100p)$$
therefore
$$\frac{dR}{dp} = 2000 - 200p$$

which is zero if $p = 10$. Moreover, $R = 0$ at the endpoints of the relevant interval $0 \le p \le 20$. Thus, monthly revenue is greatest when selling price p is 10.

31. Now monthly profit P (to be maximized) is given by

$$
\begin{aligned}
P &= \text{revenue} - \text{cost} \\
&= px - (500 + 10x) \\
&= 3000p - 100p^2 - 20500
\end{aligned}
$$

so

$$\frac{dP}{dp} = 3000 - 200p$$

which is zero if $p = 15$. In addition, $P = 0$ at the endpoints of the appropriate interval $\left[15 - 2\sqrt{5},\ 15 + 2\sqrt{5}\right]$. So: monthly profits are maximized when $p = 15$.

32. Let r be the number of daily rentals and f the daily rental fee. We are given that $\frac{dr}{df} = -10$ and that $r = 500$ when $f = 30$, so

$$
\begin{aligned}
r - 500 &= -10(f - 30) \\
r &= 800 - 10f.
\end{aligned}
$$

therefore

Now, daily revenue R is the function to maximize, given by

$$
\begin{aligned}
R &= fr = f(800 - 10f) \\
\frac{dR}{df} &= 800 - 20f
\end{aligned}
$$

therefore

which is zero when $f = 40$. Examining the endpoints of the relevant interval $0 \le f \le 80$ as usual reveals that this rental fee of \$40 maximizes revenue.

33. Let the number of rooms rented per day be y and the daily rate be \$$x$ per room. We are told that

$$
\begin{aligned}
y - 200 &= (-4)(x - 40) \\
y &= 360 - 4x.
\end{aligned}
$$

therefore

Now daily revenue R (to be maximized) is given by

$$
\begin{aligned}
R &= xy = x(360 - 4x) \\
\tfrac{dR}{dx} &= 360 - 8x
\end{aligned}
$$

therefore

which is zero when $x = 45$. Checking endpoints of the appropriate interval $0 \le x \le 90$ as usual, we see that this rental rate of \$45 maximizes revenue.

34.

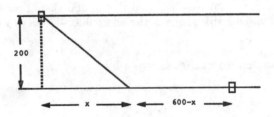

Exercise 34

With the notation of the figure above, if the cable hits shore a distance of $600 - x$ (meters) from the boathouse then it travels a distance of $\sqrt{200^2 + x^2}$ underwater and $600 - x$ on land, incurring a total cost of C (dollars) where

$$
\begin{aligned}
C &= 30(600 - x) + 50\sqrt{200^2 + x^2} \\
\frac{dC}{dx} &= -30 + \frac{50x}{\sqrt{200^2 + x^2}}
\end{aligned}
$$

therefore

which is zero when $x = 150$. Now, $x = 150 \Rightarrow C = 13850$, $x = 0$ (one end of the relevant interval) $\Rightarrow C = 28000$ and $x = 600$ (the other end) $\Rightarrow C = 10000\sqrt{10}$; so cost is minimized if $x = 150$, so that the cable hits shore a distance of $450\,\text{m}$ from the boathouse.

35. If S is the survival rate at distance x then $S = kdp$ for some constant of proportionality k. With the given data, this function to maximize is

216

$$S = \frac{1}{10} k \frac{x}{1 + \left(\frac{x}{5}\right)^2}$$

therefore

$$\frac{dS}{dx} = \frac{1}{10} k \frac{\left(1 - \frac{x^2}{25}\right)}{\left(1 + \frac{x^2}{25}\right)}$$

which is zero when $x = 5$. Now, $x = 5 \Rightarrow S = \frac{k}{4}$, $x = 0$ (one end of the given interval) $\Rightarrow S = 0$ and $x = 10$ (the other end) $\Rightarrow S = \frac{k}{5}$; so survival rate is a maximum when $x = 5$: at a distance of $5\,m$ from the trunk.

36. Here, the function to maximize is

$$S = kdp = \frac{k}{x(x - 10)^2}$$

therefore

$$\frac{dS}{dx} = \frac{k(10 - 3x)}{x^2(x - 10)^3}$$

which is zero when $x = \frac{10}{3}$. Now $x = \frac{10}{3} \Rightarrow S = \frac{27k}{4000}$, $x = 1$ (one endpoint) $\Rightarrow S = \frac{k}{81}$ and $x = 9$ (the other) $\Rightarrow S = \frac{k}{9}$; of these values for S, the least is $\frac{27k}{4000}$. Thus, survival rate is a minimum at $\frac{10}{3}\,m$ from the trunk.

37. a) $v = \dfrac{(\rho - 100)^2}{100} \Rightarrow \dfrac{dv}{d\rho} = \dfrac{\rho - 100}{50}$

   which is zero when $\rho = 100$. Since $\rho = 0 \Rightarrow v = 100$ and $\rho = 100 \Rightarrow v = 0$, it follows that velocity will be a maximum when $\rho = 0$ (!).

   b) $q = \dfrac{\rho(\rho - 100)^2}{100} \Rightarrow \dfrac{dq}{d\rho} = \dfrac{3\rho^2 - 400\rho + 10000}{100}$

   which is zero when $\rho = 100$ and when $\rho = \frac{100}{3}$. Now, $\rho = \frac{100}{3} \Rightarrow q = \frac{40000}{27}$, $\rho = 100 \Rightarrow q = 0$ and $\rho = 0 \Rightarrow q = 0$; of these values for q, the greatest is $\frac{40000}{27}$. So: maximum flow rate occurs when $\rho = \frac{100}{3}$, so that there are 100 automobiles in every 3 km.

38. This problem requires an interpretation, of which the following is a suitable one. We consider a ray of light from A to B reflecting at P, where distances are as in the accompanying figure, and seek to minimize the distance D travelled, where (by Pythagoras)

$$D = \sqrt{a^2 + x^2} + \sqrt{b^2 + (l - x)^2}.$$

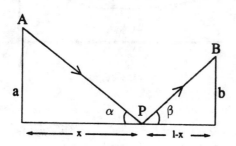

Exercise 38

217

Now

$$\frac{dD}{dx} = 2\frac{x\sqrt{b^2+(l-x)^2}+(l-x)\sqrt{a^2+x^2}}{\sqrt{a^2+x^2}\sqrt{b^2+(l-x)^2}}$$

which is zero when

$$x\sqrt{b^2+(l-x)^2} = (x-l)\sqrt{a^2+x^2}$$

which upon squaring and cancelling yields

$$x^2 b^2 = a^2(l-x)^2$$

so that

$$\tan\alpha = \frac{a}{x} = \frac{b}{l-x} = \tan\beta$$

And therefore $\alpha = \beta$ (since α and β are acute). In this case,

$$D = \sqrt{(a+b)^2 + l^2}$$

whilst if $x=0$ then $D = a+\sqrt{b^2+l^2}$ and if $x=l$ then $D = \sqrt{a^2+l^2}+b$. The least of these values of D is indeed $\sqrt{(a+b)^2+l^2}$. [We remark that there are other (shorter) solutions to this problem of a more geometrical nature: find one, or ask the instructor].

39. a) From the diagram $\tan\theta = \frac{8-h}{1.5}$, so $h = 8 - 1.5\tan\theta$. Also $\cos\theta = \frac{1.5}{l}$ so $l = 1.5\sec\theta$. Therefore,

$$\begin{aligned} C(\theta) &= 275(1.5\sec\theta) + 185(8 - 1.5\tan\theta) \\ &= 412.5\sec\theta - 277.5\tan\theta + 1480. \end{aligned}$$

b) The domain is $0 \le \theta \le \theta_0$ where $\tan\theta_0 = \frac{8}{1.5}$. So $\theta = 1.385448$ radians or $79.380345°$.

c) Graph using the range $[0,1.4] \times [1600,2400]$. The minimum is at $\theta = 0.7368$ with $C(\theta) = 1785.21$.

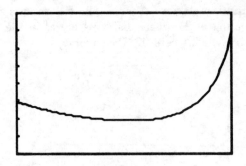

Exercise 39. c)

d) $$\begin{aligned} C'(\theta) &= 412.5\sec\theta\tan\theta - 277.5\sec^2\theta \\ &= \sec\theta(412.5\tan\theta - 277.5\sec\theta). \end{aligned}$$

Zooming in we find $C'(\theta) = 0$ for $\theta = 0.737889$. For this value of θ the cost is $C(0.737889) = 1785.20$. A more precise value of θ can be found by solving

$$412.5\tan\theta - 277.5\sec\theta = 0$$

or

$$\sin \theta = \frac{277.5}{412.5} = 0.6727272727.$$

Therefore $\theta_* = 0.7378886885$ radians or $42.27791°$. This gives a distance of $l = 1.5 \sec \theta_* = 2.02733$ miles under water and $h = 8 - 1.5 \tan \theta_* = 6.63616$ miles on land.

40. a) Let s be the cost of steel and a be the cost of aluminum. Therefore $s = 2.5a$. Now the area of the top $= \pi r^2$ so the cost of the top $= s\pi r^2 = 2.5\pi a r^2$. The area of the sides and bottom $= \pi r^2 + 2\pi r h$ and so the cost of the sides and bottom $= \pi a r(r + 2h)$. The volume of the can is $12 \, oz$ so $V = 12(1.80469) \, in^3 = 21.65628 \, in^2$. Also $V = \pi r^2 h$ so $h = \dfrac{21.65628}{\pi r^2} = \dfrac{6.89341}{r^2}$. Therefore the cost as function of a and r is

$$
\begin{aligned}
C(r, a) &= \pi a r \left(r + \frac{2(6.89341)}{r^2} \right) \\
&= \pi a \left(r^2 + \frac{13.78682}{r} \right).
\end{aligned}
$$

b) Below are the graphs for $a = 0.5, 1, 2, 3.5$ using the range: $[0, 6] \times [0, 180]$. Note that all have an absolute minimum at about $r = 1.9$.

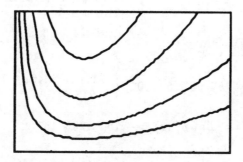

Exercise 40. b)

To find a precise value for the absolute minimum we must solve:

$$\frac{dC}{dr} = \pi a \left(2r - \frac{13.8682}{r^2} \right) = 0.$$

This gives the equation: $2r^3 - 13.78682 = 0$ so $r = \sqrt[3]{13.78682} = 1.90317$.

c) Dimensions: $r = 1.90317 \, in$ and $h = \dfrac{689341}{(1.90317)^3} = 1.90317 \, in$.

41. a) Below is the graph of the function using the range: $[0, 6] \times [-0.00025, 0]$.

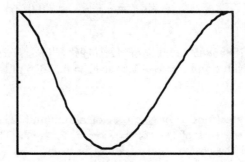

Exercise 41. a)

From the graph the maximum deflection is 0.000241 at about $x = 2.53$.

b) Note the $y(3) = -0.0002278125$ and $\lim\limits_{x \to 3^+} y(x) = -0.0002278125$ so $y(x)$ is continuous at $x = 3$. Also note the following computations:

$$y'(x) = 3.125 \times 10^{-7} \begin{cases} 20x^4 - 240x^3 + 972x^2 - 1242x & \text{if } x \leq 3 \\ -108x^2 + 918x - 1620 & \text{if } x > 3. \end{cases}$$

$y'(3) = 0.000050625 = \lim\limits_{x \to 3^+} y'(x).$

$$y''(x) = 3.125 \times 10^{-7} \begin{cases} 80x^3 - 720x^2 + 1944x - 1242 & \text{if } x \leq 3 \\ -216x + 918 & \text{if } x > 3. \end{cases}$$

$y''(3) = 0.000084375 = \lim\limits_{x \to 3^+} y''(x).$

$$y^{(3)}(x) = 3.125 \times 10^{-7} \begin{cases} 240x^2 - 1440x + 1944 & \text{if } x \leq 3 \\ -216 & \text{if } x > 3. \end{cases}$$

$y^{(3)}(3) = -0.0000675 = \lim\limits_{x \to 3^+} y'''(x) = -216(3.125 \times 10^{-7}).$

$$y^{(4)}(x) = 3.125 \times 10^{-7} \begin{cases} 480x - 1440 & \text{if } x \leq 3 \\ 0 & \text{if } x > 3. \end{cases}$$

$y^{(4)}(3) = 0 = \lim\limits_{x \to 3^+} y^{(4)}(x).$

$$y^{(5)}(x) = 3.125 \times 10^{-7} \begin{cases} 480 & \text{if } x \leq 3 \\ 0 & \text{if } x > 3. \end{cases}$$

$y^{(5)}(3) = 480 \neq \lim\limits_{x \to 3^+} y^{(5)}(x) = 0.$ These computations show that $y(x)$, $y'(x)$, $y^{(3)}(x)$, and $y^{(4)}(x)$ are all continuous, but $y^{(5)}(x)$ is not continuous.

4.3 The Mean Value Theorem

1. $f(x) = x^2 - 2x + 1$ on $[0, 2]$. Clearly, f is continuous on $[0, 2]$ and differentiable on $(0, 2)$. Also, $f(0) = f(2) = 1$. $f'(x) = 2x - 2$; $f'(c) = 2c - 2 = 0$ when $c = 1$.

2. $f(x) = 1 - x^2$ on $[-1, 1]$. f is continuous on $[-1, 1]$ and differentiable on $(-1, 1)$. $f(-1) = f(1) = 0$. $f'(x) = -2x$; $f'(c) = -2c = 0$ when $c = 0$.

3. $f(x) = \sin 3x$ on $[0, 2\pi]$. Clearly, f is continuous on $[0, 2\pi]$ and differentiable on $(0, 2\pi)$.

$$f(0) = f(2\pi) = 0$$
$$f'(x) = 3\cos 3x;$$
$$f'(c) = 3\cos 3c = 0 \text{ when } \cos 3c = 0$$
$$\Rightarrow 3c = \frac{\pi}{2}, \frac{3\pi}{2}, \frac{5\pi}{2}, \frac{7\pi}{2}, \frac{9\pi}{2}, \frac{11\pi}{2} \Rightarrow c = \frac{\pi}{6}, \frac{\pi}{2}, \frac{5\pi}{6}, \frac{7\pi}{6}, \frac{3\pi}{2}, \frac{11\pi}{6}.$$

4. $f(x) = 1 + \cos 2x$ on $[0, \pi]$. Clearly, f is continuous on $[0, \pi]$ and differentiable on $(0, \pi)$. $f(0) = f(\pi) = 2$. $f'(x) = -2\sin 2x$; $f'(c) = -2\sin 2c = 0$ when $\sin 2c = 0 \Rightarrow 2c = 0,\ \pi \Rightarrow c = \frac{\pi}{2}$.

5. $f(x) = 3x + 1$, $x \in [0, 3]$. Clearly, f is continuous on $[0, 3]$ and differentiable on $(0, 3)$.

$$f'(x) = 3;\ f'(c) = 3.$$
$$\frac{f(3) - f(0)}{3 - 0} = \frac{9}{3} = 3 = f'(c)$$

Thus all numbers in $(0, 3)$ will satisfy the equation $f'(c) = \dfrac{f(b) - f(a)}{b - a}$.

6. $f(x) = 1 - 2x$, $x \in [-1, 2]$. Clearly, f is continuous on $[-1, 2]$ and differentiable on $(-1, 2)$.

$$f'(x) = -2,\ f'(c) = -2$$
$$\frac{f(2) - f(-1)}{2 - (-1)} = \frac{-3 - 3}{3} = -2 = f'(c)$$

Thus all numbers in $(-1, 2)$ will satisfy the equation $f'(c) = \dfrac{f(b) - f(a)}{b - a}$.

7. $f(x) = \cos x$, $x \in \left[\dfrac{\pi}{4}, \dfrac{7\pi}{4}\right]$. Clearly, f is continuous on $\left[\dfrac{\pi}{4}, \dfrac{7\pi}{4}\right]$ and differentiable on $\left(\dfrac{\pi}{4}, \dfrac{7\pi}{4}\right)$.

$$f'(x) = -\sin x,\ f'(c) = -\sin c$$
$$-\sin c = \frac{f\left(\frac{7\pi}{4}\right) - f\left(\frac{\pi}{4}\right)}{\frac{7\pi}{4} - \frac{\pi}{4}} = \frac{\frac{\sqrt{2}}{2} - \frac{\sqrt{2}}{2}}{\frac{3\pi}{2}} = 0$$

Thus $c = \pi$.

8. $f(x) = 4 - x^2$, $x \in [0, 2]$. Clearly, f is continuous on $[0, 2]$ and differentiable on $(0, 2)$.

$$f'(x) = -2x,\ f'(c) = -2c$$
$$-2c = \frac{f(2) - f(0)}{2 - 0} = \frac{0 - 4}{2} = -2$$

Thus $c = 1$.

9. $f(x) = x^2 + 2x$, $x \in [0, 4]$. Clearly, f is continuous on $[0, 4]$ and differentiable on $(0, 4)$.

$$f'(x) = 2x + 2;\ f'(c) = 2c + 2$$
$$2c + 2 = \frac{f(4) - f(0)}{4 - 0} = \frac{24 - 0}{4} = 6$$

Thus $c = 2$.

10. $f(x) = 3\sin 2x$, $x \in \left[0, \frac{\pi}{2}\right]$. f is continuous on $\left[0, \frac{\pi}{2}\right]$ and differentiable on $\left(0, \frac{\pi}{2}\right)$.

$$f'(x) = 6\cos 2x; \ f'(c) = 6\cos 2c$$

$$6\cos 2c = \frac{f\left(\frac{\pi}{2}\right) - f(0)}{\frac{\pi}{2} - 0} = \frac{0 - 0}{\frac{\pi}{2}} = 0$$

Thus $2c = \frac{\pi}{2} \Rightarrow c = \frac{\pi}{4}$.

11. $f(x) = x^2 + 2x - 3$, $x \in [-3, 0]$. f is continuous on $[-3, 0]$ and differentiable on $(-3, 0)$.

$$f'(x) = 2x + 2, \ f'(c) = 2c + 2$$

$$2c + 2 = \frac{f(0) - f(-3)}{0 - (-3)} = \frac{-3 - 0}{3} = -1$$

Thus $c = -\frac{3}{2}$.

12. $f(x) = x^3 - 2x + 4$, $x \in [0, 2]$. f is continuous on $[0, 2]$ and differentiable on $(0, 2)$.

$$f'(x) = 3x^2 - 2, \ f'(c) = 3c^2 - 2$$

$$3c^2 - 2 = \frac{f(2) - f(0)}{2 - 0} = \frac{8 - 4}{2} = 2$$

Thus $c = \frac{2\sqrt{3}}{3}$.

13. $f(x) = x^3 + x^2 + x$, $x \in [1, 3]$. f is continuous on $[1, 3]$ and differentiable on $(1, 3)$.

$$f'(c) = 3c^2 + 2c + 1 = \frac{f(3) - f(1)}{3 - 1} = \frac{39 - 3}{2} = 18$$

Thus $c = \frac{-1 + 2\sqrt{13}}{3}$.

14. $f(x) = x^3 + x^2$, $x \in [-1, 1]$. f is continuous on $[-1, 1]$ and differentiable on $(-1, 1)$.

$$f'(c) = 3c^2 + 2c = \frac{f(1) - f(-1)}{1 - (-1)} = \frac{2 - 0}{2} = 1$$

Thus $c = \frac{1}{3}$.

15.

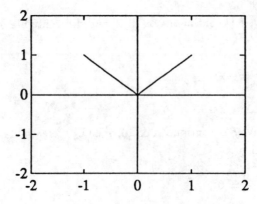

Graph of $f(x) = |x|$ on $[-1, 1]$.

The Mean Value Theorem does not apply for f on $[-1, 1]$ since f is not differentiable at $x = 0$.

16. f is not continuous on $[0, 3]$. [f is discontinuous at $x = 3$.]

17. Let $f(x) = x^3 + 3x^2 + 6x + 1$. Then f is continuous on $[-1, 0]$. $f(-1) = -3$ and $f(0) = 1$. By the Intermediate Value Theorem, there exists a $c \in (-1, 0)$ with $f(c) = 0$. Thus $x^3 + 3x^2 + 6x + 1 = 0$ has a solution between -1 and 0.
$$\begin{aligned} f'(x) &= 3x^2 + 6x + 6 = 3(x^2 + 2x + 1) + 3 \\ &= 3(x+1)^2 + 3 > 0 \text{ for all } x. \end{aligned}$$
By Rolle's Theorem, c is the only solution to $x^3 + 3x^2 + 6x + 1 = 0$ since if there were a d such that $f(d) = 0 = f(c)$, then $f'(z)$ must equal to zero for some z between c and d, but we have shown that $f'(x) > 0$ for all x.

18.
a) If $0 < x < \frac{\pi}{2}$ then $f'(x) = \sec^2 x = \frac{1}{\cos^2 x} > 1 = g'(x)$ since $0 < \cos^2 x < 1$.

b) If $0 < x < \frac{\pi}{2}$, then $(f - g)'(x) = f'(x) - g'(x) > 0$ using part a) and **Theorem 1** of Chapter 3.

c) We know $(f - g)(0) = 0$. If there were an x in $\left(0, \frac{\pi}{2}\right)$ such that $\tan x = x$ then we would have $(f - g)(x) = 0$. Rolle's Theorem applied to $f - g$ in $[0, x]$ would then produce a $c \in (0, x)$ such that $(f - g)'(c) = 0$, contrary to part a).

d) $\tan \frac{\pi}{4} = 1 > \frac{\pi}{4}$ since $\pi < 4$.

e) Suppose there exists x in $\left(0, \frac{\pi}{2}\right)$ such that $\tan x \not> x$. Part c) excludes the possibility that $\tan x = x$, so $\tan x < x$. Now, since $\tan \frac{\pi}{4} > \frac{\pi}{4}$ by part d), it follows that $(f - g)\left(\frac{\pi}{4}\right) > 0$, and since (by supposition) $\tan x < x$, it follows that $(f - g)(x) < 0$. The I.V.T. applied to $f - g$ on the closed interval with endpoints x and $\frac{\pi}{4}$ yields a y between x and $\frac{\pi}{4}$ such that $(f - g)(y) = 0$, whence $\tan y = y$; this is absurd, in view of part c).

19. According to the MVT, there is a c in (a, b) such that $f(b) - f(a) = f'(c)(b - a)$ so
$$|f(b) - f(a)| = |f'(c)|\,|b - a| \leq M(b - a)$$
since it is given that $|f'(c)| \leq M$.

20. If $x = a$ then there is nothing to prove: both sides of the purported inequality are actually equal. If $x > a$ then the MVT provides us with a $c \in (a, x)$ such that $f(x) - f(a) = f'(c)(x - a)$ and then (as in Exercise 19)
$$|f(x) - f(a)| \leq M(x - a).$$
In particular, $f(x) - f(a) \leq M(x - a)$ and so $f(x) \leq f(a) + M(x - a)$ as required.

21. Take $f(x) = \sin x$ so that $f'(x) = \cos x$ and certainly $|f'(x)| \leq 1$ for all x. Since $|f'(t)| \leq 1$ for all t in the interval (x, y) (assuming, for the sake of argument, that $x < y$) it follows at once from Exercise 19 (with $M = 1$) that $|\sin y - \sin x| \leq |y - x|$.

22. Apply the MVT: there is a $c \in (a, b)$ such that
$$f(b) - f(a) = f'(c)(b - a).$$
Since it is given that $m \leq f'(c) \leq M$, we may conclude from $b - a \geq 0$ that
$$m(b - a) \leq f'(c)(b - a) \leq M(b - a)$$
so
$$m(b - a) \leq f(b) - f(a) \leq M(b - a).$$

23. Let $f(t)$ be the distance (in km) traveled after t hours so that $f'(t)$ is the speed (in km/h) after t hours. The total distance traveled after 6 hours is $f(6) - f(0)$ and we are given that $f'(t)$ lies betweeen 75 and 90 for $0 < t < 6$. Applying Exercise 22 with $a = 0$, $b = 6$, $m = 75$, $M = 90$, we deduce that

$$450 = 75 \times 6 \ \leq \ f(6) - f(0) \ \leq \ 90 \times 6 \ = \ 540.$$

Thus: the automobile travels between 450 km and 540 km.

24. Take $f(x) = \sqrt{x}$ on the interval $[a, b] = [36, 36.2]$ say. From $f'(x) = \frac{1}{2\sqrt{x}}$ we see that if $x \geq 36$ then $\sqrt{x} \geq 6$ and $0 < f'(x) \leq \frac{1}{2 \times 6} = \frac{1}{12}$. Setting $M = \frac{1}{12}$ in Exercise 20, it follows that

$$\begin{aligned} \sqrt{36.2} \ &= \ f(36.2) \ \leq \ f(36) + M(36.2 - 36) \\ &= \ 6 + \frac{1}{12}(0.2) \ < \ 6 + \frac{1}{10}(0.2) \\ &= \ 6.02 \end{aligned}$$

and the inequality $6 < \sqrt{36.2}$ is clear. [Alternatively, use Exercise 22.]

25. Let $f(x) = x^{1/3}$ on the interval $[a, b] = [1, 1 + \Delta x]$ say. From $f'(x) = \frac{1}{3} x^{-2/3}$ it follows that if $x \geq 1$ then $0 < f'(x) \leq \frac{1}{3}$. Putting $M = \frac{1}{3}$ in Exercise 20, we have

$$\begin{aligned} \sqrt[3]{1 + \Delta x} \ &= \ f(1 + \Delta x) \ \leq \ f(1) + M(1 + \Delta x - 1) \\ &= \ 1 + \frac{1}{3}\Delta x. \end{aligned}$$

Notice that here the inequality is weak ($\leq$) rather than strong ($<$). To see that $\sqrt[3]{1 + \Delta x} < 1 + \frac{1}{3}\Delta x$, we must see that $\sqrt[3]{1 + \Delta x} \neq 1 + \frac{1}{3}\Delta x$. However, if $\sqrt[3]{1 + \Delta x} = 1 + \frac{1}{3}\Delta x$ then (cube both sides) $1 + \Delta x = 1 + \Delta x + \frac{1}{3}(\Delta x)^2 + \frac{1}{27}(\Delta x)^3$ so that $0 = \frac{1}{3}(\Delta x)^2 + \frac{1}{27}(\Delta x)^3$ which is absurd since $\Delta x > 0$. [Remark: This last step, getting from $\leq$ to $<$, seems easier than trying to modify Exercise 20's conclusion. Notice also that with a little work, the same closing argument gives a direct proof that $\sqrt[3]{1 + \Delta x} < 1 + \frac{1}{3}\Delta x$.]

26. No! Simply let f be any nonzero constant function: since f is constant, $f'(x) = 0$ for all x; but $f(x) \neq 0$ for all x.

27. Let $h = g - f$. According to the MVT, there exists a c in (a, b) such that

$$h(b) - h(a) \ = \ h'(c)(b - a).$$

Now, $h'(c) = g'(c) - f'(c) > 0$ since $g'(c) > f'(c)$; also, $h(a) = g(a) - f(a) = 0$ since $g(a) = f(a)$. So

$$h(b) \ = \ h(a) + h'(c)(b - a) \ > \ 0.$$

But $h(b) = g(b) - f(b)$; so $g(b) > f(b)$.

28. $f(x) = x^4 - 3x^2 + 2 \Rightarrow f'(x) = 4x^3 - 3x$ and $f(1) = 0$, $f(4) = 210$. So we wish to find $c \in (1, 4)$ so that

$$\begin{aligned} 210 \ &= \ f(4) - f(1) = f'(c)(4 - 1) = 3f'(c) \\ \text{or} \qquad 70 \ &= \ 4c^3 - 3c. \end{aligned}$$

A reasonable first guess for the root c is $x_1 = 3$. Write $g(x) = 4x^3 - 3x - 70$ so $g'(x) = 12x^2 - 3$ and use

$$x_{n+1} = x_n - \frac{g(x_n)}{g'(x_n)}.$$

Tabulate as in Example 1 of Section 3.10:

n	x_n	$g(x_n)$	$g'(x_n)$	x_{n+1}
1	3.00000	29.00000	105.00000	2.72381
2	2.72381	2.66185	86.02966	2.69287
3	2.69287	0.03117	84.01849	2.69250
4	2.69250	0.00004	83.99451	2.69250

Since $x_4 = x_5 = 2.69250$, an approximation to c is $c \approx 2.6925$ (to five significant figures).

29. $f(x) = \tan x \Rightarrow f'(x) = \sec^2 x$, $f(0) = 0$, $f\left(\frac{\pi}{4}\right) = 1$. So: we're looking for a $c \in \left(0, \frac{\pi}{4}\right)$ such that

$$1 = f\left(\frac{\pi}{4}\right) - f(0) = f'(c)\left(\frac{\pi}{4} - 0\right) = \frac{\pi}{4} f'(c)$$

or $\qquad 1 = \frac{\pi}{4} \sec^2 c$ or $\cos^2 c = \frac{\pi}{4}$.

Let $g(x) = \cos^2 x - \frac{\pi}{4}$ so $g'(x) = -2\cos x \sin x = -\sin 2x$.

n	x_n	$g(x_n)$	$g'(x_n)$	x_{n+1}
1	0.50000	−0.01525	−0.84147	0.48188
2	0.48188	−0.00018	−0.82134	0.48166
3	0.48166	0.00000		

(The initial guess was made noting that $\cos^2 c \approx \frac{3}{4} \Rightarrow \cos c \approx \frac{\sqrt{3}}{2} \Rightarrow c \approx \frac{\pi}{6} \approx \frac{1}{2}$). So: an approximation to c is $c \approx 0.48166$.

30. $f(x) = \sin \sqrt{x} \Rightarrow f'(x) = \frac{\cos \sqrt{x}}{2\sqrt{x}}$, $f\left(\frac{\pi^2}{16}\right) = \sin \frac{\pi}{4} = \frac{1}{\sqrt{2}}$ and $f\left(\frac{\pi^2}{4}\right) = \sin \frac{\pi}{2} = 1$. So: we want c such that

$$1 - \frac{1}{\sqrt{2}} = f\left(\frac{\pi^2}{4}\right) - f\left(\frac{\pi^2}{16}\right) = f'(c)\left(\frac{\pi^2}{4} - \frac{\pi^2}{16}\right) = \frac{3\pi^2}{16} f'(c)$$

or

$$\frac{\cos \sqrt{c}}{2\sqrt{c}} = \frac{16}{3\pi^2}\left(1 - \frac{1}{\sqrt{2}}\right) \approx 0.15827.$$

Let $\qquad g(x) = \frac{\cos \sqrt{x}}{2\sqrt{x}} - \frac{16}{3\pi^2}\left(1 - \frac{1}{\sqrt{2}}\right)$

$$\approx \frac{\cos \sqrt{x}}{2\sqrt{x}} - 0.15827$$

so $\qquad g'(x) = -\left(\frac{\sqrt{x}\sin \sqrt{x} + \cos \sqrt{x}}{4x^{3/2}}\right).$

As an initial guess, try $x_1 = 1$, between $\frac{\pi^2}{16}$ and $\frac{\pi^2}{4}$.

n	x_n	$g(x_n)$	$g'(x_n)$	x_{n+1}
1	1.00000	0.11188	−0.34544	1.32387
2	1.32387	0.01900	−0.23937	1.40325
3	1.40325	0.00072	−0.22169	1.40649
4	1.40649	0.00001	−0.22101	1.40649

So: an approximation to c is 1.40649.

31. $f(x) = ax^2 + bx + c \Rightarrow f'(x) = 2ax + b$. Now if we wish $d \in (x_1, x_2)$ to satisfy

$$f(x_2) - f(x_1) = f'(d)(x_2 - x_1)$$

then

$$ax_2^2 + bx_2 - ax_1^2 - bx_1 = (2ad + b)(x_2 - x_1)$$

or

$$a(x_1 + x_2)(x_2 - x_1) + b(x_2 - x_1) = (2ad + b)(x_2 - x_1)$$

or (cancelling $x_2 - x_1 \neq 0$ and subtracting b)

$$a(x_1 + x_2) = 2ad$$

whece $d = \frac{x_1 + x_2}{2}$ so long as $a \neq 0$. [If $a = 0$ then any point d between x_1 and x_2 works in the MVT].

For Exercises 32–33, the two lines given are parallel.

32. Slope: $m = 2$
    Point of intersection: $c = 0$
    l_1 : $y = 2x + 3$,
    l_2 : $y = 2x - 1$.

33. Slope: $m = 1 - \dfrac{2}{3\pi} = 0.7877934092$
    Point of intersection: $c = -1.070795768$
    l_1 : $y = 0.7877934092 \left(x + \dfrac{\pi}{2} \right) - \dfrac{\pi}{2}$
    l_2 : $y = 0.7877934092 \, (x + 1.070795768) - 2.068290875$.

34. Slope: $m = \dfrac{5 + \sqrt[3]{4}}{3} = 2.195800351$
    Point of intersection: $c = 0.0338803751$
    l_1: $y = 2.195800351(x + 1) + 3$
    l_2: $y = 2.195800351(x - 0.0338803751) + 0.10700029$. The Mean-Value Theorem gives sufficient conditions when $f'(x)$ intersects the line $y = m = \dfrac{f(b) - f(a)}{b - a}$. It does not say that these are the only conditions under which this could occur.

35. Slope: $m = -0.1958003507 = \dfrac{1 - \sqrt[3]{4}}{3}$. There is no point of intersection. The Mean-Value Theorem requires $f'(x)$ to exist on the open interval (a, b) in order to guarantee that $f'(c) = m$ for some number c between a and b. If $f'(x)$ fails to exist then there may not be such a number.

4.4   Increasing and Decreasing Functions

1. $f(x) = x^2 - 4x + 6$
   $f'(x) = 2x - 4 = 0$ at $x = 2$.

Therefore the only critical number is 2.

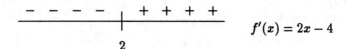

Thus f is decreasing on $(-\infty, 2]$ and increasing on $[2, \infty)$. f has a relative minimum and absolute minimum at $(2, 2)$.

2. $f(x) = x(x - 4)$
   $f'(x) = 2x - 4 = 0$ at $x = 2$.
   Therefore the only critical number is 2.

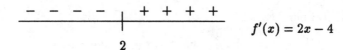

Thus f is decreasing on $(-\infty, 2]$ and increasing on $[2, \infty)$. f has a relative minimum and absolute minimum at $(2, -4)$.

3. $f(x) = 2x^3 - 3x^2 - 12x$
   $f'(x) = 6x^2 - 6x - 12 = 6(x - 2)(x + 1) = 0$ at $x = 2, -1$.
   Therefore the critical numbers are 2 and -1.

$$
\begin{array}{ccccccc}
+ & + & + & + & - & - & - & + & + & + & + \\
\end{array}
$$

$$-1 \qquad 2 \qquad\qquad f'(x) = 6(x - 2)(x + 1)$$

Thus f is increasing on $(-\infty, -1]$ and $[2, \infty)$ and decreasing on $[-1, 2]$. f has a relative minimum at $(2, -20)$ and a relative maximum at $(-1, 7)$.

4. $f(x) = (x - 3)(x + 5)$
   $f'(x) = (x + 5) + (x - 3) = 2x + 2 = 0$ at $x = -1$.
   Therefore the only critical number is -1.

$$
\begin{array}{ccccccc}
- & - & - & - & + & + & + & + \\
\end{array}
$$

$$-1 \qquad\qquad f'(x) = 2x + 2$$

Thus f is decreasing on $(-\infty, -1]$ and increasing on $[-1, \infty)$. f has a relative minimum and absolute minimum at $(-1, -16)$.

5. $f(x) = 4 + x^{2/3}$

   $f'(x) = \dfrac{2}{3x^{1/3}} \neq 0$ for all x. $f'(x)$ is undefined when $3x^{1/3} = 0 \Rightarrow x = 0$.

   Therefore the only critical number is 0.

$$
\begin{array}{ccccccc}
- & - & - & - & + & + & + & + \\
\end{array}
$$

$$0 \qquad\qquad f'(x) = \dfrac{2}{3x^{1/3}}$$

Thus f is decreasing on $(-\infty, 0]$ and increasing on $[0, \infty)$. f has a relative minimum and absolute minimum at $(0, 4)$.

6. $f(x) = 7 - 2x + x^2$
   $f'(x) = -2 + 2x = 0$ at $x = 1$.
   Therefore the only critical number is 1.

$$\underset{1}{\underbrace{\quad - \quad - \quad - \quad - \quad \Big| \quad + \quad + \quad + \quad + \quad}} \qquad f'(x) = -2 + 2x$$

   Thus f is decreasing on $(-\infty, 1]$ and increasing on $[1, \infty)$. f has a relative minimum and absolute minimum at $(1, 6)$.

7. $f(x) = x^3 + 4$
   $f'(x) = 3x^2 = 0$ at $x = 0$.
   Therefore the only critical number is 0.

$$\underset{0}{\underbrace{\quad + \quad + \quad + \quad + \quad \Big| \quad + \quad + \quad + \quad + \quad}} \qquad f'(x) = 3x^2$$

   Thus f is increasing on $(-\infty, \infty)$. f has no relative extrema.

8. $f(x) = 9 - x^2$
   $f'(x) = -2x = 0$ at $x = 0$.
   Therefore the only critical number is 0.

$$\underset{0}{\underbrace{\quad + \quad + \quad + \quad + \quad + \quad \Big| \quad - \quad - \quad - \quad - \quad - \quad}} \qquad f'(x) = -2x$$

   Thus f is increasing on $(-\infty, 0]$ and decreasing on $[0, \infty)$. f has a relative maximum and absolute maximum at $(0, 9)$.

9. $f(x) = |4 - x^2|$. Note that $f(x) = \begin{cases} 4 - x^2 & \text{if } -2 \le x \le 2 \\ x^2 - 4 & \text{if } -\infty < x \le -2 \text{ and } 2 \le x < \infty. \end{cases}$

$$\lim_{x \to -2^-} \frac{f(x) - f(-2)}{x + 2} = \lim_{x \to -2^-} \frac{x^2 - 4 - 0}{x + 2} = \lim_{x \to -2^-} (x - 2) = -4$$

$$\lim_{x \to -2^+} \frac{f(x) - f(-2)}{x + 2} = \lim_{x \to -2^+} \frac{4 - x^2 - 0}{x + 2} = \lim_{x \to -2^+} (2 - x) = 4$$

Since $\lim_{x \to -2^-} \frac{f(x) - f(-2)}{x + 2} \ne \lim_{x \to -2^+} \frac{f(x) - f(-2)}{x + 2}$, $f'(-2)$ does not exist.

$$\lim_{x \to 2^-} \frac{f(x) - f(2)}{x - 2} = \lim_{x \to 2^-} \frac{4 - x^2 - 0}{x - 2} = \lim_{x \to 2^-} -(2 + x) = -4$$

$$\lim_{x \to 2^+} \frac{f(x) - f(2)}{x - 2} = \lim_{x \to 2^+} \frac{x^2 - 9 - 0}{x - 2} = \lim_{x \to 2^+} (x + 2) = 4$$

Since $\lim_{x \to 2^+} \frac{f(x) - f(2)}{x - 2} \ne \lim_{x \to 2^-} \frac{f(x) - f(2)}{x - 2}$, $f'(2)$ does not exist.

Therefore we have: $f'(x) = \begin{cases} -2x & \text{if } -2 < x < 2 \\ 2x & \text{if } -\infty < x < -2 \text{ and } 2 < x < \infty \\ \text{does not exist} & \text{if } x = \pm 2. \end{cases}$

The critical numbers are $0, -2, 2$.

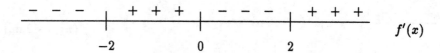

Thus f is increasing on $[-2, 0]$ and $[2, \infty)$ and decreasing on $(-\infty, -2]$ and $[0, 2]$. f has a relative maximum at $(0, 4)$ and a relative (absolute) minima at $(\pm 2, 0)$.

10. $f(x) = |x^2 - 9|$. Note that $f(x) = \begin{cases} x^2 - 9 & \text{if } -\infty < x \le -3 \text{ and } 3 \le x < \infty \\ 9 - x^2 & \text{if } -3 \le x \le 3. \end{cases}$

$\displaystyle \lim_{x \to -3^-} \frac{f(x) - f(-3)}{x - (-3)} = \lim_{x \to -3^-} \frac{x^2 - 9 - 0}{x - (-3)} = -6$ and $\displaystyle \lim_{x \to -3^+} \frac{f(x) - f(-3)}{x - (-3)} = \lim_{x \to -3^+} \frac{9 - x^2 - 0}{x - (-3)} = 6.$

Thus $f'(-3)$ does not exist.

$\displaystyle \lim_{x \to 3^-} \frac{f(x) - f(3)}{x - 3} = \lim_{x \to 3^-} \frac{9 - x^2 - 0}{x - 3} = -6$ and $\displaystyle \lim_{x \to 3^+} \frac{f(x) - f(3)}{x - 3} = \lim_{x \to 3^+} \frac{x^2 - 9 - 0}{x - 3} = 6.$

Thus $f'(3)$ does not exist.

Therefore we have $f'(x) = \begin{cases} 2x & \text{if } -\infty < x < -3 \text{ and } 3 < x < \infty \\ -2x & \text{if } -3 < x < 3 \\ \text{does not exist} & \text{if } x = \pm 3. \end{cases}$

The critical numbers are $0, \pm 3$.

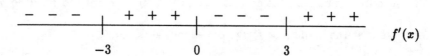

f is decreasing on $(-\infty, -3]$ and $[0, 3]$ and increasing on $[-3, 0]$ and $[3, \infty)$. f has a relative maximum at $(0, 9)$ and relative (absolute) minima at $(\pm 3, 0)$.

11. $f(x) = \sin\left(x + \frac{\pi}{4}\right)$, $0 \le x \le 2\pi$

$f'(x) = \cos\left(x + \frac{\pi}{4}\right) = 0$ when $x + \frac{\pi}{4} = \frac{\pi}{2}, \frac{3\pi}{2} \Rightarrow x = \frac{\pi}{4}, \frac{5\pi}{4}$.

Therefore the critical numbers are $\frac{\pi}{4}$ and $\frac{5\pi}{4}$.

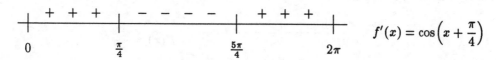

f is increasing on $\left[0, \frac{\pi}{4}\right]$ and $\left[\frac{5\pi}{4}, 2\pi\right]$ and decreasing on $\left[\frac{\pi}{4}, \frac{5\pi}{4}\right]$. f has a relative maximum at $\left(2\pi, \frac{\sqrt{2}}{2}\right)$, an absolute maximum at $\left(\frac{\pi}{4}, 1\right)$, a relative minimum at $\left(0, \frac{\sqrt{2}}{2}\right)$, and an absolute minimum at $\left(\frac{5\pi}{4}, -1\right)$.

12. $f(x) = \cos\left(2x - \frac{\pi}{2}\right)$, $0 \le x \le \pi$

$f'(x) = -2\sin\left(2x - \frac{\pi}{2}\right) = 0$ when $2x - \frac{\pi}{2} = 0, \pi \Rightarrow 2x = \frac{\pi}{2}, \frac{3\pi}{2} \Rightarrow x = \frac{\pi}{4}, \frac{3\pi}{4}$.

Therefore the critical numbers are $\frac{\pi}{4}$ and $\frac{3\pi}{4}$.

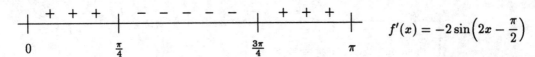

$$f'(x) = -2\sin\left(2x - \frac{\pi}{2}\right)$$

f is increasing on $\left[0, \frac{\pi}{4}\right]$ and on $\left[\frac{3\pi}{4}, \pi\right]$ and decreasing on $\left[\frac{\pi}{4}, \frac{3\pi}{4}\right]$. f has a relative maximum at $(\pi, 0)$, an absolute maximum at $\left(\frac{\pi}{4}, 1\right)$, a relative minimum at $(0, 0)$, and an absolute minimum at $\left(\frac{3\pi}{4}, -1\right)$.

13. $f(x) = x^4 - 1$
$f'(x) = 4x^3 = 0$ at $x = 0$.
Therefore the only critical number is 0.

$$f'(x) = 4x^3$$

Thus f is decreasing on $(-\infty, 0]$ and increasing on $[0, \infty)$. f has a relative minimum and absolute minimum at $(0, -1)$.

14. $f(x) = \dfrac{3}{x-2}$. Note that the domain of f is $(-\infty, 2) \cup (2, \infty)$.

$f'(x) = \dfrac{-3}{(x-2)^2} < 0$ for all x in the domain.
Therefore f has no critical numbers, f is decreasing on $(-\infty, 2) \cup (2, \infty)$ and f has no relative extrema.

15. $f(x) = \dfrac{2}{x+1}$. Note that the domain of f is $(-\infty, -1) \cup (-1, \infty)$.

$f'(x) = \dfrac{-2}{(x+1)^2} < 0$ for all x in the domain.
Thus f is decreasing on $(-\infty, -1) \cup (-1, \infty)$ and f has no relative extrema.

16. $f(x) = \dfrac{1}{x^2}$. Note that the domain of f is $(-\infty, 0) \cup (0, \infty)$.

$f'(x) = \dfrac{-2}{x^3} \neq 0$ for all x in the domain.

Therefore f has no critical numbers.

$$f'(x) = \dfrac{-2}{x^3}$$

Thus f is increasing on $(-\infty, 0)$ and decreasing on $(0, \infty)$. f has no relative extrema.

17. $f(x) = \begin{cases} 4 - x^2 & \text{if } -\infty < x \le 1 \\ x + 2 & \text{if } 1 < x < \infty. \end{cases}$

$\displaystyle \lim_{x \to 1^-} \frac{f(x) - f(1)}{x - 1} = \lim_{x \to 1^-} \frac{4 - x^2 - 3}{x - 1} = \lim_{x \to 1^-} -(1 + x) = -2$

$\displaystyle \lim_{x \to 1^+} \frac{f(x) - f(1)}{x - 1} = \lim_{x \to 1^+} \frac{x + 2 - 3}{x - 1} = 1$

230

Since $\lim\limits_{x \to 1^-} \dfrac{f(x) - f(1)}{x - 1} \neq \lim\limits_{x \to 1^+} \dfrac{f(x) - f(1)}{x - 1}$, $f'(1)$ does not exist.

We have $f'(x) = \begin{cases} -2x & \text{if } -\infty < x < 1 \\ 1 & \text{if } 1 < x < \infty \\ \text{does not exist} & \text{if } x = 1. \end{cases}$

Therefore the critical numbers are 0 and 1.

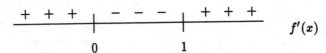

Thus f is increasing on $(-\infty, 0]$ and $[1, \infty)$ and decreasing on $[0, 1]$. f has a relative maximum at $(0, 4)$ and a relative minimum at $(1, 3)$.

18. $f(x) = |3x - x^3|$
$3x - x^3 = x(3 - x^2) = 0$ at $x = 0, \pm\sqrt{3}$.

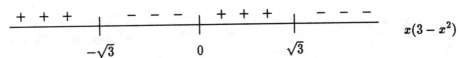

Therefore $f(x) = \begin{cases} 3x - x^3 & \text{if } -\infty < x \leq -\sqrt{3} \text{ and } 0 \leq x \leq \sqrt{3} \\ x^3 - 3x & \text{if } -\sqrt{3} \leq x \leq 0 \text{ and } \sqrt{3} \leq x < \infty. \end{cases}$

a) $\lim\limits_{x \to -\sqrt{3}^-} \dfrac{f(x) - f\left(-\sqrt{3}\right)}{x - \left(-\sqrt{3}\right)} = \lim\limits_{x \to -\sqrt{3}^-} \dfrac{3x - x^3 - 0}{x + \sqrt{3}} = -6$ and

$\lim\limits_{x \to -\sqrt{3}^+} \dfrac{f(x) - f\left(-\sqrt{3}\right)}{x - \left(-\sqrt{3}\right)} = \lim\limits_{x \to -\sqrt{3}^+} \dfrac{x^3 - 3x - 0}{x + \sqrt{3}} = 6.$

Thus $f'\left(-\sqrt{3}\right)$ does not exist.

b) $\lim\limits_{x \to 0^-} \dfrac{f(x) - f(0)}{x - 0} = \lim\limits_{x \to 0^-} \dfrac{x^3 - 3x - 0}{x - 0} = -3$ and

$\lim\limits_{x \to 0^+} \dfrac{f(x) - f(0)}{x - 0} = \lim\limits_{x \to 0^+} \dfrac{3x - x^3 - 0}{x - 0} = 3.$

Thus $f'(0)$ does not exist.

c) $\lim\limits_{x \to \sqrt{3}^-} \dfrac{f(x) - f\left(\sqrt{3}\right)}{x - \sqrt{3}} = \lim\limits_{x \to \sqrt{3}^-} \dfrac{3x - x^3 - 0}{x - \sqrt{3}} = -6$ and

$\lim\limits_{x \to \sqrt{3}^+} \dfrac{f(x) - f\left(\sqrt{3}\right)}{x - \sqrt{3}} = \lim\limits_{x \to \sqrt{3}^+} \dfrac{x^3 - 3x - 0}{x - \sqrt{3}} = 6.$

Thus $f'(\sqrt{3})$ does not exist.

We have $f'(x) = \begin{cases} 3 - 3x^2 & \text{if } -\infty < x < -\sqrt{3} \text{ and } 0 < x < \sqrt{3} \\ 3x^2 - 3 & \text{if } -\sqrt{3} < x < 0 \text{ and } \sqrt{3} < x < \infty \\ \text{does not exist} & \text{if } x = 0, \pm\sqrt{3}. \end{cases}$

$3 - 3x^2 = 0$ at $x = 1$ for $x \in \left(-\infty, -\sqrt{3}\right) \cup \left(0, \sqrt{3}\right)$.

$3x^2 - 3 = 0$ at $x = -1$ for $x \in \left(-\sqrt{3}, 0\right) \cup \left(\sqrt{3}, \infty\right)$.

Therefore the critical numbers are $0, \pm 1, \pm\sqrt{3}$.

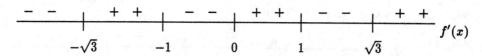

f is decreasing on $\left(-\infty, -\sqrt{3}\right]$, $[-1, 0]$ and $\left[1, \sqrt{3}\right]$ and increasing on $\left[-\sqrt{3}, -1\right], [0, 1]$ and $\left[\sqrt{3}, \infty\right)$. f has relative (absolute) minima at $(0, 0)$ and $\left(\pm\sqrt{3}, 0\right)$ and relative maxima at $(\pm 1, 2)$.

19. $f(x) = 4x^3 + 9x^2 - 12x + 7$
$f'(x) = 12x^2 + 18x - 12 = 6(2x - 1)(x + 2) = 0$ at $x = \frac{1}{2}, -2$.
Therefore the critical numbers are $\frac{1}{2}$ and -2.

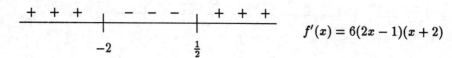

f is increasing on $(-\infty, -2]$ and $\left[\frac{1}{2}, \infty\right)$ and decreasing on $\left[-2, \frac{1}{2}\right]$. f has a relative maximum at $(-2, 35)$ and a relative minimum at $\left(\frac{1}{2}, \frac{15}{4}\right)$.

20. $f(x) = x^3 + x^2 - 8x + 8$
$f'(x) = 3x^2 + 2x - 8 = (3x - 4)(x + 2) = 0$ at $x = \frac{4}{3}, -2$.
Therefore the critical numbers are $\frac{4}{3}$ and -2.

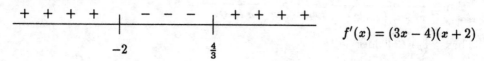

f is increasing on $(-\infty, -2]$ and on $\left[\frac{4}{3}, \infty\right)$ and decreasing on $\left[-2, \frac{4}{3}\right]$. f has a relative maximum at $(-2, 20)$ and a relative minimum at $\left(\frac{4}{3}, \frac{40}{27}\right)$.

21. $f(x) = x + \sin x$
$f'(x) = 1 + \cos x = 0$ when $\cos x = -1 \Rightarrow x = \pi + 2n\pi$ for $n \in \mathcal{Z}$.
Therefore f has infinitely many critical numbers.

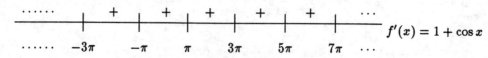

f is increasing on $(-\infty, \infty)$ and f has no relative extrema.

22. $f(x) = \tan^2 x = \dfrac{\sin^2 x}{\cos^2 x}$. Note that the domain of f is all $x \neq \left(\frac{\pi}{2} + n\pi\right)$.
$f'(x) = 2\tan x \sec^2 x = \dfrac{2\sin x}{\cos^3 x} = 0$ when $\sin x = 0 \Rightarrow x = n\pi$ for $n \in \mathcal{Z}$.

Therefore the critical numbers are $(n\pi)$ for $n \in \mathbb{Z}$.

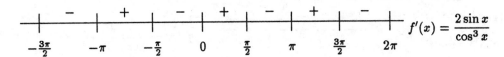

Thus f is increasing on each of the intervals $\left[n\pi, \frac{\pi}{2} + n\pi\right)$ for $n \in \mathbb{Z}$ and decreasing on each of the intervals $\left(\frac{\pi}{2} + n\pi, (n+1)\pi\right]$ for $n \in \mathbb{Z}$. f has relative (absolute) minima at $(n\pi, 0)$ for $n \in \mathbb{Z}$.

23. $f(x) = \dfrac{x}{1+x}$. Note that the domain of f is $(-\infty, -1) \cup (-1, \infty)$.

$f'(x) = \dfrac{(1+x) - x}{(1+x)^2} = \dfrac{1}{(1+x)^2} > 0$ for all x in the domain.

Thus f is increasing for all x in the domain and f has no relative extrema.

24. $f(x) = \sqrt{x+2}$. Note that the domain of f is $[-2, \infty)$.

$f'(x) = \dfrac{1}{2\sqrt{x+2}} > 0$ for all x in the domain.

Thus f is increasing for all x in the domain and f has an absolute minimum at $(-2, 0)$.

25. $f(x) = \dfrac{1}{1+x^2}$. Note that the domain of f is $(-\infty, \infty)$.

$f'(x) = \dfrac{-2x}{(1+x^2)^2} = 0$ when $x = 0$.

Therefore the only critical number is 0.

$$+\ +\ +\ \Big|\ -\ -\ - \qquad f'(x) = \dfrac{-2x}{(1+x^2)^2}$$
$$0$$

Thus f is increasing on $(-\infty, 0]$ and decreasing on $[0, \infty)$. f has a relative (absolute) maximum at $(0, 1)$.

26. $f(x) = \dfrac{x}{1+x^2}$. Note that the domain of f is $(-\infty, \infty)$.

$f'(x) = \dfrac{(1+x^2) - x(2x)}{(1+x^2)^2} = \dfrac{1-x^2}{(1+x^2)^2} = 0$ at $x = \pm 1$.

Therefore the critical numbers are ± 1.

$$-\ -\ -\ \Big|\ +\ +\ +\ \Big|\ -\ -\ - \qquad f'(x) = \dfrac{1-x^2}{(1+x^2)^2}$$
$$-1 \qquad\qquad 1$$

Thus f is decreasing on $(-\infty, -1]$ and on $[1, \infty)$ and increasing on $[-1, 1]$. f has a relative (absolute) maximum at $\left(1, \frac{1}{2}\right)$ and a relative (absolute) minimum at $\left(-1, -\frac{1}{2}\right)$.

27. $f(x) = (x+3)^{2/3}$

$f'(x) = \dfrac{2}{3(x+3)^{1/3}} \neq 0$ for all x.

233

$f'(x)$ is undefined when $3(x+3)^{1/3} = 0 \Rightarrow x = -3$.
Therefore the only critical number is -3.

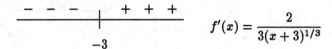

$$f'(x) = \frac{2}{3(x+3)^{1/3}}$$

Thus f is decreasing on $(-\infty, -3]$ and increasing on $[-3, \infty)$. f has a relative (absolute) minimum at $(-3, 0)$.

28. $f(x) = 1 - x^{2/3}$

$f'(x) = \dfrac{-2}{3x^{1/3}} \neq 0$ for all x.

$f'(x)$ is undefined when $3x^{1/3} = 0 \Rightarrow x = 0$.
Therefore the only critical number is 0.

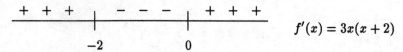

$$f'(x) = \frac{-2}{3x^{1/3}}$$

Thus f is increasing on $(-\infty, 0]$ and decreasing on $[0, \infty)$. f has a relative (absolute) maximum at $(0, 1)$.

29. $f(x) = \frac{1}{3}x^3 - 3x^2 - 7x + 5$
$f'(x) = x^2 - 6x - 7 = (x-7)(x+1) = 0$ at $x = -1, 7$.
Therefore the critical numbers are -1 and 7.

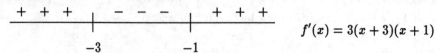

$$f'(x) = (x-7)(x+1)$$

Thus f is increasing on $(-\infty, -1]$ and on $[7, \infty)$ and decreasing on $[-1, 7]$. f has a relative maximum at $\left(-1, \frac{26}{3}\right)$ and a relative minimum at $\left(7, \frac{-230}{3}\right)$.

30. $f(x) = x^3 + 3x^2 + 10$
$f'(x) = 3x^2 + 6x = 3x(x+2) = 0$ at $x = 0, -2$.
Therefore the critical numbers are 0 and -2.

$$\begin{array}{ccc} + + + & - - - & + + + \\ \hline & \underset{-2}{\big|} & \underset{0}{\big|} \end{array} \qquad f'(x) = 3x(x+2)$$

Thus f is increasing on $(-\infty, -2]$ and on $[0, \infty)$ and decreasing on $[-2, 0]$. f has a relative maximum at $(-2, 14)$ and a relative minimum at $(0, 10)$.

31. $f(x) = x^3 + 6x^2 + 9x + 1$
$f'(x) = 3x^2 + 12x + 9 = 3(x+3)(x+1) = 0$ at $x = -1, -3$.
Therefore the critical numbers are -1 and -3.

$$\begin{array}{ccc} + + + & - - - & + + + \\ \hline & \underset{-3}{\big|} & \underset{-1}{\big|} \end{array} \qquad f'(x) = 3(x+3)(x+1)$$

Thus f is increasing on $(-\infty, -3]$ and on $[-1, \infty)$ and decreasing on $[-3, -1]$. f has a relative maximum at $(-3, 1)$ and a relative minimum at $(-1, -3)$.

32. $f(x) = 2x^3 + 9x^2 - 12x + 2$

$f'(x) = 6x^2 + 18x - 12 = 6(x^2 + 3x - 2) = 0$ when $x = \dfrac{-3 \pm \sqrt{17}}{2} \approx .562$ and -3.56.

Therefore the critical numbers are $\dfrac{-3 \pm \sqrt{17}}{2}$.

$f'(x) = 6(x^2 + 3x - 2)$

Thus f is increasing on $\left(-\infty, \dfrac{-3 - \sqrt{17}}{2}\right]$ and on $\left[\dfrac{-3 + \sqrt{17}}{2}, \infty\right)$ and

decreasing on $\left[\dfrac{-3 - \sqrt{17}}{2}, \dfrac{-3 + \sqrt{17}}{2}\right]$.

f has a relative maximum at $\left(\dfrac{-3 - \sqrt{17}}{2}, 68.55\right)$ and a relative minimum at $\left(\dfrac{-3 + \sqrt{17}}{2}, -1.55\right)$.

33. $f(x) = x^3 + 3x^2 + 6x - 1$

$f'(x) = 3x^2 + 6x + 6 = 3(x^2 + 2x + 2) = 3\left[(x + 1)^2 + 1\right] > 0$ for all x.

Thus f is increasing on $(-\infty, \infty)$ and f has no relative extrema.

34. $f(x) = x^3 + 6x^2 + 15x + 2$

$f'(x) = 3x^2 + 12x + 15 = 3(x^2 + 4x + 5) = 3\left[(x + 2)^2 + 1\right] > 0$ for all x.

Thus f is increasing on $(-\infty, \infty)$ and f has no relative extrema.

35. $f(x) = x^4 + 4x^3 + 2x^2 + 1$

$f'(x) = 4x^3 + 12x^2 + 4x = 4x(x^2 + 3x + 1) = 0$ when $x = 0, \dfrac{-3 \pm \sqrt{5}}{2}.$ $\left(\dfrac{-3 \pm \sqrt{5}}{2} \approx -.38, -2.62\right)$

Therefore the critical numbers are $0, \dfrac{-3 \pm \sqrt{5}}{2}$.

$f'(x) = 4x(x^2 + 3x + 1)$

Thus f is decreasing on $\left(-\infty, \dfrac{-3 - \sqrt{5}}{2}\right]$ and on $\left[\dfrac{-3 + \sqrt{5}}{2}, 0\right]$ and

increasing on $\left[\dfrac{-3 - \sqrt{5}}{2}, \dfrac{-3 + \sqrt{5}}{2}\right]$ and on $[0, \infty)$.

f has a relative maximum at $\left(\dfrac{-3 + \sqrt{5}}{2}, 1.09\right)$ and a relative minimum at $(0, 1)$ and a relative

(absolute) minimum at $\left(\dfrac{-3 - \sqrt{5}}{2}, -10.09\right)$.

36. $f(x) = 3x^4 - 16x^3 + 30x^2 - 24x + 5$
$f'(x) = 12x^3 - 48x^2 + 60x - 24 = 12(x^3 - 4x^2 + 5x - 2) = 12(x-1)^2(x-2) = 0$ at $x = 1, 2$.
Therefore the critical numbers are 1 and 2.

$$\begin{array}{ccccccc} - & - & - & - & - & - & + & + & + \\ \hline & & & | & & & | & & \\ & & & 1 & & & 2 & & \end{array} \qquad f'(x) = 12(x-1)^2(x-2)$$

Thus f is decreasing on $(-\infty, 2]$ and increasing on $[2, \infty)$. f has a relative (absolute) minimum at $(2, -3)$.

37. $f(x) = 6x^{5/2} - 70x^{3/2} + 15$. Note that the domain of f is $[0, \infty)$.
$f'(x) = 15x^{3/2} - 105x^{1/2} = 15x^{1/2}(x - 7) = 0$ when $x = 0, 7$.

$$\begin{array}{ccccccc} & - & - & - & - & + & + & + & + \\ \hline | & & & & | & & & \\ 0 & & & & 7 & & & \end{array} \qquad f'(x) = 15x^{1/2}(x-7)$$

Thus f is decreasing on $[0, 7]$ and increasing on $[7, \infty)$. f has a relative (absolute) minimum at $(7, -503.6)$.

38. $f(x) = \sqrt[3]{x}\,(x-7)^2$
$f'(x) = \dfrac{1}{3x^{2/3}}\,(x-7)^2 + \sqrt[3]{x}\,(2)(x-7) = (x-7)\left(\dfrac{x-7+6x}{3x^{2/3}}\right) = 7(x-7)\left(\dfrac{x-1}{3x^{2/3}}\right) = 0$ when $x = 1, 7$.

$f'(x)$ is undefined when $3x^{2/3} = 0 \Rightarrow x = 0$.
Therefore the critical numbers are $0, 1, 7$.

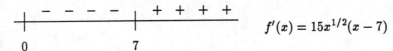

$$\begin{array}{ccccccc} + & + & + & + & + & + & - & - & - & + & + & + \\ \hline & & | & & & | & & & | & & \\ & & 0 & & & 1 & & & 7 & & \end{array} \qquad f'(x) = \dfrac{7(x-7)(x-1)}{3x^{2/3}}$$

Thus f is increasing on $(-\infty, 1]$ and on $[7, \infty)$ and decreasing on $[1, 7]$. f has a relative maximum at $(1, 36)$ and a relative minimum at $(7, 0)$.

39. $f(x) = \dfrac{x^2}{1+x^2}$. Note that the domain of f is $(-\infty, \infty)$.

$f'(x) = \dfrac{2x(1+x^2) - x^2(2x)}{(1+x^2)^2} = \dfrac{2x}{(1+x^2)^2} = 0$ at $x = 0$.

$$\begin{array}{ccccccc} - & - & - & - & + & + & + & + \\ \hline & & & | & & & \\ & & & 0 & & & \end{array} \qquad f'(x) = \dfrac{2x}{(1+x^2)^2}$$

Thus f is decreasing on $(-\infty, 0]$ and increasing on $[0, \infty)$. f has a relative (absolute) minimum at $(0, 0)$.

40. $f(x) = \dfrac{\sqrt{x}}{4-x}$. Note that the domain of f is $[0, 4) \cup (4, \infty)$.

$f'(x) = \dfrac{\frac{1}{2\sqrt{x}}(4-x) - \sqrt{x}(-1)}{(4-x^2)} = \dfrac{4+x}{2\sqrt{x}\,(4-x)^2} \neq 0$ for all x in the domain.

$$\begin{array}{ccccccc} & + & + & + & + & + & + & + & + \\ \hline | & & & & | & & & \\ 0 & & & & 4 & & & \end{array} \qquad f'(x) = \dfrac{4+x}{2\sqrt{x}\,(4-x)^2}$$

Thus f is increasing on $[0,4)$ and on $(4,\infty)$. f has a relative minimum at $(0,0)$.

41. $f(x) = \dfrac{x^{1/3}}{x^{2/3} - 4}$. Note that $x^{2/3} - 4 = 0$ when $x = \pm 8$.
Therefore the domain of f is $(-\infty, -8) \cup (-8, 8) \cup (8, \infty)$.

$$f'(x) = \frac{\frac{1}{3x^{2/3}}(x^{2/3} - 4) - x^{1/3}(\frac{2}{3x^{1/3}})}{(x^{2/3} - 4)^2} = \frac{x^{2/3} - 4 - 2x^{2/3}}{3x^{2/3}(x^{2/3} - 4)^2} = \frac{-(x^{2/3} + 4)}{3x^{2/3}(x^{2/3} - 4)^2} \neq 0$$

for all x in the domain.

$f'(x)$ is undefined at $x = 0$. Therefore, the only critical number is 0.

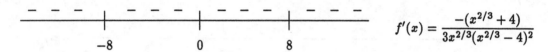

$$f'(x) = \frac{-(x^{2/3} + 4)}{3x^{2/3}(x^{2/3} - 4)^2}$$

Thus f is decreasing on $(-\infty, -8) \cup (-8, 8) \cup (8, \infty)$. f has no relative extrema.

42. $f(x) = x^{2/3}(x + 8)^2$
$$f'(x) = \frac{2}{3x^{1/3}}(x + 8)^2 + x^{2/3}(2)(x + 8) = \frac{8(x + 8)(x + 2)}{3x^{1/3}} = 0 \text{ at } x = -8, -2.$$

$f'(x)$ is undefined when $x = 0$.
Therefore the critical numbers are $0, -8$ and -2.

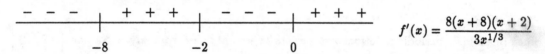

$$f'(x) = \frac{8(x + 8)(x + 2)}{3x^{1/3}}$$

Thus f is decreasing on $(-\infty, -8]$ and on $[-2, 0]$ and increasing on $[-8, -2]$ and on $[0, \infty)$. f has a relative maximum at $\left(-2, 36\sqrt[3]{4}\right)$ and a relative (absolute) minima at $(0, 0)$ and $(-8, 0)$.

43. $f(x) = \sec^2 x$, $0 \le x \le 2\pi$. Note that the domain of f is $\left[0, \frac{\pi}{2}\right) \cup \left(\frac{\pi}{2}, \frac{3\pi}{2}\right) \cup \left(\frac{3\pi}{2}, 2\pi\right]$.
$$f'(x) = 2\sec x \sec x \tan x = \frac{2\sin x}{\cos^3 x} = 0 \text{ when } \sin x = 0 \Rightarrow x = \pi \text{ for } x \in (0, 2\pi).$$

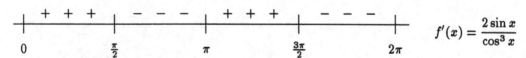

$$f'(x) = \frac{2\sin x}{\cos^3 x}$$

Thus f is increasing on $\left[0, \frac{\pi}{2}\right)$ and $\left[\pi, \frac{3\pi}{2}\right)$ and decreasing on $\left(\frac{\pi}{2}, \pi\right]$ and $\left(\frac{3\pi}{2}, 2\pi\right]$. f has a relative (absolute) minima at $(0, 1)$, $(\pi, 1)$ and $(2\pi, 1)$.

44. $f(x) = \sin^2 x + \cos x$, $0 \le x \le 2\pi$
$f'(x) = 2\sin x \cos x - \sin x = \sin x(2\cos x - 1) = 0$ when $\sin x = 0 \Rightarrow x = \pi$ for $x \in (0, 2\pi)$ or when $2\cos x - 1 = 0 \Rightarrow x = \frac{\pi}{3}, \frac{5\pi}{3}$.

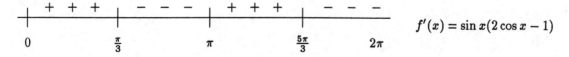

$$f'(x) = \sin x(2\cos x - 1)$$

Thus f is increasing on $\left[0, \frac{\pi}{3}\right]$ and $\left[\pi, \frac{5\pi}{3}\right]$ and decreasing on $\left[\frac{\pi}{3}, \pi\right]$ and $\left[\frac{5\pi}{3}, 2\pi\right]$. f has relative minima at $(0, 1)$ and $(2\pi, 1)$, an absolute minimum at $(\pi, -1)$ and relative (absolute) maxima at $\left(\frac{\pi}{3}, \frac{5}{4}\right)$ and $\left(\frac{5\pi}{3}, \frac{5}{4}\right)$.

45. $f(x) = |\sin x|$

Since $(\sin x)$ is a periodic function with period 2π we only need to analyze $f(x)$ on $[0, 2\pi]$.

$$f(x) = \begin{cases} \sin x & \text{if } 0 \le x \le \pi \\ -\sin x & \text{if } \pi \le x \le 2\pi. \end{cases} \quad \text{Then } f'(x) = \begin{cases} \cos x & \text{if } 0 < x < \pi \\ -\cos x & \text{if } \pi < x < 2\pi \\ \text{does not exist} & \text{if } x = \pi, 0, 2\pi. \end{cases}$$

$f'(x) = 0$ when $x = \frac{\pi}{2}, \frac{3\pi}{2}$.

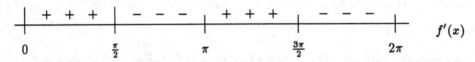

Thus f is increasing on $\left[0, \frac{\pi}{2}\right]$ and $\left[\pi, \frac{3\pi}{2}\right]$ and decreasing on $\left[\frac{\pi}{2}, \pi\right]$ and $\left[\frac{3\pi}{2}, 2\pi\right]$. f has a relative (absolute) minimum at $(\pi, 0)$ and relative (absolute) maxima at $\left(\frac{\pi}{2}, 1\right)$ and $\left(\frac{3\pi}{2}, 1\right)$.

In general, f is increasing on $\left[n\pi, (2n+1)\frac{\pi}{2}\right]$ and decreasing on $\left[(2n-1)\frac{\pi}{2}, n\pi\right]$ for $n \in \mathcal{Z}$. f has relative (absolute) minima at $(n\pi, 0)$ and relative (absolute) maxima at $\left((2n+1)\frac{\pi}{2}, 1\right)$ for $n \in \mathcal{Z}$.

46. $f(x) = |\cos x|$

Since $(\cos x)$ is a periodic function with period 2π we only need to analyze $f(x)$ on $\left[-\frac{\pi}{2}, \frac{3\pi}{2}\right]$.

$$f(x) = \begin{cases} \cos x & \text{if } -\frac{\pi}{2} \le x \le \frac{\pi}{2} \\ -\cos x & \text{if } \frac{\pi}{2} \le x \le \frac{3\pi}{2}. \end{cases} \quad \text{Then } f'(x) = \begin{cases} -\sin x & \text{if } -\frac{\pi}{2} < x < \frac{\pi}{2} \\ \sin x & \text{if } \frac{\pi}{2} < x < \frac{3\pi}{2} \\ \text{does not exist} & \text{if } x = -\frac{\pi}{2}, \frac{\pi}{2}, \frac{3\pi}{2}. \end{cases}$$

$f'(x) = 0$ when $x = 0, \pi$.

Thus f is increasing on $\left[-\frac{\pi}{2}, 0\right]$ and $\left[\frac{\pi}{2}, \pi\right]$ and decreasing on $\left[0, \frac{\pi}{2}\right]$ and $\left[\pi, \frac{3\pi}{2}\right]$. f has a relative (absolute) minimum at $\left(\frac{\pi}{2}, 0\right)$ and a relative (absolute) maxima at $(0, 1)$ and $(\pi, 1)$.

In general, f is increasing on $\left[(2n-1)\frac{\pi}{2}, n\pi\right]$ and decreasing on $\left[n\pi, (2n+1)\frac{\pi}{2}\right]$ for $n \in \mathcal{Z}$. f has relative (absolute) minima at $\left((2n+1)\frac{\pi}{2}, 0\right)$ and a relative (absolute) maxima at $(n\pi, 1)$ for $n \in \mathcal{Z}$.

47. $f(x) = x^{5/3} - 5x^{2/3} + 3$

$f'(x) = \frac{5}{3}x^{2/3} - \frac{10}{3x^{1/3}} = \frac{5(x-2)}{3x^{1/3}} = 0$ at $x = 2$.

$f'(x)$ is undefined when $3x^{1/3} = 0 \Rightarrow x = 0$.
Therefore the critical numbers are 0 and 2.

$$+ \ + \ + \ \Big| \ - \ - \ - \ \Big| \ + \ + \ +$$
$$ 0 2 $$
$$f'(x) = \frac{5(x-2)}{3x^{1/3}}$$

Thus f is increasing on $(-\infty, 0]$ and $[2, \infty)$ and decreasing on $[0, 2]$. f has a relative maximum at $(0, 3)$ and a relative minimum at $(2, -1.76)$.

48. $f(x) = \dfrac{1-x}{1+x^2}$. Note that the domain of f is $(-\infty, \infty)$.

$f'(x) = \dfrac{(-1)(1+x^2) - (1-x)(2x)}{(1+x^2)^2} = \dfrac{x^2 - 2x - 1}{(1+x^2)^2} = 0$ when $x = 1 \pm \sqrt{2} \approx 2.41, \ -.41$.

Therefore the critical numbers are $1 \pm \sqrt{2}$.

$$+ \ + \ + \ \Big| \ - \ - \ - \ \Big| \ + \ + \ +$$
$$ 1 - \sqrt{2} 1 + \sqrt{2} $$
$$f'(x) = \frac{x^2 - 2x - 1}{(1+x^2)^2}$$

Thus f is increasing on $\left(-\infty, \ 1-\sqrt{2}\right]$ and $\left[1+\sqrt{2}, \ \infty\right)$ and decreasing on $\left[1-\sqrt{2}, \ 1+\sqrt{2}\right]$. f has a relative (absolute) maximum at $\left(1-\sqrt{2}, \ 1.21\right)$ and a relative (absolute) minimum at $\left(1+\sqrt{2}, \ -.207\right)$.

49. $f(x) = \frac{1}{4}x^4 - 2x^3 + \frac{3}{2}x^2 + 10x - 8$
    $f'(x) = x^3 - 6x^2 + 3x + 10 = (x-2)(x-5)(x+1) = 0$ when $x = -1, 2, 5$.

$$- \ - \ - \ \Big| \ + \ + \ + \ \Big| \ - \ - \ - \ \Big| \ + \ + \ +$$
$$ -1 2 5 $$
$$f'(x) = (x-2)(x-5)(x+1)$$

Thus f is decreasing on $(-\infty, -1]$ and $[2, 5]$ and increasing on $[-1, 2]$ and $[5, \infty)$. f has a relative maximum at $(2, 6)$ and a relative (absolute) minima at $\left(-1, \ \frac{-57}{4}\right)$ and $\left(5, \ \frac{-57}{4}\right)$.

50. $f(x) = \sqrt[3]{8 - x^3}$

$f'(x) = \dfrac{-x^2}{(8-x^3)^{2/3}} = 0$ at $x = 0$.

$f'(x)$ is undefined when $(8 - x^3) = 0 \Rightarrow x = 2$.
Therefore the critical numbers are 0 and 2.

$$- \ - \ - \ \Big| \ - \ - \ - \ \Big| \ - \ - \ -$$
$$ 0 2 $$
$$f'(x) = \frac{-x^2}{(8-x^3)^{2/3}}$$

Thus f is decreasing on $(-\infty, \infty)$. f has no relative extrema.

51. $f(x) = x + |\sin x|, \ 0 \leq x \leq 2\pi$

$$f(x) = \begin{cases} x + \sin x & \text{if } 0 \leq x \leq \pi \\ x - \sin x & \text{if } \pi \leq x \leq 2\pi. \end{cases}$$

Then $f'(x) = \begin{cases} 1 + \cos x & \text{if } 0 < x < \pi \\ 1 - \cos x & \text{if } \pi < x < 2\pi \\ \text{does not exist} & \text{if } x = \pi, 0, 2\pi. \end{cases}$

$f'(x) \neq 0$ for $x \in (0, 2\pi)$.

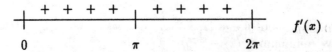

Thus f is increasing on $[0, 2\pi]$. f has a relative minimum at $(0, 0)$ and a relative maximum at $(2\pi, 2\pi)$.

52. $f(x) = \cos x + |\cos x|, \quad -2\pi \leq x \leq 2\pi$

$$f(x) = \begin{cases} \cos x + \cos x = 2\cos x & \text{if } -2\pi \leq x \leq \dfrac{-3\pi}{2} \text{ and } -\dfrac{\pi}{2} \leq x \leq \dfrac{\pi}{2} \text{ and } \dfrac{3\pi}{2} \leq x \leq 2\pi \\ \cos x - \cos x = 0 & \text{if } \dfrac{-3\pi}{2} \leq x \leq -\dfrac{\pi}{2} \text{ and } \dfrac{\pi}{2} \leq x \leq \dfrac{3\pi}{2}. \end{cases}$$

$$f'(x) = \begin{cases} -2\sin x & \text{if } -2\pi < x < \dfrac{-3\pi}{2} \text{ and } -\dfrac{\pi}{2} < x < \dfrac{\pi}{2} \text{ and } \dfrac{3\pi}{2} < x < 2\pi \\ 0 & \text{if } \dfrac{-3\pi}{2} < x < -\dfrac{\pi}{2} \text{ and } \dfrac{\pi}{2} < x < \dfrac{3\pi}{2} \\ \text{does not exist} & \text{if } x = -2\pi, \dfrac{-3\pi}{2}, -\dfrac{\pi}{2}, \dfrac{\pi}{2}, \dfrac{3\pi}{2}, 2\pi. \end{cases}$$

$f'(x) = 0$ also at $x = 0$.

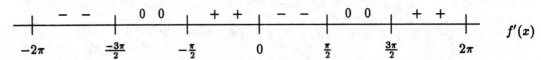

Thus f is decreasing on $\left[-2\pi, \frac{-3\pi}{2}\right]$ and $\left[0, \frac{\pi}{2}\right]$ and increasing on $\left[-\frac{\pi}{2}, 0\right]$ and $\left[\frac{3\pi}{2}, 2\pi\right]$. Also, f is constant on $\left[\frac{-3\pi}{2}, -\frac{\pi}{2}\right]$ and on $\left[\frac{\pi}{2}, \frac{3\pi}{2}\right]$. f has a relative (absolute) maxima at $(0, 2)$, $(-2\pi, 2)$ and $(2\pi, 2)$ and the absolute minimum is 0 for all $x \in \left[\frac{-3\pi}{2}, -\frac{\pi}{2}\right] \cup \left[\frac{\pi}{2}, \frac{3\pi}{2}\right]$.

53. $f(x) = (x^2 - 4x)^3$
$f'(x) = 3(x^2 - 4x)^2(2x - 4) = 6(x^2)(x - 4)^2(x - 2) = 0$ when $x = 0, 4$ and 2.
Therefore the critical numbers are $0, 4$ and 2.

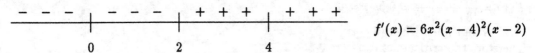

Thus f is decreasing on $(-\infty, 2]$ and increasing on $[2, \infty)$. f has a relative (absolute) minimum at $(2, -64)$.

54. $f(x) = x\sqrt{3 - x}$. Note that the domain of f is $(-\infty, 3]$.
$f'(x) = \sqrt{3 - x} + x\left(\frac{1}{2}\right)(3 - x)^{-1/2}(-1) = \dfrac{3(2 - x)}{2(3 - x)^{1/2}} = 0$ when $x = 2$.

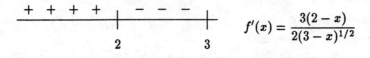

Thus f is inceasing on $(-\infty, 2]$ and decreasing on $[2, 3]$. f has a relative (absolute) maximum at $(2, 2)$ and a relative minimum at $(3, 0)$.

55. $f(x) = 2x^3 - 3ax^2 + 6$

$f'(x) = 6x^2 - 6ax = 6x(x - a) < 0$ if and only if x lies between 0 and a. Thus $a = 3$.

56. $f(x) = 2x^2 + qx + 5$

$f'(x) = 4x + q > 0$ if and only if $x > -\dfrac{q}{4}$.

Therefore $-\dfrac{q}{4} = -3 \Rightarrow q = 12$.

57. $f(x) = \frac{1}{3}x^3 - x^2 + qx + 10 \Rightarrow f'(x) = x^2 - 2x + q$. We are given that f should be increasing on $(-\infty, -3]$ and $[5, \infty)$ and decreasing on $[-3, 5]$. In particular, $f'(-3)$ and $f'(5)$ should be zero. Since $f'(-3) = 15 + q = f'(5)$ it is clear that we should take $q = -15$, so that

$$f'(x) = x^2 - 2x - 15 = (x - 5)(x + 3)$$

which is indeed positive on $(-\infty, -3)$ and $(5, \infty)$ and negative on $(-3, 5)$.

58. $f'(x) = 2x - 6$ is positive if $x > 3$ and negative if $x < 3$. Assuming f to be defined on $(-\infty, \infty)$ it follows that f is increasing on $[3, \infty)$.

59. $f(x) = x^2 + bx + c \Rightarrow f'(x) = 2x + b$. Notice that $f'\left(-\frac{b}{2}\right) = 0$ and that $f'(x) < 0$ if $x < -\frac{b}{2}$ whilst $f'(x) > 0$ if $x > -\frac{b}{2}$; so f has a relative minimum at $x = -\frac{b}{2}$. Of course, f cannot have a relative maximum at any point, since f has only one critical number: a (relative) minimum.

60. If $f(x)$ is a polynomial of degree n then $f'(x)$ is a polynomial of degree $n - 1$ and so has at most $n - 1$ zeros; thus $f(x)$ has at most $n - 1$ relative extrema. It can have fewer: for example, $f(x) = x^{2n}$ has only one relative extremum, at $x = 0$. [Strictly speaking, we exclude the case of a constant polynomial, corresponding to $n = 0$: in such a case, every point is a critical point.]

61. $f(x) = x^2 - ax + b \Rightarrow f'(x) = 2x - a$. Given that f has a relative minimum at $x = 2$ we deduce that

$$0 = f'(2) = 4 - a$$

so $a = 4$.

62. $f(x) = x^3 + ax^2 + bx + 7 \Rightarrow f'(x) = 3x^2 + 2ax + b$.

a) Given that f has relative extrema at $x = 1$ and $x = -3$ we deduce

$$\begin{aligned} 0 &= f'(1) = 3 + 2a + b \\ 0 &= f'(-3) = 27 - 6a + b \end{aligned}$$

so (by subtraction) $a = 3$ and (by substitution) $b = -9$.

b) Now

$$\begin{aligned} f'(x) &= 3x^2 + 6x - 9 = 3(x^2 + 2x - 3) \\ &= 3(x + 3)(x - 1) \end{aligned}$$

so that $f'(x) > 0$ if $x < -3$ or $x > 1$ and $f'(x) < 0$ if $-3 < x < 1$. It follows that f has a relative maximum at $x = -3$ and a relative minimum at $x = 1$.

63. $f(x) = a(x^2 - bx + 16) \Rightarrow f'(x) = a(2x - b)$.

    a) Since f has a relative extremum at $x = 5$, we have $0 = f'(5) = a(10 - b)$ so that $b = 10$. Of course, we exclude the possibility that $a = 0$!

    b) In order that $(5, f(5))$ be a relative minimum, we ask that $f'(x)$ be negative if $x < 5$ and positive if $x > 5$. Since $f'(x) = a(2x - 10)$ it is clear that a should be positive.

64. $f(x) = \sin ax \Rightarrow f'(x) = a \cos ax$, so $0 = f'\left(\frac{\pi}{6}\right) \Rightarrow 0 = a \cos\left(a\frac{\pi}{6}\right)$. If $a = 0$ then of course f has critical points everywhere, so we set aside this case. If $a \neq 0$ then $\cos\left(a\frac{\pi}{6}\right) = 0$ so that $a\frac{\pi}{6}$ is an odd multiple of $\frac{\pi}{2}$: say $a\frac{\pi}{6} = (2n + 1)\frac{\pi}{2}$ for some integer n; then $a = 3(2n + 1)$. Thus: f has a critical number at $x = \frac{\pi}{6}$ precisely when a is either zero or an odd multiple of 3.

65.   a) if the length of the rectangle is l, then its perimeter is $2x + 2l$; given that this equals 24, it follows that $l = 12 - x$. The area A of the rectangle is now $xl = x(12 - x) = A(x)$.

    b) $y = x(12 - x) = 12x - x^2$ implies that $\frac{dy}{dx} = 12 - 2x$; this is positive if $x < 6$ and negative if $x > 6$. So: the largest subinterval of $[0, 12]$ on which y is increasing is $[0, 6]$.

    c) At the endpoints $x = 0$ and $x = 12$, y is zero. From $\frac{dy}{dx} = 12 - 2x$ we see that y has a local extremum at $x = 6$, where $y = 6(12 - 6) = 36$. So: the maximum value of y on $[0, 12]$ is 36.

66. Let the carton have height h and base radius r: its volume is then (base area) $\times$ (height) $= \pi r^2 h$, which we are given as 500, and its surface area (counting curved surface and both ends) is

$$A = 2\pi rh + 2\pi r^2.$$

From $500 = \pi r^2 h$ we deduce that $h = \frac{500}{\pi r^2}$ so that A is given by

$$A = \frac{1000}{r} + 2\pi r^2$$

as a function of r. Differentiating,

$$\frac{dA}{dr} = -\frac{1000}{r^2} + 4\pi r = \frac{4}{r^2}(\pi r^3 - 250)$$

which equals zero (for a relative extremum) when $r = \sqrt[3]{250/\pi}$. If r is greater than this value then $\frac{dA}{dr} > 0$ whilst if r is less than $\frac{dA}{dr} < 0$; so A has a minimum when $r = \sqrt[3]{250/\pi}$, the value of A being then

$$100\left(\sqrt[3]{4\pi} + \sqrt[3]{\pi/2}\right) \approx 348.7$$

after simplification. Thus: for the area of the container to be minimized, $r = \sqrt[3]{250/\pi} \approx 4.3013$ and $h = \frac{500}{\pi r^2} \approx 8.6025$.

67.   a) $s(t) = -t^2 + 6t - 8 \Rightarrow s'(t) = -2t + 6$; so $s'(t) > 0$ for $t < 3$ and $s'(t) < 0$ for $t > 3$, from which we deduce the largest interval on which s is increasing to be $[0, 3]$. During this time, the motion is to the right of the horizontal number line.

    b) At time $t = 3$, the direction of motion changes: the particle comes to a halt $(s'(3) = 0)$ and then moves to the left $(s'(t) < 0$ for $t > 3)$.

68. If x is one of the numbers, then $50 - x$ is the other.

    a) If the product of the numbers is $p(x)$ then $p(x) = x(50 - x) = 50x - x^2$ so $p'(x) = 50 - 2x$; so $p(x)$ is increasing for $x < 25$, and decreasing for $x > 25$.

242

b) The maximum value of the product is $p(25) = 625$.

69. Let the box have height y and square base of side x; its (fixed) volume V is then given by $V = x^2 y$ and its surface area is $A = 2x^2 + 4xy$ (counting four sides, top and bottom). Solve for y in terms of x and V to get $y = \frac{V}{x^2}$ and substitute into A to get $A = 2x^2 + \frac{4V}{x}$. Now

$$\frac{dA}{dx} = 4x - \frac{4V}{x^2} = \frac{4(x^3 - V)}{x^2}$$

which is zero when $x^3 = V$ so that $x = \sqrt[3]{V}$ and $y = \frac{V}{x^2} = \sqrt[3]{V}$ also. If $x < \sqrt[3]{V}$ then $\frac{dA}{dx} < 0$ whilst if $x > \sqrt[3]{V}$ then $\frac{dA}{dx} > 0$; so A has a minimum value when x and y equal $\sqrt[3]{V}$. This shows that the box has least surface area when it is actually a cube.

70. The distance D from (x, x^2) to $(-3, 0)$ is

$$D = \sqrt{(x+3)^2 + (x^2 - 0)^2} = \sqrt{(x+3)^2 + x^4}.$$

As in **Example 10**, we attempt to minimize not D itself but rather

$$D^2 = x^4 + x^2 + 6x + 9.$$

Now

$$\begin{aligned} \frac{d(D^2)}{dx} &= 4x^3 + 2x + 6 \\ &= 2(2x^3 + x + 3) \\ &= 2(x+1)(2x^2 - 2x + 3) \end{aligned}$$

which equals zero when $x = -1$ (Note: the quadratic factor $2x^2 - 2x + 3$ has no real zeros). Since $x < -1 \Rightarrow \frac{d(D^2)}{dx} < 0$ and $x > -1 \Rightarrow \frac{d(D^2)}{dx} > 0$, it follows that D^2 (and hence D) is minimized when $x = -1$. Thus: $(-1, 1)$ is the point on $y = x^2$ nearest to $(-3, 0)$.

71. Let the bunker have height y and square base of side x; its volume is then given by $96 = x^2 y$. The base has area x^2 and so costs $10x^2$; the top has area x^2 and so costs $6x^2$; and the four sides have total area $4xy$ and so cost $24xy$. The total cost is therefore

$$C = 16x^2 + 24xy.$$

Solving for y in terms of x from $96 = x^2 y$ and substituting into C gives

$$C = 16x^2 + \frac{24 \times 96}{x}.$$

Now

$$\begin{aligned} \frac{dC}{dx} &= 32x - \frac{24 \times 96}{x^2} \\ &= \frac{32(x^3 - 72)}{x^2} \end{aligned}$$

which is zero when $x = 2\sqrt[3]{9}$, negative when $x < 2\sqrt[3]{9}$ and positive when $x > 2\sqrt[3]{9}$. Thus the cost of the bunker is least when $x = 2\sqrt[3]{9}$ and $y = \frac{96}{x^2} = \frac{8}{3}\sqrt[3]{9}$.

72. Take a right circular cylinder of height h and base radius r; its volume is then $\pi r^2 h = V$ (fixed) and its surface area is $A = 2\pi rh + 2\pi r^2$ (including both ends). The volume equation gives $h = \frac{V}{\pi r^2}$ which when substituted into the area equation yields

$$A = \frac{2V}{r} + 2\pi r^2.$$

Now

$$\frac{dA}{dr} = -\frac{2V}{r^2} + 4\pi r = \frac{2}{r^2}(2\pi r^3 - V)$$

which is zero when $r = \sqrt[3]{V/2\pi}$, negative when $r < \sqrt[3]{V/2\pi}$ and positive when $r > \sqrt[3]{V/2\pi}$. Thus: the surface area of the cylinder is least when $r = \sqrt[3]{V/2\pi}$ which implies $V = 2\pi r^3$. In this case, the equations $2\pi r^3 = V$ and $V = \pi r^2 h$ at once imply that $h = 2r$: the height of the cylinder equals the diameter of its base.

73. The idea here is to minimize the distance D from (x_0, y_0) to an arbitrary point (x, y) on the given line; as in **Example 10** and **Remark 3**, it is convenient to minimize $S = D^2$ first. If $b = 0$ then the given line is $x = -\frac{c}{a}$, and the solution is obvious: the least distance from (x_0, y_0) to the line $x = -\frac{c}{a}$ is the horizontal one. So suppose $b \neq 0$ and substitute $y = -\dfrac{c + ax}{b}$ into

$$S = (x - x_0)^2 + (y - y_0)^2$$

to get

$$S = (x - x_0)^2 + \left(\frac{c + ax}{b} + y_0\right)^2$$

whence

$$\frac{dS}{dx} = 2(x - x_0) + 2\frac{a}{b}\left(\frac{c + ax}{b} + y_0\right)$$

which is zero when (and only when)

$$(a^2 + b^2)x = b(bx_0 - ay_0) - ac$$

and so (from $y = -\dfrac{c + ax}{b}$ or from symmetry)

$$(a^2 + b_2)y = a(ay_0 - bx_0) - bc.$$

Checking the sign of S' shows that these values of x and y produce a minimum value for S. To calculate this minimum value, note that

$$
\begin{aligned}
(a^2 + b^2)(x - x_0) &= (a^2 + b^2)x - (a^2 + b^2)x_0 \\
&= b(bx_0 - ay_0) - ac - a^2 x_0 - b^2 x_0 \\
&= -a(ax_0 + by_0 + c)
\end{aligned}
$$

and similarly

$$(a^2 + b^2)(y - y_0) = -b(ax_0 + by_0 + c).$$

244

Thus

$$(a^2 + b^2)^2 S = \left((a^2 + b^2)(x - x_0)\right)^2 + \left((a^2 + b^2)^2(y - y_0)\right)^2$$
$$= (a^2 + b^2)(ax_0 + by_0 + c)$$

whence cancellation of a factor $a^2 + b^2$ and taking the square root yields

$$D = \frac{|ax_0 + by_0 + c|}{\sqrt{a^2 + b^2}}.$$

74. Let the width of the printed region be x and its height y (in centimetres) so that $xy = 800$. The total page area is

$$A = (x+4)(y+8)$$
$$= xy + 8x + 4y + 32$$

since the page has width $x + 4$ and height $y + 8$. Substitute $y = \frac{800}{x}$ in A to obtain

$$A = 832 + 8x + \frac{3200}{x}$$

so

$$\frac{dA}{dx} = 8 - \frac{3200}{x^2}.$$

Thus $\frac{dA}{dx} = 0$ when $8x^2 = 3200$ so that $x = 20$ ($x = -20$ is of course absurd); since $x < 20 \Rightarrow \frac{dA}{dx} < 0$ and $x > 20 \Rightarrow \frac{dA}{dx} > 0$, it follows that A is minimized when $x = 20$ and $y = 40$, so that the pages are 24 cm wide and 48 cm high.

75. The blue lines in the figure indicated the length L available to the rod when it makes an angle θ with the South wall of the corridor; from the figure, this available length is given by

$$L = 4 \csc \theta + 4 \sec \theta.$$

Now

$$\frac{dL}{d\theta} = -4 \csc \theta \cot \theta + 4 \sec \theta \tan \theta = 4\left(-\frac{\cos \theta}{\sin^2 \theta} + \frac{\sin \theta}{\cos^3 \theta}\right)$$

which is zero when $\cos^3 \theta = \sin^3 \theta$ or $\cos \theta = \sin \theta$ or $\theta = \frac{\pi}{4}$ (note that $0 \le \theta \le \frac{\pi}{2}$ from the figure). If $0 < \theta < \frac{\pi}{4}$ then $\frac{dL}{d\theta} < 0$ whilst if $\frac{\pi}{4} < \theta < \frac{\pi}{2}$ then $\frac{dL}{d\theta} > 0$: thus, the available length L is least when $\theta = \frac{\pi}{4}$; of course, this agrees with intuition. Since the rod is required to negotiate the corner completely, its length must be at most

$$L\left(\frac{\pi}{4}\right) = 4 \csc \frac{\pi}{4} + 4 \sec \frac{\pi}{4} = 8\sqrt{2},$$

which is therefore the length (in feet) of the longest rod which can be carried horizontally around the corner.

76. Regard the figure for Exercise 75 as a bird's eye view of this problem. Imagine carrying a rod of variable length L around the corner so that (for the sake of argument) one end is against the foot of the South wall and the other against the top of the West wall. The blue lines indicate the rod's length as seen from above, which (by Pythagoras' Theorem) is $\sqrt{L^2 - 8^2}$ since the corridor is 8 ft. high. Now

$$\sqrt{L^2 - 8^2} = 4 \csc \theta + 4 \sec \theta$$

so

$$\frac{1}{2}\left(L^2 - 8^2\right)^{-1/2} \cdot 2L\frac{d}{d\theta} = -4\csc\theta\cot\theta + 4\sec\theta\tan\theta$$

or

$$\frac{L}{\sqrt{L^2 - 8^2}}\frac{dL}{d\theta} = 4\left(-\frac{\cos\theta}{\sin^2\theta} + \frac{\sin\theta}{\cos^2\theta}\right).$$

It follows as before that the available length L is again least when $\theta = \frac{\pi}{4}$, so that

$$\sqrt{L^2 - 8^2} = 8\sqrt{2} \Rightarrow L^2 = 3 \cdot 8^2$$

whence the longest rod that will pass the corner is $8\sqrt{3}$ feet in length.

77. The value of the conical cup is

$$36\pi = \frac{1}{3}\left(\pi r^2\right)h$$

so that

$$108 = r^2 h.$$

Substituting $h = \frac{108}{r^2}$ into the formula for the surface area yields

$$S = \pi r\sqrt{r^2 + h^2} = \pi r\sqrt{r^2 + \frac{(108)^2}{r^4}}.$$

Bearing in mind **Remark 3**, we find it convenient to minimize S^2, where

$$S^2 = \pi^2 r^2\left(r^2 + \frac{(108)^2}{r^4}\right)$$

$$= \pi^2\left(r^4 + \frac{(108)^2}{r^2}\right).$$

From

$$\frac{d(S^2)}{dr} = \pi^2\left(4r^3 - \frac{2(108)^2}{r^3}\right)$$

we see that $\frac{d(S^2)}{dr} = 0$ when

$$4r^6 = 2(108)^2 \quad \text{therefore} \quad r^6 = 2^3\,3^6 \quad \text{therefore} \quad r = 3\sqrt{2}$$

and that $r < 3\sqrt{2} \Rightarrow \frac{d(S^2)}{dr} < 0$ whereas $r > 3\sqrt{2} \Rightarrow \frac{d(S^2)}{dr} > 0$. So the area of the conical cup is a minimum when the cup has radius $r = 3\sqrt{2}$ and height $h = 6$ (in centimetres).

78. According to **Remark 3** we can choose to minimize d^2 rather than d. Now

$$d^2 = (x - 2)^2 + (y - 0)^2$$
$$= x^2 - 4x + 4 + y^2$$
$$= x^2 - 3x + 4$$

since $y = \sqrt{x}$, so

$$\frac{d}{dx}(d^2) = 2x - 3$$

which is zero if $x = \frac{3}{2}$, negative if $x < \frac{3}{2}$ and positive if $x > \frac{3}{2}$. The least value of d^2 (hence of d) is therefore achieved when $x = \frac{3}{2}$, so that the closest point on $y = \sqrt{x}$ to $(2, 0)$ is $\left(\frac{3}{2}, \sqrt{3/2}\right)$.

79. $f(x) = x + \frac{1}{x} \Rightarrow f'(x) = 1 - \frac{1}{x^2}$

   a) $f'(x) > 0 \Leftrightarrow 1 > \frac{1}{x^2} \Leftrightarrow x^2 > 1 \Leftrightarrow x < -1$ or $x > 1$ so: f is increasing on $(-\infty, -1]$ and $[1, \infty)$.

   b) On $[1, \infty)$ the function f is increasing, so surely $f(x) \geq f(1) = 2$ for $x \geq 1$; on $(0, 1]$ the function f is decreasing, so also $f(x) \geq f(1) = 2$ if $0 < x \leq 1$. Consequently, $x + \frac{1}{x} \geq 2$ for $x > 0$.

   c) From $(x - 1)^2 \geq 0$ we deduce that $x^2 - 2x + 1 \geq 0$ so $x^2 + 1 \geq 2x$ and therefore (dividing by positive x) $1 + \frac{1}{x} \geq 2$.

   If $x < 0$ then $x + \frac{1}{x} = f(x) \leq f(-1) = -2$. This can be seen either by modifying the above argument or noting that f is an odd function.

80. $E = mc^2 \left(1 - \frac{v^2}{c^2}\right)^{-1/2}$ implies that $\frac{dE}{dv} = mc^2 \left(-\frac{1}{2}\right) \left(-\frac{2v}{c^2}\right) \left(1 - \frac{v^2}{c^2}\right)^{-3/2} = mv \left(1 - \frac{v^2}{c^2}\right)^{-3/2}$. Since this is positive when $0 < v < c$, it follows that E is an increasing function of v.

81. $r = \dfrac{1/wC}{\sqrt{R^2 + \left(1/wC\right)^2}} = \dfrac{1}{\sqrt{R^2 w^2 C^2 + 1}}$ implies that $\frac{dr}{dw} = -\frac{1}{2}(2R^2 C^2 w)(1 + R^2 w^2 C^2)^{-3/2}$

   $= -R^2 C^2 w (1 + R^2 C^2 w^2)^{-3/2}$, which is negative if $w > 0$; so r is a decreasing function of w on $(0, \infty)$.

82. $S = \sqrt{GM_e}\,(R_e + h)^{-1/2}$ implies that $\frac{dS}{dh} = \sqrt{GM_e}\left(-\frac{1}{2}\right)(R_e + h)^{-3/2}$, which is plainly negative; so orbital speed S decreases as altitude h increases.

83. Let x_1 and x_2 lie in $[a, c]$ with $x_1 < x_2$. If both x_1 and x_2 lie in $[a, b]$ then $f(x_1) < f(x_2)$ since f is increasing on $[a, b]$; likewise, $f(x_1) < f(x_2)$ if both x_1 and x_2 lie in $[b, c]$. So we may suppose $x_1 \in [a, b]$ and $x_2 \in [b, c]$. If in fact $x_1 < b$ and $b \leq x_2$ then $f(x_1) < f(b)$ and $f(b) \leq f(x_2)$ so that $f(x_1) < f(x_2)$; here, we again use the fact that f increases on $[a, b]$ and $[b, c]$. If instead $x_1 \leq b$ and $b < x_2$ then, similarly, $f(x_1) \leq f(b)$ and $f(b) < f(x_2)$ whence $f(x_1) < f(x_2)$ again. Having covered all contingencies, the proof is complete.

84. To show f is not decreasing on $[0, \infty)$ all we need do is find x_1 and x_2 in $[0, \infty)$ for which $x_1 < x_2$ but $f(x_1) \not> f(x_2)$. For example, take $x_1 = 0$ and $x_2 = 1$: then $f(x_1) = f(0) = 0 < 1 = f(1) = f(x_2)$. Similarly, $-1 < 0$ and yet $f(-1) = -1 < 0 = f(0)$, so f is not decreasing on $(-\infty, 0]$.

85. (ii) Either: proceed exactly as in the text's proof of (i) but reverse all inequalities and interchange the words 'increasing' and 'decreasing;' or apply part (i) to the function $g = -f$.

   (iii) Since $f'(x) > 0$ for all $x \in (a, c)$ and f is continuous on $[a, c]$ we conclude that f is increasing on $[a, c]$ by Therorem 6; similarly, f is increasing on $[c, b]$. Now apply the result of Exercise 83: since f is increasing on $[a, c]$ and $[c, b]$, it is increasing on $[a, b]$.

   (iv) Let $g = -f$; since $f' < 0$ on $(a, c) \cup (c, b)$ it follows that $g' = -f' > 0$ on $(a, c) \cup (c, b)$; according to part (iii) it follows that g is increasing on $[a, b]$, therefore $f = -g$ is decreasing on $[a, b]$.

86. a) $f' = \sec^2 x = \frac{1}{\cos^2 x}$ and $g'(x) = 1$. Now, if $0 < x < \frac{\pi}{2}$ then $1 > \cos x > 0$ so $\frac{1}{\cos^2 x} > 1$ and therefore $f'(x) > g'(x)$. It follows that if $x \in \left(0, \frac{\pi}{2}\right)$ then $(f - g)'(x) = f'(x) - g'(x) > 0$.

b) Let I be the interval $\left(0, \frac{\pi}{2}\right)$. Having seen above that $(f - g)'(x) > 0$ for all x in I it follows at once from **Theorem 6** that $(f - g)(x) > (f - g)(0) = 0$ for all x in I.

c) From $(f - g)(x) > 0$ it of course follows that $f(x) > g(x)$ or, in other words, $\tan x > x$ if $x \in \left(0, \frac{\pi}{2}\right)$.

This argument rests on the Mean Value Theorem (used in **Theorem 6**); the approach in Exercise 18 of 4.3 rests on Rolle's Theorem.

87. It will be enough to use the technique of Exercise 86 for $0 < x < \frac{\pi}{2}$ since $|\sin x| \leq 1 < \frac{\pi}{2}$ for all x. Let $f(x) = x$ and $g(x) = \sin x$, and note that $f(0) = 0 = g(0)$. If $0 < x < \frac{\pi}{2}$ then

$$g'(x) = \cos x < 1 = f'(x)$$

so that $(f - g)'(x) > 0$. According to **Theorem 6** (applied to $I = \left[0, \frac{\pi}{2}\right)$) it follows that if $0 < x < \frac{\pi}{2}$ then $(f - g)(x) > (f - g)(0) = 0$, whence $\sin x = g(x) < f(x) = x$ for such x.

88. Let $f(x) = \cos x$ and $g(x) = 1 - \frac{x^2}{2}$, so that $f'(x) = -\sin x$ and $g'(x) = -x$. From Exercise 87 we know that if $x > 0$ then $(f - g)'(x) = f'(x) - g'(x) = x - \sin x > 0$ so that by **Theorem 6**

$$
\begin{aligned}
(f - g)(x) \; &> \; (f - g)(0) \\
&= \; f(0) - g(0) \; = \; 0
\end{aligned}
$$

and therefore $f(x) > g(x)$ as required: in other words,

$$x > 0 \Rightarrow \cos x > 1 - \frac{x^2}{2}.$$

89. Exercise 87 tells us that if $x > 0$ then $\sin x < x$. To handle the inequality $x - \frac{x^3}{6} < \sin x$ for $x > 0$, let $f(x) = \sin x$ and $g(x) = x - \frac{x^3}{6}$ so that $f'(x) = \cos x$ and $g'(x) = 1 - \frac{x^2}{2}$. Clearly $f(0) = g(0) = 0$, whilst Exercise 88 implies that if $x > 0$ then

$$(f - g)'(x) \; = \; f'(x) - g'(x) \; = \; \cos x - \left(1 - \frac{x^2}{2}\right) \; > \; 0$$

so that (by **Theorem 6**) if $x > 0$ then

$$(f - g)(x) \; > \; (f - g)(0) \; = \; 0$$

whence

$$x \; > \; 0 \; \Rightarrow \; \sin x \; > \; x - \frac{x^3}{6}.$$

[Incidentally, if we divide through the inequalities $x - \frac{x^3}{6} < \sin x < x$ by $x > 0$ we obtain $1 - \frac{x^2}{6} < \frac{\sin x}{x} < 1$; these last inequalitites also hold for $x < 0$ since sine is odd. By an application of the "Pinching Theorem" it follows that $\lim\limits_{x \to 0} \frac{\sin x}{x} = 1$ once again.

90. Example: $f(x) = \begin{cases} \frac{1}{|x|} & \text{if } x \neq 0 \\ 0 & \text{if } x = 0. \end{cases}$

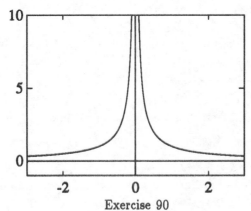

Exercise 90

91. a) $s(x) = x^2 \Rightarrow s'(x) = 2x$, which is positive if $x > 0$; by **Theorem 6**, it follows that s is increasing on $[0, \infty)$.

  b) To say that $f(c)$ is a (relative) minimum for f is to say that $f(c) \leq f(x)$ for all x (near c). Since s is increasing on $[0, \infty)$ and f is nonnegative, the inequality $f(c) \leq f(x)$ is equivalent to $f(c)^2 \leq f(x)^2$ and this in turn means that $f(c)^2$ is a minimum for f^2.

  c) In the notation of **Example 10** and **Remark 3**,

$$S(x) = x^2 - 7x + 16$$
$$\Rightarrow S'(x) = 2x - 7$$

  which is zero when $x = \frac{7}{2}$. As in **Example 10**, this is the only critical number for S in $[0, \infty)$ and in fact yields a minimum for S, hence a minimum for D by part b).

  d) As suggested in Remark 3, the difference between this approach and that of **Example 10** is that this approach circumvents the unsightly fractional powers.

92. If $f(x) = 3x^3 + x^2 + x - 3$ then

$$f'(x) = 9x^2 + 2x + 1$$
$$= 9\left(x + \frac{1}{9}\right)^2 + \frac{8}{9},$$

completing the square. So $f'(x) \geq \frac{8}{9} > 0$ for all x, whence f is increasing on $(-\infty, \infty)$. Since $f(0) = -3$ and $f(1) = 2$, we know that f has exactly one root, and this root lies between 0 and 1. In fact, the root is approximately 0.80286.

93. $f(x) = 4x^5 + 5x^4 + 2 \Rightarrow f'(x) = 20x^3(x + 1)$. Now, $f' < 0$ on $(-1, 0)$ and $f' > 0$ on $(-\infty, -1)$ and $(0, \infty)$; thus f is decreasing on $[-1, 0]$ and increasing on the intervals $(-\infty, -1]$ and $[0, \infty)$. We have $f(-2) = -46$, $f(-1) = 3$, and $f(0) = 2$. Therefore, f has exactly one root, and it lies between -2 and -1. In fact, the root is approximately -1.38564.

94. $f(x) = 3x^5 - 10x^3 + 15x - 2 \Rightarrow f'(x) = 15x^4 - 30x^2 + 15 = 15(x^2 - 1)^2$. Since $f'(x) > 0$ on $(-\infty, -1)$, $(-1, 1)$ and $(1, \infty)$ it follows that f is increasing on $(-\infty, -1]$, $[-1, 1]$ and $[1, \infty)$: that is, f is increasing on $(-\infty, \infty)$. Since $f(0) = -2$ and $f(1) = 6$, f has exactly one root, and it lies between 0 and 1. It is approximately 0.13496.

95. $$\begin{aligned} f(x) &= 40x^7 - 140x^6 + 168x^5 - 70x^4 + 1 \\ \Rightarrow f'(x) &= 280x^6 - 840x^5 + 840x^4 - 280x^3 \\ &= 280x^3(x^3 - 3x^2 + 3x - 1) \\ &= 280x^3(x-1)^3. \end{aligned}$$

Here, $f' > 0$ on $(-\infty, 0)$ and $(1, \infty)$ whilst $f' < 0$ on $(0, 1)$. Thus, f is increasing on $(-\infty, 0]$ and on $[1, \infty)$ and decreasing on $[0, 1]$. Since $f(-1) = -417$, $f(0) = 1$, $f(1) = -1$, and $f(2) = 417$, we know that f has exactly three roots—one in $(-1, 0)$, one in $(0, 1)$, and one in $(1, 2)$. Their approximate values are -0.29465, 0.5, and 1.29465.

96. a) If $v_0 = 72$ feet/sec and $h_0 = 3.5$ feet, then the equation becomes

$$y = (-0.0030864198 \sec^2\theta)x^2 + \tan\theta x + 3.5.$$

Graphs for $\theta = 25°, 30°, 45°, 55°$, and $65°$ are shown below. The range is $[0, 175] \times [0, 75]$. These angles must first be converted to radian measure. Among these graphs it appears that $\theta = 45°$ yields the maximum horizontal distance.

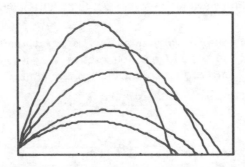

Exercise 96. a)

b) Table:

θ	25°	30°	40°	45°	50°	55°	60°	65°
height	17.97	23.75	36.97	44	51.03	57.85	64.25	70.03

c) The maximum horizontal distance travelled by the projectile is given by the positive root of the quadratic equation:

$$\frac{-16\sec^2\theta}{v_0^2}x^2 + \tan\theta x + h_0 = 0.$$

The positive root is given by

$$x = \frac{\tan\theta + \sqrt{\tan^2\theta + 64h_0\sec^2\theta/v_0^2}}{32\sec^2\theta/v_0^2}.$$

Therefore the distance $D(\theta)$ is given by

$$\begin{aligned} D(\theta) &= \frac{v_0^2\tan\theta + v_0\sqrt{v_0^2\tan^2\theta + 64h_0\sec^2\theta}}{32\sec^2\theta} \\ &= \frac{1}{32}\left(v_0^2\sin\theta\cos\theta + v_0\cos\theta\sqrt{v_0^2\sin^2\theta + 64h_0}\right). \end{aligned}$$

If $v_0 = 72$ and $h_0 = 3.5$ then

$$D(\theta) = 162\sin\theta\cos\theta + 18\cos\theta\sqrt{81\sin^2\theta + 3.5}.$$

Graph using the range: $\left[0, \frac{\pi}{2}\right] \times [0, 170]$.

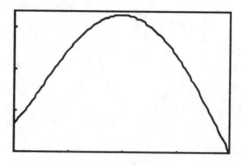

Exercise 96. c)

d) Graph with the range: $\left[0, \frac{\pi}{2}\right] \times [-330, 300]$. The value of θ where $\dfrac{D(\theta + \Delta\theta) - D(\theta)}{\Delta\theta}$ is zero is approximately the value of θ that maximizes the horizontal distance travelled by the projectile. The value of θ is approximately $\theta = 0.7748234$.

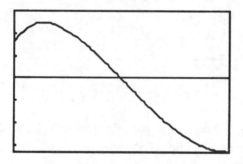

Exercise 96. d)

e) For the parabola given by $y = ax^2 + bx + c$, the vertex is given by the formulas $x = -\frac{b}{2a}$ and $y = \frac{c - b^2}{4x}$. Thus

$$
\begin{aligned}
M(\theta) &= \frac{h_0 \tan^2 \theta}{\frac{-64 \sec^2 \theta}{v_0^2}} \\
&= h_0 + \frac{v_0^2}{64} \sin^2 \theta \\
&= 3.5 + 81.5 \sin^2 \theta.
\end{aligned}
$$

The maximum occurs when $\sin \theta = 1$ or $\theta = \frac{\pi}{2}$ (see part f). $M\left(\frac{\pi}{2}\right) = 3.5 + 81.5 = 84.5$ feet.

f) Graph with the range: $\left[0, \frac{\pi}{2}\right] \times [0, 85]$.

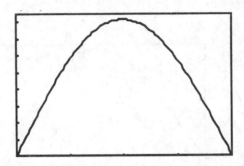

Exercise 96. f)

Note that $\dfrac{M(\theta + \Delta\theta) - M(\theta)}{\Delta\theta} \geq 0$. So $M(\theta)$ is increasing on $\left[0, \frac{\pi}{2}\right]$. Therefore the maximum occurs at the right endpoint of this interval, namely $\frac{\pi}{2}$.

4.5 Signficance of the Second Derivative: Concavity

1. a) The intervals on which the function is increasing: $[1, 2]$, $[4, 6]$ and $[8, 9]$.

  b) The intervals on which the graph is concave up: $[3, 6]$ and $[6, 9]$.

  c) The relative minima: $x = 1$, $x = 4$ and $x = 8$.

  d) The relative maxima: $x = 2$, $x = 6$ and $x = 9$.

  e) The inflection point: $x = 3$.

2. $f(x) = 3x + 1$
$f'(x) = 3$, $f''(x) = 0$.
According to Definition 4 of Section 4.4 and Definition 5 of Section 4.5, f is neither concave up nor concave down. Therefore, f has no inflection points.

3. $f(x) = x^2 - 4x - 5$
$f'(x) = 2x - 4$
$f''(x) = 2 > 0$ for all x.
Thus f is concave up on $(-\infty, \infty)$ and f has no inflection points.

4. $f(x) = 9 - x^3$
$f'(x) = -3x^2$
$f''(x) = -6x = 0$ at $x = 0$.

$$+ \; + \; + \; \big| \; - \; - \; - \qquad f''(x) = -6x$$
$$0$$

f is concave up on $(-\infty, 0]$ and concave down on $[0, \infty)$. f has an inflection point at $(0, 9)$.

5. $f(x) = |1 - x^2| = \begin{cases} 1 - x^2 & \text{if } -1 \leq x \leq 1 \\ x^2 - 1 & \text{if } x \leq -1 \text{ and } x \geq 1. \end{cases}$

$f'(x) = \begin{cases} -2x & \text{if } -1 < x < 1 \\ 2x & \text{if } x < -1 \text{ and } x > 1 \\ \text{does not exist} & \text{if } x = \pm 1. \end{cases}$ $f''(x) = \begin{cases} -2 & \text{if } -1 < x < 1 \\ 2 & \text{if } x < -1 \text{ and } x > 1 \\ \text{does not exist} & \text{if } x = \pm 1. \end{cases}$

Thus f is concave up on $(-\infty, -1]$ and $[1, \infty)$ and concave down on $[-1, 1]$. f has inflection points at $(\pm 1, 0)$.

6. $f(x) = |x^2 - 4| = \begin{cases} x^2 - 4 & \text{if } x \leq -2 \text{ and } x \geq 2 \\ 4 - x^2 & \text{if } -2 \leq x \leq 2. \end{cases}$

$f'(x) = \begin{cases} 2x & \text{if } x < -2 \text{ and } x > 2 \\ -2x & \text{if } -2 < x < 2 \\ \text{does not exist} & \text{if } x = \pm 2. \end{cases}$ $\qquad f''(x) = \begin{cases} 2 & \text{if } x < -2 \text{ and } x > 2 \\ -2 & \text{if } -2 < x < 2 \\ \text{does not exist} & \text{if } x = \pm 2. \end{cases}$

f is concave up on $(-\infty, -2]$ and $[2, \infty)$ and concave down on $[-2, 2]$. f has inflection points at $(\pm 2, 0)$.

7. $f(x) = \cos x$, $0 \leq x \leq 2\pi$
$f'(x) = -\sin x$
$f''(x) = -\cos x = 0$ at $x = \frac{\pi}{2}, \frac{3\pi}{2}$.

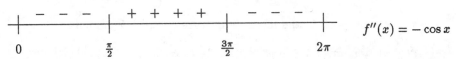

Thus f is concave down on $\left[0, \frac{\pi}{2}\right]$ and $\left[\frac{3\pi}{2}, 2\pi\right]$ and concave up on $\left[\frac{\pi}{2}, \frac{3\pi}{2}\right]$. f has inflection points at $\left(\frac{\pi}{2}, 0\right)$ and $\left(\frac{3\pi}{2}, 0\right)$.

8. $f(x) = \tan x$, $-\frac{\pi}{2} < x < \frac{\pi}{2}$
$f'(x) = \sec^2 x$
$f''(x) = 2 \sec^2 x \tan x = \dfrac{2 \sin x}{\cos^3 x} = 0$ at $x = 0$.

$f''(x) = \dfrac{2 \sin x}{\cos^3 x}$

Thus f is concave down on $\left(-\frac{\pi}{2}, 0\right]$ and concave up on $\left[0, \frac{\pi}{2}\right)$. f has an inflection point at $(0, 0)$.

9. $f(x) = (x + 3)^3$
$f'(x) = 3(x + 3)^2$
$f''(x) = 6(x + 3) = 0$ at $x = -3$.

$f''(x) = 6(x + 3)$

f is concave down on $(-\infty, -3]$ and concave up on $[-3, \infty)$. f has an inflection point at $(-3, 0)$.

10. $f(x) = \dfrac{4}{x - 2}$. (**Note:** domain of f: $(-\infty, 2) \cup (2, \infty)$)

$f'(x) = \dfrac{-4}{(x - 2)^2}$; $\quad f''(x) = \dfrac{8}{(x - 2)^3} \neq 0$ for all x in the domain.

$f''(x) = \dfrac{8}{(x - 2)^3}$

f is concave down on $(-\infty, 2)$ and concave up on $(2, \infty)$ and f has no inflection points.

11. $f(x) = \sqrt{x + 2}$. (**Note:** domain of f: $[-2, \infty)$)

$f'(x) = \frac{1}{2}(x + 2)^{-1/2}$

$f''(x) = \frac{-1}{4(x + 2)^{3/2}} < 0$ for all x in $(-2, \infty)$.

Thus f is concave down on $[-2, \infty)$ and f has no inflection points.

12. $f(x) = x^{2/3}$

$f'(x) = \frac{2}{3} x^{-1/3}$

$f''(x) = \frac{-2}{9x^{4/3}} \neq 0$ for all x. $f''(x)$ is undefined when $x = 0$.

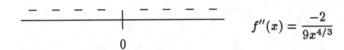

Thus f is concave down on $(-\infty, 0]$ and $[0, \infty)$ and f has no inflection points.

13. $f(x) = \frac{x}{x + 1}$. (**Note:** domain of f: $(-\infty, -1) \cup (-1, \infty)$)

$f'(x) = \frac{(x + 1) - x}{(x + 1)^2} = \frac{1}{(x + 1)^2}$

$f''(x) = \frac{-2}{(x + 1)^3} \neq 0$ for all x in the domain.

$$\begin{array}{ccc} + \ + \ + & | & - \ - \ - \\ & -1 & \end{array} \qquad f''(x) = \frac{-2}{(x + 1)^3}$$

f is concave up on $(-\infty, -1)$ and concave down on $(-1, \infty)$ and f has no inflection points.

14. $f(x) = (x + 2)^{1/3}$

$f'(x) = \frac{1}{3}(x + 2)^{-2/3}$

$f''(x) = \frac{-2}{9(x + 2)^{5/3}} \neq 0$ for all x. $f''(x)$ is undefined when $x = -2$.

$$\begin{array}{ccc} + \ + \ + & | & - \ - \ - \\ & -2 & \end{array} \qquad f''(x) = \frac{-2}{9(x + 2)^{5/3}}$$

f is concave up on $(-\infty, -2]$ and concave down on $[-2, \infty)$. f has an inflection point at $(-2, 0)$.

15. $f(x) = (2x + 1)^3$

$f'(x) = 6(2x + 1)^2$

$f''(x) = 24(2x + 1) = 0$ at $x = -\frac{1}{2}$.

$$\begin{array}{ccc} - \ - \ - & | & + \ + \ + \\ & -\frac{1}{2} & \end{array} \qquad f''(x) = 24(2x + 1)$$

f is concave down on $\left(-\infty, -\frac{1}{2}\right]$ and concave up on $\left[-\frac{1}{2}, \infty\right)$. f has an inflection point at $\left(-\frac{1}{2}, 0\right)$.

16. $f(x) = (1 - 4x)^3$
$f'(x) = -12(1 - 4x)^2$
$f''(x) = 96(1 - 4x) = 0$ when $x = \frac{1}{4}$.

+ + + | − − −

$\frac{1}{4}$

$f''(x) = 96(1 - 4x)$

f is concave up on $\left(-\infty, \frac{1}{4}\right]$ and concave down on $\left[\frac{1}{4}, \infty\right)$. f has an inflection point at $\left(\frac{1}{4}, 0\right)$.

17. $f(x) = 2x^3 - 3x^2 + 18x - 12$
$f'(x) = 6x^2 - 6x + 18$
$f''(x) = 12x - 6 = 6(2x - 1) = 0$ at $x = \frac{1}{2}$

− − − | + + +

$\frac{1}{2}$

$f''(x) = 6(2x - 1)$

f is concave down on $\left(-\infty, \frac{1}{2}\right]$ and concave up on $\left[\frac{1}{2}, \infty\right)$. f has an inflection point at $\left(\frac{1}{2}, -\frac{7}{2}\right)$.

18. $f(x) = 2x^3 + 12x^2 + 18x + 12$
$f'(x) = 6x^2 + 24x + 18$
$f''(x) = 12x + 24 = 12(x + 2) = 0$ at $x = -2$.

− − − | + + +

-2

$f''(x) = 12(x + 2)$

f is concave down on $(-\infty, -2]$ and concave up on $[-2, \infty)$. f has an inflection point at $(-2, 8)$.

19. $f(x) = x^4 + 2x^3 - 36x^2 + 24x - 6$
$f'(x) = 4x^3 + 6x^2 - 72x + 24$
$f''(x) = 12x^2 + 12x - 72 = 12(x^2 + x - 6) = 12(x + 3)(x - 2) = 0$ at $x = -3, 2$.

+ + + | − − − | + + +

-3 2

$f''(x) = 12(x + 3)(x - 2)$

f is concave up on $(-\infty, -3]$ and $[2, \infty)$ and concave down on $[-3, 2]$. f has inflection points at $(-3, -375)$ and $(2, -70)$.

20. $f(x) = \frac{1}{12}x^4 - \frac{2}{3}x^3 + 2x^2 + 5x - 8$
$f'(x) = \frac{1}{3}x^3 - 2x^2 + 4x + 5$
$f''(x) = x^2 - 4x + 4 = (x - 2)^2 = 0$ at $x = 2$.

+ + + | + + +

2

$f''(x) = (x - 2)^2$

Thus f is concave up on $(-\infty, \infty)$ and f has no inflection points.

21. $f(x) = x^4 + 6x^3 + 6x^2 + 12x + 1$

    $f'(x) = 4x^3 + 18x^2 + 12x + 12$

    $f''(x) = 12x^2 + 36x + 12 = 12(x^2 + 3x + 1) = 0$ where $x = \dfrac{-3 \pm \sqrt{5}}{2} \approx -.382, -2.62$.

    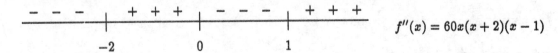

    $$f''(x) = 12(x^2 + 3x + 1)$$

    Thus f is concave up on $\left(-\infty, \frac{-3-\sqrt{5}}{2}\right]$ and $\left[\frac{-3+\sqrt{5}}{2}, \infty\right)$ and concave down on $\left[\frac{-3-\sqrt{5}}{2}, \frac{-3+\sqrt{5}}{2}\right]$.

    f has inflection points at $\left(\frac{-3-\sqrt{5}}{2}, -50.04\right)$ and $\left(\frac{-3+\sqrt{5}}{2}, -3.02\right)$.

22. $f(x) = 3x^5 + 5x^4 - 20x^3 + 10x + 4$

    $f'(x) = 15x^4 + 20x^3 - 60x^2 + 10$

    $f''(x) = 60x^3 + 60x^2 - 120x = 60x(x^2 + x - 2) = 60x(x + 2)(x - 1) = 0$ when $x = 0, -2, 1$.

    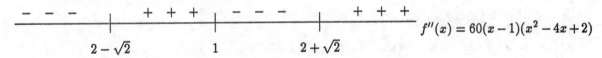

    $$f''(x) = 60x(x + 2)(x - 1)$$

    Thus f is concave down on $(-\infty, -2]$ and $[0, 1]$ and concave up on $[-2, 0]$ and $[1, \infty)$.

    f has inflection points at $(-2, 128), (0, 4)$ and $(1, 2)$.

23. $f(x) = 3x^5 - 25x^4 + 60x^3 - 60x^2 + 2x + 1$

    $f'(x) = 15x^4 - 100x^3 + 180x^2 - 120x + 2$

    $f''(x) = 60x^3 - 300x^2 + 360x - 120 = 60(x^3 - 5x^2 + 6x - 2) = 60(x - 1)(x^2 - 4x + 2) = 0$ when

    $x = 1, 2 \pm \sqrt{2}$. $[2 \pm \sqrt{2} \approx 3.41, .586]$.

    ```
     - - -        + + +        - - -           + + +
    ─────────┼─────────┼──────────────┼──────────    f''(x) = 60(x − 1)(x² − 4x + 2)
        2 − √2           1           2 + √2
    ```

    Thus f is concave down on $\left(-\infty, 2-\sqrt{2}\right]$ and $\left[1, 2+\sqrt{2}\right]$ and concave up on $\left[2-\sqrt{2}, 1\right]$ and $\left[2+\sqrt{2}, \infty\right)$.

    f has inflection points at $x = 2 - \sqrt{2}, 1$ and $2 + \sqrt{2}$.

24. $f(x) = 3x^5 + 10x^4 + 20x^3 + 4x + 2$

    $f'(x) = 15x^4 + 40x^3 + 60x^2 + 4$

    $f''(x) = 60x^3 + 120x^2 + 120x = 60x(x^2 + 2x + 2) = 0$ when $x = 0$.

    ```
          - - -    + + +
    ──────────────┼──────────    f''(x) = 60x(x² + 2x + 2)
                  0
    ```

    f is concave down on $(-\infty, 0]$ and concave up on $[0, \infty)$. f has an inflection point at $(0, 2)$.

25. $f(x) = x^{2/3}(119 - x^5)$

    $f'(x) = \frac{2}{3}x^{-1/3}(119 - x^5) + x^{2/3}(-5x^4) = \dfrac{17(14 - x^5)}{3x^{1/3}}$

$$f''(x) = \frac{17}{3}\left[\frac{-5x^4(x^{1/3}) - (14 - x^5)\frac{1}{3}(x^{-2/3})}{x^{2/3}}\right] = \frac{-238}{9}\left(\frac{x^5 + 1}{x^{4/3}}\right) = 0 \text{ at } x = -1.$$

$f''(x)$ is undefined when $x = 0$.

$$f''(x) = \frac{-238}{9}\left(\frac{x^5 + 1}{x^{4/3}}\right)$$

Thus f is concave up on $(-\infty, -1]$ and concave down on $[-1, 0]$ and $[0, \infty)$. f has an inflection point at $(-1, 120)$.

26. $f(x) = \dfrac{x}{1 + x^2}$

$$f'(x) = \frac{1 + x^2 - x(2x)}{(1 + x^2)^2} = \frac{1 - x^2}{(1 + x^2)^2}$$

$$f''(x) = \frac{-2x(1 + x^2)^2 - (1 - x^2)\,2\,(1 + x^2)(2x)}{(1 + x^2)^4} = \frac{2x^3 - 6x}{(1 + x^2)^3} = \frac{2x(x^2 - 3)}{(1 + x^2)^3} = 0 \text{ when } x = 0, \pm\sqrt{3}.$$

$$f''(x) = \frac{2x(x^2 - 3)}{(1 + x^2)^3}$$

f is concave down on $\left(-\infty, -\sqrt{3}\right]$ and $\left[0, \sqrt{3}\right]$ and concave up on $\left[-\sqrt{3}, 0\right]$ and $\left[\sqrt{3}, \infty\right)$.

f has inflection points at $\left(-\sqrt{3}, -\frac{\sqrt{3}}{4}\right)$, $(0, 0)$ and $\left(\sqrt{3}, \frac{\sqrt{3}}{4}\right)$.

27. $f(x) = x^{5/3} - 5x^{2/3} + 3$

$$f'(x) = \frac{5}{3}x^{2/3} - \frac{10}{3}x^{-1/3}$$

$$f''(x) = \frac{10}{9}x^{-1/3} + \frac{10}{9}x^{-4/3} = \frac{10}{9}\left(\frac{x + 1}{x^{4/3}}\right) = 0 \text{ at } x = -1. \quad f''(x) \text{ is undefined when } x = 0.$$

$$f''(x) = \frac{10}{9}\left(\frac{x + 1}{x^{4/3}}\right)$$

f is concave down on $(-\infty, -1]$ and concave up on $[-1, 0]$ and $[0, \infty)$. f has an inflection point at $(-1, -3)$.

28. $f(x) = \dfrac{\sqrt[3]{x}}{1 - x}$. (**Note:** domain of f: $(-\infty, 1) \cup (1, \infty)$)

$$f'(x) = \frac{(\frac{1}{3}x^{-2/3})(1 - x) + \sqrt[3]{x}}{(1 - x)^2} = \frac{1 + 2x}{3x^{2/3}(1 - x)^2}$$

$$f''(x) = \frac{2(3x^{2/3})(1 - x)^2 - (1 + 2x)[2x^{-1/3}(1 - x)^2 + 3x^{2/3}(2)(1 - x)(-1)]}{9x^{4/3}(1 - x)^4} = \frac{2(5x^2 + 5x - 1)}{9x^{5/3}(1 - x)^3} = 0 \text{ at}$$

$$x = \frac{-5 \pm 3\sqrt{5}}{10} \approx .171, -1.17$$

$f''(x)$ is undefined at $x = 0$.

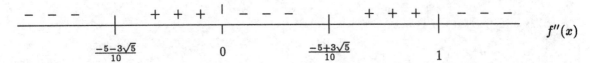

Thus f is concave down on $\left(-\infty, \frac{-5-3\sqrt{5}}{10}\right]$, $\left[0, \frac{-5+3\sqrt{5}}{10}\right]$ and $(1, \infty)$ and concave up on $\left[\frac{-5-3\sqrt{5}}{10}, 0\right]$ and $\left[\frac{-5+3\sqrt{5}}{10}, 1\right)$. f has inflection points at $x = 0, \dfrac{-5 \pm 3\sqrt{5}}{10}$.

29. $f(x) = \sin^2 x$, $x = \frac{\pi}{2}$

$f'(x) = 2 \sin x \cos x = \sin 2x$; $f'\left(\frac{\pi}{2}\right) = 0$

$f''(x) = 2 \cos 2x$; $f''\left(\frac{\pi}{2}\right) = 2 \cos \pi = -2 < 0$. Thus f has a relative maximum at $x = \frac{\pi}{2}$.

30. $f(x) = x^2 + \frac{2}{x}$, $x = 1$

$f'(x) = 2x - \frac{2}{x^2}$, $f'(1) = 0$

$f''(x) = 2 + \frac{4}{x^3}$; $f''(1) = 2 + 4 = 6 > 0$. Thus f has a relative minimum at $x = 1$.

31. $f(x) = x^4 - 4x^3 - 48x^2 + 24x + 20$, $x = 4$
$f'(x) = 4x^3 - 12x^2 - 96x + 24$; $f'(4) = -296 \neq 0$
Thus $x = 4$ is not a critical number. Therefore f does not have a relative extremum at $x = 4$.

32. $f(x) = 2x^3 - 3x^2$, $x = 1$
$f'(x) = 6x^2 - 6x$; $f'(1) = 0$
$f''(x) = 12x - 6$; $f''(1) = 12 - 6 = 6 > 0$
Thus f has a relative minimum at $x = 1$.

33. $f(x) = x^4 - 4x^3 + 6x^2 - 4x$, $x = 1$
$f'(x) = 4x^3 - 12x^2 + 12x - 4$; $f'(1) = 0$
$f''(x) = 12x^2 - 24x + 12$; $f''(1) = 0$. Thus the Second Derivative Test fails. We'll use the First Derivative Test.

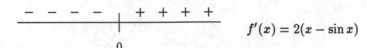

Thus f has a relative (absolute) minimum at $x = 1$.

34. $f(x) = x^2 + 2 \cos x$, $x = 0$
$f'(x) = 2x - 2 \sin x$; $f'(0) = 0$
$f''(x) = 2 - 2 \cos x$; $f''(0) = 0$. Thus the Second Derivative Test fails. We'll use the First Derivative Test. Note that $x = 0$ is the only critical number of f.

$$\begin{array}{ccc} - \ - \ - \ - & \Big| & + \ + \ + \ + \\ & 0 & \end{array} \qquad f'(x) = 2(x - \sin x)$$

Thus f has a relative (absolute) minimum at $x = 0$.

35. $f'(x) = \dfrac{x-1}{x+1}$, $x = 1$

    $f'(1) = 0$

$$\begin{array}{ccc} & - \; - \; - & + \; + \; + \\ \hline & | & | \\ & -1 & 1 \end{array}$$

$f'(x) = \dfrac{x-1}{x+1}$

Thus f has a relative minimum at $x = 1$.

36. $f'(x) = \dfrac{1 - \sin 2x}{\cos x}$, $x = \dfrac{\pi}{4}$

    $f'\left(\dfrac{\pi}{4}\right) = 0$

$$\begin{array}{ccc} + \; + \; + & + \; + \; + \\ \hline | & | & | \\ -\frac{\pi}{2} & \frac{\pi}{4} & \frac{\pi}{2} \end{array}$$

$f'(x) = \dfrac{1 - \sin 2x}{\cos x}$

Thus $x = \dfrac{\pi}{4}$ is not a relative extremum of f.

37. $f'(x) = (x-1)(x+2)$, $x = -2$.
    $f'(-2) = 0$

$$\begin{array}{ccc} + \; + \; + & - \; - \; - \\ \hline | & | \\ -2 & 1 \end{array}$$

$f'(x) = (x-1)(x+2)$

Thus f has a relative maximum at $x = -2$.

38. $f'(x) = (x-2)^3(x+1)$, $x = 2$

$$\begin{array}{ccc} - \; - \; - & + \; + \; + \\ \hline | & | \\ -1 & 2 \end{array}$$

$f'(x) = (x-2)^3(x+1)$

Thus f has a relative minimum at $x = 2$.

39. $f(x) = x^3 - 3x^2 + 6$
    $f'(x) = 3x^2 - 6x = 3x(x-2) = 0$ at $x = 0, 2$.
    $f''(x) = 6x - 6 = 6(x-1) = 0$ at $x = 1$.
    $f''(0) = -6 < 0$. Thus f has a relative maximum at $(0, 6)$.
    $f''(2) = 6 > 0$. Thus f has a relative minimum at $(2, 2)$.

f is concave down on $(-\infty, 1]$ and concave up on $[1, \infty)$. f has an inflection point at $(1, 4)$.

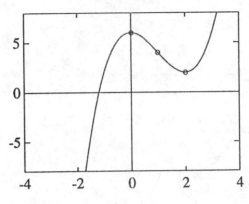

Exercise 39

40. $y = x^4 + 4x^3 - 8x^2 - 48x + 9$

$y' = 4x^3 + 12x^2 - 16x - 48 = 4(x+3)(x+2)(x-2) = 0$ at $x = -3, -2, 2$.

$y'' = 12x^2 + 24x - 16 = 4(3x^2 + 6x - 4) = 0$ at $x = -1 \pm \frac{\sqrt{21}}{3} \approx .528, -2.53$

$y''(-3) = 20 > 0$. Thus y has a relative minimum at $(-3, 54)$.

$y''(-2) = -16 < 0$. Thus y has a relative maximum at $(-2, 57)$.

$y''(2) = 80 > 0$. Thus y has a relative (absolute) minimum at $(2, -71)$.

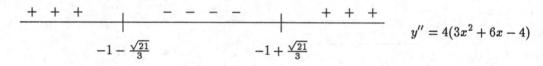

$$y'' = 4(3x^2 + 6x - 4)$$

Thus y is concave up on $\left(-\infty, -1 - \frac{\sqrt{21}}{3}\right]$ and $\left[-1 + \frac{\sqrt{21}}{3}, \infty\right)$ and concave down on $\left[-1 - \frac{\sqrt{21}}{3}, -1 + \frac{\sqrt{21}}{3}\right]$.

y has inflection points at $\left(-1 - \frac{\sqrt{21}}{3}, 55.4\right)$ and $\left(-1 + \frac{\sqrt{21}}{3}, -17.9\right)$.

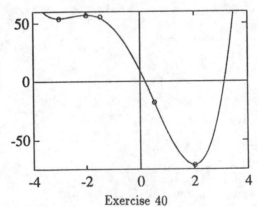

Exercise 40

41. $y = |9 - x^2| = \begin{cases} 9 - x^2 & \text{if } -3 \le x \le 3 \\ x^2 - 9 & \text{if } x \le -3 \text{ and } x \ge 3 \end{cases}$

$y' = \begin{cases} -2x & \text{if } -3 < x < 3 \\ 2x & \text{if } x < -3 \text{ and } x > 3 \\ \text{does not exist} & \text{if } x = \pm 3. \end{cases}$

$y' = 0$ at $x = 0$.

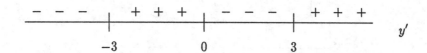

Thus y has a relative maximum at $(0, 9)$ and relative (absolute) minima at $(\pm 3, 0)$.

$$y'' = \begin{cases} -2 & \text{if } -3 < x < 3 \\ 2 & \text{if } x < -3 \text{ and } x > 3 \\ \text{does not exist} & \text{if } x = \pm 3. \end{cases}$$

Thus y is concave up on $(-\infty, -3]$ and $[3, \infty)$ and concave down on $[-3, 3]$. y has inflection points at $(\pm 3, 0)$.

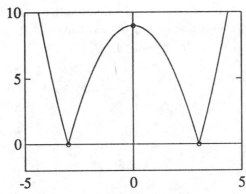

Exercise 41

42. $f(x) = (x - 8)^{2/3}$

$f'(x) = \dfrac{2}{3(x-8)^{1/3}} \neq 0$ for all x.

$f'(x)$ is undefined at $x = 8$.

$$f'(x) = \dfrac{2}{3(x-8)^{1/3}}$$

Thus f has a relative (absolute) minimum at $(8, 0)$.

$f''(x) = \dfrac{-2}{9(x-8)^{4/3}} < 0$ for all $x \neq 8$.

Thus f is concave down on $(-\infty, 8]$ and $[8, \infty)$. f has no inflection points.

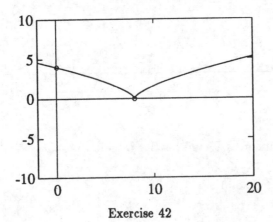

Exercise 42

43. Let the distance from P_1 to Q be x (in meters). Let the lamp at P_2 be twice as strong as that at P_1. The illumination at Q due to P_1 is $I_1 = \frac{k}{x^2}$ and the illumination at Q due to P_2 is $\frac{2k}{(20-x)^2}$ where $k > 0$ is the constant of proportionality multiplied by the intensity of the lamp at P_1. Thus: the total illumination at Q is

$$I = I_1 + I_2 = \frac{k}{x^2} + \frac{2k}{(20-x)^2}$$

therefore

$$\frac{dI}{dx} = -k\left(\frac{2}{x^3} + \frac{4}{(x-20)^3}\right)$$

and

$$\frac{d^2 I}{dx^2} = k\left(\frac{6}{x^4} + \frac{12}{(x-20)^4}\right).$$

Now, $\frac{dI}{dx} = 0$ when $\left(\frac{20-x}{x}\right)^3 = 2$ therefore $\frac{20}{x} - 1 = \sqrt[3]{2}$ so $x = \frac{20}{1+\sqrt[3]{2}}$. Since $\frac{d^2 I}{dx^2}$ is clearly positive (for $0 < x < 20$) it follows that I is a minimum when $x = \frac{20}{1+\sqrt[3]{2}}$.

44. Let the can have height h and base radius r; its volume is then

$$2\pi = \pi r^2 h$$

(so that $h = \frac{2}{r^2}$) and its surface area is

$$A = 2\pi r^2 + 2\pi r h$$
$$= 2\pi r^2 + \frac{4\pi}{r}$$

therefore

$$\frac{dA}{dr} = 4\pi r - \frac{4\pi}{r^2}$$

and

$$\frac{d^2 A}{dr^2} = 4\pi + \frac{8\pi}{r^3}.$$

Thus $\frac{dA}{dr} = 0$ when $r = 1$, and this minimizes A since $\frac{d^2 A}{dr^2} > 0$ (for $r > 0$). The most economical can therefore has radius 1 inch and height 2 inches.

45. f is increasing on $[1, 3]$ and $[5, 6]$. f is concave down on $[2, 4]$ and $[7, 8]$. The function f has an endpoint relative maximum at $x = 0$ and an endpoint relative minimum at $x = 8$. It also has relative minima at $x = 1$ and $x = 5$ and relative maxima at $x = 3$ and $x = 6$.

46. We offer an informal proof which can easily be made rigorous. **Remark 1** and **Theorem 8** imply that the graph of f lies above its tangent at any point, since $f'' > 0$. Now, let $f(a) = b < 0$ and $f'(a) = \alpha > 0$. The tangent to the graph of f at $\big(a, f(a)\big) = (a, b)$ is $y - b = \alpha(x - a)$ and crosses the x-axis when $y = 0$ so that $x = a - \frac{b}{\alpha} > a$. Since the graph of f lies above the tangent, it follows that $f\big(a - \frac{b}{\alpha}\big) > 0$. The Intermediate Value Theorem now tells us that f is zero at some point between $x = a$ (where $f < 0$) and $x = a - \frac{b}{\alpha}$ (where $f > 0$). This is absurd, given that f is everywhere negative.

47. $f(x) = ax^2 + bx + c \Rightarrow f'(x) = 2ax + b \Rightarrow f''(x) = 2a$. In order that f be concave up for all x, it is necessary and sufficient that a be positive: for if $a > 0$ then $f''(x) > 0$ for all x and we can apply **Theorem 8**; whilst if $a \leq 0$ then $f''(x) \leq 0$ for all x so that f' is not increasing.

48. Let $f(x) = -x^4$ for all real x. Then surely $f(0)$ is a relative (indeed, absolute) maximum, since $x \neq 0 \Rightarrow x^4 > 0$. However, $f''(x) = -12x^2$ so that $f''(0) = 0$.

49. Let $f(x) = x^4$ for all real x. Then $x \neq 0 \Rightarrow f(x) = x^4 > 0 = f(0)$, so that $f(0)$ is a relative (in fact, absolute) minimum, but $f''(x) = 12x^2$ so that $f''(0) = 0$.

50. Let $f(x) = x^3$; then $f'(x) = 3x^2$ and $f''(x) = 6x$, so $f'(0) = f''(0) = 0$. However, $f(0)$ is not a relative extremum, since $x < 0 \Rightarrow f(x) < f(0)$ and $x > 0 \Rightarrow f(x) > f(0)$.

51. If $n = 2$ then f can have no inflection points, since f'' is a constant and so cannot change sign. If $n = 3$ then f has one inflection point: f'' is a linear function, and so crosses the x-axis exactly once. If $n = k$ then f'' is a polynomial of degree $k - 2$ and so has at most $k - 2$ real roots; thus, f has at most $k - 2$ inflection points (but may have fewer: for example, $f(x) = x^4$ has none).

52. An example is

$$f(x) = \frac{1}{3 + (x - 1)^2}$$

for which

$$f'(x) = \frac{2(1 - x)}{\big(3 + (x - 1)^2\big)^2}$$

and

$$f''(x) = \frac{6x(x - 2)}{\big(3 + (x - 1)^2\big)^3}.$$

53. An example: $f(x) = -x^{1/3}$ (for which $f'(0)$ and $f''(0)$ are undefined). A sketch:

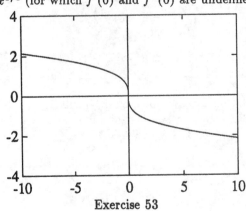

Exercise 53

54. False. Consider $f(x) = x - x^2$. Here, $f(0) = 0$ and $f''(x) = -2 < 0$ for all x; but $f(\frac{1}{2}) = \frac{1}{4} > 0$.

55.   56.

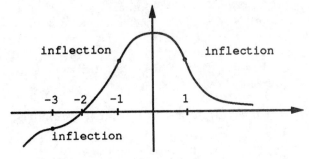

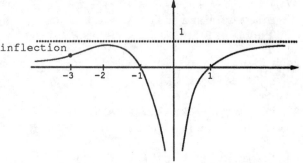

Exercise 55   Exercise 56

57. Let $a < b$ be points of I. Since the graph of f lies above the tangent at $(a, f(a))$ we have

$$f(b) > f(a) + f'(a)(b - a)$$

whilst since the graph of f lies above the tangent at $(b, f(b))$ we have

$$f(a) > f(b) + f'(b)(a - b).$$

The first of these inequalities implies

$$f'(a) < \frac{f(b) - f(a)}{b - a}$$

whilst the second implies

$$f'(b) > \frac{f(b) - f(a)}{b - a}$$

in view of $a < b$. It follows at once that $f'(a) < f'(b)$. Thus: f' is increasing on I.

58. Assume f' is increasing on I and let $a \in I$. Since the tangent to the graph of f at $(a, f(a))$ has equation

$$y = f(a) + f'(a)(x - a)$$

we consider the function

$$d(x) = f(x) - \{f(a) + f'(a)(x - a)\}.$$

Since

$$d'(x) = f'(x) - f'(a)$$

we see (from f' increasing) that $x > a \Rightarrow d'(x) > 0$ whilst $x < a \Rightarrow d'(x) < 0$. Now, if $b \in I$ and $b > a$ then (MVT) there exists $x \in (a, b)$ such that $d(b) - d(a) = d'(x)(b - a) > 0$ since $d'(x) > 0$; thus $d(b) > d(a)$. If $b \in I$ and $b < a$ then (MVT) there exists $x \in (b, a)$ such that $d(b) - d(a) = d'(x)(b - a) > 0$ since $d'(x) < 0$; thus $d(b) > d(a)$. It now follows that if $b \in I$ and $b \neq a$ then $d(b) > d(a) = 0$. Looking back to the definition of d, this establishes that the graph of f lies above its tangents, as required.

59.   a) As suggested, let $f''(c) = L < 0$.

b) Since $f'(c) = 0$ we have

$$\frac{f'(x)}{x - c} = \frac{f'(x) - f'(c)}{x - c}$$

which (by definition) approaches $f''(c) = L$ as $x \to c$.

c) In the ϵ-δ definition of limit, let $\epsilon = \frac{1}{2}|L| = -\frac{1}{2}L$; there exists a $\delta > 0$ such that $|x - c| < \delta$ implies $\left|\frac{f'(x)}{x - c} - L\right| < \frac{1}{2}|L|$, whence in fact $\frac{f'(x)}{x - c} < \frac{1}{2}L < 0$. Thus we may take $(a, b) = (c - \delta,\ c + \delta)$.

d) If $x \in (a, c)$ then $x - c < 0$ and $\frac{f'(x)}{x - c} < 0$ imply $f'(x) > 0$.

e) Similarly, $x \in (c, b)$ implies $x - c > 0$, so $\frac{f'(x)}{x - c} < 0$ implies $f'(x) < 0$.

f) According to the First Derivative Test, **Theorem 7** (especially part (i)) in Section 4.4, it follows at once from d) and e) above that $(c, f(c))$ is a relative maximum as desired.

60. $T'(x) = \dfrac{x}{v_1\sqrt{a^2 + x^2}} - \dfrac{c - x}{v_2\sqrt{(c - x)^2 + b^2}}.$

Now,

$$\frac{d}{dx}\left(\frac{x}{v_1\sqrt{a^2 + x^2}}\right) = \frac{1}{v_1}\left\{\frac{\sqrt{a^2 + x^2} - x \cdot \frac{x}{\sqrt{a^2+x^2}}}{a^2 + x^2}\right\}$$

$$= \frac{a^2}{v_1(a^2 + x^2)^{3/2}}$$

and

$$\frac{d}{dx}\left(\frac{c - x}{v_2\sqrt{(c - x)^2 + b^2}}\right) = \frac{1}{v_2}\left\{\frac{\sqrt{(c - x)^2 + b^2}\,(-1) - (c - x) \cdot \frac{-(c-x)}{\sqrt{(c-x)^2+b^2}}}{(c - x)^2 + b^2}\right\}$$

$$= \frac{-b^2}{v_2\left((c - x)^2 + b^2\right)^{3/2}}.$$

Taking our two expressions over a common denominator yields

$$T''(x) = \frac{a^2 v_2\left((c - x)^2 + b^2\right)^{3/2} + b^2 v_1(a^2 + x^2)^{3/2}}{v_1 v_2(a^2 + x^2)^{3/2}\left((c - x)^2 + b^2\right)^{3/2}}.$$

4.6   Large-Scale Behavior: Asymptotes

1. $\displaystyle\lim_{x\to\infty}\frac{3x^2+2}{10x^2-3x}=\lim_{x\to\infty}\frac{3x^2+2}{10x^2-3x}\frac{\left(1/x^2\right)}{\left(1/x^2\right)}=\lim_{x\to\infty}\frac{3+\frac{2}{x^2}}{10-\frac{3}{x}}=\frac{3+0}{10-0}=\frac{3}{10}$

2. $\displaystyle\lim_{x\to\infty}\frac{6x^4-8x}{7-3x^4}=\lim_{x\to\infty}\frac{6x^4-8x}{7-3x^4}\frac{\left(1/x^4\right)}{\left(1/x^4\right)}=\lim_{x\to\infty}\frac{6-\frac{8}{x^3}}{\frac{7}{x^4}-3}=\frac{6-0}{0-3}=-2$

3. $\displaystyle\lim_{x\to\infty}\frac{x(4-x^3)}{3x^4+2x^2}=\lim_{x\to\infty}\frac{x(4-x^3)}{3x^4+2x^2}\frac{\left(1/x^4\right)}{\left(1/x^4\right)}=\lim_{x\to\infty}\frac{\frac{4}{x^3}-1}{3+\frac{2}{x^2}}=\frac{0-1}{3+0}=-\frac{1}{3}$

4. $\displaystyle\lim_{x\to\infty}\left(\frac{1}{x}\right)^5=\lim_{x\to\infty}\frac{1}{x^5}=0$

5. $\displaystyle\lim_{x\to-\infty}x^{-1/3}=\lim_{x\to-\infty}\frac{1}{x^{1/3}}=0$

6. $\displaystyle\lim_{x\to-\infty}\frac{2x+6}{x^2+1}=\lim_{x\to-\infty}\frac{2x+6}{x^2+1}\frac{\left(1/x^2\right)}{\left(1/x^2\right)}=\lim_{x\to-\infty}\frac{\frac{2}{x}+\frac{6}{x^2}}{1+\frac{1}{x^2}}=\frac{0+0}{1+0}=0$

7. $\displaystyle\lim_{x\to\infty}\frac{3x^2+7x}{1-x^4}=\lim_{x\to\infty}\frac{3x^2+7x}{1-x^4}\frac{\left(1/x^4\right)}{\left(1/x^4\right)}=\lim_{x\to\infty}\frac{\frac{3}{x^2}+\frac{7}{x^3}}{\frac{1}{x^4}-1}=\frac{0+0}{0-1}=0$

8. $\displaystyle\lim_{x\to\infty}\frac{x^4-4}{x^3+7x^2}=\lim_{x\to\infty}\frac{x^4-4}{x^3+7x^2}\frac{\left(1/x^3\right)}{\left(1/x^3\right)}=\lim_{x\to\infty}\frac{x-\frac{4}{x^3}}{1+\frac{7}{x}};$

   numerator: $\displaystyle\lim_{x\to\infty}\left(x-\frac{4}{x^3}\right)=\left(\lim_{x\to\infty}x\right)-0=\infty$

   denominator: $\displaystyle\lim_{x\to\infty}\left(1+\frac{7}{x}\right)=1+0=1$

   Therefore $\displaystyle\lim_{x\to\infty}\frac{x-\frac{4}{x^3}}{1+\frac{7}{x}}=+\infty.$

9. $\displaystyle\lim_{x\to-\infty}\frac{\sin x}{x}.$ Note that $-\dfrac{1}{x}\le\dfrac{\sin x}{x}\le\dfrac{1}{x}.$

   Since $\displaystyle\lim_{x\to-\infty}-\frac{1}{x}=\lim_{x\to-\infty}\frac{1}{x}=0,\ \lim_{x\to-\infty}\frac{\sin x}{x}=0.$

10. $\displaystyle\lim_{x\to\infty}\frac{x^3-2x^2+5x-4}{x-3}=\lim_{x\to\infty}\frac{x^3-2x^2+5x-4}{x-3}\frac{\left(1/x\right)}{\left(1/x\right)}=\lim_{x\to\infty}\frac{x^2-2x+5-\frac{4}{x}}{1-\frac{3}{x}};$

    numerator: $\displaystyle\lim_{x\to\infty}\left(x^2-2x+5-\frac{4}{x}\right)=\left[\lim_{x\to\infty}(x^2-2x)\right]+5-0=\infty$

    denominator: $\displaystyle\lim_{x\to\infty}\left(1-\frac{3}{x}\right)=1-0=1$

    Therefore $\displaystyle\lim_{x\to\infty}\frac{x^2-2x+5-\frac{4}{x}}{1-\frac{3}{x}}=+\infty.$

11. $\lim\limits_{x \to -\infty} \dfrac{x^4 - 3\sin x}{3x + 5x^5} = \lim\limits_{x \to -\infty} \left(\dfrac{x^4 - 3\sin x}{3x + 5x^5} \right) \dfrac{(1/x^5)}{(1/x^5)} = \lim\limits_{x \to -\infty} \dfrac{\frac{1}{x} - \frac{3\sin x}{x^5}}{\frac{3}{x^4} + 5}$

$$= \lim\limits_{x \to -\infty} \dfrac{\frac{1}{x} - 3\left(\frac{\sin x}{x} \right)\left(\frac{1}{x^4} \right)}{\frac{3}{x^4} + 5} = \dfrac{0 - 3(0)(0)}{0 + 5} = 0$$

Note that $\lim\limits_{x \to -\infty} \dfrac{\sin x}{x} = 0$ (See problem 9.)

12. $\lim\limits_{x \to \infty} \dfrac{x^2 \left| 2 + x^2 \right|}{4 - x^4} = \lim\limits_{x \to \infty} \dfrac{x^2 \left(2 + x^2 \right)}{(2 - x^2)(2 + x^2)} = \lim\limits_{x \to \infty} \dfrac{x^2}{2 - x^2} \dfrac{\left(1/x^2 \right)}{\left(1/x^2 \right)} = \lim\limits_{x \to \infty} \dfrac{1}{\frac{2}{x^2} - 1} = \dfrac{1}{0 - 1} = -1$

13. $\lim\limits_{x \to \infty} \dfrac{\left| 2 - x^2 \right|}{x^2 + 1} = \lim\limits_{x \to \infty} \dfrac{\left| 2 - x^2 \right|}{x^2 + 1} \dfrac{\left(1/x^2 \right)}{\left(1/x^2 \right)} = \lim\limits_{x \to \infty} \dfrac{\left| \frac{2}{x^2} - 1 \right|}{1 + \frac{1}{x^2}} = \dfrac{|0 - 1|}{1 + 0} = \dfrac{1}{1} = 1$

14. $\lim\limits_{x \to -\infty} \dfrac{(x - 2)^3}{\left| x^3 - x \right|}$

Note that $x^3 = -|x|^3$ for $x < 0$.

Therefore $\lim\limits_{x \to -\infty} \dfrac{(x - 2)^3}{\left| x^3 - x \right|} = \lim\limits_{x \to -\infty} \dfrac{(x - 2)^3}{\left| x^3 - x \right|} \dfrac{\left(1/x^3 \right)}{\left(1/x^3 \right)} = \lim\limits_{x \to -\infty} \dfrac{\frac{(x-2)^3}{x^3}}{-\frac{1}{|x|^3} \left| x^3 - x \right|}$

$$= \lim\limits_{x \to -\infty} \dfrac{\left(1 - \frac{2}{x} \right)^3}{-\left| 1 - \frac{1}{x^2} \right|} = \dfrac{1}{-1} = -1$$

15. $\lim\limits_{x \to \infty} \dfrac{\sqrt{x - 1}}{x^2} = \lim\limits_{x \to \infty} \dfrac{\sqrt{x - 1}}{x^2} \dfrac{\left(1/x^2 \right)}{\left(1/x^2 \right)} = \lim\limits_{x \to \infty} \dfrac{\sqrt{\frac{x-1}{x^4}}}{1} = \lim\limits_{x \to \infty} \dfrac{\sqrt{\frac{1}{x^3} - \frac{1}{x^4}}}{1} = \dfrac{0}{1} = 0$

16. $\lim\limits_{x \to \infty} \dfrac{x^{2/3} + x^{4/3}}{x^2} = \lim\limits_{x \to \infty} \dfrac{x^{2/3} + x^{4/3}}{x^2} \dfrac{\left(1/x^2 \right)}{\left(1/x^2 \right)} = \lim\limits_{x \to \infty} \dfrac{\frac{1}{x^{4/3}} + \frac{1}{x^{2/3}}}{1} = \dfrac{0 + 0}{1} = 0$

17. $\lim\limits_{x \to \infty} \dfrac{x^{2/3} + x}{1 + x^{3/4}} = \lim\limits_{x \to \infty} \dfrac{x^{2/3} + x}{1 + x^{3/4}} \dfrac{\left(1/x^{3/4} \right)}{\left(1/x^{3/4} \right)} = \lim\limits_{x \to \infty} \dfrac{\frac{1}{x^{1/12}} + x^{1/4}}{\frac{1}{x^{3/4}} + 1}$

numerator: $0 + \lim\limits_{x \to \infty} x^{1/4} = \infty$

denominator: $0 + 1 = 1$

Therefore $\lim\limits_{x \to \infty} \dfrac{\frac{1}{x^{1/12}} + x^{1/4}}{\frac{1}{x^{3/4}} + 1} = +\infty.$

18. $\lim\limits_{x \to \infty} \dfrac{\sqrt{x} + 7}{1 - \sqrt[3]{x}} = \lim\limits_{x \to \infty} \dfrac{x^{1/2} + 7}{1 - x^{1/3}} \dfrac{\left(1/x^{1/3}\right)}{\left(1/x^{1/3}\right)} = \lim\limits_{x \to \infty} \dfrac{x^{1/6} + \frac{7}{x^{1/3}}}{\frac{1}{x^{1/3}} - 1}$

numerator: $\left(\lim\limits_{x \to \infty} x^{1/6}\right) + 0 = \infty$

denominator: $0 - 1 = -1$

Therefore $\lim\limits_{x \to \infty} \dfrac{x^{1/6} + \frac{7}{x^{1/3}}}{\frac{1}{x^{1/3}} - 1} = -\infty.$

19. $\lim\limits_{x \to \infty} x \sin x$

As x runs through the odd multiples of $\frac{\pi}{2}$, $(\sin x)$ takes the values $+1$ and -1 alternately. So at these values, $(x \sin x)$ is alternately $+x$ and $-x$. Since these values of x can be arbitrarily large, $(x \sin x)$ assumes arbitrarily large positive and negative values as $x \to \infty$. Thus $\lim\limits_{x \to \infty} (x \sin x)$ does not exist.

(See Figure 1.5 on page 193.)

20. $\lim\limits_{x \to -\infty} (x + \sin x)$

Note that $x + (-1) \le x + \sin x \le x + 1$. Since $(x + (-1))$ and $(x + 1)$ both approach $-\infty$ as $x \to -\infty$, $\lim\limits_{x \to -\infty} (x + \sin x) = -\infty$.

21. $\lim\limits_{x \to 2^+} \dfrac{1}{x - 2} = \infty$

22. $\lim\limits_{x \to -2^-} \dfrac{1}{(x + 2)^2} = \infty$

23. $\lim\limits_{x \to \frac{\pi}{2}^-} \tan x = \lim\limits_{x \to \frac{\pi}{2}^-} \dfrac{\sin x}{\cos x} = \infty$ (since $\sin x \to 1$ as $x \to \frac{\pi}{2}^-$ and $cos x \to 0^+$ as $x \to \frac{\pi}{2}^-$)

24. $\lim\limits_{x \to \frac{\pi}{2}^+} \tan x = \lim\limits_{x \to \frac{\pi}{2}^+} \dfrac{\sin x}{\cos x} = -\infty$ (since as $x \to \frac{\pi}{2}^+$, $\sin x \to 1$ and $\cos x \to 0^-$)

25. $\lim\limits_{x \to \frac{\pi}{2}} \tan x$: does not exist (See problems 23 and 24.)

26. $\lim\limits_{x \to -\frac{\pi}{2}^+} x \tan x = \lim\limits_{x \to -\frac{\pi}{2}^+} \dfrac{x \sin x}{\cos x} = \infty$ (since as $x \to -\frac{\pi}{2}^+$, $x \sin x \to \left(-\frac{\pi}{2}\right)(-1) = \frac{\pi}{2}$ and $\cos x \to 0^+$)

27. $\lim\limits_{x \to 0^-} \dfrac{|x|}{x} = \lim\limits_{x \to 0^-} \dfrac{-x}{x} = \lim\limits_{x \to 0^-} (-1) = -1$

28. $\lim\limits_{x \to 0^+} \dfrac{|x|}{x} = \lim\limits_{x \to 0^+} \dfrac{x}{x} = \lim\limits_{x \to 0^+} (1) = 1$

29. $\lim\limits_{x \to 5^-} \dfrac{x^2 + 1}{x + 5} = \dfrac{25 + 1}{5 + 5} = \dfrac{26}{10} = \dfrac{13}{5}$

30. $\lim\limits_{x \to 3} \dfrac{x^2 - 4}{x + 3} = \dfrac{9 - 4}{3 + 3} = \dfrac{5}{6}$

31. $\lim\limits_{x \to 5^-} \dfrac{x^2 - 25}{x - 5} = \lim\limits_{x \to 5^-} \dfrac{(x + 5)(x - 5)}{x - 5} = \lim\limits_{x \to 5^-} (x + 5) = 10$

32. $\lim\limits_{x \to 3} \dfrac{x^2 - 6x + 9}{x - 3} = \lim\limits_{x \to 3} \dfrac{(x-3)^2}{x-3} = \lim\limits_{x \to 3}(x-3) = 0$

33. $\lim\limits_{x \to 5^-} \dfrac{x - 7}{x - 5} = \infty$ (since as $x \to 5^-$, $x - 7 \to -2$ and $x - 5 \to 0^-$).

34. $\lim\limits_{x \to 3} \dfrac{10 - x^2}{x - 3}$: does not exist (since $\lim\limits_{x \to 3^+} \dfrac{10 - x^2}{x - 3} = \infty$ and $\lim\limits_{x \to 3^-} \dfrac{10 - x^2}{x - 3} = -\infty$).

35. $\lim\limits_{x \to 1^+} \dfrac{1}{(x - 1)^{1/3}} = \infty$ (since as $x \to 1^+$, $(x - 1)^{1/3} \to 0^+$).

36. $\lim\limits_{x \to 1^-} \dfrac{1}{(x - 1)^{1/3}} = -\infty$ (since as $x \to 1^-$, $(x - 1)^{1/3} \to 0^-$)

37. $\lim\limits_{x \to 1^+} \dfrac{1}{(x - 1)^{2/3}} = \infty$

38. $\lim\limits_{x \to 1^-} \dfrac{1}{(x - 1)^{2/3}} = \infty$. (Note that the fractional power has an **even numerator**.)

39. $y = \dfrac{1}{1 + x}$

$\lim\limits_{x \to \infty} \dfrac{1}{1 + x} = \lim\limits_{x \to \infty} \dfrac{1}{1 + x} \dfrac{(1/x)}{(1/x)} = 0$

$\lim\limits_{x \to -\infty} \dfrac{1}{1 + x} = 0$

Therefore $y = 0$ is the horizontal asymptote.

40. $y = \dfrac{1 + x}{3 - x}$

$\lim\limits_{x \to \infty} \dfrac{1 + x}{3 - x} = \lim\limits_{x \to \infty} \dfrac{1 + x}{3 - x} \dfrac{(1/x)}{(1/x)} = \dfrac{1}{-1} = -1$

$\lim\limits_{x \to -\infty} \dfrac{1 + x}{3 - x} = -1$

Therefore $y = -1$ is the horizontal asymptote.

41. $y = \dfrac{2x^2}{x^2 + 1}$

$\lim\limits_{x \to \infty} \dfrac{2x^2}{x^2 + 1} = 2;\ \lim\limits_{x \to -\infty} \dfrac{2x^2}{x^2 + 1} = 2$

Therefore $y = 2$ is the horizontal asymptote.

42. $y = \dfrac{x^2}{1 - x}$

$\lim\limits_{x \to \infty} \dfrac{x^2}{1 - x} = \lim\limits_{x \to \infty} \dfrac{x^2}{1 - x} \dfrac{(1/x)}{(1/x)} = \lim\limits_{x \to \infty} \dfrac{x}{\frac{1}{x} - 1} = -\infty$

Similarly, $\displaystyle\lim_{x\to-\infty}\frac{x^2}{1-x}=\lim_{x\to-\infty}\frac{x}{\frac{1}{x}-1}=\infty.$

Thus $y=\dfrac{x^2}{1-x}$ has no horizontal asymptotes.

43. $y=\dfrac{2x^2}{(x^2+1)^2}$

$$\lim_{x\to\infty}\frac{2x^2}{(x^2+1)^2}=\lim_{x\to\infty}\frac{2x^2}{(x^2+1)^2}\frac{\left(1/x^4\right)}{\left(1/x^4\right)}=\lim_{x\to\infty}\frac{\frac{2}{x^2}}{\left(1+\frac{1}{x^2}\right)^2}=0$$

Similarly, $\displaystyle\lim_{x\to-\infty}\frac{2x^2}{(x^2+1)^2}=0.$

Therefore $y=0$ is the horizontal asymptote.

44. $y=\dfrac{x^2+3}{1-3x^2}$

$$\lim_{x\to\infty}\frac{x^2+3}{1-3x^2}=\lim_{x\to\infty}\frac{x^2+3}{1-3x^2}\frac{\left(1/x^2\right)}{\left(1/x^2\right)}=\lim_{x\to\infty}\frac{1+\frac{3}{x^2}}{\frac{1}{x^2}-3}=-\frac{1}{3}$$

Similarly, $\displaystyle\lim_{x\to-\infty}\frac{x^2+3}{1-3x^2}=-\frac{1}{3}.$

Therefore $y=-\frac{1}{3}$ is the horizontal asymptote.

45. $y=3+\dfrac{\sin x}{x}$

$$\lim_{x\to\infty}\left(3+\frac{\sin x}{x}\right)=3+\lim_{x\to\infty}\frac{\sin x}{x}=3+0=3$$

Similarly, $\displaystyle\lim_{x\to-\infty}\left(3+\frac{\sin x}{x}\right)=3.$

Therefore $y=3$ is the horizontal asymptote.

46. $y=\dfrac{4x-\sqrt{x}}{x^{2/3}+x}$

$$\lim_{x\to\infty}\frac{4x-\sqrt{x}}{x^{2/3}+x}=\lim_{x\to\infty}\frac{4x-x^{1/2}}{x^{2/3}+x}\frac{\left(1/x\right)}{\left(1/x\right)}=\lim_{x\to\infty}\frac{4-\frac{1}{x^{1/2}}}{\frac{1}{x^{1/3}}+1}=4$$

Similarly, $\displaystyle\lim_{x\to-\infty}\frac{4x-\sqrt{x}}{x^{2/3}+x}=4$

Therefore $y=4$ is the horizontal asymptote.

47. $y=\dfrac{4-3x}{\sqrt{1+2x^2}}$

Note that for $x>0$, $\sqrt{x^2}=x$ and for $x<0$, $\sqrt{x^2}=-x$.

$$\lim_{x\to\infty}\frac{4-3x}{\sqrt{1+2x^2}}=\lim_{x\to\infty}\frac{4-3x}{\sqrt{1+2x^2}}\frac{\left(1/x\right)}{\left(1/x\right)}=\lim_{x\to\infty}\frac{\frac{4}{x}-3}{\sqrt{\frac{1}{x^2}+2}}=-\frac{3}{\sqrt{2}}=\frac{-3\sqrt{2}}{2}$$

$$\lim_{x\to-\infty}\frac{4-3x}{\sqrt{1+2x^2}}=\lim_{x\to-\infty}\frac{4-3x}{\sqrt{1+2x^2}}\frac{\left(-1/x\right)}{\left(-1/x\right)}=\lim_{x\to-\infty}\frac{-\frac{4}{x}+3}{\sqrt{\frac{1}{x^2}+2}}=\frac{3}{\sqrt{2}}=\frac{3\sqrt{2}}{2}$$

Thus $y = \pm\dfrac{3\sqrt{2}}{2}$ are the horizontal asymptotes.

48. $y = x \sin\left(\dfrac{1}{x}\right)$

Note that $y = x \sin\dfrac{1}{x} = \dfrac{\sin\frac{1}{x}}{\frac{1}{x}}$. Now, let $z = \dfrac{1}{x}$. Then $z \to 0$ as $x \to \pm\infty$.

Therefore $\displaystyle\lim_{x \to \pm\infty} x \sin\left(\dfrac{1}{x}\right) = \lim_{z \to 0}\dfrac{\sin z}{z} = 1$

Thus $y = 1$ is the horizontal asymptote.

49. $3 = \displaystyle\lim_{x \to \pm\infty}\dfrac{ax + 7}{4 - x} = \lim_{x \to \pm\infty}\dfrac{ax + 7}{4 - x}\dfrac{(1/x)}{(1/x)} = \lim_{x \to \pm\infty}\dfrac{a + \frac{7}{x}}{\frac{4}{x} - 1} = -a$

Thus $a = -3$.

50. $y = \dfrac{1}{x - 2}$

$\displaystyle\lim_{x \to 2+}\dfrac{1}{x - 2} = \infty$ and $\displaystyle\lim_{x \to 2-}\dfrac{1}{x - 2} = -\infty$

Thus $x = 2$ is the vertical asymptote.

51. $y = \dfrac{1}{(x + 2)^2}$

$\displaystyle\lim_{x \to -2+}\dfrac{1}{(x + 2)^2} = \infty$ and $\displaystyle\lim_{x \to -2-}\dfrac{1}{(x + 2)^2} = \infty$

Thus $x = -2$ is the vertical asymptote.

52. $y = \dfrac{1}{(x - 1)^{1/3}}$

$\displaystyle\lim_{x \to 1+}\dfrac{1}{(x - 1)^{1/3}} = \infty$ and $\displaystyle\lim_{x \to 1-}\dfrac{1}{(x - 1)^{1/3}} = -\infty$.

Thus $x = 1$ is the vertical asymptote.

53. $y = \dfrac{1}{(x - 1)^{2/3}}$; $\displaystyle\lim_{x \to 1+}\dfrac{1}{(x - 1)^{2/3}} = \lim_{x \to 1-}\dfrac{1}{(x - 1)^{2/3}} = \infty$.

Thus $x = 1$ is the vertical asymptote.

54. $y = \dfrac{1 + x}{3 - x}$

$\displaystyle\lim_{x \to 3+}\dfrac{1 + x}{3 - x} = -\infty$ and $\displaystyle\lim_{x \to 3-}\dfrac{1 + x}{3 - x} = \infty$

Thus $x = 3$ is the vertical asymptote.

55. $y = \dfrac{x^2}{1 - x}$

$\displaystyle\lim_{x \to 1+}\dfrac{x^2}{1 - x} = -\infty$ and $\displaystyle\lim_{x \to 1-}\dfrac{x^2}{1 - x} = \infty$.

Thus $x = 1$ is the vertical asymptote.

271

56. $\lim\limits_{x \to \pm\infty} \dfrac{x^r + 3x}{7 - x^{4/3}} = -1$ if and only if $\dfrac{x^r}{-x^{4/3}} = -1$ which implies $r = \frac{4}{3}$.

57. $\lim\limits_{x \to \pm\infty} \dfrac{\pi + ax^r}{1 - 3x^{2/3}} = -2$ if and only if $\dfrac{ax^r}{-3x^{2/3}} = -2$ which implies $r = \frac{2}{3}$ and $\frac{a}{-3} = -2$. Hence $a = 6$.

58. In order for a rational function to have a nonzero horizontal asymptote, the denominator and the numerator must have equal degree. It follows that $r \leq 3$. Clearly, $r = 3$ works. If $r < 3$ then
$\lim\limits_{x \to \pm\infty} \dfrac{3x^r + 2x^3}{r\,x^3} = \dfrac{2}{r}$ which equals $\dfrac{5}{3}$ when $r = \dfrac{6}{5}$. Thus r can be either 3 or $\dfrac{6}{5}$.

59. $y = (x^2 + ax + b)^{-1} = \dfrac{1}{x^2 + ax + b}$ has vertical asymptotes of $x = 3$ and $x = 5$ if and only if $x^2 + ax + b = (x-3)(x-5) = x^2 - 8x + 15$.
Thus $a = -8$ and $b = 15$.

60.   a) Assume $n < m$. Then $m - n > 0$.
$$\lim_{x \to \infty} \frac{a_n x^n + a_{n-1} x^{n-1} + \cdots + a_1 x + a_0}{b_m x^m + b_{m-1} x^{m-1} + \cdots + b_1 x + b_0} = \lim_{x \to \infty} \frac{a_n x^n + a_{n-1} x^{n-1} + \cdots + a_1 x + a_0}{b_m x^m + b_{m-1} x^{m-1} + \cdots + b_1 x + b_0} \frac{(1/x^m)}{(1/x^m)}$$
$$= \lim_{x \to \infty} \frac{\frac{a_n}{x^{m-n}} + \frac{a_{n-1}}{x^{m-n+1}} + \cdots + \frac{a_1}{x^{m-1}} + \frac{a_0}{x^m}}{b_m + \frac{b_{m-1}}{x} + \cdots + \frac{b_1}{x^{m-1}} + \frac{b_0}{x^m}} = \frac{0 + 0 + \cdots + 0 + 0}{b_m + 0 + \cdots + 0 + 0} = 0$$

  b) Assume $n > m$ and a_n and b_m have like signs. (Note that $n - m > 0$.)
$$\text{Then } \lim_{x \to \infty} \frac{a_n x^n + a_{n-1} x^{n-1} + \cdots + a_1 x + a_0}{b_m x^m + b_{m-1} x^{m-1} + \cdots + b_1 x + b_0} \frac{(1/x^m)}{(1/x^m)}$$
$$= \lim_{x \to \infty} \frac{a_n x^{n-m} + a_{n-1} x^{n-1-m} + \cdots + a_1 x^{1-m} + a_0 x^{-m}}{b_m + \frac{b_{m-1}}{x} + \cdots + \frac{b_1}{x^{m-1}} + \frac{b_0}{x^m}} = \infty$$

since as $x \to \infty$, the numerator approaches infinity and the denominator approaches b_m.

  c) Assume $n = m$. Then $\lim\limits_{x \to \infty} \dfrac{a_n x^n + a_{n-1} x^{n-1} + \cdots + a_1 x + a_0}{b_m x^m + b_{m-1} x^{m-1} + \cdots + b_1 x + b_0} \dfrac{(1/x^m)}{(1/x^m)}$
$$= \lim_{x \to \infty} \frac{a_n + \frac{a_{n-1}}{x} + \cdots + \frac{a_1}{x^{m-1}} + \frac{a_0}{x^m}}{b_m + \frac{b_{m-1}}{x} + \cdots + \frac{b_1}{x^{m-1}} + \frac{b_0}{x^m}} = \frac{a_n}{b_m}.$$

61.   a)

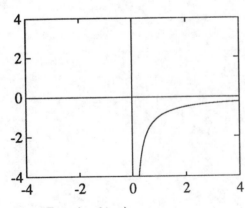

Exercise 61. a)

  b) $f(x) = -\frac{1}{x}$ for $x > 0$

c) No.

   Counterexample: $f(x) = -1 - \frac{1}{x}$

   $\lim\limits_{x \to \infty} f(x) = \lim\limits_{x \to \infty} \left(-1 - \frac{1}{x} \right) = -1 \neq 0$

62. a) $\lim\limits_{x \to \infty} f(x) = \lim\limits_{x \to \infty} \frac{2x^3 + 1}{x - 1} = \lim\limits_{x \to \infty} \frac{2x^3 + 1}{x - 1} \frac{\left(1/x \right)}{\left(1/x \right)} = \lim\limits_{x \to \infty} \frac{2x^2 + \frac{1}{x}}{1 - \frac{1}{x}} = \infty$

   b) $\lim\limits_{x \to \infty} [f(x) - g(x)] = \lim\limits_{x \to \infty} \left[\frac{2x^3 + 1}{x - 1} - (2x^2 + 2x + 2) \right] = \lim\limits_{x \to \infty} \frac{2x^3 + 1 - 2(x^2 + x + 1)(x - 1)}{x - 1}$

   $$= \lim\limits_{x \to \infty} \frac{2x^3 + 1 - 2(x^3 - 1)}{x - 1} = \lim\limits_{x \to \infty} \frac{3}{x - 1} = 0$$

   c) The calculation in part (b) says that the vertical distance $|f(x) - g(x)|$ between the graphs of $f(x)$ and $g(x)$ approaches zero as $x \to \infty$.

63.

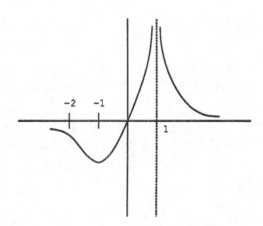

-2   -1

1

Exercise 63

64. a) Let $\epsilon > 0$ be given. Choose any integer N such that $N > \frac{1}{\epsilon}$. If $x > N$, then $|f(x) - L| = \left| \frac{1}{x} - 0 \right| = \left| \frac{1}{x} \right| < \frac{1}{N} < \epsilon$.

   b) Note that here $|f(x) - L| = \left| \frac{3x + 1}{x} - 3 \right| = \left| \frac{1}{x} \right|$. Let $\epsilon > 0$ be given. Choose N as in part (a). The same argument shows $x > N \Rightarrow |f(x) - L| < \epsilon$ in this case.

   c) There $|f(x) - L| = \left| \frac{\sin x}{x + 2} - 0 \right| = \left| \frac{\sin x}{x + 2} \right| \leq \left| \frac{1}{x + 2} \right| < \left| \frac{1}{x} \right|$ if $x > 0$. Let $\epsilon > 0$ be given. Choose N as in part a). The same argument shows $x > N \Rightarrow |f(x) - L| < \left| \frac{1}{x} \right| < \frac{1}{N} < \epsilon$.

65. A convenient definition of " $\lim\limits_{x \to -\infty} f(x) = L$" is "given $\epsilon > 0$ there is an integer N so that if $x < -N$ then $|f(x) - L| < \epsilon$."

   a) $f(x) = \frac{1}{x}$, $L = 0$. Given $\epsilon > 0$ pick any integer $N > \frac{1}{\epsilon}$. If $x < -N$ then $|x| = -x > N$ so that $|f(x) - L| = \left| \frac{1}{x} - 0 \right| = \frac{1}{|x|} < \frac{1}{N} < \epsilon$.

b) $f(x) = \dfrac{x}{2x+1}$, $L = \frac{1}{2}$. Here, $|f(x) - L| = \left|\dfrac{x}{2x+1} - \frac{1}{2}\right| = \left|\dfrac{-1}{4x+2}\right| = \dfrac{1}{|4x+2|}$. For $x < -\frac{1}{2}$ we have $4x + 2 < 0$ so $|4x+2| = -4x - 2$. Now, given $\epsilon > 0$ pick any integer N so that $4N - 2 > \frac{1}{\epsilon}$ (that is, $N > \frac{1}{2} + \frac{1}{4\epsilon}$). If $x < -N$ then $4x + 2 < -4N + 2$ so $-4x - 2 > 4N - 2 > \frac{1}{\epsilon}$ and therefore $|f(x) - L| = \dfrac{1}{|4x+2|} = -\dfrac{1}{4x+2} < \epsilon$.

c) $f(x) = \dfrac{6x - 2}{x + 1}$, $L = 6$. Here, $|f(x) - L| = \left|\dfrac{6x-2}{x+1} - 6\right| = \left|\dfrac{-8}{x+1}\right| = \dfrac{8}{|x+1|}$. For $x < -1$ we have $x + 1 < 0$ so $|x + 1| = -(x+1)$. Now, given $\epsilon > 0$ select any integer N such that $N > 1 + \frac{8}{\epsilon}$. If $x < -N$ then $x + 1 < -N + 1$ so $-(x+1) > N - 1 > \frac{8}{\epsilon}$ so that $|f(x) - L| = \dfrac{-8}{x+1} < \epsilon$.

66. Note first that in the definition it may be supposed that N is positive.

 a) $f(x) = \dfrac{1}{|x|}$, $a = 0$. Given $N > 0$, set $\delta = \frac{1}{N}$. If $0 < |x| < \delta = \frac{1}{N}$ then $\dfrac{1}{|x|} > \frac{1}{\delta} = N$: so here, if $0 < |x - a| < \delta$ then $f(x) > N$ as required.

 b) $f(x) = \dfrac{1}{x^2}$, $a = 0$. Given $N > 0$, set $\delta = \frac{1}{\sqrt{N}}$. If $0 < |x| < \frac{1}{\sqrt{N}}$ then $0 < x^2 < \frac{1}{N}$ so $\dfrac{1}{x^2} > N$ as required.

 c) $f(x) = |\tan x|$, $a = \frac{\pi}{2}$. Note first that if $\frac{\pi}{4} < x < \frac{3\pi}{4}$ then $\sin x > \frac{\sqrt{2}}{2}$. Also, cosine is continuous and is zero at $\frac{\pi}{2}$, so if $\epsilon > 0$ is given then we can find $\delta > 0$ so that $\left|x - \frac{\pi}{2}\right| < \delta \Rightarrow |cos x| < \epsilon$. Now, let $N > 0$ be given and put $\epsilon = \frac{\sqrt{2}}{2N}$. If we insist (as we may) that $\delta < \frac{\pi}{4}$ then whenever $0 < \left|x - \frac{\pi}{2}\right| < \delta$ we have $|\tan x| = \left|\dfrac{\sin x}{\cos x}\right| > \dfrac{\sqrt{2}/2}{\epsilon} = N$. This proves that $\lim\limits_{x \to \pi/2} |\tan x| = \infty$.

67. "The limit of $f(x)$ as $x \to a^-$ is ∞" means that if N is any real number then there is a $\delta > 0$ so that $f(x) > N$ whenever $a - \delta < x < a$. Here, N may be assumed to be positive without harm. Now let $f(x) = \tan x$ and $a = \frac{\pi}{2}$. As in the previous question, $\sin x > \frac{\sqrt{2}}{2}$ whenever $x \in \left(-\frac{\pi}{2}, \frac{\pi}{2}\right)$. If $N > 0$ is given then pick $\delta > 0$ so that $\left|x - \frac{\pi}{2}\right| < \delta$ implies $|\cos x| < \frac{\sqrt{2}}{2N}$; this is again possible since cosine is continuous and $\cos \frac{\pi}{2} = 0$. If we once more stipulate that δ is at most $\frac{\pi}{4}$, then $\frac{\pi}{2} - \delta < x < \frac{\pi}{2}$ implies that $\sin x > \frac{\sqrt{2}}{2}$ and $0 < \cos x < \frac{\sqrt{2}}{2N}$, whence

$$\tan x = \frac{\sin x}{\cos x} > N$$

as required.

68. "The limit of $f(x)$ as $x \to a^-$ is $-\infty$" means that given any N there is a $\delta > 0$ so that if $a - \delta < x < a$ then $f(x) < N$. Here, it may be supposed that N is negative. Now let $f(x) = \frac{1}{x}$ and $a = 0$. Given $N < 0$ select $\delta = -\frac{1}{N} > 0$. If $-\delta < x < 0$ then $\frac{1}{x} < -\frac{1}{\delta} = N$, as required.

69. Let $\lim\limits_{x \to \infty} f(x) = L$ and $\lim\limits_{x \to \infty} g(x) = M$.

 i) Given $\epsilon > 0$ choose N large enough so that both $|f(x) - L| < \frac{1}{2}\epsilon$ and $|g(x) - M| < \frac{1}{2}\epsilon$ if $x > N$; this ensures that if $x > N$ then

$$|(f(x) + g(x)) - (L + M)| = |(f(x) - L) + (g(x) - M)| \le |f(x) - L| + |g(x) - M| < \epsilon$$

so that $\lim\limits_{x \to \infty} (f(x) + g(x)) = L + M$.

ii) If $c = 0$ then both sides of the purported equality are surely zero. Let $c \neq 0$ and, given $\epsilon > 0$, choose N so large that if $x > N$ then $|f(x) - L| < \frac{\epsilon}{|c|}$. Then

$$x > N \Rightarrow |c\,f(x) - cL| = |c|\,|f(x) - L| < \epsilon$$

so that $\lim_{x \to \infty} (c\,f(x)) = cL$.

iii) This is less straightforward. Note that

$$f(x)\,g(x) - LM = (f(x) - L)\,g(x) + L(g(x) - M).$$

By hypothesis, large values of x will make both $|f(x) - L|$ and $|g(x) - M|$ small: so the right side will be small if we can control the term $g(x)$ multiplying $(f(x) - L)$; we can first pick N_1 so that $x > N_1 \Rightarrow |g(x) - M| < 1$ (say: any positive number would do in place of 1); this is possible since $\lim_{x \to \infty} g(x) = M$. Note now that if $x > N_1$ then

$$|g(x)| = |g(x) - M + M| \leq |g(x) - M| + |M| < 1 + |M|.$$

Now, given $\epsilon > 0$ we pick N_2 so that $x > N_2 \Rightarrow |f(x) - L| < \frac{\epsilon}{2(1+|M|)}$ and pick N_3 so that $x > N_3 \Rightarrow |g(x) - M| < \frac{\epsilon}{2(1+|L|)}$. Finally, let N be any number greater than N_1, N_2 and N_3. If $x > N$ then

$$
\begin{aligned}
|f(x)\,g(x) - LM| &= |(f(x) - L)\,g(x) + L(g(x) - M)| \\
&\leq |f(x) - L|\,|g(x)| + |L|\,|g(x) - M| \\
&\leq \frac{\epsilon}{2(1 + |M|)}\,(1 + |M|) + |L|\frac{\epsilon}{2(1 + |L|)} \\
&< \epsilon \qquad \text{as required.}
\end{aligned}
$$

iv) Suppose we can show $\lim_{x \to \infty} \frac{1}{g(x)} = \frac{1}{M}$. Then applying (iii) with $\frac{1}{g}$ in place of g will give

$$
\begin{aligned}
\lim_{x \to \infty} \left(\frac{f(x)}{g(x)} \right) &= \lim_{x \to \infty} \left(f(x) \cdot \frac{1}{g(x)} \right) \\
&= L \cdot \frac{1}{M} = \frac{L}{M}.
\end{aligned}
$$

So we need only show $\lim_{x \to \infty} \frac{1}{g(x)} = \frac{1}{M}$, for which the calculations are less involved. Note that

$$* \qquad \left| \frac{1}{g(x)} - \frac{1}{M} \right| = \left| \frac{M - g(x)}{M\,g(x)} \right| = \frac{|g(x) - M|}{|M|\,|g(x)|}.$$

Since $M \neq 0$ we can find an N_1 so that $x > N_1 \Rightarrow |g(x) - M| < \frac{1}{2}|M| \Rightarrow |g(x)| > \frac{1}{2}|M|$; this ensures that the denominator of $(*)$ does not cause difficulties by being zero. Now, given $\epsilon > 0$, choose N_2 so that $x > N_2 \Rightarrow |g(x) - M| < \frac{1}{2}|M|^2\epsilon$. Letting N be any number greater than N_1 and N_2, if $x > N$ then $|M|\,|g(x)| > \frac{1}{2}|M|^2$ and $|g(x) - M| < \frac{1}{2}|M|^2\,\epsilon$, so

$$\left| \frac{1}{g(x)} - \frac{1}{M} \right| < \frac{\frac{1}{2}|M|^2\,\epsilon}{\frac{1}{2}|M|^2} = \epsilon$$

so that $\lim_{x \to \infty} \frac{1}{g(x)} = \frac{1}{M}$ as required.

Remark: These proofs are of course similar to the proofs of **Theorem 1**; see Section 2.4 and Appendix II of the text. Also, part (ii) is a special case of part (iii) given by letting $g(x) = c$ for all x.

For Exercises 70–73, HA denotes horizontal asymptote and VA denotes vertical asymptote.

70. VA: $x = -1$, $x = -\frac{1}{2}$
    HA: $y = \frac{5}{2}$
    Relative Maximum: $\left(-\frac{2}{3}, -20\right)$
    Relative Minimum: $(0, 0)$

71. VA: $x = \pm 3$, $x = 0$
    HA: $y = 0$
    Relative Maximum: $(1.26229, -0.59827)$
    Relative Minimum: $(-1.26229, 0.59827)$

72. VA: $x = 0.201314$, $x = 1$
    HA: $y = 0$
    Relative Maximum: $(0.56610, -2.92384)$
    Relative Minimum: $(-0.67420, -0.36873)$

73. VA: $x = -0.222102$
    HA: $y = \frac{1}{2}$
    This function has an infinite number of relative extrema.
    In the interval $[-2, 2]$ we have:
    Relative Maxima: $(-1.16199, 1.31915)$
                     $(0.55592, 8.22225)$
                     $(1.71356, 0.51648)$
    Relative Minima: $(-0.61225, 0.69910)$
                     $(1.43590, 0.48692)$

74. Oblique asymptote: $y = -2x - 4$.

75. Oblique asymptote: $y = x$.

76. For $|x|$ large, $f(x)$ looks like the parabola $y = 2x^2 - 2x - 1$.

77. For $x > 0$ and large $f(x)$ looks like the horizontal line $y = 7.3891$.

78. Consider the figure given below.

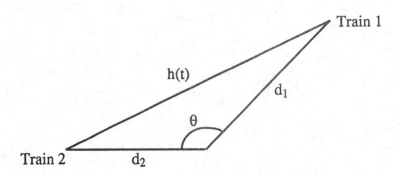

Exercise 78

a) By the Law of Cosines we have

$$h^2 = d_1^2 + d_2^2 - 2d_1 d_2 \cos 135°$$
$$= d_1^2 + d_2^2 + \frac{2d_2 d_2}{\sqrt{2}}.$$

Let $t = 0$ be 2:00 A.M. when the second train leaves Kansas City. In that case $d_2 = 65t$ and $d_1 = 50(t + 1)$. So at $t = 0$ train 2 is in the station, but train 1 is 50 miles away. Therefore

$$h^2 = [50(t + 1)]^2 + (65t)^2 + \frac{2(50)(t + 1)65t}{\sqrt{2}}$$
$$= 2500t^2 + 5000t + 2500 + 4225t^2 + \frac{6500t^2 + 6500t}{\sqrt{2}}$$
$$= 11.321.19408t^2 + 9596.194078t + 2500.$$

276

Thus $h(t) = \sqrt{11321.19408t^2 + 9596.194078t + 2500}$.

b) Graph with the range: $[0,5] \times [50, 600]$.

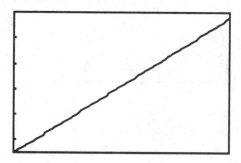

Exercise 78. b)

It does appear that this graph approaches an oblique asymptote since it seems to become a straight line for t large. To show this, suppose $a = 11321.194078$, $b = 9596.194078$, and $c = 2500$. Then the graph is given by $y = \sqrt{at^2 + bt + c}$. This implies the following:

$$y^2 - at^2 - bt = c$$

$$y^2 - a\left(t^2 + \frac{b}{a}t \right) = c$$

$$y^2 - a\left(t^2 + \frac{b}{a}t + \frac{b^2}{4a^2}\right) = c + \frac{b^2}{4a}$$

$$y^2 - a\left(t + \frac{b}{2a}\right)^2 = c + \frac{b^2}{4a}$$

An equation of this type produces the graph of a hyperbola, centered at $\left(\frac{-b}{2a}, 0\right) = (-0.423815, 0)$. With asymptotes $y = \pm\sqrt{a}\left(t + \frac{b}{2a}\right) = 106.4011(t - 0.423815)$. To see this numerically note that: $h(5) - f(5) = 0.40403$, $h(10) - f(10) = 0.210281$, and $h(20) - f(20) = 0.107330$.

c) $h'(t) = \dfrac{1}{2}(at^2 + bt + c)^{-1/2}(2at + b)$

$$= \frac{22643.38816t + 9596.194078}{2\sqrt{11321.19408t^2 + 9596.194078t + 2500}}.$$

Graph with the range: $[0,5] \times [50, 150]$. Note that the graph of $h'(t)$ becomes horizontal as t increases. This indicates that the tangent lines to the curve eventually have constant slope, i.e., $h(t)$ is nearly a straight line for t large.

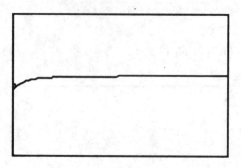

Exercise 78. c)

4.7 Curve Sketching

1. $f(x) = x^2 - 2x - 8$

    1) Domain: $(-\infty, \infty)$

    2) $x^2 - 2x - 8 = (x - 4)(x + 2) = 0$ when $x = 4$, $x = -2$.

    3) f has no vertical or horizontal asymptotes.

    4) $f'(x) = 2x - 2 = 0$ when $x = 1$.

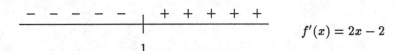

$$f'(x) = 2x - 2$$

    Thus f is decreasing on $(-\infty, 1]$ and increasing on $[1, \infty)$. The point $(1, -9)$ is the minimum.

    5) $f''(x) = 2 > 0$. Thus f is concave up on $(-\infty, \infty)$.

    6) $f(0) = -8$

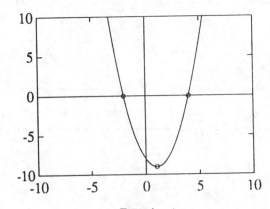

Exercise 1

2. $f(x) = 9x - x^2$

    1) Domain: $(-\infty, \infty)$

    2) $9x - x^2 = x(9 - x) = 0$ when $x = 0$ and $x = 9$.

3) f has no vertical or horizontal asymptotes.

4) $f'(x) = 9 - 2x = 0$ when $x = \frac{9}{2}$.

$$f'(x) = 9 - 2x$$

f is increasing on $\left(-\infty, \frac{9}{2}\right]$ and decreasing on $\left[\frac{9}{2}, \infty\right)$. The point $\left(\frac{9}{2}, \frac{81}{4}\right)$ is the maximum.

5) $f''(x) = -2 < 0$. Thus f is concave down on $(-\infty, \infty)$.

6)

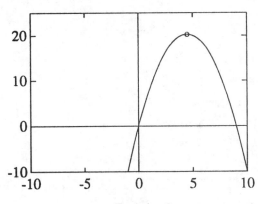

Exercise 2

3. $f(x) = 2x^3 - 3x^2$

1) Domain: $(-\infty, \infty)$

2) $2x^3 - 3x^2 = x^2(2x - 3) = 0$ when $x = 0$ and $x = \frac{3}{2}$.

3) f has no vertical or horizontal asymptotes.

4) $f'(x) = 6x^2 - 6x = 6x(x - 1) = 0$ when $x = 0, 1$.

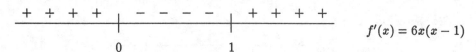

$$f'(x) = 6x(x - 1)$$

f is increasing on $(-\infty, 0]$ and $[1, \infty)$ and decreasing on $[0, 1]$. The point $(0, 0)$ is the relative maximum and $(1, -1)$ is the relative minimum.

5) $f''(x) = 12x - 6 = 0$ when $x = \frac{1}{2}$.

$$f''(x) = 12x - 6$$

f is concave down on $\left(-\infty, \frac{1}{2}\right]$ and concave up on $\left[\frac{1}{2}, \infty\right)$. The point $\left(\frac{1}{2}, -\frac{1}{2}\right)$ is the inflection point.

6)

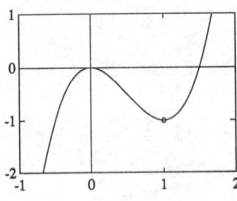

Exercise 3

4. $f(x) = 3x^4 - 4x^3$

1) Domain: $(-\infty, \infty)$

2) $3x^4 - 4x^3 = x^3(3x - 4) = 0$ when $x = 0, \frac{4}{3}$.

3) f has no vertical or horizontal asymptotes.

4) $f'(x) = 12x^3 - 12x^2 = 12x^2(x - 1) = 0$ when $x = 0, 1$.

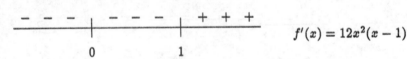

$f(x)$ is decreasing on $(-\infty, 1]$ and increasing on $[1, \infty)$. The point $(1, -1)$ is the minimum.

5) $f''(x) = 36x^2 - 24x = 12x(3x - 2) = 0$ when $x = 0, \frac{2}{3}$.

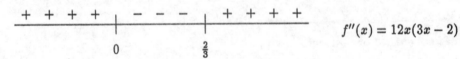

f is concave up on $(-\infty, 0]$ and $\left[\frac{2}{3}, \infty\right)$ and concave down on $\left[0, \frac{2}{3}\right]$. The points $(0, 0)$ and $\left(\frac{2}{3}, \frac{-16}{27}\right)$ are inflection points.

6)

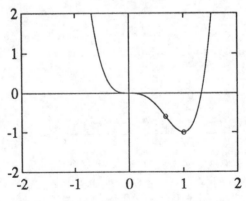

Exercise 4

280

5. $f(x) = x^4 - 8x^3 + 24x^2 - 32x + 19$

   1) Domain: $(-\infty, \infty)$

   2) f has no vertical or horizontal asymptotes.

   3) $f'(x) = 4x^3 - 24x^2 + 48x - 32 = 4(x^3 - 6x^2 + 12x - 8) = 4(x-2)^3 = 0$ at $x = 2$.

$$\underset{\textstyle 2}{\underline{\begin{array}{ccccccc} - & - & - & & + & + & + \end{array}}} \qquad f'(x) = 4(x-2)^3$$

   f is decreasing on $(-\infty, 2]$ and increasing on $[2, \infty)$. f has a minimum at $(2, 3)$.

   4) $f''(x) = 12(x-2)^2 = 0$ at $x = 2$.

$$\underset{\textstyle 2}{\underline{\begin{array}{cccccccc} + & + & + & + & + & + & + & + \end{array}}} \qquad f''(x) = 12(x-2)^2$$

   f is concave up on $(-\infty, \infty)$.

   5) $f(0) = 19$ and $f(4) = 19$.

   6)

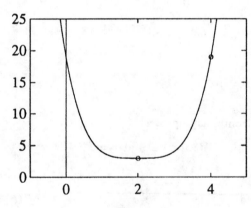

Exercise 5

6. $f(x) = 3x^5 - 15x^4 + 20x^3$

   1) Domain: $(-\infty, \infty)$

   2) $3x^5 - 15x^4 + 20x^3 = x^3(3x^2 - 15x + 20) = 0$ when $x = 0$.
      Note that we have no real zeros from $(3x^2 - 15x + 20)$.

   3) f has no vertical or horizontal asymptotes.

   4) $f'(x) = 15x^4 - 60x^3 + 60x^2 = 15x^2(x-2)^2 = 0$ when $x = 0, 2$.

$$\underset{\textstyle 0 \qquad 2}{\underline{\begin{array}{ccccccccc} + & + & + & & + & + & & + & + & + \end{array}}} \qquad f'(x) = 15x^2(x-2)^2$$

   f is increasing on $(-\infty, \infty)$. Therefore, f has no extrema on $(-\infty, \infty)$.

281

5) $f''(x) = 60x^3 - 180x^2 + 120x = 60x(x-2)(x-1) = 0$ when $x = 0, 1, 2$.

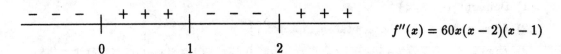

$$f''(x) = 60x(x-2)(x-1)$$

f is concave up on $[0, 1]$ and $[2, \infty)$ and concave down on $(-\infty, 0]$ and $[1, 2]$. The points $(0, 0)$, $(1, 8)$ and $(2, 16)$ are inflection points.

6)

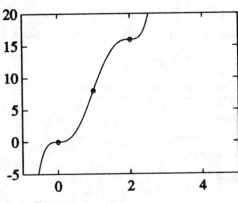

Exercise 6

7. $f(x) = x^3 + x^2 - 8x + 8$

1) Domain: $(-\infty, \infty)$

2) f has no vertical or horizontal asymptotes.

3) $f'(x) = 3x^2 + 2x - 8 = (3x - 4)(x + 2) = 0$ at $x = \frac{4}{3}, -2$.

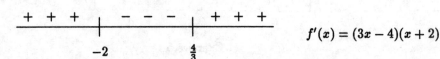

$$f'(x) = (3x - 4)(x + 2)$$

f is increasing on $(-\infty, -2]$ and $\left[\frac{4}{3}, \infty\right)$ and decreasing on $\left[-2, \frac{4}{3}\right]$. f has a relative maximum at $(-2, 20)$ and a relative minimum at $\left(\frac{4}{3}, \frac{40}{27}\right)$.

4) $f''(x) = 6x + 2 = 2(3x + 1) = 0$ at $x = -\frac{1}{3}$.

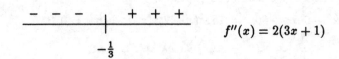

$$f''(x) = 2(3x + 1)$$

f is concave down on $\left(-\infty, -\frac{1}{3}\right]$ and concave up on $\left[-\frac{1}{3}, \infty\right)$. The inflection point is $\left(-\frac{1}{3}, \frac{290}{27}\right)$.

5) $f(0) = 8$ and $f(-4) = -8$

6)

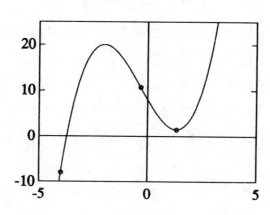

Exercise 7

8. $f(x) = x^3 + 2x^2 - x - 2$

1) Domain: $(-\infty, \infty)$

2) $x^3 + 2x^2 - x - 2 = x^2(x+2) - (x+2) = (x+2)(x-1)(x+1) = 0$ when $x = -2, -1, 1$.

3) f has no vertical or horizontal asymptotes.

4) $f'(x) = 3x^2 + 4x - 1 = 0$ when $x = \dfrac{-2 \pm \sqrt{7}}{3}$ (using the quadratic formula) $\approx .215, -1.55$.

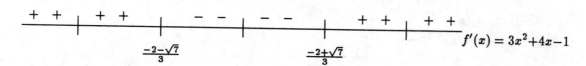

f is decreasing on $\left[\frac{-2-\sqrt{7}}{3}, \frac{-2+\sqrt{7}}{3}\right]$ and increasing on $\left(-\infty, \frac{-2-\sqrt{7}}{3}\right]$ and $\left[\frac{-2+\sqrt{7}}{3}, \infty\right)$. The point $\left(\frac{-2-\sqrt{7}}{3}, .631\right)$ is a relative maximum and $\left(\frac{-2+\sqrt{7}}{3}, -2.11\right)$ is a relative minimum.

5) $f''(x) = 6x + 4 = 2(3x + 2) = 0$ at $x = -\frac{2}{3}$

$$
\begin{array}{ccc}
-\ -\ -\ - & + + + + \\
\hline
& -\frac{2}{3} &
\end{array}
\qquad f''(x) = 2(3x+2)
$$

f is concave up on $\left[-\frac{2}{3}, \infty\right)$ and concave down on $\left(-\infty, -\frac{2}{3}\right]$. $\left(-\frac{2}{3}, -\frac{20}{27}\right)$ is an inflection point.

6.) $f(0) = -2$

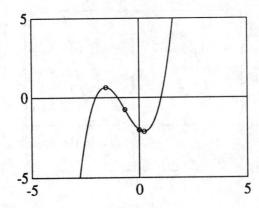

Exercise 8

9. $f(x) = x^3 - 7x + 6$

1) Domain: $(-\infty, \infty)$

2) $x^3 - 7x + 6 = (x-1)(x+3)(x-2) = 0$ at $x = -3, 1, 2.$

3) f has no vertical or horizontal asymptotes.

4) $f'(x) = 3x^2 - 7 = 0$ at $x = \pm\frac{\sqrt{21}}{3}.$

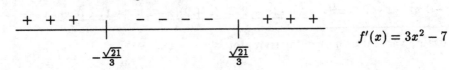

f is increasing on $\left(-\infty, -\frac{\sqrt{21}}{3}\right]$ and $\left[\frac{\sqrt{21}}{3}, \infty\right)$ and decreasing on $\left[-\frac{\sqrt{21}}{3}, \frac{\sqrt{21}}{3}\right]$. f has a relative maximum at $\left(-\frac{\sqrt{21}}{3}, 13.13\right)$ and a relative minimum at $\left(\frac{\sqrt{21}}{3}, -1.13\right).$

5) $f''(x) = 6x = 0$ at $x = 0.$

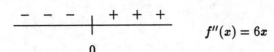

f is concave up on $[0, \infty)$ and concave down on $(-\infty, 0]$. $(0, 6)$ is an inflection point.

6)

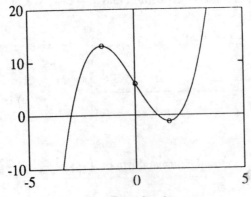

Exercise 9

284

10. $f(x) = \dfrac{x-1}{x+1}$

    1) $x+1 = 0$ at $x = -1$; therefore Domain: $(-\infty, -1) \cup (-1, \infty)$.

    2) $f(x) = 0$ when $x - 1 = 0 \Rightarrow x = 1$.

    3) $\lim\limits_{x \to \pm\infty} f(x) = 1$, therefore $y = 1$ is the horizontal asymptote.

        $\lim\limits_{x \to -1^-} f(x) = +\infty$ and $\lim\limits_{x \to -1^+} f(x) = -\infty$; therefore $x = -1$ is the vertical asymptote.

    4) $f'(x) = \dfrac{(x+1) - (x-1)}{(x+2)^2} = \dfrac{2}{(x+1)^2} > 0$ for all x in the domain.
    Thus f is increasing for all x in the domain.

    5) $f''(x) = \dfrac{-4}{(x+1)^3}$

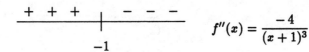

    f is concave up on $(-\infty, -1)$ and concave down on $(-1, \infty)$.

    6) $f(0) = -1$ and $f(-2) = 3$

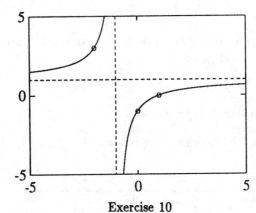

Exercise 10

11. $f(x) = \dfrac{x+4}{x-4}$

    1) $x - 4 = 0$ at $x = 4$, therefore Domain: $(-\infty, 4) \cup (4, \infty)$.

    2) $f(x) = 0$ when $x + 4 = 0 \Rightarrow x = -4$.

    3) $\lim\limits_{x \to \pm\infty} f(x) = 1$, therefore $y = 1$ is the horizontal asymptote.

        $\lim\limits_{x \to 4^-} f(x) = -\infty$ and $\lim\limits_{x \to 4^+} f(x) = \infty$; therefore $x = 4$ is the vertical asymptote.

    4) $f'(x) = \dfrac{(x-4) - (x+4)}{(x-4)^2} = \dfrac{-8}{(x-4)^2} < 0$ for all x in the domain.

    Therefore f is decreasing for all x in the domain.

    5) $f''(x) = \dfrac{16}{(x-4)^3}$

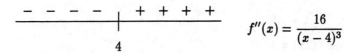

f is concave up on $(4, \infty)$ and concave down on $(-\infty, 4)$.

6) $f(0) = -1$ and $f(5) = 9$.

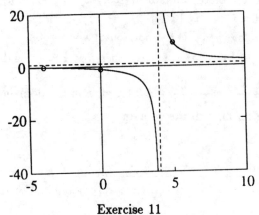

Exercise 11

12. $f(x) = \sin x + \cos x$

1) Domain: $(-\infty, \infty)$. Note that f is a periodic function with period 2π. Therefore we only need to sketch f over the interval $[0, 2\pi]$.

2) $f(x) = 0$ when $\sin x = -\cos x \Rightarrow x = \frac{3\pi}{4}, \frac{7\pi}{4}$ on $[0, 2\pi]$.

3) f has no vertical or horizontal asymptotes.

4) $f'(x) = \cos x - \sin x = 0$ when $\cos x = \sin x \Rightarrow x = \frac{\pi}{4}, \frac{5\pi}{4}$ on $[0, 2\pi]$.

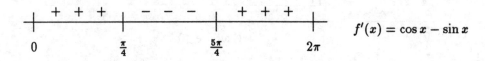

f is increasing on $\left[0, \frac{\pi}{4}\right]$ and $\left[\frac{5\pi}{4}, 2\pi\right]$ and decreasing on $\left[\frac{\pi}{4}, \frac{5\pi}{4}\right]$. f has a maximum at $\left(\frac{\pi}{4}, \sqrt{2}\right)$ and a minimum at $\left(\frac{5\pi}{4}, -\sqrt{2}\right)$.

5) $f''(x) = -\sin x - \cos x = 0$ when $-\sin x = \cos x \Rightarrow x = \frac{3\pi}{4}, \frac{7\pi}{4}$ on $[0, 2\pi]$.

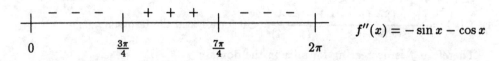

f is concave down on $\left[0, \frac{3\pi}{4}\right]$ and $\left[\frac{7\pi}{4}, 2\pi\right]$ and concave up on $\left[\frac{3\pi}{4}, \frac{7\pi}{4}\right]$. f has inflection points at $\left(\frac{3\pi}{4}, 0\right)$ and $\left(\frac{7\pi}{4}, 0\right)$.

6) $f(0) = 1$ and $f(2\pi) = 1$.

286

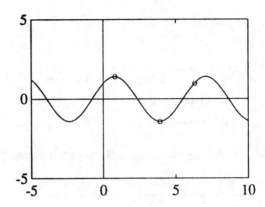

Exercise 12

[**Remark**: It might help to notice that in this example, the addition rules for sine and cosine tell us that

$$\sin x + \cos x = \begin{cases} \sqrt{2} \;\; \cos\left(x - \frac{\pi}{4}\right) \\ \text{or} \\ \sqrt{2} \;\; \sin\left(x + \frac{\pi}{4}\right) \end{cases}]$$

13. $f(x) = 9x - x^{-1} = 9x - \dfrac{1}{x} = \dfrac{9x^2 - 1}{x}$

1) Domain: $(-\infty, 0) \cup (0, \infty)$

2) $f(x) = 0$ when $9x^2 - 1 = 0 \Rightarrow x = \pm\frac{1}{3}$.

3) $f(x)$ has no horizontal asymptote.
$\lim\limits_{x \to 0^-} f(x) = +\infty$ and $\lim\limits_{x \to 0^+} f(x) = -\infty$; thus, $x = 0$ is the vertical asymptote. Also, note that $y = 9x$ is an oblique asymptote of $f(x)$.

4) $f'(x) = 9 + x^{-2} = 9 + \dfrac{1}{x^2} = \dfrac{9x^2 + 1}{x^2} > 0$ for all x in the domain. Thus f is increasing for all x in the domain.

$$+ \; + \; + \; + \quad - \; - \; - \\ \underline{\hspace{6cm}} \\ 0$$

$f''(x) = \dfrac{-2}{x^3}$

f is concave up on $(-\infty, 0)$ and concave down on $(0, \infty)$.

6)

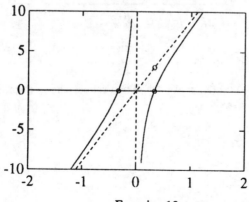

Exercise 13

287

14. $f(x) = x + 3x^{2/3} = x^{2/3}(x^{1/3} + 3)$

1) Domain: $(-\infty, \infty)$

2) $f(x) = 0$ when $x^{2/3} = 0 \Rightarrow x = 0$ or when $x^{1/3} + 3 = 0 \Rightarrow x = -27$.

3) f has no vertical or horizontal asymptotes.

4) $f'(x) = 1 + \dfrac{2}{x^{1/3}} = \dfrac{x^{1/3} + 2}{x^{1/3}} = 0$ at $x = -8$.
   $f'(x)$ is undefined at $x = 0$. Therefore, the critical numbers are 0 and -8.

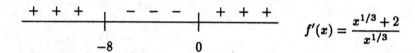

f is increasing on $(-\infty, -8]$ and $[0, \infty)$ and decreasing on $[-8, 0]$. f has a relative maximum at $(-8, 4)$ and a relative minimum at $(0, 0)$.

5) $f''(x) = \dfrac{-2}{3x^{4/3}}$. $f''(x)$ is undefined at $x = 0$.

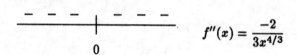

f is concave down on $(-\infty, 0]$ and $[0, \infty)$.

6) $f(1) = 4$

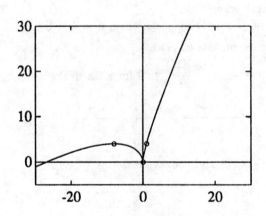

Exercise 14

15. $f(x) = |4 - x^2|$
    To graph $f(x)$, first we will graph $g(x) = 4 - x^2$. Then reflect the portion of the graph of $g(x)$ extending below the x-axis through the x-axis.

- $g(x) = 4 - x^2$

1) Domain: $(-\infty, \infty)$

2) $g(x) = 0$ at $x = \pm 2$.

3) $g(x)$ has no vertical or horizontal asymptotes.

288

4) $g'(x) = -2x = 0$ at $x = 0$.

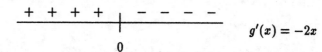

$g'(x) = -2x$

0

g is increasing on $(-\infty, 0]$ and decreasing on $[0, \infty)$. g has a maximum at $(0, 4)$.

5) $g''(x) = -2 < 0$. Thus g is concave down for all x.

6)

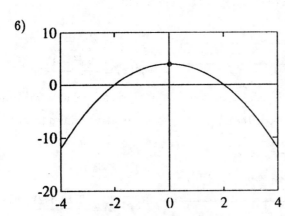

The graph of $g(x) = 4 - x^2$
Exercise 15

For the graph of $f(x)$, see Figure 5.6 on page 240 of the textbook.

16. $f(x) = \sqrt{x + 4}$

1) $x + 4 \geq 0$ when $x \geq -4$. Domain: $[-4, \infty)$

2) $f(x) = 0$ when $x + 4 = 0 \Rightarrow x = -4$.

3) f has no vertical or horizontal asymptote.

4) $f'(x) = \dfrac{1}{2\sqrt{x + 4}} > 0$ for all x in the domain. Thus f is increasing for all x in the domain.

5) $f''(x) = \dfrac{-1}{4(x + 4)^{3/2}} < 0$ for all x in the domain. Thus f is concave down for all x in the domain.

6) $f(0) = 2$

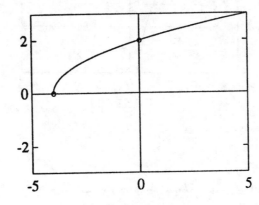

Exercise 16

289

17. $f(x) = x^2 + \frac{2}{x}$

   1) Domain: $(-\infty, 0) \cup (0, \infty)$

   2) $f(x) = x^2 + \dfrac{2}{x} = \dfrac{x^3 + 2}{x} = 0$ when $x^3 + 2 = 0 \Rightarrow x = \sqrt[3]{-2}$.

   3) $\displaystyle\lim_{x \to \pm\infty} f(x) = \infty$, therefore f has no horizontal asymptotes.

       $\displaystyle\lim_{x \to 0^-} f(x) = -\infty$ and $\displaystyle\lim_{x \to 0^+} f(x) = \infty$, therefore $x = 0$ is the vertical asymptote.

   4) $f'(x) = 2x - \dfrac{2}{x^2} = \dfrac{2(x^3 - 1)}{x^2} = 0$ at $x = 1$.

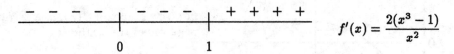

$$f'(x) = \frac{2(x^3 - 1)}{x^2}$$

   f is increasing on $[1, \infty)$ and decreasing on $(-\infty, 0)$ and $(0, 1]$. f has a relative minimum at $(1, 3)$.

   5) $f''(x) = 2 + \dfrac{4}{x^3} = \dfrac{2(x^3 + 2)}{x^3} = 0$ at $x = \sqrt[3]{-2}$.

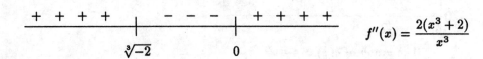

$$f''(x) = \frac{2(x^3 + 2)}{x^3}$$

   f is concave up on $\left(-\infty, \sqrt[3]{-2}\,\right]$ and $(0, \infty)$ and concave down on $\left[\sqrt[3]{-2},\, 0\right)$. $\left(\sqrt[3]{-2},\, 0\right)$ is an inflection point.

   6) Note that $y = x^2$ is an oblique asymptote for f.

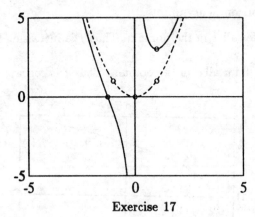

Exercise 17

18. $f(x) = (x + 1)^{5/3} - (x + 1)^{2/3}$

   1) Domain: $(-\infty, \infty)$

   2) $(x + 1)^{5/3} - (x + 1)^{2/3} = (x + 1)^{2/3}(x) = 0$ at $x = -1$ and 0.

   3) f has no vertical or horizontal asymptotes.

4) $f'(x) = \frac{5}{3}(x+1)^{2/3} - \frac{2}{3}(x+1)^{-1/3} = \frac{5x+3}{3(x+1)^{1/3}} = 0$ at $x = -\frac{3}{5}$.

$f'(x)$ is undefined when $(x+1)^{1/3} = 0 \Rightarrow x = -1$. Therefore the critical numbers are -1 and $-\frac{3}{5}$.

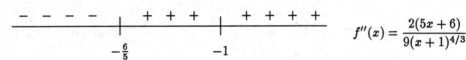

$$f'(x) = \frac{5x+3}{3(x+1)^{1/3}}$$

f is increasing on $(-\infty, -1]$ and $\left[-\frac{3}{5}, \infty\right)$ and decreasing on $\left[-1, -\frac{3}{5}\right]$. f has a relative maximum at $(-1, 0)$ and a relative minimum at $\left(-\frac{3}{5}, -.326\right)$.

5) $f''(x) = \frac{10}{9}(x+1)^{-1/3} + \frac{2}{9}(x+1)^{-4/3} = \frac{2(5x+6)}{9(x+1)^{4/3}} = 0$ at $x = -\frac{6}{5}$.

$f''(x)$ is undefined when $(x+1)^{4/3} = 0 \Rightarrow x = -1$.

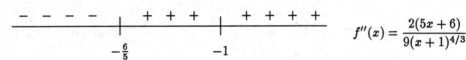

$$f''(x) = \frac{2(5x+6)}{9(x+1)^{4/3}}$$

f is concave up on $\left[-\frac{6}{5}, -1\right]$ and concave down on $\left(-\infty, -\frac{6}{5}\right]$. $\left(-\frac{6}{5}, -.41\right)$ is an inflection point.

6)

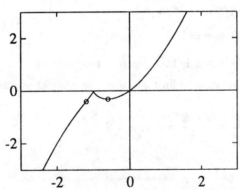

Exercise 18

19. $f(x) = |x^2 - 3x + 2|$

To graph $f(x)$, first we will graph $g(x) = x^2 - 3x + 2$. Then reflect the portion of the graph extending below the x-axis through the x-axis. (See problem 38 in Section 4.1.) Consider $g(x) = x^2 - 3x + 2$.

1) Domain: $(-\infty, \infty)$

2) $g(x) = x^2 - 3x + 2 = (x-1)(x-2) = 0$ at $x = 1, 2$.

3) g has no vertical or horizontal asymptote.

4) $g'(x) = 2x - 3 = 0$ at $x = \frac{3}{2}$.

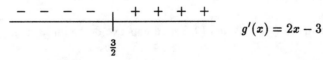

$$g'(x) = 2x - 3$$

g is decreasing on $\left(-\infty, \frac{3}{2}\right]$ and increasing on $\left[\frac{3}{2}, \infty\right)$. g has a minimum at $\left(\frac{3}{2}, -\frac{1}{4}\right)$.

5) $g''(x) = 2 > 0$. Thus g is concave up on $(-\infty, \infty)$.

6) $g(0) = 2$

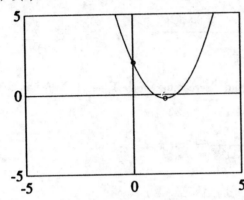

The graph of $g(x)$

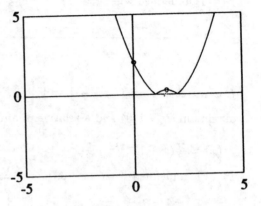

The graph of $f(x)$

20. $f(x) = x^{-2} - x^{-1} = \dfrac{1 - x}{x^2}$

1) Domain: $(-\infty, 0) \cup (0, \infty)$

2) $f(x) = 0$ when $1 - x = 0 \Rightarrow x = 1$.

3) $\displaystyle\lim_{x \to \pm\infty} f(x) = 0$; therefore $y = 0$ is the horizontal asymptote.

$\displaystyle\lim_{x \to 0^-} f(x) = \lim_{x \to 0^+} f(x) = \infty$; therefore $x = 0$ is the vertical asymptote.

4) $f'(x) = \dfrac{-x^2 - 2x(1 - x)}{x^4} = \dfrac{x - 2}{x^3} = 0$ at $x = 2$

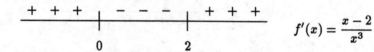

$f'(x) = \dfrac{x - 2}{x^3}$

f is increasing on $(-\infty, 0)$ and $[2, \infty)$ and decreasing on $(0, 2]$. f has a minimum at $\left(2, -\frac{1}{4}\right)$.

5) $f''(x) = \dfrac{x^3 - 3x^2(x - 2)}{x^6} = \dfrac{6 - 2x}{x^4} = 0$ at $x = 3$.

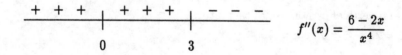

$f''(x) = \dfrac{6 - 2x}{x^4}$

f is concave up on $(-\infty, 0)$ and $(0, 3]$ and concave down on $[3, \infty)$. f has a point of inflection at $\left(3, -\frac{2}{9}\right)$.

6) $f(-1) = 2$

292

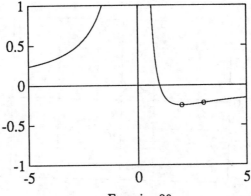

Exercise 20

21. $f(x) = \dfrac{1}{x(x-4)} = \dfrac{1}{x^2 - 4x}$

   1) $x(x-4) = 0$ at $x = 0, 4$; therefore Domain: $(-\infty, 0) \cup (0, 4) \cup (4, \infty)$

   2) $f(x) \neq 0$ for all x in the domain.

   3) $\lim\limits_{x \to \pm\infty} f(x) = 0$; therefore $y = 0$ is the horizontal asymptote.
   $\lim\limits_{x \to 0^-} f(x) = \infty$ and $\lim\limits_{x \to 0^+} f(x) = -\infty$, therefore $x = 0$ is a vertical asymptote.
   $\lim\limits_{x \to 4^-} f(x) = -\infty$ and $\lim\limits_{x \to 4^+} f(x) = \infty$, therefore $x = 4$ is a vertical asymptote.

   4) $f'(x) = \dfrac{-(2x-4)}{x^2(x-4)^2} = \dfrac{2(2-x)}{x^2(x-4)^2} = 0$ at $x = 2$.

$$
\begin{array}{ccccccc}
+ \ + \ + & | & + \ + \ + & | & - \ - \ - & | & - \ - \ - \\
\hline
 & 0 & & 2 & & 4 &
\end{array}
\qquad f'(x) = \dfrac{2(2-x)}{x^2(x-4)^2}
$$

   f is increasing on $(-\infty, 0)$ and $(0, 2]$ and decreasing on $[2, 4)$ and $(4, \infty)$. f has a relative maximum at $\left(2, -\frac{1}{4}\right)$.

   5) $f''(x) = \dfrac{-2(x^2)(x-4)^2 - (2)(2-x)[2x(x-4)^2 + 2(x-4)x^2]}{x^4(x-4)^4}$

   $= \dfrac{-2x(x-4) + 2(x-2)[2(x-4) + 2x]}{x^3(x-4)^3} = \dfrac{2(3x^2 - 12x + 16)}{x^3(x-4)^3}$

   $= \dfrac{2[3(x-2)^2 + 4]}{x^3(x-4)^3} \neq 0$ for all x in the domain.

$$
\begin{array}{ccccc}
+ \ + \ + & | & - \ - \ - & | & + \ + \ + \\
\hline
 & 0 & & 4 &
\end{array}
\qquad f''(x) = \dfrac{2[3(x-2)^2 + 4]}{x^3(x-4)^3}
$$

   f is concave up on $(-\infty, 0)$ and $(4, \infty)$ and concave down on $(0, 4)$. f has no inflection points.

6) $f(-1) = \frac{1}{5}$ and $f(5) = \frac{1}{5}$.

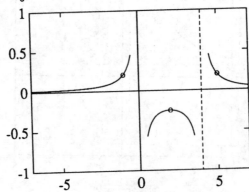

Exercise 21

22. $f(x) = x^2 - \dfrac{9}{x^2} = \dfrac{x^4 - 9}{x^2}$

1) Domain: $(-\infty, 0) \cup (0, \infty)$

2) $f(x) = 0$ when $x^4 - 9 = 0 \Rightarrow x = \pm 9^{1/4}$.

3) f has no horizontal asymptotes.
$\displaystyle\lim_{x \to 0^-} f(x) = \lim_{x \to 0^+} f(x) = -\infty$. Thus $x = 0$ is the vertical asymptote of $f(x)$. Note also that $y = x^2$ is an oblique asymptote.

4) $f'(x) = 2x + \dfrac{18}{x^3} = \dfrac{2x^4 + 18}{x^3} \neq 0$ for all x in the domain.

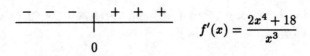

$$f'(x) = \frac{2x^4 + 18}{x^3}$$

f is decreasing on $(-\infty, 0)$ and increasing on $(0, \infty)$.

5) $f''(x) = 2 - \dfrac{54}{x^4} = \dfrac{2(x^4 - 27)}{x^4} = 0$ at $x = \pm 27^{1/4}$

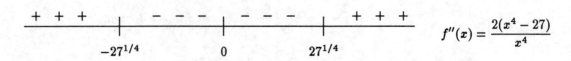

$$f''(x) = \frac{2(x^4 - 27)}{x^4}$$

f is concave up on $(-\infty, -27^{1/4}]$ and $[27^{1/4}, \infty)$ and concave down on $[-27^{1/4}, 0)$ and $(0, 27^{1/4}]$. The points of inflection are $\left(\pm 27^{1/4}, \dfrac{18}{27^{1/2}} \right)$. Note: $\dfrac{18}{27^{1/2}} \approx 3.464$.

6)

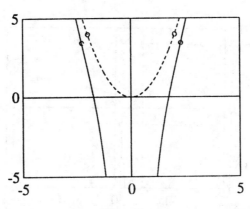

Exercise 22

23. $f(x) = \dfrac{x}{(2x+1)^2}$

1) $2x + 1 = 0$ at $x = -\frac{1}{2}$; therefore Domain: $\left(-\infty, -\frac{1}{2}\right) \cup \left(-\frac{1}{2}, \infty\right)$.

2) $f(x) = 0$ when $x = 0$.

3) $\displaystyle\lim_{x \to \pm\infty} f(x) = 0$; therefore $y = 0$ is the horizontal asymptote.

$\displaystyle\lim_{x \to -\frac{1}{2}^-} f(x) = \lim_{x \to -\frac{1}{2}^+} f(x) = -\infty$; therefore $x = -\frac{1}{2}$ is the vertical asymptote.

4) $f'(x) = \dfrac{(2x+1)^2 - x(2)(2x+1)(2)}{(2x+1)^4} = \dfrac{1 - 2x}{(2x+1)^3} = 0$ at $x = \frac{1}{2}$.

$$- \quad - \quad - \quad | \quad + \quad + \quad + \quad | \quad - \quad - \quad -$$
$$ -\tfrac{1}{2} \tfrac{1}{2}$$

$f'(x) = \dfrac{1 - 2x}{(2x+1)^3}$

f is decreasing on $\left(-\infty, -\frac{1}{2}\right)$ and $\left[\frac{1}{2}, \infty\right)$ and increasing on $\left(-\frac{1}{2}, \frac{1}{2}\right]$.
f has a maximum at $\left(\frac{1}{2}, \frac{1}{8}\right)$.

5) $f''(x) = \dfrac{-2(2x+1)^3 - (1 - 2x)\,3\,(2x+1)^2(2)}{(2x+1)^6}$

$ = \dfrac{-2(2x+1) + 6(2x-1)}{(2x+1)^4} = \dfrac{8(x-1)}{(2x+1)^4} = 0$ at $x = 1$

$$- \quad - \quad - \quad | \quad - \quad - \quad - \quad | \quad + \quad + \quad +$$
$$ -\tfrac{1}{2} 1$$

$f''(x) = \dfrac{8(x-1)}{(2x+1)^4}$

f is concave down on $\left(-\infty, -\frac{1}{2}\right)$ and $\left(-\frac{1}{2}, 1\right]$ and concave up on $[1, \infty)$. f has an inflection point at $\left(1, \frac{1}{9}\right)$.

6.) $f(-1) = -1$

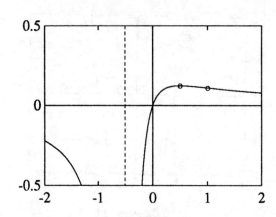

Exercise 23

24. $f(x) = \sqrt{2x - x^2}$

1) $2x - x^2 = x(2 - x) = 0$ at $x = 0, 2$.

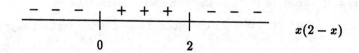

$$x(2 - x)$$

Thus $x(2 - x) \geq 0$ on $[0, 2]$.
Therefore the domain is $[0, 2]$.

2) $f(x) = 0$ at $x = 0, 2$.

3) f has no vertical or horizontal asymptotes.

4) $f'(x) = \dfrac{2 - 2x}{2\sqrt{2x - x^2}} = \dfrac{1 - x}{\sqrt{2x - x^2}} = 0$ at $x = 1$

$$f'(x) = \dfrac{1 - x}{\sqrt{2x - x^2}}$$

f is increasing on $[0, 1]$ and decreasing on $[1, 2]$. f has a maximum at $(1, 1)$.

5) $f''(x) = -1(2x - x^2)^{-1/2} + (1 - x)\left(-\frac{1}{2}\right)(2x - x^2)^{-3/2}(2 - 2x)$

$$= \dfrac{-1}{(2x - x^2)^{3/2}} < 0 \quad \text{for all } x \text{ in the domain.}$$

Therefore f is concave down for all x in $[0, 2]$.

6)

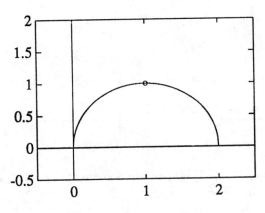

Exercise 24

25. $f(x) = (x - 3)^{2/3} + 1$

1) Domain: $(-\infty, \infty)$

2) $f(x) > 0$ for all x since $(x - 3)^{2/3} > 0$ for all x.

3) f has no vertical or horizontal asymptote.

4) $f'(x) = \frac{2}{3}(x - 3)^{-1/3} \neq 0$ for all x.
   $f'(x)$ is undefined when $x - 3 = 0 \Rightarrow x = 3$.
   Therefore $x = 3$ is a critical number.

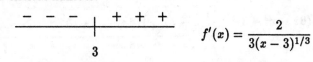

$$f'(x) = \frac{2}{3(x - 3)^{1/3}}$$

f is decreasing on $(-\infty, 3]$ and increasing on $(3, \infty)$. f has a minimum at $(3, 1)$.

5) $f''(x) = -\frac{2}{9}(x - 3)^{-4/3} = \frac{-2}{9(x - 3)^{4/3}} \neq 0$ for all x.
   $f''(x)$ is undefined at $x = 3$.

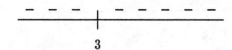

$$f''(x) = \frac{-2}{9(x - 3)^{4/3}}$$

f is concave down on $(-\infty, 3]$ and $[3, \infty)$.

6) $f(0) \approx 3.08$

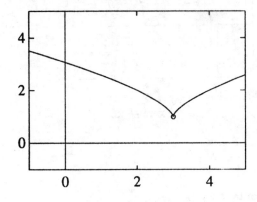

Exercise 25

26. $f(x) = \sqrt{6x - x^2 - 8}$

1) $6x - x^2 - 8 = -(x - 4)(x - 2) = 0$ at $x = 2, 4$.

$$- \quad - \quad - \quad | \quad + \quad + \quad + \quad | \quad - \quad - \quad -$$
$$2 \qquad\qquad 4$$

$$- (x - 4)(x - 2)$$

$-(x - 4)(x - 2) \geq 0$ on $[2, 4]$. Therefore the domain of $f(x)$ is $[2, 4]$.

2) $f(x) = 0$ at $x = 2, 4$.

3) f has no vertical or horizontal asymptotes.

4) $f'(x) = \dfrac{6 - 2x}{2\sqrt{6x - x^2 - 8}} = \dfrac{3 - x}{\sqrt{6x - x^2 - 8}} = 0$ at 3.

$$+ \quad + \quad + \quad | \quad - \quad - \quad -$$
$$2 \qquad\qquad 3 \qquad\qquad 4$$

$$f'(x) = \dfrac{3 - x}{\sqrt{6x - x^2 - 8}}$$

f is increasing on $[2, 3]$ and decreasing on $[3, 4]$. f has a maximum at $(3, 1)$.

5) $f''(x) = \dfrac{-1(6x - x^2 - 8)^{1/2} - (3 - x)\left(\frac{1}{2}\right)(6x - x^2 - 8)^{-1/2}(6 - 2x)}{6x - x^2 - 8}$

$$= \dfrac{-(6x - x^2 - 8) - (3 - x)^2}{(6x - x^2 - 8)^{3/2}} = \dfrac{-1}{(6x - x^2 - 8)^{3/2}} < 0$$

for all x in the domain. Thus f is concave down on $[2, 4]$.

6)

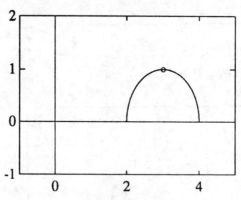

Exercise 26

27. $f(x) = \frac{1}{3}x^3 - x^2 - 3x + 4$

1) Domain: $(-\infty, \infty)$

2) f has no vertical or horizontal asymptotes.

298

3) $f'(x) = x^2 - 2x - 3 = (x-3)(x+1) = 0$ at $x = -1, 3$.

$$+\ +\ +\ \Big|\ -\ -\ -\ \Big|\ +\ +\ + \qquad f'(x) = (x-3)(x+1)$$
$$-1 3$$

f is increasing on $(-\infty, -1]$ and $[3, \infty)$ and decreasing on $[-1, 3]$. f has a relative maximum at $(-1, 6^{1/3})$ and a relative minimum at $(3, -5)$.

4) $f''(x) = 2x - 2 = 2(x-1) = 0$ at $x = 1$.

$$-\ -\ -\ \Big|\ +\ +\ + \qquad f''(x) = 2(x-1)$$
$$1$$

f is concave down on $(-\infty, 1]$ and concave up on $[1, \infty)$. f has an inflection point at $\left(1, \frac{1}{3}\right)$.

5) $f(0) = 4$

6) (See Figure 4.9 on p.226 of the textbook.)

28. $f(x) = (x^2 + 2)^{-1} = \dfrac{1}{x^2 + 2}$. Note that f is an even function; therefore, the graph of f is symmetric with respect to the y-axis.

1) Since $(x^2 + 2) > 0$ for all real numbers x, the domain is $(-\infty, \infty)$.

2) $f(x) \neq 0$ for all x in $(-\infty, \infty)$.

3) $\lim\limits_{x \to \pm\infty} f(x) = 0$; therefore $y = 0$ is the horizontal asymptote.

4) $f'(x) = \dfrac{-2x}{(x^2 + 2)^2} = 0$ at $x = 0$.

$$+\ +\ +\ \Big|\ -\ -\ - \qquad f'(x) = \dfrac{-2x}{(x^2 + 2)^2}$$
$$0$$

f is increasing on $(-\infty, 0]$ and decreasing on $[0, \infty)$. f has a maximum at $\left(0, \frac{1}{2}\right)$.

5) $f''(x) = -2(x^2 + 2)^{-2} + (-2x)(-2)(x^2 + 2)^{-3}(2x) = \dfrac{2(3x^2 - 2)}{(x^2 + 2)^3} = 0$ at $x = \pm\dfrac{\sqrt{6}}{3}$

$$+\ +\ +\ \Big|\ -\ -\ -\ \cdot\ \Big|\ +\ +\ + \qquad f''(x) = \dfrac{2(3x^2 - 2)}{(x^2 + 2)^3}$$
$$-\tfrac{\sqrt{6}}{3} \tfrac{\sqrt{6}}{3}$$

f is concave up on $\left(-\infty, -\frac{\sqrt{6}}{3}\right]$ and $\left[\frac{\sqrt{6}}{3}, \infty\right)$ and concave down on $\left[-\frac{\sqrt{6}}{3}, \frac{\sqrt{6}}{3}\right]$. The points of inflection are $\left(\pm\frac{\sqrt{6}}{3}, \frac{3}{8}\right)$.

6)

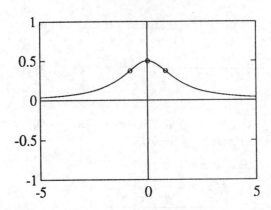

Exercise 28

29. $f(x) = \dfrac{x^2 - 4x + 5}{x - 2} = \dfrac{(x^2 - 4x + 4) + 1}{x - 2} = \dfrac{(x - 2)^2 + 1}{x - 2}$

1) $x - 2 = 0$ at $x = 2$; therefore Domain: $(-\infty, 2) \cup (2, \infty)$.

2) $(x - 2)^2 + 1 > 0$ for all x in the domain. Thus $f(x) \neq 0$ for all x in the domain.

3) f has no horizontal asymptotes.
   $\lim\limits_{x \to 2^-} f(x) = -\infty$ and $\lim\limits_{x \to 2^+} f(x) = \infty$; thus $x = 2$ is the vertical asymptote.

4) $f'(x) = \dfrac{(2x - 4)(x - 2) - (x^2 - 4x + 5)}{(x - 2)^2} = \dfrac{(x - 3)(x - 1)}{(x - 2)^2} = 0$

$$+ \ + \ + \quad | \quad - \ - \ - \quad | \quad - \ - \ - \quad | \quad + \ + \ + \qquad f'(x) = \dfrac{(x - 3)(x - 1)}{(x - 2)^2}$$
$$ 1 2 3$$

f is increasing on $(-\infty, 1]$ and $[3, \infty)$ and decreasing on $[1, 2)$ and $(2, 3]$. f has a relative maximum at $(1, -2)$ and a relative minimum at $(3, 2)$.

5) $f''(x) = \dfrac{(2x - 4)(x - 2)^2 - (x^2 - 4x + 3)\, 2\,(x - 2)}{(x - 2)^4} = \dfrac{2}{(x - 2)^3} \neq 0$ for all x in the domain.

$$- \ - \ - \quad | \quad + \ + \ + \qquad f''(x) = \dfrac{2}{(x - 2)^3}$$
$$ 2$$

f is concave down on $(-\infty, 2)$ and concave up on $(2, \infty)$.

6) $f(x) = \dfrac{(x - 2)^2 + 1}{x - 2} = (x - 2) + \dfrac{1}{x - 2}$
   Note that $y = x - 2$ is an oblique asymptote for $f(x)$.

7) $f(0) = -\frac{5}{2}$

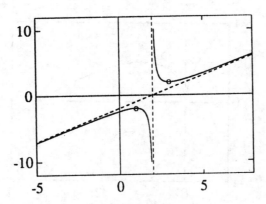

Exercise 29

30. $f(x) = \frac{3}{2} x^{2/3} - x = x^{2/3}\left(\frac{3}{2} - x^{1/3}\right)$

1) Domain: $(-\infty, \infty)$

2) $f(x) = 0$ when $x^{2/3} = 0 \Rightarrow x = 0$ or when $\frac{3}{2} - x^{1/3} = 0 \Rightarrow x = \frac{27}{8}$.

3) f has no vertical or horizontal asymptotes.

4) $f'(x) = x^{-1/3} - 1 = \dfrac{1 - x^{1/3}}{x^{1/3}} = 0$ at $x = 1$.
   $f'(x)$ is undefined when $x^{1/3} = 0 \Rightarrow x = 0$.
   Therefore the critical numbers are 0 and 1.

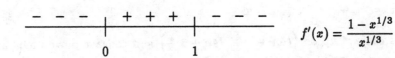

f is decreasing on $(-\infty, 0]$ and $[1, \infty)$ and increasing on $[0, 1]$. f has a relative minimum at $(0, 0)$ and a relative maximum at $\left(1, \frac{1}{2}\right)$.

5) $f''(x) = \dfrac{-1}{3x^{4/3}} \neq 0$ for all x. $f''(x)$ is undefined at 0.

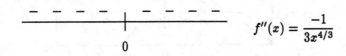

f is concave down on $(-\infty, 0]$ and $[0, \infty)$.

301

6)

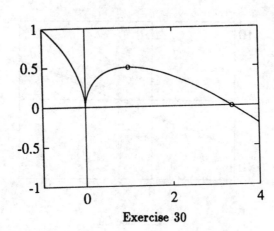

Exercise 30

31. $f(x) = x + \cos x$.

1)–5) This function is defined on $(-\infty, \infty)$ and has neither horizontal nor vertical asymptotes.

$$f'(x) = 1 - \sin x$$
$$f'(x) = 0 \Rightarrow \sin x = 1 \Rightarrow x = (4n+1)\tfrac{\pi}{2} \text{(for any integer } n)$$
$$= \ldots\ldots, -\frac{3\pi}{2}, \frac{\pi}{2}, \frac{5\pi}{2}, \ldots\ldots.$$
$$f'(x) > 0 \text{ for all other values of } x.$$

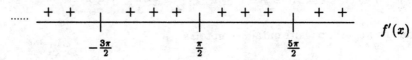

So f is increasing on $(-\infty, \infty)$ with critical numbers $x = (4n+1)\tfrac{\pi}{2}$ and no relative extrema.

$f''(x) = -\cos x$, so

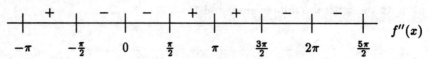

So f is concave up on $\left((4n+1)\tfrac{\pi}{2}, (4n+3)\tfrac{\pi}{2}\right)$ and concave down on $\left((4n-1)\tfrac{\pi}{2}, (4n+1)\tfrac{\pi}{2}\right)$ with inflection points where x is any odd multiple of $\tfrac{\pi}{2}$.

6)

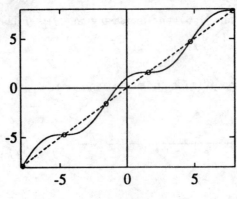

Exercise 31

302

(Dashed line $y = x$ helps sketching: graph of $y = f(x)$ crosses this line when x is an odd multiple of $\frac{\pi}{2}$ since then $\cos x = 0$.)

32. $f(x) = x + \sin(-2x) = x - \sin 2x$

1)–5) Domain: $(-\infty, \infty)$

f has no vertical or horizontal asymptotes.

$$f'(x) = 1 - 2\cos 2x$$
$$f'(x) = 0 \Rightarrow \cos 2x = \tfrac{1}{2}$$
$$\Rightarrow 2x = \left(2n \pm \tfrac{1}{3}\right)\pi$$
$$\Rightarrow x = \left(n \pm \tfrac{1}{6}\right)\pi \text{ (any integer } n)$$

Thus

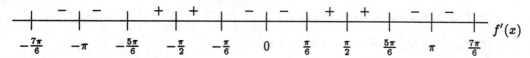

so that f is increasing on the intervals $\left[\left(n + \tfrac{1}{6}\right)\pi, \left(n + \tfrac{5}{6}\right)\pi\right]$ and decreasing on the intervals $\left[\left(n - \tfrac{1}{6}\right)\pi, \left(n + \tfrac{1}{6}\right)\pi\right]$, with relative maxima when $x = \left(n - \tfrac{1}{6}\right)\pi$ and relative minima when $x = \left(n + \tfrac{1}{6}\right)\pi$, for any integer n. It has no absolute extrema.

$f''(x) = 4\sin 2x$ is familiar:

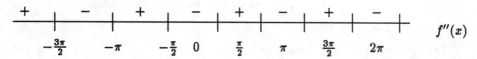

So f is concave up on $\left[n\pi, \left(n + \tfrac{1}{2}\right)\pi\right]$ and concave down on $\left[\left(n - \tfrac{1}{2}\right)\pi, n\pi\right]$ with points of inflection where $x = \frac{n\pi}{2}$, for any integer n.

6)

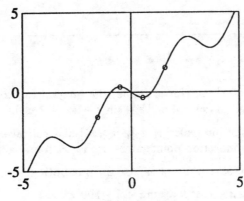

Exercise 32

33. $f(x) = \sin x + \cos^2 x$

1)–5) This function is defined on $(-\infty, \infty)$ with no asymptotes of either kind. Since it is periodic of period 2π we confine attention to the interval $[0, 2\pi]$.

$$f'(x) = \cos x - 2\sin x \cos x$$
$$= \cos x(1 - 2\sin x)$$
$$f'(x) = 0 \Rightarrow \cos x = 0 \text{ or } \sin x = \tfrac{1}{2}$$
$$\Rightarrow x = \tfrac{\pi}{2}, \tfrac{3\pi}{2}, \tfrac{\pi}{6}, \tfrac{5\pi}{6}$$

Since $\cos x > 0$ for x in $\left[0, \tfrac{\pi}{2}\right) \cup \left(\tfrac{3\pi}{2}, 2\pi\right]$ and $\cos x < 0$ for x in $\left(\tfrac{\pi}{2}, \tfrac{3\pi}{2}\right)$ while $\sin x > \tfrac{1}{2}$ for x in $\left(\tfrac{\pi}{6}, \tfrac{5\pi}{6}\right)$ and $\sin x < \tfrac{1}{2}$ for x in $\left[0, \tfrac{\pi}{6}\right) \cup \left(\tfrac{5\pi}{6}, 2\pi\right)$ we see:

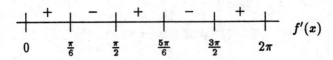

So f is increasing on the intervals $\left[0, \tfrac{\pi}{6}\right]$, $\left[\tfrac{\pi}{2}, \tfrac{5\pi}{6}\right]$, $\left[\tfrac{3\pi}{2}, 2\pi\right]$ and decreasing on the intervals $\left[\tfrac{\pi}{6}, \tfrac{\pi}{2}\right]$, $\left[\tfrac{5\pi}{6}, \tfrac{3\pi}{2}\right]$, and has relative extrema at

$$\left(\tfrac{\pi}{6}, \tfrac{5}{4}\right) \quad \longleftarrow \text{ absolute maximum}$$

$$\left(\tfrac{\pi}{2}, 1\right) \quad \longleftarrow \text{ relative minimum}$$

$$\left(\tfrac{5\pi}{6}, \tfrac{5}{4}\right) \quad \longleftarrow \text{ absolute maximum}$$

$$\left(\tfrac{3\pi}{2}, -1\right) \quad \longleftarrow \text{ absolute minimum}$$

$$f''(x) = -\sin x - 2\cos^2 x + 2\sin^2 x$$
$$= 4\sin^2 x - \sin x - 2$$

By the quadratic formula, $f''(x) = 0 \Rightarrow \sin x = \frac{1 \pm \sqrt{33}}{8}$.

Approximate solutions (in degrees): $x_1 \approx 57\tfrac{1}{2}^\circ$, $x_2 \approx 122\tfrac{1}{2}^\circ$, $x_3 \approx 216\tfrac{1}{3}^\circ$, $x_4 \approx 323\tfrac{2}{3}^\circ$.

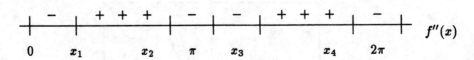

[To determine the sign of f'' on (for example) the interval (x_2, x_3) simply substitute a specific value of x such as π: then $f''(\pi) = 4\sin^3 \pi - \sin \pi - 2 = -2 < 0$.]

So f is concave up on the intervals $[x_1, x_2]$, $[x_3, x_4]$ and concave down on the intervals $[0, x_1]$, $[x_2, x_3]$, $[x_4, 2\pi]$ and has inflection points at x_1, x_2, x_3, x_4 found above (approximately).

To see where the graph crosses the x-axis, use the quadratic formula again:

$$f(x) = \sin x + \cos^2 x = \sin x + 1 - \sin^2 x.$$

Therefore $f(x) = 0 \Rightarrow \sin x = \frac{1 \pm \sqrt{5}}{2}$.

Of these, only $\sin x = \frac{1 - \sqrt{5}}{2}$ is possible, yielding (approximately, in degrees)

$$x \approx 218^\circ \text{ and } x \approx 322^\circ.$$

Exercise Set 4.7

6)

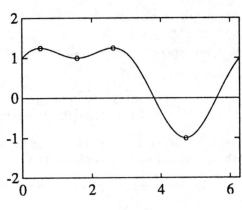

Exercise 33

34. $f(x) = \sin^2 x + \sin x$

1)–5) Again this function is periodic of period 2π on $(-\infty, \infty)$ with neither vertical nor horizontal asymptotes; we focus on the interval $[0, 2\pi]$.

$$\begin{aligned} f'(x) &= 2\sin x \cos x + \cos x \\ &= (2\sin x + 1)\cos x \\ f'(x) &= 0 \Rightarrow \sin x = -\tfrac{1}{2} \text{ or } \cos x = 0 \\ &\Rightarrow x = \tfrac{7\pi}{6}, \tfrac{11\pi}{6}, \tfrac{\pi}{2}, \tfrac{3\pi}{2} \end{aligned}$$

Arguing along lines similar to those in problem 33, we arrive at

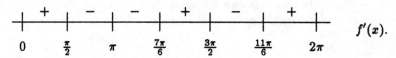

So f is increasing on $\left[0, \tfrac{\pi}{2}\right]$, $\left[\tfrac{7\pi}{6}, \tfrac{3\pi}{2}\right]$, $\left[\tfrac{11\pi}{6}, 2\pi\right]$ and decreasing on $\left[\tfrac{\pi}{2}, \tfrac{7\pi}{6}\right]$, $\left[\tfrac{3\pi}{2}, \tfrac{11\pi}{6}\right]$, and has relative extrema at

$$\left(\tfrac{\pi}{2}, 2\right) \quad \leftarrow \text{ absolute maximum}$$
$$\left(\tfrac{7\pi}{6}, -\tfrac{1}{4}\right) \quad \leftarrow \text{ absolute minimum}$$
$$\left(\tfrac{3\pi}{2}, 0\right) \quad \leftarrow \text{ relative maximum}$$
$$\left(\tfrac{11\pi}{6}, -\tfrac{1}{4}\right) \quad \leftarrow \text{ absolute minimum}$$

$$\begin{aligned} f''(x) &= 2\cos^2 x - 2\sin^2 x - \sin x \\ &= 2 - \sin x - 4\sin^2 x \\ f''(x) &= 0 \Rightarrow \sin x = \frac{-1 \pm \sqrt{33}}{2} \end{aligned}$$

Approximate solution (in degrees): $x_1 \approx 36\tfrac{1}{3}^\circ$, $x_2 \approx 143\tfrac{2}{3}^\circ$, $x_3 \approx 237\tfrac{1}{2}^\circ$, $x_4 \approx 302\tfrac{1}{2}^\circ$

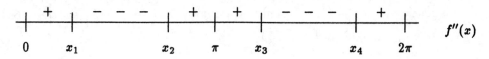

305

So f is concave up on $[0, x_1]$, $[x_2, x_3]$, $[x_4, 2\pi]$ and concave down on $[x_1, x_2]$, $[x_3, x_4]$ and has inflection points at x_1, x_2, x_3, x_4.

The graph crosses the x-axis where $0 = \sin^2 x + \sin x = \sin x(\sin x + 1)$ so that $x = 0, \pi, \frac{3\pi}{2}, 2\pi$.

[Remark: Notice that (by trigonometry)

$$
\begin{aligned}
f(-x) &= \sin^2 x - \sin x \\
&= 1 - (\sin x + \cos^2 x).
\end{aligned}
$$

This means that the graph of $f(x) = \sin^2 x + \sin x$ can be obtained by reflecting the graph of problem 33 through the origin and shifting vertically by one unit.]

6)

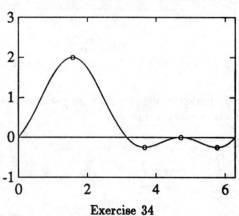

Exercise 34

35. $f(x) = 4x^2(1 - x^2)$. Note that f is an even function.
Therefore, f is symmetric with respect to the y-axis.

1) Domain: $(-\infty, \infty)$

2) $f(x) = 0$ when $x = 0, \pm 1$.

3) f has no vertical or horizontal asymptotes.

4) $f'(x) = 8x(1 - x^2) + 4x^2(-2x) = 8x(1 - 2x^2) = 0$ at $x = 0, \pm\frac{\sqrt{2}}{2}$

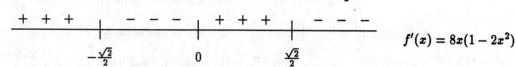

f is increasing on $\left(-\infty, -\frac{\sqrt{2}}{2}\right]$ and $\left[0, \frac{\sqrt{2}}{2}\right]$ and decreasing on $\left[-\frac{\sqrt{2}}{2}, 0\right]$ and $\left[\frac{\sqrt{2}}{2}, \infty\right)$.

f has a relative minimum at $(0, 0)$ and maxima at $\left(\pm\frac{\sqrt{2}}{2}, 1\right)$.

5) $f''(x) = 8(1 - 2x^2) + 8x(-4x) = 8(1 - 6x^2) = 0$ at $x = \pm\frac{\sqrt{6}}{6}$.

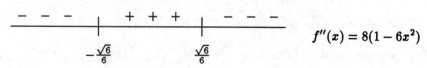

f is concave up on $\left[-\frac{\sqrt{6}}{6}, \frac{\sqrt{6}}{6}\right]$ and concave down on $\left(-\infty, -\frac{\sqrt{6}}{6}\right]$ and $\left[\frac{\sqrt{6}}{6}, \infty\right)$.

The inflection points are $\left(\pm\frac{\sqrt{6}}{6}, \frac{5}{9}\right)$.

6)

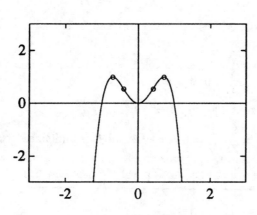

Exercise 35

36. $f(x) = x^4 - 3x^2 + 2$. Note that f ia an even function $\Rightarrow f$ is symmetric with respect to the y-axis.

1) Domain: $(-\infty, \infty)$

2) $x^4 - 3x^2 + 2 = (x^2 - 1)(x^2 - 2) = 0$ at $x = \pm 1, \pm\sqrt{2}$.

3) f has no vertical or horizontal asymptotes.

4) $f'(x) = 4x^3 - 6x = 2x(2x^2 - 3) = 0$ at $x = 0, \pm\frac{\sqrt{6}}{2}$.

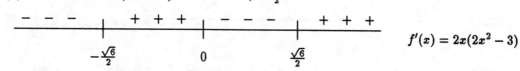

f is increasing on $\left[-\frac{\sqrt{6}}{2}, 0\right]$ and $\left[\frac{\sqrt{6}}{2}, \infty\right)$ and decreasing on $\left(-\infty, -\frac{\sqrt{6}}{2}\right]$ and $\left[0, \frac{\sqrt{6}}{2}\right]$.

f has a minima at $\left(-\frac{\sqrt{6}}{2}, -\frac{1}{4}\right)$ and $\left(\frac{\sqrt{6}}{2}, -\frac{1}{4}\right)$ and a relative maximum at $(0, 2)$.

5) $f''(x) = 12x^2 - 6 = 6(2x^2 - 1) = 0$ at $x = \pm\frac{\sqrt{2}}{2}$.

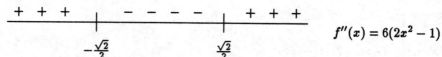

f is concave up on $\left(-\infty, -\frac{\sqrt{2}}{2}\right]$ and $\left[\frac{\sqrt{2}}{2}, \infty\right)$ and concave down on $\left[-\frac{\sqrt{2}}{2}, \frac{\sqrt{2}}{2}\right]$.

The inflection points are $\left(-\frac{\sqrt{2}}{2}, \frac{3}{4}\right)$ and $\left(\frac{\sqrt{2}}{2}, \frac{3}{4}\right)$.

6)

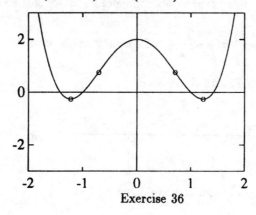

Exercise 36

307

37. $f(x) = x^{2/3} - \frac{1}{5} x^{5/3}$

1) Domain: $(-\infty, \infty)$

2) $x^{2/3} - \frac{1}{5}x^{5/3} = x^{2/3}\left(1 - \frac{1}{5}x\right) = 0$ at $x = 0, 5$.

3) f has no vertical or horizontal asymptotes.

4) $f'(x) = \frac{2}{3}x^{-1/3} - \frac{1}{3}x^{2/3} = \frac{2-x}{3x^{1/3}} = 0$ at $x = 2$ and f' is undefined at $x = 0$. The critical numbers are $0, 2$.

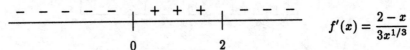

f is increasing on $[0, 2]$ and decreasing on $(-\infty, 0]$ and $[2, \infty)$. f has a relative minimum at $(0, 0)$ and a relative maximum at $\left(2, \frac{3\sqrt[3]{4}}{5}\right)$.

5) $f''(x) = -\frac{2}{9}x^{-4/3} - \frac{2}{9}x^{-1/3} = -\frac{2}{9}\left(\frac{1+x}{x^{4/3}}\right) = 0$ at $x = -1$. $f''(x)$ is undefined at $x = 0$.

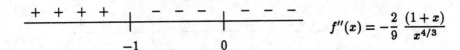

f is concave up on $(-\infty, -1]$ and concave down on $[-1, 0)$ and $[0, \infty)$. $\left(-1, \frac{6}{5}\right)$ is an inflection point.

6) See Figure 5.8 on page 241 of the textbook.

38. $f(x) = \dfrac{2x+1}{x+2}$

1) Domain: $(-\infty, -2) \cup (-2, \infty)$

2) $f(x) = 0$ when $2x + 1 = 0 \Rightarrow x = -\frac{1}{2}$.

3) $\lim\limits_{x \to \pm\infty} f(x) = 2$; therefore $y = 2$ is the horizontal asymptote.
   $\lim\limits_{x \to -2^-} f(x) = +\infty$ and $\lim\limits_{x \to -2^+} f(x) = -\infty$; therefore $x = -2$ is the vertical asymptote.

4) $f'(x) = \dfrac{2(x+2) - (2x+1)}{(x+2)^2} = \dfrac{3}{(x+2)^2} > 0$ for all x in the domain.

5) $f''(x) = \dfrac{-6}{(x+2)^3} \neq 0$ for all x in the domain.

$$\begin{array}{ccc} + + + + & & - - - - \\ \hline & -2 & \end{array} \qquad f''(x) = \dfrac{-6}{(x+2)^3}$$

f is concave up on $(-\infty, -2)$ and concave down on $(-2, \infty)$.

6) $f(0) = \frac{1}{2}$

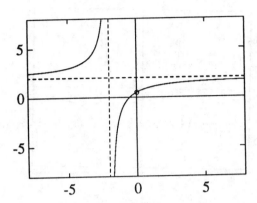

Exercise 38

39. $f(x) = \dfrac{x^2}{9 - x^2}$

1) $9 - x^2 = 0$ when $x = \pm 3$; therefore Domain: $(-\infty, -3) \cup (-3, 3) \cup (3, \infty)$.

2) $f(x) = 0$ when $x = 0$.

3) $\lim\limits_{x \to \pm\infty} f(x) = -1$; therefore $y = -1$ is the horizontal asymptote.

    $\lim\limits_{x \to -3^-} f(x) = -\infty$ and $\lim\limits_{x \to -3^+} f(x) = +\infty$; therefore $x = -3$ is a vertical asymptote.

    $\lim\limits_{x \to 3^-} f(x) = +\infty$ and $\lim\limits_{x \to 3^+} f(x) = -\infty$; therefore $x = 3$ is a vertical asymptote.

4) $f'(x) = \dfrac{2x(9 - x^2) - x^2(-2x)}{(9 - x^2)^2} = \dfrac{18x}{(9 - x^2)^2} = 0$ at $x = 0$.

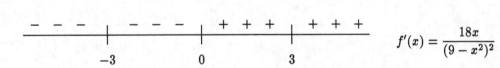

$f'(x) = \dfrac{18x}{(9 - x^2)^2}$

f is decreasing on $(-\infty, -3)$ and $(-3, 0]$ and increasing on $[0, 3)$ and $(3, \infty)$. f has a relative minimum at $(0, 0)$.

5) $f''(x) = \dfrac{18(9 - x^2)^2 - 18x(2)(9 - x^2)(-2x)}{(9 - x^2)^4} = \dfrac{18(9 - x^2 + 4x^2)}{(9 - x^2)^3} = \dfrac{54(3 + x^2)}{(9 - x^2)^3} \neq 0$ for all x in the domain.

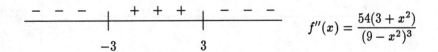

$f''(x) = \dfrac{54(3 + x^2)}{(9 - x^2)^3}$

f is concave down on $(-\infty, -3)$ and $(3, \infty)$ and concave up on $(-3, 3)$.

6)

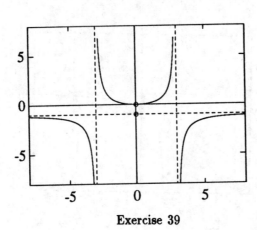

Exercise 39

40. $f(x) = \dfrac{x^2}{x^2 - 16}$. Note that f is an even function $\Rightarrow$ f is symmetric with respect to the y-axis.

1) $x^2 - 16 = 0$ at $x = \pm 4$; therefore Domain: $(-\infty, -4) \cup (-4, 4) \cup (4, \infty)$.

2) $f(x) = 0$ when $x^2 = 0 \Rightarrow x = 0$.

3) $\lim\limits_{x \to \pm\infty} f(x) = 1$, therefore $y = 1$ is the horizontal asymptote.
   $\lim\limits_{x \to -4^+} f(x) = -\infty$, $\lim\limits_{x \to -4^-} f(x) = -\infty$, therefore $x = -4$ is a vertical asymptote.
   $\lim\limits_{x \to 4^+} f(x) = \infty$, $\lim\limits_{x \to 4^-} f(x) = -\infty$, therefore $x = 4$ is a vertical asymptote.

4) $f'(x) = \dfrac{2x(x^2 - 16) - x^2(2x)}{(x^2 - 16)^2} = \dfrac{-32x}{(x^2 - 16)^2} = 0$ at $x = 0$.

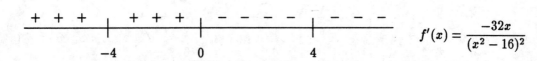

$f'(x) = \dfrac{-32x}{(x^2 - 16)^2}$

f is increasing on $(-\infty, -4)$ and $(-4, 0]$ and decreasing on $[0, 4)$ and $(4, \infty)$. f has a relative maximum at $(0, 0)$.

5) $f''(x) = \dfrac{-32(x^2 - 16)^2 + 32x(2)(x^2 - 16)(2x)}{(x^2 - 16)^4} = \dfrac{32(3x^2 + 16)}{(x^2 - 16)^3}$ which is always different from 0.

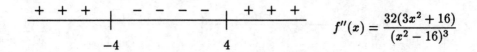

$f''(x) = \dfrac{32(3x^2 + 16)}{(x^2 - 16)^3}$

f is concave up on $(-\infty, -4)$ and $(4, \infty)$ and concave down on $(-4, 4)$. f has no inflection points.

6)

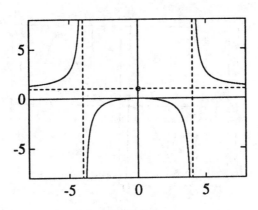

Exercise 40

41. $f(x) = 5x^3 - x^5$. Note that f is an odd function. Therefore the graph of f is symmetric with respect to the origin.

1) Domain: $(-\infty, \infty)$

2) $5x^3 - x^5 = x^3(5 - x^2) = 0$ at $x = 0, \pm\sqrt{5}$.

3) f has no vertical or horizontal asymptotes.

4) $f'(x) = 15x^2 - 5x^4 = 5x^2(3 - x^2) = 0$ at $x = 0, \pm\sqrt{3}$.

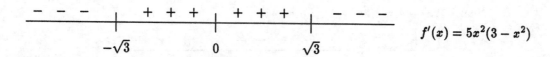

f is increasing on $\left[-\sqrt{3}, \sqrt{3}\right]$ and decreasing on $\left(-\infty, -\sqrt{3}\right]$ and $\left[\sqrt{3}, \infty\right)$. f has a relative minimum at $\left(-\sqrt{3}, -6\sqrt{3}\right)$ and a relative maximum at $\left(\sqrt{3}, 6\sqrt{3}\right)$.

5) $f''(x) = 30x - 20x^3 = 10x(3 - 2x^2) = 0$ at $x = 0, \pm\frac{\sqrt{6}}{2}$.

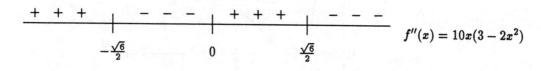

f is concave up on $\left(-\infty, -\frac{\sqrt{6}}{2}\right]$ and $\left[0, \frac{\sqrt{6}}{2}\right]$ and concave down on $\left[-\frac{\sqrt{6}}{2}, 0\right]$ and $\left[\frac{\sqrt{6}}{2}, \infty\right)$. The inflection points are $\left(-\frac{\sqrt{6}}{2}, \frac{-21\sqrt{6}}{8}\right)$, $(0,0)$ and $\left(\frac{\sqrt{6}}{2}, \frac{21\sqrt{6}}{8}\right)$.

311

6)

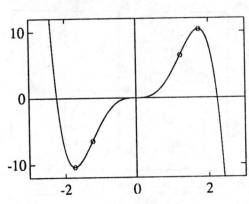

Exercise 41

42. $f(x) = x^4 - 2x^2 + 1$. Note that f is an even function. Therefore, the graph of f is symmetric with respect to the y-axis.

1) Domain: $(-\infty, \infty)$

2) $x^4 - 2x^2 + 1 = (x^2 - 1)(x^2 - 1) = 0$ at $x = \pm 1$.

3) f has no vertical or horizontal asymptotes.

4) $f'(x) = 4x^3 - 4x = 4x(x^2 - 1) = 0$ at $x = 0, \pm 1$.

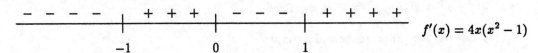

f is increasing on $[-1, 0]$ and $[1, \infty)$ and decreasing on $(-\infty, -1]$ and $[0, 1]$. f has minima at $(-1, 0)$ and $(1, 0)$ and a relative maximum at $(0, 1)$.

5) $f''(x) = 12x^2 - 4 = 4(3x^2 - 1) = 0$ at $x = \pm\frac{\sqrt{3}}{3}$.

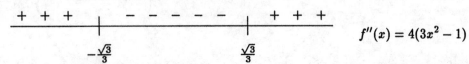

f is concave up on $\left(-\infty, -\frac{\sqrt{3}}{3}\right]$ and $\left[\frac{\sqrt{3}}{3}, \infty\right)$ and concave down on $\left[-\frac{\sqrt{3}}{3}, \frac{\sqrt{3}}{3}\right]$. The inflection points are $\left(-\frac{\sqrt{3}}{3}, \frac{4}{9}\right)$ and $\left(\frac{\sqrt{3}}{3}, \frac{4}{9}\right)$.

6)

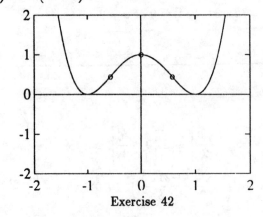

Exercise 42

312

43. $f(x) = 16 - 20x^3 + 3x^5$

1) Domain: $(-\infty, \infty)$

2) $f(x) = 0$ when $x \approx -2.6377, .9774, 2.5163$

3) f has no vertical or horizontal asymptotes.

4) $f'(x) = -60x^2 + 15x^4 = 15x^2(x^2 - 4) = 0$ at $x = 0, \pm 2$.

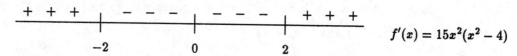

$$f'(x) = 15x^2(x^2 - 4)$$

f is increasing on $(-\infty, -2]$ and $[2, \infty)$ and decreasing on $[-2, 2]$. f has a relative maximum at $(-2, 80)$ and a relative minimum at $(2, -48)$.

5) $f''(x) = -120x + 60x^3 = 60x(x^2 - 2) = 0$ at $x = 0, \pm\sqrt{2}$.

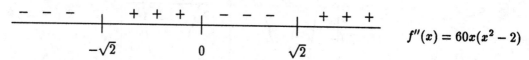

$$f''(x) = 60x(x^2 - 2)$$

f is concave down on $\left(-\infty, -\sqrt{2}\right]$ and $\left[0, \sqrt{2}\right]$ and concave up on $\left[-\sqrt{2}, 0\right]$ and $\left[\sqrt{2}, \infty\right)$. The inflection points are $\left(-\sqrt{2},\ 16 + 28\sqrt{2}\right)$, $(0, 16)$ and $\left(\sqrt{2},\ 16 - 28\sqrt{2}\right)$.

6)

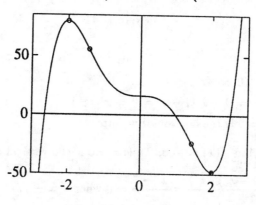

Exercise 43

44. $f(x) = x^4 - 18x^2 + 32$. Note that f is an even function. Therefore, the graph of f is symmetric with respect to the y-axis.

1) Domain: $(-\infty, \infty)$

2) $x^4 - 18x^2 + 32 = (x^2 - 16)(x^2 - 2) = 0$ when $x = \pm 4, \pm\sqrt{2}$.

3) f has no vertical or horizontal asymptotes.

4) $f'(x) = 4x^3 - 36x = 4x(x^2 - 9) = 0$ at $x = 0, \pm 3$.

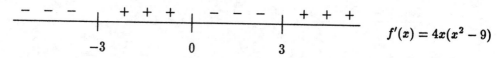

$$f'(x) = 4x(x^2 - 9)$$

f is increasing on $[-3, 0]$ and $[3, \infty)$ and decreasing on $(-\infty, -3]$ and $[0, 3]$. f has minima at $(-3, -49)$ and $(3, -49)$ and a relative maximum at $(0, 32)$.

5) $f''(x) = 12x^2 - 36 = 12(x^2 - 3) = 0$ at $x = \pm\sqrt{3}$.

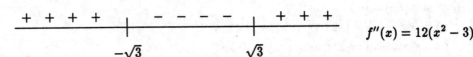

$$f''(x) = 12(x^2 - 3)$$

f is concave up on $\left(-\infty, -\sqrt{3}\right]$ and $\left[\sqrt{3}, \infty\right)$ and concave down on $\left[-\sqrt{3},\ \sqrt{3}\right]$. The inflection points are $\left(-\sqrt{3}, -13\right)$ and $\left(\sqrt{3}, -13\right)$.

6)

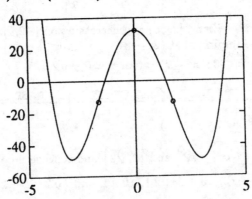

Exercise 44

45. $f(x) = \dfrac{1 - x^2}{x^3}$

1) Domain: $(-\infty, 0) \cup (0, \infty)$

2) $f(x) = 0$ when $1 - x^2 = (1 - x)(1 + x) = 0 \Rightarrow x = \pm 1$.

3) $\lim\limits_{x \to \pm\infty} f(x) = 0 \Rightarrow y = 0$ is the horizontal asymptote.

   $\lim\limits_{x \to 0^+} f(x) = \infty$ and $\lim\limits_{x \to 0^-} f(x) = -\infty$. Thus $x = 0$ is the vertical asymptote.

4) $f'(x) = \dfrac{-2x(x^3) - (1 - x^2)(3x^2)}{x^6} = \dfrac{x^2 - 3}{x^4} = 0$ when $x = \pm\sqrt{3}$.

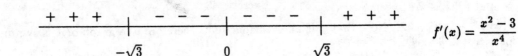

$$f'(x) = \frac{x^2 - 3}{x^4}$$

f is increasing on $\left(-\infty, -\sqrt{3}\right]$ and $\left[\sqrt{3}, \infty\right)$ and decreasing on $\left[-\sqrt{3}, 0\right)$ and $\left(0, \sqrt{3}\right]$. f has a relative minimum at $\left(\sqrt{3}, \dfrac{-2\sqrt{3}}{9}\right)$ and a relative maximum at $\left(-\sqrt{3}, \dfrac{2\sqrt{3}}{9}\right)$.

5) $f''(x) = \dfrac{2x(x^4) - 4x^3(x^2 - 3)}{x^8} = \dfrac{2(6 - x^2)}{x^5} = 0$ at $x = \pm\sqrt{6}$.

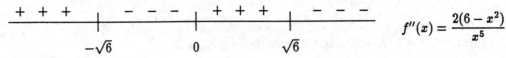

$$f''(x) = \frac{2(6 - x^2)}{x^5}$$

314

f is concave up on $\left(-\infty, -\sqrt{6}\right]$ and $\left(0, \sqrt{6}\right]$ and concave down on $\left[-\sqrt{6}, 0\right)$ and $\left[\sqrt{6}, \infty\right)$. $\left(-\sqrt{6}, \dfrac{5\sqrt{6}}{36}\right)$ and $\left(\sqrt{6}, \dfrac{-5\sqrt{6}}{36}\right)$ are inflection points.

6)

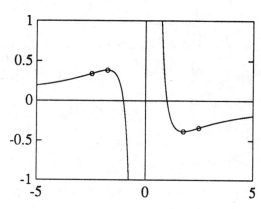

Exercise 45

46. $f(x) = 3x^{5/3} - 15x^{2/3} + 5$

1) Domain: $(-\infty, \infty)$

2) f has no vertical or horizontal asymptotes.

3) $f'(x) = 5x^{2/3} - 10x^{-1/3} = \dfrac{5(x - 2)}{x^{1/3}} = 0$ at $x = 2$. $f'(x)$ is undefined when $x = 0$. Thus the critical numbers are 0 and 2.

$$
\begin{array}{ccccccccc}
+ & + & + & & - & - & - & & + & + & + \\
\hline
& & & 0 & & & & 2 & & &
\end{array}
\qquad f'(x) = \dfrac{5(x - 2)}{x^{1/3}}
$$

f is increasing on $(-\infty, 0]$ and $[2, \infty)$ and decreasing on $[0, 2]$. f has a relative maximum at $(0, 5)$ and a relative minimum at $(2, -9.29)$.

4) $f''(x) = \dfrac{5x^{1/3} - 5(x - 2)\left(\frac{1}{3}\right)x^{-2/3}}{x^{2/3}} = \dfrac{15x - 5x + 10}{3x^{4/3}} = \dfrac{10(x + 1)}{3x^{4/3}} = 0$ at $x = -1$.

$f''(x)$ is undefined at $x = 0$.

$$
\begin{array}{ccccccccc}
- & - & - & & + & + & + & & + & + & + \\
\hline
& & & -1 & & & & 0 & & &
\end{array}
\qquad f''(x) = \dfrac{10(x + 1)}{3x^{4/3}}
$$

f is concave down on $(-\infty, -1]$ and concave up on $[-1, 0]$ and $[0, \infty)$. f has an inflection point at $(-1, -13)$.

5) $f(1) = -7$, $f(4) \approx -2.56$, $f(5) = 5$

6)

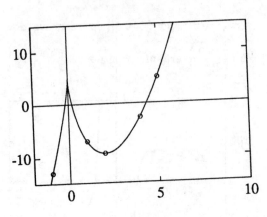

Exercise 46

47. $f(x) = \dfrac{x}{x^2+3}$.   Note that f is an odd function. Therefore, the graph of f is symmetric with respect to the origin.

1) (x^2+3) is always positive. Therefore Domain: $(-\infty, \infty)$.

2) $f(x) = 0$ when $x = 0$.

3) $\lim\limits_{x \to \pm\infty} f(x) = 0 \Rightarrow y = 0$ is the horizontal asymptote. $f(x)$ has no vertical asymptotes.

4) $f'(x) = \dfrac{x^2+3-x(2x)}{(x^2+3)^2} = \dfrac{3-x^2}{(x^2+3)^2} = 0$ when $x = \pm\sqrt{3}$.

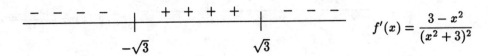

f is decreasing on $\left(-\infty, -\sqrt{3}\right]$ and $\left[\sqrt{3}, \infty\right)$ and increasing on $\left[-\sqrt{3}, \sqrt{3}\right]$.
The point $\left(-\sqrt{3}, \dfrac{-\sqrt{3}}{6}\right)$ is a relative minimum and $\left(\sqrt{3}, \dfrac{\sqrt{3}}{6}\right)$ is a relative maximum.

5) $f''(x) = \dfrac{-2x(x^2+3)^2+(x^2-3)(2)(x^2+3)(2x)}{(x^2+3)^4} = \dfrac{2x(x+3)(x-3)}{(x^2+3)^3} = 0$ when $x = 0, -3, 3$.

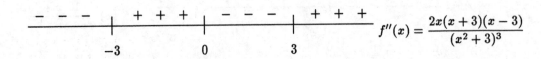

f is concave up on $[-3, 0]$ and $[3, \infty)$ and concave down on $(-\infty, -3]$ and $[0, 3]$.
The points $(0, 0)$, $\left(-3, -\dfrac{1}{4}\right)$ and $\left(3, \dfrac{1}{4}\right)$ are inflection points.

316

6)

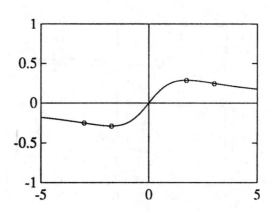

Exercise 47

48. $f(x) = \dfrac{x^2 + 3x}{x^2 + 2x + 1}$

1) $x^2 + 2x + 1 = (x + 1)^2 = 0$ when $x = -1$.
   Therefore Domain: $(-\infty, -1) \cup (-1, \infty)$.

2) $f(x) = 0$ when $x^2 + 3x = x(x + 3) = 0 \Rightarrow x = 0, -3$.

3) $\lim\limits_{x \to \pm\infty} f(x) = 1 \Rightarrow y = 1$ is the horizontal asymptote.
   $\lim\limits_{x \to -1^+} f(x) = \lim\limits_{x \to -1^-} f(x) = -\infty$. Thus $x = -1$ is the vertical asymptote.

4) $f'(x) = \dfrac{(2x + 3)(x^2 + 2x + 1) - (x^2 + 3x)(2x + 2)}{(x^2 + 2x + 1)^2} = \dfrac{-x^2 + 2x + 3}{(x^2 + 2x + 1)^2}$

   $= \dfrac{-(x - 3)(x + 1)}{(x + 1)^4} = \dfrac{-(x - 3)}{(x + 1)^3} = 0$ when $x = 3$

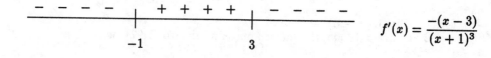

f is decreasing on $(-\infty, -1)$ and $[3, \infty)$ and increasing on $(-1, 3]$. The point $\left(3, \frac{9}{8}\right)$ is an absolute maximum.

5) $f''(x) = \dfrac{(x + 1)^3(-1) - 3(x + 1)^2(3 - x)}{(x + 1)^6} = \dfrac{2(x - 5)}{(x + 1)^4}$

   Therefore, $f''(x) < 0$ on $(-\infty, -1)$ and $(-1, 5)$, and $f''(x) > 0$ on $(5, \infty)$. So the graph of f is concave down on $(-\infty, -1)$ and on $(-1, 5)$, and it is concave up on $(5, \infty)$. The point $\left(5, f(5)\right) = \left(5, \frac{10}{9}\right)$ is an inflection point.

6) To see whether $f(x)$ crosses the horizontal asymptote, let $f(x) = 1 \Rightarrow \dfrac{x^2 + 3x}{x2 + 2x + 1} = 1 \Rightarrow x = 1$.

317

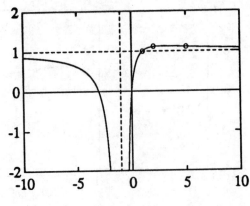

Exercise 48

4.8 Linear Approximation Revisited

In Exercises 1 through 4, we take $f(x) = x^2$ so that $f'(x) = 2x$ and $f''(x) = 2$; consequently, we take $M = 2$ in each case. [Note: in each case, the estimated error is the actual error; this is typical for $f(x) = x^2$, as can be seen from the binomial expansion of $(x_0 + \Delta x)^2$.]

1. Here, with $x_0 = 2$ and $\Delta x = 0.02$, the linear approximation to $(2.02)^2$ is

$$f(2.02) \approx f(2) + f'(2)(0.02) = 4.08$$

with error at most

$$\frac{1}{2}M(\Delta x)^2 = 0.0004.$$

2. With $x_0 = 7$ and $\Delta x = 0.47$, the linear approximation to $(7.47)^2$ is

$$f(7.47) \approx f(7) + f'(7)(0.47) = 55.58$$

with error at most

$$\frac{1}{2}(2)(0.47)^2 = 0.2209$$

3. With $x_0 = 2$ and $\Delta x = -0.03$, the linear approximation to $(1.97)^2$ is

$$f(1.97) \approx f(2) + f'(2)(-0.03) = 3.88$$

with error at most

$$\frac{1}{2}(2)(-0.03)^2 = 0.0009.$$

4. With $x_0 = 7$ and $\Delta x = -0.5$, the linear approximation to $(6.5)^2$ is

$$f(6.5) \approx f(7) + f'(7)(-0.5) = 42$$

with error at most

$$\frac{1}{2}(2)(-0.5)^2 = 0.25.$$

318

5. $f(x) = x^4 \Rightarrow f'(x) = 4x^3 \Rightarrow f''(x) = 12x^2$. The maximum value of f'' on $[2, 2.03]$ is assumed at the right endpoint, so we take $M = 12(2.03)^2 = 49.4508$. With $x_0 = 2$ and $\Delta x = 0.03$, the linear approximation to $(2.03)^4$ is

$$
\begin{aligned}
f(2.03) &\approx f(2) + f'(2)(0.03) \\
&= 16 + (32)(0.03) = 16.96
\end{aligned}
$$

with error at most

$$
\frac{1}{2}(49.4508)(0.03)^2 = 0.02225286.
$$

Remark: It might be more in keeping with the level of calculations in the earlier problems to handle the error a little differently. The maximum value of f'' on $[2, 2.03]$ is surely less than $f''(3) = (12)(9) = 108$, so the error in problem 5 is surely less than $\frac{1}{2}(108)(0.03)^2 = 0.0486$; this error estimate is easier to calculate than the (closer) estimate provided by taking $M = f''(2.03)$.]

6. $f(x) = x^7 \Rightarrow f'(x) = 7x^6 \Rightarrow f''(x) = 42x^5$. The maximum value of f'' on $[1, 1.11]$ is $f''(1.11) = 42(1.11)^5$; rather than calculate this exactly, note that it is less than $42(1.2)^5 = 104.50944$. The linear approximation to $(1.11)^7$ is then (with $x_0 = 1$ and $\Delta x = 0.11$) given by

$$
f(1.11) \approx f(1) + f'(1)(0.11) = 1.77
$$

with error at most

$$
\frac{1}{2}(104.50944)(0.11)^2 = 0.632282112.
$$

[Taking $M = 42(1.11)^5$ shows that the error is less than this; in fact, $(1.11)^7 \approx 2.076 \cdots \cdots$].

7. $f(x) = \sin x \Rightarrow f'(x) = \cos x \Rightarrow f''(x) = -\sin x$ (with x in radians). Using the approximation $|\sin x| \le |x|$ we see that if x lies in $[\frac{\pi}{4}, \frac{23\pi}{90}]$ then $|f''(x)| \le \frac{23\pi}{90}$; here, of course, $\frac{23\pi}{90}$ radians corresponds to $46°$. With $x_0 = \frac{\pi}{4}$ and $\Delta x = \frac{\pi}{180}$, the linear approximation to $\sin 46° = \sin \frac{23\pi}{90}$ is

$$
\begin{aligned}
f\left(\frac{23\pi}{90}\right) &\approx f\left(\frac{\pi}{4}\right) + f'\left(\frac{\pi}{4}\right)\left(\frac{\pi}{180}\right) \\
&= \sin\frac{\pi}{4} + \left(\cos\frac{\pi}{4}\right)\left(\frac{\pi}{180}\right) \\
&\approx 0.71945
\end{aligned}
$$

with error at most

$$
\frac{1}{2}\left(\frac{23\pi}{90}\right)\left(\frac{\pi}{180}\right)^2 \approx 0.00012.
$$

8. $f(x) = \cos x \Rightarrow f'(x) = -\sin x \Rightarrow f''(x) = -\cos x$. With $x_0 = \frac{\pi}{4}$ and $\Delta x = \frac{\pi}{180}$, the linear approximation is $\cos 46° = \cos \frac{23\pi}{90}$ is

$$
\begin{aligned}
f\left(\frac{23\pi}{90}\right) &\approx f\left(\frac{\pi}{4}\right) + f'\left(\frac{\pi}{4}\right)\left(\frac{\pi}{180}\right) \\
&= \cos\frac{\pi}{4} - \left(\sin\frac{\pi}{4}\right)\left(\frac{\pi}{180}\right) \\
&\approx 0.69477.
\end{aligned}
$$

To estimate the error, note that $|f''(x)| = |\cos x| \le 1$ so (taking $M = 1$ for convenience) the error is at most

$$
\frac{1}{2}(1)\left(\frac{\pi}{180}\right)^2 \approx 0.00015.
$$

9. $f(x) = \sin x \Rightarrow f'(x) = \cos x \Rightarrow f''(x) = -\sin x$. Here, $|f''(x)| = |\sin x| \le |x|$ so that $|f''(x)| \le \frac{5\pi}{180} = \frac{\pi}{36} = M$ for x in $\left[0, \frac{5\pi}{180}\right]$. Consequently, if we let $x_0 = 0$ and $\Delta x = \frac{5\pi}{180}$ then the corresponding linear approximation to $\sin 5° = \sin \frac{5\pi}{180}$ is

$$f\left(\frac{\pi}{36}\right) \approx f(0) + f'(0)\frac{\pi}{36}$$
$$= \frac{\pi}{36} \approx 0.08727$$

with error at most

$$\frac{1}{2}M(\Delta x)^2 = \frac{1}{2}\left(\frac{\pi}{36}\right)\left(\frac{\pi}{36}\right)^2 \approx 0.00033.$$

10. $f(x) = \cos x \Rightarrow f'(x) = -\sin x \Rightarrow f''(x) = -\cos x$. Here, $|f''(x)| = |\cos x| \le 1 = M$ for all x. Consequently, if we let $x_0 = \frac{\pi}{2}$ and $\Delta x = -\frac{\pi}{180}$ (in radians; equal to -1 in degrees) then the corresponding linear approximation to $\cos 89°$ is

$$\cos 89° \approx f\left(\frac{\pi}{2}\right) + f'\left(\frac{\pi}{2}\right)\left(-\frac{\pi}{180}\right)$$
$$= \frac{\pi}{180} \approx 0.01745$$

with error at most

$$\frac{1}{2}M(\Delta x)^2 = \frac{1}{2}(1)\left(-\frac{\pi}{180}\right)^2$$
$$\approx 0.00015.$$

11. Let $f(x) = x^{1/2}$, so $f'(x) = \frac{1}{2}x^{-1/2}$ and $f''(x) = -\frac{1}{4}x^{-3/2}$. With $x_0 = 36$ and $\Delta x = 1$, the linear approximation to $\sqrt{37} = f(37)$ is

$$f(37) \approx f(36) + f'(36)(1)$$
$$= 6\frac{1}{12} = 6.0833.$$

To estimate the error, note this $|f''(x)| = -|\frac{1}{4}x^{-3/2}|$ is decreasing for $x \in [36, 37]$; thus, we may take $M = |f''(36)| = \frac{1}{864}$ and find that the error is at most

$$\frac{1}{2}M(\Delta x)^2 \approx 0.000579.$$

12. $f(x) = \frac{1}{x} \Rightarrow f'(x) = -\frac{1}{x^2} \Rightarrow f''(x) = \frac{2}{x^3}$. With $x_0 = 2$ and $\Delta x = 0.1$, the linear approximation to $\frac{1}{2.1}$ is

$$f(2.1) \approx f(2) + f'(2)(0.1)$$
$$= 0.475.$$

To estimate the error, note that $f''(x)$ decreases for $x \in (0, \infty)$; thus we may take $M = f''(2) = 0.25$ and find that the error is at most

$$\frac{1}{2}(0.25)(0.1)^2 = 0.00125.$$

320

13. Let $f(x) = \sqrt{x}$ so that $f'(x) = \frac{1}{2\sqrt{x}}$ and $f''(x) = -\frac{1}{4\sqrt{x^3}}$. With $x_0 = 49$ we get

$$f(50) \approx f(49) + f'(49)(1) = 7\frac{1}{14} \approx 7.07143$$

and with $x_0 = 50.41 (= (7.1)^2)$ we get

$$\begin{aligned} f(50) &\approx 7.1 + f'(50.41)(-0.41) \\ &\approx 7.07113. \end{aligned}$$

Since $|f''(x)| = \frac{1}{4\sqrt{x^3}}$ is decreasing on $[49, 50]$ we estimate the error (according to **Theorem 11**) in the second approximation as being at most

$$\begin{aligned} \frac{1}{2} |f''(49)| (-0.41)^2 &= \frac{1}{2} \left(\frac{1}{4 \times 343} \right) (-0.41)^2 \\ &= 0.00006; \end{aligned}$$

this makes it clear that 7.1 is a better approximation to $\sqrt{50}$ than is 7. [Alternatively, note that

$$\left(\sqrt{50} - 7 \right) \left(\sqrt{50} + 7 \right) = 50 - 49 = 1$$

so

$$\sqrt{50} - 7 = \frac{1}{\sqrt{50} + 7}$$

whereas

$$\left(7.1 - \sqrt{50} \right) \left(7.1 + \sqrt{50} \right) = 50.41 - 50 = 0.41$$

so

$$7.1 - \sqrt{50} = \frac{0.41}{\sqrt{50} + 7.1}.$$

Since $0.41 < 1$ and $\sqrt{50} + 7.1 > \sqrt{50} + 7$ it follows that $7.1 - \sqrt{50} < \sqrt{50} - 7$.]

14. Let $f(x) = \sqrt{x}$ so that $f'(x) = \frac{1}{2\sqrt{x}}$ and $f''(x) = -\frac{1}{4\sqrt{x^3}}$. With $x_0 = 1$ we get the approximation

$$f(2.5) \approx f(1) + f'(1)(1.5) = 1.75$$

and with $x_0 = 4$ we get

$$f(25) \approx f(4) + f'(4)(-1.5) = 1.625.$$

In each case $|\Delta x| = 1.5$, so the upper bound on the error made (according to **Theorem 11**) will be least when $x_0 = 4$ since $|f''(x)| = \frac{1}{4\sqrt{x^3}}$ is decreasing on $[1, 4]$. [However, note that this refers to the bound on the error rather than to the error itself.]

15. $f(x) = \sin x \Rightarrow f''(x) = -\sin x$. Since $|\sin x| \le 1$ we shall take $M = 1$ in **Theorem 11**; see the naïve estimation in **Example 3** of the text. According to **Theorem 11**, the error made in the linear approximation

$$f(x) \approx f(0) + f'(0)x = x$$

is at most

$$\frac{1}{2} M x^2 \;=\; \frac{1}{2}\, x^2.$$

If we wish to make this error at most 0.005 (see **Remark 2** of the text) then we take $x^2 \leq 0.01$ so that $|x| \leq 0.1$. An interval as required is therefore $[-0.1, 0.1]$. [Needless to say, this is not the largest such interval.]

16. Take $f(x) = \sqrt{x}$ so that $f'(x) = \frac{1}{2\sqrt{x}}$ and $f''(x) = -\frac{1}{4\sqrt{x^3}}$. With $x_0 = 1$ we obtain the linear approximation

$$f(1 + \Delta x) \;\approx\; f(1) + f'(1)\Delta x$$

or

$$\sqrt{1 + \Delta x} \;\approx\; 1 + \frac{1}{2}\Delta x.$$

To estimate the error, note that $|f''(x)|$ is a decreasing function of positive x. Had the exercise asked for an interval with 1 as its left endpoint then we could take $M = \left|f''(1)\right| = \frac{1}{4}$. As it stands, the left endpoint is undetermined. To make matters simple we shall look for an interval contained in $\left[\frac{1}{4}, \frac{7}{4}\right]$ since we can easily compute $\left|f''\left(\frac{1}{4}\right)\right| = 2$, which we take to be M. Now according to **Theorem 11** the error in $\sqrt{1 + \Delta x} \approx 1 + \frac{1}{2}\Delta x$ is at most $\frac{1}{2}M\Delta x^2 = (\Delta x)^2$. If this is (see **Remark 2**) to be at most 0.05 then we take $\left|\Delta x\right| \leq \sqrt{0.05}$. We cannot compute this square-root exactly, but surely $\left|\Delta x\right| \leq \sqrt{0.04}$ also works, and $\sqrt{0.04} = 0.2$. Thus: on the interval $\left[0.8, 1.2\right]$ the linear approximation $\sqrt{1 + \Delta x} \approx 1 + \frac{1}{2}\Delta x$ is accurate to one decimal place. [Again, this interval could be enlarged by better estimates on M].

17. Graph with the range: $[-3, 2] \times [-3, 6]$.

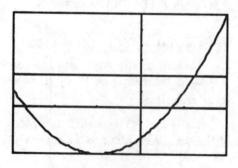

Exercise 17

Since $f''(x) = 2$, the linear approximation has the same accuracy at all points.

18. Graph with the range: $[-2, 2] \times [-8, 8]$.

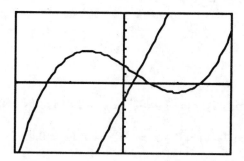

Exercise 18

Since $f''(x) = 12x - 2$, the linear approximation is most accurate at $x = \frac{1}{6}$ where $f''(x) = 0$. It is less accurate as $|x|$ increases.

19. Graph with the range: $[0, \pi] \times [-1, 2.5]$.

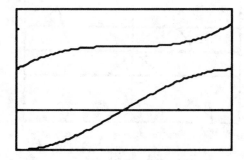

Exercise 19

Since $f''(x) = -\cos x$, the linear approximation is most accurate at $x = \frac{\pi}{2}$ where $f''(x) = 0$ and least accurate at $x = 0$ and $x = \pi$ where $|f''(x)|$ is the largest.

20. Graph with the range: $[-1, 3] \times [-10, 35]$.

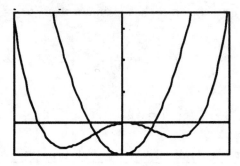

Exercise 20

Since $f''(x) = 12x^2 - 10$, the linear approximation is most accurate at $x = \pm\sqrt{5/6}$ where $f''(x) = 0$ and least accurate at $x = \pm 3$ where $|f''(x)|$ is largest.

21. $f'(x) = \dfrac{3x^2}{8} - 1$

   a) $f(-2) = 1$ and $f'(-2) = \frac{1}{2}$ so $T_1(x) = 1 + \frac{x+2}{2} = \frac{x}{2} + 2$. Take $h_1 = 0.11$.

   b) $f(0) = 0$ and $f'(0) = -1$ so $T_2(x) = 0 - x = -x$. Take $h_2 = 0.42$.

   c) $f''(x) = \frac{3x}{4}$ so $|f''(-2)| = \frac{3}{2}$ and $|f''(0)| = 0$. The error terms $|f(x) - T_1(x)|$ and $|f(x) - T_2(x)|$ depend on the size of $f''(x)$ near $x = -2$ and $x = 0$. This explains why the value for h_1 in part a is larger than the value in part b.

   d) Graph with the range: $[-5, 5] \times [-3, 3]$.

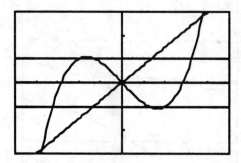

Exercise 21. d)

Graph $f(x)$, $f''(x)$, and $y = \pm 1$. From the graph we see that $|f''(x)| \le 1$ for $-\frac{4}{3} \le x \le \frac{4}{3}$. For any point, a, in this interval we have

$$|f(x) - T(x)| = \frac{|f''(z_x)|}{2} |x - a|^2$$

where $T(x) = f(a) + f'(a)(x - a)$. Since the maximum of $|f''(x)| = 1$ on the interval $-\frac{4}{3} \le x \le \frac{4}{3}$ then

$$|f(x) - T(x)| \le \frac{1}{2} |x - a|^2.$$

22. Graph with the range: $[-3, 3] \times [-18, 18]$, $h = 0.01$.

$$\begin{aligned} S_h(x) &= \frac{-(x+h)^3 + 4(x+h) - 2(-x^3 + 4x) + (x-h)^3 - 4(x-h)^3}{h^2} \\ &= -6x. \end{aligned}$$

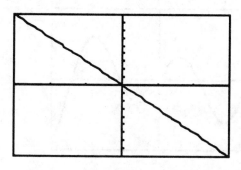

Exercise 22

23. Graph with the range: $[-\pi, \pi] \times [-4, 4]$, $h = 0.01$.

$$S_h(x) = \frac{\sin(2(x+h)) - 2\sin 2x + \sin(2(x-h))}{h^2}$$
$$= \frac{8\sin x \cos x(\cos^2 h - 1)}{h^2}$$

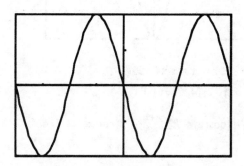

Exercise 23

24. Graph with the range: $[-2\pi, 2\pi] \times [-7, 7]$, $h = 0.01$.

$$S_h(x) = \frac{2x\cos x(\cos h - 1) - 2h\sin x \sin h}{h^2}.$$

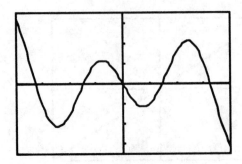

Exercise 24

25. Graph with the range: $[-2, 2] \times [-2, 1]$, $h = 0.01$.

$$S_h(x) = \frac{1}{h^2} \left[\frac{1}{1 + (x+h)^2} - \frac{2}{1 + x^2} + \frac{1}{1 + (x-h)^2} \right].$$

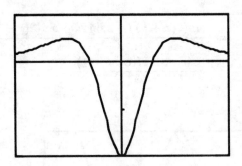

Exercise 25

26. Graph with the range: $\left[-\frac{\pi}{4}, \frac{\pi}{2}\right] \times [-2, 5]$, $h = 0.01$.

$$S_h(x) = \frac{1}{h^2} \left[\left(\cos(x+h)\right)^{x+h} - 2(\cos x)^x + \left(\cos(x-h)\right)^{x-h} \right].$$

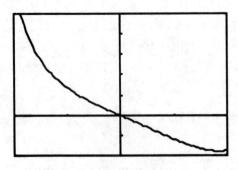

Exercise 26

326

Review Exercises—Chapter 4

1. $f(x) = 4x - x^2$

   1) $f'(x) = 4 - 2x = 0$ at $x = 2$.

   $$\begin{array}{ccc} + \ + \ + & | & - \ - \ - \\ & 2 & \end{array} \qquad f'(x) = 4 - 2x$$

   f is increasing on $(-\infty, 2]$ and decreasing on $[2, \infty)$. f has an absolute maximum at $(2, 4)$.

   2) $f''(x) = -2 < 0$. Thus f is concave down on $(-\infty, \infty)$. f has no inflection points.

   3) $f(x) = 0$ at $x = 0, 4$.

   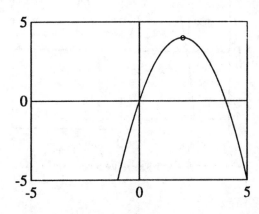

   Exercise 1

2. $f(x) = x(x-1)(x+3)$

   1) $f'(x) = (x-1)(x+3) + x(x+3) + x(x-1) = 3x^2 + 4x - 3 = 0$

   $$\text{when } x = \frac{-2 \pm \sqrt{13}}{3} \approx .535 \text{ and } -1.869.$$

   $$\begin{array}{ccccc} + \ + \ + & | & - \ - \ - & | & + \ + \ + \\ & \frac{-2-\sqrt{13}}{3} & & \frac{-2+\sqrt{13}}{3} & \end{array} \qquad f'(x) = 3x^2 + 4x - 3$$

   f is increasing on $\left(-\infty, \dfrac{-2-\sqrt{13}}{3}\right]$ and on $\left[\dfrac{-2+\sqrt{13}}{3}, \infty\right)$ and

   $$\text{decreasing on } \left[\frac{-2-\sqrt{13}}{3}, \frac{-2+\sqrt{13}}{3}\right].$$

   f has a relative maximum at $\left(\dfrac{-2-\sqrt{13}}{3}, 6.06\right)$ & a relative minimum at $\left(\dfrac{-2+\sqrt{13}}{3}, -.879\right)$.

2) $f''(x) = 6x + 4 = 2(3x + 2) = 0$ at $x = -\frac{2}{3}$.

$$\overset{\textstyle - \ - \ - \quad\mid\quad + \ + \ +}{\underset{\textstyle -\frac{2}{3}}{\rule{7cm}{0.4pt}}} \qquad f''(x) = 2(3x + 2)$$

f is concave up on $\left[-\frac{2}{3}, \infty\right)$ and concave down on $\left(-\infty, -\frac{2}{3}\right]$. f has an inflection point at $\left(-\frac{2}{3}, \frac{70}{27}\right)$.

3) $f(x) = 0$ at $x = 0, 1, -3$.

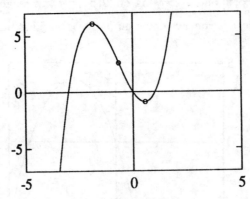

Exercise 2

3. $f(x) = x^2 - 2x + 3$

1) $f'(x) = 2x - 2 = 0$ at $x = 1$.

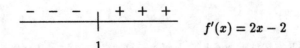

f is decreasing on $(-\infty, 1]$ and increasing on $[1, \infty)$. f has an absolute minimum at $(1, 2)$.

2) $f''(x) = 2 > 0$. Thus f is concave up on $(-\infty, \infty)$. f has no inflection points.

3) $f(0) = 3$

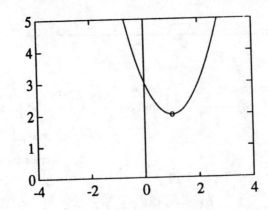

Exercise 3

4. $f(x) = \sin 4x$. Since $\sin 4x$ has period $\frac{\pi}{2}$, we will analyze $f(x)$ on $\left[0, \frac{\pi}{2}\right]$ only.

   1) $f'(x) = 4\cos 4x = 0$ when $\cos 4x = 0 \Rightarrow 4x = \frac{\pi}{2}, \frac{3\pi}{2} \Rightarrow x = \frac{\pi}{8}, \frac{3\pi}{8}$.

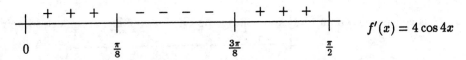

   f is increasing on $\left[0, \frac{\pi}{8}\right]$ and $\left[\frac{3\pi}{8}, \frac{\pi}{2}\right]$ and decreasing on $\left[\frac{\pi}{8}, \frac{3\pi}{8}\right]$. f has an absolute maximum at $\left(\frac{\pi}{8}, 1\right)$ and an absolute minimum at $\left(\frac{3\pi}{8}, -1\right)$.

   2) $f''(x) = -16\sin 4x = 0$ when $\sin 4x = 0 \Rightarrow 4x = \pi \Rightarrow x = \frac{\pi}{4}$.

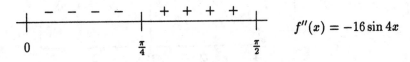

   f is concave down on $\left[0, \frac{\pi}{4}\right]$ aand concave up on $\left[\frac{\pi}{4}, \frac{\pi}{2}\right]$. f has an inflection point at $\left(\frac{\pi}{4}, 0\right)$.

   3) $f(0) = f\left(\frac{\pi}{2}\right) = 0$

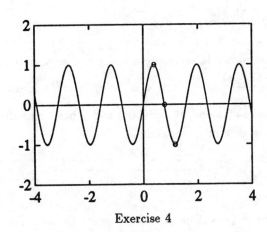

Exercise 4

5. $f(x) = 2\sin(\pi x)$. Since $\sin \pi x$ has period 2, we will need to analyze $f(x)$ on $[0, 2]$ only.

   1) $f(0) = f(2) = 0$

   2) $f'(x) = 2\pi \cos(\pi x) = 0$ when $\cos(\pi x) = 0 \Rightarrow \pi x = \frac{\pi}{2}, \frac{3\pi}{2} \Rightarrow x = \frac{1}{2}, \frac{3}{2}$.

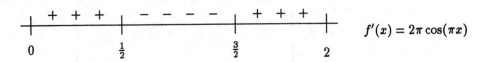

   f is increasing on $\left[0, \frac{1}{2}\right]$ and $\left[\frac{3}{2}, 2\right]$ and decreasing on $\left[\frac{1}{2}, \frac{3}{2}\right]$. f has an absolute maximum at $\left(\frac{1}{2}, 2\right)$ and an absolute minimum at $\left(\frac{3}{2}, -2\right)$.

3) $f''(x) = -2\pi^2 \sin \pi x = 0$ when $\sin \pi x = 0 \Rightarrow \pi x = \pi \Rightarrow x = 1$.

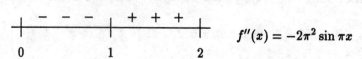

$f''(x) = -2\pi^2 \sin \pi x$

f is concave down on $[0, 1]$ and concave up on $[1, 2]$. f has an inflection point at $(1, 0)$.

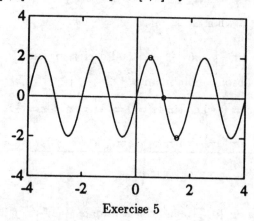

Exercise 5

6. $f(x) = \dfrac{1 - x}{x}$

1) Domain: $(-\infty, 0) \cup (0, \infty)$

2) $f(x) = 0$ when $1 - x = 0 \Rightarrow x = 1$.

3) f has $y = -1$ as a horizontal asymptote and $x = 0$ as a vertical asymptote.

4) $f'(x) = \dfrac{-1(x) - (1 - x)}{x^2} = \dfrac{-1}{x^2} < 0$ for all x in the x^2 domain.

Thus f is decreasing on $(-\infty, 0) \cup (0, \infty)$ and f has no relative extrema.

5) $f''(x) = \dfrac{2}{x^3} \neq 0$ for all x in the domain.

$f''(x) = \dfrac{2}{x^3}$

f is concave down on $(-\infty, 0)$ and concave up on $(0, \infty)$. f has no inflection points.

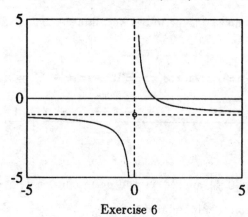

Exercise 6

330

7. $f(t) = 2\cos^2(2t) = 2\left[\frac{1}{2}(1 + \cos 4t)\right] = 1 + \cos 4t$. Since $\cos 4t$ has period $\frac{\pi}{2}$, we'll need to analyze $f(t)$ on $\left[0, \frac{\pi}{2}\right]$ only.

1) $f(t) = 0$ when $\cos 4t = -1 \Rightarrow 4t = \pi \Rightarrow t = \frac{\pi}{4}$.

2) $f'(t) = -4\sin 4t = 0$ when $\sin 4t = 0 \Rightarrow 4t = \pi \Rightarrow t = \frac{\pi}{4}$.

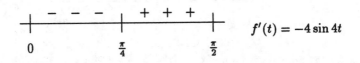

f is decreasing on $\left[0, \frac{\pi}{4}\right]$ and increasing on $\left[\frac{\pi}{4}, \frac{\pi}{2}\right]$. f has an absolute minimum at $\left(\frac{\pi}{4}, 0\right)$. f has absolute maxima at $(0, 2)$ and $\left(\frac{\pi}{2}, 2\right)$.

3) $f''(t) = -16\cos 4t = 0$ when $\cos 4t = 0 \Rightarrow 4t = \frac{\pi}{2}, \frac{3\pi}{2} \Rightarrow t = \frac{\pi}{8}, \frac{3\pi}{8}$.

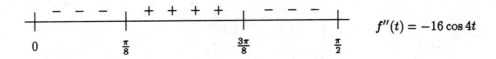

f is concave down on $\left[0, \frac{\pi}{8}\right]$ and $\left[\frac{3\pi}{8}, \frac{\pi}{2}\right]$ and concave up on $\left[\frac{\pi}{8}, \frac{3\pi}{8}\right]$. f has points of inflection at $\left(\frac{\pi}{8}, 1\right)$ and $\left(\frac{3\pi}{8}, 1\right)$.

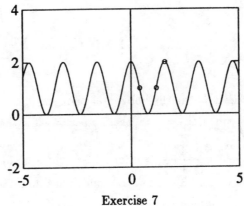

Exercise 7

8. $f(x) = \sqrt{1 - \sin^2 x} = \sqrt{\cos^2 x} = |\cos x|$

Since $\cos x$ has period 2π, we'll limit our analysis of $f(x)$ on $\left[-\frac{\pi}{2}, \frac{3\pi}{2}\right]$.

1) See problem 46 of Section 4.4 for the analysis of $f'(x)$.

2) $f''(x) = \begin{cases} -\cos x & \text{if } -\dfrac{\pi}{2} < x < \dfrac{\pi}{2} \\[2mm] \cos x & \text{if } \dfrac{\pi}{2} < x < \dfrac{3\pi}{2} \\[2mm] \text{does not exist} & \text{if } x = -\dfrac{\pi}{2},\ \dfrac{\pi}{2},\ \dfrac{3\pi}{2} \end{cases}$

Note that $f''(x) \neq 0$ for $x \in \left[-\frac{\pi}{2}, \frac{3\pi}{2}\right]$.

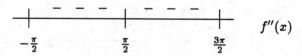

Thus f is concave down on $\left[-\frac{\pi}{2}, \frac{\pi}{2}\right]$ and $\left[\frac{\pi}{2}, \frac{3\pi}{2}\right]$. f has no inflection points.

3) $f\left(-\frac{\pi}{2}\right) = f\left(\frac{\pi}{2}\right) = f\left(\frac{3\pi}{2}\right) = 0$

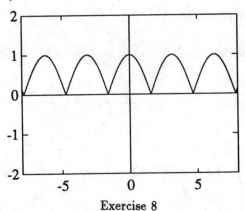

Exercise 8

9. $f(x) = \dfrac{x-3}{x+3}$

1) Domain: $(-\infty, -3) \cup (-3, \infty)$

2) $f(x) = 0$ when $x = 3$.

3) $\displaystyle\lim_{x \to \pm\infty} f(x) = 1$, therefore $y = 1$ is the horizontal asymptote.

$\displaystyle\lim_{x \to -3^-} f(x) = \infty$ and $\displaystyle\lim_{x \to -3^+} f(x) = -\infty$, therefore $x = -3$ is the vertical asymptote.

4) $f'(x) = \dfrac{(x+3) - (x-3)}{(x+3)^2} = \dfrac{6}{(x+3)^2} > 0$ for all x in the domain.

Thus f is increasing on $(-\infty, -3) \cup (-3, \infty)$ and f has no relative extrema.

5) $f''(x) = \dfrac{-12}{(x+3)^3} \neq 0$ for all x in the domain.

$$+\ +\ +\ \ |\ \ -\ -\ -\ - \qquad f''(x) = \dfrac{-12}{(x+3)^3}$$
$$-3$$

f is concave up on $(-\infty, -3)$ and concave down on $(-3, \infty)$. f has no inflection points.

6) $f(0) = -1$

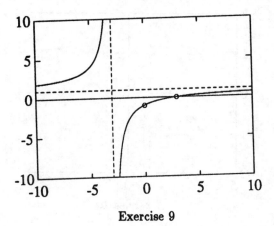

Exercise 9

10. $f(t) = \dfrac{t^2}{t^2 - 1}$

1) Domain: $(-\infty, -1) \cup (-1, 1) \cup (1, \infty)$

2) $f(t) = 0$ when $t = 0$.

3) $\lim\limits_{t \to \pm\infty} f(t) = 1$; therefore $y = 1$ is the horizontal asymptote.

   $\lim\limits_{t \to -1^-} f(t) = \infty$ and $\lim\limits_{t \to -1^+} f(t) = -\infty$; therefore $x = -1$ is a vertical asymptote.

   $\lim\limits_{t \to 1^-} f(t) = -\infty$ and $\lim\limits_{t \to 1^+} f(t) = \infty$; therefore $x = 1$ is a vertical asymptote.

4) $f'(t) = \dfrac{2t(t^2 - 1) - t^2(2t)}{(t^2 - 1)^2} = \dfrac{-2t}{(t^2 - 1)^2} = 0$ at $t = 0$.

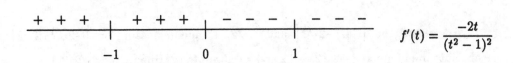

$f'(t) = \dfrac{-2t}{(t^2 - 1)^2}$

f is increasing on $(-\infty, -1)$ and $(-1, 0]$ and decreasing on $[0, 1)$ and $(1, \infty)$. f has a relative maximum at $(0, 0)$.

5) $f''(t) = \dfrac{-2(t^2 - 1)^2 + 2t\,(2)\,(t^2 - 1)(2t)}{(t^2 - 1)^4} = \dfrac{6t^2 + 2}{(t^2 - 1)^3} \neq 0$ for all x in the domain.

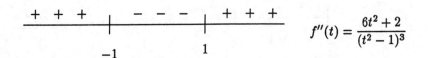

$f''(t) = \dfrac{6t^2 + 2}{(t^2 - 1)^3}$

333

f is concave up on $(-\infty, -1)$ and $(1, \infty)$ and concave down on $(-1, 1)$. f has no inflection points.

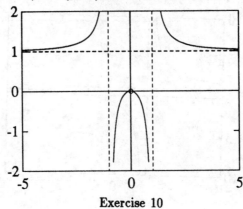

Exercise 10

11. $f(x) = x\sqrt{16 - x^2}$

1) Domain: $[-4, 4]$

2) $f(x) = 0$ when $x = 0, \pm 4$.

3) $f'(x) = (16 - x^2)^{1/2} + x\left(\frac{1}{2}\right)(16 - x^2)^{-1/2}(-2x) = \dfrac{2(8 - x^2)}{(16 - x^2)^{1/2}} = 0$ when $x = \pm 2\sqrt{2}$.

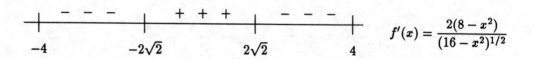

$f'(x) = \dfrac{2(8 - x^2)}{(16 - x^2)^{1/2}}$

f is decreasing on $\left[-4, -2\sqrt{2}\right]$ and $\left[2\sqrt{2}, 4\right]$ and increasing on $\left[-2\sqrt{2}, 2\sqrt{2}\right]$. f has an absolute minimum at $\left(-2\sqrt{2}, -8\right)$ and an absolute maximum at $\left(2\sqrt{2}, 8\right)$.

4) $f''(x) = \dfrac{-4x(16 - x^2)^{1/2} - 2(8 - x^2)\left(\frac{1}{2}\right)(16 - x^2)^{-1/2}(-2x)}{(16 - x^2)} = \dfrac{2x(x^2 - 24)}{(16 - x^2)^{3/2}} = 0$ when $x = 0$.

Note that $x^2 - 24 \neq 0$ on $[-4, 4]$.

$f''(x) = \dfrac{2x(x^2 - 24)}{(16 - x^2)^{3/2}}$

334

f is concave up on $[-4, 0]$ and concave down on $[0, 4]$. f has an inflection point at $(0, 0)$.

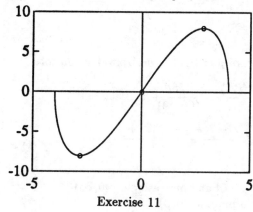

Exercise 11

12. $f(x) = \dfrac{\sqrt{x}}{1 + \sqrt{x}}$

1) Domain: $[0, \infty)$

2) $f(x) = 0$ when $x = 0$.

3) $\lim\limits_{x \to \infty} f(x) = 1$; therefore $y = 1$ is the horizontal asymptote.

4) $f'(x) = \dfrac{\frac{1}{2}x^{-1/2}(1 + \sqrt{x}) - (\sqrt{x}\,\frac{1}{2}\,x^{-1/2})}{(1 + \sqrt{x})^2} = \dfrac{1}{2\sqrt{x}(1 + \sqrt{x})^2} \neq 0$ for all x in the domain.

$f'(x) > 0$ on $(0, \infty)$. Thus f is increasing on $[0, \infty)$. f has no relative extrema.

5) $f''(x) = \dfrac{-\left[2\left(\frac{1}{2}\right)x^{-1/2}\left(1 + \sqrt{x}\right)^2 + 2\sqrt{x}\,(2)\left(1 + \sqrt{x}\right)\left(\frac{1}{2}\right)x^{-1/2}\right]}{4x\left(1 + \sqrt{x}\right)^4}$

$$= \dfrac{-\left(1 + 3\sqrt{x}\right)}{4x^{3/2}\left(1 + \sqrt{x}\right)^3} < 0 \text{ on } (0, \infty).$$

Thus f is concave down on $[0, \infty)$. f has no inflection points.

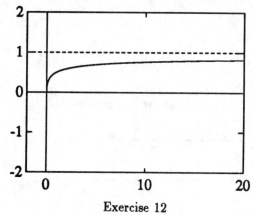

Exercise 12

335

13. $f(t) = \dfrac{t^2}{t^2 + 9}$

1) Domain: $(-\infty, \infty)$

2) $f(t) = 0$ when $t = 0$.

3) $\displaystyle\lim_{t \to \pm\infty} f(t) = 1$; therefore $y = 1$ is the horizontal asymptote.

4) $f'(t) = \dfrac{2t(t^2 + 9) - t^2(2t)}{(t^2 + 9)^2} = \dfrac{18t}{(t^2 + 9)^2} = 0$ at $t = 0$.

$$\underline{\quad -\ -\ -\ \underset{0}{|}\ +\ +\ +\quad} \qquad f'(t) = \dfrac{18t}{(t^2 + 9)^2}$$

f is decreasing on $(-\infty, 0]$ and increasing on $[0, \infty)$.
f has an absolute minimum at $(0, 0)$.

5) $f''(t) = \dfrac{18(t^2 + 9)^2 - (18t)(2)(t^2 + 9)(2t)}{(t^2 + 9)^4} = \dfrac{54(3 - t^2)}{(t^2 + 9)^3} = 0$ when $t = \pm\sqrt{3}$.

$$\underline{\quad -\ -\ -\ \underset{-\sqrt{3}}{|}\ +\ +\ +\ \underset{\sqrt{3}}{|}\ -\ -\ -\quad} \qquad f''(t) = \dfrac{54(3 - t^2)}{(t^2 + 9)^3}$$

f is concave down on $(-\infty, -\sqrt{3}]$ and $[\sqrt{3}, \infty)$ and concave up on $[-\sqrt{3}, \sqrt{3}]$. f has points of inflection at $\left(\pm\sqrt{3}, \frac{1}{4}\right)$.

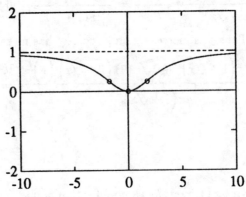

Exercise 13

14. $f(x) = \dfrac{2}{\sqrt{x}} + \dfrac{\sqrt{x}}{2} = \dfrac{4 + x}{2\sqrt{x}}$

1) Domain: $(0, \infty)$

2) $f(x) \neq 0$ for all x in $(0, \infty)$.

3) $\displaystyle\lim_{x \to 0^+} f(x) = \infty$; therefore $x = 0$ is the vertical asymptote.

4) $f'(x) = \dfrac{2\sqrt{x} - (4 + x)(2)(\frac{1}{2})x^{-1/2}}{4x} = \dfrac{x - 4}{4x^{3/2}} = 0$ at $x = 4$.

$$\underline{\quad \underset{0}{|}\ -\ -\ -\ \underset{4}{|}\ +\ +\ +\quad} \qquad f'(x) = \dfrac{x - 4}{4x^{3/2}}$$

f is decreasing on $(0, 4]$ and increasing on $[4, \infty)$. f has an absolute minimum at $(4, 2)$.

5) $f''(x) = \dfrac{4x^{3/2} - (x - 4)(4)\left(\frac{3}{2}\right)x^{1/2}}{16x^3} = \dfrac{12 - x}{8x^{5/2}} = 0$ when $x = 12$.

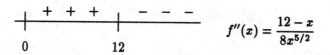

$$f''(x) = \dfrac{12 - x}{8x^{5/2}}$$

f is concave up on $(0, 12]$ and concave down on $[12, \infty)$. f has an inflection point at $\left(12, \frac{4\sqrt{3}}{3}\right)$.

6) Note that $y = \frac{1}{2}\sqrt{x}$ is an oblique asymptote of f.

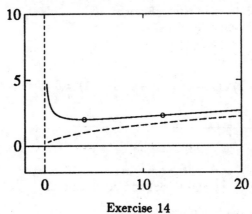

Exercise 14

15. $f(x) = x^4 - 2x^2$

1) $f(x) = x^2(x^2 - 2) = 0$ when $x = 0, \pm\sqrt{2}$.

2) $f'(x) = 4x^3 - 4x = 4x(x^2 - 1) = 0$ when $x = 0, \pm 1$.

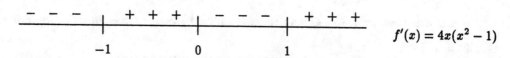

$$f'(x) = 4x(x^2 - 1)$$

$f(x)$ is decreasing on $(-\infty, -1]$ and $[0, 1]$ and increasing on $[-1, 0]$ and $[1, \infty)$. f has a relative maximum at $(0, 0)$ and absolute minima at $(\pm 1, -1)$.

3) $f''(x) = 12x^2 - 4 = 4(3x^2 - 1) = 0$ at $x = \pm\frac{\sqrt{3}}{3}$.

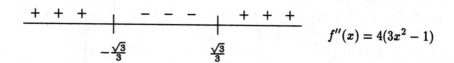

$$f''(x) = 4(3x^2 - 1)$$

f is concave up on $\left(-\infty, -\frac{\sqrt{3}}{3}\right]$ and on $\left[\frac{\sqrt{3}}{3}, \infty\right)$ and concave down on $\left[-\frac{\sqrt{3}}{3}, \frac{\sqrt{3}}{3}\right]$. f has inflection points at $\left(\pm\frac{\sqrt{3}}{3}, -\frac{5}{9}\right)$.

337

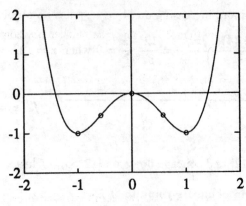

Exercise 15

16. $f(x) = (x+1)(x-1)^2$

    1) $f(x) = 0$ when $x = \pm 1$.

    2) $f'(x) = (x-1)^2 + (x+1)\,2\,(x-1) = (x-1)(3x+1) = 0$ when $x = 1, -\frac{1}{3}$

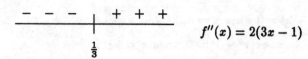

    f is increasing on $\left(-\infty, -\frac{1}{3}\right]$ and $[1, \infty)$ and decreasing on $\left[-\frac{1}{3}, 1\right]$. f has a relative maximum at $\left(-\frac{1}{3}, \frac{32}{27}\right)$ and a relative minimum at $(1, 0)$.

    3) $f''(x) = (3x+1) + 3(x-1) = 6x - 2 = 2(3x-1) = 0$ when $x = \frac{1}{3}$.

$$\begin{array}{ccc} - \ - \ - & + \ + \ + \\ \hline & \frac{1}{3} \end{array} \qquad f''(x) = 2(3x-1)$$

    f is concave down on $\left(-\infty, \frac{1}{3}\right]$ and concave up on $\left[\frac{1}{3}, \infty\right)$. f has an inflection point at $\left(\frac{1}{3}, \frac{16}{27}\right)$.

    4) $f(0) = 1$

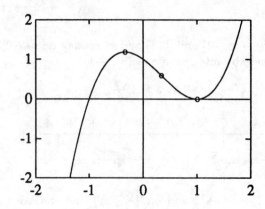

Exercise 16

338

17. $f(x) = x - \cos x$

This function is defined on the whole real line and has neither horizontal nor vertical asymptotes.

1) $f'(x) = 1 + \sin x$

$f'(x) = 0 \Rightarrow \sin x = -1 \Rightarrow x = (4n+3)\frac{\pi}{2}$ (any integer n) $= \cdots\cdots, -\frac{\pi}{2}, \frac{3\pi}{2}, \frac{7\pi}{2}, \cdots\cdots$.

$f'(x) > 0$ for all other values of x.

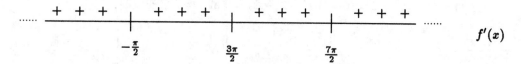

So f is increasing on $(-\infty, \infty)$ with critical numbers $x = (4n+3)\frac{\pi}{2}$ and no relative extrema.

2) $f''(x) = \cos x$

This being a familiar function, we see:

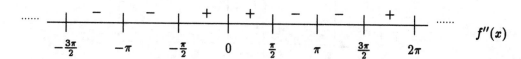

So f is concave up on the intervals $\left((4n-1)\frac{\pi}{2}, (4n+1)\frac{\pi}{2}\right)$ and concave down on the intervals $\left((4n+1)\frac{\pi}{2}, (4n+3)\frac{\pi}{2}\right)$ and has inflection points where x is any odd multiple of $\frac{\pi}{2}$.

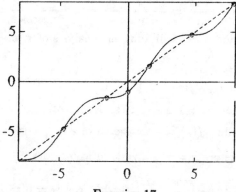

Exercise 17

(Dashed line helps to draw graph: the graph of $f(x)$ crosses $y = x$ when x is an odd multiple of $\frac{\pi}{2}$ since then $\cos x = 0$).

18. $f(t) = t^3 - 3t^2 + 2$

1) $f'(t) = 3t^2 - 6t = 3t(t-2) = 0$ when $t = 0, 2$.

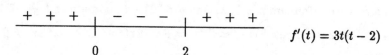

f is increasing on $(-\infty, 0]$ and $[2, \infty)$ and decreasing on $[0, 2]$. f has a relative maximum at $(0, 2)$ and a relative minimum at $(2, -2)$.

2) $f''(t) = 6t - 6 = 6(t - 1) = 0$ at $t = 1$.

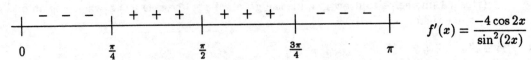

$$f''(t) = 6(t - 1)$$

f is concave down on $(-\infty, 1]$ and concave up on $[1, \infty)$. f has an inflection point at $(1, 0)$.

3) $f(t) = t^3 - 3t^2 + 2 = (t - 1)(t^2 - 2t - 2) = 0$ at $t = 1$, $1 \pm \sqrt{3}$.

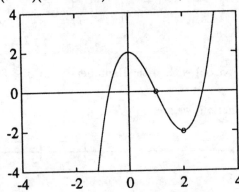

Exercise 18

19. $f(x) = x^2 + \frac{2}{x}$. (See problem (17) in Section 4.7.)

20. $f(x) = \tan x + \cot x = \dfrac{\sin x}{\cos x} + \dfrac{\cos x}{\sin x} = \dfrac{\sin^2 x + \cos^2 x}{\sin x \cos x} = \dfrac{1}{\sin x \cos x} = \dfrac{2}{\sin 2x} = 2 \csc 2x.$

Note: Since $\csc 2x$ has period π, we will limit our analysis of f on $[0, \pi]$.

1) Domain: $\left(0, \frac{\pi}{2}\right) \cup \left(\frac{\pi}{2}, \pi\right)$.

2) $f(x) \neq 0$ for all x in the domain.

3) $\lim\limits_{x \to 0^+} f(x) = \infty$, therefore $x = 0$ is a vertical asymptote.

$\lim\limits_{x \to \frac{\pi}{2}^-} f(x) = \infty$ and $\lim\limits_{x \to \frac{\pi}{2}^+} f(x) = -\infty$; therefore $x = \frac{\pi}{2}$ is a vertical asymptote.

$\lim\limits_{x \to \pi^-} f(x) = -\infty$; therefore $x = \pi$ is a vertical asymptote.

4) $f'(x) = -4 \csc 2x \cot 2x = \dfrac{-4 \cos(2x)}{\sin^2(2x)} = 0$ when $\cos 2x = 0 \Rightarrow 2x = \frac{\pi}{2}, \frac{3\pi}{2} \Rightarrow x = \frac{\pi}{4}, \frac{3\pi}{4}$.

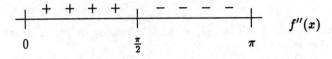

$$f'(x) = \dfrac{-4 \cos 2x}{\sin^2(2x)}$$

f is decreasing on $\left(0, \frac{\pi}{4}\right]$ and $\left[\frac{3\pi}{4}, \pi\right)$ and increasing on $\left[\frac{\pi}{4}, \frac{\pi}{2}\right)$ and $\left(\frac{\pi}{2}, \frac{3\pi}{4}\right]$.

f has a relative minimum at $\left(\frac{\pi}{4}, 2\right)$ and a relative maximum at $\left(\frac{3\pi}{4}, -2\right)$.

5) $f''(x) = -4(-2 \csc 2x \cot^2 2x - 2 \csc^3 2x) = 8 \csc 2x(\cot^2 2x + \csc^2 2x) \neq 0$ for all x in the domain.

$$f''(x)$$

f is concave up on $\left(0, \frac{\pi}{2}\right)$ and concave down on $\left(\frac{\pi}{2}, \pi\right)$. f has no inflection points.

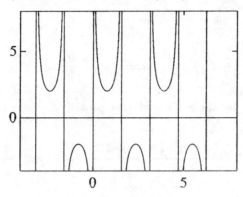

Exercise 20

21. $f(u) = u\sqrt{u^2 + 1}$

   1) Domain: $(-\infty, \infty)$

   2) $f(u) = 0$ when $u = 0$.

   3) $f'(u) = \sqrt{u^2 + 1} + u\left(\frac{1}{2}\right)(u^2 + 1)^{-1/2}(2u) = \dfrac{2u^2 + 1}{(u^2 + 1)^{1/2}} > 0$ for all x.

   Thus f is increasing on $(-\infty, \infty)$ and f has no relative extrema.

   4) $f''(u) = \dfrac{4u(u^2 + 1)^{1/2} - (2u^2 + 1)\left(\frac{1}{2}\right)(u^2 + 1)^{-1/2}(2u)}{u^2 + 1}$ $= \dfrac{u[4(u^2 + 1) - (2u^2 + 1)]}{(u^2 + 1)^{3/2}}$

   $= \dfrac{u(2u^2 + 3)}{(u^2 + 1)^{3/2}} = 0$ at $u = 0$.

$$\begin{array}{ccccccc} - & - & - & | & + & + & + \\ & & & 0 & & & \end{array} \qquad f''(u) = \dfrac{u(2u^2 + 3)}{(u^2 + 1)^{3/2}}$$

f is concave down on $(-\infty, 0]$ and concave up on $[0, \infty)$. f has an inflection point at $(0, 0)$.

   5) $f(1) = \sqrt{2}$ and $f(-1) = -\sqrt{2}$.

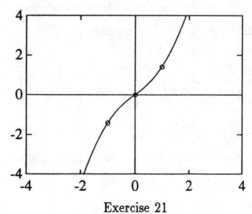

Exercise 21

22. $f(x) = \dfrac{3x - 1}{\sqrt{2x^2 + 1}}$

1) Domain: $(-\infty, \infty)$

2) $f(x) = 0$ when $3x - 1 = 0 \Rightarrow x = \frac{1}{3}$.

3) $\displaystyle\lim_{x \to \infty} \frac{3x - 1}{\sqrt{2x^2 + 1}} = \frac{3\sqrt{2}}{2}$ and $\displaystyle\lim_{x \to -\infty} \frac{3x - 1}{\sqrt{2x^2 + 1}} = -\frac{3\sqrt{2}}{2}$.

Therefore $y = \pm\dfrac{3\sqrt{2}}{2}$ are the horizontal asymptotes.

4) $f'(x) = \dfrac{3(2x^2 + 1)^{1/2} - (3x - 1)(\frac{1}{2})(2x^2 + 1)^{-1/2}(4x)}{2x^2 + 1} = \dfrac{3(2x^2 + 1) - (3x - 1)(2x)}{(2x^2 + 1)^{3/2}}$

$= \dfrac{2x + 3}{(2x^2 + 1)^{3/2}} = 0$ at $x = -\dfrac{3}{2}$.

$$- \quad - \quad - \quad \Big|_{-\frac{3}{2}} \quad + \quad + \quad +$$

$f'(x) = \dfrac{2x + 3}{(2x^2 + 1)^{3/2}}$

f is decreasing on $\left(-\infty, -\frac{3}{2}\right]$ and increasing on $\left[-\frac{3}{2}, \infty\right)$. f has an absolute minimum at $\left(-\frac{3}{2}, \frac{-\sqrt{22}}{2}\right)$.

5) $f''(x) = \dfrac{2(2x^2 + 1)^{3/2} - (2x + 3)(\frac{3}{2})(2x^2 + 1)^{1/2}(4x)}{(2x^2 + 1)^3} = \dfrac{2[(2x^2 + 1) - (2x + 3)(3x)]}{(2x^2 + 1)^{5/2}}$

$= \dfrac{-2(4x^2 + 9x - 1)}{(2x^2 + 1)^{5/2}} = 0$ when $x = \dfrac{-9 \pm \sqrt{97}}{8} \approx .106, -2.36$.

$$- \quad - \quad - \quad \Big|_{\frac{-9 - \sqrt{97}}{8}} \quad + \quad + \quad + \quad + \quad \Big|_{\frac{-9 + \sqrt{97}}{8}} \quad - \quad - \quad -$$

$f''(x) = \dfrac{-2(4x^2 + 9x - 1)}{(2x^2 + 1)^{5/2}}$

f is concave down on $\left(-\infty, \dfrac{-9 - \sqrt{97}}{8}\right]$ and on $\left[\dfrac{-9 + \sqrt{97}}{8}, \infty\right)$

and concave up on $\left[\dfrac{-9 - \sqrt{97}}{8}, \dfrac{-9 + \sqrt{97}}{8}\right]$.

f has points of inflection at $\left(\dfrac{-9 - \sqrt{97}}{8}, -2.32\right)$ and $\left(\dfrac{-9 + \sqrt{97}}{8}, -.675\right)$.

6) $f(0) = -1$

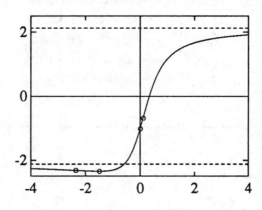

Exercise 22

23. $f(x) = \cos x + \sin^2 x$

This can be treated along the same lines as Exercise Set 4.7 number 33 $\cdots\cdots$ or reduced to this very problem by two simple trigonometric substitutions. From $\cos\left(x - \frac{\pi}{2}\right) = \cos x \cos\frac{\pi}{2} + \sin x \sin\frac{\pi}{2} = \sin x$ and similarly $\sin\left(x - \frac{\pi}{2}\right) = -\cos x$, it follows that

$$
\begin{aligned}
f\left(x - \frac{\pi}{2}\right) &= \sin x + (-\cos x)^2 \\
&= \sin x + \cos^2 x
\end{aligned}
$$

which is the very function of number 33 in Exercise Set 4.7. This means that in order to obtain the graph of $f(x) = \cos x + \sin^2 x$, all we need do is shift the graph in 4.7 number 33 by $\frac{\pi}{2}$ units to the left. Notice that the resulting graph is symmetric about the vertical line $x = \pi$. If we had drawn it on all of $(-\infty, \infty)$ we would have seen that the graph is also symmetric about the y-axis (after all, $f(x) = \cos x + \sin^2 x$ is an even function).

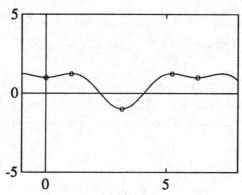

Exercise 23

By shifting the solution to 4.7 number 33 a total of $\frac{\pi}{2}$ radians (or 90°) to the left, we find:

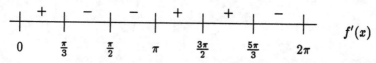

So: f increases on $\left[0, \frac{\pi}{3}\right]$, $\left[\pi, \frac{5\pi}{3}\right]$ and decreases on $\left[\frac{\pi}{3}, \pi\right]$, $\left[\frac{5\pi}{3}, 2\pi\right]$ and has relative extrema at

343

$$(0,1) \longleftarrow \text{relative minimum}$$
$$\left(\tfrac{\pi}{3}, \tfrac{5}{4}\right) \longleftarrow \text{absolute maximum}$$
$$(\pi, -1) \longleftarrow \text{absolute minimum}$$
$$\left(\tfrac{5\pi}{3}, \tfrac{5}{4}\right) \longleftarrow \text{absolute maximum}$$
$$(2\pi, 1) \longleftarrow \text{relative minimum}$$

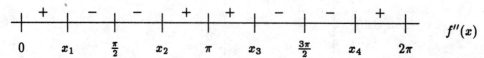

Here (approximately, in degrees):

$$x_1 \approx 32\tfrac{1}{2}^\circ \ , \ x_2 \approx 126\tfrac{1}{3}^\circ \ , \ x_3 \approx 233\tfrac{2}{3}^\circ \ , \ x_4 \approx 302\tfrac{1}{2}^\circ$$

So f is concave up on $[0, x_1]$, $[x_2, x_3]$, $[x_4, 2\pi]$ and concave down on $[x_1, x_2]$, $[x_3, x_4]$, with inflection points at x_1, x_2, x_3, x_4 found above.

24. $f(x) = x^{2/3} - x^{8/3}$

   1) Domain: $(-\infty, \infty)$

   2) $f(x) = x^{2/3} - x^{8/3} = x^{2/3}(1 - x^2) = 0$ when $x = 0, 1, -1$.

   3) $f'(x) = \tfrac{2}{3}x^{-1/3} - \tfrac{8}{3}x^{5/3} = \dfrac{2(1 - 4x^2)}{3x^{1/3}} = 0$ when $1 - 4x^2 = 0 \Rightarrow x = \pm\tfrac{1}{2}$.

      f' is undefined when $x = 0$.

$$f'(x) = \dfrac{2(1 - 4x^2)}{3x^{1/3}}$$

f is increasing on $\left(-\infty, -\tfrac{1}{2}\right]$ and on $\left[0, \tfrac{1}{2}\right]$ and decreasing on $\left[-\tfrac{1}{2}, 0\right]$ and $\left[\tfrac{1}{2}, \infty\right)$. f has a relative minimum at $(0, 0)$ and absolute maxima at $\left(\pm\tfrac{1}{2}, \ .472\right)$.

   4) $f''(x) = -\tfrac{2}{9}x^{-4/3} - \tfrac{40}{9}x^{2/3} = \dfrac{-2(1 + 20x^2)}{9x^{4/3}} < 0$ for all x.

      Thus f is concave down on $(-\infty, 0]$ and $[0, \infty)$.

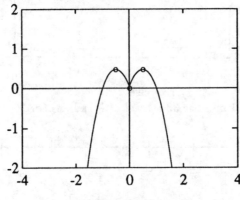

Exercise 24

344

25. $f(x) = \dfrac{1}{x^2 - 4}, \; x \in [-1, 1]$

$f'(x) = \dfrac{-2x}{(x^2 - 4)^2} = 0$ at $x = 0$. Therefore the only critical number of f on $[-1, 1]$ is 0.

$$f(-1) \;=\; -\tfrac{1}{3} \quad \longleftarrow \text{ minimum value of } f \text{ on } [-1, 1]$$

$$f(0) \;=\; -\tfrac{1}{4} \quad \longleftarrow \text{ maximum value of } f \text{ on } [-1, 1]$$

$$f(1) \;=\; -\tfrac{1}{3} \quad \longleftarrow \text{ minimum value of } f \text{ on } [-1, 1]$$

26. $f(x) = \sin x + \cos x, \; x \in [-\pi, \pi]$

$f'(x) = \cos x - \sin x = 0$ when $\cos x = \sin x \Rightarrow x = -\dfrac{3\pi}{4}, \; \dfrac{\pi}{4}$.

Therefore the critical numbers of f on $[-\pi, \pi]$ are $-\dfrac{3\pi}{4}, \; \dfrac{\pi}{4}$.

$$f(-\pi) \;=\; -1$$

$$f(-\tfrac{3\pi}{4}) \;=\; -\sqrt{2} \quad \longleftarrow \text{ minimum value of } f \text{ on } [-\pi, \pi]$$

$$f(\tfrac{\pi}{4}) \;=\; \sqrt{2} \quad \longleftarrow \text{ maximum value of } f \text{ on } [-\pi, \pi]$$

$$f(\pi) \;=\; -1$$

27. $y = x^3 - 3x^2 + 1, \; x \in [-1, 1]$
$y' = 3x^2 - 6x = 3x(x - 2) = 0$ when $x = 0$. Therefore the only critical number of f on $[-1, 1]$ is 0.
$$y(-1) \;=\; -3 \quad \longleftarrow \text{ minimum value of } y \text{ on } [-1, 1]$$
$$y(0) \;=\; 1 \quad \longleftarrow \text{ maximum value of } y \text{ on } [-1, 1]$$
$$y(1) \;=\; -1$$

28. $f(x) = x\sqrt{1 - x}, \; x \in [-3, 0]$
$f'(x) = \sqrt{1 - x} + x\left(\tfrac{1}{2}\right)(1 - x)^{-1/2}(-1) = \dfrac{2 - 3x}{2(1 - x)^{1/2}} \neq 0$ on $[-3, 0]$.

Therefore f has no critical numbers on $[-3, 0]$.
$$f(-3) \;=\; -6 \quad \longleftarrow \text{ minimum value of } f \text{ on } [-3, 0]$$
$$f(0) \;=\; 0 \quad \longleftarrow \text{ maximum value of } f \text{ on } [-3, 0]$$

29. $f(x) = x + x^{2/3}, \; x \in [-1, 1]$

$f'(x) = 1 + \dfrac{2}{3x^{1/3}} = \dfrac{3x^{1/3} + 2}{3x^{1/3}} = 0$ at $x = -\tfrac{8}{27}$.

$f'(x)$ is undefined when $x = 0$. Therefore the critical numbers of f are 0 and $-\tfrac{8}{27}$.

$$f(-1) \;=\; 0 \quad \longleftarrow \text{ minimum value of } f \text{ on } [-1, 1]$$

$$f(-\tfrac{8}{27}) \;=\; \tfrac{4}{27}$$

$$f(0) \;=\; 0 \quad \longleftarrow \text{ minimum value of } f \text{ on } [-1, 1]$$

$$f(1) \;=\; 2 \quad \longleftarrow \text{ maximum value of } f \text{ on } [-1, 1]$$

30. $y = x - 2|x|$, $x \in [-3, 2]$

$$y = \begin{cases} x - 2x = -x & \text{if } 0 \le x \le 2 \\ x + 2x = 3x & \text{if } -3 \le x \le 0 \end{cases} \qquad y' = \begin{cases} -1 & \text{if } 0 < x < 2 \\ 3 & \text{if } -3 < x < 0 \\ \text{does not exist} & \text{if } x = -3, 2, 0 \end{cases}$$

Therefore the only critical number of y is 0.

$$\begin{aligned} y(-3) &= -9 \quad \longleftarrow \text{ minimum value of } y \text{ on } [-3, 2] \\ y(0) &= 0 \quad \longleftarrow \text{ maximum value of } y \text{ on } [-3, 2] \\ y(2) &= -2 \end{aligned}$$

31. $y = x - \sqrt{1 - x^2}$, $x \in [-1, 1]$

$$y' = 1 - \tfrac{1}{2}(1 - x^2)^{1/2}(-2x) = \frac{(1 - x^2)^{1/2} + x}{(1 - x^2)^{1/2}} = 0 \text{ when } x = -\frac{\sqrt{2}}{2}.$$

$$\begin{aligned} y(-1) &= -1 \\ y\left(-\tfrac{\sqrt{2}}{2}\right) &= -\sqrt{2} \quad \longleftarrow \text{ minimum value of } y \text{ on } [-1, 1] \\ y(1) &= 1 \quad \longleftarrow \text{ maximum value of } y \text{ on } [-1, 1] \end{aligned}$$

32. $f(x) = \sqrt[3]{x} - x$, $x \in [-1, 1]$

$$f'(x) = \frac{1}{3x^{2/3}} - 1 = \frac{1 - 3x^{2/3}}{3x^{2/3}} = 0 \text{ when } x = \pm\frac{\sqrt{3}}{9}. \quad f'(x) \text{ is undefined when } x = 0.$$

$$\begin{aligned} f(-1) &= 0 \\ f\left(-\tfrac{\sqrt{3}}{9}\right) &\approx -.385 \quad \longleftarrow \text{ minimum value of } f \text{ on } [-1, 1] \\ f(0) &= 0 \\ f\left(\tfrac{\sqrt{3}}{9}\right) &\approx .385 \quad \longleftarrow \text{ maximum value of } f \text{ on } [-1, 1] \\ f(1) &= 0 \end{aligned}$$

33. $f(x) = x - 4x^{-2}$, $x \in [-3, -1]$

$$f'(x) = 1 + 8x^{-3} = \frac{x^3 + 8}{x^3} = 0 \text{ when } x = -2.$$

$$\begin{aligned} f(-3) &= -\tfrac{31}{9} \\ f(-2) &= -3 \quad \longleftarrow \text{ maximum value of } f \text{ on } [-3, -1] \\ f(-1) &= -5 \quad \longleftarrow \text{ minimum value of } f \text{ on } [-3, -1] \end{aligned}$$

34. $y = \dfrac{x}{1 + x^2}$, $x \in [-2, 2]$

$$y' = \frac{(1 + x^2) - x(2x)}{(1 + x^2)^2} = \frac{1 - x^2}{(1 + x^2)^2} = 0 \text{ at } x = \pm 1.$$

$$y(-2) = -\tfrac{2}{5}$$

$$y(-1) = -\tfrac{1}{2} \quad \longleftarrow \text{ minimum value of } y \text{ on } [-2, 2]$$

$$y(1) = \tfrac{1}{2} \quad \longleftarrow \text{ maximum value of } y \text{ on } [-2, 2]$$

$$y(2) = \tfrac{2}{5}$$

35. $y = x^4 - 2x^2, \ x \in [-2, 2]$
$y' = 4x^3 - 4x = 4x(x^2 - 1) = 0$ when $x = 0, \pm 1$.

$$y(-2) = 8 \quad \longleftarrow \text{ maximum value of } y \text{ on } [-2, 2]$$
$$y(-1) = -1 \quad \longleftarrow \text{ minimum value of } y \text{ on } [-2, 2]$$
$$y(0) = 0$$
$$y(1) = -1 \quad \longleftarrow \text{ minimum value of } y \text{ on } [-2, 2]$$
$$y(2) = 8 \quad \longleftarrow \text{ maximum value of } f \text{ on } [-2, 2]$$

36. $f(x) = |x + \sin x|, \ x \in \left[-\tfrac{\pi}{2}, \tfrac{\pi}{2}\right]$
[See Example 8 on page 243 of the textbook.]

$$f(x) = \begin{cases} x + \sin x & \text{if } 0 \le x \le \tfrac{\pi}{2} \\ -(x + \sin x) & \text{if } -\tfrac{\pi}{2} \le x \le 0 \end{cases} \qquad f'(x) = \begin{cases} 1 + \cos x & \text{if } 0 < x < \tfrac{\pi}{2} \\ -(1 + \cos x) & \text{if } -\tfrac{\pi}{2} < x < 0 \\ \text{does not exist} & \text{if } 0, \pm \tfrac{\pi}{2} \end{cases}$$

$f'(x) \ne 0$ for all x in the domain.

$$f\left(-\tfrac{\pi}{2}\right) = \tfrac{\pi + 2}{2} \quad \longleftarrow \text{ maximum value of } f \text{ on } \left[-\tfrac{\pi}{2}, \tfrac{\pi}{2}\right]$$

$$f(0) = 0 \quad \longleftarrow \text{ minimum value of } f \text{ on } \left[-\tfrac{\pi}{2}, \tfrac{\pi}{2}\right]$$

$$f\left(\tfrac{\pi}{2}\right) = \tfrac{\pi + 2}{2} \quad \longleftarrow \text{ maximum value of } f \text{ on } \left[-\tfrac{\pi}{2}, \tfrac{\pi}{2}\right]$$

37. $f(x) = \dfrac{2 + \cos x}{2 - \cos x}, \ x \in [-\pi, \pi]$

$$f'(x) = \frac{-\sin x(2 - \cos x) - (2 + \cos x)(\sin x)}{(2 - \cos x)^2} = \frac{-4\sin x}{(2 - \cos x)^2} = 0 \text{ when } -4\sin x = 0 \Rightarrow x = 0.$$

Therefore the only critical number of f on $[-\pi, \pi]$ is 0.

$$f(-\pi) = \tfrac{1}{3} \quad \longleftarrow \text{ minimum value of } f \text{ on } [-\pi, \pi]$$
$$f(0) = 3 \quad \longleftarrow \text{ maximum value of } f \text{ on } [-\pi, \pi]$$
$$f(\pi) = \tfrac{1}{3} \quad \longleftarrow \text{ maximum value of } f \text{ on } [-\pi, \pi]$$

38. $y = \dfrac{x - 1}{x^2 + 3}, \ x \in [-2, 4]$

$$y' = \frac{(x^2 + 3) - (x - 1)(2x)}{(x^2 + 3)^2} = \frac{-x^2 + 2x + 3}{(x^2 + 3)^2} = 0 \text{ when } x = -1, 3.$$

$$y(-2) = -\tfrac{3}{7}$$

$$y(-1) = -\tfrac{1}{2} \quad \longleftarrow \text{ minimum value of } y \text{ on } [-2, 4]$$

$$y(3) = \tfrac{1}{6} \quad \longleftarrow \text{ maximum value of } y \text{ on } [-2, 4]$$

$$y(4) = \tfrac{3}{19}$$

39. $y = \dfrac{x + 2}{\sqrt{x^2 + 1}}, \ x \in [0, 2]$

347

$$y' = \frac{(x^2+1)^{1/2} - (x+2)(\frac{1}{2})(x^2+1)^{-1/2}(2x)}{x^2+1} = \frac{1-2x}{(x^2+1)^{3/2}} = 0 \text{ when } x = \frac{1}{2}.$$

$$y(0) = 2$$

$$y\left(\frac{1}{2}\right) = \sqrt{5} \quad \longleftarrow \text{ maximum value of } y \text{ on } [0,2]$$

$$y(2) = \frac{4\sqrt{5}}{5} \quad \longleftarrow \text{ minimum value of } y \text{ on } [0,2]$$

40. $f(x) = |x^2 - 1|, \; x \in [-3,3]$

$$f(x) = \begin{cases} x^2 - 1 & \text{if } -3 \le x \le -1 \text{ and } 1 \le x \le 3 \\ 1 - x^2 & \text{if } -1 \le x \le 1 \end{cases}$$

$$f'(x) = \begin{cases} 2x & \text{if } -3 < x < -1 \text{ and } 1 < x < 3 \\ -2x & \text{if } -1 < x < 1 \\ \text{does not exist} & \text{if } x = -3, -1, 1, 3 \end{cases}$$

$f'(x) = 0$ when $x = 0$.

$$\begin{aligned} f(-3) &= 8 \quad \longleftarrow \text{ maximum value of } f \text{ on } [-3,3] \\ f(-1) &= 0 \quad \longleftarrow \text{ minimum value of } f \text{ on } [-3,3] \\ f(0) &= 1 \\ f(1) &= 0 \quad \longleftarrow \text{ minimum value of } f \text{ on } [-3,3] \\ f(3) &= 8 \quad \longleftarrow \text{ maximum value of } f \text{ on } [-3,3] \end{aligned}$$

41. $f(x) = |x^3 + x^2 - x|, \; x \in \left[-\frac{3}{2}, 1\right]$

$$f(x) = \begin{cases} x^3 + x^2 - x & \text{if } -\frac{3}{2} \le x \le 0 \text{ and } \frac{-1+\sqrt{5}}{2} \le x \le 1 \\ -(x^3 + x^2 - x) & \text{if } 0 \le x \le \frac{-1+\sqrt{5}}{2} \end{cases}$$

$$f'(x) = \begin{cases} 3x^2 + 2x - 1 & \text{if } -\frac{3}{2} < x < 0 \text{ and } \frac{-1+\sqrt{5}}{2} < x < 1 \\ -(3x^2 + 2x - 1) & \text{if } 0 < x < \frac{-1+\sqrt{5}}{2} \\ \text{does not exist} & \text{if } x = -\frac{3}{2}, \; 0, \; \frac{-1+\sqrt{5}}{2}, \; 1 \end{cases}$$

$f'(x) = 0$ when $x = -1, \frac{1}{3}$.

$$f\left(-\frac{3}{2}\right) = \frac{3}{8}$$

$$f(-1) = 1 \quad \longleftarrow \text{ maximum value of } f \text{ on } \left[-\frac{3}{2}, 1\right]$$

$$f(0) = 0 \quad \longleftarrow \text{ minimum value of } f \text{ on } \left[-\frac{3}{2}, 1\right]$$

$$f\left(\frac{1}{3}\right) = \frac{5}{27}$$

$$f\left(\frac{-1+\sqrt{5}}{2}\right) = 0 \quad \longleftarrow \text{ minimum value of } f \text{ on } \left[-\frac{3}{2}, 1\right]$$

$$f(1) = 1 \quad \longleftarrow \text{ maximum value of } f \text{ on } \left[-\frac{3}{2}, 1\right]$$

42. $\displaystyle \lim_{x \to \infty} \frac{x^4 + 2x + 1}{2x^5 + 3x^2 + 2} = \lim_{x \to \infty} \frac{\frac{1}{x} + \frac{2}{x^4} + \frac{1}{x^5}}{2 + \frac{3}{x^3} + \frac{2}{x^5}} = 0$

348

43. $\lim\limits_{x \to -\infty} \dfrac{3x^2 + x + 1}{x^2 + 1} = \lim\limits_{x \to -\infty} \dfrac{3 + \frac{1}{x} + \frac{1}{x^2}}{1 + \frac{1}{x^2}} = 3$

44. $\lim\limits_{x \to 3+} \dfrac{x^2 + 3x - 4}{x^2 - 2x - 3} = \lim\limits_{x \to 3+} \dfrac{(x+4)(x-1)}{(x-3)(x+1)} = \infty$

45. $\lim\limits_{x \to -1+} \dfrac{x^2 + x + 4}{x^2 - 2x - 3} = \lim\limits_{x \to -1+} \dfrac{x^2 + x + 4}{(x-3)(x+1)} = -\infty$

46. $\lim\limits_{x \to 7-} \dfrac{x - 8}{x^2 - 8x + 7} = \lim\limits_{x \to 7-} \dfrac{x-8}{(x-7)(x-1)} = \infty$

47. $\lim\limits_{x \to 7} \dfrac{x-8}{x^2 - 8x + 7} =$ does not exist since $\lim\limits_{x \to 7+} \dfrac{x-8}{x^2 - 8x + 7} = \lim\limits_{x \to 7+} \dfrac{x-8}{(x-7)(x-1)} = -\infty$

$$\text{and } \lim\limits_{x \to 7-} \dfrac{x-8}{x^2 - 8x + 7} = \infty.$$

48. $\lim\limits_{x \to 7-} \dfrac{x - 8}{x^2 - 14x + 49} = \lim\limits_{x \to 7-} \dfrac{x-8}{(x-7)^2} = -\infty.$

49. $\lim\limits_{x \to 7} \dfrac{x - 8}{x^2 - 14x + 49} = \lim\limits_{x \to 7} \dfrac{x-8}{(x-7)^2} = -\infty$

50. $\lim\limits_{x \to 0+} \dfrac{\sin\left(\frac{1}{x}\right)}{x}$. Let $y = \frac{1}{x}$. Then $x = \frac{1}{y}$.

As $x \to 0^+$, $y \to \infty$. Thus $\lim\limits_{x \to 0+} \dfrac{\sin\frac{1}{x}}{x} = \lim\limits_{y \to \infty} \dfrac{\sin y}{\frac{1}{y}} = \lim\limits_{y \to \infty} y \sin y$ which does not exist. (See problem 19 of Section 4.6.)

51. $\lim\limits_{x \to \infty} \cos\left(\dfrac{1}{x}\right) = \cos 0 = 1$

52. $\lim\limits_{x \to 0+} \cos\left(\dfrac{1}{x}\right)$. Let $y = \frac{1}{x}$, then as $x \to 0^+$, $y \to \infty$.

Then $\lim\limits_{x \to 0+} \cos\left(\dfrac{1}{x}\right) = \lim\limits_{y \to \infty} \cos y$ which does not exist since as y runs through the multiples of π, $(\cos y)$ takes the values $+1$ and -1 alternately.

53. $\lim\limits_{\theta \to \frac{\pi}{2}-} \dfrac{\sec \theta}{\tan \theta} = \lim\limits_{\theta \to \frac{\pi}{2}-} \dfrac{1}{\sin \theta} = 1$

54. Let x and y be the two numbers. Let D be the desired difference.

We want to maximize $D = x^2 - y^2$ (1)
subject to the condition that $x + y = 10$.   (2)
From (2), we have $y = 10 - x$.   (3)

Substituting from (3) into (1), we have $D(x) = x^2 - (10 - x)^2$. Equation (2) together with the fact that x and y are positive numbers implies that $0 < x < 10$.
We have: $\quad D(x) = x^2 - (10 - x)^2, \quad 0 < x < 10$
$\qquad\qquad D'(x) = 2x + 2(10 - x) = 20 > 0.$

Thus $D(x)$ is increasing on $(0, 10)$. Since x cannot be 10, $D(x)$ does not attain a maximum value on $(0, 10)$. However, if x and y were nonnegative and not strictly positive, $D(x)$ would attain a maximum value of 100 when $x = 10$ and $y = 0$.

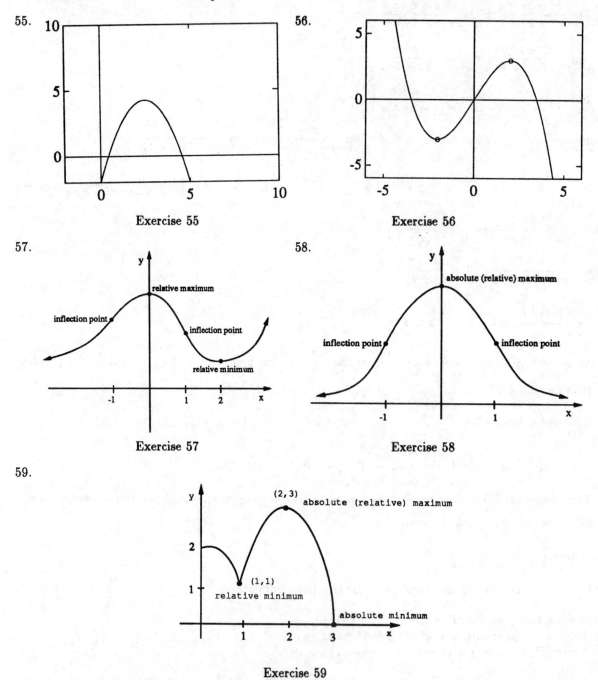

55.

Exercise 55

56.

Exercise 56

57.

Exercise 57

58.

Exercise 58

59.

Exercise 59

60. The properties of f in this case contradict one another. Apply Rolle's Theorem to $f'(x)$. Since $f'(x)$ is continuous on $[1, 2]$ and differentiable on $(1, 2)$ and $f'(1) = f'(2) = 0$, there exists a $c \in (1, 2)$ such that $f''(c) = 0$. However, property (4) says that $f''(x) > 0$ for all $x \in (0, 2)$. We have a contradiction.

61.

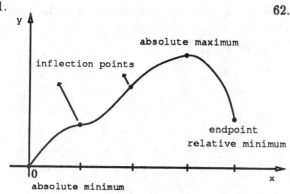

Exercise 61

62.

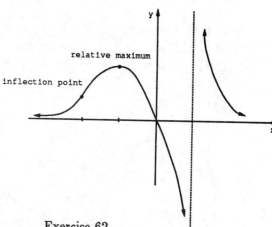

Exercise 62

63. Let the foot of the ladder be a distance x (in meters) from the foot of the fence, so (Pythagoras) the distance along the ladder from its foot to the top of the fence is $\sqrt{x^2+9}$. An application of similar triangles shows that the length L of the ladder is given by

$$\frac{L}{\sqrt{x^2+9}} = \frac{x+1}{x}$$

therefore

$$L = \left(1+\frac{1}{x}\right)\sqrt{x^2+9}$$

therefore

$$\frac{dL}{dx} = \frac{x^3-9}{x^2\sqrt{x^2+9}}$$

which is zero if $x=\sqrt[3]{9}$, negative if $x<\sqrt[3]{9}$ and positive if $x>\sqrt[3]{9}$. The shortest ladder for the job therefore has length $(1+9^{-1/3})(9^{2/3}+9)^{1/2}$ meters.

64. Since f has a (continuous) derivative on $[a,b]$ it is surely continuous on $[a,b]$. Because $f(a)$ and $f(b)$ have opposite signs, there is a $p \in (a,b)$ such that $f(p)=0$ (Intermediate Value Theorem). Suppose there is another point, say q, in (a,b) at which f is zero. Then (Rolle's Theorem) there exists an x between p and q such that $f'(x)=0$, contrary to what is given. This shows that p is the unique point such that $f(p)=0$.

65. Let $s=$ number of subscriptions sold per year and let $r=$ yearly subscription rate. The marketing department predicts $\frac{ds}{dr}=-50$, so the (linear!) equation relating r and s is

$$s-2000 = (-50)(r-40)$$

therefore

$$s = 4000-50r.$$

Now, total revenue R is

$$R = rs = r(4000-50r)$$

therefore

$$\frac{dR}{dr} = 4000-100r$$

which is zero if $r = 40$, < 0 if $r > 40$ and > 0 if $r < 40$. So: the subscription rate to maximize revenue is \$40 per year.

66.　a)　$f(x) = \sin x$
　　　　$a = 0 \quad b = 2\pi$

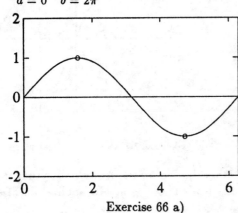

Exercise 66 a)

　b)　$f(x) = \cos x$
　　　$a = -\frac{3\pi}{2} \quad b = \frac{3\pi}{2}$

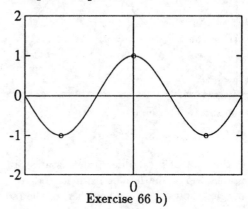

Exercise 66 b)

67.　　　$f(t) \;=\; t^3 - 6t^2 + 14t + 5$

$\Rightarrow \; f'(t) \;=\; 3t^2 - 12t + 14$

$$= \; 3\left(t^2 - 4t + \frac{14}{3}\right)$$

$$= \; 3\left((t-2)^2 + \frac{2}{3}\right) \;\geq\; 2 \text{ for all } t.$$

Since $f'(t) > 0$ for all real t, it follows that f is increasing on $(-\infty, \infty)$ by Theorem 6.

68.　$f(x) = x^2 + bx + c \Rightarrow f'(x) = 2x + b$, so $f'(x) = 0$ when $x = -\frac{1}{2}b$. Now, f assumes a minimum value when $x = -\frac{1}{2}b$: in fact, $f'(x) < 0$ if $x < -\frac{1}{2}b$ and $f'(x) > 0$ if $x > -\frac{1}{2}b$. This minimum value is

$$f\left(-\frac{1}{2}b\right) = c - \frac{b^2}{4},$$

so $f(x) > 0$ for all x if and only if $4c > b^2$. [Alternatively, complete the square:

$$f(x) \;=\; x^2 + bx + c \;=\; \left(x + \frac{1}{2}b\right)^2 + \left(c - \frac{b^2}{4}\right)$$

$$\geq \; c - \frac{b^2}{4} \qquad \Big].$$

69.　　　$f(x) \;=\; \dfrac{x}{1 + ax^2}$

$\Rightarrow \; f'(x) \;=\; \dfrac{1 - ax^2}{(1 + ax^2)^2}.$

Now, since f is to assume its minima at the interior points $x = \pm 3$, we must have $f'(\pm 3) = 0$; thus $9a = 1$ and so $a = \frac{1}{9}$.

70. We wish to find the value of x for which $f(x) = \sqrt{x} - x$ is maximized; of course, we suppose $x \geq 0$. Now if $x > 0$ then $f'(x) = \frac{1}{2\sqrt{x}} - 1$, which is zero when $2\sqrt{x} = 1$ so that $x = \frac{1}{4}$; further, $x < \frac{1}{4} \Rightarrow f'(x) > 0$ and $x > \frac{1}{4} \Rightarrow f'(x) < 0$. So the number most exceeded by its square root is $x = \frac{1}{4}$.

352

71. $f(x) = x^n + ax + b \Rightarrow f'(x) = n\,x^{n-1} + a$. If n is even then f' is increasing (look at the sign of f''). In particular, $f'(x) = 0$ for just one value of x. Now, if $p < q < r$ are three zeros of f then (by Rolle's Theorem) there exist $s \in (p, q)$ and $t \in (q, r)$ such that $f'(s) = 0$ and $f'(t) = 0$: this contradicts the finding of the previous sentence, and shows that f has at most two zeros. If n is odd then f can have three zeros [for example, look at $f(x) = x^3 - x = (x + 1)\,x\,(x - 1)$].

72. Let y be the yield per tree, t be the number of trees per acre and p be the total number of peaches per acre in the crop; so $p = yt$. We are told that $\frac{dy}{dt} = -10$, so the (linear) equation relating y and t is

$$y - 300 = -10(t - 25)$$

therefore
$$y = 550 - 10t$$

therefore
$$p = (550 - 10t)t$$

therefore
$$\frac{dp}{dt} = 550 - 20t$$

So $\frac{dp}{dt} = 0$ when $t = 27\frac{1}{2}$, $\frac{dp}{dt} < 0$ when $t > 27\frac{1}{2}$ and $\frac{dp}{dt} > 0$ when $t < 27\frac{1}{2}$. Thus the largest peach crop (per acre) will be produced by planting $27\frac{1}{2}$ trees (per acre).

73. The surface area of the pool is $S = \frac{1}{2}r^2\theta$ (constant) and its perimeter P is given by $P = 2r + r\theta$. Solve $S = \frac{1}{2}r^2\theta$ for $\theta = \frac{2S}{r^2}$ and substitute into P to get

$$P = 2r + \frac{2S}{r}$$

therefore
$$\frac{dP}{dr} = 2 - \frac{2S}{r^2}$$

which is 0 if $r = \sqrt{S}$, < 0 if $r < \sqrt{S}$ and > 0 if $r > \sqrt{S}$. Thus the perimeter P is minimized when $r = \sqrt{S}$ and $\theta = 2$ (radians).

74. Assume that the rectangle has width x and height y and is situated inside the equilateral triangle as in the figure; the rectangle of course has area $A = xy$. The length AP is $5\sqrt{3}$ units, since $\angle ABP$ is 60°. Since the triangles APB and QRB are similar, $\dfrac{5\sqrt{3}}{5} = \dfrac{y}{5 - \frac{x}{2}}$ so that $y = 5\sqrt{3} - \dfrac{\sqrt{3}}{2}\,x$. Now

$$A = \sqrt{3}x\left(5 - \frac{x}{2}\right)$$

therefore
$$\frac{dA}{dx} = \sqrt{3}\,(5 - x)$$

which is 0, < 0 or > 0 according to whether $x = 5$, $x > 5$ or $x < 5$. Since $A = 0$ when $x = 0$ or $x = 10$ (the smallest and largest values of x to consider) it follows that A is maximized when $x = 5$ and $y = \frac{5\sqrt{3}}{2}$.

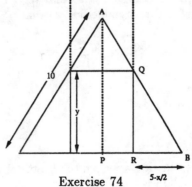

Exercise 74

75. Let $S = D^2$ where D is the distance from $(0,0)$ to the point (x, y) on the hyperbola $x^2 - x + \frac{5}{4} - y^2 = 0$, so that

$$S = x^2 + y^2 = 2x^2 - x + \frac{5}{4}.$$

Then $\frac{dS}{dx} = 4x - 1$, which is zero if $x = \frac{1}{4}$, negative if $x < \frac{1}{4}$ and positive if $x > \frac{1}{4}$; so S is a minimum when $x = \frac{1}{4}$ and so (Remark 3 of Section 4.4) D is a minimum when $x = \frac{1}{4}$. From $y^2 = x^2 - x + \frac{5}{4}$ we deduce that $y = \pm\frac{\sqrt{17}}{4}$. Thus: the sought for points are $\left(\frac{1}{4}, \pm\frac{\sqrt{17}}{4}\right)$.

76. We suppose the rectangle to be so situated that its center is at the origin and its four vertices on the ellipse. If (x, y) denotes the vertex in the positive quadrant of the plane, then the rectangle has width $2x$ and height $2y$, hence area $A = 4xy$; also, $x^2 + \frac{y^2}{4} = 1$, so that $y = 2\sqrt{1 - x^2}$. Now

$$A = 8x\sqrt{1 - x^2}$$

therefore

$$\frac{dA}{dx} = 8\sqrt{1 - x^2} - \frac{8x^2}{\sqrt{1 - x^2}}$$

$$= \frac{8(1 - 2x^2)}{\sqrt{1 - x^2}}$$

which is zero if $x = \frac{1}{\sqrt{2}}$, > 0 if $0 < x < \frac{1}{\sqrt{2}}$ and < 0 if $\frac{1}{\sqrt{2}} < x < 1$. Since $A = 0$ when $x = 0$ as $x = 1$ (the appropriate limits) it follows that A is maximized when $x = \frac{1}{\sqrt{2}}$ and $y = \sqrt{2}$, so that the rectangle has width $\sqrt{2}$ and height $2\sqrt{2}$ (hence area 4 square units).

77. Let the log have diameter D (given as fixed) and let the beam have width w and depth d, so (Pythagoras) $D^2 = w^2 + d^2$. The strength S of the beam is given by $S = kwd^2$ where k is some constant of proportionality; by substitution, it follows that

$$S = kw(D^2 - w^2)$$

therefore

$$\frac{dS}{dw} = kD^2 - 3kw^2$$

which is zero when $w = \frac{D}{\sqrt{3}}$. An examination of the sign of $\frac{dS}{dw}$ on either side of $w = \frac{D}{\sqrt{3}}$ shows that $w = \frac{D}{\sqrt{3}}$ maximizes S. Now, $w = \frac{D}{\sqrt{3}} \Rightarrow d = \sqrt{D^2 - w^2} = D\sqrt{2/3}$ so that the optimal ratio of d to w is $\frac{d}{w} = \sqrt{2}$.

78. Let the distance from light to student at time t be f, so that $\frac{df}{dt} = 1.2$. Let the student's shadow have length s at time t. An application of similar triangles shows that

$$\frac{s + f}{s} = \frac{8}{1.6} = 5$$

so that

$$s = \frac{1}{4}f$$

and therefore $\frac{ds}{dt} = \frac{1}{4}\frac{df}{dt} = 0.3$. Thus, the rate at which the length of the student's shadow is increasing is a constant $0.3 \ m/s$ (and the 20 m data is irrelevant).

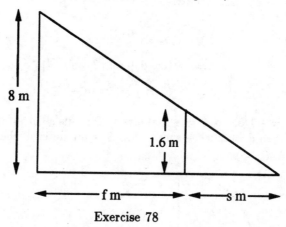

Exercise 78

79. Let the cylindrical portion of the tank have length l and radius r. If the cost of the material for this portion is k per unit area then the total construction cost C is the sum of $2\pi r l k$ (for cylinder) and $(4\pi r^2)(3k)$ (for caps) so

$$C = 2\pi r l k + 12\pi r^2 k.$$

The total volume of the tank is fixed at

$$V \;=\; \pi r^2 h + \frac{4}{3}\pi r^3$$

so

$$l \;=\; \frac{V - \frac{4}{3}\pi r^3}{\pi r^2}.$$

Thus

$$C \;=\; 2k\left[\frac{V}{r} + \frac{14}{3}\pi r^2\right]$$

so

$$\frac{dC}{dr} \;=\; 2k\left[\frac{-V}{r^2} + \frac{28}{3}\pi r\right]$$

which is zero when $r = \sqrt[3]{\dfrac{3V}{28\pi}}$. Checking the sign of $\frac{dC}{dr}$ for r less than or greater than this value shows that cost C is minimized when

$$r = \sqrt[3]{\frac{3V}{28\pi}} \quad \text{and} \quad l = 8r.$$

80. See figure for notation. If the given circle has center O then θ is the angle between AO and either OC or OB. Working in triangle ODC we have $\frac{w}{2} = r\sin\theta$ and $h = r\cos\theta$, so $\triangle ABC$ has width $w = 2r\sin\theta$ and height $r + r\cos\theta$, whence its area S is given by

$$\begin{aligned}
S \;&=\; \frac{1}{2}(\text{width}) \times (\text{height}) \\
&=\; r\sin\theta(r + r\cos\theta) \\
&=\; r^2(\sin\theta + \sin\theta\cos\theta).
\end{aligned}$$

Now

$$\frac{dS}{d\theta} = r^2(\cos\theta + \cos^2\theta - \sin^2\theta)$$
$$= r^2(2\cos^2\theta + \cos\theta - 1)$$

which is zero when $\cos\theta = -1$ or $\cos\theta = \frac{1}{2}$ (quadratic formula). In $[0, \pi]$ this has solutions $\theta = \pi$ and $\theta = \frac{\pi}{3}$. Checking S when $\theta = 0$, π, $\frac{\pi}{3}$ shows that S is maximized when $\theta = \frac{\pi}{3}$ (so the triangle is equilateral).

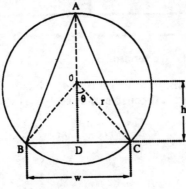

Exercise 80

81. Given $x_1, \cdots\cdots, x_n$ let us write

$$S = (\overline{x} - x_1)^2 + \cdots\cdots + (\overline{x} - x_n)^2.$$

Then

$$\frac{dS}{d\overline{x}} = 2(\overline{x} - x_1) + \cdots\cdots + 2(\overline{x} - x_n)$$

which is zero when

$$\overline{x} = \frac{x_1 + \cdots\cdots + x_n}{n};$$

checking the sign of $\frac{dS}{d\overline{x}}$ as usual shows that S is minimized.

82. If $E = (mx_1 - y_1)^2 + \cdots\cdots + (mx_n - y_n)^2$ then

$$\frac{dE}{dm} = 2x_1(mx_1 - y_1) + \cdots\cdots + 2x_n(mx_n - y_n)$$
$$= 2(x_1^2 m - x_1 y_1 + \cdots\cdots + x_n^2 m - x_n y_n)$$
$$= 2((x_1^2 + \cdots\cdots + x_n^2)m - (x_1 y_1 + \cdots\cdots + x_n y_n))$$

which is zero when

$$m = \frac{x_1 y_1 + \cdots\cdots + x_n y_n}{x_1^2 + \cdots\cdots + x_n^2};$$

checking the sign of $\frac{dE}{dm}$ as usual on either side of this value for m shows that E is minimized.

356

83. Let the box have width w, length $2w$, height h (in centimeters). Its volume is then

$$1024 = w \times 2w \times h$$

so
$$h = \frac{512}{w^2}.$$

Now, the total cost of construction (in cents) is

$$
\begin{aligned}
C &= \underset{\text{(top)}}{12 \times 2w^2} + \underset{\text{(bottom)}}{6 \times 2w^2} + \underset{\text{(sides)}}{6 \times 6wh} \\
&= 36(w^2 + wh) \\
&= 36\left(w^2 + \frac{512}{w}\right).
\end{aligned}
$$

Thus

$$\frac{dC}{dw} = 36\left(2w - \frac{512}{w^2}\right)$$

which is zero when $w^3 = 256$ so $w = 4\sqrt[3]{4}$. If $w > 4\sqrt[3]{4}$ then $\frac{dC}{dw} > 0$ whilst if $0 < w < 4\sqrt[3]{4}$ then $\frac{dC}{dw} < 0$. Thus, cost C is least when $w = 4\sqrt[3]{4}$ and $h = 2\sqrt[3]{256}$.

84.   a) $(x+3)(x-5) = x^2 - 2x - 15 = 0$ when $x = -3$ and $x = 5$; so Rolle's Theorem applies to the interval $[-3, 5]$.

   b) $f'(x) = 2x - 2$, which is zero when $x = 1$; the value of c required is thus $c = 1$.

85. $$
\begin{aligned}
f(x) &= x^3 + qx^2 + 5x - 6 \\
\Rightarrow \quad f'(x) &= 3x^2 + 2qx + 5.
\end{aligned}
$$

With $c = 2$, $a = 0$ and $b = 2$, the Mean Value Theorem applied to f yields

$$
\begin{aligned}
f(2) - f(0) &= 2f'(2) \\
\end{aligned}
$$
or
$$(12 + 4q) - (-6) = 2(17 + 4q)$$
or
$$q = -4.$$

86. At a speed of s km/h, the time taken to travel 1 km is $\frac{1}{s}$ hr; the total cost per km is thus C given by

$$C = \frac{200}{s} + \frac{s}{100} \quad \text{(in dollars)}.$$

From $\dfrac{dC}{ds} = \dfrac{-200}{s^2} + \dfrac{1}{100}$ we see that $\frac{dC}{ds} = 0$ when $s^2 = 20{,}000$ so that $s = 100\sqrt{2}$; checking signs as usual shows that C is minimized. Thus: cost per km is least when the aircraft travels at $100\sqrt{2} \approx 141.4$ km/h.

87. Let x be the number of times per year that parts are ordered, and we assume that all orders are equal in size. Then the cost function (in dollars per year) is

$$
\begin{aligned}
C(x) &= (.05)\,3000 + 20x + \left(\frac{.48}{x}\right)\left(\frac{1500}{x}\right)x \\
&= 150 + 20x + \frac{720}{x}.
\end{aligned}
$$

Thus

$$C'(x) = 20 - \frac{720}{x^2}.$$

To find the critical numbers, we solve $C'(x) = 0$. That is,

$$\frac{720}{x^2} = 20$$
$$36 = x^2,$$

and $x = 6$ is the critical number in the interval $[1, \infty)$. Note that $x = 6$ corresponds to a minimum because

$$C''(x) = \frac{1440}{x^3},$$

which implies that the graph of C is concave up on $[1, \infty)$.

88. Let P be the number of people signing up, C the cost per person, and $R = CP$ the total revenue. Since we are given that $\frac{dC}{dP} = -1.5$ and that $C = 400$ when $P = 100$, it follows that

$$C - 400 = (-1.5)(P - 100)$$
or
$$C = 550 - 1.5P.$$

Consequently

$$R = (550 - 1.5P)P$$
so
$$\frac{dR}{dP} = 550 - 3P$$

which is < 0 if $P > \frac{550}{3}$ and > 0 if $P < \frac{550}{3}$. Since $\frac{550}{3}$ is not a whole number, we look to the whole numbers on either side: 183 and 184. Now, if $P = 183$ then $R = 50416.5$ whilst if $P = 184$ then $R = 50416$; in view of our remarks on the sign of $\frac{dR}{dP}$, it follows that revenue is greatest when 183 sign up.

89. This problem is similar to Exercise 87. Let x be the number of times per year that shirts are ordered. Then the cost function is

$$C(x) = (.20)\,2000 + 50x + \left(\frac{3}{x}\right)\left(\frac{1000}{x}\right)x$$
$$= 400 + 50x + \frac{3000}{x}.$$

Thus

$$C'(x) = 50 - \frac{3000}{x^2},$$

and the critical numbers are obtained by solving $C'(x) = 0$. We obtain

$$x^2 = 60,$$

which yields the critical number $\sqrt{60}$. If we want an integral number for x, we use the fact that $\sqrt{60}$ lies between 7 and 8, and we calculate

$$C(7) = 1178.59$$
$$C(8) = 1175.$$

Note that $C''(x) > 0$ on $[1, \infty)$, so the graph of C is concave upward on $[1, \infty)$. So shirts should be ordered 8 times per year.

Chapter 5

Antidifferentiation

5.1 Antiderivatives

In Exercises 1–18, find the indefinite integral.

1. $\int \left(2x^2 + 1\right) dx = \dfrac{2x^3}{3} + x + C$

2. $\int \left(1 - x^3\right) dx = x - \dfrac{x^4}{4} + C$

3. $\int \left(x^{2/3} + x^{5/2}\right) dx = \dfrac{3}{5} x^{5/3} + \dfrac{2}{7} x^{7/2} + C$

4. $\int \left(\sqrt{x} + 1\right)^2 dx \;=\; \int \left(x + 2\sqrt{x} + 1\right) dx = \int \left(x + 2x^{1/2} + 1\right) dx$

   $\qquad\qquad\quad = \dfrac{x^2}{2} + 2 \cdot \dfrac{2}{3} x^{3/2} + x + C = \dfrac{x^2}{2} + \dfrac{4}{3} x^{3/2} + x + C$

5. $\int \left(2\cos t - \sin t\right) dt = 2\sin t + \cos t + C$

6. $\int 4\sec^2 \theta \, d\theta = 4 \int \sec^2 \theta \, d\theta = 4\tan \theta + C$

7. $\int \left(\sin x + \sec^2 x\right) dx = \int \sin x \, dx + \int \sec^2 x \, dx = -\cos x + \tan x + C$

8. $\int \sqrt{t}\,(t^2 + 1)\, dt = \int t^{1/2}\,(t^2 + 1)\, dt = \int t^{5/2} + t^{1/2}\, dt = \dfrac{2}{7} t^{7/2} + \dfrac{2}{3} t^{3/2} + C$

9. $\int \dfrac{t^2 + 1}{\sqrt{t}}\, dt = \int \left(t^2 + 1\right) \cdot t^{-1/2}\, dt = \int t^{3/2} + t^{-1/2}\, dt = \dfrac{2}{5} t^{5/2} + 2t^{1/2} + C$

10. $\int (t - 1)(t + 1)\, dt = \int t^2 - 1\, dt = \dfrac{t^3}{3} - t + C$

11. $\int (t^2 + 1)(t + 2)\, dt = \int \left(t^3 + t + 2t^2 + 2\right) dt = \dfrac{t^4}{4} + \dfrac{2t^3}{3} + \dfrac{t^2}{2} + 2t + C$

359

12. $\int \left(\sqrt{t} + 2\right)\left(t^3 + 1\right) dt = \int \left(t^{1/2} + 2\right)\left(t^3 + 1\right) dt$

$= \int \left(t^{7/2} + 2t^3 + t^{1/2} + 2\right) dt = \frac{2}{9} t^{9/2} + \frac{1}{2} t^4 + \frac{2}{3} t^{3/2} + 2t + C$

13. $\int \frac{1}{\sqrt[3]{x}} dx = \int x^{-1/3} dx = \frac{3}{2} x^{2/3} + C$

14. $\int \frac{1}{1 - \sin^2 x} dx = \int \frac{1}{\cos^2 x} dx = \int \sec^2 x \, dx = \tan x + C$

15. $\int \left(\tan^2 x + 1\right) dx = \int \left((\sec^2 x - 1) + 1\right) dx = \int \sec^2 x \, dx = \tan x + C$

16. $\int \left(5 \cos x + \sqrt{x}\right) dx = 5 \int \cos x \, dx + \int x^{1/2} dx = 5 \sin x + \frac{2}{3} x^{3/2} + C$

17. $\int \left(3x^2 - \sin x + 2 \sec^2 x\right) dx = 3 \int x^2 \, dx - \int \sin x \, dx + 2 \int \sec^2 x \, dx$

$= 3 \frac{x^3}{3} - - \cos x + 2 \tan x + C = x^3 + \cos x + 2 \tan x + C$

18. $\int (6x + 5 - 3 \cot x \csc x) dx = \int (6x + 5) dx - 3 \int \csc x \cot x \, dx$

$= 6 \frac{x^2}{2} + 5x - 3(- \csc x) + C = 3x^2 + 5x + 3 \csc x + C$

In Exercises 19–26, find the position function s corresponding to the given velocity function and initial condition.

19. $v(t) = \cos t, \ s(0) = 2$

$$\begin{aligned} s(t) &= \int \cos t \, dt, & s(0) &= \sin(0) + C = 2 \\ s(t) &= \sin t + C & & 0 + C = 2 \\ & & & C = 2 \\ & & s(t) &= \sin t + 2 \end{aligned}$$

20. $v(t) = 1 + 2t, \ s(0) = 0$

$$\begin{aligned} s(t) &= \int (1 + 2t) \, dt, & s(0) &= 0 + 0^2 + C = 0 \\ s(t) &= t + t^2 + C & & C = 0 \\ & & s(t) &= t + t^2 \end{aligned}$$

21. $v(t) = 2t^2, \ s(0) = 4$

$$\begin{aligned} s(t) &= \int 2t^2 \, dt, & s(0) &= 0 + C = 4 \\ s(t) &= \frac{2t^3}{3} + C & & C = 4 \\ & & s(t) &= \frac{2t^3}{3} + 4 \end{aligned}$$

22. $v(t) = t(t + 2), \ s(0) = 2$

$$s(t) = \int t(t+2)\,dt \qquad s(0) = 0 + 0 + C = 2$$
$$= \int (t^2 + 2t)\,dt \qquad\qquad C = 2$$
$$s(t) = \frac{t^3}{3} + t^2 + C \qquad s(t) = \frac{t^3}{3} + t^2 + 2$$

23. $v(t) = \sqrt{t}(t+4),\ s(0) = 0$

$$s(t) = \int \sqrt{t}(t+4)\,dt \qquad s(0) = 0 + C = 0$$
$$= \int t^{1/2}(t+4)\,dt \qquad\qquad C = 0$$
$$= \int \left(t^{3/2} + 4t^{1/2} \right)\,dt \qquad s(t) = \frac{2}{5}t^{5/2} + \frac{8}{3}t^{3/2}$$
$$s(t) = \frac{2}{5}t^{5/2} + \frac{8}{3}t^{3/2} + C$$

24. $v(t) = \dfrac{\left(\sqrt{t}+3\right)^2}{\sqrt{t}},\ s(1) = \dfrac{2}{3}$

$$s(t) = \int \frac{\left(\sqrt{t}+3\right)^2}{\sqrt{t}}\,dt \qquad s(t) = \frac{2}{3}t^{3/2} + 6t + 18t^{1/2} + C$$
$$= \int \left((\sqrt{t})^2 + 6\sqrt{t} + 9 \right)t^{-1/2}\,dt \qquad s(1) = \frac{2}{3} + 6 + 18 + C = \frac{2}{3}$$
$$= \int (t + 6t^{1/2} + 9)t^{-1/2}\,dt \qquad\qquad 24\tfrac{2}{3} + C = \frac{2}{3}$$
$$= \int (t^{1/2} + 6 + 9t^{-1/2})\,dt \qquad\qquad C = -24$$
$$s(t) = \frac{2}{3}t^{3/2} + 6t + 9 \cdot 2\,t^{1/2} + C \qquad s(t) = \frac{2}{3}t^{3/2} + 6t + 18t^{1/2} - 24$$

25. $v(t) = \left(\sqrt{t}+4\right)\left(\sqrt{t}-4\right),\ s(2) = 3$

$$s(t) = \int \left(\sqrt{t}+4\right)\left(\sqrt{t}-4\right)\,dt \qquad s(2) = \frac{4}{2} - 16 \cdot 2 + C = 3$$
$$= \int (t - 16)\,dt \qquad\qquad 2 - 32 + C = 3$$
$$s(t) = \frac{t^2}{2} - 16t + C \qquad\qquad -30 + C = 3$$
$$\qquad\qquad C = 33$$
$$s(t) = \frac{t^2}{2} - 16t + 33$$

26. $v(t) = \sin t,\ s(\pi) = 1$

$$s(t) = \int \sin t\,dt \qquad s(\pi) = -\cos(\pi) + C = 1$$
$$s(t) = -\cos t + C \qquad\qquad -(-1) + C = 1$$
$$\qquad\qquad C = 0$$
$$s(t) = -\cos t$$

In Exercises 27–37, find the velocity function v and position function s corresponding to the given acceleration function and initial conditions.

27. $a(t) = 2$, $v(0) = 3$, $s(0) = 0$

$$v(t) = \int a(t)\, dt = \int 2\, dt = 2t + C_1$$
$$v(t) = 2t + C_1$$
$$v(0) = 0 + C_1 = 3, \ C_1 = 3$$
$$v(t) = 2t + 3$$

$$s(t) = \int v(t)\, dt$$
$$= \int (2t + 3)\, dt = t^2 + 3t + C_2$$
$$s(t) = t^2 + 3t + C_2$$
$$s(0) = 0 + C_2 = 0$$
$$C_2 = 0$$
$$s(t) = t^2 + 3t$$

28. $a(t) = -2$, $v(0) = 6$, $s(0) = 10$

$$v(t) = \int -2\, dt$$
$$v(t) = -2t + C_1$$
$$v(0) = 0 + C_1 = 6$$
$$C_1 = 6$$
$$v(t) = -2t + 6$$

$$s(t) = \int (-2t + 6)\, dt$$
$$s(t) = -t^2 + 6t + C_2$$
$$s(0) = 0 + 0 + C_2 = 10$$
$$C_2 = 10$$
$$s(t) = -t^2 + 6t + 10$$

29. $a(t) = 3t$, $v(0) = 0$, $s(0) = 20$

$$v(t) = \int 3t\, dt$$
$$v(t) = \frac{3t^2}{2} + C_1$$
$$v(0) = 0 + C_1 = 0$$
$$C_1 = 0$$
$$v(t) = \frac{3t^2}{2}$$

$$s(t) = \int \frac{3t^2}{2}\, dt$$
$$s(t) = \frac{3t^3}{6} + C_2$$
$$s(t) = \frac{t^3}{2} + C_2$$
$$s(0) = 0 + C_2 = 20$$
$$C_2 = 20$$
$$s(t) = \frac{t^3}{2} + 20$$

30. $a(t) = 200$, $v(0) = 100$, $s(0) = 200$

$$v(t) = \int 200\, dt$$
$$v(t) = 200t + C_1$$
$$v(0) = 0 + C_1 = 100$$
$$C_1 = 100$$
$$v(t) = 200t + 100$$

$$s(t) = \int (200t + 100)\, dt$$
$$s(t) = \frac{200t^2}{2} + 100t + C_2$$
$$s(t) = 100t^2 + 100t + C_2$$
$$s(0) = 0 + 0 + C_2 = 200$$
$$C_2 = 200$$
$$s(t) = 100t^2 + 100t + 200$$

31. $a(t) = 4t + 4$, $v(0) = 8$, $s(0) = 12$

$$v(t) = \int (4t + 4)\, dt$$
$$v(t) = 2t^2 + 4t + C_1$$
$$v(0) = 0 + 0 + C_1 = 8$$
$$C_1 = 8$$
$$v(t) = 2t^2 + 4t + 8$$
$$s(t) = \int (2t^2 + 4t + 8)\, dt$$

$$s(t) = \frac{2t^3}{3} + \frac{4t^2}{2} + 8t + C_2$$
$$s(t) = \frac{2t^3}{3} + 2t^2 + 8t + C_2$$
$$s(0) = 0 + 0 + 0 + C_2 = 12$$
$$C_2 = 12$$
$$s(t) = \frac{2t^3}{3} + 2t^2 + 8t + 12$$

32. $a(t) = \cos t,\; v(0) = 1,\; s(0) = 9$

$$v(t) = \int \cos t\, dt$$
$$v(t) = \sin t + C_1$$
$$v(0) = \sin(0) + C_1 = 1$$
$$0 + C_1 = 1$$
$$C_1 = 1$$
$$v(t) = \sin t + 1$$

$$s(t) = \int (\sin t + 1)\, dt$$
$$s(t) = -\cos t + t + C_2$$
$$s(0) = -\cos(0) + 0 + C_2 = 9$$
$$-1 + C_2 = 9$$
$$C_2 = 10$$
$$s(t) = -\cos t + t + 10$$

33. $a(t) = \sin t,\; v(0) = 0,\; s(0) = 2$

$$v(t) = \int \sin t\, dt$$
$$v(t) = -\cos t + C_1$$
$$v(0) = -\cos(0) + C_1 = 0$$
$$-1 + C_1 = 0$$
$$C_1 = 1$$
$$v(t) = -\cos t + 1$$

$$s(t) = \int (-\cos t + 1)\, dt$$
$$s(t) = -\sin t + t + C_2$$
$$s(0) = -\sin(0) + 0 + C_2 = 2$$
$$0 + 0 + C_2 = 2$$
$$C_2 = 2$$
$$s(t) = -\sin t + t + 2$$

34. $a(t) = 2t,\; v(1) = 1,\; s(1) = 1$

$$v(t) = \int 2t\, dt$$
$$v(t) = t^2 + C_1$$
$$v(1) = 1 + C_1 = 1$$
$$C_1 = 0$$
$$v(t) = t^2$$

$$s(t) = \int t^2\, dt$$
$$s(t) = \frac{t^3}{3} + C_2$$
$$s(1) = \frac{1}{3} + C_2 = 1$$
$$C_2 = \frac{2}{3}$$
$$s(t) = \frac{t^3}{3} + \frac{2}{3}$$

35. $a(t) = -2\sin t,\; v(\pi) = -3,\; s(\pi) = -2\pi$

$$v(t) = \int -2\sin t\, dt$$
$$v(t) = 2\cos t + C_1$$
$$v(\pi) = 2\cos(\pi) + C_1 = -3$$
$$2(-1) + C_1 = -3$$
$$-2 + C_1 = -3$$
$$C_1 = -1$$

$$v(t) = 2\cos t - 1$$
$$s(t) = \int (2\cos t - 1)\, dt$$
$$s(t) = 2\sin t - t + C_2$$
$$s(\pi) = 2\sin \pi - \pi + C_2 = -2\pi$$
$$0 - \pi + C_2 = -2\pi$$
$$C_2 = -\pi$$

36. $a(t) = -2$, $v(2) = -5$, $s(1) = -2$

$$
\begin{aligned}
v(t) &= \int -2\, dt & s(t) &= \int (-2t - 1)\, dt \\
v(t) &= -2t + C_1 & s(t) &= -t^2 - t + C_2 \\
v(2) &= -2(2) + C_1 = -5 & s(1) &= -1 - 1 + C_2 = -2 \\
&\quad -4 + C_1 = -5 & & \quad C_2 = 0 \\
&\quad C_1 = -1 & s(t) &= -t^2 - t - 0 \\
v(t) &= -2t - 1 & s(t) &= -t^2 - t
\end{aligned}
$$

37. $a(t) = 0$, $v(1) = 1$, $s(2) = 3$

$$
\begin{aligned}
v(t) &= \int 0\, dt & s(t) &= \int 1\, dt \\
v(t) &= C_1 & s(t) &= t + C_2 \\
v(1) &= C_1 = 1 & s(2) &= 2 + C_2 = 3 \\
&\quad C_1 = 1 & & \quad C_2 = 1 \\
v(t) &= 1 & s(t) &= t + 1
\end{aligned}
$$

38. $a(t) = 2 - t$, $v(t) = 2t - \dfrac{1}{2}t^2 + C_1$, $s(t) = t^2 - \dfrac{1}{6}t^3 + C_1 t + C_2$

We can pick $C_1 = C_2 = 0$. Then $s(t) = t^2 - \dfrac{t^3}{6}$ and $v(t) = 2t - \dfrac{t^2}{2} = 2t\left(1 - \dfrac{t}{4}\right)$ and $2t\left(1 - \dfrac{t}{4}\right) = 0$ when $t = 0, 4$. So $v(t) > 0$ for t in $(0, 4)$. Thus for the above position function, the particle does not change direction for t in the interval $(0, 4)$.

39. $v(0) = 15$ m/s

$$
\begin{aligned}
a &= -9.8 \text{ m/s}^2 & C_1 &= 15 \\
v(t) &= \int -9.8\, dt & v(t) &= -9.8t + 15 \\
v(t) &= -9.8t + C_1 & \text{velocity} &= 0 \text{ when } -9.8t + 15 = 0 \\
v(0) &= 0 + C_1 = 15 & 15 &= 9.8t \\
& & \frac{15}{9.8} &= t
\end{aligned}
$$

   (a) Stops rising at $t = \dfrac{15}{9.8} \approx 1.53$ secs.

   (b) Strikes the ground at $t = 2 \cdot \dfrac{15}{9.8} \approx 3.06$ secs.

40. The rocket reaches maximum height when velocity $= 0$.

$$
\begin{aligned}
s(t) &= -4.9t^2 + 45t \\
v(t) &= s'(t) = -9.8t + 45 \\
&\quad -9.8t + 45 = 0 \\
&\quad 45 = 9.8t \\
&\quad \frac{45}{9.8} = t \approx 4.59 \text{ secs.}
\end{aligned}
$$

41. A particle moves along a line with constant acceleration $a = 3m/s^2$.

   (a) How fast is it moving after $6\,s$ if its initial velocity is $v(0) = 10m/s$?

$$v(t) = \int 3\,dt$$
$$= 3t + C_1$$
$$v(0) = 0 + C_1 = 10$$
$$C_1 = 10$$

$$v(t) = 3t + 10\,m/s$$
$$v(6) = 3.6 + 10$$
$$= 28\,m/s$$

(b) What is its initial velocity $v(0)$ if its speed after $3\,s$ is $15m/s$?

$$v(3) = 3 \cdot 3 + C_1 = 15$$
$$9 + C_1 = 15$$
$$C_1 = 6$$
$$v(t) = 3t + 6\,m/s$$

Find $v(0)$,

$$v(0) = 3 \cdot 0 + 6 = 6\,m/s$$

42. $s = 150m$, $a = -9.8\,m/s^2$, $v(0) = 0$, , $s(0) = 0$

$$v(t) = \int -9.8\,dt$$
$$v(t) = -9.8t + C_1$$
$$v(0) = 0 + C_1 = 0$$
$$C_1 = 0$$
$$v(t) = -9.8t$$

$$s(t) = \int -9.8t\,dt$$
$$s(t) = \frac{-9.8t^2}{2} + C_2$$
$$s(0) = 0 + C_2 = 0$$
$$C_2 = 0$$
$$s(t) = -4.9t^2$$

(a)

Find t when $s = -150m$
$$-4.9t^2 = -150$$
$$t^2 = \frac{150}{4.9}$$
$$t = \sqrt{\frac{150}{4.9}} \approx 5.53 \text{ secs to hit the ground.}$$

(b)

$$v\left(\sqrt{\frac{150}{4.9}}\right) = -9.8\sqrt{\frac{150}{4.9}}$$

$$v \approx -54.22\,m/\text{secs}$$

54.22 m/sec speed at which it hits the ground.

43. What constant acceleration will enable the driver of an automobile to increase its speed from $20\,m/s$ to $25\,m/s$ in $10\,s$? Find $a(t) = K$.

if $\left.\begin{array}{rcl} v(0) &=& 20\,m/s \\ v(10) &=& 25\,m/s \end{array}\right\}$ $\begin{array}{rcl} v(t) &=& \int K\,dt \\ v(t) &=& Kt + C_1 \end{array}$

365

$$v(0) \;=\; 0 + C_1 = 20 \qquad\qquad v(10) \;=\; 10K + 20 = 25$$
$$C_1 = 20 \qquad\qquad\qquad\qquad\quad 10K = 5$$
$$v(t) \;=\; Kt + 20 \qquad\qquad\qquad\qquad K = \frac{1}{2}$$

$$\text{Therefore } a(t) \;=\; \frac{1}{2}\ m/s^2.$$

44. What constant negative acceleration (deceleration) is required to bring an automobile travelling at a rate of 72 km/h to a full stop in 100 m? Find $a(t) = -K$.

$$v(0) \;=\; 72\ Km/hr \;=\; \frac{72000m}{3600\ \text{sec}} \;=\; 20\ m/s$$

$$v(t) \;=\; \int -K\,dt$$

$$=\; -Kt + C_1$$

$$v(0) \;=\; 0 + C_1 = 20$$

$$C_1 = 20$$

$$v(t) \;=\; -Kt + 20$$

$$s(t) \;=\; \int -Kt + 20\,dt$$

$$s(t) \;=\; \frac{-Kt^2}{2} + 20t + C_2$$

$$s(0) = 0$$
$$C_2 = 0$$
$$\text{when } s = 100$$
$$v = 0 \Rightarrow$$
$$-Kt + 20 = 0$$
$$-Kt = -20$$
$$t = \frac{20}{K}.$$

So
$$\frac{-Kt^2}{2} + 20t = 100$$

$$-K \cdot \frac{400}{2K^2} + 20 \cdot \frac{20}{K} = 100$$

$$\frac{-200}{K} + \frac{400}{K} = 100$$

$$\frac{200}{K} = 100$$

$$K = 2$$

$$\Rightarrow a = \; -2\ m/s^2.$$

45. Figures a–d illustrate the graphs $y = f(x)$ of four particular functions, and Figures i–iv represent the slope portraits for their antiderivatives F. Match each function f with the slope portrait of its antiderivative F. Then, on each of the four slope portraits, illustrate the antidifferentiation process by

sketching the graphs $y = F(x)$ of three different antiderivatives.

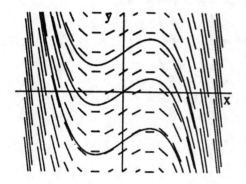

Exercise 45(a)
Matches ii

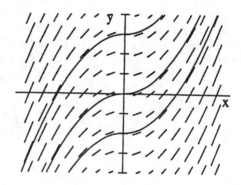

Exercise 45(b)
Matches i

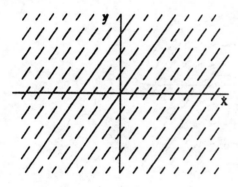

Exercise 45(c)
Matches iv

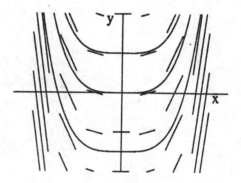

Exercise 45(d)
Matches iii

In each of Exercises 46–49, sketch the slope portrait for the antiderivatives F of the given function f. Then, on each portrait, sketch the graphs $y = F(x)$ of three antiderivatives of f. Finally, using the formula for $f(x)$ calculate the indefinite integral of f, and reconcile the graphs $y = F(x)$ with the integral.

46. $f(x) = x$

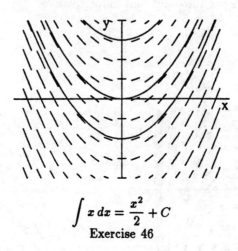

$$\int x\,dx = \frac{x^2}{2} + C$$
Exercise 46

47. $f(x) = 1 - x$

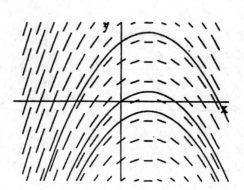

m	
0	$1 - x = 0,\ x = 1$
1	$1 - x = 1,\ x = 0$
2	$1 - x = 2,\ x = -1$
-1	$1 - x = -1,\ x = 2$

$$\int (1 - x)\,dx = x - \frac{x^2}{2} + C$$
Exercise 47

48. $f(x) = 1 - x^2$

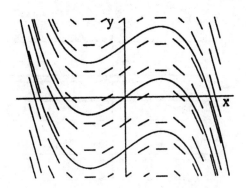

m	
0	$1 - x^2 = 0, \; x = \pm 1$
1	$1 - x^2 = 1, \; x = 0$
-3	$1 - x^2 = -3, \; x = \pm 2$

$$\int (1 - x^2)\, dx = x - \frac{x^3}{3} + C$$

Exercise 48

49. $f(x) = \sin x$

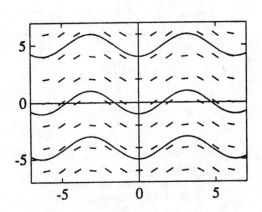

m	
0	$\sin x = 0$
	$x = 0, \pi, 2\pi, \ldots$
	$x = -\pi, -2\pi, \ldots$
1	$\sin x = 1$
	$x = \pi/2, 5\pi/2, 9\pi/2, \ldots$
-1	$\sin x = -1$
	$x = 3\pi/2, 7\pi/2, \ldots$

$$\int \sin x \, dx = -\cos x + C$$

Exercise 49

50. Plot the slope portrait that illustrates Theorem 2. The slope portrait is 0 everywhere. Therefore the only solutions are horizontal lines.

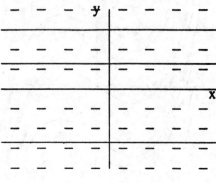

Exercise 50

51. Let $f(x) = 1/x$. From the sketch, the $\lim\limits_{x \to 0} \dfrac{1}{x}$ appears to be $-\infty$.

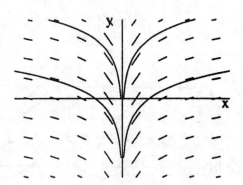

m	
0	$1/x = 0$ undefined
1	$1/x = 1$, $x = 1$
2	$1/x = 2$, $x = 1/2$
3	$1/x = 3$, $x = 1/3$
1/3	$1/x = 1/3$, $x = 3$
1/2	$1/x = 1/2$, $x = 2$

Exercise 51

52. For each f, determine which of the graphs i–ix are graphs of antiderivatives of f.

(a) Graphs vi and ix match Graph a.

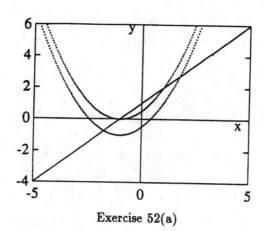

Exercise 52(a)

(b) Graph iv matches Graph b.

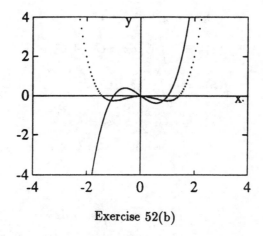

Exercise 52(b)

(c) Graph vii matches Graph c.

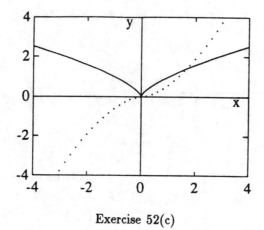

Exercise 52(c)

53. $f''(x) = 0$ $f(0) = -2$ $f(1) = 1$

$$
\begin{aligned}
f'(x) &= \int 0 \, dx \\
&= C_1 \\
f(x) &= \int C_1 \, dx \\
&= C_1 x + C_2 \\
\text{Now } f(0) &= 0 + C_2 = -2, \ C_2 = -2 \\
f(x) &= C_1 x - 2 \\
\text{and } f(1) &= C_1 - 2 = 1 \\
C_1 &= 3. \\
\text{Therefore } f(x) &= 3x - 2.
\end{aligned}
$$

Only one such function that satisfies the two given initial conditions.

54. $f^{(3)}(x) = 0$

$$
\begin{aligned}
f''(x) &= \int 0 \, dx = C_1 \\
f'(x) &= \int C_1 \, dx = C_1 x + C_2 \\
f(x) &= \int (C_1 x + C_2) \, dx = \frac{C_1 x^2}{2} + C_2 x + C_3 \\
f(x) &= C_1 \frac{x^2}{2} + C_2 x + C_3 \\
&\text{or just } C_1 x^2 + C_2 x + C_3.
\end{aligned}
$$

55. $f''(x) = 2 \sin x$

$$
\begin{aligned}
f'(x) &= 2 \int \sin x \, dx \\
&= -2 \cos x + C_1 \\
f(x) &= (-2 \cos x + C_1) \, dx \\
&= -2 \sin x + C_1 x + C_2
\end{aligned}
$$

56. Find cubic $p(x)$ such that

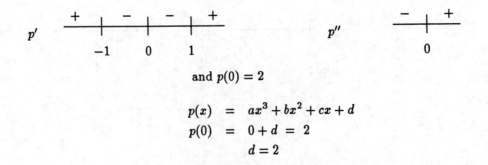

and $p(0) = 2$

$$
\begin{aligned}
p(x) &= ax^3 + bx^2 + cx + d \\
p(0) &= 0 + d = 2 \\
d &= 2
\end{aligned}
$$

$$p(x) = ax^3 + bx^2 + cx + 2$$
$$p'(x) = 3ax^2 + 2bx + c$$
$$p''(x) = 6ax + 2b$$

If $p''(x)$ changes sign at 0, then $p''(0) = 0 + 2b = 0$
$$b = 0.$$
$$\text{Therefore } p''(x) = 6ax$$
$$p'(x) = 3ax^2 + c$$

If $p'(x)$ changes sign at ± 1, then ± 1 must be factors of $p'(x)$.
$$p'(1) = 3a + c = 0$$
$$c = -3a$$
$$\text{Therefore, } p'(x) = 3ax^2 - 3a$$
$$\text{and } p(x) = ax^3 + bx^2 + cx + d$$
$$= ax^3 + 0\,x^2 - 3ax + 2$$
$$p(x) = ax^3 - 3ax + 2$$

57. F is continuous on $[-2, 12]$; $F(0) = 0$; and F is the antiderivative of f.

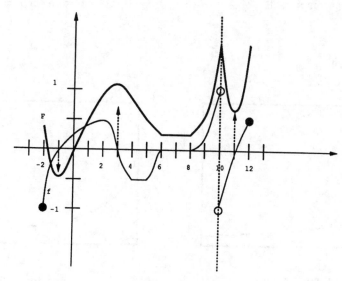

Exercise 57

58. Range: $[-1, 1] \times [-2, 2]$.

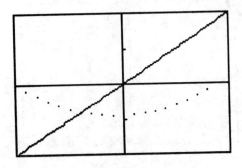

Graph for $n = 20$

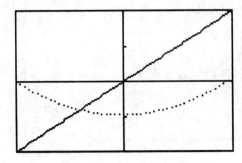

Graph for $n = 50$

59. Range: $[-2\pi, 2\pi] \times [-1.5, 1.5]$.

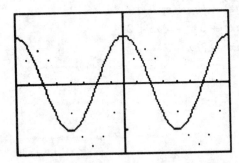

Graph for $n = 20$

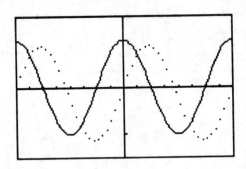

Graph for $n = 50$

60. Range: $\left[\frac{1}{2}, 5\right] \times [0, 3]$.

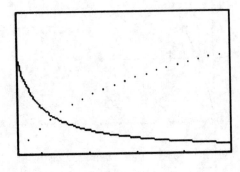

Graph for $n = 20$

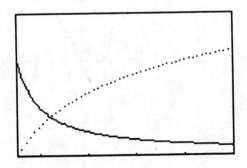

Graph for $n = 50$

61. Range: $[-2, 2] \times [0, 3]$.

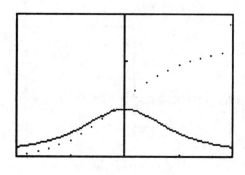 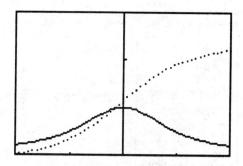

Graph for $n = 20$ Graph for $n = 50$

62. Graphs given below have the range $[-1, 1] \times [-4, 4]$. The graph on the left has $m = 1$ and $C = -1, 0$, and 2 and the graph on the right has $m = -2$ and $C = -1, 0$, and 2. For each graph $n = 50$.

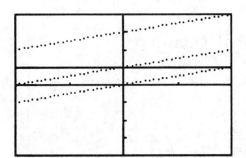 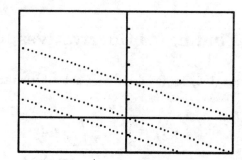

The graphs produced are always the form $y = mx + C$. This is because the tangent line approximation will be exact if $F''(x) = 0$, which is the case if $F(x) = mx + C$.

63. a) The graph to the left is that of the points $(t, r(t))$ using the range $[0, 30] \times [0, 35]$. The graph to the right is the prediction of the shape of $n(t)$.

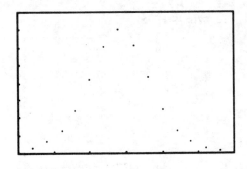 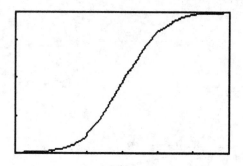

b) Use the tangent line approximation with $h = 2$ and $n(0) = 1$. This takes the form of $n(t_i) = n(t_{i-1}) + hr(t_{i-1})$ where $t_0 = 0$, $t_1 = 2$, $t_2 = 4,\ldots$, etc. The values of $n(t_i)$ are given in the following table.

t	$n(t)$
0	1.0000
2	2.4798
4	5.5572
6	11.8760
8	24.5174
10	48.5335
12	89.9374
14	150.4876
16	220.4014
18	281.9954
20	324.6750
22	349.6395
24	362.8414
26	369.4562
28	372.6816
30	374.2334

After 30 days all 375 persons have the disease.

5.2 Finding Antiderivatives by Substitution

In Exercises 1–6, use the designated substitution to find the specified antiderivatives.

1. $\displaystyle \int x^5 \sin\left(x^6 + 2\right) dx = \frac{1}{6} \int \sin u \, du \qquad\qquad u = x^6 + 2$
    $du = 6x^5 \, dx$

    $\displaystyle = \frac{1}{6}(-\cos u) + C$

    $\displaystyle = \frac{-\cos(x^6 + 2)}{6} + C$

2. $\displaystyle \int (x+1)\sqrt{x^2 + 2x}\, dx = \frac{1}{2} \int \sqrt{u}\, du \qquad\qquad u = x^2 + 2x$
    $du = 2x + 2\, dx$

    $\displaystyle \phantom{\int (x+1)\sqrt{x^2+2x}\,dx} = \frac{1}{2} \int u^{1/2}\, du \qquad\qquad\; = 2(x+1)\, dx$

    $\displaystyle \phantom{\int (x+1)\sqrt{x^2+2x}\,dx} = \frac{1}{2} \cdot \frac{2}{3} u^{3/2} + C$

    $\displaystyle \phantom{\int (x+1)\sqrt{x^2+2x}\,dx} = \frac{1}{3}\left(x^2 + 2x\right)^{3/2} + C$

3. $\displaystyle \int \frac{\cos\sqrt{x}}{\sqrt{x}}\, dx = 2 \int \cos u \, du \qquad u = \sqrt{x} = x^{1/2}$

    $\displaystyle \phantom{\int \frac{\cos\sqrt{x}}{\sqrt{x}}\, dx} = 2\sin u + C \qquad du = \frac{1}{2} x^{-1/2}\, dx$

    $\displaystyle \phantom{\int \frac{\cos\sqrt{x}}{\sqrt{x}}\, dx} = 2\sin\sqrt{x} + C$

    $\displaystyle \phantom{\int \frac{\cos\sqrt{x}}{\sqrt{x}}\, dx = 2\sin\sqrt{x} + C \qquad} du = \frac{1}{2\sqrt{x}}\, dx$

376

4. $\displaystyle \int \frac{\sin x \cos x}{\sqrt{1+\cos^2 x}}\, dx$ $= -\dfrac{1}{2}\displaystyle\int \frac{1}{\sqrt{u}}\, du$ $u = 1+\cos^2 x$

$\qquad\qquad\qquad\qquad\quad = -\dfrac{1}{2}\displaystyle\int u^{-1/2}\, du$ $du = 2\cos x(-\sin x)\, dx$

$\qquad\qquad\qquad\qquad\qquad\qquad\qquad\qquad\qquad\qquad du = -2\sin x\cos x\, dx$

$\qquad\qquad\qquad\qquad\quad = -\dfrac{1}{2}\cdot 2u^{1/2}+C$

$\qquad\qquad\qquad\qquad\quad = -\sqrt{u}+C$

$\qquad\qquad\qquad\qquad\quad = -\sqrt{1+\cos^2 x}+C$

5. $\displaystyle \int x\sqrt{x+2}\, dx$ $= \displaystyle\int (u-2)u^{1/2}\, du$ $u = x+2$

$\qquad\qquad\qquad\quad = \displaystyle\int \left(u^{3/2}-2u^{1/2}\right) du$ $du = dx$

$\qquad\qquad\qquad\qquad\qquad\qquad\qquad\qquad\qquad\quad x = u-2$

$\qquad\qquad\qquad\quad = \dfrac{2}{5}u^{5/2}-2\cdot\dfrac{2}{3}u^{3/2}+C$

$\qquad\qquad\qquad\quad = \dfrac{2}{5}(x+2)^{5/2}-\dfrac{4}{3}(x+2)^{3/2}+C$

6. $\displaystyle \int x\sin x^2 \cos x^2\, dx$ $= \dfrac{1}{2}\displaystyle\int u\, du$ $u = \sin x^2$

$\qquad\qquad\qquad\qquad\qquad = \dfrac{1}{2}\cdot\dfrac{u^2}{2}+C$ $du = \cos x^2 \cdot 2x\, dx$

$\qquad\qquad\qquad\qquad\qquad = \dfrac{u^2}{4}+C$

$\qquad\qquad\qquad\qquad\qquad = \dfrac{\sin^2 x^2}{4}+C$

In Exercises 7–14, use the method of substitution to find the antiderivative. Check your answer by differentiating.

7. $\displaystyle \int x\sqrt{x^2+1}\, dx$ $= \dfrac{1}{2}\displaystyle\int u^{1/2}\, du$ $u = x^2+1$

$\qquad\qquad\qquad\qquad = \dfrac{1}{2}\cdot\dfrac{2}{3}u^{3/2}+C$ $du = 2x\, dx$

$\qquad\qquad\qquad\qquad = \dfrac{1}{3}(x^2+1)^{3/2}+C$

8. $\displaystyle \int x\cos x^2\, dx$ $= \dfrac{1}{2}\displaystyle\int \cos u\, du$ $u = x^2$

$\qquad\qquad\qquad\quad = \dfrac{1}{2}(\sin u)+C$ $du = 2x\, dx$

$\qquad\qquad\qquad\quad = \dfrac{1}{2}\sin x^2 +C$

9. $\displaystyle \int \sec 2\theta \tan 2\theta\, d\theta$ $= \dfrac{1}{2}\displaystyle\int \sec u\tan u\, du$ $u = 2\theta$

$\qquad\qquad\qquad\qquad\quad = \dfrac{1}{2}\sec u + C$ $du = 2\, d\theta$

$\qquad\qquad\qquad\qquad\quad = \dfrac{1}{2}\sec 2\theta + C$

10. $\displaystyle\int \csc(\pi x)\cot(\pi x)\,dx \;=\; \frac{1}{\pi}\int \csc u \cot u\,du$ 　　　$u = \pi x$
$$du = \pi\,dx$$
$$= \frac{1}{\pi}(-\csc u) + C$$
$$= -\frac{1}{\pi}\csc(\pi x) + C$$

11. $\displaystyle\int x\csc^2(x^2)\,dx \;\Rightarrow\; \frac{1}{2}\int \csc^2 u\,du$ 　　　$u = x^2$
$$du = 2x\,dx$$
$$= -\frac{1}{2}\cot u + C$$
$$= -\frac{1}{2}\cot(x^2) + C$$

12. $\displaystyle\int \frac{\sin\sqrt{x}\,\cos\sqrt{x}}{\sqrt{x}}\,dx \;=\; 2\int u\,du$ 　　　$u = \sin\sqrt{x}$
$$du = \cos\sqrt{x}\cdot\frac{1}{2}x^{-1/2}\,dx$$
$$= 2\frac{u^2}{2} + C$$
$$du = \frac{\cos\sqrt{x}}{2\sqrt{x}}\,dx$$
$$= u^2 + C$$
$$= \sin^2\sqrt{x} + C$$

13. $\displaystyle\int\left(1-t^2\right)\sqrt{3t^3-9t+9}\,dt \;=\; -\frac{1}{9}\int u^{1/2}\,du$ 　　　$u = 3t^3 - 9t + 9$
$$du = 9t^2 - 9\,dt$$
$$= -\frac{1}{9}\cdot\frac{2}{3}u^{3/2} + C \qquad\qquad = 9(t^2-1)\,dt$$
$$= -9(1-t^2)\,dt$$
$$= -\frac{2}{27}(3t^3-9t+9)^{3/2} + C$$

14. $\displaystyle\int (1-x)^3\,dx \;=\; -\int u^3\,du$ 　　　$u = 1-x$
$$du = -dx$$
$$= -\frac{u^4}{4} + C$$
$$= \frac{-(1-x)^4}{4} + C$$

In Exercises 15–24, find the antiderivative.

15. $\displaystyle\int \sec^2(x-\pi)\,dx \;=\; \int \sec^2 u\,du$ 　　　$u = x-\pi$
$$du = dx$$
$$= \tan u + C$$
$$= \tan(x-\pi) + C \text{ or } \tan x + C$$

16. $\displaystyle\int x\sec^2 x^2\,dx \;=\; \frac{1}{2}\int \sec^2 u\,du$ 　　　$u = x^2$
$$du = 2x\,dx$$
$$= \frac{1}{2}\tan u + C$$
$$= \frac{1}{2}\tan x^2 + C$$

17. $\displaystyle\int \frac{(2\sqrt{x}+3)^2}{\sqrt{x}}\,dx \;=\; \int u^2\,du$ 　　　$u = 2\sqrt{x}+3$
$$du = 2\cdot\frac{1}{2}x^{-1/2}\,dx$$
$$= \frac{u^3}{3} + C$$
$$= \frac{(2\sqrt{x}+3)^3}{3} + C \qquad du = \frac{1}{\sqrt{x}}\,dx$$

18. $\displaystyle\int (t^3 - 6t + 7)^5 (2 - t^2)\, dt \;=\; -\frac{1}{3}\int u^5\, du$

$u \;=\; t^3 - 6t + 7$
$du \;=\; (3t^2 - 6)\, dt$
$ \;=\; -3(-t^2 + 2)\, dt$

$\displaystyle \;=\; -\frac{1}{3}\frac{u^6}{6} + C$

$\displaystyle \;=\; -\frac{u^6}{18} + C$

$\displaystyle \;=\; \frac{-(t^3 - 6t + 7)^6}{18} + C$

19. $\displaystyle\int (x^2 + 1)^2\, dx \;=\; \int (x^4 + 2x^2 + 1)\, dx$

$\displaystyle \;=\; \frac{x^5}{5} + \frac{2x^3}{3} + x + C$

20. $\displaystyle\int \sqrt{t}(t + 1)\, dt \;=\; \int t^{3/2} + t^{1/2}\, dt$

$\displaystyle \;=\; \frac{2}{5}t^{5/2} + \frac{2}{3}t^{3/2} + C$

21. $\displaystyle\int x\sqrt{x - 1}\sqrt{x + 1}\, dx \;=\; \int x\sqrt{x^2 - 1}\, dx$

$u \;=\; x^2 - 1$
$du \;=\; 2x\, dx$

$\displaystyle \;=\; \frac{1}{2}\int u^{1/2}\, du$

$\displaystyle \;=\; \frac{1}{2}\cdot\frac{2}{3}u^{3/2} + C$

$\displaystyle \;=\; \frac{1}{3}(x^2 - 1)^{3/2} + C$

22. $\displaystyle\int \sqrt{1 - \sin x}\,\cos x\, dx \;=\; -\int u^{1/2}\, du$

$u \;=\; 1 - \sin x$
$du \;=\; -\cos x\, dx$

$\displaystyle \;=\; -\frac{2}{3}u^{3/2} + C$

$\displaystyle \;=\; -\frac{2}{3}(1 - \sin x)^{3/2} + C$

23. $\displaystyle\int x\sec(\pi - x^2)\tan(\pi - x^2)\, dx \;=\; -\frac{1}{2}\int \sec u \tan u\, du$

$u \;=\; \pi - x^2$
$du \;=\; -2x\, dx$

$\displaystyle \;=\; -\frac{1}{2}\sec u + C$

$\displaystyle \;=\; -\frac{1}{2}\sec(\pi - x^2) + C$

24. $\displaystyle\int \frac{\sec^2\sqrt{2x + 1}}{\sqrt{2x + 1}}\, dx \;=\; \int \sec^2 u\, du$

$u \;=\; \sqrt{2x + 1} \;=\; (2x + 1)^{1/2}$
$du \;=\; \frac{1}{2}(2x + 1)^{-1/2}\cdot 2\, dx$

$\displaystyle \;=\; \tan u + C$

$\displaystyle \;=\; \tan \sqrt{2x + 1} + C$

$\displaystyle \;=\; \frac{1}{\sqrt{2x + 1}}\, dx$

25. $\displaystyle\int (x^5 - 2x^3)(x^6 - 3x^4)^{5/2}\, dx \;=\; \frac{1}{6}\int u^{5/2}\, du$

$u \;=\; x^6 - 3x^4$
$du \;=\; 6x^5 - 12x^3\, dx$
$ \;=\; 6(x^5 - 2x^3)\, dx$

$\displaystyle \;=\; \frac{1}{6}\cdot\frac{2}{7}u^{7/2} + C$

$\displaystyle \;=\; \frac{1}{21}(x^6 - 3x^4)^{7/2} + C$

26. $\displaystyle\int x^3(1-x^4)^{10}\,dx \;=\; -\frac{1}{4}\int u^{10}\,du$ $\qquad u \;=\; 1-x^4$
$\qquad\qquad\qquad\qquad du \;=\; -4x^3\,dx$

$\displaystyle\qquad\qquad\qquad\quad =\; -\frac{1}{4}\frac{u^{11}}{11}+C$

$\displaystyle\qquad\qquad\qquad\quad =\; \frac{-(1-x^4)^{11}}{44}+C$

27. $\displaystyle\int x^2(1-x^4)^2\,dx \;=\; \int x^2(1-2x^4+x^8)\,dx$

$\displaystyle\qquad\qquad\qquad\quad =\; \int x^2-2x^6+x^{10}\,dx$

$\displaystyle\qquad\qquad\qquad\quad =\; \frac{x^3}{3}-\frac{2x^7}{7}+\frac{x^{11}}{11}+C$

28. $\displaystyle\int \frac{x^3+2x}{\sqrt[3]{x^4+4x^2}}\,dx \;=\; \frac{1}{4}\int \frac{1}{\sqrt[3]{u}}\,du$ $\qquad u \;=\; x^4+4x^2$
$\qquad\qquad\qquad\qquad\qquad\quad du \;=\; 4x^3+8x\,dx$
$\qquad\qquad\qquad\qquad\qquad\quad du \;=\; 4(x^3+2x)\,dx$

$\displaystyle\qquad\qquad\qquad\qquad\quad =\; \frac{1}{4}\int u^{-1/3}\,du$

$\displaystyle\qquad\qquad\qquad\qquad\quad =\; \frac{1}{4}\cdot\frac{3}{2}u^{2/3}+C$

$\displaystyle\qquad\qquad\qquad\qquad\quad =\; \frac{3}{8}(x^4+4x^2)^{2/3}+C$

29. $\displaystyle\int \frac{t^2+2}{\sqrt{t^3+6t}}\,dt \;=\; \frac{1}{3}\int \frac{1}{\sqrt{u}}\,du$ $\qquad u \;=\; t^3+6t$
$\qquad\qquad\qquad\qquad\qquad du \;=\; 3t^2+6\,dt$
$\qquad\qquad\qquad\qquad\qquad\quad =\; 3(t^2+2)\,dt$

$\displaystyle\qquad\qquad\qquad\quad =\; \frac{1}{3}\int u^{-1/2}\,du$

$\displaystyle\qquad\qquad\qquad\quad =\; \frac{1}{3}\cdot 2u^{1/2}+C$

$\displaystyle\qquad\qquad\qquad\quad =\; \frac{2}{3}\sqrt{t^3+6t}+C$

30. $\displaystyle\int \frac{\cos^3\sqrt{x}\sin\sqrt{x}}{\sqrt{x}}\,dx \;=\; -2\int u^3\,du$ $\qquad u \;=\; \cos\sqrt{x}$
$\qquad\qquad\qquad\qquad\qquad\qquad du \;=\; -\sin\sqrt{x}\cdot\frac{1}{2}x^{-1/2}\,dx$

$\displaystyle\qquad\qquad\qquad\qquad\quad =\; -\frac{2u^4}{4}+C$
$\qquad\qquad\qquad\qquad\qquad\qquad\quad =\; -\frac{\sin\sqrt{x}}{2\sqrt{x}}\,dx$

$\displaystyle\qquad\qquad\qquad\qquad\quad =\; -\frac{u^4}{2}+C$

$\displaystyle\qquad\qquad\qquad\qquad\quad =\; -\frac{\cos^4\sqrt{x}}{2}+C$

31. $\displaystyle\int \frac{1}{(1-4x)^{2/3}}\,dx \;=\; -\frac{1}{4}\int u^{-2/3}\,du$ $\qquad u \;=\; 1-4x$
$\qquad\qquad\qquad\qquad\qquad\quad du \;=\; -4\,dx$

$\displaystyle\qquad\qquad\qquad\qquad\quad =\; -\frac{1}{4}\cdot 3u^{1/3}+C$

$\displaystyle\qquad\qquad\qquad\qquad\quad =\; -\frac{3}{4}(1-4x)^{1/3}+C$

32. $\displaystyle\int\left(\frac{1+\tan^2\sqrt{x}}{\sqrt{x}}\right)dx = 2\int 1+\tan^2 u\,du$ $u = \sqrt{x}$

$\displaystyle = 2\int 1+(\sec^2 u-1)\,du$ $du = \frac{1}{2}x^{-1/2}\,dx$

$\displaystyle = 2\int\sec^2 u\,du$ $= \frac{1}{2\sqrt{x}}\,dx$

$\displaystyle = 2\tan u + C$

$\displaystyle = 2\tan\sqrt{x} + C$

33. $\displaystyle\int x^3\sqrt{x^2+1}\,dx = \frac{1}{2}\int(u-1)u^{1/2}\,du$ $u = x^2+1$

$\displaystyle = \frac{1}{2}\int u^{3/2}-u^{1/2}\,du$ $du = 2x\,dx$

                               $x^2 = u-1$

$\displaystyle = \frac{1}{2}\cdot\left(\frac{2}{5}u^{5/2}-\frac{2}{3}u^{3/2}\right)+C$

$\displaystyle = \frac{1}{5}(x^2+1)^{5/2}-\frac{1}{3}(x^2+1)^{3/2}+C$

34. $\displaystyle\int\frac{x}{\sqrt[3]{(1+x^2)}}\,dx = \int\frac{(u-1)}{u^{2/3}}\,du$ $u = 1+x$

$\displaystyle = \int(u-1)u^{-2/3}\,du$ $du = dx$

                               $x = u-1$

$\displaystyle = \int\left(u^{1/3}-u^{-2/3}\right)du$

$\displaystyle = \frac{3}{4}u^{4/3}-3u^{1/3}+C$

$\displaystyle = \frac{3}{4}(1+x)^{4/3}-3(1+x)^{1/3}+C$

In Exercises 35–37, state a substitution appropriate to each part and find the antiderivative.

35. $u = 3x+4$
       $du = 3\,dx$

(a) $\displaystyle\int(3x+4)^{10}\,dx = \frac{1}{3}\int u^{10}\,du = \frac{1}{3}\cdot\frac{u^{11}}{11}+C = \frac{1}{33}(3x+4)^{11}+C$

(b) $\displaystyle\int\sqrt{3x+4}\,dx = \frac{1}{3}\int u^{1/2}\,du = \frac{1}{3}\cdot\frac{2}{3}u^{3/2}+C = \frac{2}{9}(3x+4)^{3/2}+C$

(c) $\displaystyle\int\sin(3x+4)\,dx = \frac{1}{3}\int\sin u\,du = \frac{1}{3}(-\cos u)+C = -\frac{1}{3}\cos(3x+4)+C$

(d) $\displaystyle\int\sin^2(3x+4)\cos(3x+4)\,dx = \int\sin^2 u\cos u\,du$ $w = \sin u$

                                                       $dw = \cos u\,du$

$\displaystyle = \frac{1}{3}\int w^2\,dw$

$\displaystyle = \frac{1}{3}\frac{w^3}{3}+C$

$\displaystyle = \frac{\sin^3 u}{9}+C$

$\displaystyle = \frac{\sin^3(3x+4)}{9}+C$

36. (a) $\displaystyle\int \sin(3x)\,dx \;=\; \frac{1}{3}\int \sin u\,du \qquad \begin{aligned} u &= 3x \\ du &= 3dx \end{aligned}$

$$= \; -\frac{1}{3}\cos u + C$$

$$= \; -\frac{1}{3}\cos(3x) + C$$

(b) $\displaystyle\int \sin(3x)\cos(3x)\,dx = \frac{1}{3}\int u\,du \;=\; \frac{1}{3}\frac{u^2}{2} + C$ or do substitution in 2 steps

$$\begin{aligned} u &= \sin(3x) \\ du &= 3\cos(3x)\,dx \end{aligned}$$

$$= \; \frac{u^2}{6} + C$$

$$= \; \frac{\sin^2(3x)}{6} + C$$

(c) Substitution same as in (b) or do in 2 steps.

$$\int \sin^4(3x)\cos(3x)\,dx = \frac{1}{3}\int u^4\,du = \frac{1}{3}\frac{u^5}{5} + C = \frac{u^5}{15} + C = \frac{\sin^5(3x)}{15} + C$$

37. $\begin{aligned} u &= x^2 + 6 \\ du &= 2x\,dx \end{aligned}$

(a) $\displaystyle\int (x^2+6)^{21}x\,dx = \frac{1}{2}\int u^{21}\,du = \frac{1}{2}\frac{u^{22}}{22} + C = \frac{(x^2+6)^{22}}{44} + C$

(b) $\displaystyle\int \frac{x}{\sqrt{x^2+6}}\,dx = \frac{1}{2}\int u^{-1/2}\,du = \frac{1}{2}\cdot 2u^{1/2} + C = \sqrt{x^2+6} + C$

(c) $\displaystyle\int x\sec^2(x^2+6)\,dx = \frac{1}{2}\int \sec^2 u\,du = \frac{1}{2}\tan u + C = \frac{1}{2}\tan(x^2+6) + C$

38. $\displaystyle\int \sin nx\,dx \;=\; \frac{1}{n}\int \sin u\,du \qquad \begin{aligned} u &= nx \\ du &= n\,dx \end{aligned}$

$$= \; \frac{1}{n}(-\cos u) + C$$

$$= \; -\frac{\cos nx}{n} + C$$

39.   I. $\displaystyle\int \sin x \cos x\,dx \;=\; \int u\,du \qquad \begin{aligned} u &= \sin x \\ du &= \cos x\,dx \end{aligned}$

$$= \; \frac{u^2}{2} + C_1$$

$$= \; \frac{\sin^2 x}{2} + C_1$$

II. $\sin 2x = 2\sin x \cos x$

$$\int \sin x \cos x\,dx \;=\; \frac{1}{2}\int \sin 2x\,dx \qquad \begin{aligned} u &= 2x \\ du &= 2\,dx \end{aligned}$$

$$= \; \frac{1}{4}\int \sin u\,du$$

$$= \; \frac{1}{4}(-\cos u) + C_2$$

$$= \; -\frac{\cos(2x)}{4} + C_2$$

III. Use $\cos 2x = \cos^2 x - \sin^2 x$ and $\cos^2 x = 1 - \sin^2 x$ in answer II.

$$
\begin{aligned}
-\frac{\cos(2x)}{4} + C_2 &= \frac{-(\cos^2 x - \sin^2 x)}{4} + C_2 \\
&= \frac{-\cos^2 x + \sin^2 x}{4} + C_2 \\
&= \frac{-(1 - \sin^2 x) + \sin^2 x}{4} + C_2 \\
&= \frac{-1 + 2\sin^2 x}{4} + C_2 \\
&= \frac{\sin^2 x}{2} - \frac{1}{4} + C_2
\end{aligned}
$$

Yes, answers are the same if $C_1 = -\dfrac{1}{4} + C_2$.

40. (a) Using the substitution $u = x + 1$, calculate the antiderivatives $\displaystyle\int x\sqrt{x+1}\,dx$

$$\text{and } \int \sqrt{x+1}\,dx.$$

$$
\begin{aligned}
u &= x + 1 \\
du &= dx \\
x &= u - 1
\end{aligned}
$$

I. $\displaystyle\int x\sqrt{x+1}\,dx = \int (u-1)u^{1/2}\,du = \int u^{3/2} - u^{1/2}\,du = \frac{2}{5}u^{5/2} - \frac{2}{3}u^{3/2} + C$

$$= \frac{2}{5}(x+1)^{5/2} - \frac{2}{3}(x+1)^{3/2} + C$$

II. $\displaystyle\int \sqrt{x+1}\,dx = \int u^{1/2}\,du = \frac{2}{3}u^{3/2} + C = \frac{2}{3}(x+1)^{3/2} + C$

(b) Explain why

$$\int x\sqrt{x+1}\,dx \neq x\int \sqrt{x+1}\,dx.$$

They are not equal since

$$
\begin{aligned}
\frac{d}{dx}\int x\sqrt{x+1}\,dx &= x\sqrt{x+1} \neq \frac{d}{dx}\left(x\int \sqrt{x+1}\,dx\right) \\
&= x\frac{d}{dx}\int \sqrt{x+1}\,dx + \left(\frac{dx}{dx}\right)\int \sqrt{x+1}\,dx = x\sqrt{x+1} + \frac{2}{3}(x+1)^{3/2} + C
\end{aligned}
$$

41. Four of the six antiderivatives can be calculated using the methods discussed in Sections 5.1 and 5.2.

I. Which are they? (a, c, d, and e). Three of these four can be determined using a substitution. Determine which three and indicate the substitution.

(a) $\displaystyle\int \sec^2(2\theta)\,d\theta$ $\quad$ $\begin{aligned} u &= 2\theta \\ du &= 2\,d\theta \end{aligned}$ $\quad$ substitution

(d) $\displaystyle\int u(u^2+1)^{10}\, du$ $w = u^2+1$ substitution

$\qquad\qquad\qquad\qquad\qquad dw = 2u\, du$

(e) $\displaystyle\int \frac{t+1}{\sqrt{t^2+2t}}\, dt$ $u = t^2+2t$ substitution

$\qquad\qquad\qquad\qquad\qquad du = (2t+2)\, dt$

$\qquad\qquad\qquad\qquad\qquad\quad = 2(t+1)\, dt$

II. How do you calculate the fourth?

$\qquad$ [(c)] $\displaystyle\int (u^2+1)^{10}\, du$

$\qquad$ Since substitution won't work, we can multiply it out and then integrate.

III. One of the three amenable to the method of substitution can also be determined without using substitution. Which one? How can this antiderivative be determined if substitution is not used?

$\qquad$ [(d)] $\displaystyle\int u(u^2+1)^{10}\, du$

$\qquad$ It can be multiplied out.

5.3   Differential Equations

In Exercises 1–14, find the general solution of the differential equation.

1. $\displaystyle\frac{dy}{dx} = x-1$

$\quad y = \displaystyle\int (x-1)\, dx$

$\quad y = \dfrac{x^2}{2} - x + C$

2. $\displaystyle\frac{dy}{dx} = 2+\sec^2 x$

$\quad y = \displaystyle\int (2+\sec^2 x)\, dx$

$\quad y = 2x + \tan x + C$

3. $\displaystyle\frac{dy}{dx} = x^2 - \frac{1}{x^2}$

$\quad y = \displaystyle\int \left(x^2 - x^{-2}\right) dx$

$\qquad = \dfrac{x^3}{3} - \dfrac{x^{-1}}{-1} + C$

$\quad y = \dfrac{x^3}{3} + \dfrac{1}{x} + C$

4. $\displaystyle\frac{dy}{dx} = 4\cos 2x$ $u = 2x$

$\qquad\qquad\qquad\qquad du = 2\, dx$

$\quad y = 4\displaystyle\int \cos 2x\, dx$

$\quad y = 4 \cdot \dfrac{1}{2}\displaystyle\int \cos u\, du$

$\quad y = 2\sin u + C$

$\quad y = 2\sin 2x + C$

5. $\displaystyle\frac{dy}{dx} = \sin x \cos x \sqrt{1+\sin^2 x}$ $u = 1+\sin^2 x$

$\qquad\qquad\qquad\qquad\qquad\qquad du = 2\sin x \cos x\, dx$

$\quad y = \displaystyle\int \sin x \cos x \sqrt{1+\sin^2 x}\, dx$

$\quad y = \dfrac{1}{2}\displaystyle\int u^{1/2}\, du$

$\quad y = \dfrac{1}{2}\cdot\dfrac{2}{3}\, u^{3/2} + C$

$\quad y = \dfrac{1}{3}(1+\sin^2 x)^{3/2} + C$

6. $\dfrac{dy}{dx} = (x+1)\sqrt{x^2+2x+2}$ $\quad\quad u = x^2+2x+2$

$\quad\quad y = \displaystyle\int (x+1)\sqrt{x^2+2x+2}\,dx \quad\quad du = 2x+2\,dx$

$\quad\quad\quad\quad\quad\quad\quad\quad\quad\quad\quad\quad\quad\quad\quad\quad\quad\quad du = 2(x+1)\,dx$

$\quad\quad\quad = \dfrac{1}{2}\displaystyle\int u^{1/2}\,du$

$\quad\quad y = \dfrac{1}{2}\cdot\dfrac{2}{3}\,u^{3/2} + C$

$\quad\quad y = \dfrac{1}{3}(x^2+2x+2)^{3/2} + C$

7. $dy = (x - \sqrt{x})\,dx$

$\quad y = \displaystyle\int x - x^{1/2}\,dx$

$\quad y = \dfrac{x^2}{2} - \dfrac{2}{3}\,x^{3/2} + C$

8. $dy + 3\,dx = 0$

$\quad y = -3\displaystyle\int dx$

$\quad y = -3x + C$

9. $\quad dy = \dfrac{x}{y}\,dx$

$\quad\displaystyle\int y\,dy = \displaystyle\int x\,dx$

$\quad\quad \dfrac{y^2}{2} = \dfrac{x^2}{2} + C$

$\dfrac{y^2}{2} - \dfrac{x^2}{2} = C$

$\quad y^2 - x^2 = C \text{ or } y^2 = x^2 + C$

Note: C is an arbitrary constant.
$2C$ is just another constant!

10. $\quad \dfrac{dy}{dx} = \dfrac{x}{y^2}$

$\quad \displaystyle\int y^2\,dy = \displaystyle\int x\,dx$

$\quad\quad \dfrac{y^3}{3} = \dfrac{x^2}{2} + C$

$\quad\quad y = \left(\dfrac{3}{2}x^2 + C\right)^{1/3}$

11. $\quad dy = -4xy^2\,dx$

$\quad \dfrac{1}{y^2}\,dy = -4x\,dx$

$\quad \displaystyle\int y^{-2}\,dy = -4\displaystyle\int x\,dx$

$\quad\quad \dfrac{y^{-1}}{-1} = -\dfrac{4x^2}{2} + C$

$\quad\quad -\dfrac{1}{y} = -2x^2 + C$

$\quad\quad\quad y = \dfrac{1}{2x^2 + C}$

12. $\quad dy = \dfrac{\sqrt{x}}{\sqrt{y}}\,dx$

$\quad \sqrt{y}\,dy = \sqrt{x}\,dx$

$\quad \displaystyle\int y^{1/2}\,dy = \displaystyle\int x^{1/2}\,dx$

$\quad\quad \dfrac{2y^{3/2}}{3} = \dfrac{2}{3}x^{3/2} + C$

$\quad\quad y^{3/2} = x^{3/2} + C$

$\quad\quad y = \left(x^{3/2} + C\right)^{2/3}$

13. $\quad \dfrac{dy}{dx} = -\dfrac{x}{y}$

$\quad \displaystyle\int y\,dy = -\displaystyle\int x\,dx$

$\quad\quad \dfrac{y^2}{2} = -\dfrac{x^2}{2} + C$

$\dfrac{y^2}{2} + \dfrac{x^2}{2} = C$

$\quad x^2 + y^2 = C \text{ or } y^2 = C - x^2$

14. $\quad dy = y^2\,dx$

$\quad \displaystyle\int y^{-2}\,dy = \displaystyle\int dx$

$\quad\quad \dfrac{y^{-1}}{-1} = x + C$

$\quad\quad -\dfrac{1}{y} = x + C$

$\quad\quad y = -\dfrac{1}{x + C}$

In Exercises 15–26, find the solution of the initial value problem.

15. $\dfrac{dy}{dx} = \dfrac{x}{y}, \quad y(1) = 4$

$$
\begin{aligned}
y\, dy &= x\, dx \\
\frac{y^2}{2} &= \frac{x^2}{2} + C \\
y^2 &= x^2 + C \\
y &= \pm\sqrt{x^2 + C}
\end{aligned}
$$

Now $y(1) = \pm\sqrt{1^2 + C} = 4$ implies that we must use the positive radical.

$$
\begin{aligned}
16 &= 1^2 + C \\
15 &= C \\
y &= \sqrt{x^2 + 15}
\end{aligned}
$$

16. $\dfrac{dy}{dx} = \dfrac{x}{y}, \quad y(1) = -4$

$$
\begin{aligned}
\int y\, dy &= \int x\, dx \\
\frac{y^2}{2} &= \frac{x^2}{2} + C \\
y^2 &= x^2 + C \\
y &= \pm\sqrt{x^2 + C}
\end{aligned}
$$

Now $y(1) = \pm\sqrt{1^2 + C} = -4$ implies we must use the negative radical.

$$
\begin{aligned}
16 &= 1^2 + C \\
15 &= C \\
y &= -\sqrt{x^2 + 15}
\end{aligned}
$$

17. $\dfrac{dy}{dx} = -\dfrac{x}{y}, \quad y\!\left(\sqrt{3}\right) = 1$

$$
\begin{aligned}
\int y\, dy &= -\int x\, dx \\
\frac{y^2}{2} &= -\frac{x^2}{2} + C \\
y^2 &= -x^2 + C \\
(1)^2 &= -\left(\sqrt{3}\right)^2 + C
\end{aligned}
$$

Now $y\!\left(\sqrt{3}\right) = \pm\sqrt{-3 + C} = 1$ implies we must use the positive radical.

$$
\begin{aligned}
C &= 4 \\
y &= \sqrt{-x^2 + 4}
\end{aligned}
$$

18. $2y\, dy = \sqrt{x}\, dx, \quad y(0) = 2$

$$
\begin{aligned}
2\int y\, dy &= \int x^{1/2}\, dx \\
\frac{2y^2}{2} &= \frac{2}{3}\, x^{3/2} + C \\
y^2 &= \frac{2}{3}\, x^{3/2} + C \\
y &= \pm\sqrt{\frac{2}{3}\, x^{3/2} + C}
\end{aligned}
$$

Now $y(0) = \pm\sqrt{0 + C} = 2$ implies we must use the positive radical.

$$
\begin{aligned}
C &= 2^2 = 4 \\
y &= \sqrt{\frac{2}{3}\, x^{3/2} + 4}
\end{aligned}
$$

19. $dy = \dfrac{x^2}{\left(1 + x^3\right)^2}\, dx, \quad y(0) = \dfrac{1}{2}$

$$y = \int \frac{x^2}{(1+x^3)^2}\,dx \qquad u = 1+x^3$$
$$du = 3x^2\,dx$$

$$y = \frac{1}{3}\int \frac{1}{u^2}\,du$$

$$y = \frac{1}{3}\int u^{-2}\,du$$

$$y = \frac{1}{3}\frac{u^{-1}}{-1} + C$$

$$y = -\frac{1}{3u} + C$$

$$y = \frac{-1}{3(1+x^3)} + C$$

$$\frac{1}{2} = \frac{-1}{3(1+0)} + C$$

$$\frac{1}{2} = \frac{-1}{3} + C$$

$$\frac{5}{6} = C$$

$$y = \frac{-1}{3(1+x^3)} + \frac{5}{6}$$

20. $dy = \dfrac{x}{y\sqrt{1+x^2}}\,dx, \quad y(0) = 2$

$$\int y\,dy = \int \frac{x}{\sqrt{1+x^2}}\,dx \qquad u = 1+x^2$$
$$du = 2x\,dx$$

$$\frac{y^2}{2} = \frac{1}{2}\int u^{-1/2}\,du$$

$$\frac{y^2}{2} = \frac{1}{2}\cdot 2u^{1/2} + C$$

$$y^2 = 2\sqrt{1+x^2} + C$$

$$y = \pm\sqrt{2(1+x^2)^{1/2} + C}$$

Now $y(0) = \pm\sqrt{2(1+0)^{1/2} + C} = 2$

implies we must
use the positive radical.

$$\sqrt{2+C} = 2$$
$$2+C = 2^2$$
$$C = 2$$
$$y = \sqrt{2(1+x^2)^{1/2} + 2}$$

21. $dy = y^2(1+x^2)\,dx, \quad y(0) = 1$

$$\int y^{-2}\,dy = \int (1+x^2)\,dx$$

$$\frac{y^{-1}}{-1} = x + \frac{x^3}{3} + C$$

$$-\frac{1}{y} = \frac{3x + x^3 + C}{3}$$

$$y = -\frac{3}{x^3 + 3x + C}$$

$$1 = -\frac{3}{0+0+C}$$

$$C = -3$$

$$y = -\frac{3}{x^3 + 3x - 3}$$

22. $\dfrac{dy}{dx} = 2x^2y^2, \quad y(0) = -3$

$$y^{-2}\,dy = 2x^2\,dx$$

$$\frac{y^{-1}}{-1} = \frac{2x^3}{3} + C$$

$$-\frac{1}{y} = \frac{2x^3 + C}{3}$$

$$y = -\frac{3}{2x^3 + C}$$

$$-3 = -\frac{3}{0+C}$$

$$-3C = -3$$

$$C = 1$$

$$y = -\frac{3}{2x^3 + 1}$$

23. $\dfrac{d^2y}{dx^2} = x^2, \quad y(0) = 2, \quad y'(0) = 4$

$$y' = \frac{dy}{dx} = \frac{x^3}{3} + C_1 \Rightarrow \begin{array}{rcl} 4 &=& 0 + C_1 \\ 4 &=& C_1 \end{array} \qquad \begin{array}{rcl} 2 &=& 0 + 0 + C_2 \\ 2 &=& C_2 \end{array}$$

$$y' = \frac{x^3}{3} + 4 \qquad\qquad y = \frac{x^4}{12} + 4x + 2$$

$$y = \frac{x^4}{12} + 4x + C_2$$

24. $\dfrac{d^2y}{dx^2} = \sin x, \quad y(0) = \pi, \quad y'(0) = 2$

$$\begin{array}{rcl} y' &=& -\cos x + C_1 \\ 2 &=& -\cos 0 + C_1 \\ 2 &=& -1 + C_1 \\ 3 &=& C_1 \\ y' &=& -\cos x + 3 \end{array} \qquad\qquad \begin{array}{rcl} y &=& -\sin x + 3x + C_2 \\ \pi &=& -\sin 0 + 0 + C_2 \\ \pi &=& C_2 \\ y &=& -\sin x + 3x + \pi \end{array}$$

25. $\dfrac{d^2y}{dx^2} = -16\sin 4x, \quad y(\pi) = \pi + 1, \quad y'(\pi) = 1$

$$\begin{array}{rcl} y' &=& -16\displaystyle\int \sin 4x\, dx \qquad u = 4x \\ & & \qquad\qquad\qquad\quad du = 4dx \\ y' &=& -4\displaystyle\int \sin u\, du \\ y' &=& -4(-\cos u) + C_1 \\ y' &=& 4\cos u + C_1 \\ y' &=& 4\cos(4x) + C_1 \\ 1 &=& 4\cos 4\pi + C_1 \\ 1 &=& 4 + C_1 \\ -3 &=& C_1 \\ y' &=& 4\cos(4x) - 3 \end{array}$$

$$\begin{array}{rcl} y' &=& 4\cos 4x - 3 \\ y &=& 4\displaystyle\int \cos 4x\, dx - \int 3\, dx \\ y &=& 4 \cdot \dfrac{1}{4}\displaystyle\int \cos u\, du - 3x + C_2 \\ y &=& \sin u - 3x + C_2 \\ y &=& \sin 4x - 3x + C_2 \\ & & \pi + 1 = \sin 4\pi - 3\pi + C_2 \\ & & \pi + 1 = 0 - 3\pi + C_2 \\ & & 4\pi + 1 = C_2 \end{array}$$

Therefore $y = \sin 4x - 3x + 4\pi + 1$.

26. $\dfrac{d^2y}{dx^2} = \sqrt{x}, \quad y(1) = \dfrac{1}{3}, \quad y'(1) = 1$

$$\begin{array}{rcl} y' &=& \displaystyle\int x^{1/2}\, dx \\ y' &=& \dfrac{2}{3}x^{3/2} + C_1 \\ 1 &=& \dfrac{2}{3}(1)^{3/2} + C_1 \\ \dfrac{1}{3} &=& C_1 \\ y' &=& \dfrac{2}{3}x^{3/2} + \dfrac{1}{3} \\ y &=& \displaystyle\int\left(\dfrac{2}{3}x^{3/2} + \dfrac{1}{3}\right) dx \end{array}$$

$$\begin{array}{rcl} y &=& \dfrac{2}{3}\cdot\dfrac{2}{5}x^{5/2} + \dfrac{1}{3}x + C_2 \\ y &=& \dfrac{4}{15}x^{5/2} + \dfrac{1}{3}x + C_2 \\ \dfrac{1}{3} &=& \dfrac{4}{15} + \dfrac{1}{3} + C_2 \\ C_2 &=& -\dfrac{4}{15} \\ y &=& \dfrac{4}{15}x^{5/2} + \dfrac{1}{3}x - \dfrac{4}{15} \end{array}$$

27. $\dfrac{dy}{dx} = x - \sqrt{x}, \quad y(1) = 2$

$$y = \int \left(x - x^{1/2}\right) dx$$

$$y = \frac{x^2}{2} - \frac{2}{3} x^{3/2} + C$$

$$2 = \frac{1}{2} - \frac{2}{3} + C$$

$$2 = -\frac{1}{6} + C$$

$$\frac{13}{6} = C$$

$$y = \frac{x^2}{2} - \frac{2}{3} x^{3/2} + \frac{13}{6}$$

28. Normal slope $= \dfrac{y}{x}$, $y(2) = 1$

Tan slope $= \dfrac{dy}{dx} = -\dfrac{x}{y}$

$$\int y \, dy = -\int x \, dx$$

$$\frac{y^2}{2} = -\frac{x^2}{2} + C$$

$$y^2 = -x^2 + C$$

Now $y(2) = \pm\sqrt{C - 2^2} = 1$

implies we must use the positive radical

$$1 = -2^2 + C$$

$$5 = C$$

$$y = \sqrt{5 - x^2}$$

29. $\dfrac{dy}{dx} = 2$ for $x \le 1$, $\dfrac{dy}{dx} = x + 1$ for $x > 1$

$$y = \int 2 \, dx$$

$$y = 2x + C_1 \text{ for } x \le 1$$

$$y(0) = 1$$

$$y(0) = 1 = C_1$$

$$y = \int (x + 1) \, dx$$

$$y = \frac{x^2}{2} + x + C_2 \text{ for } x > 1$$

Therefore $y = 2x + 1$ for $x \le 1$.

If the curve is continuous then $\displaystyle\lim_{x \to 1+} \frac{x^2}{2} + x + C_2 = f(1) = 2(1) + 1 = 3$

$$\frac{1}{2} + 1 + C_2 = 3$$

$$C_2 = \frac{3}{2}$$

$$y = \frac{x^2}{2} + x + \frac{3}{2} \text{ for } x > 1$$

and $y(2) = \dfrac{2^2}{2} + 2 + \dfrac{3}{2}$

$$y(2) = \frac{11}{2}.$$

30. Find V if $\dfrac{dV}{dt} = \sqrt{t + 1}$, $V(0) = 5$

$$V = \int (t+1)^{1/2}\, dt \qquad\qquad u = t+1$$
$$\qquad\qquad\qquad\qquad\qquad\qquad\qquad du = dt$$
$$V = \int u^{1/2}\, du$$
$$V(t) = \frac{2}{3} u^{3/2} + C$$
$$V(t) = \frac{2}{3}(t+1)^{3/2} + C$$
$$V(0) = 5 = \frac{2}{3}(0+1)^{3/2} + C,\ C = \frac{13}{3}$$
$$V(t) = \frac{2}{3}(t+1)^{3/2} + \frac{13}{3}$$

31. $\dfrac{dP}{dt} = \dfrac{900}{P^2},\quad P(0) = 10$

$$\int P^2\, dP = 900 \int dt \qquad\qquad \frac{P^3}{3} = 900t + \frac{10^3}{3}$$
$$\frac{P^3}{3} = 900t + C_1 \qquad\qquad P^3 = 2700t + 10^3$$
$$\frac{10^3}{3} = 0 + C_1,\ C_1 = \frac{10^3}{3} \qquad\qquad P(t) = (2700t + 1000)^{1/3}$$

32. $\dfrac{dv}{dt} = Kv^{-1/2},\quad v(0) = 9\ m/s$

$$\int v^{1/2}\, dv = K \int dt \qquad\qquad \frac{2}{3} v^{3/2} = Kt + 18$$
$$\frac{2}{3} v^{3/2} = Kt + C \qquad\qquad v^{3/2} = \frac{3}{2} Kt + 27$$
$$\frac{2}{3} \cdot 9^{3/2} = 0 + C \qquad\qquad v(t) = \left(\frac{3}{2} Kt + 27 \right)^{2/3}$$
$$18 = C$$

33. $\dfrac{dP}{dt} = 4\sqrt{P},\quad P(0) = 10$

$$\int P^{-1/2}\, dP = 4 \int dt$$
$$2\sqrt{P} = 4t + C$$
$$\sqrt{P} = 2t + C$$
$$P = (2t + C)^2$$
$$P(0) = 10 = C^2,\ C = \sqrt{10}$$
$$P(t) = \left(2t + \sqrt{10} \right)^2$$

34. Range: $[0, 4] \times [0, 4]$.

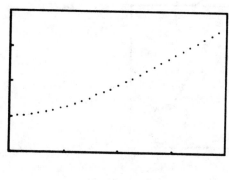 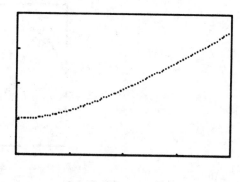

Graph for $n = 30$ Graph for $n = 50$

35. Use the initial condition $y(-2) = -2$. Range: $[-2, 2] \times [-2, 0]$. For $n = 10$ and $n = 30$ the value of y_i eventually overflows due to round-off error. If you try $n = 60$ or $n = 120$ then a graph is produced.

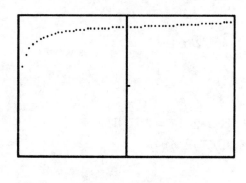 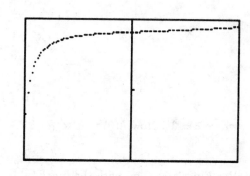

Graph for $n = 60$ Graph for $n = 120$

36. Range: $[0, 6] \times [-1, 7]$. Graphs for all values of n.

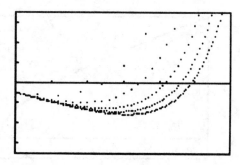

37. Range: $[0, 2\pi] \times [0, 1]$. Graphs for all values of n.

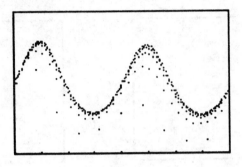

38.   a) Since

$$\frac{dh}{dt} = \frac{-r^2\sqrt{2g}}{h(2r-h)}$$

then

$$\int (2hR - h^2)\, dh = \int -r^2\sqrt{2g}\, dt.$$

So

$$Rh^2 - \frac{1}{3}h^3 = -r^2\sqrt{2h}\,t + c$$

or

$$h^2(3R - h) = -3r^2\sqrt{2g}\,t + \bar{c}$$

where $\bar{c} = 3c$. Now $h(0) = R$ so $R^2(3R - R) = \bar{c}$. Therefore $\bar{c} = 2R^3$. Thus

$$h^2(3R - h) = -3r^2\sqrt{2g}\,t + 2R^3.$$

b) Use Euler's Method with the function

$$\frac{-0.25\sqrt{2(9.8)}}{h(12-h)} = \frac{-1.106797181}{h(12-h)}.$$

With the calculator you would use **(-1.106797181)/(Y(12-Y))**. Taking the range to be $[0, 135] \times [-1, 6]$ and $n = 5(135) = 657$ the following graph is produced.

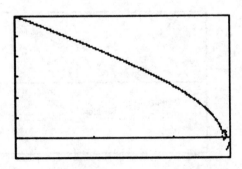

392

The graph crosses the axis at approximately $t = 130.1$.

c. The time taken for the tank to empty is when $h = 0$ so from part a)

$$-3r^2\sqrt{2g}\,t + 2R^3 = 0$$

so $t = \dfrac{2R^3}{3r^2\sqrt{2g}}$. The exact time for part b) is then $t = \dfrac{2(6^3)}{3(0.5)^2\sqrt{2g}} = 130.105138$.

39.   a) Now

$$\frac{dv}{dt} = -kv^2$$

so

$$\int \frac{dv}{v^2}\, dt = -k \int dt.$$

Therefore

$$\frac{-1}{v} = kt + c$$

or

$$v = \frac{-1}{kt + c}.$$

Now $v(0) = 75$ so $75 = -\frac{1}{c}$, i.e., $c = -\dfrac{1}{75}$. Therefore

$$v(t) = \frac{-1}{kt - \frac{1}{75}} = \frac{75}{1 - 75kt}.$$

b) Since $v(2) = 45$ we have that $45 = \dfrac{75}{1 - 75k(2)}$ and so $1 - 150k = \dfrac{75}{45}$ implying that $k = -\dfrac{1}{225}$.
So

$$v(t) = \frac{75}{1 + \frac{75}{225}t} = \frac{225}{3 + t}.$$

Graph with the range: $[0, 100] \times [0, 75]$.

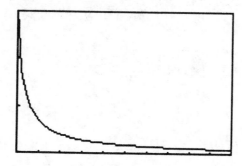

Now 5% of 75 is 3.75 and so we must solve $\dfrac{225}{3 + t} = 3.75$ so $t = 57$.

c) Now $\dfrac{ds}{dt} = \dfrac{225}{3+t}$. So use Euler's Method with the function $\mathbf{225/(3+X)}$. The graph below has the range $[0, 57] \times [0, 700]$ and was produced with $n = 300$.

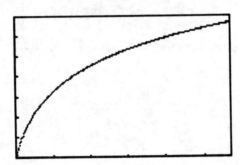

The last value computed by the programs is $s(57)$ and this value is 667.35 feet. More accurate values of this can be found by increasing the value of n. For example, if $n = 1000$, $s(57) \approx 672.02$ and if $n = 2000$, $s(57) \approx 673.03$.

Review Exercises—Chapter 5

In Exercises 1–18, find the indefinite integral for the given function.

1. $f(x) = 6x^2 - 2x + 1$

$$\int \left(6x^2 - 2x + 1\right) dx = 2x^3 - x^2 + x + C$$

2. $f(x) = (x^2 - 6x)^2$

$$\int (x^2 - 6x)^2 \, dx = \int x^4 - 12x^3 + 36x^2 \, dx = \frac{x^5}{5} - 3x^4 + 12x^3 + C$$

3. $y = \dfrac{\sin\sqrt{t}}{\sqrt{t}}$

$$
\begin{aligned}
\int \frac{\sin\sqrt{t}}{\sqrt{t}} \, dt &= 2\int \sin u \, du & u &= \sqrt{t} \\
&= 2(-\cos u) + C & du &= \frac{1}{2}t^{-1/2} \, dt \\
&= -2\cos\sqrt{t} + C & &= \frac{1}{2\sqrt{t}} \, dt
\end{aligned}
$$

4. $f(x) = x\sqrt{9x^4}$

$$\int x\sqrt{9x^4} \, dx = \int x \cdot 3x^2 \, dx = \int 3x^3 \, dx = \frac{3x^4}{4} + C.$$

5. $f(x) = 3\sqrt{x} + \dfrac{3}{\sqrt{x}}$

$$\int \left(3\sqrt{x} + \frac{3}{\sqrt{x}}\right) dx = \int \left(3x^{1/2} + 3x^{-1/2}\right) dx = 3 \cdot \frac{2}{3} x^{3/2} + 3 \cdot 2x^{1/2} + C = 2x^{3/2} + 6x^{1/2} + C$$

6. $y = \left(t + \sqrt{t}\right)^3$

$$\int \left(t + \sqrt{t}\,\right)^3 dt = \int \left[t^3 + 3t^2\left(\sqrt{t}\,\right) + 3t\left(\sqrt{t}\,\right)^2 + \left(\sqrt{t}\,\right)^3\right] dt = \int \left(t^3 + 3t^{5/2} + 3t^2 + t^{3/2}\right) dt$$

$$= \frac{t^4}{4} + \frac{6}{7}\,t^{7/2} + t^3 + \frac{2}{5}\,t^{5/2} + C$$

7. $y = x \sec^2 x^2$

$$\begin{aligned}
\int x \sec^2 x^2\, dx &= \frac{1}{2}\int \sec^2 u\, du &\qquad u &= x^2 \\
&= \frac{1}{2}\tan u + C & du &= 2x\, dx \\
&= \frac{1}{2}\tan x^2 + C
\end{aligned}$$

8. $f(s) = s\sqrt{9 - s^2}$

$$\begin{aligned}
\int s\sqrt{9 - s^2}\, ds &= -\frac{1}{2}\int u^{1/2}\, du &\qquad f(s) &= s\sqrt{9 - s^2} \\
&= -\frac{1}{2}\cdot\frac{2}{3}\,u^{3/2} + C & u &= 9 - s^2 \\
&= -\frac{1}{3}(9 - s^2)^{3/2} + C & du &= -2s\, ds
\end{aligned}$$

9. $f(x) = \dfrac{x^3 - 7x^2 + 6x}{x}$

$$\int \frac{x^3 - 7x^2 + 6x}{x}\, dx = \int x^2 - 7x + 6\, dx = \frac{x^3}{3} - \frac{7x^2}{2} + 6x + C$$

10. $y = \dfrac{x^3 + x^2 - x + 2}{x + 2}$

$$\int \frac{x^3 + x^2 - x + 2}{x + 2}\, dx = \int \frac{(x + 2)(x^2 - x + 1)}{x + 2}\, dx = \int x^2 - x + 1\, dx\,(x \neq -2) = \frac{x^3}{3} - \frac{x^2}{2} + x + C$$

11. $f(t) = t^5\sqrt{t^2 + 1}$

$$\begin{aligned}
\int t^5\sqrt{t^2 + 1}\, dt &= \int t^4\sqrt{t^2 + 1}\cdot t\, dt &\qquad u &= t^2 + 1 \\
&= \frac{1}{2}\int (u - 1)^2 u^{1/2}\, du & du &= 2t\, dt \\
&= \frac{1}{2}\int \left(u^2 - 2u + 1\right)u^{1/2}\, du & t^2 &= u - 1 \\
&= \frac{1}{2}\int \left(u^{5/2} - 2u^{3/2} + u^{1/2}\right) du \\
&= \frac{1}{2}\left(\frac{2}{7}u^{7/2} - 2\cdot\frac{2}{5}\,u^{5/2} + \frac{2}{3}\,u^{3/2}\right) + C \\
&= \frac{1}{7}\,u^{7/2} - \frac{2}{5}\,u^{5/2} + \frac{1}{3}\,u^{3/2} + C \\
&= \frac{1}{7}(t^2 + 1)^{7/2} - \frac{2}{5}(t^2 + 1)^{5/2} + \frac{1}{3}(t^2 + 1)^{3/2} + C
\end{aligned}$$

12. $y = x^4\left(1 + x^3\right)^2$

$$\int x^4 \left(1 + x^3\right)^2 dx = \int x^4 \left(x^6 + 2x^3 + 1\right) dx = \int \left(x^{10} + 2x^7 + x^4\right) dx = \frac{x^{11}}{11} + \frac{2x^8}{8} + \frac{x^5}{5} + C$$

$$= \frac{x^{11}}{11} + \frac{x^8}{4} + \frac{x^5}{5} + C$$

13. $f(x) = x \cos(1 + x^2)$

$$\begin{aligned}
\int x \cos(1 + x^2) \, dx &= \frac{1}{2} \int \cos u \, du \\
&= \frac{1}{2} \sin u + C \\
&= \frac{1}{2} \sin(1 + x^2) + C
\end{aligned}$$

$$\begin{aligned}
u &= 1 + x^2 \\
du &= 2x \, dx
\end{aligned}$$

14. $f(x) = \dfrac{\sec x}{1 + \tan^2 x}$

$$\int \frac{\sec x}{1 + \tan^2 x} \, dx = \int \frac{\sec x}{1 + (\sec^2 x - 1)} \, dx = \int \frac{\sec x}{\sec^2 x} \, dx = \int \frac{1}{\sec x} \, dx = \int \cos x \, dx = \sin x + C$$

15. $y = (2x - 1)(2x + 1)$

$$\int (2x - 1)(2x + 1) \, dx = \int (4x^2 - 1) \, dx = \frac{4}{3} x^3 - x + C$$

16. $y = \dfrac{x}{4x^4 + 4x^2 + 1}$

$$\begin{aligned}
\int \frac{x}{4x^4 + 4x^2 + 1} \, dx &= \int \frac{x \, dx}{\left(2x^2 + 1\right)^2} \\
&= \frac{1}{4} \int u^{-2} \, du \\
&= \frac{1}{4} \frac{u^{-1}}{-1} + C \\
&= \frac{-1}{4} u + C \\
&= \frac{-1}{4\left(2x^2 + 1\right)} + C
\end{aligned}$$

$$\begin{aligned}
u &= 2x^2 + 1 \\
du &= 4x \, dx
\end{aligned}$$

17. $f(x) = |x|$

$$\int |x| \, dx \Rightarrow \begin{cases} \int x \, dx & \text{if } x \geq 0 \\ \int -x \, dx & \text{if } x < 0 \end{cases} \qquad f(x) = \begin{cases} \dfrac{x^2}{2} + C & x \geq 0 \\ -\dfrac{x^2}{2} + C & x < 0 \end{cases}$$

18. $y = \left|1 - t^2\right|$

$$\int \left|1 - t^2\right| \, dt = \begin{cases} \int (1 - t^2) \, dt, & -1 \leq t \leq 1 \\ \int (t^2 - 1) \, dt, & t < -1 \text{ or } t > 1 \end{cases} = \begin{cases} t - \dfrac{t^3}{3} + C & -1 \leq t \leq 1 \\ \dfrac{t^3}{3} - t + C & t < -1 \text{ or } t > 1 \end{cases}$$

In Exercises 19–24, find the particular function satisfying the stated conditions.

19. $f'(x) = 1 + \cos x, \quad f(0) = 3$

$$f(x) = \int (1 + \cos x)\, dx$$
$$f(x) = x + \sin x + C$$
$$f(0) = 0 + 0 + C = 3$$
$$C = 3$$
$$f(x) = x + \sin x + 3$$

20. $f'(x) = \dfrac{x}{\sqrt{1+x^2}}, \quad f(0) = 4$

$$f(x) = \int \frac{x}{\sqrt{1+x^2}}\, dx \qquad u = 1 + x^2$$
$$du = 2x\, dx$$
$$= \frac{1}{2} \int u^{-1/2}\, du$$
$$= \frac{1}{2} \cdot 2u^{1/2} + C$$
$$= \sqrt{u} + C$$
$$= \sqrt{1+x^2} + C$$

$$f(0) = \sqrt{1+0} + C = 4$$
$$1 + C = 4$$
$$C = 3$$
$$f(x) = \sqrt{1+x^2} + 3$$

21. $f''(x) = 3, \quad f'(1) = 6, \quad f(0) = 4$

$$f'(x) = \int 3\, dx$$
$$f'(x) = 3x + C_1$$
$$f'(1) = 3 + C_1 = 6$$
$$C_1 = 3$$
$$f'(x) = 3x + 3$$
$$f(x) = \int (3x + 3)\, dx$$

$$f(x) = \frac{3x^2}{2} + 3x + C_2$$
$$f(0) = 0 + 0 + C_2 = 4$$
$$C_2 = 4$$
$$f(x) = \frac{3x^2}{2} + 3x + 4$$

22. $f''(x) = \sin x - \cos x, \quad f'(0) = 3, \quad f(0) = 0$

$$f'(x) = \int (\sin x - \cos x)\, dx$$
$$f'(x) = -\cos x - \sin x + C$$
$$f'(0) = -\cos 0 - \sin 0 + C = 3$$
$$-1 - 0 + C = 3$$
$$C = 4$$
$$f'(x) = -\cos x - \sin x + 4$$

$$f(x) = \int (-\cos x - \sin x + 4)\, dx$$
$$= -\sin x + \cos x + 4x + C_2$$
$$f(0) = -\sin 0 + \cos 0 + 0 + C_2 = 0$$
$$1 + C_2 = 0$$
$$C_2 = -1$$
$$f(x) = -\sin x + \cos x + 4x - 1$$

23. $f'(x) = x \sec x^2 \tan x^2, \quad f\left(\sqrt{\pi/4}\right) = 1$

$$f(x) = \int x \sec x^2 \tan x^2\, dx \qquad u = x^2$$
$$du = 2x\, dx$$
$$f(x) = \frac{1}{2} \int \sec u \tan u\, du$$
$$= \frac{1}{2} \sec u + C$$
$$= \frac{1}{2} \sec x^2 + C$$

$$f\left(\sqrt{\frac{\pi}{4}}\right) = \frac{1}{2} \sec \left(\sqrt{\frac{\pi}{4}}\right)^2 + C = 1$$
$$= \frac{1}{2} \sec \frac{\pi}{4} + C = 1$$
$$= \frac{1}{2} \cdot \sqrt{2} + C = 1$$
$$C = 1 - \frac{\sqrt{2}}{2} = \frac{2 - \sqrt{2}}{2}$$
$$f(x) = \frac{1}{2} \sec(x^2) + \frac{2 - \sqrt{2}}{2}$$

24. $f''(x) = 2\sec^2 x \tan x$, $f'(0) = 2$, $f(0) = 2$

$$f'(x) = 2\int \sec^2 x \tan x\, dx = 2\int u\, du \qquad \begin{aligned} u &= \sec x \\ du &= \sec x \tan x\, dx \end{aligned}$$

$$= 2\frac{u^2}{2} + C_1$$

$$\begin{aligned} f'(x) &= \sec^2 x + C_1 \\ f'(0) &= \sec^2(0) + C_1 = 2 \\ & \quad 1 + C_1 = 2 \\ & \quad C_1 = 1 \\ f'(x) &= \sec^2 x + 1 \end{aligned} \qquad \begin{aligned} f(x) &= \int (\sec^2 x + 1)\, dx \\ f(x) &= \tan x + x + C_2 \\ f(0) &= \tan(0) + 0 + C_2 = 2 \\ & \quad 0 + C_2 = 2 \\ & \quad C_2 = 2 \\ f(x) &= \tan x + x + 2 \end{aligned}$$

25. $\dfrac{dy}{dx} = 2x$, $y(2) = 9$

$$\begin{aligned} y &= 2\int x\, dx \\ y &= x^2 + C \\ y(2) &= 2^2 + C = 9 \end{aligned} \qquad \begin{aligned} 4 + C &= 9 \\ C &= 5 \\ y &= x^2 + 5 \end{aligned}$$

26. $\dfrac{dy}{dx} = \dfrac{x}{y}$, $y(0) = 1$

$$\begin{aligned} \int y\, dy &= \int x\, dx \\ \frac{y^2}{2} &= \frac{x^2}{2} + C \\ y^2 &= x^2 + C \end{aligned}$$

Now $y(0) = \pm\sqrt{0^2 + C} = 1$
implies we must
use the positive radical.

$$\begin{aligned} \sqrt{0 + C} &= 1 \\ C &= 1 \\ y &= \sqrt{x^2 + 1} \end{aligned}$$

27. Find s if $a = v'(t) = s''(t) = -9.8 m/s^2$, $v(0) = 0$, $s(0) = 0$.

$$\begin{aligned} v(t) &= -\int 9.8\, dt \\ &= -9.8t + C_1 \\ v(0) &= 0 + C_1 = 0 \\ & \quad C_1 = 0 \\ v(t) &= -9.8t \\ -9.8t &= -49 \\ t &= \frac{49}{9.8} = 5 \end{aligned} \qquad \begin{aligned} s(t) &= \int -9.8t\, dt \\ s(t) &= -\frac{9.8t^2}{2} + C_2 \\ s(0) &= 0 + C_2 = 0 \\ & \quad C_2 = 0 \\ s(t) &= -4.9t^2 \\ s(5) &= -4.9(5)^2 \\ &\approx -122.5m \end{aligned}$$

Therefore the initial height is $\approx 122.5m$.

In Exercises 28–31, find the solution of the initial value problem.

28. $\dfrac{dy}{dx} = \sec^2 x$, $y(0) = 1$

$$y = \int \sec^2 x \, dx \qquad\qquad y(0) = \tan 0 + C = 1$$
$$y = \tan x + C \qquad\qquad\qquad 0 + C = 1$$
$$\qquad\qquad\qquad\qquad\qquad\qquad C = 1$$
$$\qquad\qquad\qquad\qquad y = \tan x + 1$$

29. $\dfrac{dy}{dx} = \dfrac{x+1}{y}, \quad y(1) = 2$

$$\int y \, dy = \int (x+1) \, dx \qquad\qquad 2^2 = 1^2 + 2(1) + C$$
$$\frac{y^2}{2} = \frac{x^2}{2} + x + C \qquad\qquad 4 = 1 + 2 + C$$
$$y^2 = x^2 + 2x + C \qquad\qquad 1 = C$$
$$y(1) = 2 \qquad\qquad\qquad y^2 = x^2 + 2x + 1$$
$$\qquad\qquad\qquad\qquad\qquad y^2 = (x+1)^2$$
$$\qquad\qquad\qquad\qquad\qquad y = x + 1$$

30. $dy = 2xy^2 \, dx, \quad y(0) = -\dfrac{1}{2}$

$$\int y^{-2} \, dy = 2x \, dx \qquad\qquad y(0) = -\frac{1}{0+C} = -\frac{1}{2}$$
$$\frac{y^{-1}}{-1} = x^2 + C \qquad\qquad\qquad C = 2$$
$$-\frac{1}{y} = x^2 + C \qquad\qquad\qquad y = -\frac{1}{x^2 + 2}$$
$$y = -\frac{1}{x^2 + C}$$

31. $dy = \dfrac{\sqrt{x+1}}{\sqrt{y}} \, dx, \quad y(0) = 1$

$$\int y^{1/2} \, dy = \int (x+1)^{1/2} \, dx \qquad u = x + 1 \qquad\qquad y(0) = 1$$
$$\qquad\qquad\qquad\qquad\qquad\qquad du = dx \qquad\qquad (0+1)^{3/2} + C = 1$$
$$\frac{2}{3} y^{3/2} = \int u^{1/2} \, du \qquad\qquad\qquad\qquad\qquad C = 0$$
$$\frac{2}{3} y^{3/2} = \frac{2}{3} u^{3/2} + C \qquad\qquad\qquad\qquad y^{3/2} = (x+1)^{3/2}$$
$$y^{3/2} = u^{3/2} + C \qquad\qquad\qquad\qquad\qquad y = x + 1$$
$$y^{3/2} = (x+1)^{3/2} + C$$

32. $v(t) = 2t - (t+1)^{-2}$

(a) $\quad s(t) = \displaystyle\int v(t) \, dt \qquad\quad u = t + 1 \qquad\qquad s(t) = t^2 + \dfrac{1}{t+1} + C$

$$\qquad\qquad\qquad\qquad\qquad\qquad du = dt \qquad\qquad s(0) = 0 + \frac{1}{0+1} + C$$
$$= \int 2t - (t+1)^{-2} \, dt \qquad\qquad\qquad\qquad C = -1$$
$$= \int 2t \, dt - \int u^{-2} \, du$$
$$= t^2 - \frac{u^{-1}}{-1} + C \qquad\qquad\qquad\qquad s(t) = t^2 + \frac{1}{t+1} - 1$$

(b) $\quad a(t) \;=\; v'(t)$ where $v(t) = 2t - (t+1)^{-2}$

$\qquad\qquad = \; 2 + 2(t+1)^{-3}$

$\quad\; a(t) \;=\; 2 + \dfrac{2}{(t+1)^3}$

33. $s(t) = t^3 - 6t^2 + 9t - 4$

(a) $\quad s(0) \;=\; 0 - 0 + 0 - 4$

$\qquad s(0) \;=\; -4$ or 4 units to the left of the origin.

(b) $\quad v(t) \;=\; 3t^2 - 12t + 9$

$\qquad v(0) \;=\; 0 - 0 + 9 = 9$ units moving in a positive direction, i.e. right.

(c) Changes direction when velocity $= 0$.

$\qquad v(t) \;=\; 0, \; 3t^2 - 12t + 9 = 0$

$\qquad\qquad\qquad 3(t^2 - 4t + 3) = 0$

$\qquad\qquad\qquad 3(t-3)(t-1) = 0$

$\qquad\qquad\qquad t = 1, \; 3$

It changes direction at time $t = 1$ and at $t = 3$.

(d) It changes direction twice.

34. $\quad V'(t) \;=\; \sqrt{36+t}$

$\qquad V(t) \;=\; \displaystyle\int \sqrt{36+t}\,dt \qquad u \;=\; 36+t$

$\qquad\qquad\qquad\qquad\qquad\qquad du \;=\; dt$

$\qquad\qquad = \; \displaystyle\int u^{1/2}\,du$

$\qquad V(t) \;=\; \dfrac{2}{3}u^{3/2} + C$

$\qquad V(t) \;=\; \dfrac{2}{3}(36+t)^{3/2} + C$

In 1985 $t = 0$, $\qquad V(0) = 200.$

$\qquad V(0) \;=\; \dfrac{2}{3}(36+0)^{3/2} + C \;=\; 200$

$\qquad\qquad\quad \dfrac{2 \cdot 216}{3} + C \;=\; 200$

$\qquad\qquad\quad 144 + C = 200$

$\qquad\qquad\quad C = 56$

$\qquad V(t) \;=\; \dfrac{2}{3}(36+t)^{3/2} + 56$

Chapter 6

The Definite Integral

6.1 Area and Summation

1. $\displaystyle\sum_{j=1}^{10} j = 1 + 2 + \cdots + 9 + 10$

   [Use special formula] [Eq (4)]

   $= \dfrac{10(10+1)}{2} = 55.$

2. $\displaystyle\sum_{j=1}^{8} j^2 = 1^2 + 2^2 + \cdots + 7^2 + 8^2$

   [Use special formula] [Eq(5)]

   $= \dfrac{8(8+1)(2 \cdot 8 + 1)}{6}$

   $= 204.$

3. $\displaystyle\sum_{j=1}^{12} j^3 = 1^3 + 2^3 + \cdots + 11^3 + 12^3$

   [Use special formula] [Eq(6)]

   $= (12)^2 \dfrac{(12+1)^2}{4}$

   $= 6084.$

4. $\displaystyle\sum_{j=3}^{13} 4 = 4 + 4 + \cdots + 4 + 4$

   $= 11(4) = 44.$

   [Use special formula] [Eq(3)]

5. $\displaystyle\sum_{j=1}^{5} \sin\left(\dfrac{j\pi}{2}\right) = \sin\left(\dfrac{1\pi}{2}\right) + \sin\left(\dfrac{2\pi}{2}\right) + \sin\left(\dfrac{3\pi}{2}\right) + \sin\left(\dfrac{4\pi}{2}\right) + \sin\left(\dfrac{5\pi}{2}\right) = 1 + 0 - 1 + 0 + 1 = 1.$

6. $\displaystyle\sum_{j=1}^{6} \cos\left(\dfrac{j\pi}{3}\right) = \cos\left(\dfrac{1\pi}{3}\right) + \cos\left(\dfrac{2\pi}{3}\right) + \cos\left(\dfrac{3\pi}{3}\right) + \cos\left(\dfrac{4\pi}{3}\right) + \cos\left(\dfrac{5\pi}{3}\right) + \cos\left(\dfrac{6\pi}{3}\right)$

   $= \dfrac{1}{2} - \dfrac{1}{2} - 1 - \dfrac{1}{2} + \dfrac{1}{2} + 1 = 0.$

7. $\displaystyle\sum_{j=2}^{200} (-1)^j = (-1)^2 + (-1)^3 + \cdots + (-1)^{198} + (-1)^{199} + (-1)^{200} = (+1 - 1) + \cdots + (+1 - 1) + (1) =$
   $0 + \cdots + 0 + 1 = 1.$ Method: group the terms in pairs, each of which sums to zero.

8. $\displaystyle\sum_{j=3}^{147} j$

   [Use special formula] [Eq(4)]

$$= \sum_{j=3}^{147} j + \left(\sum_{j=1}^{2} j - \sum_{j=1}^{2} j \right) = \left(\sum_{j=3}^{147} + \sum_{j=1}^{2} \right) - \left(\sum_{j=1}^{2} \right)$$

$$= \left(\sum_{j=1}^{147} j \right) - \left(\sum_{j=1}^{2} j \right) = \left[\frac{(147)(147+1)}{2} \right] - [1+2] = 10,875.$$

9. $2 + 4 + 6 + 8 + 10$
   $$= 2(1) + 2(2) + 2(3) + 2(4) + 2(5)$$
   $$= 2(1 + 2 + 3 + 4 + 5)$$
   $$= 2\sum_{j=1}^{5} j = \sum_{j=1}^{5} 2j$$

10. $5 + 8 + 11 + 14 + 17 + 20$
    $$= (5 + 0 \cdot 3) + (5 + 1 \cdot 3) + \cdots + (5 + 5 \cdot 3)$$
    $$= \sum_{j=0}^{5} (5 + 3j)$$

    Note that there are many ways to express the same value. For example:

    $$30 + \sum_{j=0}^{5} (3j) = 30 + 3\sum_{j=0}^{5} j \text{ or } \sum_{j=1}^{6} (3j + 2)$$

    and so on.

11. $4 + 9 + 16 + 25 + 36$
    $$= 2^2 + 3^2 + 4^2 + 5^2 + 6^2$$
    $$= \sum_{j=2}^{6} j^2$$

12. $8 + 18 + 32 + 50 + 72$
    $$= 2(4) + 2(9) + 2(16) + 2(25) + 2(36)$$
    $$= 2(2^2) + 2(3^2) + 2(4^2) + 2(5^2) + 2(6^2)$$
    $$= \sum_{j=2}^{6} 2j^2$$

13. Since $\Delta x = \frac{b-a}{n} = \frac{4-0}{4} = 1$ then $x_0 = 0$, $x_1 = 1$, $x_2 = 2$, $x_3 = 3$, $x_4 = 4$. Since $f(x) = x$ increases, then for the lower sum, the c_j are the left endpoints. So

$$\underline{S}_4 = \sum_{j=1}^{4} f(c_j)\Delta x$$
$$= 0(1) + 1(1) + 2(1) + 3(1)$$
$$= 6.$$

14. Since $\Delta x = \frac{b-a}{n} = \frac{3-1}{4} = \frac{1}{2}$; then $x_0 = 1$, $x_1 = \frac{3}{2}$, $x_2 = \frac{4}{2}$, $x_3 = \frac{5}{2}$, $x_4 = \frac{6}{2}$. Since $f(x) = 2x + 5$ increases, then the c_j are the left endpoints for a lower sum. So

$$\underline{S}_4 = \sum_{j=1}^{4} f(c_j)\Delta x$$
$$= [2(1) + 5] \left(\frac{1}{2}\right) + \left[2\left(\frac{3}{2}\right) + 5\right] \left(\frac{1}{2}\right) + \left[2\left(\frac{4}{2}\right) + 5\right] \left(\frac{1}{2}\right) + \left[2\left(\frac{5}{2}\right) + 5\right] \left(\frac{1}{2}\right)$$
$$= 17.$$

15. Since $\Delta x = \frac{b-a}{n} = \frac{4-0}{4} = 1$, then $x_0 = 0$, $x_1 = 1$, $x_2 = 2$, $x_3 = 3$, $x_4 = 4$. Since $f(x) = 2x^2 + 3$ increases on $[0,4]$, then the c_j are the left endpoints for a lower sum. So

$$
\begin{aligned}
\underline{S}_4 &= \sum_{j=1}^{4} f(c_j)\Delta x \\
&= [2(0^2) + 3](1) + [2(1)^2 + 3](1) + [2(2)^2 + 3](1) + [2(3)^2 + 3](1) \\
&= 40.
\end{aligned}
$$

16. Since $\Delta x = \frac{b-a}{n} = \frac{0+3}{3} = 1$, then $x_0 = -3$, $x_1 = -2$, $x_2 = -1$, $x_3 = 0$. Since $f(x) = 9 - x^2$ increases on $[-3,0]$, the c_j are the left endpoints for a lower sum. So

$$
\begin{aligned}
\underline{S}_3 &= \sum_{j=1}^{3} f(c_j)\Delta x \\
&= [9 - (-3)^2](1) + [9 - (-2)^2](1) + [9 - (-1)^2]^2(1) \\
&= 13.
\end{aligned}
$$

17. Since $\Delta x = \frac{b-a}{n} = \frac{\pi-0}{4} = \frac{\pi}{4}$ then $x_0 = 0$, $x_1 = \frac{\pi}{4}$, $x_2 = \frac{2\pi}{4}$, $x_3 = \frac{3\pi}{4}$, $x_4 = \frac{4\pi}{4}$. Since $f(x) = \sin(x)$ increases on $\left[0, \frac{\pi}{2}\right]$ and decreases on $\left[\frac{\pi}{2}, \pi\right]$, then for a lower sum the endpoints are $c_1 = 0$, $c_2 = \frac{\pi}{4}$, $c_3 = \frac{3\pi}{4}$, $c_4 = \frac{4\pi}{4}$. So

$$
\begin{aligned}
\underline{S}_4 &= \sum_{j=1}^{4} f(c_j)\Delta x \\
&= [\sin(0)]\frac{\pi}{4} + \left[\sin\left(\frac{\pi}{4}\right)\right]\frac{\pi}{4} + \left[\sin\left(\frac{3\pi}{4}\right)\right]\frac{\pi}{4} + \left[\sin\left(\frac{4\pi}{4}\right)\right]\frac{\pi}{4} \\
&= 0 + \left(\sqrt{\frac{1}{2}}\right)\frac{\pi}{4} + \left(\sqrt{\frac{1}{2}}\right)\frac{\pi}{4} + 0 = \frac{\pi\sqrt{2}}{4}.
\end{aligned}
$$

18. Since $\Delta x = \frac{b-a}{n} = \frac{6-0}{3} = 2$, then $x_0 = 0$, $x_1 = 2$, $x_2 = 4$, $x_3 = 6$. Since $f(x) = x^3 + x + 1$, then $f(0) = 1$, $f(2) = 11$, $f(4) = 69$, $f(6) = 223$, and f increases on $[0,6]$. So

$$
\begin{aligned}
\underline{S}_3 &= \sum_{j=1}^{3} f(c_j)\Delta x \\
&= 1(2) + 11(2) + 69(2) \\
&= 162.
\end{aligned}
$$

19. Since $\Delta x = \frac{b-a}{n} = \frac{3-1}{200} = \frac{1}{100}$, then $x_0 = 1$, $x_1 = 1.01, \cdots$, $x_{199} = 2.99$, $x_{200} = 3$. Since $f(x) = 4x + 1$ increases on $[1,3]$, then the c_j are the left endpoints for a lower sum:

$$
\begin{aligned}
c_j &= 1 + \left(\frac{j}{100}\right) - \left(\frac{1}{100}\right) \\
&= \frac{99}{100} + \frac{j}{100}.
\end{aligned}
$$

So

$$
\underline{S}_{200} = \sum_{j=1}^{200} f(c_j)\Delta x = \sum_{j=1}^{200} \left[f\left(\frac{99}{100} + \frac{j}{100}\right)\right]\frac{1}{100}
$$

See 29 a) b).

$$= \frac{1}{100}\left\{\sum_{j=1}^{200} 4\left(\frac{99}{100}+\frac{j}{100}\right)+1\right\} = \frac{1}{100}\left\{\sum_{j=1}^{200}\left(\frac{99}{25}+1\right)+\frac{j}{25}\right\}$$

$$= \frac{1}{100}\left\{\frac{124}{25}\left(\sum_{j=1}^{200} 1\right)+\frac{1}{25}\left(\sum_{j=1}^{200} j\right)\right\}$$

[Use special formula for $\sum n$ and $\sum j$.] [Eq(3)] [Eq(4)]

$$= \frac{1}{100}\left\{\left(\frac{124}{25}\right)(200)+\frac{1}{25}\left(\frac{(200)(200+1)}{2}\right)\right\} = 17.96.$$

20. Since $\Delta x = \frac{b-a}{n} = \frac{2-0}{100} = \frac{1}{50}$, then $x_0 = 0$, $x_1 = \frac{1}{50}$, $x_2 = \frac{2}{50}, \cdots$, $x_{99} = \frac{99}{50}$, $x_{100} = \frac{100}{50}$. Since $f(x) = 2 - x$ decreases on $[0,2]$, then for a lower sum, the c_j are the right endpoints. So

$$S_{100} = \sum_{j=1}^{100} f(c_j)\Delta x \;=\; \left[2-\frac{1}{50}\right]\left(\frac{1}{50}\right)+\left[2-\frac{2}{50}\right]\left(\frac{1}{50}\right)+\cdots$$

$$+\left[2-\frac{99}{50}\right]\left(\frac{1}{50}\right)+\left[2-\frac{100}{50}\right]\left(\frac{1}{50}\right)$$

$$= \sum_{j=1}^{100}\left[2-\frac{j}{50}\right]\left(\frac{1}{50}\right) = \left(\frac{1}{50}\right)\sum_{j=1}^{100}\left[2-\frac{j}{50}\right]$$

$$= \frac{1}{50}\left\{\left(2\sum_{j=1}^{100} 1\right)-\frac{1}{50}\left(\sum_{j=1}^{100} j\right)\right\}$$

[Use special formulas: Eq(3) for $\sum n$ and Eq(4) for $\sum j$.]

$$= \frac{1}{50}\left\{100(2)-\frac{1}{50}\frac{(100)(100+1)}{2}\right\} = 1.98.$$

21. Since $\Delta x = \frac{b-a}{n} = \frac{3-0}{3} = 1$, then $x_0 = 0$, $x_1 = 1$, $x_2 = 2$, $x_3 = 3$. Since $f(x) = 2x$ increases on $[0,3]$, then the c_j for an upper sum are the right endpoints. So

$$\overline{S}_3 = \sum_{j=1}^{3} f(c_j)\Delta x = 2(1)(1)+2(2)(1)+2(3)(1) = 12.$$

22. Since $\Delta x = \frac{b-a}{n} = \frac{3-0}{6} = \frac{1}{2}$, then $x_0 = 0$, $x_1 = \frac{1}{2}, \cdots, x_5 = \frac{5}{2}$, $x_6 = \frac{6}{2}$. Since $f(x) = 6 - x$ decreases on $[0,3]$ then the c_j for an upper sum are the left endpoints. So

$$\overline{S}_6 = \sum_{j=1}^{6} f(c_j)\Delta x = (6-0)\frac{1}{5}+\left(6-\frac{1}{2}\right)\frac{1}{5}+\cdots+\left(6-\frac{5}{2}\right)\frac{1}{5} = 14.25.$$

23. Since $\Delta x = \frac{b-a}{n} = \frac{3+1}{4} = 1$, then $x_0 = -1$, $x_1 = 0$, $x_2 = 1$, $x_3 = 2$, $x_4 = 3$. Since $f(x) = 3x^2 + 10$ decreases on $[-1,0]$ and increases on $[0,3]$, then

$$\overline{S}_4 = f(-1)(1)+f(1)(1)+f(2)(1)+f(3)(1) = 13+13+22+37 = 85.$$

24. Since $\Delta x = \frac{b-a}{n} = \frac{2-0}{4} = \frac{1}{2}$, then $x_0 = 0$, $x_1 = \frac{1}{2}$, $x_2 = \frac{2}{2}$, $x_3 = \frac{3}{2}$, $x_4 = \frac{4}{2}$. Since $f(x) = 4 - x^2$ decreases on $[0, 2]$, then the c_j for an upper sum are the left endpoints. So

$$\overline{S}_4 = \sum_{j=1}^{4} f(c_j)\Delta x = [4 - 0^2]\frac{1}{2} + \left[4 - \left(\frac{1}{2}\right)^2\right]\frac{1}{2} + \left[4 - \left(\frac{2}{2}\right)^2\right]\frac{1}{2} + \left[4 - \left(\frac{3}{2}\right)^2\right]\frac{1}{2} = 6.25.$$

25. Since $\Delta x = \frac{b-a}{n} = \left(\frac{\pi}{2} + \frac{\pi}{2}\right)/4 = \frac{\pi}{4}$, then $x_0 = -\frac{2\pi}{4}$, $x_1 = -\frac{\pi}{4}$, $x_2 = 0$, $x_3 = \frac{\pi}{4}$, $x_4 = \frac{2\pi}{4}$. Since $f(x) = \cos(x)$ increases on $\left[-\frac{\pi}{2}, 0\right]$ and decreases on $\left[0, \frac{\pi}{2}\right]$, then

$$\begin{aligned}
\overline{S}_4 &= \sum_{j=1}^{4} f(c_j)\Delta x = \left[\cos\left(-\frac{\pi}{4}\right)\right]\left(\frac{\pi}{4}\right) + [\cos(0)]\left(\frac{\pi}{4}\right) + [\cos(0)]\left(\frac{\pi}{4}\right) + \left[\cos\left(\frac{\pi}{4}\right)\right] \\
&= \left(\sqrt{\frac{1}{2}}\right)\frac{\pi}{4} + \frac{\pi}{4} + \frac{\pi}{4} + \left(\sqrt{\frac{1}{2}}\right)\frac{\pi}{4} = \frac{\pi}{2} + \frac{\pi\sqrt{2}}{4} = \pi\left(\frac{2 + \sqrt{2}}{4}\right).
\end{aligned}$$

26. Since $\Delta x = \frac{b-a}{n} = \frac{6-0}{3} = 2$, then $x_0 = 0$, $x_1 = 2$, $x_2 = 4$, $x_3 = 6$. Since $f(0) = 1$, $f(2) = 11$, $f(4) = 69$, $f(6) = 223$, and f increases on $[0, 6]$, then the c_j for an upper sum are the right endpoints. So

$$\overline{S}_3 = \sum_{j=1}^{3} f(c_j)\Delta x = [f(2)]\,2 + [f(4)]\,2 + [f(6)]\,2 = (11)2 + (69)\,2 + (223)\,2 = 606.$$

27. Since $\Delta x = \frac{b-a}{n} = \frac{4-1}{300} = \frac{1}{100}$, then $x_0 = 1$, $x_1 = 1.01, \cdots$, $x_{299} = 3.99$, $x_{300} = 4$. Since $f(x) = 4x - 1$ increases, then the c_j for an upper sum are the right endpoints: $c_j = 1 + \frac{j}{100}$. So

$$\begin{aligned}
\overline{S}_{300} &= \sum_{j=1}^{300} f(c_j)\Delta x = \sum_{j=1}^{300} f\left(1 + \frac{j}{100}\right)\cdot\Delta x = \sum_{j=1}^{300}\left[4\left(1 + \frac{j}{100}\right) - 1\right]\left(\frac{1}{100}\right) \\
&= \sum_{j=1}^{300}\left[3 + \frac{j}{25}\right]\left(\frac{1}{100}\right) = \frac{1}{100}\left\{3\left(\sum_{j=1}^{300} 1\right) + \frac{1}{25}\left(\sum_{j=1}^{300} j\right)\right\}
\end{aligned}$$

[Use special formula: Eq (3) for $\sum n$ and Eq (4) for $\sum j$.]

$$= \frac{1}{100}\left\{300(3) + \frac{1}{25}\frac{300(300 + 1)}{2}\right\} = 27.06.$$

28. Since $\Delta x = \frac{b-a}{n} = \frac{3-1}{200} = \frac{1}{100}$, then $x_0 = 1$, $x_1 = 1.01, \cdots$, $x_{199} = 2.99$, $x_{200} = 3$. Since $f(x) = 2x - 1$ increases, then for an upper sum, the c_j are the right endpoints: $c_j = 1 + \frac{j}{100}$. So

$$\begin{aligned}
\overline{S}_{200} &= \sum_{j=1}^{200} f(c_j)\Delta x = \sum_{j=1}^{200} f\left(1 + \frac{j}{100}\right)\Delta x \\
&= \sum_{j=1}^{200}\left[2\left(1 + \frac{j}{100}\right) - 1\right]\left(\frac{1}{100}\right) \\
&= \sum_{j=1}^{200}\left[1 + \frac{j}{50}\right]\left(\frac{1}{100}\right) = \left(\frac{1}{100}\right)\left\{\left(\sum_{j=1}^{200} 1\right) + \frac{1}{50}\left(\sum_{j=1}^{200} j\right)\right\}
\end{aligned}$$

[Use special formulas Eq(3) for $\sum n$ and Eq(4) for $\sum j$.]

$$= \frac{1}{100}\left\{200(1) + \frac{1}{50}\frac{(200)(200+1)}{2}\right\} = 6.02.$$

29. a) $\displaystyle\sum_{j=1}^{n}(x_j + y_j)$. The summation notation means $(x_1 + y_1) + (x_2 + y_2) + \cdots + (x_n + y_n)$. Since addition is associative and commutative, this can be commuted and associated (regrouped) as $(x_1 + x_2 + \cdots + x_n) + (y_1 + y_2 + \cdots + y_n)$. The summation notation for this is $\displaystyle\sum_{j=1}^{n} x_j + \sum_{j=1}^{n} y_j$.

b) The summation notation $\displaystyle\sum_{j=1}^{n} cx_j$ means $cx_1 + cx_2 + \cdots + cx_n$. By the distributive law, this is

$c(x_1 + x_2 + \cdots + x_n)$. The summation notation for this is $c\displaystyle\sum_{j=1}^{n} x_j$.

c) This is the same as a), where we replace y_j with $-y_j$ as follows:

$$\sum_{j=1}^{n} x_j - y_j = \sum_{j=1}^{n} x_j + (-1)y_j.$$

By 29 a), b) this is equal to

$$\sum x_j + \sum(-1)(y_j) = \left[\sum x_j\right] + (-1)\left[\sum y_j\right]$$
$$= \sum x_j - \sum y_j.$$

d) $\displaystyle\sum_{j=1}^{n} c$ means $c + c + \cdots + c$ where c is summed n times. By the distributive law, this equals nc.

30. $\displaystyle\sum_{j=1}^{10} 2j = 2\sum_{j=1}^{10} j$ [29 b)]

$$= 2\frac{10(10+1)}{2} \quad [\text{Eq}(4)]$$
$$= 110.$$

31. $\displaystyle\sum_{j=1}^{7} 3j^2 = 3\sum_{j=1}^{7} j^2$ [29 b)]

$$= 3\frac{7(7+1)(2\cdot 7+1)}{6} \quad [\text{Eq}(4)]$$
$$= 420.$$

32. $\displaystyle\sum_{j=1}^{8} j^2 + j + 1 = \sum_{j=1}^{8} j^2 + \sum_{j=1}^{8} j + \sum_{j=1}^{8} 1$ [29 a)] $= 204 + 36 + 8 = 248.$

33. $\displaystyle\sum_{j=1}^{6} 2j^2 + 3j + 5 = \sum_{j=1}^{6} 2j^2 + \sum_{j=1}^{6} 3j + \sum_{j=1}^{6} 5$ [29 a)]

$$= 2\sum_{j=1}^{6} j^2 + 3\sum_{j=1}^{6} j + \sum_{j=1}^{6} 5 \quad [29\text{ b)}]$$
$$= 2\frac{(6)(6+1)(2\cdot 6+1)}{6} + 3\frac{(6)(6+1)}{2} + 6\cdot 5 \quad \text{By [Eq(3),(4),(5)]}$$
$$= 275.$$

406

34. $\displaystyle\sum_{j=1}^{n} j^2 + 2j + 3 \;=\; \sum_{j=1}^{n} j^2 + 2\sum_{j=1}^{n} j + \sum_{j=1}^{n} 3$ [By 29 a), b)]

$$=\; \frac{(n)(n+1)(2n+1)}{6} + 2\left[\frac{n(n+1)}{2}\right] + n\cdot 3 \qquad \text{[By Eq(3),(4),(5)]}$$

$$=\; \frac{2n^3 + 9n^2 + 25n}{6}.$$

35. $\displaystyle\sum_{j=1}^{n} (6j^2 - 4j) \;=\; 6\cdot\sum_{j=1}^{n} j^2 - 4\cdot\sum_{j=1}^{n} j$ [By 29 b), c)]

$$=\; 6\left[\frac{n(n+1)(2n+1)}{6}\right] - 4\left[\frac{n(n+1)}{2}\right] \qquad \text{[By Eq (4), (5)]}$$

$$=\; (2n^3 + 3n^2 + n) - (2n^2 + 2n)$$

$$=\; 2n^3 + n^2 - n.$$

36. If $55 = \displaystyle\sum_{j=1}^{n} j$ then $55 = \dfrac{n(n+1)}{2}$ by Eq. (4).   Solve for n :

$$\begin{aligned} n(n+1) &= 2(55) \\ n^2 + n - 110 &= 0 \\ (n-10)(n+11) &= 0 \\ n &= 10, -11. \\ \text{Solution: } n &= 10. \end{aligned}$$

37. If $120 = \displaystyle\sum_{j=4}^{n} 2j$ then

$$120 = 2\left[\sum_{j=4}^{n}(j)\right] = 2\left[\sum_{j=1}^{n} j - \sum_{j=1}^{3} j\right] = 2\left[\frac{n(n+1)}{2} - (1+2+3)\right] = n(n+1) - 2(6) = n^2 + n - 12.$$

Solve $120 = n^2 + n - 12$ for n:

$$\begin{aligned} n^2 + n - 132 &= 0 \\ (n-11)(n+12) &= 0 \\ n &= 11, -12. \\ \text{Solution: } n &= 11. \end{aligned}$$

38. If $\displaystyle\sum_{j=0}^{n}(3j - 1) = 441$, then

$$\begin{aligned} 441 &= 3\left[\sum_{j=0}^{n} j\right] - \left[\sum_{j=0}^{n} 1\right] = 3\left[0 + \sum_{j=1}^{n} j\right] - \left[1 + \sum_{j=1}^{n} 1\right] \\ &= 3\left[\frac{n(n+1)}{2}\right] - [1 + n\cdot 1] = \frac{3(n^2 + n) - 2(n+1)}{2}. \end{aligned}$$

Solve for n

$$\begin{aligned} 2(441) &= (3n^2 + 3n) - (2n + 2) \\ 882 &= 3n^2 + n - 2 \\ 0 &= 3n^2 + n - 884 \end{aligned}$$

By quadratic formula,
Our solution is

$$\begin{aligned} n &= 17, -17\tfrac{1}{3} \\ n &= 17. \end{aligned}$$

39. If $\displaystyle\sum_{j=n}^{15}(1-2j) = -209$ then

$$-209 = \sum_{j=1}^{15}(1-2j) - \sum_{j=1}^{n-1}(1-2j) = \left[\sum_{j=1}^{15}1 - 2\sum_{j=1}^{15}j\right] - \left[\sum_{j=1}^{n-1}1 - 2\sum_{j=1}^{n-1}j\right]$$

$$= \left[15 - 2\left(\frac{15(15+1)}{2}\right)\right] - \left[(n-1) - 2\frac{(n-1)(n)}{2}\right]$$

$$= [15 - 240] - [-n^2 + 2n - 1] = n^2 - 2n - 224.$$

Solve $-209 = n^2 - 2n - 224$ for n: $0 = n^2 - 2n - 15$
$0 = (n+3)(n-5)$ Our solution is $n = 5$.

40. If $\displaystyle\sum_{j=n}^{12} 6j^2 = 3870$, then

$$3870 = 6\sum_{j=n}^{12} j^2 = 6\left[\sum_{j=n}^{12} j^2\right] = 6\left[\sum_{j=1}^{12} j^2 - \sum_{j=1}^{n-1} j^2\right]$$

$$= 6\left[\frac{12(12+1)(2\cdot 12+1)}{6} - \frac{(n-1)(n)(2n-1)}{6}\right]$$

$$= 12(13)(25) - (2n^3 - 3n^2 + n) = -2n^3 + 3n^2 - n + 3900.$$

Solve for n:

$$0 = -2n^3 + 3n^2 - n + 30$$

By trial and error,   n = 3.

Probably better just to begin adding: $12^2 + 11^2 + 10^2 + \cdots$ until getting a sum of $\frac{3870}{6} = 645$.
Answer: $12^2 + 11^2 + \cdots + 4^2 + 3^2 = 645$.

41. $13 + 15 + \cdots + 241 + 243 = \displaystyle\sum_{j=1}^{116}(11+2j) = \sum 11 + 2\sum j$

$$= [116(11)] + 2\left[\frac{(116)(116+1)}{2}\right]\quad [\text{By Eq(4)}]$$

$$= 1276 + (116)(117) = 14,848.$$

42. Total number of logs, if there are n on the bottom row, is $1 + 2 + 3 + \cdots + n = \frac{n(n+1)}{2}$.
So solve $105 = \frac{n(n+1)}{2}$ for n: $210 = n(n+1)$
$0 = n^2 + n - 210 = (n+15)(n-14)$.
Our solution is $n = 14$ logs on bottom.

43. $\left(\dfrac{1}{n^2}\right)\displaystyle\sum_{j=1}^{n}(2j+1) = \dfrac{1}{n^2}\left[2\left(\sum_{j=1}^{n} j\right) + \left(\sum_{j=1}^{n} 1\right)\right] = \dfrac{1}{n^2}\left[2\left(\frac{(n)(n+1)}{2}\right) + n\cdot 1\right]$

$$= \frac{1}{n^2}[n^2 + 2n] = 1 + \frac{2}{n} \to 1 \text{ as } n \to \infty.$$

44. $\left(\dfrac{1}{n^2}\right) \sum\limits_{j=1}^{n} (3j-2) = \left(\dfrac{1}{n^2}\right)\left[3\sum\limits_{j=1}^{n} j - \sum\limits_{j=1}^{n} 2\right] = \left(\dfrac{1}{n^2}\right)\left[3\left(\dfrac{n(n+1)}{2}\right) - n\cdot 2\right]$

$= \dfrac{1}{n^2}\dfrac{[3n^2+3n-4n]}{2} = \dfrac{3n^2-n}{2n^2} = \dfrac{3}{2} - \dfrac{1}{2n} \to \dfrac{3}{2}$ as $n \to \infty$.

45. $\left(\dfrac{1}{n^3}\right) \sum\limits_{j=1}^{n} 6j^2 = \left(\dfrac{1}{n^3}\right) 6\sum j^2 = \left(\dfrac{6}{n^3}\right)\dfrac{(n(n+1)(2n+1))}{6} = \dfrac{n(n+1)(2n+1)}{n\cdot n\cdot n}$

$= (1)\left(1+\tfrac{1}{n}\right)\left(2+\tfrac{1}{n}\right) \to 1\cdot 1\cdot 2 = 2$ as $n \to \infty$.

46. $\left(\dfrac{1}{n^4}\right) \sum\limits_{j=1}^{n}(3j^3+1) = \left(\dfrac{1}{n^4}\right)\left(3\sum j^3 + \sum 1\right)$

$= \left(\dfrac{1}{n^4}\right)\left[3\left(\dfrac{n^2(n+1)^2}{4}\right) + (n\cdot 1)\right]$ [Eq (3)] [Eq (6)]

$= \dfrac{3n^2(n+1)^2}{4\,n^2\,n^2} + \dfrac{n}{n^4} = \dfrac{3}{4}(1)\left(1+\dfrac{1}{n}\right)^2 + \dfrac{1}{n^3} \to \dfrac{3}{4}(1)(1)+0$

$= \dfrac{3}{4}$ as $n \to \infty$.

47. a) Since $\Delta x = \dfrac{b-a}{n} = \dfrac{2-0}{4} = \dfrac{1}{2}$, then $x_0 = 0$, $x_1 = \dfrac{1}{2}$, $x_2 = \dfrac{2}{2}$, $x_3 = \dfrac{3}{2}$, $x_4 = \dfrac{4}{2}$. Since f increases, then for a lower sum, the c_j are the left endpoints. So

$\underline{S}_4 = \sum\limits_{j=1}^{4} f(c_j)\Delta x = \left(\dfrac{1}{2}\right)\sum f(c_j) = \left(\dfrac{1}{2}\right)\left[f(0) + f\left(\dfrac{1}{2}\right) + f(1) + f\left(\dfrac{3}{2}\right)\right]$

$= \left(\dfrac{1}{2}\right)\left(2 + \dfrac{7}{2} + \dfrac{10}{2} + \dfrac{13}{2}\right) = 8.5.$

b) $\Delta x = \dfrac{2}{100} = \dfrac{1}{50}$ and $c_j = 0 + \dfrac{j-1}{50}$. So

$\underline{S}_{100} = \sum\limits_{j=1}^{100} f(c_j)\Delta x = \left(\dfrac{1}{50}\right)\sum f\left(\dfrac{j-1}{50}\right) = \left(\dfrac{1}{50}\right)\sum\left[3\left(\dfrac{j-1}{50}\right)+2\right]$

$= \dfrac{1}{50}\left\{\dfrac{3}{50}\left(\sum j\right) + \left(2 - \dfrac{3}{50}\right)\left(\sum 1\right)\right\}$

$= \dfrac{1}{50}\left\{\dfrac{3}{50}\left(\dfrac{100(100+1)}{2}\right) + \dfrac{97}{50}(100)\right\} = 9.94$

c) $\Delta x = \dfrac{2-0}{n} = \dfrac{2}{n}$ and $c_j = 0 + (\Delta x)(j-1) = \dfrac{2(j-1)}{n}$. So

$\underline{S}_n = \sum f(c_j)\Delta x = \sum\limits_{j=1}^{n}\left[3\left(\dfrac{2(j-1)}{n}\right)+2\right]\left(\dfrac{2}{n}\right)$

$= \sum\left[\dfrac{6}{n}(j-1)\left(\dfrac{2}{n}\right) + 2\left(\dfrac{2}{n}\right)\right] = \sum\left[\dfrac{12}{n^2}(j-1) + \dfrac{4}{n}\right]$

$= \dfrac{12}{n^2}\sum j - \dfrac{12}{n^2}\sum 1 + \dfrac{4}{n}\sum 1 = \dfrac{12}{n^2}\cdot\dfrac{n(n+1)}{2} - \dfrac{12}{n^2}\cdot\dfrac{n}{1} + \dfrac{4n}{n\,1}$

$= \dfrac{12}{1}\left(\dfrac{n}{n}\right)\dfrac{1}{2}\left(\dfrac{n+1}{n}\right) - \dfrac{12}{n} + 4 = \dfrac{6}{1}\left(1+\dfrac{1}{n}\right) - \dfrac{12}{n} + 4 = 10 - \dfrac{6}{n}.$

d) $\underline{S} = 10 - \frac{6}{n} \to 10$ as $n \to \infty$.

e) Use the right endpoints: $c_j = 0 + \frac{1}{50} j$.

$$
\begin{aligned}
\overline{S}_{100} &= \sum f(c_j)\Delta x = \sum_{j=1}^{100} \left[3\left(\frac{j}{50}\right) + 2\right]\left(\frac{1}{50}\right) \\
&= \sum \left(\frac{1}{50} \cdot \frac{3}{50}\right) j + \sum \frac{2}{50} \\
&= \frac{3}{50(50)} \left(\sum j\right) + \frac{2}{50} \left(\sum 1\right) \\
&= \frac{1}{50}\left(\frac{3}{50}\right)\frac{(100)(100+1)}{2} + \frac{2}{50}(100) = 10.06.
\end{aligned}
$$

f) $\Delta x = \frac{2-0}{n}$,

   $c_j = 0 + \left(\frac{2}{n}\right) j$.

$$
\begin{aligned}
\text{So} \quad \overline{S}_n &= \sum f(c_j)\Delta x = \sum_{j=1}^{100} \left[3\left(\frac{2}{n}j\right) + 2\right]\left(\frac{2}{n}\right) \\
&= 3\left(\frac{2}{n}\right)\left(\frac{2}{n}\right)\sum j + 2\left(\frac{2}{n}\right)\sum 1 = \left(\frac{12}{n^2}\right)\frac{n(n+1)}{2} + \left(\frac{4}{n}\right) \cdot n \\
&= \frac{12}{2}\left(\frac{n}{n}\right)\left(\frac{n+1}{n}\right) + 4 = (6)\left(1 + \frac{1}{n}\right) + 4 = 10 + \frac{6}{n}.
\end{aligned}
$$

g) $\overline{S}_n = 10 + \frac{6}{n} \to 10$ as $n \to \infty$.

h) The area is $10 = \lim \overline{S}_n = \lim \underline{S}_n$ by **Definition 1**.

48.   a) Since $\Delta x = \frac{b-a}{n} = \frac{4-0}{4} = 1$, then $x_0 = 0$, $x_1 = 1$, $x_2 = 2$, $x_3 = 3$, $x_4 = 4$. Since f decreases, then the c_j are the right endpoints for a lower sum. So

$$
\begin{aligned}
\underline{S}_4 &= \sum_{j=1}^{4} f(c_j)\Delta x = \sum f(c_j)(1) \\
&= f(1) + f(2) + f(3) + f(4) \\
&= 6 + 4 + 2 + 0 = 12.
\end{aligned}
$$

b) Since $\Delta x = \frac{4}{100} = \frac{1}{25}$, then the $c_j = 0 + (\Delta x)j = \frac{j}{25}$. So

$$
\begin{aligned}
\underline{S}_{100} &= \sum f(c_j)\Delta x = \sum_{j=1}^{100} \left[8 - 2\left(\frac{j}{25}\right)\right]\left(\frac{1}{25}\right) = \left\{8\left(\sum 1\right) - \frac{2}{25}\left(\sum j\right)\right\}\left(\frac{1}{25}\right) \\
&= \left\{100(8) - \frac{2}{25}\frac{(100)(100+1)}{2}\right\}\left(\frac{1}{25}\right) = 15.84.
\end{aligned}
$$

c) $\Delta x = \frac{4}{n}$, and $c_j = (\Delta x)j = \frac{4j}{n}$. So

$$
\underline{S}_n = \sum f(c_j)\Delta x = \sum_{j=1}^{n} \left[8 - 2\left(\frac{4j}{n}\right)\right]\left(\frac{4}{n}\right) = \sum 8\left(\frac{4}{n}\right) - \sum 2\frac{4(4)j}{n^2}
$$

$$= \frac{32}{n}\left(\sum 1\right) - \frac{32}{n^2}\left(\sum j\right) = \frac{32}{n}(n) - \frac{32}{n^2}\left(\frac{n(n+1)}{2}\right)$$

$$= 32 - 16\left(\frac{n}{n}\right)\left(\frac{n+1}{n}\right) = 16 - \frac{16}{n}.$$

d) $\underline{S}_n = 16 - \frac{16}{n} \to 16$ as $n \to \infty$.

e) Now $\Delta x = \frac{1}{25}$, and c_j are the right endpoints: $c_j = 0 + \left(\frac{j}{25}\right) - \left(\frac{1}{25}\right)$. So

$$\overline{S}_{100} = \sum f(c_j)\Delta x = \sum_{j=1}^{100}\left[8 - 2\left(\frac{j}{25} - \frac{1}{25}\right)\right]\left(\frac{1}{25}\right)$$

$$= \left\{\sum\left[\left(8 + \frac{2}{25}\right) - \frac{2}{25}j\right]\right\}\left(\frac{1}{25}\right) = \left\{\frac{202}{25}\left(\sum 1\right) - \frac{2}{25}\left(\sum j\right)\right\}\left(\frac{1}{25}\right)$$

$$= \left[\left(\frac{202}{25}\right) - \frac{2}{25}\frac{(100)(100+1)}{2}\right]\left(\frac{1}{25}\right) = 16.16.$$

f) Now $\Delta x = \frac{4}{n}$ and $c_j = (\Delta x)(j - 1) = \left(\frac{4}{n}\right)j - \left(\frac{4}{n}\right)$. So

$$\overline{S}_n = \sum f(c_j)\Delta x = \sum_{j=1}^{n}\left[8 - 2\left(\frac{4j}{n} - \frac{4}{n}\right)\right]\left(\frac{4}{n}\right)$$

$$= \left(8 + \frac{8}{n}\right)\left(\frac{4}{n}\right)\sum 1 - 2\frac{4}{n}\left(\frac{4}{n}\right)\sum j$$

$$= \left[\left(8 + \frac{8}{n}\right)\left(\frac{4}{n}\right)(n)\right] - \left[\frac{2}{1}\frac{4}{n}\frac{4}{n}\frac{(n)(n+1)}{2}\right]$$

$$= \left[32 + \frac{32}{n}\right] - \left[16 + \frac{16}{n}\right] = 16 + \frac{16}{n}.$$

g) $\overline{S}_n = 16 + \frac{16}{n} \to 16$ as $n \to \infty$.

h) Since $\lim \overline{S}_n = \lim \underline{S}_n = 16$, then by **Definition 1**, the area is 16.

49. a) Since $\Delta x = \frac{b-0}{n} = \frac{b}{n}$ and $f(x) = x^2$ increases on $[0, b]$, then c_j is the left endpoint for a lower sum: $c_j = 0 + (\Delta x)j - \Delta x = \left(\frac{b}{n}\right)j - \left(\frac{b}{n}\right)$. So

$$\underline{S}_n = \sum_{j=1}^{n} f(c_j)\Delta x = \sum\left[\frac{b}{n}(j-1)\right]^2\left(\frac{b}{n}\right) = \sum\left(\frac{b}{n}\right)^2(j^2 - 2j + 1)\left(\frac{b}{n}\right)$$

$$= \left(\frac{b}{n}\right)^3\left\{\sum j^2 - 2\sum j + \sum 1\right\} = \frac{b^3}{n^3}\left\{\frac{(n)(n+1)(2n+1)}{6} - \frac{2}{1}\frac{n(n+1)}{2} + n\right\}$$

$$= b^3\left\{\left(\frac{n}{n}\right)\left(\frac{n+1}{n}\right)\left(\frac{2n+1}{n}\right)\frac{1}{6} - \left(\frac{n}{n}\right)\left(\frac{n+1}{n}\right)\left(\frac{1}{n}\right) + \frac{n}{n}\frac{1}{n^2}\right\}$$

$$= b^3\left\{\frac{1}{6}\left(1 + \frac{1}{n}\right)\left(2 + \frac{1}{n}\right) - \left(1 + \frac{1}{n}\right)\left(\frac{1}{n}\right) + \frac{1}{n^2}\right\} = b^3\left\{\frac{1}{3} - \frac{1}{2}\frac{1}{n} + \frac{1}{6}\frac{1}{n^2}\right\}.$$

b) $\underline{S}_n \to b^3\left\{\frac{1}{3} - 0 + 0\right\} = \frac{b^3}{3}$ as $n \to \infty$.

c) $\Delta x = \frac{b-a}{n}$, and $c_j = a + (\Delta x)j - \Delta x$. So

$$\underline{S}_n = \sum_{j=1}^{n} f(c_j)\Delta x = \sum \left[a + \left(\frac{b-a}{n} \right)(j-1) \right]^2 \left(\frac{b-a}{n} \right)$$

$$= \sum \left[a^2 + 2a\left(\frac{b-a}{n} \right)(j-1) + \left(\frac{b-a}{n} \right)^2 (j-1)^2 \right] \left(\frac{b-a}{n} \right)$$

$$= \left\{ \sum \left[a^2 - 2a\left(\frac{b-a}{n} \right) + \left(\frac{b-a}{n} \right)^2 \right] + \left[2a\left(\frac{b-a}{n} \right) - 2\left(\frac{b-a}{n} \right)^2 \right] j \right.$$

$$\left. + \left[\left(\frac{b-a}{n} \right)^2 \right] j^2 \right\} \left(\frac{b-a}{n} \right)$$

$$= \left\{ \left[a^2 - \frac{2a(b-a)}{n} + \left(\frac{b-a}{n} \right)^2 \right] n + \left[2a\left(\frac{b-a}{n} \right) - 2\left(\frac{b-a}{n} \right)^2 \right] \frac{n(n+1)}{2} \right.$$

$$\left. + \left[\left(\frac{b-a}{n} \right)^2 \right] \frac{n(n+1)(2n+1)}{6} \right\} \left(\frac{b-a}{n} \right)$$

$$= \left[(b-a)a^2 \frac{n}{n} \right] + \left[\frac{-2a(b-a)2n}{n^2} \right] + \left[(b-a)^3 \left(\frac{n}{n^3} \right) \right] + \left[\frac{2a(b-a)^2}{2} \frac{n(n+1)}{n \cdot n} \right]$$

$$+ \left[\frac{-2(b-a)^3}{2} \frac{(n)}{n} \frac{(n+1)}{n} \frac{1}{n} \right] + \left[\frac{(b-a)^3}{6} \frac{(n)}{n} \frac{(n+1)}{n} \frac{(2n+1)}{n} \right]$$

This expression is convenient for taking the limit, in 49 d).

d) Take the limit of each term of $\underline{S}_n$ from 49 c):

$$\underline{S}_n \rightarrow (b-a)a^2 + 0 + 0 + a(b-a)^2 + 0 + \frac{(b-a)^3}{6} \left(\frac{2}{1} \right)$$

$$= a^2b - a^3 + a^3 - 2a^2b + ab^2 + \frac{b^3 - 3b^2a + 3ba^2 - a^3}{3} = \frac{-a^3}{3} + \frac{b^3}{3}.$$

e) The area A0b from 0 to b is equal to the sum of the area A0a from 0 to a and the area Aab from a to b: A0b = A0a + Aab. So Aab = A0b − A0a. And A0a = $\frac{a^3}{3}$ from 49 b). So Aab = $\frac{b^3}{3} - \frac{a^3}{3}$.

f) The area from 1 to 5 is the area from 0 to 5 minus the area from 0 to 1: $\frac{5^3}{3} - \frac{1^3}{3} = \frac{125-1}{3} = \frac{124}{3}$.

50. a) $\Delta x = \frac{2-1}{4} = \frac{1}{4}$, and f increases, so the c_j are the left endpoints of the intervals $x_0 = \frac{4}{4}, x_1 = \frac{5}{4}, x_2 = \frac{6}{4}, x_3 = \frac{7}{4}, x_4 = \frac{8}{4}$. So

$$\underline{S}_4 = \sum_{j=1}^{4} f(c_j)\Delta x = f\left(\frac{4}{4} \right) \cdot \frac{1}{4} + f\left(\frac{5}{4} \right) \cdot \frac{1}{4} + f\left(\frac{6}{4} \right) \cdot \frac{1}{4} + f\left(\frac{7}{4} \right) \cdot \frac{1}{4}$$

$$= \left\{ \left(\frac{12}{4} + 4 \right) + \left(\frac{15}{4} + 4 \right) + \left(\frac{18}{4} + 4 \right) + \left(\frac{21}{4} + 4 \right) \right\} \left(\frac{1}{4} \right)$$

$$= \left\{ \frac{66}{4} + 4.4 \right\} \left(\frac{1}{4} \right) = \frac{66}{16} + 4 = \frac{65}{8}.$$

b) $\Delta x = \frac{2-1}{100} = \frac{1}{100}$ and $c_j = 1 + \Delta x(j-1) = \left(\frac{j}{100} \right) + \left(\frac{99}{100} \right)$. So

$$\underline{S}_{100} = \sum_{j=1}^{100} f(c_j)\Delta x = \sum \left[3\left(\frac{j}{100} + \frac{99}{100} \right) + 4 \right] \left(\frac{1}{100} \right)$$

412

$$= \left\{ \left(\frac{3}{100}\right) \sum j + \left(\frac{3(99)}{100} + 4\right) \sum 1 \right\} \left(\frac{1}{100}\right)$$

$$= \left\{ \left(\frac{3}{100}\right) \left(\frac{100(100+1)}{2}\right) + \left(\frac{3(99)}{100} + \frac{4}{1}\right)(100) \right\} \left(\frac{1}{100}\right) = 8.485.$$

c) Now $\Delta x = \frac{2-1}{n} = \frac{1}{n}$ and $c_j = 1 + (\Delta x)(j-1) = 1 + \left(\frac{1}{n}\right)(j-1) = 1 - \frac{1}{n} + \frac{j}{n}$. So

$$\begin{aligned}
\underline{S}_n &= \sum_{j=1}^{n} f(c_j)\Delta x = \sum \left[3\left(1 - \frac{1}{n} + \frac{j}{n}\right) + 4\right]\left(\frac{1}{n}\right) \\
&= \sum \left[\left(7 - \frac{3}{n}\right) + \left(\frac{3}{n}\right)j\right]\frac{1}{n} = \left\{\left(7 - \frac{3}{n}\right)\sum 1 + \left(\frac{3}{n}\right)\sum j\right\}\left(\frac{1}{n}\right) \\
&= \left\{\left(7 - \frac{3}{n}\right)n + \left(\frac{3}{n}\right)\frac{n(n+1)}{2}\right\}\left(\frac{1}{n}\right) = \left(7 - \frac{3}{n}\right) + \left(\frac{3}{2}\left(1 + \frac{1}{n}\right)\right) = \frac{17}{2} - \frac{3}{2}\frac{1}{n}.
\end{aligned}$$

d) $S_n = \frac{17}{2} - \frac{3}{2}\frac{1}{n} \to \frac{17}{2}$ as $n \to \infty$.

e) $\Delta x = \frac{2-1}{100} = \frac{1}{100}$ and the c_j are now the right endpoints: $c_j = 1 + j(\Delta x) = 1 + \frac{j}{100}$. So

$$\begin{aligned}
\overline{S}_{100} &= \sum_{j=1}^{100} f(c_j)\Delta x = \sum \left[3\left(1 + \frac{j}{100}\right) + 4\right]\left(\frac{1}{100}\right) \\
&= \sum \left[3 + \frac{3}{100}j + 4\right]\frac{1}{100} = \left\{7\left(\sum 1\right) + \frac{3}{100}\left(\sum j\right)\right\}\left(\frac{1}{100}\right) \\
&= \left\{7(100) + \frac{3}{100}\left(\frac{100(100+1)}{2}\right)\right\}\left(\frac{1}{100}\right) = 8.515.
\end{aligned}$$

f) Now $\Delta x = \frac{2-1}{n} = \frac{1}{n}$ and $c_j = 1 + (\Delta x)j = 1 + \frac{j}{n}$. So

$$\begin{aligned}
\overline{S}_n &= \sum_{j=1}^{n} f(c_j)\Delta x = \sum \left[3\left(1 + \frac{j}{n}\right) + 4\right]\left(\frac{1}{n}\right) = \sum \left[7 + \frac{3}{n}j\right]\frac{1}{n} \\
&= \frac{7}{n}\left(\sum 1\right) + \frac{3}{n^2}\left(\sum j\right) = \frac{7}{n}(n) + \frac{3}{n^2}\left(\frac{n(n+1)}{2}\right) \\
&= 7 + \frac{3}{2}\left(\frac{n}{n}\right)\left(\frac{n+1}{n}\right) = 7 + \frac{3}{2}\left(1 + \frac{1}{n}\right) = \frac{17}{2} + \frac{3}{2}\frac{1}{n}.
\end{aligned}$$

g) $\overline{S}_n = \frac{17}{2} + \frac{3}{2}\frac{1}{n} \to \frac{17}{2}$ as $n \to \infty$.

h) Since $\lim \overline{S}_n = \lim \underline{S}_n = \frac{17}{2}$, then the area is $\frac{17}{2}$, by **Definition 1**.

51. Use the method pointed out in 49 e): find the area from $x = 0$ to $x = 3$, then from $x = 0$ to $x = 2$, then subtract the second area.

(1) The function increases, so use left endpoints for a lower sum: $c_j = 0 + \left(\frac{3}{n}\right)(j-1)$.

$$\begin{aligned}
\underline{S}_n &= \sum_{j=1}^{n} f(c_j)\Delta x = \sum \left[3\left(\frac{3}{n}(j-1)\right)^2 + 7\right]\left(\frac{3}{n}\right) \\
&= \sum \left[3\left\{\left(\frac{3}{n}\right)^2(j^2 - 2j + 1)\right\} + 7\right]\left(\frac{3}{n}\right)
\end{aligned}$$

413

$$= 3\left(\frac{3}{n}\right)^2\left(\frac{3}{n}\right)\left(\sum j^2\right) - 3\left(\frac{3}{n}\right)^2\left(\frac{3}{n}\right)(2)\left(\sum j\right) + \left[3\left(\frac{3}{n}\right)^2 + 7\right]\left(\frac{3}{n}\right)\left(\sum 1\right)$$

$$= 3\left(\frac{3^3}{n^3}\right)\left(\frac{(n)(n+1)(2n+1)}{6}\right) - (2)(3)\left(\frac{3}{n}\right)^3\left(\frac{n(n+1)}{2}\right) + \left[3\left(\frac{3}{n}\right)^3 + 7\left(\frac{3}{n}\right)\right]\left(\frac{n}{1}\right)$$

$$= \left(\frac{3}{6}\right)\frac{3^3}{1}\left(\frac{n}{n}\right)\left(\frac{n+1}{n}\right)\left(\frac{2n+1}{n}\right) - (3)\frac{3^3}{n}\left(\frac{n}{n}\frac{n+1}{n}\right) + \left(3\frac{3^3}{n^2} + \frac{3(7)}{1}\right)$$

which approaches $\frac{3^3}{2}\left(\frac{2}{1}\right) - (0) + (0 + 21) = 48$ as $n \to \infty$.

(2) Replacing 3 by 2 above, we get.

$$\underline{S}_n = \left(\frac{3}{6}\right)2^3\left(\frac{n}{n}\frac{n+1}{n}\frac{2n+1}{n}\right) - 3\frac{2^3}{n}\frac{n}{n}\frac{n+1}{n} + \left(3\frac{2^3}{n^2} + 2(7)\right)$$

which approaches $22 = 8 - 0 + 14$ as $n \to \infty$. The difference between the areas is $48 - 22 = 26$.

52. The only difference between this and Ex. 51 is that now $c_j = 0 + j(\Delta x) = \left(\frac{3}{n}\right)j$; so

$$\overline{S}_n = \sum_{j=1}^{n} f(c_j)\Delta x = \sum\left(3\left(\frac{3}{n}j\right)^2 + 7\right)\left(\frac{3}{n}\right)$$

$$= 3\left(\frac{3}{n}\right)^3\left(\sum j^2\right) + 7\left(\frac{3}{n}\right)\left(\sum 1\right) = 3\left(\frac{3}{n}\right)^3\left(\frac{n(n+1)(2n+1)}{6}\right) + 7\left(\frac{3}{n}\right)\left(\frac{n}{1}\right)$$

$$= 3\left(\frac{3}{1}\right)^3\left(\frac{1}{6}\right)\left(\frac{n}{n}\frac{n+1}{n}\frac{2n+1}{n}\right) + 7\left(\frac{3}{1}\right)\left(\frac{n}{n}\right) \to 3\left(\frac{3}{1}\right)^3\left(\frac{1}{6}\right)(1)(1)\left(\frac{2}{1}\right) + 7\left(\frac{3}{1}\right)(1)$$

$$= 27 + 21 = 48 \text{ as } n \to \infty.$$

To do the area from $x = 0$ to $x = 2$, above replace $\frac{3}{n}$ by $\frac{2}{n}$. This gives

$$\overline{S}_n \to 3\left(\frac{2}{1}\right)^3\left(\frac{1}{6}\right)(1)(1)\left(\frac{2}{1}\right) + 7\left(\frac{2}{1}\right)(1) = 8 + 14 = 22 \text{ as } n \to \infty.$$

The difference between the two areas is $48 - 22 = 26$.

53. Now $\Delta x = \frac{1-0}{n} = \frac{1}{n}$, and since f increases on $[0, 1]$, then the c_j are the left endpoints for a lower sum: $c_j = 0 + (\Delta x)(j-1) = \left(\frac{j}{n}\right) - \left(\frac{1}{n}\right)$. So

$$\underline{S}_n = \sum_{j=1}^{n} f(c_j)\Delta x = \sum\left[3\left[\frac{j-1}{n}\right]^2 + 6\right]\left(\frac{1}{n}\right) = \sum\left[3(j^2 - 2j + 1)\left(\frac{1}{n}\right)^2 + 6\right]\left(\frac{1}{n}\right)$$

$$= 3\left(\frac{1}{n}\right)^3\left(\sum j^2\right) - 3(2)\left(\frac{1}{n}\right)^3\left(\sum j\right) + \left[3\left(\frac{1}{n}\right)^3 + 6\frac{1}{n}\right]\left(\sum 1\right)$$

$$= 3\frac{1}{n^3}\frac{n(n+1)(2n+1)}{6} - 3(2)\frac{1}{n^3}\cdot\frac{n(n+1)}{2} + \left[3\frac{1}{n^3} + 6\frac{1}{n}\right]\frac{n}{1}$$

$$= \frac{1}{2}\left(\frac{n}{n}\right)\left(1 + \frac{1}{n}\right)\left(2 + \frac{1}{n}\right) - \left(\frac{3}{1}\right)\frac{n}{n}\left(1 + \frac{1}{n}\right)\left(\frac{1}{n}\right) + \left[\frac{3}{n^2} + 6\right]$$

which approaches $\frac{1}{2}(1)(1)(2) - 3(1)(1)(0) + (0) + 6 = 7$ as $n \to \infty$.

54. Now $\Delta x = \frac{2-0}{n} = \frac{2}{n}$, and since f increases on $[0,2]$ (the vertex is at $x = -\frac{1}{4}$), then the c_j for a lower sum are the left endpoints: $c_j = 0 + \Delta x(j-1) = \left(\frac{2}{n}\right)(j-1)$. So

$$
\begin{aligned}
\underline{S}_n &= \sum_{j=1}^{n} f(c_j)\Delta x = \sum \left\{ 2\left[\left(\frac{2}{n}\right)(j-1)\right]^2 + \left[\left(\frac{2}{n}\right)(j-1)\right] + 3 \right\}\left(\frac{2}{n}\right) \\
&= \sum \left\{ 2\left(\frac{2}{n}\right)^3 (j^2 - 2j + 1) + \left(\frac{2}{n}\right)^2 (j-1) + \left(\frac{2}{n}\right)3 \right\} \\
&= 2\left(\frac{2}{n}\right)^3 \left(\sum j^2\right) + \left[-2\left(\frac{2}{n}\right)^3 2 + \left(\frac{2}{n}\right)^2\right]\left(\sum j\right) + \left[2\left(\frac{2}{n}\right)^3 - \left(\frac{2}{n}\right)^2 + \left(\frac{2}{n}\right)3\right]\left(\sum 1\right) \\
&= \frac{2}{1}\left(\frac{2^3}{n^3}\right)\frac{n(n+1)(2n+1)}{6} + \left[\frac{-2^5}{n^3} + \frac{2^2}{n^2}\right]\frac{n(n+1)}{2} + \left[\frac{2^4}{n^3} - \frac{2^2}{n^2} + \frac{2(3)}{n}\right]\frac{n}{1} \\
&= \left[\frac{8}{3}\left(\frac{n}{n}\right)\left(\frac{n+1}{n}\right)\left(\frac{2n+1}{n}\right)\right] + \left[\frac{-2^4}{1}\frac{n}{n}\frac{n+1}{n}\frac{1}{n}\right] + \left[\frac{2}{1}\frac{n}{n}\frac{n+1}{n}\right] + \left[\frac{2^4}{n^2}\right] - \left[\frac{2^2}{n}\right] + \left[\frac{2(3)}{1}\right]
\end{aligned}
$$

which approaches $\frac{8}{3} \cdot \frac{2}{1} - 2^4(0) + 2 + 0 - 0 + 6 = \frac{40}{3}$ as $n \to \infty$.

55. Since $f(x) = \text{ABS}(1 - x^2)$ decreases from 1 to 0 on $[0,1]$ and increases from 0 to 1 on $[1,2]$, we will divide the sum into two parts, each with n terms, each covering one of those intervals.

In the first sum, the c_j are the right endpoints: $c_j = 0 + (\Delta x)j = \frac{j}{n}$, and $1 - x^2 \geq 0$.

$$
\begin{aligned}
\underline{S}_n \text{ on}[0,1] &= \sum_{j=1}^{n} f(c_j)\Delta x = \sum \text{ABS}\left(1 - \left(\frac{j}{n}\right)^2\right)\left(\frac{1}{n}\right) = \frac{-1}{n^3}\sum j^2 + \frac{1}{n}\sum 1 \\
&= \left(\frac{-1}{n^3}\right)\frac{n(n+1)(2n+1)}{6} + \left(\frac{1}{n}\right)n = -\frac{1}{6}\frac{n}{n}\frac{n+1}{n}\frac{2n+1}{n} + 1 \to -\frac{1}{6}\frac{2}{1} + 1 \\
&= \frac{2}{3} \text{ as } n \to \infty.
\end{aligned}
$$

So

In the second sum, on $[1,2]$, the c_j are the left endpoints, $c_j = 1 + (\Delta x)(j-1)$, and $1 - x^2 \leq 0$.

$$
\begin{aligned}
\underline{S}_n \text{ on } [1,2] &= \sum_{j=1}^{n} f(c_j)\Delta x = \sum \left\{ \text{ABS}\left(1 - \left[1 + \frac{j-1}{n}\right]^2\right)\right\}\left(\frac{1}{n}\right) \\
&= \sum \left\{\left[1 + 2\left(\frac{j-1}{n}\right) + \frac{j^2 - 2j + 1}{n^2}\right] - 1\right\}\left(\frac{1}{n}\right) \\
&= \left(\frac{1}{n^3}\right)\left(\sum j^2\right) + \left(\frac{2}{n^2} - \frac{2}{n^3}\right)\left(\sum j\right) + \left(-\frac{2}{n^2} + \frac{1}{n^3}\right)\sum 1 \\
&= \left[\frac{n(n+1)(2n+1)}{n \cdot n \cdot n \cdot 6}\right] + \left[\left(\frac{2}{n^2} - \frac{2}{n^3}\right)\frac{n(n+1)}{2}\right] + \left[\left(-\frac{2}{n^2} + \frac{1}{n^3}\right)\frac{n}{1}\right] \\
&\to \left[\frac{2}{1}\frac{1}{6}\right] + [1 - 0] + [0 + 0] \\
&= \frac{4}{3} \text{ as } n \to \infty.
\end{aligned}
$$

So the entire sum approaches $\frac{2}{3} + \frac{4}{3} = 2$.

56. a) Width $= \frac{b}{n}$ for each rectangle.

   b) The right side of the kth rectangle is $\left(\frac{k}{n}\right)b$ away from the left vertex, $k = 1, 2, \cdots, n$. By similar triangles, its height x is such that

   $$\frac{x}{\left(\frac{k}{n}\right)b} = \frac{h}{b} \text{ as } x = \frac{hk}{n}.$$

   c) Area of rectangle is base $\times$ height $= \frac{b}{n} \times \frac{hk}{n} = \frac{bhk}{n^2}$.

   d) The total area $A_n = \displaystyle\sum_{k=1}^{n}$ areas kth rectangle

   $$= \sum \frac{bhk}{n^2} = \frac{bh}{n^2} \sum k = \frac{bh}{n^2} \cdot \frac{n(n+1)}{2} = \frac{bh}{1}\left(\frac{n+1}{2n}\right) = bh\left(\frac{1}{2} + \frac{1}{2n}\right).$$

   e) $A_n \to bh\left(\frac{1}{2} + 0\right) = \frac{bh}{2}$ as $n \to \infty$.

57. a) If $\frac{x^2}{9} + \frac{y^2}{4} = 1$, then

   $$\frac{y^2}{4} = 1 - \frac{x^2}{9}, \quad \frac{y}{2} = \pm\sqrt{1 - \frac{x^2}{9}}, \quad \text{and} \quad y = \pm 2\sqrt{1 - \frac{x^2}{9}}.$$

   b) The graph is symmetric about the x-axis and y-axis, so we can do only the part in Quadrant I, then multiply by 4 to get the total area.

   c) The function decreases in Quadrant I, so use the right endpoints for a lower sum.
   $\underline{S}_3 : c_1 = 1, \ c_2 = 2, \ c_3 = 3$. Also $\Delta x = \frac{3-0}{3} = 1$. So

   $$\underline{S}_3 = \sum_{j=1}^{3} f(c_j)\Delta x = 2\sqrt{1 - \frac{1^2}{9}} + 2\sqrt{1 - \frac{2^2}{9}} + 2\sqrt{1 - \frac{3^2}{9}} \approx 3.37633.$$

   So $\underline{S}_3 \cdot 4 \approx 13.50532.$

   d) If $n = 6$, $\Delta x = \frac{3-0}{6} = \frac{1}{2}$ then $c_1 = \frac{1}{2}, c_2 = \frac{2}{2}, \ c_3 = \frac{3}{2}, \cdots, \ c_6 = \frac{6}{2}$, and

   $$\underline{S}_6 = \sum_{j=1}^{6} f(c_j)\Delta x \approx 4.09297.$$

   So $\underline{S}_6 \cdot 4 \approx 16.37190.$

58. a) The unit circle is $x^2 + y^2 = 1$. For the top half, we want $y > 0$. Solve for y first. $y^2 = 1 - x^2$, $y = \pm\sqrt{1 - x^2}$. So $y = +\sqrt{1 - x^2}$ for the top half.

   b) For $\underline{S}_3$ in Quadrant I, the c_j are the right endpoints: $\frac{1}{3}, \frac{2}{3}, \frac{3}{3}$.

   $$\underline{S}_3 = \left(\sqrt{1 - \left(\frac{1}{3}\right)^2}\right)\left(\frac{1}{3}\right) + \left(\sqrt{1 - \left(\frac{2}{3}\right)^2}\right)\left(\frac{1}{3}\right) + \left(\sqrt{1 - \left(\frac{3}{3}\right)^2}\right)\left(\frac{1}{3}\right) \approx 0.56272.$$

c) For $\underline{S}_6$ in Quadrant I, the c_j are $\frac{1}{6}$, $\frac{2}{6}$, $\cdots$, $\frac{6}{6}$, and $\Delta x = \frac{1}{6}$. So

$$\underline{S}_6 = \sum_{j=1}^{6} f(c_j)\Delta x \approx 0.68216.$$

d) Exact value is $\frac{\pi r^2}{4} = \frac{\pi 1^2}{4} \approx 0.78540$.

59. a) We have divided the entire circle, measuring $2\pi(rad)$, into $2n$ equal angles, so each measures $\frac{2\pi}{2n} = \frac{\pi}{n}(rad)$.

   b) As shown in the figure, the base of each triangle is $r\cos\left(\frac{\pi}{n}\right)$ and the height is $r\sin\left(\frac{\pi}{n}\right)$, from basic trigonometry. So the area is $\left(\frac{1}{2}\right)$ (base) (height) $= \left(\frac{r^2}{2}\right)\sin\left(\frac{\pi}{n}\right)\cos\left(\frac{\pi}{n}\right) = \left(\frac{r^2}{4}\right)\left[2\sin\left(\frac{\pi}{n}\right)\cos\left(\frac{\pi}{n}\right)\right] = \left(\frac{r^2}{4}\right)\sin\left(\frac{2\pi}{n}\right)$ by the double-angle identity for sin.

   c) To get A_n, sum the areas of the triangles:

$$A_n = \sum_{t=1}^{2n} \frac{r^2}{4}\sin\left(\frac{2\pi}{n}\right)$$

$$= \frac{r^2}{4}\sin\left(\frac{2\pi}{n}\right)\left(\sum 1\right)$$

$$= \frac{r^2}{4}\left[\sin\left(\frac{2\pi}{n}\right)\right](2n)$$

$$= \frac{nr^2}{2}\sin\left(\frac{2\pi}{n}\right).$$

   d) Table 1.2

r	n	A_n	$A = \pi r^2$
1	5		
1	20	3.09017	
1	50		
4	5	38.04226	50.26548
4	20		
4	50		
4	200	50.25721	
10	5		314.15927
10	20		
10	50		
10	200	314.10759	
10	500	314.15100	
10	1000		

60. a) There are $2n$ equal angles around the center of the circle, so each measures $\frac{2\pi}{2n} = \frac{\pi}{n}$ (rad). The height of each triangle is r, and the base is $r\tan\left(\frac{\pi}{n}\right)$ by basic trigonometry. Hence the area is

$$\frac{1}{2}\left(r\tan\left(\frac{\pi}{n}\right)\right)(r) = \frac{r^2}{2}\tan\left(\frac{\pi}{n}\right).$$

   b) There are $2n$ such triangles, so the total area of them is

$$\frac{2n}{1}\frac{r^2}{2}\tan\left(\frac{\pi}{n}\right) = nr^2\tan\left(\frac{\pi}{n}\right).$$

c)

r	n	A_n	A
1	5	3.63271	3.14159
1	20	3.16769	
1	50	3.14573	
4	5	58.12340	50.26548
4	20	50.68302	
4	50	50.33173	
4	200	50.26962	
10	5	363.27126	314.15927
10	20	316.76888	
10	50	314.57334	
10	200	314.18511	
10	500	314.16340	
10	1000	314.16030	

61. The triangles in Ex. 59 are inscribed in the circle, so the sum of their areas, $\left(\frac{nr^2}{2}\right)\sin\left(\frac{2\pi}{n}\right)$, is a lower sum for the area. The triangles in Ex. 60 are circumscribed, so $nr^2\tan\left(\frac{\pi}{n}\right)$ is an upper sum for the area. Take the limits of the two sums:

$$\frac{nr^2}{2}\sin\left(\frac{2\pi}{n}\right) = \pi r^2\left(\frac{n}{2\pi}\right)\sin\left(\frac{2\pi}{n}\right) = \pi r^2\frac{\sin(u)}{u}\left[u = \frac{2\pi}{n}\right]$$

which approaches $\pi r^2 \cdot 1$ as $n \to \infty$ since $\frac{2\pi}{n} = u \to 0$ as $n \to \infty$.

Also $nr^2 \tan\left(\frac{\pi}{n}\right) = \pi r^2\left(\frac{n}{\pi}\right)\tan\left(\frac{\pi}{n}\right) = \pi r^2 \frac{\tan(w)}{w}$ $\left[w = \frac{\pi}{n}\right]$ which approaches $\pi r^2 \cdot 1$ as $n \to \infty$, since $w = \frac{\pi}{n} \to 0$ as $n \to \infty$.

62. $j^2 - (j-1)^2 = j^2 - (j^2 - 2j + 1) = j^2 - j^2 + 2j - 1 = 2j - 1$. Following the hint,

$$\sum_{j=1}^{4}(2j-1) = \sum_{j=1}^{4} j^2 - (j-1)^2 = [1^2 - (0)^2] + [2^2 - (1)^2] + [3^2 - (2)^2] + [4^2 - (3)^2]$$
$$= 1^2 - 1^2 + 2^2 - 2^2 + 3^2 - 3^2 + 4^2 = 4^2.$$

The middle terms "telescope", leaving only $4^2 - 0^2 = 4^2$. In general,

$$\sum_{j=1}^{4}(2j-1) = \sum_{j=1}^{n} j^2 - (j-1)^2 = \sum_{j=1}^{n} j^2 - \sum_{j=1}^{n}(j-1)^2$$
$$= 1^2 + 2^2 + \cdots + (n-1)^2 + n^2 - [0^2 + 1^2 + 2^2 + \cdots + (n-1)^2] = -0^2 + n^2 = n^2.$$

63. Assume Eq(3) and Ex. 62. Then

$$\sum_{j=1}^{n} j = \sum\left(\frac{1}{2}(2j-1) + \frac{1}{2}\right) = \frac{1}{2}\sum(2j-1) + \frac{1}{2}\sum 1$$
$$= \frac{1}{2}n^2 + \frac{1}{2}n = \frac{1}{2}(n^2 + n) = \frac{1}{2}n(n+1).$$

64. $j^3 - (j-1)^3 = j^3 - (j^3 - 3j^2 + 3j - 1) = 3j^2 - 3j + 1$. Also

$$\sum_{j=1}^{n} j^3 - (j-1)^3 = (1^3 - 0^3) + (2^3 - 1^3) + (3^3 - 2^3) + \cdots + [(n-1)^3 - (n-2)^3] + [n^3 - (n-1)^3]$$
$$= (-0)^3 + (1^3 - 1^3) + (2^3 - 2^3) + \cdots + [(n-1)^3 - (n-1)^3] + n^3$$
$$= 0 + 0 + 0 + \cdots + 0 + n^3 = n^3.$$

Assuming Eq(3) and Eq(4), we get

$$\sum_{j=1}^{n} j^2 = \sum\frac{1}{3}(3j^2 - 3j + 1) + j - \frac{1}{3} = \frac{1}{3}\left(\sum 3j^2 - 3j + 1\right) + \left(\sum j\right) - \frac{1}{3}\left(\sum 1\right)$$
$$= \frac{1}{3}n^3 + \frac{1}{2}n(n+1) - \frac{1}{3}n = \frac{2n^3 + 3n(n+1) - 2n}{6} = \frac{1}{6}(2n^3 + 3n^2 + n)$$
$$= \frac{1}{6}(n)(2n^2 + 3n + 1) = \frac{1}{6}(n)(n+1)(2n+1).$$

65. a) Using addition identitites: $\sin(\alpha + \beta) - \sin(\alpha - \beta) = [\sin\alpha\cos\beta + \cos\alpha\sin\beta] - [\sin\alpha\cos\beta - \cos\alpha\sin\beta] = 2\cos\alpha\sin\beta$.

 b) To show

$$\sum_{j=0}^{n-1}\cos(j\gamma) = \frac{1}{2} + \frac{\sin\left(n - \frac{1}{2}\right)\gamma}{2\sin\left(\frac{1}{2}\gamma\right)}$$

for all n, show it for (1) $n = 1$, then show (2) that if it is true for n, then it is also true for $n+1$.

(1) $\displaystyle\sum_{j=0}^{1-1} \cos(j\gamma) = \cos(0 \cdot \gamma) = 1$ and

$$\frac{1}{2}\,\frac{\sin\left(1-\frac{1}{2}\right)\gamma}{2\sin\left(\frac{1}{2}\gamma\right)} = \frac{1}{2} + \frac{\sin\left(\frac{1}{2}r\right)}{2\sin\left(\frac{1}{2}r\right)} = \frac{1}{2} + \frac{1}{2} = 1.$$

(2) Assume the equation is true for n. Then

$$\sum_{j=0}^{(n+1)-1} \cos j\gamma = \left[\sum_{j=0}^{n-1} \cos(j\gamma)\right] + \cos(n\gamma) = \left[\frac{1}{2} + \frac{1}{2}\,\frac{\sin\left(n-\frac{1}{2}\right)\gamma}{\sin\left(\frac{1}{2}\right)\gamma}\right] + \cos(n\gamma).$$

We want to show that this is equal to $\left[\dfrac{1}{2} + \dfrac{1}{2}\,\dfrac{\sin\left(n+1-\frac{1}{2}\right)\gamma}{\sin\left(\frac{1}{2}\gamma\right)}\right]$. This is equivalent to showing that $\cos(n\gamma)$ is equal to the difference of the last two expressions in brackets. So take the difference and simplify it:

$$\left[\frac{1}{2} + \frac{1}{2}\,\frac{\sin\left(n+\frac{1}{2}\right)\gamma}{\sin\left(\frac{1}{2}\gamma\right)}\right] - \left[\frac{1}{2} + \frac{1}{2}\,\frac{\sin\left(n-\frac{1}{2}\right)\gamma}{\sin\left(\frac{1}{2}\gamma\right)}\right] =$$

$$\left[\frac{1}{2} - \frac{1}{2}\right] + \left[\frac{1}{2\sin\left(\frac{1}{2}\gamma\right)}\right]\left[\sin\left(n+\frac{1}{2}\right)\gamma - \sin\left(n-\frac{1}{2}\right)\gamma\right]$$

$$= 0 + \frac{1}{2\sin\left(\frac{1}{2}\gamma\right)}\left[\sin\left(n\gamma + \frac{\gamma}{2}\right) - \sin\left(n\gamma - \frac{\gamma}{2}\right)\right]$$

[Use part a) now, where $\alpha = n\gamma$ and $\beta = \dfrac{\gamma}{2}$.]

$$= \frac{1}{2\sin\left(\frac{1}{2}\gamma\right)}\left[2\cos(n\gamma)\sin\left(\frac{\gamma}{2}\right)\right] = \cos(n\gamma)$$

after cancelling.

66. We note that $\left(\dfrac{1}{j}\right) - \dfrac{1}{j+1} = \dfrac{j+1}{j(j+1)} - \dfrac{1(j)}{(j+1)j} = \dfrac{(j+1)-1}{j(j+1)} = \dfrac{1}{j(j+1)}.$
    So

$$\sum_{j=1}^{n} \frac{1}{j(j+1)} = \sum \frac{1}{j} - \frac{1}{j+1}$$

[which will "telescope" as in Ex. 62 and Ex. 64.]

$$= \left(\frac{1}{1} - \frac{1}{2}\right) + \left(\frac{1}{2} - \frac{1}{3}\right) + \cdots + \left(\frac{1}{n-1} - \frac{1}{n}\right) + \left(\frac{1}{n} - \frac{1}{n+1}\right)$$

[regroup]

$$= 1 + \left(-\frac{1}{2} + \frac{1}{2}\right) + \left(-\frac{1}{3} + \frac{1}{3}\right) + \cdots + \left(-\frac{1}{n} + \frac{1}{n}\right) - \frac{1}{n+1} = 1 - \frac{1}{n+1}.$$

67. Graph using the range $[-1, 2] \times [0, 2]$ and $n = 16$.

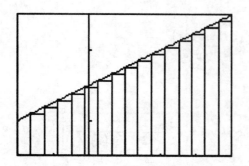 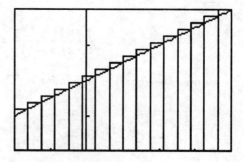

Table:

n	4	8	16	64
$\underline{S}_n$	3.1875	3.46875	3.609375	3.71484375
$\overline{S}_n$	4.3125	4.03125	3.890625	3.78515625

68. Graph using the range $[0, 3] \times [0, 9]$ and $n = 16$.

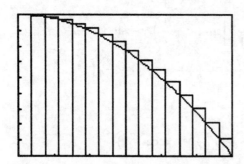 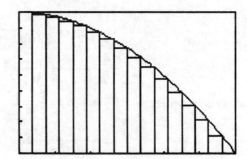

Table:

n	4	8	16	64
$\underline{S}_n$	14.34375	16.2421875	17.138671875	17.787963867
$\overline{S}_n$	21.09375	19.6171875	18.826171875	18.209838867

69. Graph using the range $[0, 1] \times [0, 1]$ and $n = 16$.

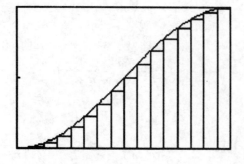 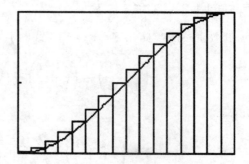

Table:

n	4	8	16	64
$\underline{S}_n$	0.375	0.4375	0.46875	0.4921875
$\overline{S}_n$	0.625	0.5625	0.53125	0.5078125

70. Graph using the range $[-1, 3] \times [0, 2]$ and $n = 16$.

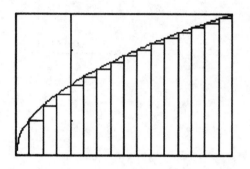 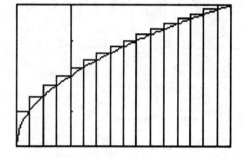

Table:

n	4	8	16	64
$\underline{S}_n$	4.14626437	4.765041773	5.05864457	5.26766649
$\overline{S}_n$	6.14626437	5.765041773	5.55864957	5.39266649

71. Table:

n	$\overline{S}_n$	$\underline{S}_n$	$\overline{S}_n - \underline{S}_n$
1	4	2	2
2	3.6	2.6	1
4	3.3811765	2.8811765	0.5
8	3.2639885	3.0139885	0.25
16	3.20344161	3.07844161	0.125
32	3.17267989	3.11017989	0.0625
64	3.15717696	3.12592696	0.3125

From the table it appears that $\overline{S}_n - \underline{S}_n = \frac{1}{2}^{m-1}$ where m is chosen so that $2^m = n$. So for $n = 1$, $m = 0$; $n = 2$, $m = 1$; $n = 4$, $m = 2$; $n = 8$, $m = 3$; etc. Therefore if $\overline{S}_n - \underline{S}_n < 10^{-5}$ then we must choose m so that $10^{-5} < \frac{1}{2}^{m-1}$. So solving the equation $2^{1-m} = 10^{-5}$ gives $m = 17.6096$. Thus take $m = 18$. Then $\frac{1}{2}^{17} = 7.6295 \times 10^{-6} < 10^{-5}$. Thus $n = 2^{18} = 262144$ yields the prescribed error.

6.2   Riemann Sums: The Definite Integral

1. $f(x) = x + 2$, $P_3 = \{1, 2, 3, 5\}$

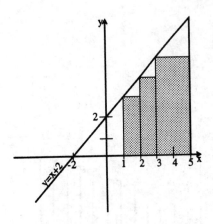

$$R_3 = \sum_{j=1}^{3} f(t_j)\Delta x_j$$

$$= f(1)(2-1) + f(2)(3-2)$$

$$+ f(3)(5-3)$$

$$= 3(1) + 4(1) + 5(2) = 17.$$

Exercise 1

2. $f(x) = 2x + 3$, $P_4 = \{-1, 0, 2, 3, 4\}$

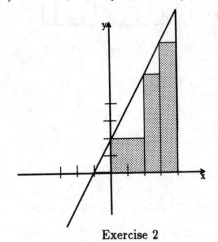

$$R_4 = \sum_{j=1}^{4} f(t_j)\Delta x_j$$

$$= f(-1)(0-(-1)) + f(0)(2-0)$$

$$+ f(2)(3-2) + f(3)(4-3)$$

$$= 1(1) + 3(2) + 7(1) + 9(1)$$

$$= 23.$$

Exercise 2

3. $f(x) = 3 - x$, $P_4 = \left\{0, 1, \frac{3}{2}, 2, 4\right\}$

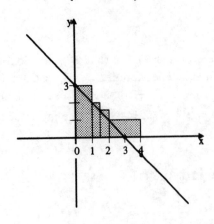

$$R_4 = \sum_{j=1}^{4} f(t_j)\Delta x_j$$

$$= f(0)(1-0) + f(1)\left(\frac{3}{2}-1\right)$$

$$+ f\left(\frac{3}{2}\right)\left(2-\frac{3}{2}\right) + f(2)(4-2)$$

$$= 3(1) + 2\left(\frac{1}{2}\right)$$

$$+ \frac{3}{2}\left(\frac{1}{2}\right) + 1(2)$$

$$= 6\frac{3}{4}.$$

Exercise 3

422

4. $f(x) = 4 - 2x$, $P_3 = \left\{0, \frac{1}{2}, 1, 2, 3\right\}$.

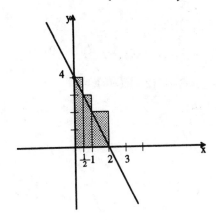

Exercise 4

$$
\begin{aligned}
R_4 &= \sum_{j=1}^{4} f(t_j)\Delta x_j \\
&= f(0)\left(\tfrac{1}{2} - 0\right) + f\left(\tfrac{1}{2}\right)\left(1 - \tfrac{1}{2}\right) \\
&\quad + f(1)(2 - 1) + f(2)(3 - 2) \\
&= 4\left(\tfrac{1}{2}\right) + 3\left(\tfrac{1}{2}\right) + 2(1) + 0(1) \\
&= 5\tfrac{1}{2}.
\end{aligned}
$$

5. $f(x) = x^2 - x$, $P_3 = \{0, 1, 2, 4\}$.

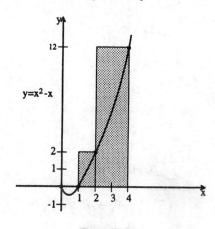

Exercise 5

$$
\begin{aligned}
R_3 &= \sum_{j=1}^{3} f(t_j)\Delta x_j \\
&= f(1)(1 - 0) + f(2)(2 - 1) \\
&\quad + f(4)(4 - 2) \\
&= 0(1) + 2(1) + 12(2) \\
&= 26.
\end{aligned}
$$

6. $f(x) = 4 - x^2$, $P_3 = \{-1, 0, 2, 3\}$.

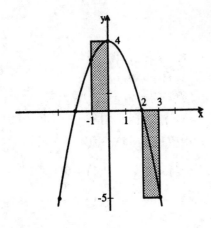

$$R_3 = \sum_{j=1}^{3} f(t_j)\Delta x_j$$

$$= f(0)\big(0 - (-1)\big)$$

$$+ f(2)(2 - 0) + f(3)(3 - 2)$$

$$= 4(1) + 0(2) + (-5)(1)$$

$$= -1.$$

Exercise 6

7. $f(x) = \cos x, \; P_4 = \left\{0, \; \dfrac{\pi}{2}, \; \pi, \; 2\pi, \; \dfrac{5\pi}{2}\right\}.$

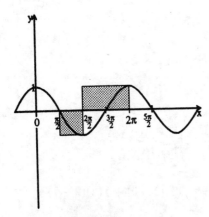

$$R_4 = \sum_{j=1}^{4} f(t_j)\Delta x_j$$

$$= f\big(\tfrac{\pi}{2}\big)\big(\tfrac{\pi}{2} - 0\big) + f(\pi)\big(\pi - \tfrac{\pi}{2}\big)$$

$$+ f(2\pi)(2\pi - \pi)$$

$$+ f\big(\tfrac{5\pi}{2}\big)\big(\tfrac{5\pi}{2} - 2\pi\big)$$

$$= 0\big(\tfrac{\pi}{2}\big) + (-1)\big(\tfrac{\pi}{2}\big)$$

$$+ 1(\pi) + 0\big(\tfrac{\pi}{2}\big)$$

$$= \tfrac{\pi}{2}.$$

Exercise 7

8. $f(x) = \sin x, \; P_3 = \left\{0, \dfrac{\pi}{2}, \dfrac{3\pi}{2}, 2\pi\right\}.$

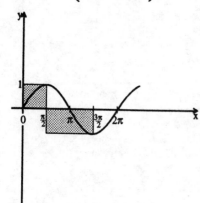

$$R_3 = \sum_{j=1}^{3} f(t_j)\Delta x_j$$

$$= f\big(\tfrac{\pi}{2}\big)\big(\tfrac{\pi}{2} - 0\big) + f\big(\tfrac{3\pi}{2}\big)\big(\tfrac{3\pi}{2} - \tfrac{\pi}{2}\big)$$

$$+ f(2\pi)\big(2\pi - \tfrac{3\pi}{2}\big)$$

$$= 1\big(\tfrac{\pi}{2}\big) + (-1)(\pi) + 0\big(\tfrac{\pi}{2}\big)$$

$$= -\tfrac{\pi}{2}.$$

Exercise 8

9. $f(x) = \sqrt{x}$, $P_3 = \{0, 2, 6, 12\}$.

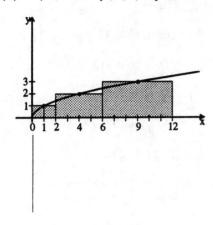

$$
\begin{aligned}
R_3 &= \sum_{j=1}^{3} f(t_j)\Delta x_j \\
&= f(1)(2-0) + f(4)(6-2) \\
&\quad + f(9)(12-6) \\
&= 1(2) + 2(4) + 3(6) \\
&= 28.
\end{aligned}
$$

Exercise 9

10. $f(x) = \sqrt{1-x}$, $P_3 = \{-11, -5, -1, 1\}$

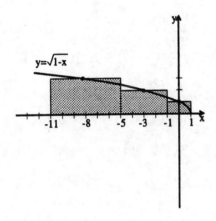

$y = \sqrt{1-x}$

$$
\begin{aligned}
R_3 &= \sum_{j=1}^{3} f(t_j)\Delta x_j \\
&= f(-8)\bigl(-5 - (-11)\bigr) \\
&\quad + f(-3)\bigl(-1 - (-5)\bigr) \\
&\quad + f(0)\bigl(1 - (-1)\bigr) \\
&= 3(6) + 2(4) + 1(2) \\
&= 28.
\end{aligned}
$$

Exercise 10

11. $f(x) = \sin x$, $P_4 = \left\{0, \dfrac{\pi}{2}, \pi, \dfrac{3\pi}{2}, 2\pi\right\}$.

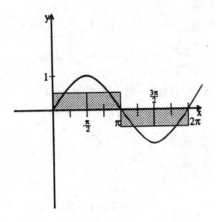

$$R_4 = \sum_{j=1}^{4} f(t_j)\Delta x_j$$

$$= f\left(\tfrac{\pi}{4}\right)\left(\tfrac{\pi}{2}-0\right) + f\left(\tfrac{3\pi}{4}\right)\left(\pi - \tfrac{\pi}{2}\right)$$

$$+ f\left(\tfrac{5\pi}{4}\right)\left(\tfrac{3\pi}{2}-\pi\right)$$

$$+ f\left(\tfrac{7\pi}{4}\right)\left(2\pi - \tfrac{3\pi}{2}\right)$$

$$+ \sqrt{\tfrac{1}{2}}\left(\tfrac{\pi}{2}\right) + \sqrt{\tfrac{1}{2}}\left(\tfrac{\pi}{2}\right)$$

$$+ \left(-\sqrt{\tfrac{1}{2}}\right)\left(\tfrac{\pi}{2}\right) + \left(-\sqrt{\tfrac{1}{2}}\right)\left(\tfrac{1}{2}\right)$$

$$= 0.$$

Exercise 11

12. $f(x) = \cos x,\ P_3 = \left\{0, \dfrac{\pi}{2}, \dfrac{3\pi}{2}, 2\pi\right\}$

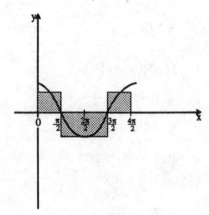

$$R_3 = \sum_{j=1}^{3} f(t_j)\Delta x_j$$

$$= f\left(\tfrac{\pi}{4}\right)\left(\tfrac{\pi}{2}-0\right) + f(\pi)\left(\tfrac{3\pi}{2} - \tfrac{\pi}{2}\right)$$

$$+ f\left(\tfrac{7\pi}{4}\right)\left(2\pi - \tfrac{3\pi}{2}\right)$$

$$= \sqrt{\tfrac{1}{2}}\left(\tfrac{\pi}{2}\right) + (-1)(\pi) + \sqrt{\tfrac{1}{2}}\left(\tfrac{\pi}{2}\right)$$

$$= \pi\left(\sqrt{\tfrac{1}{2}} - 1\right).$$

Exercise 12

13. $f(x) = 4 - x^2,\ [a, b] = [-2, 2]$
    $P_4 = \{-2, -1, 0, 1, 2\}$

426

$$\text{largest } R_n \quad = \quad f(-1)(-1-(-2))$$
$$+ f(0)(0-(-1))$$
$$+ f(0)(1-0)$$
$$+ f(1)(2-1)$$
$$= \quad 3(1) + 4(1) + 4(1) + 3(1)$$
$$= \quad 14.$$
$$\text{smallest } R_n \quad = \quad f(-2)(1) + f(-1)(1)$$
$$+ f(1)(1) + f(2)(1)$$
$$= \quad 0 + 3 + 3 + 0$$
$$= \quad 6.$$

Exercise 13

14. $f(x) = x^2 - 2x = (x-1)^2 - 1$, $[a,b] = [0,4]$
$P_4 = \{0, 1, 2, 3, 4\}$.

$$\text{largest } R_n \quad = \quad f(0)(1-0) + f(2)(2-1) + f(3)(3-2) + f(4)(4-3)$$
$$= \quad 0(1) + 0(1) + 3(1) + 8(1) = 11.$$
$$\text{smallest } R_n \quad = \quad f(1)(1-0) + f(1)(2-1) + f(2)(3-2) + f(3)(4-3)$$
$$= \quad (-1) + (-1) + (0) + (3) = 1.$$

15. $f(x) = \sin x$, $[a,b] = [0, 2\pi]$
$P_4 = \left\{0, \dfrac{\pi}{2}, \pi, \dfrac{3\pi}{2}, 2\pi\right\}$, $\Delta x_j = \dfrac{\pi}{2}$.

$$\text{largest } R_n \quad = \quad f\left(\frac{\pi}{2}\right) \cdot \frac{\pi}{2} + f\left(\frac{\pi}{2}\right) \cdot \frac{\pi}{2} + f(\pi)\frac{\pi}{2} + f(2\pi)\frac{\pi}{2}$$
$$= \quad (1 + 1 + 0 + 0)\left(\frac{\pi}{2}\right) = \pi.$$
$$\text{smallest } R_n \quad = \quad \left[f(0) + f(\pi) + f\left(\frac{3\pi}{2}\right) + f\left(\frac{3\pi}{2}\right)\right]\left(\frac{\pi}{2}\right)$$
$$= \quad (0 + 0 - 1 - 1)\frac{\pi}{2} = -\pi.$$

16. $f(x) = \sin x$, $[a,b] = [-\pi, \pi]$
$P_4 = \left\{-\pi, -\frac{\pi}{2}, 0, \frac{\pi}{2}, \pi\right\}$, $\Delta x_j = \frac{\pi}{2}$

$$\text{largest } R_n \quad = \quad \left[f(-\pi) + f(0) + f\left(\frac{\pi}{2}\right) + f\left(\frac{\pi}{2}\right)\right]\left(\frac{\pi}{2}\right)$$
$$= \quad (0 + 0 + 1 + 1)\left(\frac{\pi}{2}\right) = \pi.$$

$$\text{smallest } R_n = \left[f\left(-\frac{\pi}{2}\right) + f\left(-\frac{\pi}{2}\right) + f(0) + f(\pi) \right] \frac{\pi}{2}$$
$$= (-1 - 1 + 0 + 0)\left(\frac{\pi}{2}\right) = -\pi.$$

17. $\displaystyle\int_0^5 6\,dx = 6(5-0) = 30$ by Example 5.

18. $\displaystyle\int_0^3 2x + 5\,dx = \int_0^3 2x\,dx + \int_0^3 5\,dx = 2\int_0^3 x\,dx + 5\int_0^3 1\,dx$

$$= 2\left[\frac{1}{2}(3^2 - 0^2)\right] + 5[(3-0)] = 9 + 15 = 24.$$

19. $\displaystyle\int_{-2}^4 (3x - 2)\,dx = 3\int_{-2}^4 x\,dx - 2\int_{-2}^4 1\,dx$

$$= 3\left[\frac{1}{2}(4^2 - (-2)^2)\right] - 2[4 - (-2)] = 3[6] - 2[6] = 6.$$

20. $\displaystyle\int_5^2 (5x + 7)\,dx = 5\int_5^2 x\,dx + 7\int_5^2 1\,dx = 5\left[\frac{1}{2}(2^2 - 5^2)\right] + 7(2 - 5)$

$$= 5\left[\frac{-21}{2}\right] + 7[-3] = \frac{-105 - 42}{2} = \frac{-147}{2}.$$

21. $\displaystyle\int_4^0 (9 - 2x)\,dx = 9\int_4^0 1\,dx - 2\int_4^0 x\,dx = 9(0 - 4) - 2\left[\frac{1}{2}(0^2 - 4^2)\right]$

$$= -36 - 2[-8] = -20.$$

22. $\displaystyle\int_2^5 2x^2\,dx = 2\int_2^5 x^2\,dx$

$$= 2\left[\frac{1}{3}(5^3 - 2^3)\right] = 2\left[\frac{117}{3}\right] = 78.$$

23. $\displaystyle\int_3^1 (6 - x^2)\,dx = \int_3^1 6\,dx - \int_3^1 x^2\,dx = 6(1 - 3) - \left[\frac{1}{3}(1^3 - 3^3)\right]$

$$= -12 - \left(-\frac{26}{3}\right) = \frac{-36 + 26}{3} = -\frac{10}{3}.$$

24. $\displaystyle\int_{-4}^{-1} (3x^2 - 1)\,dx = \int_{-4}^{-1} 3x^2\,dx - \int_{-4}^{-1} 1\,dx = 3\int_{-4}^{-1} x^2\,dx - \int_{-4}^{-1} 1\,dx$

$$= 3\left\{\frac{1}{3}[(-1)^3 - (-4)^3]\right\} - [-1 - (-4)] = 3\left\{\frac{1}{3}63\right\} - [+3] = 60.$$

25. $\displaystyle\int_0^1 (x^2 - x + 2)\,dx = \int_0^1 x^2\,dx - \int_0^1 x\,dx + 2\int_0^1 1\,dx$

$$= \frac{1}{3}(1^3 - 0^3) - \frac{1}{2}(1^2 - 0^2) + 2(1 - 0) = \frac{1}{3} - \frac{1}{2} + 2 = \frac{11}{6}.$$

26. $\displaystyle\int_{-2}^{2}(2x^2-3x+2)\,dx \;=\; 2\int_{-2}^{2}x^2\,dx-3\int_{-2}^{2}x\,dx+2\int_{-2}^{2}1\,dx$

$$= \; 2\frac{1}{3}(2^3-(-2)^3)-3\frac{1}{2}(2^2-(-2)^2)+2(2-(-2))$$

$$= \; \frac{2}{3}(16)-\frac{3}{2}(0)+2(4) \;=\; \frac{32+24}{3} \;=\; \frac{56}{3}.$$

27. $\displaystyle\int_{0}^{4}x(2+x)\,dx \;=\; \int_{0}^{4}(x^2+2x)\,dx \;=\; \int_{0}^{4}x^2\,dx+2\int_{0}^{4}x\,dx$

$$= \; \frac{1}{3}(4^3-0^3)+2\frac{1}{2}(4^2-0^2) \;=\; \frac{1}{3}64+16 \;=\; \frac{64+48}{3} \;=\; \frac{112}{3}.$$

28. $\displaystyle\int_{3}^{-1}(x-2)(x+2)\,dx \;=\; \int_{3}^{-1}(x^2-4)\,dx \;=\; \int_{3}^{-1}x^2\,dx-4\int_{3}^{-1}1\,dx$

$$= \; \frac{1}{3}\left((-1)^3-(3)^3\right)-4(-1-3) \;=\; \frac{1}{3}(-28)-4(-4)$$

$$= \; \frac{-28+48}{3} \;=\; \frac{20}{3}.$$

29. $\displaystyle\int_{-4}^{-1}2x(1-2x)\,dx \;=\; \int(-4x^2+2x)\,dx \;=\; -4\int x^2\,dx+2\int x\,dx$

$$= \; (-4)\frac{1}{3}\left[(-1)^3-(-4)^3\right]+2\left[\frac{1}{2}\left((-1)^2-(-4)^2\right)\right]$$

$$= \; \left(-\frac{4}{3}\right)(63)+(-15) \;=\; -84-15 \;=\; -99.$$

30. $\displaystyle\int_{-1}^{1}(ax^2+bx+c)\,dx \;=\; a\int x^2\,dx+b\int x\,dx+c\int 1\,dx$

$$= \; a\frac{1}{3}(1^3-(-1)^3)+b\frac{1}{2}(1^2-(-1)^2)+c(1-(-1)) \;=\; \frac{2}{3}a+0\cdot b+2c.$$

31. a) $\displaystyle\int_{2}^{0}f(x)\,dx=-\int_{0}^{2}f(x)\,dx=-(3).$

   b) $\displaystyle\int_{0}^{5}f(x)\,dx=\int_{0}^{2}f(x)\,dx+\int_{2}^{5}f(x)\,dx=3-2=1.$

   c) $\displaystyle\int_{2}^{8}f(x)\,dx=\int_{2}^{5}+\int_{5}^{8}=-2+5=3.$

   d) $\displaystyle\int_{0}^{8}f(x)\,dx=\int_{0}^{2}+\int_{2}^{5}+\int_{5}^{8}=3-2+5=6.$

   e) $\displaystyle\int_{5}^{0}f(x)\,dx=-\int_{0}^{5}=-1$ from (b).

   f) $\displaystyle\int_{8}^{2}f(x)\,dx=-\int_{2}^{8}=-3$ from (c).

32. $\displaystyle\int_0^{2\pi} \sin^2(x)\,dx + \int_0^{2\pi} \cos^2(x)\,dx = \int_0^{2\pi} \sin^2(x) + \cos^2(x)\,dx = \int_0^{2\pi} 1\,dx = 2\pi - 0 = 2\pi.$

33. a) $\displaystyle\int_1^3 [f(x) + g(x)]\,dx = \int_1^3 f(x)\,dx + \int_1^3 g(x)\,dx = 5 - 2 = 3.$

b) $\displaystyle\int_1^3 [6g(x) - 2f(x)]\,dx = 6\int_1^3 g(x)\,dx - 2\int_1^3 f(x)\,dx = 6(-2) - 2(5) = -22.$

c) $\displaystyle\int_3^1 2f(x)\,dx - \int_1^3 g(x)\,dx = 2\left(-\int_1^3 f(x)\,dx\right) - (-2) = 2(-5) + 2 = -8.$

d) $\displaystyle\int_1^3 [2f(x) + 3]\,dx = 2\int_1^3 f(x)\,dx + 3\int_1^3 1\,dx = 2(5) + 3(3 - 1) = 16.$

e) $\displaystyle\int_3^1 [g(x) - 4f(x) + 5]\,dx = -\int_1^3 g(x)\,dx - 4\left(-\int_1^3 f(x)\,dx\right) + 5\left(-\int_1^3 1\,dx\right)$
$$= -(-2) - 4(-5) + 5(-2) = 12.$$

f) $\displaystyle\int_1^3 [1 - 4g(x) + 3f(x)]\,dx = \int_1^3 1\,dx - 4\int_1^3 g(x)\,dx + 3\int_1^3 f(x)\,dx$
$$= 2 - 4(-2) + 3(5) = 25.$$

34.

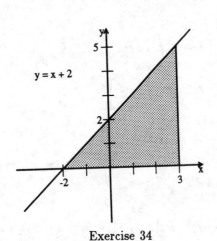

$y = x + 2$

Exercise 34

$$\int_{-2}^3 (x + 2)\,dx = \text{area of triangle}$$
$$= \tfrac{1}{2}bh$$
$$= \tfrac{1}{2}(5)(5)$$
$$= \tfrac{25}{2}.$$

35.

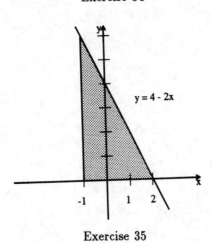

$y = 4 - 2x$

Exercise 35

$$\int_{-1}^2 (4 - 2x)\,dx = \text{area of triangle}$$
$$= \tfrac{1}{2}bh$$
$$= \tfrac{1}{2}(3)(6)$$
$$= 9.$$

36.

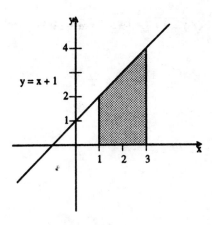

Exercise 36

$$\int_1^3 (x+1)\,dx \;=\; \text{area of trapezoid}$$
$$=\; \tfrac{1}{2}(b_1 + b_2)(h)$$
$$=\; \tfrac{1}{2}(2+4)(2)$$
$$=\; 6.$$

37.

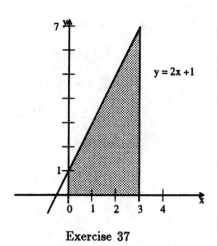

Exercise 37

$$\int_0^3 (2x+1)\,dx \;=\; \text{area of trapezoid}$$
$$=\; \tfrac{1}{2}(b_1 + b_2)(h)$$
$$=\; \tfrac{1}{2}(1+7)(3)$$
$$=\; 12.$$

38.

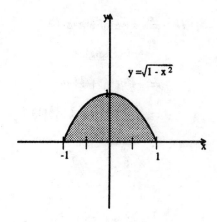

Exercise 38

$$\int_{-1}^{1} \sqrt{1-x^2}\,dx \;=\; \text{area semicircle}$$
$$=\; \tfrac{1}{2}\pi r^2$$
$$=\; \tfrac{1}{2}\pi 1^2$$
$$=\; \tfrac{\pi}{2}.$$

39.

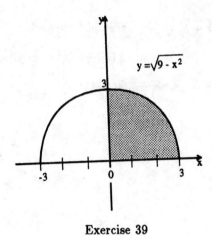

Exercise 39

$$\int_{+0}^{3} \sqrt{9 - x^2}\, dx \;=\; \text{area quarter circle}$$
$$=\; \tfrac{1}{4}\pi r^2$$
$$=\; \tfrac{9\pi}{4}.$$

40.

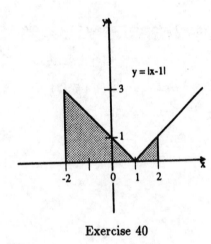

Exercise 40

$$\int_{-2}^{2} |x - 1|\, dx \;=\; \text{sum of areas}$$
$$\text{of triangles}$$
$$=\; \tfrac{1}{2} b_1 h_1 + \tfrac{1}{2} b_2 h_2$$
$$=\; \tfrac{1}{2}(3)(3) + \tfrac{1}{2}(1)(1)$$
$$=\; \tfrac{10}{2}$$
$$=\; 5.$$

41.

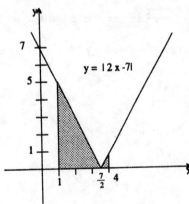

Exercise 41

$$\int_{1}^{4} |2x - 7|\, dx \;=\; \text{sum of areas}$$
$$\text{of triangles}$$
$$=\; \tfrac{1}{2} b_1 h_1 + \tfrac{1}{2} b_2 h_2$$
$$=\; \tfrac{1}{2}\left(\tfrac{5}{2}\right)(5) + \tfrac{1}{2}\left(\tfrac{1}{2}\right)(1)$$
$$=\; \tfrac{25+1}{4}$$
$$=\; \tfrac{13}{2}.$$

432

42.

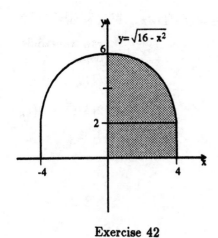

Exercise 42

$$\int_0^4 \left(\sqrt{16 - x^2} + 2 \right) dx = \text{area}$$

quarter circle
+ area

rectangle

$$= \tfrac{1}{4} \pi r^2 + lw$$

$$= \tfrac{1}{4} \pi 4^2 + 4 \cdot 2$$

$$= 4\pi + 8.$$

43.

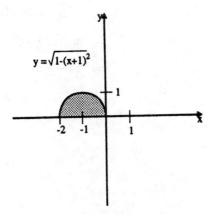

Exercise 43

$$\int_{-2}^0 \sqrt{1 - (x+1)^2} \, dx = \text{area semicircle}$$

$$= \tfrac{1}{2} \pi r^2$$

$$= \tfrac{1}{2} \pi 1^2$$

$$= \tfrac{\pi}{2}.$$

44.

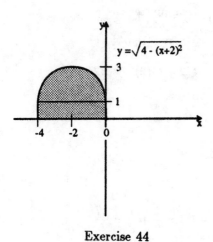

Exercise 44

$$\int_{-4}^0 \left(\sqrt{4 - (x+2)^2} + 1 \right) dx$$

$$= \text{area semicircle}$$

$$+ \text{area rectangle}$$

$$= \tfrac{1}{2} \pi r^2 + lw$$

$$= \tfrac{1}{2} \pi 2^2 + 4(1)$$

$$= 2\pi + 4.$$

45.

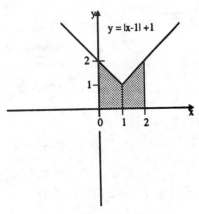

$$\int_0^2 (|x-1|+1)\, dx \;=\; \text{area of}$$
$$\text{two trapezoids}$$
$$=\; \text{twice}$$
$$\left(\tfrac{1}{2}(b_1+b_2)h\right)$$
$$=\; 2(\tfrac{1}{2})(2+1)(1)$$
$$=\; 3.$$

Exercise 45

46.

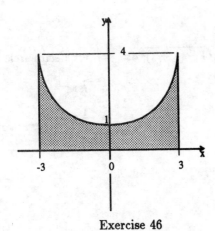

$$\int_{-3}^{3} \left(4 - \sqrt{9-x^2}\right) dx \;=\; \text{area rectangle} -$$
$$\text{area semicircle}$$
$$=\; 6(4) - \tfrac{1}{2}\pi 3^2$$
$$=\; 24 - \tfrac{9}{2}\pi.$$

Exercise 46

47.

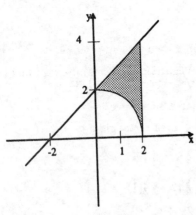

$$\int_0^2 \left(x + 2 - \sqrt{4-x^2}\right) dx$$
$$=\; \text{area trapezoid} -$$
$$\text{area quarter circle}$$
$$=\; \tfrac{1}{2}(b_1+b_2)h - \tfrac{1}{4}\pi r^2$$
$$=\; \tfrac{1}{2}(2+4)(2) - \tfrac{1}{4}\pi 2^2$$
$$=\; 6 - \pi.$$

Exercise 47

48. $\displaystyle \int_0^2 f(x)\, dx = \int_0^1 x\, dx + \int_1^2 x^2\, dx = \frac{1}{2}\left(1^2 - 0^2\right) + \frac{1}{3}\left(2^3 - 1^3\right) = \frac{1}{2} + \frac{7}{3} = \frac{17}{6}.$

49.

$$\int_0^2 f(x)\,dx \;=\; \int_0^1 (1-x)\,dx + \int_1^2 (x^2-1)\,dx$$

$$= \int_0^1 1\,dx - \int_0^1 x\,dx + \int_1^2 x^2\,dx - \int_1^2 1\,dx \;=\; (1-0) - \frac{1}{2}(1^2 - 0^2)$$

$$+ \frac{1}{3}(2^3 - 1^3) - (2-1) \;=\; 1 - \frac{1}{2} + \frac{7}{3} - 1 \;=\; \frac{14-3}{6} \;=\; \frac{11}{6}.$$

50.

$$\int_0^2 f(x)\,dx \;=\; \int_0^1 (-x^2)\,dx + \int_1^2 (x-2)\,dx \;=\; -\int_0^1 x^2\,dx + \int_1^2 x\,dx - 2\int_1^2 1\,dx$$

$$= -\frac{1}{3}(1^3 - 0^3) + \frac{1}{2}(2^2 - 1^2) - 2(2-1) \;=\; -\frac{1}{3} + \frac{3}{2} - 2 \;=\; \frac{-2+9-12}{6} \;=\; -\frac{5}{6}.$$

51.

$$\int_0^2 f(x)\,dx \;=\; \int_0^1 (x^2-1)\,dx + \int_1^2 (x-1)\,dx \;=\; \int_0^1 x^2\,dx - \int_0^1 1\,dx + \int_1^2 x\,dx - \int_1^2 1\,dx$$

$$= \frac{1}{3}(1^3 - 0^3) - (1-0) + \frac{1}{2}(2^2 - 1^2) - (2-1) \;=\; \frac{1}{3} - 1 + \frac{3}{2} - 1$$

$$= \frac{2-6+9-6}{6} \;=\; -\frac{1}{6}.$$

52. a) Suppose $f(x)$ and $g(x)$ are continuous on $[a,b]$ and $f(x) \leq g(x)$ on $[a,b]$.

Then $h(x) = g(x) - f(x)$ is nonnegative, and $\int_a^b g(x)\,dx = \int_a^b [h(x) + f(x)]\,dx = \int_a^b h(x)\,dx + \int_a^b f(x)\,dx$. The proof is finished if we show that $\int_a^b h(x)\,dx \geq 0$. The general Riemann sum for this definite integral is

$$R_n = \sum_{j=1}^n h(t_j)\Delta x_j.$$

Since $\Delta x_j > 0$ (assuming $a < b$) and $h(t_j) \geq 0$ for all j, then also $R_n \geq 0$.

Since $R_n \to \int_a^b h(x)\,dx$ and all the R_n are non-negative, then the integral must be nonnegative also.

b) Since the range of $\sin(x)$ is $[-1,1]$, then

$$-1 \leq \sin(x^2) \leq 1.$$

Also

$$1 \;\leq\; 1 + x^2$$

so

$$\int_0^1 \sin(x^2)\,dx \;\leq\; \int_0^1 (x^2 + 1)\,dx \quad \text{by } \#\,52\ a).$$

53.   a) Suppose f is continuous on $[a, b]$. Then it has a max M and a min m on $[a, b]$; so $m \leq f(x) \leq M$ for $x \in [a, b]$. Hence for any Riemann sum R_n,

$$\sum_{j=1}^{n} m \Delta x_j \ \leq \ \sum_{j=1}^{n} f(t_j) \Delta x_j \ \leq \ \sum_{j=1}^{n} M \Delta x_j.$$

Factor m and M out of each sum:

$$m \left(\sum \Delta x_j \right) \ \leq \ R_n \ \leq \ M \left(\sum \Delta x_j \right).$$

So

$$m(b - a) \leq R_n \ \leq \ M(b - a),$$

since $\sum_{j=1}^{n} \Delta x_j = b - a$. So each R_n is sandwiched between the constants $m(b - a)$ and $M(b - a)$, and therefore its limit, $\int_a^b f(x)\, dx$, must also lie in the interval $[m(b - a), \ M(b - a)]$.

  b) If $1 \leq x \leq 4$, then $1 \leq \sqrt{x} \leq 2$, so $\frac{1}{1} \geq \frac{1}{\sqrt{x}} \geq \frac{1}{2}$, and the min and max of $f(x) = \frac{1}{\sqrt{x}}$ on $[1, 4]$ are $\frac{1}{2}$ and 1. So by # 53 a),

$$\frac{1}{2}(4 - 1) \leq \int_1^4 \frac{1}{\sqrt{x}}\, dx \ \leq \ 1(4 - 1), \quad \text{or} \quad \frac{3}{2} \leq \int_1^4 \frac{1}{\sqrt{x}}\, dx \leq 3.$$

54.   a) Suppose f is continuous on $[a, b]$. Then $|f|$ is also continuous. And by the triangle inequality,

$$\left| \sum_{j=1}^{n} f(t_j) \Delta x_j \right| \leq \sum_{j=1}^{n} |f(t_j) \Delta x_j| = \sum_{j=1}^{n} |f(t_j)| \Delta x_j$$

where we assume $a < b$, so that $\Delta x_j > 0$. As $n \to \infty$ (and $\|P_n\| \to 0$), the sums on the left side approach $\left| \int_a^b f(x)\, dx \right|$, and the sums on the right side approach $\int_a^b |f(x)|\, dx$. Hence the inequality holds between those two values also,

$$\left| \int_a^b f(x)\, dx \right| \leq \int_a^b |f(x)|\, dx.$$

(This is a sort of generalized triangle inequality.)

  b) $\left| \int_0^3 (x - 2)\, dx \right| = \left| \int_0^3 x\, dx - 2 \int_0^3 1\, dx \right| = \left| \frac{1}{2}(3^2 - 0^2) - 2(3 - 0) \right| = \left| \frac{9}{2} - 6 \right| = \frac{3}{2}.$

And $\int_0^3 |x - 2|\, dx = \text{area of triangles} = \frac{1}{2}(2)(2) + \frac{1}{2}(1)(1) = \frac{5}{2}.$

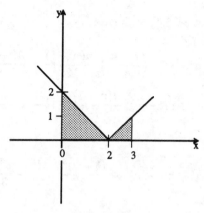

Exercise 54 b)

55. By **Theorem 2**, we can use the partitions with equal-length, since the limit is independent of which we use, so long as $\|P_n\| \to 0$. Then $\Delta x_j = \frac{b-a}{n} = \Delta x$, and $x_j = a + j\Delta x$ is the jth point of the partition $P_n = \{x_0, x_1, \cdots, x_n\}$. Choose the right endpoint of each subinterval:

$$t_j = x_j = a + j\Delta x \text{ for } j = 1, 2, \cdots, n.$$

Then

$$R_n = \sum_{j=1}^{n} f(t_j)\Delta x_j = \sum_{j=1}^{n}(a + j\Delta x)\Delta x = \sum a\Delta x + \sum j(\Delta x)^2$$

$$= a\Delta x \sum 1 + (\Delta x)^2 \sum j = a\Delta x \cdot n + (\Delta x)^2 \frac{n(n+1)}{2}$$

$$= \frac{a(b-a)}{n}\frac{n}{1} + \frac{(b-a)^2}{n^2}\frac{n(n+1)}{2} = a(b-a) + \frac{1}{2}(b-a)^2 \left(\frac{n}{n}\right)\left(\frac{n+1}{n}\right).$$

As $n \to \infty$, this approaches

$$a(b-a) + \frac{1}{2}(b-a)^2(1)(1) = ab - a^2 + \frac{1}{2}(b^2 - 2ab + a^2)$$

$$= ab - a^2 + \frac{1}{2}b^2 - ab + \frac{1}{2}a^2 = \frac{1}{2}b^2 - \frac{1}{2}a^2 = \frac{1}{2}(b^2 - a^2).$$

56. (See the solution to Ex. 55 for more details.)

Using P_n as in Ex. 55, we have

$$R_n = \sum_{j=1}^{n} f(t_j)\Delta x_j = \sum_{j=1}^{n}(a + j\Delta x)^2\Delta x$$

$$= \Delta x \sum_{j=1}^{n}(a^2 + 2aj\Delta x + j^2\Delta x^2)$$

$$= \frac{b-a}{n}\left[\sum_{j=1}^{n}a^2 + \sum_{j=1}^{n}2aj\Delta x + \sum_{j=1}^{n}j^2\Delta x^2\right]$$

$$= \frac{b-a}{n}\left[a^2\sum_{j=1}^{n}1 + 2a\left(\frac{b-a}{n}\right)\sum_{j=1}^{n}j + \left(\frac{b-a}{n}\right)^2\sum_{j=1}^{n}j^2\right].$$

437

Using Formulas (3)–(5) in Section 6.1, we have

$$R_n = \frac{b-a}{n}\left[a^2 n + 2a\left(\frac{b-a}{n}\right)\frac{n(n+1)}{2} + \left(\frac{b-a}{n}\right)^2 \frac{n(n+1)(2n+1)}{6}\right]$$

$$= (b-a)a^2 + a(b-a)^2 \frac{n(n+1)}{n^2} + (b-a)^3 \frac{n(n+1)(2n+1)}{6n^3}.$$

As $n \to \infty$,

$$R_n \to (b-a)a^2 + a(b-a)^2 + \frac{(b-a)^3}{3} = \frac{b^3}{3} - \frac{a^3}{3}.$$

57. a) $\displaystyle\sum_{j=1}^{n}\frac{j}{n^2} = \sum_{j=1}^{n}\left(\frac{j}{n}\right)\left(\frac{1}{n}\right)$. If we partition $[0,1]$ into n equal parts, then

$$\Delta x = \frac{1}{n}, \text{ and } x_j = j\Delta x, \ j = 0,1,2,\cdots,n.$$

If we then let t_j be the right endpoint of the jth subintervals,

$$t_j = x_j = \frac{j}{n}, \ j = 1,2,\cdots,n,$$

then

$$\sum_{j=1}^{n}\left(\frac{j}{n}\right)\left(\frac{1}{n}\right) = \sum_{j=1}^{n}t_j\Delta x = \sum_{j=1}^{n}f(t_j)\Delta x_j$$

which is a Riemann sum R_n for the function $f(x) = x$ over the interval $[0,1]$.

b) Since $R_n \to \int_0^1 f(x)\,dx$ as $n \to \infty$, then

$$\lim_{n\to\infty}\sum_{j=1}^{n}\frac{j}{n^2} = \lim_{n\to\infty}(R_n) = \int_0^1 f(x)\,dx = \frac{1}{2}(1^2 - 0^2) = \frac{1}{2}.$$

58. (See Ex. 57 for more explanation.) $\displaystyle\lim_{n\to\infty}\sum_{j=1}^{n}\frac{j^2}{n^3} = \lim\sum_{j=1}^{n}\left(\frac{j}{n}\right)^2\left(\frac{1}{n}\right)$

$$= \lim_{n\to\infty}R_n = \int_0^1 x^2\,dx = \frac{1}{3}(1^3 - 0^3) = \frac{1}{3}.$$

59. $\displaystyle\lim_{n\to\infty}\sum_{j=1}^{n}\frac{\sqrt{n^2 - j^2}}{n^2} = \lim_{n\to\infty}\sum\frac{\sqrt{n^2 - j^2}}{n}\cdot\left(\frac{1}{n}\right) = \lim_{n\to\infty}\sum\sqrt{1 - \left(\frac{j}{n}\right)^2}\left(\frac{1}{n}\right)$

$$= \lim_{n\to\infty}R_n = \int_0^1\left(\sqrt{1 - x^2}\right)dx$$

$$= \text{area of a quarter circle} = \frac{1}{4}\pi 1^2 = \frac{1}{4}\pi.$$

438

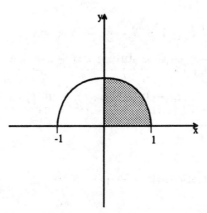

Exercise 59

60.

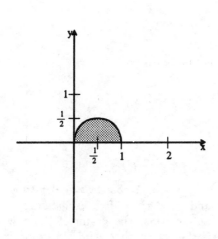

$$\lim_{n\to\infty} \sum \frac{\sqrt{jn - j^2}}{n^2}$$

$$= \lim_{n\to\infty} \sum \frac{\sqrt{jn - j^2}}{n} \cdot \frac{1}{n}$$

$$= \lim_{n\to\infty} \sum_{j=1}^{n} \left(\sqrt{\left(\frac{j}{n}\right) - \left(\frac{j}{n}\right)^2} \right) \left(\frac{1}{n}\right)$$

$$= \lim_{n\to\infty} \sum_{j=1}^{\infty} \left[f\left(\frac{j}{n}\right) \right] \left(\frac{1}{n}\right)$$

$$= \lim_{n\to\infty} R_n = \int_0^1 \left(\sqrt{x - x^2} \right) dx$$

[Now complete the square.]

$$= \int_0^1 \sqrt{\left(\frac{1}{2}\right)^2 - \left(x - \frac{1}{2}\right)^2} \, dx$$

$$= \text{area of semicircle}$$

$$= \frac{1}{2}\pi r^2 = \frac{1}{2}\pi \left(\frac{1}{2}\right)^2 = \frac{\pi}{8}.$$

Exercise 60

61. For $\int_0^b \cos x \, dx$, we use the partition $P_n = \{x_j\}$ where $x_j = j\left(\frac{b}{n}\right)$, $j = 0, 1, \cdots, n$, and the left endpoints $t_j = x_{j-1} = (j-1)\left(\frac{b}{n}\right)$ for the Riemann sum:

$$R_n = \sum_{j=1}^{n} f(t_j)\Delta x_j = \sum_{j=1}^{n} \cos\left(\frac{(j-1)b}{n}\right)\left(\frac{b}{n}\right)$$

$$= \left(\frac{b}{n}\right) \sum_{j=1}^{n} \cos\left[(j-1)\left(\frac{b}{n}\right)\right] = \left(\frac{b}{n}\right) \sum_{k=0}^{n-1} \cos\left(k\left(\frac{b}{n}\right)\right)$$

where we substitute $k = j - 1$ in the summation for convenience. So

$$R_n = \left(\frac{b}{2n}\right) + \left[\frac{\left(\frac{b}{2n}\right)}{\sin\left(\frac{b}{2n}\right)}\right] \sin\left[\left(n - \frac{1}{2}\right)\left(\frac{b}{n}\right)\right]$$

as $n \to \infty$, $\left(\frac{b}{2n}\right) \to 0$, and $\left(n - \frac{1}{2}\right)\left(\frac{b}{n}\right) = b - \frac{b}{2n} \to b$ so $\sin\left[\left(n - \frac{1}{2}\right)\frac{b}{n}\right] \to \sin(b)$ since $\sin x$ is continuous. Also $\frac{\sin x}{x} \to 1$ as $x \to 0$, so substituting $x = \frac{b}{2n}$, we get

$$R_n \to 0 + [1][\sin(b)] = \sin(b).$$

So $\int_0^b \cos(x)\,dx = \sin(b)$.

62. a)

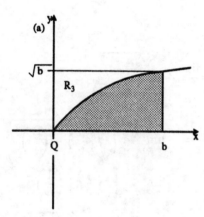

Exercise 62 a)

$$\int_0^b \sqrt{x}\,dx = \text{area shaded region}$$
$$= R_1.$$
$$R_2 = \text{area of the rectangle.}$$
$$R_3 = R_2 - R_1.$$

b)

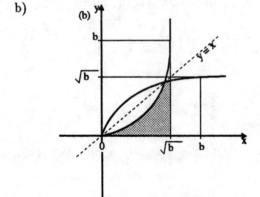

Exercise 62 b)

In $y = \sqrt{x}$, exchange x and y to get $x = \sqrt{y}$ and $y = x^2$. After exchanging x and y in an equation, the graph of the new equation is the reflection of the old graph in the line $y = x$. Reflecting a graph does not change lengths and areas. Hence the shaded region in Figure 6.2 b) has the same area as R_3. So $R_3 = \int_0^{\sqrt{b}} x^2\,dx$ [which we can write as $\int_0^{\sqrt{b}} y^2\,dy$ since any variable can be used.]

b) **Alternate solution.** To compute R_3, the area shaded, integrate along the y-axis, not the x-axis. Since $y = \sqrt{x}$, then $x = y^2 = f(y)$ now gives the function we integrate, from $y = 0$ to $y = \sqrt{b}$ so

$$R_3 = \int_0^{\sqrt{b}} y^2\,dy = \frac{1}{3}\left(\left(\sqrt{b}\right)^3 - 0^3\right) = \frac{1}{3}b^{3/2}.$$

c) Hence $R_3 = \frac{1}{3}((\sqrt{b})^3 - 0^3) = \frac{1}{3}b^{3/2}$. So $R_1 = R_2 - R_3 = (b)\sqrt{b} - \frac{1}{3}b^{3/2} = \frac{2}{3}b^{3/2}$.

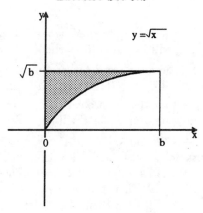

$$y = \sqrt{x}$$

Exercise 62 c)

63. The text shows that if C is between A and B, then $\int_A^B = \int_A^C + \int_C^A$. Now we use that fact to show that $\int_a^b = \int_a^c + \int_c^b$, no matter what the order among a, b, c.

    The six possible orders are

    (i) $a < c < b$

    (ii) $c < a < b$ (already discussed in the text)

    (iii) $a < b < c$

    (iv) $b < a < c$

    (v) $b < c < a$ and

    (vi) $c < b < a$.

We use the same method for each order, starting with the fact that the equation above is true for any $A < C < B$, and **Definition 4**.

   (iii) We know from the proof, $\displaystyle\int_a^c = \int_a^b + \int_b^c$, since b is between a and c. [Hence $a = A$, $b = C$, $c = B$]. Hence $\displaystyle\int_a^b = +\int_a^c - \int_b^c = \int_a^c + \int_c^b$ by **Definition 4**.

   (iv) a is between b and c, so $\displaystyle\int_b^c = \int_b^a + \int_a^c$. Hence $-\displaystyle\int_b^a = \int_a^c - \int_b^c$, so $\displaystyle\int_a^b = \int_a^c + \int_c^b$ by **Definition 4**.

   (v) c is between b and a, so $\displaystyle\int_b^a = \int_b^c + \int_c^a$. Hence $-\displaystyle\int_b^a = -\int_c^a - \int_b^c$, so $\displaystyle\int_a^b = \int_a^c + \int_c^b$.

   (vi) b is between c and a, so $\displaystyle\int_c^a = \int_c^b + \int_b^a$. Hence $-\displaystyle\int_b^a = -\int_c^a + \int_c^b$, so $\displaystyle\int_a^b = \int_a^c + \int_c^b$.

64. If $S_n = \displaystyle\sum_{j=1}^n c\, f(t_j)\Delta x_j$ and $R_n = \displaystyle\sum_{j=1}^n f(t_j)\Delta x_j$ are Riemann sums for $c\, f(x)$ and $f(x)$, respectively, on the interval $[a, b]$ for the same partition P and the same t_j, then by the distributive law, the first sum is c times as large as the second: $S_n = c\, R_n$. Hence $\lim S_n = c \lim R_n$. Since $S_n \to \int_a^b c\, f(x)\,dx$ and $R_n \to \int_a^b f(x)\,dx$ as $n \to \infty$, then we conclude that

$$\int_a^b c\, f(x)\,dx = c \int_a^b f(x)\,dx.$$

65. a) [As with several problems in this set, the statement of the problem leaves nothing else to be said.]

b) $f(x) = g(x)$ for $0 \le x < 1$. So if we never choose 1 for one of the t_j in a Riemann sum for g, then the sums for f and g over $[0, 1]$ will be equal if we choose the same P and t_j.

c) **Theorem 2** tells us that the Riemann sums for f approach a limit, the same limit, regardless of our choice of t_j, so long as $||P|| \to 0$. Hence we can avoid $t_n = 1$ in our sums for f over the interval $[0, 1]$, and still be sure that they approach $\int_0^1 f(x)\,dx$ as $n \to \infty$; $||P|| \to 0$. By (b), if we use the same P and t_j for our sums for g, they will be the same as the sums for f. Hence those sums for g approach the same limit: $\int_0^1 x\,dx = \frac{1}{2}(1^2 - 0^2) = \frac{1}{2}$.

But to show g is integrable, we also must show that we get the same limit in case t_n can be equal to 1 in the sums. We argue as follows. If $t_n = 1$ in some R_n, then we compute R'_n with the same P and t_j, except $t'_n \ne 1$. Then the difference between R_n and R'_n is no larger than $(g(1) - g(t'_n))\Delta x_n \le 2||P||$. As $||P|| \to 0$, then, this difference approaches zero. So the sums R_n with $t_n = 1$ are larger than the R'_n sum, but they all approach the same limit, which we now can call $\int_0^1 g(x)\,dx$.

66. In the Riemann sum R_n for g over $[a, c]$, we can include b as one of the points in partition $P = \{x_j\}$, $j = 0, 1, \cdots, n$, as in **Theorem 5** in the text. Since $f_2(x) = \hat{f}_2(x)$ except for $x = b$, then so long as $t_j \ne b$ in the subinterval whose left endpoint is b, we have

$R_n = S_m + T_p$ where $m + p = n$, and S_m and T_p are Riemann sums for f_1 over $[a, b]$ and $\hat{f}_2$ over $[b, c]$, respectively. Since f_1 and $\hat{f}_2$ are both continuous, then by **Theorem 2**, $S_m \to \int_a^b f_1(x)\,dx$ and $T_p \to \int_b^c \hat{f}_2(x)\,dx$ as $||P|| \to 0$. So for such partitions, $R_n \to \int_a^b f_1 + \int_b^c \hat{f}_2$.

As in Ex. 65, we must consider what happens if $t_{m+1} = b$, in the first subinterval of T_p, before we conclude that **all** possible Riemann sums will approach the same limit as $||P|| \to 0$. The argument is basically the same as in Ex. 65. If R_n is a sum with $t_{m+1} = b$, then form another sum R'_n identical to it except that $t'_{m+1} \ne b$. Then $|R_n - R'_n| = |\hat{f}_2(t'_{m+1}) - \hat{f}_2(b)|(\Delta x_{m+1}) \le (M - m)||P||$ where M and m are the max and min of the continuous function $\hat{f}_2$ on the interval $[b, c]$.

So $R_n - R'_n \to 0$ as $||P|| \to 0$, and **all** the sums R_n will approach the same limit as the R'_n, which is $\int_a^b f_1 + \int_b^c \hat{f}_2$.

Now we conclude that $g(x)$ is integrable and

$$\int_a^c g = \int_a^b f_1 + \int_b^c \hat{f}_2.$$

67. The same procedure as used in Ex. 66 can be applied to any finite number of discontinuities—but not necessarily an infinite number. If g is continuous on $[a, b]$, except at the points c_j, $j = 1, 2, \cdots, n$. $a = c_0 < c_1 < c_2 < \cdots < c_n < c_{n+1} = b$ then on each subinterval $[c_j, c_{j+1}]$, $j = 0, 1, \cdots, n$ we define

$$f_j(x) = \begin{cases} \lim_{x \to c_j^+} g(x) & \text{if } x = c_j \\ g(x) & \text{if } c_j < x < c_{j+1} \\ \lim_{x \to c_j^-} g(x) & \text{if } x = c_{j+1}. \end{cases}$$

Then each f_j is continuous, and a Riemann sum $(R\,g)_m$ for g over $[a, b]$ is equal to the sum of n different sums

$$(R\,g)_m = (R\,f_1)_{m_1} + \cdots (R\,f_n)_{m_n}$$

where $m_1 + m_2 + \cdots + m_n = m$ and $(R\,f_j)_{m_j}$ is a sum for f_j over $[c_j, c_{j+1}]$, and where we do **not** allow any t_k to be equal to any c_j, and where each c_j is in the partition P.

Such sums approach $\int_a^{c_1} f_0 + \int_{c_1}^{c_2} f_1 + \cdots + \int_{c_n}^b f_n$ as $\|P\| \to 0$. If any $t_k = c_j$, then as before (Ex. 66) we form a sum $(Rg)'_m$ such that in $(Rg)'_m$ no t_k is equal to any c_j, but

$$|(Rg)'_m - (Rg)_m| \le B\|P\|,$$

where now B equals the **finite** sum of $(\text{Max}_1 - \min_1) + \cdots + (\text{Max}_n - \min_n)$ where Max_j, $\min_j$ are the max and min of f_j on $[c_j, c_{j+1}]$. So, as before, all Riemann sums for g over $[a, b]$ approach the same limit, so g is integrable over $[a, b]$, and

$$\int_a^b g = \int_a^{c_1} f_0 + \cdots + \int_{c_n}^b f_n.$$

[NOTE: If there were an infinite number of points c_j, we could not conclude that necessarily the sum $(\text{Max}_1 - \min_1 + \cdots) = B$ was finite.]

68. Evaluate each individual integral, over which the function is continuous, then add up those values.

69. $h(x) = \begin{cases} \frac{1}{x} & \text{if } x \ne 0 \\ 0 & \text{if } x = 0. \end{cases}$ If we can show that any particular sequence of sums R_n does not approach a (finite) limit, then we have shown that it is not true all sums R_n approach a definite limit, and so we have shown that h is not integrable.

Let $P = \{0, \frac{1}{n}, \frac{2}{n}, \cdots, 1\}$ and choose $t_1 = \frac{1}{n^2}$, $t_2 = \frac{1}{n}, t_3 = \frac{2}{n}, \cdots, t_n = \frac{n-1}{n}$. Then

$$\begin{aligned} R_n &= \sum h(t_j)\Delta x_j = \left[\frac{1}{\left(\frac{1}{n^2}\right)} + \frac{1}{\frac{1}{n}} + \cdots + \frac{1}{\frac{n-1}{n}}\right]\left(\frac{1}{n}\right) \\ &= \left[n^2 + n + \frac{1}{2}n + \frac{1}{3}n + \cdots + \left(\frac{1}{n-1}\right)n\right]\left(\frac{1}{n}\right) \\ &= n + \left(1 + \frac{1}{2} + \frac{1}{3} + \cdots + \frac{1}{n-1}\right) > n. \end{aligned}$$

So as $n \to \infty$, $R_n \to \infty$, and h is not integrable over $[0, 1]$.

70. $\displaystyle\int_1^3 \frac{1}{x}\,dx \approx \sum_{j=1}^3 f(m_j)\Delta x = \frac{1}{\frac{3}{2}}(1) + \frac{1}{\frac{5}{2}}(1) = \frac{2}{3} + \frac{2}{5} = \frac{16}{15}.$

71. $\displaystyle\int_1^4 \frac{1}{x}\,dx \approx \sum_{j=1}^4 f(m_j)\Delta x_j = \frac{1}{\frac{3}{2}}(1) + \frac{1}{\frac{5}{2}}(1) + \frac{1}{\frac{7}{2}}(1) = \frac{2}{3} + \frac{2}{5} + \frac{2}{7} = \frac{70 + 42 + 30}{3 \cdot 5 \cdot 7} = \frac{142}{105}.$

72. $\displaystyle\int_0^6 \sqrt{x^3 + 1}\,dx \approx \sum_{j=1}^3 f(x_j)\Delta x_j = \sqrt{1^3 + 1}\,(2) + \sqrt{(3^3) + 1}\,(2) + \sqrt{(5^3) + 1}\,(2) \approx 35.86137668.$

73. $\displaystyle\int_0^4 \sqrt{x^3 + 1}\,dx \approx \sum_{j=1}^4 f(m_j)\Delta x_j = f\left(\frac{1}{2}\right)(1) + f\left(\frac{3}{2}\right)(1) + f\left(\frac{5}{2}\right)(1) + f\left(\frac{7}{2}\right)(1) \approx 13.85350749.$

74. $\displaystyle\int_0^1 \frac{1}{1 + x^2}\,dx \approx \sum_{j=1}^2 f(m_j)\Delta x_j = \left(\frac{1}{1 + \left(\frac{1}{4}\right)^2}\right)\left(\frac{1}{2}\right) + \frac{1}{1 + \left(\frac{3}{4}\right)^2}\left(\frac{1}{2}\right) \approx 0.790588235.$

75. $\displaystyle\int_0^4 \frac{1}{1 + x^2}\,dx \approx \sum_{j=1}^4 f(m_j)\Delta x_j = f\left(\frac{1}{2}\right) \cdot (1) + f\left(\frac{3}{2}\right) \cdot (1) + f\left(\frac{5}{2}\right) \cdot (1) + f\left(\frac{7}{2}\right) \cdot (1) \approx 1.321095040.$

76. $\displaystyle\int_0^3 \frac{1}{\sqrt{1+x^3}}\,dx \approx \sum_{j=1}^{6} f(m_j)\Delta x_j \left[\{m_j\} = \left\{\frac{1}{4},\frac{3}{4},\frac{5}{4},\frac{7}{4},\frac{9}{4},\frac{11}{4}\right\} \Delta x_j = \frac{1}{2}\right] \approx 1.653822774.$

77. $\displaystyle\int_0^1 \sqrt{x^4+1}\,dx \;\approx\; \sum_{j=1}^{6} f(m_j)\Delta x_j \approx 1.087792562$

$\quad$ where $\{m_j\} \;=\; \left\{\dfrac{1}{12},\dfrac{3}{12},\dfrac{5}{12},\dfrac{7}{12},\dfrac{9}{12},\dfrac{11}{12}\right\}$ and $\Delta x_j \;=\; \dfrac{1}{6}.$

78. $\displaystyle\int_1^2 \frac{1}{x}\,dx = \sum_{j=1}^{6} f(m_j)\Delta x_j \approx 0.692284321$ where $\{m_j\} = \left\{\dfrac{13}{12},\dfrac{15}{12},\dfrac{17}{12},\dfrac{19}{12},\dfrac{21}{12},\dfrac{23}{12}\right\}$ and $\Delta x_j = \dfrac{1}{6}.$

79. $\displaystyle\int_0^3 \sin\sqrt{x}\,dx \approx \sum_{j=1}^{6} f(m_j)\Delta x_j \approx 2.551760163$ where $\{m_j\} = \left\{\dfrac{1}{4},\dfrac{3}{4},\dfrac{5}{4},\dfrac{7}{4},\dfrac{9}{4},\dfrac{11}{4}\right\}$ and $\Delta x_j = \dfrac{1}{2}.$

80.

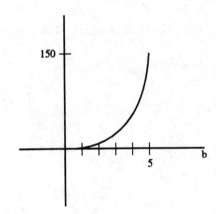

b	$\displaystyle\int_0^b x^3\,dx$ midpoint, $n=5$
0.5	0.0153125
1	0.245
1.5	1.2403125
2	3.920
2.5	9.5703125
3	19.845
3.5	36.7653125
4	62.720
4.5	100.4653125
5	153.125

Exercise 80

81.

$y=\int_0^b \sin x^2\,dx$

b	$\displaystyle\int_0^b \sin(x^2)\,dx$ midpoint rule, $n=5$
0.25	0.005154896
0.5	0.041077014
0.75	0.136284354
1.0	0.308439081
1.25	0.545833991
1.5	0.785324320
1.75	0.912511721
2.0	0.824278992
2.25	0.557850284
2.5	0.367701578
2.75	0.529437490
3.0	0.889646736

Exercise 81

82.

$$\int_0^1 2x\,dx \approx \sum_{j=1}^n f(m_j)\Delta x_j = \sum_{j=1}^n f\left(\frac{1}{2n}+\frac{j-1}{n}\right)\cdot\left(\frac{1}{n}\right)$$

$$= \sum 2\left[\frac{1}{2n}+\frac{j-1}{n}\right]\left(\frac{1}{n}\right) = \sum\frac{1}{n^2}+\frac{2(j-1)}{n^2}$$

$$= \left[\left(-\frac{1}{n^2}\right)\sum 1+\left(\frac{2}{n^2}\right)\sum j\right]$$

$$= \left(-\frac{1}{n^2}\right)\frac{n}{1}+\left(\frac{2}{n^2}\right)\frac{(n)(n+1)}{2}$$

$$= -\frac{1}{n}+\left(\frac{n}{n}\cdot\frac{n+1}{n}\right) = -\frac{1}{n}+1+\frac{1}{n} = 1.$$

The estimate $\sum f(m_j)\Delta x_j$ is exact if f is a linear function, because then each term

$$f(m_j)\Delta x_j = f\left(\frac{x_{j-1}+x_j}{2}\right)(x_j-x_{j-1})$$

$$= \frac{f(x_{j-1})+f(x_j)}{2}(x_j-x_{j-1})$$

$$= \text{area of the shaded trapezoid}$$

which is exactly the integral of f

between x_{j-1} and x_j.

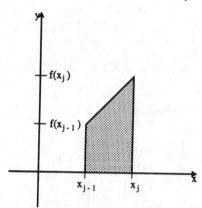

Exercise 82

83. Graph using the range $[-4,5]\times[-1.5,3]$ and $n=16$.

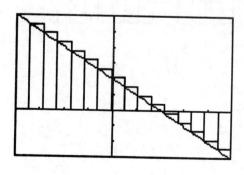 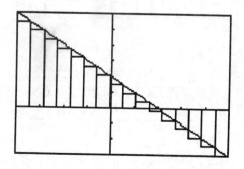

Table:

n	4	8	16	32	64
L_n	11.8125	9.28125	8.015625	7.3828125	7.06640625
R_n	1.6875	4.21875	5.484375	6.1171875	6.43359375

84. Graph using the range $[-1, 2] \times [-3.5, 3]$ and $n = 16$.

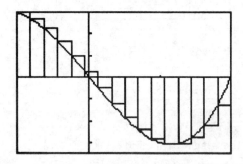 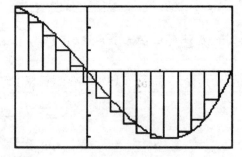

Table:

n	4	8	16	32	64
L_n	-0.703125	-1.58203125	-1.9423828	-2.1027832	-2.1780396
R_n	-2.953125	-2.70703125	-2.5048828	-2.3840332	-2.3186646

85. Graph using the range $[0, 4] \times [0, 5]$ and $n = 16$.

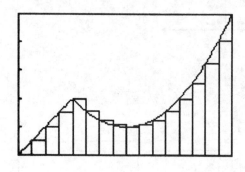 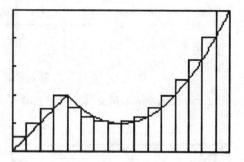

Table:

n	4	8	16	32	64
L_n	5	5.875	6.40625	6.6953125	6.8457031
R_n	10	8.375	7.65625	7.3203125	7.1582031

86. Graph using the range $[0.125, 3] \times [0, 8]$ and $n = 16$.

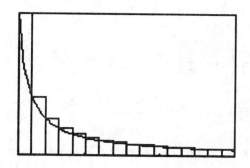 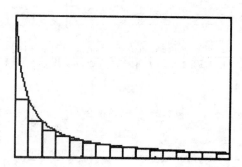

Table:

n	4	8	16	32	64
L_n	7.3769203	5.0521835	4.0173083	3.5635851	3.3608668
R_n	1.8665037	2.2969752	2.6397041	2.8747830	3.0164658

In Exercises 87–90, we approximate to 3 significant digits:

87. $L_n = 3.1425795$, $R_n = 3.1405795$, $n = 2000$.

88. $L_n = 8.3813570$, $R_n = 8.3838004$, $n = 5000$.

89. $L_n = 1.8500817$, $R_n = 1.8472228$, $n = 1000$.

90. $L_n = -1.5707955$, $R_n = -1.5707955$, $n = 2500$.

6.3 The Fundamental Theorem of Calculus

1. $\displaystyle\int_{-2}^{2} (x^3 - 1)\, dx = \frac{x^4}{4} - x\bigg]_{-2}^{2} = \left(\frac{16}{4} - 2\right) - \left(\frac{16}{4} - (-2)\right) = -4.$

2. $\displaystyle\int_{0}^{4} (x^2 - 2x + 3)\, dx = \frac{x^3}{3} - x^2 + 3x\bigg]_{0}^{4} = \frac{64}{3} - 16 + 12 = \frac{52}{3}.$

3. $\displaystyle\int_{1}^{2} \left(\frac{2}{x^3} + 5x\right) dx = -\frac{1}{x^2} + \frac{5}{2} x^2\bigg]_{1}^{2} = \left(-\frac{1}{4} + 10\right) - \left(-\frac{1}{1} + \frac{5}{2}\right) = \frac{33}{4}.$

4. $\displaystyle\int_{0}^{3} t\left(\sqrt[3]{t} - 2\right) dt = \int_{0}^{3} (t^{4/3} - 2t)\, dt = \frac{3}{7} t^{7/3} - t^2\bigg]_{0}^{3} = \frac{3}{7} 3^{7/3} - 9 = \frac{3^{10/3}}{7} - 9 = -3.4370.$

5. $\displaystyle\int_{1}^{3} (t^2 + 2)^2\, dt = \int_{1}^{3} (t^4 + 4t^2 + 4)\, dt = \frac{1}{5} t^5 + \frac{4}{3} t^3 + 4t\bigg]_{1}^{3} = \frac{243}{5} + 36 + 12 - \left(\frac{1}{5} + \frac{4}{3} + 4\right) = 91\frac{1}{15}.$

6. $\displaystyle\int_{0}^{1} (x^{3/5} - x^{5/3})\, dx = \frac{5}{8} x^{8/5} - \frac{3}{8} x^{8/3}\bigg]_{0}^{1} = \frac{5}{8} - \frac{3}{8} = \frac{1}{4}.$

7. $\displaystyle\int_{1}^{2} \frac{1 - x}{x^3}\, dx = -\frac{1}{2x^2} + \frac{1}{x}\bigg]_{1}^{2} = -\frac{1}{8} + \frac{1}{2} - \left(-\frac{1}{2} + \frac{1}{1}\right) = -\frac{1}{8}.$

8. $\int_4^9 \left(\sqrt{x} - \frac{1}{\sqrt{x}}\right) dx = \frac{2}{3} x^{3/2} - 2\sqrt{x}\Big]_4^9 = \frac{2}{3}(27-8) - 2(3-2) = \frac{38}{3} - 2 = \frac{32}{3}.$

9. $\int_2^0 x\left(\sqrt{x} - 1\right) dx = \int_2^0 (x^{3/2} - x) dx = \frac{2}{5} x^{5/2} - \frac{x^2}{2}\Big]_2^0 = -\frac{2}{5}\left(4\sqrt{2}\right) + \frac{4}{2} = 2 - \frac{8}{5}\sqrt{2} = -0.26274.$

10. $\int_9^4 \left(\sqrt{x} + \frac{1}{\sqrt{x}}\right) dx = \frac{2}{3} x^{3/2} + 2\sqrt{x}\Big]_9^4 = \frac{2}{3}(8-27) + 2(2-3) = -\frac{38}{3} - 2 = -\frac{44}{3}.$

Alternatively, we may write $-\left(\frac{2}{3} x^{3/2} + 2\sqrt{x}\right)\Big]_4^9 = -\left(\frac{2}{3}(27-8) + 2(3-2)\right)$

$$= -\left(\frac{38}{3} + 2\right) = -\frac{44}{3}.$$

11. $\int_0^4 |x-3|\, dx = -\int_0^3 (x-3)\, dx + \int_3^4 (x-3)\, dx = -\left[\frac{x^2}{2} - 3x\right]_0^3 + \left[\frac{x^2}{2} - 3x\right]_3^4$

$$= -\left(\frac{9}{2} - 9\right) + \left[\left(\frac{16}{2} - \frac{9}{2}\right) - 3(4-3)\right] = \frac{9}{2} + \frac{7}{2} - 3 = 5.$$

12. $\int_{-2}^2 |1-x^2|\, dx = -\int_{-2}^{-1} (1-x^2)\, dx + \int_{-1}^1 (1-x^2)\, dx - \int_1^2 (1-x^2)\, dx$

$$= -\left[x - \frac{x^3}{3}\right]_{-2}^{-1} + \left[x - \frac{x^3}{3}\right]_{-1}^1 - \left[x - \frac{x^3}{3}\right]_1^2$$

$$= -\left[(-1-(-2)) - \frac{1}{3}(-1-(-2)^3)\right] + \left[(1-(-1)) - \frac{1}{3}(1-(-1))\right]$$

$$- \left[\left(2 - \frac{8}{3}\right) - \left(1 - \frac{1}{3}\right)\right]$$

$$= -\left[1 - \frac{7}{3}\right] + \left[2 - \frac{2}{3}\right] - \left[-\frac{2}{3} - \frac{2}{3}\right] = \frac{12}{3} = 4.$$

13. $\int_0^4 |x - \sqrt{x}|\, dx = -\int_0^1 (x - \sqrt{x})\, dx + \int_1^4 (x - \sqrt{x})\, dx$

$$= -\left[\frac{x^2}{2} - \frac{2}{3} x^{3/2}\right]_0^1 + \left[\frac{x^2}{2} - \frac{2}{3} x^{3/2}\right]_1^4$$

$$= -\left(\frac{1}{2} - \frac{2}{3}\right) + \left(\left(\frac{16}{2} - \frac{1}{2}\right) - \frac{2}{3}(8-1)\right) = 3.$$

14. $\int_1^4 (u^3 - \sqrt{u})\, du = \frac{1}{4} u^4 - \frac{2}{3} u^{3/2}\Big]_1^4 = \left(64 - \frac{1}{4}\right) - \frac{2}{3}(8-1) = \frac{709}{12} = 59.0833.$

15. $\int_4^9 x^{1/2}(1 - x^{3/2})\, dx = \int_4^9 (x^{1/2} - x^2)\, dx = \frac{2}{3} x^{3/2} - \frac{x^3}{3}\Big]_4^9 = \frac{2}{3}(27-8) - \frac{1}{3}(729-64) = -209.$

16. $\displaystyle\int_0^4 |9 - x^2|\, dx = \int_0^3 (9 - x^2)\, dx - \int_3^4 (9 - x^2)\, dx = 9x - \dfrac{x^3}{3}\Big]_0^3 - \left(9x - \dfrac{x^3}{3}\right)\Big]_3^4$

$\qquad\qquad = 27 - 9 - \left(36 - \dfrac{64}{3}\right) + (27 - 9) = \dfrac{64}{3}.$

17. $\displaystyle\int_{-1}^4 |x^2 - x - 2|\, dx = x^2 - x - 2 = (x - 2)(x + 1) \Rightarrow x^2 - x - 2 = 0 \text{ at } x = -1,\ x = 2$

$\qquad -\displaystyle\int_{-1}^2 (x^2 - x - 2)\, dx + \int_2^4 (x^2 - x - 2)\, dx = -\left[\dfrac{x^3}{3} - \dfrac{x^2}{2} - 2x\right]_{-1}^2 + \left[\dfrac{x^3}{3} - \dfrac{x^2}{2} - 2x\right]_2^4$

$\qquad\qquad\qquad\qquad = -\left(\dfrac{8}{3} - \dfrac{4}{2} - 4\right) + \left(-\dfrac{1}{3} - \dfrac{1}{2} + 2\right)$

$\qquad\qquad\qquad\qquad\quad + \left(\dfrac{64}{3} - \dfrac{16}{2} - 8\right) - \left(\dfrac{8}{3} - \dfrac{4}{2} - 4\right) = \dfrac{79}{6}.$

18. $\displaystyle\int_0^{\pi/2} \cos t\, dt = \sin t \Big]_0^{\pi/2} = 1 - 0 = 1.$

19. $\displaystyle\int_0^{\pi} \sin u\, du = -\cos u \Big]_0^{\pi} = -(-1 - 1) = 2.$

20. $\displaystyle\int_0^{\pi/4} \sec x \tan x\, dx = \sec x \Big]_0^{\pi/4} = \sqrt{2} - 1.$

21. $\displaystyle\int_{-\pi/3}^{\pi/3} \sec^2 \theta\, d\theta = \tan \theta \Big]_{-\pi/3}^{\pi/3} = \sqrt{3} - (-\sqrt{3}) = 2\sqrt{3}.$

22. $\displaystyle\int_0^{\pi/6} \tan^2 x\, dx = \int_0^{\pi/6} (\sec^2 x - 1)\, dx = \tan x - x \Big]_0^{\pi/6} = \dfrac{\sqrt{3}}{3} - \dfrac{\pi}{6} \approx 0.05375.$

23. $\displaystyle\int_{-2}^3 f(x)\, dx$, where $f(x) = \begin{cases} x^2 - 1 & \text{for } x \leq 1 \\ x - 1 & \text{for } x > 1 \end{cases}$

$\qquad\qquad = \displaystyle\int_{-2}^1 (x^2 - 1)\, dx + \int_1^3 (x - 1)\, dx$

$\qquad\qquad = \dfrac{x^3}{3} - x \Big]_{-2}^1 + \dfrac{x^2}{2} - x \Big]_1^3 = \dfrac{1}{3}(1 + 8) - (1 + 2) + \dfrac{1}{2}(9 - 1) - (3 - 1)$

$\qquad\qquad = 3 - 3 + 4 - 2 = 2.$

24. $\displaystyle\int_1^5 f(x)\,dx$, where $f(x) = \begin{cases} \sqrt{x} & \text{for } 0 \le x \le 4 \\ 2x^2 - 7x - 2 & \text{for } x > 4 \end{cases}$

$$= \int_1^4 \sqrt{x}\,dx + \int_4^5 (2x^2 - 7x - 2)\,dx = \frac{2}{3} x^{3/2} \Big]_1^4 + \frac{2}{3} x^3 - \frac{7}{2} x^2 - 2x \Big]_4^5$$

$$= \frac{2}{3}(8-1) + \frac{2}{3}(125-64) - \frac{7}{2}(25-16) - 2(5-4)$$

$$= \frac{14}{3} + \frac{122}{3} - \frac{63}{2} - 2 = \frac{71}{6}.$$

25. $\displaystyle\int_0^\pi \sin 1\,dx = (\sin 1)x \Big]_0^\pi = \pi \sin 1 \approx 2.6436.$

26. $\displaystyle\int_0^{\pi/3} \sec^2 1\,dx = (\sec^2 1)x \Big]_0^{\pi/3} = \frac{\pi}{3} \sec^2 1 \approx 3.5872.$

27. $f(x) = x^2 + x + 1$, $[0,1]$

$$\bar{f} = \frac{1}{1-0} \int_0^1 (x^2 + x + 1)\,dx = \frac{x^3}{3} + \frac{x^2}{2} + x \Big]_0^1 = \frac{1}{3} + \frac{1}{2} + 1 = \frac{11}{6}.$$

28. $f(x) = \sin x$, $[0, 2\pi]$

$$\bar{f} = \frac{1}{2\pi - 0} \int_0^{2\pi} \sin x\,dx = -\frac{1}{2\pi} \cos x \Big]_0^{2\pi} = -\frac{1}{2\pi}(1-1) = 0.$$

29. $f(x) = \sec^2 x$, $\left[0, \dfrac{\pi}{4}\right]$

$$\bar{f} = \frac{1}{\frac{\pi}{4} - 0} \int_0^{\pi/4} \sec^2 x\,dx = \frac{4}{\pi} \tan x \Big]_0^{\pi/4} = \frac{4}{\pi}(1-0) = \frac{4}{\pi}.$$

30. $f(x) = \sec x \tan x$, $\left[0, \dfrac{\pi}{4}\right]$.

$$\bar{f} = \frac{1}{\frac{\pi}{4} - 0} \int_0^{\pi/4} \sec x \tan x\,dx = \frac{4}{\pi} \sec x \Big]_0^{\pi/4} = \frac{\pi}{4}(\sqrt{2} - 1) \approx 0.3253.$$

31. $\displaystyle F(x) = \int_1^x t\,dt = \frac{t^2}{2} \Big]_1^x = \frac{x^2 - 1}{2}. \qquad F'(x) = \frac{2x}{2} = x.$

32. $\displaystyle F(x) = \int_{-2}^x (t^2 + 2t)\,dt = \frac{t^3}{3} + t^2 \Big]_{-2}^x = \frac{x^3}{3} + x^2 + \frac{8}{3} - 4 = \frac{x^3}{3} + x^2 - \frac{4}{3}.$

$F'(x) = x^2 + 2x.$

33. $\displaystyle F(x) = \int_0^x \cos t\,dt = \sin t \Big]_0^x = \sin x - 0 = \sin x.$

$F'(x) = \cos x.$

34. $F(x) = \int_0^x (at^2 + bt + c)\, dt = \dfrac{a}{3} t^3 + \dfrac{b}{2} t^2 + ct \Big]_0^x = \dfrac{a}{3} x^3 + \dfrac{b}{2} x^2 + cx.$

   $F'(x) = ax^2 + bx + c.$

35. $F(x) = \int_2^x t^2 \sin t^2\, dt = \int_2^x f(t)\, dt.$ $\qquad F'(x) = f(x) = x^2 \sin x^2.$

36. $F(x) = \int_0^x \sqrt{t^4 + 1}\, dt = \int_0^x f(t)\, dt.$ $\qquad F'(x) = f(x) = \sqrt{x^4 + 1}.$

37. $F(x) = \int_x^1 t^3 \cos^2 t\, dt = -\int_1^x t^3 \cos^2 t\, dt = -\int_1^x f(t)\, dt.$

   $F'(x) = -f(x) = -x^3 \cos^2 x.$

38. $F(x) = \int_x^5 \sqrt{4 + t^3}\, dt = -\int_5^x \sqrt{4 + t^3}\, dt = -\int_5^x f(t)\, dt.$

   $F'(x) = -f(x) = -\sqrt{4 + x^3}.$

39. $F(x) = \int_0^{3x} \sqrt{1 + \sin t}\, dt = \int_0^{g(x)} f(t)\, dt.$ $\qquad F'(x) = f[g(x)]\, g'(x) = 3\sqrt{1 + \sin 3x}.$

40. $F(x) = \int_2^{x^2} \dfrac{1}{1 + t^3}\, dt = \int_0^{g(x)} f(t)\, dt.$ $\qquad F'(x) = f[g(x)]g'(x) = \dfrac{2x}{1 + x^6}.$

41. $F(x) = \int_{x^2}^x \cos^3(t + 1)\, dt = \int_{g(x)}^x f(t)\, dt.$

   $F'(x) = f(x) - f[g(x)]g'(x) = \cos^3(x + 1) - 2x\left(\cos^3(x^2 + 1)\right).$

42. $F(x) = \int_{-x}^x \sqrt{t^2 + 1}\, dt = \int_{g(x)}^x f(t)\, dt.$

   $F'(x) = f(x) - f[g(x)]g'(x) = \sqrt{x^2 + 1} - \left(-\sqrt{x^2 + 1}\right) = 2\sqrt{x^2 + 1}.$

43. $F(x) = x \int_3^{x^2} (t^2 + 1)^{-3}\, dt = x \int_3^{g(x)} f(t)\, dt.$

   $F'(x) = \int_3^{g(x)} f(t)\, dt + x\, f[g(x)]g'(x) = \int_3^{x^2} (t^2 + 1)^{-3}\, dt + 2x^2 (x^4 + 1)^{-3}.$

44. $F(x) = \cos x \int_{\sin x}^{\pi} \sqrt{t^4 + 1}\, dt = -\cos x \int_{\pi}^{\sin x} \sqrt{t^4 + 1}\, dt = -\cos x \int_{\pi}^{g(x)} f(t)\, dt.$

   $F'(x) = \sin x \int_{\pi}^{\sin x} \sqrt{t^4 + 1}\, dt - \cos^2 x \sqrt{\sin^4 x + 1}.$

45. a) $\dfrac{d}{dx}\left[\int_a^x f(t)\, dt\right] = f(x).$ $\qquad\qquad$ b) $\dfrac{d}{dx}\left[\int_x^b f(t)\, dt\right] = -f(x).$

   c) $\dfrac{d}{dx}\left[\int_a^b f(t)\, dt\right] = 0.$ $\qquad\qquad$ d) $\dfrac{d}{dx}\left[\int f(x)\, dx\right] = f(x).$

e) $\displaystyle\int_a^b f'(x)\,dx = f(b) - f(a).$ f) $\displaystyle\int f'(x)\,dx = f(x) + C.$

46. a) $f(x) = 2x + 5,\ [4,6]$

$$\int_4^6 (2x + 5)\,dx = (6 - 4)f(c)$$

$$x^2 + 5x\Big]_4^6 = (36 - 16) + 5(6 - 4) = 30$$

$$30 = 2\,f(c);\ f(c) = 15;\ 2c + 5 = 15 \Rightarrow c = 5.$$

b) $f(x) = 3x^2 - 4x + 1,\ [1,4].$

$$\int_1^4 (3x^2 - 4x + 1)\,dx = (4 - 1)f(c) = 3(3c^2 - 4c + 1).$$

$$x^3 - 2x^2 + x\Big]_1^4 = (64 - 1) - 2(16 - 1) + (4 - 1) = 36.$$

$$36 = 3(3c^2 - 4c + 1);\ 3c^2 - 4c - 11 = 0.$$

$$c = \frac{4 \pm \sqrt{16 + 132}}{6} = \frac{4 \pm \sqrt{148}}{6} = \frac{2 \pm \sqrt{37}}{3}.$$

Use plus sign. $c = \dfrac{2 + \sqrt{37}}{3} = \dfrac{1}{3}(2 + \sqrt{37}) \approx 2.6943.$

47. a) $\displaystyle\int_1^4 |x - 2|\,dx = (4 - 1)f(c) = 3|c - 2|$

$$\int_1^2 (2 - x)\,dx + \int_2^4 (x - 2)\,dx = 3|c - 2|$$

$$2x - \frac{x^2}{2}\Big]_1^2 + \frac{x^2}{2} - 2x\Big]_2^4 = 3|c - 2|$$

$$\frac{1}{2} + 2 = 3|c - 2|$$

So we want c such that $|c - 2| = \dfrac{5}{6}.$ Thus $c = \dfrac{7}{6}$ or $\dfrac{17}{6}.$

b) $f(x) = 4 - x^2,\ [-2,3].$

$$\int_{-2}^3 (4 - x^2)\,dx = (3 + 2)f(c) = 5(4 - c^2).$$

$$4x - \frac{x^3}{3}\Big]_{-2}^3 = 4(3 + 2) - \frac{1}{3}(27 + 8) = 20 - \frac{35}{3} = \frac{25}{3}.$$

$$\frac{25}{3} = 5(4 - c^2);\ 4 - c^2 = \frac{5}{3};\ c^2 = 4 - \frac{5}{3} = \frac{7}{3}.$$

$$c = \pm\sqrt{\frac{7}{3}} \approx 1.5275.$$

48. $f(x) = x^2 + ax + 3,\ [0,6];\ \bar{f} = 27.$

$$\bar{f} = \frac{1}{6 - 0}\int_0^6 (x^2 + ax + 3)\,dx = \frac{1}{6}\left[\frac{x^3}{3} + \frac{a}{2}x^2 + 3x\right]_0^6 = \frac{1}{6}\left[\frac{216}{3} + 18a + 18\right] = 27;\ 90 + 18a = 162.$$

$18a = 72;\ a = 4.$

49. $f(x) = 2x + 3$, $[1, b]$; $\bar{f} = 11$.

$$\bar{f} = \frac{1}{b-1} \int_1^b (2x+3)\, dx = \frac{1}{b-1} \left[x^2 + 3x \right]_1^b = \frac{1}{b-1}(b^2 + 3b - 4) = 11.$$

$b^2 + 3b - 4 = 11b - 11$; $b^2 - 8b + 7 = 0$.
$(b-1)(b-7) = 0$; choose $b = 7$.

50. $f(x) = 4x - x^2$, $[0, b]$, $\bar{f} = 1$.

$$\bar{f} = \frac{1}{b-0} \int_0^b (4x - x^2)\, dx = \frac{1}{b}\left[2x^2 - \frac{x^3}{3} \right]_0^b = \frac{1}{b}\left(2b^2 - \frac{b^3}{3} \right) = 2b - \frac{b^2}{3}.$$

$2b - \dfrac{b^2}{3} = 1$; $b^2 - 6b + 3 = 0$; $b = \dfrac{6 \pm \sqrt{36 - 12}}{2}$.

$b = 3 \pm \sqrt{6} \approx 0.55051,\ 5.44949.$

51. $A(x) = \displaystyle\int_0^x f(t)\, dt$, $0 \le x \le 8$

a) $A'(x) = f(x)$, which is greater than zero on $[0, 4]$ and $[7, 8]$.

b) A is constant on the interval $[6, 7]$.

c) $A'(x) = 0 \Rightarrow f(x) = 0$, which occurs at $x = 4$, and $6 \le x \le 7$.
   A relative maximum of A occurs at $x = 4$ and at the endpoint $x = 8$.

d) $A(0) = \displaystyle\int_0^0 f(t)\, dt = 0.$

52. $\bar{f} = \dfrac{1}{2\pi} \displaystyle\int_0^{2\pi} \sin x\, dx = -\dfrac{1}{2\pi} \cos x \bigg]_0^{2\pi} = -\dfrac{1}{2\pi}(1 - 1) = 0$

$\bar{f} = \dfrac{1}{2\pi} \displaystyle\int_0^{2\pi} \cos x\, dx = \dfrac{1}{2\pi} \sin x \bigg]_0^{2\pi} = \dfrac{1}{2\pi}(0 - 0) = 0.$

The average value for $\sin x$ occurs at $x = 0, \pi$, and 2π.
The average value for $\cos x$ occurs at $x = \frac{\pi}{2}$ and $x = \frac{3\pi}{2}$.

53. $\bar{f} = \dfrac{1}{b-a} \displaystyle\int_a^b f(x)\, dx$; $\bar{g} = \dfrac{1}{b-a} \displaystyle\int_a^b g(x)\, dx.$

$\bar{f} - \bar{g} = \dfrac{1}{b-a} \displaystyle\int_a^b [f(x) - g(x)]\, dx.$ If $f(x) - g(x) \ge 0$ on $[a, b]$, then $\displaystyle\int_a^b [f(x) - g(x)]\, dx \ge 0$ and $\bar{f} - \bar{g} \ge 0.$

54. Let $f(x) = x$ and $g(x) = 0.8$, on $[a, 1]$.

$$\bar{f} = \int_0^1 x\, dx = \frac{x^2}{2} \bigg]_0^1 = \frac{1}{2}.$$

$$\bar{g} = \int_0^1 0.8\, dx = 0.8.$$

$\bar{f} < \bar{g}$, but $f(x) > g(x)$ for $x \in (0.8, 1]$.

55. $\bar{f} = \dfrac{1}{b-a} \displaystyle\int_a^b f(x)\,dx \le \dfrac{1}{b-a}\displaystyle\int_a^b M\,dx.$

$\bar{f}(b-a) = M(b-a) \Rightarrow \bar{f} = M.$

If $\bar{f} = M$, $\bar{f}(b-a) = \displaystyle\int_a^b f(x)\,dx \le \displaystyle\int_a^b M\,dx \Rightarrow f(x) = M$ for all $x \in [a,b]$.

56. $\displaystyle\lim_{n\to\infty} \left(\dfrac{\pi}{2n}\right) \sum_{j=1}^n \left(\cos \dfrac{j\pi}{2n}\right) = \lim_{n\to\infty} \sum_{j=1}^n \cos x_j \Delta x$, where $x_j = \dfrac{\pi}{2n}j$ and

$$\Delta x = \dfrac{\pi}{2n}$$

The limit of the sum is $\displaystyle\int_0^{\pi/2} \cos x\,dx = \sin x \Big]_0^{\pi/2} = 1 - 0 = 1.$

57. $\displaystyle\lim_{n\to\infty} \left(\dfrac{x}{n}\right) \sum_{j=1}^n \left(\sin \dfrac{jx}{n}\right) = \lim_{n\to\infty} \sum_{j=1}^n \sin y_j \Delta y$, where $y_j = \dfrac{x}{n}j$ and $\Delta y = \dfrac{x}{n}$.

$$\displaystyle\int_0^x \sin y\,dy = -\cos y \Big]_0^x = -\cos x + 1 = 1 - \cos x.$$

58. a) Graph using the range $[0,3] \times [-1.5, 1.5]$.

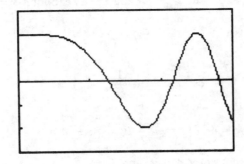

$A(x)$ is increasing in the intervals $\left[0, \sqrt{\pi/2}\right]$ and $\left[\sqrt{3\pi/2},\ \sqrt{5\pi/2}\right]$.

$A(x)$ is deceasing in the intervals $\left[\sqrt{\pi/2},\ \sqrt{3\pi/2}\right]$ and $\left[\sqrt{5\pi/2},\ 3\right]$.

$A(x)$ is nonnegative since the total area below the x-axis is less than the total area above the x-axis.

b) Use $n = 10$ on each of the intervals $[0, 0.5]$, $[0.5, 1]$, $[1, 1.5]$, $[1.5, 2]$, $[2, 2.5]$, $[2.5, 3]$ and add the results. To see this note that

$$A(1.5) = \int_0^{0.5} \cos x^2\,dx + \int_{0.5}^1 \cos x^2\,dx + \int_1^{1.5} \cos x^2\,dx.$$

So use the Midpoint Rule with $n = 10$ on each of the 3 integrals to approximate $A(1.5)$. The results are summarized in the following table.

n	0.5	1.0	1.5	2	2.5	3
$A(x)$	0.4969	0.9047	0.8994	0.4611	0.6053	0.7031

c) Graph of $f(x) = \cos x^2$ and the values from the table using the range $[0, 3] \times [-1, 1]$. Also on this graph are indicated the relative extrema of $A(x)$. (See Appendix V.)

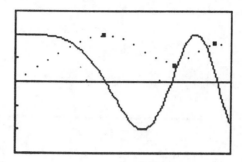

d) The derivative of $A'(x)$ is $f(x) = \cos x^2$. So on the interval $[0, 3]$, $A(x)$ has critical points where $f(x) = 0$. The values are at $x = \sqrt{\pi/2}$, $\sqrt{3\pi/2}$, and $\sqrt{5\pi/2}$. From the graph of $f(x)$ $(A'(x))$ we see that on the interval $\left[0, \sqrt{\pi/2}\right]$, $A(x)$ is increasing and on the interval $\left[\sqrt{\pi/2}, \sqrt{3\pi/2}\right]$, $A(x)$ is decreasing so $x = \sqrt{\pi/2}$ is a relative maximum. On the interval $\left[\sqrt{3\pi/2}, \sqrt{5\pi/2}\right]$, $A(x)$ is increasing so $x = \sqrt{3\pi/2}$ is a relative minimum and since $A(x)$ is decreasing on the interval $\left[\sqrt{5\pi/2}, 3\right]$, $x = \sqrt{5\pi/2}$ is a relative maximum.

59. a) Graph using the range $[0, 10] \times [-50, 50]$.

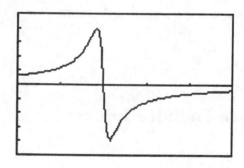

Since $\frac{dp}{dt} = 0$ when $t = 4$ and $\frac{dp}{dt} > 0$ for $t < 4$, $\frac{dp}{dt} < 0$ for $t > 4$, then at $t = 4$, $p(t)$ has a maximum.

b) Using the Midpoint Rule programs for drawing an antiderivative (Appendix V) with **K=32, N=20**, and the range $[0, 8] \times [0, 135]$ the following graph is drawn.

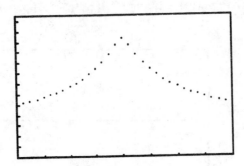

c) From the graph in part b) it appears that the maximum number of rabbits is approximately 114.

d) Using the Midpoint Rule programs for drawing an antiderivative (Appendix V) with **K=50** and **N=6**, and the range $[0, 50] \times [-20, 135]$ the following graph is drawn.

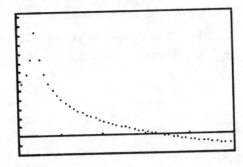

The graph crosses the axis at approximately $t = 34$. So the rabbits will die out after 34 years.

6.4 Substitution in Definite Integrals

1. $\int_0^{\pi/4} \cos 2x \, dx, \ u = 2x$

   $du = 2x \, dx$

   if $x = 0, \ u = 0$; if $x = \frac{\pi}{4}, \ u = \frac{\pi}{2}$

   $\frac{1}{2} \int_0^{\pi/2} \cos u \, du \ = \ \frac{1}{2} \sin u \Big]_0^{\pi/2}$

   $= \ \frac{1}{2}(1 - 0) = \frac{1}{2}$.

2. $\int_0^1 x(x^2 + 1)^{100} \, dx, \ u = x^2 + 1$

   $du = 2x \, dx$

   if $x = 0, \ u = 1$; if $x = 1, \ u = 2$

   $\frac{1}{2} \int_1^2 u^{100} \, du \ = \ \frac{1}{2(101)} u^{101} \Big]_1^2$

   $= \ \frac{1}{202}(2^{101} - 1)$.

3. $\int_0^{\pi/6} \sec(2x)\tan(2x)\,dx, \quad u = 2x$

$\quad du = 2\,dx$

$\quad$ if $x = 0$, $u = 0$; if $x = \frac{\pi}{6}$, $u = \frac{\pi}{3}$

$\quad \frac{1}{2}\int_0^{\pi/3} \sec u\tan u\,du \;=\; \frac{1}{2}\sec u\Big]_0^{\pi/3}$

$\qquad\qquad\qquad\qquad\quad = \;\frac{1}{2}(2-1)$

$\qquad\qquad\qquad\qquad\quad = \;\frac{1}{2}.$

4. $\int_0^{\pi/12} \sec^2 3x\,dx, \quad u = 3x$

$\quad du = 3\,dx$

$\quad$ if $x = 0$, $u = 0$; if $x = \frac{\pi}{12}$, $u = \frac{\pi}{4}$

$\quad \frac{1}{3}\int_0^{\pi/4} \sec^2 u\,du \;=\; \frac{1}{3}\tan u\Big]_0^{\pi/4}$

$\qquad\qquad\qquad\qquad\quad = \;\frac{1}{3}(1-0)$

$\qquad\qquad\qquad\qquad\quad = \;\frac{1}{3}.$

5. $\int_1^2 (2-x^2)(x^3-6x+7)^{50}\,dx, \quad u = x^3-6x+7$

$\quad du = (3x^2-6)\,dx = -3(2-x^2)\,dx$

$\quad$ if $x = 1$, $u = 2$; if $x = 2$, $u = 3$

$\quad -\frac{1}{3}\int_2^3 u^{50}\,du \;=\; -\frac{1}{3(51)}u^{51}\Big]_2^3$

$\qquad\qquad\qquad\qquad\quad = \;-\frac{1}{153}\left(3^{51}-2^{51}\right)$

$\qquad\qquad\qquad\qquad\quad \approx \;-1.40764\times 10^{22}.$

6. $\int_0^3 x\sqrt{9-x^2}\,dx, \quad u = 9-x^2$

$\quad du = -2x\,dx$

$\quad$ if $x = 0$, $u = 9$; if $x = 3$, $u = 0$

$\quad -\frac{1}{2}\int_9^0 \sqrt{u}\,du \;=\; \frac{1}{3}u^{3/2}\Big]_0^9$

$\qquad\qquad\qquad\qquad\quad = \;\frac{1}{3}(9^{3/2}) = \frac{1}{3}(3^3) = 9.$

7. $\int_0^{\pi/4}(1-\cos 2x)\,dx \;=\; x - \frac{1}{2}\sin 2x\Big]_0^{\pi/4}$

$\qquad\qquad\qquad\qquad\quad = \;\frac{\pi}{4}-\frac{1}{2} \approx 0.2854.$

8. $\int_0^{2\pi/3}\sin(2x)\,dx \;=\; -\frac{1}{2}\cos 2x\Big]_0^{2\pi/3}$

$\qquad\qquad\qquad\qquad\quad = \;-\frac{1}{2}\left(-\frac{1}{2}-1\right) = \frac{3}{4}.$

9. $\int_0^1 2x(x^2+1)^2\,dx$

$\quad$ Let $u = x^2+1$, $du = 2x\,dx$

$\quad$ if $x = 0$, $u = 1$; if $x = 1$, $u = 2$

$\quad \int_1^2 u^2\,du \;=\; \frac{1}{3}u^3\Big]_1^2$

$\qquad\qquad\qquad = \;\frac{1}{3}(8-1) = \frac{7}{3}.$

10. $\int_0^1 x\sqrt{1-x^2}\,dx$

$\quad$ Let $u = 1-x^2$, $du = -2x\,dx$

$\quad$ if $x = 0$, $u = 1$; if $x = 1$, $u = 0$

$\quad -\frac{1}{2}\int_1^0 \sqrt{u}\,du \;=\; \frac{1}{2}\left(\frac{2}{3}\right)u^{3/2}\Big]_0^1$

$\qquad\qquad\qquad\qquad\quad = \;\frac{1}{3}.$

11. $\int_1^2 x\sqrt{4-x^2}\,dx$

$\quad$ Let $u = 4-x^2$, $du = -2x\,dx$

$\quad$ if $x = 1$, $u = 3$; if $x = 2$, $u = 0$

$\quad -\frac{1}{2}\int_3^0 \sqrt{u}\,du \;=\; \frac{1}{2}\left(\frac{2}{3}\right)u^{3/2}\Big]_0^3$

$\qquad\qquad\qquad\qquad\quad = \;\frac{1}{3}(3^{3/2}) = \sqrt{3}.$

12. $\int_{-\pi/6}^0 \sec^2(2x)\,dx \;=\; \frac{1}{2}\tan 2x\Big]_{-\pi/6}^0$

$\qquad\qquad\qquad\qquad\quad = \;\frac{1}{2}(0+\sqrt{3}) = \frac{\sqrt{3}}{2}.$

13. $\displaystyle\int_0^{\sqrt{\pi/2}} t\sin(\pi - t^2)\,dt$

Let $u = \pi - t^2$, $du = -2t\,dt$

if $t = 0$, $u = \pi$; if $t = \frac{\sqrt{\pi}}{2}$, $u = \frac{3\pi}{4}$

$\displaystyle -\frac{1}{2}\int_\pi^{3\pi/4} \sin u\,du \;=\; \frac{1}{2}\cos u\Big]_\pi^{3\pi/4}$

$\displaystyle \qquad\qquad = \frac{1}{2}\left(-\frac{\sqrt{2}}{2} + 1\right)$

$\displaystyle \qquad\qquad = \frac{2-\sqrt{2}}{4}.$

14. $\displaystyle\int_0^{\pi/2} \cos^3 t \sin t\,dt \;=\; -\frac{1}{4}\cos^4 t\Big]_0^{\pi/2}$

$\displaystyle \qquad\qquad\qquad\qquad = -\frac{1}{4}(0 - 1) = \frac{1}{4}.$

15. $\displaystyle\int_{-\pi/4}^{\pi/4} \frac{\sin x}{\cos^2 x}\,dx \;=\; \int_{-\pi/4}^{\pi/4} (\cos x)^{-2}\sin x\,dx$

$\displaystyle \qquad\qquad = \frac{1}{\cos x}\Big]_{-\pi/4}^{\pi/4}$

$\displaystyle \qquad\qquad = \sec x\Big]_{-\pi/4}^{\pi/4}$

$\displaystyle \qquad\qquad = \sqrt{2} - \sqrt{2} = 0.$

16. $\displaystyle\int_0^{\pi/4} \tan\theta \sec^2\theta\,d\theta \;=\; \frac{1}{2}\tan^2\theta\Big]_0^{\pi/4}$

$\displaystyle \qquad\qquad\qquad\qquad = \frac{1}{2}(1 - 0) = \frac{1}{2}.$

17. $\displaystyle\int_0^4 \frac{dt}{\sqrt{2t+1}}$

Let $u = 2t + 1$, $du = 2\,dt$

if $t = 0$, $u = 1$; if $t = 4$, $u = 9$.

$\displaystyle \frac{1}{2}\int_1^9 \frac{du}{\sqrt{u}} \;=\; \sqrt{u}\Big]_1^9 = 3 - 1 = 2.$

18. $\displaystyle\int_0^2 \frac{x}{\sqrt{16 + x^2}}\,dx$

Let $u = 16 + x^2$, $du = 2x\,dx$

if $x = 0$, $u = 16$; if $x = 2$, $u = 20$

$\displaystyle \frac{1}{2}\int_{16}^{20} \frac{du}{\sqrt{u}} \;=\; \sqrt{u}\Big]_{16}^{20}$

$\displaystyle \qquad\qquad = 2\sqrt{5} - 4 = 2(\sqrt{5} - 2).$

19. $\displaystyle\int_{-1}^1 \frac{x}{(1 + x^2)^3}\,dx$

Let $u = 1 + x^2$, $du = 2x\,dx$

if $x = -1$, $u = 2$; if $x = 1$, $u = 2$

$\displaystyle \frac{1}{2}\int_2^2 \frac{du}{u^3} \;=\; 0.$

20. $\displaystyle\int_0^3 x\sqrt{9 - x^2}\,dx$

Let $u = 9 - x^2$, $du = -2x\,dx$

if $x = 0$, $u = 9$; if $x = 3$, $u = 0$

$\displaystyle -\frac{1}{2}\int_9^0 \sqrt{u}\,du \;=\; -\frac{1}{2}\left(\frac{2}{3}\right)u^{3/2}\Big]_9^0$

$\displaystyle \qquad\qquad = \frac{1}{3}(27) = 9.$

21. $\displaystyle\int_{\pi^2/4}^{\pi^2} \frac{\sin\sqrt{x}}{\sqrt{x}}\,dx$

Let $u = \sqrt{x}$, $du = \frac{1}{2\sqrt{x}}\,dx$

if $x = \frac{\pi^2}{4}$, $u = \frac{\pi}{2}$; if $x = \pi^2$, $u = \pi$

$\displaystyle 2\int_{\pi/2}^{\pi}\sin u\,du \;=\; -2\cos u\Big]_{\pi/2}^{\pi}$

$= -2(-1 - 0) = 2.$

22. $\displaystyle\int_0^{\pi/4}\sin x\sqrt{\cos x}\,dx$

Let $u = \cos x$, $du = -\sin x\,dx$

if $x = 0$, $u = 1$; if $x = \frac{\pi}{4}$, $u = \frac{\sqrt{2}}{2}$

$\displaystyle -\int_1^{\sqrt{2}/2}\sqrt{u}\,du \;=\; -\frac{2}{3}u^{3/2}\Big]_1^{\sqrt{2}/2}$

$= -\frac{2}{3}\left(\frac{2^{1/4}}{2} - 1\right)$

$= \frac{1}{3}(2 - 2^{1/4}) \approx 0.2703.$

23. $\displaystyle\int_1^2 \frac{x}{(2x^2 - 1)^3}\,dx$

Let $u = 2x^2 - 1$, $du = 4x\,dx$

if $x = 1$, $u = 1$; if $x = 2$, $u = 7$

$\displaystyle \frac{1}{4}\int_1^7 \frac{du}{u^3} \;=\; -\frac{1}{8}\frac{1}{u^2}\Big]_1^7$

$= \frac{1}{8}\left(1 - \frac{1}{49}\right) = \frac{6}{49}.$

24. $\displaystyle\int_0^2 t|t^2 - 1|^{10}\,dt = \int_0^2 t(t^2 - 1)^{10}\,dt$

Let $u = t^2 - 1$, $du = 2t\,dt$

if $t = 0$, $u = -1$; if $t = 2$, $u = 3$

$\displaystyle \frac{1}{2}\int_{-1}^3 u^{10}\,du \;=\; \frac{1}{22}u^{11}\Big]_{-1}^3$

$= \frac{1}{22}(3^{11} + 1)$

$= \frac{88{,}574}{11} \approx 8052.2.$

25. $\displaystyle\int_1^4 s|2 - s|^{12}\,ds = \int_1^4 s(2 - s)^{12}\,ds$

Let $u = 2 - s$, $du = -ds$, $s = 2 - u$. If $s = 1$, $u = 1$; if $s = 4$, $u = -2$.

$\displaystyle -\int_1^{-2}(2 - u)u^{12}\,du \;=\; \int_{-2}^1 2\,u^{12}\,du - \int_{-2}^1 u^{13}\,du$

$= \frac{2}{13}u^{13}\Big]_{-2}^1 - \frac{1}{14}u^{14}\Big]_{-2}^1 = \frac{2}{13}(1 + 2^{13}) - \frac{1}{14}(1 - 2^{14})$

$= \frac{2}{13} - \frac{1}{14} + 2^{14}\left(\frac{1}{13} + \frac{1}{14}\right) = \frac{15}{182} + \frac{27\cdot 2^{14}}{182} = \frac{442{,}383}{182} \approx 2430.68.$

26. $\displaystyle\int_0^2 x\sqrt{|1 - x^2|}\,dx = \int_0^1 x\sqrt{1 - x^2}\,dx + \int_1^2 x\sqrt{x^2 - 1}\,dx$

Let $u = 1 - x^2$, $du = -2x\,dx$ 　　　Let $u = x^2 - 1$, $du = 2x\,dx$

if $x = 0$, $u = 1$; if $x = 1$, $u = 0$. 　　if $x = 1$, $u = 0$; if $x = 2$, $u = 3$.

$\displaystyle -\frac{1}{2}\int_1^0 \sqrt{u}\,du + \frac{1}{2}\int_0^3 \sqrt{u}\,du = \frac{1}{3}u^{3/2}\Big]_0^1 + \frac{1}{3}u^{3/2}\Big]_0^3 = \frac{1}{3}(1 - 0) + \frac{1}{3}(3\sqrt{3} - 0) = \frac{1 + 3\sqrt{3}}{3} \approx 2.0654.$

27. $\displaystyle\int_0^1 \frac{1}{(4 - x)^2}\,dx = \frac{1}{4 - x}\Big]_0^1 = \frac{1}{3} - \frac{1}{4} = \frac{1}{12}.$

28. $\displaystyle\int_0^{\pi/4}\sin^2 x\,dx = \int_0^{\pi/4}\left(\frac{1}{2} - \frac{1}{2}\cos 2x\right)dx = \frac{x}{2} - \frac{1}{4}\sin 2x\Big]_0^{\pi/4} = \frac{\pi}{8} - \frac{1}{4} = \frac{\pi - 2}{8} \approx 0.1427.$

29. $\displaystyle\int_0^{\pi}\cos^2 x\,dx = \int_0^{\pi}\left(\frac{1}{2} + \frac{1}{2}\cos 2x\right)dx = \frac{x}{2} + \frac{1}{4}\sin 2x\Big]_0^{\pi} = \frac{\pi}{2} + \frac{1}{4}(0) = \frac{\pi}{2}.$

30. $\displaystyle\int_0^{\sqrt{\pi/4}} u \sec(u^2)\tan(u^2)\, du$

Let $z = u^2$, $dz = 2u\, du$

if $u = 0$, $z = 0$; if $u = \sqrt{\frac{\pi}{4}}$, $z = \frac{\pi}{4}$

$\displaystyle\frac{1}{2}\int_0^{\pi/4} \sec z \tan z\, dz$

$\displaystyle = \frac{1}{2}\sec z \Big]_0^{\pi/4} = \frac{1}{2}(\sqrt{2}-1)$

$\approx 0.2071.$

31. $\displaystyle\int_0^{\pi/4} \sec^2 u \sqrt{\tan u}\, du = \frac{2}{3}(\tan u)^{3/2}\Big]_0^{\pi/4}$

$\displaystyle = \frac{2}{3}.$

(Let $z = \tan u$, etc.)

32. $\displaystyle\int_{-\pi/4}^{\pi/4} \tan^2\theta\, d\theta = \int_{-\pi/4}^{\pi/4}(\sec^2\theta - 1)\, d\theta$

$\displaystyle = \tan\theta - \theta\Big]_{-\pi/4}^{\pi/4}$

$\displaystyle = 1 - (-1) - \frac{\pi}{4} + \left(-\frac{\pi}{4}\right)$

$\displaystyle = 2 - \frac{\pi}{2} \approx 0.4292.$

33. $\displaystyle\int_0^{\sqrt{\pi/2}} t\cos^3(t^2)\sin(t^2)\, dt$

Let $u = t^2$, $du = 2t\, dt$

if $t = 0$, $u = 0$; if $t = \sqrt{\frac{\pi}{2}}$, $u = \frac{\pi}{2}$

$\displaystyle\frac{1}{2}\int_0^{\pi/2} \cos^3 u \sin u\, du$

$\displaystyle = -\frac{1}{8}\cos^4(u)\Big]_0^{\pi/2}$

$\displaystyle = \frac{1}{8}.$

(Let $z = \cos u$, etc.)

34. $\displaystyle\int_0^{\pi/2} \cos t \sqrt{1 - \sin^2 t}\, dt.$

a) $\sin^2 t + \cos^2 t = 1 \ \Rightarrow\ \sqrt{1 - \sin^2 t} = \cos t$

$\displaystyle\int_0^{\pi/2} \cos^2 t\, dt = \frac{1}{2}\int_0^{\pi/2}(1 + \cos 2t)\, dt = \frac{1}{2}\left[t + \frac{1}{2}\sin 2t\right]_0^{\pi/2}$

$\displaystyle = \frac{1}{2}\left(\frac{\pi}{2} + 0\right) = \frac{\pi}{4} \approx 0.7854.$

b) Let $u = \sin t$, $du = \cos t\, dt$

if $t = 0$, $u = 0$; if $t = \frac{\pi}{2}$, $u = 1$

$\displaystyle\int_0^1 \sqrt{1 - u^2}\, du = $ area of $\frac{1}{4}$ of circle of radius 1

$\displaystyle = \frac{\pi}{4} \approx 0.7854.$

35. $\displaystyle\int_0^{\pi/4} \tan^2 x \sec^2 x\, dx$

Let $u = \sec x$, $du = \sec x \tan x\, dx$; $\tan x = \sqrt{\sec^2 x - 1} = \sqrt{u^2 - 1}$

if $x = 0$, $u = 1$; if $x = \frac{\pi}{4}$, $x = \sqrt{2}$

$\displaystyle\int_1^{\sqrt{2}} \sqrt{u^2 - 1}\, u\, du = \frac{1}{2}\cdot\frac{2}{3}(u^2 - 1)^{3/2}\Big]_1^{\sqrt{2}}$

$\displaystyle = \frac{1}{3}(1 - 0) = \frac{1}{3}.$

(Let $z = u^2 - 1$, $dz = 2u\, du$, etc.)

36. $f(x) = \sqrt{x+2}, \; [2,7]$

$$\bar{f} = \frac{1}{7-2} \int_2^7 \sqrt{x+2}\, dx$$

$$= \frac{1}{5} \cdot \frac{2}{3}(x+2)^{3/2} \Big]_2^7$$

$$= \frac{2}{15}(27-8) = \frac{38}{15}.$$

37. $f(x) = x\sqrt{x^2+1}, \; [0,2]$

$$\bar{f} = \frac{1}{2-0} \int_0^2 x\sqrt{x^2+1}\, dx$$

$$= \frac{1}{2}\left(\frac{1}{3}\right)(x^2+1)^{3/2} \Big]_0^2$$

$$= \frac{1}{6}(5^{3/2}-1) \approx 1.6967.$$
$$(\text{Let } u = x^2+1, \; du = 2x\, dx, \; \text{etc;})$$

38. $f(x) = (x+4)^{2/3}, \; [-4,4]$

$$\bar{f} = \frac{1}{4-(-4)} \int_{-4}^4 (x+4)^{2/3}\, dx$$

$$= \frac{1}{8} \cdot \frac{3}{5}(x+4)^{5/3} \Big]_{-4}^4$$

$$= \frac{3}{40}(32-0) = \frac{12}{5} = 2.40.$$

39. $f(x) = x\cos x^2, \; \left[0, \dfrac{\sqrt{\pi}}{2}\right]$

$$\bar{f} = \frac{1}{\frac{\sqrt{\pi}}{2}-0} \int_0^{\sqrt{\pi}/2} x\cos x^2\, dx$$

Let $u = x^2$, $du = 2x\, dx$
if $x = 0$, $u = 0$; if $x = \frac{\sqrt{\pi}}{2}$, $u = \frac{\pi}{4}$

$$\bar{f} = \frac{2}{\sqrt{\pi}} \cdot \frac{1}{2} \int_0^{\pi/4} \cos u\, du$$

$$= \frac{1}{\sqrt{\pi}} \sin u \Big]_0^{\pi/4} = \frac{1}{\sqrt{\pi}} \frac{\sqrt{2}}{2}$$

$$= \frac{1}{\sqrt{2\pi}} \approx 0.3989.$$

40. $F(x) = \displaystyle\int_2^x \sqrt{1+t}\, dt, \; x \geq -1$

$$= \frac{2}{3}(1+t)^{3/2} \Big]_2^x$$

$$= \frac{2}{3}\left[(1+x)^{3/2} - 3^{3/2}\right].$$

$$F'(x) = \frac{2}{3} \cdot \frac{3}{2}(1+x)^{1/2} = \sqrt{1+x}.$$

41. $F(x) = \displaystyle\int_1^x \frac{t}{\sqrt{1+t^2}}\, dt = \sqrt{1+t^2} \Big]_1^x$

$$= \sqrt{1+x^2} - \sqrt{2}.$$

$$F'(x) = \frac{2x}{2\sqrt{1+x^2}} = \frac{x}{\sqrt{1+x^2}}.$$

42. $F(x) = \displaystyle\int_0^x t\sin t^2\, dt$

$$= \frac{1}{2} \int_0^x (\sin t^2)(2t\, dt)$$

$$= -\frac{1}{2}\cos t^2 \Big]_0^x = -\frac{1}{2}\cos x^2 + \frac{1}{2}.$$

$$F'(x) = -\frac{1}{2}(-\sin x^2)(2x)$$

$$= x\sin x^2.$$

43. $F(x) = \displaystyle\int_x^0 \cos^2 t \sin t\, dt$

$$= -\frac{1}{3}\cos^3 t \Big]_x^0 = -\frac{1}{3} + \frac{1}{3}\cos^3 x.$$

$$F'(x) = \frac{1}{3}(3\cos^2 x)(-\sin x)$$

$$= -\cos^2 x \sin x.$$

6.5   Finding Areas By Integration

1. $f(x) = 2x + 5,\ a = 0,\ b = 2.$

$$A = \int_0^2 (2x + 5)\, dx = x^2 + 5x \Big]_0^2 = 4 + 10 = 14.$$

2. $f(x) = 9 - x^2,\ a = -3,\ b = 3.$

$$A = \int_{-3}^3 (9 - x^2)\, dx = 9x - \frac{x^3}{3} \Big]_{-3}^3 = 2\left(9x - \frac{x^3}{3}\right)\Big]_0^3 = 2(27 - 9) = 36.$$

3. $f(x) = \dfrac{1}{\sqrt{x - 2}},\ a = 3,\ b = 5.$

$$A = \int_3^5 \frac{1}{\sqrt{x - 2}}\, dx = 2\sqrt{x - 2} \Big]_3^5 = 2(\sqrt{3} - 1) \approx 1.4641.$$

4. $f(x) = \sin 2x,\ a = 0,\ b = \dfrac{\pi}{2}.$

$$A = \int_0^{\pi/2} \sin 2x\, dx = -\frac{1}{2}\cos 2x \Big]_0^{\pi/2} = -\frac{1}{2}(-1 - 1) = 1.$$

5. $f(x) = \dfrac{x + 1}{(x^2 + 2x)^2},\ a = 1,\ b = 2.$

$$A = \int_1^2 \frac{x + 1}{(x^2 + 2x)^2}\, dx = -\frac{1}{2}\left(\frac{1}{x^2 + 2x}\right)\Big]_1^2 = -\frac{1}{2}\left(\frac{1}{8} - \frac{1}{3}\right) = \frac{5}{48}.$$

6. $f(x) = x\sqrt{1 + x^2},\ a = 0,\ b = 1.$

$$A = \int_0^1 x\sqrt{1 + x^2}\, dx = \frac{1}{2}\left(\frac{2}{3}\right)(1 + x^2)^{3/2}\Big]_0^1 = \frac{1}{3}(2^{3/2} - 1) \approx 0.6095.$$

7. $f(x) = \sqrt{5 + x},\ a = -4,\ b = 4.$

$$A = \int_{-4}^4 \sqrt{5 + x}\, dx = \frac{2}{3}(5 + x)^{3/2}\Big]_{-4}^4 = \frac{2}{3}(27 - 1) = \frac{52}{3}.$$

8. $f(x) = \dfrac{x}{\sqrt{9 + x^2}},\ a = 0,\ b = 4.$

$$A = \int_0^4 \frac{x}{\sqrt{9 + x^2}}\, dx = \sqrt{9 + x^2}\Big]_0^4 = 5 - 3 = 2.$$

9. $f(x) = x + 1,\ g(x) = -2x + 1,\ 0 \le x \le 2.$

$$A = \int_0^2 [x + 1 - (-2x + 1)]\, dx$$

$$= \int_0^2 3x\, dx$$

$$= \left. \frac{3}{2} x^2 \right]_0^2$$

$$= 6.$$

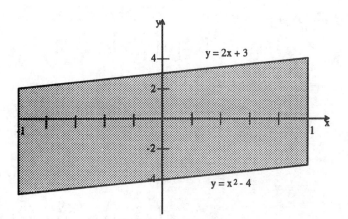

Exercise 9

10. $f(x) = 2x + 3$, $g(x) = x^2 - 4$, $-1 \le x \le 1$.

$$A = \int_{-1}^1 [2x + 3 - (x^2 - 4)]\, dx$$

$$= \left. -\frac{x^3}{3} + x^2 + 7x \right]_{-1}^1$$

$$= -\frac{1}{3}(1 + 1) + (1 - 1) + 7(1 + 1)$$

$$= \frac{40}{3}.$$

Exercise 10

11. $f(x) = \sqrt{x}$, $g(x) = -x^2$, $0 \le x \le 4$.

$$A = \int_0^4 [\sqrt{x} - (-x^2)]\, dx$$

$$= \left. \frac{2}{3} x^{3/2} + \frac{x^3}{3} \right]_0^4$$

$$= \frac{2}{3}(8) + \frac{64}{3}$$

$$= \frac{80}{3}.$$

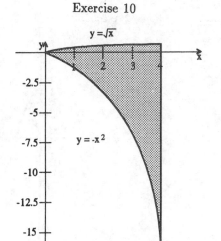

Exercise 11

12. $f(x) = \dfrac{1}{x^2}$, $g(x) = x^{2/3}$, $1 \le x \le 8$.

$$A = \int_1^8 \left(x^{2/3} - \frac{1}{x^2} \right) dx$$

$$= \left. \frac{3}{5} x^{5/3} + \frac{1}{x} \right]_1^8$$

$$= \frac{3}{5}(32-1) + \left(\frac{1}{8} - 1 \right)$$

$$= \frac{709}{40}$$

$$= 17.725.$$

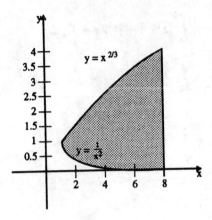

Exercise 12

13. $f(x) = \sin x$, $g(x) = \cos x$, $0 \le x \le 2\pi$.

$\cos x \ge \sin x$ on $\left[0, \dfrac{\pi}{4} \right] \cup \left[\dfrac{5\pi}{4}, 2\pi \right]$

$\sin x \ge \cos x$ on $\left[\dfrac{\pi}{4}, \dfrac{5\pi}{4} \right]$

$$A = \int_0^{\pi/4} (\cos x - \sin x)\, dx + \int_{\pi/4}^{5\pi/4} (\sin x - \cos x)\, dx + \int_{5\pi/4}^{2\pi} (\cos x - \sin x)\, dx$$

$$= \left. \sin x + \cos x \right]_0^{\pi/4} + \left. (-\cos x - \sin x) \right]_{\pi/4}^{5\pi/4} + \left. \sin x + \cos x \right]_{5\pi/4}^{2\pi}$$

$$= \frac{\sqrt{2}}{2} + \left(\frac{\sqrt{2}}{2} - 1 \right) + \left(\frac{\sqrt{2}}{2} + \frac{\sqrt{2}}{2} \right) + \left(\frac{\sqrt{2}}{2} + \frac{\sqrt{2}}{2} \right) + \left(0 + \frac{\sqrt{2}}{2} + 1 + \frac{\sqrt{2}}{2} \right)$$

$$= 4\sqrt{2} \approx 5.6569.$$

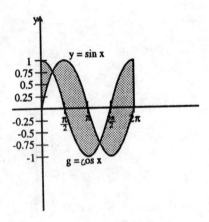

Exercise 13

14. $f(x) = 1$, $g(x) = \cos x$, $0 \le x \le 2\pi$.

$$A = \int_0^{2\pi} (1 - \cos x)\, dx$$

$$= x - \sin x \Big]_0^{2\pi}$$

$$= 2\pi.$$

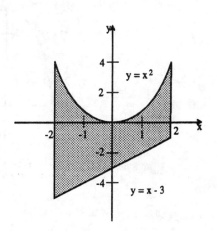

Exercise 14

15. $f(x) = x - 3$, $g(x) = x^2$, $-2 \le x \le 2$.

$$A = \int_{-2}^{2} [x^2 - (x - 3)]\, dx$$

$$= \frac{x^3}{3} - \frac{x^2}{2} + 3x \Big]_{-2}^{2}$$

$$= \frac{1}{3}(8 + 8) - \frac{1}{2}(4 - 4) + 3(2 + 2)$$

$$= \frac{52}{3}.$$

Exercise 15

16. $f(x) = x\sqrt{9 - x^2}$, $g(x) = -x$, $-3 \le x \le 3$.

Using symmetry,

$$A = 2\int_0^3 \left[x\sqrt{9 - x^2} - (-x) \right] dx$$

$$= -\frac{2}{3}(9 - x^2)^{3/2} + x^2 \Big]_0^3$$

$$= -\frac{2}{3}(0 - 27) + 9$$

$$= 27.$$

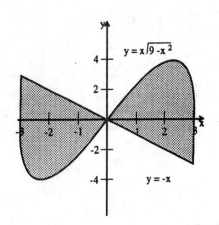

Exercise 16

465

17. $f(x) = |4 - x^2|$, $g(x) = 5$, $-3 \le x \le 3$.

$$A = \int_{-3}^{3} (5 - |4 - x^2|)\, dx = \int_{-3}^{-2} (5 - x^2 + 4)\, dx + \int_{-2}^{2} (5 - 4 + x^2)\, dx + \int_{2}^{3} (5 - x^2 + 4)\, dx$$

$$= 9x - \frac{x^3}{3}\Big]_{-3}^{-2} + x + \frac{x^3}{3}\Big]_{-2}^{2} + 9x - \frac{x^3}{3}\Big]_{2}^{3}$$

$$= 9(-2 + 3) - \frac{1}{3}(-8 + 27) + (2 + 2) + \frac{1}{3}(8 + 8) + 9(3 - 2) - \frac{1}{3}(27 - 8)$$

$$= \frac{44}{3}.$$

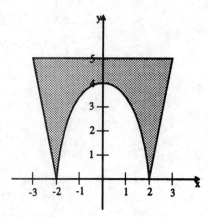

Exercise 17

18. $f(x) = \sin x$, $g(x) = x$, $0 \le x \le \pi$.

$$A = \int_{0}^{\pi} (x - \sin x)\, dx$$

$$= \frac{x^2}{2} + \cos x\Big]_{0}^{\pi}$$

$$= \frac{\pi^2}{2} - 1 - 1$$

$$= \frac{\pi^2}{2} - 2$$

$$\approx 2.9348.$$

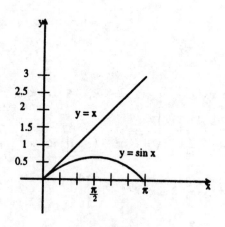

Exercise 18

19. $f(x) = \dfrac{x^2 - 1}{x^2}$, $g(x) = \dfrac{1 - x^2}{x^2}$, $1 \le x \le 2$.

$$A = \int_1^2 \left(\frac{x^2-1}{x^2} - \frac{1-x^2}{x^2} \right) dx$$

$$= \int_1^2 \left(2 - \frac{2}{x^2} \right) dx$$

$$= 2x + \frac{2}{x} \Big]_1^2$$

$$= 2(2-1) + 2 \left(\frac{1}{2} - 1 \right)$$

$$= 1.$$

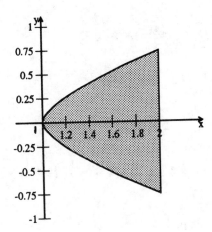

Exercise 19

20. $f(x) = x^{2/3}$, $g(x) = x^{1/3}$, $-1 \le x \le 1$.

   $f(x) \ge g(x)$ on $[-1, 0]$; $g(x) \ge f(x)$ on $[0, 1]$

$$A = \int_{-1}^0 (x^{2/3} - x^{1/3}) \, dx + \int_0^1 (x^{1/3} - x^{2/3}) \, dx$$

$$= \frac{3}{5} x^{5/3} - \frac{3}{4} x^{4/3} \Big]_{-1}^0 + \frac{3}{4} x^{4/3} - \frac{3}{5} x^{5/3} \Big]_0^1$$

$$= \frac{3}{5}(0+1) - \frac{3}{4}(0-1) + \frac{3}{4}(1-0) - \frac{3}{5}(1-0)$$

$$= \frac{6}{4}$$

$$= \frac{3}{2}.$$

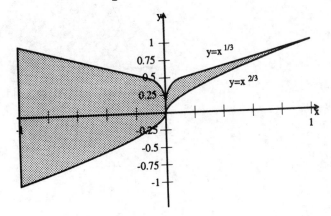

Exercise 20

21. $y = 4 - x^2$, $y = x - 2$.

   $4 - x^2 = x - 2 \Rightarrow x^2 + x - 6 = 0$;

   $(x + 3)(x - 2) = 0$; $x = 2, -3$

467

$$A = \int_{-3}^{2} [4 - x^2 - (x-2)]\, dx$$

$$= \int_{-3}^{2} (-x^2 - x + 6)\, dx$$

$$= -\frac{x^3}{3} - \frac{x^2}{2} + 6x \Big]_{-3}^{2}$$

$$= -\frac{1}{3}(8 + 27) - \frac{1}{2}(4 - 9) + 6(2 + 3)$$

$$= \frac{125}{6}$$

$$= 20.833.$$

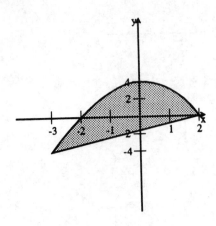

Exercise 21

22. $y = 9 - x^2, \; 9y - x^2 + 9 = 0.$
    $9(9 - x^2) - x^2 + 9 = 0 \Rightarrow$
    $-10x^2 + 90 = 0 \Rightarrow x = \pm 3$

$$A = \int_{-3}^{3} \left(9 - x^2 - \frac{x^2 - 9}{9}\right) dx$$

$$= 2\int_{0}^{3} \left(10 - \frac{10}{9}x^2\right) dx$$

$$= 2\left(10x - \frac{10}{27}x^3\right)\Big]_{0}^{3}$$

$$= 20(3 - 1)$$

$$= 40.$$

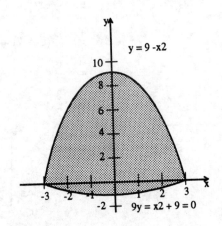

Exercise 22

23. $y = x^2, \; y = x^3.$

$x^2 = x^3 \Rightarrow x^2(x - 1) = 0 \Rightarrow x = 0, 1$

468

$$A = \int_0^1 (x^2 - x^3)\, dx$$

$$= \left. \frac{x^3}{3} - \frac{x^4}{4} \right]_0^1$$

$$= \frac{1}{3} - \frac{1}{4}$$

$$= \frac{1}{12}.$$

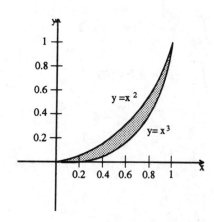

Exercise 23

24. $y = x^3$, $y = x$.

$x^3 = x \Rightarrow x(x^2 - 1) = 0 \Rightarrow x = 0, \pm 1$

$x^3 \geq x$ on $[-1, 0]$; $x \geq x^3$ on $[0, 1]$

$$A = \int_{-1}^0 (x^3 - x)\, dx + \int_0^1 (x - x^3)\, dx$$

$$= \left. \frac{x^4}{4} - \frac{x^2}{2} \right]_{-1}^0 + \left. \frac{x^2}{2} - \frac{x^4}{4} \right]_0^1$$

$$= \frac{1}{4}(0 - 1) - \frac{1}{2}(0 - 1) + \frac{1}{2} - \frac{1}{4}$$

$$= \frac{2}{4}$$

$$= \frac{1}{2}.$$

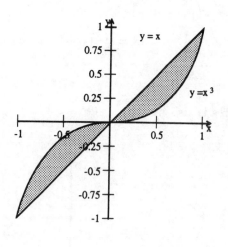

Exercise 24

25. $x + y^2 = 4$, $y = x + 2$.

$4 - y^2 = y - 2 \Rightarrow y^2 + y - 6 = 0 \Rightarrow (y + 3)(y - 2) = 0$

$$y = 2, -3$$

469

$$A = \int_{-3}^{2} [4 - y^2 - (y - 2)] \, dy$$

$$= \int_{-3}^{2} (-y^2 - y + 6) \, dy$$

$$= -\frac{y^3}{3} - \frac{y^2}{2} + 6y \Big]_{-3}^{2}$$

$$= -\frac{1}{3}(8 + 27) - \frac{1}{2}(4 - 9) + 6(2 + 3)$$

$$= \frac{125}{6}$$

$$\approx 20.833.$$

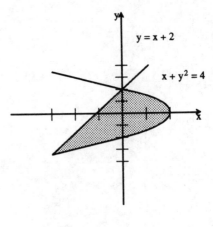

Exercise 25

26. $x = \sqrt{y}, \quad x = \sqrt[3]{y}.$
$\quad x^2 = x^3 \Rightarrow x(x^2 - 1) = 0$
$\qquad\qquad \Rightarrow x = 0, 1$

Note that the graphs do not intersect at $x = -1$, since the values of y are $+1$ and -1, respectively.

$$A = \int_0^1 \left(\sqrt[3]{y} - \sqrt{y} \right) dy = \frac{3}{4} y^{4/3} - \frac{2}{3} y^{3/2} \Big]_0^1 = \frac{3}{4} - \frac{2}{3} = \frac{1}{12}$$

The area is also given by $\int_0^1 (x^2 - x^3) \, dx = \frac{x^3}{3} - \frac{x^4}{4} \Big]_0^1 = \frac{1}{12}.$

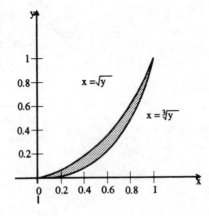

27. $2x = y^2, \quad x + 2 = y^2.$
$\quad 2x = x + 2 \Rightarrow x = 2; \ y = \pm 2$

$$A = \int_{-2}^{2} \left[\frac{1}{2} y^2 - (y^2 - 2) \right] dy$$

$$= 2 \int_{0}^{2} \left(-\frac{y^2}{2} + 2 \right) dy$$

$$= 2 \left[-\frac{y^3}{6} + 2y \right]_{0}^{2}$$

$$= 2 \left(-\frac{8}{6} + 4 \right)$$

$$= \frac{16}{3}.$$

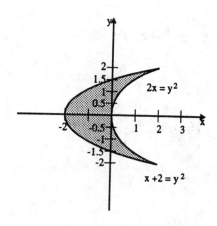

Exercise 27

28. $y = x^{2/3}, \; y = 2 - x^2.$

At $x = \pm 1, \; y = 1$

$$A = \int_{-1}^{1} (2 - x^2 - x^{2/3}) \, dx$$

$$= 2 \int_{0}^{1} (-x^2 - x^{2/3} + 2) \, dx$$

$$= 2 \left[-\frac{x^3}{3} - \frac{3}{5} x^{5/3} + 2x \right]_{0}^{1}$$

$$= 2 \left(-\frac{1}{3} - \frac{3}{5} + 2 \right)$$

$$= \frac{32}{15}$$

$$\approx 2.1333.$$

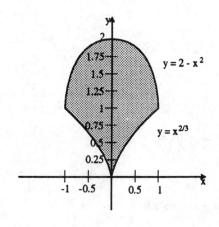

Exercise 28

29. $y = x^{2/3}, \; y = x^2.$
$x^{2/3} = x^2 \Rightarrow x = 0, \pm 1$

$$A = \int_{-1}^{1} (x^{2/3} - x^2)\,dx$$

$$= 2\int_{0}^{1} (x^{2/3} - x^2)\,dx$$

$$= 2\left[\frac{3}{5}x^{5/3} - \frac{x^3}{3}\right]_{0}^{1}$$

$$= 2\left(\frac{3}{5} - \frac{1}{3}\right)$$

$$= \frac{8}{15}$$

$$\approx 0.5333.$$

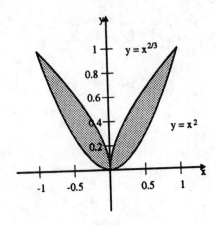

Exercise 29

30. $f(x) = 2x + 1$, $[-1, 2]$.

a) $\displaystyle\int_{-1}^{2} (2x + 1)\,dx = x^2 + x\Big]_{-1}^{2} = (4 - 1) + (2 + 1) = 6.$

b) $2x + 1 = 0 \Rightarrow x = -\dfrac{1}{2}$

$$A = \int_{-1}^{-1/2} -(2x + 1)\,dx + \int_{-1/2}^{2} (2x + 1)\,dx$$

$$= -(x^2 + x)\Big]_{-1}^{-1/2} + x^2 + x\Big]_{-1/2}^{2}$$

$$= -\left(\frac{1}{4} - \frac{1}{2}\right) + (1 - 1) + (4 + 2) - \left(\frac{1}{4} - \frac{1}{2}\right) = \frac{13}{2}.$$

31. $f(x) = 3 - x$, $[0, 4]$.

a) $\displaystyle\int_{0}^{4} (3 - x)\,dx = 3x - \frac{x^2}{2}\Big]_{0}^{4} = 12 - 8 = 4.$

b) $3 - x = 0 \Rightarrow x = 3$

$$A = \int_{0}^{3} (3 - x)\,dx - \int_{3}^{4} (3 - x)\,dx$$

$$= 3x - \frac{x^2}{2}\Big]_{0}^{3} - \left(3x - \frac{x^2}{2}\right)\Big]_{3}^{4}$$

$$= 9 - \frac{9}{2} - (12 - 8) + \left(9 - \frac{9}{2}\right) = 5.$$

32. $f(x) = 3 - x^2$, $[0, 2]$.

a) $\displaystyle\int_{0}^{2} (3 - x^2)\,dx = 3x - \frac{x^3}{3}\Big]_{0}^{2} = 6 - \frac{8}{3} = \frac{10}{3}.$

b) $3 - x^2 = 0 \Rightarrow x = \sqrt{3}$

$$
\begin{aligned}
A &= \int_0^{\sqrt{3}} (3 - x^2)\, dx - \int_{\sqrt{3}}^2 (3 - x^2)\, dx \\
&= 3x - \frac{x^3}{3}\bigg]_0^{\sqrt{3}} - \left(3x - \frac{x^3}{3}\right)\bigg]_{\sqrt{3}}^2 \\
&= 3\sqrt{3} - \sqrt{3} - \left(6 - \frac{8}{3}\right) + (3\sqrt{3} - \sqrt{3}) \\
&= 4\sqrt{3} - \frac{10}{3} \approx 3.5949.
\end{aligned}
$$

33. $f(x) = 2x^2 - 1$, $[-1, 1]$.

a) $\int_{-1}^1 (2x^2 - 1)\, dx = \frac{2}{3}x^3 - x\bigg]_{-1}^1 = \frac{2}{3}(1 + 1) - (1 + 1) = -\frac{2}{3}.$

b) $2x^2 - 1 = 0 \Rightarrow x = \pm\dfrac{\sqrt{2}}{2}$

$$
\begin{aligned}
A &= \int_{-1}^{-\sqrt{2}/2} (2x^2 - 1)\, dx - \int_{-\sqrt{2}/2}^{\sqrt{2}/2} (2x^2 - 1)\, dx + \int_{\sqrt{2}/2}^1 (2x^2 - 1)\, dx \\
&= \frac{2}{3}x^3 - x\bigg]_{-1}^{-\sqrt{2}/2} - \left(\frac{2}{3}x^3 - x\right)\bigg]_{-\sqrt{2}/2}^{\sqrt{2}/2} + \frac{2}{3}x^3 - x\bigg]_{\sqrt{2}/2}^1 \\
&= \frac{2}{3}\left(-\frac{\sqrt{2}}{4} + 1\right) - \left(-\frac{\sqrt{2}}{2} + 1\right) - \left(\frac{2}{3}\cdot\frac{\sqrt{2}}{4} - \frac{\sqrt{2}}{2} + \frac{2}{3}\cdot\frac{\sqrt{2}}{4} - \frac{\sqrt{2}}{2}\right) \\
&\quad + \frac{2}{3}\left(1 - \frac{\sqrt{2}}{4}\right) - \left(1 - \frac{\sqrt{2}}{2}\right) \\
&= \frac{2}{3}(2\sqrt{2} - 1) \approx 1.2190.
\end{aligned}
$$

34. $f(x) = 2x^2 + 6x + 1$, $[1, 3]$.

a) $\int_1^3 (2x^2 + 6x + 1)\, dx = \frac{2}{3}x^3 + 3x^2 + x\bigg]_1^3 = \frac{2}{3}(27 - 1) + 3(9 - 1) + (3 - 1) = \frac{130}{3} \approx 43.333.$

b) $2x^2 + 6x + 1 = 0 \Rightarrow x = \dfrac{-6 \pm \sqrt{36 - 8}}{4} = \dfrac{-3 \pm \sqrt{7}}{2}$

Since $2x^2 + 6x + 1 > 0$ on $[1, 3]$, $A \approx 43.333$.

35. $f(x) = x^2 + 6$, $[0, 3]$.

a) $\int_0^3 (x^2 + 6)\, dx = \frac{x^3}{3} + 6x\bigg]_0^3 = 9 + 18 = 27.$

b) Since $x^2 + 6 > 0$ on $[0, 3]$, $A = 27$.

36. $f(x) = x(x + 1)$, $[-2, 1]$.

a) $\int_{-2}^{1} x(x+1)\,dx = \dfrac{x^3}{3} + \dfrac{x^2}{2}\Big]_{-2}^{1} = \dfrac{1}{3}(1+8) + \dfrac{1}{2}(1-4) = 3 - \dfrac{3}{2} = \dfrac{3}{2}.$

b) $x(x+1) = 0 \Rightarrow x = -1, 0$
$x(x+1) \geq 0$ on $[-2, -1]$ and $[0, 1]$

$$A = \int_{-2}^{-1} x(x+1)\,dx - \int_{-1}^{0} x(x+1)\,dx + \int_{0}^{1} x(x+1)\,dx$$

$$= \dfrac{x^3}{3} + \dfrac{x^2}{2}\Big]_{-2}^{-1} - \left(\dfrac{x^3}{3} + \dfrac{x^2}{2}\right)\Big]_{-1}^{0} + \dfrac{x^3}{3} + \dfrac{x^2}{2}\Big]_{0}^{1}$$

$$= \dfrac{1}{3}(-1+8) + \dfrac{1}{2}(1-4) + \left(-\dfrac{1}{3} + \dfrac{1}{2}\right) + \left(\dfrac{1}{3} + \dfrac{1}{2}\right) = \dfrac{11}{6}.$$

37. $y = x^{2/3}, \; y = 1$
$x^{2/3} = 1 \Rightarrow x = \pm 1$

$A = \int_{-1}^{1} (1 - x^{2/3})\,dx = x - \dfrac{3}{5}x^{5/3}\Big]_{-1}^{1} = 1 + 1 - \dfrac{3}{5}(1+1) = \dfrac{4}{5}.$

38. $y = x^{2/3}, \; y = 1$
$x = \pm y^{3/2}$

$A = \int_{0}^{1} \left[y^{3/2} - (-y^{3/2}) \right] dy = \dfrac{4}{5} y^{5/2}\Big]_{0}^{1} = \dfrac{4}{5}.$

39. $y = x^3, \; y = -x^3, \; y = 1, \; y = -1$
$y = x^3 \Rightarrow x = y^{1/3}$
By symmetry the area is
$4 \times$ area in quadrant I.

$A = 4 \int_{0}^{1} (y^{1/3} - 0)\,dy$

$= \dfrac{4 \cdot 3}{4} y^{4/3}\Big]_{0}^{1}$

$= 3.$

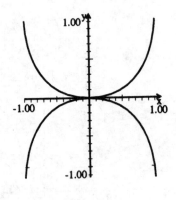

Exercise 39

40. $y = x^3, \; y = -x^3, \; y = 1, \; y = -1$

By symmetry the area is $4 \times$ area in quadrant I.

$A = 4 \int_{0}^{1} (1 - x^3)\,dx = 4 \left(x - \dfrac{x^4}{4} \right)\Big]_{0}^{1} = 4 \left(1 - \dfrac{1}{4} \right) = 3.$

41. $y = 4 - x^2$, an inverted parabola with x-intercepts ± 2. For the areas above and below $y = a$ to be equal,

474

$$\int_0^a \sqrt{4-y}\,dy = \int_a^4 \sqrt{4-y}\,dy$$

$$-\frac{2}{3}(4-y)^{3/2}\Big]_0^a = -\frac{2}{3}(4-y)^{3/2}\Big]_a^4$$

$$(4-a)^{3/2} - 8 = 0 - (4-a)^{3/2}$$

$$(4-a)^{3/2} = 4$$

$$4-a = 4^{2/3}$$

$$a = 4 - 4^{2/3} \approx 1.4802.$$

Exercise 41

42. $y = 4x - 4x^2 = (-1 + 4x - 4x^2) + 1 = -(2x-1)^2 + 1$
$y - 1 = -(2x-1)^2$, an inverted parabola with vertex at $\left(\frac{1}{2}, 1\right)$.
x-intercepts are 0 and 1. $x = \frac{1}{2}(1 \pm \sqrt{1-y})$

$$\int_0^a \left[\frac{1}{2}(1+\sqrt{1-y}) - \frac{1}{2}\right]\,dy = \int_a^1 \left[\frac{1}{2} - \frac{1}{2}(1-\sqrt{1-y})\right]\,dy$$

$$-\frac{2}{3}(1-y)^{3/2}\Big]_0^a = -\frac{2}{3}(1-y)^{3/2}\Big]_a^1$$

$$(1-a)^{3/2} - 1 = 0 - (1-a)^{3/2}$$

$$(1-a)^{3/2} = \frac{1}{2}$$

$$1-a = \left(\frac{1}{2}\right)^{2/3}$$

$$a = 1 - \left(\frac{1}{2}\right)^{2/3} \approx 0.3700.$$

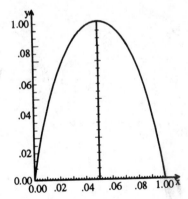

Exercise 42

43.   a) $y = x$, $x = 3$, $y = -1$
$$A = \int_{-1}^3 [x - (-1)]\,dx = \frac{x^2}{2} + x\Big]_{-1}^3 = \frac{1}{2}(9-1) + 3 + 1 = 8.$$

475

The figure is a triangle with base of length 4 and altitude of length 4. Its area $= \frac{1}{2}(4 \cdot 4) = 8$.

b) $A = \int_1^3 [(2x + 2) - (-2x - 1)]\, dx = 2x^2 + 3x \Big]_1^3 = 2(9 - 1) + 3(3 - 1) = 22.$

The figure is a parallelogram with base at $x = 3$ of length $8 - (-7) = 15$ and base at $x = 1$ of length $4 - (-3) = 7$. The altitude is $3 - 1 = 2$. Area $= \left(\dfrac{b_1 + b_2}{2}\right) \times$ altitude $= \left(\dfrac{15 + 7}{2}\right) \cdot 2 = 22.$

44. $f(x) = \sin(x^2),\ [0, \sqrt{2\pi}\,]$

$\Delta x = \dfrac{\sqrt{2\pi}}{8} \approx 0.31333$

i	x_i	$\sin(x_i^2)$
1	0.15670	0.02454
2	0.46999	0.21910
3	0.78332	0.57581
4	1.09665	0.93299
5	1.40998	0.91421
6	1.72331	0.17096
7	2.03664	−0.84485
8	2.34996	−0.68954

Area $\approx \Delta x \displaystyle\sum_{i=1}^{8} |\sin(x_i^2)|$

$\approx 0.31333(4.372)$

$\approx 1.3699.$

If $n = 50$, we obtain the approximation: area $\approx 1.3594.$

45. $f(x) = \sin \sqrt{x},\ [0, 4\pi^2]$

$\Delta x = \dfrac{4\pi^2}{8} = \dfrac{\pi^2}{2} \approx 4.9348$

i	x_i	$\sin \sqrt{x_i}$
1	2.4674	1.0000
2	7.4022	0.4086
3	12.3370	−0.3624
4	17.2718	−0.8491
5	22.2066	−1.0000
6	27.1414	−0.8788
7	32.0762	−0.5807
8	37.0110	−0.1982

Area $\approx \Delta x \displaystyle\sum_{i=1}^{8} |\sin \sqrt{x_i}|$

$\approx 4.9348(5.2778)$

$\approx 26.045.$

If $n = 50$, we obtain the approximation: area $\approx 25.159.$

6.6 Numerical Approximation of the Definite Integral

1. $\displaystyle\int_1^5 \frac{1}{x}\, dx \approx \left\{\sum_{j=1}^{8} f(m_j)\right\} \Delta x = M_8$

$= \left\{ f\left(\dfrac{5}{4}\right) + f\left(\dfrac{7}{4}\right) + f\left(\dfrac{9}{4}\right) + \cdots + f\left(\dfrac{19}{4}\right) \right\} \left(\dfrac{5 - 1}{8}\right)$

$= \left(\dfrac{4}{5} + \dfrac{4}{7} + \dfrac{4}{9} + \cdots + \dfrac{4}{19} \right) \left(\dfrac{1}{2}\right) \approx 1.599844394.$

2. $\displaystyle\int_5^8 \frac{1}{x}\, dx \approx M_6 = \left\{\sum_{j=1}^{6} f(m_j)\right\} \Delta x$

$= \left\{ f\left(\dfrac{21}{4}\right) + f\left(\dfrac{23}{4}\right) + f\left(\dfrac{25}{4}\right) + f\left(\dfrac{27}{4}\right) + f\left(\dfrac{29}{4}\right) + f\left(\dfrac{31}{4}\right) \right\} \left(\dfrac{8 - 5}{6}\right)$

$= \left\{ \dfrac{1}{21} + \dfrac{1}{23} + \cdots + \dfrac{1}{31} \right\} \left(\dfrac{4}{1}\right)\left(\dfrac{3}{6}\right) \approx 0.469750337.$

3. $\int_0^8 \sqrt{1+x^3}\,dx \;\approx\; M_4 \;=\; \left\{\sum_{j=1}^4 f(m_j)\right\}\Delta x$

$\qquad\qquad = \left\{\sqrt{1+1^3} + \sqrt{1+3^3} + \sqrt{1+5^3} + \sqrt{1+7^3}\right\}\left(\dfrac{8-0}{4}\right) \;\approx\; 72.95585067.$

4. $\int_\pi^{4\pi} \dfrac{\sin x}{x}\,dx \approx M_3 \;=\; \left\{\sum_{j=1}^3 f(m_j)\right\}\Delta x \;=\; \left\{f\left(\dfrac{3\pi}{2}\right) + f\left(\dfrac{5\pi}{2}\right) + f\left(\dfrac{7\pi}{2}\right)\right\}\left(\dfrac{4\pi-\pi}{3}\right)$

$\qquad\qquad = \left\{\dfrac{2}{3\pi}(-1) + \dfrac{2}{5\pi}(1) + \dfrac{2}{7\pi}(-1)\right\}(\pi) \;=\; \left\{-\dfrac{1}{3} + \dfrac{1}{5} - \dfrac{1}{7}\right\}\left(\dfrac{2}{\pi}\right)\left(\dfrac{\pi}{1}\right)$

$\qquad\qquad \approx\; -0.552380952.$

5. $\int_0^4 \sqrt{x^3+2}\,dx \;\approx\; T_4 \;=\; \left(\dfrac{4-0}{4}\right)\left(\dfrac{1}{2}\right)[f(0) + 2f(1) + 2f(2) + 2f(3) + f(4)]$

$\qquad\qquad = (1)\left(\dfrac{1}{2}\right)\left[\sqrt{2} + 2\sqrt{1+2} + 2\sqrt{8+2} + 2\sqrt{27+2} + \sqrt{64+2}\right] \;\approx\; 15.04861925.$

6. $\int_1^5 \sqrt{x^3+1}\,dx \;\approx\; T_4$

$\qquad\qquad = \left(\dfrac{5-1}{4}\right)\left(\dfrac{1}{2}\right)\left[\sqrt{1^3+1} + 2\sqrt{2^3+1} + 2\sqrt{3^3+1} + 2\sqrt{4^3+1} + \sqrt{5^3+1}\right]$

$\qquad\qquad = (1)\left(\dfrac{1}{2}\right)\left[\sqrt{2} + 2\sqrt{9} + 2\sqrt{28} + 2\sqrt{65} + \sqrt{126}\right] \;\approx\; 22.67335323.$

7. $\int_1^5 \dfrac{1}{x}\,dx \;\approx\; T_8$

$\qquad\qquad = \left(\dfrac{5-1}{8}\right)\left(\dfrac{1}{2}\right)\left[\left(\dfrac{2}{2}\right) + 2\left(\dfrac{2}{3}\right) + 2\left(\dfrac{2}{4}\right) + 2\left(\dfrac{2}{5}\right) + 2\left(\dfrac{2}{6}\right) + 2\left(\dfrac{2}{7}\right) + \left(\dfrac{2}{8}\right)\right]$

$\qquad\qquad \approx\; 1.628968254.$

8. $\int_1^3 \dfrac{1}{x}\,dx \;\approx\; T_6$

$\qquad\qquad = \left(\dfrac{3-1}{6}\right)\dfrac{1}{2}\left[\left(\dfrac{3}{3}\right) + 2\left(\dfrac{3}{4}\right) + 2\left(\dfrac{3}{5}\right) + 2\left(\dfrac{3}{6}\right) + 2\left(\dfrac{3}{7}\right) + 2\left(\dfrac{3}{8}\right) + \left(\dfrac{3}{9}\right)\right]$

$\qquad\qquad \approx\; 1.106746032.$

9. $\int_0^8 \sqrt{1+x^3}\,dx \;\approx\; T_4$

$\qquad\qquad = \left(\dfrac{8-0}{4}\right)\left(\dfrac{1}{2}\right)\left[\sqrt{1} + 2\sqrt{1+2^3} + 2\sqrt{1+4^3} + 2\sqrt{1+6^3} + \sqrt{1+8^3}\right]$

$\qquad\qquad \approx\; 75.23585852.$

10. $\int_0^3 \sqrt{x^4 + 1} \, dx \approx T_6$

$$= \left(\frac{3-0}{6}\right)\left(\frac{1}{2}\right)\left[\sqrt{1} + 2\sqrt{\left(\frac{1}{2}\right)^4 + 1} + \cdots + 2\sqrt{\left(\frac{5}{2}\right)^4 + 1} + \sqrt{\left(\frac{6}{2}\right)^4 + 1}\right]$$

$$\approx 10.19374853.$$

11. $\int_1^5 \frac{1}{x} \, dx \approx S_8$

$$= \left(\frac{5-1}{8}\right)\left(\frac{1}{3}\right)\left[\left(\frac{2}{2}\right) + 4\left(\frac{2}{3}\right) + 2\left(\frac{2}{4}\right) + 4\left(\frac{2}{5}\right) + 2\left(\frac{2}{6}\right) + 4\left(\frac{2}{7}\right)\right.$$

$$\left. + 2\left(\frac{2}{8}\right) + 4\left(\frac{2}{9}\right) + \left(\frac{2}{10}\right)\right]$$

$$\approx 1.610846561.$$

12. $\int_1^4 \frac{1}{x} \, dx \approx S_6$

$$= \left(\frac{4-1}{6}\right)\left(\frac{1}{3}\right)\left[\left(\frac{2}{2}\right) + 4\left(\frac{2}{3}\right) + 2\left(\frac{2}{4}\right) + 4\left(\frac{2}{5}\right) + 2\left(\frac{2}{6}\right) + 4\left(\frac{2}{7}\right) + \left(\frac{2}{8}\right)\right]$$

$$\approx 1.387698413.$$

13. $\int_1^5 \sqrt{x^3 + 1} \, dx \approx S_4$

$$= \left(\frac{5-1}{4}\right)\left(\frac{1}{3}\right)\left[\sqrt{1^3 + 1} + 4\sqrt{2^3 + 1} + 2\sqrt{3^3 + 1} + 4\sqrt{4^3 + 1} + \sqrt{5^3 + 1}\right]$$

$$\approx 22.49040732.$$

14. $\int_0^3 \sqrt{x^4 + 1} \, dx \approx S_6$

$$= \left(\frac{3-0}{6}\right)\frac{1}{3}\left[\sqrt{1} + 4\sqrt{\left(\frac{1}{2}\right)^4 + 1} + 2\sqrt{\left(\frac{2}{2}\right)^4 + 1} + 4\sqrt{\left(\frac{3}{2}\right)^4 + 1}\right.$$

$$\left. + 2\sqrt{\left(\frac{4}{2}\right)^4 + 1} + 4\sqrt{\left(\frac{5}{2}\right)^4 + 1} + \sqrt{\left(\frac{6}{2}\right)^4 + 1}\right]$$

$$\approx 10.069994.$$

15. $\displaystyle\int_0^2 \sin(x^2)\,dx \approx T_{10}$

$$= \left(\frac{2-0}{10}\right)\left(\frac{1}{2}\right)\left[\sin(0) + 2\sin\left(\frac{1}{5}\right)^2 + \cdots + 2\sin\left(\frac{9}{5}\right)^2 + \sin\left(\frac{10}{5}\right)\right]$$

$$\approx 0.795924733.$$

16. $\displaystyle\int_0^1 \frac{1}{\sqrt{x^3+1}}\,dx \approx T_{10}$

$$= \left(\frac{1-0}{10}\right)\left(\frac{1}{2}\right)\left[\frac{1}{\sqrt{0^3+1}} + 2\frac{1}{\sqrt{\left(\frac{1}{10}\right)^3+1}} + \cdots\right.$$

$$\left. +2\frac{1}{\sqrt{\left(\frac{9}{10}\right)^3+1}} + \frac{1}{\sqrt{\left(\frac{10}{10}\right)^3+1}}\right]$$

$$\approx 0.909161659.$$

17. $\displaystyle\int_0^1 \frac{1}{1+x^2}\,dx \approx T_{10} = \left(\frac{1-0}{10}\right)\left(\frac{1}{2}\right)\left[\frac{1}{1+0^2} + \frac{2}{1+(.1)^2} + \cdots + \frac{2}{1+(.9)^2} + \frac{1}{1+(1.0)^2}\right]$

$$\approx 0.784981497.$$

18. $\displaystyle\int_0^3 \cos\sqrt{x}\,dx \approx T_{10} = \left(\frac{3-0}{10}\right)\left(\frac{1}{2}\right)\left[\cos(0) + 2\cos\sqrt{\frac{3}{10}} + \cdots + 2\cos\sqrt{\frac{27}{10}} + \cos\sqrt{\frac{30}{10}}\right]$

$$\approx 1.099660524.$$

19. $\displaystyle\int_0^2 \frac{1}{\sqrt{1+x^3}}\,dx \approx S_{10} = \left(\frac{2-0}{10}\right)\left(\frac{1}{3}\right)\left[\frac{1}{\sqrt{1}} + \frac{4}{\sqrt{1+\left(\frac{1}{5}\right)^3}} + \frac{2}{\sqrt{1+\left(\frac{2}{5}\right)^3}} + \cdots\right.$

$$\left. +\frac{4}{\sqrt{1+\left(\frac{9}{5}\right)^3}} + \frac{1}{\sqrt{1+2^3}}\right] \approx 1.402206274.$$

20. $\displaystyle\int_0^1 \sqrt{1+x^4}\,dx \approx S_{10}$

$$= \left(\frac{1-0}{10}\right)\left(\frac{1}{3}\right)\left[\sqrt{1+0^4} + 4\sqrt{1+\left(\frac{1}{10}\right)^4} + 2\sqrt{1+\left(\frac{2}{10}\right)^4} + \cdots\right.$$

$$\left. +4\sqrt{1+\left(\frac{9}{10}\right)^4} + \sqrt{1+\left(\frac{10}{10}\right)^4}\right] \approx 1.089429384.$$

21. $\displaystyle\int_1^2 \frac{1}{x}\,dx \approx S_{10} = \left(\frac{2-1}{10}\right)\left(\frac{1}{3}\right)\left[\frac{1}{1} + \frac{4}{1.1} + \frac{2}{1.2} + \cdots + \frac{4}{1.9} + \frac{1}{2}\right]$

$$\approx 0.693150231.$$

22. $\int_0^4 \sin\sqrt{x}\, dx \approx S_{10}$

$$= \left(\frac{4-0}{10}\right)\left(\frac{1}{3}\right)\left[\sin 0 + 4\sin\sqrt{0.4} + 2\sin\sqrt{0.8} + \cdots + 4\sin\sqrt{3.6} + \sin\sqrt{4}\right]$$

$$\approx 3.462402876.$$

23. $\int_0^1 \frac{4}{1+x^2}\, dx \approx M_{10} = 3.142425985 \approx T_{10} = 3.139925989$

$$M_{100} \approx 3.141600987$$

$$T_{100} \approx 3.141575987.$$

24. Since $f(x) = \frac{1}{x}$ is concave up and positive between $x = 1$ and $x = 5$, then by **Theorem 10**,

$$M_8 \le \int_1^5 \frac{1}{x}\, dx \le T_8$$

$$1.599844394 \le \int_1^5 \le 1.628968254.$$

25. Since $f(x) = \sqrt{x^3+1}$ is positive and $f''(x) = \frac{3}{4}\frac{x}{(x^3+1)^{3/2}}[x^3+1] = \frac{3}{4}\left(\frac{x}{\sqrt{x^3+1}}\right)$ which is positive for $0 < x < 8$, then $y = f(x)$ is concave up between 0 and 8. So by **Theorem 10**,

$$M_4 \le \int_0^8 f(x)\, dx \le T_4,$$

so

$$72.9558 \le \int_0^8 \le 75.2359$$

26. Assume that T approximates the "actual" total rainfall. There are 4 intervals over 8 hours.

$$T_4 = \left(\frac{8-0}{4}\right)\left(\frac{\text{hr}}{1}\right)\left(\frac{1}{2}\right)[(0.2) + 2(0.6) + 2(0.5) + 2(1.2) + (0.3)]\left(\frac{\text{in}}{\text{hr}}\right) = 5.1 \text{ in.}$$

27. Since no measurement is shown in Fig 6.12 to the left of the line measuring 12 ft, we will use 0 ft as $f(x_0)$ and 13 ft as $f(x_5)$. Then the area is approximately

$$T_5 = (5 \text{ ft})\left(\frac{1}{2}\right)[0 + 2(12) + 2(15) + 2(15) + 2(20) + 2(18) + 13]\,(\text{ft})$$

$$= \frac{5}{2}[173]\,(\text{ft}^2) = 432.5 \text{ ft}^2$$

[Note: if we use 0 for $f(x_0)$ and 13 for $f(x_6)$, then we get $T_6 = \left(\frac{5}{2}\right)[0+2(12)+\cdots+2(18)+13] = 432.5$.]

28. We want to estimate the average temperature,

$$\frac{1}{24-0}\int_0^{24} T(h)\, dh \approx \frac{1}{24}S_8 = \left(\frac{1}{24\,\text{hr}}\right)\left(\frac{24-0}{8}\right)\left(\frac{\text{hr}}{1}\right)\left(\frac{1}{3}\right)$$

$$[(19) + 4(16) + 2(18) + 4(26) + 2(31) + 4(31) + 2(27) + 4(23) + (19)]\,(^\circ C)$$

$$= \frac{1}{24}[574]\,(^\circ C) = 23.91\overline{6}\,(^\circ C).$$

29. We will use trapezoids to estimate the number of autos in each subinterval. Note that the 5 subintervals are not equal in length.

$$A_1 = [(9-5)(\text{hrs})]\left(\frac{1}{2}\right)\left[(10+100)\left(\frac{\text{auto}}{\text{hr}}\right)\right] = 220 \text{ autos}$$

$$A_2 = [4 \text{ hrs}]\left(\frac{1}{2}\right)\left[(100+50)\left(\frac{\text{auto}}{\text{hr}}\right)\right] = 300 \text{ autos}$$

$$A_3 = [2 \text{ hrs}]\left(\frac{1}{2}\right)\left[(50+60)\left(\frac{\text{auto}}{\text{hr}}\right)\right] = 110 \text{ autos}$$

$$A_4 = [2 \text{ hrs}]\left(\frac{1}{2}\right)\left[(60+110)\left(\frac{\text{auto}}{\text{hr}}\right)\right] = 170 \text{ autos}$$

$$A_5 = [6 \text{ hrs}]\left(\frac{1}{2}\right)\left[(110+20)\left(\frac{\text{auto}}{\text{hr}}\right)\right] = 390 \text{ autos}$$

$$A_6 = [(11-5)(\text{hrs})]\left(\frac{1}{2}\right)\left[(20+10)\left(\frac{\text{auto}}{\text{hr}}\right)\right] = 90 \text{ autos}$$

$$\sum A_j = 1280 \text{ autos.}$$

30. We want to estimate $\int_a^b f(x)\,dx$. If $\{s_0, s_1, \cdots, s_{2n}\}$ are the $2n+1$ points used for S_{2n}; $\{m_1, m_2, \cdots, m_n\}$ are the n points used for M_n; and $\{t_0, t_1, \cdots, t_n\}$ are the $n+1$ points used for T_n; then $s_0 = t_0$, $s_1 = m_1$, $s_2 = t_1$, $s_3 = m_2, \cdots, s_{2n-1} = m_n$, and $s_{2n} = t_n$. That is, the odd-numbered s_j are midpoints, and the even-numbered s_j are the endpoints, for the partition of n equal subintervals. Note that the subintervals for S_{2n} are $\frac{b-a}{2n}$ in length.

$$
\begin{aligned}
\frac{2}{3}M_n + \frac{1}{3}T_n &= \frac{2}{3}\cdot\left(\frac{b-a}{n}\right)[f(m_1)+f(m_2)+\cdots+f(m_n)] \\
&\quad + \frac{1}{3}\cdot\left(\frac{b-a}{n}\right)\frac{1}{2}[f(t_0)+2f(t_1)+\cdots+2f(t_{n-1})+f(t_n)] \\
&= \left(\frac{b-a}{n}\right)\left(\frac{1}{2}\right)\frac{1}{3}\Big\{4[f(s_1)+f(s_3)+\cdots+f(s_{2n-1})] \\
&\quad + \left(\frac{2}{1}\right)\left(\frac{1}{2}\right)[f(s_0)+2f(s_2)+\cdots+2f(s_{2n-2})+f(s_{2n})]\Big\} \\
&= \left(\frac{b-a}{2n}\right)\left(\frac{1}{3}\right)\Big\{[4f(s_1)+\cdots+4f(s_{2n-1})] \\
&\quad + [f(s_0)+2f(s_2)+\cdots+2f(s_{2n-2})+f(s_{2n})]\Big\} \\
&= \left(\frac{b-a}{2n}\right)\left(\frac{1}{3}\right)\Big\{f(s_0)+4f(s_1)+2f(s_2)+\cdots+4f(s_{2n-1})+F(s_{2n})\Big\} \\
&= S_{2n}.
\end{aligned}
$$

31. Graph the antiderivative with the range $[1,5]\times[0,2]$. Take $F(1)=0$, $k=50$, and $n=6$. So using the programs set **FA=0, K=50**, and **N=6**. Typical graph will look like the following.

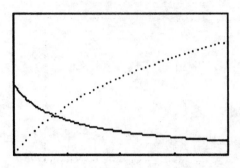

Exercise 31

32. Graph the antiderivative with the range $[-\pi, \pi] \times [0, 3.5]$. Take $F(-\pi) = 0$, $k = 50$, and $n = 6$. So using the programs set **FA=0**, **K=50**, and **N=6**. Typical graph will look like the following.

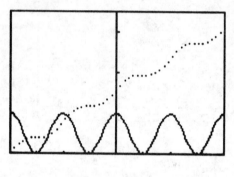

Exercise 32

33. Graph the antiderivative with the range $[-4, 4] \times [0, 3]$. Take $F(-4) = 0$, $k = 50$, and $n = 6$. So using the programs set **FA=0**, **K=50**, and **N=6**. Typical graph will look like the following.

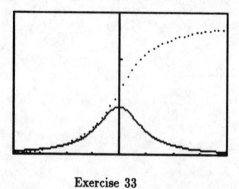

Exercise 33

34. Graph the antiderivative with the range $[0, 8] \times [0, 35]$. Take $F(0) = 0$, $k = 50$, and $n = 6$. So using the programs set **FA=0**, **K=50**, and **N=6**. Typical graph will look like the following.

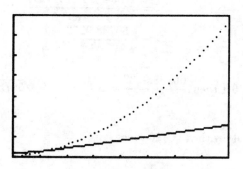

Exercise 34

35.   a) Graph with the **range** $[1, 365] \times [-10, 100]$.

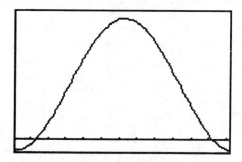

Exercise 35

Hottest months: June, July, and August.

Coldest months: January and December.

Now the minimum is -6.884 and the maximum is 92.983 so

$$-6.884 < \frac{1}{364} \int_1^{365} T(t)\, dt < 92.983.$$

Therefore the approximate average value is $\dfrac{92.983 - 6.884}{2} = 43.0495$.

b) Now $T'(t) = \dfrac{102\pi}{365} \cos\left(\dfrac{2\pi}{365}(t - 95.2)\right)$, and $T''(t) = \dfrac{-204\pi^2}{133225} \sin\left(\dfrac{2\pi}{365}(t - 95.2)\right)$.

So the maximum of $|T''|$ occurs where $\sin\left(\dfrac{2\pi}{365}(t - 95.2)\right) = 1$ or when

$$\begin{aligned}
\frac{2\pi}{365}(t - 95.2) &= \frac{\pi}{2} \\
t - 95.2 &= \frac{365}{4} \\
t &= 186.45.
\end{aligned}$$

Therefore the maximum of $|T''| = \dfrac{204\pi^2}{133225} = 0.01511277$. The error for the Trapezoidal Rule is then

$$|\text{error}| \quad \leq \quad \frac{(b-a)^3 M}{12 n^2}$$

483

$$= \frac{(364)^2(0.01511277)}{12n^2}$$

$$= \frac{60738.9233}{n^2}.$$

So if we require $|\text{error}| < 0.1$ then $\dfrac{60738.9233}{n^2} < 0.1$. So $n > \sqrt{607389.233} = 779.352$. Thus take $n = 780$.

c) Using the Trapezoidal Rule with $n = 780$ gives

$$\int_1^{365} T(t)\, dt \approx 15702.91.$$

So the average value is approximately $\dfrac{15702.91}{364} = 43.14$.

36. a) Table:

e	P
0	6.2832
0.1	6.2674
0.2	6.2199
0.3	6.1393
0.4	6.0238
0.5	5.8698
0.6	5.6723
0.7	5.4226
0.8	5.1054
0.9	4.6868
1.0	4.0000

b) Graph the pairs (e, P) using the range $[0, 1] \times [4, 6.5]$.

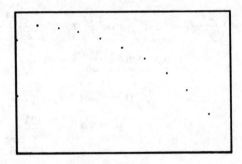

Exercise 36 b)

c) The values of $r = 0.8$ and $s = 0.85$ produce a fairly good fit. Graph is given below with the same range as in part b).

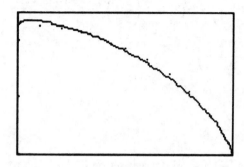

Exercise 36. c)

Review Exercises—Chapter 6

1. $f(x) = 3x + 1$, $x \in [0, 3]$, $n = 6$.

Note that f is increasing on $[0, 3]$ and $\Delta x = \frac{3-0}{6} = \frac{1}{2}$. Therefore, the partition is $\left\{ 0, \frac{1}{2}, 1, \frac{3}{2}, 2, \frac{5}{2}, 3 \right\}$.

a) $\underline{S}_6 = \frac{1}{2} \left(f(0) + f\left(\frac{1}{2}\right) + f(1) + f\left(\frac{3}{2}\right) + f(2) + f\left(\frac{5}{2}\right) \right)$

$= \frac{1}{2} \left(1 + \frac{5}{2} + 4 + \frac{11}{2} + 7 + \frac{17}{2} \right)$

$= \frac{57}{4}.$

b) $\overline{S}_6 = \frac{1}{2} \left(f\left(\frac{1}{2}\right) + f(1) + f\left(\frac{3}{2}\right) + f(2) + f\left(\frac{5}{2}\right) + f(3) \right)$

$= \frac{1}{2} \left(\frac{5}{2} + 4 + \frac{11}{2} + 7 + \frac{17}{2} + 10 \right)$

$= \frac{75}{4}.$

2. $f(x) = \frac{1}{x}$, $x \in [1, 3]$, $n = 4$.

Note that f is decreasing on $[1, 3]$ and that $\Delta x = \frac{3-1}{4} = \frac{1}{2}$. Therefore, the partition is $\left\{ 1, \frac{3}{2}, 2, \frac{5}{2}, 3 \right\}$.

a) $\underline{S}_4 = \frac{1}{2} \left(f\left(\frac{3}{2}\right) + f(2) + f\left(\frac{5}{2}\right) + f(3) \right)$

$= \frac{1}{2} \left(\frac{2}{3} + \frac{1}{2} + \frac{2}{5} + \frac{1}{3} \right)$

$= \frac{19}{20}.$

b) $\overline{S}_4 \;=\; \dfrac{1}{2}\left(f(1) + f\left(\dfrac{3}{2}\right) + f(2) + f\left(\dfrac{5}{2}\right)\right)$

$\qquad = \dfrac{1}{2}\left(1 + \dfrac{2}{3} + \dfrac{1}{2} + \dfrac{2}{5}\right)$

$\qquad = \dfrac{77}{60}.$

3. $f(x) = \dfrac{1}{1 + x^2}$, $x \in [-1, 1]$, $n = 6$.

Note that f is increasing on $[-1, 0]$ and decreasing on $[0, 1]$. We have $\Delta x = \dfrac{1-(-1)}{6} = \dfrac{1}{3}$. The partition is

$$
\begin{aligned}
x_i &= -1 + i(\Delta x) \\
&= -1 + \dfrac{i}{3}.
\end{aligned}
$$

That is, the partition is $\left\{-1, -\dfrac{2}{3}, -\dfrac{1}{3}, 0, \dfrac{1}{3}, \dfrac{2}{3}, 1\right\}$.

a) $\underline{S}_6 \;=\; \Delta x(f(x_0) + f(x_1) + f(x_2) + f(x_4) + f(x_5) + f(x_6))$

$\qquad = \dfrac{1}{3}\left(\dfrac{1}{2} + \dfrac{9}{13} + \dfrac{9}{10} + \dfrac{9}{10} + \dfrac{9}{13} + \dfrac{1}{2}\right)$

$\qquad = \dfrac{272}{195}.$

b) $\overline{S}_6 \;=\; \Delta x(f(x_1) + f(x_2) + f(x_3) + f(x_3) + f(x_4) + f(x_5))$

$\qquad = \dfrac{1}{3}\left(\dfrac{9}{13} + \dfrac{9}{10} + 1 + 1 + \dfrac{9}{10} + \dfrac{9}{13}\right)$

$\qquad = \dfrac{337}{195}.$

4. $f(x) = \sin \pi x$, $x \in [0, 1]$, $n = 6$.

Note that f is increasing on $\left[0, \dfrac{1}{2}\right]$ and decreasing on $\left[\dfrac{1}{2}, 1\right]$. We have $\Delta x = \dfrac{1-0}{6} = \dfrac{1}{6}$. The partition is

$$
\begin{aligned}
x_i &= 0 + i(\Delta x) \\
&= \dfrac{i}{6}.
\end{aligned}
$$

In other words, the partition is $\left\{0, \dfrac{1}{6}, \dfrac{1}{3}, \dfrac{1}{2}, \dfrac{2}{3}, \dfrac{5}{6}, 0\right\}$.

a) $\underline{S}_6 \;=\; \Delta x(f(x_0) + f(x_1) + f(x_2) + f(x_4) + f(x_5) + f(x_6))$

$\qquad = \dfrac{1}{6}\left(0 + \dfrac{1}{2} + \dfrac{\sqrt{3}}{2} + \dfrac{\sqrt{3}}{2} + \dfrac{1}{2} + 0\right)$

$\qquad = \dfrac{1 + \sqrt{3}}{6}.$

b) $\overline{S}_6 = \Delta x(f(x_1) + f(x_2) + f(x_3) + f(x_3) + f(x_4) + f(x_5))$

$$= \frac{1}{6}\left(\frac{1}{2} + \frac{\sqrt{3}}{2} + 1 + 1 + \frac{\sqrt{3}}{2} + \frac{1}{2}\right)$$

$$= \frac{3 + \sqrt{3}}{6}.$$

5. $f(x) = \cos \pi x$, $x \in \left[-\frac{1}{2}, \frac{1}{2}\right]$, $n = 4$.

Note that f is increasing on $\left[-\frac{1}{2}, 0\right]$ and decreasing on $\left[0, \frac{1}{2}\right]$. We have $\Delta x = \frac{\frac{1}{2}-(-\frac{1}{2})}{4} = \frac{1}{4}$. The partition is

$$x_i = -\frac{1}{2} + i(\Delta x)$$
$$= -\frac{1}{2} + \frac{i}{4}.$$

In other words, the partition is $\left\{-\frac{1}{2}, -\frac{1}{4}, 0, \frac{1}{4}, \frac{1}{2}\right\}$.

a) $\underline{S}_4 = \Delta x(f(x_0) + f(x_1) + f(x_3) + f(x_4))$.

$$= \frac{1}{4}\left(0 + \frac{\sqrt{2}}{2} + \frac{\sqrt{2}}{2} + 0\right)$$

$$= \frac{\sqrt{2}}{4}.$$

b) $\overline{S}_4 = \Delta x(f(x_1) + f(x_2) + f(x_2) + f(x_3))$

$$= \frac{1}{4}\left(\frac{\sqrt{2}}{2} + 1 + 1 + \frac{\sqrt{2}}{2}\right)$$

$$= \frac{1}{2} + \frac{\sqrt{2}}{4}.$$

6. $f(x) = \sec x$, $x \in \left[-\frac{\pi}{3}, \frac{\pi}{3}\right]$, $n = 4$.

Note that f is decreasing on $\left[-\frac{\pi}{3}, 0\right]$ and increasing on $\left[0, \frac{\pi}{3}\right]$. We have $\Delta x = \frac{\frac{\pi}{3}-(-\frac{\pi}{3})}{4} = \frac{\pi}{6}$. The partition is

$$x_i = -\frac{\pi}{3} + i(\Delta x)$$
$$= -\frac{\pi}{3} + i\frac{\pi}{6}.$$

In other words, the partition is $\left\{-\frac{\pi}{3}, -\frac{\pi}{6}, 0, \frac{\pi}{6}, \frac{\pi}{3}\right\}$.

a) $\underline{S}_4 = \frac{\pi}{6}\left(f(x_1) + f(x_2) + f(x_2) + f(x_3)\right)$

$= \frac{\pi}{6}\left(\frac{2}{\sqrt{3}} + 1 + 1 + \frac{2}{\sqrt{3}}\right)$

$= \pi\left(\frac{1}{3} + \frac{2\sqrt{3}}{9}\right).$

b) $\overline{S}_4 = \frac{\pi}{6}\left(f(x_0) + f(x_1) + f(x_3) + f(x_4)\right)$

$= \frac{\pi}{6}\left(2 + \frac{2}{\sqrt{3}} + \frac{2}{\sqrt{3}} + 2\right)$

$= \pi\left(\frac{2}{3} + \frac{2\sqrt{3}}{9}\right).$

7. $f(x) = 3 - x$, $x \in [0,2]$, $n = 100$.

Note that f is decreasing on $[0,2]$. Also, we have $\Delta x = \frac{2-0}{100} = \frac{1}{50}$. The partition is

$$x_i = 0 + i(\Delta x)$$
$$= \frac{i}{50}.$$

In other words, the partition is $\left\{0, \frac{1}{50}, \frac{1}{25}, \ldots, \frac{49}{25}, \frac{99}{50}, 2\right\}$.

a) $\underline{S}_{100} = \Delta x\left(\sum_{i=1}^{100} f(x_i)\right)$

$= \frac{1}{50}\left(\sum_{i=1}^{100}(3 - x_i)\right)$

$= \frac{1}{50}\left(\sum_{i=1}^{100}\left(3 - \frac{i}{50}\right)\right)$

$= \frac{1}{50}\left(300 - \frac{1}{50}\sum_{i=1}^{100} i\right)$

$= 6 - \frac{1}{50^2}\left(\frac{100(101)}{2}\right)$

$= 6 - \frac{101}{50} = 3\frac{49}{50} = 3.98.$

b) $\overline{S}_{100} = \Delta x\left(\sum_{i=0}^{99} f(x_i)\right)$

$= \frac{1}{50}\left(\sum_{i=0}^{99}(3 - x_i)\right)$

$= \frac{1}{50}\left(\sum_{i=0}^{99}\left(3 - \frac{i}{50}\right)\right)$

$= \frac{1}{50}\left(300 - \frac{1}{50}\sum_{i=0}^{99} i\right)$

$= 6 - \frac{1}{50^2}\left(\frac{99(100)}{2}\right)$

$= 6 - \frac{99}{50} = 4\frac{1}{50} = 4.02.$

8. $f(x) = x^2 + 1$, $x \in [-1,1]$, $n = 200$.

Note that f is decreasing on $[-1,0]$ and increasing on $[0,1]$. We have $\Delta x = \frac{1-(-1)}{200} = \frac{1}{100}$. The partition is

$$x_i = -1 + \frac{i}{100}.$$

In other words, the partition is $\left\{-1, \frac{-99}{100}, \ldots, \frac{99}{100}, 1\right\}$.

a) $\underline{S}_{200} = \Delta x(f(x_1) + \ldots + f(x_{100}) + f(x_{100}) + \ldots + f(x_{199}))$. Using the fact that $f(-x) = f(x)$, we calculate

$$
\begin{aligned}
\underline{S}_{200} &= 2\Delta x \sum_{i=1}^{100} f(x_i) \\
&= 2\left(\frac{1}{100}\right) \sum_{i=1}^{100} (x_i^2 + 1) \\
&= \frac{1}{50} \sum_{i=1}^{100} \left(\left(-1 + \frac{i}{100}\right)^2 + 1\right) \\
&= \frac{1}{50} \sum_{i=1}^{100} \left(2 - \frac{i}{50} + \frac{i^2}{10000}\right) \\
&= 4 - \frac{1}{50^2} \sum_{i=1}^{100} i + \frac{1}{500,000} \sum_{i=1}^{100} i^2 \\
&= 4 - \frac{1}{50^2}\left(\frac{100(101)}{2}\right) + \frac{1}{500,000}\frac{100(101)(201)}{6} \\
&= \frac{26567}{10000} = 2.6567.
\end{aligned}
$$

b) $\overline{S}_{200} = \Delta x(f(x_0) + \ldots + f(x_{99}) + f(x_{101}) + \ldots + f(x_{200}))$. Using the fact that $f(-x) = f(x)$, we calculate

$$
\begin{aligned}
\overline{S}_{200} &= 2\Delta x \left(\sum_{i=0}^{99} f(x_i)\right) \\
&= 2\left(\frac{1}{100}\right) \sum_{i=0}^{99} (x_i^2 + 1) \\
&= \frac{1}{50} \sum_{i=0}^{99} \left(\left(-1 + \frac{i}{100}\right)^2 + 1\right) \\
&= \frac{1}{50} \sum_{i=0}^{99} \left(2 - \frac{i}{50} + \frac{i^2}{10000}\right) \\
&= 4 - \frac{1}{50^2} \sum_{i=0}^{99} i + \frac{1}{500,000} \sum_{i=0}^{99} i^2 \\
&= 4 - \frac{1}{50^2}\left(\frac{99(100)}{2}\right) + \frac{1}{500,000}\frac{99(100)(199)}{6} \\
&= \frac{26767}{10000} = 2.6767.
\end{aligned}
$$

9. $f(x) = \cos x$, $x \in \left[-\frac{\pi}{2}, \pi\right]$, $n = 6$

Note that f is positive on $\left[-\frac{\pi}{2}, \frac{\pi}{2}\right]$ and negative on $\left[\frac{\pi}{2}, \pi\right]$. Also, f is increasing on $\left[-\frac{\pi}{2}, 0\right]$ and decreasing on $[0, \pi]$. We have $\Delta x = \frac{\pi - \left(\frac{\pi}{2}\right)}{6} = \frac{\pi}{4}$. So the partition is

$$
x_i = -\frac{\pi}{2} + i\frac{\pi}{4},
$$

which is $\left\{-\frac{\pi}{2}, -\frac{\pi}{4}, 0, \frac{\pi}{4}, \frac{\pi}{2}, \frac{3\pi}{4}, \pi\right\}$.

i	x_i	$f(x_i)$
0	$-\pi/2$	0
1	$-\pi/4$	$\sqrt{2}/2$
2	0	1
3	$\pi/4$	$\sqrt{2}/2$
4	$\pi/2$	0
5	$3\pi/4$	$-\sqrt{2}/2$
6	π	-1

The smallest Riemann sum for $n = 6$ is

$$\Delta x(f(x_0) + f(x_1) + f(x_3) + f(x_4) + f(x_5) + f(x_6)) = \frac{\pi}{4}\left(0 + \frac{\sqrt{2}}{2} + \frac{\sqrt{2}}{2} + 0 - \frac{\sqrt{2}}{2} - 1\right)$$
$$= \frac{\pi}{8}\left(\sqrt{2} - 2\right).$$

The largest Riemann sum for $n = 6$ is

$$\Delta x(f(x_1) + f(x_2) + f(x_2) + f(x_3) + f(x_4) + f(x_5)) = \frac{\pi}{4}\left(\frac{\sqrt{2}}{2} + 1 + 1 + \frac{\sqrt{2}}{2} + 0 - \frac{\sqrt{2}}{2}\right)$$
$$= \frac{\pi}{8}\left(\sqrt{2} + 4\right).$$

10. $f(x) = 9 - x^2$, $x \in [0, 3]$, $n = 6$.

Note that f is decreasing and positive on $[0, 3]$. We have $\Delta x = \frac{3-0}{6} = \frac{1}{2}$. So the partition is

$$x_i = 0 + i(\Delta x) = \frac{i}{2}.$$

In other words, the partition is $\left\{0, \frac{1}{2}, 1, \frac{3}{2}, 2, \frac{5}{2}, 3\right\}$.

The smallest Riemann sum is

$$\begin{aligned}
\underline{S}_6 &= \frac{1}{2}\left(f(x_1) + f(x_2) + f(x_3) + f(x_4) + f(x_5) + f(x_6)\right) \\
&= \frac{1}{2}\left(\frac{35}{4} + 8 + \frac{27}{4} + 5 + \frac{11}{4} + 0\right) \\
&= \frac{125}{8}.
\end{aligned}$$

The largest Riemann sum is

$$\begin{aligned}
\overline{S}_6 &= \frac{1}{2}\left(f(x_0) + f(x_1) + f(x_2) + f(x_3) + f(x_4) + f(x_5)\right) \\
&= \frac{1}{2}\left(9 + \frac{35}{4} + 8 + \frac{27}{4} + 5 + \frac{11}{4}\right) \\
&= \frac{161}{8}.
\end{aligned}$$

11. $f(x) = x^2 + 1$, $x \in [-1, 2]$, $n = 6$

Note that f is positive on $[-1, 2]$. Moreover, it is decreasing on $[-1, 0]$ and increasing on $[0, 2]$. We have $\Delta x = \frac{2-(-1)}{6} = \frac{1}{2}$. So the partition is

$$
\begin{aligned}
x_i &= -1 + i(\Delta x) \\
&= -1 + \frac{i}{2}.
\end{aligned}
$$

In other words, the partition is $\left\{-1, -\frac{1}{2}, 0, \frac{1}{2}, 1, \frac{3}{2}, 2\right\}$.

The smallest Riemann sum is

$$
\begin{aligned}
\underline{S}_6 &= \frac{1}{2}\left(f(x_1) + f(x_2) + f(x_2) + f(x_3) + f(x_4) + f(x_5)\right) \\
&= \frac{1}{2}\left(\frac{5}{4} + 1 + 1 + \frac{5}{4} + 2 + \frac{13}{4}\right) \\
&= \frac{39}{8}.
\end{aligned}
$$

The largest Riemann sum is

$$
\begin{aligned}
\overline{S}_6 &= \frac{1}{2}\left(f(x_0) + f(x_1) + f(x_3) + f(x_4) + f(x_5) + f(x_6)\right) \\
&= \frac{1}{2}\left(2 + \frac{5}{4} + \frac{5}{4} + 2 + \frac{13}{4} + 5\right) \\
&= \frac{59}{8}.
\end{aligned}
$$

12. $f(x) = 2x - x^2$, $x \in [-1, 3]$, $n = 4$.

Note that f is increasing on $[-1, 1]$ and decreasing on $[1, 3]$. Moreover, it is positive on $[0, 2]$ and negative otherwise. We have $\Delta x = \frac{3-(-1)}{4} = 1$. So the partition is

$$
x_i = -1 + i(\Delta x) = -1 + i.
$$

In other words, the partition is $\{-1, 0, 1, 2, 3\}$.

The smallest Riemann sum is

$$
\begin{aligned}
1(f(x_0) + f(x_1) + f(x_3) + f(x_4)) &= -3 + 0 + 0 + -3 \\
&= -6.
\end{aligned}
$$

The largest Riemann sum is

$$
\begin{aligned}
1(f(x_1) + f(x_2) + f(x_2) + f(x_3)) &= 0 + 1 + 1 + 0 \\
&= 2.
\end{aligned}
$$

13. $y = 2x - 2$, $[2, 4]$

Let $\Delta x = \frac{4-2}{n} = \frac{2}{n}$, $x_i = 2 + \left(\frac{2}{n}\right)i$

f is increasing on $[2, 4]$.

$$\begin{aligned}
\underline{S}_n &= \Delta x \sum_{i=0}^{n-1} f(x_i) = \frac{2}{n} \sum_{i=0}^{n-1} 4 + \left(\frac{4}{n}\right) i - 2 \\
&= \frac{2}{n} \sum_{i=0}^{n-1} 2 + \frac{4i}{n} = \frac{2}{n}(2n) + \frac{8}{n^2} \sum_{i=0}^{n-1} i \\
&= 4 + \frac{8}{n^2} \frac{(n-1)(n)}{2}. \\
\lim_{n \to \infty} \underline{S}_n &= 4 + 4 = 8.
\end{aligned}$$

14.
$$\begin{aligned}
\overline{S}_n &= \Delta x \sum_{i=1}^{n} f(x_i) = \frac{2}{n} \sum_{i=1}^{n} 4 + \left(\frac{4}{n}\right) i - 2 = \frac{2}{n} \sum_{i=1}^{n} 2 + \frac{4i}{n} = \frac{2}{n}(2n) + \frac{8}{n^2} \sum_{i=1}^{n} i \\
&= 4 + \frac{8}{n^2} \frac{(n)(n+1)}{2}
\end{aligned}$$

$$\lim_{n \to \infty} \overline{S}_n = 4 + 4 = 8.$$

15. $\displaystyle \int_0^9 \sqrt{x}\, dx = \frac{2}{3} x^{3/2} \Big]_0^9 = \frac{2}{3} \cdot 27 = 18$

16. $\displaystyle \int_0^3 3\sqrt{x+1}\, dx = 3 \cdot \frac{2}{3}(x+1)^{3/2} \Big]_0^3 = 2(8-1) = 14$

17. $\displaystyle \int_0^1 (x^{2/3} - x^{1/2})\, dx = \frac{3}{5} x^{5/3} - \frac{2}{3} x^{3/2} \Big]_0^1 = \frac{3}{5} - \frac{2}{3} = -\frac{1}{15}$

18. $\displaystyle \int_1^2 \frac{1-t}{t^3}\, dt = -\frac{1}{2t^2} + \frac{1}{t} \Big]_1^2 = -\frac{1}{8} + \frac{1}{2} + \frac{1}{2} - 1 = -\frac{1}{8}$

19. $\displaystyle \int_0^1 x^3(x+1)\, dx = \frac{x^5}{5} + \frac{x^4}{4} \Big]_0^1 = \frac{1}{5} + \frac{1}{4} = \frac{9}{20}$

20. $\displaystyle \int_0^\pi \sin^2 x\, dx = \frac{1}{2} \int_0^\pi (1 - \cos 2x)\, dx = \frac{1}{2}\left[x - \frac{1}{2} \sin 2x \right]_0^\pi = \frac{1}{2}(\pi) = \frac{\pi}{2}$

21. $\displaystyle \int_0^1 (3x + 2)\, dx = \frac{3}{2} x^2 + 2x \Big]_0^1 = \frac{3}{2} + 2 = \frac{7}{2}$

22. $\displaystyle \int_1^3 (7 + 3x)\, dx = 7x + \frac{3}{2} x^2 \Big]_1^3 = 7(3-1) + \frac{3}{2}(9-1) = 14 + 12 = 26$

23. $\displaystyle \int_{-3}^5 (x^2 + 2)\, dx = \frac{x^3}{3} + 2x \Big]_{-3}^5 = \frac{1}{3}(125 + 27) + 2(5 + 3) = \frac{152}{3} + 16 \approx 66.667$

24. $\displaystyle \int_1^5 (3x^2 - 2)\, dx = x^3 - 2x \Big]_1^5 = 125 - 1 - 10 + 2 = 116$

25. $\int_1^4 \sqrt{x}\,dx = \frac{2}{3}\,x^{3/2}\Big]_1^4 = \frac{2}{3}(8-1) = \frac{14}{3}$

26. $\int_0^8 \sqrt{x+1}\,dx = \frac{2}{3}(x+1)^{3/2}\Big]_0^8 = \frac{2}{3}(27-1) = \frac{52}{3}$

27. $\int_1^4 \left(\sqrt{x} - \frac{1}{\sqrt{x}}\right)dx = \frac{2}{3}\,x^{3/2} - 2\sqrt{x}\Big]_1^4 = \frac{2}{3}(8-1) - 2(2-1) = \frac{14}{3} - 2 = \frac{8}{3}$

28. $\int_1^8 (x^{1/3} - 1)\,dx = \frac{3}{4}\,x^{4/3} - x\Big]_1^8 = \frac{3}{4}(16-1) - (8-1) = \frac{45}{4} - 7 = \frac{17}{4}$

29. $\int_0^2 (x+7)(2x+2)\,dx = \int_0^2 (2x^2 + 16x + 14)\,dx = \frac{2}{3}\,x^3 + 8x^2 + 14x\Big]_0^2 = \frac{16}{3} + 32 + 28 = \frac{196}{3} \approx 65.333$

30. $\int_1^4 (x^2 - 1)(x + 2)\,dx = \int_1^4 (x^3 + 2x^2 - x - 2)\,dx = \frac{x^4}{4} + \frac{2}{3}\,x^3 - \frac{x^2}{2} - 2x\Big]_1^4$

$$= \frac{1}{4}(256 - 1) + \frac{2}{3}(64 - 1) - \frac{1}{2}(16 - 1) - 2(4 - 1)$$

$$= \frac{225}{4} + 36 = \frac{369}{4} = 92.25$$

31. $\int_0^{\pi/4} \sin x\,dx = -\cos x\Big]_0^{\pi/4} = -\frac{\sqrt{2}}{2} + 1 = \frac{2 - \sqrt{2}}{2} \approx 0.29289$

32. $\int_{-\pi/4}^{\pi/4} \cos x\,dx = 2\int_0^{\pi/4} \cos x\,dx = 2\,\sin x\Big]_0^{\pi/4} = \sqrt{2}$

33. $\int_0^{\pi/4} \sin(2x)\,dx = -\frac{1}{2}\cos 2x\Big]_0^{\pi/4} = -\frac{1}{2}(0 - 1) = \frac{1}{2}$

34. $\int_{\pi/4}^{\pi/2} \cos(\pi - 2x)\,dx = -\frac{1}{2}\sin(\pi - 2x)\Big]_{\pi/4}^{\pi/2} = -\frac{1}{2}(0 - 1) = \frac{1}{2}$

35. $\int_1^2 (x + 4)^{10}\,dx = \frac{1}{11}(x + 4)^{11}\Big]_1^2 = \frac{1}{11}(6^{11} - 5^{11}) \approx 28.543 \times 10^6$

36. $\int_0^4 (\sqrt{a} + \sqrt{x})^2\,dx = \int_0^4 (a + 2\sqrt{a}\sqrt{x} + x)\,dx = ax + \frac{4}{3}\sqrt{a}\,x^{3/2} + \frac{x^2}{2}\Big]_0^4$

$$= 4a + \frac{4}{3}\sqrt{a}(8) + 8 = 4a + \frac{32}{3}\sqrt{a} + 8$$

37. $\int_2^4 \frac{t^2 - 2t}{5}\,dt = \frac{t^3}{15} - \frac{t^2}{5}\Big]_2^4 = \frac{1}{15}(64 - 8) - \frac{1}{5}(16 - 4) = \frac{56}{15} - \frac{12}{5} = \frac{4}{3}$

38. $\int_1^2 \frac{1 + t}{t^3}\,dt = -\frac{1}{2t^2} - \frac{1}{t}\Big]_1^2 = -\frac{1}{2}\left(\frac{1}{4} - 1\right) - \frac{1}{2} + 1 = \frac{3}{8} + \frac{1}{2} = \frac{7}{8}$

39. $\displaystyle\int_0^3 \frac{dt}{(t+1)^2} = -\left.\frac{1}{t+1}\right]_0^3 = -\left(\frac{1}{4}-1\right) = \frac{3}{4}$

40. $\displaystyle\int_3^8 \frac{1}{\sqrt{x+1}}\,dx = \left.2\sqrt{x+1}\right]_3^8 = 2(3-2) = 2$

41. $\displaystyle\int_0^{\pi/4} \frac{1}{\cos^2 x}\,dx = \int_0^{\pi/4} \sec^2 x\,dx = \left.\tan x\right]_0^{\pi/4} = 1$

42. $\displaystyle\int_0^{\pi/3} \frac{\sin x}{\cos^2 x}\,dx = \left.\frac{1}{\cos x}\right]_0^{\pi/3} = 2-1 = 1$

43. $\displaystyle\int_0^1 \left(x-\sqrt{x}\right)^2\,dx = \int_0^1 \left(x^2 - 2x\sqrt{x} + x\right)\,dx = \left.\frac{x^3}{3} - \frac{4}{5}x^{5/2} + \frac{x^2}{2}\right]_0^1 = \frac{1}{3} - \frac{4}{5} + \frac{1}{2} = \frac{1}{30}$

44. $\displaystyle\int_1^4 x^{1/2}(1+x^{3/2})^5\,dx$

    Let $u = 1 + x^{3/2}$, $du = \frac{3}{2}x^{1/2}\,dx$
    if $x = 1$, $u = 2$; if $x = 4$, $u = 9$.

    $$\frac{2}{3}\int_2^9 u^5\,du = \left.\frac{1}{9}u^6\right]_2^9 = 59{,}049 - \frac{64}{9} \approx 59{,}041.89$$

45. $\displaystyle\int_1^8 t\left(\sqrt[3]{t} - 2t\right)\,dt = \int_1^8 (t^{4/3} - 2t^2)\,dt = \left.\frac{3}{7}t^{7/3} - \frac{2}{3}t^3\right]_1^8 = \frac{3}{7}(128-1) - \frac{2}{3}(512-1) = -\frac{6011}{21} \approx -286.238$

46. $\displaystyle\int_1^4 \frac{x^2+2x+4}{\sqrt{x}}\,dx = \int_1^4 \left(x^{3/2} + 2\sqrt{x} + \frac{4}{\sqrt{x}}\right)\,dx = \left.\frac{2}{5}x^{5/2} + \frac{4}{3}x^{3/2} + 8\sqrt{x}\right]_1^4$

    $\qquad = \frac{2}{5}(32-1) + \frac{4}{3}(8-1) + 8(2-1) = \frac{446}{15} \approx 29.733$

47. $\displaystyle\int_1^2 \frac{1}{(1-2x)^3}\,dx$

    Let $u = 1-2x$, $du = -2\,dx$
    if $x = 1$, $u = -1$; if $x = 2$, $u = -3$.

    $$-\frac{1}{2}\int_{-1}^{-3} \frac{du}{u^3} = \left.\frac{1}{4}\cdot\frac{1}{u^2}\right]_{-1}^{-3} = \frac{1}{4}\left(\frac{1}{9}\cdot\frac{1}{1}\right) = -\frac{2}{9} \approx -0.2223$$

48. $\displaystyle\int_{-\pi/4}^{\pi/4} \sec^2 t\,dt = \left.\tan t\right]_{-\pi/4}^{\pi/4} = 1-(-1) = 2$

49. $\displaystyle\int_0^1 \frac{x^3+8}{x+2}\,dx = \int_0^1 (x^2 - 2x + 4)\,dx = \left.\frac{x^3}{3} - x^2 + 4x\right]_0^1 = \frac{1}{3} - 1 + 4 = \frac{10}{3}$

50. $\displaystyle\int_0^2 x^2\sqrt{x^3+1}\,dx = \frac{1}{3}\int_0^2 3x^2\sqrt{x^3+1}\,dx = \left.\frac{2}{9}(x^3+1)^{3/2}\right]_0^2 = \frac{2}{9}(27-1) = \frac{52}{9} \approx 5.7778$

494

51. $\displaystyle\int_0^{\pi/2} \sec\left(\frac{u}{2}\right)\tan\left(\frac{u}{2}\right)\,du = 2\sec\left(\frac{u}{2}\right)\Big]_0^{\pi/2} = 2(\sqrt{2}-1) \approx 0.8284$

52. $\displaystyle\int_{-\pi}^{\pi} \sec^2\left(\frac{v}{3}\right)\,dv = 3\tan\left(\frac{v}{3}\right)\Big]_{-\pi}^{\pi} = 3\left(\sqrt{3}+\sqrt{3}\right) = 6\sqrt{3} \approx 10.392$

53. $\displaystyle\int_{-1}^{3}|x-x^2|\,dx = -\int_{-1}^{0}(x-x^2)\,dx + \int_0^1 (x-x^2)\,dx - \int_1^3 (x-x^2)\,dx = -\left(\frac{x^2}{2}-\frac{x^3}{3}\right)\Big]_{-1}^{0} + \frac{x^2}{2} - \frac{x^3}{3}\Big]_0^1 -$

$\left(\frac{x^2}{2}-\frac{x^3}{3}\right)\Big]_1^3 = -\left(0-\frac{1}{2}-0-\frac{1}{3}\right) + \left(\frac{1}{2}-\frac{1}{3}\right) - \left(\frac{9}{2}-\frac{1}{2}-9+\frac{1}{3}\right) = \frac{5}{6}+\frac{1}{6}+\frac{14}{3} = \frac{17}{3} \approx 5.6667$

54. $\displaystyle\int_0^{2\pi}|\sin 2x|\,dx \;=\; 2\left[\int_0^{\pi/2}\sin 2x\,dx - \int_{\pi/2}^{\pi}\sin 2x\,dx\right] = 2\left\{\left[-\frac{1}{2}\cos 2x\right]_0^{\pi/2} + \frac{1}{2}\cos 2x\Big]_{\pi/2}^{\pi}\right\}$

$$= 2\left[\left(\frac{1}{2}+\frac{1}{2}\right)+\left(\frac{1}{2}+\frac{1}{2}\right)\right] = 4$$

55. $\displaystyle\int_0^1 x\sqrt{x+1}\,dx$

Let $u = x+1$, $du = dx$, $x = u-1$
if $x = 0$, $u = 1$; if $x = 1$, $u = 2$.

$$\int_1^2 (u-1)\sqrt{u}\,du \;=\; \int_1^2 (u^{3/2}-\sqrt{u})\,du = \frac{2}{5}u^{5/2} - \frac{2}{3}u^{3/2}\Big]_1^2 = \frac{2}{5}(4\sqrt{2}-1) - \frac{2}{3}(2\sqrt{2}-1)$$

$$= \frac{4}{15}\sqrt{2} + \frac{4}{15} = \frac{4}{15}(\sqrt{2}+1) \approx 0.6438$$

56. $\displaystyle\int_0^2 x^3\sqrt{x^2+2}\,dx$

Let $u = x^2+2$. Then $du = 2x\,dx$ and $x^2 = u-2$. Also, if $x = 0$, $u = 2$; if $x = 2$, $u = 6$. We have

$$\int_0^2 x^3\sqrt{x^2+2}\,dx \;=\; \int_2^6 (u-2)\sqrt{u}\left(\frac{1}{2}\right)\,du$$

$$= \frac{1}{2}\int_2^6 u^{3/2} - 2u^{1/2}\,du$$

$$= \frac{u^{5/2}}{5} - \frac{2u^{3/2}}{3}\Big]_2^6$$

$$= \frac{16}{5}\sqrt{6} + \frac{8}{15}\sqrt{2} \approx 8.59261.$$

57. $\displaystyle 2\sum_{i=1}^{m} i - 2\sum_{i=1}^{10} i = 1782.$

$$\frac{2(m)(m+1)}{2} - 2\frac{(10)(11)}{2} \;=\; 1782$$
$$m(m+1) \;=\; 1782 + 110 \;=\; 1892$$

$$m^2 + m - 1892 = 0$$

$$m = \frac{-1 \pm \sqrt{1 + 7568}}{2} = \frac{-1 \pm 87}{2} = \frac{86}{2} = 43$$

$$n = 2m = 86$$

58. Number of push-ups $= \left(\sum_{i=1}^{30} i \right) + 30 \cdot 9 = \frac{30 \cdot 31}{2} + 270 = 735.$

59. $f(x) = \sqrt{x - 1},\ 1 \le x \le 5$

$$A = \int_1^5 \sqrt{x - 1}\, dx = \frac{2}{3}(x - 1)^{3/2} \Big]_1^5 = \frac{2}{3}(8) = \frac{16}{3}$$

60. $f(x) = \dfrac{1}{\sqrt{x - 1}},\ 2 \le x \le 10.$

$$A = \int_2^{10} \frac{1}{\sqrt{x - 1}}\, dx = 2\sqrt{x - 1} \Big]_2^{10} = 2(3 - 1) = 4$$

61. $f(x) = (x^2 + 2)^2,\ 0 \le x \le 1$

$$A = \int_0^1 (x^2 + 2)^2\, dx = \int_0^1 (x^4 + 4x^2 + 4)\, dx = \frac{x^5}{5} + \frac{4x^3}{3} + 4x \Big]_0^1 = \frac{1}{5} + \frac{4}{3} + 4 = \frac{83}{15} \approx 5.533$$

62. $f(x) = (x - 1)(x + 2),\ 0 \le x \le 2$

$$A = -\int_0^1 (x^2 + x - 2)\, dx + \int_1^2 (x^2 + x - 2)\, dx = -\left(\frac{x^3}{3} + \frac{x^2}{2} - 2x \right) \Big]_0^1 + \frac{x^3}{3} + \frac{x^2}{2} - 2x \Big]_1^2$$

$$= -\left(\frac{1}{3} + \frac{1}{2} - 2 \right) + \left(\frac{8}{3} + 2 - 4 \right) - \left(\frac{1}{3} + \frac{1}{2} - 2 \right) = 3$$

63. $f(x) = 4x - x^2,\ 4 \le x \le 5$

$$A = -\int_4^5 (4x - x^2)\, dx = -\left(2x^2 - \frac{x^3}{3} \right) \Big]_4^5 = -\left(50 - \frac{125}{3} \right) + \left(32 - \frac{64}{3} \right) = -18 + \frac{61}{3} = \frac{7}{3}$$

64. $f(x) = \sin \pi x,\ 0 \le x \le 2$

$$A = \int_0^1 \sin \pi x\, dx - \int_1^2 \sin \pi x\, dx = -\frac{1}{\pi} \cos \pi x \Big]_0^1 + \frac{1}{\pi} \cos \pi x \Big]_1^2 = -\frac{1}{\pi}(-1 - 1) + \frac{1}{\pi}(1 + 1) = \frac{4}{\pi}$$

65. $f(x) = 3 + 2x - x^2,\ 1 \le x \le 4$

$f \ge 0$ on $[1, 3]$; $f \le 0$ on $[3, 4]$.

$$A = \int_1^3 (3 + 2x - x^2)\, dx - \int_3^4 (3 + 2x - x^2)\, dx = 3x + x^2 - \frac{x^3}{3} \Big]_1^3 - \left(3x + x^2 - \frac{x^3}{3} \right) \Big]_3^4$$

$$= 9 + 9 - 9 - \left(3 + 1 - \frac{1}{3} \right) - \left(12 + 16 - \frac{64}{3} \right) + (9 + 9 - 9) = \frac{23}{3} \approx 7.667$$

66. $f(x) = x^3 - x,\ 1 \le x \le 3$

$$A = \int_1^3 (x^3 - x)\, dx = \frac{x^4}{4} - \frac{x^2}{2} \Big]_1^3 = \frac{81}{4} - \frac{1}{4} - \frac{9}{2} + \frac{1}{2} = 16$$

67. $f(x) = x^2 - x - 2, \; 2 \le x \le 4$

$$A = \int_2^4 (x^2 - x - 2)\, dx = \frac{x^3}{3} - \frac{x^2}{2} - 2x \Big]_2^4 = \frac{64}{3} - \frac{8}{3} - \frac{16}{2} + \frac{4}{2} - 8 + 4 = \frac{26}{3} \approx 8.667$$

68. $f(x) = \sin(2x), \; 0 \le x \le \dfrac{\pi}{4}$

$$A = \int_0^{\pi/4} \sin(2x)\, dx = -\frac{1}{2}\cos(2x) \Big]_0^{\pi/4} = -\frac{1}{2}(0 - 1) = \frac{1}{2}$$

69. $f(x) = \sqrt{1 - x}, \; -1 \le x \le 1$

$$A = \int_{-1}^1 \sqrt{1 - x}\, dx = -\frac{2}{3}(1 - x)^{3/2} \Big]_{-1}^1 = -\frac{2}{3}(0 - 2^{3/2}) = \frac{4\sqrt{2}}{3} \approx 1.8856$$

70. $f(x) = \dfrac{1}{(x - 2)^2}, \; -2 \le x \le 1$

$$A = \int_{-2}^1 \frac{1}{(x-2)^2}\, dx = -\frac{1}{x-2} \Big]_{-2}^1 = -\left(\frac{1}{-1} - \frac{1}{-4} \right) = \frac{3}{4}$$

71. $f(x) = (ax^2 + 3)^2, \; 0 \le x \le 3.$

$$A = \int_0^3 (ax^2 + 3)^2\, dx = \int_0^3 (a^2 x^4 + 6ax^2 + 9)\, dx$$

$$= \frac{1}{5}a^2 x^5 + 2ax^3 + 9x \Big]_0^3 = \frac{1}{5}a^2(243) + 2a(27) + 27 = \frac{243}{5}a^2 + 54a + 27$$

72. $f(x) = 9 - x^2, \; -3 \le x \le 3$

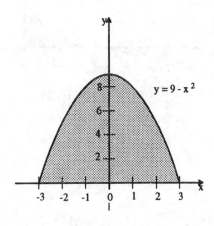

Exercise 72

$$A = \int_{-3}^3 (9 - x^2)\, dx = 2\int_0^3 (9 - x^2)\, dx = 2\left[9x - \frac{x^3}{3} \right]_0^3 = 2(27 - 9) = 36$$

73. $f(x) = x^2 + 2x - 3, \; -3 \le x \le 1$

497

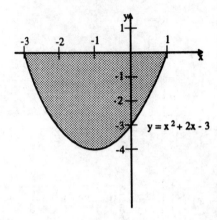

Exercise 73

$$A = -\int_{-3}^{1} (x^2 + 2x - 3)\, dx = -\left[\frac{x^3}{3} + x^2 - 3x\right]_{-3}^{1} = -\left(\frac{1}{3} + 1 - 3\right) + (-9 + 9 + 9) = \frac{32}{3}$$

74. $f(x) = (x-1)^3,\ 0 \leq x \leq 2$

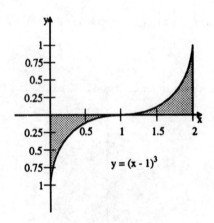

Exercise 74

$$A = -\int_{0}^{1} (x-1)^3\, dx + \int_{1}^{2} (x-1)^3\, dx = -\left[\frac{1}{4}(x-1)^4\right]_{0}^{1} + \left[\frac{1}{4}(x-1)^3\right]_{1}^{2} = -\frac{1}{4}(0-1) + \frac{1}{4}(1-0) = \frac{1}{2}$$

75. $f(x) = 2x - 4,\ 0 \leq x \leq 4$

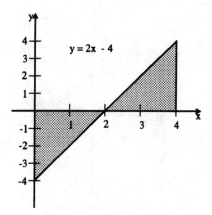

Exercise 75

$$A = -\int_0^2 (2x-4)\,dx + \int_2^4 (2x-4)\,dx = -\left[x^2 - 4x\right]_0^2 + \left[x^2 - 4x\right]_2^4 = -(4-8)+(16-16)-(4-8) = 8$$

76. $f(x) = \sqrt{x} - 2,\ 0 \le x \le 9$

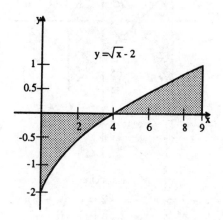

Exercise 76

$$\begin{aligned}
A &= -\int_0^4 \left(\sqrt{x} - 2\right)\,dx + \int_4^9 \left(\sqrt{x} - 2\right)\,dx = -\left[\frac{2}{3}\,x^{3/2} - 2x\right]_0^4 + \left[\frac{2}{3}\,x^{3/2} - 2x\right]_4^9 \\
&= -\left(\frac{16}{3} - 8\right) + (18 - 18) - \left(\frac{16}{3} - 8\right) = \frac{16}{3}
\end{aligned}$$

77. $y = x^3,\ y = 0,\ x = -2,\ x = 2$

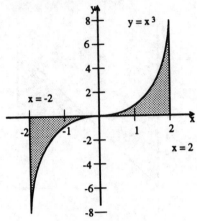

Exercise 77

$$A = -\int_{-2}^{0} x^3\,dx + \int_{0}^{2} x^3\,dx = 2\int_{0}^{2} x^3\,dx = \frac{1}{2}x^4\Big]_{0}^{2} = 8$$

78. $y = 1 - x^2$, $y = -x - 1$, $x = -2$, $x = 1$

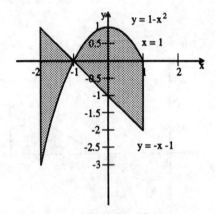

Exercise 78

Find intersections: $1 - x^2 = -x - 1 \Rightarrow x^2 - x - 2 = 0$; $(x - 2)(x + 1) = 0$; $x = -1, 2$
$-x - 1 \geq 1 - x^2$ on $[-2, -1]$; $1 - x^2 \geq -x - 1$ on $[-1, 1]$.

$$
\begin{aligned}
A &= \int_{-2}^{-1}\left[-x - 1 - (1 - x^2)\right]dx + \int_{-1}^{1}\left[1 - x^2 - (-x - 1)\right]dx \\
&= -\frac{x^2}{2} - 2x + \frac{x^3}{3}\Big]_{-2}^{-1} + 2x - \frac{x^3}{3} + \frac{x^2}{2}\Big]_{-1}^{1} \\
&= -\frac{1}{2}(1 - 4) - 2(-1 + 2) + \frac{1}{3}(-1 + 8) + 2(1 + 1) - \frac{1}{3}(1 + 1) + \frac{1}{2}(1 - 1) \\
&= \frac{31}{6} \approx 5.1667
\end{aligned}
$$

79. $y = \sqrt{x}$, $y = -\sqrt{x}$, $x = 0$, $x = 4$

500

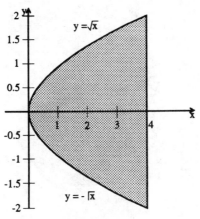

Exercise 79

$$A = \int_0^4 \left[\sqrt{x} - (-\sqrt{x})\right] dx = \frac{4}{3} x^{3/2} \Big]_0^4 = \frac{4}{3}(8) = \frac{32}{3}$$

80. $y = x^2$, $y = 2 - x$, $x = -2$, $x = 1$

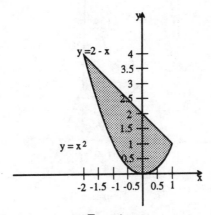

Exercise 80

Find intersections: $x^2 = 2 - x \Rightarrow x^2 + x - 2 = 0$; $(x + 2)(x - 1) = 0$; $x = -2, 1$.

$$A = \int_{-2}^1 (2 - x - x^2) dx = 2x - \frac{x^2}{2} - \frac{x^3}{3} \Big]_{-2}^1 = 2(1 + 2) - \frac{1}{2}(1 - 4) - \frac{1}{3}(1 + 8) = \frac{9}{2}$$

81. $y = x^2 - 4x + 2$, $x + y = 6$

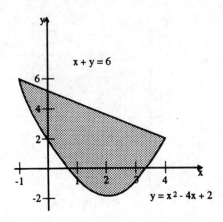

$x + y = 6$

$y = x^2 - 4x + 2$

Exercise 81

Find intersections: $x^2 - 4x + 2 = 6 - x \Rightarrow x^2 - 3x - 4 = 0;\ (x+1)(x-4) = 0;\ x = -1, 4.$

$$A = \int_{-1}^{4} [6 - x - (x^2 - 4x + 2)]\, dx = 4x + \frac{3}{2} x^2 - \frac{x^3}{3} \Big]_{-1}^{4} = 4(4+1) + \frac{3}{2}(16-1) - \frac{1}{3}(64+1) = \frac{125}{6} \approx 20.833$$

82. $y = 2 - 2x - x^2,\ x = -y$

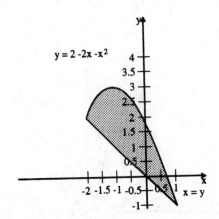

$y = 2 - 2x - x^2$

$x = y$

Exercise 82

Find intersections: $2 - 2x - x^2 = -x \Rightarrow x^2 + x - 2 = 0;\ (x+2)(x-1) = 0;\ x = -2, 1.$

$$A = \int_{-2}^{1} [2 - 2x - x^2 - (-x)]\, dx = 2x - \frac{x^2}{2} - \frac{x^3}{3} \Big]_{-2}^{1} = 2(1+2) - \frac{1}{2}(1-4) - \frac{1}{3}(1+8) = \frac{9}{2}.$$

83. $y = \dfrac{x-2}{\sqrt{x^2 - 4x + 8}},\ x = 0,\ y = 0;\ y = f(x) < 0$ on $[0, 2].$

502

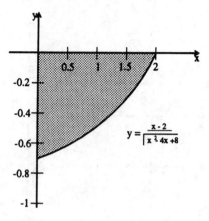

Exercise 83

$$A = \int_0^2 -\frac{(x-2)}{\sqrt{x^2 - 4x + 8}}\, dx$$

Let $u = x^2 - 4x + 8$, $du = (2x - 4)\, dx$
if $x = 0$, $u = 8$; if $x = 2$, $u = 4$.

$$A = \int_8^4 \frac{-du}{2\sqrt{u}} = \left. -\sqrt{u}\,\right]_8^4 = \sqrt{8} - 2 = 2\left(\sqrt{2} - 1\right) \approx 0.8284$$

84. $x - y^2 + 3 = 0$, $x - 2y = 0$

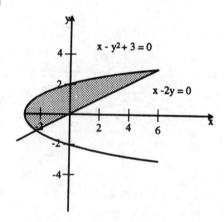

Exercise 84

Find intersections: $y^2 - 3 = 2y \Rightarrow y^2 - 2y - 3 = 0$; $(y - 3)(y + 1) = 0$; $y = -1, 3$.

$$A = \int_{-1}^3 [2y - (y^2 - 3)]\, dy = \left. y^2 - \frac{y^3}{3} + 3y\,\right]_{-1}^3 = 9 - 1 - \frac{1}{3}(27 + 1) + 3(3 + 1) = \frac{32}{3} \approx 10.667$$

85. $x + y^2 = 0$, $x + y + 2 = 0$

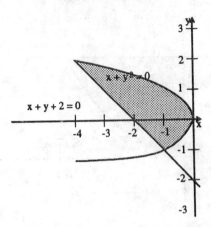

Exercise 85

Find intersections: $-y^2 = -y - 2 \Rightarrow y^2 - y - 2 = 0; (y-2)(y+1) = 0; y = -1, 2.$

$$A = \int_{-1}^{2} [-y^2 - (-y-2)] \, dy = -\frac{y^3}{3} + \frac{y^2}{2} + 2y \Big]_{-1}^{2} = -\frac{1}{3}(8+1) + \frac{1}{2}(4-1) + 2(2+1) = \frac{9}{2}$$

86. $x = 2y^2 - 3, \ x = y^2 + 1$

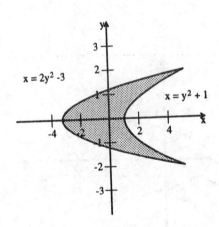

Exercise 86

Find intersections: $2y^2 - 3 = y^2 + 1 \Rightarrow y^2 - 4 = 0 \Rightarrow y = \pm 2.$

$$A = \int_{-2}^{2} [y^2 + 1 - (2y^2 - 3)] \, dy = -\frac{y^3}{3} + 4y \Big]_{-2}^{2} = 2 \left[\frac{-y^3}{3} + 4y \right]_{0}^{2} = 2\left(-\frac{8}{3} + 8 \right) = \frac{32}{3} \approx 10.6667$$

87. $x = y^{2/3}, \ x = y^2$

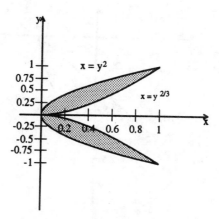

Exercise 87

Find intersections: $y^{2/3} = y^2 \Rightarrow y^{2/3}(1 - y^{4/3}) = 0 \Rightarrow y = 0, -1, 1$

$$A = \int_{-1}^{1} (y^{2/3} - y^2)\, dy = 2\left[\frac{3}{5} y^{5/3} - \frac{y^3}{3}\right]_0^1 = 2\left(\frac{3}{5} - \frac{1}{3}\right) = \frac{8}{15}.$$

88. $y = 4x^2$, $4x + y - 8 = 0$

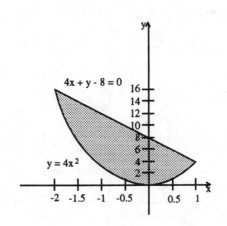

Exercise 88

Find intersections: $4x^2 = -4x + 8 \Rightarrow x^2 + x - 2 = 0$; $(x+2)(x-1) = 0$; $x = -2, 1$.

$$A = \int_{-2}^{1} (8 - 4x - 4x^2)\, dx = 8x - 2x^2 - \frac{4}{3} x^3 \bigg]_{-2}^{1} = 8(1+2) - 2(1-4) - \frac{4}{3}(1+8) = 18$$

89. $y = x^2 + x - 2$, and the line through $(-1, -2)$ and $(1, 0)$.

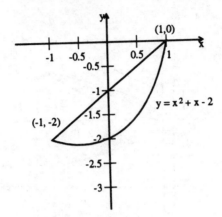

Exercise 89

Slope of line: $\quad m \;=\; \dfrac{0-(-2)}{1-(-1)} \;=\; 1$

$\dfrac{y-0}{x-1} \;=\; 1 \;\Rightarrow\; y \;=\; x-1.$

Find intersections: $x^2+x-2 \;=\; x-1 \;\Rightarrow\; x^2-1 \;=\; 0 \;\Rightarrow\; x \;=\; -1, 1.$

$$A = \int_{-1}^{1}\left[x-1-(x^2+x-2)\right]dx = x - \frac{x^3}{3}\Bigg]_{-1}^{1} = (1+1) - \frac{1}{3}(1+1) = \frac{4}{3}$$

90. $y = 2 - x - x^2$, and the line through $(-1, 2)$ and $(1, 0)$.

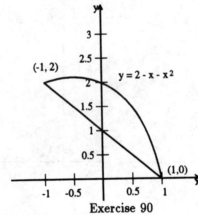

Exercise 90

Slope of line: $\quad m \;=\; \dfrac{0-2}{1-(-1)} \;=\; -1$

$\dfrac{y-0}{x-1} \;=\; -1$

$\qquad\;\Rightarrow\; y \;=\; -x+1.$

Find intersections: $2-x-x^2 \;=\; -x+1 \Rightarrow x^2-1 \;=\; 0 \;\Rightarrow\; x \;=\; -1, 1.$

$$A = \int_{-1}^{1}\left[2-x-x^2-(-x+1)\right]dx = x - \frac{x^3}{3}\Bigg]_{-1}^{1} = (1+1) - \frac{1}{3}(1+1) = \frac{4}{3}$$

91. $3x + 5y = 23, \; 5x - 2y = 28, \; 2x - 7y = -26$

506

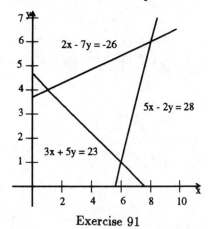

Exercise 91

Find intersections:

$$
\begin{aligned}
2x - 7y &= -26 \\
3x + 5y &= 23
\end{aligned}
\quad \rightarrow \quad
\begin{aligned}
6x - 21y &= -78 \\
6x + 10y &= 46 \\
\hline
-31y &= -124
\end{aligned}
\quad \Rightarrow \quad y = 4,\ x = 1
$$

$$
\begin{aligned}
2x - 7y &= -26 \\
5x - 2y &= 28
\end{aligned}
\quad \rightarrow \quad
\begin{aligned}
10x - 35y &= -130 \\
10x - 4y &= 56 \\
\hline
-31y &= -186
\end{aligned}
\quad \Rightarrow \quad y = 6,\ x = 8
$$

$$
\begin{aligned}
3x + 5y &= 23 \\
5x - 2y &= 28
\end{aligned}
\quad \rightarrow \quad
\begin{aligned}
15x + 25y &= 115 \\
15x - 6y &= 84 \\
\hline
31y &= 31
\end{aligned}
\quad \Rightarrow \quad y = 1,\ x = 6
$$

$$
\begin{aligned}
A &= \int_1^6 \left(\frac{2x+26}{7} - \frac{23-3x}{5} \right) dx + \int_6^8 \left(\frac{2x+26}{7} - \frac{5x-28}{2} \right) dx \\
&= x^2 \left(\frac{1}{7} + \frac{3}{10} \right) + x \left(\frac{26}{7} - \frac{23}{5} \right) \Big]_1^6 + x^2 \left(\frac{1}{7} - \frac{5}{4} \right) + x \left(\frac{26}{7} + 14 \right) \Big]_6^8 \\
&= \frac{31}{70}(36-1) - \frac{31}{35}(6-1) - \frac{31}{28}(64-36) + \frac{124}{7}(8-6) \\
&= \frac{31}{2} - \frac{31}{7} - 31 + \frac{248}{7} = \frac{217}{14} = \frac{31}{2}
\end{aligned}
$$

92. $y = x\sqrt{1-x^2}$ on $[-1, 1]$.

$$
\bar{f} = \frac{1}{1-(-1)} \int_{-1}^1 x\sqrt{1-x^2}\, dx = \frac{1}{2} \left(-\frac{1}{2} \right) \left(\frac{2}{3} \right) (1-x^2)^{3/2} \Big]_{-1}^1 = -\frac{1}{6}(0-0) = 0.
$$

93. $y = \sin x \cos x$ on $\left[0, \frac{\pi}{2} \right]$.

$$
\bar{f} = \frac{1}{\frac{\pi}{2} - 0} \int_0^{\pi/2} \sin x \cos x\, dx = \frac{2}{\pi} \cdot \frac{1}{2} \sin^2 x \Big]_0^{\pi/2} = \frac{1}{\pi}(1-0) = \frac{1}{\pi}
$$

507

94. $y = x^3 \sin x^2$ on $[-\pi, \pi]$.

Since $y = f(x)$ is odd, $\int_{-\pi}^{\pi} f(x)\, dx = 0$. Therefore $\overline{f} = 0$.

95. $y = \dfrac{x}{\sqrt{1 + x^2}}$ on $[0, 2]$.

$$\overline{f} = \frac{1}{2 - 0} \int_0^2 \frac{x}{\sqrt{1 + x^2}}\, dx = \frac{1}{2} \sqrt{1 + x^2}\Big]_0^2 = \frac{1}{2}\left(\sqrt{5} - 1\right) \approx 0.6180$$

96. $y = \sqrt{x} + \sqrt[3]{x}$ on $[0, 1]$.

$$\overline{f} = \frac{1}{1 - 0} \int_0^1 \left(\sqrt{x} + \sqrt[3]{x}\right) dx = \frac{2}{3} x^{3/2} + \frac{3}{4} x^{4/3}\Big]_0^1 = \frac{2}{3} + \frac{3}{4} = \frac{17}{12} \approx 1.4167.$$

97. $y = \sin x$ on $[0, 10\pi]$.

$$\overline{f} = \frac{1}{10\pi - 0} \int_0^{10\pi} \sin x\, dx = -\frac{1}{10\pi} \cos x\Big]_0^{10\pi} = -\frac{1}{10\pi}(1 - 1) = 0$$

98. If $f(x) = \sin x$, $\overline{f} = \dfrac{1}{b - a} \int_a^b \sin x\, dx$.
If $a = 2n\pi$ and $b = 2m\pi$,

$$\overline{f} = \frac{1}{2\pi(m - n)} \int_{2n\pi}^{2m\pi} \sin x\, dx = -\frac{1}{2\pi(m - n)} \cos x\Big]_{2n\pi}^{2m\pi} = -\frac{1}{2\pi(m - n)}(1 - 1) = 0.$$

If $a = x_0 + 2n\pi$ and $b = x_0 + 2m\pi$,

$$\overline{f} = \frac{1}{2\pi(m - n)} \int_{x_0 + 2n\pi}^{x_0 + 2m\pi} \sin x\, dx = -\frac{1}{2\pi(m - n)} \cos x\Big]_{x_0 + 2n\pi}^{x_0 + 2m\pi}$$
$$= -\frac{1}{2\pi(m - n)}[\cos(x_0 + 2m\pi) - \cos(x_0 + 2n\pi)]$$
$$= -\frac{1}{2\pi(m - n)}(\cos x_0 \cos 2m\pi - \sin x_0 \sin 2m\pi - \cos x_0 \cos 2n\pi + \sin x_0 \sin 2n\pi)$$
$$= -\frac{1}{2\pi(m - n)}(\cos x_0 - 0 - \cos x_0 + 0) = 0.$$

99. $f(x) = \sin^2 x$ on $[0, n\pi]$.

$$\overline{f} = \frac{1}{n\pi} \int_0^{n\pi} \sin^2 x\, dx = \frac{1}{2n\pi} \int_0^{n\pi} (1 - \cos 2x)\, dx$$
$$= \frac{1}{2n\pi}\left[x - \frac{1}{2} \sin 2x\right]_0^{n\pi} = \frac{1}{2n\pi}\left(n\pi - \frac{1}{2}(0)\right) = \frac{1}{2}$$

100. $\overline{f} = \dfrac{1}{L - 0} \int_0^L \sin^2 x\, dx = \dfrac{1}{2L} \int_0^L (1 - \cos 2x)\, dx$
$$= \frac{1}{2L}\left[x - \frac{1}{2} \sin 2x\right]_0^L = \frac{1}{2L}\left(L - \frac{1}{2} \sin 2L\right).$$

$$\lim_{L \to \infty} \overline{f} = \frac{1}{2} - \frac{1}{4} \lim_{L \to \infty} \frac{\sin 2L}{L} = \frac{1}{2}. \quad \left(\text{Remember: } \lim_{x \to \infty} \frac{\sin x}{x} = 0 \right)$$

$$\lim_{L \to \infty} \frac{1}{L} \int_0^L \cos^2 x \, dx = \lim_{L \to \infty} \frac{1}{2L} \int_0^L (1 + \cos 2x) \, dx$$

$$= \lim_{L \to \infty} \frac{1}{2L} \left[x + \frac{1}{2} \sin 2x \right]_0^L = \lim_{L \to \infty} \frac{1}{2L} \left(L + \frac{1}{2} \sin 2L \right)$$

$$= \frac{1}{2} + \frac{1}{4} \lim_{L \to \infty} \frac{\sin L}{L} = \frac{1}{2}.$$

101. $F(x) = \int_0^x t^2 \sqrt{1+t} \, dt$ for $x > -1$.

    a) $F(0) = \int_0^0 t^2 \sqrt{1+t} \, dt = 0$

    b) $F'(x) = x^2 \sqrt{1+x}$

    c) $F'(3) = 9\sqrt{4} = 18$

    d) $F(2x) = \int_0^{2x} t^2 \sqrt{1+t} \, dt$

        $F'(2x) = (2x)^2 \sqrt{1+2x} = 4x^2 \sqrt{1+2x}$

102. $F(x) = \int_0^{2x} \frac{1}{1+t^2} \, dt$

    a) $F'(x) = \frac{1}{1+(2x)^2} \frac{d}{dx}(2x) = \frac{2}{1+4x^2}$

    b) $F'(1) = \frac{2}{1+4} = \frac{2}{5}$

    c) $F'(x^2) = \frac{2}{1+4(x^2)^2} = \frac{2}{1+4x^4}$

103. $\frac{d}{dx} \left[\int_a^x f(t) \, dt \right] = f(x)$

    $\int_a^x \frac{d}{dt} [f(t)] \, dt = \int_a^x f'(t) \, dt$

                    $= f(x) - f(a)$.

So the equality does not hold if $f(a) \neq 0$.

104. False; consider the function $f(x) = \sin x$ over the interval $[0, 2\pi]$. We have

$$\int_0^{2\pi} \sin x \, dx = -\cos x \Big]_0^{2\pi} = 0.$$

Nevertheless, $f(x) = \sin x$ is not identically zero on $[0, 2\pi]$.

105. If $a \neq b$, then we know that there exists a number c between a and b such that

$$f(c) = \frac{\int_a^b f(x)\,dx}{b-a} = 0$$

by the Mean Value Theorem for Integrals. So the statement is true if $a \neq b$. If $a = b$, then $\int_a^b f(x)\,dx = 0$ regardless of the value of $f(a)$.

106. If $a \neq b$ and $\int_a^b |f(x)|\,dx = 0$, then $f(x)$ must be identically zero on $[a, b]$. That is, if $f(c) = 0$ for some $c \in [a, b]$, then there exists an interval $[c_1, c_2]$ containing c and a minimum value $m > 0$ for $|f|$ on $[c_1, c_2]$. In this case

$$
\begin{aligned}
0 < m(c_2 - c_1) &\leq \int_{c_1}^{c_2} |f(x)|\,dx \\
&\leq \int_a^b |f(x)|\,dx.
\end{aligned}
$$

107. $\dfrac{d}{dx}\left[\displaystyle\int_{2x}^{0} \sec t\,dt\right] = -\dfrac{d}{dx}\left[\displaystyle\int_{0}^{2x} \sec t\,dt\right] = -\sec(2x)\dfrac{d}{dx}(2x) = -2\sec(2x)$

108. $\dfrac{d}{dx}\left[\displaystyle\int_{x}^{x^2} \sqrt{1+t^2}\,dt\right] = \dfrac{d}{dx}\left[\displaystyle\int_{a}^{x^2} \sqrt{1+t^2}\,dt\right] - \dfrac{d}{dx}\left[\displaystyle\int_{a}^{x} \sqrt{1+t^2}\,dt\right]$

$$= \sqrt{1+(x^2)^2}\,\dfrac{d}{dx}(x^2) - \sqrt{1+x^2} = 2x\sqrt{1+x^4} - \sqrt{1+x^2}$$

109. $\displaystyle\int_a^b [f(x)]^n f'(x)\,dx$; assume that $f'(x)$ is continuous on $[a, b]$. Let $f(x) = u$, $f'(x)\,dx = du$.

$$\int u^n\,du = \frac{1}{n+1}u^{n+1}; \quad \frac{1}{n+1}\left[f(x)^{n+1}\right]_a^b = \frac{1}{n+1}\left[f(b)^{n+1} - f(a)^{n+1}\right] \quad (n \neq -1)$$

110. $\displaystyle\int_1^4 (2x+1)\,dx$

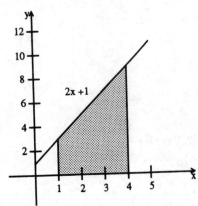

Exercise 110

$$
\begin{aligned}
A &= \int_1^4 (2x+1)\,dx \\
&= \left. x^2 + x \right]_1^4 \\
&= (16-1) + (4-1) \\
&= 18 \\
&= \text{area of square} \\
&\quad + \text{area of triangle} \\
&= 3 \cdot 3 + \frac{3 \cdot 6}{2} \\
&= 9 + 9 = 18.
\end{aligned}
$$

510

111. $\displaystyle\int_{-1}^{1} \sqrt{1 - x^2}\, dx$ = area of semicircle of radius 1 = $\dfrac{\pi}{2}$.

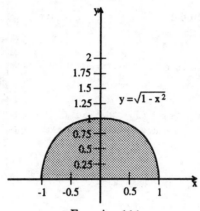

Exercise 111

112. $\displaystyle\int_{0}^{3} \left(1 + \sqrt{9 - x^2}\right) dx$

    =   area of $\frac{1}{4}$ of circle
        of radius 3

    = $\frac{1}{4}(9\pi)$

    = $\frac{9}{4}\pi$
        plus area of rectangle A

        over $[0, 3]$

    = $3 \cdot 1$

    =   3.

Total area = $\frac{9}{4}\pi + 3$

           $\approx$ 10.069.

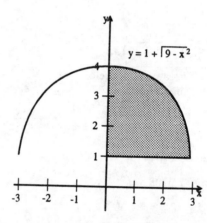

Exercise 112

113. $\displaystyle\int_{-3}^{3} |3 - x|\, dx$

    =   area of triangle of base 6
        and altitude 6.

Area = $\frac{1}{2}(6 \cdot 6)$

      = 18.

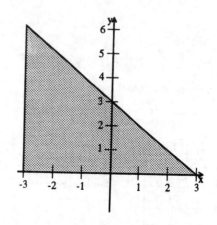

Exercise 113

511

114. $\displaystyle\int_0^4 \sqrt{16-x^2}\,dx$

  = $\frac{1}{4}$ area of circle of radius 4

  = $\frac{1}{4}(\pi)(16)$

  = 4π.

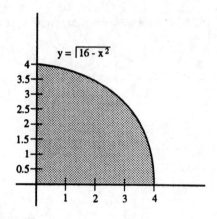

Exercise 114

115. $\displaystyle\int_{-1}^1 (3+\sqrt{1-x^2})\,dx$

  = area of semicircle
     of radius 1
     plus area of rectangle
     (2×3)

  = $\frac{1}{2}\pi + 6$

  ≈ 7.5708.

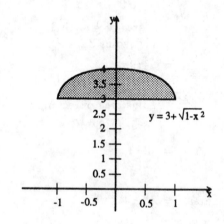

Exercise 115

116. $\displaystyle\int_{-2}^2 \frac{1}{1+x^2}\,dx,\ n=4.$

  $\Delta x = \dfrac{2-(-2)}{4} = 1,\ x_i = -2+i$

i	x_i	$f(x_i)$
0	−2	0.2
1	−1	0.5
2	0	1.0
3	1	0.5
4	2	0.2

$\displaystyle\int_{-2}^2 f(x)\,dx \approx \frac{1}{2}[0.2 + 2(0.5)$

  $+2(1.0) + 2(0.5) + 0.2]$

  = 2.2.

512

117. $\int_0^4 \sqrt{1+x^2}\,dx, \; n=4$

$\Delta x = \dfrac{4-0}{4} = 1, \; x_i = i$

i	x_i	$f(x_i)$
0	0	1.0000
1	1	1.4142
2	2	2.2361
3	3	3.1623
4	4	4.1231

$\int_0^4 f(x)\,dx \approx \tfrac{1}{2}[1.0000 + 2(1.4142)$
$+ 2(2.2361) + 2(3.1623)$
$+ 4.1231]$
$= 9.3742.$

118. $\int_{-1}^2 \dfrac{1}{x+2}\,dx, \; n=6$

$\Delta x = \dfrac{2-(-1)}{6} = 0.5, \; x_i = -1 + 0.5i$

i	x_i	$f(x_i)$
0	-1.0	1.0000
1	-0.5	0.6667
2	0.0	0.5000
3	0.5	0.4000
4	1.0	0.3333
5	1.5	0.2857
6	2.0	0.2500

$\int_{-1}^2 f(x)\,dx \approx \dfrac{0.5}{2}[1.0000 + 2(0.6667)$
$+ 2(0.5000) + 2(0.4000)$
$+ 2(0.3333) + 2(0.2857)$
$+ 0.2500]$
$\approx 1.4054.$

119. $\overline{f} = \dfrac{1}{10-0}\int_0^{10} f(x)\,dx$

i	x_i	$f(x_i)$
0	0	72
1	2	82
2	4	90
3	6	94
4	8	91
5	10	85

$\int_0^{10} f(x)\,dx = \dfrac{\Delta x}{2}[72 + 2(82) + 2(90)$
$+ 2(94) + 2(91) + 85]$
$= \tfrac{2}{2}(871)$
$= 871$
$\overline{f} = \text{average temperature}$
$= \tfrac{1}{10}(871)$
$= 87.1.$

120. $\int_0^{2\pi} \sqrt{|\sin x|}\,dx, \; n=4$

$\Delta x = \dfrac{2\pi - 0}{4} = \dfrac{\pi}{2}, \; x_i = \dfrac{\pi}{4} + \dfrac{\pi}{2}i - \Delta x$

i	x_i	$f(x_i)$
1	$\pi/4$	0.8409
2	$3\pi/4$	0.8409
3	$5\pi/4$	0.8409
4	$7\pi/4$	0.8409

$\int_0^{2\pi} f(x)\,dx \approx \dfrac{\pi}{2}(3.3636)$
$= 1.6818\pi$
≈ 5.2835

121. $\displaystyle\int_0^{2\pi} \frac{\sin x}{x}\,dx$, $n = 4$

$\Delta x = \dfrac{2\pi - 0}{4} = \dfrac{\pi}{2}$, $x_i = \dfrac{\pi}{4} + \dfrac{\pi}{2}i - \Delta x$

i	x_i	$f(x_i)$
1	$\pi/4$	0.9003
2	$3\pi/4$	0.3001
3	$5\pi/4$	-0.1801
4	$7\pi/4$	-0.1286

$$\int_0^{2\pi} f(x)\,dx \approx \frac{\pi}{2}\,(0.9003 + 0.3001$$
$$- 0.1801 - 0.1286)$$
$$= 0.4458\pi$$
$$\approx 1.4007$$

Chapter 7

Applications of the Definite Integral

7.1   Calculating Volumes by Slicing

1. $f(x) = 2x + 1,\ 1 \le x \le 4$

   a) $V = \displaystyle\int_1^4 \pi(2x+1)^2\,dx$

   b) $V = \dfrac{\pi}{6}\,(2x+1)^3\Big]_1^4 = \dfrac{\pi}{6}\,(729-27) = 117\pi$

2. $f(x) = \sqrt{4x-1},\ 1 \le x \le 5$

   a) $V = \displaystyle\int_1^5 \pi(4x-1)\,dx$

   b) $V = \pi(2x^2 - x)\Big]_1^5 = \pi(45-1) = 44\pi$

3. $f(x) = \sin x,\ 0 \le x \le \pi$

   a) $V = \displaystyle\int_0^\pi \pi\sin^2 x\,dx$

   b) $V = \dfrac{\pi}{2}\displaystyle\int_0^\pi (1-\cos 2x)\,dx = \dfrac{\pi}{2}\left(x - \dfrac{\sin 2x}{2}\right)\Big]_0^\pi = \dfrac{\pi}{2}(\pi) = \dfrac{\pi^2}{2}$

4. $f(x) = \sin x\sqrt{\cos x},\ 0 \le x \le \dfrac{\pi}{2}$

   a) $V = \displaystyle\int_0^{\pi/2} \pi\sin^2 x\cos x\,dx$

   b) $V = \dfrac{\pi}{3}\,\sin^3 x\Big]_0^{\pi/2} = \dfrac{\pi}{3}$

5. $f(x) = x(x^3 - 1)^2,\ 0 \le x \le 1$

   a) $V = \displaystyle\int_0^1 \pi x^2(x^3-1)^4\,dx$

b) $V = \dfrac{\pi}{15}\,(x^3-1)^5\Big]_0^1 = \dfrac{\pi}{15}\,(0+1) = \dfrac{\pi}{15}$

6. $f(x) = 2x^2(x^5+1)^5,\ -1 \le x \le 0$

   a) $V = \displaystyle\int_{-1}^0 4\pi x^4(x^5+1)^{10}\,dx$

   b) $V = \dfrac{4\pi}{55}\,(x^5+1)^{11}\Big]_{-1}^0 = \dfrac{4\pi}{55}\,(1-0) = \dfrac{4\pi}{55}$

7. $f(x) = \sqrt{4-x^2},\ 1 \le x \le 2$

   a) $V = \displaystyle\int_1^2 \pi(4-x^2)\,dx$

   b) $V = \pi\left(4x - \dfrac{x^3}{3}\right)\Big]_1^2 = \pi\left[8 - \dfrac{8}{3} - \left(4 - \dfrac{1}{3}\right)\right] = \dfrac{5\pi}{3}$

8. $f(x) = \dfrac{\sqrt{1+x}}{x^{3/2}},\ 1 \le x \le 2$

   a) $V = \displaystyle\int_1^2 \pi\dfrac{1+x}{x^3}\,dx$

   b) $V = \pi\left(-\dfrac{1}{2x^2} - \dfrac{1}{x}\right)\Big]_1^2 = \pi\left[-\dfrac{1}{8} - \dfrac{1}{2} + \left(\dfrac{1}{2}+1\right)\right] = \dfrac{7\pi}{8}$

9. $f(x) = \tan x,\ 0 \le x \le \dfrac{\pi}{4}$

$$V = \int_0^{\pi/4} \pi\tan^2 x\,dx = \pi\int_0^{\pi/4}(\sec^2 x - 1)\,dx = \pi(\tan x - x)\Big]_0^{\pi/4} = \pi\left(1 - \dfrac{\pi}{4}\right) = \dfrac{\pi}{4}\,(4-\pi)$$

10. $f(x) = \sin x,\ 0 \le x \le \pi$

$$V = \int_0^{\pi} \pi\sin^2 x\,dx = \dfrac{\pi}{2}\int_0^{\pi}(1-\cos 2x)\,dx = \dfrac{\pi}{2}\left(x - \dfrac{\sin 2x}{2}\right)\Big]_0^{\pi} = \dfrac{\pi}{2}\,(\pi) = \dfrac{\pi^2}{2}$$

11. $f(x) = \sec x,\ 0 \le x \le \dfrac{\pi}{4}$

$$V = \int_0^{\pi/4} \pi\sec^2 x\,dx = \pi\tan x\Big]_0^{\pi/4} = \pi$$

12. $f(x) = |x-1|,\ 0 \le x \le 3$

$$V = \int_0^3 \pi(x-1)^2\,dx = \dfrac{\pi}{3}\,(x-1)^3\Big]_0^3 = \dfrac{\pi}{3}\,(8+1) = 3\pi$$

13. $f(x) = |3x-6|,\ 0 \le x \le 5$

$$V = \int_0^5 \pi(3x-6)^2\,dx = \dfrac{\pi}{9}\,(3x-6)^3\Big]_0^5 = \dfrac{\pi}{9}\,(729+216) = 105\pi$$

14. $f(x) = \sqrt{\cos x}$, $g(x) = 1$, $0 \le x \le \dfrac{\pi}{2}$

    a) $V = \displaystyle\int_0^{\pi/2} \pi(1^2 - \cos x)\,dx$

    b) $V = \pi(x - \sin x)\Big]_0^{\pi/2} = \pi\left(\dfrac{\pi}{2} - 1\right)$

15. $f(x) = -x^2 + 6$, $g(x) = 2$

    a) Find intersections: $-x^2 + 6 = 2 \Rightarrow x^2 = 4 \Rightarrow x = \pm 2$.

$$V = \int_{-2}^{2} \pi[(-x^2 + 6)^2 - 4]\,dx = 2\pi \int_0^2 [(-x^2 + 6)^2 - 4]\,dx$$

    b) $V = 2\pi\left[\dfrac{x^5}{5} - \dfrac{12}{3}x^3 + 36x - 4x\right]_0^2 = 2\pi\left(\dfrac{32}{5} - 32 + 72 - 8\right) = \dfrac{384}{5}\pi \approx 241.274$

16. $f(x) = x^2$, $g(x) = x^3$

    a) Find intersections: $x^2 = x^3 \Rightarrow x(x^2 - 1) = 0 \Rightarrow x = 0, 1$.

$$V = \int_0^1 \pi[x^4 - x^6]\,dx$$

    b) $V = \pi\left[\dfrac{x^5}{5} - \dfrac{x^7}{7}\right]_0^1 = \pi\left(\dfrac{1}{5} - \dfrac{1}{7}\right) = \dfrac{2\pi}{35}$

17. $f(x) = \dfrac{x^2}{4}$, $g(x) = x$

    a) Find intersections: $\dfrac{x^2}{4} = x \Rightarrow x\left(\dfrac{x}{4} - 1\right) = 0 \Rightarrow x = 0, 4$

$$V = \int_0^4 \pi\left[x^2 - \dfrac{x^4}{16}\right]dx$$

    b) $V = \pi\left[\dfrac{x^3}{3} - \dfrac{x^5}{80}\right]_0^4 = \pi\left(\dfrac{64}{3} - \dfrac{1024}{80}\right) = \pi\left(\dfrac{64}{3} - \dfrac{64}{5}\right) = \dfrac{128\pi}{15}$

18. $f(x) = \frac{1}{x}$, $g(x) = \sqrt{x}$, $1 \le x \le 4$

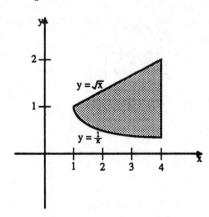

$$\begin{aligned} V &= \int_1^4 \pi\left(x - \dfrac{1}{x^2}\right)dx \\[2mm] &= \pi\left[\dfrac{x^2}{2} + \dfrac{1}{x}\right]_1^4 \\[2mm] &= \pi\left[8 + \dfrac{1}{4} - \dfrac{1}{2} - 1\right] \\[2mm] &= \dfrac{27\pi}{4} \end{aligned}$$

Exercise 18

19. $f(x) = \sin x$, $g(x) = \cos x$, $0 \le x \le \dfrac{\pi}{2}$

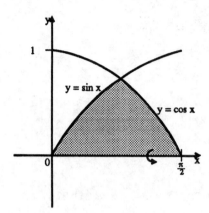

Exercise 19

$$
\begin{aligned}
V &= \int_0^{\pi/4} \pi(\cos^2 x - \sin^2 x)\, dx \\
&\quad + \int_{\pi/4}^{\pi/2} \pi(\sin^2 x - \cos^2 x)\, dx \\
&= 2\pi \int_0^{\pi/4} (\cos^2 x - \sin^2 x)\, dx \\
&= 2\pi \int_0^{\pi/4} \cos 2x\, dx \\
&= \pi(\sin 2x)\Big]_0^{\pi/4} \\
&= \pi
\end{aligned}
$$

20. $x + y = 4$, $x = 0$, $0 \le y \le 4$

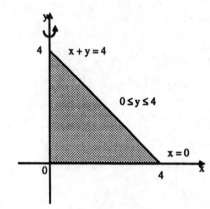

Exercise 20

a) $V = \displaystyle\int_0^4 \pi(4-y)^2\, dy$

b) $V = -\dfrac{\pi}{3}(4-y)^3\Big]_0^4$

$ = -\dfrac{\pi}{3}(0 - 64)$

$ = \dfrac{64\pi}{3}$

21. $y = x^2$, $y = 0$, $0 \le x \le 2$

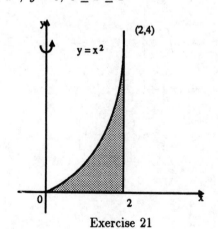

Exercise 21

a) $x = 2 \Rightarrow y = 2^2 = 4$

$ V = \displaystyle\int_0^4 \pi[4 - y]\, dy$

b) $V = \pi\left[4y - \dfrac{y^2}{2}\right]_0^4$

$ = (16 - 8)\pi$

$ = 8\pi$

22. $y = 4$, $y = x^2$

518

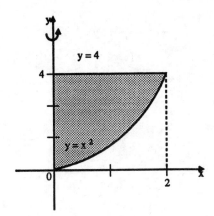

a) $V = \int_0^4 \pi y \, dy$

b) $V = \left. \dfrac{\pi y^2}{2} \right]_0^4$

$= 8\pi$

Exercise 22

23. $y = x^3, x = 2, \ y = 0$

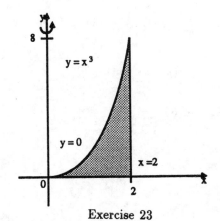

a) $x = 2 \Rightarrow y = 2^3 = 8$

$V = \int_0^8 \pi[4 - y^{2/3}] \, dy$

b) $V = \pi\left[4y - \dfrac{3}{5} y^{5/3}\right]_0^8$

$= \pi\left(32 - \dfrac{96}{5}\right)$

$= \dfrac{64\pi}{5}$

Exercise 23

24. $y = x^2, \ y = x^3$

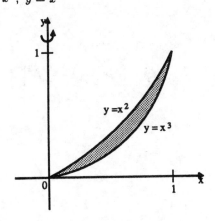

a) $V = \int_0^1 \pi(y^{2/3} - y) \, dy$

b) $V = \pi\left[\dfrac{3}{5} y^{5/3} - \dfrac{y^2}{2}\right]_0^1$

$= \pi\left(\dfrac{3}{5} - \dfrac{1}{2}\right)$

$= \dfrac{\pi}{10}$

Exercise 24

25. $y = x^4, \ y = 0, \ x = 2$

$x = 2 \Rightarrow y = 2^4 = 16$

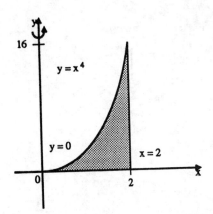

$$V = \int_0^{16} \pi(4 - \sqrt{y})\, dy$$

$$= \pi \left[4y - \frac{2}{3} y^{3/2} \right]_0^{16}$$

$$= \pi \left(64 - \frac{128}{3} \right)$$

$$= \frac{64}{3} \pi$$

Exercise 25

26. $y = |x - 2|$, $y = 2$

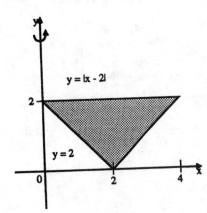

$$V = \int_0^2 \pi[(y+2)^2 - (2-y)^2]\, dy$$

$$= \frac{\pi}{3} \left[(y+2)^3 + (2-y)^3 \right]_0^2$$

$$= \frac{\pi}{3}[(64-8) + (0-8)]$$

$$= 16\pi$$

Exercise 26

27. $y = (x - 2)^2$, $y = 4$
$x - 2 = \pm\sqrt{y} \ \Rightarrow \ x = 2 + \sqrt{y}$,
$\qquad\qquad\qquad\quad x = 2 - \sqrt{y}$

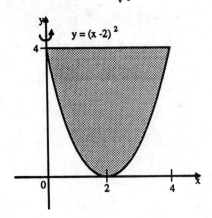

$$V = \int_0^4 \pi \left[(2+\sqrt{y})^2 - (2-\sqrt{y})^2 \right] dy$$

$$= \pi \int_0^4 8\sqrt{y}\, dy$$

$$= \frac{16\pi}{3} y^{3/2} \Big]_0^4$$

$$= \frac{16\pi}{3} (8)$$

$$= \frac{128\pi}{3}$$

Exercise 27

28. Side dimension $= 10 \left(1 - \frac{y}{8}\right)$

$$V = \int_0^8 \left[10 - \left(1 - \frac{y}{8}\right)\right]^2 dy = 100 \int_0^8 \left(1 - \frac{y}{8}\right)^2 dy = -\frac{800}{3} \left(1 - \frac{y}{8}\right)^3 \Big]_0^8 = -\frac{800}{3} (0 - 1) = \frac{800}{3}$$

29. Consider rotating a semicircle of radius r about the y-axis.

$$V = 2 \int_0^r \pi(r^2 - y^2) \, dy = 2\pi \left[r^2 y - \frac{y^3}{3}\right]_0^r = 2\pi \left(r^3 - \frac{r^3}{3}\right) = \frac{4\pi r^3}{3}$$

30. Let the x-axis be perpendicular to the bases of the triangles. For $x \in [-1, 1]$, $y = \sqrt{16 - x^2} =$ one-half the base of the triangle. The altitude of the triangle $= 2y \sin 60° = \sqrt{3} y$. The area of the triangle $= \frac{1}{2}(2y)(\sqrt{3}y) = \sqrt{3}y^2 = \sqrt{3}(16 - x^2)$.

$$V = 2 \int_0^4 \sqrt{3}(16 - x^2) \, dx = 2\sqrt{3} \left(16x - \frac{x^3}{3}\right)\Big]_0^4 = 2\sqrt{3}\left(64 - \frac{64}{3}\right) = \frac{256}{3}\sqrt{3} \approx 147.802$$

31. $A(x) = \frac{1}{2}(2y)^2 = 2y^2 = 2(16 - x^2)$

$$V = 2 \int_0^4 2(16 - x^2) \, dx = 4\left(16x - \frac{x^3}{3}\right)\Big]_0^4 = 4\left(64 - \frac{64}{3}\right) = \frac{512}{3} \approx 170.667$$

32. Assume that the hemisphere is placed with the circle of radius $10\,\text{m}$ at the top, along the y-axis. Cross-sections perpendicular to the y-axis are circles of radius x, where $x^2 + (10 - y^2) = 10^2 = 100$.

$$\begin{aligned} V &= \int_0^6 A(y) \, dy = \int_0^6 \pi[100 - (10 - y)^2] \, dy = \int_0^6 \pi(20y - y^2) \, dy \\ &= \pi\left(10y^2 - \frac{y^3}{3}\right)\Big]_0^6 = \pi(360 - 72) = 288\pi \approx 904.779 \, m^3 \end{aligned}$$

33. $f(x) = x^2$, $g(x) = 8 - x^2$.
    Intersections: $x^2 = 8 - x^2 \Rightarrow 2x^2 = 8 \Rightarrow x = \pm 2$.
    At any value of $x \in [-2, 2]$, the base of the square has length $8 - x^2 - x^2 = 8 - 2x^2$. $A(x) = (8 - 2x^2)^2$.

$$\begin{aligned} V &= \int_{-2}^2 (8 - 2x^2)^2 \, dx = 2\int_0^2 (64 - 32x^2 + 4x^4) \, dx = 2\left[64x - \frac{32}{3}x^3 + \frac{4}{5}x^5\right]_0^2 \\ &= 2\left(128 - \frac{32}{3}(8) + \frac{4}{5}(32)\right) = \frac{2048}{15} \approx 136.533 \end{aligned}$$

34. At any value of $x \in [-2, 2]$, the diameter of the circle is $8 - 2x^2$. $A(x) = \frac{\pi}{8}(8 - 2x^2)^2$.

$$\begin{aligned} V &= \int_{-2}^2 \frac{\pi}{8}(8 - 2x^2)^2 \, dx = \frac{\pi}{4}\int_0^2 (64 - 32x^2 + 4x^4) \, dx = \frac{\pi}{4}\left[64x - \frac{32}{3}x^3 + \frac{4}{5}x^5\right]_0^2 \\ &= \frac{\pi}{4}\left(128 - \frac{256}{3} + \frac{128}{5}\right) = \frac{256}{15}\pi \approx 53.617 \end{aligned}$$

35. Consider circles perpendicular to the y-axis. The radius at y is $R - \frac{y}{h}(R - r)$, and the area is
$\pi \left[R - y \frac{(R - r)}{h} \right]^2$.

$$V = \int_0^h \pi \left[R - y \frac{(R - r)}{h} \right]^2 dy = -\frac{\pi}{3} \frac{h}{(R - r)} \left[R - y \frac{(R - r)}{h} \right]^3 \Bigg]_0^h$$

$$= -\frac{\pi}{3} \frac{h}{(R - r)} [r^3 - R^3] = \frac{\pi h}{3} (R^2 + rR + r^2)$$

36. For the line from $(0, 0)$ to $(2, 5)$, $y = \frac{5}{2} x$. For the line from $(2, 5)$ to $(5, 0)$, $m = \frac{5 - 0}{2 - 5} = -\frac{5}{3}$
$\frac{y - 0}{x - 5} = -\frac{5}{3}$; $y = -\frac{5}{3} x + \frac{25}{3}$

$$V = \int_0^2 \pi \left(\frac{5}{2} x \right)^2 dx + \int_2^5 \pi \left(-\frac{5}{3} x + \frac{25}{3} \right)^2 dx$$

$$= \frac{25}{4} \pi \frac{x^3}{3} \Bigg]_0^2 - \frac{25}{9} \frac{\pi}{3} (-x + 5)^3 \Bigg]_2^5 = \frac{25}{12} \pi (8) - \frac{25}{27} \pi (0 - 27) = \frac{125}{3} \pi \approx 130.9$$

37. From Exercise 32, $\quad V = \pi \int_0^h (20y - y^2)\, dy$

$$\frac{dV}{dt} = \pi(20h - h^2) \frac{dh}{dt}; \quad 10 = \pi(20 \cdot 4 - 16) \frac{dh}{dt}$$

$$\frac{dh}{dt} = \frac{10}{\pi(64)} \approx 0.04974 \text{ m/min}$$

38. The material removed from the sphere consists of a cylinder and two end caps. Orient the sphere as shown in the two-dimensional sketch.

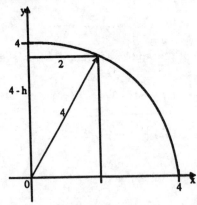

Exercise 38

The length of the cylinder is $2\sqrt{16 - 4} = 2\sqrt{12} = 4\sqrt{3}$. Its volume is $\pi r^2 h = \pi(4)(4\sqrt{3}) = 16\sqrt{3}\pi$.

The thickness of the cap, h, is given by $4 + (4 - h)^2 = 16$, or $4 - h = \sqrt{12}$; $h = 4 - \sqrt{12}$.
Cross-sections of the cap perpendicular to the y-axis are circles of radius x, where $x^2 + (4 - h)^2 = 16$,

as h varies from 0 to $4 - \sqrt{12}$.

$$V_{\text{cap}} \;=\; \int_0^{4-\sqrt{12}} \pi(16 - (4-h)^2)\,dh = \pi \int_0^{4-\sqrt{12}} (8h - h^2)\,dh = \pi \left[4h^2 - \frac{h^3}{3}\right]_0^{4-\sqrt{12}}$$

$$\;=\; 4\pi(16 - 8\sqrt{12} + 12) - \frac{\pi}{3}(64 - 48\sqrt{12} + 144 - 12\sqrt{12}) = \pi\left(\frac{128}{3} - 24\sqrt{3}\right)$$

$$\text{Total volume removed} \;=\; \pi\left(16\sqrt{3} + \frac{256}{3} - 48\sqrt{3}\right)$$

$$\;=\; \pi\left(\frac{256}{3} - 32\sqrt{3}\right) \approx 93.9578 \text{ cm}^3.$$

$$\text{So, volume of the remaining solid} \;=\; \frac{256}{3}\pi - \pi\left(\frac{256}{3} - 32\sqrt{3}\right)$$

$$\;=\; 32\sqrt{3}\,\pi.$$

39. $y = 1 - x^2$ and x-axis about the line $y = -3$.

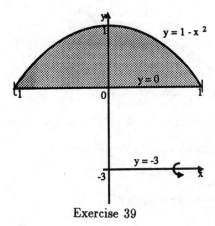

Exercise 39

$$V \;=\; \int_{-1}^{1} \pi[(1 - x^2 - (-3))^2 - 9]\,dx = \pi \int_{-1}^{1} (16 - 8x^2 + x^4 - 9)\,dx$$

$$\;=\; \pi\left[7x - \frac{8}{3}x^3 + \frac{x^5}{5}\right]_{-1}^{1} = 2\pi\left(7 - \frac{8}{3} + \frac{1}{5}\right) = \frac{136}{15}\pi \approx 28.4838$$

40. $y = \sqrt{x}$ and $y = \frac{x}{2}$ about the line $y = 4$.

Intersections: $\sqrt{x} = \frac{x}{2} \Rightarrow (2 - \sqrt{x})\sqrt{x} = 0 \Rightarrow x = 0, 4.$

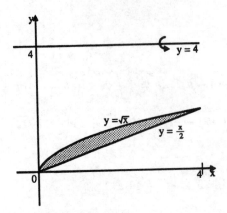

Exercise 40

$$V = \int_0^4 \pi \left[\left(4 - \frac{x}{2}\right)^2 - \left(4 - \sqrt{x}\right)^2 \right] dx = \pi \int_0^4 \left(-4x + \frac{x^2}{4} + 8\sqrt{x} - x\right) dx$$

$$= \pi \left[-\frac{5x^2}{2} + \frac{x^3}{12} + \frac{16}{3} x^{3/2} \right]_0^4 = \pi \left(-40 + \frac{16}{3}(1+8)\right) = 8\pi$$

41. $y = \sqrt{x}$, $y = 0$, and $x = 9$ about the line $y = -2$.

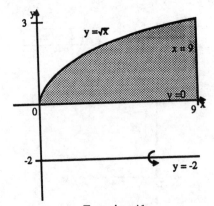

Exercise 41

$$V = \int_0^9 \pi[(\sqrt{x} + 2)^2 - 4] \, dx = \pi \int_0^9 (x + 4\sqrt{x}) \, dx$$

$$= \pi \left[\frac{x^2}{2} + \frac{8}{3} x^{3/2} \right]_0^9 = \pi \left(\frac{81}{2} + \frac{8}{3} \cdot 27\right) = \frac{225}{2}\pi \approx 353.429$$

42. $\dfrac{x^2}{a^2} + \dfrac{y^2}{b^2} = 1$ about the x-axis.

$$V = \int_{-a}^a \pi y^2 \, dx = 2\pi \int_0^a b^2 \left(1 - \frac{x^2}{a^2}\right) dx = 2\pi b^2 \left[x - \frac{x^3}{3a^2} \right]_0^a = 2\pi b^2 \left(a - \frac{a^3}{3a^2}\right) = \frac{4\pi ab^2}{3}$$

43. $\dfrac{x^2}{a^2} + \dfrac{y^2}{b^2} = 1$ about the y-axis.

$$V = \int_{-b}^{b} \pi x^2 \, dy = 2\pi \int_{0}^{b} a^2 \left(1 - \frac{y^2}{b^2}\right) dy = 2\pi a^2 \left[y - \frac{y^3}{b^2}\right]_{0}^{b} = \frac{4\pi a^2 b}{3}$$

44. If $a = b = r$ in Exercise 42 and 43, $V = \frac{4}{3}\pi r^3$, the formula for the volume of a sphere.

45. Cross-sections taken perpendicular to the x-axis are rectangles of height $x \tan \theta$ and width $2y$, where y is given by $x^2 + y^2 = r^2$.

$$V = 2 \int_{0}^{r} (x \tan \theta)\sqrt{r^2 - x^2} \, dx = -\frac{2}{3} \tan \theta (r^2 - x^2)^{3/2} \Big]_{0}^{r} = \frac{2}{3} r^3 \tan \theta$$

Alternatively, cross-sections taken perpendicular to the y-axis are right triangles of base x and altitude $x \tan \theta$.

$$V = 2 \int_{0}^{r} \frac{1}{2} x \, (x \tan \theta) \, dy = \tan \theta \int_{0}^{r} (r^2 - y^2) \, dy = \tan \theta \left[r^2 y - \frac{y^3}{3}\right]_{0}^{r} = \frac{2}{3} r^3 \tan \theta$$

46 $f(x) = \dfrac{1}{\sqrt{1 + x^2}}$, $y = 0$; $0 \le x \le 4$

$$V = \int_{0}^{4} \pi \left(\frac{1}{1 + x^2}\right) dx = \pi \int_{0}^{4} \frac{dx}{1 + x^2} = \pi \int_{0}^{4} g(x) \, dx$$

$\Delta x = \frac{4}{8} = 0.5$

i	x_i	$g(x_i)$
0	0.0	1.0000
1	0.5	0.8000
2	1.0	0.5000
3	1.5	0.3077

i	x_i	$g(x_i)$
4	2.0	0.2000
5	2.5	0.1379
6	3.0	0.1000
7	3.5	0.0755
8	4.0	0.0588

$$V = \pi \left(\frac{0.5}{2}\right) \left(f(x_0) + 2 \sum_{i=1}^{7} f(x_i) + f(x_8)\right) = 0.25\pi(5.301) \approx 4.1634$$

47. $f(x) = \dfrac{1}{\sqrt{1 + x^2}}$, $y = 0$, $0 \le x \le 4$

$$V = \int_{0}^{4} \pi \left(\frac{1}{1 + x^2}\right) dx = \pi \int_{0}^{4} \frac{dx}{1 + x^2} = \pi \int_{0}^{4} g(x) \, dx$$

$\Delta x = \frac{4}{12} = \frac{1}{3}$; $x_i = 0 + \frac{1}{3} i$

i	x_i	$g(x_i)$
0	0.0000	1.0000
1	0.3333	0.9000
2	0.6667	0.6923
3	1.0000	0.5000
4	1.3333	0.3600
5	1.6667	0.2647
6	2.0000	0.2000

i	x_i	$g(x_i)$
7	2.3333	0.1552
8	2.6667	0.1233
9	3.0000	0.1000
10	3.3333	0.0826
11	3.6667	0.0692
12	4.0000	0.0588

$$V = \pi\left(\frac{1}{9}\right)[f(x_0) + 4f(x_1) + 2f(x_2) + 4f(x_3) + \cdots + 2f(x_{10}) + 4f(x_{11}) + f(x_{12})]$$

$$= \frac{\pi}{9}(11.9316) \approx 4.1649$$

48. $y = x^{-1/2}$ and the x-axis about the x-axis for $1 \le x \le 4$.

$$V = \int_1^4 \pi\left(\frac{1}{x}\right) dx = \pi \ln|x|\Big]_1^4 = \pi \ln 4 \approx 4.3552$$

49. $\dfrac{x^2}{a^2} + \dfrac{y^2}{b^2} = 1$ about the line $y = -2$, with $a = 3$ and $b = 1$.

$$y^2 = \frac{9 - x^2}{9} \Rightarrow y = \pm\frac{1}{3}\sqrt{9 - x^2}$$

$$V = \int_{-3}^3 \pi\left[\left(2 + \frac{1}{3}\sqrt{9 - x^2}\right)^2 - \left(2 - \frac{1}{3}\sqrt{9 - x^2}\right)^2\right] dx = \frac{\pi}{9}\int_{-3}^3 24\sqrt{9 - x^2}\,dx = \frac{8\pi}{3}\int_{-3}^3 \sqrt{9 - x^2}\,dx$$

The integral $\displaystyle\int_{-3}^3 \sqrt{9 - x^2}\,dx$ equals the area of a semicircle of radius 3, or $\frac{\pi}{2}(9)$.

$$\text{Volume} = \frac{8\pi}{3} \cdot \frac{9\pi}{2} = 12\pi^2.$$

7.2 Calculating Volumes by the Method of Cylindrical Shells

1. $x + y = 1$, $x = 0$, $y = 0$

   a)

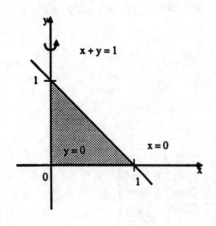

   b) $\displaystyle V = \int_0^1 2\pi x(1 - x)\,dx$

   $$= 2\pi\int_0^1 (x - x^2)\,dx$$

   c) $\displaystyle V = 2\pi\left[\frac{x^2}{2} - \frac{x^3}{3}\right]_0^1$

   $$= 2\pi\left(\frac{1}{2} - \frac{1}{3}\right)$$

   $$= \frac{\pi}{3}$$

Exercise 1

526

2. $y = \sqrt{x}$, $y = 0$, $1 \le x \le 4$

   a)

   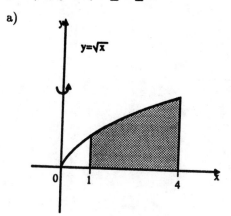

   Exercise 2

   b) $V = \int_{1}^{4} 2\pi x (\sqrt{x})\, dx$

   $= 2\pi \int_{1}^{4} x^{3/2}\, dx$

   c) $V = 2\pi \left(\dfrac{2}{5}\right) x^{5/2} \Big]_{1}^{4}$

   $= \dfrac{4\pi}{5}(32 - 1)$

   $= \dfrac{124\pi}{5}$

   ≈ 77.9115

3. $y = x^3$, $y = 0$, $1 \le x \le 3$

   a)

   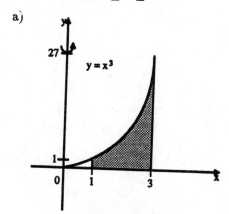

   Exercise 3

   b) $V = \int_{1}^{3} 2\pi x (x^3)\, dx$

   $= 2\pi \int_{1}^{3} x^4\, dx$

   c) $V = \dfrac{2\pi}{5} x^5 \Big]_{1}^{3}$

   $= \dfrac{2\pi}{5}(243 - 1)$

   $= \dfrac{484\pi}{5}$

   ≈ 304.106

4. $y = \sqrt{1 + x^2}$, $y = 0$, $0 \le x \le 3$

   a)

   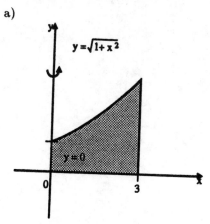

   Exercise 4

   b) $V = \int_{0}^{3} 2\pi x \sqrt{1 + x^2}\, dx$

   $= \pi \int_{0}^{3} 2x \sqrt{1 + x^2}\, dx$

   c) $V = \dfrac{2\pi}{3}(1 + x^2)^{3/2} \Big]_{0}^{3}$

   $= \dfrac{2\pi}{3}(10^{3/2} - 1)$

   ≈ 64.1362

5. $y = \sqrt{x}$, $y = -x$, $x = 4$

a)

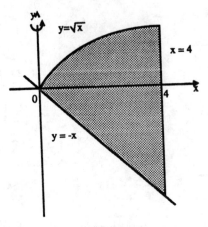

Exercise 5

b) $V = \int_0^4 2\pi x(\sqrt{x} + x)\, dx = 2\pi \int_0^4 (x^{3/2} + x^2)\, dx$

c) $V = 2\pi \left[\dfrac{2}{5} x^{5/2} + \dfrac{x^3}{3} \right]_0^4 = \dfrac{4\pi}{5}(32) + \dfrac{2\pi}{3}(64) = \dfrac{1024\pi}{15} \approx 214.467$

6. $y = 1 + x + x^2$, $y = -2$, $1 \le x \le 3$

a)

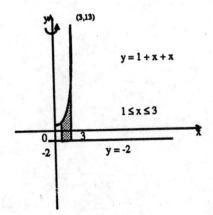

$y = x^2 + x + 1$

$\quad = \left(x + \dfrac{1}{2} \right)^2 + \dfrac{3}{4}$

Exercise 6

b) $V = \int_1^3 2\pi x(1 + x + x^2 + 2)\, dx = 2\pi \int_1^3 (3x + x^2 + x^3)\, dx$

c) $V = 2\pi \left[\dfrac{3x^2}{2} + \dfrac{x^3}{3} + \dfrac{x^4}{4} \right]_1^3 = 2\pi \left[\dfrac{3}{2}(9-1) + \dfrac{1}{3}(27-1) + \dfrac{1}{4}(81-1) \right]$

$\quad = 2\pi \left(\dfrac{122}{3} \right) = \dfrac{244\pi}{3} \approx 255.52$

7. $y = -x^2 - 4x - 3$, $y = 0$

a) $y = -(x^2 + 4x + 3) = -(x+2)^2 + 1$

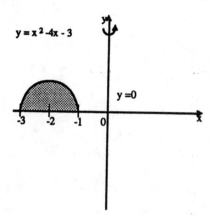

$$y = 0 \Rightarrow (x+2)^2 = 1 \quad \Rightarrow \quad x+2 = \pm 1$$

$$\Rightarrow \quad x = -3, -1$$

Exercise 7

b) Since $x < 0$ on $[-3, -1]$, use $|x| = -x$ as the radius of the shell.

$$V = \int_{-3}^{-1} 2\pi(-x)(-x^2 - 4x - 3)\,dx = 2\pi \int_{-3}^{-1} (x^3 + 4x^2 + 3x)\,dx$$

c) $V = 2\pi\left[\dfrac{x^4}{4} + \dfrac{4}{3}x^3 + \dfrac{3}{2}x^2\right]_{-3}^{-1} = 2\pi\left[\dfrac{1}{4}(1-81) + \dfrac{4}{3}(-1+27) + \dfrac{3}{2}(1-9)\right]$

$$= 2\pi\left(\dfrac{8}{3}\right) = \dfrac{16\pi}{3} \approx 16.7552$$

8. $y = \dfrac{1}{x},\ y = 1,\ x = 4$

a)

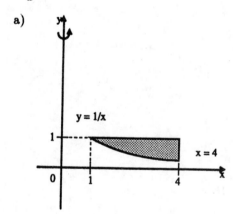

Exercise 8

b) $V = \displaystyle\int_1^4 2\pi x\left(1 - \dfrac{1}{x}\right) dx = 2\pi \int_1^4 (x-1)\,dx$

c) $V = 2\pi\left[\dfrac{x^2}{2} - x\right]_1^4 = 2\pi\left[(8-4) - \left(\dfrac{1}{2} - 1\right)\right] = 2\pi\left(\dfrac{9}{2}\right) = 9\pi$

9. $y = \sin x^2,\ y = 1,\ 0 \le x \le \sqrt{\dfrac{\pi}{2}}$

a)

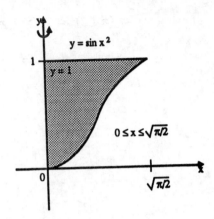

Exercise 9

b) $V = \int_0^{\sqrt{\pi/2}} 2\pi x (1 - \sin x^2)\, dx = \pi \int_0^{\sqrt{\pi/2}} 2x(1 - \sin x^2)\, dx$

c) $V = \pi(x^2 + \cos x^2) \Big]_0^{\sqrt{\pi/2}} = \pi \left(\dfrac{\pi}{2} - 1 \right)$

10. $y = 2,\ y = \sqrt[3]{x},\ x = 0$

a)

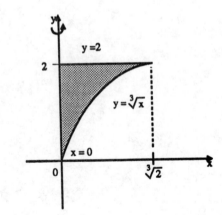

Exercise 10

b) $V = \int_0^8 2\pi x \left(2 - \sqrt[3]{x}\right) dx = 2\pi \int_0^8 (2x - x^{4/3})\, dx$

c) $V = 2\pi \left[x^2 - \dfrac{3}{7} x^{7/3} \right]_0^8 = 2\pi \left(64 - \dfrac{3}{7} 8^{7/3} \right) = \dfrac{128\pi}{7}$

11. $y = -x \cos x^3,\ \ 0 \le x \le \sqrt[3]{\dfrac{\pi}{2}}$

530

a)

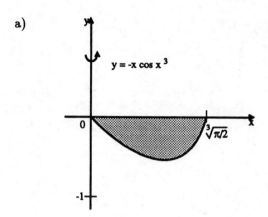

Exercise 11

b) $V = \int_0^{\sqrt[3]{\pi/2}} 2\pi x(0 + x\cos x^3)\,dx = 2\pi \int_0^{\sqrt[3]{\pi/2}} x^2 \cos x^3\,dx = \frac{2\pi}{3}\sin x^3 \Big]_0^{\sqrt[3]{\pi/2}} = \frac{2\pi}{3}$

12. $y = \dfrac{1}{\sqrt{4-x^2}}$, $y = 0$, $0 \le x \le 1$

a)

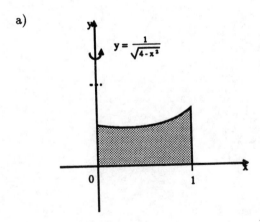

Exercise 12

b) $V = \int_0^1 2\pi x \left(\dfrac{1}{\sqrt{4-x^2}}\right)\,dx = \pi \int_0^1 \dfrac{2x\,dx}{\sqrt{4-x^2}}$

$= -2\pi\sqrt{4-x^2}\Big]_0^1 = -2\pi\left(\sqrt{3}-2\right) = 2\pi\left(2-\sqrt{3}\right) \approx 1.6836$

13. $y = \sqrt{9-x^2}$, $y = 0$, $0 \le x \le 3$

531

a)

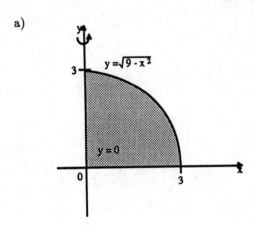

Exercise 13

b) $V = \int_0^3 2\pi x \sqrt{9 - x^2} \, dx = -\frac{2\pi}{3}(9 - x^2)^{3/2} \Big]_0^3 = -\frac{2\pi}{3}(0 - 27) = 18\pi$

14. $y = \frac{1}{\sqrt{x}} + \sqrt{x}$, $y = 0$, $1 \le x \le 4$

a)

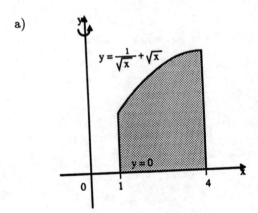

Exercise 14

b) $V = \int_1^4 2\pi x \left(\frac{1}{\sqrt{x}} + \sqrt{x}\right) dx = 2\pi \int_1^4 \left(\sqrt{x} + x^{3/2}\right) dx = 2\pi \left[\frac{2}{3} x^{3/2} + \frac{2}{5} x^{5/2}\right]_1^4$

$= 2\pi \left[\frac{2}{3}(8 - 1) + \frac{2}{5}(32 - 1)\right] = \frac{512\pi}{15} \approx 107.233$

15. $y = \frac{\sqrt{1 + x^{3/2}}}{\sqrt{x}}$, $y = 0$, $1 \le x \le 4$

a)

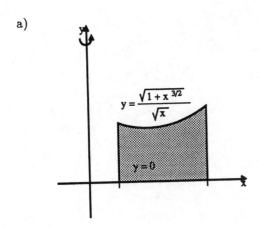

$$y = \frac{\sqrt{1 + x^{3/2}}}{\sqrt{x}}$$

$y = 0$

Exercise 15

b) $\quad V = \displaystyle\int_1^4 2\pi x \left(\frac{\sqrt{1 + x^{3/2}}}{\sqrt{x}} \right) dx = 2\pi \int_1^4 \sqrt{1 + x^{3/2}} \, x^{1/2} \, dx$

$\qquad = \dfrac{8\pi}{9} (1 + x^{3/2})^{3/2} \Big]_1^4 = \dfrac{8\pi}{9} \left(27 - 2\sqrt{2} \right) \approx 67.500$

16. $y = \sin x^2$, $y = -x$, $0 \le x \le \sqrt{\pi}$, about the y-axis.

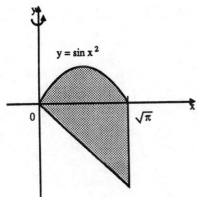

$y = \sin x^2$

$0 \qquad \sqrt{\pi}$

Exercise 16

$V = \displaystyle\int_0^{\sqrt{\pi}} 2\pi x (\sin x^2 + x) \, dx = 2\pi \int_0^{\sqrt{\pi}} (x \sin x^2 + x^2) \, dx$

$\qquad 2\pi \left[-\dfrac{1}{2} \cos x^2 + \dfrac{x^3}{3} \right]_0^{\sqrt{\pi}} = -\pi(-1 - 1) + \dfrac{2\pi}{3} (\pi)^{3/2} = \dfrac{2\pi}{3} (3 + \pi^{3/2}) \approx 17.945$

17. $x = y^2$, $x = 4$, about the line $x = -1$.

533

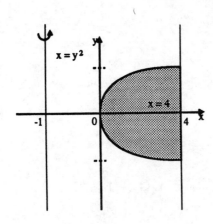

Exercise 17

$$V = \int_0^4 2\pi(x+1)\left(\sqrt{x} - (-\sqrt{x})\right) dx = 4\pi \int_0^4 \left(x^{3/2} + \sqrt{x}\right) dx = 4\pi \left[\frac{2}{5} x^{5/2} + \frac{2}{3} x^{3/2}\right]_0^4$$

$$= 4\pi \left(\frac{64}{5} + \frac{16}{3}\right) = \frac{1088\pi}{15} \approx 227.870$$

18. $x = \sqrt{4+y}$, $x = 0$, $y = 0$, about the line $x = -2$.
     $x^2 = 4+y \Rightarrow y = x^2 - 4$

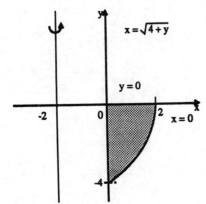

Exercise 18

$$V = \int_0^2 2\pi(x+2)\left(-(x^2-4)\right) dx = 2\pi \int_0^2 (-x^3 - 2x^2 + 4x + 8)\, dx$$

$$= 2\pi \left[-\frac{x^4}{4} - \frac{2}{3} x^3 + 2x^2 + 8x\right]_0^2 = 2\pi \left(-4 - \frac{16}{3} + 8 + 16\right) = \frac{88\pi}{3} \approx 92.153$$

19. $y = x^2$, $y = \sqrt{x}$, about the line $y = -2$.

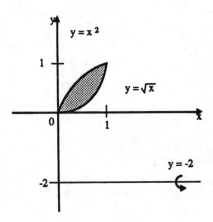

Exercise 19

$$x^2 = \sqrt{x} \Rightarrow x = 0, 1; \quad y = x^2 \Rightarrow x = \sqrt{y}; \quad y = \sqrt{x} \Rightarrow x = y^2$$

$$
\begin{aligned}
V &= \int_0^1 2\pi (y+2)(\sqrt{y} - y^2) \, dy = 2\pi \int_0^1 (-y^3 - 2y^2 + y^{3/2} + 2y^{1/2}) \, dy \\
&= 2\pi \left[-\frac{y^4}{4} - \frac{2}{3} y^3 + \frac{2}{5} y^{5/2} + \frac{4}{3} y^{3/2} \right]_0^1 = 2\pi \left(-\frac{1}{4} - \frac{2}{3} + \frac{2}{5} + \frac{4}{3} \right) \\
&= \frac{49\pi}{30} \approx 5.1313
\end{aligned}
$$

20. Triangle with vertices $(1,0)$, $(2,4)$, and $(4,0)$ about the y-axis.

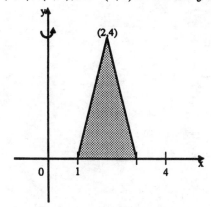

Exercise 20

line 1: $\quad m = \dfrac{4-0}{2-1} = 4; \qquad \dfrac{y-0}{x-1} = 4; \qquad y = 4x - 4$

line 2: $\quad m = \dfrac{4-0}{2-4} = -2; \qquad \dfrac{y-0}{x-4} = -2; \quad y = -2x + 8$

$$
\begin{aligned}
V &= \int_0^4 \pi \left[\left(\frac{8-y}{2} \right)^2 - \left(\frac{y+4}{4} \right)^2 \right] dy = \frac{\pi}{16} \int_0^4 (3y^2 - 72y + 240) \, dy \\
&= \frac{\pi}{16} \left[y^3 - 36y^2 + 240y \right]_0^4 = \frac{\pi}{16} (64 - 576 + 960) = 28\pi
\end{aligned}
$$

21. $y = x$, $y = x^3$, about the y-axis.

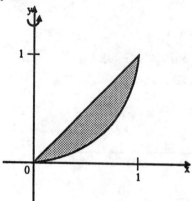

Exercise 21

$$V \;=\; 2\int_0^1 2\pi x(x - x^3)\,dx = 4\pi \int_0^1 (x^2 - x^4)\,dx = 4\pi \left[\frac{x^3}{3} - \frac{x^5}{5}\right]_0^1 = 4\pi\left(\frac{1}{3} - \frac{1}{5}\right) = \frac{8\pi}{15}$$

22. $y = 4$, $y = x^2$, about the line $y = -2$.

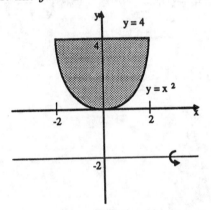

Exercise 22

Use cylinders parallel to the x-axis. $y = x^2 \Rightarrow x = \pm\sqrt{y}$

$$V \;=\; \int_0^4 2\pi(y + 2)\left(\sqrt{y} - (-\sqrt{y})\right) = 4\pi\int_0^4 (y^{3/2} + 2y^{1/2})\,dy = 4\pi\left[\frac{2}{5}y^{5/2} + \frac{4}{3}y^{3/2}\right]_0^4$$

$$\;=\; 4\pi\left(\frac{64}{5} + \frac{32}{3}\right) = \frac{1408\pi}{15} \approx 294.891$$

Alternatively, disks parallel to the y-axis may be used.

$$V \;=\; 2\pi\int_0^2 [6^2 - (x^2 + 2)^2]\,dx = 2\pi\int_0^2 (32 - 4x^2 - x^4)\,dx = 2\pi\left[32x - \frac{4}{3}x^3 - \frac{x^5}{5}\right]_0^2$$

$$\;=\; 2\pi\left(64 - \frac{32}{3} - \frac{32}{5}\right) = \frac{1408\pi}{15}$$

23. Triangle with vertices $(1, 3)$, $(1, 7)$, and $(4, 7)$ about the line $y = 1$.

536

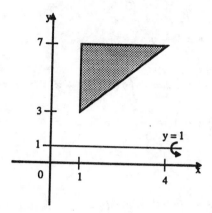

Exercise 23

Use cylinders parallel to the x-axis.

Equation of line: $\quad m = \dfrac{7-3}{4-1} = \dfrac{4}{3}$

$$\frac{y-3}{x-1} = \frac{4}{3}; \quad y = \frac{4}{3}x + \frac{5}{3} \Rightarrow x = \frac{3y-5}{4}$$

$$V = \int_3^7 2\pi(y-1)\left(\frac{3y-5}{4} - 1\right) dy = \frac{3\pi}{2}\int_3^7 (y-1)(y-3)\,dy = \frac{3\pi}{2}\int_3^7 (y^2 - 4y + 3)\,dy$$

$$= \frac{3\pi}{2}\left[\frac{y^3}{3} - 2y^2 + 3y\right]_3^7 = \frac{3\pi}{2}\left[\frac{1}{3}(343 - 27) - 2(49 - 9) + 3(7 - 3)\right] = \frac{3\pi}{2}\left(\frac{112}{3}\right)$$

$$= 56\pi \approx 175.929$$

Alternatively, disks parallel to the y-axis may be used.

$$V = \int_1^4 \pi\left[6^2 - \left(\frac{4}{3}x + \frac{5}{3} - 1\right)^2\right] dx = \frac{16\pi}{9}\int_1^4 (20 - x - x^2)\,dx = \frac{16\pi}{9}\left[20x - \frac{x^2}{2} - \frac{x^3}{3}\right]_1^4$$

$$= \frac{16\pi}{9}\left[20(4-1) - \frac{1}{2}(16-1) - \frac{1}{3}(64-1)\right] = \frac{504\pi}{9} = 56\pi$$

24. Triangles with vertices $(0,0)$, $(0,2)$, and $(2,0)$ about

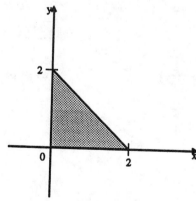

Exercise 24

a) x-axis.

Equation of line $y = 2 - x$ or $x = 2 - y$.

$$V = \int_0^2 \pi(2-x)^2 \, dx = \pi \int_0^2 (4 - 4x + x^2) \, dx = \pi \left[4x - 2x^2 + \frac{x^3}{3} \right]_0^2$$

$$= \pi \left(8 - 8 + \frac{8}{3} \right) = \frac{8\pi}{3}$$

b) y-axis.

$$V = \int_0^2 \pi(2-y)^2 \, dy = \frac{8\pi}{3}, \quad \text{as in part a).}$$

c) The line $x = -1$.

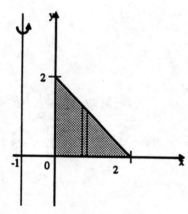

$$V = \int_0^2 2\pi(x+1)(2-x) \, dx$$

$$= 2\pi \int_0^2 (-x^2 + x + 2) \, dx$$

$$= 2\pi \left[-\frac{x^3}{3} + \frac{x^2}{2} + 2x \right]_0^2$$

$$= 2\pi \left(-\frac{8}{3} + 2 + 4 \right)$$

$$= \frac{20\pi}{3}$$

Exercise 24. c)

d) The line $y = -1$.

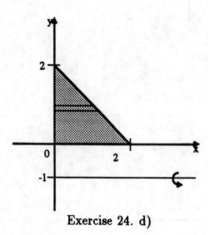

$$V = \int_0^2 2\pi(y+1)(2-y) \, dy$$

$$= 2\pi \int_0^2 (-y^2 + y + 2) \, dy$$

$$= \frac{20\pi}{3},$$

as in part c).

Exercise 24. d)

25. $\dfrac{x^2}{a^2} + \dfrac{y^2}{b^2} = 1$ about the y-axis.

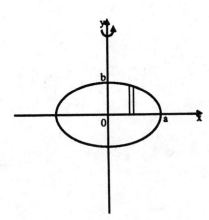

$$y = \frac{b}{a}\sqrt{a^2 - x^2}$$

Exercise 25

$$V = 2\int_0^a 2\pi x \left(\frac{b}{a}\sqrt{a^2 - x^2}\right) dx = \frac{4\pi b}{a}\int_0^a \left(\sqrt{a^2 - x^2}\right) x\, dx$$

$$= -\frac{4\pi b}{3a}(a^2 - x^2)^{3/2}\Big]_0^a = -\frac{4\pi b}{3a}(0 - a^3) = \frac{4\pi a^2 b}{3}$$

26. $x^2 + (y - 2)^2 = 1$ about the x-axis.
For the upper half of the circle, $y = 2 + \sqrt{1 - x^2}$, and for the lower half, $y = 2 - \sqrt{1 - x^2}$.

a) Using disks parallel to the y-axis, we have

$$V = \int_{-1}^1 \pi\left[\left(2 + \sqrt{1 - x^2}\right)^2 - \left(2 - \sqrt{1 - x^2}\right)^2\right] dx$$

$$= \pi\int_{-1}^1 \left[2 + \sqrt{1 - x^2}\right]^2 dx - \pi\int_{-1}^1 \left[2 - \sqrt{1 - x^2}\right]^2 dx.$$

b) $V = \pi\int_{-1}^1 \left(4 + 4\sqrt{1 - x^2} + 1 - x^2 - 4 + 4\sqrt{1 - x^2} - 1 + x^2\right) dx = 8\pi\int_{-1}^1 \sqrt{1 - x^2}\, dx$

c) $\int_{-1}^1 \sqrt{1 - x^2}\, dx = $ area of a semicircle of radius $1 = \frac{\pi}{2}$.

d) $V = 8\pi\int_{-1}^1 \sqrt{1 - x^2}\, dx = 8\pi\left(\frac{\pi}{2}\right) = 4\pi^2$

27. $y = \sin x$, x-axis, about the y-axis, for $0 \le x \le \pi$.

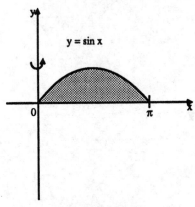

Exercise 27

$$V = \int_0^\pi 2\pi x (\sin x)\, dx = 2\pi \int_0^\pi x \sin x\, dx$$

With $n = 8$, $\Delta x = \frac{\pi}{8}$ and $x_i = \frac{\pi}{8} i$. Let $f(x_i) = x_i \sin x_i$.

i	x_i	$f(x_i)$
0	0	0.0000
1	$\pi/8$	0.1503
2	$\pi/4$	0.5554
3	$3\pi/8$	1.0884
4	$\pi/2$	1.5708
5	$5\pi/8$	1.8140
6	$3\pi/4$	1.6661
7	$7\pi/8$	1.0520
8	π	0.0000

$$
\begin{aligned}
V &= 2\pi \left(\frac{\pi}{16}\right)\left[f(x_0) + f(x_8)\right.\\
&\qquad \left. + 2\sum_{i=1}^{7} f(x_i)\right]\\
&= \frac{\pi^2}{8}(15.794)\\
&\approx 19.485
\end{aligned}
$$

28. By Simpson's Rule, with $n = 8$,

$$V = 2\pi \left(\frac{\pi}{24}\right)\left[f(x_0) + 4f(x_1) + 2f(x_2) + \cdots + 4f(x_7) + f(x_8)\right] \approx 19.742$$

With $n = 16$, $V \approx 19.739$. The exact value is $2\pi \left[\sin x - x \cos x\right]_0^\pi = 2\pi[-\pi(-1)] = 2\pi^2 \approx 19.7392$.

29. $y = \sqrt{1 + x^4}$, $y = 0$, $x = 0$, $x = 2$, about the line $y = -1$.

540

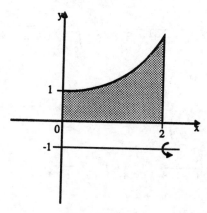

Exercise 29

$$V = \pi \int_0^2 \left[\left(1 + \sqrt{1 + x^4} \right)^2 - 1^2 \right] dx = \pi \int_0^2 \left(2\sqrt{1 + x^4} + 1 + x^4 \right) dx \approx 15.707\pi \approx 49.345$$

Simpson's Rule with $n = 16$ was used.

7.3   Arc Length and Surface Area

1. $y = 2x + 3$, from $(-3, -3)$ to $(2, 7)$.

  a) $L = \displaystyle\int_{-3}^2 \sqrt{1 + (y')^2}\, dx = \int_{-3}^2 \sqrt{1 + 4}\, dx = \sqrt{5} \int_{-3}^2 dx$

  b) $L = \sqrt{5}\, x \Big]_{-3}^2 = 5\sqrt{5}$

2. $y = x^{3/2}$, from $(0, 0)$ to $(4, 8)$.

  a) $L = \displaystyle\int_0^4 \sqrt{1 + (y')^2}\, dx = \int_0^4 \sqrt{1 + \left(\frac{3}{2}\sqrt{x} \right)^2}\, dx = \frac{1}{2} \int_0^4 \sqrt{4 + 9x}\, dx$

  b) $L = \dfrac{1}{2} \cdot \dfrac{2}{27} (4 + 9x)^{3/2} \Big]_0^4 = \dfrac{1}{27} (40^{3/2} - 8) = \dfrac{8}{27} (10^{3/2} - 1) \approx 9.0734$

3. $y - 2 = (x + 2)^{3/2}$, from $(-2, 2)$ to $(2, 10)$.

  a) $L = \displaystyle\int_{-2}^2 \sqrt{1 + (y')^2}\, dx = \int_{-2}^2 \sqrt{1 + \left(\frac{3}{2}(x + 2)^{1/2} \right)^2}\, dx = \frac{1}{2} \int_{-2}^2 \sqrt{4 + 9(x + 2)}\, dx$

  $ = \dfrac{1}{2} \displaystyle\int_{-2}^2 \sqrt{22 + 9x}\, dx$

  b) $L = \dfrac{1}{2} \cdot \dfrac{2}{27} (22 + 9x)^{3/2} \Big]_{-2}^2 = \dfrac{1}{27} (40^{3/2} - 4^{3/2}) = \dfrac{8}{27}(10^{3/2} - 1) \approx 9.0734$

  Note that the graph in Exercise 3 is the same as that in Exercise 2, translated from $(0, 0)$ to $(-2, 2)$ at its origin.

4. $3y = (x^2 + 2)^{3/2}$, for $0 \le x \le 2$

a) $L = \int_0^2 \sqrt{1 + (y')^2}\, dx = \int_0^2 \sqrt{1 + x^2(x^2 + 2)}\, dx = \int_0^2 \sqrt{1 + 2x^2 + x^4}\, dx = \int_0^2 (1 + x^2)\, dx$

b) $L = x + \dfrac{x^3}{3} \Big]_0^2 = 2 + \dfrac{8}{3} = \dfrac{14}{3}$

5. $3y = 2(x + 1)^{3/2}$, from $\left(0, \dfrac{2}{3}\right)$ to $\left(3, \dfrac{16}{3}\right)$

a) $L = \int_0^3 \sqrt{1 + (y')^2}\, dx = \int_0^3 \sqrt{1 + x + 1}\, dx = \int_0^3 \sqrt{2 + x}\, dx$

b) $L = \dfrac{2}{3}(2 + x)^{3/2} \Big]_0^3 = \dfrac{2}{3}\left[5^{3/2} - 2^{3/2}\right] \approx 5.568$

6. $y = \dfrac{x^3}{3} + \dfrac{1}{4x}$, for $1 \le x \le 3$

a) $L = \int_1^3 \sqrt{1 + (y')^2}\, dx = \int_1^3 \sqrt{1 + \left(x^2 - \dfrac{1}{4x^2}\right)^2}\, dx = \int_1^3 \dfrac{\sqrt{16x^8 + 8x^4 + 1}}{4x^2}\, dx$

$= \dfrac{1}{4} \int_1^3 \dfrac{4x^4 + 1}{x^2}\, dx$

b) $L = \dfrac{1}{4}\left[\dfrac{4x^3}{3} - \dfrac{1}{x}\right]_1^3 = \left(\dfrac{x^3}{3} - \dfrac{1}{4x}\right)\Big]_1^3 = \left(\dfrac{27}{3} - \dfrac{1}{12}\right) - \left(\dfrac{1}{3} - \dfrac{1}{4}\right) = \dfrac{53}{6}$

7. $(3y - 6)^2 = (x^2 + 2)^3$, for $0 \le x \le 3$

$(y - 2)^2 = \dfrac{1}{9}(x^2 + 2)^3; \quad y - 2 = \pm\dfrac{1}{3}(x^2 + 2)^{3/2}$

$y' = \pm x(x^2 + 2)^{1/2}; \quad (y')^2 = x^2(x^2 + 2)$

$L = \int_0^3 \sqrt{1 + (y')^2}\, dx = \int_0^3 \sqrt{1 + 2x^2 + x^4}\, dx = \int_0^3 (1 + x^2)\, dx = x + \dfrac{x^3}{3} \Big]_0^3 = 3 + 9 = 12$

Note that the graph has two branches, symmetrical about the line $y = 2$. Each arc has length 12.

8. $y = \dfrac{x^4}{4} + \dfrac{1}{8x^2}$, for $1 \le x \le 2$

$y' = x^3 - \dfrac{1}{4x^3}; \quad (y')^2 = x^6 - \dfrac{1}{2} + \dfrac{1}{16x^6}$

$L = \int_1^2 \sqrt{1 + (y')^2}\, dx = \int_1^2 \sqrt{x^6 + \dfrac{1}{2} + \dfrac{1}{16x^6}}\, dx = \int_1^2 \left(x^3 + \dfrac{1}{4x^3}\right) dx = \dfrac{x^4}{4} - \dfrac{1}{8x^2}\Big]_1^2$

$= \dfrac{1}{4}(16 - 1) - \dfrac{1}{8}\left(\dfrac{1}{4} - 1\right) = \dfrac{123}{32} \approx 3.8438$

9. $y = \dfrac{x^3}{6} + \dfrac{1}{2x}$, from $\left(1, \dfrac{2}{3}\right)$ to $\left(4, \dfrac{259}{24}\right)$.

$$y' = \frac{x^2}{2} - \frac{1}{2x^2}; \ (y')^2 = \frac{1}{4}\left(x^4 - 2 + \frac{1}{x^4}\right)$$

$$\begin{aligned} L &= \int_1^4 \sqrt{1 + (y')^2}\, dx = \frac{1}{2}\int_1^4 \sqrt{x^4 + 2 + \frac{1}{x^4}}\, dx = \frac{1}{2}\int_1^4 \left(x^2 + \frac{1}{x^2}\right) dx \\ &= \frac{1}{2}\left[\frac{x^3}{3} - \frac{1}{x}\right]_1^4 = \frac{1}{2}\left[\frac{1}{3}(64 - 1) - \left(\frac{1}{4} - 1\right)\right] = \frac{87}{8} = 10.875 \end{aligned}$$

10. $y = \frac{2}{3}x^{3/2} - \frac{1}{2}x^{1/2}$, from $\left(1, \frac{1}{6}\right)$ to $\left(3, \frac{3\sqrt{3}}{2}\right)$.

$$y' = x^{1/2} - \frac{1}{4}x^{-1/2}; \ (y')^2 = x - \frac{1}{2} + \frac{1}{16x}$$

$$\begin{aligned} L &= \int_1^3 \sqrt{1 + (y')^2}\, dx = \int_1^3 \sqrt{x + \frac{1}{2} + \frac{1}{16x}}\, dx = \int_1^3 \left(x^{1/2} + \frac{1}{4}x^{-1/2}\right) dx \\ &= \frac{2}{3}x^{3/2} + \frac{1}{2}x^{1/2}\bigg]_1^3 = \frac{2}{3}(3^{3/2} - 1) + \frac{1}{2}(3^{1/2} - 1) \approx 3.1635 \end{aligned}$$

11. $y = \frac{x^4 + 6x + 3}{6x}$, for $2 \le x \le 4$

$$y' = \frac{x^2}{2} + 0 - \frac{1}{2x^2}; \ (y')^2 = \frac{1}{4}\left(x^4 - 2 + \frac{1}{x^4}\right)$$

$$\begin{aligned} L &= \int_2^4 \sqrt{1 + (y')^2}\, dx = \frac{1}{2}\int_2^4 \sqrt{x^4 + 2 + \frac{1}{x^4}}\, dx = \frac{1}{2}\int_2^4 \left(x^2 + \frac{1}{x^2}\right) dx = \frac{1}{2}\left[\frac{x^3}{3} - \frac{1}{x}\right]_2^4 \\ &= \frac{1}{2}\left[\frac{1}{3}(64 - 8) - \left(\frac{1}{4} - \frac{1}{2}\right)\right] = \frac{227}{24} \approx 9.4583 \end{aligned}$$

12. $y = \frac{3x^5}{5} + \frac{1}{36x^3}$, for $1 \le x \le 2$

$$y' = 3x^4 - \frac{1}{12x^4}; \ (y')^2 = 9x^{16} - \frac{1}{2} + \frac{1}{144x^{16}}$$

$$\begin{aligned} L &= \int_1^2 \sqrt{1 + (y')^2}\, dx = \int_1^2 \sqrt{9x^{16} + \frac{1}{2} + \frac{1}{144x^{16}}}\, dx = \int_1^2 \left(3x^4 + \frac{1}{12x^4}\right) dx \\ &= \frac{3x^5}{5} - \frac{1}{36x^3}\bigg]_1^2 = \frac{3}{5}(32 - 1) - \frac{1}{36}\left(\frac{1}{8} - 1\right) = \frac{26,819}{1440} \approx 18.6243 \end{aligned}$$

13. $y = 2x + 3$, x-axis, for $1 \le x \le 3$ about the x-axis.

$$y' = 2; \ (y')^2 = 4$$

$$\begin{aligned} S &= \int_1^3 2\pi y\sqrt{1 + (y')^2}\, dx = 2\pi\int_1^3 (2x + 3)\sqrt{5}\, dx = 2\pi\sqrt{5}\left[x^2 + 3x\right]_1^3 \\ &= 2\pi\sqrt{5}\left[(9 - 1) + 3(3 - 1)\right] = 2\pi\sqrt{5}(14) = 28\pi\sqrt{5} \approx 196.695 \end{aligned}$$

14. $y = x^3$, x-axis, for $0 \leq x \leq 3$ about the x-axis.

$y' = 3x^2$; $(y')^2 = 9x^4$

$$S = \int_0^3 2\pi y \sqrt{1+(y')^2}\, dx = 2\pi \int_0^3 x^3 \sqrt{1+9x^4}\, dx = \frac{2\pi}{36} \cdot \frac{2}{3}\left(1+9x^4\right)^{3/2}\Big]_0^3$$

$$= \frac{\pi}{27}\left(730^{3/2} - 1\right) \approx 2294.819$$

15. $y = x^{1/3}$, y-axis, for $0 \leq y \leq 2$ about the y-axis.

$x = y^3$; $\dfrac{dx}{dy} = 3y^2$; $\left(\dfrac{dx}{dy}\right)^2 = 9y^4$

$$S = \int_0^2 2\pi x \sqrt{1+(x')^2}\, dy = 2\pi \int_0^2 y^3 \sqrt{1+9y^4}\, dy = \frac{2\pi}{36} \cdot \frac{2}{3}\left(1+9y^4\right)^{3/2}\Big]_0^2$$

$$= \frac{\pi}{27}\left(145^{3/2} - 1\right) \approx 203.044$$

16. $x = 2\sqrt{y}$, y-axis, for $1 \leq y \leq 4$ about the y-axis.

$\dfrac{dx}{dy} = \dfrac{1}{\sqrt{y}}$; $\left(\dfrac{dx}{dy}\right)^2 = \dfrac{1}{y}$

$$S = \int_1^4 2\pi x \sqrt{1+(x')^2}\, dy = 2\pi \int_1^4 2\sqrt{y}\sqrt{1+\frac{1}{y}}\, dy = 4\pi \int_1^4 \sqrt{y+1}\, dy = \frac{8\pi}{3}(y+1)^{3/2}\Big]_1^4$$

$$= \frac{8\pi}{3}\left(5^{3/2} - 2^{3/2}\right) \approx 69.969$$

17. $y = \sqrt{x}$, x-axis, for $1 \leq x \leq 2$ about the x-axis.

$y' = \dfrac{1}{2\sqrt{x}}$; $(y')^2 = \dfrac{1}{4x}$

$$S = \int_1^2 2\pi y \sqrt{1+(y')^2}\, dx = 2\pi \int_1^2 \sqrt{x}\sqrt{1+\frac{1}{4x}}\, dx = \pi \int_1^2 \sqrt{4x+1}\, dx = \frac{\pi}{6}(4x+1)^{3/2}\Big]_1^2$$

$$= \frac{\pi}{6}\left(9^{3/2} - 5^{3/2}\right) = \frac{\pi}{6}\left(27 - 5^{3/2}\right) \approx 8.2832$$

18. $y = \sqrt{2x - 1}$, x-axis, for $2 \leq x \leq 8$ about the x-axis.

$y' = \dfrac{1}{\sqrt{2x-1}}$; $(y')^2 = \dfrac{1}{2x-1}$

$$S = \int_2^8 2\pi y \sqrt{1+(y')^2}\, dx = 2\pi \int_2^8 \sqrt{2x-1}\sqrt{1+\frac{1}{2x-1}}\, dx = 2\pi \int_2^8 \sqrt{2x}\, dx$$

$$= 2\sqrt{2}\pi \left(\frac{2}{3}\right) x^{3/2}\Big]_2^8 = \frac{4\sqrt{2}\pi}{3}\left(8^{3/2} - 2^{3/2}\right) = \frac{16\pi}{3}\left(4^{3/2} - 1\right) = \frac{112\pi}{3} \approx 117.286$$

19. $12xy - 3y^4 = 4$, y-axis, for $1 \leq y \leq 3$ about the y-axis.

544

$$x = \frac{1}{3y} + \frac{y^3}{4}; \quad \frac{dx}{dy} = -\frac{1}{3y^2} + \frac{3y^2}{4}$$

$$\left(\frac{dx}{dy}\right)^2 = \frac{1}{9y^4} - \frac{1}{2} + \frac{9y^4}{16}$$

$$\begin{aligned}
S &= \int_1^3 2\pi x \sqrt{1 + (x')^2}\, dy = 2\pi \int_1^3 \left(\frac{1}{3y} + \frac{y^3}{4}\right)\sqrt{\frac{1}{9y^4} + \frac{1}{2} + \frac{9y^4}{16}}\, dy \\
&= 2\pi \int_1^3 \left(\frac{1}{3y} + \frac{y^3}{4}\right)\left(\frac{1}{3y^2} + \frac{3y^2}{4}\right) dy = 2\pi \int_1^3 \left(\frac{1}{9y^3} + \frac{y}{3} + \frac{3y^5}{16}\right) dy \\
&= 2\pi \left[-\frac{1}{18y^2} + \frac{y^2}{6} + \frac{3y^6}{96}\right]_1^3 = 2\pi \left[-\frac{1}{18}\left(\frac{1}{9} - 1\right) + \frac{1}{6}(9 - 1) + \frac{1}{32}(729 - 1)\right] \\
&= \frac{7819\pi}{162} \approx 151.630
\end{aligned}$$

20. $y = \dfrac{x^3}{4} + \dfrac{1}{3x}$, x-axis, for $1 \le x \le 3$ about the x-axis.

$$y' = \frac{3x^2}{4} - \frac{1}{3x^2}; \quad (y')^2 = \frac{9x^4}{16} - \frac{1}{2} + \frac{1}{9x^4}$$

$$\begin{aligned}
S &= \int_1^3 2\pi y \sqrt{1 + (y')^2}\, dx = 2\pi \int_1^3 \left(\frac{x^3}{4} + \frac{1}{3x}\right)\sqrt{\frac{9x^4}{16} + \frac{1}{2} + \frac{1}{9x^4}}\, dx \\
&= 2\pi \int_1^3 \left(\frac{x^3}{4} + \frac{1}{3x}\right)\left(\frac{3x^2}{4} + \frac{1}{3x^2}\right) dx = \frac{7819\pi}{162} \approx 151.630,
\end{aligned}$$

as in Exercise 19.

21. $f(x) = \displaystyle\int_0^x \sqrt{2t^2 + t^4}\, dt$, $1 \le x \le 2$

$$L = \int_1^2 \sqrt{1 + (y')^2}\, dx; \quad y' = \sqrt{2x^2 + x^4}$$

$$L = \int_1^2 \sqrt{1 + 2x^2 + x^4}\, dx = \int_1^2 (1 + x^2)\, dx = x + \frac{x^3}{3}\bigg]_1^2 = (2 - 1) + \frac{1}{3}(8 - 1) = \frac{10}{3}$$

22. $f(x) = \displaystyle\int_1^x \sqrt{t^2 - 1}\, dt$, $2 \le x \le 10$

$$f'(x) = \sqrt{x^2 - 1},\ [f'(x)]^2 = x^2 - 1$$

$$L = \int_2^{10} \sqrt{1 + (x^2 - 1)}\, dx = \int_2^{10} x\, dx = \frac{x^2}{2}\bigg]_2^{10} = \frac{1}{2}(100 - 4) = 48$$

23. $y = \dfrac{rx}{\sqrt{l^2 - r^2}}$, $0 \le x \le \sqrt{l^2 - r^2}$, about the x-axis.

$$y' = \frac{r}{\sqrt{l^2 - r^2}},\ 1 + (y')^2 = 1 + \frac{r^2}{l^2 - r^2} = \frac{l^2}{l^2 - r^2}$$

$$S = \int_0^{\sqrt{l^2 - r^2}} 2\pi y \sqrt{1 + (y')^2}\, dx = 2\pi \int_0^{\sqrt{l^2 - r^2}} \frac{rx}{\sqrt{l^2 - r^2}} \cdot \frac{l}{\sqrt{l^2 - r^2}}\, dx$$

$$= \frac{2\pi r l}{l^2 - r^2} \left(\frac{x^2}{2}\right)\Bigg]_0^{\sqrt{l^2 - r^2}} = \frac{\pi r l}{l^2 - r^2}(l^2 - r^2) = \pi r l$$

24. $x^2 + (y-1)^2 = 25$ from $(3,5)$ to $(4,4)$ about the x-axis.

$(y-1)^2 = 25 - x^2$; $y = 1 + \sqrt{25 - x^2}$; $y' = \dfrac{-x}{\sqrt{25 - x^2}}$

$$S = \int_3^4 2\pi y \sqrt{1 + (y')^2}\, dx = 2\pi \int_3^4 \left(1 + \sqrt{25 - x^2}\right)\sqrt{1 + \frac{x^2}{25 - x^2}}\, dx$$

$$= 2\pi \int_3^4 \left(1 + \sqrt{25 - x^2}\right)\left(\frac{5}{\sqrt{25 - x^2}}\right) dx = 10\pi \int_3^4 \left(\frac{1}{\sqrt{25 - x^2}} + 1\right) dx \approx 40.3316$$

Answers will vary, depending on the value of n used in the Trapezoidal Rule. The exact answer is $10\pi \left(\sin^{-1}\dfrac{4}{5} - \sin^{-1}\dfrac{3}{5} + 1\right)$.

25. Circle centered at $(1,1)$ with radius 5. Arc joining $(4,5)$ and $(5,4)$ revolved about x-axis.

$(x-1) + (y-1)^2 = 25$; $y = 1 + \sqrt{25 - (x-1)^2}$

$y' = \dfrac{-(x-1)}{\sqrt{25 - (x-1)^2}}$; $1 + (y')^2 = 1 + \dfrac{(x-1)^2}{25 - (x-1)^2} = \dfrac{25}{25 - (x-1)^2}$

$$S = \int_4^5 2\pi y \sqrt{1 + (y')^2}\, dx = 2\pi \int_4^5 \left[1 + \sqrt{25 - (x-1)^2}\right]\left(\frac{5}{\sqrt{25 - (x-1)^2}}\right) dx$$

$$= 10\pi \int_4^5 \left(\frac{1}{\sqrt{25 - (x-1)^2}} + 1\right) dx$$

Let $u = x - 1$, $du = dx$; $x = 4 \Rightarrow u = 3$, $x = 5 \Rightarrow u = 4$

$$S = 10\pi \int_3^4 \left(\frac{1}{\sqrt{25 - u^2}} + 1\right) du \approx 40.3316 \text{ as in Exercise 24.}$$

26. $f(x) = \sin x$ between $(0,0)$ and $\left(\dfrac{\pi}{2}, 1\right)$.

$f'(x) = \cos x$

$$L = \int_0^{\pi/2} \sqrt{1 + \cos^2 x}\, dx$$

$n = 10 \Rightarrow \Delta x = \dfrac{\pi}{20}$, $x_i = \left(\dfrac{\pi}{20}\right)i$, $f(x_i) = \sqrt{1 + \cos^2 x_i}$

i	x_i	$f(x_i)$
0	0	1.4142
1	$\pi/20$	1.4055
2	$\pi/10$	1.3800
3	$3\pi/20$	1.3394
4	$\pi/5$	1.2863
5	$\pi/4$	1.2247

i	x_i	$f(x_i)$
6	$3\pi/10$	1.1600
7	$7\pi/20$	1.0982
8	$2\pi/5$	1.0467
9	$9\pi/20$	1.0122
10	$\pi/2$	1.0000

$$L \approx \frac{h}{2}\left[f(x_0) + f(x_{10}) + 2\sum_{i=1}^{9} f(x_i)\right] \approx 1.9101$$

27. $f(x) = \sin x$ between $(0,0)$ and $(\pi, 0)$ about the x-axis.

$f'(x) = \cos x$

$$S = \int_0^{\pi} 2\pi y \sqrt{1 + (y')^2}\, dx = 2\pi \int_0^{\pi} \sqrt{1 + \cos^2 x}\, \sin x\, dx$$

Let $n = 10$, $\Delta x = \dfrac{\pi}{10}$, $x_i = \dfrac{\pi}{20} + \left(\dfrac{\pi}{10}\right)(i-1)$, $f(x_i) = \sqrt{1 + \cos^2 x_i}\, \sin x_i$

i	x_i	$f(x_i)$
1	$\pi/20$	0.2199
2	$3\pi/20$	0.6081
3	$5\pi/20$	0.8660
4	$7\pi/20$	0.9785
5	$9\pi/20$	0.9997

i	x_i	$f(x_i)$
6	$11\pi/20$	0.9997
7	$13\pi/20$	0.9785
8	$15\pi/20$	0.8660
9	$17\pi/20$	0.6081
10	$19\pi/20$	0.2199

$$S \approx 2\pi h \sum_{i=1}^{10} f(x_i) \approx 14.497$$

28. $f(x) = x^2$ between $(-1, 1)$ and $(1, 1)$.

$f'(x) = 2x$

$$L = \int_{-1}^{1} \sqrt{1 + 4x^2}\, dx$$

Let $n = 20$, $\Delta x = 0.2$, $x_i = -1 + 0.2i$, $f(x_i) = \sqrt{1 + 4x_i^2}$

i	x_i	$f(x_i)$
0	-1.0	2.2361
1	-0.8	1.8868
2	-0.6	1.5620
3	-0.4	1.2806
4	-0.2	1.0770
5	0.0	1.0000

i	x_i	$f(x_i)$
6	0.2	1.0770
7	0.4	1.2806
8	0.6	1.5620
9	0.8	1.8868
10	1.0	2.2361

By Simpson's Rule, $L \approx \dfrac{h}{3}[f(x_0) + 4f(x_1) + 2f(x_2) + \cdots + 4f(x_9) + f(x_{10})] \approx 2.958$.

29. $\dfrac{x^2}{a^2} + \dfrac{y^2}{b^2} = 1$, for $-\dfrac{a}{2} \le x \le \dfrac{a}{2}$ about the x-axis.

$y = \dfrac{b}{a}\sqrt{a^2 - x^2}$; $\quad y' = \dfrac{-bx}{a\sqrt{a^2 - x^2}}$

$1 + (y')^2 = 1 + \dfrac{b^2 x^2}{a^2(a^2 - x^2)}$

Let $a = 4$, $b = 3$.

$$S = \int_{-2}^{2} 2\pi \left(\frac{3}{4}\sqrt{16 - x^2}\right) \sqrt{1 + \frac{9x^2}{16(16 - x^2)}}\, dx = \frac{3\pi}{8} \int_{-2}^{2} \sqrt{256 - 7x^2}\, dx$$

547

With $n = 10$, $\Delta x = 0.4$, $x_i = -2 + 0.4i$, $f(x_i) = \sqrt{256 - 7x_i^2}$

i	x_i	$f(x_i)$
0	-2.0	15.0997
1	-1.6	15.4298
2	-1.2	15.6818
3	-0.8	15.8594
4	-0.4	15.9650
5	0.0	16.0000

i	x_i	$f(x_i)$
6	0.4	15.9650
7	0.8	15.8594
8	1.2	15.6818
9	1.6	15.4298
10	2.0	15.0997

$$S \approx \frac{3\pi}{8} \left(\frac{h}{3}\right) [f(x_0) + 4f(x_1) + 2f(x_2) + \cdots + 4f(x_9) + f(x_{10})] \approx 74.000$$

7.4 Distance and Velocity

1. $v(t) = 9 - t^2$, $a = 0$, $b = 3$
$v(t) \geq 0$ on $[0, 3]$.

$$D = \int_0^3 v(t)\, dt = \int_0^3 (9 - t^2)\, dt = 9t - \frac{t^3}{3}\Big]_0^3 = 27 - 9 = 18$$

2. $v(t) = \sqrt{t + 1}$, $a = 0$, $b = 8$

$$D = \int_0^8 \sqrt{t + 1}\, dt = \frac{2}{3}(t + 1)^{3/2}\Big]_0^8 = \frac{2}{3}(9^{3/2} - 1) = \frac{2}{3}(27 - 1) = \frac{52}{3} \approx 17.333$$

3. $v(t) = \sin t$, $a = 0$, $b = \pi$

$$D = \int_0^\pi \sin t\, dt = -\cos t\Big]_0^\pi = -(-1 - 1) = 2$$

4. $v(t) = t - t^3$, $a = 0$, $b = 1$

$$D = \int_0^1 (t - t^3)\, dt = \frac{t^2}{2} - \frac{t^4}{4}\Big]_0^1 = \frac{1}{2} - \frac{1}{4} = \frac{1}{4}$$

5. $v(t) = 2 + t^2$, $a = 0$, $b = 5$

$$D = \int_0^5 (2 + t^2)\, dt = 2t + \frac{t^3}{3}\Big]_0^5 = 10 + \frac{125}{3} = \frac{155}{3} \approx 51.667$$

6. $v(t) = t^2 + t - 6$, $a = 0$, $b = 2$
$v(t) = (t + 3)(t - 2) \Rightarrow v(t) = 0$ at $t = -3, 2$.

$$D = \int_0^2 |v(t)|\, dt = \int_0^2 (-t^2 - t + 6)\, dt = -\frac{t^3}{3} - \frac{t^2}{2} + 6t\Big]_0^2 = -\frac{8}{3} - 2 + 12 = \frac{22}{3}$$

7. $v(t) = \cos \pi t$, $a = \dfrac{1}{2}$, $b = \dfrac{3}{2}$

$\cos \pi t \;=\; 0$ at $t = \dfrac{1}{2}$, $t = \dfrac{3}{2}$

$\cos \pi t \;\geq\; 0$ on $\left[0, \dfrac{1}{2}\right]$

$\cos \pi t \;\leq\; 0$ on $\left[\dfrac{1}{2}, \dfrac{3}{2}\right]$

$$D = -\int_{1/2}^{3/2} \cos \pi t \, dt = -\left. \frac{1}{\pi} \sin \pi t \right]_{1/2}^{3/2} = -\frac{1}{\pi}(-1-1) = \frac{2}{\pi}$$

8. $v(t) = \dfrac{1-t}{\sqrt{t}}$, $a = 2$, $b = 4$

$$D = \int_2^4 |v(t)| \, dt = \int_2^4 \left|\frac{1-t}{\sqrt{t}}\right| \, dt = \int_2^4 \left(\sqrt{t} - \frac{1}{\sqrt{t}}\right) dt = \left. \frac{2}{3} t^{3/2} - 2\sqrt{t} \right]_2^4$$

$$= \frac{2}{3}(8 - 2^{3/2}) - 2\left(2 - \sqrt{2}\right) = \frac{4}{3}\left(4 - \sqrt{2}\right) - 2\left(2 - \sqrt{2}\right) = \frac{4}{3} + \frac{2\sqrt{2}}{3} \approx 2.2761$$

9. $v(t) = \sin^2 t \cos t$, $a = 0$, $b = \dfrac{\pi}{2}$

$$D = \int_0^{\pi/2} \sin^2 t \cos t \, dt = \left. \frac{1}{3} \sin^3 t \right]_0^{\pi/2} = \frac{1}{3}$$

10. $v(t) = \sqrt{t}\,(t+2)$, $a = 0$, $b = 1$

$$D = \int_0^1 \sqrt{t}\,(t+2)\, dt = \int_0^1 \left(t^{3/2} + 2\sqrt{t}\right) dt$$

$$= \left[\frac{2}{5} t^{5/2} + \frac{4}{3} t^{3/2}\right]_0^1 = \frac{2}{5} + \frac{4}{3} = \frac{26}{15}$$

11. $v(t) = (2t - 5)(t^2 - 5t + 6)^{1/3}$, $a = 3$, $b = 4$
$t^2 - 5t + 6 = (t - 2)(t - 3) \Rightarrow$ zeros at $t = 2, 3$.

$$D = \int_3^4 (t^2 - 5t + 6)^{1/3}(2t - 5)\, dt = \left. \frac{3}{4}(t^2 - 5t + 6)^{4/3} \right]_3^4 = \frac{3}{4}[2^{4/3} - 0] = \frac{3}{2^{2/3}} \approx 1.8899$$

12. $v(t) = t - 4$, $t = 0$ to $t = 6$
$v(t) \leq 0$ on $[0, 4]$; $v(t) \geq 0$ on $[4, 6]$.

$$D = -\int_0^4 (t - 4)\, dt + \int_4^6 (t - 4)\, dt = \left. -\frac{t^2}{2} + 4t \right]_0^4 + \left. \frac{t^2}{2} - 4t \right]_4^6 = -8 + 16 + \frac{1}{2}(36 - 16) - 4(6 - 4) = 10$$

13. $v(t) = 2 - t^2$, $t = 0$ to $t = 4$.
    $2 - t^2 = 0 \Rightarrow t = \sqrt{2}$

$$D = \int_0^{\sqrt{2}} (2 - t^2)\, dt + \int_{\sqrt{2}}^4 (t^2 - 2)\, dt = 2t - \frac{t^3}{3}\Big]_0^{\sqrt{2}} + \frac{t^3}{3} - 2t\Big]_{\sqrt{2}}^4$$

$$= 2\sqrt{2} - \frac{2\sqrt{2}}{3} + \frac{64}{3} - 8 - \frac{2\sqrt{2}}{3} + 2\sqrt{2} = \frac{8\sqrt{2}}{3} + \frac{40}{3} \approx 17.105$$

14. $v(t) = \sin(2t)$, $t = 0$ to $t = \frac{\pi}{2}$.
    $\sin(2t) = 0 \Rightarrow 2t = 0,\ \pi \Rightarrow t = 0,\ \frac{\pi}{2}$

$$D = \int_0^{\pi/2} \sin(2t)\, dt = -\frac{1}{2}\cos(2t)\Big]_0^{\pi/2} = -\frac{1}{2}(-1 - 1) = 1$$

15. $v(t) = \cos t \sin t$, $t = 0$ to $t = \frac{\pi}{2}$.
    $\cos t \sin t \geq 0$ on $\left[0, \frac{\pi}{2}\right]$

$$D = \int_0^{\pi/2} \cos t \sin t\, dt = \frac{1}{2}\sin^2 t\Big]_0^{\pi/2} = \frac{1}{2}$$

16. $v(t) = t^{1/3} - t^{2/3}$, $t = 0$ to $t = 8$.
    $t^{1/3} - t^{2/3} = 0 \Rightarrow t(t - 1) = 0 \Rightarrow t = 0,\ 1$

$$D = \int_0^1 (t^{1/3} - t^{2/3})\, dt + \int_1^8 (t^{2/3} - t^{1/3})\, dt = \frac{3}{4}t^{4/3} - \frac{3}{5}t^{5/3}\Big]_0^1 + \frac{3}{5}t^{5/3} - \frac{3}{4}t^{4/3}\Big]_1^8$$

$$= \frac{3}{4} - \frac{3}{5} + \frac{3}{5}(32 - 1) - \frac{3}{4}(16 - 1) = \frac{3}{20} + \frac{93}{5} - \frac{45}{4} = \frac{150}{20} = 7.5$$

17. $v(t) = (t - 1)^5$, $t = 0$ to $t = 2$.
    $(t - 1)^5 = 0 \Rightarrow t = 1$

$$D = \int_0^1 -(t - 1)^5\, dt + \int_1^2 (t - 1)^5\, dt = -\frac{1}{6}(t - 1)^6\Big]_0^1 + \frac{1}{6}(t - 1)^6\Big]_1^2 = -\frac{1}{6}(-1) + \frac{1}{6}(1) = \frac{1}{3}$$

18. $v(t) = \dfrac{t^2 - 9}{t + 3}$, $t = 0$ to $t = 4$.
    $t^2 - 9 = 0 \Rightarrow t = 3$

$$D = -\int_0^3 \left(\frac{t^2 - 9}{t + 3}\right) dt + \int_3^4 \left(\frac{t^2 - 9}{t + 3}\right) dt = -\int_0^3 (t - 3)\, dt + \int_3^4 (t - 3)\, dt$$

$$= -\left(\frac{t^2}{2} - 3t\right)\Big]_0^3 + \left(\frac{t^2}{2} - 3t\right)\Big]_3^4 = -\left(\frac{9}{2} - 9\right) + \left[\left(8 - \frac{9}{2}\right) - (12 - 9)\right] = 5$$

19. $v(t) = t(t^2 - 9)^5$, $t = 2$ to $t = 4$.
    $t^2 - 9 = 0 \Rightarrow t = 3$

$$D = -\int_2^3 t(t^2 - 9)^5\, dt + \int_3^4 t(t^2 - 9)^5\, dt = -\frac{1}{12}(t^2 - 9)^6\Big]_2^3 + \frac{1}{12}(t^2 - 9)^6\Big]_3^4$$

$$= -\frac{1}{12}(0 - 15{,}625) + \frac{1}{12}(117{,}649) = \frac{133{,}274}{12} = \frac{66{,}637}{6} \approx 11{,}106.17$$

20. $v(t) = t^2(t^3 - 1)^3$, $t = 0$ to $t = 2$.
$t^3 - 1 = 0 \Rightarrow t = 1$

$$D = -\int_0^1 t^2(t^3-1)^3 \, dt + \int_1^2 t^2(t^3-1)^3 \, dt = -\frac{1}{12}(t^3-1)^4 \Big]_0^1 + \frac{1}{12}(t^3-1)^4 \Big]_1^2$$

$$= -\frac{1}{12}(0-1) + \frac{1}{12}(7^4-0) = \frac{7^4+1}{12} \approx 200.167$$

21.   a) $v(t) = t - 4$, $t = 0$ to $t = 6$.
Distance = 10, calculated in Exercise 12.
Displacement $= \dfrac{t^2}{2} - 4t \Big]_0^6 = 18 - 24 = -6$.

    b) $v(t) = 2 - t^2$, $t = 0$ to $t = 4$.
Distance $= \dfrac{8\sqrt{2} + 40}{3}$, calculated in Exercise 13.
Displacement $= 2t - \dfrac{t^3}{3} \Big]_0^4 = 8 - \dfrac{64}{3} = -\dfrac{40}{3}$.

    c) $v(t) = \sin(2t)$, $t = 0$ to $t = \frac{\pi}{2}$.
Since $v(t) \geq 0$ on $\left[0, \frac{\pi}{2}\right]$, the distance and the displacement are the same. The value is 1, calculated in Exercise 14.

    d) $v(t) = \cos t \sin t$, $t = 0$ to $t = \frac{\pi}{2}$.
Since $v(t) \geq 0$ on $\left[0, \frac{\pi}{2}\right]$, the distance and the displacement are the same. The value is $\frac{1}{2}$, calculated in Exercise 15.

22. $s(t) = t^2 - 7t + 10$

    a) $v(t) = s'(t) = 2t - 7$
    b) $a(t) = v'(t) = 2$
    c) $v(t) = 0 \Rightarrow 2t - 7 = 0 \Rightarrow t = \frac{7}{2}$
    d) $v(t) > 0$ when $2t - 7 > 0 \Rightarrow t > \frac{7}{2}$
    e) $v(t) < 0$ when $2t - 7 < 0 \Rightarrow t < \frac{7}{2}$

    f) $\quad D = -\int_1^{7/2}(2t-7)\,dt + \int_{7/2}^8(2t-7)\,dt = -(t^2-7t)\Big]_1^{7/2} + (t^2-7t)\Big]_{7/2}^8$

$$= -\left[\left(\frac{49}{4}-1\right)-7\left(\frac{7}{2}-1\right)\right] + \left[\left(64-\frac{49}{4}\right)-7\left(8-\frac{7}{2}\right)\right] = \frac{25}{4} + \frac{81}{4} = \frac{53}{2}$$

    g) Displacement $= t^2 - 7t \Big]_1^8 = (64 - 56) - (1 - 7) = 14$.

23. $v(t) = -t^2 + 5t - 6$, $s(0) = 4$

    a) $\quad s(t) = \int(-t^2 + 5t - 6)\,dt = -\frac{t^3}{3} + \frac{5}{2}t^2 - 6t + C$

$\qquad s(0) = 4 \Rightarrow C = 4$

$\qquad s(t) = -\frac{t^3}{3} + \frac{5}{2}t^2 - 6t + 4$

b) $a(t) = v'(t) = -2t + 5$

c) $v(t) = 0 \Rightarrow (t-3)(t-2) = 0 \Rightarrow t = 2, 3$

d) $v(t) > 0$ for $t \in (2,3)$

e) $v(t) < 0$ for $0 \le t \le 2$, $3 \le t < \infty$

f) $\quad D \;=\; -\int_0^2 (-t^2 + 5t - 6)\, dt + \int_2^3 (-t^2 + 5t - 6)\, dt$

$$= \left. \frac{t^3}{3} - \frac{5}{2} t^2 + 6t \right]_0^2 - \left(\frac{t^3}{3} - \frac{5}{2} t^2 + 6t \right) \Big]_2^3$$

$$= \frac{8}{3} - 10 + 12 - \left(9 - \frac{45}{2} + 18 \right) + \left(\frac{8}{3} - 10 + 12 \right) = \frac{29}{6}$$

g) Displacement $= -\dfrac{t^3}{3} + \dfrac{5}{2} t^2 - 6t \Big]_0^3 = -9 + \dfrac{45}{2} - 18 = -4.5.$

24. $a(t) = 2t - 4$, $v(0) = 3$, $s(0) = 5$

a) $\quad v(t) \;=\; \int a(t)\, dt = \int (2t - 4)\, dt = t^2 - 4t + C_1$

$\qquad v(0) \;=\; 3 \Rightarrow C_1 = 3$

$\qquad v(t) \;=\; t^2 - 4t + 3$

b) $\quad s(t) \;=\; \int v(t)\, dt = \dfrac{t^3}{3} - 2t^2 + 3t + C_2$

$\qquad \overset{\bullet}{s}(0) \;=\; 5 \Rightarrow C_2 = 5$

$\qquad s(t) \;=\; \dfrac{t^3}{3} - 2t^2 + 3t + 5$

c) $v(t) = 0 \Rightarrow (t-1)(t-3) = 0 \Rightarrow t = 1, 3$

d) $v(t) > 0 \Rightarrow t \in [0,1) \cup (3, \infty)$

e) $v(t) < 0 \Rightarrow t \in (1,3)$

f) $\quad$ Distance $\;=\; \int_0^1 (t^2 - 4t + 3)\, dt - \int_1^3 (t^2 - 4t + 3)\, dt$

$$= \left. \frac{t^3}{3} - 2t^2 + 3t \right]_0^1 - \left(\frac{t^3}{3} - 2t^2 + 3t \right) \Big]_1^3$$

$$= \frac{1}{3} - 2 + 3 - \left[(9 - 18 + 9) - \left(\frac{1}{3} - 2 + 3 \right) \right] = \frac{8}{3}.$$

g) Displacement $= \displaystyle\int_0^3 (t^2 - 4t + 3)\, dt = \dfrac{t^3}{3} - 2t^2 + 3t \Big]_0^3 = 9 - 18 + 9 = 0.$

25. $a(t) = -9.8 \, \text{m/s}^2$, $v(0) = 12 \, \text{m/s}$

a) $\quad v(t) \;=\; \int -9.8\, dt = -9.8t + C_1$

$\qquad v(0) \;=\; 12 \Rightarrow C_1 = 12; \; v(t) = -9.8t + 12$

b) $\quad s(t) \;=\; \displaystyle\int (-9.8t + 12)\, dt = -4.9t^2 + 12t + C_2$

$\qquad s(0) \;=\; 10 \Rightarrow C_2 = 10; \; s(t) = -4.9t^2 + 12t + 10$

c) $v(t) = 0 \Rightarrow -9.8t + 12 = 0 \Rightarrow t = \frac{12}{9.8} \approx 1.2245$ seconds.

d) $\quad s(t) = 0 \;\Rightarrow\; -4.9t^2 + 12t + 10 = 0$

$\qquad\qquad\quad \Rightarrow \; t \approx 3.106$ seconds

e) Displacement $= -10$ meters.

$$\text{Distance} \;=\; \int_0^{1.2245} (-9.8t + 12)\, dt - \int_{1.2245}^{3.106} (-9.8t + 12)\, dt$$

$$=\; (-4.9t^2 + 12t) \Big]_0^{1.2245} - (-4.9t^2 + 12t) \Big]_{1.2245}^{3.106}$$

$$=\; 24.694.$$

f) $v\left(\frac{12}{4.9}\right) = -v(0) = -12 \,\text{m/s}$

The minus sign indicates that the ball is falling.

26. $a = 0.5 \,\text{m/s}^2$ until $v = 36 \,\text{km/h}$; the train decelerates at $a = -0.5 \,\text{m/s}^2$; total time $= 2$ minutes.

$v(t) = \displaystyle\int a(t)\, dt = 0.5t + v_0 = 0.5t$, since $v_0 = 0$.

$36 \,\text{km/h} = 36{,}000 \,\text{m/h} = \frac{36{,}000}{60^2} \,\text{m/s} = 10 \,\text{m/s}$

$v(T) = 0.5T = 10 \,\text{m/s} \Rightarrow T = \frac{10}{0.5} = 20 \,\text{s}$

Therefore $v(t) = \begin{cases} 0.5t & 0 \le t \le 20 \\ 10 & 20 < t \le 100 \\ -0.5(t - 100) + 10 = -0.5t + 60 & 100 < t \le 120. \end{cases}$

$$D \;=\; \int_0^{20} 0.5t\, dt + \int_{20}^{100} 10\, dt + \int_{100}^{120} (60 - 0.5t)\, dt = 0.25t^2 \Big]_0^{20} + 10t \Big]_{20}^{100} + (60t - 0.25t^2) \Big]_{100}^{120}$$

$$=\; 100 + 800 + 60(20) - 0.25(14400 - 10000) = 1000 \,\text{m}$$

27. $a_1(t) = 1.0 \,\text{m/s}^2, \; a_2(t) = -1.0 \,\text{m/s}^2, \; v_{\max} = 108 \,\text{km/h}$

$v(t) = \displaystyle\int a(t)\, dt = 1.0t + v_0 = t$, since $v_0 = 0$

$108 \,\text{km/h} = \frac{108{,}000}{3600} \,\text{m/s} = 30 \,\text{m/s}$

$v(T) = T = 30 \,\text{m/s} \Rightarrow T = 30 \,\text{s}$

Therefore $v(t) = \begin{cases} t & 0 \le t \le 30 \\ 30 & 30 < t < 10(60) \end{cases}$

$$D = \int_0^{30} t\, dt + \int_{30}^{600} 30\, dt = \frac{t^2}{2} \Big]_0^{30} + 30t \Big]_{30}^{600} = 450 + 30(600 - 30) = 17{,}550 \,\text{m} = 17.55 \,\text{km}$$

28. The train accelerates for $30\,\text{s}$, travels at constant speed for $9\frac{1}{2}$ minutes and decelerates for 30 seconds.

$$v(t) = \begin{cases} t & 0 \le t \le 30 \\ 30 & 30 < t \le 570 \\ -t + 600 & 570 < t \le 600 \end{cases}$$

$$\begin{aligned} D &= \int_0^{30} t\, dt + \int_{30}^{570} 30\, dt + \int_{570}^{600} (-t + 600)\, dt = \left. \frac{t^2}{2} \right]_0^{30} + 30t \Big]_{30}^{570} + 600t - \left. \frac{t^2}{2} \right]_{570}^{600} \\ &= 450 + 30(540) + 600(30) - \frac{1}{2}(360,000 - 324,900) = 17,100\,\text{m} = 17.1\,\text{km} \end{aligned}$$

29. If $v(t) \le 0$ for all $t \in [a, b]$, then

$$\int_a^b |v(t)|\, dt = \int_a^b -v(t)\, dt = \int_b^a v(t)\, dt$$

$= s(a) - s(b)$ if s is an antiderivative of v, that is, $s'(t) = v(t)$.

30. $\int_a^b |v(t)|\, dt = $ distance traveled for time $\in [a, b]$.

$s(b) - s(a) = $ change in position for time $\in [a, b]$, or $s(b) - s(a) = \int_a^b v(t)\, dt$. Therefore, $|s(b) - s(a)| < \int_a^b |v(t)|\, dt$.

7.5 Hydrostatic Pressure

1.

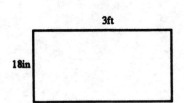

Exercise 1

$$\begin{aligned} F &= \int_0^{1.5} (62.4)\, h\, (3)\, dh \\ &= 187.2 \left. \frac{h^2}{2} \right]_0^{1.5} \\ &= 187.2 \left(\frac{2.25}{2} \right) \\ &= 210.6\,\text{lb} \end{aligned}$$

on each end panel

or $2(210.6) = 421.2$ lb

on both ends.

2.

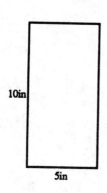

10in

5in

$$F = \int_0^{10/12} (62.4) h \left(\frac{5}{12}\right) dh$$

$$= 26 \left.\frac{h^2}{2}\right]_0^{10/12}$$

$$= 13 \left(\frac{100}{144}\right)$$

$$\approx 9.028 \text{ lb}$$

Exercise 2

3.

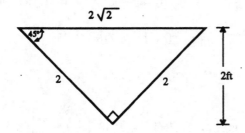

$2\sqrt{2}$

45°

2

2

2ft

Exercise 3

$$F = \int_0^2 (62.4)h(2)(2-h)\, dh = 124.8 \left[h^2 - \frac{h^3}{3}\right]_0^2 = 124.8 \left(\frac{4}{3}\right) = 166.4 \text{ lb}$$

4.

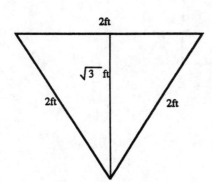

2ft

$\sqrt{3}$ ft

2ft

2ft

Exercise 4

$$F = \int_0^{\sqrt{3}} (62.4)\, h \left[2\left(1 - \frac{h}{\sqrt{3}}\right)\right] dh = 124.8 \left[\frac{h^2}{2} - \frac{h^3}{3\sqrt{3}}\right]_0^{\sqrt{3}} = 124.8 \left(\frac{3}{2} - \frac{3\sqrt{3}}{3\sqrt{3}}\right) = \frac{124.8}{2} = 62.4 \text{ lb}$$

5.

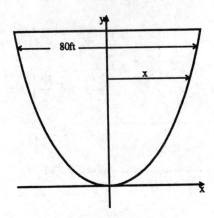

Exercise 5

$$y = \frac{x^2}{100} = \frac{40^2}{100} = 16 \text{ ft depth.}$$

$x^2 = 100y \Rightarrow x = 10\sqrt{y}$, measured from the bottom. $x = 10\sqrt{16 - y}$, measured from the top.

$$F = 2 \int_0^{16} (62.4) h \left(10\sqrt{16 - h}\right) dh = 1248 \int_0^{16} h\sqrt{16 - h} \, dh$$

Let $u = 16 - h$, $du = -dh$, $h = 16 - u$
if $h = 0$, $u = 16$; if $h = 16$, $u = 0$.

$$-1248 \int_{16}^{0} (16 - u) \sqrt{u} \, du \quad = \quad -1248 \left(\frac{32}{3}\right) u^{3/2} \Big]_{16}^{0} + 1248 \left(\frac{2}{5}\right) u^{5/2} \Big]_{16}^{0}$$

$$= \quad 851,968 - 511,180.8 = 340,787.2 \text{ lb}$$

6.

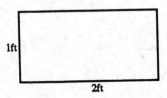

Exercise 6

$$F \quad = \quad \int_0^{x} (62.4) h \, (2) \, dh = 62.4 h^2 \Big]_0^{x} = 62.4x^2$$

If $x = 1$, force $= 62.4$ lb.

For $F = \dfrac{62.4}{2}$, $\dfrac{62.4}{2} = 62.4x^2$.

$$x^2 = \frac{1}{2} \Rightarrow x = \frac{1}{\sqrt{2}} = \frac{\sqrt{2}}{2} \approx 0.7071 \text{ ft}$$

7.

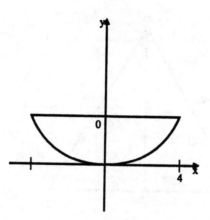

Assume that the trough is full.

Exercise 7

$$\frac{x^2}{16} + \frac{y^2}{4} = 1 \Rightarrow x = \pm 2\sqrt{4 - y^2} \text{ with coordinates centered at top of trough.}$$

$$F = 2\int_0^2 (62.4)\, h\left[2\sqrt{4 - h^2}\right] dh = 124.8 \int_0^2 2h\sqrt{4 - h^2}\, dh = -124.8 \left(\frac{2}{3}\right)(4 - h^2)^{3/2}\Big]_0^2$$

$$= -\frac{249.6}{3}(-8) = 665.6 \text{ lb}$$

8.

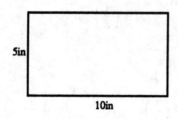

5in

10in

Exercise 8

$$F = \int_3^{30+\frac{5}{12}} (62.4)\, h\left(\frac{10}{12}\right) dh$$

$$= 26h^2\Big]_{30}^{30+\frac{5}{12}}$$

$$\approx 26\,(25.1736)$$

$$\approx 654.51 \text{ lb}$$

9.

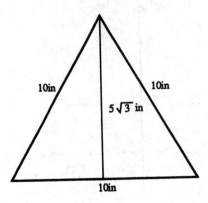

Exercise 9

$$F \ = \ \int_{30}^{30+\frac{5\sqrt{3}}{12}} (62.4)\,h \left[\frac{10}{12}\frac{h-30}{\frac{5\sqrt{3}}{12}}\right]\,dh = \frac{124.8}{\sqrt{3}}\left[\frac{h^3}{3}-15h^2\right]_{30}^{30+\frac{5\sqrt{3}}{12}}$$

$$\approx \ \frac{124.8}{\sqrt{3}}\,[(9665.269-9000)-(14,157.332-13,500)] \approx 571.89 \text{ lb}$$

10. The equation becomes $F = \displaystyle\int_{a}^{b} c\,h\,w(h)\,dh.$

11.

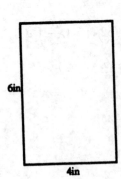

6in

4in

Exercise 11

$$F \ = \ \int_{0}^{1/2}(72)\,h\left(\frac{1}{3}\right)\,dh$$

$$= \ 12h^2\Big]_{0}^{1/2}$$

$$= \ 12\left(\frac{1}{4}\right)$$

$$= \ 3 \text{ lb}$$

12.

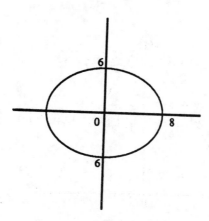

Exercise 12

$\frac{x^2}{64} + \frac{y^2}{36} = 1 \Rightarrow x = \pm\frac{4}{3}\sqrt{36 - y^2}$. The coordinates are centered at the top surface of the oil.

$$F = 2\int_0^6 (58)\, h\left(\frac{4}{3}\sqrt{36 - h^2}\right) dh = \frac{232}{3}\int_0^6 2h\sqrt{36 - h^2}\, dh - \frac{232}{3}\left(\frac{2}{3}\right)(36 - h^2)^{3/2}\Big]_0^6$$

$$= -\frac{464}{9}(-216) = 11,136 \text{ lb}$$

13. $F = \int_0^{12} 58h\left(\frac{4}{3}\sqrt{36 - (h-6)^2}\right) dh \approx 52,230$ lb with $n = 20$ in Simpson's Rule.

(Exact answer is $16,704\pi$ lb $\approx 52,477$ lb)

14. The plate does not move because there is an equal force in the opposite direction on the back face of the plate.

15. $F = \int_2^{2+\frac{10}{12}} (62.4)\, h\left(\frac{15}{12}\right) dh = 39h^2\Big]_2^{2+\frac{10}{12}} = 39\left(\frac{289}{36} - 4\right)$

$$= \frac{39(145)}{36} = \frac{1885}{12} \approx 157.083 \text{ lb}$$

Using the average depth gives $F = (62.4)\left(\frac{15.10}{144}\right)\left(2 + \frac{5}{12}\right) = 65\left(\frac{29}{12}\right) = \frac{1885}{12}$, as calculated above.

16. $F = \int_6^8 (62.4)\, h\,(2)\, dh = 62.4h^2\Big]_6^8 = 62.4(64 - 36) = 1747.2$ lb

Using the average depth gives $F = (62.4)(4)(7) = 1747.2$ lb

17. $F = \int_7^9 (62.4)\, h\,(3)\, dh = 93.6h^2\Big]_7^9 = 93.6(81 - 49) = 2995.2$ lb

Using the average depth gives $F = (62.4)(6)(7 + 1) = 2995.2$ lb.

18. Conjecture: The hydrostatic force exerted on a submerged plate equals the density of the fluid multiplied by the area of the plate and the depth of the center of the plate.

7.6   Work

1. $W = \int_0^{1/2} F(x)\, dx = \int_0^{1/2} 40x\, dx = 20x^2 \Big]_0^{1/2} = \dfrac{20}{4} = 5$ ft-lb

2. $W = \int_0^{0.80} kx\, dx = \int_0^{0.80} 5x\, dx = \dfrac{5}{2} x^2 \Big]_0^{0.80} = \dfrac{5}{2}(0.64) = 1.60$ joules

3. $W = \int_{1/2}^1 40x\, dx = 20x^2 \Big]_{1/2}^1 = 20 - 5 = 15$ ft-lb

4. $W = \int_{0.80}^{1.20} 5x\, dx = \dfrac{5}{2} x^2 \Big]_{0.80}^{1.20} = \dfrac{5}{2}(1.44 - 0.64) = 2.00$ joules

5. $F = kx;\ 50 = k\left(\frac{4}{12}\right);\ k = 150$ lb/ft

$$W = \int_{1/3}^{2/3} 150x\, dx = 75x^2 \Big]_{1/3}^{2/3} = 75\left(\frac{4}{9} - \frac{1}{9}\right) = 25 \text{ ft-lb}$$

6. $W = 40 = \int_0^{1/2} kx\, dx = \dfrac{k}{2} x^2 \Big]_0^{1/2} = \dfrac{k}{8}$

$$k = 40(8) = 320 \text{ lb/ft}$$

7. True. $W_1 = \int_A^B F(x)\, dx,\ W_2 = \int_B^C F(x)\, dx$

$$W = W_1 + W_2 = \int_A^C F(x)\, dx$$

The property of definite integrals is

$$\int_A^B f(x)\, dx + \int_B^C f(x)\, dx = \int_A^C f(x)\, dx.$$

8. $W = 2 = \int_0^6 kx\, dx = \dfrac{kx^2}{2} \Big]_0^6 = 18k$

$$k = \frac{2}{18} = \frac{1}{9} \text{ dyne/cm}$$

9.   a) $F = kx;\ 10 = k(2);\ k = 5$ lb/ft

     b) $F = 5(3) = 15$ lb; additional weight $= 5$ lb.

     c) $W = \int_0^3 kx\, dx = \int_0^3 5x\, dx = \dfrac{5}{2} x^2 \Big]_0^3 = \dfrac{45}{2}$ ft-lb

10.

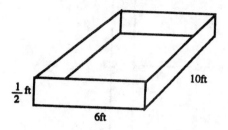

Exercise 10

$$W = \int_0^{1/2} (62.4)(6.10)h\, dh = 1872\, h^2 \Big]_0^{1/2} = 468 \text{ ft-lb}$$

11.

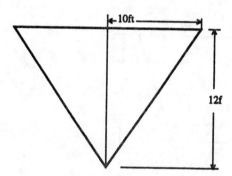

Exercise 11

r depends on h: $r = 10\left(1 - \dfrac{h}{12}\right)$

$$
\begin{aligned}
W &= \int_0^{12} (62.4)\pi \left[100\left(1 - \frac{h}{12}\right)^2 \right] h\, dh \\
&= 6240\,\pi \int_0^{12} \left(h - \frac{h^2}{6} + \frac{h^3}{144} \right) dh \\
&= 6240\pi \left[\frac{h^2}{2} - \frac{h^3}{18} + \frac{h^4}{576} \right]_0^{12} \\
&= 6240\pi(72 - 96 + 36) \\
&= 74,880\pi \text{ ft-lb} \approx 235,242 \text{ ft-lb}
\end{aligned}
$$

12.

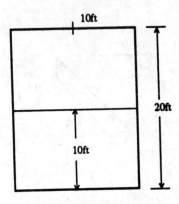

Exercise 12

$$W = \int_{10}^{20} \pi r^2 \rho h\, dh = 62.4(100)\pi \int_{10}^{20} h\, dh = 3120\pi\, h^2 \Big]_{10}^{20}$$
$$= 3120\pi(400 - 100) = 936,000\pi \text{ ft-lb} \approx 2,940,531 \text{ ft-lb}$$

13. $\displaystyle W = \int_{0}^{20} \pi r^2 \rho h\, dh = 3120\pi h^2 \Big]_{0}^{20} = 1,248,000\pi \text{ ft-lb}$
$\approx 3,920,708 \text{ ft-lb}$

14.

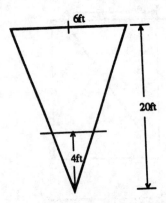

Exercise 14

At a depth h, $r = 6\left(1 - \dfrac{h}{20}\right)$.

$$W = \int_{16}^{20} \pi r^2 \rho h\, dh = 62.4\pi \int_{16}^{20} 36\left(1 - \frac{h}{20}\right)^2 h\, dh$$
$$= 2246.4\pi \int_{16}^{20} \left(h - \frac{h^2}{10} + \frac{h^3}{400}\right) dh$$
$$= 2246.4\pi \left[\frac{h^2}{2} - \frac{h^3}{30} + \frac{h^4}{1600}\right]_{16}^{20}$$
$$= 2246.4\pi \left[(200 - 128) - \frac{1}{30}(8000 - 4096) + \frac{1}{1600}(160,000 - 65,536)\right]$$

$$= 2246.4\pi \left(72 - \frac{3904}{30} + \frac{1476}{25} \right) = \frac{152,755.2}{75} \pi$$

$$\approx 2036.74\pi \text{ ft-lb} \approx 6398.6 \text{ ft-lb}$$

15. $$W = \int_0^{20} \pi r^2 \rho h \, dh = 62.4\pi \int_0^{20} 36 \left(1 - \frac{h}{20} \right)^2 h \, dh = 2246.4\pi \int_0^{20} \left(h - \frac{h^2}{10} + \frac{h^3}{400} \right) dh$$

$$= 2246.4\pi \left[\frac{h^2}{2} - \frac{h^3}{30} + \frac{h^4}{1600} \right]_0^{20} = 2246.4\pi \left(200 - \frac{8000}{30} + \frac{160,000}{1600} \right)$$

$$= 74,880\pi \text{ ft-lb} \approx 235,242 \text{ ft-lb}$$

16.
Volume of water	=	$\pi r^2 h = \pi (10^2)(10) = 1000\pi$ cu ft.
Work to raise water 12 ft	=	$(62.4)(1000\pi)(10)$ ft-lb $= 624,000\pi$ ft-lb.
Total work	=	$936,000\pi + 624,000\pi$ ft-lb
	=	$1,560,000\pi$ ft-lb
	$\approx$	$4,900,884.5$ ft-lb.

17.
Volume of water	=	$\frac{\pi r^2 h}{3}$.
r	=	$\frac{4}{20}(6) = 1.2$ ft; $V = \frac{\pi (1.2^2)(4)}{3} = 1.92\pi$ cu ft
Work to raise water 15 ft	=	$(62.4)(1.92\pi)(15)$
	=	1797.12π ft-lb.
Total work	=	$2036.74\pi + 1797.12\pi$ ft-lb
	=	3833.86π ft-lb
	$\approx$	$12,044$ ft-lb.

Alternatively, $W = \int_{16}^{20} \pi r^2 \rho (h + 15) \, dh = 3833.86\pi$ ft-lb.

18.

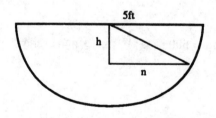

Exercise 18

$$W = \int_0^5 \pi r^2 \rho h \, dh; \quad h^2 + r^2 = 25 \Rightarrow r^2 = 25 - h^2$$

$$W = 62.4\pi \int_0^5 (25 - h^2) h \, dh = 62.4\pi \left[\frac{25}{2} h^2 - \frac{h^4}{4} \right]_0^5$$

$$= 62.4\pi \left(\frac{625}{2} - \frac{625}{4} \right) = 9750\pi \text{ ft-lb} \approx 30,631 \text{ ft-lb}$$

19. $W = \displaystyle\int_0^{20} F(x)\,dx = \int_0^{20} 2(20-x)\,dx = 40x - x^2 \Big]_0^{20} = 800 - 400 \text{ ft-lb} = 400 \text{ ft-lb}$

20. $W = \displaystyle\int_0^{30} F(x)\,dx = \int_0^{30} 40(30-x)\,dx = 1200x - 20x^2 \Big]_0^{30} = 36,000 - 18,000 \text{ joules} = 18,000 \text{ joules}$

21.

Exercise 21

$$W = \int_0^{20} F(x)\,dx$$

Leakage = 10 lb/s; $\frac{10}{5} = 2$ lb/ft.

$$W = \int_0^{20} (100 + 500 - 2x)\,dx = 600x - x^2 \Big]_0^{20} = 12,000 - 400 = 11,600 \text{ ft-lb}$$

If the cable weighs 3 lb/ft,

$$W = \int_0^{20} (100 + 500 + 60 - (3+2)x)\,dx = 660x - \frac{5x^2}{2} \Big]_0^{20} = 13,200 - 1000 = 12,200 \text{ ft-lb.}$$

Alternatively, add

$$W = \int_0^{20} 3(20-x)\,dx = 60x - \frac{3}{2}x^2 \Big]_0^{20} = 1200 - 600 = 600 \text{ ft-lb to original answer.}$$

22.

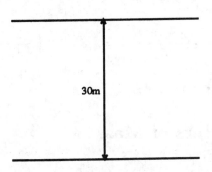

Exercise 22

$$W = \int_0^{30} F(x)\, dx$$

Leakage = 0.1 liter/s= 0.1 liter/m;   loss in weight = 1 newton/m.

$$W = \int_0^{30} [45 + 40(10) - 1(x)]\, dx = 445x - \frac{x^2}{2}\Big]_0^{30} = 13,350 - 450 = 12,900 \text{ joules}$$

23.   a) $F = K\,\dfrac{q_1 q_2}{r^2};\ 10 = \dfrac{K\,q^2}{4^2};\ K = \dfrac{160}{q^2}\ \dfrac{\text{newton-meter}^2}{\text{coulomb}^2}$

   b) $W = \displaystyle\int_0^2 F(x)\, dx = \int_0^2 \frac{160}{q^2}\,\frac{q^2}{(4-x)^2}\, dx = \frac{160}{(4-x)}\Big]_0^2 = 160\left(\frac{1}{2} - \frac{1}{4}\right) = 40 \text{ joules}$

24. $W = \displaystyle\int_0^4 \frac{160}{1^2}\,\frac{1^2}{(5-x)^2}\, dx = \frac{160}{5-x}\Big]_0^4 = 160\left(1 - \frac{1}{5}\right) = 128 \text{ joules}$

25.

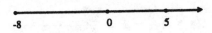

Exercise 25

$$\begin{aligned} W &= \int_0^4 \left[\frac{160}{1}\,\frac{1}{(5-x)^2} - \frac{160}{1}\,\frac{1}{(8+x)^2}\right] dx = 160\left[\frac{1}{5-x} + \frac{1}{8+x}\right]_0^4 \\ &= 160\left[\left(1 - \frac{1}{5}\right) + \left(\frac{1}{12} - \frac{1}{8}\right)\right] = 160\left(\frac{4}{5} - \frac{1}{24}\right) = \frac{364}{3} \approx 121.33 \text{ joules} \end{aligned}$$

26. a) $w(r) = \dfrac{k}{r^2};\quad 5 = \dfrac{k}{4000^2};$

 $k = 80,000,000$ ton-miles^2

   b) $W = \displaystyle\int_{4000}^{4100} \dfrac{8 \times 10^7}{r^2}\, dr = -\dfrac{8 \times 10^7}{r}\Bigg]_{4000}^{4100} = -8 \times 10^7 \left(\dfrac{1}{4100} - \dfrac{1}{4000} \right)$

   $= \dfrac{8 \times 10^7\,(100)}{(4100)(4000)} \approx 487.8$ ton-miles

7.7 Moments and Centers of Mass

1. $\bar{x} = \dfrac{\sum_{j=1}^n x_j m_j}{\sum_{j=1}^n m_j} = \dfrac{1(-4) + 31(2) + 7(6)}{1 + 31 + 7} = \dfrac{100}{39} \approx 2.5641$

2. $\bar{x} = \dfrac{2(-10) + 5(-3) + 1(1) + 10(5)}{2 + 5 + 1 + 10} = \dfrac{16}{18} = \dfrac{8}{9}$

3. $80 \cdot 3 = 50x;\ x = \dfrac{24}{5} = 4.8$ ft

4. Let $x =$ the distance of the plant from city A.

$$100x = 300(200 - x);\ \dfrac{x}{3} = 200 - x;\ x = 150 \text{ miles}$$

This problem is satisfied by equation (1) if w is interpreted to be the number of items being shipped.

5. The moment of mass m_j about the y-axis is $x_j m_j$, independent of the value of y_j. Hence,

$$\sum_{j=1}^n (x_j - \bar{x}) m_j = 0,$$

where $\bar{x}$ is the x-coordinate of the center of mass.

6. The moment of mass m_j about the x-axis is $y_j m_j$, independent of the value of x_j. Hence,

$$\sum_{j=1}^n (y_j - \bar{y}) m_j = 0,$$

where $\bar{y}$ is the y-coordinate of the center of mass.

7. $\displaystyle\sum_{j=1}^n (x_j - \bar{x}) m_j = 0 \Rightarrow \sum_{j=1}^n x_j m_j = \bar{x} \sum_{j=1}^n m_j \Rightarrow \bar{x} = \dfrac{\sum_{j=1}^n x_j m_j}{\sum_{j=1}^n m_j}$

 Similarly, $\bar{y} = \dfrac{\sum_{j=1}^n y_j m_j}{\sum_{j=1}^n m_j}.$

8. $\bar{x} = \dfrac{10(0) + 15(2) + 2(4) + 10(1)}{10 + 15 + 2 + 10} = \dfrac{48}{37}$ $\qquad \bar{y} = \dfrac{10(0) + 15(-2) + 2(-2) + 10(1)}{37} = -\dfrac{24}{37}$

9. $\bar{x} = 0 = \dfrac{2(1) + 4(-2) + 3x_3}{2 + 4 + 3} = \dfrac{-6 + 3x_3}{9}$ $\qquad \bar{y} = 1 = \dfrac{2(1) + 4(3) + 3y_3}{9} = \dfrac{14 + 3y_3}{9}$

 $x_3 = \dfrac{6}{3} = 2$ $\qquad\qquad\qquad\qquad\qquad y_3 = \dfrac{9 - 14}{3} = -\dfrac{5}{3}$

10. $y = x^3$, $y = x^{1/3}$, $0 \le x \le 1$

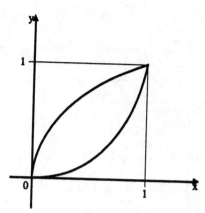

Exercise 10

$$\overline{x} = \frac{1}{A}\int_0^1 x(x^{1/3} - x^3)\,dx$$

$$A = \int_0^1 (x^{1/3} - x^3)\,dx = \frac{3}{4}x^{4/3} - \frac{1}{4}x^4\Big]_0^1 = \frac{3}{4} - \frac{1}{4} = \frac{1}{2}$$

$$\overline{x} = 2\int_0^1 (x^{4/3} - x^4)\,dx = \frac{6}{7}x^{7/3} - \frac{2}{5}x^5\Big]_0^1 = \frac{6}{7} - \frac{2}{5} = \frac{16}{35}$$

$$\overline{y} = \overline{x} = \frac{16}{35} \text{ by symmetry.}$$

$$(\overline{x}, \overline{y}) = \left(\frac{16}{35}, \frac{16}{35}\right)$$

11. $y = x^2$, $y = x^3$

Intersections: $x = 0, 1$

$$A = \int_0^1 (x^2 - x^3)\,dx = \frac{x^3}{3} - \frac{x^4}{4}\Big]_0^1 = \frac{1}{3} - \frac{1}{4} = \frac{1}{12}$$

$$\overline{x} = 6\int_0^1 x(x^2 - x^3)\,dx = 6\left[\frac{x^4}{4} - \frac{x^5}{5}\right]_0^1 = 12\left(\frac{1}{4} - \frac{1}{5}\right) = \frac{3}{5}$$

$$\overline{y} = 6\int_0^1 (x^4 - x^6)\,dx = 6\left[\frac{x^5}{5} - \frac{x^7}{7}\right]_0^1 = 6\left(\frac{1}{5} - \frac{1}{7}\right) = 6\left(\frac{2}{35}\right) = \frac{12}{35}$$

$$(\overline{x}, \overline{y}) = \left(\frac{3}{5}, \frac{12}{35}\right)$$

12. $x = 4 - y^2$, $x = 0$
    $y = 0 \Rightarrow x = 4$

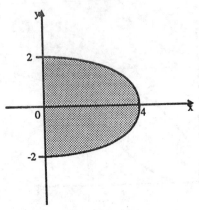

Exercise 12

$$A = 2\int_0^4 \sqrt{4-x}\,dx = -\frac{4}{3}(4-x)^{3/2}\Big]_0^4 = -\frac{4}{3}(0-8) = \frac{32}{3}$$

$$\bar{x} = \frac{3}{32}\int_0^4 x\left[\sqrt{4-x}+\sqrt{4-x}\right]dx$$

Let $u = 4-x$, $du = -dx$, $x = 4-u$
$\quad x = 0 \Rightarrow u = 4$, $x = 4 \Rightarrow u = 0$.

$$\bar{x} = -\frac{3}{16}\int_4^0 (4-u)\sqrt{u}\,du = -\frac{3}{16}\left[\frac{8}{3}u^{3/2}-\frac{2}{5}u^{5/2}\right]_4^0 = \frac{3}{16}\left[\frac{8}{3}(8)-\frac{2}{5}(32)\right] = \frac{8}{5}$$

By symmetry, $\bar{y} = 0$.

$$(\bar{x},\bar{y}) = \left(\frac{8}{5},0\right)$$

13. $y = 4-x$, $x = 0$, $y = 0$

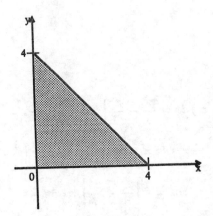

Exercise 13

$$A = \int_0^4 (4-x)\,dx = 4x-\frac{x^2}{2}\Big]_0^4 = 16-8 = 8$$

$$\bar{x} = \frac{1}{8}\int_0^4 x(4-x)\,dx = \frac{1}{8}\left[2x^2-\frac{x^3}{3}\right]_0^4 = \frac{1}{8}\left(32-\frac{64}{3}\right) = \frac{4}{3}$$

$$\bar{y} = \frac{4}{3} \text{ by symmetry.}$$

$$(\bar{x}, \bar{y}) = \left(\frac{4}{3}, \frac{4}{3}\right)$$

14. $y = x, \ y = 4 - x, \ y = 0$

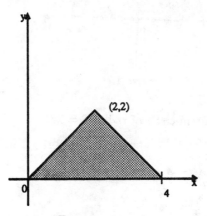

Exercise 14

$$A = \int_0^2 (4 - y - y) \, dy = 4y - y^2 \Big]_0^2 = 8 - 4 = 4$$

(or, use $A = \frac{1}{2} bh = \frac{1}{2}(4 \cdot 2) = 4$.)

By symmetry, $\bar{x} = 2$.

$$\bar{y} = \frac{1}{8}\left[\int_0^2 x^2 \, dx + \int_2^4 (4 - x)^2 \, dx\right] = \frac{1}{8}\left[\frac{x^3}{3}\right]_0^2 - \frac{1}{24}\left[(4 - x)^3\right]_2^4$$

$$= \frac{1}{8} \cdot \frac{8}{3} - \frac{1}{24}(0 - 8) = \frac{16}{24} = \frac{2}{3}$$

Alternatively,

$$\bar{y} = \frac{1}{4}\int_0^2 y(4 - 2y) \, dy = \frac{1}{4}\left[2y^2 - \frac{2}{3}y^3\right]_0^2 = \frac{1}{4}\left(8 - \frac{16}{3}\right) = \frac{2}{3}$$

$$(\bar{x}, \bar{y}) = \left(2, \frac{2}{3}\right)$$

569

15. $y = \sqrt{4 - x^2}, \; y = 0$

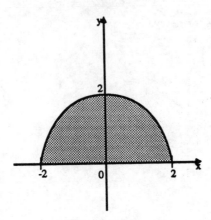

Exercise 15

$$A \;=\; \text{area of semicircle of radius } 2 = 2\pi.$$

$$\overline{x} \;=\; 0 \text{ by symmetry.}$$

$$\overline{y} \;=\; \frac{1}{4\pi}\int_{-2}^{2}(4 - x^2)\,dx = \frac{1}{4\pi}\left[4x - \frac{x^3}{3}\right]_{-2}^{2} = \frac{1}{2\pi}\left(8 - \frac{8}{3}\right) = \frac{8}{3\pi}$$

$$(\overline{x}, \overline{y}) \;=\; \left(0, \frac{8}{3\pi}\right)$$

16. $x = y^2, \; x = 4 - y^2$

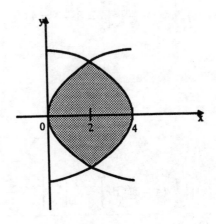

Exercise 16

Intersections: $y^2 = 4 - y^2 \Rightarrow \; y^2 = 2 \Rightarrow y = \pm\sqrt{2}$
$\qquad\qquad\qquad\qquad\quad x = 2.$

By symmetry, $\overline{x} = 2, \; \overline{y} = 0.$

$$(\overline{x}, \overline{y}) \;=\; (2, 0)$$

570

17. $y = 2x^2$, $y = x^2 + 1$

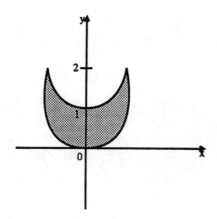

Exercise 17

Intersections: $\quad 2x^2 \;=\; x^2 + 1 \Rightarrow x^2 = 1 \Rightarrow x = \pm 1,\; y = 2$

$$A \;=\; 2\int_0^1 (x^2 + 1 - 2x^2)\, dx$$

$$=\; 2\int_0^1 (1 - x^2)\, dx = 2x - \frac{2x^3}{3}\Bigg]_0^1 = \frac{4}{3}.$$

By symmetry, $\overline{x} = 0$.

$$\overline{y} \;=\; \frac{3}{8}\int_{-1}^1 [(x^2 + 1)^2 - 4x^4]\, dx = \frac{3}{4}\int_0^1 (-3x^4 + 2x^2 + 1)\, dx$$

$$=\; \frac{3}{4}\left[-\frac{3}{5}x^5 + \frac{2}{3}x^3 + x\right]_0^1 = \frac{3}{4}\left(-\frac{3}{5} + \frac{2}{3} + 1\right) = \frac{4}{5}$$

$$(\overline{x}, \overline{y}) \;=\; \left(0, \frac{4}{5}\right)$$

18. $y = \frac{1}{3}x^2$, $y = -\frac{2}{3}x^2 + 4$

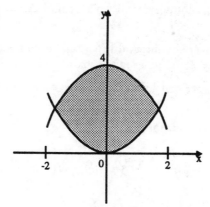

Exercise 18

Intersections: $\frac{1}{3}x^2 = -\frac{2}{3}x^2 + 4 \Rightarrow x = \pm 2,\; y = \frac{4}{3}$.

$$A \;=\; \int_{-2}^2 \left(-\frac{2}{3}x^2 + 4 - \frac{1}{3}x^2\right)\, dx = 2\int_0^2 (4 - x^2)\, dx = 2\left[4x - \frac{x^3}{3}\right]_0^2$$

$$= 2\left(8 - \frac{8}{3}\right) = \frac{32}{3}$$

By symmetry, $\bar{x} = 0$.

$$\bar{y} = \frac{3}{64}\int_{-2}^{2}\left[\left(-\frac{2}{3}x^2 + 4\right)^2 - \left(\frac{1}{3}x^2\right)^2\right]dx = \frac{3}{32}\int_{0}^{2}\left(\frac{1}{3}x^4 - \frac{16}{3}x^2 + 16\right)dx$$

$$= \frac{3}{32}\left[\frac{x^5}{15} - \frac{16x^3}{9} + 16x\right]_{0}^{2} = \frac{3}{32}\left(\frac{32}{15} - \frac{128}{9} + 32\right) = \frac{28}{15}$$

$$(\bar{x}, \bar{y}) = \left(0, \frac{28}{15}\right)$$

19.

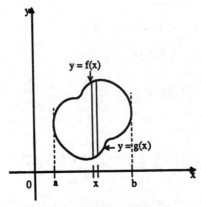

Exercise 19

a) Let R be a region bounded above by the graph of $y = f(x)$ and below by the graph of $y = g(x)$, for $a \le x \le b$. Assume that R is the face of a plate of uniform thickness c and of uniform density ρ.

b) Partition the interval $[a, b]$ into n subintervals of equal length $\Delta x = \frac{b-a}{n}$. Select one number t_j in each subinterval $[x_{j-1}, x_j]$.

c) Consider the section of R lying over the interval $[x_{j-1}, x_j]$ to be a slab of height $[f(t_j) - g(t_j)]$, width Δx, thickness c and density ρ. Its mass is approximated by $m_j \approx c\rho[f(t_j) - g(t_j)]\Delta x$.

d) The center of mass of the slab described in step c), with respect to the y-axis, must lie on the line $y = y_j = \frac{1}{2}[f(t_j) + g(t_j)]$.

e) An analogue of equation (6), for equilibrium, is

$$0 = \sum_{j=1}^{n}(y_j - \bar{y})m_j \approx \sum_{j=1}^{n}(y_j - \bar{y})c\rho[f(t_j) - g(t_j)]\Delta x$$

$$= \sum_{j=1}^{n}\left\{\frac{1}{2}[f(t_j) + g(t_j)] - \bar{y}\right\}c\rho[f(t_j) - g(t_j)]\Delta x$$

$$= c\rho\sum_{j=1}^{n}\left[\frac{1}{2}[f(t_j)^2 - g(t_j)^2] - \bar{y}[f(t_j) - g(t_j)]\right]\Delta x$$

Take the limit as $n \to \infty$, and divide through by $c\rho \ne 0$.

$$0 = \int_a^b \frac{1}{2}[f(x)^2 - g(x)^2]\,dx - \overline{y}\int_a^b [f(x) - g(x)]\,dx$$

$$\overline{y} = \frac{\frac{1}{2}\int_a^b [f(x)^2 - g(x)^2]\,dx}{\int_a^b = [f(x) - g(x)]\,dx} = \frac{1}{2A}\int_a^b [f(x)^2 - g(x)^2]\,dx$$

20. $f(x) = 1 - x^2$, $g(x) = 2$, for $-1 \le x \le 1$.

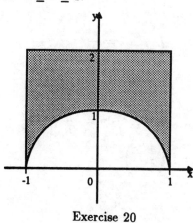

Exercise 20

$$A = \int_{-1}^1 [2 - (1 - x^2)]\,dx = \int_{-1}^1 (1 + x^2)\,dx = 2\left[x + \frac{x^3}{3}\right]_0^1 = \frac{8}{3}$$

$\overline{x} = 0$, by symmetry.

$$\overline{y} = \frac{3}{16}\int_{-1}^1 [4 - (1 - x^2)^2]\,dx = \frac{3}{8}\int_0^1 (3 + 2x^2 - x^4)\,dx = \frac{3}{8}\left[3x + \frac{2}{3}x^3 - \frac{x^5}{5}\right]_0^1$$

$$= \frac{3}{8}\left(3 + \frac{2}{3} - \frac{1}{5}\right) = \left(\frac{3}{8}\right)\left(\frac{52}{15}\right) = \frac{13}{10}$$

$$(\overline{x}, \overline{y}) = \left(0, \frac{13}{10}\right)$$

21. $f(x) = x^3$, $g(x) = 0$, for $0 \le x \le 1$.

573

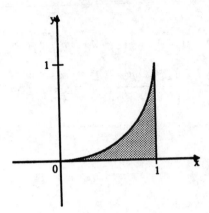

Exercise 21

$$A = \int_0^1 x^3\,dx = \frac{x^4}{4} = \frac{1}{4}$$

$$\overline{x} = 4\int_0^1 x(x^3)\,dx = \frac{4x^5}{5}\bigg]_0^1 = \frac{4}{5}$$

$$\overline{y} = 2\int_0^1 (x^6)\,dx = \frac{2x^7}{7}\bigg]_0^1 = \frac{2}{7}$$

$$(\overline{x},\overline{y}) = \left(\frac{4}{5},\frac{2}{7}\right)$$

22. $f(x) = \sqrt{x}$, $g(x) = 0$, for $0 \le x \le 4$.

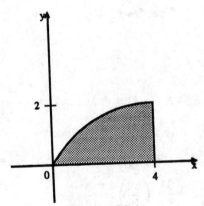

Exercise 22

$$A = \int_0^4 \sqrt{x}\,dx = \frac{2}{3}x^{3/2}\bigg]_0^4 = \frac{16}{3}$$

$$\overline{x} = \frac{3}{16}\int_0^4 x\sqrt{x}\,dx = \frac{3}{16}\cdot\frac{2}{5}x^{5/2}\bigg]_0^4 = \frac{3}{40}(32) = \frac{12}{5}$$

$$\overline{y} = \frac{3}{32}\int_0^4 x\,dx = \frac{3}{32}\frac{x^2}{2}\bigg]_0^4 = \frac{3}{4}$$

$$(\overline{x},\overline{y}) = \left(\frac{12}{5},\frac{3}{4}\right)$$

23. $f(x) = x$, $g(x) = -x$, for $0 \le x \le 4$.

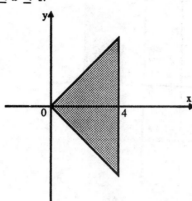

Exercise 23

$$A = \int_0^4 [x - (-x)]\, dx = x^2 \Big]_0^4 = 16$$

$$\bar{x} = \frac{1}{16} \int_0^4 x(2x)\, dx = \frac{1}{16} \cdot \frac{2}{3} x^3 \Big]_0^4 = \frac{1}{24}(64) = \frac{8}{3}$$

$$\bar{y} = 0, \text{ by symmetry.}$$

$$(\bar{x}, \bar{y}) = \left(\frac{8}{3}, 0\right)$$

24. $f(x) = \sqrt{9 - x^2}$, $g(x) = 0$, for $-3 \le x \le 3$.

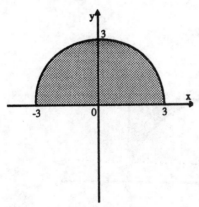

Exercise 24

$$A = \frac{\pi}{2}(9) = \frac{9\pi}{2}, \text{ the area of a semicircle of radius 3.}$$

$$\bar{x} = 0, \text{ by symmetry.}$$

$$\bar{y} = \frac{1}{9\pi} \int_{-3}^3 (9 - x^2)\, dx = \frac{2}{9\pi} \left[9x - \frac{x^3}{3}\right]_0^3 = \frac{2}{9\pi}(27 - 9) = \frac{4}{\pi}$$

$$(\bar{x}, \bar{y}) = \left(0, \frac{4}{\pi}\right)$$

25. $f(x) = x^2 + x + 1$, $g(x) = 0$, for $1 \le x \le 3$.

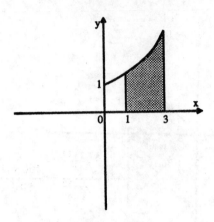

Exercise 25

$$A = \int_1^3 (x^2 + x + 1)\,dx = \frac{x^3}{3} + \frac{x^2}{2} + x\Big]_1^3 = 9 + \frac{9}{2} + 3 - \left(\frac{1}{3} + \frac{1}{2} + 1\right) = \frac{44}{3}$$

$$\overline{x} = \frac{3}{44}\int_1^3 x(x^2 + x + 1)\,dx = \frac{3}{44}\left[\frac{x^4}{4} + \frac{x^3}{3} + \frac{x^2}{2}\right]_1^3 = \frac{3}{44}\left[\frac{81}{4} + 9 + \frac{9}{2} - \left(\frac{1}{4} + \frac{1}{3} + \frac{1}{2}\right)\right] = \frac{49}{22}$$

$$\overline{y} = \frac{3}{88}\int_1^3 (x^2 + x + 1)^2\,dx = \frac{3}{88}\int_1^3 (x^4 + 2x^3 + 3x^2 + 2x + 1)\,dx$$

$$= \frac{3}{88}\left[\frac{x^5}{5} + \frac{x^4}{2} + x^3 + x^2 + x\right]_1^3 = \frac{3}{88}\left[\frac{243}{5} + \frac{81}{2} + 27 + 9 + 3 - \left(\frac{1}{5} + \frac{1}{2} + 3\right)\right] = \frac{933}{220}$$

$$(\overline{x}, \overline{y}) = \left(\frac{49}{22}, \frac{933}{220}\right)$$

26. $f(x) = 4$, $g(x) = 4 - x^2$, for $0 \le x \le 2$.

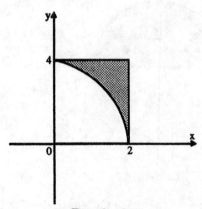

Exercise 26

$$A = \int_0^2 [4 - (4 - x^2)]\,dx = \int_0^2 x^2\,dx = \frac{x^3}{3}\Big]_0^2 = \frac{8}{3}$$

$$\overline{x} = \frac{3}{8}\int_0^2 x(x^2)\,dx = \frac{3}{32}x^4\Big]_0^2 = \frac{3}{2}$$

$$\overline{y} = \frac{3}{16}\int_0^2 [16 - (4 - x^2)^2]\,dx = \frac{3}{16}\int_0^2 (8x^2 - x^4)\,dx = \frac{3}{16}\left[\frac{8}{3}x^3 - \frac{x^5}{5}\right]_0^2$$

576

$$= \frac{3}{16}\left(\frac{64}{3} - \frac{32}{5}\right) = \frac{14}{5}$$

$$(\bar{x}, \bar{y}) = \left(\frac{3}{2}, \frac{14}{5}\right)$$

27. $\rho(x) = 2 + \left(\dfrac{12-2}{10}\right) x = 2 + x$

$$\bar{x} = \frac{\int_0^{10} x\rho(x)\,dx}{\int_0^{10} \rho(x)\,dx} = \frac{\int_0^{10} x(2+x)\,dx}{\int_0^{10}(2+x)\,dx} = \frac{x^2 + \frac{x^3}{3}\Big]_0^{10}}{2x + \frac{x^2}{2}\Big]_0^{10}} = \frac{100 + \frac{1000}{3}}{20 + 50} = \frac{1300}{210} = \frac{130}{21}$$

28. $\rho(x) = 1 + x$

$$\bar{x} = \frac{\int_0^{10} x(1+x)\,dx}{\int_0^{10}(1+x)\,dx} = \frac{\frac{x^2}{2} + \frac{x^3}{3}\Big]_0^{10}}{x + \frac{x^2}{2}\Big]_0^{10}} = \frac{50 + \frac{1000}{3}}{10 + 50} = \frac{1150}{180} = \frac{115}{18}$$

29. $\rho(x) = 1 + x^2$

$$\bar{x} = \frac{\int_0^{10} x(1+x^2)\,dx}{\int_0^{10}(1+x^2)\,dx} = \frac{\frac{x^2}{2} + \frac{x^4}{4}\Big]_0^{10}}{x + \frac{x^3}{3}\Big]_0^{10}} = \frac{50 + 2500}{10 + \frac{1000}{3}} = \frac{7650}{1030} = \frac{765}{103}$$

30. $\rho(x) = x + x^2$

$$\bar{x} = \frac{\int_0^6 x(x+x^2)\,dx}{\int_0^6(x+x^2)\,dx} = \frac{\frac{x^3}{3} + \frac{x^4}{4}\Big]_0^6}{\frac{x^2}{2} + \frac{x^3}{3}\Big]_0^6} = \frac{72 + 324}{18 + 72} = \frac{22}{5}$$

31.

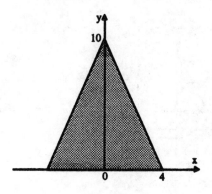

Exercise 31

$$\bar{y} = \frac{\int_0^h \rho y A(y)\,dy}{\int_0^h \rho A(y)\,dy} = \frac{\int_0^{10} y\pi r^2\,dy}{\int_0^{10} \pi r^2\,dy}$$

Use $r = \dfrac{4(10-y)}{10}$.

$$\overline{y} = \frac{\int_0^{10} y(10-y)^2\,dy}{\int_0^{10}(10-y)^2\,dy} = \frac{50y^2 - \frac{20}{3}y^3 + \frac{y^4}{4}\Big]_0^{10}}{100y - 10y^2 + \frac{y^3}{3}\Big]_0^{10}} = \frac{5000 - \frac{20,000}{3} + 2500}{1000 - 1000 + \frac{1000}{3}} = \frac{2500}{1000}$$

$$= \quad 2.5 \text{ cm from the base.}$$

$$\overline{x} = \quad 0, \text{ by symmetry.}$$

$$(\overline{x}, \overline{y}) = \quad (0, 2.5)$$

32. $y = x^3$, for $0 \le x \le 2$, about x-axis.

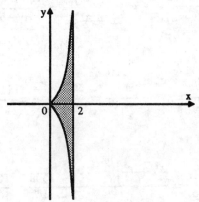

Exercise 32

$$\overline{x} = \frac{\int_0^2 x A(x)\,dx}{\int_0^2 A(x)\,dx} = \frac{\int_0^2 x(\pi x^6)\,dx}{\int_0^2 \pi x^6\,dx} = \frac{\frac{x^8}{8}\Big]_0^2}{\frac{x^7}{7}\Big]_0^2} = \frac{32}{\frac{128}{7}} = \frac{7}{4}$$

$$\overline{y} = \quad 0, \text{ by symmetry.}$$

$$(\overline{x}, \overline{y}) = \quad (1.75, 0)$$

33. $y = x^2$, $0 \le x \le 2$, about x-axis.

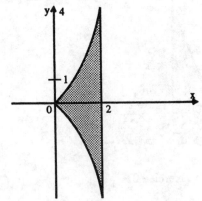

Exercise 33

$$\overline{x} = \frac{\int_0^2 x A(x)\,dx}{\int_0^2 A(x)\,dx} = \frac{\int_0^2 x \pi x^4\,dx}{\int_0^2 \pi x^4\,dx} = \frac{\frac{x^6}{6}\Big]_0^2}{\frac{x^5}{5}\Big]_0^2} = \frac{\frac{32}{3}}{\frac{32}{5}} = \frac{5}{3}$$

578

$$\overline{y} = 0, \text{ by symmetry.}$$
$$(\overline{x}, \overline{y}) = \left(\frac{5}{3}, 0\right)$$

34. $f(x) = \dfrac{1}{x^3}$, $1 \le x \le 3$, about x-axis.

$$\overline{x} = \frac{\int_1^3 x A(x)\, dx}{\int_1^3 A(x)\, dx} = \frac{\int_1^3 x\left(\frac{\pi}{x^6}\right) dx}{\int_1^3 \left(\frac{\pi}{x^6}\right) dx} = \frac{\int_1^3 \frac{dx}{x^5}}{\int_1^3 \frac{dx}{x^6}} = \frac{-\frac{1}{4x^4}\Big]_1^3}{-\frac{1}{5x^5}\Big]_1^3}$$

$$= \frac{5}{4}\frac{\left(\frac{1}{81}-1\right)}{\left(\frac{1}{243}-1\right)} = \frac{5}{4}\left(\frac{80}{81}\right)\left(\frac{243}{242}\right) = \frac{150}{121}$$

$$\overline{y} = 0, \text{ by symmetry.}$$
$$(\overline{x}, \overline{y}) = \left(\frac{150}{121}, 0\right)$$

35. A circle centered at the origin is symmetric with respect to both the x-axis and the y-axis. Therefore, $(\overline{x}, \overline{y})$ is the center of the circle.

36. $y = \sin x$, $y = -\sin x$, for $0 \le x \le \pi$.

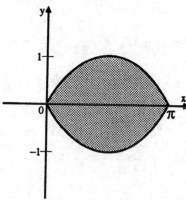

Exercise 36

The figure is symmetric with respect to the lines $y = 0$ and $x = \dfrac{\pi}{2}$. Therefore, $(\overline{x}, \overline{y}) = \left(\dfrac{\pi}{2}, 0\right)$.

37. $f(x) = 4 - x^2$, $f(x) = \begin{cases} -x - 2 & \text{for } -2 \leq x \leq 0 \\ x - 2 & \text{for } 0 \leq x \leq 2 \end{cases}$

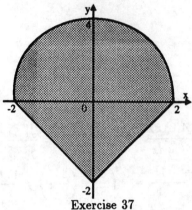

Exercise 37

$$A = 2\int_0^2 [4 - x^2 - (x - 2)]\, dx = 2\int_0^2 (6 - x - x^2)\, dx = 2\left[6x - \frac{x^2}{2} - \frac{x^3}{3}\right]_0^2$$

$$= 2\left(12 - 2 - \frac{8}{3}\right) = \frac{44}{3}$$

$$\overline{x} = 0, \text{ by symmetry.}$$

$$\overline{y} = \frac{3}{88}\int_{-2}^0 [(4 - x^2)^2 - (-x - 2)^2]\, dx + \frac{3}{88}\int_0^2 [(4 - x^2)^2 - (x - 2)^2]\, dx$$

$$= \frac{3}{88}\int_{-2}^0 (x^4 - 9x^2 - 4x + 12)\, dx + \frac{3}{88}\int_0^2 (x^4 - 9x^2 + 4x + 12)\, dx$$

$$= \frac{3}{44}\int_0^2 (x^4 - 9x^2 + 12)\, dx - \frac{3}{88}\int_{-2}^0 4x\, dx + \frac{3}{88}\int_0^2 4x\, dx$$

$$= \frac{3}{44}\left[\frac{x^5}{5} - 3x^3 + 12x\right]_0^2 - \frac{3}{44}x^2\Big]_{-2}^0 + \frac{3}{44}x^2\Big]_0^2$$

$$= \frac{3}{44}\left(\frac{32}{5} - 24 + 24\right) - \frac{3}{44}(-4) + \frac{3}{44}(4) = \frac{3}{44}\left(\frac{72}{5}\right) = \frac{54}{55}$$

$$(\overline{x}, \overline{y}) = \left(0, \frac{54}{55}\right)$$

38. $x^2 + y^2 - xy = 6$

The equation is that of an ellipse rotated $45°$. If we change coordinates by letting

$$\frac{\sqrt{2}}{2}(x' + y') = x \text{ and } \frac{\sqrt{2}}{2}(-x' + y') = y$$

we get

$$3x'^2 + y'^2 = 12 \text{ or } \frac{x'^2}{4} + \frac{y'^2}{12} = 1.$$

Thus, along $y = x$, the major axis has length $2\sqrt{12}$ and the minor axis has length 4. The centroid is $(0, 0)$, by symmetry.

39.

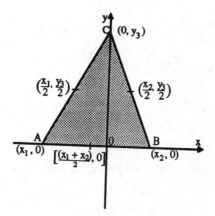

Exercise 39

Position the triangle with its base on the x-axis and vertex on the y-axis.

Area $= (x_2 - x_1) \frac{y_3}{2}$.

line AC: $\dfrac{y - 0}{x - x_1} = \dfrac{y_3 - 0}{0 - x_1}$; $y = \dfrac{-y_3(x - x_1)}{x_1}$

line BC: $\dfrac{y - 0}{x - x_2} = \dfrac{y_3 - 0}{0 - x_2}$; $y = \dfrac{-y_3(x - x_2)}{x_2}$

$$\bar{x} = \frac{2}{y_3(x_2 - x_1)}\left[\int_{x_1}^{0} \frac{-xy_3(x - x_1)}{x_1}\,dx + \int_{0}^{x_2} \frac{-xy_3(x - x_2)}{x_2}\,dx\right]$$

$$= \frac{2}{(x_2 - x_1)}\left\{\left.\frac{-x^3}{3x_1} + \frac{x^2}{2}\right]_{x_1}^{0} + \left.\frac{-x^3}{3x_2} + \frac{x^2}{2}\right]_{0}^{x_2}\right\} = \frac{2}{(x_2 - x_1)}\left(\frac{x_1^2}{3} - \frac{x_1^2}{2} - \frac{x_2^2}{3} + \frac{x_2^2}{2}\right)$$

$$= \frac{2(x_2^2 - x_1^2)}{6(x_2 - x_1)} = \frac{x_2 + x_1}{3}$$

$$\bar{y} = \frac{1}{y_3(x_2 - x_1)}\left[\int_{x_1}^{0} \frac{y_3^2(x - x_1)^2}{x_1^2}\,dx + \int_{0}^{x_2} \frac{y_3^2(x - x_2)^2}{x_2^2}\,dx\right]$$

$$= \frac{y_3}{(x_2 - x_1)}\left\{\left.\frac{(x - x_1)^3}{3x_1^2}\right]_{x_1}^{0} + \left.\frac{(x - x_2)^3}{3x_2^2}\right]_{0}^{x_2}\right\} = \frac{y_3}{(x_2 - x_1)}\left(-\frac{x_1}{3} - \frac{-x_2}{3}\right) = \frac{y_3}{3}$$

The equations of the median lines are:

From A: $\dfrac{y - 0}{x - x_1} = \dfrac{\frac{y_3}{2}}{\frac{x_2}{2} - x_1}$; $y = \dfrac{y_3(x - x_1)}{x_2 - 2x_1}$

From B: $\dfrac{y - 0}{x - x_2} = \dfrac{\frac{y_3}{2}}{\frac{x_1}{2} - x_2}$; $y = \dfrac{y_3(x - x_2)}{x_1 - 2x_2}$

From C: $\dfrac{y - y_3}{x - 0} = \dfrac{y_3}{0 - \frac{x_1 + x_2}{2}}$; $y = y_3 - \dfrac{2xy_3}{x_1 + x_2}$, $x_1 \neq -x_2$

Find intersections:

AB: $\dfrac{x - x_1}{x_2 - 2x_1} = \dfrac{x - x_2}{x_1 - 2x_2}$; $x(x_1 - 2x_2 - x_2 + 2x_1) = x_1^2 - x_2^2$

$$x = \frac{x_1^2 - x_2^2}{3(x_1 - x_2)} = \frac{x_1 + x_2}{3}$$

$$y = \frac{y_3\left(\frac{x_1 + x_2}{3} - x_1\right)}{x_2 - 2x_1} = \frac{y_3}{3}$$

AC: $\dfrac{x - x_1}{x_2 - 2x_1} = \dfrac{x_1 + x_2 - 2x}{x_1 + x_2}$; $\quad x(x_1 + x_2 + 2x_2 - 4x_1) = -x_1^2 + x_2^2$

$$x = \frac{x_2^2 - x_1^2}{3(x_2 - x_1)} = \frac{x_1 + x_2}{3}$$

BC: $\dfrac{x - x_2}{x_1 - 2x_2} = \dfrac{x_1 + x_2 - 2x}{x_1 + x_2}$; $\quad$ same as AC with $x_1 \leftrightarrow x_2$

$$x = \frac{x_1 + x_2}{3}$$

Therefore the centroid equals the point of intersection of the medians.

7.8   The Theorem of Pappus

1. $x^2 + (y - 5)^2 = 9$, about the x-axis.

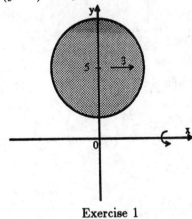

Exercise 1

$$\begin{aligned} V &= 2\pi A \overline{y} \\ &= 2\pi(9\pi)\,5 \\ &= 90\pi^2 \end{aligned}$$

2. $(x - 2)^2 + (y + 3)^2 = 4$, about the line $y = 3$.

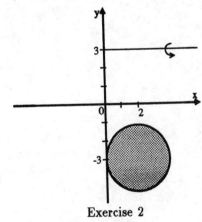

Exercise 2

$$\begin{aligned} V &= 2\pi A \overline{y} \\ &= 2\pi(4\pi)(3 + 3) \\ &= 48\pi^2 \end{aligned}$$

3. $y = x^2$, $y = \sqrt{x}$, about the y-axis.

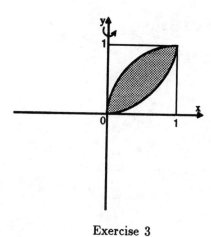

$$V = 2\pi A\overline{x}$$
$$= 2\pi \left(\frac{1}{3}\right) \left(\frac{9}{20}\right)$$
$$= \frac{3\pi}{10}$$

Exercise 3

4. $x^2 + y^2 = 16$, about the line $y = -x$.

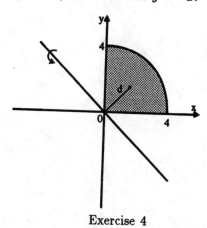

$$V = 2\pi A d$$
$$= 2\pi(4\pi)\sqrt{2\left(\frac{16}{3\pi}\right)^2}$$
$$= 8\sqrt{2}\,\pi^2 \left(\frac{16}{3\pi}\right)$$
$$= \frac{128}{3}\sqrt{2}\,\pi$$
$$\approx 189.563$$

Exercise 4

5.

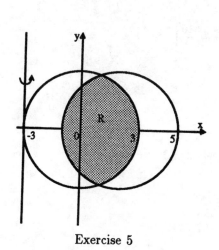

$$V = 2\pi A(\overline{x} + 3)$$
$$A = 2\pi A(1 + 3) = 8\pi A$$

Exercise 5

6.

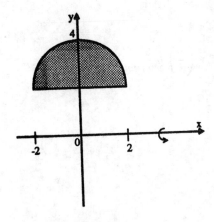

$$y = 2 + \sqrt{4 - y^2},$$
$$y = 0,$$
$$-2 \leq x \leq 2$$

Exercise 6

The area of the region is

$$A = 4(2) + \frac{4\pi}{2} = 8 + 2\pi.$$

So

$$\bar{y} = \frac{1}{2A} \int_{-2}^{2} \left(2 + \sqrt{4 - x^2}\right)^2 dx$$

$$= \frac{1}{16 + 4\pi} \int_{-2}^{2} 4 + 4\sqrt{4 - x^2} + (4 - x^2)\, dx$$

$$= \frac{1}{16 + 4\pi} \int_{-2}^{2} 8 + 4\sqrt{4 - x^2} - x^2\, dx.$$

Now

$$\int_{-2}^{2} 8\, dx = 32$$

$$\int_{-2}^{2} 4\sqrt{4 - x^2}\, dx = 4(2\pi) \quad \text{(semicircle)}$$

$$\int_{-2}^{2} x^2\, dx = \left.\frac{x^3}{3}\right]_{-2}^{2} = \frac{16}{3}.$$

Thus,

$$\bar{y} = \frac{1}{16 + 4\pi}\left[32 + 8\pi - \frac{16}{3}\right]$$

$$= \frac{20 + 6\pi}{3(4 + \pi)}.$$

By the Theorem of Pappus, we have

$$V = 2\pi \bar{y} A$$

$$= 2\pi \left(\frac{20 + 6\pi}{12 + 3\pi}\right)(8 + 2\pi)$$

$$= 2\pi(20 + 6\pi)\left(\frac{2}{3}\right)$$

$$= \frac{8\pi}{3}(10 + 3\pi) \approx 162.733.$$

584

7. $\bar{y} = \dfrac{8}{2} = 4$

force $= (10)\,(8)\,(4)\,(62.4) = 19,968$ lb

8. $\bar{y} = 30 + \dfrac{4}{12}$ ft $= \dfrac{91}{3}$ ft

force $= A\rho\bar{y} = \pi\left(\dfrac{1}{3}\right)^2 (62.4)\left(\dfrac{91}{3}\right) \approx 661$ lb

9. $\bar{y} = 30 + \dfrac{4}{12}$ ft $= \dfrac{91}{3}$ ft

$A = (6)(8) + \pi(4^2) = 48 + 16\pi$ in^2

force $= A\rho\bar{y} = \dfrac{48 + 16\pi}{144}(62.4)\left(\dfrac{91}{3}\right) \approx 1291.65$ lb

10.

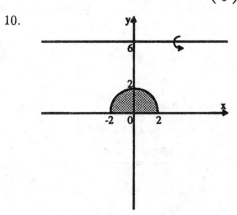

$$
\begin{aligned}
f(x) &= 6 \\
g(x) &= \sqrt{4 - x^2} \\
-2 &\le x \le 2
\end{aligned}
$$

Exercise 10

The area of the region is $A = 6(4) - 2\pi = 24 - 2\pi$.

Now

$$
\begin{aligned}
\bar{y} &= \frac{1}{2A}\int_{-2}^{2} 36 - (4 - x^2)\,dx \\
&= \frac{1}{48 - 4\pi}\int_{-2}^{2} 32 + x^2\,dx \\
&= \frac{1}{48 - 4\pi}\left[128 + \frac{16}{3}\right] \\
&= \frac{1}{48 - 4\pi}\left[\frac{400}{3}\right] = \frac{100}{36 - 3\pi}.
\end{aligned}
$$

By the Theorem of Pappus, we have

$$
\begin{aligned}
V &= 2\pi(6 - \bar{y})A \\
&= 2\pi\left(6 - \frac{100}{36 - 3\pi}\right)(24 - 2\pi) \\
&= \frac{8\pi}{3}(58 - 9\pi) \approx 249.03.
\end{aligned}
$$

11. $\quad V \;=\; \displaystyle\int_1^3 2\pi x \left[2\sqrt{1-(x-2)^2}\,\right] dx$

Let $u \;=\; x-2,\; du = dx,\; x = u+2$
$\quad x \;=\; 1 \Rightarrow u = -1;\; x = 3 \Rightarrow u = 1$
$\quad V \;=\; 4\pi \displaystyle\int_{-1}^1 (u+2)\sqrt{1-u^2}\,du = 4\pi \int_{-1}^1 u\sqrt{1-u^2}\,du + 8\pi \int_{-1}^1 \sqrt{1-u^2}\,du$

$$\;=\; -\frac{4\pi}{3}(1-u^2)^{3/2}\Big]_{-1}^1 + 8\pi \left(\frac{\pi}{2}\right) \quad \text{(area of semicircle)}$$

$$\;=\; -\frac{4\pi}{3}(0-0) + 4\pi^2 = 4\pi^2$$

Review Exercises—Chapter 7

1. Rotate the region bounded by $y = \sqrt{r^2 - x^2}$ and the x-axis about the x-axis.

$$V = \int_{-r}^r \pi(r^2 - x^2)\,dx = 2\pi\left[r^2 x - \frac{x^3}{3}\right]_0^r = 2\pi\left(r^3 - \frac{r^3}{3}\right) = \frac{4\pi r^3}{3}$$

2. $x = 4,\; y^2 = x$, about the x-axis.

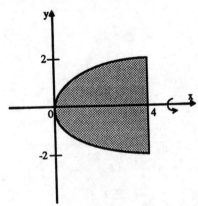

Exercise 2

$$V \;=\; \int_0^4 \pi x\,dx$$

$$\;=\; \pi \frac{x^2}{2}\Big]_0^4$$

$$\;=\; 8\pi$$

3. $y = \dfrac{1}{x},\; 1 \le x \le 4$, about the x-axis,

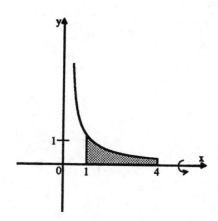

$$V = \int_1^4 \pi \left(\frac{1}{x}\right)^2 dx$$

$$= -\frac{\pi}{x}\Big]_1^4$$

$$= \pi \left(1 - \frac{1}{4}\right)$$

$$= \frac{3\pi}{4}$$

Exercise 3

4. $y = \sec\left(\dfrac{\pi x}{2}\right)$, $-\dfrac{1}{2} \le x \le \dfrac{1}{2}$, about the x-axis.

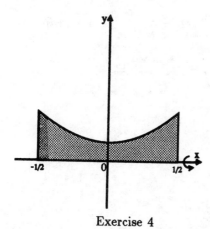

$$V = \int_{-1/2}^{1/2} \pi \sec^2\left(\frac{\pi x}{2}\right) dx$$

$$= 4\tan\left(\frac{\pi x}{2}\right)\Big]_0^{1/2}$$

$$= 4$$

Exercise 4

5. $y = \sin x$, $0 \le x \le \dfrac{3\pi}{2}$, about the x-axis.

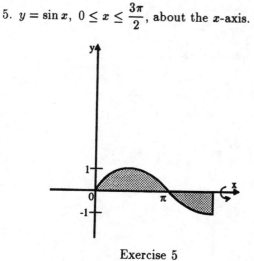

$$V = \int_0^{3\pi/2} \pi \sin^2 x \, dx$$

$$= \frac{\pi}{2} \int_0^{3\pi/2} (1 - \cos 2x) \, dx$$

$$= \frac{\pi}{2}\left[x - \frac{1}{2}\sin 2x\right]_0^{3\pi/2}$$

$$= \frac{\pi}{2}\left(\frac{3\pi}{2}\right)$$

$$= \frac{3\pi^2}{4}$$

Exercise 5

587

6. $f(x) = \sqrt{1-x^2}$, $g(x) = \dfrac{1}{2}$, about the x-axis.

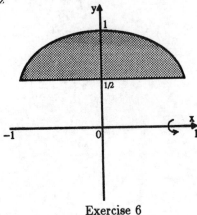

Exercise 6

Use cylinders, with axes parallel to the x-axis.

$$V = \int_{1/2}^{1} 2\pi yx\, dy = 2\pi \int_{1/2}^{1} y\left[\sqrt{1-y^2} + \sqrt{1-y^2}\right] dy = 4\pi \int_{1/2}^{1} y\sqrt{1-y^2}\, dy$$

$$= -\frac{4\pi}{3}(1-y^2)^{3/2}\Bigg]_{1/2}^{1} = -\frac{4\pi}{3}\left(0 - \left(\frac{3}{4}\right)^{3/2}\right) = \frac{4\pi}{3}\cdot\frac{3\sqrt{3}}{8} = \frac{\sqrt{3}\pi}{2}$$

7. $f(x) = x^{2/3}$, $0 \le x \le 2$, about the y-axis.

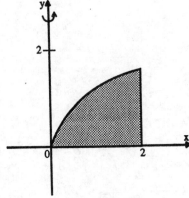

Exercise 7

$$V = \int_{0}^{2} 2\pi x(x^{2/3})\, dx = 2\pi \int_{0}^{2} x^{5/3}\, dx = \frac{3\pi}{4} x^{8/3}\Bigg]_{0}^{2} = \frac{3\pi}{4}(4 \cdot 2^{2/3})$$

$$= (3\pi)2^{2/3} \approx 14.961$$

8. $y = \sin x^2$, $0 \le \dfrac{\sqrt{\pi}}{2}$, about the y-axis.

588

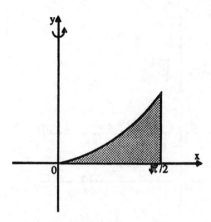

Exercise 8

$$V = \int_0^{\sqrt{\pi}/2} 2\pi x(\sin x^2)\, dx = -\pi \cos x^2 \bigg]_0^{\sqrt{\pi}/2} = -\pi\left(\frac{\sqrt{2}}{2} - 1\right) = \frac{\pi}{2}\left(2 - \sqrt{2}\right) \approx 0.9202$$

9. $y_1 = \sin x^2$, $y_2 = \cos x^2$, $0 \le x \le \dfrac{\sqrt{\pi}}{2}$, about the y-axis.

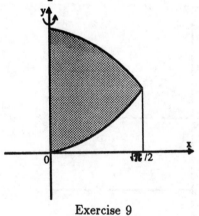

Exercise 9

$$V = \int_0^{\sqrt{\pi}/2} 2\pi x(\cos x^2 - \sin x^2)\, dx = \pi\left[\sin x^2 + \cos x^2\right]_0^{\sqrt{\pi}/2} = \pi\left(\frac{\sqrt{2}}{2} + \frac{\sqrt{2}}{2} - 1\right)$$

$$= \pi\left(\sqrt{2} - 1\right) \approx 1.3013$$

10. $y = \dfrac{1}{x^2}$, $1 \le x \le 4$, about the line $y = -2$

589

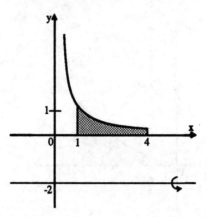

Exercise 10

$$V = \int_1^4 \pi \left[\left(\frac{1}{x^2} + 2 \right)^2 - 2^2 \right] dx = \pi \int_1^4 \left(\frac{1}{x^4} + \frac{4}{x^2} \right) dx = \pi \left[-\frac{1}{3x^3} - \frac{4}{x} \right]_1^4$$

$$= \pi \left[-\frac{1}{3} \left(\frac{1}{64} - 1 \right) - 4 \left(\frac{1}{4} - 1 \right) \right] = \frac{213}{64} \pi \approx 10.4556$$

11. $y = x^{2/3} + 1$, $0 \le x \le 2$, about $y = -3$.

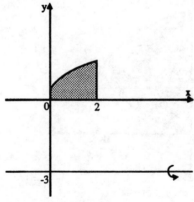

Exercise 11

$$V = \int_0^2 \pi [(x^{2/3} + 4)^2 - 3^2] dx = \pi \int_0^2 (x^{4/3} + 8x^{2/3} + 7) dx = \pi \left[\frac{3}{7} x^{7/3} + \frac{24}{5} x^{5/3} + 7x \right]_0^2$$

$$= \pi \left[\frac{12}{7} \sqrt[3]{2} + \frac{48}{5} \sqrt[3]{4} + 14 \right] \approx 98.643$$

12. $y = 3 + x^3$, $x = 0$, $y = 0$, about the line $x = 2$.

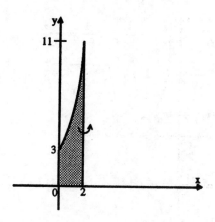

Exercise 12

$$V = \int_{\sqrt[3]{3}}^{0} 2\pi(2-x)(3+x^3)\,dx = 2\pi\int_{-\sqrt[3]{3}}^{0}(-x^4+2x^3-3x+6)\,dx$$

$$= 2\pi\left[-\frac{x^5}{5}+\frac{x^4}{2}-\frac{3}{2}x^2+6x\right]_{-\sqrt[3]{3}}^{0} = 2\pi\left(-\frac{3}{5}\sqrt[3]{9}-\frac{3}{2}\sqrt[3]{3}+\frac{3}{2}\sqrt[3]{9}+6\sqrt[3]{3}\right)$$

$$= 9\pi\sqrt[3]{3}+\frac{9}{5}\pi\sqrt[3]{9}.$$

13. $y = 1 + x^{3/2}$ from $(0,1)$ to $(4,9)$.

$$y' = \frac{3}{2}x^{1/2};\ 1+(y')^2 = 1+\frac{9}{4}x$$

$$L = \int_{0}^{4}\sqrt{1+(y')^2}\,dx = \int_{0}^{4}\sqrt{1+\frac{9}{4}x}\,dx = \frac{8}{27}\left(1+\frac{9}{4}x\right)^{3/2}\Bigg]_{0}^{4}$$

$$= \frac{8}{27}\left(10^{3/2}-1\right) \approx 9.0734$$

14. $x^2 = y^3$ from $(0,0)$ to $(1,1)$

$$2x = 3y^2 y'; \quad (y')^2 = \frac{4x^2}{9y^4} = \frac{4y^3}{9y^4} = \frac{4}{9y} = \frac{4}{9x^{2/3}}$$

$$L = \int_0^1 \sqrt{1+(y')^2}\, dx = \int_0^1 \sqrt{1 + \frac{4}{9x^{2/3}}}\, dx = \frac{1}{3}\int_0^1 \frac{1}{x^{1/3}}\sqrt{4+9x^{2/3}}\, dx$$

Let $u = 4 + 9x^{2/3}, \; du = 6x^{-1/3}\, dx$

$x = 0 \Rightarrow u = 4; \; x = 1 \Rightarrow u = 13$

$$\frac{1}{3}\int_4^{13} \frac{1}{6}\sqrt{u}\, du = \frac{1}{18}\cdot\frac{2}{3}u^{3/2}\Big]_4^{13} = \frac{1}{27}\left(13\sqrt{13}-8\right) \approx 1.4397$$

Alternatively, let $x = y^{3/2}; \; \dfrac{dx}{dy} = \dfrac{3}{2}y^{1/2}$

$$L = \int_0^1 \sqrt{1+(x')^2}\, dy = \int_0^1 \sqrt{1 + \frac{9}{4}y}\, dy = \frac{4}{9}\cdot\frac{2}{3}\left(1+\frac{9}{4}y\right)^{3/2}\Big]_0^1$$

$$= \frac{8}{27}\left(\left(\frac{13}{4}\right)^{3/2}-1\right) = \frac{8}{27}\left(\frac{13\sqrt{13}}{8}-1\right) = \frac{1}{27}\left(13\sqrt{13}-8\right)$$

15. $y^2 = (x+1)^3$ from $(0,1)$ to $(1,\sqrt{8})$

$$y = (x+1)^{3/2}; \quad y' = \frac{3}{2}(x+1)^{1/2}$$

$$L = \int_0^1 \sqrt{1+(y')^2}\, dx = \int_0^1 \sqrt{1 + \frac{9}{4}(x+1)}\, dx = \frac{4}{9}\cdot\frac{2}{3}\left[1+\frac{9}{4}(x+1)\right]^{3/2}\Big]_0^1$$

$$= \frac{8}{27}\left[\left(\frac{11}{2}\right)^{3/2}-\left(\frac{13}{4}\right)^{3/2}\right] = \frac{8}{27}\left(\frac{11\sqrt{11}}{2\sqrt{2}}-\frac{13\sqrt{13}}{8}\right) = \frac{1}{27}\left(\frac{44\sqrt{11}}{\sqrt{2}}-13\sqrt{13}\right)$$

$$= \frac{22\sqrt{22}-13\sqrt{13}}{27} \approx 2.0858$$

16. $y = \dfrac{1}{8}\left(x^4 + \dfrac{2}{x^2}\right)$ from $\left(1,\dfrac{3}{8}\right)$ to $\left(2,\dfrac{33}{16}\right)$

$$y' = \frac{x^3}{2} - \frac{1}{2x^3}$$

$$L = \int_1^2 \sqrt{1+(y')^2}\, dx = \int_1^2 \sqrt{1 + \frac{1}{4}\left(x^6 - 2 + \frac{1}{x^6}\right)}\, dx = \int_1^2 \sqrt{\frac{1}{4}\left(x^6 + 2 + \frac{1}{x^6}\right)}\, dx$$

$$= \frac{1}{2}\int_1^2 \left(x^3 + \frac{1}{x^3}\right) dx = \frac{1}{2}\left[\frac{x^4}{4} - \frac{1}{2x^2}\right]_1^2 = \frac{1}{2}\left[\left(4-\frac{1}{4}\right)-\left(\frac{1}{8}-\frac{1}{2}\right)\right] = \frac{33}{16}$$

17. $y = x^5 + \dfrac{1}{12x} + 1$ from $\left(1,\dfrac{25}{12}\right)$ to $\left(2,\dfrac{217}{24}\right)$

$$y' = 3x^2 - \frac{1}{12x^2}$$

$$L = \int_1^2 \sqrt{1+(y')^2}\,dx = \int_1^2 \sqrt{1+9x^4-\frac{1}{2}+\frac{1}{144x^4}}\,dx = \int_1^2 \sqrt{9x^4+\frac{1}{2}+\frac{1}{144x^4}}\,dx$$

$$= \int_1^2 \left(3x^2+\frac{1}{12x^2}\right)dx = x^3-\frac{1}{12x}\bigg]_1^2 = (8-1)-\frac{1}{12}\left(\frac{1}{2}-1\right) = \frac{169}{24}$$

18. $y = x^{5/2}+\dfrac{1}{5\sqrt{x}}+3$ from $\left(4,\dfrac{351}{10}\right)$ to $\left(16,\dfrac{20541}{20}\right)$.

$$y' = \frac{5}{2}x^{3/2}-\frac{1}{10x^{3/2}}$$

$$L = \int_4^{16}\sqrt{1+(y')^2}\,dx = \int_4^{16}\sqrt{1+\frac{25}{4}x^3-\frac{1}{2}+\frac{1}{100x^3}}\,dx$$

$$= \int_4^{16}\sqrt{\frac{25}{4}x^3+\frac{1}{2}+\frac{1}{100x^3}}\,dx = \int_4^{16}\left(\frac{5}{2}x^{3/2}+\frac{1}{10x^{3/2}}\right)dx$$

$$= x^{5/2}-\frac{1}{5\sqrt{x}}\bigg]_4^{16} = (1024-32)-\frac{1}{5}\left(\frac{1}{4}-\frac{1}{2}\right)$$

$$= \frac{19,841}{20} = 992.05$$

19. $A(y) = A\left(1-\dfrac{y}{h}\right)^2$

$$V = \int_0^h A(y)\,dy = \int_0^h A\left(1-\frac{y}{h}\right)^2 dy = -\frac{Ah}{3}\left(1-\frac{y}{h}\right)^3\bigg]_0^h$$

$$= -\frac{Ah}{3}(0-1) = \frac{Ah}{3}$$

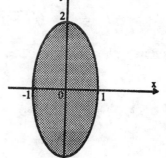

20. $x^2+\dfrac{y^2}{4} = 1$

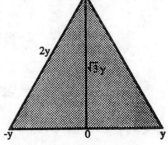

a) $V = \displaystyle\int_{-1}^1 A(x)\,dx$

Each square has side $2y$, where $y^2 = 4(1-x^2)$

$$V = \int_{-1}^1 4y^2\,dx = \int_{-1}^1 16(1-x^2)\,dx = 16\left[x-\frac{x^3}{3}\right]_{-1}^1 = 32\left[x-\frac{x^3}{3}\right]_0^1 = \frac{64}{3}$$

b) $V = \displaystyle\int_{-1}^1 A(x)\,dx$

Each triangle has base $2y$ and altitude $\sqrt{3}\,y$.

$$V = \int_{-1}^1 \frac{2y\left(\sqrt{3}\,y\right)}{2}\,dx = \sqrt{3}\int_{-1}^1 4(1-x^2)\,dx = 4\sqrt{3}\left[x-\frac{x^3}{3}\right]_{-1}^1 = 8\sqrt{3}\left[x-\frac{x^3}{3}\right]_0^1 = \frac{16\sqrt{3}}{3}$$

21.

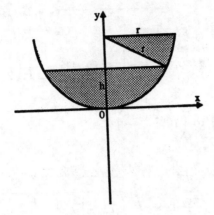

Exercise 21

At any value of h, $r^2 = (r-h)^2 + x^2$

$$V = \int_0^{r/2} \pi[r^2 - (r-h)^2]\,dh = \pi \int_0^{r/2} (2rh - h^2)\,dh = \pi\left[rh^2 - \frac{h^3}{3}\right]_0^{r/2}$$

$$= \pi\left(\frac{r^3}{4} - \frac{r^3}{24}\right) = \frac{5\pi r^3}{24}$$

$$\% \text{ of capacity} = \frac{\frac{5\pi r^3}{24}}{\frac{2\pi r^3}{3}}(100) = \frac{500}{16} = 31.25\%$$

22. $f(x) = \cos^2 x$, $0 \le x \le \pi$

$$f'(x) = -2\cos x \sin x = -\sin 2x$$
$$L = \int_0^\pi \sqrt{1 + [f'(x)]^2}\,dx = \int_0^\pi \sqrt{1 + \sin^2(2x)}\,dx$$

Using Simpson's Rule with $n = 16$ gives $L \approx 3.820$.

23. $f(x) = \sin x$, $0 \le x \le \pi$, about the x-axis.

$$f'(x) = \cos x$$
$$S = \int_0^\pi 2\pi(\sin x)\sqrt{1 + \cos^2 x}\,dx$$

Using Simpson's Rule with $n = 16$ gives $S \approx 14.424$.

24. $F = kx$

$$20 = k(10); \quad k = 2 \text{ lb/in}$$
$$W = \int_0^{15} F(x)\,dx = \int_0^{15} 2x\,dx = x^2\Big]_0^{15} = 225 \text{ lb/in} = \frac{225}{12} \text{ ft-lb} = \frac{75}{4} \text{ ft-lb}$$

25. $F = kx$

$$10 = k(20); \ k = 0.5 \text{ newton/cm}$$

$$W = \int_0^{50} F(x)\,dx = \int_0^{50} 0.5x\,dx = \left. \frac{x^2}{4} \right]_0^{50}$$

$$= 625 \text{ newton-cm} = 6.25 \text{ newton-m or joules}$$

$$W = \int_{50}^{60} 0.5x\,dx = \left. \frac{x^2}{4} \right]_{50}^{60}$$

$$= \frac{1}{4}(3600 - 2500) = 275 \text{ newton-cm} = 2.75 \text{ newton-m or joules}$$

26. $\displaystyle W = \int_0^{20} F(h)\,dh = \int_0^{20} \rho\pi r^2 h\,dh = 62.4\pi(25)\int_0^{20} h\,dh = \left. 780h^2 \right]_0^{20}$

$$= 312{,}000\pi \text{ ft-lb} \approx 980{,}177 \text{ ft-lb}$$

27.

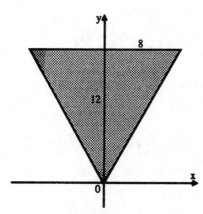

Exercise 27

From the figure, we see that

$$\frac{r}{h} = \frac{8}{12} = \frac{2}{3}.$$

So

$$\begin{aligned}
W &= \int_0^{12} F(h)\,dh = \int_0^{12} \rho\pi r^2(22 - h)\,dh \\
&= 62.4\pi \int_0^{12} \frac{4h^2}{9}(22 - h)\,dh \\
&= 62.4\pi(3328) \\
&= 207667.2\pi \text{ ft-lbs} \\
&= 652{,}405.75 \text{ ft-lbs}
\end{aligned}$$

28. Note the solution to Exercise 27. The total volume is

$$V = \frac{\pi r^2 h}{3} = \frac{\pi(64)(12)}{3} = 256\pi \text{ ft}^3.$$

First, we determine the height corresponding to one-half of V. We want

$$\frac{\pi r^2 h}{3} = 128\pi,$$

and $r = \frac{2}{3}h$. So

$$\frac{\pi 4 h^3}{27} = 128\pi,$$

and $h = 6\sqrt[3]{4}$. Therefore, the work required to pump one-half of the water is

$$
\begin{aligned}
\int_{6\sqrt[3]{4}}^{12} F(h)\, dh &= \int_{6\sqrt[3]{4}}^{12} \rho \pi r^2 (22 - h)\, dh \\
&= 62.4\pi \int_{6\sqrt[3]{4}}^{12} \frac{4h^2}{9}(22 - h)\, dh \\
&= 62.4\pi(64)\left(8 + 9\sqrt[3]{4}\right) \\
&= 279{,}613.7 \text{ ft-lbs}
\end{aligned}
$$

29. $W = \displaystyle\int_0^{20} F(x)\, dx = \int_0^{20} \frac{400}{20}x\, dx = 20\left.\frac{x^2}{2}\right]_0^{20} = 4000$ ft-lb

30. $W = 4000 + 50(20) = 5000$ ft-lb

31. $W = \displaystyle\int_0^{30} F(x)\, dx = \int_0^{30}(1000 + 6x)\, dx = 1000x + 3x^2 \Big]_0^{30} = 30{,}000 + 2700 = 32{,}700$ ft-lb

32. $W = \displaystyle\int_0^{15}(1000 + 6x)\, dx + \int_{15}^{30} 6x\, dx = 1000x\Big]_0^{15} + 3x^2\Big]_0^{30} = 15{,}000 + 2700 = 17{,}700$ ft-lb

33.

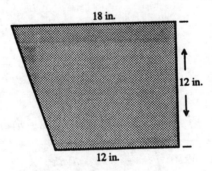

18 in.

12 in.

12 in.

Exercise 33

$$
\begin{aligned}
F &= \int_0^1 62.4(h)\,(w(h))\, dh \\
w(h) &= 1.5 - \frac{h}{2} \\
F &= 62.4 \int_0^1 h\left(1.5 - \frac{h}{2}\right) dh = 62.4\left[\frac{3}{4}h^2 - \frac{1}{6}h^3\right]_0^1 = 62.4\left(\frac{14}{24}\right) = 36.4\,\text{lb}
\end{aligned}
$$

34.

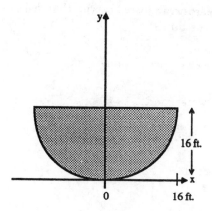

16 ft.

16 ft.

0

Exercise 34

$$F = \int_0^{16} 62.4(16 - h)w(h)\, dh$$

But $w(h) = 2x$ where $y = \frac{x^2}{16}$. That is, $x = 4\sqrt{y}$, so $w(h) = 8\sqrt{h}$.

$$\begin{aligned} F &= \int_0^{16} 62.4(16 - h)8\sqrt{h}\, dh \\ &= 136{,}315 \text{ lb} \end{aligned}$$

35.

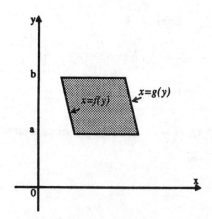

b

$x=f(y)$

$x=g(y)$

a

0

Exercise 35

Let the left-hand boundary of the face be given by $x = f(y)$ and the right-hand boundary by $x = g(y)$. Then, by integrating with respect to y,

$$\bar{y} = \frac{\int_a^b y[g(y) - f(y)]\, dy}{\int_a^b [g(y) - f(y)]\, dy} = \frac{\int_a^b hw(h)\, dh}{A}$$

where h is used in place of y and $w(h)$ represents the width of the face.

The force on the face of the plate is

$$\rho A\bar{y} = \rho \int_a^b hw(h)\, dh = \int_a^b \rho hw(h)\, dh,$$

which is the usual form for the hydrostatic force, except that h is usually measured as increasing downward. Let $h = -y$ and change the limits:

$$F = \int_b^a \rho h w(h)\, dh.$$

Usually $b = 0$, at the surface of the water, and

$$F = \int_0^a \rho h w(h)\, dh.$$

36.

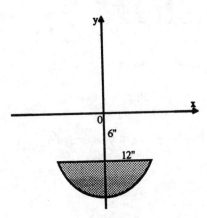

Exercise 36

$$
\begin{aligned}
F &= \int_0^1 \rho\left(h+\frac{1}{2}\right)\left[2\sqrt{1-h^2}\right]\, dh = 62.4\,(2)\int_0^1\left[h\sqrt{1-h^2}+\frac{1}{2}\sqrt{1-h^2}\right]\, dh \\[2mm]
&= 124.8\left[-\frac{1}{3}\left(1-h^2\right)^{3/2}\right]_0^1 + 62.4\int_0^1\sqrt{1-h^2}\, dh \\[2mm]
&= -\frac{124.8}{3}(0-1) + 62.4\left(\frac{\pi}{4}\right) \quad \text{(one quadrant of a circle)} \\[2mm]
&= 41.6 + 15.6\pi \approx 90.609\ \text{lb}
\end{aligned}
$$

37. $\bar{x} = \dfrac{\int_0^{20} 16\pi x \rho(x)\, dx}{\int_0^{20} 16\pi \rho(x)\, dx}$

$$
\begin{aligned}
\rho(x) &= 2 + \frac{6}{20}x \\[2mm]
\bar{x} &= \frac{\int_0^{20} x\left(2+\frac{6}{20}x\right) dx}{\int_0^{20}\left(2+\frac{6}{20}x\right) dx} = \frac{\int_0^{20}\left(2x+\frac{6}{20}x^2\right) dx}{\int_0^{20}\left(2+\frac{6}{20}x\right) dx} = \frac{x^2+\frac{2}{20}x^3\big]_0^{20}}{2x+\frac{3}{20}x^2\big]_0^{20}} = \frac{400+800}{40+60} = \frac{1200}{100} = 12\ \text{cm}
\end{aligned}
$$

38. $\bar{x} = \dfrac{\int_0^{10}\rho A(x)\, dx}{\int_0^{10}\rho A(x)\, dx} = \dfrac{\int_0^{10}(2+0.05x^2)x\, dx}{\int_0^{10}(2+0.05x^2)\, dx} = \dfrac{\int_0^{10}(2x+0.05x^3)\, dx}{\int_0^{10}(2+0.05x^2)\, dx} = \dfrac{x^2+\frac{0.05}{4}x^4\big]_0^{10}}{2x+\frac{0.05x^3}{3}\big]_0^{10}}$

$$= \frac{100+125}{20+\frac{50}{3}} = \frac{(225)(3)}{110} = 6.136\ \text{cm}$$

39. $\bar{x} = \dfrac{\int_0^3 \rho A(x)\,dx}{\int_0^3 \rho A(x)\,dx} = \dfrac{\int_0^3 \frac{20\left(10+\frac{90x}{3}\right)}{2(100^2)}\,x\,dx}{\int_0^3 \frac{20\left(10+\frac{90x}{3}\right)}{2(100^2)}\,dx} = \dfrac{\int_0^3 (10x + 30x^2)\,dx}{\int_0^3 (10 + 30x)\,dx} = \dfrac{5x^2 + 10x^3 \Big]_0^3}{10x + 15x^2 \Big]_0^3}$

   $= \dfrac{45 + 270}{30 + 135} = \dfrac{315}{165} = \dfrac{21}{11}$ m

40. $\bar{x} = \dfrac{\sum_{j=1}^3 x_j w_j}{\sum_{j=1}^3 w_j} = \dfrac{-5(2) + 1(5) + 4(7)}{2 + 5 + 7} = \dfrac{23}{14}$

41. $\bar{x} = \dfrac{\sum_{j=1}^4 x_j w_j}{\sum_{j=1}^4 w_j} = \dfrac{m(-2 - 2 + 0 + 1)}{4m} = -\dfrac{3}{4}$

   $\bar{y} = \dfrac{m(4 - 2 + 6 - 4)}{4m} = 1$

42. $\bar{x} = \dfrac{\int_0^{100} \rho(x) A x\,dx}{\int_0^{100} \rho(x) A\,dx} = \dfrac{\int_0^{100} (1 + \sqrt{x})\,x\,dx}{\int_0^{100} (1 + \sqrt{x})\,dx} = \dfrac{\frac{x^2}{2} + \frac{2}{5} x^{5/2} \Big]_0^{100}}{x + \frac{2}{3} x^{3/2} \Big]_0^{100}} = \dfrac{5000 + 40{,}000}{100 + \frac{2000}{3}}$

   $= \dfrac{135{,}000}{2300} \approx 58.696$ cm

43. Answers will vary.

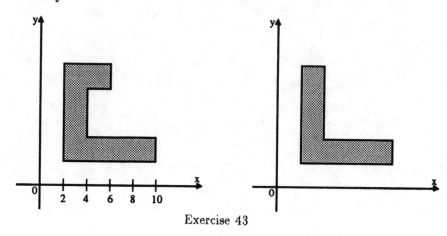

Exercise 43

44. $y = x,\ y = 0,\ x = 0,\ x = 2$

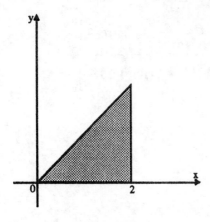

$$A = \frac{1}{2}(2 \cdot 2) = 2$$

$$\bar{x} = \frac{1}{A}\int_0^2 x(x)\,dx$$

$$= \frac{1}{6}x^3\Big]_0^2 = \frac{8}{6} = \frac{4}{3}$$

$$\bar{y} = \frac{1}{2A}\int_0^2 x^2\,dx$$

$$= \frac{1}{12}x^3\Big]_0^2 = \frac{2}{3}$$

Exercise 44

45. $y = 16 - x^2$

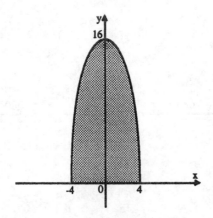

Exercise 45

$$\bar{x} = 0, \text{ by symmetry}$$

$$A = 2\int_0^4 (16 - x^2)\,dx = 2\left[16x - \frac{x^3}{3}\right]_0^4 = 2\left(64 - \frac{64}{3}\right) = \frac{256}{3}$$

$$\bar{y} = \frac{1}{2A}\int_{-4}^4 (16 - x^2)^2\,dx = \frac{3}{512}\int_{-4}^4 (256 - 32x^2 + x^4)\,dx = \frac{3}{256}\left[256x - \frac{32}{3}x^3 + \frac{x^5}{5}\right]_0^4$$

$$= \frac{3}{256}\left(1024 - \frac{2048}{3} + \frac{1024}{5}\right) = 3\left(4 - \frac{8}{3} + \frac{4}{5}\right) = \frac{32}{5}$$

$$(\bar{x}, \bar{y}) = (0, 6.4)$$

46. $y = x,\ y = x^2$

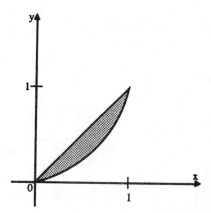

Exercise 46

Intersections: $x = x^2 \Rightarrow x = 0, 1$

$$A = \int_0^1 (x - x^2)\, dx = \frac{x^2}{2} - \frac{x^3}{3} \Big]_0^1 = \frac{1}{6}$$

$$\bar{x} = \frac{1}{A} \int_0^1 x(x - x^2)\, dx = 6 \left[\frac{x^3}{3} - \frac{x^4}{4} \right]_0^1 = 6 \left(\frac{1}{12} \right) = \frac{1}{2}$$

$$\bar{y} = \frac{1}{2A} \int_0^1 (x^2 - x^4)\, dx = 3 \left[\frac{x^3}{3} - \frac{x^5}{5} \right]_0^1 = \frac{2}{5}$$

$$(\bar{x}, \bar{y}) = \left(\frac{1}{2}, \frac{2}{5} \right)$$

47. $y = 6x - x^2 = -9 + 6x - x^2 + 9 = -(x - 3)^2 + 9$
$y - 9 = -(x - 3)^2$
Parabola with vertex at $(3, 9)$.

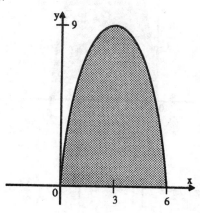

Exercise 47

$\bar{x} = 3$, by symmetry

$$A = 2 \int_0^3 (6x - x^2)\, dx = 2 \left[3x^2 - \frac{x^3}{3} \right]_0^3 = 2(27 - 9) = 36$$

$$\bar{y} = \frac{1}{2A} \int_0^6 (6x - x^2)^2\, dx = \frac{1}{72} \int_0^6 (36x^2 - 12x^3 + x^4)\, dx = \frac{1}{72} \left[12x^3 - 3x^4 + \frac{x^5}{5} \right]_0^6$$

$$= \frac{1}{72}\left(2592 - 3888 + \frac{7776}{5}\right) = 36 - 54 + \frac{108}{5} = \frac{18}{5}$$

$$(\overline{x}, \overline{y}) = (3, 3.6)$$

48. $9x^2 + 4y^2 = 36$

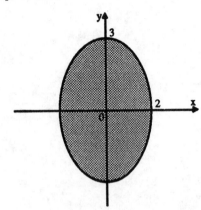

Exercise 48

$$\frac{x^2}{4} + \frac{y^2}{9} = 1$$

$$(\overline{x}, \overline{y}) = (0, 0), \text{ by symmetry}$$

49. $y = 6x - x^2$, $y = 3 - |x - 3|$
   $y = 6x - x^2 = -9 + 6x - x^2 + 9 = -(x-3)^2 + 9$
   $y - 9 = -(x-3)^2$, a parabola with vertex at $(3, 9)$

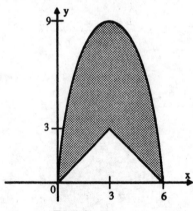

Exercise 49

$$\overline{x} = 3, \text{ by symmetry}$$

$$A = 2\int_0^3 (6x - x^2 - x)\,dx = 2\left[\frac{5}{2}x^2 - \frac{x^3}{3}\right]_0^3 = 2\left(\frac{45}{2} - 9\right) = 27$$

$$\overline{y} = \frac{1}{2A}\int_0^6 [(6x - x^2)^2 - (3 - |x-3|)^2]\,dx$$

$$= \frac{1}{54}\left[\int_0^3 (36x^2 - 12x^3 + x^4 - x^2)\,dx + \int_3^6 (36x^2 - 12x^3 + x^4 - 36 + 12x - x^2)\,dx\right]$$

$$= \frac{1}{54}\left[\frac{35}{3}x^3 - 3x^4 + \frac{x^5}{5}\right]_0^3 + \frac{1}{54}\left[\frac{35}{3}x^3 - 3x^4 + \frac{x^5}{5} - 36x + 6x^2\right]$$

$$= \frac{1}{54}\left(315 - 243 + \frac{243}{5}\right) + \frac{1}{54}\left[\frac{35}{3}(216 - 27) - 3(1296 - 81)\right.$$

$$\left. + \frac{1}{5}(7776 - 243) - 36(6 - 3) + 6(36 - 9)\right]$$

$$= \frac{1}{54}\left(\frac{603}{5}\right) + \frac{1}{54}\left(\frac{603}{5}\right) = \frac{67}{15}$$

$$(\overline{x}, \overline{y}) = \left(3, \frac{67}{15}\right)$$

50. $y = |2x|$, $y = 4$

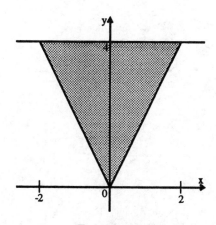

Exercise 50

$$\overline{x} = 0, \text{ by symmetry}$$

$$A = 2\left(\frac{2 \cdot 4}{2}\right) = 8$$

$$\overline{y} = \frac{1}{2A}\int_{-2}^{2}[4^2 - (|2x|)^2]\,dx = \frac{1}{16}\int_{-2}^{2}(16 - 4x^2)\,dx = \frac{1}{16}\left[16x - \frac{4}{3}x^3\right]_{-2}^{2}$$

$$= \frac{1}{8}\left(32 - \frac{32}{3}\right) = \frac{8}{3}$$

$$(\overline{x}, \overline{y}) = \left(0, \frac{8}{3}\right)$$

51. $y = |x - 2|$, $y = -\frac{1}{5}x$

603

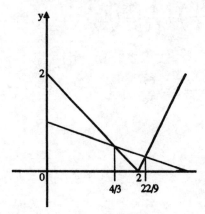

Exercise 51

To obtain the points of intersection, we solve

$$x - 2 = -\frac{x}{5} + \frac{14}{5},$$

which yields $x = 4$. We also solve

$$2 - x = -\frac{x}{5} + \frac{14}{5},$$

which yields $x = -1$.

The area A is

$$A = \int_{-1}^{2} \left(-\frac{x}{5} + \frac{14}{5}\right) - (2 - x)\, dx + \int_{2}^{4} \left(-\frac{x}{5} + \frac{14}{5}\right) - (x - 2)\, dx = 6.$$

Then

$$\overline{x} = \frac{1}{A} \int_{-1}^{2} x \left(\left(-\frac{x}{5} + \frac{14}{5}\right) - (2 - x)\right) dx + \frac{1}{A} \int_{2}^{4} x \left(\left(-\frac{x}{5} + \frac{14}{5}\right) - (x - 2)\right) dx$$

$$= \frac{1}{6}\left(\frac{18}{5}\right) + \frac{1}{6}\left(\frac{32}{5}\right) = \frac{5}{3},$$

and

$$\overline{y} = \frac{1}{2A} \int_{-1}^{2} \left(-\frac{x}{5} + \frac{14}{5}\right)^2 - (2 - x)^2\, dx + \frac{1}{2A} \int_{2}^{4} \left(-\frac{x}{5} + \frac{14}{5}\right)^2 - (x - 2)^2\, dx$$

$$= \frac{1}{12}\left(\frac{324}{25}\right) + \frac{1}{12}\left(\frac{176}{25}\right) = \frac{5}{3}.$$

52.

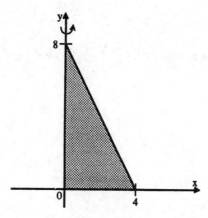

Exercise 52

$$A = \frac{4 \cdot 8}{2} = 16$$

$$\bar{x} = \frac{1}{16} \int_0^4 x(8 - 2x)\, dx = \frac{1}{16} \left[4x^2 - \frac{2}{3} x^3 \right]_0^4 = \frac{1}{16} \left(64 - \frac{128}{3} \right) = \frac{4}{3}$$

$$V = 2\pi \bar{x} A = 2\pi(16) \left(\frac{4}{3} \right) = \frac{128\pi}{3}$$

53. $y = \sqrt{4 - x^2}$

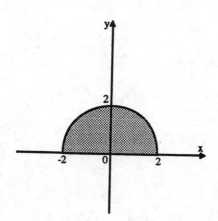

Exercise 53

$$\bar{y} = \frac{1}{2A} \int_{-2}^{2} (4 - x^2)\, dx = \frac{2}{4\pi} \left[4x - \frac{x^3}{3} \right]_0^2 = \frac{1}{2\pi} \left(8 - \frac{8}{3} \right) = \frac{8}{3\pi}$$

a) about $y = 0$

$$V = 2\pi \bar{y} A = 2\pi \left(\frac{8}{3\pi} \right) (2\pi) = \frac{32\pi}{3}$$

b) about $y = -2$

$$V = 2\pi(\bar{y} + 2)A = 2\pi \left(\frac{8}{3\pi} + 2 \right) (2\pi) = \frac{32}{3}\pi + 8\pi^2$$

605

c) about $y = 6$

$$V = 2\pi(6 - \overline{y})A = 2\pi\left(6 - \frac{8}{3\pi}\right)(2\pi) = 4\pi^2\left(6 - \frac{8}{3\pi}\right) = 24\pi^2 - \frac{32\pi}{3}$$

Chapter 8

Logarithmic and Exponential Functions

8.1 Review of Logarithms and Inverse Functions

1. $\log_{10} 1000 = \log_{10} 10^3 = 3$

2. $\log_7 343 = \log_7 7^3 = 3$

3. $\log_8 2 = \log_8 8^{1/3} = 1/3$

4. $\log_4(0.5) = \log_4 2^{-1} = \log_4 4^{-1/2} = -1/2$

5. $\log_5(1/125) = \log_5 5^{-3} = -3$

6. $\log_2 \sqrt{2} = \log_2 2^{1/2} = 1/2$

7. $\log_4 8 = \log_4 2^3 = \log_4 4^{3/2} = 3/2$

8. $\log_{343} 7 = \log_{343} 343^{1/3} = 1/3$

9. $\log_{10} 1 = \log_{10} 10^0 = 0$

10. $\log_b 1 = \log_b b^0 = 0$

11. $\log_x 16 = 4 \Longrightarrow x^4 = 16 \Longrightarrow x = 2$ since $x > 0$.

12. $\log_5 x = 3 \Longrightarrow x = 5^3 = 125$

13. $\log_{1/2} x = 2 \Longrightarrow x = (1/2)^2 = 1/4$

14. $\log_{125} x = 2/3 \Longrightarrow x = 125^{2/3} = 25$

15. $\log_3 x = 1/6 \Longrightarrow x = 3^{1/6}$

16. $\log_x 243 = 5 \Longrightarrow x^5 = 243 = 3^5 \Longrightarrow x = 3$

17. $\log_x 27 = 3/2 \Longrightarrow x^{3/2} = 27 = 9^{3/2} \Longrightarrow x = 9$

18. $\log_7 x = 1 \Longrightarrow x = 7^1 = 7$

19. $\log_{10} x = 0 \implies x = 10^0 = 1$

20. $\log_7 x = 0 \implies x = 7^0 = 1$

21. $\log_2 x = -3 \implies x = 2^{-3} = 1/8$

22. $\log_5 x = -2 \implies x = 5^{-2} = 1/25$

23. $y = f(x) = 3x - 2 \implies x = \frac{1}{3}y + \frac{2}{3}$. Hence the inverse of f is the function $f^{-1} : (-\infty, \infty) \to (-\infty, \infty)$, where $f^{-1}(x) = \frac{1}{3}x + \frac{2}{3}$. The graphs of $y = f(x)$ and $y = f^{-1}(x)$ are shown below.

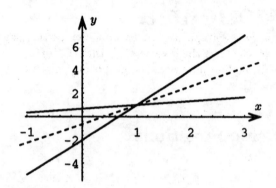

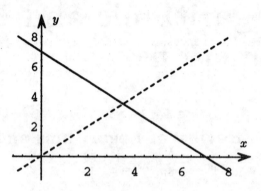

$$y = 3x - 2, \quad y = (1/3)x + 2/3$$

Exercise 23

$$y = 7 - x$$

Exercise 24

24. $y = f(x) = 7 - x \iff x = 7 - y$. Hence the inverse of f is the function $f^{-1} : (-\infty, \infty) \to (-\infty, \infty)$, where $f^{-1}(x) = 7 - x$. In this case $f^{-1}(x) = f(x)$. The graphs of $y = f(x)$ and $y = f^{-1}(x)$ are shown above. Note that the graph is symmetric about the line $y = x$, which shows, geometrically, that $f = f^{-1}$.

25. $y = \frac{1}{x+2} \iff x = \frac{1}{y} - 2$. Hence the inverse of the function $f : (-\infty, -2) \cup (-2, \infty) \to (-\infty, 0) \cup (0, \infty)$ is the function $f^{-1} : (-\infty, 0) \cup (0, \infty) \to (-\infty, -2) \cup (-2, \infty)$, where $f^{-1}(x) = (1/x) - 2$.

26. $y = \frac{2}{4-x} \iff x = 4 - 2/y$. Hence the inverse of the function $f : (-\infty, 4) \cup (4, \infty) \to (-\infty, 0) \cup (0, \infty)$ is the function $f^{-1} : (-\infty, 0) \cup (0, \infty) \to (-\infty, 4) \cup (4, \infty)$, where $f^{-1}(x) = 4 - 2/x$.

27. $y = \frac{x}{x+3} \iff x = \frac{3y}{1-y}$. Hence the inverse of the function $f : (-\infty, -3) \cup (-3, \infty) \to (-\infty, 1) \cup (1, \infty)$ is the function $f^{-1} : (-\infty, 1) \cup (1, \infty) \to (-\infty, -3) \cup (-3, \infty)$, where $f^{-1}(x) = \frac{3x}{1-x}$.

28. $y = \frac{x}{5-x} \iff x = \frac{5y}{1+y}$. Hence the inverse of the function $f : (-\infty, 5) \cup (5, \infty) \to (-\infty, -1) \cup (-1, \infty)$ is the function $f^{-1} : (-\infty, -1) \cup (-1, \infty) \to (-\infty, 5) \cup (5, \infty)$, where $f^{-1}(x) = \frac{5x}{1+x}$.

29. $y = 4 + \sqrt{x-1} \iff x = (y-4)^2 + 1$. Hence the inverse of the function $f : [1, \infty) \to [4, \infty)$ is the function $f^{-1} : [4, \infty) \to [1, \infty)$, where $f^{-1}(x) = (x-4)^2 + 1$.

30. $y = \sqrt[3]{x+3} \iff x = y^3 - 3$. Hence the inverse of the function $f : (-\infty, \infty) \to (-\infty, \infty)$ is the function $f^{-1} : (-\infty, \infty) \to (-\infty, \infty)$, where $f^{-1}(x) = x^3 - 3$.

31. $y = 3 + x^{5/3} \iff x = (y-3)^{3/5}$. Hence the inverse of the function $f : (-\infty, \infty) \to (-\infty, \infty)$ is the function $f^{-1} : (-\infty, \infty) \to (-\infty, \infty)$, where $f^{-1}(x) = (x-3)^{3/5}$.

32. $y = \frac{1}{\sqrt{x+1}} \iff x = \frac{1}{y^2} - 1$. Hence the inverse of the function $f : (-1, \infty) \to (0, \infty)$ is the function $f^{-1} : (0, \infty) \to (-1, \infty)$, where $f^{-1}(x) = \frac{1}{x^2} - 1$.

33. $f'(x) = -\frac{6}{(7-2x)^2} \implies (f^{-1})'(3/5) = 1/f'(1) = -25/6$

34. $f'(x) = 4x + 3x^2 \implies (f^{-1})'(16) = 1/f'(2) = 1/20$

35. $f'(x) = \sec^2 x \implies (f^{-1})'(1) = 1/f'(\pi/4) = \cos^2(\pi/4) = 1/2$

36. $f'(x) = -\sin x \implies (f^{-1})'(\sqrt{2}/2) = 1/f'(\pi/4) = -1/\sin(\pi/4) = -\sqrt{2}$

37. Since $f(f(x)) = 1/f(x) = 1/(1/x) = x$, it follows that $f^{-1} = f$.

38. True. For suppose a horizontal line, say $y = b$, intersects the graph of f at two distinct points, say (a, b) and (a', b), where $a' \neq a$. Then $f(a) = f(a') = b$. If f^{-1} is the inverse of f, then $f^{-1}(f(a)) = f^{-1}(f(a'))$ and hence $a = a'$, a contradiction. Therefore no line intersects the graph of f more than once.

39. (a) $y = x^2 + 3 \implies x = \pm\sqrt{y-3}$. Since $x \geq 0$, $f^{-1}(x) = \sqrt{x-3}$.

    (b) $y = (x+1)^2 \implies x = -1 \pm \sqrt{y}$. Since $x \leq -1$, $f^{-1}(x) = -1 - \sqrt{x}$.

    (c) $y = (x+2)^2 \implies x = -2 \pm \sqrt{y}$. Since $x \geq -2$, $f^{-1}(x) = -2 + \sqrt{x}$.

    (d) $y = x^2 - 2x - 3 = (x-1)^2 - 4 \implies x = 1 \pm \sqrt{y+4}$. Since $x \leq 1$, $f^{-1}(x) = 1 - \sqrt{x+4}$.

40. Let $y = ax^2 + bx + c$. Then $ax^2 + bx + c - y = 0$ and hence, using the quadratic formula to solve for x, we find that

$$x = -\frac{b}{2a} \pm \frac{\sqrt{b^2 - 4a(c-y)}}{2a}.$$

If x is restricted to the domain $[-b/2a, \infty)$, then $x \geq -b/2a$ and therefore

$$x = -\frac{b}{2a} + \frac{\sqrt{b^2 - 4a(c-y)}}{2a}.$$

Thus, if $x \geq -b/2a$, the function f has the inverse $f^{-1}(x) = (-b + \sqrt{b^2 - 4a(c-x)})/(2a)$.
Remark: As an alternate way to obtain this result, observe that the function $f(x)$ has an absolute extremum at $x = -b/2a$. Thus, if $x \geq -b/2a$, $f(x)$ is either increasing or decreasing and hence, by Theorem 2, has an inverse on the interval $[-b/2a, +\infty)$.

41. (a) Since $f'(x) = \frac{1}{3}(x+5)^{-2/3}$ and $f(3) = 2$, $f'(3) = 1/12$ and hence $(f^{-1})'(2) = 1/f'(3) = 1/(1/12) = 12$.

    (b) $y = \sqrt[3]{x+5} \implies x = y^3 - 5$. Hence $f^{-1}(x) = x^3 - 5$.

    (c) $(f^{-1})'(x) = 3x^2$.

    (d) From part (c), $(f^{-1})'(2) = 3 \cdot 2^2 = 12$.

42. The graph on the following page shows that the sine function $f(x) = \sin x$ restricted to the interval $[-\pi/2, \pi/2]$ is an increasing function and hence has an inverse when restricted to the domain $[-\pi/2, \pi/2]$.

43. If $g = f^{-1}$, where $f : [-\pi/2, \pi/2] \to [-1, 1]$ is defined by $f(x) = \sin x$ for all $x \in [-\pi/2, \pi/2]$, then $g'(b) = 1/f'(a) = 1/\cos a$, where $b = f(a) = \sin a$. Hence:

    (a) $\sin \pi/4 = \sqrt{2}/2 \implies g'(\sqrt{2}/2) = 1/\cos(\pi/4) = \sqrt{2}$;

(b) $\sin 0 = 0 \Longrightarrow g'(0) = 1/\cos 0 = 1$;

(c) $\sin \pi/3 = \sqrt{3}/2 \Longrightarrow g'(\sqrt{3}/2) = 1/\cos(\pi/3) = 2$;

(d) $\sin(-\pi/6) = -1/2 \Longrightarrow g'(-1/2) = 1/\cos(-\pi/6) = 2\sqrt{3}/3$.

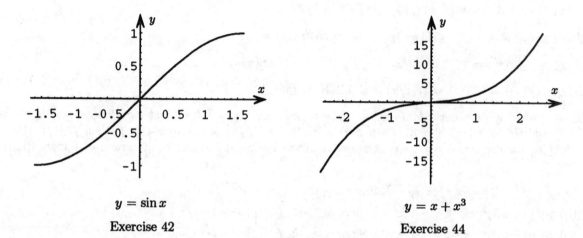

$y = \sin x$

Exercise 42

$y = x + x^3$

Exercise 44

44. Since $f : (-\infty, \infty) \to (-\infty, \infty)$ and $f'(x) = 1 + 3x^2 > 0$ for all x, f is an increasing function, as illustrated in the above figure, and hence has an inverse on $(-\infty, \infty)$.

45. (a) $f(x) = x + x^3 = 0 \Longrightarrow x = 0 \Longrightarrow g(0) = 0$.

(b) $f(x) = x + x^3 = 2 \Longrightarrow x = 1 \Longrightarrow g(2) = 1$.

(c) $g'(0) = 1/f'(0) = 1/1 = 1$.

(d) $g'(2) = 1/f'(1) = 1/4$.

46. Since $f'(x) = -35 - 5x^4 = -(35 + 5x^4) < 0$, f is decreasing on the interval $(-\infty, \infty)$ and hence has an inverse.

47. (a) $f(x) = -35x - x^5 = 0 \Longrightarrow x = 0 \Longrightarrow g(0) = 0$.

(b) $g'(0) = 1/f'(0) = -1/35$.

(c) $f(x) = -35x - x^5 = -36 \Longrightarrow x = 1 \Longrightarrow g(-36) = 1$.

(d) $g'(-36) = 1/f'(1) = -1/40$.

(e) Since $g^{-1} = f$, $g^{-1}(1) = f(1) = -36$.

48. Use induction on t. The formula $P(t) = (1 + r)^t P_0$ is valid if $t = 1$. Now, let $t > 1$ and assume as induction hypothesis that $P(t) = (1 + r)^t P_0$. The amount on deposit after $t + 1$ years is then equal to

$$(1 + r)^t P_0 + r[(1 + r)^t P_0] = (1 + r)^t P_0[1 + r] = (1 + r)^{t+1} P_0 = P(t + 1).$$

Hence the formula is valid for all positive integers t.

49. The formula for $P(t)$ obtained in Exercise 48 shows that after t compoundings, the amount on deposit is $P(t) = (1 + r)^t P_0$. If there are n compoundings per year, each compounding period has an interest rate of r/n and there are nt such compoundings in t years. Hence the amount on deposit after t years is equal to $\left(1 + \frac{r}{n}\right)^{nt} P_0$.

50. (a) If $x = \log_b m$ and $y = \log_b n$, then, by definition, $b^x = m$ and $b^y = n$.

(b) $mn = b^x b^y = b^{x+y}$.

(c) Suppose that $z = n/m$. Let $c = b^z$. Then $c^m = b^n$ and hence $\log_b c^m = \log_b b^n = n$. But $\log_b c^m = m \log_b c$. Therefore $\log_b b^z = \log_b c = n/m = z$.

(d) It follows from parts (a), (b), and (c) that $\log_b(mn) = \log_b b^{x+y} = x + y = \log_b x + \log_b y$.

51. Let $x = \log_b m$ and $y = \log_b n$. Then $b^x = m$ and $b^y = n$, and hence $m/n = b^x/b^y = b^{x-y}$. Now, if z is rational, we showed in Exercise 50 that $\log_b b^z = z$. Hence $\log_b m/n = \log_b b^{x-y} = x - y = \log_b m - \log_b n$.

52. Let $x = \log_b m$ and let r be a rational number. Then $b^x = m$ and hence $m^r = (b^x)^r = b^{xr}$. Hence $\log_b m^r = \log_b b^{xr} = xr = r \log_b x$.

53. $\log_{10}(I/I_0) = kx \implies I/I_0 = 10^{kx} \implies I = I_0 \cdot 10^{kx}$.

54. True. To say that $y = \log_5 x$ means that $5^y = x$ and, in this section, 5^y is defined only if y is rational. For the definition of 5^y when y is any real number, see Section 8.4.

55. It depends on the value of a. To see this, let $n = \log_a x$ and $m = \log_a y$. Then $a^n = x$, $a^m = y$, and hence $x/y = a^{n-m}$. Now, if $0 < x < y$ and $a > 1$, then $a^{n-m} = x/y < 1$ and hence $n - m < 0$; that is, $\log_a x < \log_a y$. But if $0 < a < 1$, then $n - m > 0$; that is, $\log_a x > \log_a y$. To illustrate this last case, consider the numbers 2 and 4. Clearly, $2 < 4$; but $\log_{1/2} 2 = -1 > \log_{1/2} 4 = -2$.

For Exercises 56–60, we use parametric plotting programs on the HP-28S or the Casio with the x and y coordinates defined by $x = f(t)$, $y = t$. On the TI-81 use the **Param** mode and set $\mathbf{X_{1T}} = \mathbf{f(T)}$ and $\mathbf{Y_{1T}} = \mathbf{T}$. With HP-48SX use the **PARA** plot type and enter **'f(t)+i*t'** for the function to be plotted. The range for the variable t should be the same as the domain of f(x). The function $y = f(x)$ can also be drawn by taking $x = t$ and $y = f(t)$.

56. Graph using the range $[-1, 4] \times [-1, 4]$; see the graph below on the left. The inverse relation is not a function since $f(x) = 0$ at $x = 0$ and $x = 1/2$. This implies that $f(x)$ is not one-to-one.

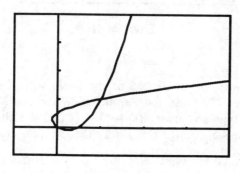
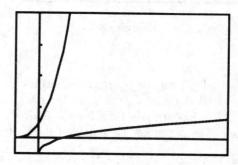

Exercise 56   Exercise 57

57. Graph using the range $[-1, 8] \times [-1, 8]$; see the graph above on the right. The inverse relation is a function since $f(x)$ is one-to-one.

58. Graph using the range:
Casio or TI-81: $[-4.7, 4.7] \times [-3.2, 3.2]$.
HP-28S: $[-13.6, 13.6] \times [-3, 3.2]$.
HP-48SX: $[-6.5, 6.5] \times [-3.1, 3.2]$.

See the graph below on the left. The inverse relation is not a functin since $f(x)$ is not one-to-one. To see this note that $f(x) = 0$ for $x = \pm 2$ and $x = 0$.

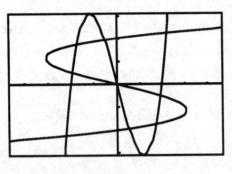

Exercise 58

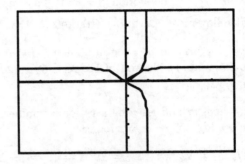

Exercise 59

59. Graph using the range $[-10, 10] \times [-10, 10]$; see above on the right. The inverse relation is not a function since $f(x)$ is not one-to-one. Notice that it fails the horizontal line test.

60. Graph using the range $[-2\pi, 2\pi] \times [-2\pi, 2\pi]$. The inverse relation is a function since $f(x)$ is one-to-one. To see this we note that $f'(x) = 1 - \sin x \geq 0$. This implies that $f(x)$ is increasing.

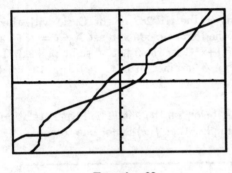

Exercise 60

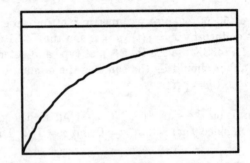

Exercise 61

For Exercises 61–63 we use parametric plotting programs on the HP-28S or the Casio with the x and y coordinates defined by $x = f(t)$, $y = 1/f'(t)$. On the TI-81 use the **Param** mode and set $\mathbf{X_{1T}} = \mathbf{f(T)}$ and $\mathbf{Y_{1T}} = \mathbf{1/f'(T)}$. With HP-48SX use the **PARA** plot type and enter **'f(t)+i*INV(f'(t))'** for the function to be plotted. The range for the variable t should be the same as the domain for the function $f(x)$.

61. Graph using the range $[1/3, 1] \times [-9, 1]$; see above on the right. Note that $f(1) = 1$ and $f(3) = 1/3$ so the domain of $f^{-1}(x)$ is $[1/3, 1]$. Now $f^{-1}(x) = 1/x$ and $(f^{-1})'(x) = -1/x^2$. So we plot the pairs defined by $x = 1/t$ and $y = -t^2$. Since $x = 1/t$ we have that $y = -1/x^2 = (f^{-1})'(x)$.

62. Graph using the range $[-1.5, 1.5] \times [0, 7]$; see below on the left. The range of $f(x)$ is $[\sqrt[3]{-2}, \sqrt[3]{2}]$ which is contained in the interval $[-1.5, 1.5]$. The inverse function is given by $f^{-1}(x) = x^3 + 1$ so $(f^{-1})'(x) = 3x^2$. Therefore take $x = \sqrt[3]{t-1}$ and $y = 1/f'(t) = 3(t-1)^{2/3}$. But $t = x^3 + 1$ so $y = 3(1 + x^3 - 1)^{2/3} = 3x^2$.

63. Graph using the range $[-1, 1] \times [0, 10]$; see below on the right. The range of $f(x)$ is $[-1, 1]$. This graph is produced by plotting the pair $x = \sin t$, $y = \sec t$.

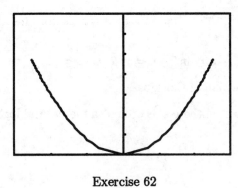

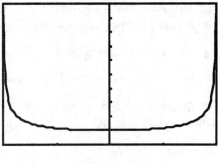

Exercise 62   Exercise 63

64. The graph on the left uses the range $[0,13] \times [1000, 2500]$ and depicts the graphs for annual, $n = 1$; quarterly, $n = 4$; and daily, $n = 365$ compounding. The graph on the right is the same except the range is $[7,9] \times [1700, 2200]$.

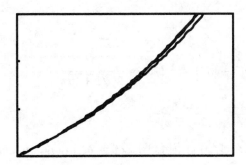

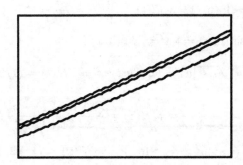

The following table gives the doubling time for each of the compounding periods.

Compounding	n	Doubling Time
Annually	1	8 years 181 days
Semi-annually	2	8 years 119 days
Quarterly	4	8 years 88 days
Monthly	12	8 years 67 days
Daily	365	8 years 57 days
Hourly	8760	8 years 56 days

8.2   The Natural Logarithm Function

1. (a) $\ln x = 0 \implies 2x = e^0 = 1 \implies x = 1/2$.

   (b) $2\ln x = \ln 2x \implies \ln x^2 = \ln 2x \implies x^2 = 2x \implies x = 0, 2$. Since $\ln x$ is only defined for $x > 0$, the only solution is $x = 2$.

   (c) $\ln(2/x) - \ln x = 0 \implies \ln(2/x) = \ln x \implies 2/x = x \implies x = \pm\sqrt{2}$. Since $\ln x$ is only defined for $x > 0$, the only solution is $x = \sqrt{2}$.

   (d) $\ln 3^{x^2} = 0 \implies x^2 \ln 3 = 0 \implies x = 0$ since $\ln 3 \neq 0$.

   (e) $\sqrt{\ln \sqrt{x}} = 1 \implies \ln \sqrt{x} = 1 \implies \sqrt{x} = e^1 = e \implies x = e^2$.

(f) $3\ln x + x = 2 + \ln x^3 \implies 3\ln x + x = 2 + 3\ln x \implies x = 2.$

(g) $\int_1^x \frac{3}{t}\,dt = 6 \implies 3\ln x = 6 \implies \ln x = 2 \implies x = e^2.$

(h) $\int_2^x \frac{1}{t}\,dt = 0 \implies \ln x = \int_1^x \frac{1}{t}\,dt = \int_1^2 \frac{1}{t}\,dt + \int_2^x \frac{1}{t}\,dt = \ln 2 + 0 = \ln 2 \implies x = 2.$

2. Because it is differentiable by the Fundamental Theorem of Calculus.

3. True. The graph of $y = \ln x$ shows that every number on the y-axis occurs as the natural logarithm of some number x.

4. $y = \ln 2x \implies y' = \frac{1}{2x}(2) = \frac{1}{x}.$

5. $y = x\ln x \implies y' = x(1/x) + (\ln x)(1) = 1 + \ln x.$

6. $y = \ln(6 - x^2) \implies y' = \frac{1}{6-x^2}(-2x) = \frac{-2x}{6-x^2}.$

7. $f(x) = \ln\sqrt{x^3 - x} = (1/2)\ln(x^3 - x) \implies y' = \frac{1}{2}\frac{1}{x^3-x}(3x^2 - 1) = \frac{3x^2 - 1}{2(x^3 - x)}.$

8. $f(t) = \ln(\ln t) \implies f'(t) = \frac{1}{\ln t}\frac{1}{t} = \frac{1}{t\ln t}.$

9. $y = \sin(\ln x) \implies y' = [\cos(\ln x)](1/x).$

10. $h(t) = \sqrt{1 + \ln t} = (1 + \ln t)^{1/2} \implies h'(t) = (1/2)(1 + \ln t)^{-1/2}(1/t) = \frac{1}{2t\sqrt{1 + \ln t}}.$

11. $f(x) = \frac{x}{1 + \ln x} \implies f'(x) = \frac{(1 + \ln x)(1) - x(1/x)}{(1 + \ln x)^2} = \frac{\ln x}{(1 + \ln x)^2}.$

12. $f(x) = x^2\ln^2 x = x^2(\ln x)^2 \implies f'(x) = (x^2)[2(\ln x)(1/x)] + [(\ln x)^2](2x) = 2x(\ln x)(1 + \ln x).$

13. $y = (3\ln\sqrt{x})^4 = (81/16)(\ln x)^4 \implies y' = \frac{81}{16}4(\ln x)^3\left(\frac{1}{x}\right) = \frac{81\ln^3 x}{4x}.$

14. $f(x) = (\sin x)\ln(1 + \sqrt{x}) \implies$

$$f'(x) = (\sin x)\left(\frac{1}{1 + \sqrt{x}}\right)\left(\frac{1}{2}x^{-1/2}\right) + (\ln(1 + \sqrt{x}))(\cos x) = \frac{\sin x}{2\sqrt{x}(1 + \sqrt{x})} + (\cos x)\ln(1 + \sqrt{x}).$$

15. $f(t) = \frac{\ln(a + bt)}{\ln(c + dt)} \implies$

$$f'(t) = \frac{(\ln(c + dt))\left[\frac{1}{a+bt}b\right] - (\ln(a + bt))\left[\frac{1}{c+dt}d\right]}{[\ln(c + dt)]^2} = \frac{\frac{b\ln(c+dt)}{a+bt} - \frac{d\ln(a+bt)}{c+dt}}{[\ln(c + dt)]^2}.$$

16. $y = \ln(\sin t - t\cos t) \implies y' = \frac{\cos t - (-t\sin t + \cos t)}{\sin t - t\cos t} = \frac{t\sin t}{\sin t - t\cos t}.$

17. $f(x) = \ln\cos x \implies f'(x) = \frac{1}{\cos x}(-\sin x) = -\frac{\sin x}{\cos x} = -\tan x.$

18. $y = x\ln(x^2 - x - 3) \implies y' = x\left(\frac{1}{x^2 - x - 3}\right)(2x - 1) + \ln(x^2 - x - 3) = \frac{x(2x - 1)}{x^2 - x - 3} + \ln(x^2 - x - 3).$

19. $u(t) = \dfrac{t}{1 + \ln^2 t} \implies u'(t) = \dfrac{(1 + \ln^2 t)(1) - t(2\ln t)(1/t)}{(1 + \ln^2 t)^2} = \dfrac{\ln^2 t - 2\ln t + 1}{(\ln^2 t + 1)^2} = \left(\dfrac{\ln t - 1}{\ln^2 t + 1}\right)^2.$

20. $y = \ln(xy) \implies y' = \dfrac{1}{xy}(xy' + y) = \dfrac{y'}{y} + \dfrac{1}{x} \implies yy'x = y'x + y \implies y'(xy - x) = y \implies y' = \frac{y}{x(y-1)}.$

21. $\ln(x+y) + \ln(x-y) = 1 \implies \ln(x^2 - y^2) = 1 \implies \dfrac{2x - 2yy'}{x^2 - y^2} = 0 \implies 2x - 2yy' = 0 \implies y' = x/y.$

22. $x\ln y = 3 \implies x(1/y)y' + \ln y = 0 \implies y' = -(y/x)\ln y.$

23. $x\ln y + y\ln x = x \implies x(1/y)y' + \ln y + y(1/x) + y'\ln x = 1 \implies x^2 y' + xy\ln y + y^2 + xyy'\ln x = xy$
    $\implies y'(x^2 + xy\ln x) = xy - xy\ln y - y^2 \implies$

$$y' = \frac{xy(1 - \ln y)}{x(x + y\ln x)} = \frac{y(1 - \ln y)}{x + y\ln x}.$$

24. $y = \displaystyle\int_5^x \frac{1}{t}\,dt = \int_1^x \frac{1}{t}\,dt - \int_1^5 \frac{1}{t}\,dt = \ln x - \ln 5 \implies y' = 1/x.$

25. $y = \displaystyle\int_x^1 \frac{1}{2t}\,dt = -\frac{1}{2}\int_1^x \frac{1}{t}\,dt = -(1/2)\ln x \implies y' = -1/(2x).$

26. $y = \displaystyle\int_1^{x^2} \ln 2t\,dt = f(g(x))$, where $f(x) = \int_1^x \ln 2t\,dt$ and $g(x) = x^2$. Using the chain rule and the fact that $f'(x) = \ln 2x$, it follows that $y' = f'(g(x))g'(x) = (\ln 2x^2)(2x)$.

27. Since $y' = 1 - 1/x = 0 \iff x = 1$, the graph of $y = x - \ln x$ has one critical number, namely $x = 1$. Moreover, $x = 1$ is a local minimum since $y'' = 1/x^2$ and $y''(1) = 1 > 0$. Hence, since $y(1) = 1 - \ln 1 = 1$, the point $(1, 1)$ is a local minimum. The graph is shown below.

28. $\ln xy = 1 \implies xy = e \implies y = e/x$. Since $y' = -e/x^2 \neq 0$, the graph of $y = e/x$, shown below, has no relative extrema.

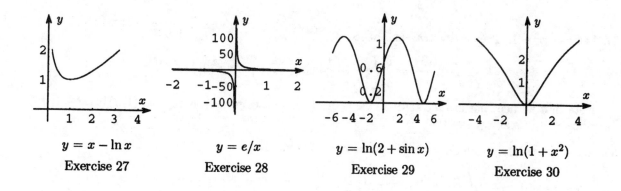

$y = x - \ln x$

Exercise 27

$y = e/x$

Exercise 28

$y = \ln(2 + \sin x)$

Exercise 29

$y = \ln(1 + x^2)$

Exercise 30

29. $y = \ln(2 + \sin x) \implies y' = (\cos x)/(2 + \sin x)$. Therefore

$$y' = 0 \quad \iff \quad \cos x = 0 \quad \iff \quad x = \ldots, -\pi/2, \pi/2, 3\pi/2, \ldots .$$

Now,

$$y'' = \frac{(2+\sin x)(-\sin x) - (\cos x)(\cos x)}{(2+\sin x)^2} = \frac{-1 - 2\sin x}{(2+\sin x)^2}.$$

Therefore

$$y''(-\pi/2) = 1 > 0 \implies x = -\pi/2 \text{ is a relative minimum}$$
$$y''(\pi/2) = -1/3 < 0 \implies x = \pi/2 \text{ is a relative maximum}$$
$$y''(3\pi/2) = 1 > 0 \implies x = 3\pi/2 \text{ is a relative minimum}$$

Thus, on the interval $[0, 2\pi]$ the graph of $y = \ln(2 + \sin x)$ has two relative extrema: a relative maximum at $(\pi/2, \ln 3)$ and a relative minimum at $(3\pi/2, 0)$. Since $\sin x$ is periodic with period 2π, the graph of $y = \ln(2 + \sin x)$ is also periodic with period 2π.

30. $y = \ln(1 + x^2) \implies y' = (2x)/(1 + x^2)$. Hence $y' = 0 \iff x = 0$. Since

$$y'' = \frac{(1+x^2)(2) - (2x)(2x)}{(1+x^2)^2} = \frac{2(1-x^2)}{(1+x^2)^2},$$

$y''(0) = 2 > 0 \implies (0, 0)$ is a relative minimum.

31. Let $u = x + 3$. Then $du = dx$ and hence

$$\int \frac{dx}{x+3} = \int \frac{du}{u} = \ln|u| + C = \ln|x+3| + C.$$

32. Let $u = 2x + 1$. Then $du = 2dx$ and hence

$$\int \frac{dx}{2x+1} = \frac{1}{2} \int \frac{du}{u} = \frac{1}{2} \ln|u| + C = \frac{1}{2} \ln|2x+1| + C.$$

33. Let $u = 1 - x$. Then $du = -dx$ and hence

$$\int \frac{dx}{1-x} = -\int \frac{du}{u} = -\ln|u| + C = -\ln|1-x| + C.$$

34. Let $u = x^2 + 1$. Then $du = 2xdx$ and hence

$$\int \frac{x\,dx}{x^2+1} = \frac{1}{2} \int \frac{du}{u} = \frac{1}{2} \ln|u| + C = \frac{1}{2} \ln(x^2 + 1) + C.$$

Note that we need not use the absolute value of $x^2 + 1$ in the antiderivative, *i.e.*, $(1/2) \ln|x^2 + 1| + C$, since $x^2 + 1$ is always positive.

35. Let $u = x^2 - 2x$. Then $du = (2x - 2)dx = 2(x - 1)dx$ and hence

$$\int \frac{x-1}{x^2-2x}\,dx = \frac{1}{2} \int \frac{du}{u} = \frac{1}{2} \ln|u| + C = \frac{1}{2} \ln|x^2 - 2x| + C.$$

36. Let $u = 1 - 3x^2$. Then $du = -6xdx$ and hence

$$\int \frac{x}{1-3x^2}\,dx = -\frac{1}{6} \int \frac{du}{u} = -\frac{1}{6} \ln|u| + C = -\frac{1}{6} \ln|1 - 3x^2| + C.$$

37. Let $u = x^3 + 9x$. Then $du = (3x^2 + 9)dx = 3(x^2 + 3)dx$ and hence

$$\int \frac{x^2 + 3}{x^3 + 9x}\, dx = \frac{1}{3} \int \frac{du}{u} = \frac{1}{3}\ln|u| + C = \frac{1}{3}\ln|x^3 + 9x| + C.$$

38. Let $u = \sin x$. Then $du = \cos x\, dx$ and hence

$$\int \cot x\, dx = \int \frac{\cos x}{\sin x}\, dx = \int \frac{du}{u} = \ln|u| + C = \ln|\sin x| + C.$$

39. Let $u = 4 + 2\cos t$. Then $du = -2\sin t\, dt$ and hence

$$\int \frac{\sin t\, dt}{4 + 2\cos t} = -\frac{1}{2} \int \frac{du}{u} = -\frac{1}{2}\ln|u| + C = -\frac{1}{2}\ln(4 + 2\cos t) + C.$$

We need not use the absolute value of $4 + 2\cos t$ in the antiderivative since $4 + 2\cos t$ is always positive.

40. Let $u = \ln x$. Then $du = \dfrac{1}{x}\, dx$ and hence

$$\int \frac{dx}{x\ln x} = \int \frac{du}{u} = \ln|u| + C = \ln|\ln x| + C.$$

41. Let $u = \ln x$. Then $du = \dfrac{1}{x}\, dx$ and hence

$$\int \frac{\ln^2 x}{x}\, dx = \int u^2\, du = \frac{1}{3}u^3 + C = \frac{1}{3}\ln^3 x + C.$$

42. Since $\dfrac{x^2}{x + 1} = x - 1 + \dfrac{1}{x + 1}$, it follows that

$$\int \frac{x^2}{x + 1}\, dx = \int \left(x - 1 + \frac{1}{x + 1}\right)dx = \frac{1}{2}x^2 - x + \ln|x + 1| + C.$$

43. Let $u = 1 - \sqrt{x}$. Then $du = -\dfrac{1}{2\sqrt{x}}\, dx$ and hence

$$\int \frac{1}{\sqrt{x}(1 - \sqrt{x})}\, dx = -2 \int \frac{du}{u} = -2\ln|u| + C = -2\ln|1 - \sqrt{x}| + C.$$

44. Since $\dfrac{1 - 2t^2}{1 - t} = 2t + 2 - \dfrac{1}{1 - t}$, it follows that

$$\int \frac{1 - 2t^2}{1 - t}\, dt = \int \left(2t + 2 - \frac{1}{1 - t}\right)dt = t^2 + 2t + \ln|1 - t| + C.$$

45. Since $\dfrac{x^4 + 3x^2 + x + 1}{x + 1} = x^3 - x^2 + 4x - 3 + \dfrac{4}{x + 1}$, it follows that

$$\begin{aligned}
\int \frac{x^4 + 3x^2 + x + 1}{x + 1}\, dx &= \int \left(x^3 - x^2 + 4x - 3 + \frac{4}{x + 1}\right)dx \\
&= \frac{1}{4}x^4 - \frac{1}{3}x^3 + 2x^2 - 3x + 4\ln|x + 1| + C.
\end{aligned}$$

46. $\int_1^e \dfrac{1}{x}\,dx = \ln x\big|_1^e = \ln e - \ln 1 = 1.$

47. Let $u = \ln x$. Then $du = \dfrac{1}{x}\,dx$ and hence

$$\int_e^{e^2} \frac{1}{x\ln x}\,dx = \int_1^2 \frac{1}{u}\,du = \ln u\big|_1^2 = \ln 2 - \ln 1 = \ln 2.$$

48. Let $u = \ln x$. Then $du = \dfrac{1}{x}\,dx$ and hence

$$\int_1^e \frac{\ln x}{x}\,dx = \int_0^1 u\,du = \frac{1}{2}u^2\Big|_0^1 = \frac{1}{2}.$$

49. Let $u = \ln(x^2) = 2\ln x$. Then $du = \dfrac{2}{x}\,dx$ and hence

$$\int_e^{e^2} \frac{1}{x\ln(x^2)}\,dx = \frac{1}{2}\int_2^4 \frac{1}{u}\,du = \frac{1}{2}\ln u\Big|_2^4 = \frac{1}{2}(\ln 4 - \ln 2) = \frac{1}{2}\ln\frac{4}{2} = \frac{1}{2}\ln 2.$$

50. Let $u = x^2 + 1$. Then $du = 2x\,dx$ and hence

$$\int_2^3 \frac{x}{x^2+1}\,dx = \frac{1}{2}\int_5^{10} \frac{du}{u} = \frac{1}{2}\ln u\Big|_5^{10} = \frac{1}{2}(\ln 10 - \ln 5) = \frac{1}{2}\ln\frac{10}{5} = \frac{1}{2}\ln 2.$$

51. $\int_1^2 \left(\dfrac{1}{1+x} - \dfrac{1}{2+x} \right)dx = (\ln(1+x) - \ln(2+x))\big|_1^2 = (\ln 3 - \ln 4) - (\ln 2 - \ln 3) = \ln(9/8).$

52. Let $u = x^3 + 3x$. Then $du = (3x^2 + 3)dx = 3(x^2+1)dx$ and hence

$$\int_1^2 \frac{x^2+1}{x^3+3x}\,dx = \frac{1}{3}\int_4^{14} \frac{du}{u} = \frac{1}{3}\ln u\Big|_4^{14} = \frac{1}{3}(\ln 14 - \ln 4) = \frac{1}{3}\ln\frac{7}{2}.$$

53. Let $u = \cos x$. Then $du = \sin x\,dx$ and hence

$$\int_0^{\pi/3} \tan x\,dx = \int_0^{\pi/3} \frac{\sin x}{\cos x}\,dx = -\int_1^{1/2} \frac{1}{u}\,du = -\ln u\big|_1^{1/2} = -(\ln\frac{1}{2} - \ln 1) = \ln 2.$$

54. If $y = x(\ln x)^2 + x/(\ln x)$, then

$$y' = x\left[2(\ln x)\frac{1}{x} \right] + (\ln x)^2 + \frac{\ln x - x(1/x)}{(\ln x)^2} = 2\ln x + (\ln x)^2 + \frac{\ln x - 1}{(\ln x)^2}$$

and hence the slope of the tangent line at $(e, 2e)$ is

$$y'(e) = 2\ln e + (\ln e)^2 + \frac{\ln e - 1}{(\ln e)^2} = 3.$$

Therefore the equation of the tangent line is $y - 2e = 3(x - e)$, or $y = 3x - e$.

618

55. If $y = x^3 \ln^2 x$, then $y' = (x^3) \, 2 \, (\ln x)(1/x) + (\ln^2 x)(3x^2)$ and hence $y'(1) = 2 \ln 1 + (\ln 1)(3) = 0$. Thus, the slope of the tangent line at $x = 1$ is 0. Since $y(1) = 1 \ln 1 = 0$, an equation of the tangent line is $y - 0 = 0(x - 1)$, or $y = 0$.

56. Area $= \displaystyle\int_3^4 \frac{1}{x - 2} \, dx = \ln(x - 2)|_3^4 = \ln 2 - \ln 1 = \ln 2.$

57. We first find the intersection points of the two graphs. Since $x + y = 6$, $y = 6 - x$. Hence $xy = 8 \implies x(6 - x) = 8 \implies x^2 - 6x + 8 = 0 \implies (x - 4)(x - 2) = 0 \implies x = 2, 4$. Thus, the area bounded by the two graphs is equal to

$$\int_2^4 \left[(6 - x) - \frac{8}{x}\right] \, dx = \left(6x - \frac{1}{2}x^2 - 8 \ln x\right)\Big|_2^4 = 6 - 8 \ln 2.$$

58. (a) Let $f(x) = \ln(1 + x)$. Then $f'(x) = 1/(1 + x)$, $f'(0) = 1$, and $f(0) = 0$. Hence the equation of the tangent line to the graph of $f(x) = \ln(1 + x)$ at $x = 0$ is $y = x$. Using the tangent line as a linear approximation to the values of $f(x)$ for x near 0, it follows that

$$\ln(1 + x) \approx x \quad \text{for} \quad x \approx 0.$$

In particular, $\ln(1 + \varepsilon) \approx \varepsilon$.

(b) Choosing $\varepsilon = -0.2, -0.05, 0.05$, and 0.2, respectively, we find that:

$$\ln 0.8 \approx -0.2$$
$$\ln 0.95 \approx -0.05$$
$$\ln 1.05 \approx 0.05$$
$$\ln 1.2 \approx 0.2$$

(c) Using a calculator, we find, to four decimal places, that $\ln 0.8 = -0.2231$, $\ln 0.95 = -0.0513$, $\ln 1.05 = 0.0488$, and $\ln 1.2 = 0.1823$.

59. True. To see this, let x be a positive real number. Then the slope of the tangent line to the graph of $y = \ln x$ is $1/x$, while the slope of the tangent line to the graph of $y = \ln ax$ is $(1/ax)(a) = 1/x$. Thus, for a given value of x, the two graphs have the same slope.

60. Let dV be the volume of a differential cross-section disk of thickness dx. Then

$$dV = \pi r^2 dx = \pi y^2 dx = \pi \, \frac{2}{x - 1} \, dx$$

and hence the volume of revolution is

$$V = 2\pi \int_3^5 \frac{1}{x - 1} \, dx = 2\pi \ln(x - 1)|_3^5 = 2\pi \ln 4 - 2\pi \ln 2 = 2\pi \ln 2.$$

61. Let dV be the volume of a differential cross-section disk of thickness dx. Then

$$dV = \pi r^2 dx = \pi y^2 dx = \pi \, \frac{\ln^2 x}{x} \, dx$$

and hence the volume of revolution is

$$V = \pi \int_1^2 \frac{\ln^2 x}{x} \, dx = \frac{\pi}{3} \ln^3 x\Big|_3^5 = \frac{\pi}{3}(\ln 2)^3 \qquad \text{(by Exercise 41)}.$$

62. $F'(x) = x(1/x) + \ln x - 1 = \ln x$. Therefore $F(x)$ is an antiderivative of $\ln x$.

63. By definition, the average value of $\ln x$ on the interval $[1, e]$ is

$$\frac{1}{e-1} \int_1^e \ln x \, dx = \frac{1}{e-1}(x \ln x - x)\Big|_1^e = \frac{1}{e-1}[(e \ln e - e) - (\ln 1 - 1)] = \frac{1}{e-1}. \qquad \text{(by Exercise 62)}$$

64. Let $s(t)$ and $v(t)$ stand for the distance and velocity of the particle at time t. Then

$$v(t) = \int a(t) \, dt = \int \frac{1}{t+2} \, dt = \ln(t+2) + C.$$

Since $v(0) = 0$, $C = -\ln 2$. Therefore $v(t) = \ln(t+2) - \ln 2$. Since $v(t) > 0$ when $t > 0$, $s(t)$ is an increasing function and hence the distance travelled during the time interval $[0, 4]$ is

$$s(4) - s(0) = \int_0^4 |v(t)| \, dt = \int_0^4 [\ln(t+2) - \ln 2] \, dt = [(t+2)\ln(t+2) - (t+2) - (\ln 2)t]|_0^4 = 6\ln 3 - 4.$$

65. $dy/dx = 1/x \Longrightarrow y = \int \frac{1}{x} \, dx = \ln x + C_1 = \ln(C_2 x)$, $x > 0$.

66. Since $(x^2 + 3)dy - 2x dx = 0$, $dy = \frac{2x}{x^2+3} \, dx \Longrightarrow y = \int \frac{2x}{x^2+3} \, dx = \ln(x^2 + 3) + C$.

67. $\cos x \, dy = \sin x \, dx \Longrightarrow dy = \tan x \, dx \Longrightarrow y = \int \tan x \, dx = \ln|\sec x| + C$. (See Exercise 53)

68. $y^{-1} dx = x \, dy \Longrightarrow y \, dy = (1/x) dx \Longrightarrow (1/2)y^2 = \ln|x| + C \Longrightarrow y^2 = 2\ln|x| + C$.

69. Since $(2x+1)dy = y^2 dx$,

$$\frac{1}{y^2} \, dy = \frac{1}{2x+1} \, dx \Longrightarrow -\frac{1}{y} = \int \frac{1}{2x+1} \, dx = \frac{1}{2} \ln|2x+1| + C \Longrightarrow y = \frac{-2}{\ln|2x+1| + C}.$$

70. (a) If $y = f(t)$, then $\frac{d}{dt} \ln(f(t)) = (1/f(t))(f'(t)) = G$. Thus, the growth of $f(t)$ is the derivative of $\ln f(t)$.

(b) Let $f(t)$ and $g(t)$ be functions. Then the growth of the product $f(t)g(t)$ is equal to

$$\frac{d}{dt} \ln(f(t)g(t)) = \frac{1}{f(t)g(t)}(f(t)g'(t) + g(t)f'(t)) = \frac{g'(t)}{g(t)} + \frac{f'(t)}{f(t)},$$

which is the sum of the growth of $f(t)$ and the growth of $g(t)$.

(c) Let $n(t)$ stand for the number of gallons of heating oil sold at a price of $p(t)$ dollars per gallon. Then the revenue function $r(t)$ is the product $p(t)n(t)$; i.e., $r(t) = n(t)p(t)$. Now, we are given that $p'(t) = 0.15p(t)$ and $n'(t) = -0.10n(t)$. Thus, the growth rate for price is $p'/p = 0.15$ and for quantity is $n/n' = -0.10$. It follows from part (b) that the growth rate of revenue is the sum: $r'/r = p'/p + n'/n = 0.05$; that is, revenue increases at an annual rate of 5%.

71. (a) Regarding P as a function of T and differentiating both sides of the equation

$$\ln P = \frac{-2000}{T} + 5.5$$

with respect to T, we obtain

$$\frac{P'}{P} = \frac{2000}{T^2} \quad \Longrightarrow \quad \frac{dP}{dT} = P' = \frac{2000}{T^2} \, P.$$

620

(b) Regarding T as a function of P and differentiating both sides of the equation

$$\ln P = \frac{-2000}{T} + 5.5$$

with respect to P, we obtain

$$\frac{1}{P} = \frac{2000}{T^2}\frac{dT}{dP} \implies \frac{dT}{dP} = \frac{1}{2000}\frac{T^2}{P}.$$

72. Let $f(x) = \ln(x/b)$. Then

$$f'(x) = \frac{1}{\frac{x}{b}}\frac{1}{b} = \frac{1}{x}.$$

Therefore $f(x) = \ln x + C$ for some constant C. Now, $f(x) = \ln(x/b) \implies f(b) = \ln 1 = 0$, while $f(x) = \ln x + C \implies f(b) = \ln b + C$. Hence $\ln b + C = 0 \implies C = -\ln b$. Therefore $\ln(x/b) = \ln x - \ln b$. Letting $x = a$, it follows that $\ln(a/b) = \ln a - \ln b$.

73. Let $f(x) = \ln x^r$, where r is a rational number. Then $f'(x) = \frac{1}{x^r}(rx^{r-1}) = \frac{r}{x} = (r\ln x)'$. Hence $f(x) = r\ln x + C$ for some constant C. Now, $f(x) = \ln x^r \implies f(1) = \ln 1 = 0$, while $f(x) = r\ln x + C \implies f(1) = r\ln 1 + C = C$. Hence $C = 0$ and therefore $\ln x^r = r\ln x$. Letting $x = a$, it follows that $\ln a^r = r\ln a$.

74. (a) If $1 \le x \le 2$, then $x \le 2 \implies 1/2 \le 1/x$, while $1 \le x \implies 1/x \le 1$. Hence $1/2 \le 1/x \le 1$.

(b) Since, by (a), the graph of $y = 1/2$ lies below that of $y = 1/x$, which lies below that of $y = 1$ on the interval $[1, 2]$, it follows that the areas under these graphs increase; that is,

$$\int_1^2 \frac{1}{2}\,dx \le \int_1^2 \frac{1}{x}\,dx \le \int_1^2 1\,dx.$$

(c) Since $\int_1^2 \frac{1}{2}\,dx = 1/2$, $\int_1^2 \frac{1}{x}\,dx = \ln 2$, and $\int_1^2 1\,dx = 1$, it follows from (b) that $1/2 \le \ln 2 \le 1$.

(d) Multiplying all terms of the inequality in (c) by 2, we obtain $1 \le 2\ln 2 \le 2$.

(e) Since $1 = \ln e$, $\ln 4 = 2\ln 2$, and $2 = \ln e^2$, we obtain $\ln e \le \ln 4 \le \ln e^2$.

(f) Since the function $f(x) = \ln x$ is an increasing function, it follows from (e) that $e \le 4 \le e^2$. For either $e = 4$, $e > 4$, or $e < 4$; if $e > 4$, then $\ln e > \ln 4$, which contradicts (e). Hence $e \le 4$. Similarly, $4 \le e^2$.

(g) Since $4 \le e^2$, taking square roots gives $2 \le e$. Hence $2 \le e \le 4$.

75. Subdividing the interval $[1, 5]$ into 8 equal subintervals of length $1/2$, we obtain the subintervals $[1, 1.5]$, $[1.5, 2], \ldots, [4.5, 5]$, whose midpoints are 1.25, 1.75, 2.25, 2.75, 3.25, 3.75, 4.25, and 4.75. Hence, using the Midpoint Rule with $f(x) = 1/x$,

$$\ln 5 = \int_1^5 \frac{1}{x}\,dx \approx 0.5[f(1.25) + f(1.75) + \cdots + f(4.75)] = 0.5\left(\frac{1}{1.25} + \frac{1}{1.75} + \cdots + \frac{1}{4.75}\right) \approx 1.5998.$$

76. Using the Trapezoidal Rule,

$$\ln 5 = \int_1^5 \frac{1}{x}\,dx \approx \left(\frac{1}{2}\right)(0.5)[f(1) + 2f(1.5) + 2f(2) + \cdots + 2f(4.5) + f(5)] \approx 1.6290.$$

77. Since the function $f(x) = 1/x$ is decreasing on the interval $[1, 5]$, the actual area under the curve is less than area of the trapezoid on each subinterval. Hence

$$\int_1^5 \frac{1}{x}\, dx < \text{Trapezoidal Rule approximation} \quad \Longrightarrow \quad \ln 5 < 1.6290.$$

Now, it was shown in Theorem 10, Chapter 6, that if the function $f(x)$ is concave up on the interval $[a, b]$, then the Midpoint Rule value is smaller than $\int_a^b f(x)\, dx$, which is smaller than the Trapezoidal Rule value:

$$\text{Midpoint Rule} < \int_a^b f(x)\, dx < \text{Trapezoidal Rule}.$$

Hence, since $f(x) = 1/x$ is concave up on the interval $[1, 5]$, it follows from Exercises 75, 76 that $1.5998 < \ln 5 < 1.6290$.

78. Let $f(x) = 1/x$. Using the Trapezoidal Rule to approximate $\int_1^{2.7} f(x)dx$, with $n = 6$, we find that

$$\ln 2.7 = \int_1^{2.7} \frac{1}{x}\, dx \approx \frac{1.7}{6}\frac{1}{2}[f(1) + 2f(1.2833) + 2f(1.5667) + 2f(1.8500)$$
$$+ 2f(2.1333) + f(2.4167)] = 0.998973.$$

Hence $\ln 2.7 < 1 \Longrightarrow 2.7 < e$. Using the Midpoint Rule to approximate $\int_1^{2.8} f(x)dx$, with $n = 6$, we find that

$$\ln 2.8 = \int_1^{2.8} \frac{1}{x}\, dx = \frac{1.8}{6}[f(1.15) + f(1.45) + f(1.75) + f(2.05) + f(2.35) + f(2.65)] = 1.0264.$$

Hence $\ln 2.8 > 1 \Longrightarrow 2.8 > e$. Therefore $2.7 < e < 2.8$.

79. Graph using the range $[0, 10] \times [-2, 1]$.

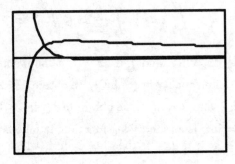

$f'(x) = (1 - \ln x)/x^2$.
Domain: $(0, +\infty)$. Range: $(-\infty, e^{-1})$. Absolute maximum: (e, e^{-1}).

80. Graph using the range $[0, 10] \times [-3, 5]$.

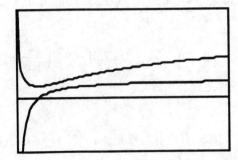

$f'(x) = (x^2 - 1)/(x^2 + x)$.
Domain: $(0, +\infty)$. Range: $[\ln 2, +\infty)$. Absolute maximum: $(1, \ln 2)$.

81. Graph using the range $[-4, 4] \times [-1.5, 1.5]$.

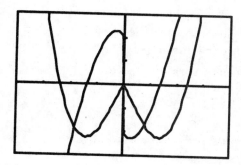

$$f'(x) = \begin{cases} 2x\ln(1-x) + 1 - x & \text{if } x < 0 \\ 2x\ln(1+x) - 1 - x & \text{if } x \geq 0. \end{cases}$$

Domain: $(-\infty, +\infty)$. Range: $[-1.12024037, +\infty)$. Absolute minima: $(\pm 1.37301614, -1.12024037)$.

82. From Exercise 39, Section 5.3 we have $v(t) = 225/(3 + t)$. So

$$\frac{ds}{dt} = \frac{225}{3 + t}.$$

This implies that

$$\int ds = \int \frac{225}{3 + t}\, dt.$$

Thus $s(t) = 225\ln(3 + t) + C$. Since $s(0) = 0$, choose C so that $225\ln 3 + C = 0$, i.e., $C = -225\ln 3$. Therefore

$$s(t) = 225\ln(3 + t) - 225\ln 3 = 225\ln\left(\frac{3 + t}{3}\right) = 225\ln(1 + t/3).$$

The graph on the left is the graph of $s(t)$ and that from Euler's Method using $n = 300$ and using the range of $[0, 57] \times [0, 700]$. The graph on the right is the same but using the range $[30, 57] \times [520, 700]$.

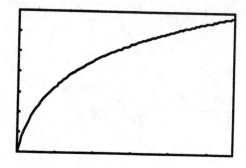

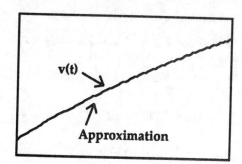

8.3   The Natural Exponential Function

1. (a) $e^{\ln 2} = 2$

   (b) $e^{-\ln 4} = e^{\ln 1/4} = 1/4$

   (c) $e^{(\ln x - \ln y)} = e^{\ln x}/e^{\ln y} = x/y$

   (d) $\ln e^{-x^2} = -x^2$

   (e) $\ln xe^{\sqrt{x}} - \ln x = \ln x + \ln e^{\sqrt{x}} - \ln x = \sqrt{x}$

   (f) $e^{x \ln 2} = e^{\ln 2^x} = 2^x$

   (g) $e^{\ln(1/x)} = 1/x$

   (h) $e^{4 \ln x} = e^{\ln x^4} = x^4$

   (i) $\ln xe^{x^2} = \ln x + \ln e^{x^2} = \ln x + x^2$

   (j) $e^{x - \ln x} = e^x/e^{\ln x} = e^x/x$

2. (a) $\ln x = 2 \Longrightarrow x = e^2.$

   (b) $\ln x^2 = 9 \Longrightarrow x^2 = e^9 \Longrightarrow x = \pm\sqrt{e^9} = \pm e^{9/2}.$

   (c) $e^{x^2} = 5 \Longrightarrow x^2 = \ln 5 \Longrightarrow x = \pm\sqrt{\ln 5}.$

   (d) $e^{2x} - 2e^x + 1 = 0 \Longrightarrow (e^x - 1)^2 = 0 \Longrightarrow e^x = 1 \Longrightarrow x = \ln 1 = 0.$

3. $e^{x-y} = x + 3 \Longrightarrow \ln e^{x-y} = \ln(x+3) \Longrightarrow x - y = \ln(x+3) \Longrightarrow y = x - \ln(x+3).$

4. $e^{(y-1)^2} = x^4 + 1 \Longrightarrow (y-1)^2 = \ln(x^4+1) \Longrightarrow y - 1 = \pm\sqrt{\ln(x^4+1)} \Longrightarrow y = 1 \pm \sqrt{\ln(x^4+1)}.$ Hence, if $y > 1$, $y = 1 + \sqrt{\ln(x^4+1)}.$

5. $y = e^{3x} \Longrightarrow y' = e^{3x}(3) = 3e^{3x}.$

6. $f(x) = xe^{-x} \Longrightarrow f'(x) = x(-e^{-x}) + e^{-x} = e^{-x}(1-x).$

7. $f(t) = e^{\sqrt{t}} \Longrightarrow f'(t) = e^{\sqrt{t}}(1/2)t^{-1/2} = e^{\sqrt{t}}/(2\sqrt{t}).$

8. $y = \dfrac{e^x + 1}{e^x} = 1 + e^{-x} \Longrightarrow y' = -e^{-x}.$

9. $y = e^{x^2 - x} \Longrightarrow y' = (2x-1)e^{x^2-x}.$

10. $f(t) = e^{\sin t} \Longrightarrow f'(t) = e^{\sin t} \cos t.$

11. $f(x) = e^x \sin x \Longrightarrow y' = e^x \cos x + e^x \sin x = e^x(\cos x + \sin x).$

12. $y = xe^{-3 \ln x} = xx^{-3} = x^{-2} \Longrightarrow y' = -2x^{-3}.$

13. $f(x) = \ln \dfrac{e^x + 1}{x + 1} = \ln(e^x + 1) - \ln(x+1) \Longrightarrow f'(x) = \dfrac{e^x}{e^x + 1} - \dfrac{1}{x+1}.$

14. $f(x) = (e^x - 1)/(e^x + 1) \Longrightarrow f'(x) = \dfrac{(e^x + 1)(e^x) - (e^x - 1)(e^x)}{(e^x + 1)^2} = \dfrac{2e^x}{(e^x + 1)^2}.$

15. $f(x) = (2 - e^{x^2})^3 \Longrightarrow f'(x) = 3(2 - e^{x^2})^2(-2xe^{x^2}) = -6xe^{x^2}(2 - e^{x^2})^2.$

16. $y = x\ln(e^x + x) \Longrightarrow y' = x\dfrac{1}{e^x + x}(e^x + 1) + \ln(e^x + x) = \dfrac{x(e^x + 1)}{e^x + x} + \ln(e^x + x).$

17. $y = (1/2)(e^x + e^{-x}) \Longrightarrow y' = (1/2)(e^x - e^{-x}).$

624

18. $f(x) = (x^2 + x - 1)e^{x^2+3} \implies f'(x) = (x^2 + x - 1)e^{x^2+3}(2x) + e^{x^2+3}(2x + 1) = e^{x^2+3}(2x^3 + 2x^2 + 1).$

19. $y = e^{\sqrt{x}} \ln \sqrt{x} = (1/2)\, e^{\sqrt{x}} \ln x \implies$

$$y' = \frac{1}{2}e^{\sqrt{x}}\frac{1}{x} + \frac{1}{2}e^{\sqrt{x}}\frac{1}{2\sqrt{x}} \ln x = \frac{e^{\sqrt{x}}(2 + \sqrt{x} \ln x)}{4x}.$$

20. $f(t) = te^{(1/t)^2} = te^{t^{-2}} \implies f'(t) = te^{t^{-2}}(-2t^{-3}) + e^{t^{-2}} = e^{t^{-2}}(1 - 2t^{-2}).$

21. $e^{xy} = x \implies e^{xy}(xy' + y) = 1 \implies xy' + y = e^{-xy} \implies y' = \dfrac{e^{-xy} - y}{x} = \dfrac{1/x - y}{x} = \dfrac{1 - xy}{x^2}.$

22. $e^{x-y} = xy \implies e^{x-y}(1 - y') = xy' + y \implies y'(-e^{x-y} - x) = y - e^{x-y} \implies y' = \frac{e^{x-y} - y}{e^{x-y} + x}.$

23. $\ln(x + 2y) = e^y \implies \dfrac{1}{x + 2y}(1 + 2y') = e^y y' \implies y'(e^y(x + 2y) - 2) = 1 \implies y' = \dfrac{1}{e^y(x + 2y) - 2}.$

24. $y^2 e^x + y \ln x = 2 \implies y^2 e^x + 2yy'e^x + y/x + y' \ln x = 0 \implies y' = \dfrac{-y^2 e^x - y/x}{2ye^x + \ln x}$

25. True; if $y = e^{kx}$ for some constant k, then $y' = ke^{kx} = ky$ and hence y' is proportional to y.

26. False; if $y = 0$, there is no x such that $e^x = 0$.

27. True; e^x is never zero.

28. Let $u = 2x$. Then $du = 2dx$ and hence

$$\int e^{2x}\, dx = \frac{1}{2}\int e^u\, du = \frac{1}{2}e^u + C = \frac{1}{2}e^{2x} + C.$$

29. Let $u = -x$. Then $du = -dx$ and hence

$$\int e^{-x}\, dx = -\int e^u\, du = -e^u + C = -e^{-x} + C.$$

30. Let $u = 2x + 6$. Then $du = 2dx$ and hence

$$\int e^{2x+6}\, dx = \frac{1}{2}\int e^u\, du = \frac{1}{2}e^u + C = \frac{1}{2}e^{2x+6} + C.$$

31. Let $u = x^2 + 3$. Then $du = 2xdx$ and hence

$$\int xe^{x^2+3}\, dx = \frac{1}{2}\int e^u\, du = \frac{1}{2}e^u + C = \frac{1}{2}e^{x^2+3} + C.$$

32. Let $u = 1 - x^3$. Then $du = -3x^2 dx$ and hence

$$\int x^2 e^{1-x^3}\, dx = -\frac{1}{3}\int e^u\, du = -\frac{1}{3}e^u + C = -\frac{1}{3}e^{1-x^3} + C.$$

33. Let $u = 1 + e^{2x}$. Then $du = 2e^{2x} dx$ and hence

$$\int e^{2x}(1 + e^{2x})^3\, dx = \frac{1}{2}\int u^3\, du = \left(\frac{1}{2}\right)\left(\frac{1}{4}\right)u^4 + C = \frac{1}{8}(1 + e^{2x})^4 + C.$$

34. Let $u = \sqrt{x}$. Then $du = 1/(2\sqrt{x})\,dx$ and hence

$$\int \frac{e^{\sqrt{x}}}{\sqrt{x}}\,dx = 2\int e^u\,du = 2e^u + C = 2e^{\sqrt{x}} + C.$$

35. Let $u = 1/x$. Then $du = -(1/x^2)dx$ and hence

$$\int \frac{e^{1/x}}{x^2}\,dx = -\int e^u\,du = -e^u + C = -e^{1/x} + C.$$

36. Let $u = e^x + 1$. Then $du = e^x dx$ and hence

$$\int \frac{e^x}{\sqrt{e^x + 1}}\,dx = \int \frac{1}{\sqrt{u}}\,du = 2\sqrt{u} + C = 2\sqrt{e^x + 1} + C.$$

37. Let $u = 1 + e^x$. Then $du = e^x dx$ and hence

$$\int \frac{e^x}{1 + e^x}\,dx = \int \frac{1}{u}\,du = \ln|u| + C = \ln(1 + e^x) + C.$$

Note that the absolute value in the antiderivative is not needed, *i.e.* $\ln|1 + e^x|$, since $1 + e^x$ is always positive.

38. Let $u = \sin x$. Then $du = \cos x\,dx$ and hence

$$\int (\cos x)e^{\sin x}\,dx = \int e^u\,du = e^u + C = e^{\sin x} + C.$$

39. Let $u = 3 + e^x$. Then $du = e^x dx$ and hence

$$\int \frac{e^x}{(3 + e^x)^2}\,dx = \int \frac{1}{u^2}\,du = -\frac{1}{u} + C = -\frac{1}{3 + e^x} + C.$$

40. First observe that

$$\int \frac{1 + e^{-ax}}{1 - e^{-ax}}\,dx = \int \left(\frac{1}{1 - e^{-ax}} + \frac{e^{-ax}}{1 - e^{-ax}} \right) dx = \int \underbrace{\frac{e^{ax}}{e^{ax} - 1}}_{u = e^{ax}}\,dx + \int \underbrace{\frac{e^{-ax}}{1 - e^{-ax}}}_{v = e^{-ax}}\,dx.$$

For the first integral, let $u = e^{ax}$. Then $du = ae^{ax}dx$ and hence

$$\int \frac{e^{ax}}{e^{ax} - 1}\,dx = \frac{1}{a}\int \frac{1}{u - 1}\,du = \frac{1}{a}\ln|u - 1| + C = \frac{1}{a}\ln|e^{ax} - 1| + C$$

$$= \frac{1}{a}\ln\left| \frac{1 - e^{-ax}}{e^{-ax}} \right| + C = \frac{1}{a}\ln|1 - e^{-ax}| + x + C.$$

For the second integral, let $v = e^{-ax}$. Then $dv = -ae^{-ax}dx$ and hence

$$\int \frac{e^{-ax}}{1 - e^{-ax}}\,dx = -\frac{1}{a}\int \frac{1}{1 - v}\,dv = \frac{1}{a}\ln|1 - v| + C = \frac{1}{a}\ln|1 - e^{-ax}| + C.$$

Therefore

$$\int \frac{1 + e^{-ax}}{1 - e^{-ax}}\,dx = \frac{1}{a}\ln|1 - e^{-ax}| + x + \frac{1}{a}\ln|1 - e^{-ax}| + C = x + \frac{2}{a}\ln|1 - e^{-ax}| + C.$$

626

41. Let $u = 1 + e^{\sqrt{x}}$. Then $du = e^{\sqrt{x}}/(2\sqrt{x})\,dx$ and hence

$$\int \frac{(1 + e^{\sqrt{x}})e^{\sqrt{x}}}{\sqrt{x}}\,dx = 2\int u\,du = u^2 + C = (1 + e^{\sqrt{x}})^2 + C.$$

42. Referring to Exercise 28, we find that

$$\int_0^1 e^{2x}\,dx = \frac{1}{2}e^{2x}\Big|_0^1 = \frac{1}{2}e^2 - \frac{1}{2}e^0 = \frac{1}{2}(e^2 - 1).$$

43. Referring to Exercise 29, we find that

$$\int_1^{\ln 4} e^{-x}\,dx = -e^{-x}\Big|_1^{\ln 4} = -e^{-\ln 4} + e^{-1} = \frac{1}{e} - \frac{1}{4}.$$

44. Let $u = 1 - x^2$. Then $du = -2x\,dx$ and hence

$$\int_0^1 xe^{1-x^2}\,dx = -\frac{1}{2}\int_1^0 e^u\,du = -\frac{1}{2}\Big|_1^0 = -\frac{1}{2}e^0 + \frac{1}{2}e = \frac{1}{2}(e - 1).$$

45. Referring to Exercise 38, we find that

$$\int_0^{\pi/2} (\cos x)e^{\sin x}\,dx = e^{\sin x}\Big|_0^{\pi/2} = e^1 - e^0 = e - 1.$$

46. Let $u = 2 + e^x$. Then $du = e^x\,dx$ and hence

$$\int_0^{\ln 2} \frac{e^x}{2 + e^x}\,dx = \int_3^4 \frac{1}{u}\,du = \ln u\Big|_3^4 = \ln 4 - \ln 3 = \ln\frac{4}{3}.$$

47. Let $u = \ln x$. Then $du = (1/x)\,dx$ and hence

$$\int_1^{e^2} \frac{\ln x}{x}\,dx = \int_0^2 u\,du = \frac{1}{2}u^2\Big|_0^2 = 2.$$

48. $\displaystyle\int_0^2 \frac{e^x + e^{-x}}{2}\,dx = \frac{1}{2}(e^x - e^{-x})\Big|_0^2 = \frac{1}{2}(e^2 - e^{-2}) - \frac{1}{2}(e^0 - e^0) = \frac{1}{2}(e^2 - e^{-2}).$

49. $\displaystyle\int_0^2 e^x\,dx = e^x\Big|_0^2 = e^2 - 1.$

50. Let $u = \tan x$. Then $du = \sec^2 x\,dx$ and hence

$$\int_0^{\pi/4} (\sec^2 x)e^{\tan x}\,dx = \int_0^1 e^u\,du = e^u\Big|_0^1 = e - 1.$$

51. Let $u = \sin^2 x$. Then $du = 2\sin x \cos x\,dx$ and hence

$$\int_0^{\pi/3} \sin x \cos x\, e^{\sin^2 x}\,dx = \frac{1}{2}\int_0^{3/4} e^u\,du = \frac{1}{2}e^u\Big|_0^{3/4} = \frac{1}{2}(e^{3/4} - 1).$$

52. $y = (3 - x^2)e^x \implies y' = (3 - x^2)e^x + e^x(-2x) = e^x(-x^2 - 2x + 3) = e^x(-x + 1)(x + 3) = 0 \iff x = 1, -3$. Thus, the graph of $y = (3 - x^2)e^x$ has two critical points: $(1, 2e)$ and $(-3, -6e^{-3})$. Now,

$$y'' = e^x(-2x - 2) + (-x^2 - 2x + 3)e^x = e^x(-x^2 - 4x + 1).$$

Thus $y''(1) = -4e < 0 \implies (1, 2e)$ is a relative maximum, while $y''(-3) = 4e^{-3} > 0 \implies (-3, -6e^{-3})$ is a relative minimum, as shown below. Since the curve crosses the x-axis only at the points $x = \pm\sqrt{3}$, it follows that $2e$ is the maximum value of the function.

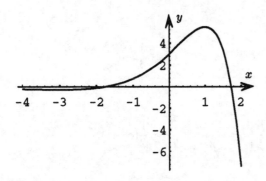

$$y = (3 - x^2)e^x$$

Exercise 52

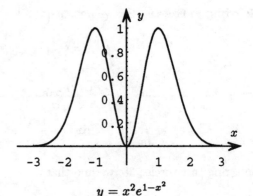

$$y = x^2 e^{1 - x^2}$$

Exercise 53

53. $y = x^2 e^{1-x^2} \implies y' = x^2 e^{1-x^2}(-2x) + e^{1-x^2}(2x) = 2xe^{1-x^2}(1 - x^2) = 0 \implies x = 0, \pm 1$. Thus, the graph of $y = x^2 e^{1-x^2}$ has three critical points: $(0, 0)$, $(1, 1)$, and $(-1, 1)$. Now,

$$y'' = 2xe^{1-x^2}(-2x) + (1 - x^2)[2xe^{1-x^2}(-2x) + 2e^{1-x^2}] = e^{1-x^2}(4x^4 - 10x^2 + 2).$$

Thus, $y''(0) = 2e > 0 \implies (0, 0)$ is a relative minimum, while $y''(\pm 1) = -4 < 0 \implies (1, 1)$ and $(-1, 1)$ are relative maximums, as shown above.

54. Area $= \displaystyle\int_0^1 x \, dy = \int_0^1 e^y \, dy = e^y \big|_0^1 = e - 1.$

55. We find that

$$f'(x) = xe^{2x}(2) + e^{2x} = e^{2x}(2x + 1)$$

and

$$f''(x) = e^{2x}(2) + (2x + 1)e^{2x}(2) = e^{2x}(4x + 4).$$

(a) Since $f'(x) > 0 \iff x > -1/2$, $f(x)$ is increasing on the interval $[-1/2, +\infty)$.

(b) Since $f'(x) < 0 \iff x < -1/2$, $f(x)$ is decreasing on the interval $(-\infty, -1/2]$.

(c) $f'(x) = 0 \iff x = -1/2$. Hence $f(x)$ has a relative extremum at the point $(-1/2, -1/2 \ e^{-1})$. Since $f''(-1/2) = e^{-1}(2) > 0$, this is a relative minimum.

(d) Since $f''(x) > 0 \iff x > -1$, the graph is concave up on the interval $[-1, +\infty)$.

(e) Since $f''(x) < 0 \iff x < -1$, the graph is concave down on the interval $(-\infty, -1]$.

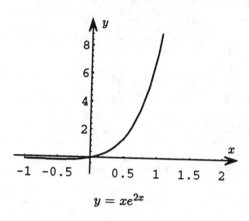

$y = xe^{2x}$

Exercise 55

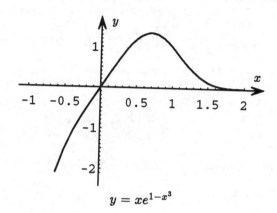

$y = xe^{1-x^3}$

Exercise 56

(f) Since the concavity changes only at the point $x = -1$, $(-1, -e^{-2})$ is the only inflection point. The graph is shown below.

56. We find that

$$f'(x) = xe^{1-x^3}(-3x^2) + e^{1-x^3} = e^{1-x^3}(1 - 3x^3)$$

and

$$f''(x) = e^{1-x^3}(-9x^2) + (1 - 3x^3)e^{1-x^3}(-3x^2) = -3x^2e^{1-x^3}(4 - 3x^3).$$

(a) Since $f'(x) > 0 \iff x < \sqrt[3]{1/3}$, $f(x)$ is increasing on the interval $(-\infty, \sqrt[3]{1/3}]$.

(b) Since $f'(x) < 0 \iff x > \sqrt[3]{1/3}$, $f(x)$ is decreasing on the interval $[\sqrt[3]{1/3}, \infty)$.

(c) $f'(x) = 0 \iff x = \sqrt[3]{1/3}$. Hence $f(x)$ has a relative extremum at the point $(\sqrt[3]{1/3}, e^{2/3}\sqrt[3]{1/3})$. Since $f''(\sqrt[3]{1/3}) = -3(3^{-2/3})e^{2/3} < 0$, this is a relative maximum.

(d) Since $f''(x) > 0 \iff x > \sqrt[3]{4/3}$, the graph is concave up on the interval $[\sqrt[3]{4/3}, +\infty)$.

(e) Since $f''(x) < 0 \iff x < \sqrt[3]{4/3}$, the graph is concave down on the interval $(-\infty, \sqrt[3]{4/3}]$.

(f) $f''(x) = 0 \iff x = 0$, $\sqrt[3]{4/3}$. Since the concavity changes at $x = \sqrt[3]{4/3}$, $(\sqrt[3]{4/3}, \sqrt[3]{4/3}\, e^{-1/3})$ is an inflection point. The concavity does not change at $x = 0$, however, since $f''(x) < 0$ for values of x slightly smaller than and greater than $x = 0$. The graph is shown above.

57. $y = (1/2)(e^x + e^{-x}) \implies y' = (1/2)(e^x - e^{-x})$. Hence the length of graph from $(0, 1)$ to $(\ln 2, 5/4)$ is equal to

$$\int_0^{\ln 2} \sqrt{1 + (y')^2}\, dx = \int_0^{\ln 2} \sqrt{1 + \frac{1}{4}(e^x - e^{-x})^2}\, dx = \frac{1}{2}\int_0^{\ln 2} \sqrt{4 + e^{2x} - 2 + e^{-2x}}\, dx$$

$$= \frac{1}{2}\int_0^{\ln 2} \sqrt{e^{2x} + 2 + e^{-2x}}\, dx = \frac{1}{2}\int_0^{\ln 2} (e^x + e^{-x})\, dx = \frac{1}{2}(e^x - e^{-x})\Big|_0^{\ln 2} = \frac{1}{2}\left(2 - \frac{1}{2}\right) - \frac{1}{2}(1 - 1) = \frac{3}{4}.$$

58. Since, from Exercise 57, $y = 1/2\,(e^x + e^{-x})$ and $y' = 1/2\,(e^x - e^{-x})$, the surface area is equal to

$$\int_0^{\ln 2} 2\pi y\sqrt{1+(y')^2}\,dx = 2\pi \int_0^{\ln 2} \frac{1}{2}(e^x + e^{-x})\sqrt{1+\frac{1}{4}(e^x - e^{-x})^2}\,dx = \frac{\pi}{2}\int_0^{\ln 2}(e^x + e^{-x})^2\,dx$$

$$= \frac{\pi}{2}\int_0^{\ln 2}(e^{2x}+2+e^{-2x})\,dx = \frac{\pi}{2}\left(\frac{1}{2}e^{2x}+2x-\frac{1}{2}e^{-2x}\right)\Big|_0^{\ln 2}$$

$$= \frac{\pi}{2}\left[\frac{1}{2}(4)+2\ln 2-\frac{1}{2}\left(\frac{1}{4}\right)\right] - \frac{\pi}{2}\left[\frac{1}{2}-\frac{1}{2}\right] = \frac{\pi}{2}\left(\frac{15}{8}+2\ln 2\right).$$

59. $\frac{dy}{dx} = 2xy \implies \frac{dy}{y} = 2x\,dx \implies \ln|y| = x^2 + C \implies y = Ce^{x^2}$, where C is any constant.

60. If $c > 4$,

$$\int_4^c \frac{1}{x-3}\,dx = 1 \implies (\ln|x-3|)\big|_4^c = 1 \implies \ln(c-3) = 1 \implies c-3 = e \implies c = 3+e.$$

61. Since $y' = xe^x + e^x = e^x(x+1) = 0 \iff x = -1$, the graph of $y = xe^x$ has only one point at which the tangent line is horizontal, namely $(-1, -1/e)$. Hence $P = (-1, -1/e)$.

62. $y = (e^x+1)/e^{3x} = e^{-2x}+e^{-3x} \implies y' = -2e^{-2x}-3e^{-3x}$. Therefore $y(\ln 2) = (e^{\ln 2}+1)/e^{3\ln 2} = 3/8$ and $y'(\ln 2) = -2e^{-2\ln 2} - 3e^{-3\ln 2} = -2(1/4)-3(1/8) = -7/8$. Thus, the equation of the tangent line at $x = \ln 2$ is $y - 3/8 = -7/8\,(x-\ln 2)$.

63. First observe in the figure on the following page that the region is symmetric with respect to the y-axis. Hence the area of the region is

$$2\int_0^1 (e^x - e^{-x})\,dx = 2(e^x + e^{-x})\big|_0^1 = 2(e+e^{-1}) - 2(1+1) = 2(e-2)+2/e.$$

64. Let dV be the volume of a differential cross-section disk of thickness dx. Then

$$dV = \pi r^2\,dx = \pi y^2\,dx = \pi e^{4x}\,dx$$

and hence the volume of revolution is

$$V = \int_0^2 \pi e^{4x}\,dx = \frac{\pi}{4}e^{4x}\Big|_0^2 = \frac{\pi}{4}(e^8-1).$$

65. $y = Ae^{kt} \implies y' = Ake^{kt} = ky$ and $y(0) = Ae^0 = A$.

   (a) $y' = y \implies k = 1$. Since $y(0) = 1 = A$, $y = e^t$.
   (b) $y' = \pi y \implies k = \pi$. Since $y(0) = -2 = A$, $y = -2e^{\pi t}$.
   (c) $y' = -3y \implies k = -3$. Since $y(0) = 2 = A$, $y = 2e^{-3t}$.

66. The average value of $y = xe^{1-x^2}$ on $[-2, 2]$ is equal to

$$\frac{1}{4}\int_{-2}^2 \underbrace{xe^{1-x^2}}_{\substack{u = 1-x^2 \\ du = -2x\,dx}}\,dx = -\frac{1}{8}\int_{-3}^{-3} e^u\,du = 0.$$

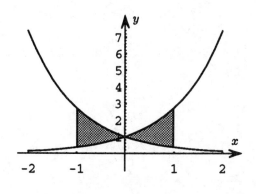

$$y = e^{-x} \qquad y = e^x$$

Exercise 63

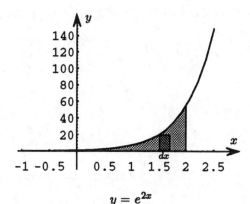

$$y = e^{2x}$$

Exercise 64

67. $f(-x) = (e^{-x} + e^x)/2 = f(x) \implies f(x)$ is an even function.
$g(-x) = (e^{-x} - e^x)/2 = -g(x) \implies g(x)$ is an odd function.

68. (a) $e^{x-y} = ye^x \implies e^{x-y}(1 - y') = ye^x + e^x y' \implies y'(e^x + e^{x-y}) = e^{x-y} - ye^x \implies y' = \frac{e^{x-y} - ye^x}{e^x + e^{x-y}} = \frac{e^{-y} - y}{1 + e^{-y}}$.

(b) If $e^{x-y} = ye^x$, then $e^x(e^{-y} - y) = 0$ and hence $e^{-y} - y = 0$ since $e^x \neq 0$. Therefore $y' = \frac{e^{-y} - y}{1 + e^{-y}} = 0$. It follows that $y = c$ for some constant c.

(c) As we noted in part (b), for any number x, $e^{x-y} = ye^x \implies e^{-y} = y$. Now, there is only one number y such that $e^{-y} = y$. To see this, let $f(x) = xe^x$. The graph of the function $f(x)$ is basically the same as the graph of the function $y = xe^{2x}$ shown in Exercise 55. Since $f(x)$ is continuous and increasing, it follows from the intermediate value theorem that there is one and only one value of x such that $f(x) = xe^x = 1$. Thus, the function y defined implicitly by the equation $e^{x-y} = ye^x$ is in fact a constant function.

Remark: The equation $xe^x = 1$ cannot be solved in closed form. Using Newton's method discussed in Chapter 3, one can show that $x \approx 0.567143$.

69. $I = I_0 e^{-kl} \implies dI/dl = I_0 e^{-kl}(-k) = -kI$. Hence, if $dI/dl = -kI = 4I$, then $k = -4$.

70. (a) Let x be a real number. Then

$$f(\mu + x) = \frac{1}{\sigma\sqrt{2\pi}} e^{-x^2/2\sigma^2} \quad \text{and} \quad f(\mu - x) = \frac{1}{\sigma\sqrt{2\pi}} e^{-x^2/2\sigma^2}.$$

Hence $f(\mu + x) = f(\mu - x)$ for all x and therefore the graph of f is symmetric about the line $x = \mu$.

(b) Let us first find $f'(x)$ and $f''(x)$:

$$f'(x) = \frac{1}{\sigma\sqrt{2\pi}} e^{-(x-\mu)^2/2\sigma^2} [-2(x - \mu)/2\sigma^2] = \frac{-1}{\sigma^3\sqrt{2\pi}} e^{-(x-\mu)^2/2\sigma^2} (x - \mu)$$

and

$$f''(x) = \frac{-1}{\sigma^3\sqrt{2\pi}} \left[e^{-(x-\mu)^2/2\sigma^2} (1) + (x - \mu)e^{-(x-\mu)^2/2\sigma^2} [-2(x - \mu)/2\sigma^2] \right]$$

631

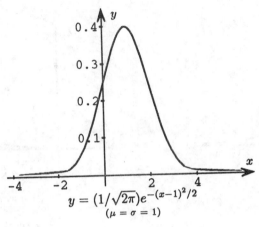

$$y = (1/\sqrt{2\pi})e^{-(x-1)^2/2}$$
$$(\mu = \sigma = 1)$$

Exercise 70

$$= \frac{-1}{\sigma^5\sqrt{2\pi}}e^{-(x-\mu)^2/2\sigma^2}[\sigma^2 - (x-\mu)^2].$$

Therefore $f''(x) = 0 \iff x - \mu = \pm\sigma \iff x = \mu \pm \sigma$. Since the sign of $f''(x)$ clearly changes at both $\mu \pm \sigma$, it follows that $\mu \pm \sigma$ are inflection points for the graph of $f(x)$.

(c) $f'(x) = 0 \implies x = \mu$. Since $f''(\mu) = (-1/\sigma^5\sqrt{2\pi})\sigma^2 < 0$, $x = \mu$ is a relative maximum for the graph, as shown above.

71. Since $-b_1(t - c_1) < 0$ and $-b_2(t - c_2) < 0$ as $t \to \infty$, $e^{-b_1(t-c_1)} \to 0$ and $e^{-b_2(t-c_2)} \to 0$. Therefore

$$\lim_{t\to\infty} y(t) = a \lim_{t\to\infty} \frac{1}{1 + e^{-b_1(t-c_1)}} + (f - a) \lim_{t\to\infty} \frac{1}{1 + e^{-b_2(t-c_2)}} = a(1) + (f - a)(1) = f.$$

72. (a) $E = E_+ + E_- = -1/r + 10e^{-r} \implies dE/dr = 1/r^2 - 10e^{-r}$. Thus, to find the relative extrema of E, we must solve the equation

$$\frac{1}{r^2} - 10e^{-r} = 0.$$

This equation cannot be solved in closed form. Let us use Newton's Method to approximate its roots. The table below on the left gives values of the function $f(r) = 1/r^2 - 10e^{-r}$ for selected values of r.

r	$1/r^2 - 10e^{-r}$
0.1	90.95
1	−2.68
2	−1.10
3	−0.39
4	−0.12
5	−0.03
6	2.99

$r_{n+1} = r_n - f(r_n)/f'(r_n)$	
0.2	5
0.2695	5.5329
0.3344	5.7801
0.3729	5.8265
0.3825	5.8279
0.3830	5.8279
0.3830	5.8279

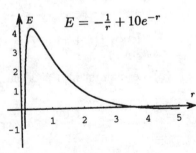

$$E = -\frac{1}{r} + 10e^{-r}$$

It shows that there are two roots, one between 0.1 and 1, the other between 5 and 6. Using 0.2 and 5 as initial values, the table in the center lists the successive approximations to the roots using Newton's formula. To within four decimal places, the roots are 0.3830 and 5.8279. Now, $E''(r) = -2/r^3 + 10e^{-r} \implies E''(0.3830) = -28.78 < 0 \implies$ maximum potential $= E(0.383) \approx 4.2072$, and $E''(5.8279) = 0.0193 > 0 \implies$ minimum potential $= E(5.8279) \approx -0.1421$.

(b) The graph of E is shown above on the right.

73. Let $f(x) = e^{2x} - 8x + 1$. The graph of $y = f(x)$, drawn by computer and shown below on the right, indicates that $f(x)$ has two zeros, one between 0 and 0.5, the other between 0.5 and 1. Using $r_0 = 0.4$ and $r_0 = 0.9$ as initial values in Newton's Method and setting

$$r_{n+1} = r_n - \frac{f(r_n)}{f'(r_n)},$$

the table below on the left summarizes the successive values of r_n and shows that the roots, to within four decimal places, are 0.4073 and 0.9333.

$r_{n+1} = r_n - f(r_n)/f'(r_n)$	
0.4	0.9
0.407197	0.936678
0.407263	0.933355
0.407263	0.933326
0.407263	0.933326
0.407263	0.933326

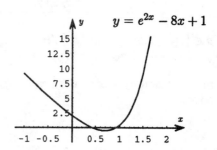

74. Let $f(x) = e^x + x - 5$. The graph of $y = f(x)$, drawn by computer and shown below on the right, indicates that $f(x)$ has one zero, between 1 and 2. Using $r_0 = 1.1$ as initial value in Newton's Method and setting

$$r_{n+1} = r_n - \frac{f(r_n)}{f'(r_n)},$$

the table below on the left summarizes the successive values of r_n and shows that the root, to within four decimal places, is 1.3066.

$r_{n+1} = r_n - f(r_n)/f'(r_n)$
1.1
1.32373
1.30667
1.30656
1.30656

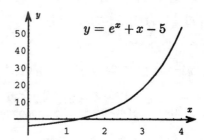

75. The values of the polynomials are most easily calculated using a spreadsheet. For example, let the first column contain the values of x. In the second column, cell B1, enter the formula EXP(A1), which means to raise e to the number in cell A1; in the third column, cell C1, enter the formula A1 + 1, which adds one to the number in cell A1 and gives $P_1(x)$; the fourth column the formula

```
C1 + (A1)^2/FACT(2),
```

which gives the value of the polynomial $P_2(x)$ when x equals the value in cell A1. Similarly, in the fifth and sixth columns enter the formulas

```
D1 + (A1)^3/FACT(3)
E1 + (A1)^4/FACT(4)
```

which give $P_3(x)$ and $P_4(x)$. Now select cells B1 through D6 and do a FILL DOWN. The table below is the result of these calculations.

x	e^x	$P_1(x)$	$P_2(x)$	$P_3(x)$	$P_4(x)$
1	2.71828183	2	2.5	2.66666667	2.70833333
0.5	1.64872127	1.5	1.625	1.64583333	1.6484375
2	7.3890561	3	5	6.33333333	7
−2	0.13533528	−1	1	−0.3333333	0.33333333

76. Let $y \in [f(a), f(b)]$. Then $(f^{-1})'(y) = 1/f'(x)$, where $y = f(x)$. Since f is increasing on the interval $[a, b]$, $f'(x) > 0$. Hence $(f^{-1})'(y) > 0$ and therefore f^{-1} is increasing on the interval $[f(a), f(b)]$.

77. Let $y_1 = e^{x_1}$ and $y_2 = e^{x_2}$. Then $x_1 = \ln y_1$, $x_2 = \ln y_2$ and hence $x_1 - x_2 = \ln y_1 - \ln y_2 = \ln(y_1/y_2)$. Therefore
$$e^{x_1 - x_2} = e^{\ln(y_1/y_2)} = \frac{y_1}{y_2} = \frac{e^{x_1}}{e^{x_2}}.$$

78. Let $y_1 = e^{x_1}$. Then $x_1 = \ln y_1$ and hence $r x_1 = r \ln y_1 = \ln y_1^r$ for any rational number r. Therefore
$$e^{r x_1} = e^{\ln y_1^r} = y_1^r = (e^{x_1})^r.$$

8.4 Exponentials and Logs to other Bases

1. (a) $2x = e^{\ln 2x}$

   (b) $\pi^3 = e^{\ln \pi^3} = e^{3 \ln \pi}$

   (c) $7^{\ln x} = e^{\ln 7^{\ln x}} = e^{(\ln x)(\ln 7)}$

   (d) $4^{\sqrt{2}} = e^{\ln 4^{\sqrt{2}}} = e^{\sqrt{2} \ln 4}$

   (e) $3^{\sin x} = e^{\ln 3^{\sin x}} = e^{(\sin x) \ln 3}$

   (f) $2^x 4^{1-x} = 2^x 2^{2-2x} = 2^{2-x} = e^{\ln 2^{2-x}} = e^{(2-x) \ln 2}$

2. $\pi^3 = e^{3 \ln \pi} = e^{3.434189658} = 31.00627669$
   $4^{\sqrt{2}} = e^{\sqrt{2} \ln 4} = e^{1.960516287} = 7.102993302.$

3. For any real number x,
$$10^x = e^{\ln 10^x} = e^{x \ln 10}$$

   and
$$e^x = 10^{\log_{10} e^x} = 10^{x \log_{10} e} = 10^{x/\ln 10}.$$

   Note that $\log_{10} e = \ln e / \ln 10 = 1/\ln 10$.

4. False; by definition, $e^{x \ln a} = a^x$ for all real x but only for $a > 0$.

5. $y = 3^x \implies y' = 3^x \ln 3$.

6. $f(x) = \pi^{x^2-1} \implies f'(x) = \pi^{x^2-1}(2x \ln \pi)$.

7. $f(x) = \log_{10}(2x - 1) = \frac{\ln(2x-1)}{\ln 10} \implies f'(x) = \frac{2}{(\ln 10)(2x-1)}$.

8. $y = x10^{x-1} \implies y' = x(10^{x-1} \ln 10) + 10^{x-1} = 10^{x-1}(x \ln 10 + 1)$.

9. $y = \log_{10}(\ln x) = \frac{\ln(\ln x)}{\ln 10} \implies y' = \frac{1}{x(\ln 10)(\ln x)}$.

10. $f(x) = 2^{1-x} \log_2 \sqrt{x} = 2^{1-x} \frac{\ln \sqrt{x}}{\ln 2} = 2^{1-x} \frac{(1/2)\ln x}{\ln 2} = \frac{2^{-x} \ln x}{\ln 2} \implies$

$$f'(x) = \frac{1}{\ln 2}\left(2^{-x}\frac{1}{x} + (\ln x)2^{-x}(-1)\ln 2\right) = \frac{2^{-x}}{\ln 2}\left(\frac{1}{x} - (\ln x)(\ln 2)\right).$$

11. $g(t) = t^2 \log_2 t = t^2 \frac{\ln t}{\ln 2} \implies$

$$g'(t) = \frac{1}{\ln 2}\left(t^2\left(\frac{1}{t}\right) + (\ln t)(2t)\right) = \frac{t}{\ln 2}(1 + 2\ln t).$$

12. $y = \log_{10} e^{\sqrt{t^2+1}} = (t^2 + 1)^{1/2} \log_{10} e \implies$

$$y' = \frac{1}{2}(t^2 + 1)^{-1/2}(2t)\log_{10} e = \frac{t \log_{10} e}{\sqrt{t^2 + 1}}.$$

13. $f(x) = \pi^x + x^\pi \implies f'(x) = \pi^x \ln \pi + \pi x^{\pi-1}$.

14. $y = \log_{10} 2^{x^2} = x^2 \log_{10} 2 \implies y' = 2x \log_{10} 2$.

15. $y = x^x = e^{x \ln x} \implies y' = e^{x \ln x}(x\frac{1}{x} + \ln x) = x^x(1 + \ln x)$.

16. $y = x^{\cos x} = e^{(\cos x)\ln x} \implies y' = e^{(\cos x)\ln x}(\cos x\frac{1}{x} - (\sin x)\ln x) = x^{\cos x}(\frac{\cos x}{x} - (\sin x)\ln x)$.

17. $y = x^{\sqrt{x}} = e^{\sqrt{x}\ln x} \implies$

$$y' = e^{\sqrt{x}\ln x}\left(\sqrt{x}\,\frac{1}{x} + (\ln x)\frac{1}{2\sqrt{x}}\right) = \frac{x^{\sqrt{x}}}{\sqrt{x}}\left(1 + \frac{1}{2}\ln x\right).$$

18. $y = x^{\ln x} = e^{(\ln x)^2} \implies y' = e^{(\ln x)^2}2(\ln x)\frac{1}{x} = 2x^{(\ln x)-1}\ln x$.

19. $y = (\cos x)^{\sin x} = e^{(\sin x)\ln \cos x} \implies$

$$y' = e^{(\sin x)\ln \cos x}\left(\sin x\frac{1}{\cos x}(-\sin x) + (\ln \cos x)(\cos x)\right) = (\cos x)^{\sin x}\left(\frac{-\sin^2 x}{\cos x} + (\ln \cos x)(\cos x)\right).$$

20. $y = (\ln x)^x = e^{x \ln \ln x} \implies y' = e^{x \ln \ln x}(x\frac{1}{\ln x}\frac{1}{x} + \ln \ln x) = (\ln x)^x(\frac{1}{\ln x} + \ln \ln x)$.

21. $\int 5^x \, dx = \int e^{x \ln 5} \, dx = \frac{1}{\ln 5} 5^x + C$.

22. Let $u = 1 - x^2$. Then $du = -2x\,dx$ and hence

$$\int x2^{1-x^2} \, dx = -\frac{1}{2}\int 2^u \, du = -\frac{1}{2}\frac{1}{\ln 2} 2^u + C = -\frac{1}{2\ln 2} 2^{1-x^2} + C.$$

23. Let $u = \sqrt{x}$. Then $du = \frac{1}{2\sqrt{x}}dx$ and hence

$$\int \frac{\pi^{\sqrt{x}}}{\sqrt{x}} \, dx = 2 \int \pi^u \, du = \frac{2}{\ln \pi} \, \pi^u + C = \frac{2}{\ln \pi} \, \pi^{\sqrt{x}} + C.$$

24. $\int 2^{1-x} 2^{1+x} \, dx = \int 2^2 \, dx = 4x + C.$

25. Since $a^{2\ln x} = (e^{\ln a})^{2\ln x} = x^{2\ln a}$,

$$\int a^{2\ln x} \, dx = \int x^{2\ln a} \, dx = \frac{1}{1 + 2\ln a} \, x^{1+2\ln a} + C.$$

26. Let $u = ax^2 + bx$. Then $du = (2ax + b)dx$ and hence

$$\int 7^{ax^2+bx} \left(ax + \frac{b}{2} \right) \, dx = \frac{1}{2} \int 7^u \, du = \frac{1}{2} \frac{1}{\ln 7} 7^u + C = \frac{1}{2\ln 7} 7^{ax^2+bx} + C.$$

27. Let $u = x^2$. Then $du = 2xdx$ and hence

$$\int_0^2 x3^{x^2} \, dx = \frac{1}{2} \int_0^4 3^u \, du = \frac{1}{2\ln 3} 3^u \Big|_0^4 = \frac{1}{2\ln 3}(3^4 - 1) = \frac{40}{\ln 3}.$$

28. Let $u = 3t - 1$. Then $du = 3dt$ and hence

$$\int_0^2 2^{3t-1} \, dt = \frac{1}{3} \int_{-1}^5 2^u \, du = \frac{1}{3\ln 2} 2^u \Big|_{-1}^5 = \frac{1}{3\ln 2}(2^5 - 2^{-1}) = \frac{21}{2\ln 2}.$$

29. Let $u = e^{2x}$. Then $du = 2e^{2x}dx$ and hence

$$\int_0^{\ln 2} 3^{e^{2x}} e^{2x} \, dx = \frac{1}{2} \int_1^4 3^u \, du = \frac{1}{2} \frac{1}{\ln 3} 3^u \Big|_1^4 = \frac{39}{\ln 3}.$$

30. Let $u = \cos 2x$. Then $du = -2\sin 2xdx$ and hence

$$\int_0^{\pi/4} (\sin 2x)7^{\cos 2x} \, dx = -\frac{1}{2} \int_1^0 7^u \, du = -\frac{1}{2} \frac{1}{\ln 7} 7^u \Big|_1^0 = -\frac{1}{2\ln 7}(1 - 7) = \frac{3}{\ln 7}.$$

31. True.

32. For all real numbers x and y, $a^x a^y = e^{x\ln a} e^{y\ln a} = e^{(x+y)\ln a} = a^{x+y}.$

33. For all real numbers x and y, $a^x/a^y = e^{x\ln a}/e^{y\ln a} = e^{(x-y)\ln a} = a^{x-y}.$

34. (a) $y = x2^{1-x} = xe^{(1-x)\ln 2} \implies$

$$y' = x \left(e^{(1-x)\ln 2}(-\ln 2) \right) + 2^{1-x} = 2^{1-x}(1 - x\ln 2)$$

and

$$y'' = 2^{1-x}(-\ln 2) + (1 - \ln 2)2^{1-x}(-\ln 2) = 2^{1-x}(\ln 2)(x\ln 2 - 2).$$

Hence $y' = 0 \iff x = 1/\ln 2$. Since $y''(1/\ln 2) = 2^{1-1/\ln 2}(-\ln 2) < 0$, $(1/\ln 2, 2^{1-1/\ln 2}/\ln 2)$ is a relative maximum.

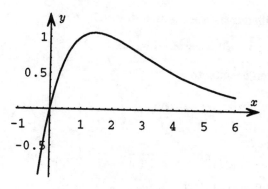

$$y = x2^{1-x}$$

Exercise 34

(b) $y' > 0 \iff x < 1/\ln 2$. Hence y is increasing on the interval $(-\infty, 1/\ln 2)$.

(c) Clearly, y'' changes sign at $x = 2/\ln 2$. Hence $x = 2/\ln 2$ is an inflection point.

(d) $y'' > 0 \iff x > 2/\ln 2$. Hence y is concave up on the interval $(2/\ln 2, \infty)$ and concave down on $(-\infty, 2/\ln 2)$.

35. $y = a^x \implies y' = a^x \ln a$. Hence $y' - 2y = 0 \implies a^x \ln a - 2a^x = a^x(\ln a - 2) = 0 \implies \ln a = 2 \implies a = e^2$.

36. Let $f_n(x) = \log_{10}(10^n + x) = \frac{\ln(10^n + x)}{\ln 10}$. Then

$$f_n'(x) = \frac{1}{\ln 10} \frac{1}{10^n + x}.$$

Now, for small values of h,

$$f_n(x + h) \approx f_n(x) + f_n'(x)h.$$

Thus, setting $x = 0$, it follows that

$$f_n(h) \approx f_n(0) + f_n'(0)h = \log_{10} 10^n + \frac{1}{\ln 10} \frac{1}{10^n} h,$$

or simply,

$$f_n(h) \approx n + \frac{h}{(\ln 10)10^n}, \quad \text{if} \quad h \approx 0.$$

(a) $\log_{10} 10.2 = \log_{10}(10^1 + 0.2) = f_1(0.2) \approx 1 + \frac{0.2}{(\ln 10)10} \approx 1.00868589$.
Calculator value: $\log_{10} 10.2 = 1.00860017$.

(b) $\log_{10} 98 = \log_{10}(10^2 - 2) = f_2(-2) \approx 2 - \frac{2}{(\ln 10)10^2} \approx 1.99131411$.
Calculator value: $\log_{10} 98 = 1.99122608$.

(c) $\log_{10} 995 = \log_{10}(10^3 - 5) = f_3(-5) \approx 3 - \frac{5}{(\ln 10)10^3} \approx 2.9978285$.
Calculator value: $\log_{10} 995 = 2.99782308$.

(d) $\log_{10} 0.09 = \log_{10}(10^{-1} - 0.01) = f_{-1}(-0.01) \approx -1 - \frac{0.01}{(\ln 10)10^{-1}} \approx -1.043429$.
Calculator value: $\log_{10} 0.09 = -1.04575749$.

37. Let dV be the volume of a differential cross-section disk. Then

$$dV = \pi r^2\,dx = \pi y^2\,dx = \pi 3^{2x}\,dx.$$

Hence the volume of revolution is equal to

$$V = \int_0^1 \pi 3^{2x}\,dx = \pi \int_0^1 e^{2x\ln 3}\,dx$$

$$= \frac{\pi}{2\ln 3}3^{2x}\Big|_0^1 = \frac{\pi}{2\ln 3}(3^2 - 3^0) = \frac{4\pi}{\ln 3}.$$

38. Let dV be the volume of a differential shell. Then

$$dV = 2\pi r h\,dx = 2\pi x(16 - 2^{x^2}).$$

Hence, since $x = 2$ when $y = 16$, the volume of revolution is equal to

$$\int_0^2 2\pi x(16 - 2^{x^2})\,dx = 2\pi \int_0^2 16x\,dx - 2\pi\int_0^2 x 2^{x^2}\,dx = 2\pi(16)\frac{1}{2}x^2\Big|_0^2 - \frac{2\pi}{2\ln 2}2^{x^2}\Big|_0^2 = 64\pi - \frac{15\pi}{\ln 2}.$$

39. $y = \log_{10} x = (\ln x)/(\ln 10) \implies y' = 1/(x\ln 10) \implies y'(1) = 1/\ln 10$. Hence the equation of the tangent line to the graph at $x = 1$ is $y = (1/\ln 10)(x - 1)$.

40. $(1.10)^n P_0 = e^{nr}P_0 \implies e^{nr} = (1.10)^n \implies nr = n\ln 1.10 \implies r = \ln 1.10 \approx 0.0953$. Thus an interest rate of 10% compounded annually is equivalent to an interest rate of 9.53% compounded continuously.

41. $V(t) = (5000)2^{\sqrt{t}} = 5000e^{\sqrt{t}\ln 2} \implies V'(t) = 5000e^{\sqrt{t}\ln 2}\left(\frac{1}{2\sqrt{t}}\ln 2\right) = (5000)2^{\sqrt{t}}\left(\frac{\ln 2}{2\sqrt{t}}\right)$. Therefore

$$V'(4) = (5000)(2^2)\left(\frac{\ln 2}{4}\right) \approx 3465.74.$$

Hence after 4 years the value of the painting is increasing at the rate of \$3,465.74 per year.

42. $y = \dfrac{x(x + 1)(x + 2)}{(x + 3)(x + 4)(x + 5)}$

$$\implies \ln y = \ln x + \ln(x + 1) + \ln(x + 2) - \ln(x + 3) - \ln(x + 4) - \ln(x + 5)$$

$$\implies \frac{1}{y}\,y' = \frac{1}{x} + \frac{1}{x + 1} + \frac{1}{x + 2} - \frac{1}{x + 3} - \frac{1}{x + 4} - \frac{1}{x + 5}$$

$$\implies y' = y\left(\frac{1}{x} + \frac{1}{x + 1} + \frac{1}{x + 2} - \frac{1}{x + 3} - \frac{1}{x + 4} - \frac{1}{x + 5}\right).$$

43. $y = \sqrt[3]{\dfrac{x + 2}{x + 3}}$

$$\implies \ln y = \frac{1}{3}(\ln(x + 2) - \ln(x + 3))$$

$$\implies \frac{1}{y}\,y' = \frac{1}{3}\left(\frac{1}{x + 2} - \frac{1}{x + 3}\right)$$

$$\implies y' = \frac{1}{3}y\left(\frac{1}{x + 2} - \frac{1}{x + 3}\right).$$

44. $y = \dfrac{\sqrt[3]{x^2+1}}{(x+1)(x+2)^2}$

$$\implies \ln y = \frac{1}{3}\ln(x^2+1) - \ln(x+1) - 2\ln(x+2)$$

$$\implies \frac{1}{y}y' = \frac{2x}{3(x^2+1)} - \frac{1}{x+1} - \frac{2}{x+2}$$

$$\implies y' = y\left(\frac{2x}{3(x^2+1)} - \frac{1}{x+1} - \frac{2}{x+2}\right).$$

45. $y = \dfrac{(x^2+2)^3(x-1)^5}{x\sqrt{x+1}\sqrt{x+2}}$

$$\implies \ln y = 3\ln(x^2+2) + 5\ln(x-1) - \ln x - \frac{1}{2}\ln(x+1) - \frac{1}{2}\ln(x+2)$$

$$\implies \frac{1}{y}y' = \frac{3(2x)}{x^2+2} + \frac{5}{x-1} - \frac{1}{x} - \frac{1}{2}\frac{1}{x+1} - \frac{1}{2}\frac{1}{x+3}$$

$$\implies y' = y\left(\frac{6x}{x^2+2} + \frac{5}{x-1} - \frac{1}{x} - \frac{1}{2(x+1)} - \frac{1}{2(x+2)}\right).$$

46. $y = x^{\sqrt{x+1}}$

$$\implies \ln y = \sqrt{x+1}\,\ln x$$

$$\implies \frac{1}{y}y' = \frac{1}{2}(x+1)^{-1/2}\ln x + \sqrt{x+1}\,\frac{1}{x}$$

$$\implies y' = y\left(\frac{\ln x}{2\sqrt{x+1}} + \frac{\sqrt{x+1}}{x}\right).$$

47. $y = (x^2+1)^x$

$$\implies \ln y = x\ln(x^2+1)$$

$$\implies \frac{1}{y}y' = x\frac{2x}{x^2+1} + \ln(x^2+1)$$

$$\implies y' = y\left(\frac{2x^2}{x^2+1} + \ln(x^2+1)\right).$$

48. Let r be any real number. Then $x^r = e^{r\ln x}$ and hence

$$\frac{d}{dx}(x^r) = \frac{d}{dx}(e^{r\ln x}) = e^{r\ln x}\left(\frac{r}{x}\right) = \frac{rx^r}{x} = rx^{r-1}.$$

49. (a) Since $x \to \infty \iff h \to 0^+$,

$$y = \lim_{x\to\infty}\left(1 + \frac{r}{x}\right)^x = \lim_{h\to 0^+}(1+rh)^{1/h}.$$

(b) $f(x) = \ln(1+rx) \implies$

$$f'(0) = \lim_{h\to 0^+}\frac{f(h)-f(0)}{h}$$

$$= \lim_{h \to 0^+} \frac{\ln(1 + rh)}{h}$$

$$= \lim_{h \to 0^+} \ln(1 + rh)^{1/h}$$

$$= \ln \lim_{h \to 0^+} (1 + rh)^{1/h} \quad \text{since ln is continuous}$$

$$= \ln y.$$

(c) Since $f'(x) = r/(1 + rx)$, $f'(0) = r$. Hence, by part (b), $\ln y = f'(0) = r$. Therefore $y = e^r$; *i.e.*,

$$\lim_{x \to \infty} \left(1 + \frac{r}{x}\right)^x = e^r.$$

50. (a) Casio: $f(10^{11}) = 2.718281828$
                         $f(1.3 \times 10^{11}) = 2.829217014$
                         $f(1.5 \times 10^{11}) = 1$
                         If $x > 1.5 \times 10^{11}$, then $f(x)$ is evaluated as 1.

       TI-81: $f(10^{11}) = 2.718281828$
                         $f(1.3 \times 10^{11}) = 2.829217014$
                         $f(1.5 \times 10^{11}) = 2.85765118$
                         $f(2 \times 10^{11}) = 2.718281828$
                         $f(2 \times 10^{12}) = 7.389056099$
                         If $x > 2 \times 10^{12}$, then $f(x)$ is evaluated as 1.

       HP-28S: $f(10^{11}) = 2.71828182845$
                         $f(1.3 \times 10^{11}) = 3.6692966676$
                         $f(1.5 \times 10^{11}) = 4.4816890703$
                         $f(2 \times 10^{11}) = 1$
                         If $x \geq 2 \times 10^{11}$, then $f(x)$ is evaluated as 1.

The HP-48SX is the same as the HP-28S.

(b) Casio: $f(10^{-11}) = 2.718281828$
                         $f(8 \times 10^{-12}) = 2.829217014$
                         $f(7.5 \times 10^{-12}) = 2.905677747$
                         If $x < 7.5 \times 10^{-12}$, then $f(x)$ is evaluated as 1.

       TI-81: $f(10^{-11}) = 2.718281828$
                           $f(8 \times 10^{-12}) = 2.718281828$
                         $f(7.5 \times 10^{-12}) = 2.905677747$
                         $f(10^{-12}) = 2.718281828$
                         $f(7.5 \times 10^{-13}) = 3.793667895$
                         If $x < 5 \times 10^{-13}$, then $f(x)$ is evaluated as 1.

       HP-28S: $f(10^{-11}) = 2.71828182845$
                         $f(8 \times 10^{-12}) = 3.49034295744$
                         $f(6 \times 10^{-12}) = 5.29449005044$
                         If $x < 5 \times 10^{-12}$, then $f(x)$ is evaluated as 1.

The HP-48SX is the same as the HP-28S.

51. (a) The intersection points are summarized in the table below on the left:

b	Intersection Points
1.5	1.5, 7.408765
2	2,4
2.5	2.5, 2.970287
3	2.478053, 3
4	2, 4
5	1.764922, 5

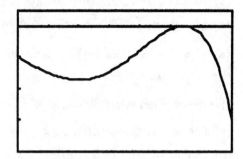

Notice that for each value of b there are 2 points of intersection. Also notice that the points get close between 2.5 and 3 and as b gets larger one of the points moves towards zero.

(b) Find b so that $f(x) = g(x)$ and $f'(x) = g'(x)$, i.e.,

$$x^b - b^x = 0, \text{ and } bx^{x-1} - b^x \ln b = 0.$$

Note that in part a the points are closer if b is taken between 2.5 and 3. A good value between 2.5 and 3 to check is $x = e$. If x is set equal to e we have $x^e - e^x = 0$ if and only if $x = e$ and $ex^{e-1} - e^x = x^e - e^x = 0$. So if $b = e$ then $f(x) = g(x)$ and $f'(x) = g'(x)$. Below is the graph of $F(x) = x^e - e^x$ using the range $[0, 3.5] \times [-4, .5]$.

8.5 Exponential Growth and Decay

1. $y = e^{2t}$.

2. $y = 2e^{-4t}$.

3. $y = y(0)e^{5t} \implies y(1) = y(0)e^5 = 1 \implies y(0) = e^{-5}$. Hence $y = e^{-5}e^{5t} = e^{5t-5}$.

4. $y = 2e^{kt} \implies y(2) = 2e^{2k} = 6 \implies k = (1/2)\ln 3$. Hence $y = 2e^{(t/2)\ln 3} = 2(3^{t/2})$.

5. $y = e^{kt} \implies y(1) = e^k = e^{-2} \implies k = -2$. Hence $y = e^{-2t}$.

6. $y = 0$.

7. False. To see this, let y_0 be the amount of isotope present at some time $t = 0$. Then $y = y_0e^{kt}$ and hence, if $t_{1/2}$ is the half-life of the isotope,

$$y(t_{1/2}) = y_0 e^{kt_{1/2}} = \frac{1}{2}y_0.$$

Therefore $e^{kt_{1/2}} = 1/2$. The solution of this equation for the half-life is $t_{1/2} = -(1/k)\ln 2$, which is independent of y_0. Thus the half-life is independent of the amount present.

8. Let y_0 be the intial amount of the isotope, $y(t)$ the amount after t days. Then $y(t) = y_0e^{kt}$. Since the half-life is 20 days, $y(20) = (1/2)y_0$. Therefore

$$y_0e^{20k} = \frac{1}{2}y_0 \implies 20k = \ln\frac{1}{2} = -\ln 2 \implies k = -\frac{\ln 2}{20} \implies y(t) = y_0e^{-t\ln 2/20}.$$

Moreover,

$$y(10) = 50 \implies y_0e^{-10\ln 2/20} = 50 \implies y_0 = 50e^{\ln 2/2} = 50\sqrt{2}.$$

Hence $y(t) = 50\sqrt{2}e^{-t\ln 2/20}$.

(a) Initial amount $= y_0 = 50\sqrt{2} \approx 70.71$ mg.

(b) Amount remaining after 30 days $= y(30) = 50\sqrt{2}e^{-30\ln 2/20} = 25$ mg.

(c) $y(t) = 50\sqrt{2}e^{-t\ln 2/20} = 0.1(50\sqrt{2}) \Longrightarrow -t(\ln 2)/20 = \ln 0.1 \Longrightarrow t \approx 66.4$ days.

9. (a) $v' = (-c/m)v \Longrightarrow v(t) = v_0 e^{-(c/m)t}$.

 (b) $v(t) = 10e^{-(40/10)t} = 10e^{-4t}$.

 (c) yes; $\lim_{t\to\infty} v(t) = \lim_{t\to\infty} 10e^{-4t} = 0$.

10. (a) $[A]' = -k[A] \Longrightarrow [A(t)] = [A_0]e^{-kt}$.

 (b) $[A_0] = 10 \Longrightarrow [A(t)] = 10e^{-kt}$. Moreover,

$$[A(20)] = 6 \Longrightarrow 10e^{-20k} = 6 \Longrightarrow -20k = \ln(6/10) \Longrightarrow k = -(1/20)\ln(3/5).$$

Hence $[A(t)] = 10e^{(t/20)\ln(3/5)}$. Thus, after 1 hour (60 seconds) the concentration is

$$[A(60)] = 10e^{(60/20)\ln(3/5)} \approx 2.16 \text{ moles/liter}.$$

11. Let $y(t)$ stand for the number of bacteria after t hours. Then $y(t) = 50e^{kt}$. Since $y(12) = 400$,

$$50e^{12k} = 400 \quad\Longrightarrow\quad 12k = \ln 8 \quad\Longrightarrow\quad k = \frac{1}{12}\ln 8.$$

Hence $y(t) = 50e^{(t/12)\ln 8}$.

 (a) $y(t) = 2(50) \Longrightarrow 50e^{(t/12)\ln 8} = 2(50) \Longrightarrow (t/12)\ln 8 = \ln 2 \Longrightarrow t = 12\frac{\ln 2}{\ln 8} = 4$ hours.

 (b) $y(16) = 50e^{(16/12)\ln 8} = 800$ bacteria.

12. $dp/dh = -0.116p \Longrightarrow p(h) = p_0 e^{-0.116h}$. If $p_0 = 1$, the pressure 8 km above sea level is $p(8) = e^{-0.116(8)}$ ≈ 0.395 atmospheres.

13. Let $R(t)$ be the present value of the revenue received from selling the land t years from now. Then

$$R(t) = V(t)e^{-0.1t} = 10000e^{\sqrt{t}\ln 1.2 - 0.1t}.$$

To maximize this, we set $R'(t) = 0$ and solve for t:

$$R'(t) = 10000e^{\sqrt{t}\ln 1.2 - 0.1t}\left(\frac{\ln 1.2}{2\sqrt{t}} - 0.1\right) = 0$$

$$\Longrightarrow \quad \frac{\ln 1.2}{2\sqrt{t}} - 0.1 = 0 \quad\Longrightarrow\quad \sqrt{t} = \frac{\ln 1.2}{2(0.1)} \approx 0.9116 \quad\Longrightarrow\quad t \approx 0.831 \text{ years} \approx 10 \text{ months}.$$

The original price of the land, \$10,000, does not affect this calculation.

14. (a) $dN/dt = bN - dN = (b-d)N \Longrightarrow N(t) = N_0 e^{(b-d)t}$.

 (b) If $b = 20/100 = 0.2$, $d = 12/100 = 0.12$, and $N_0 = 200$, then $N(t) = 200e^{0.08t}$. Thus, in 20 years the size of the population will be $N(20) = 200e^{0.08(20)} \approx 991$ individuals.

15. Let $y(t)$ be the amount of the substance present after t years. Then $y(t) = y_0 e^{kt}$. If 30% of the substance is excreted after 12 hours, then 70% remains and hence $y(12) = 0.7y_0$. Therefore

$$y_0 e^{12k} = 0.7y_0 \implies k = \frac{1}{12} \ln 0.7.$$

Hence $y(t) = y_0 e^{(t/12)\ln 0.7}$. To find the half-life $t_{1/2}$ of the substance, set $y(t_{1/2}) = \frac{1}{2}y_0$ and solve for $t_{1/2}$:

$$y_0 e^{(t_{1/2}/12)\ln 0.7} = \frac{1}{2}y_0 \implies \frac{t_{1/2}}{12}\ln 0.7 = \ln\frac{1}{2} \implies t_{1/2} = 12\frac{\ln 0.5}{\ln 0.7} \approx 23.32 \text{ hours.}$$

16. Let $y(t)$ be the amount of ^{14}C in the organism t years after it dies. Then $y(t) = y_0 e^{kt}$. Since the half-life of ^{14}C is 5760 years,

$$y(5760) = (1/2)y_0 \implies y_0 e^{5760k} = \frac{1}{2}y_0 \implies k = \frac{\ln 0.5}{5760}.$$

Hence $y(t) = y_0 e^{(t/5760)\ln 0.5}$.

(a) $y(t) = 0.8y_0 \implies y_0 e^{(t/5760)\ln 0.5} = 0.8y_0 \implies (t/5760)\ln 0.5 = \ln 0.8 \implies t = 5760(\ln 0.8)/(\ln 0.5)$
    ≈ 1854.31 years.

(b) $y(t) = 0.1y_0 \implies y_0 e^{(t/5760)\ln 0.5} = 0.1y_0 \implies (t/5760)\ln 0.5 = \ln 0.1 \implies t = 5760(\ln 0.1)/(\ln 0.5)$
    ≈ 19134.31 years.

17. Let $y(t)$ be the amount of the iodine isotope present in the body after t days. Then $y(t) = y_0 e^{kt}$. If the half-life is 8 days, then $y(8) = (1/2)y_0$ and hence

$$y_0 e^{8k} = \frac{1}{2}y_0 \implies 8k = \ln 0.5 \implies k = \frac{\ln 0.5}{8}.$$

Therefore $y(t) = y_0 e^{(t/8)\ln 0.5}$. Thus, if $y_0 = 50$, the amount remaining after 3 weeks (21 days) is $y(21)$ $= 50e^{(21/8)\ln 0.5} \approx 8.11$ micrograms.

18. Let $y(t)$ be the amount of undissolved chemical after t minutes. Then $y(t) = y_0 e^{kt}$. If $y_0 = 20$ and $y(5) = 10$, then

$$20e^{5k} = 10 \implies 5k = \ln 0.5 \implies k = \frac{\ln 0.5}{5}.$$

Hence $y(t) = 20e^{(t/5)\ln 0.5}$. When 90% of the chemical is dissolved, 10% remains. Therefore $y(t) = 0.1(20) = 2$ and hence

$$20e^{(t/5)\ln 0.5} = 2 \implies \frac{t}{5}\ln 0.5 = \ln 0.1 \implies t = 5\frac{\ln 0.1}{\ln 0.5} \approx 16.61 \text{ minutes.}$$

19. $T(t) = T_e + (T_0 - T_e)e^{-kt} \implies dT/dt = (T_0 - T_e)e^{-kt}(-k) = -k(T - T_e)$ and $T(0) = T_e + (T_0 - T_e)e^0$ $= T_0$. As $t \to \infty$,

$$\lim_{t\to\infty} T(t) = \lim_{t\to\infty} \left[T_e + (T_0 - T_e)e^{-kt}\right] = T_e.$$

20. If $T(t_1)$ is known, $T(t_1) = T_e + (T_0 - T_e)e^{-kt_1}$ and hence

$$e^{-kt_1} = \frac{T(t_1) - T_e}{T_0 - T_e} \implies -kt_1 = \ln\frac{T(t_1) - T_e}{T_0 - T_e} \implies k = -\frac{1}{t_1}\ln\frac{T(t_1) - T_e}{T_0 - T_e}.$$

21. $T_e = 20$, $T_0 = 100$, $t_1 = 10$, $T(10) = 50 \implies$

$$k = -\frac{1}{10} \ln \frac{50 - 20}{100 - 20} = -\frac{1}{10} \ln \frac{3}{8} \implies T(t) = 20 + 80 e^{(t/10)(\ln 3/8)}.$$

To find the time to cool to $30°C$, set $T(t) = 30$ and solve for t:

$$20 + 80 e^{(t/10)(\ln 3/8)} = 30 \implies \frac{t}{10} \ln \frac{3}{8} = \ln \frac{1}{8} \implies t = 10 \frac{\ln 1/8}{\ln 3/8} \approx 21.2 \text{ minutes.}$$

22. $T(60) = 20 + 80 e^{(60/10)(\ln 3/8)} \approx 20.22°C.$

23. $T_e = 0$, $T_0 = 100$, $T(5) = 50 \implies k = -(1/5)\ln 0.5 \implies T(t) = 100 e^{(t/5)\ln 0.5}$. To find the time to cool to $30°C$, set $T(t) = 30$ and solve for t:

$$100 e^{(t/5)\ln 0.5} = 30 \implies \frac{t}{5} \ln 0.5 = \ln \frac{3}{10} \implies t = 5 \frac{\ln 0.3}{\ln 0.5} \approx 8.68 \text{ minutes.}$$

After 1 hour, the temperature will be $T(60) = 100 e^{(60/5)\ln 0.5} \approx 0.024°C.$

24. $T_e = 100$, $T_0 = 20$, $T(10) = 30 \implies k = -(1/10)(\ln 7/8) \implies T(t) = 100 - 80 e^{(t/10)(\ln 7/8)}$. To find the time to warm to $50°C$, set $T(t) = 50$ and solve for t:

$$100 - 80 e^{(t/10)(\ln 7/8)} = 50 \implies \frac{t}{10} \ln \frac{7}{8} = \ln \frac{5}{8} \implies t = 10 \frac{\ln 5/8}{\ln 7/8} \approx 35.2 \text{ minutes.}$$

25. Let $R(t)$ be the present value of the revenue received from selling the wine t years from now. Then

$$R(t) = V(t) e^{-0.125t} = 100(1.5)^{\sqrt{t}} e^{-0.125t} = 100 e^{\sqrt{t}\ln 1.5 - 0.125t}.$$

To maximize revenue, set $R'(t) = 0$ and solve for t:

$$R'(t) = 100 e^{\sqrt{t}\ln 1.5 - 0.125t} \left(\frac{\ln 1.5}{2\sqrt{t}} - 0.125 \right) = 0$$

$$\implies \frac{\ln 1.5}{2\sqrt{t}} - 0.125 = 0 \implies \sqrt{t} = \frac{\ln 1.5}{2(0.125)} \approx 1.622 \implies t \approx 2.63 \text{ years.}$$

26.   (a) $P(1) = 100 e^{(0.1)(1)} \approx \$100.52.$

    (b) If we invest \$100 at 15% interest compounded continuously, the yield after one year is $P(1) = 100 e^{(0.15)(1)} \approx \$116.18$. Thus, the effective annual yield is 16.18%.

27. $P(1) = P_0 e^{r(1)} = 0.08 P_0 \implies r = \ln 1.08 \approx .077$, or 7.7% compounded continuously.

28. Let P_0 be the amount of money. Then

$$P(8) = P_0 e^{(0.1)(8)} = 2500 \implies P_0 = 2500 e^{-0.8} \approx \$1123.32.$$

29.   (a) $P(1) = P_0 \left(1 + \frac{0.1}{2}\right)^2 = P_0(1 + r) \implies r = 0.1025$, or 10.25%.

    (b) $P(1) = P_0 \left(1 + \frac{0.1}{4}\right)^4 = P_0(1 + r) \implies r = 0.1038$, or 10.38%.

    (c) $P(1) = P_0 \left(1 + \frac{0.1}{365}\right)^{365} = P_0(1 + r) \implies r = 0.105156$, or 10.5156%.

    (d) $P(1) = P_0 e^{0.1} = P_0(1 + r) \implies r = e^{0.1} - 1 = 0.10517$ or 10.517%.

30. Let t be the time to double the investment. Then $P(t) = P_0 e^{0.1t} = 2P_0 \Longrightarrow 0.1t = \ln 2 \Longrightarrow t = 6.93$ years.

31. $P(1) = P_0 e^r = P_0(1 + i) \Longrightarrow i = e^r - 1$.

32. Present value at 10% interest compounded continuously $= 1000e^{-0.1(5)} \approx \$606.53$.
    Present value at 10% effective annual yield $= 1000(1.1)^{-5} = \$620.92$

33. False. The equation $P'(t) = 0.05P(t)$ means that at each moment the growth rate is 5%, which is false. The 5% growth rate refers to the effective yearly growth rate. To find the continuous growth rate r which yields an annual 5% growth, set $P(t) = P_0 e^{rt}$. Then $P(1) = P_0 e^r = (1 + 0.05)P_0 \Longrightarrow r = \ln 1.05 \approx 0.0488$. Hence $P(t) = P_0 e^{r \ln 1.05} = P_0(1.05)^r$ and therefore the correct equation is $P'(t) = P_0 e^{r \ln 1.05}(\ln 1.05) = (\ln 1.05)P(t)$. In particular, a continuous growth rate of 4.88% is equivalent to a yearly growth rate of 5%.

34. $k = \ln 1.05$; see Exercise 33.

35. (a) If $y = f(t)$ is a differentiable solution of the equation $y' = ky$, then $f'(t) = kf(t)$ for all t.
    (b) $g'(t) = f(t)(-ke^{-kt}) + e^{-kt}f'(t) = -kye^{-kt} + (ky)e^{-kt} = 0$.
    (c) Since $g'(t) = 0$ for all t, $g(t) = C$ for some constant C. Hence $f(t) = Ce^{kt}$.

36. After 1 year, \$1 at 100% interest compounded continuously will yield

$$e = \lim_{x \to \infty} \left(1 + \frac{1}{x}\right)^x$$

dollars.

37. (a) Graph of the pairs $(\ln a, \ln A)$ using the range $[-1, 2] \times [0, 4]$.

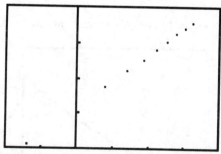

The graph of the pairs $(\ln a, \ln A)$ appears to be a straight line, therefore $\ln A = m \ln a + b$. Exponentiating gives $A = e^{m \ln a + b} = e^b e^{\ln a^m} = ca^m$ or $A = ca^r$ where $c = e^b$ and $m = r$.
(b) Slope: $m = \frac{\ln 35.1241 - \ln 1.1107}{\ln 5 - \ln 0.5} = 1.500001$
    Intercept: $b = \ln 1.1107 - 1.500001 \ln 0.5 = 1.44711911$.
    Hence $A = e^b a^{1.500001}$ or $A = 3.1415a^{1.5}$.

38. If $A = P(1 + r/n)^{nt}$ then when $A = 2P$ we have $(1 + r/n)^{nt} = 2$. Taking logarithms gives $nt \ln(1 + r/n) = \ln 2$. Hence

$$t = \frac{\ln 2}{n \ln(1 + r/n)}.$$

(a) Graph $Y = \ln 2/(N \ln (1+X/N))$ using the range $[0.05, .13] \times [0, 14]$ which is the range for the most common interest rates and $n = 1, 4,$ and 365. Little difference is seen between the graphs.

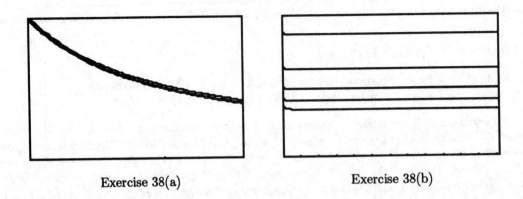

Exercise 38(a)   Exercise 38(b)

(b) Graph **Y=ln 2/(X ln (1+R/X))** using the range $[1, 365] \times [0, 16]$ and $r =0.05, 0.07, 0.09, 0.11,$ and 0.13. We see from these graphs that the doubling time is affected significantly by the interest rate no matter what the compounding period. The effect though decreases as the interest rate r increases.

(c) It appears that the interest rate has more effect on the doubling time than the compounding period.

39. (a) If $N(t) = N_0 e^{kt}$ then $N(0) = 75.995 = N_0 e^{k(0)}$. So $N_0 = 75.995$ and so $N(t) = 75.995 e^{kt}$. Since $N(8) = 226.505$, then choose k so that $226.505 = 75.995 e^{8k}$. So $e^{8k} = 2.980525$ and thus $k = \frac{1}{8} \ln(2.980525) = 0.1365124$. Therefore

$$N(t) = 75.995 e^{0.1365124t}.$$

Graph this function and the values in the table using the range $[0, 10] \times [75, 350]$.

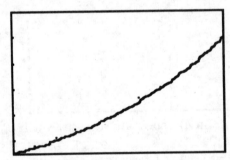

For the year 1990 this model predicts the population to be $N(9) = 259.663$ and for the year 2000, $N(10) = 297.612$. These appear to be fairly resonable estimates.

(b) If $N(0) = 5.308$ then $N_0 = 5.308$ and so $N(t) = 5.308 e^{kt}$. At $t = 11$ (the year 1900) we choose k so than $75.995 = 5.308 e^{11k}$. Thus $k = \frac{1}{11} \ln(75.995/5.308) = 0.2419502$. Therefore $N(t) = 5.308 e^{0.2419502t}$. The graph of this function along with the values in the table using the range $[11, 21] \times [75, 35]$ are shown below on the left. It appears that this model is worse than the one in part a. If instead we choose k so that at $t = 19$ (the year 1980) $5.308 e^{19k} = 226.505$, then

$$k = \frac{1}{19} \ln(226.505/5.308) = 0.1975554.$$

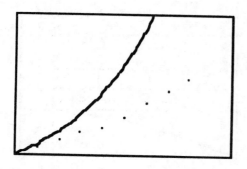 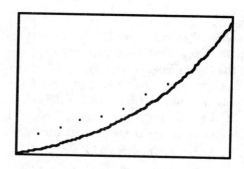

The graph of $N(t) = 5.308e^{0.1975554t}$ using the range $[11, 21] \times [45, 350]$ is shown above on the right. This model doesn't appear to be a very good predictor either, except for the later part of this century.

40. (a) Substituting this expression for b into equation (2), we find that

$$\frac{N_4 N_0 (N_0 - N_2 e^{-2k})}{N_0 N_2 (1 - e^{-2k})} + N_4 e^{-4k} - \frac{N_4 N_0 (N_0 - N_2 e^{-2k})}{N_0 N_2 (1 - e^{-2k})} = N_0.$$

Now, set $x = e^{-2k}$. Then $x^2 = e^{-4k}$ and hence

$$\frac{N_4 N_0 (N_0 - N_2 x)}{N_0 N_2 (1-x)} + N_4 x^2 - \frac{N_4 N_0 (N_0 - N_2 x)}{N_0 N_2 (1-x)} = N_0$$

$$\implies \quad N_4 (N_0 - N_2 x) + N_4 x^2 [N_2 (1-x)] - N_4 (N_0 - N_2 x) x^2 = N_0 N_2 (1-x)$$

$$\implies \quad N_4 (N_0 - N_2 x)(1 - x^2) + N_2 N_4 x^2 (1-x) = N_0 N_2 (1-x)$$

$$\implies \quad (1-x) \left[N_4 (N_0 - N_2 x)(1+x) + N_2 n_4 x^2 - N_0 N_2 \right] = 0$$

$$\implies \quad (1-x) \left[N_0 (N_4 - N_2) - x N_4 (N_2 - N_0) \right] = 0.$$

Finally, if $k > 0$, then $1 - x = 1 - e^{-2k} \neq 0$ and hence

$$N_0 (N_4 - N_2) - x N_4 (N_2 - N_0) = 0 \quad \implies \quad x = e^{-2k} = \frac{N_0 (N_4 - N_2)}{N_4 (N_2 - N_0)}.$$

Since, from the population data in Exercise 24, Chapter 3, Section 3, $N_0 = 75.995$, $N_2 = 105.711$ and $N_4 = 131.669$,

$$e^{-2k} = \frac{(75.995)(131.669 - 105.711)}{131.669(105.711 - 75.995)} = 0.504176$$

$$\implies \quad k = -\tfrac{1}{2} \ln(0.504176) = 0.342415$$

$$\implies \quad b = \frac{N_0 - N_2 e^{-2k}}{N_0 N_2 (1 - e^{-2k})} = \frac{75.995 - 105.711(0.504176)}{(75.995)(105.711)(1 - 0.504176)} = 0.005698.$$

Therefore

$$N(t) = \frac{75.995}{0.433053 + 0.566947 e^{-0.342415t}}.$$

The graph of $N(t)$ with range $[0, 10] \times [75, 350]$ and the population data values, shown to the right, indicates that this model is not a good predictor of the U.S. population for this century. Note that $N(9) = 165.543$ and $N(10) = 168.309$. These values are smaller than the populations of 1960, 1970, and 1980. These numbers are not realistic for the population in the years 1990 and 2000.

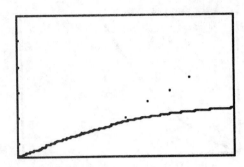

(b) Since $k > 0$, $\lim\limits_{t \to +\infty} e^{-kt} = 0$ and hence

$$\lim_{t \to +\infty} N(t) = \frac{75.995}{0.433053} = 175.487.$$

This value is again smaller than the populations in 1960, 1970, and 1980. We conclude that this model is not a plausible description of the growth of the population of the United States for this or any other century.

(c) Let $N_0 = 75.995$, $N_6 = 179.323$, and $N_8 = 225.505$. Then, as in part (a),

$$b = \frac{N_0 - N_6 e^{-6k}}{N_0 N_6 (1 - e^{-6k})}.$$

If we set $x = e^{-6k}$, then $x^{4/3} = e^{-8k}$ and hence

$$\frac{N_8(N_0 - N_6 x)}{N_6(1 - x)} + N_8 x^{4/3} - \frac{N_8(N_0 - N_6 x) x^{4/3}}{N_6(1 - x)} = N_0$$

or

$$N_8(N_0 - N_6 x) + N_8 x^{4/3}(N_6(1 - x) - N_8(N_0 - N_6 x) x^{4/3} - N_0 N_6(1 - x) = 0.$$

Simplifying, we obtain the equation

$$N_8(N_0 - N_6 x)(1 - x^{4/3}) + N_6 N_8 x^{4/3}(1 - x) - N_0 N_6(1 - x) = 0.$$

The graph of the function

$$\textbf{C(A-BX)(1-X\^{} (4/3))+BCX\^{} (4/3)(1-X)-AB(1-X)}$$

with **A**=75.995, **B**=179.323, and **C** =226.505 using the range $[0, 2] \times [-1000, 7000]$ is shown below on the left.

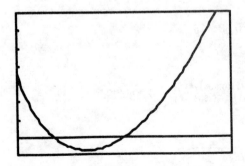
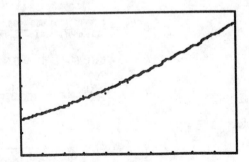

We see that $x = 1$ is a root, but this implies that $e^{-6k} = 1$ so $k = 0$. If $k = 0$ then b is not defined. There is also another root in the interval $[0, 1]$. Magnifying the graph at this point we see that $x = 0.33299263$. So $e^{-6k} = 0.33299263$ and therefore $k = -\frac{1}{6}\ln(0.33299263) = 0.1832725$ and

$$b = \frac{75.995 - 179.323(0.33299263)}{(75.995)(179.323)(1 - 0.33299263)} = 0.00179122.$$

Therefore

$$N(t) = \frac{75.995}{0.1361240 + 0.8638760e^{-0.1832725x}}.$$

The graph of $N(t)$ and the values from the data table using the range $[0, 10] \times [0, 300]$ is shown above on the right. This appears to be quite a good model. For this model, $N(9) = 251.537$ and $N(10) = 277.026$ which seem to be quite plausible. The carrying capacity for this model is

$$\lim_{t \to +\infty} N(t) = \frac{75.995}{0.1361240} = 558.278.$$

If a person is 20 years old in 1991 then in in the year 2036, when the person is 65 years old, this model predicts the population to be $N(13.6) = 366.120$ million. (When $t = 13.6$ this means 136 years from 1900 which is 2036.) The value of 366.120 is 65.6% of the carrying capacity. For

$$N(t) = \frac{N_0}{N_0 b + e^{-kt}(1 - N_0 b)}$$

$$N'(t) = \frac{N_0(1 - N_0 b)ke^{-kt}}{(N_0 b + e^{-kt}(1 - N_0 b))^2}$$

and

$$N''(t) = \frac{2N_0(1 - N_0 b)^2 k^2 e^{-2kt}}{(N_0 b + e^{-kt}(1 - N_0 b))^3} + \frac{N_0(1 - N_0 b)k^2 e^{-kt}}{(N_0 b + e^{-kt}(1 - N_0 b))^2}.$$

Below are the graphs of $N'(t)$ and $N''(t)$. On the left is N' using the range $[0, 40] \times [10, 30]$. On the right is the graph of N'' using the range $[0, 40] \times [-2, 2]$.

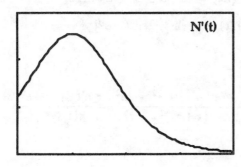
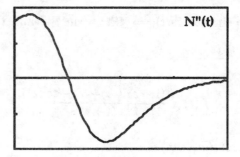

We see that N' is at a maximum for $t = 10.11$, or in the year 2001. So the fastest population growth predicted by this model will be in the year 2001.

Review Exercises – Chapter 8

1. (a) $8^{2/3} = 4$

   (b) $36^{-5/2} = 6^{-5} = 1.286 \times 10^{-4}$

(c) $16^{5/4} = 32$

(d) $4^{-3/4} = 2^{-3/2} = 1/(2\sqrt{2})$

2. (a) $\log_2 16 = 4$

   (b) $\log_8 16 = \log_8 8 + \log_8 2 = 1 + \log_8 8^{1/3} = 4/3$

   (c) $\ln e^2 = 2$

   (d) $\log_2 3\sqrt{2} = \log_2 3 + 1/2 = \ln 3/\ln 2 + 0.5 = 2.08$

3. (a) $x = y + 2 \Longrightarrow y = f^{-1}(x) = x - 2$

   (b) $x = y^2 \Longrightarrow y = g^{-1}(x) = \sqrt{x}$

   (c) $x = 2y + 1 \Longrightarrow y = (1/2)x - 1/2$

   (d) $x = \sqrt{y} \Longrightarrow y = x^2, \; x \geq 0$

4. (a) $3\ln x - \ln 3x = 0 \Longrightarrow \ln(x^3/3x) = 0 \Longrightarrow \frac{1}{3}x^2 = 1 \Longrightarrow x = \sqrt{3}$

   (b) $\ln x^3 = 3 \Longrightarrow 3\ln x = 3 \Longrightarrow \ln x = 1 \Longrightarrow x = e$

   (c) $\int_1^x \frac{2}{t}\,dt = 4 \Longrightarrow 2\ln x = 4 \Longrightarrow \ln x = 2 \Longrightarrow x = e^2$

   (d) $\int_2^x \frac{1}{t}\,dt = 3 + \ln 2 \Longrightarrow \int_1^x \frac{1}{t}\,dt - \int_1^2 \frac{1}{t}\,dt = 3 + \ln 2 \Longrightarrow \ln x - \ln 2 = 3 + \ln 2 \Longrightarrow \ln(x/4) = 3 \Longrightarrow$
   $x/4 = e^3 \Longrightarrow x = 4e^3$

5. $y = x^2 \ln(x - a) \Longrightarrow y' = x^2 \dfrac{1}{x - a} + 2x\ln(x - a)$

6. $f(x) = xe^{1-x} \Longrightarrow f'(x) = xe^{1-x}(-1) + e^{1-x} = e^{1-x}(1 - x)$

7. $f(x) = \ln(\ln^2 x) = 2\ln(\ln x) \Longrightarrow f'(x) = 2\dfrac{1}{\ln x}\dfrac{1}{x} = \dfrac{2}{x\ln x}$

8. $y = (2\ln\sqrt{x})^3 = (\ln x)^3 \Longrightarrow y' = 3(\ln x)^2(1/x)$

9. $y = e^{x^2}\tan 2x \Longrightarrow y' = e^{x^2}(\sec 2x)(2) + (\tan 2x)e^{x^2}(2x) = 2e^{x^2}[\sec^2 2x + x\tan 2x]$

10. $f(t) = \sin^2 t\cos(\ln t) \Longrightarrow f'(t) = \sin^2 t[-\sin(\ln t)](1/t) + \cos(\ln t)[2\sin t\cos t]$

11. $f(t) = \dfrac{te^t}{1 + e^t} \Longrightarrow$

$$f'(t) = \frac{(1 + e^t)(te^t + e^t) - te^t(e^t)}{(1 + e^t)^2} = \frac{te^t + te^{2t} + e^t + e^{2t} - te^{2t}}{(1 + e^t)^2} = \frac{e^t(t + 1 + e^t)}{(1 + e^t)^2}$$

12. $y = x\ln(\sqrt{x} - e^{-x}) \Longrightarrow$

$$y' = x\,\frac{1}{\sqrt{x} - e^{-x}}\left(\frac{1}{2\sqrt{x}} + e^{-x}\right) + \ln(\sqrt{x} - e^{-x})$$

13. $y = \dfrac{\ln(t - a)}{\ln(t^2 + b)} \Longrightarrow$

$$y' = \frac{[\ln(t^2 + b)]\left(\frac{1}{t-a}\right) - [\ln(t - a)]\left(\frac{1}{t^2+b}\right)(2t)}{[\ln(t^2 + b)]^2}$$

14. $f(t) = \ln\cos t \implies f'(t) = \dfrac{1}{\cos t}(-\sin t) = -\tan t$

15. $f(t) = e^{\sqrt{t}-\ln t} \implies f'(t) = e^{\sqrt{t}-\ln t}\left(\dfrac{1}{2\sqrt{t}} - \dfrac{1}{t}\right) = e^{\sqrt{t}-\ln t}\left(\dfrac{\sqrt{t}-2}{2t}\right) = \left(\dfrac{\sqrt{t}-2}{2t^2}\right)e^{\sqrt{t}}$

16. $y = \dfrac{1}{a}\ln\dfrac{a+bx}{x} = \dfrac{1}{a}[\ln(a+bx) - \ln x] \implies y' = \dfrac{1}{a}\left[\dfrac{b}{a+bx} - \dfrac{1}{x}\right] = \dfrac{-1}{x(a+bx)}$

17. $y = \dfrac{1}{3x} + \dfrac{1}{4}\ln\left(\dfrac{1-2x}{\sqrt{x}}\right) = \dfrac{1}{3}x^{-1} + \dfrac{1}{4}\ln(1-2x) - \dfrac{1}{8}\ln x$

$$\implies \quad y' = -\dfrac{1}{3}x^{-2} + \dfrac{1}{4}\dfrac{1}{1-2x}(-2) - \dfrac{1}{8x} = -\dfrac{1}{3x^2} - \dfrac{1}{2(1-2x)} - \dfrac{1}{8x} = \dfrac{-6x^2 + 13x - 8}{24x^2(1-2x)}$$

18. $f(t) = \sin(\ln t + te^{\sqrt{t}})$

$$\implies \quad f'(t) = [\cos(\ln t + te^{\sqrt{t}})]\left[\dfrac{1}{t} + te^{\sqrt{t}}\dfrac{1}{2\sqrt{t}} + e^{\sqrt{t}}\right] = \left(\dfrac{1}{t} + \dfrac{\sqrt{t}e^{\sqrt{t}}}{2} + e^{\sqrt{t}}\right)\cos(\ln t + te^{\sqrt{t}})$$

19. $x\ln y^2 + y\ln x = 1 \implies 2x\ln y + y\ln x = 1$

$$\implies (2x)\left(\dfrac{1}{y}\,y'\right) + (\ln y)(2) + y\left(\dfrac{1}{x}\right) + (\ln x)y' = 0$$

$$\implies y'\left(\dfrac{2x}{y} + \ln x\right) = -2\ln y - \dfrac{y}{x}$$

$$\implies y' = \dfrac{-2\ln y - \frac{y}{x}}{\frac{2x}{y} + \ln x} = \dfrac{-2xy\ln y - y^2}{2x^2 + xy\ln x}$$

20. $y = e^{4x}(e^{-x} + \ln x)^2$

$$\implies y' = e^{4x}\left[2(e^{-x} + \ln x)\left(-e^{-x} + \dfrac{1}{x}\right)\right] + (e^{-x} + \ln x)^2 e^{4x}(4)$$

$$= e^{4x}(e^{-x} + \ln x)\left[2\left(-e^{-x} + \dfrac{1}{x}\right) + 4(e^{-x} + \ln x)\right]$$

$$= 2e^{4x}(e^{-x} + \ln x)\left(e^{-x} + \dfrac{1}{x} + 2\ln x\right)$$

21. $x^2 + e^{xy} - y^2 = 2 \implies 2x + e^{xy}(xy' + y) - 2yy' = 0 \implies y'(xe^{xy} - 2y) = -2x - ye^{xy} \implies y' = \dfrac{2x + ye^{xy}}{2y - xe^{xy}}$

22. $y = xe^{1-\sqrt{x}\ln x}$

$$\implies y' = xe^{1-\sqrt{x}\ln x}\left[-\sqrt{x}\dfrac{1}{x} - (\ln x)\dfrac{1}{2\sqrt{x}}\right] + e^{1-\sqrt{x}\ln x}$$

$$= e^{1-\sqrt{x}\ln x}\left[x\left(-\dfrac{1}{\sqrt{x}} - \dfrac{\ln x}{2\sqrt{x}}\right) + 1\right]$$

$$= e^{1-\sqrt{x}\ln x}\left(-\sqrt{x} - \dfrac{1}{2}\sqrt{x}\ln x + 1\right)$$

23. $e^{xy} = \sqrt{xy}$

$$\implies \quad e^{xy}(xy' + y) = \sqrt{x}\,\frac{1}{2\sqrt{y}}\,y' + \sqrt{y}\,\frac{1}{2\sqrt{x}}$$

$$\implies \quad y'\left[xe^{xy} - \frac{\sqrt{x}}{2\sqrt{y}}\right] = \frac{\sqrt{y}}{2\sqrt{x}} - ye^{xy}$$

$$\implies \quad y' = \frac{\frac{\sqrt{y}}{2\sqrt{x}} - ye^{xy}}{xe^{xy} - \frac{\sqrt{x}}{2\sqrt{y}}} = \frac{y - 2y\sqrt{x}\sqrt{y}e^{xy}}{2x\sqrt{x}\sqrt{y}e^{xy} - x} = \frac{y(1 - 2e^{2xy})}{x(2e^{2xy} - 1)} = -\frac{y}{x}$$

24. $y = \sqrt{e^{-x} + e^x} \implies y' = \frac{1}{2}(e^{-x} + e^x)^{-1/2}(-e^{-x} + e^x) = \frac{e^x - e^{-x}}{2\sqrt{e^x + e^{-x}}}$

25. $y = \ln\sqrt{e^{2x} + \sin\sqrt{x}} = (1/2)\ln(e^{2x} + \sin\sqrt{x}) \implies$

$$y' = \frac{1}{2}\,\frac{1}{e^{2x} + \sin\sqrt{x}}\left(2e^{2x} + \frac{\cos\sqrt{x}}{2\sqrt{x}}\right)$$

26. $y = \int_1^{\sin x} \ln(t+1)\,dt \implies y' = [\ln(\sin x + 1)]\cos x$

27. Let $u = 1 - x^2$. Then $du = -2x\,dx$ and hence

$$\int \frac{x\,dx}{1 - x^2} = -\frac{1}{2}\int \frac{du}{u} = -\frac{1}{2}\ln|u| + C = -\frac{1}{2}\ln|1 - x^2| + C.$$

28. Let $u = 1 - x^2$. Then $du = -2x\,dx$ and hence

$$\int xe^{1-x^2}\,dx = -\frac{1}{2}\int e^u\,du = -\frac{1}{2}e^u + C = -\frac{1}{2}e^{1-x^2} + C.$$

29. Let $u = 4x + 2x^2$. Then $du = (4 + 4x)dx = 4(1 + x)dx$ and hence

$$\int \frac{x + 1}{4x + 2x^2}\,dx = \frac{1}{4}\int \frac{du}{u} = \frac{1}{4}\ln|u| + C = \frac{1}{4}\ln|4x + 2x^2| + C.$$

30. Let $u = \ln(x + 1)$. Then $du = \dfrac{1}{x + 1}\,dx$ and hence

$$\int \frac{dx}{(x + 1)\ln^3(x + 1)} = \int \frac{du}{u^3} = -\frac{1}{2}\frac{1}{u^2} + C = -\frac{1}{2}\frac{1}{\ln^2(x + 1)} + C.$$

31. $\int \sqrt{e^x}\,dx = \int e^{\frac{1}{2}x}\,dx = 2e^{\frac{1}{2}x} + C = 2\sqrt{e^x} + C.$

32. $\int (e^x + 1)^2\,dx = \int (e^{2x} + 2e^x + 1)\,dx = (1/2)e^{2x} + 2e^x + x + C.$

33. Let $u = \ln x$. Then $du = (1/x)dx$ and hence

$$\int_e^{e^2} \frac{1}{x\sqrt{\ln x}}\,dx = \int_1^2 u^{-1/2}\,du = 2\sqrt{u}\Big|_1^2 = 2\sqrt{2} - 2.$$

34. Let $u = e^x + e^{-x}$. Then $du = (e^x - e^{-x})dx$ and hence

$$\int \frac{e^x - e^{-x}}{e^x + e^{-x}}\, dx = \int \frac{du}{u} = \ln|u| + C = \ln(e^x + e^{-x}) + C.$$

Note that we need not use the absolute value in the antiderivative, *i.e.*, $\ln|e^x + e^{-x}|$, since $e^x + e^{-x}$ is always positive.

35. $\displaystyle\int_0^1 \frac{x^3 - 1}{x + 1}\, dx = \int_0^1 \left(x^2 - x + 1 - \frac{2}{x+1}\right)\, dx = \left[\frac{1}{3}x^3 - \frac{1}{2}x^2 + x - 2\ln(x+1)\right]\Big|_0^1 = \frac{5}{6} - 2\ln 2.$

36. Let $u = \sqrt{x}$. Then $du = 1/(2\sqrt{x})dx$ and hence

$$\int_1^4 \frac{e^{\sqrt{x}}}{\sqrt{x}}\, dx = 2\int_1^2 e^u\, du = 2e^u\big|_1^2 = 2e(e - 1).$$

37. Let $u = x^2$. Then $du = 2x\,dx$ and hence

$$\int_0^2 x(e^{x^2} + 1)\, dx = \frac{1}{2}\int_0^4 (e^u + 1)\, du = \frac{1}{2}(e^u + u)\Big|_0^4 = \frac{1}{2}e^4 + \frac{3}{2}.$$

38. Let $u = 1 + \sqrt{x}$. Then $du = (1/2\sqrt{x})dx$ and hence

$$\int_4^9 \frac{1}{\sqrt{x}(1 + \sqrt{x})}\, dx = 2\int_3^4 \frac{1}{u} = 2\ln u\big|_3^4 = 2\ln 4 - 2\ln 3 = 2\ln\frac{4}{3}.$$

39. $\int e^x(1 - e^{2x})^3\, dx = \int e^x(1 - 3e^{2x} + 3e^{4x} - e^{6x})\, dx = \int(e^x - 3e^{3x} + 3e^{5x} - e^{7x})\, dx = e^x - e^{3x} + \frac{3}{5}e^{5x} - \frac{1}{7}e^{7x} + C.$

40. Let $u = \sin\sqrt{x}$. Then $du = (\cos\sqrt{x})/(2\sqrt{x})\, dx$ and hence

$$\int \frac{\cot\sqrt{x}}{\sqrt{x}}\, dx = \int \frac{\cos\sqrt{x}}{\sqrt{x}\sin\sqrt{x}}\, dx = 2\int \frac{du}{u} = 2\ln|u| + C = 2\ln|\sin\sqrt{x}| + C.$$

41. Let $u = x^3 + x^2 - 7$. Then $du = (3x^2 + 2x)dx$ and hence

$$\int \frac{2x + 3x^2}{x^3 + x^2 - 7}\, dx = \int \frac{du}{u} = \ln|u| + C = \ln|x^3 + x^2 - 7| + C.$$

42. $\int_1^{\ln 2}(e^x + 1)(e^x - 1)\, dx = \int_1^{\ln 2}(e^{2x} - 1)\, dx = \frac{1}{2}e^{2x} - x)\big|_1^{\ln 2} = 3 - \ln 2 - \frac{1}{2}e^2.$

43. Let $u = x^2 + 1$. Then $du = 2x\,dx$ and hence

$$\int_{-3}^{-2} \frac{x}{x^2 + 1}\, dx = \frac{1}{2}\int_{10}^5 \frac{du}{u} = \frac{1}{2}\ln u\Big|_{10}^5 = \frac{1}{2}\ln 5 - \frac{1}{2}\ln 10 = \frac{1}{2}\ln\frac{1}{2} = -\frac{\ln 2}{2}.$$

44. $\int_0^1 \left(\frac{1}{x+1} - \frac{1}{x+2}\right)\, dx = [\ln(x+1) - \ln(x+2)]\big|_0^1 = \ln\frac{4}{3}.$

45. $y = x^{\sqrt{x}} \Longrightarrow \ln y = \sqrt{x}\ln x \Longrightarrow \frac{1}{y}y' = \sqrt{x}\left(\frac{1}{x}\right) + (\ln x)\left(\frac{1}{2\sqrt{x}}\right) = \frac{1}{\sqrt{x}} + \frac{\ln x}{2\sqrt{x}} \Longrightarrow y' = y\left(\frac{1}{\sqrt{x}} + \frac{\ln x}{2\sqrt{x}}\right).$

46. $y = \sqrt{\dfrac{(x-1)^2(x+1)}{x(x+3)^3}}$

$$\implies \ln y = \frac{1}{2}[2\ln(x-1) + \ln(x+1) - \ln x - 3\ln(x+3)]$$

$$\implies \frac{1}{y}\,y' = \frac{1}{2}\left[\frac{2}{x-1} + \frac{1}{x+1} - \frac{1}{x} - \frac{3}{x+3}\right]$$

$$\implies y' = \frac{y}{2}\left[\frac{1}{x-1} + \frac{1}{x+1} - \frac{1}{x} - \frac{3}{x+3}\right].$$

47. $y = x^{\sin^2 x}$

$$\implies \ln y = (\sin^2 x)(\ln x)$$

$$\implies \frac{1}{y}\,y' = (\sin^2 x)\left(\frac{1}{x}\right) + (\ln x)2\sin x\cos x$$

$$\implies y' = y\left[\frac{\sin^2 x}{x} + 2(\ln x)(\sin x)(\cos x)\right].$$

48. $y = (\tan x)^{\cos x}$

$$\implies \ln y = (\cos x)\ln\tan x$$

$$\implies \frac{1}{y}\,y' = (\cos x)\frac{1}{\tan x}\,\sec^2 x + (\ln\tan x)(-\sin x)$$

$$= \cos x\,\frac{\cos x}{\sin x}\,\frac{1}{\cos^2 x} - (\ln\tan x)(\sin x)$$

$$= \frac{1}{\sin x} - (\ln\tan x)\sin x$$

$$\implies y' = (\tan x)^{\cos x}[\csc x - (\ln\tan x)\sin x].$$

49. $y = x^2\ln x \implies y' = x^2\left(\frac{1}{x}\right) + (\ln x)(2x) = x(1 + 2\ln x)$. Hence $y' = 0 \iff x = 0$ or $\ln x = -1/2 \implies x = e^{-1/2}$. Now, $x = 0$ is not in the domain of the function and hence is not a critical number. Since $y'' = x\left(\frac{2}{x}\right) + (1 + 2\ln x) = 3 + 2\ln x$, $y''(e^{-1/2}) = 3 + 2\ln e^{-1/2} = 2 > 0$ and hence $x = e^{-1/2}$ is a local minimum. Thus $y = x^2\ln x$ has only one relative extremum, a local minimum at $(e^{-1/2}, -\frac{1}{2}e^{-1})$. The graph is shown below on the left.

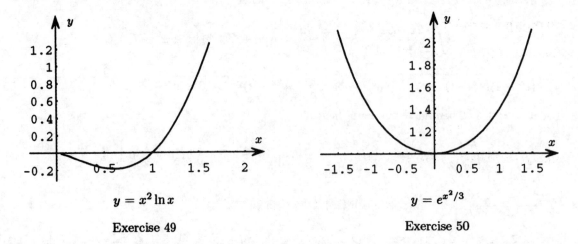

$y = x^2\ln x$

$y = e^{x^2/3}$

Exercise 49

Exercise 50

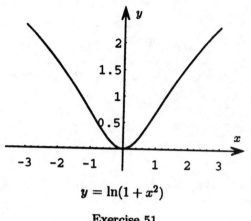

$$y = \ln(1 + x^2)$$

Exercise 51

$$y = x^2 - \ln x^2$$

Exercise 52

50. $f(x) = e^{x^2/3} \implies f'(x) = e^{x^2/3}(2x/3) = (2/3)xe^{x^2/3}$. Hence $f'(x) = 0 \iff x = 0$. Since $f''(x) = \frac{2}{3}e^{x^2/3}\left(1 + \frac{2}{3}x^2\right)$, $f''(0) = 2/3 > 0$ and hence $x = 0$ is a relative minimum. Thus $f(x) = e^{x^2/3}$ has only one relative extremum, a relative minimum at $(0, 1)$. It follows that the maximum value on $[-1, \sqrt{8}]$ $= f(\sqrt{8}) = e^{8/3}$, while the minimum value on $[-1, \sqrt{8}] = f(0) = 1$. Note that the graph is symmetric about the y-axis since $f(-x) = f(x)$.

51. $y = \ln(1 + x^2) \implies y' = \frac{2x}{1+x^2}$ and $y'' = \frac{2-2x^2}{(1+x^2)^2}$. Hence $y'' = 0 \iff x = \pm 1$. Since $y'' > 0$ when $-1 < x < 1$, the graph is concave up on the interval $[-1, 1]$ and concave down otherwise.

52. $y = x^2 - \ln x^2 \implies y' = 2x - \frac{2}{x}$ and $y'' = 2 + \frac{2}{x^2}$. Hence $y' = 0 \iff 2x = \frac{2}{x} \iff x^2 = 1 \iff x = \pm 1$. Both of the points $(1, 1)$ and $(-1, 1)$ are local minima since $y''(1) = y''(-1) = 4 > 0$. The graph is shown above. Note that the graph is symmetric about the y-axis since $f(-x) = f(x)$.

53. The area of T is $A(x) = \frac{1}{2}x\ln x$. Thus the rate of change of the area is equal to

$$\frac{dA}{dt} = \left[\left(\frac{1}{2}x\right)\frac{1}{x} + (\ln x)\frac{1}{2}\right]\frac{dx}{dt} = \frac{1}{2}(1 + \ln x)(4) = 2(1 + \ln x).$$

Hence, when $x = 5$ the area is increasing at a rate of $2(1 + \ln 5)$ square units/sec.

54. The area of the circle is $A(r) = \pi r^2$ and hence its rate of change is equal to

$$\frac{dA}{dt} = 2\pi r\frac{dr}{dt} = 2\pi r e^r.$$

Hence, when $r = 2$ inches the area is increasing at a rate of $4\pi e^2$ square inches/sec.

55. Since $f(-x) = -f(x)$, the graph is symmetric about the origin and hence the area is equal to

$$2\int_0^2 \underbrace{\frac{x}{x^2 + 1}}_{u = x^2 + 1}\, dx = \int_1^5 \frac{du}{u} = \ln u\Big|_1^5 = \ln 5.$$

655

56. We first find the intersection points of the two graphs:

$$\frac{10x - 21}{4x - 10} = \frac{1}{x} \quad \implies \quad 10x^2 - 21x = 4x - 10 \quad \implies \quad (2x - 1)(x - 2) = 0 \quad \implies \quad x = \frac{1}{2}, 2.$$

Now, on the interval $[1/2, 2]$ the graph of $y = \frac{10x-21}{4x-10}$ lies above the graph of $y = \frac{1}{x}$. Hence, since

$$\frac{10x - 21}{4x - 10} = \frac{10}{4} + \frac{4}{4x - 10},$$

the area between the graphs is equal to

$$\int_{1/2}^{2} \left(\frac{10}{4} + \frac{4}{4x - 10} - \frac{1}{x} \right) dx = \left(\frac{5}{2}x + \ln|4x - 10| - \ln x \right)\Big|_{1/2}^{2} = \frac{15}{4} - \ln 16.$$

57. Let dV be the volume of a differential cross-section disk of thickness dx. Then

$$dV = \pi r^2 dx = \pi y^2 dx = \pi \frac{2}{x - 1} \, dx$$

and hence the volume of revolution is

$$V = \int_{3}^{5} \frac{2\pi}{x - 1} \, dx = 2\pi \ln(x - 1)\big|_{3}^{5} = 2\pi \ln 2.$$

58. (a) $P\{0 \le x \le 1\} = \int_0^1 e^{-x} \, dx = -e^{-x}\big|_0^1 = 1 - 1/e \approx 0.632.$

(b) $P\{0 \le x \le \ln 2\} = \int_0^{\ln 2} e^{-x} \, dx = -e^{-x}\big|_0^{\ln 2} = 1/2.$

(c) $P\{0 \le x \le 4\} = \int_0^4 e^{-x} \, dx = -e^{-x}\big|_0^4 = 1 - 1/e^4 \approx 0.982.$

59. For $x > 0$, the graph is the decreasing exponential function $y = e^{-x}$, while for $x < 0$ it is identically zero, as illustrated below.

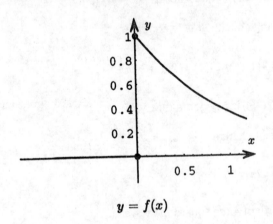

$$y = f(x)$$

Exercise 59

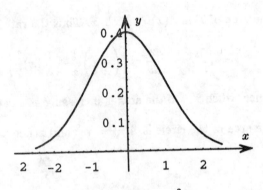

$$y = (1/\sqrt{2\pi})\, e^{-x^2/2}$$

Exercise 61

60. (a) Using the Midpoint Rule with n subdivisions, we find that

$$P\{-1 \le x \le 1\} = \int_{-1}^{1} f(x) \, dx \approx \frac{2}{n} \sum_{i=0}^{n-1} f\left(-1 + \frac{2i + 1}{n} \right).$$

The table below, column 2, summarizes the values obtained by letting $n = 20, 40, 60, 80$. Hence, to three decimal places, $\int_{-1}^{1} f(x)\, dx \approx 0.683$.

n	$\frac{2}{n}\sum_{i=0}^{n-1} f\left(-1+\frac{2i+1}{n}\right)$	$\frac{2}{n}\sum_{i=0}^{n-1} f\left(\frac{2i+1}{n}\right)$
20	0.682891	0.477295
40	0.682740	0.477261
60	0.682712	0.477255
80	0.682702	0.477253

(b) Using the Midpoint Rule with n subdivisions, we find that

$$P\{0 \le x \le 2\} = \int_0^2 f(x)\, dx \approx \frac{2}{n}\sum_{i=0}^{n-1} f\left(\frac{2i+1}{n}\right).$$

The table above, column 3, summarizes the values obtained by letting $n = 20, 40, 60, 80$. Hence, to three decimal places, $\int_0^2 f(x)\, dx \approx 0.477$.

61. We find that

$$f(x) = \frac{1}{\sqrt{2\pi}} e^{-x^2/2}$$

$$f'(x) = \frac{1}{\sqrt{2\pi}} e^{-x^2/2}(-x)$$

$$f''(x) = \frac{-x}{\sqrt{2\pi}} e^{-x^2/2}(-x) + e^{-x^2/2}\left(\frac{-1}{\sqrt{2\pi}}\right) = \frac{e^{-x^2/2}}{\sqrt{2\pi}}(x^2 - 1).$$

Hence $f'(x) = 0 \iff x = 0$. Since $f''(0) = -1/\sqrt{2\pi} < 0$, $x = 0$ is a relative maximum. Moreover, $f''(x) = 0 \iff x = \pm 1$. Since the sign of f'' changes at both ± 1, $x = \pm 1$ are inflection points.

62. $y = xe^{2x} \implies y' = xe^{2x}(2) + e^{2x} = e^{2x}(2x+1) \implies y'(\ln 2) = e^{2\ln 2}(2\ln 2 + 1) = 4(2\ln 2 + 1)$. Hence the equation of the tangent line at $(\ln 2, 4\ln 2)$ is $y - 4\ln 2 = 4(2\ln 2 + 1)(x - \ln 2)$.

63. $y = x^2 \ln x \implies y' = x^2\left(\frac{1}{x}\right) + 2x\ln x = x + 2x\ln x \implies y'(e) = 3e$. Hence the slope of the line perpendicular to the graph of $y = x^2 \ln x$ at (e, e^2) is $-\frac{1}{3e}$ and therefore its equation is $y - e^2 = -\frac{1}{3e}(x - e)$.

64. $y = e^{-3x} + 1 \implies y' = -3e^{-3x} \implies y'(\ln 3) = -3e^{-3\ln 3} = -1/9$. Since $y(\ln 3) = 28/27$, the equation of the tangent line is $y - \frac{28}{27} = -\frac{1}{9}(x - \ln 3)$.

65. $y = x\ln x^2 = 2x\ln x \implies y' = 2x\left(\frac{1}{x}\right) + 2\ln x = 2 + 2\ln x \implies y'(e) = 2 + 2\ln e = 4$. Hence

$$y \approx y(e) + y'(e)(x - e) = e\ln e^2 + 4(x - e) = 4x - 2e \quad \text{for } x \approx e.$$

If $x = 2.7$, $y(2.7) \approx 4(2.7) - 2e \approx 5.363436$. (*Calculator value:* $y(2.7) = 5.36355957$.)

66. Let $f(x) = e^{\sqrt{x}}$. Then $f'(x) = e^{\sqrt{x}}/2\sqrt{x} \implies f'(4) = e^2/4$. Hence

$$f(x) \approx f(4) + f'(4)(x - 4) = e^2 + \frac{e^2}{4}(x - 4) = \frac{e^2}{4}x \quad \text{for } x \approx 4.$$

If $x = 4.1$, $f(4.1) = e^{\sqrt{4.1}} \approx \frac{e^2}{4}(4.1) = 7.573782$. (*Calculator value:* $e^{\sqrt{4.1}} \approx 7.574942$.)

67. (a) $f'(x) = (1/2)(e^x - e^{-x}) = g(x)$, $g'(x) = (1/2)(e^x + e^{-x}) = f(x)$.

657

(b) Since $f''(x) = g'(x) = f(x)$ and $g''(x) = f'(x) = g(x)$, both $y = f(x)$ and $y = g(x)$ satisfy the differential equation $y'' - y = 0$.

(c) Let $f(x) = (1/2)(e^{kx} + e^{-kx})$ and $g(x) = (1/2)(e^{kx} - e^{-kx})$. Then $f''(x) = (1/2)k^2(e^{kx} + e^{-kx}) = k^2 f(x)$ and $g''(x) = (1/2)k^2(e^{kx} - e^{-kx}) = k^2 g(x)$. Hence $y = (1/2)(e^{kx} + e^{-kx})$ and $y = (1/2)(e^{kx} - e^{-kx})$ are two solutions of the differential equation $y'' - k^2 y = 0$.

68. (a) $v(t) = \int a(t)\, dt = \int (t+2)^{-2}\, dt = -\frac{1}{t+2} + C$. Since $v(0) = -\frac{1}{2} + C = 0$, $C = \frac{1}{2}$ and hence $v(t) = \frac{1}{2} - \frac{1}{t+2}$.

(b) $\int_0^{10} v(t)\, dt = \int_0^{10} \left(\frac{1}{2} - \frac{1}{t+2}\right) dt = [(1/2)t - \ln(t+2)]\big|_0^{10} = 5 - \ln 6$.

69. Using the Midpoint Rule with $f(x) = e^{-x^2}$ and $n = 20$, we find that

$$\int_0^5 e^{-x^2}\, dx \approx \frac{5}{20} \sum_{i=0}^{19} f\left(\frac{5}{2n} + \frac{5i}{n}\right) = 0.886227.$$

70. $2^x = e^{rx} \implies x \ln 2 = rx \implies r = \ln 2$.

71. Set $f(x) = 1/x$. Using the Midpoint Rule with $n = 10$, we find that:

(a) $\ln 4 = \int_1^4 \frac{1}{x}\, dx \approx \frac{3}{10} \sum_{i=1}^{10} f\left(1 + \left(i + \frac{1}{2}\right)\frac{3}{10}\right) = 1.38284$

(b) $\ln 5 = \int_1^5 \frac{1}{x}\, dx \approx \frac{4}{10} \sum_{i=1}^{10} f\left(1 + \left(i + \frac{1}{2}\right)\frac{4}{10}\right) = 1.60321$

(c) $\ln \frac{1}{2} = \int_1^{1/2} \frac{1}{x}\, dx = -\int_{1/2}^1 \frac{1}{x}\, dx \approx -\frac{1}{20} \sum_{i=1}^{10} f\left(1 + \left(i + \frac{1}{2}\right)\frac{1}{20}\right) = -0.692835$

72. Let $f(x) = \ln x + 2 - x$. The table below on the left gives values of $f(x)$ for selected values of x and shows that there are two roots, one between 0.1 and 1, the other between 3 and 4. Using 0.1 and 3.5 as initial values, the table in the center lists the successive approximations to the roots using Newton's formula:

$$r_{n+1} = r_n - \frac{f(r_n)}{f'(r_n)}.$$

Thus, to within four decimal places, the roots are 0.1586 and 3.1462.

x	$\ln x + 2 - x$
0.1	−0.4
1	1
2	0.69
3	0.1
4	−0.6

$r_{n+1} = r_n - f(r_n)/f'(r_n)$	
0.1	3.5
0.144732	3.15387
0.157864	3.1462
0.158592	3.14619
0.158594	3.14619
0.158594	3.14619

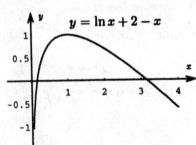

$y = \ln x + 2 - x$

73. $\frac{dy}{dx} = 2y \implies \frac{dy}{y} = 2dx \implies \ln|y| = 2x + x \implies y = Ce^{2x}$. Since $y(0) = C = 1$, $y = e^{2x}$.

74. $\frac{dy}{dx} + y = 0 \implies \frac{dy}{y} = -dx \implies \ln|y| = -x + C \implies y = Ce^{-x}$. Since $y(\ln 2) = Ce^{-\ln 2} = \frac{1}{2}C = 2$, $C = 4$ and hence $y = 4e^{-x}$.

75. $2y' + 4y = 0 \Rightarrow 2\frac{dy}{dx} = -4y \Rightarrow \frac{dy}{y} = -2dx \Rightarrow \ln|y| = -2x + C \Rightarrow y = Ce^{-2x}$. Since $y(0) = C$
$= \pi$, $y = \pi e^{-2x}$.

76. $y' + y = 2 \Rightarrow \frac{dy}{dx} = 2 - y \Rightarrow \frac{dy}{2-y} = dx \Rightarrow -\ln|2-y| = x + C \Rightarrow 2 - y = Ce^{-x} \Rightarrow y = 2 - Ce^{-x}$.
Since $y(0) = 2 - C = 1$, $C = 1$ and hence $y = 2 - e^{-x}$.

77. $4y' - 4y + 4 = 0 \Rightarrow \frac{dy}{dx} = y - 1 \Rightarrow \frac{dy}{y-1} = dx \Rightarrow \ln|y-1| = x + C \Rightarrow y - 1 = Ce^x \Rightarrow y = 1 + Ce^x$.
Since $y(0) = 1 + C = 1$, $C = 0$ and hence $y = 1$.

78. $(y' + y)^2 = 0 \Rightarrow y' + y = 0 \Rightarrow \frac{dy}{y} = -dx \Rightarrow \ln|y| = -x + C \Rightarrow y = Ce^{-x}$. Since $y(0) = C = 1$,
$y = e^{-x}$.

79. The average value is equal to

$$\frac{1}{\sqrt{\ln 2}} \int_0^{\sqrt{\ln 2}} x e^{x^2+1} \, dx.$$

Let $u = x^2 + 1$. Then $du = 2x\,dx$ and hence

$$\frac{1}{\sqrt{\ln 2}} \int_0^{\sqrt{\ln 2}} x e^{x^2+1} \, dx = \frac{1}{\sqrt{\ln 2}} \frac{1}{2} \int_1^{1+\ln 2} e^u \, du = \frac{1}{2\sqrt{\ln 2}} e^u \Big|_1^{1+\ln 2} = \frac{1}{2\sqrt{\ln 2}} (e^{1+\ln 2} - e) = \frac{e}{2\sqrt{\ln 2}}.$$

80. The area A of the region is

$$A = \int_e^{e^2} \frac{1}{x} \, dx = \ln x \Big|_e^{e^2} = 1.$$

Hence the centroid $(\overline{x}, \overline{y})$ of the region is given by

$$\overline{x} = \frac{1}{A} \int_e^{e^2} x \left[\frac{1}{x} - 0\right] \, dx = \int_e^{e^2} dx = x \Big|_e^{e^2} = e^2 - e$$

and

$$\overline{y} = \frac{1}{2A} \int_e^{e^2} \left[\left(\frac{1}{x}\right)^2 - 0\right] \, dx = -\frac{1}{2x} \Big|_e^{e^2} = \frac{e-1}{2e^2}.$$

81. The curves $y = 1/x$ and $y = x^2$ intersect when $1/x = x^2 \Rightarrow x^3 = 1 \Rightarrow x = 1$. Now, on the interval
$[1, 3]$, $x^2 > \frac{1}{x}$. Hence the area A of the region is

$$A = \int_1^3 \left(x^2 - \frac{1}{x}\right) \, dx = \left(\frac{1}{3}x^3 - \ln x\right) \Big|_1^3 = \frac{26}{3} - \ln 3.$$

Therefore the centroid $(\overline{x}, \overline{y})$ is given by

$$\overline{x} = \frac{1}{A} \int_1^3 x \left(x^2 - \frac{1}{x}\right) \, dx = \frac{1}{A} \int_1^3 (x^3 - 1) \, dx = \frac{1}{A} \left(\frac{1}{4}x^4 - x\right) \Big|_1^3 = \frac{18}{A} = \frac{54}{26 - 3\ln 3}$$

and

$$\overline{y} = \frac{1}{2A} \int_1^3 \left[(x^2)^2 - \left(\frac{1}{x}\right)^2\right] \, dx = \frac{1}{2A} \int_1^3 (x^4 - x^{-2}) \, dx = \frac{1}{2A} \left(\frac{1}{5}x^5 + \frac{1}{x}\right) \Big|_1^3 = \frac{358}{15A} = \frac{358}{5(26 - 3\ln 3)}.$$

82. Let $f(x) = e^x + 2x$. Then $f(0) = 1$ and $f(-1) = 1/e - 2 < 0$. Hence the root of $f(x)$ lies between -1 and 0. Let $r_0 = -0.5$ and

$$r_{n+1} = r_n - \frac{f(r_n)}{f'(r_n)} = r_n - \frac{e^{r_n} + 2r_n}{e^{r_n} + 2}, \quad \text{for } n = 1, 2, \dots .$$

Then

$$r_0 = -0.5, r_1 = -0.349045, r_2 = -0.351733, r_3 = -0.351734, r_4 = -0.351734.$$

Hence, to five decimal places, the root of $f(x)$ is -0.35173.

83. Let $P(t)$ be the population t years after 1980. Then $P(t) = P_0 e^{kt}$, where $P_0 = 203$. Since $P(10) = 227$,

$$203 e^{10k} = 227 \quad \Longrightarrow \quad 10k = \ln\frac{227}{203} \quad \Longrightarrow \quad k = \frac{1}{10}\ln\frac{227}{203}.$$

Hence $P(t) = 203 e^{(t/10)\ln(227/203)}$. Thus, the population in 2000 will be

$$P(20) = 203 e^{\frac{20}{10}\ln\frac{227}{203}} \approx 203(1.25) = 254 \text{ million people,}$$

and in the year 2010 will be

$$P(30) = 203 e^{\frac{30}{10}\ln\frac{227}{203}} \approx 203(1.4) = 284 \text{ million people.}$$

84. Let $y(t)$ be the amount of raw sugar remaining after t hours. Then $y(t) = y_0 e^{kt} = 100 e^{kt}$. Since $y(6) = 75$,

$$100 e^{6k} = 75 \quad \Longrightarrow \quad 6k = \ln\frac{3}{4} \quad \Longrightarrow \quad k = \frac{1}{6}\ln\frac{3}{4}.$$

Hence $y(t) = 100 e^{(t/6)\ln(3/4)}$.

(a) When half the raw sugar has been inverted, $y(t) = 50$. Hence

$$100 e^{\frac{t}{6}\ln\frac{3}{4}} = 50 \quad \Longrightarrow \quad \frac{t}{6}\ln\frac{3}{4} = \ln\frac{1}{2} \quad \Longrightarrow \quad t = 6\frac{\ln(1/2)}{\ln(3/4)} \approx 14.5 \text{ hours.}$$

(b) When 90% of the raw sugar has been inverted, 10% remains and hence

$$100 e^{\frac{t}{6}\ln\frac{3}{4}} = 0.1(100) \quad \Longrightarrow \quad \frac{t}{6}\ln\frac{3}{4} = \ln 0.1 \quad \Longrightarrow \quad t = 6\frac{\ln 0.1}{\ln 0.75} \approx 48 \text{ hours.}$$

85. Let $y(t)$ be the number of fruit flies after t days. Then $y(t) = 100 e^{kt}$. Since $y(10) = 500$,

$$100 e^{10k} = 500 \quad \Longrightarrow \quad 10k = \ln 5 \quad \Longrightarrow \quad k = \frac{1}{10}\ln 5.$$

Hence $y(t) = 100 e^{(t/10)\ln 5}$. Thus, after 4 days the number of fruit flies is equal to

$$y(4) = 100 e^{(4/10)\ln 5} \approx 190.$$

86. Let V and r stand for the volume and radius of the snowball after t minutes. Then $V = \frac{4}{3}\pi r^3$.

(a) Since the snowball melts at a rate proportional to the surface area and the surface area is equal to $4\pi r^2$, it follows that $\frac{dV}{dt} = k(4\pi r^2)$. But

$$V = \frac{4}{3}\pi r^3 \quad \Longrightarrow \quad \frac{dV}{dt} = \frac{4}{3}\pi(3r^2)\frac{dr}{dt} = 4\pi r^2 \frac{dr}{dt}.$$

Hence

$$\frac{dV}{dt} = k(4\pi r^2) \quad \Longrightarrow \quad 4\pi r^2 \frac{dr}{dt} = k(4\pi r^2) \quad \Longrightarrow \quad \frac{dr}{dt} = k.$$

(b) $\frac{dr}{dt} = k \implies dr = k\,dt \implies r = kt + C$, where k and C are constants.

(c) Since $r(0) = C = 10$ and $r(20) = 20k + C = 8 \implies k = -1/10$, it follows that $r(t) = -\frac{t}{10} + 10$.

(d) $r(60) = 10 - \frac{60}{10} = 4$ cm.

87. Let $y(t)$ be the amount of the substance after t hours. Then $y(t) = 100e^{kt}$. Since $y(6) = 40$,

$$100e^{6k} = 40 \quad \implies \quad 6k = \ln\frac{4}{10} \quad \implies \quad k = \frac{1}{6}\ln\frac{2}{5}.$$

Hence $y(t) = 100e^{(t/6)\ln(2/5)}$. If $t_{1/2}$ is the half-life of the substance, then

$$y(t_{1/2}) = 100e^{(t_{1/2}/6)\ln(2/5)} = 50 \quad \implies \quad \frac{t_{1/2}}{6}\ln\frac{2}{5} = \ln\frac{1}{2} \quad \implies \quad t_{1/2} = 6\,\frac{\ln 0.5}{\ln 0.4} \approx 4.54 \text{ hours.}$$

88. $f(x) = Ce^{kx} \implies f'(x) = kCe^{kx}$. Hence $f'(x+a) = kCe^{k(x+a)} = kf(x+a)$.

Chapter 9

Trigonometric and Inverse Trigonometric Functions

9.1 Integrals of the Trigonometric Functions

1. $\displaystyle\int \cos 3x\, dx = \frac{1}{3}\sin 3x + C$

2. $\displaystyle\int_0^{\pi/8} \sec^2 2x\, dx = \frac{1}{2}\tan 2x\Big|_0^{\pi/8} = \frac{1}{2}$

3. $\displaystyle\int_0^{\pi/8} \sec 2x \tan 2x\, dx = \frac{1}{2}\sec 2x\Big|_0^{\pi/8} = \frac{1}{2}\sqrt{2} - \frac{1}{2}$

4. Let $u = x^2$. Then $du = 2x\,dx$ and hence
$$\int x\csc x^2\, dx = \frac{1}{2}\int \csc u\, du = \frac{1}{2}\ln|\csc u - \cot u| + C = \frac{1}{2}\ln|\csc x^2 - \cot x^2| + C.$$

5. $\displaystyle\int (\tan^2 x + 1)\, dx = \int \sec^2 x\, dx = \tan x + C$

6. Let $u = \sqrt{x}$. Then $du = \dfrac{1}{2\sqrt{x}}\, dx$ and hence
$$\int \frac{\tan \sqrt{x}}{\sqrt{x}}\, dx = 2\int \tan u\, du = 2\ln|\sec u| + C = 2\ln|\sec \sqrt{x}| + C.$$

7. Let $u = x^2$. Then $du = 2x\,dx$ and hence
$$\int \frac{x}{\cos x^2}\, dx = \frac{1}{2}\int \sec u\, du = \frac{1}{2}\ln|\sec u + \tan u| + C = \frac{1}{2}\ln|\sec x^2 + \tan x^2| + C.$$

8. Let $u = \sqrt{x}$. Then $du = \dfrac{1}{2\sqrt{x}}\, dx$ and hence
$$\int \frac{1}{\sqrt{x}\sec \sqrt{x}}\, dx = 2\int \cos u\, du = 2\sin u + C = 2\sin \sqrt{x} + C.$$

662

9. Let $u = 3x^2 - 1$. Then $du = 6xdx$ and hence

$$\int x \tan(3x^2 - 1)\, dx = \frac{1}{6} \int \tan u\, du = \frac{1}{6} \ln|\sec u| + C = \frac{1}{6} \ln|\sec(3x^2 - 1)| + C.$$

10. Let $u = 1 + \cos x$. Then $du = -\sin x dx$ and hence

$$\int_0^{\pi/2} (\sin x)\sqrt{1 + \cos x}\, dx = -\int_2^1 \sqrt{u}\, du = -\frac{2}{3} u^{3/2} \Big|_2^1 = \frac{2}{3} (2\sqrt{2} - 1).$$

11. Let $u = \tan x$. Then $du = \sec^2 x dx$ and hence

$$\int_{\pi/4}^{\pi/3} \tan^3 x \sec^2 x\, dx = \int_1^{\sqrt{3}} u^3\, du = \frac{1}{4} u^4 \Big|_1^{\sqrt{3}} = 2.$$

12. Let $u = \ln(\sin x)$. Then $du = \dfrac{1}{\sin x} \cos x dx = \cot x dx$ and hence

$$\int_{\pi/4}^{\pi/2} (\cot x) \ln(\sin x)\, dx = \int_{\ln(\sqrt{2}/2)}^0 u\, du = \frac{1}{2} u^2 \Big|_{\ln(\sqrt{2}/2)}^0 = -\frac{1}{2} \left(\ln \frac{\sqrt{2}}{2}\right)^2 = -\frac{1}{8}(\ln 2)^2.$$

13. Let $u = \tan x$. Then $du = \sec^2 x dx$ and hence

$$\int_{\pi/6}^{\pi/4} \sec^2 x\, e^{\tan x}\, dx = \int_{1/\sqrt{3}}^1 e^u\, du = e^u \big|_{1/\sqrt{3}}^1 = e - e^{1/\sqrt{3}}.$$

14. Let $u = 2x - 1$. Then $du = 2dx$ and hence

$$\int \frac{\cos^2(2x - 1)}{\sin(2x - 1)}\, dx = \frac{1}{2} \int \frac{\cos^2 u}{\sin u}\, du = \frac{1}{2} \int \frac{1 - \sin^2 u}{\sin u}\, du = \frac{1}{2} \int (\csc u - \sin u)\, du$$

$$= \frac{1}{2} [\ln|\csc u - \cot u| + \cos u] + C = \frac{1}{2} [\ln|\csc(2x - 1) - \cot(2x - 1)| + \cos(2x - 1)] + C.$$

15. Let $u = x + \tan x$. Then $du = (1 + \sec^2 x)dx$ and hence

$$\int \frac{\sec^2 x + 1}{x + \tan x}\, dx = \int \frac{du}{u} = \ln|u| + C = \ln|x + \tan x| + C.$$

16. Let $u = x + \pi/4$. Then $du = dx$ and hence

$$\int \frac{\csc^2(x + \pi/4)}{\cot^3(x + \pi/4)}\, dx = \int \frac{\csc^2 u}{\cot^3 u}\, du = \int \frac{1}{\sin^2 u} \frac{\sin^3 u}{\cos^3 u}\, du$$

$$= \int \underbrace{\frac{\sin u}{\cos^3 u}}_{v = \cos u}\, du = -\int \frac{dv}{v^3} = \frac{1}{2} \frac{1}{v^2} + C = \frac{1}{2} \frac{1}{\cos^2 u} + C = \frac{1}{2\cos^2(x + \pi/4)} + C.$$

17. Let $u = 1 - x^2$. Then $du = -2xdx$ and hence

$$\int x \cot^2(1 - x^2)\, dx = -\frac{1}{2} \int \cot^2 u\, du = -\frac{1}{2} \int (\csc^2 u - 1)\, du$$

$$= -\frac{1}{2}(-\cot u - u) + C = \frac{1}{2} \cot(1 - x^2) + \frac{1}{2} (1 - x^2) + C.$$

18. Let $u = \tan x$. Then $du = \sec^2 x dx$ and hence $\displaystyle\int \frac{\sec^2 x}{\sqrt{\tan x}}\,dx = \int u^{-1/2}\,du = 2\sqrt{u} + C = 2\sqrt{\tan x} + C.$

19. $\displaystyle\int (\sec x + \tan x)^2\,dx = \int (\sec^2 x + 2\sec x \tan x + \underbrace{\tan^2 x}_{\sec^2 x - 1})\,dx = \int (2\sec^2 x + 2\sec x \tan x - 1)\,dx =$

$2\tan x + 2\sec x - x + C$

20. Let $u = \sin x - \cos x$. Then $du = (\cos x + \sin x)dx$ and hence

$$\int \frac{\sin x + \cos x}{\sin x - \cos x}\,dx = \int \frac{du}{u} = \ln|u| + C = \ln|\sin x - \cos x| + C.$$

21. Let $u = x^2 - 2x$. Then $du = (2x - 2)dx = 2(x - 1)dx$ and hence

$$\int (1 - x)\sec(x^2 - 2x)\,dx = -\frac{1}{2}\int \sec u\,du$$

$$= -\frac{1}{2}\ln|\sec u + \tan u| + C = -\frac{1}{2}\ln|\sec(x^2 - 2x) + \tan(x^2 - 2x)| + C.$$

22. Let $u = \sec\sqrt{x}$. Then $du = \dfrac{\sec\sqrt{x}\tan\sqrt{x}}{2\sqrt{x}}\,dx$ and hence

$$\int \frac{\sec^5\sqrt{x}\tan\sqrt{x}}{\sqrt{x}}\,dx = 2\int u^4\,du = \frac{2}{5}u^5 + C = \frac{2}{5}\sec^5\sqrt{x} + C.$$

23. Let $u = \sin x$. Then $du = \cos x dx$ and hence

$$\int \sqrt{1 - \sin^2 x}\,e^{\sin x}\,dx = \int |\cos x| e^{\sin x}\,dx = \begin{cases} \int e^u\,du = e^u + C = e^{\sin x} + C, & \cos x \geq 0 \\ -\int e^u\,du = -e^u + C = -e^{\sin x} + C, & \cos x < 0. \end{cases}$$

24. Let $u = \ln x$. Then $du = (1/x)dx$ and hence

$$\int \frac{\sec\ln x}{x}\,dx = \int \sec u\,du = \ln|\sec u + \tan u| + C = \ln|\sec\ln x + \tan\ln x| + C.$$

25. Let $u = \sqrt{x}$. Then $du = \dfrac{1}{2\sqrt{x}}\,dx$ and hence

$$\int \frac{dx}{\sqrt{x}(1 + \cos\sqrt{x})} = 2\int \frac{du}{1 + \cos u} \cdot \frac{1 - \cos u}{1 - \cos u} = 2\int \frac{1 - \cos u}{\sin^2 u}\,du$$

$$= 2\int (\csc^2 u - \cot u \csc u)\,du = 2(-\cot u + \csc u) + C = 2(-\cot\sqrt{x} + \csc\sqrt{x}) + C.$$

26. Let $u = \cot 2x$. Then $du = -2\csc^2 2x\,dx$ and hence

$$\int \csc^4 2x\,dx = \int (\csc^2 2x)(1 + \cot^2 2x)\,dx$$

$$= -\frac{1}{2}\int (1 + u^2)\,du = -\frac{1}{2}\left(u + \frac{1}{3}u^3\right) = -\frac{1}{2}\left(\cot 2x + \frac{1}{3}\cot^3 2x\right) + C.$$

27. Let $u = 1 + \sin 4x$. Then $du = 4\cos 4x\,dx$ and hence

$$\int \frac{\cos 4x}{\sqrt{1 + \sin 4x}}\,dx = \frac{1}{4}\int u^{-1/2}\,du = \frac{1}{2}\sqrt{u} + C = \frac{1}{2}\sqrt{1 + \sin 4x} + C.$$

28. Let $u = \cos 2\theta$. Then $du = -2\sin 2\theta\,d\theta$ and hence

$$\int \sin^3 2\theta\,d\theta = \int (\sin 2\theta)(1 - \cos^2 2\theta)\,d\theta$$

$$= -\frac{1}{2}\int (1 - u^2)\,du = -\frac{1}{2}\left(u - \frac{1}{3}u^3\right) + C = -\frac{1}{2}\left(\cos 2\theta - \frac{1}{3}\cos^3 2\theta\right) + C.$$

29. Let $u = \tan\theta + 1$. Then $du = \sec^2\theta\,d\theta$ and hence

$$\int \frac{\sec^2\theta\,d\theta}{\sqrt{\tan\theta + 1}} = \int \frac{du}{\sqrt{u}} = 2\sqrt{u} + C = 2\sqrt{\tan\theta + 1} + C.$$

30. $\int \dfrac{\tan x}{1 + \tan^2 x}\,dx = \int \dfrac{\tan x}{\sec^2 x}\,dx = \int \underbrace{\sin x\cos x\,dx}_{u=\sin x} = \int u\,du = \dfrac{1}{2}u^2 + C = \dfrac{1}{2}\sin^2 x + C$

31. Area $= \displaystyle\int_0^{\sqrt{\pi/3}} \underbrace{x\sec x^2\,dx}_{u=x^2} = \frac{1}{2}\int_0^{\pi/3}\sec u\,du = \frac{1}{2}\ln|\sec u + \tan u|\Big|_0^{\pi/3} = \frac{1}{2}\ln(2 + \sqrt{3})$

32. Area $= 2\displaystyle\int_0^{\pi/3}\sec x\tan x\,dx = 2\sec x\big|_0^{\pi/3} = 2$

33. Average value on $[\pi/6, \pi/3] =$

$$\frac{1}{\pi/3 - \pi/6}\int_{\pi/6}^{\pi/3}\csc x\,dx = \frac{6}{\pi}\ln|\csc x - \cot x|\Big|_{\pi/6}^{\pi/3} = \frac{6}{\pi}\ln\left|\frac{2}{\sqrt{3}} - \frac{1}{\sqrt{3}}\right| - \frac{6}{\pi}\ln|2 - \sqrt{3}| = -\frac{6}{\pi}\ln(2\sqrt{3} - 3)$$

34. Method 1: $\int \underbrace{\sin x\cos x\,dx}_{u=\sin x} = \int u\,du = \frac{1}{2}u^2 + C = \frac{1}{2}\sin^2 x + C$

Method 2: $\int \sin x\cos x\,dx = \frac{1}{2}\int \sin 2x\,dx = -\frac{1}{4}\cos 2x + C$

35. Let $u = \tan x$. Then $\int \tan x\sec^2 x\,dx = \int u\,du = \frac{1}{2}u^2 + C = \frac{1}{2}\tan^2 x + C$. Alternatively,

$$\int \tan x\sec^2 x\,dx = \int \frac{\sin x}{\cos^3 x}\,dx = -\int \frac{1}{u^3}\,du = \frac{1}{2}\frac{1}{u^2} + C = \frac{1}{2\cos^2 x} + C = \frac{1}{2}\sec^2 x + C.$$

36. Let $u = \csc x - \cot x$. Then $du = (-\csc x\cot x + \csc^2 x)dx = (\csc x)(\csc x - \cot x)dx$ and hence

$$\int \csc x\,dx = \int \frac{(\csc x)(\csc x - \cot x)}{\csc x - \cot x}\,dx = \int \frac{du}{u} = \ln|u| + C = \ln|\csc x - \cot x| + C.$$

665

37. Since $y' = \dfrac{1}{\sin x}(\cos x) = \cot x$,

$$\text{length} = \int_{\pi/6}^{\pi/3} \sqrt{1 + \cot^2 x}\; dx = \int_{\pi/6}^{\pi/3} \csc x\; dx$$

$$= \ln|\csc x - \cot x|\big|_{\pi/6}^{\pi/3} = \ln\left|\frac{2}{\sqrt{3}} - \frac{1}{\sqrt{3}}\right| - \ln(2 - \sqrt{3}) = -\ln(2\sqrt{3} - 3).$$

38. Since $y' = \dfrac{1}{\cos x}(-\sin x) = -\tan x$,

$$\text{length} = \int_0^{\pi/4} \sqrt{1 + \tan^2 x}\; dx = \int_0^{\pi/4} \sec x\; dx = \ln|\sec x + \tan x|\big|_0^{\pi/4} = \ln(1 + \sqrt{2}).$$

39. Let dV be the volume of a differential cross-section disk of thickness dx. Then $dV = \pi y^2 dx = \pi \sec^2 x \tan^2 x\, dx$. Since the solid of revolution is symmetric about the y-axis, the total volume of revolution is

$$V = 2\int_0^{\pi/4} \pi \underbrace{\sec^2 x \tan^2 x}_{u = \tan x}\; dx = 2\pi \int_0^1 u^2\; du = \frac{2\pi}{3} u^3 \bigg|_0^1 = \frac{2\pi}{3}.$$

40. Let dV be the volume of a differential cross-section disk of thickness dx. Since $\sec x > 1$, the graph of $y = \sec x \tan x$ lies above that of $y = \tan x$ for $0 \le x \le \pi/4$. Hence the cross-section is a washer of outer radius $\sec x \tan x$ and inner radius $\tan x$. Therefore $dV = \pi(\tan^2 x \sec^2 x - \tan^2 x)dx$. Since the solid of revolution is symmetric about the y-axis, it follows that the volume of revolution is

$$V = 2\int_0^{\pi/4} \pi(\tan^2 x \sec^2 x - \tan^2 x)\; dx = 2\pi \left[\int_0^{\pi/4} \tan^2 x \sec^2 x\; dx - \int_0^{\pi/4} (\sec^2 x - 1)\; dx \right]$$

$$= 2\pi \frac{\tan^3 x}{3} \bigg|_0^{\pi/4} - 2\pi(\tan x - x)\big|_0^{\pi/4} = \frac{\pi^2}{2} - \frac{4\pi}{3}.$$

41. The line $y = -2$ intersects the graph of $y = \sec x$ when $\sec x = -2 \implies \cos x = -1/2 \implies x = 2\pi/3,$ $4\pi/3$. Since this region is symmetric about the line $x = \pi$, the area of the region is equal to

$$2\int_\pi^{4\pi/3} (\sec x + 2)\; dx = (2\ln|\sec x + \tan x| + 4x)\big|_\pi^{4\pi/3} = 2\ln(2 - \sqrt{3}) + \frac{4\pi}{3} = \frac{4\pi}{3} - 2\ln(2 + \sqrt{3}).$$

42. Let dV be the differential volume of a shell of thickness dx revolved about the y-axis. Then $dV = 2\pi r h dx = 2\pi x y dx = 2\pi x \sec x^2 dx$ and hence the volume of revolution is

$$V = 2\pi \int_0^{\sqrt{\pi/4}} \underbrace{x \sec x^2}_{u = x^2}\; dx = \pi \int_0^{\pi/4} \sec u\; du = \pi \ln|\sec u + \tan u|\big|_0^{\pi/4} = \pi \ln(\sqrt{2} + 1).$$

43. Let dV be the differential volume of a shell of thickness dx revolved about the y-axis. Then $dV = 2\pi r h dx = 2\pi x y dx = 2\pi x \tan 2x^2 dx$ and hence the volume of revolution is

$$V = 2\pi \int_0^{\sqrt{\pi/8}} \underbrace{x \tan 2x^2}_{u = 2x^2}\; dx = \frac{\pi}{2} \int_0^{\pi/4} \tan u\; du = \frac{\pi}{2} \ln|\sec u|\big|_0^{\pi/4} = \frac{\pi}{2} \ln\sqrt{2}.$$

44. Since $y' = -\sin x$, the length of the graph over the interval $[0, 2\pi]$ is $\int_0^{2\pi} \sqrt{1 + \sin^2 x} \; dx$. Using Simpson's Rule with $f(x) = \sqrt{1 + \sin^2 x}$ and $n = 6$, we find that

$$\int_0^{2\pi} \sqrt{1 + \sin^2 x} \; dx \approx \frac{2\pi}{3(6)} \left[f(0) + 4f\left(\frac{2\pi}{6}\right) + 2f\left(\frac{4\pi}{6}\right) + 4f\left(\frac{6\pi}{6}\right) \right.$$
$$\left. + 2f\left(\frac{8\pi}{6}\right) + 4f\left(\frac{10\pi}{6}\right) + f(2\pi) \right] = 7.6356$$

9.2 Integrals Involving Products of Trigonometric Functions

1. $\displaystyle\int \sin^2 2x \; dx = \frac{1}{2} \int (1 - \cos 4x) \; dx = \frac{1}{2}\left(x - \frac{1}{4}\sin 4x\right) + C$

2. $\displaystyle\int_0^{\pi/4} \sec^2 x \; dx = \tan x \big|_0^{\pi/4} = 1$

3. Let $u = \cos x$. Then $du = -\sin x dx$ and hence

$$\int_{\pi/4}^{\pi/2} \sin x \cos^2 x \; dx = -\int_{\sqrt{2}/2}^0 u^2 \; du = -\frac{1}{3}u^3 \Big|_{\sqrt{2}/2}^0 = \frac{\sqrt{2}}{12}.$$

4. Let $u = \cos x$. Then $du = -\sin x dx$ and hence

$$\int \sin x \cos^3 x \; dx = -\int u^3 \; du = -\frac{1}{4}u^4 + C = -\frac{1}{4}\cos^4 x + C.$$

5. Let $u = \cos x$. Then $du = -\sin x dx$ and hence

$$\int \sin^5 x \; dx = \int \sin^4 x (\sin x \; dx) = \int (1 - \cos^2 x)^2 (\sin x \; dx) = -\int (1 - u^2)^2 \; du$$

$$= -\int (1 - 2u^2 + u^4) \; du = -\left(u - \frac{2}{3}u^3 + \frac{1}{5}u^5\right) + C = -\cos x + \frac{2}{3}\cos^3 x - \frac{1}{5}\cos^5 x + C.$$

6. Let $u = \tan x$. Then $du = \sec^2 x dx$ and hence

$$\int \tan x \sec^2 x \; dx = \int u \; du = \frac{1}{2}u^2 + C = \frac{1}{2}\tan^2 x + C.$$

7. Let $u = \sin x$. Then $du = \cos x dx$ and hence

$$\int_0^{\pi/2} \sin^2 x \cos^3 x \; dx = \int_0^{\pi/2} \sin^2 x(1 - \sin^2 x)\cos x \; dx = \int_0^1 u^2(1 - u^2) \; du = \left(\frac{1}{3}u^3 - \frac{1}{5}u^5\right)\Big|_0^1 = \frac{2}{15}.$$

8. $\displaystyle\int_0^{\pi/4} \sin^2 x \cos^2 x \; dx = \frac{1}{4}\int_0^{\pi/4} \sin^2 2x \; dx = \frac{1}{4}\int_0^{\pi/4} \frac{1}{2}(1 - \cos 4x) \; dx = \frac{1}{8}\left(x - \frac{1}{4}\sin 4x\right)\Big|_0^{\pi/4} = \pi/32$

9. Let $u = \sec x$. Then $du = \sec x \tan x \; dx$ and hence

$$\int \sec^3 x \tan^3 x \; dx = \int \sec^2 x \underbrace{(\sec^2 x - 1)}_{\tan^2 x}(\sec x \tan x \; dx)$$

$$= \int u^2(u^2 - 1) \; du = \frac{1}{5}u^5 - \frac{1}{3}u^3 + C = \frac{1}{5}\sec^5 x - \frac{1}{3}\sec^3 x + C.$$

10. Let $u = \tan(2x - 1)$. Then $du = 2\sec^2(2x - 1)dx$ and hence

$$\int \sec^4(2x - 1)\,dx = \int \sec^2(2x - 1)\underbrace{(1 + \tan^2(2x - 1))}_{\sec^2(2x-1)}\,dx = \frac{1}{2}\int(1 + u^2)\,du$$

$$= \frac{1}{2}\left(u + \frac{1}{3}u^3\right) + C = \frac{1}{2}\tan(2x - 1) + \frac{1}{6}\tan^3(2x - 1) + C.$$

11. $\displaystyle\int_0^\pi \sin^6 x\,dx = \int_0^\pi (\sin^2 x)^3\,dx = \int_0^\pi \left[\frac{1}{2}(1 - \cos 2x)\right]^3\,dx$

$$= \frac{1}{8}\int_0^\pi (1 - 3\cos 2x + 3\cos^2 2x - \cos^3 2x)\,dx$$

$$= \frac{1}{8}\int_0^\pi \left(1 - 3\cos 2x + \frac{3}{2}(1 + \cos 4x) - (\cos 2x)(1 - \sin^2 2x)\right)\,dx$$

$$= \frac{1}{8}\int_0^\pi \left(\frac{5}{2} - 3\cos 2x + \frac{3}{2}\cos 4x\right)\,dx - \frac{1}{8}\int_0^\pi \underbrace{(1 - \sin^2 2x)\cos 2x}_{u=\sin 2x}\,dx$$

$$= \frac{1}{8}\left[\frac{5}{2}\,x - \frac{3}{2}\sin 2x + \frac{3}{8}\sin 4x\right]\Big|_0^\pi - \frac{1}{8}\cdot\frac{1}{2}\int_0^0 (1 - u^2)\,du$$

$$= \frac{1}{8}\left(\frac{5}{2}\,\pi\right)$$

$$= \frac{5\pi}{16}.$$

12. Let $u = \tan x$. Then $du = \sec^2 x\,dx$ and hence

$$\int_0^{\pi/4} \tan^3 x \sec^4 x\,dx = \int_0^{\pi/4} \tan^3 x\underbrace{(1 + \tan^2 x)}_{\sec^2 x}(\sec^2 x\,dx)$$

$$= \int_0^1 (u^3 + u^5)\,du = \left(\frac{1}{4}\,u^4 + \frac{1}{6}\,u^6\right)\Big|_0^1 = \frac{5}{12}.$$

13. Let $u = \cot x$. Then $du = -\csc^2 x\,dx$ and hence

$$\int \cot^3 x\,dx = \int \cot x\underbrace{(\csc^2 x - 1)}_{\cot^2 x}\,dx = \int \underbrace{\cot x \csc^2 x}_{u=\cot x}\,dx - \int \cot x\,dx$$

$$= -\int u\,du - \ln|\sin x| + C = -\frac{1}{2}\cot^2 x - \ln|\sin x| + C.$$

14. Let $u = \cos x$. Then $du = -\sin x\,dx$ and hence

$$\int \sin^3 x\sqrt{\cos x}\,dx = \int (\sin x)\underbrace{(1 - \cos^2 x)}_{\sin^2 x}\sqrt{\cos x}\,dx = -\int(1 - u^2)\sqrt{u}\,du$$

$$= -\int(u^{1/2} - u^{5/2})\,du = -\frac{2}{3}\,u^{3/2} + \frac{2}{7}\,u^{7/2} + C = -\frac{2}{3}\cos^{3/2}x + \frac{2}{7}\cos^{7/2}x + C.$$

15. $\displaystyle\int_0^{\pi/4} \sin x \sin 3x \, dx = \frac{1}{2}\int_0^{\pi/4} (\cos 2x - \cos 4x) \, dx = \frac{1}{2}\left(\frac{1}{2}\sin 2x - \frac{1}{4}\sin 4x\right)\Big|_0^{\pi/4} = \frac{1}{4}$

16. $\displaystyle\int_0^{\pi/2} \sin 5x \cos 3x \, dx = \frac{1}{2}\int_0^{\pi/2} (\sin 8x + \sin 2x) \, dx = \frac{1}{2}\left(-\frac{1}{8}\cos 8x - \frac{1}{2}\cos 2x\right)\Big|_0^{\pi/2} = \frac{1}{2}$

17. $\displaystyle\int_{\pi/2}^{\pi} (\sin x + \cos x)^2 \, dx = \int_{\pi/2}^{\pi} (\sin^2 x + 2\sin x \cos x + \cos^2 x) \, dx = \int_{\pi/2}^{\pi} (1 + \sin 2x) \, dx$

$$= \left(x - \frac{1}{2}\cos 2x\right)\Big|_{\pi/2}^{\pi} = \frac{\pi}{2} - 1$$

18. Let $u = \sin x$. Then $du = \cos x \, dx$ and hence

$$\int_0^{\pi/2} \sin^2 x \cos^5 x \, dx = \int_0^{\pi/2} \sin^2 x \cos^4 x \, (\cos x) \, dx = \int_0^1 u^2(1 - u^2)^2 \, du$$

$$= \int_0^1 (u^2 - 2u^4 + u^6) \, du = \left(\frac{1}{3}u^3 - \frac{2}{5}u^5 + \frac{1}{7}u^7\right)\Big|_0^1 = \frac{8}{105}.$$

19. Let $u = \tan x$. Then $du = \sec^2 x \, dx$ and hence

$$\int \tan^4 x \sec^4 x \, dx = \int \tan^4 x \underbrace{(1 + \tan^2 x)}_{\sec^2 x} \sec^2 x \, dx$$

$$= \int u^4(1 + u^2) \, du = \frac{1}{5}u^5 + \frac{1}{7}u^7 + C = \frac{1}{5}\tan^5 x + \frac{1}{7}\tan^7 x + C.$$

20. Let $u = \csc x$. Then $du = -\csc x \cot x \, dx$ and hence

$$\int \cot^5 x \, dx = \int \underbrace{(\csc^2 x - 1)^2}_{\cot^2 x} \cot x \, dx = \int (\csc^4 x - 2\csc^2 x + 1)\cot x \, dx$$

$$= \int \csc^4 x \cot x \, dx - 2\int \csc^2 x \cot x \, dx + \int \cot x \, dx = -\int u^3 \, du + 2\int u \, du + \int \cot x \, dx$$

$$= -\frac{1}{4}u^4 + u^2 + \ln|\sin x| + C = -\frac{1}{4}\csc^4 x + \csc^2 x + \ln|\sin x| + C.$$

21. Let $u = \sec x$. Then $du = \sec x \tan x \, dx$ and hence

$$\int \tan^5 x \sec^3 x \, dx = \int \underbrace{(\sec^2 x - 1)^2}_{\tan^2 x} \sec^2 x (\sec x \tan x) \, dx = \int (u^2 - 1)^2 u^2 \, du$$

$$= \int (u^6 - 2u^4 + u^2) \, du = \frac{1}{7}u^7 - \frac{2}{5}u^5 + \frac{1}{3}u^3 + C = \frac{1}{7}\sec^7 x - \frac{2}{5}\sec^5 x + \frac{1}{3}\sec^3 x + C.$$

22. Let $u = 1 + \tan x$. Then $du = \sec^2 x \, dx$ and hence

$$\int \frac{\sec^2 x}{1 + \tan x} \, dx = \int \frac{du}{u} = \ln|u| + C = \ln|1 + \tan x| + C.$$

23. $\displaystyle\int \sin x \sin 2x \, dx = \frac{1}{2}\int (\cos x - \cos 3x) \, dx = \frac{1}{2}\sin x - \frac{1}{6}\sin 3x + C$

24. $\displaystyle\int \sin 4x \cos 3x \, dx = \frac{1}{2}\int (\sin 7x + \sin x) \, dx = = -\frac{1}{14}\cos 7x - \frac{1}{2}\cos x + C$

25. Let $u = \sin x$. Then $du = \cos x dx$ and hence

$$\int \frac{\cos^3 x}{\sqrt{\sin x}} \, dx = \int \frac{\overbrace{1 - \sin^2 x}^{\cos^2 x}}{\sqrt{\sin x}} \cos x \, dx = \int \frac{1 - u^2}{\sqrt{u}} \, du$$

$$= \int (u^{-1/2} - u^{3/2}) \, du = 2u^{1/2} - \frac{2}{5}u^{5/2} + C = 2\sqrt{\sin x} - \frac{2}{5}\sin^{5/2} x + C.$$

26. $\displaystyle\int (\tan x + \cot x)^2 \, dx = \int (\underbrace{\tan^2 x}_{\sec^2 x - 1} + 2 + \underbrace{\cot^2 x}_{\csc^2 x - 1}) \, dx = \int (\sec^2 x + \csc^2 x) \, dx = \tan x - \cot x + C$

27. Let $u = 1 + \tan x$. Then $du = \sec^2 x dx$ and hence

$$\int \frac{\sec^2 x}{(1 + \tan x)^4} \, dx = \int u^{-4} \, du = -\frac{1}{3}\frac{1}{u^3} + C = -\frac{1}{3}\frac{1}{(1 + \tan x)^3} + C.$$

28. Let $u = \sin x$. Then $du = \cos x dx$ and hence

$$\int \csc x \cot x \, dx = \int \frac{\cos x}{\sin^2 x} \, dx = \int u^{-2} \, du = -\frac{1}{u} + C = -\csc x + C.$$

29. $\displaystyle\int \cos 5x \cos 3x \, dx = \frac{1}{2}\int (\cos 8x + \cos 2x) \, dx = \frac{1}{16}\sin 8x + \frac{1}{4}\sin 2x + C$

30. $\displaystyle\int \cos 7x \sin 2x \, dx = \frac{1}{2}\int (\sin 9x - \sin 5x) \, dx = -\frac{1}{18}\cos 9x + \frac{1}{10}\cos 5x + C$

31. Let $u = \cos \theta$. Then $du = -\sin \theta d\theta$ and hence

$$\int \frac{\sin^3 \theta}{\cos \theta} \, d\theta = \int \frac{1 - \cos^2 \theta}{\cos \theta} \sin \theta \, d\theta = -\int \frac{1 - u^2}{u} \, du$$

$$= -\int \left(\frac{1}{u} - u\right) \, du = -\ln|u| + \frac{1}{2}u^2 + C = -\ln|\cos \theta| + \frac{1}{2}\cos^2 \theta + C.$$

32. Let $u = \sec \theta$. Then $du = \sec \theta \tan \theta \, d\theta$ and hence

$$\int \frac{\tan^3 \theta}{\sec \theta} \, d\theta = \int \frac{\tan^2 \theta}{\sec^2 \theta} \sec \theta \tan \theta \, d\theta$$

$$= \int \frac{u^2 - 1}{u^2} \, du = \int \left(1 - \frac{1}{u^2}\right) \, du = u + \frac{1}{u} + C = \sec \theta + \cos \theta + C.$$

33. Let $u = \tan x$. Then $du = \sec^2 x dx$ and hence

$$\int \sec^2 x \sqrt{\tan x} \, dx = \int \sqrt{u} \, du = \frac{2}{3}u^{3/2} + C = \frac{2}{3}\tan^{3/2} x + C.$$

34. Since $\sin ax \sin bx = \dfrac{1}{2} [\cos(b-a)x - \cos(b+a)x]$,

$$\sin ax \sin bx \cos cx = \frac{1}{2}[\cos(b-a)x \cos cx - \cos(b+a)x \cos cx]$$

$$= \frac{1}{4}[\cos(b-a+c)x + \cos(b-a-c)x - \cos(b+a+c)x - \cos(b+a-c)x].$$

Hence

$$\int \sin ax \sin bx \cos cx \, dx = \frac{1}{4}\int [\cos(b-a+c)x + \cos(b-a-c)x] \, dx$$

$$-\frac{1}{4}\int [\cos(b+a+c)x + \cos(b+a-c)x] \, dx$$

$$= \frac{\sin(b-a+c)x}{4(b-a+c)} + \frac{\sin(b-a-c)x}{4(b-a-c)}$$

$$-\frac{\sin(b+a+c)x}{4(b+a+c)} - \frac{\sin(b+a-c)x}{4(b+a-c)} + C.$$

35. Since $\sin ax \cos bx = \frac{1}{2}[\sin(a+b)x + \sin(a-b)x]$,

$$\sin ax \cos bx \cos cx = \frac{1}{2}[\sin(a+b)x \cos cx + \sin(a-b)x \cos cx]$$

$$= \frac{1}{4}[\sin(a+b+c)x + \sin(a+b-c)x + \sin(a-b+c)x - \sin(a-b-c)x].$$

Hence

$$\int \sin ax \cos bx \cos cx \, dx = \frac{1}{4}\int [\sin(a+b+c)x + \sin(a+b-c)x] \, dx$$

$$+\frac{1}{4}\int [\sin(a-b+c)x + \sin(a-b-c)x] \, dx$$

$$= -\frac{\cos(a+b+c)x}{4(a+b+c)} - \frac{\cos(a+b-c)x}{4(a+b-c)}$$

$$-\frac{\cos(a-b+c)x}{4(a-b+c)} - \frac{\cos(a-b-c)x}{4(a-b-c)} + C.$$

36. $\displaystyle \int \sin(ax+b)\cos(cx+d) \, dx = \frac{1}{2}\int [\sin((a+c)x+b+d) + \sin((a-c)x+b-d) \, dx$

$$= -\frac{\cos((a+c)x+b+d)}{2(a+c)} - \frac{\cos((a-c)x+b-d)}{2(a-c)} + C$$

37. $\displaystyle \frac{1}{\pi}\int_{-\pi}^{\pi} \cos n\theta \, d\theta = \frac{1}{n\pi} \sin n\theta \Big|_{-\pi}^{\pi} = 0$

38. $\displaystyle \frac{2}{\pi}\int_{0}^{\pi} \sin n\theta \, d\theta = -\frac{2}{n\pi} \cos n\theta \Big|_{0}^{\pi} = \begin{cases} 0, & n \text{ even} \\ \frac{4}{n\pi}, & n \text{ odd} \end{cases}$

39. Let $u = \dfrac{n\pi x}{L}$. Then $du = \dfrac{n\pi}{L} dx$ and hence

$$\int_{-L}^{L} \cos^2\left(\frac{n\pi x}{L}\right) dx = \frac{L}{n\pi}\int_{-n\pi}^{n\pi} \cos^2 u \, du$$

671

$$= \frac{L}{n\pi} \int_{-n\pi}^{n\pi} \frac{1}{2} \left(1 + \cos 2u\right) du = \frac{L}{2n\pi} \left[u + \frac{1}{2} \sin 2u \right]\Big|_{-n\pi}^{n\pi} = L.$$

40. $\int_0^{2\pi} \sin^2 nx \, dx = \frac{1}{2} \int_0^{2\pi} (1 - \cos 2nx) \, dx = \frac{1}{2} \left(x - \frac{1}{2n} \sin 2nx \right)\Big|_0^{2\pi} = \pi$

41. $\int_0^{2\pi} \sin nx \cos mx \, dx = \frac{1}{2} \int_0^{2\pi} [\sin(n+m)x + \sin(n-m)x] \, dx$

$$= \frac{1}{2} \left[-\frac{\cos(n+m)x}{n+m} - \frac{\cos(n-m)x}{n-m} \right]\Big|_0^{2\pi} = 0$$

42. $\int_0^{2\pi} \cos nx \cos mx \, dx = \frac{1}{2} \int_0^{2\pi} [\cos(n+m)x + \cos(n-m)x] \, dx$

$$= \frac{1}{2} \left[\frac{\sin(n+m)x}{n+m} + \frac{\sin(n-m)x}{n-m} \right]\Big|_0^{2\pi} = 0$$

43. Since $\tan^n x = \tan^{n-2} x \tan^2 x = \tan^{n-2} x (\sec^2 x - 1)$, it follows that

$$\int \tan^n x \, dx = \int \underbrace{\tan^{n-2} x \sec^2 x}_{u = \tan x} \, dx - \int \tan^{n-2} x \, dx$$

$$= \int u^{n-2} \, du - \int \tan^{n-2} x \, dx = \frac{\tan^{n-1} x}{n-1} - \int \tan^{n-2} x \, dx.$$

44. Since $\cot^n x = \cot^{n-2} x \cot^2 x = \cot^{n-2} x (\csc^2 x - 1)$, it follows that

$$\int \cot^n x \, dx = \int \underbrace{\cot^{n-2} x \csc^2 x}_{u = \cot x} \, dx - \int \cot^{n-2} x \, dx$$

$$= -\int u^{n-2} \, du - \int \cot^{n-2} x \, dx = -\frac{\cot^{n-1} x}{n-1} - \int \cot^{n-2} x \, dx.$$

45. Adding the two identities, we find that

$$\sin(\theta_1 + \theta_2) + \sin(\theta_1 - \theta_2) = 2 \sin \theta_1 \cos \theta_2 \quad \Longrightarrow \quad \sin \theta_1 \cos \theta_2 = \frac{1}{2} \left[\sin(\theta_1 + \theta_2) + \sin(\theta_1 - \theta_2) \right].$$

46. Subtracting the two identities, we find that

$$\cos(\theta_1 - \theta_2) - \cos(\theta_1 + \theta_2) = 2 \sin \theta_1 \sin \theta_2 \quad \Longrightarrow \quad \sin \theta_1 \sin \theta_2 = \frac{1}{2} \left[\cos(\theta_1 - \theta_2) - \cos(\theta_1 + \theta_2) \right].$$

47. Adding the two identities in Exercise 46, we find that

$$\cos(\theta_1 - \theta_2) + \cos(\theta_1 + \theta_2) = 2 \cos \theta_1 \cos \theta_2 \quad \Longrightarrow \quad \cos \theta_1 \cos \theta_2 = \frac{1}{2} \left[\cos(\theta_1 + \theta_2) + \cos(\theta_1 - \theta_2) \right].$$

48. $\dfrac{1}{\pi/2} \int_{-\pi/4}^{\pi/4} \underbrace{\tan x \sec^2 x}_{u = \tan x} \, dx = \dfrac{2}{\pi} \int_{-1}^{1} u \, du = \dfrac{1}{\pi} u^2 \Big|_{-1}^{1} = 0$

49. $\dfrac{1}{\pi}\displaystyle\int_0^\pi \sin^2 x\,dx = \dfrac{1}{2\pi}\displaystyle\int_0^\pi (1-\cos 2x)\,dx = \dfrac{1}{2\pi}\left(x-\dfrac{1}{2}\sin 2x\right)\Big|_0^\pi = 1/2.$

50. Let dV be the volume of a differential cross-section disk of thickness dx. Then $dV = \pi y^2 dx = \pi \tan^2 x \sec^2 x\,dx$ and hence the volume of revolution is

$$V = \int_0^{\pi/4} \pi \underbrace{\tan^2 x \sec^2 x}_{u=\tan x}\,dx = = \pi \int_0^1 u^2\,du = \dfrac{\pi}{3}\,u^3\Big|_0^1 = \dfrac{\pi}{3}.$$

51. Let dV be the volume of a differential cross-section disk of thickness dx. Then $dV = \pi y^2 dx = \pi \tan^4 x \sec^2 x\,dx$ and hence the volume of revolution is

$$V = \int_0^{\pi/4} \pi \underbrace{\tan^4 x \sec^2 x}_{u=\tan x}\,dx = \pi \int_0^1 u^4\,du = \dfrac{\pi}{5}\,u^5\Big|_0^1 = \dfrac{\pi}{5}.$$

52. Area $= \displaystyle\int_0^{\pi/4} 2\tan^2 x\,dx = 2\int_0^{\pi/4}(\sec^2 x - 1)\,dx = 2(\tan x - x)|_0^{\pi/4} = 2 - \dfrac{\pi}{2}$

53. Area $= \displaystyle\int_0^{\pi/4} \sec x\,dx = \ln|\sec x + \tan x|\big|_0^{\pi/4} = \ln(1+\sqrt{2})$

54. (a) $v(t) = \displaystyle\int a(t)\,dt = \int \cos^2 t\,dt = \dfrac{1}{2}\int(1+\cos 2t)\,dt = \dfrac{1}{2}\left(t+\dfrac{1}{2}\sin 2t\right)+C.$ Since $v(0)=C=0$,

$v(t)=\dfrac{1}{2}t+\dfrac{1}{4}\sin 2t.$

(b) $s(t) = \displaystyle\int v(t)\,dt = \int\left(\dfrac{1}{2}t+\dfrac{1}{4}\sin 2t\right)dt = \dfrac{1}{4}t^2 - \dfrac{1}{8}\cos 2t + C.$ Since $s(0) = -\dfrac{1}{8}+C=4$, $C=$

$\dfrac{33}{8}$ and hence $s(t) = \dfrac{1}{4}t^2 - \dfrac{1}{8}\cos 2t + \dfrac{33}{8}.$

55. $s = \displaystyle\int_0^{\pi/6} v(t)\,dt = \int_0^{\pi/6}\sin 3t\cos 2t\,dt = \dfrac{1}{2}\int_0^{\pi/6}(\sin 5t + \sin t)\,dt = \left(-\dfrac{1}{10}\cos 5t - \dfrac{1}{2}\cos t\right)\Big|_0^{\pi/6} =$

$\dfrac{1}{5}(3-\sqrt{3})$

56. Since $v(t+\pi/2) = -v(t)$ for $0 \le t \le \pi/2$,

$$s = 2\int_0^{\pi/2} v(t)\,dt = 2\int_0^{\pi/2}\sin^2 t\cos^3 t\,dt = 2\int_0^{\pi/2}\underbrace{\sin^2 t\,(1-\sin^2 t)\cos t}_{u=\sin t}\,dt$$

$$= 2\int_0^1 u^2(1-u^2)\,dt = 2\left(\dfrac{1}{3}u^3 - \dfrac{1}{5}u^5\right)\Big|_0^1 = \dfrac{4}{15}.$$

9.3 The Inverse Trigonometric Functions

1. $\text{Sin}^{-1}(1/2) = \pi/6$

2. $\text{Cos}^{-1}(-\sqrt{3}/2) = 5\pi/6$

3. $\text{Tan}^{-1}(0) = 0$

4. $\text{Sec}^{-1}(2) = \pi/3$

5. $\text{Cos}^{-1}(-1) = \pi$

6. $\text{Cot}^{-1}(-\sqrt{3}) = -\pi/6$

7. $\text{Sin}^{-1}(1) + \text{Cos}^{-1}(1) = \pi/2$

8. $\text{Tan}^{-1}(-1/\sqrt{3}) = -\pi/6$

9. $\tan(\text{Sin}^{-1}(1/2)) = \tan(\pi/2) = \sqrt{3}/3$

10. $\text{Cos}^{-1}(\sin\pi/4) = \pi/4$

11. $\sec(\text{Tan}^{-1}(\sqrt{3})) = 2$

12. $\text{Cot}^{-1}(\tan(2\pi/3)) = 5\pi/6$

13. $\cos(2\text{Tan}^{-1}(\sqrt{3})) = 2\cos^2(\text{Tan}^{-1}(\sqrt{3})) - 1 = -1/2$

14. $\sin(2\text{Cot}^{-1}(3/4)) = 2\sin(\text{Cot}^{-1}(3/4))\cos(\text{Cot}^{-1}(3/4)) = 2(4/5)(3/5) = 24/25$

15. $\tan^2(\pi - \text{Sin}^{-1}(\sqrt{2}/2)) = \tan^2(\pi - \pi/4) = 1$

16. $\text{Sec}^{-1}(2\tan(\pi/4)) = \text{Sec}^{-1}2 = \pi/3$

17. $\text{Cos}^{-1}(2 - \sqrt{2}\sin(\pi/4)) = \text{Cos}^{-1}1 = 0$

18. $\sec(2\text{Tan}^{-1}1) = \sec(\pi/2)$ is undefined

19. $\tan(\text{Sec}^{-1}2) = \tan\pi/3 = \sqrt{3}$

20. $y = \cos(\text{Sin}^{-1}x) = \sqrt{1-x^2}$

21. $y = \tan(\text{Sin}^{-1}x) = x/\sqrt{1-x^2}$

22. $y = \tan(\text{Tan}^{-1}x) = x$

23. $y = \sin(2\text{Sin}^{-1}(x)) = 2\sin(\text{Sin}^{-1}x)\cos(\text{Sin}^{-1}x) = 2x\sqrt{1-x^2}$

24. $y = \sin(\text{Tan}^{-1}x) = x/\sqrt{1+x^2}$

25. $y = \sin(\text{Sec}^{-1}x) = \sqrt{x^2-1}/x$

26. $y = \cos(\text{Tan}^{-1}x) = 1/\sqrt{1+x^2}$

27. $y = \tan(\text{Cot}^{-1}x) = 1/x$

28. $y = \tan(\text{Sec}^{-1}x) = \sqrt{x^2-1}$

29. $y = \cos(\text{Csc}^{-1}x) = \sqrt{x^2-1}/x$

30. For $-1 \leq x \leq 1$, $\text{Sin}^{-1}x$ is the angle whose sine is equal to x; *i.e.*, $\sin(\text{Sin}^{-1}x) = x$. Recall that since the sine of an angle lies between -1 and $+1$, $\text{Sin}^{-1}x$ is defined only when $-1 \leq x \leq 1$. Similar identities are true for the remaining trig functions:

$$\begin{aligned}
\cos(\text{Cos}^{-1}x) &= x, & &-1 \leq x \leq 1 \\
\tan(\text{Tan}^{-1}x) &= x, & &\text{for all real } x \\
\sec(\text{Sec}^{-1}x) &= x, & &x \leq -1 \text{ or } x \geq 1 \\
\csc(\text{Csc}^{-1}x) &= x, & &x \leq -1 \text{ or } x \geq 1 \\
\cot(\text{Cot}^{-1}x) &= x, & &\text{for all real } x
\end{aligned}$$

31. No. $\text{Sec}^{-1}(\cos \pi/4) = \text{Sec}^{-1}(\sqrt{2}/2)$, which is undefined since $\sqrt{2}/2 < 1$.

32. (a) If y lies in the interval $[-1, 1]$, there is one and only one value of x in the interval $[\pi/2, 3\pi/2]$ such that $y = \sin x$. Hence the sine function has a legitimate inverse when x is restricted to the interval $[\pi/2, 3\pi/2]$.

    (b) If x is restricted to the interval $[0, \pi/2]$, only non-negative values of $y = \sin x$ are obtained and hence the inverse, $\text{Sin}^{-1}x$ would only be defined for $0 \leq x \leq 1$.

    (c) It should be defined for all possible values of the function; *i.e.*, its domain should equal the range of the function.

33. $\tan \theta = x/100 \implies \theta = \text{Tan}^{-1}(x/100)$.

34. $\tan \theta = 9/30 \implies \theta = \text{Tan}^{-1}(3/10) \approx 16.7°$. In general, $\theta = \text{Tan}^{-1}(9/l)$.

35. $\sin \theta = 1/2 \implies \theta = \text{Sin}^{-1}(1/2) = \pi/6$.

36. $\dfrac{\sin \theta_1}{n_1} = \dfrac{\sin \theta_2}{n_2} \implies \sin \theta_2 = \dfrac{n_2}{n_1} \sin \theta_1 \implies \theta_2 = \text{Sin}^{-1}\left(\dfrac{n_2}{n_1} \sin \theta_1\right)$.

37. $\tan \theta = x/20 \implies \theta = \text{Tan}^{-1}(x/20)$.

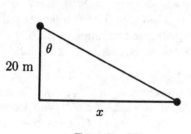

Exercise 37

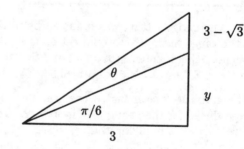

Exercise 38

38. $\tan \pi/6 = y/3 \implies y = 3\tan \pi/6 = \sqrt{3}$. Hence $\tan(\theta + \pi/6) = (3 - \sqrt{3} + \sqrt{3})/3 = 1 \implies \theta + \pi/6 = \pi/4 \implies \theta = \pi/12$.

39. Let θ_1 be the angle between the kite and the horizontal. Then $\sin \theta_1 = 100/(100\sqrt{2}) = \sqrt{2}/2 \implies \theta_1 = \pi/4$. Now, let θ_2 be the new angle between the kite and the horizontal after it has lost half its altitude. Then $\sin \theta_2 = 50/(100\sqrt{2}) = 1/(2\sqrt{2}) \implies \theta_2 = \text{Sin}^{-1}(1/(2\sqrt{2})) \approx 20.705°$. Hence the change in angle is $\theta_1 - \theta_2 = \pi/4 - \text{Sin}^{-1}(1/(2\sqrt{2})) \approx 24.295°$, or 0.478 radians.

40. The graphs are shown below.

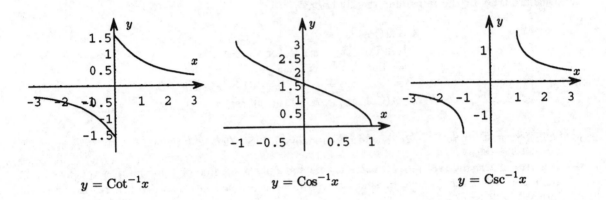

$$y = \text{Cot}^{-1}x \qquad\qquad y = \text{Cos}^{-1}x \qquad\qquad y = \text{Csc}^{-1}x$$

41. The graph on the left is $f(x) = \text{Cos}^{-1}(1/x)$ using the range $[1, \pi] \times [0, 2]$. The graph on the right is that of $f(x) = \text{Cos}^{-1}(1/x)$, $g(x) = \sec x$, and $y = x$ using the range $[-1.4, 3.35] \times [-.6, 2.55]$.

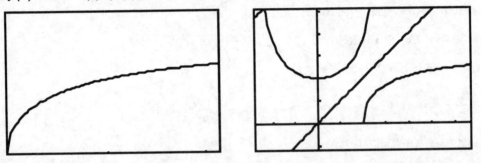

Remark: This range setting was obtained by using the **RANGE** program with **X=Y=1** and **S=0.05**. This is the range setting you would get on the TI-81. The Casio would be similar. For the HP-28S a good range setting is entering **X=Y=1** and **S=0.075**. This will give the range $[-4.1, 6.1] \times [-.125, 2.2]$. For the HP-48SX use **X=Y=1** and **S=0.05**. This will give the range setting $[-2.25, 4.25] \times [-.55, 2.6]$.

From the graph it is seen that if the graph of $f(x)$ is reflected across the line $y = x$, the graph of $g(x)$ is produced for the values of x in $[0, \pi/2]$. To show this set $y = \text{Cos}^{-1}(1/x)$. Then $\cos y = 1/x$. Therefore $\sec y = x$ implying that $y = \text{Sec}^{-1}x$. A similar statement can be made for the functions $f(x) = \text{Sin}^{-1}(1/x)$ and $f(x) = \text{Tan}^{-1}(1/x)$. The function $f(x) = \text{Sin}^{-1}(1/x)$ is the inverse of $\csc x$ and the function $f(x) = \text{Tan}^{-1}(1/x)$ is the inverse of $\cot x$.

42. From the figure we see that

$$\tan \theta = \frac{x}{50}, \quad s = 50\alpha, \quad \text{and} \quad 2\theta + \pi - \alpha = \pi.$$

Therefore $\alpha = 2\theta$. So $s = 100\theta$ and $\theta = \text{Tan}^{-1}(x/50)$. This implies that $s = 100\text{Tan}^{-1}(x/50)$. Now the domain of $s = 100\theta$ is $[0, \pi/4]$ with range $[0, 25\pi]$. The domain for $s = 100\text{Tan}^{-1}(x/50)$ is $[0, 50]$ with range $[0, 25\pi]$. The graph below on the left is $s = 100\theta$ using the range $[0, \pi/4] \times [0, 80]$ and the graph below on the right is $s = 100\text{Tan}^{-1}(x/50)$ using the range $[0, 50] \times [0, 80]$.

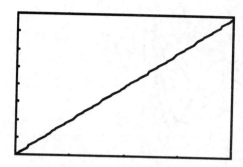 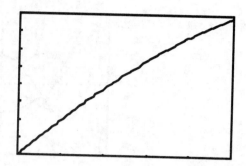

43. (a) Since $-1 \le \cos\theta \le 1$ the range of $T_n(x)$ is $[-1, 1]$. Since the domain of $y = \text{Cos}^{-1}x$ is also $[-1, 1]$, then the domain of $T_n(x)$ is $[-1, 1]$. This explains why the range for the graphs in this exercise all use the range $[-1, 1] \times [-1, 1]$. The graph below is that of $T_n(x)$, $n = 1, 2, 3, 4,$ and 5.

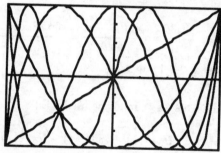

Let $\theta = \text{Cos}^{-1}x$, so $x = \cos\theta$. Then

$$T_2(x) = \cos 2\theta = 2\cos^2\theta - 1 = 2x^2 - 1$$

and

$$
\begin{aligned}
T_3(x) \;=\; \cos 3\theta \;&=\; \cos(2\theta + \theta) \\
&=\; \cos 2\theta \cos\theta - \sin 2\theta \sin\theta \\
&=\; (2\cos^2\theta - 1)\cos\theta - (2\sin\theta\cos\theta)\sin\theta \\
&=\; 2\cos^3\theta - \cos\theta - 2\sin^2\theta\cos\theta \\
&=\; 2\cos^3\theta - \cos\theta - 2(1 - \cos^2\theta)\cos\theta \\
&=\; 2\cos^3\theta - \cos\theta - 2\cos\theta + 2\cos^3\theta \\
&=\; 4\cos^3\theta - 3\cos\theta \\
&=\; 4x^3 - 3x.
\end{aligned}
$$

Notice that $T_n(x)$ defined in this way cannot be used if $|x| > 1$. For these values of x the following formula is used:

$$T_{n+1}(x) = 2xT_n(x) - T_{n-1}(x),$$

where $T_0(x) = 1$ and $T_1(x) = x$.

(b) Since $-1 \le \sin\theta \le 1$ the range of $U_n(x)$ is $[-1, 1]$. Since the domain of $y = \text{Cos}^{-1}x$ is also $[-1, 1]$, then the domain of $U_n(x)$ is $[-1, 1]$. This explains why the range for the graphs in this exercise all use the range $[-1, 1] \times [-2.5, 2.5]$. The graph below is that of $U_n(x)$, $n = 1, 2, 3, 4,$ and 5.

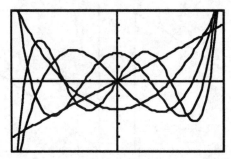

Let $\theta = \text{Cos}^{-1}x$, so $x = \cos\theta$. Then

$$U_1(x) = \frac{\sin 2\theta}{\sin\theta} = \frac{2\sin\theta\cos\theta}{\sin\theta} = 2\cos\theta = 2x$$

and

$$U_2(x) = \frac{\sin 3\theta}{\sin\theta} = \frac{\sin(2\theta + \theta)}{\sin\theta} = \frac{\sin 2\theta\cos\theta + \cos 2\theta\sin\theta}{\sin\theta}$$

$$= \frac{2\sin\theta\cos^2\theta + (2\cos^2\theta - 1)\sin\theta}{\sin\theta} = 2\cos^2\theta + 2\cos^2\theta - 1 = 4\cos^2\theta - 1 = 4x^2 - 1.$$

Notice that $U_n(x)$ defined in this way cannot be used if $|x| > 1$. For these values of x the following formula is used:

$$U_{n+1}(x) = 2xU_n(x) - U_{n-1}(x),$$

where $U_0(x) = 1$ and $U_1(x) = 2x$.

9.4   Derivatives of the Inverse Trigonometric Functions

1. $y = \text{Sin}^{-1}3x \Longrightarrow y' = \dfrac{3}{\sqrt{1 - 9x^2}}$

2. $f(x) = x\text{Tan}^{-1}(x - 1) \Longrightarrow f'(x) = x\,\dfrac{1}{1 + (x-1)^2} + \text{Tan}^{-1}(x - 1) = \dfrac{x}{x^2 - 2x + 2} + \text{Tan}^{-1}(x - 1)$

3. $f(t) = \text{Sin}^{-1}\sqrt{t} \Longrightarrow f'(t) = \dfrac{1}{\sqrt{1 - t}}\,\dfrac{1}{2\sqrt{t}} = \dfrac{1}{2\sqrt{t}\sqrt{1 - t}}$

4. $y = x^3\text{Csc}^{-1}(1 + x) \Longrightarrow$

$$y' = x^3\,\frac{-1}{(1 + x)\sqrt{(1 + x)^2 - 1}} + (\text{Csc}^{-1}(1 + x))3x^2 = \frac{-x^3}{(1 + x)\sqrt{x^2 + 2x}} + 3x^2\text{Csc}^{-1}(1 + x)$$

5. $f(x) = \text{Sin}^{-1}e^{-x} \Longrightarrow f'(x) = \dfrac{1}{\sqrt{1 - (e^{-x})^2}}\,(-e^{-x}) = \dfrac{-e^{-x}}{\sqrt{1 - e^{-2x}}}$

6. $y = \text{Tan}^{-1}\sqrt{x^2 - 1} \Longrightarrow y' = \dfrac{1}{1 + (x^2 - 1)}\,\dfrac{1}{2}\,(x^2 - 1)^{-1/2}(2x) = \dfrac{1}{x\sqrt{x^2 - 1}}$

7. $y = \sqrt{\text{Cos}^{-1}x} \Longrightarrow y' = \dfrac{1}{2}\,(\text{Cos}^{-1}x)^{-1/2}\,\dfrac{-1}{\sqrt{1 - x^2}} = \dfrac{-1}{2\sqrt{(\text{Cos}^{-1}x)(1 - x^2)}}$

8. $y = \text{Sin}^{-1}x^2 - x\text{Sec}^{-1}(x+3) \implies$

$$
\begin{aligned}
f'(x) &= \frac{1}{\sqrt{1-(x^2)^2}} - \left[x \frac{1}{(x+3)\sqrt{(x+3)^2-1}} + \text{Sec}^{-1}(x+3) \right] \\
&= \frac{2x}{\sqrt{1-x^4}} - \frac{x}{(x+3)\sqrt{x^2+6x+8}} - \text{Sec}^{-1}(x+3)
\end{aligned}
$$

9. $f(x) = \ln\text{Tan}^{-1}x \implies f'(x) = \frac{1}{\text{Tan}^{-1}x} \frac{1}{1+x^2}$

10. $y = \text{Tan}^{-1}\left(\frac{1-x}{1+x}\right) \implies y' = \frac{1}{1+\left(\frac{1-x}{1+x}\right)^2} \frac{(1+x)(-1)-(1-x)(1)}{(1+x)^2} = \frac{-1}{1+x^2}$

11. $y = \text{Sec}^{-1}(1/x) \implies y' = \frac{1}{\frac{1}{x}\sqrt{\left(\frac{1}{x}\right)^2-1}}\left(-\frac{1}{x^2}\right) = \frac{-1}{\sqrt{1-x^2}}$

12. $y = e^{\text{Tan}^{-1}\sqrt{x}} \implies y' = e^{\text{Tan}^{-1}\sqrt{x}} \frac{1}{1+(\sqrt{x})^2}\left(\frac{1}{2\sqrt{x}}\right) = \frac{e^{\text{Tan}^{-1}\sqrt{x}}}{2\sqrt{x}(1+x)}$

13. $f(x) = \frac{\text{Tan}^{-1}x}{1+x^2} \implies f'(x) = \frac{(1+x^2)\frac{1}{1+x^2}-(\text{Tan}^{-1}x)(2x)}{(1+x^2)^2} = \frac{1-2x\text{Tan}^{-1}x}{(1+x^2)^2}$

14. $f(x) = x^2\text{Sin}^{-1}2x \implies f'(x) = x^2\frac{1}{\sqrt{1-(2x)^2}}(2) + (\text{Sin}^{-1}2x)(2x) = \frac{2x^2}{\sqrt{1-4x^2}} + 2x\text{Sin}^{-1}2x$

15. Let $u = x/2$. Then $du = (1/2)dx$ and hence

$$
\int \frac{1}{4+x^2}\, dx = \frac{1}{4}\int \frac{1}{1+\left(\frac{x}{2}\right)^2}\, dx = \frac{1}{2}\int \frac{1}{1+u^2}\, du = \frac{1}{2}\text{Tan}^{-1}u + C = \frac{1}{2}\text{Tan}^{-1}\frac{x}{2} + C.
$$

16. $\int_0^{\sqrt{3}/2} \frac{dx}{\sqrt{1-x^2}} = \text{Sin}^{-1}x\Big|_0^{\sqrt{3}/2} = \pi/3$

17. Let $u = 3x$. Then $du = 3dx$ and hence

$$
\int \frac{1}{1+9x^2}\, dx = \frac{1}{3}\int \frac{1}{1+u^2}\, du = \frac{1}{3}\text{Tan}^{-1}u + C = \frac{1}{3}\text{Tan}^{-1}3x + C.
$$

18. Let $u = 2x$. Then $du = 2dx$ and hence

$$
\int \frac{5}{\sqrt{1-4x^2}}\, dx = 5\int \frac{1}{\sqrt{1-(2x)^2}}\, dx = \frac{5}{2}\int \frac{1}{\sqrt{1-u^2}}\, du = \frac{5}{2}\text{Sin}^{-1}u + C = \frac{5}{2}\text{Sin}^{-1}2x + C.
$$

19. Let $u = x/4$. Then $du = (1/4)dx$ and hence

$$
\int_{8\sqrt{3}/3}^8 \frac{1}{2x\sqrt{x^2-16}}\, dx = \frac{1}{8}\int_{8\sqrt{3}/3}^8 \frac{(1/4)\, dx}{\frac{x}{4}\sqrt{\left(\frac{x}{4}\right)^2-1}} = \frac{1}{8}\int_{2\sqrt{3}/2}^2 \frac{1}{u\sqrt{u^2-1}}\, du = \frac{1}{8}\text{Sec}^{-1}u\Big|_{2\sqrt{3}/3}^2 = \frac{\pi}{48}.
$$

20. Let $u = \sin x$. Then $du = \cos x\,dx$ and hence

$$\int \frac{\cos x}{1 + \sin^2 x}\,dx = \int \frac{1}{1 + u^2}\,du = \text{Tan}^{-1}u + C = \text{Tan}^{-1}\sin x + C.$$

21. Let $u = 1 - x^2$. Then $du = -2x\,dx$ and hence

$$\int \frac{x}{\sqrt{1 - x^2}}\,dx = -\frac{1}{2}\int u^{-1/2}\,du = -\sqrt{u} + C = -\sqrt{1 - x^2} + C.$$

22. $\displaystyle\int_{\sqrt{2}}^{2} \frac{1}{x\sqrt{x^2 - 1}}\,dx = \text{Sec}^{-1}x\Big|_{\sqrt{2}}^{2} = \pi/12$

23. Let $u = e^x$. Then $du = e^x\,dx$ and hence

$$\int \frac{1}{\sqrt{e^{2x} - 1}}\,dx = \int \frac{1}{u\sqrt{u^2 - 1}}\,du = \text{Sec}^{-1}u + C = \text{Sec}^{-1}e^x + C.$$

24. Let $u = x^2$. Then $du = 2x\,dx$ and hence

$$\int \frac{x}{\sqrt{1 - x^4}}\,dx = \frac{1}{2}\int \frac{1}{\sqrt{1 - u^2}}\,du = \frac{1}{2}\text{Sin}^{-1}u + C = \frac{1}{2}\text{Sin}^{-1}x^2 + C.$$

25. Let $u = e^{\sqrt{x}}$. Then $du = \dfrac{e^{\sqrt{x}}}{2\sqrt{x}}\,dx$ and hence

$$\int \frac{e^{\sqrt{x}}}{\sqrt{x}(1 + e^{2\sqrt{x}})}\,dx = 2\int \frac{du}{1 + u^2} = 2\text{Tan}^{-1}u + C = 2\text{Tan}^{-1}e^{\sqrt{x}} + C.$$

26. Let $u = x^3$. Then $du = 3x^2\,dx$ and hence

$$\int \frac{x^2}{\sqrt{1 - x^6}}\,dx = \frac{1}{3}\int \frac{1}{\sqrt{1 - u^2}}\,du = \frac{1}{3}\text{Sin}^{-1}u + C = \frac{1}{3}\text{Sin}^{-1}x^3 + C.$$

27. Let $u = x^3$. Then $du = 3x^2\,dx$ and hence

$$\int \frac{x^2}{1 + x^6}\,dx = \frac{1}{3}\int \frac{1}{1 + u^2}\,du = \frac{1}{3}\text{Tan}^{-1}u + C = \frac{1}{3}\text{Tan}^{-1}x^3 + C.$$

28. Let $u = \cos x$. Then $du = \sin x\,dx$ and hence

$$\int \frac{\sin x}{1 + \cos^2 x}\,dx = -\int \frac{1}{1 + u^2}\,du = -\text{Tan}^{-1}u + C = -\text{Tan}^{-1}\cos x + C.$$

29. Let $u = \text{Tan}^{-1}x$. Then $du = \dfrac{1}{1 + x^2}\,dx$ and hence

$$\int_0^1 \frac{\text{Tan}^{-1}x}{1 + x^2}\,dx = \int_0^{\pi/4} u\,du = \frac{1}{2}u^2\Big|_0^{\pi/4} + C = \frac{\pi^2}{32}.$$

30. Let $u = \text{Tan}^{-1}3x$. Then $du = \dfrac{3}{1 + 9x^2}\,dx$ and hence

$$\int \frac{\text{Tan}^{-1}3x}{1 + 9x^2}\,dx = \frac{1}{3}\int u\,du = \frac{1}{6}u^2 + C = \frac{1}{6}(\text{Tan}^{-1}3x)^2 + C.$$

31. Let $u = \text{Cos}^{-1}2x$. Then $du = \dfrac{-2}{\sqrt{1-4x^2}}\,dx$ and hence

$$\int \frac{(\text{Cos}^{-1}2x)^3}{\sqrt{1-4x^2}}\,dx = -\frac{1}{2}\int u^3\,du = -\frac{1}{8}u^4 + C = -\frac{1}{8}\left(\text{Cos}^{-1}2x\right)^4 + C.$$

32. Let $u = \text{Sin}^{-1}x$. Then $du = \dfrac{1}{\sqrt{1-x^2}}\,dx$ and hence

$$\int \frac{1}{(\text{Sin}^{-1}x)\sqrt{1-x^2}}\,dx = \int \frac{1}{u}\,du = \ln|u| + C = \ln\left|\text{Sin}^{-1}x\right| + C.$$

33. Let $u = (1/2)\sin x$. Then $du = (1/2)\cos x\,dx$ and hence

$$\int_0^{\pi/2} \frac{\cos x}{\sqrt{4-\sin^2 x}}\,dx = \frac{1}{2}\int_0^{\pi/2} \frac{\cos x}{\sqrt{1-\left(\frac{\sin x}{2}\right)^2}}\,dx = \int_0^{1/2} \frac{du}{\sqrt{1-u^2}} = \text{Sin}^{-1}u\Big|_0^{1/2} = \pi/6.$$

34. Let $u = \dfrac{x}{\sqrt{2}}$. Then $du = \dfrac{1}{\sqrt{2}}\,dx$ and hence

$$\int_1^2 \frac{dx}{x^2 + 2} = \frac{1}{2}\int_1^2 \frac{dx}{\left(\frac{x}{\sqrt{2}}\right)^2 + 1} = \frac{\sqrt{2}}{2}\int_{1/\sqrt{2}}^{2/\sqrt{2}} \frac{du}{1+u^2} = \frac{\sqrt{2}}{2}\text{Tan}^{-1}u\Big|_{1/\sqrt{2}}^{2/\sqrt{2}} \approx 0.24.$$

35. Let $u = 4x/5$. Then $du = (4/5)dx$ and hence

$$\int_2^3 \frac{dx}{x\sqrt{16x^2 - 25}} = \frac{4}{25}\int_2^3 \frac{dx}{\left(\frac{4x}{5}\right)\sqrt{\left(\frac{4x}{5}\right)^2 - 1}} = \frac{1}{5}\int_{8/5}^{12/5} \frac{du}{u\sqrt{u^2-1}} = \frac{1}{5}\text{Sec}^{-1}u\Big|_{8/5}^{12/5} \approx 0.05.$$

36. Let $u = 2x$. Then $du = 2dx$ and hence

$$\int \frac{dx}{2x\sqrt{4x^2-1}} = \frac{1}{2}\int \frac{du}{u\sqrt{u^2-1}} = \frac{1}{2}\text{Sec}^{-1}u + C = \frac{1}{2}\text{Sec}^{-1}(2x) + C.$$

37. Let $u = \ln x$. Then $du = (1/x)dx$ and hence

$$\int_1^{\sqrt{e}} \frac{1}{x\sqrt{1-\ln^2 x}}\,dx = \int_0^{1/2} \frac{1}{\sqrt{1-u^2}}\,du = \text{Sin}^{-1}u\Big|_0^{1/2} = \pi/6.$$

38. $\dfrac{dy}{dx} = 4 + y^2 \implies \dfrac{dy}{4+y^2} = dx \implies \displaystyle\int \frac{dy}{4+y^2} = \int dx = x + C.$ Now, let $u = y/2$. Then $du = (1/2)dy$ and hence

$$\int \frac{dy}{4+y^2} = \frac{1}{4}\int \frac{dy}{1+\left(\frac{y}{2}\right)^2} = \frac{1}{2}\int \frac{du}{1+u^2} = \frac{1}{2}\text{Tan}^{-1}u = \frac{1}{2}\text{Tan}^{-1}\frac{y}{2}.$$

Hence

$$\frac{1}{2}\text{Tan}^{-1}\frac{y}{2} = x + C \implies \text{Tan}^{-1}\frac{y}{2} = 2x + C \implies y = 2\tan(2x + C).$$

Since $y(0) = 2\tan C = 1$, $C = \text{Tan}^{-1}(1/2)$. Hence

$$y = 2\tan\left(2x + \text{Tan}^{-1}\frac{1}{2}\right).$$

39. $\frac{dy}{dx} = \sqrt{1-y^2} \implies \frac{dy}{\sqrt{1-y^2}} = dx \implies \text{Sin}^{-1}y = x + C \implies y = \sin(x+C)$. Since $y(0) = \sin C = 1$,

$C = \text{Sin}^{-1}1 = \pi/2$ and hence $y = \sin(x + \pi/2) = \cos x$.

40. First note that $dx/dt = 120$ km/hr $= 100/3$ m/s. Now, $\theta = \text{Tan}^{-1}(x/20) \implies$

$$\frac{d\theta}{dt} = \frac{1}{1 + \left(\frac{x}{20}\right)^2} \left(\frac{1}{20}\right) \left(\frac{dx}{dt}\right).$$

Hence

$$\left.\frac{d\theta}{dt}\right|_{x=40} = \frac{1}{1+2^2} \left(\frac{1}{20}\right) \left(\frac{100}{3}\right) = \frac{1}{3} \text{ rad/s}.$$

In other words, the radar gun is rotating at the rate of 1/3 radians/s when the car is 40 meters away.

41. Let x be the horizontal distance of the eye from the wall. Then

$$\theta = \text{Tan}^{-1}\left(\frac{4}{x}\right) - \text{Tan}^{-1}\left(\frac{1}{x}\right)$$

$$\implies \frac{d\theta}{dx} = \left[\frac{1}{1 + \left(\frac{4}{x}\right)^2} \left(-\frac{4}{x^2}\right) - \frac{1}{1 + \left(\frac{1}{x}\right)^2} \left(-\frac{1}{x^2}\right)\right] = \frac{12 - 3x^2}{(x^2 + 16)(x^2 + 1)}.$$

It follows that $d\theta/dx = 0 \iff x = \pm 2$. We reject the value $x = -2$. Since $d\theta/dx > 0$ for $x < 2$ and $d\theta/dx < 0$ for $x > 2$, $x = 2$ is a relative maximum for θ. Thus, the maximum visibility occurs when the observer stands 2 meters from the wall.

42. Let x and y stand for the horizontal and vertical displacements of the ladder, as illustrated in the figure below, and θ the angle between the ladder and the wall. Then $\theta = \text{Sin}^{-1}(x/10)$. Therefore

$$\frac{d\theta}{dt} = \frac{1}{\sqrt{1 - \left(\frac{x}{10}\right)^2}} \left(\frac{1}{10}\right) \frac{dx}{dt}.$$

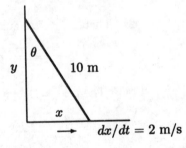

Now, when $y = 8$, $x = \sqrt{100 - 8^2} = 6$ and hence

$$\left.\frac{d\theta}{dt}\right|_{y=8} = \frac{1}{\sqrt{1 - \left(\frac{6}{10}\right)^2}} \left(\frac{1}{10}\right) (2) = 0.25 \text{ radians/s}.$$

In other words, when the top of the ladder is 8 meters above the ground, the angle between the ladder and wall is increasing at the rate of 0.25 radians/s.

43. Let x stand for the horizontal distance of the dog from the pole, as illustrated in the figure below, and θ the angle between the pole and the leash. Then $\theta = \text{Tan}^{-1}(x/10)$. Therefore

$$\frac{d\theta}{dt} = \frac{1}{1 + \left(\frac{x}{10}\right)^2} \left(\frac{1}{10}\right) \frac{dx}{dt}.$$

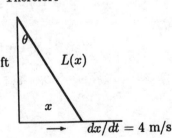

Now, when $x = 10$,

$$\left.\frac{d\theta}{dt}\right|_{x=10} = \frac{1}{1 + \left(\frac{10}{10}\right)^2} \left(\frac{1}{10}\right) (4) = 0.2 \text{ radians/s}.$$

In other words, when the dog is 10 feet from the pole, the angle between the leash and pole is increasing at the rate of 0.2 radians/s. Now, let $L(x)$ stand for the length of the leash when the dog is x feet from the pole. Then $L(x) = \sqrt{x^2 + 100}$. Hence

$$\frac{dL}{dt} = \frac{1}{2\sqrt{x^2 + 100}} \, (2x) \, \frac{dx}{dt}.$$

When $x = 10$,

$$\frac{dL}{dt}\bigg|_{x=10} = \frac{1}{2\sqrt{10^2 + 100}} \, (20)(4) = 2\sqrt{2} \text{ ft/s.}$$

Thus, when the dog is 10 feet from the pole its leash is increasing at the rate of $2\sqrt{2}$ ft/s.

44. Let x stand for half of the horizontal distance between the ends of the boards, y the vertical distance from the top of the boards to the ground, and θ the angle between the vertical and one of the boards. Then

$$\frac{\theta}{2} = \text{Tan}^{-1}(x/y) = \text{Tan}^{-1}\left(\frac{\sqrt{100 - y^2}}{y}\right) = \text{Tan}^{-1}\left(\sqrt{\frac{100}{y^2} - 1}\right)$$

$$\implies \quad \theta = 2\text{Tan}^{-1}\left(\sqrt{\frac{100}{y^2} - 1}\right).$$

Therefore

$$\frac{d\theta}{dt} = \frac{2}{1 + \left(\frac{100}{y^2} - 1\right)} \left(\frac{1}{2}\right) \frac{1}{\sqrt{\frac{100}{y^2} - 1}} \left(-\frac{200}{y^3}\right) \frac{dy}{dt}$$

$$= \frac{-2}{\sqrt{100 - y^2}} \frac{dy}{dt}.$$

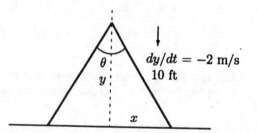

Hence, when $y = 5$,

$$\frac{d\theta}{dt}\bigg|_{y=5} = \frac{-2}{\sqrt{100 - 5^2}}(-2) = 0.46 \text{ radians/s.}$$

In other words, when the hinge is 5 meters above the ground the angle between the boards is increasing at the rate of 0.46 radians/s.

45. $\sin\theta = 1.22\lambda/d \implies \theta = \text{Sin}^{-1}(1.22\lambda/d) \implies$

$$\theta' = \frac{1}{\sqrt{1 - (1.22\lambda/d)^2}} \left(\frac{-1.22\lambda}{d^2}\right).$$

Since $\theta' < 0$, θ is a decreasing function of d.

46. $y = \text{Sin}^{-1}x^2 \implies y' = \frac{1}{\sqrt{1 - x^4}} \, (2x) \implies y'(\sqrt{2}/2) = 2\sqrt{2}/\sqrt{3}$. Hence an equation for the tangent line to the graph at $(\sqrt{2}/2, \pi/6)$ is

$$y - \frac{\pi}{6} = \frac{2\sqrt{2}}{\sqrt{3}} \left(x - \frac{\sqrt{2}}{2}\right).$$

47. Area $= \int_0^1 \dfrac{x^2}{1+x^6}\, dx$. From Exercise 27 in this section,

$$\int_0^1 \frac{x^2}{1+x^6}\, dx = \frac{1}{3}\mathrm{Tan}^{-1}x^3\Big|_0^1 = \frac{\pi}{12}.$$

48. The two graphs intersect when

$$\frac{2}{1+x^2} = x \quad\Longrightarrow\quad 2 = x + x^3 \quad\Longrightarrow\quad x = 1.$$

For $1 \le x \le \sqrt{3}$, $\dfrac{2}{1+x^2} \le x$. Hence the area bounded by the graphs is equal to

$$\int_1^{\sqrt{3}} \left(x - \frac{2}{1+x^2}\right) dx = \left(\frac{1}{2}x^2 - 2\mathrm{Tan}^{-1}x\right)\Big|_1^{\sqrt{3}} = 1 - \frac{\pi}{6}.$$

49. Let dV stand for the volume of a differential shell of thickness dx. Then $dV = 2\pi rh\,dx = 2\pi xy\,dx = (2\pi x/\sqrt{1-x^4})dx$ and hence the volume of revolution is

$$V = \int_0^{\sqrt{2}/2} \frac{2\pi x}{\sqrt{1-x^4}}\, dx.$$

Let $u = x^2$. Then $du = 2x\,dx$ and hence

$$V = \pi \int_0^{1/2} \frac{du}{\sqrt{1-u^2}} = \pi\mathrm{Sin}^{-1}u\Big|_0^{1/2} = \frac{\pi^2}{6}.$$

50. Let dV stand for the volume of a differential shell of thickness dx. Then $dV = 2\pi rh\,dx = 2\pi xy\,dx = (2\pi x/x^2\sqrt{x^2-1})dx$ and hence the volume of revolution is

$$V = \int_{\sqrt{2}}^2 \frac{2\pi x}{x^2\sqrt{x^2-1}}\, dx = 2\pi \int_{\sqrt{2}}^2 \frac{1}{x\sqrt{x^2-1}}\, dx = 2\pi\mathrm{Sec}^{-1}x\Big|_{\sqrt{2}}^2 = \frac{\pi^2}{6}.$$

51. Let dV be the differential volume of a cross-section disk of thickness dx. Then $dV = \pi r^2 dx = 4\pi/(1+4x^2)\,dx$ and hence the volume of revolution is

$$V = \int_0^{\sqrt{3}} \frac{4\pi}{1+4x^2}\, dx.$$

Let $u = 2x$. Then $du = 2dx$ and hence

$$V = 2\pi \int_0^{2\sqrt{3}} \frac{du}{1+u^2} = 2\pi\mathrm{Tan}^{-1}u\Big|_0^{2\sqrt{3}} = 2\pi\mathrm{Tan}^{-1}2\sqrt{3} \approx 8.10.$$

52. On the interval $[\sqrt{\ln 2}, 1]$, the average value of f is equal to

$$\frac{1}{1-\sqrt{\ln 2}} \int_{\sqrt{\ln 2}}^1 \frac{xe^{-x^2}}{\sqrt{1-e^{-2x^2}}}\, dx.$$

Let $u = e^{-x^2}$. Then $du = -2xe^{-x^2}dx$ and hence the average is equal to

$$\frac{-1}{2}\frac{1}{1-\sqrt{\ln 2}} \int_{1/2}^{1/e} \frac{du}{\sqrt{1-u^2}} = \frac{-1}{2(1-\sqrt{\ln 2})}\mathrm{Sin}^{-1}u\Big|_{1/2}^{1/e} = \frac{-1}{2(1-\sqrt{\ln 2})}\left(\mathrm{Sin}^{-1}(1/e) - \frac{\pi}{6}\right).$$

53. $y = \text{Cos}^{-1}x \implies \cos y = x \implies -(\sin y)y' = 1 \implies y' = -\dfrac{1}{\sin y}$. Since $0 \le y \le \pi$, $\sin y > 0$. Therefore $\sin y = +\sqrt{1 - \cos^2 y} = \sqrt{1 - x^2}$ and hence

$$y' = \frac{-1}{\sqrt{1 - x^2}}.$$

54. $y = \text{Cot}^{-1}x \implies \cot y = x \implies -(\csc^2 y)y' = 1 \implies y' = \dfrac{-1}{\csc^2 y} = \dfrac{-1}{1 + \cot^2 y} = \dfrac{-1}{1 + x^2}.$

55. $y = \text{Sec}^{-1}x \implies \sec y = x \implies (\sec y \tan y)y' = 1 \implies y' = \dfrac{1}{(\sec y)(\tan y)}$. Since $y \in [0, \pi/2) \cup [\pi, 3\pi/2)$, $\tan y > 0$. Therefore $\tan y = +\sqrt{\sec^2 y - 1} = \sqrt{x^2 - 1}$ and hence

$$y' = \frac{1}{x\sqrt{x^2 - 1}}.$$

56. $y = \text{Csc}^{-1}x \implies \csc y = x \implies -(\csc y)\cot y)y' = 1 \implies y' = \dfrac{-1}{(\csc y)(\cot y)}$. Since $y \in (0, \pi/2] \cup (\pi, 3\pi/2]$, $\cot y > 0$. Therefore $\cot y = +\sqrt{\csc^2 y - 1} = \sqrt{x^2 - 1}$ and hence

$$y' = \frac{-1}{x\sqrt{x^2 - 1}}.$$

57. Since $\dfrac{d}{dx}(\text{Cos}^{-1}x) = -1/\sqrt{1 - x^2} = \dfrac{d}{dx}(-\text{Sin}^{-1}x)$, the functions $\text{Cos}^{-1}x$ and $-\text{Sin}^{-1}x$ must differ by a constant, say $\text{Cos}^{-1}x + \text{Sin}^{-1}x = C$. Setting $x = 0$, $C = \text{Cos}^{-1}0 + \text{Sin}^{-1}0 = \pi/2$. Hence

$$\text{Cos}^{-1}x = \frac{\pi}{2} - \text{Sin}^{-1}x.$$

58. (a) From the figure we see that

$$\frac{h(t)}{250 + s(t)} = \tan(\theta + \alpha).$$

   Therefore

$$\theta(t) = \text{Tan}^{-1}\left(\frac{h(t)}{250 + s(t)}\right) - \alpha.$$

   Since $\tan \alpha = 26/250$, $\alpha = \text{Tan}^{-1}(26/250) = 0.10362746$. Hence

$$\theta(t) = \text{Tan}^{-1}\left(\frac{39 + 480t - 24t^2}{8.5t^2 + 250}\right) - 0.10362746.$$

   (b) Graph of $\theta(t)$ and $\theta'(t)$ using the range $[0, 20] \times [-0.25, 1.5]$.

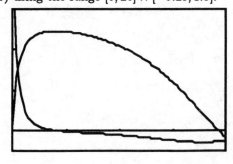

By magnifying the graph of $\theta'(t)$ we find that $\theta'(t) = 0$ for $t = 4.089081$. At this value of t, $\theta(t) = 1.226895$ radians or $70.296°$. The altitude of the missile at this time is $h(4.089081) = 1600.465$ feet. The distance from the launch pad is $s(4.089081) = 142.125$ feet.

(c) The hill that the launch pad is on will block the radar when $\theta(t) = 0$. By magnifying the graph of $\theta(t)$ we find that $\theta(t) = 0$ for $t = 19.316549$. At this time, $h(19.316549) = 355.846$ feet and $s(19.316549) = 3171.597$ feet.

(d) The missile will hit the ground when $h(t) = 0$. Using the quadratic formula we find that $h(t) = 0$ for $t = 20.080923$. At this time $s(20.080923) = 3427.569$. Hence, after the radar is blocked the missile will be in the air for only 0.764374 seconds more and travel another 255.972 feet.

59. (a) Since $\theta = \text{Tan}^{-1}(x/50)$, it follows that

$$\frac{d\theta}{dt} = \frac{1/50}{1 + (x/50)^2}\frac{dx}{dt} = \frac{50}{2500 + x^2}\frac{dx}{dt}.$$

If $dx/dt = 1.5$ then $d\theta/dt = 50(1.5)/(2500 + x^2) = 75/(2500 + x^2)$. Graph using the range $[0, 50] \times [0, 0.05]$.

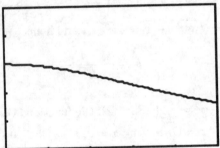

We see from the graph that the maximum of $d\theta/dt$ is when $x = 0$. For $x = 0$, $d\theta/dt = 75/2500 = 0.03$ radians/s.

(b) Since $s = 100\text{Tan}^{-1}(x/50)$,

$$\frac{ds}{dt} = \frac{100(1/50)}{1 + (x/50)^2}\frac{dx}{dt} = \frac{5000}{2500 + x^2}\frac{dx}{dt}.$$

If $dx/dt = 1.5$ then $d\theta/dt = 5000(1.5)/(2500 + x^2) = 7500/(2500 + x^2)$. As in part (a), ds/dt is a maximum when $x = 0$. For $x = 0$, $ds/dt = 3$ feet/s.

9.5   The Hyperbolic Functions

1. $\sinh 0 = (1/2)(e^0 - e^0) = 0$

2. $\sinh 1 = (1/2)(e - 1/e)$

3. $\cosh(\ln 2) = (1/2)(e^{\ln 2} + e^{-\ln 2}) = 5/4$

4. $\sinh(\ln 4) = (1/2)(e^{\ln 4} - e^{-\ln 4}) = 15/8$

5. $\tanh(\ln 2) = \dfrac{e^{\ln 2} - e^{-\ln 2}}{e^{\ln 2} + e^{-\ln 2}} = 3/5$

6. $\text{sech}\,(1) = \dfrac{2}{e + 1/e}$

7. $\coth(\ln 4) = \dfrac{e^{\ln 4} + e^{-\ln 4}}{e^{\ln 4} - e^{-\ln 4}} = 17/15$

8. $\operatorname{csch}(\ln \pi^3) = \dfrac{2}{e^{\ln \pi^3} - e^{-\ln \pi^3}} = \dfrac{2}{\pi^3 - \pi^{-3}}$

9. $y = \sinh^{-1}(1) = \ln(1 + \sqrt{2})$

10. $\tanh^{-1}(1/2) = \dfrac{1}{2} \ln\left(\dfrac{1 + 1/2}{1 - 1/2}\right) = (1/2)\ln 3$

11. $y = \sinh 2x \implies y' = 2\cosh 2x$

12. $f(x) = x\cosh x \implies f'(x) = x\sinh x + \cosh x$

13. $f(x) = \sinh x \tanh x \implies f'(x) = (\sinh x)(\operatorname{sech}^2 x) + (\tanh x)\cosh x = (\sinh x)(1 + \operatorname{sech}^2 x).$

14. $y = \sinh^3(1 - x^2) \implies y' = 3\sinh^2(1 - x^2)\cosh(1 - x^2)(-2x)$

15. $f(x) = \sqrt{\cosh 4x} = (\cosh 4x)^{1/2} \implies f'(x) = (1/2)(\cosh 4x)^{-1/2}(\sinh 4x)(4) = 2(\sinh 4x)/\sqrt{\cosh 4x}$

16. $y = \dfrac{1 - \cosh x}{1 + \cosh x} \implies y' = \dfrac{(1 + \cosh x)(-\sinh x) - (1 - \cosh x)(\sinh x)}{(1 + \cosh x)^2} = \dfrac{-2\sinh x}{(1 + \cosh x)^2}$

17. $y = 1/\cosh x = (\cosh x)^{-1} \implies y' = -(\cosh x)^{-2}(\sinh x) = -(\sinh x)/\cosh^2 x$

18. $f(x) = \ln(\tanh x^2) \implies f'(x) = \dfrac{1}{\tanh x^2}(\operatorname{sech}^2 x^2)(2x)$

19. $f(x) = e^x \operatorname{csch} x^2 \implies f'(x) = e^x(-\operatorname{csch} x^2 \coth x^2)(2x) + (\operatorname{csch} x^2)e^x$

20. $f(x) = \cosh^{-1}(1 + x^2) \implies f'(x) = \dfrac{1}{\sqrt{(1 + x^2)^2 - 1}}(2x) = \dfrac{2x}{\sqrt{x^4 + 2x^2}} = \dfrac{2x}{|x|\sqrt{x^2 + 2}}$

21. $y = \sinh^{-1} 2x \implies y' = \dfrac{1}{\sqrt{(2x)^2 + 1}}\,2 = \dfrac{2}{\sqrt{4x^2 + 1}}$

22. $y = \sqrt{\sinh^{-1}} = (\sinh^{-1} x)^{1/2} \implies y' = \dfrac{1}{2}(\sinh^{-1} x)^{-1/2}\dfrac{1}{\sqrt{x^2 + 1}}$

23. $f(s) = \tanh^{-1} s^2 \implies f'(s) = \dfrac{1}{1 - (s^2)^2}(2s) = \dfrac{2s}{1 - s^4}$

24. $y = \cosh^{-1} e^{x^2} \implies y' = \dfrac{1}{\sqrt{(e^{x^2})^2 - 1}}(e^{x^2})(2x) = \dfrac{2xe^{x^2}}{\sqrt{e^{2x^2} - 1}}$

25. $f(x) = \ln \cosh^{-1} \pi x \implies f'(x) = \dfrac{1}{\cosh^{-1} \pi x}\dfrac{1}{\sqrt{(\pi x)^2 - 1}}\pi$

26. $f(x) = \ln(1 + \cosh \pi x) \implies f'(x) = \dfrac{1}{1 + \cosh \pi x}(\sinh \pi x)(\pi)$

27. $y = \cosh x \cosh x^2 \implies y' = (\cosh x)(\sinh x^2)(2x) + (\cosh x^2)(\sinh x)$

28. $y = \sqrt{1 + \cosh^2 3x} \implies y' = \dfrac{1}{2}(1 + \cosh^2 3x)^{-1/2}(2\cosh 3x)(\sinh 3x)(3) = \dfrac{3(\sinh 3x)(\cosh 3x)}{\sqrt{1 + \cosh^2 3x}}$

687

29. $y = \sinh \ln x^2 \Longrightarrow y' = (\cosh \ln x^2)\left(\dfrac{1}{x^2} 2x\right) = \dfrac{2}{x} \cosh \ln x^2 = \dfrac{2}{x}\dfrac{1}{2}\left(x^2 + \dfrac{1}{x^2}\right) = x + \dfrac{1}{x^3}$

30. $y = x \sinh^{-1}(1 + x^2) \Longrightarrow y' = x \dfrac{1}{\sqrt{(1+x^2)^2 + 1}}(2x) + \sinh^{-1}(1 + x^2)$

31. $y = \tanh \sqrt{1 + x^2} \Longrightarrow y' = (\mathrm{sech}^2 \sqrt{1 + x^2})\dfrac{1}{2}(1 + x^2)^{-1/2}(2x) = \dfrac{x \, \mathrm{sech}^2 \sqrt{1 + x^2}}{\sqrt{1 + x^2}}$

32. $x + \cosh xy = y \Longrightarrow 1 + (\sinh xy)(xy' + y) = y' \Longrightarrow$

$$y'(x \sinh xy - 1) = -1 - y \sinh xy \Longrightarrow y' = \dfrac{1 + y \sinh xy}{1 - x \sinh xy}$$

33. $\displaystyle\int_1^2 \dfrac{1}{\sqrt{1 + x^2}}\, dx = \sinh^{-1} x \Big|_1^2 = \sinh^{-1} 2 - \sinh^{-1} 1 = \ln(2 + \sqrt{5}) - \ln(1 + \sqrt{2})$

34. Let $u = x^2$. Then $du = 2x\,dx$ and hence

$$\int_1^2 x \sinh x^2 \, dx = \dfrac{1}{2}\int_1^4 \sinh u \, du = \dfrac{1}{2}\cosh u \Big|_1^4 = \dfrac{1}{2}(\cosh 4 - \cosh 1) = \dfrac{1}{2}\left(\dfrac{e^4 + e^{-4}}{2} - \dfrac{e + e^{-1}}{2}\right)$$

35. $\displaystyle\int \tanh^2(2x)\, dx = \int (1 - \mathrm{sech}^2 2x)\, dx = x - \dfrac{1}{2}\tanh 2x + C$

36. $\displaystyle\int e^x \sinh x \, dx = \dfrac{1}{2}\int e^x(e^x - e^{-x})\, dx = \dfrac{1}{2}\int(e^{2x} - 1)\, dx = \dfrac{1}{4}e^{2x} - \dfrac{1}{2}x + C$

37. Let $u = \sinh x$. Then $du = \cosh x\,dx$ and hence

$$\int_0^2 \sinh x \cosh x \, dx = \int_{\sinh 0}^{\sinh 2} u \, du = \dfrac{1}{2}u^2 \Big|_{\sinh 0}^{\sinh 2} = \dfrac{1}{2}\left(\dfrac{e^2 - e^{-2}}{2}\right)^2.$$

38. Let $u = x/2$. Then $du = (1/2)dx$ and hence

$$\int_0^1 \dfrac{1}{4 - x^2}\, dx = \dfrac{1}{4}\int_0^1 \dfrac{1}{1 - \left(\frac{x}{2}\right)^2}\, dx = \dfrac{1}{2}\int_0^{1/2} \dfrac{1}{1 - u^2}\, du$$

$$= \dfrac{1}{2}\tanh^{-1} u \Big|_0^{1/2} = \dfrac{1}{2}(\tanh^{-1} 1/2 - \tanh^{-1} 0) = \dfrac{1}{4}\ln\dfrac{1 + 1/2}{1 - 1/2} - \dfrac{1}{4}\ln 1 = \dfrac{1}{4}\ln 3.$$

39. Let $u = x/2$. Then $du = (1/2)dx$ and hence

$$\int_4^6 \dfrac{dx}{\sqrt{x^2 - 4}} = \dfrac{1}{2}\int_4^6 \dfrac{dx}{\sqrt{\left(\frac{x}{2}\right)^2 - 1}} = \int_2^3 \dfrac{du}{\sqrt{u^2 - 1}} = \cosh^{-1} u \Big|_2^3 = \ln(3 + 2\sqrt{2}) - \ln(2 + \sqrt{3}).$$

40. Let $u = \dfrac{x}{\sqrt{2}}$. Then $du = \dfrac{1}{\sqrt{2}}dx$ and hence

$$\int_2^4 \dfrac{dx}{2 - x^2} = \dfrac{1}{2}\int_2^4 \dfrac{dx}{1 - \left(\frac{x}{\sqrt{2}}\right)^2} = \dfrac{\sqrt{2}}{2}\int_{\sqrt{2}}^{2\sqrt{2}} \dfrac{du}{1 - u^2} = \dfrac{\sqrt{2}}{2}\coth^{-1} u \Big|_{\sqrt{2}}^{2\sqrt{2}}$$

$$= \dfrac{\sqrt{2}}{2}\left(\coth^{-1} 2\sqrt{2} - \coth^{-1} \sqrt{2}\right) = \dfrac{\sqrt{2}}{4}\left(\ln\dfrac{2\sqrt{2} + 1}{2\sqrt{2} - 1} - \ln\dfrac{\sqrt{2} + 1}{\sqrt{2} - 1}\right) = \dfrac{\sqrt{2}}{4}\ln\dfrac{3 - \sqrt{2}}{3 + \sqrt{2}}.$$

41. Let $u = 3x/5$. Then $du = (3/5)dx$ and hence

$$\int_0^1 \frac{1}{\sqrt{9x^2 + 25}}\, dx = \frac{1}{5}\int_0^1 \frac{1}{\sqrt{\left(\frac{3x}{5}\right)^2 + 1}}\, dx = \frac{1}{3}\int_0^{3/5} \frac{du}{\sqrt{u^2 + 1}} = \frac{1}{3}\sinh^{-1} u \Big|_0^{3/5} = \frac{1}{3}\ln\left(\frac{3}{5} + \frac{\sqrt{34}}{5}\right).$$

42. Let $u = x/2$. Then $du = (1/2)dx$ and hence

$$\int_1^2 \frac{dx}{x\sqrt{4 + x^2}} = \frac{1}{2}\int_1^2 \frac{dx}{x\sqrt{1 + \left(\frac{x}{2}\right)^2}} = \int_{1/2}^1 \frac{du}{u\sqrt{1 + u^2}} = \operatorname{csch}^{-1} u \Big|_{1/2}^1 = \ln(1 + \sqrt{2}) - \ln(2 + \sqrt{5}).$$

43. Let $u = \cosh x$. Then $du = \sinh x\, dx$ and hence

$$\int \frac{\sinh x}{\cosh x}\, dx = \int \frac{du}{u} = \ln|u| + C = \ln|\cosh x| + C = \ln(\cosh x) + C.$$

44. $\displaystyle \int \frac{1}{\sinh^2 x}\, dx = \int \operatorname{csch}^2 x\, dx = -\coth x + C.$

45. Let $u = \sinh x$. Then $du = \cosh x\, dx$ and hence

$$\int \frac{\cosh x}{\sqrt{\sinh x}}\, dx = \int \frac{du}{\sqrt{u}} = 2\sqrt{u} + C = 2\sqrt{\sinh x} + C.$$

46. Let $u = \cosh x$. Then $du = \sinh x\, dx$ and hence

$$\int \frac{\sinh x}{1 + \cosh^2 x}\, dx = \int \frac{du}{1 + u^2} = \operatorname{Tan}^{-1}(\cosh x) + C.$$

47. Let $u = \tanh x$. Then $du = \operatorname{sech}^2 x\, dx$ and hence

$$\int \tanh^2 x \operatorname{sech}^2 x\, dx = \int u^2\, du = \frac{1}{3}u^3 + C = \frac{1}{3}\tanh^3 x + C.$$

48. Let $u = \tanh x$. Then $du = \operatorname{sech}^2 x\, dx$ and hence

$$\int \frac{1 - \tanh^2 x}{1 + \tanh^2 x}\, dx = \int \frac{\operatorname{sech}^2 x}{1 + \tanh^2 x}\, dx = \int \frac{du}{1 + u^2} = \operatorname{Tan}^{-1} u + C = \operatorname{Tan}^{-1}(\tanh x) + C.$$

49. False. For example, $\sinh 0 = 0$ but $\sinh 2\pi = \dfrac{e^{2\pi} - e^{-2\pi}}{2} \neq 0$.

50. $1 - \tanh^2 x = 1 - \dfrac{\sinh^2 x}{\cosh^2 x} = \dfrac{\cosh^2 x - \sinh^2 x}{\cosh^2 x} = \dfrac{1}{\cosh^2 x} = \operatorname{sech}^2 x$

51. $\cosh x \pm \sinh x = \dfrac{e^x + e^{-x}}{2} \pm \dfrac{e^x - e^{-x}}{2} = e^{\pm x}$

52. $\sinh 2x = \dfrac{e^{2x} - e^{-2x}}{2} = \dfrac{(e^x - e^{-x})(e^x + e^{-x})}{2} = 2\dfrac{e^x - e^{-x}}{2}\dfrac{e^x + e^{-x}}{2} = 2\sinh x \cosh x$

53. $\cosh 2x = \dfrac{e^{2x} + e^{-2x}}{2} = \dfrac{(e^x)^2 - 2 + (e^{-x})^2 + 2}{2} = 2\left(\dfrac{e^x - e^{-x}}{2}\right)^2 + 1 = 2\sinh^2 x + 1$

54. $\sinh x \cosh y + \sinh y \cosh x$

$$= \frac{e^x - e^{-x}}{2} \frac{e^y + e^{-y}}{2} + \frac{e^y - e^{-y}}{2} \frac{e^x + e^{-x}}{2}$$

$$= \frac{e^{x+y} - e^{y-x} + e^{x-y} - e^{-x-y}}{4} + \frac{e^{x+y} - e^{x-y} + e^{y-x} - e^{-x-y}}{4}$$

$$= \frac{e^{x+y} - e^{-(x+y)}}{2} = \sinh(x+y)$$

55. $\cosh x \cosh y + \sinh x \sinh y$

$$= \frac{e^x + e^{-x}}{2} \frac{e^y + e^{-y}}{2} + \frac{e^x - e^{-x}}{2} \frac{e^y - e^{-y}}{2}$$

$$= \frac{e^{x+y} + e^{y-x} + e^{x-y} + e^{-x-y}}{4} + \frac{e^{x+y} - e^{-x+y} - e^{x-y} + e^{-x-y}}{4}$$

$$= \frac{e^{x+y} + e^{-(x+y)}}{2}$$

$$= \cosh(x+y)$$

56. Since $y' = \sinh x$, length $= \int_0^b \sqrt{1 + \sinh^2 x} \, dx = \int_0^b \cosh x \, dx = \sinh x \big|_0^b = \sinh b.$

57. $g(y) = c_1 \sinh(ky) + c_2 \cosh(ky) \implies g'(y) = kc_1 \cosh(ky) + kc_2 \sinh(ky) \implies g''(y) = k^2 c_1 \sinh(ky) + k^2 c_2 \cosh(ky) = k^2 g(y) \implies g''(y) - k^2 g(y) = 0.$

58. $y = c_1 \sinh 2x + c_2 \cosh 2x = C_1 e^{2x} + C_2 e^{-2x}$

59. $y = c_1 \sinh k\pi x + c_2 \cosh k\pi x = C_1 e^{k\pi x} + C_2 e^{-k\pi x}$

60. $y = A \sin x + B \cos x + C \sinh x + D \cosh x \implies$

$$d^4 y / dx^4 = A d^4 (\sin x)/dx^4 + B d^4 (\cos x)/dx^4 + C d^4 (\sinh x)/dx^4 + D d^4 (\cosh x)/dx^4$$

$$= A \sin x + B \cos x + C \sinh x + D \cosh x = y.$$

61. Since $\coth x > \tanh x$,

$$\text{Area} = \int_{\ln 2}^{\ln 4} (\coth x - \tanh x) \, dx = \int_{\ln 2}^{\ln 4} \left(\frac{\cosh x}{\sinh x} - \frac{\sinh x}{\cosh x} \right) dx$$

$$= (\ln |\sinh x| - \ln |\cosh x|) \big|_{\ln 2}^{\ln 4} = \ln |\tanh x| \big|_{\ln 2}^{\ln 4} = \ln \frac{75}{51} = \ln \frac{25}{17}.$$

62. Let dV be the differential volume of a cross-section disk of thickness dx. Then $dV = \pi r^2 dx = \pi y^2 dx$ $= \pi \cosh^2 x \, dx$ and hence the volume of revolution is

$$V = \int_{-\ln 2}^{\ln 2} \pi \cosh^2 x \, dx = \frac{\pi}{4} \int_{-\ln 2}^{\ln 2} (e^{2x} + 2 + e^{-2x}) \, dx = \frac{\pi}{4} \left(\frac{1}{2} e^{2x} + 2x - \frac{1}{2} e^{-2x} \right) \Big|_{-\ln 2}^{\ln 2}$$

$$= \frac{\pi}{8} (e^{2\ln 2} + 4 \ln 2 - e^{-2\ln 2}) - \frac{\pi}{8} (e^{-2\ln 2} - 4 \ln 2 - e^{2\ln 2}) = \frac{\pi}{8} \left(\frac{15}{2} + 8 \ln 2 \right).$$

63. $f(x) = \cosh x - \sinh x = (e^x + e^{-x})/2 - (e^x - e^{-x})/2 = e^{-x}$ has no relative extrema.

64. $f(x) = 2\cosh x + 5\sinh x = e^x + e^{-x} + (5/2)(e^x - e^{-x}) = (7/2)e^x - (3/2)e^{-x} \implies f'(x) = (7/2)e^x + (3/2)e^{-x}$. Since $f'(x) \neq 0$ for all x, $f(x)$ has no relative extrema.

65. $f(x) = 5\cosh x - 2\sinh x = (5/2)(e^x + e^{-x}) - (e^x - e^{-x}) = (3/2)e^x + (7/2)e^{-x} \implies f'(x) = (3/2)e^x - (7/2)e^{-x} = (1/2)e^{-x}(3e^{2x} - 7) = 0 \implies e^{2x} = 7/3 \implies 2x = \ln(7/3) \implies x = (1/2)\ln(7/3)$. Since $f''(x) = (3/2)e^x + (7/2)e^{-x} > 0$, $x = (1/2)\ln(7/3)$ is a relative minimum for $f(x)$.

66. $f(x) = \cosh 2x \implies f'(x) = 2\sinh 2x \implies f''(x) = 4\cosh 2x = 4f(x)$.

67. $f(x) = \cosh 3x - 2\sinh 3x \implies f'(x) = 3\sinh 3x - 6\cosh 3x \implies f''(x) = 9\cosh 3x - 18\sinh 3x = 9f(x)$.

68. $f(x) = A\cosh 6x + B\sinh 6x \implies f'(x) = 6A\sinh 6x + 6B\cosh 6x \implies f''(x) = 36A\cosh 6x + 36B\sinh 6x = 36f(x)$.

69. $\sinh(-x) = (e^{-x} - e^x)/2 = -(e^x - e^{-x})/2 = -\sinh x \implies \sinh x$ is odd
$\cosh(-x) = (e^{-x} + e^x)/2 = \cosh x \implies \cosh x$ is even
$\tanh(-x) = \sinh(-x)/\cosh(-x) = -(\sinh x)/(\cosh x) = -\tanh x \implies \tanh x$ is odd
$\mathrm{sech}(-x) = 1/\cosh(-x) = 1/\cosh x = \mathrm{sech}\, x \implies \mathrm{sech}\, x$ is even
$\mathrm{csch}(-x) = 1/\sinh(-x) = -1/\sinh x = -\mathrm{csch}\, x \implies \mathrm{csch}\, x$ is odd
$\coth(-x) = 1/\tanh(-x) = -1/\tanh x = -\coth x \implies \coth x$ is odd

70. $\dfrac{d}{dx}(\tanh x) = \dfrac{d}{dx}\left(\dfrac{\sinh x}{\cosh x}\right) = \dfrac{\cosh^2 x - \sinh^2 x}{\cosh^2 x} = \dfrac{1}{\cosh^2 x} = \mathrm{sech}^2 x$

$\dfrac{d}{dx}(\coth x) = \dfrac{d}{dx}(\tanh x)^{-1} = -(\tanh x)^{-2}(\mathrm{sech}\, x)^2 = -\dfrac{\cosh^2 x}{\sinh^2 x}\dfrac{1}{\cosh^2 x} = -\dfrac{1}{\sinh^2 x} = -\mathrm{csch}^2 x$

$\dfrac{d}{dx}(\mathrm{sech}\, x) = \dfrac{d}{dx}(\cosh x)^{-1} = -(\cosh x)^{-2}(\sinh x) = -\dfrac{\sinh x}{\cosh x}\dfrac{1}{\cosh x} = -\mathrm{sech}\, x \tanh x$

$\dfrac{d}{dx}(\mathrm{csch}\, x) = \dfrac{d}{dx}(\sinh x)^{-1} = -(\sinh x)^{-2}(\cosh x) = -\dfrac{\cosh x}{\sinh x}\dfrac{1}{\sinh x} = -\mathrm{csch}\, x \coth x$

71. $\cosh^2 x - \sinh^2 x = 1 \implies \dfrac{\cosh^2 x}{\sinh^2 x} - 1 = \dfrac{1}{\sinh^2 x} \implies \coth^2 x - 1 = \mathrm{csch}^2 x$.

72. $\dfrac{d}{dx}(\cosh^{-1} x) = \dfrac{d}{dx}(\ln(x + \sqrt{x^2 - 1})) = \dfrac{1}{x + \sqrt{x^2 - 1}}\left(1 + \dfrac{2x}{2\sqrt{x^2 - 1}}\right) = \dfrac{1}{\sqrt{x^2 - 1}}$

$\dfrac{d}{dx}(\tanh^{-1} x) = \dfrac{d}{dx}\left(\dfrac{1}{2}\ln\dfrac{1+x}{1-x}\right) = \dfrac{1}{2}\dfrac{d}{dx}(\ln(1+x) - \ln(1-x)) = \dfrac{1}{2}\left(\dfrac{1}{1+x} + \dfrac{1}{1-x}\right) = \dfrac{1}{1-x^2}$

$\dfrac{d}{dx}(\coth^{-1} x) = \dfrac{d}{dx}\left(\dfrac{1}{2}\ln\left(\dfrac{x+1}{x-1}\right)\right) = \dfrac{1}{2}\dfrac{d}{dx}(\ln(x+1) - \ln(x-1)) = \dfrac{1}{2}\left(\dfrac{1}{x+1} - \dfrac{1}{x-1}\right) = \dfrac{1}{1-x^2}$

$\dfrac{d}{dx}(\mathrm{sech}^{-1} x) = \dfrac{d}{dx}\left(\ln\dfrac{1 + \sqrt{1 - x^2}}{x}\right) = \dfrac{d}{dx}(\ln(1 + \sqrt{1 - x^2}) - \ln x)$

$= \dfrac{1}{1 + \sqrt{1 - x^2}}\dfrac{-2x}{2\sqrt{1 - x^2}} - \dfrac{1}{x} = \dfrac{-x^2 - (\sqrt{1 - x^2})(1 + \sqrt{1 - x^2})}{x(\sqrt{1 - x^2})(1 + \sqrt{1 - x^2})} = \dfrac{-1}{x\sqrt{1 - x^2}}$

691

$$\frac{d}{dx}(\operatorname{csch}^{-1} x) = \frac{d}{dx} \ln\left(\frac{1}{x} + \frac{\sqrt{1+x^2}}{|x|}\right) = \frac{d}{dx} \ln\left(\frac{1}{x} + \sqrt{\frac{1}{x^2} + 1}\right)$$

$$= \frac{1}{\frac{1}{x} + \sqrt{\frac{1}{x^2} + 1}} \left[-\frac{1}{x^2} + \frac{-\frac{2}{x^3}}{2\sqrt{\frac{1}{x^2} + 1}}\right] = \frac{x}{1 + x\sqrt{\frac{1}{x^2} + 1}} (-1) \frac{x\sqrt{\frac{1}{x^2} + 1} + 1}{x^3 \sqrt{\frac{1}{x^2} + 1}}$$

$$= \frac{-1}{\sqrt{x^2 + x^4}} = \frac{-1}{x^2 \sqrt{\frac{1}{x^2} + 1}} = \frac{-1}{|x|\sqrt{1+x^2}}$$

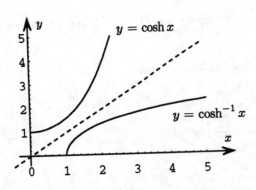

Exercise 74

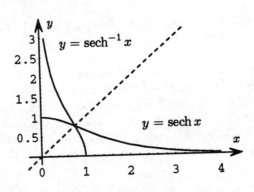

Exercise 75

73. Since $y = \cosh^{-1} x = \ln(x + \sqrt{x^2 - 1}) \geq 0$ when $x \geq 1$, the range of $\cosh^{-1}$ is $[0, \infty)$, which means that the domain of $\cosh$, that is, its principal branch, is the interval $[0, \infty)$.

74. $y = \cosh x \implies y' = \sinh x$. Since $\sinh x > 0$ for $x > 0$, it follows that $\cosh x$ is an increasing function on the interval $[0, \infty)$. The graph of $y = \cosh^{-1} x$ is shown above on the left.

75. The principal branch of sech is the interval $[0, \infty)$; hence,

$$\operatorname{sech} : [0, \infty) \to (0, 1] \implies \operatorname{sech}^{-1} : (0, 1] \to [0, \infty).$$

The graph of $y = \operatorname{sech}^{-1} x$ is shown above on the right. The principal branch of tanh is the interval $(-\infty, \infty)$; hence,

$$\tanh : (-\infty, \infty) \to (-1, 1) \implies \tanh^{-1} : (-1, 1) \to (-\infty, \infty).$$

The graph of $y = \tanh^{-1} x$ is shown below on the left. The principal branch of csch is the interval $(-\infty, 0) \cup (0, \infty)$; hence,

$$\operatorname{csch} : (-\infty, 0) \cup (0, \infty) \to (-\infty, 0) \cup (0, \infty) \implies \operatorname{csch}^{-1} : (-\infty, 0) \cup (0, \infty) \to (-\infty, 0) \cup (0, \infty).$$

The graph of $y = \operatorname{csch}^{-1} x$ is shown below on the right.

The principal branch of coth is the interval $(-\infty, 0) \cup (0, \infty)$; hence,

$$\coth : (-\infty, 0) \cup (0, \infty) \to (-\infty, -1) \cup (1, \infty) \implies \coth^{-1} : (-\infty, -1) \cup (1, \infty) \to (-\infty, 0) \cup (0, \infty).$$

The graph of $y = \coth^{-1} x$ is shown below.

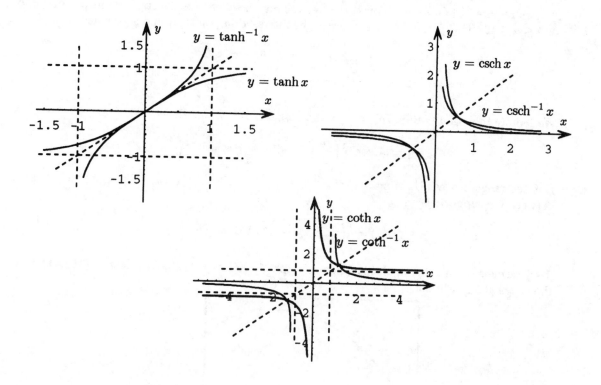

76. The graphs are shown above in Exercise 75.

77. (a) Graph $y = 10\cosh(x/10)$, $y = 16$ and $y = 28$ using the range $[-20, 20] \times [0, 30]$.

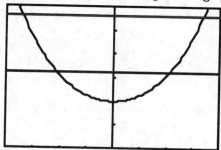

Note that $y = 16$ and $y = 28$ intersect the graph of $y = 10\cosh(x/10)$ twice. If we assume the 16 foot support is left of the lowest point then $x = -10.469679$ is the point of intersection. If the 28 foot support is right of the lowest point the $x = 16.892355$ is the point of intersection. So the distance between the supports is $16.892355 + 10.469679 = 27.362035$ feet.

(b) The length along the curve is given by

$$L = \int_{-10.469679}^{16.892355} \sqrt{1 + \left[10\left(\frac{1}{10}\sinh\left(\frac{x}{10}\right)\right)\right]^2}\, dx = \int_{-10.469679}^{16.892355} \sqrt{1 + \sinh^2\left(\frac{x}{10}\right)}\, dx$$

$$= \int_{-10.469679}^{16.892355} \cosh\left(\frac{x}{10}\right) dx = 10\sinh\left(\frac{x}{10}\right)\Big|_{-10.469679}^{16.892355} = 10(3.864339) = 38.64339 \text{ feet.}$$

(c) We are required to find a function that gives the distance between the supports as a function of a, $0 \le a \le 16$. This is the distance between the intersection points x_1 and x_2 defined by

$$a \cosh\left(\frac{x_1}{a}\right) = 16, \quad x_1 < 0$$
$$a \cosh\left(\frac{x_2}{a}\right) = 28, \quad x_2 > 0.$$

We require then that $\cosh(x_1/a) = 16/a$ and $\cosh(x_2/a) = 28/a$. Hence

$$x_1 = a\mathrm{Cosh}^{-1}\left(\frac{16}{a}\right) \quad \text{and} \quad x_2 = a\mathrm{Cosh}^{-1}\left(\frac{28}{a}\right).$$

But the range of $\mathrm{Cosh}^{-1}t$ is positive if $t > 0$ so $x_1 = -a\mathrm{Cosh}^{-1}(16/a)$. Therefore the distance between the supports is

$$D(a) = a\left(\mathrm{Cosh}^{-1}\left(\frac{28}{a}\right) + \mathrm{Cosh}^{-1}\left(\frac{16}{a}\right)\right).$$

To determine the maximum of this function graph $\mathbf{Y = X(Cosh^{-1}(26/X) + Cosh^{-1}(16/X))}$ using the range $[0, 16] \times [0, 30]$.

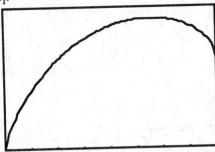

Magnifying the graph at the maximum point we find that if the lowest point on the cable is $a = 11.119420$ feet then the supports will be a maximum of $D(11.119420) = 27.579904$ feet apart.

78. (a) Since

$$m\frac{dv}{dt} = -mg + kv^2$$

then

$$\int \frac{m\, dv}{kv^2 - mg} = \int dt$$

or

$$-\frac{m}{k} \int \frac{dv}{\frac{mg}{k} - v^2} = t + c. \tag{1}$$

Now $\int \frac{du}{a^2 - u^2} = \frac{1}{a}\mathrm{Tanh}^{-1}\left(\frac{u}{a}\right) + C$. So if $u = v$ and $a = \sqrt{mg/k}$, then applying the above formula to the left hand side of equation (1) we get

$$-\frac{m}{k}\left(\sqrt{\frac{k}{mg}}\,\mathrm{Tanh}^{-1}\left(v\,\sqrt{\frac{k}{mg}}\right)\right) = t + c.$$

Thus

$$-\sqrt{\frac{m}{kg}}\,\mathrm{Tanh}^{-1}\left(v\,\sqrt{\frac{k}{mg}}\right) = t + c.$$

694

If $v(0) = 0$ and since $\text{Tanh}^{-1}(0) = 0$ we have $c = 0$ so

$$-\sqrt{\frac{m}{kg}}\,\text{Tanh}^{-1}\left(v\,\sqrt{\frac{k}{mg}}\right) = t.$$

Solving this equation for v gives

$$v = \sqrt{\frac{mg}{k}}\,\tanh\left(-t\,\sqrt{\frac{kg}{m}}\right) = -\sqrt{\frac{mg}{k}}\,\tanh\left(t\,\sqrt{\frac{kg}{m}}\right)$$

since $\tanh(-x) = -\tanh(x)$.

(b) If $m = 80$, $g = 9.8$ and $k = 0.086$, $v(t) = -95.479208\,\tanh(0.102640t)$. Graph this function using the range $[0, 50] \times [-100, 0]$.

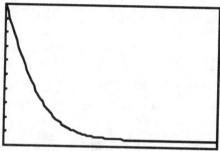

Since $\lim\limits_{x\to+\infty}\tanh x = 1$ we have that $\lim\limits_{x\to+\infty} v(t) = -\sqrt{mg/k} = -95.479208$. This implies that an 80kg (176.4lbs) person in free fall will not fall faster than 95.48 meters/second. This is 313.25 feet/second or 213.58 miles/hour.

(c) Now,

$$\frac{ds}{dt} = v(t) = -\sqrt{\frac{m}{kg}}\,\tanh\left(t\,\sqrt{\frac{k}{mg}}\right) \implies s(t) = -\sqrt{\frac{m}{kg}}\int \tanh\left(t\,\sqrt{\frac{k}{mg}}\right)\,dt.$$

Since $[\sinh(au)]' = a\cosh(au)$ it follows that

$$\int \tanh(au)\,du = \frac{1}{a}\int \frac{\sinh(au)}{\cosh(au)}\,a\,du = \frac{1}{a}\ln\cosh(au) + C.$$

Thus

$$s(t) = -\sqrt{\frac{m}{kg}}\left[\sqrt{\frac{mg}{k}}\ln\cosh\left(t\,\sqrt{\frac{kg}{m}}\right)\right] + C.$$

If $s(0) = s_0$ and $\cosh(0) = 1$ we have $C = s_0$. Hence, using the values of m, g, and k from part (b) we find that

$$s(t) = -930.232558\,\ln\cosh(0.102640t) + s_0.$$

Now 95% of the terminal velocity is $0.95(-95.479208) = -90.705248$ meters/second. This occurs at time t_0, where $v(t_0) = -95.479208\,\tanh(0.102640t_0) = -90.705248$. Solving this equation for t_0 gives $t_0 = \text{Tanh}^{-1}(0.95)/0.102640 = 17.846631$ seconds. We then choose s_0 so that $s(t_0) = -1000$. Note that $t = 0$ is where free fall begins. Substituting the value of t_0 into the expression for $s(t)$ gives $s(t_0) = -1082.745535 + s_0 = 1000$. Therefore $s_0 = 2082.745535$. So free fall should begin at an altitude of 2082.74 meters (6833.17 feet).

Review Exercises – Chapter 9

1. $\text{Sin}^{-1}(\sqrt{3}/2) = \pi/3$

2. $\text{Cos}^{-1}(-1/2) = 2\pi/3$

3. $\text{Tan}^{-1}\sqrt{3} = \pi/3$

4. $\text{Sin}^{-1}(\sin \pi/4) = \pi/4$

5. $\text{Cos}^{-1}\big(\cos(-\pi/4)\big) = \pi/4$

6. $\text{Tan}^{-1}\big(\sin(\pi/2)\big) = \pi/4$

7. $\cot(\text{Sin}^{-1}(\sqrt{2}/2)) = 1$

8. $\tan\big(\text{Sec}^{-1}(-2)\big) = \sqrt{3}$

9. $\sinh(\ln 2) = (1/2)(e^{\ln 2} - e^{-\ln 2}) = 3/4.$

10. $y = \text{Tan}^{-1}3x \Longrightarrow y' = \dfrac{1}{1 + (3x)^2}\,(3) = \dfrac{3}{1 + 9x^2}$

11. $y = \text{Sin}^{-1}\sqrt{x} \Longrightarrow y' = \dfrac{1}{\sqrt{1-x}}\,\dfrac{1}{2\sqrt{x}} = \dfrac{1}{2\sqrt{x(1-x)}}$

12. $y = \text{Tan}^{-1}\left(\dfrac{2-x}{2+x}\right) \Longrightarrow y' = \dfrac{1}{1 + \left(\frac{2-x}{2+x}\right)^2}\,\dfrac{(2+x)(-1) - (2-x)(1)}{(2+x)^2} = \dfrac{-2}{4 + x^2}$

13. $f(t) = \text{Cot}^{-1}(1 - t^2) \Longrightarrow f'(t) = \dfrac{-1}{1 + (1 - t^2)^2}\,(-2t) = \dfrac{2t}{t^4 - 2t^2 + 2}$

14. $y = \cosh(\ln x) = \dfrac{1}{2}\left(e^{\ln x} + e^{-\ln x}\right) = \dfrac{1}{2}\left(x + \dfrac{1}{x}\right) \Longrightarrow y' = \dfrac{1}{2} - \dfrac{1}{2x^2}$

15. $f(x) = x^2 \sinh(1 - x) \Longrightarrow$

$$f'(x) = x^2[\cosh(1-x)](-1) + [\sinh(1-x)](2x) = 2x\sinh(1-x) - x^2\cosh(1-x)$$

16. $y = \text{Tan}^{-1}\dfrac{\sqrt{x}}{2} \Longrightarrow y' = \dfrac{1}{1 + \left(\frac{\sqrt{x}}{2}\right)^2}\,\dfrac{1}{4\sqrt{x}} = \dfrac{1}{\sqrt{x}(4 + x)}$

17. $y = \csc(\cot 6x) \Longrightarrow y' = -\csc(\cot 6x)\cot(\cot 6x)(-\csc^2 6x)(6) = 6\csc(\cot 6x)\cot(\cot 6x)(\csc^2 6x)$

18. $y = \ln^2(\cos^2 x^2) \Longrightarrow y' = 2\ln(\cos^2 x^2)\,\dfrac{1}{\cos^2 x^2}(2\cos x^2)(-\sin x^2)(2x) = -16x\ln(\cos x^2)\tan x^2$

19. $f(x) = \text{Sin}^{-1}[\ln(2x + 1)] \Longrightarrow f'(x) = \dfrac{1}{\sqrt{1 - \ln^2(2x+1)}}\,\dfrac{1}{2x+1}\,(2)$

20. $y = \text{Sin}^{-1}(\cos e^{-x}) \Longrightarrow y' = \dfrac{1}{\sqrt{1 - \cos^2 e^{-x}}}\left(-\sin e^{-x}\right)\left(-e^{-x}\right) = \dfrac{e^{-x}\sin e^{-x}}{\sqrt{1 - \cos^2 e^{-x}}}$

21. $f(x) = \dfrac{\text{Tan}^{-1}x}{1+x^2} \Longrightarrow f'(x) = \dfrac{(1+x^2)\left(\frac{1}{1+x^2}\right) - (\text{Tan}^{-1}x)(2x)}{(1+x^2)^2} = \dfrac{1 - 2x\text{Tan}^{-1}x}{(1+x^2)^2}$

22. $y = \dfrac{\text{Sin}^{-1}e^x}{1+e^x} \Longrightarrow y' = \dfrac{(1+e^x)\frac{1}{\sqrt{1-e^{2x}}}(e^x) - (\text{Sin}^{-1}e^x)e^x}{(1+e^x)^2}$

23. $f(x) = \text{Sec}^{-1}\sqrt{x^2+4} \Longrightarrow f'(x) = \dfrac{1}{\sqrt{x^2+4}\sqrt{(x^2+4)-1}}\dfrac{1}{2}\dfrac{1}{\sqrt{x^2+4}}(2x) = \dfrac{x}{(x^2+4)\sqrt{x^2+3}}$

24. $y = \sqrt{\cosh x^2} \Longrightarrow y' = \dfrac{1}{2}(\cosh x^2)^{-1/2}(\sinh x^2)(2x) = \dfrac{x\sinh x^2}{\sqrt{\cosh x^2}}$

25. $y = 1/(\pi + \tanh x) \Longrightarrow y' = -(\pi + \tanh x)^{-2}(\text{sech}^2 x) = \dfrac{-\text{sech}^2 x}{(\pi + \tanh x)^2}$

26. $f(x) = \ln^2(\sinh x) \Longrightarrow f'(x) = 2[\ln(\sinh x)]\dfrac{1}{\sinh x}\cosh x = 2[\ln(\sinh x)]\coth x$

27. $y = (\sinh x + \text{Cos}^{-1}2x)^{1/5} \Longrightarrow y' = \dfrac{1}{5}(\sinh x + \text{Cos}^{-1}2x)^{-4/5}\left(\cosh x - \dfrac{2}{\sqrt{1-4x^2}}\right)$

28. $y = \ln|\tan x| \Longrightarrow y' = \dfrac{1}{|\tan x|}\dfrac{|\tan x|}{\tan x}(\sec^2 x) = \cot x \sec^2 x = \csc x \sec x$

29. $y = \ln(\text{Sin}^{-1}x) \Longrightarrow y' = \dfrac{1}{\text{Sin}^{-1}x}\dfrac{1}{\sqrt{1-x^2}}$

30. $f(x) = \dfrac{\tan 2x}{2+\sec 2x} \Longrightarrow$

$$f'(x) = \dfrac{(2+\sec 2x)(\sec^2 2x)(2) - (\tan 2x)(\sec 2x \tan 2x)(2)}{(2+\sec 2x)^2} = \dfrac{2(\sec 2x)(1+2\sec 2x)}{(2+\sec 2x)^2}$$

31. $f(x) = x\tanh^{-1}(\ln x) \Longrightarrow f'(x) = x\dfrac{1}{1-\ln^2 x}\dfrac{1}{x} + \tanh^{-1}(\ln x) = \dfrac{1}{1-\ln^2 x} + \tanh^{-1}(\ln x)$

32. $y = x^2\sinh^{-1}(e^x) \Longrightarrow y' = x^2\dfrac{1}{\sqrt{e^{2x}+1}}(e^x) + 2x\sinh^{-1}(e^x)$

33. $y = \ln\sqrt{\tanh^{-1}(x^2)} = \frac{1}{2}\ln(\tanh^{-1}(x^2)) \Longrightarrow y' = \dfrac{1}{2}\dfrac{1}{\tanh^{-1}(x^2)}\dfrac{1}{1-x^4}(2x) = \dfrac{x}{(1-x^4)\tanh^{-1}(x^2)}$

34. Let $u = 3x$. Then $du = 3dx$ and hence

$$\int\dfrac{dx}{\sqrt{1+9x^2}} = \dfrac{1}{3}\int\dfrac{du}{\sqrt{1+u^2}} = \dfrac{1}{3}\sinh^{-1}u + C = \dfrac{1}{3}\sinh^{-1}3x + C.$$

35. Let $u = 2x$. Then $du = 2dx$ and hence

$$\int\dfrac{dx}{\sqrt{4x^2-1}} = \dfrac{1}{2}\int\dfrac{du}{\sqrt{u^2-1}}$$

$$= \dfrac{1}{2}\cosh^{-1}u + C = \dfrac{1}{2}\cosh^{-1}u + C = \dfrac{1}{2}\cosh^{-1}2x + C = \dfrac{1}{2}\ln\left|2x + \sqrt{4x^2-1}\right| + C.$$

36. Let $u = x/2$. Then $du = (1/2)dx$ and hence

$$\int \frac{dx}{x\sqrt{4+x^2}} = \frac{1}{2}\int \frac{dx}{x\sqrt{1+\left(\frac{x}{2}\right)^2}} = \frac{1}{2}\int \frac{du}{u\sqrt{1+u^2}} = -\frac{1}{2}\operatorname{csch}^{-1}u + C = -\frac{1}{2}\operatorname{csch}^{-1}\frac{x}{2} + C.$$

37. Let $u = \sin 2x$. Then $du = 2\cos 2x\,dx$ and hence

$$\int \sin^4 2x \cos 2x\, dx = \frac{1}{2}\int u^4\, du = \frac{1}{10}u^5 + C = \frac{1}{10}\sin^5 2x + C.$$

38. Let $u = \sec x$. Then $du = \sec x \tan x\,dx$ and hence

$$\int \sec^5 x \tan x\, dx = \int u^4\, du = \frac{1}{5}u^5 + C = \frac{1}{5}\sec^5 x + C.$$

39. Let $u = \sqrt{x}$. Then $du = \frac{1}{2\sqrt{x}}dx$ and hence

$$\int_{\pi^2/16}^{\pi^2/4} \frac{\sin^2\sqrt{x}}{\sqrt{x}}\, dx = 2\int_{\pi/4}^{\pi/2} \sin^2 u\, du = \int (1 - \cos 2u)\, du = \left.\left(u - \frac{1}{2}\sin 2u\right)\right|_{\pi/4}^{\pi/2} = \frac{\pi}{4} + \frac{1}{2}.$$

40. Let $u = \sin x$. Then $du = \cos x\,dx$ and hence

$$\int_0^{\pi/2} \sin^{5/2} x \cos x\, dx = \int_0^1 u^{5/2}\, du = \left.\frac{2}{7}u^{7/2}\right|_0^1 = \frac{2}{7}.$$

41. Let $u = \frac{x}{\sqrt{2}}$. Then $du = \frac{1}{\sqrt{2}}dx$ and hence

$$\int_{-1}^1 \frac{dx}{\sqrt{2-x^2}} = \frac{1}{\sqrt{2}}\int_{-1}^1 \frac{dx}{\sqrt{1-\left(\frac{x}{\sqrt{2}}\right)^2}} = \int_{-1/\sqrt{2}}^{1/\sqrt{2}} \frac{du}{\sqrt{1-u^2}} = \left.\operatorname{Sin}^{-1}u\right|_{-1/\sqrt{2}}^{1/\sqrt{2}} = \frac{\pi}{2}.$$

42. Let $u = x^2$. Then $du = 2x\,dx$ and hence

$$\int x\csc^2 x^2\, dx = \frac{1}{2}\int \csc^2 u\, du = -\frac{1}{2}\cot u + C = -\frac{1}{2}\cot x^2 + C.$$

43. Let $u = \tan x$. Then $du = \sec^2 x\,dx$ and hence

$$\int_0^{\pi/4} \sqrt{\tan x}\sec^2 x\, dx = \int_0^1 \sqrt{u}\, du = \left.\frac{2}{3}u^{3/2}\right|_0^1 = \frac{2}{3}.$$

44. Let $u = \tan\sqrt{x}$. Then $du = \frac{\sec^2\sqrt{x}}{2\sqrt{x}}dx$ and hence

$$\int \frac{\sec^2\sqrt{x}\tan\sqrt{x}}{\sqrt{x}}\, dx = 2\int u\, du = u^2 + C = \tan^2\sqrt{x} + C.$$

45. $\displaystyle\int \frac{1+\sin^2 2x}{\cos^2 2x}\, dx = \int \frac{2-\cos^2 2x}{\cos^2 2x}\, dx = 2\int \sec^2 2x\, dx - \int dx = \tan 2x - x + C.$

46. Let $u = \sec x$. Then $du = \sec x \tan x \, dx$ and hence

$$\int \tan(\sec x) \sec x \tan x \, dx = \int \tan u \, du = \ln|\sec u| + C = \ln|\sec(\sec x)| + C.$$

47. Let $u = \tan 2x$. Then $du = 2 \sec^2 2x \, dx$ and hence

$$\int_0^{\pi/8} \tan^2(2x) \sec^2(2x) \, dx = \frac{1}{2} \int_0^1 u^2 \, du = \frac{1}{6} u^3 \Big|_0^1 = \frac{1}{6}.$$

48. Let $u = x^2$. Then $du = 2x \, dx$ and hence

$$\int \frac{x}{x^4 + 1} \, dx = \frac{1}{2} \int \frac{du}{1 + u^2} = \frac{1}{2} \operatorname{Tan}^{-1} u + C = \frac{1}{2} \operatorname{Tan}^{-1} x^2 + C.$$

49. Let $u = x/2$. Then $du = (1/2) dx$ and hence

$$\int \frac{dx}{\sqrt{4 - x^2}} = \frac{1}{2} \int \frac{dx}{\sqrt{1 - \left(\frac{x}{2}\right)^2}} = \int \frac{du}{\sqrt{1 - u^2}} = \operatorname{Sin}^{-1} u + C = \operatorname{Sin}^{-1} \frac{x}{2} + C.$$

50. Let $u = \dfrac{e^{-x}}{\sqrt{2}}$. Then $du = -\dfrac{e^{-x}}{\sqrt{2}} \, dx$ and hence

$$\int \frac{e^{-x} \, dx}{2 + e^{-2x}} = \frac{1}{2} \int \frac{e^{-x} \, dx}{1 + \left(\frac{e^{-x}}{\sqrt{2}}\right)^2} = -\frac{1}{2} \sqrt{2} \int \frac{du}{1 + u^2} = -\frac{\sqrt{2}}{2} \operatorname{Tan}^{-1} u + C = -\frac{\sqrt{2}}{2} \operatorname{Tan}^{-1} \frac{e^{-x}}{\sqrt{2}} + C.$$

51. Let $u = \operatorname{Cos}^{-1} x$. Then $du = \dfrac{-1}{\sqrt{1 - x^2}} \, dx$ and hence

$$\int \frac{\operatorname{Cos}^{-1} x}{\sqrt{1 - x^2}} \, dx = -\int u \, du = -\frac{1}{2} u^2 + C = -\frac{1}{2} (\operatorname{Cos}^{-1} x)^2 + C.$$

52. Let $u = x^{3/2}/2$. Then $du = (3/4) x^{1/2} \, dx$ and hence

$$\int \frac{\sqrt{x}}{x^3 + 4} \, dx = \frac{1}{4} \int \frac{\sqrt{x}}{1 + \left(\frac{x^{3/2}}{2}\right)^2} = \frac{1}{3} \int \frac{du}{1 + u^2} = \frac{1}{3} \operatorname{Tan}^{-1} u + C = \frac{1}{3} \operatorname{Tan}^{-1} \frac{x^{3/2}}{2} + C.$$

53. Let $u = e^{\sqrt{x}}$. Then $du = \dfrac{e^{\sqrt{x}}}{2\sqrt{x}} \, dx$ and hence

$$\int \frac{e^{\sqrt{x}}}{\sqrt{x}(1 + e^{2\sqrt{x}})} \, dx = \int \frac{e^{\sqrt{x}}}{\sqrt{x}(1 + (e^{\sqrt{x}})^2)} \, dx = 2 \int \frac{du}{1 + u^2} = 2 \operatorname{Tan}^{-1} u + C = 2 \operatorname{Tan}^{-1} e^{\sqrt{x}} + C.$$

54. Let $u = x^3$. Then $du = 3x^2 \, dx$ and hence

$$\int_1^2 \frac{x^2}{x^6 + 9} \, dx = \frac{1}{3} \int_1^8 \frac{du}{u^2 + 9} = \frac{1}{9} \operatorname{Tan}^{-1} \frac{u}{3} \Big|_1^8 = \frac{1}{9} \left(\operatorname{Tan}^{-1} \frac{8}{3} - \operatorname{Tan}^{-1} \frac{1}{3} \right).$$

55. Let $u = \text{Sin}^{-1}x$. Then $du = dx/\sqrt{1-x^2}$ and hence

$$\int \frac{\sqrt{\text{Sin}^{-1}x}}{\sqrt{1-x^2}}\,dx = \int \sqrt{u}\,du = \frac{2}{3}\,u^{3/2} + C = \frac{2}{3}\,(\text{Sin}^{-1}x)^{3/2} + C.$$

56. Let $u = \text{Tan}^{-1}2x$. Then $du = 2dx/(1+4x^2)$ and hence

$$\int \frac{(\text{Tan}^{-1}2x)^4}{1+4x^2}\,dx = \frac{1}{2}\int u^4\,du = \frac{1}{10}\,u^5 + C = \frac{1}{10}\,(\text{Tan}^{-1}2x)^5 + C.$$

57. Let $u = \text{Sec}^{-1}3x$. Then $du = \dfrac{3}{3x\sqrt{9x^2-1}}\,dx$ and hence

$$\int \frac{\text{Sec}^{-1}3x}{x\sqrt{36x^2-4}}\,dx = \frac{1}{2}\int \frac{\text{Sec}^{-1}3x}{x\sqrt{9x^2-1}}\,dx = \frac{1}{2}\int u\,du = \frac{1}{4}\,u^2 + C = \frac{1}{4}\,(\text{Sec}^{-1}3x)^2 + C.$$

58. Let $u = \dfrac{\cos x}{\sqrt{2}}$. Then $du = \dfrac{-\sin x}{\sqrt{2}}\,dx$ and hence

$$\int \frac{\sin x}{\sqrt{2-\cos^2 x}}\,dx = \frac{1}{\sqrt{2}}\int \frac{\sin x}{\sqrt{1-\left(\frac{\cos x}{\sqrt{2}}\right)^2}}\,dx = -\int \frac{du}{\sqrt{1-u^2}} = -\text{Sin}^{-1}u + C = -\text{Sin}^{-1}\left(\frac{\cos x}{\sqrt{2}}\right) + C.$$

59. Let $u = \sec 2x$. Then $du = 2\sec 2x\tan 2x\,dx$ and hence

$$\int_0^{\pi/9} \sec^3 2x\tan 2x\,dx = \frac{1}{2}\int_1^{\sec 2\pi/9} u^2\,du = \frac{1}{6}\,u^3\Big|_1^{\sec 2\pi/9} = \frac{1}{6}\,\sec^3\frac{2\pi}{9} - \frac{1}{6}.$$

60. Let $u = x/2$. Then $du = (1/2)dx$ and hence

$$\int_0^{\pi/2} \tan^3(x/2)\,dx = 2\int_0^{\pi/4} \tan u\underbrace{(\sec^2 u - 1)}_{\tan^2 u}\,du$$

$$= 2\int_0^{\pi/4} \tan u\sec^2 u\,du - 2\int_0^{\pi/4} \tan u\,du = (\tan u)^2\Big|_0^{\pi/4} - 2\ln|\sec u|\Big|_0^{\pi/4} = 1 - \ln 2.$$

61. Let $u = \tan x$. Then $du = \sec^2 x\,dx$ and hence

$$\int_0^{\pi/3} \sec^4 x\,dx = \int_0^{\pi/3} (1+\tan^2 x)\sec^2 x\,dx = \int_0^{\sqrt{3}} (1+u^2)\,du = \left(u + \frac{1}{3}\right)\Big|_0^{\sqrt{3}} = 2\sqrt{3}.$$

62. Let $u = \sec x$. Then $du = \sec x\tan x\,dx$ and hence

$$\int \sec^3 x\tan^3 x\,dx = \int \sec^2 x\underbrace{(\sec^2 x - 1)}_{\tan^2 x}(\sec x\tan x)\,dx = \int u^2(u^2-1)\,du$$

$$= \frac{1}{5}\,u^5 - \frac{1}{3}\,u^3 + C = \frac{1}{5}\,\sec^5 x - \frac{1}{3}\,\sec^3 x + C.$$

63. $\displaystyle\int_0^{\pi/3} \cos x \cos 5x \, dx = \frac{1}{2} \int_0^{\pi/3} (\cos 6x + \cos 4x) \, dx = \frac{1}{2} \left(\frac{1}{6} \sin 6x + \frac{1}{4} \sin 4x \right)\Big|_0^{\pi/3} = -\frac{\sqrt{3}}{16}$

64. $\displaystyle\int_0^{\pi/4} \sin x \cos 2x \, dx = \frac{1}{2} \int_0^{\pi/4} (\sin 3x - \sin x) \, dx = \frac{1}{2} \left(-\frac{1}{3} \cos 3x + \cos x \right)\Big|_0^{\pi/4} = \frac{1}{3}\sqrt{2} - \frac{1}{3}$

65. Let $u = \cos x$. Then $du = -\sin x \, dx$ and hence

$$\int \sqrt{\cos x}\, \sin^3 x \, dx = -\int u^{1/2}(1 - u^2) \, du = -\left(\frac{2}{3} u^{3/2} - \frac{2}{7} u^{7/2} \right) + C = -\frac{2}{3} \cos^{3/2} x + \frac{2}{7} \cos^{7/2} x + C.$$

66. Let $u = \cot 2x$. Then $du = -2\csc^2 2x \, dx$ and hence

$$\int \cot^4 2x \, dx = \int (\cot^2 2x)(\csc^2 2x - 1) \, dx = -\frac{1}{2} \int u^2 \, du - \int \cot^2 2x \, dx$$

$$= -\frac{1}{6} u^3 - \int (\csc^2 2x - 1) \, dx = -\frac{1}{6} \cot^3 2x + \frac{1}{2} \cot 2x + x + C.$$

67. Let $u = \cos x$. Then $du = -\sin x \, dx$ and hence

$$\int \frac{\sin^3 x}{\cos^2 x} \, dx = \int \frac{1 - \cos^2 x}{\cos^2 x} (\sin x) \, dx = -\int \frac{1 - u^2}{u^2} \, du$$

$$= -\int (u^{-2} - 1) \, du = u^{-1} + u + C = \sec x + \cos x + C.$$

68. $\displaystyle\int_0^{\pi/2} \sin^2 x \cos^4 x \, dx = \int_0^{\pi/2} (\sin x \cos x)^2 \cos^2 x \, dx = \frac{1}{4} \int_0^{\pi/4} \sin^2 2x \cos^2 x \, dx$

$$= \frac{1}{4} \int_0^{\pi/2} (\sin^2 2x) \left(\frac{1 + \cos 2x}{2} \right) dx = \frac{1}{8} \int_0^{\pi/2} \frac{1 - \cos 4x}{2} \, dx + \frac{1}{8} \int_0^{\pi/2} \sin^2 2x \cos 2x \, dx$$

$$= \frac{1}{16} \left(x - \frac{1}{4} \sin 4x \right) + \frac{1}{48} \sin^3 2x \Big|_0^{\pi/2} = \frac{\pi}{32}.$$

69. Let $u = 1 + \sinh x$. Then $du = \cosh x \, dx$ and hence

$$\int \frac{\cosh x}{1 + \sinh x} \, dx = \int \frac{du}{u} = \ln |u| + C = \ln |1 + \sinh x| + C.$$

70. Let $u = \sinh x^2$. Then $du = 2x \cosh x^2 \, dx$ and hence

$$\int x \coth x^2 \, dx = \int x \, \frac{\cosh x^2}{\sinh x^2} \, dx = \frac{1}{2} \int \frac{1}{u} \, du = \frac{1}{2} \ln |u| + C = \frac{1}{2} \ln |\sinh x^2| + C.$$

71. Let $u = \cosh x$. Then $du = \sinh x \, dx$ and hence

$$\int \sinh^3 x \, dx = \int (\sinh x)(\underbrace{\cosh^2 - 1}_{\sinh^2 x}) \, dx = \int u^2 \, du - \cosh x = \frac{1}{3} \cosh^3 x - \cosh x + C.$$

72. Let $u = e^{3x}/3$. Then $du = e^{3x} dx$ and hence

$$\int \frac{e^{3x}}{\sqrt{9 + e^{6x}}} \, dx = \frac{1}{3} \int \frac{e^{3x}}{\sqrt{1 + \left(\frac{e^{3x}}{3}\right)^2}} \, dx = \frac{1}{3} \int \frac{du}{\sqrt{1 + u^2}} = \frac{1}{3} \sinh^{-1} u + C = \frac{1}{3} \sinh^{-1} \frac{e^{3x}}{3} + C.$$

73. Area $= \int_0^\pi \sin^2 x \, dx = \frac{1}{2} \int_0^\pi (1 - \cos 2x) \, dx = \frac{1}{2} \left(x - \frac{1}{2} \sin 2x\right) \Big|_0^\pi = \pi/2.$

74. First note that the curves $y = \sin^2 x$ and $y = \cos^2 x$ intersect when $\sin^2 x = \cos^2 x \implies \tan^2 x = 1 \implies x = \pi/4$ if $0 \le x \le \pi/2$. Hence

$$\text{Area} = \int_0^{\pi/4} (\cos^2 x - \sin^2 x) \, dx + \int_{\pi/4}^{\pi/2} (\sin^2 x - \cos^2 x) \, dx$$

$$= \int_0^{\pi/4} \cos 2x \, dx + \int_{\pi/4}^{\pi/2} (-\cos 2x) \, dx = \frac{1}{2} \sin 2x \Big|_0^{\pi/4} - \frac{1}{2} \sin 2x \Big|_{\pi/4}^{\pi/2} = 1.$$

75. Let dV be the differential volume of a cross-section disk of thickness dx. Then $dV = \pi r^2 dx = \pi y^2 dx = \pi \tan^2 x \, dx$ and hence the volume of revolution is

$$V = \int_0^{\pi/4} \pi \tan^2 x \, dx = \pi \int_0^{\pi/4} (\sec^2 x - 1) \, dx = \pi (\tan x - x)|_0^{\pi/4} = \pi \left(1 - \frac{\pi}{4}\right).$$

76. Let dV be the differential volume of a cross-section disk of thickness dx. Then

$$dV = \pi r^2 dx = \pi y^2 dx = \frac{\pi}{1 + x^2} \, dx$$

and hence the volume of revolution is

$$V = \int_0^1 \frac{\pi}{1 + x^2} \, dx = \pi \text{Tan}^{-1} x \Big|_0^1 = \frac{\pi^2}{4}.$$

77. Let dV be the volume of a differential shell of thickness dx. Then

$$dV = 2\pi r h \, dx = 2\pi x y \, dx = \frac{2\pi x}{1 + x^4} \, dx$$

and hence the volume of revolution is

$$V = 2\pi \int_0^2 \underbrace{\frac{x}{1 + x^4}}_{u = x^2} \, dx = \pi \int_0^4 \frac{du}{1 + u^2} = \pi \text{Tan}^{-1} u \Big|_0^4 = \pi \text{Tan}^{-1} 4.$$

78. Average value of $\cosh x$ on the interval $[-\ln 2, \ln 2] =$

$$\frac{1}{2 \ln 2} \int_{-\ln 2}^{\ln 2} \cosh x \, dx = \frac{1}{2 \ln 2} \sinh x \Big|_{-\ln 2}^{\ln 2} = \frac{1}{2 \ln 2} (2 \sinh \ln 2) = \frac{3}{4 \ln 2}.$$

79. $y = \cosh^{-1} x^2 y$

$$\Rightarrow \quad y' = \frac{1}{\sqrt{(x^2 y)^2 - 1}} \, (x^2 y' + 2xy)$$
$$\Rightarrow \quad y' \sqrt{x^4 y^2 - 1} = x^2 y' + 2xy$$
$$\Rightarrow \quad y'(\sqrt{x^4 y^2 - 1} - x^2) = 2xy$$
$$\Rightarrow \quad y' = \frac{2xy}{\sqrt{x^4 y^2 - 1} - x^2}$$

80. $y = \mathrm{Cos}^{-1} x^2 y$

$$\Rightarrow \quad y' = \frac{-1}{\sqrt{1 - x^4 y^2}} \, (x^2 y' + 2xy)$$
$$\Rightarrow \quad y' \sqrt{1 - x^4 y^2} = -x^2 y' - 2xy$$
$$\Rightarrow \quad y'(\sqrt{1 - x^4 y^2} + x^2) = -2xy$$
$$\Rightarrow \quad y' = \frac{-2xy}{\sqrt{1 - x^4 y^2} + x^2}$$

81. Area $= \displaystyle\int_{-1}^{1} \cosh x \, dx = \sinh x \big|_{-1}^{1} = 2 \sinh 1 = e - 1/e$

82. $y = \sinh x \cosh x \Longrightarrow y' = \sinh^2 x + \cosh^2 x = 2 \sinh^2 x + 1$. Since $2 \sinh^2 + 1 \neq 0$, the graph has no relative extrema. Now, $y'' = 4 \sinh x \cosh x > 0 \Longleftrightarrow x > 0$. Therefore the graph is concave up on $[0, \infty)$ and concave down on $(-\infty, 0]$.

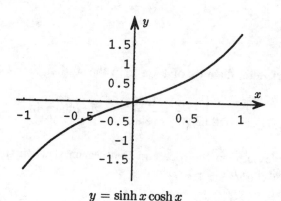

$y = \sinh x \cosh x$

Since concavity changes at $x = 0$, the graph has an inflection point at $x = 0$.

83. $\displaystyle\int_0^4 v(t) \, dt = \int_0^4 \sin^2 \pi t \, dt = \frac{1}{2} \int_0^4 (1 - \cos 2\pi t) \, dt = \frac{1}{2} \left(t - \frac{1}{2\pi} \sin 2\pi t \right) \Big|_0^4 = 2$

84. Let $f(x) = \mathrm{Sin}^{-1} x$. Then $f'(x) = 1/\sqrt{1 - x^2}$ and hence

$$f(x) \approx f(0.5) + f'(0.5)(x - 0.5) = \frac{\pi}{6} + \frac{2}{\sqrt{3}} \left(x - \frac{1}{2} \right) \quad \text{for } x \approx 0.5.$$

In particular, $f(0.48) = \text{Sin}^{-1}0.48 \approx \dfrac{\pi}{6} + \dfrac{2}{\sqrt{3}}\,(-0.02) \approx 0.5$.

85. Let x be any real number. Then

$$\cosh x - \sinh x = \frac{e^x + e^{-x}}{2} - \frac{e^x - e^{-x}}{2} = e^{-x} > 0$$

and hence $\cosh x > \sinh x$.

86. $y = \text{Sin}^{-1}x^2 \Longrightarrow$

$$y' = \frac{1}{\sqrt{1-x^4}}\,(2x) \quad \Longrightarrow \quad y'(\sqrt{2}/2) = \frac{1}{\sqrt{1-\frac{4}{16}}}\,(\sqrt{2}) = \frac{2\sqrt{2}}{\sqrt{3}}.$$

Since $y(\sqrt{2}/2) = \text{Sin}^{-1}(1/2) = \pi/6$, the equation of the tangent line at $x = \sqrt{2}/2$ is

$$y - \frac{\pi}{6} = \frac{2\sqrt{2}}{\sqrt{3}}\left(x - \frac{\sqrt{2}}{2}\right).$$

87. Let θ stand for the angle of elevation of the plane when the plane is a horizontal distance of x km from the station. Then $\theta = \text{Tan}^{-1}(3/x)$ and $dx/dt = 400$. Hence

$$\frac{d\theta}{dt} = \frac{1}{1 + \left(\frac{3}{x}\right)^2}\left(-\frac{3}{x^2}\right)\frac{dx}{dt}$$

$$= -\frac{1200}{x^2 + 9}.$$

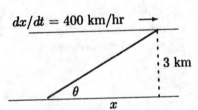

$dx/dt = 400$ km/hr

3 km

θ

x

When the plane is a horizontal distance of 4 km from the station, $x = 4$ and hence

$$\left.\frac{d\theta}{dt}\right|_{x=4} = -\frac{1200}{25} = -48 \text{ radians/hr.}$$

88. Let θ stand for the angle between the rope and the horizontal when the length of rope is x meters. Then $\theta = \text{Sin}^{-1}(10/x)$ and $dx/dt = -1$. Hence

$$\frac{d\theta}{dt} = \frac{1}{\sqrt{1 - \left(\frac{10}{x}\right)^2}}\left(-\frac{10}{x^2}\right)\frac{dx}{dt}$$

$$= \frac{10}{x\sqrt{x^2 - 100}}.$$

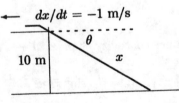

$dx/dt = -1$ m/s

θ

10 m

x

When 20 meters of rope is still out, $x = 20$ and hence

$$\left.\frac{d\theta}{dt}\right|_{x=20} = \frac{10}{(20)\sqrt{300}} = \frac{1}{20\sqrt{3}} \text{ radians/s} \approx 0.029 \text{ radians/s.}$$

89. $y = \text{Cot}^{-1}x \Longrightarrow y' = -1/(1+x^2)$. Now, the slope of the line $x + 5y - 10 = 0$ is $-1/5$. Hence

$$y' = \frac{-1}{1+x^2} = -\frac{1}{5} \quad \Longrightarrow \quad 1 + x^2 = 5 \quad \Longrightarrow \quad x = \pm 2.$$

Thus, there are two points on the graph of $y = \text{Cot}^{-1}x$ at which the tangent line is parallel to $x + 5y - 10 = 0$, namely $(2, \text{Cot}^{-1}2)$ and $(-2, \text{Cot}^{-1}(-2))$.

90. $y = \text{Sin}^{-1}x \Longrightarrow y' = 1/\sqrt{1-x^2} \Longrightarrow y'(\sqrt{2}/2) = 1/\sqrt{1-1/2} = \sqrt{2}$. Hence the tangent line at $(\sqrt{2}/2, \pi/4)$ has equation

$$y - \frac{\pi}{4} = \sqrt{2}\left(x - \frac{\sqrt{2}}{2}\right).$$

To find the x-intercept, set $y = 0$ and solve for x:

$$-\frac{\pi}{4} = \sqrt{2}\left(x - \frac{\sqrt{2}}{2}\right) \quad \Longrightarrow \quad x = \frac{\sqrt{2}}{2} - \frac{\pi}{4\sqrt{2}} = \frac{4-\pi}{4\sqrt{2}}.$$

91. For the velocity $v(t)$, we find that

$$v(t) = \int a(t)\, dt = \int \left[-1 - \frac{2t}{(1+t^2)^2}\right] dt = -t + \frac{1}{1+t^2} + C.$$

Since $v(0) = 1 + C = 4$, $C = 3$ and hence

$$v(t) = -t + \frac{1}{1+t^2} + 3.$$

Therefore

$$s(t) = \int v(t)\, dt = \int \left(-t + \frac{1}{1+t^2} + 3\right) dt = -\frac{1}{2}t^2 + \text{Tan}^{-1}t + 3t + C.$$

Since $s(0) = C = 0$,

$$s(t) = -\frac{1}{2}t^2 + \text{Tan}^{-1}t + 3t.$$

92. Because $\sinh x = (e^x - e^{-x})/2$ is the sum of two continuous functions defined for all real x.

Chapter 10

Techniques of Integration

10.1 Integration by Parts

1. Let $u = x$, $dv = e^x dx$. Then $du = dx$, $v = e^x$, and hence
$$\int xe^x \, dx = xe^x - \int e^x \, dx = xe^x - e^x + C.$$

2. Let $u = \ln x$, $dv = x dx$. Then $du = \dfrac{1}{x} \, dx$, $v = \dfrac{1}{2} x^2$, and hence
$$\int x \ln x \, dx = \frac{1}{2} x^2 \ln x - \frac{1}{2} \int \left(\frac{1}{x}\right) x^2 \, dx = \frac{1}{2} x^2 \ln x - \frac{1}{4} x^2 + C.$$

3. Let $u = x$, $dv = \cos x dx$. Then $du = dx$, $v = \sin x$, and hence
$$\int_{\pi/2}^{\pi} x \cos x \, dx = \left[x \sin x - \int \sin x \, dx \right]\Big|_{\pi/2}^{\pi} = (x \sin x + \cos x)|_{\pi/2}^{\pi} = -1 - \frac{\pi}{2}.$$

4. Let $u = \ln 2x$, $dv = x^2 dx$. Then $du = \dfrac{1}{x} \, dx$, $v = \dfrac{1}{3} x^3$, and hence
$$\int_{1/2}^{e/2} x^2 \ln 2x \, dx = \left[\frac{1}{3} x^3 \ln 2x - \frac{1}{3} \int x^2 \, dx \right]\Big|_{1/2}^{e/2} = \left(\frac{1}{3} x^3 \ln 2x - \frac{1}{9} x^3 \right)\Big|_{1/2}^{e/2} = \frac{e^3}{36} + \frac{1}{72}.$$

5. Let $u = \text{Tan}^{-1} x$, $dv = dx$. Then $du = \dfrac{1}{1+x^2} \, dx$, $v = x$, and hence
$$\int \text{Tan}^{-1} x \, dx = x\text{Tan}^{-1} x - \int \underbrace{\frac{x}{1+x^2}}_{w=1+x^2} \, dx = x\text{Tan}^{-1} x - \frac{1}{2} \int \frac{1}{w} \, dw$$
$$= x\text{Tan}^{-1} x - \frac{1}{2} \ln |w| + C = x\text{Tan}^{-1} x - \frac{1}{2} \ln(1 + x^2) + C.$$

6. Let $u = x$, $dv = \sec x \tan x dx$. Then $du = dx$, $v = \sec x$, and hence
$$\int x \sec x \tan x \, dx = x \sec x - \int \sec x \, dx = x \sec x - \ln |\sec x + \tan x| + C.$$

706

7. Let $u = x$, $dv = \sec^2 \pi x\, dx$. Then $du = dx$, $v = (1/\pi)\tan \pi x$ and hence

$$\int x\sec^2 \pi x\, dx = \frac{x}{\pi}\tan x - \frac{1}{\pi}\int \tan \pi x\, dx = \frac{x}{\pi}\tan \pi x - \frac{1}{\pi^2}\ln|\sec \pi x| + C.$$

8. Let $u = e^{ax}$, $dv = \sin x\, dx$. Then $du = ae^{ax}$, $v = -\cos x\, dx$, and hence

$$\int e^{ax}\sin x\, dx = -e^{ax}\cos x + a\underbrace{\int e^{ax}\cos x\, dx}.$$

For the last integral, let $u = e^{ax}$, $dv = \cos x\, dx$. Then $du = ae^{ax}$, $v = \sin x$, and hence

$$\int e^{ax}\cos x\, dx = e^{ax}\sin x - a\int e^{ax}\sin x\, dx.$$

Therefore

$$\int e^{ax}\sin x\, dx = -e^{ax}\cos x + a\left[e^{ax}\sin x - a\int e^{ax}\sin x\, dx\right] = -e^{ax}\cos x + ae^{ax}\sin x - a^2\int e^{ax}\sin x\, dx$$

$$\implies (1+a^2)\int e^{ax}\sin x\, dx = e^{ax}(a\sin x - \cos x) + C$$

$$\implies \int e^{ax}\sin x\, dx = \frac{e^{ax}(a\sin x - \cos x)}{1+a^2} + C.$$

9. Let $u = \sin(\ln x)$, $dv = dx$. Then $du = \dfrac{\cos(\ln x)}{x}\, dx$, $v = x$, and hence

$$\int \sin(\ln x)\, dx = x\sin(\ln x) - \underbrace{\int \cos(\ln x)\, dx}.$$

For the last integral, let $u = \cos(\ln x)$, $dv = dx$. Then $du = -\dfrac{\sin(\ln x)}{x}\, dx$, $v = x$, and hence

$$\int \cos(\ln x)\, dx = x\cos(\ln x) + \int \sin(\ln x)\, dx.$$

Therefore

$$\int \sin(\ln x)\, dx = x\sin(\ln x) - \left[x\cos(\ln x) + \int \sin(\ln x)\, dx\right]$$

$$\implies 2\int \sin(\ln x)\, dx = x[\sin(\ln x) - \cos(\ln x)] + C$$

$$\implies \int \sin(\ln x)\, dx = \frac{x}{2}[\sin(\ln x) - \cos(\ln x)] + C.$$

10. Let $u = \sin x$, $dv = \sinh x\, dx$. Then $du = \cos x\, dx$, $v = \cosh x$, and hence

$$\int \sin x\sinh x\, dx = \sin x\cosh x - \underbrace{\int \cos x\cosh x\, dx}.$$

For the last integral, let $u = \cos x$, $dv = \cosh x\, dx$. Then $du = -\sin x\, dx$, $v = \sinh x$, and hence

$$\int \cos x\cosh x\, dx = \cos x\sinh x + \int \sin x\sinh x\, dx.$$

Therefore

$$\int \sin x \sinh x \, dx = \sin x \cosh x - \left[\cos x \sinh x + \int \sin x \sinh x \, dx \right]$$

$$\implies \quad 2 \int \sin x \sinh x \, dx = \sin x \cosh x - \cos x \sinh x + C$$

$$\implies \quad \int \sin x \sinh x \, dx = \frac{1}{2} \left(\sin x \cosh x - \cos x \sinh x \right) + C.$$

11. Let $u = x$, $dv = (x+2)^8 dx$. Then $du = dx$, $v = (1/9)(x+2)^9$, and hence

$$\int_0^1 x(x+2)^8 \, dx = \left[\frac{1}{9} x(x+2)^9 - \frac{1}{9} \int (x+2)^9 \, dx \right]\Big|_0^1$$

$$= \left[\frac{1}{9} x(x+2)^9 - \frac{1}{90} (x+2)^{10} \right]\Big|_0^1 = \frac{27761}{18} \approx 1542.28.$$

12. Let $u = x$, $dv = \sqrt{x+1}\,dx$. Then $du = dx$, $v = (2/3)(x+1)^{3/2}$, and hence

$$\int_0^4 x\sqrt{x+1} \, dx = \left[\frac{2}{3} x(x+1)^{3/2} - \frac{2}{3} \int (x+1)^{3/2} \, dx \right]\Big|_0^4$$

$$= \left[\frac{2}{3} x(x+1)^{3/2} - \frac{4}{15} (x+1)^{5/2} \right]\Big|_0^4 = \frac{20\sqrt{5}}{3} + \frac{4}{15}.$$

13. Let $u = (\ln x)^2$, $dv = dx$. Then $du = \dfrac{2\ln x}{x} \, dx$, $v = x$, and hence

$$\int_e^{e^3} (\ln x)^2 \, dx = \left[x(\ln x)^2 - 2 \underbrace{\int \ln x \, dx} \right]\Big|_e^{e^3}.$$

For the last integral, let $u = \ln x$, $dv = dx$. Then $du = (1/x)dx$, $v = x$, and hence

$$\int \ln x \, dx = x \ln x - \int dx = x \ln x - x.$$

Hence

$$\int_e^{e^3} (\ln x)^2 \, dx = \left[x(\ln x)^2 - 2x \ln x + 2x \right]\Big|_e^{e^3} = 5e^3 - e.$$

14. Let $u = \ln x^2$, $dv = xdx$. Then $du = (2/x)dx$, $v = (1/2)x^2$, and hence

$$\int_1^e x \ln x^2 \, dx = \left[\frac{1}{2} x^2 \ln x^2 - \int x \, dx \right]\Big|_1^e = \left(\frac{1}{2} x^2 \ln x^2 - \frac{1}{2} x^2 \right)\Big|_1^e = \frac{1}{2} e^2 + \frac{1}{2}.$$

15. Let $u = x$, $dv = e^{2x}dx$. Then $du = dx$, $v = (1/2)e^{2x}$, and hence

$$\int_0^{\ln 2} xe^{2x} \, dx = \left[\frac{1}{2} xe^{2x} - \frac{1}{2} \int e^{2x} \, dx \right]\Big|_0^{\ln 2} = \left(\frac{1}{2} xe^{2x} - \frac{1}{4} e^{2x} \right)\Big|_0^{\ln 2} = 2\ln 2 - \frac{3}{4}.$$

16. Let $u = \ln x$, $dv = x^3 dx$. Then $du = (1/x)dx$, $v = (1/4)x^4$, and hence

$$\int_1^e x^3 \ln x \, dx = \left[\frac{1}{4} x^4 \ln x - \frac{1}{4}\int x^3 \, dx\right]\Big|_1^e = \left(\frac{1}{4} x^4 \ln x - \frac{1}{16} x^4\right)\Big|_1^e = \frac{3}{16} e^4 + \frac{1}{16}.$$

17. Let $u = x^3$, $dv = e^{2x} dx$. Then $du = 3x^2 dx$, $v = (1/2)e^{2x}$, and hence

$$\int_0^1 x^3 e^{2x} \, dx = \left[\frac{1}{2} x^3 e^{2x} - \frac{3}{2}\underbrace{\int x^2 e^{2x} \, dx}\right]\Big|_0^1.$$

For the last integral, let $u = x^2$, $dv = e^{2x} dx$. Then $du = 2x dx$, $v = (1/2)e^{2x}$, and hence, using Exercise 15,

$$\int x^2 e^{2x} \, dx = \frac{1}{2} x^2 e^{2x} - \int x e^{2x} \, dx = \frac{1}{2} x^2 e^{2x} - \frac{1}{2} x e^{2x} + \frac{1}{4} e^{2x}.$$

Therefore

$$\int_0^1 x^3 e^{2x} \, dx = \left[\frac{1}{2} x^3 e^{2x} - \frac{3}{2}\left(\frac{1}{2} x^2 e^{2x} - \frac{1}{2} x e^{2x} + \frac{1}{4} e^{2x}\right)\right]\Big|_0^1 = \frac{1}{8} e^2 + \frac{3}{8}.$$

18. Let $u = \sec^3 x$, $dv = \sec^2 x dx$. Then $du = 3\sec^3 x \tan x dx$, $v = \tan x$, and hence

$$\int \sec^5 x \, dx = \sec^3 x \tan x - 3 \int \sec^3 x \underbrace{\tan^2 x}_{\sec^2 x - 1} \, dx = \sec^3 x \tan x - 3 \int \sec^5 x \, dx + 3 \int \sec^3 x \, dx$$

$$\Longrightarrow 4 \int \sec^5 x \, dx = \sec^3 x \tan x + 3 \left[\frac{1}{2} \sec x \tan x + \frac{1}{2} \ln|\sec x + \tan x|\right] + C \quad \text{(see Example 7)}$$

$$\Longrightarrow \int \sec^5 x \, dx = \frac{1}{4} \sec^3 x \tan x + \frac{3}{8} \sec x \tan x + \frac{3}{8} \ln|\sec x + \tan x| + C.$$

19. Let $u = 3x^2 - 2x + 1$, $dv = e^{2x} dx$. Then $du = (6x - 2)dx$, $v = (1/2)e^{2x}$, and hence

$$\int_0^1 (3x^2 - 2x + 1)e^{2x} \, dx = \left[\frac{1}{2}(3x^2 - 2x + 1)e^{2x} - \int (3x - 1)e^{2x} \, dx\right]\Big|_0^1$$

$$= \left[\frac{1}{2}(3x^2 - 2x + 1)e^{2x} - 3\int x e^{2x} \, dx + \int e^{2x} \, dx\right]\Big|_0^1$$

$$= \left[\frac{1}{2}(3x^2 - 2x + 1)e^{2x} - 3\left(\frac{1}{2} x e^{2x} - \frac{1}{4} e^{2x}\right) + \frac{1}{2} e^{2x}\right]\Big|_0^1$$

$$= \frac{3}{4} e^2 - \frac{7}{4}.$$

20. Let $u = 4x^2 + 2x$, $dv = \sin(x/2)dx$. Then $du = (8x + 2)dx$, $v = -2\cos(x/2)$, and hence

$$\int_0^\pi (4x^2 + 2x)\sin(x/2) \, dx = \left[-2(4x^2 + 2x)\cos(x/2) + 2\underbrace{\int (8x + 2)\cos(x/2) \, dx}\right]\Big|_0^\pi.$$

For the last integral, let $u = 8x + 2$, $dv = \cos(x/2)dx$. Then $du = 8dx$, $v = 2\sin(x/2)$, and hence

$$\int (8x + 2) \cos(x/2)\ dx = 2(8x + 2)\sin(x/2) - 16\int \sin(x/2)\ dx = 2(8x + 2)\sin(x/2) + 32\cos(x/2).$$

Therefore

$$\int_0^\pi (4x^2 + 2x)\sin(x/2)\ dx = \left[-2(4x^2)\cos(x/2) + 2[2(8x + 2)\sin(x/2) + 32\cos(x/2)]\right]\Big|_0^\pi$$

$$= [(-8x^2 - 4x + 64)\cos(x/2) + 4(8x + 2)\sin(x/2)]\big|_0^\pi = 32\pi - 56.$$

21. Let $u = x$, $dv = \sinh x\,dx$. Then $du = dx$, $v = \cosh x$, and hence

$$\int x \sinh x\ dx = x\cosh x - \int \cosh x\ dx = x\cosh x - \sinh x + C.$$

22. Let $u = x^2$, $dv = (x/\sqrt{x^2 + 1})dx$. Then $du = 2x\,dx$, $v = \sqrt{x^2 + 1}$, and hence

$$\int_0^1 \frac{x^3}{\sqrt{x^2 + 1}}\ dx = \left[x^2\sqrt{x^2 + 1} - 2\int x\sqrt{x^2 + 1}\ dx\right]\Big|_0^1 = \left[x^2\sqrt{x^2 + 1} - \frac{2}{3}(x^2 + 1)^{3/2}\right]\Big|_0^1 = \frac{2}{3} - \frac{1}{3}\sqrt{2}.$$

23. Let $u = e^{2x}$, $dv = \cos x\,dx$. Then $du = 2e^{2x}$, $v = \sin x$, and hence

$$\int e^{2x}\cos x\ dx = e^{2x}\sin x - 2\underbrace{\int e^{2x}\sin x\ dx}.$$

For the last integral, let $u = e^{2x}$, $dv = \sin x\,dx$. Then $du = 2e^{2x}$, $v = -\cos x$, and hence

$$\int e^{2x}\sin x\ dx = -e^{2x}\cos x + 2\int e^{2x}\cos x\ dx.$$

Therefore

$$\int e^{2x}\cos x\ dx = e^{2x}\sin x - 2\left[-e^{2x}\cos x + 2\int e^{2x}\cos x\ dx\right]$$

$$\Longrightarrow \quad 5\int e^{2x}\cos x\ dx = e^{2x}\sin x + 2e^{2x}\cos x$$

$$\Longrightarrow \quad \int e^{2x}\cos x\ dx = \frac{e^{2x}}{5}(\sin x + 2\cos x)$$

$$\Longrightarrow \quad \int_0^\pi e^{2x}\cos x\ dx = -\frac{2e^{2\pi}}{5} - \frac{2}{5}.$$

24. Let $u = \text{Cos}^{-1}2x$, $dv = dx$. Then $du = \dfrac{-1}{\sqrt{1 - 4x^2}}(2)dx$, $v = x$, and hence

$$\int \text{Cos}^{-1}2x\ dx = x\text{Cos}^{-1}2x + 2\int \underbrace{\frac{x}{\sqrt{1 - 4x^2}}}_{w = 1 - 4x^2}\ dx = x\text{Cos}^{-1}2x - \frac{1}{4}\int \frac{dw}{\sqrt{w}}$$

$$= x\text{Cos}^{-1}2x - \frac{1}{2}\sqrt{w} + C = x\text{Cos}^{-1}2x - \frac{1}{2}\sqrt{1 - 4x^2} + C.$$

25. Let $u = (\text{Cos}^{-1}x)^2$, $dv = dx$. Then $du = 2(\text{Cos}^{-1}x)\dfrac{-1}{\sqrt{1-x^2}}\,dx$, $v = x$, and hence

$$\int (\text{Cos}^{-1}x)^2\,dx = x(\text{Cos}^{-1}x)^2 + 2\underbrace{\int \frac{x\text{Cos}^{-1}x}{\sqrt{1-x^2}}\,dx}.$$

For the last integral, let $u = \text{Cos}^{-1}x$, $dv = \dfrac{x}{\sqrt{1-x^2}}\,dx$. Then $du = \dfrac{-1}{\sqrt{1-x^2}}\,dx$, $v = -\sqrt{1-x^2}$, and hence

$$\int \frac{x\text{Cos}^{-1}x}{\sqrt{1-x^2}}\,dx = -\sqrt{1-x^2}\text{Cos}^{-1}x - \int dx = -\sqrt{1-x^2}\text{Cos}^{-1}x - x.$$

Therefore

$$\int (\text{Cos}^{-1}x)^2\,dx = x(\text{Cos}^{-1}x)^2 - 2\sqrt{1-x^2}\,\text{Cos}^{-1}x - 2x + C.$$

26. Let $u = \ln x$, $dv = \sqrt{x}\,dx$. Then $du = (1/x)dx$, $v = (2/3)x^{3/2}$, and hence

$$\int \sqrt{x}\ln x\,dx = \frac{2}{3}\,x^{3/2}\ln x - \frac{2}{3}\int x^{1/2}\,dx = \frac{2}{3}\,x^{3/2}\ln x - \frac{4}{9}\,x^{3/2} + C.$$

27. Let $u = x^2$, $dv = x\sqrt{1+x^2}\,dx$. Then $du = 2xdx$, $v = (1/3)(1+x^2)^{3/2}$, and hence

$$\int x^3\sqrt{1+x^2}\,dx = \frac{1}{3}\,x^2(1+x^2)^{3/2} - \frac{2}{3}\int x(1+x^2)^{3/2}\,dx$$

$$= \frac{1}{3}\,x^2(1+x^2)^{3/2} - \frac{2}{15}\,(1+x^2)^{5/2} + C = \frac{1}{15}\,(1+x^2)^{3/2}(3x^2 - 2) + C.$$

28. Let $u = x^2$, $dv = x\sqrt{9-x^2}\,dx$. Then $du = 2xdx$, $v = (-1/3)(9-x^2)^{3/2}$, and hence

$$\int x^3\sqrt{9-x^2}\,dx = -\frac{x^2}{3}\,(9-x^2)^{3/2} + \frac{2}{3}\int x(9-x^2)^{3/2}\,dx$$

$$= -\frac{x^2}{3}\,(9-x^2)^{3/2} - \frac{2}{15}\,(9-x^2)^{5/2} + C = -\frac{1}{15}\,(9-x^2)^{3/2}(23 - 2x^2) + C.$$

29. Let $u = x^2$, $dv = xe^{ax^2}dx$. Then $du = 2xdx$, $v = (1/2a)e^{ax^2}$, and hence

$$\int x^3e^{ax^2}\,dx = \frac{x^2}{2a}\,e^{ax^2} - \frac{1}{a}\int xe^{ax^2}\,dx = \frac{x^2e^{ax^2}}{2a} - \frac{1}{2a^2}\,e^{ax^2} + C = \frac{e^{ax^2}}{2a^2}\,(ax^2 - 1) + C.$$

30. Let $u = \tanh^{-1}x$, $dv = dx$. Then $du = \dfrac{1}{1-x^2}\,dx$, $v = x$, and hence

$$\int \tanh^{-1}x\,dx = x\tanh^{-1}x - \int \frac{x}{1-x^2}\,dx = x\tanh^{-1}x + \frac{1}{2}\,\ln|1-x^2| + C.$$

31. First observe that

$$\int \sec^3 x\tan^2 x\,dx = \int \sec^3 x(\sec^2 x - 1)\,dx = \int \sec^5 dx - \int \sec^3 x\,dx.$$

Hence, by Exercise 18 and Example 7,

$$\int \sec^3 x \tan^2 x \, dx = \frac{1}{4} \sec^3 x \tan x + \frac{3}{8} \sec x \tan x + \frac{3}{8} \ln|\sec x + \tan x|$$

$$- \left[\frac{1}{2} \sec x \tan x + \frac{1}{2} \ln|\sec x + \tan x| \right] + C$$

$$= \frac{1}{4} \sec^3 x \tan x - \frac{1}{8} \sec x \tan x - \frac{1}{8} \ln|\sec x + \tan x| + C.$$

32. Let $u = \tanh^{-1} x$, $dv = x \, dx$. Then $du = \dfrac{1}{1-x^2} \, dx$, $v = \dfrac{1}{2} x^2$, and hence

$$\int x \tanh^{-1} x \, dx = \frac{1}{2} x^2 \tanh^{-1} x - \frac{1}{2} \int \frac{x^2}{1-x^2} \, dx$$

$$= \frac{1}{2} x^2 \tanh^{-1} x - \frac{1}{2} \int \left(-1 + \frac{1}{1-x^2} \right) dx = \frac{1}{2} x^2 \tanh^{-1} x + \frac{1}{2} x - \frac{1}{2} \tanh^{-1} x + C.$$

33. Let $w = \sinh^{-1} x$. Then $x = \sinh w$, $dx = \cosh w \, dw$, and hence

$$\int x \sinh^{-1} x \, dx = \int (\sinh w) w \cosh w \, dw = \frac{1}{2} \int w \sinh 2w \, dw,$$

since $\sinh 2w = 2 \sinh w \cosh w$. Now, let $u = w$, $dv = \sinh 2w \, dw$. Then $du = dw$, $v = (1/2) \cosh 2w$, and hence

$$\int w \sinh 2w \, dw = \frac{1}{2} w \cosh 2w - \frac{1}{2} \int \cosh 2w \, dw = \frac{1}{2} w \cosh 2w - \frac{1}{4} \sinh 2w + C.$$

Therefore

$$\int x \sinh^{-1} x \, dx = \frac{1}{4} w \cosh 2w - \frac{1}{8} \sinh 2w + C$$

$$= \frac{1}{4} w (2 \sinh^2 w + 1) - \frac{1}{8} (2 \sinh w \cosh w) + C = \frac{2x^2+1}{4} \sinh^{-1} x - \frac{1}{4} x \sqrt{1+x^2} + C.$$

34. Let $w = \sqrt{x} \, dx$. Then $x = w^2$, $dx = 2w \, dw$, and hence

$$\int \sqrt{x} e^{-\sqrt{x}} \, dx = \int w e^{-w} (2w \, dw) = 2 \int \underbrace{w^2 e^{-w} \, dw}_{u = w^2, \, dv = e^{-w} dw} = 2 \left[-w^2 e^{-w} + 2 \int \underbrace{w e^{-w} \, dw}_{u = w, \, dv = e^{-w} dw} \right]$$

$$= -2w^2 e^{-w} + 4(-w e^{-w} - 2e^{-w}) + C = -2x e^{-\sqrt{x}} - 4\sqrt{x} e^{-\sqrt{x}} - 4e^{-\sqrt{x}} + C.$$

35. Let $u = 2x^{3/2}$, $dv = \dfrac{e^{-\sqrt{x}}}{\sqrt{x}} \, dx$. Then $du = 3x^{1/2} dx$, $v = -2e^{-\sqrt{x}}$, and hence

$$\int 2x e^{-\sqrt{x}} \, dx = -4x^{3/2} e^{-\sqrt{x}} + 6 \int \sqrt{x} e^{-\sqrt{x}} \, dx$$

$$= -4x^{3/2} e^{-\sqrt{x}} + 6 \left[-2x e^{-\sqrt{x}} - 4\sqrt{x} e^{-\sqrt{x}} - 4e^{-\sqrt{x}} \right] + C \quad \text{by Exercise 34}$$

$$= -4x^{3/2} e^{-\sqrt{x}} - 12x e^{-\sqrt{x}} - 24\sqrt{x} e^{-\sqrt{x}} - 24e^{-\sqrt{x}} + C.$$

36. Let $u = \sinh^{-1} x$, $dv = dx$. Then $du = \dfrac{1}{\sqrt{1+x^2}} \, dx$, $v = x$, and hence

$$\int \sinh^{-1} x \, dx = x \sinh^{-1} x - \int \frac{x}{\sqrt{1+x^2}} \, dx = x \sinh^{-1} x - \sqrt{1+x^2} + C.$$

37. Area $= \displaystyle\int_0^1 \underbrace{x \sin \pi x \, dx}_{u=x, dv=\sin \pi x} = \left(-\frac{x}{\pi} \cos \pi x + \frac{1}{\pi} \int \cos \pi x \, dx \right)\Big|_0^1 = \left(-\frac{x}{\pi} \cos \pi x + \frac{1}{\pi^2} \sin \pi x \right)\Big|_0^1 = \frac{1}{\pi}.$

38. Since $y = x$ is the equation of the tangent line to the graph of $y = \ln(1+x)$ at $x = 0$, $\ln(1+x) < x$. Hence the area is equal to

$$\int_0^1 [x - \ln(1+x)] \, dx = \left[\int x \, dx - \int \underbrace{\ln(1+x) \, dx}_{u=\ln(1+x), dv=dx} \right]\Bigg|_0^1 = \left[\frac{1}{2} x^2 - x \ln(1+x) + \int \frac{x}{1+x} \, dx \right]\Bigg|_0^1$$

$$= \left[\frac{1}{2} x^2 - x \ln(1+x) + \int \left(1 - \frac{1}{1+x} \right) dx \right]\Bigg|_0^1 = \left[\frac{1}{2} x^2 - x \ln(1+x) + x - \ln(1+x) \right]\Bigg|_0^1 = \frac{3}{2} - 2\ln 2.$$

39. Let dV be the differential volume of a cross-section disk of thickness dx. Then $dV = \pi r^2 dx = \pi y^2 dx = \pi(\ln x)^2 dx$ and hence the volume of revolution is

$$V = \int_1^e \pi(\ln x)^2 \, dx.$$

Let $u = (\ln x)^2$, $dv = dx$. Then $du = \dfrac{2(\ln x)}{x} \, dx$, $v = x$, and hence

$$V = \pi \left[x(\ln x)^2 - 2 \int \ln x \, dx \right]\Bigg|_1^e = \pi \left[x(\ln x)^2 - 2x \ln x + 2x \right]\big|_1^e = \pi(e - 2).$$

40. Let dV be the differential volume of a shell of thickness dx revolved about the y-axis. Then $dV = 2\pi rh dx = 2\pi xy dx = 2\pi x \ln x dx$ and hence the volume of revolution is

$$V = \int_1^e 2\pi \underbrace{x \ln x \, dx}_{u=\ln x, dv=x dx} = 2\pi \left[\frac{1}{2} x^2 \ln x - \frac{1}{2} \int x \, dx \right]\Bigg|_1^e = 2\pi \left[\frac{1}{2} x^2 \ln x - \frac{1}{4} x^2 \right]\Bigg|_1^e = \frac{\pi}{2} (e^2 + 1).$$

41. Let dV be the differential volume of a shell of thickness dx revolved about the y-axis. Then $dV = 2\pi rh dx = 2\pi xy dx = 2\pi x \sin x dx$ and hence the volume of revolution is

$$V = 2\pi \int_0^\pi \underbrace{x \sin x \, dx}_{u=x, dv=\sin x \, dx} = 2\pi \left(-x \cos x + \sin x \right)\big|_0^\pi = 2\pi^2.$$

42. Let dV be the differential volume of a shell of thickness dx revolved about the y-axis. Then $dV = 2\pi rh dx = 2\pi xy dx = 2\pi x^2 e^{2x} dx$ and hence the volume of revolution is

$$V = 2\pi \int_0^{\ln 2} \underbrace{x^2 e^{2x} \, dx}_{u=x^2, dv=e^{2x} dx} = 2\pi \left[\frac{x^2}{2} e^{2x} - \int \underbrace{xe^{2x} \, dx}_{u=x, dv=e^{2x} dx} \right]\Bigg|_0^{\ln 2}$$

$$= 2\pi \left[\frac{x^2}{2} e^{2x} - \frac{x}{2} e^{2x} + \frac{1}{2} \int e^{2x} \, dx \right]\Big|_0^{\ln 2}$$

$$= 2\pi \left[\frac{x^2}{2} e^{2x} - \frac{x}{2} e^{2x} + \frac{1}{4} e^{2x} \right]\Big|_0^{\ln 2}$$

$$= 2\pi \left[2(\ln 2)^2 - 2\ln 2 + \frac{3}{4} \right].$$

43. The average value on $[0, 1]$ is equal to $\int_0^1 e^{-x} \sin \pi x \, dx$. Now,

$$\int \underbrace{e^{-x} \sin \pi x \, dx}_{u=\sin \pi x, dv=e^{-x} \, dx} = -e^{-x} \sin \pi x + \pi \int \underbrace{e^{-x} \cos \pi x \, dx}_{u=\cos \pi x, dv=e^{-x} \, dx}$$

$$= -e^{-x} \sin \pi x + \pi \left[-e^{-x} \cos \pi x - \pi \int e^{-x} \sin \pi x \, dx \right]$$

$$= -e^{-x} \sin \pi x - \pi e^{-x} \cos \pi x - \pi^2 \int e^{-x} \sin \pi x \, dx$$

$$\Longrightarrow \quad (1 + \pi^2) \int e^{-x} \sin \pi x \, dx = -e^{-x}(\sin \pi x + \pi \cos \pi x) + C$$

$$\Longrightarrow \quad \int e^{-x} \sin \pi x \, dx = -\frac{e^{-x}}{1+\pi^2} (\sin \pi x + \pi \cos \pi x) + C.$$

Therefore

$$\text{Average value} = -\frac{e^{-x}}{1 + \pi^2} (\sin \pi x + \pi \cos \pi x)\Big|_0^1 = \frac{\pi(1 + e)}{e(1 + \pi^2)}.$$

44. The average value on $[0, 1]$ is

$$\int_0^1 \text{Sin}^{-1}x \, dx = \left(x\text{Sin}^{-1}x + \sqrt{1 - x^2} \right)\Big|_0^1 = \frac{\pi}{2} - 1. \quad \text{(see Example 5)}$$

45. Since $v(t) > 0$ for $0 < t < 1$ and $v(t) < 0$ for $1 < t < 2$, the distance travelled from $t = 0$ to $t = 2$ is

$$s = \int_0^2 |v(t)| \, dt = \int_0^1 t \sin \pi t \, dt + \int_1^2 (-t \sin \pi t) \, dt.$$

Now,

$$\int \underbrace{t \sin \pi t \, dt}_{u=t, dv=\sin \pi t \, dt} = -\frac{t}{\pi} \cos \pi t + \frac{1}{\pi} \int \cos \pi t \, dt = -\frac{t}{\pi} \cos \pi t + \frac{1}{\pi^2} \sin \pi t + C.$$

Therefore

$$s = \left(-\frac{t}{\pi} \cos \pi t + \frac{1}{\pi^2} \sin \pi t \right)\Big|_0^1 + \left(\frac{t}{\pi} \cos \pi t - \frac{1}{\pi^2} \sin \pi t \right)\Big|_1^2 = \frac{4}{\pi} \text{ meters.}$$

46. Since $v(t) < 0$ for $0 < t < 1$ and $v(t) > 0$ for $1 < t < 2$, the distance travelled is

$$s = \int_0^2 |v(t)| \, dt = \int_0^1 [-(t^2 - 1)e^{-t}] \, dt + \int_1^2 (t^2 - 1)e^{-t} \, dt.$$

Now,

$$\int \underbrace{t^2 e^{-t}\, dt}_{u=t^2,\,dv=e^{-t}\,dt} = -t^2 e^{-t} + 2\int \underbrace{t e^{-t}\, dt}_{u=t,\,dv=e^{-t}\,dt}$$

$$= -t^2 e^{-t} + 2\left[-t e^{-t} + \int e^{-t}\, dt\right]$$

$$= -t^2 e^{-t} - 2t e^{-t} - 2 e^{-t} + C.$$

Therefore

$$s = -\int_0^1 t^2 e^{-t}\, dt + \int_0^1 e^{-t}\, dt + \int_1^2 t^2 e^{-t}\, dt - \int_1^2 e^{-t}\, dt$$

$$= (t^2 e^{-t} + 2t e^{-t} + 2e^{-t})\Big|_0^1 - e^{-t}\Big|_0^1 + (-t^2 e^{-t} - 2t e^{-t} - 2e^{-t})\Big|_1^2 + e^{-t}\Big|_1^2$$

$$= \frac{8}{e} - \frac{9}{e^2} - 1.$$

47. Work $= \int_1^5 (62.4)\pi x^2 (5-y)\, dy = \int_1^5 (62.4)\pi (\ln y)^2 (5-y)\, dy$. Now,

$$\int \underbrace{(5-y)(\ln y)^2\, dy}_{u=(\ln y)^2,\,dv=(5-y)\,dy} = (\ln y)^2\left(5y - \frac{1}{2}y^2\right) - 2\int \underbrace{(\ln y)\left(5 - \frac{1}{2}y\right)\, dy}_{u=\ln y,\,dv=(5-\frac{1}{2}y)\,dy}$$

$$= (\ln y)^2\left(5y - \frac{1}{2}y^2\right) - 2\left[(\ln y)\left(5y - \frac{1}{4}y^2\right) - \int\left(5 - \frac{1}{4}y\right)\, dy\right]$$

$$= (\ln y)^2\left(5y - \frac{1}{2}y^2\right) - 2(\ln y)\left(5y - \frac{1}{4}y^2\right) + 2\left(5y - \frac{1}{8}y^2\right) + C.$$

Hence

$$\text{Work} = (62.4)\pi\left[(\ln y)^2\left(5y - \frac{1}{2}y^2\right) - 2(\ln y)\left(5y - \frac{1}{4}y^2\right) + 2\left(5y - \frac{1}{8}y^2\right)\right]\Bigg|_1^5$$

$$= (62.4)\pi\left[\frac{25}{2}(\ln 5)^2 - \frac{75}{2}(\ln 5) + 34\right].$$

48. The centroid of the region is the point $(\overline{x}, \overline{y})$, where

$$\overline{x} = \frac{1}{A}\int_0^{\ln 2} x e^x\, dx = \frac{1}{A}(x e^x - e^x)\Bigg|_0^{\ln 2} = \frac{1}{A}(2\ln 2 - 1),$$

$$\overline{y} = \frac{1}{2A}\int_0^{\ln 2} e^{2x}\, dx = \frac{1}{4A}e^{2x}\Bigg|_0^{\ln 2} = \frac{3}{4A},$$

and

$$A = \int_0^{\ln 2} e^x\, dx = e^x\Big|_0^{\ln 2} = 1.$$

49. The centroid of the region is the point $(\overline{x}, \overline{y})$, where

$$\overline{x} = \frac{1}{A}\int_1^e x(x - \ln x)\, dx = \frac{1}{A}\int_1^e \underbrace{(x^2 - x\ln x)\, dx}_{u=\ln x,\,dv=x\,dx} = \frac{1}{A}\left[\frac{1}{3}x^3 - \left(\frac{1}{2}x^2\ln x - \frac{1}{2}\int x\, dx\right)\right]\Bigg|_1^e$$

$$= \frac{1}{A}\left[\frac{1}{3}x^3 - \frac{1}{2}x^2\ln x + \frac{1}{4}x^2\right]\Big|_1^e = \frac{1}{A}\left(\frac{1}{3}e^3 - \frac{1}{4}e^2 - \frac{7}{12}\right),$$

$$\bar{y} = \frac{1}{2A}\int_1^e [x^2 - (\ln x)^2]\, dx = \frac{1}{2A}\left[\frac{1}{3}x^3 - (x(\ln x)^2 - 2x\ln x + 2x)\right]\Big|_1^e = \frac{1}{2A}\left(\frac{1}{3}e^3 - e + \frac{5}{3}\right),$$

and

$$A = \int_1^e (x - \ln x)\, dx = \left[\frac{1}{2}x^2 - (x\ln x - x)\right]\Big|_1^e = \frac{1}{2}e^2 - \frac{3}{2}.$$

50. Let $u = \sin^{n-1}x$, $dv = \sin x\, dx$. Then $du = (n-1)\sin^{n-2}x\, dx$, $v = -\cos x$, and hence

$$\int \sin^n x\, dx \quad = \quad -\sin^{n-1}x\cos x + (n-1)\int \sin^{n-2}x\, \underbrace{\cos^2 x}_{1-\sin^2 x}\, dx$$

$$= \quad -\sin^{n-1}x\cos x + (n-1)\int \sin^{n-2}x\, dx - (n-1)\int \sin^n x\, dx$$

$$\Longrightarrow \quad n\int \sin^n x\, dx \quad = \quad -\sin^{n-1}x\cos x + (n-1)\int \sin^{n-2}x\, dx$$

$$\Longrightarrow \quad \int \sin^n x\, dx \quad = \quad -\frac{1}{n}\sin^{n-1}x\cos x + \frac{n-1}{n}\int \sin^{n-2}x\, dx.$$

51. $\displaystyle \int \sin^2 x\, dx = -\frac{1}{2}\sin x\cos x + \frac{1}{2}\int dx = -\frac{1}{2}\sin x\cos x + \frac{1}{2}x + C.$

52. $\displaystyle \int \sin^4 x\, dx = -\frac{1}{4}\sin^3 x\cos x + \frac{3}{4}\int \sin^2 x\, dx = -\frac{1}{4}\sin^3 x\cos x + \frac{3}{4}\left[-\frac{1}{2}\sin x\cos x + \frac{1}{2}x\right] + C =$
$-\frac{1}{4}\sin^3 x\cos x - \frac{3}{8}\sin x\cos x + \frac{3}{8}x + C.$

53. Let $u = x^n$, $dv = e^x dx$. Then $du = nx^{n-1}dx$, $v = e^x$, and hence

$$\int x^n e^x\, dx = x^n e^x - n\int x^{n-1}e^x\, dx.$$

54. $\displaystyle \int x^4 e^x\, dx = x^4 e^x - 4\int x^3 e^x\, dx = x^4 e^x - 4\left[x^3 e^x - 3\int x^2 e^x\, dx\right]$
$= x^4 e^x - 4x^3 e^x + 12\left[x^2 e^x - 2\int xe^x\, dx\right] = x^4 e^x - 4x^3 e^x + 12x^2 e^x - 24\left[xe^x - \int e^x\, dx\right]$
$= x^4 e^x - 4x^3 e^x + 12x^2 e^x - 24xe^x + 24e^x + C.$

55. Let $u = x^n$, $dv = \sin ax\, dx$. Then $du = nx^{n-1}dx$, $v = (-1/a)\cos ax$, and hence

$$\int x^n \sin ax\, dx = -\frac{1}{a}x^n\cos ax + \frac{n}{a}\int x^{n-1}\cos ax\, dx.$$

56. $\displaystyle \int x\sin 3x\, dx = -\frac{x}{3}\cos 3x + \frac{1}{3}\int \cos 3x\, dx = -\frac{x}{3}\cos 3x + \frac{1}{9}\sin 3x + C.$

57. Let $u = \rho(x)$, $dv = e^x dx$. Then $du = \rho'(x)dx$, $v = e^x$, and hence

$$\int \rho(x)e^x\, dx = \rho(x)e^x - \int \rho'(x)e^x\, dx.$$

Since $\rho(x)$ is a polynomial of degree n, $\rho'(x)$ is a polynomial of degree at most $n-1$. Hence, after at most $n-1$ further applications of this reduction formula one obtains $\int \rho(x)e^x\, dx$ in terms of $\int e^x\, dx$, which is equal to $e^x + C$, and hence gives a complete formula for $\int \rho(x)e^x\, dx$.

10.2 Trigonometric Substitutions

1. Let $x = 2\sin\theta$. Then $dx = 2\cos\theta\,d\theta$ and hence

$$\int \sqrt{4 - x^2}\,dx = \int (2\cos\theta)(2\cos\theta\,d\theta) = 2\int(1 + \cos 2\theta)\,d\theta = 2\theta + \sin 2\theta + C$$

$$= 2\theta + 2\sin\theta\cos\theta + C = 2\text{Sin}^{-1}\frac{x}{2} + 2\left(\frac{x}{2}\right)\frac{\sqrt{4 - x^2}}{2} + C = 2\text{Sin}^{-1}\frac{x}{2} + \frac{x}{2}\sqrt{4 - x^2} + C.$$

2. Let $x = 4\tan\theta$. Then $dx = 4\sec^2\theta\,d\theta$ and hence

$$\int \frac{1}{\sqrt{x^2 + 16}}\,dx = \int \frac{1}{4\sec\theta}4\sec^2\theta\,d\theta = \int \sec\theta\,d\theta$$

$$= \ln|\sec\theta + \tan\theta| + C = \ln\left|\frac{\sqrt{x^2 + 16}}{4} + \frac{x}{4}\right| + C = \ln(\sqrt{x^2 + 16} + x) + C.$$

Note that the missing constant, $\ln 4$, is absorbed as part of C.

3. Let $x = 3\sin\theta$. Then $dx = 3\cos\theta\,d\theta$ and hence

$$\int_0^{3/2} \frac{dx}{\sqrt{9 - x^2}} = \int_0^{\pi/6} \frac{3\cos\theta}{3\cos\theta}\,d\theta = \theta\big|_0^{\pi/6} = \frac{\pi}{6}.$$

4. Let $x = \sec\theta$. Then $dx = \sec\theta\tan\theta\,d\theta$ and hence

$$\int_1^2 \sqrt{x^2 - 1}\,dx = \int_0^{\pi/3} \tan\theta\sec\theta\tan\theta\,d\theta$$

$$= \int_0^{\pi/3} \sec\theta(\sec^2\theta - 1)\,d\theta$$

$$= \int_0^{\pi/3} \sec^3\theta\,d\theta - \int_0^{\pi/3} \sec\theta\,d\theta$$

$$= \left[\frac{1}{2}\sec\theta\tan\theta + \frac{1}{2}\ln|\sec\theta + \tan\theta| - \ln|\sec\theta + \tan\theta|\right]\Big|_0^{\pi/3}$$

$$= \frac{1}{2}\left(\sec\theta\tan\theta - \ln|\sec\theta + \tan\theta|\right)\big|_0^{\pi/3}$$

$$= \frac{1}{2}(2\sqrt{3} - \ln(2 + \sqrt{3})).$$

5. Let $u = x^2 - 4$. Then $du = 2x\,dx$ and hence

$$\int \frac{x^3}{\sqrt{x^2 - 4}}\,dx = \frac{1}{2}\int \frac{u + 4}{\sqrt{u}}\,du = \frac{1}{2}\int(u^{1/2} + 4u^{-1/2})\,du$$

$$= \frac{1}{3}u^{3/2} + 4u^{1/2} + C = \frac{1}{3}(x^2 - 4)^{3/2} + 4\sqrt{x^2 - 4} + C.$$

6. Let $x = 3\sin\theta$. Then $dx = 3\cos\theta\,d\theta$ and hence

$$\int \frac{x^2}{\sqrt{9 - x^2}}\,dx = \int \frac{9\sin^2\theta}{3\cos\theta}3\cos\theta\,d\theta = 9\int \sin^2\theta\,d\theta = \frac{9}{2}\int(1 - \cos 2\theta)\,d\theta = \frac{9}{2}\left(\theta - \frac{1}{2}\sin 2\theta\right) + C$$

$$= \frac{9}{2}(\theta - \sin\theta\cos\theta) = \frac{9}{2}\left(\text{Sin}^{-1}\frac{x}{3} - \frac{x}{3}\frac{\sqrt{9 - x^2}}{3}\right) + C = \frac{9}{2}\text{Sin}^{-1}\frac{x}{3} - \frac{1}{2}x\sqrt{9 - x^2} + C.$$

7. Let $x = \sin\theta$. Then $dx = \cos\theta\,d\theta$ and hence

$$\int \frac{\sqrt{1-x^2}}{x}\,dx = \int \frac{\cos\theta}{\sin\theta}\,\cos\theta\,d\theta = \int \frac{1-\sin^2\theta}{\sin\theta}\,d\theta = \int (\csc\theta - \sin\theta)\,d\theta$$

$$= \ln|\csc\theta - \cot\theta| + \cos\theta + C = \ln\left|\frac{1}{x} - \frac{\sqrt{1-x^2}}{x}\right| + \sqrt{1-x^2} + C.$$

Therefore

$$\int_{1/2}^{1} \frac{\sqrt{1-x^2}}{x}\,dx = \left(\ln\left|\frac{1-\sqrt{1-x^2}}{x}\right| + \sqrt{1-x^2}\right)\Bigg|_{1/2}^{1} = -\ln\left(2-\sqrt{3}\right) - \frac{1}{2}\sqrt{3}.$$

8. Let $x = 4\tan\theta$. Then $dx = 4\sec^2\theta\,d\theta$ and hence

$$\int \frac{1}{x^2\sqrt{x^2+16}}\,dx = \int \frac{1}{16\tan^2\theta\sec\theta}\,4\sec^2\theta\,d\theta = \frac{1}{16}\int \underbrace{\frac{\cos\theta}{\sin^2\theta}\,d\theta}_{u=\sin\theta}$$

$$= \frac{1}{16}\int \frac{du}{u^2} = \frac{1}{16}\left(-\frac{1}{u}\right) + C = -\frac{1}{16\sin\theta} + C = -\frac{\sqrt{x^2+16}}{16x} + C.$$

Therefore

$$\int_{3}^{5} \frac{1}{x^2\sqrt{x^2+16}}\,dx = -\frac{\sqrt{x^2+16}}{16x}\Bigg|_{3}^{5} = -\frac{1}{16}\left[\frac{\sqrt{41}}{5} - \frac{5}{3}\right].$$

9. Let $x = 2\sec\theta$. Then $dx = 2\sec\theta\tan\theta\,d\theta$ and hence

$$\int \frac{1}{x^3\sqrt{x^2-4}}\,dx = \int \frac{1}{8\sec^3\theta\,2\tan\theta}(2\sec\theta\,\tan\theta)\,d\theta = \frac{1}{8}\int \cos^2\theta\,d\theta = \frac{1}{16}\int (1+\cos 2\theta)\,d\theta$$

$$= \frac{1}{16}\left(\theta + \frac{1}{2}\sin 2\theta\right) + C = \frac{1}{16}(\theta + \sin\theta\cos\theta) + C = \frac{1}{16}\left(\text{Sec}^{-1}(x/2) + \frac{\sqrt{x^2-4}}{x}\frac{2}{x}\right) + C.$$

10. Let $u = x^2 + 4$. Then $du = 2x\,dx$ and hence

$$\int \frac{x}{\sqrt{x^2+4}}\,dx = \frac{1}{2}\int \frac{du}{\sqrt{u}} = \sqrt{u} + C = \sqrt{x^2+4} + C.$$

11. Let $x = 4\tan\theta$. Then $dx = 4\sec^2\theta\,d\theta$ and hence

$$\int_{0}^{4} \frac{1}{(16+x^2)^2}\,dx = \int_{0}^{\pi/4} \frac{1}{256\sec^4\theta}\,4\sec^2\theta\,d\theta$$

$$= \frac{1}{64}\int_{0}^{\pi/4} \cos^2\theta\,d\theta = \frac{1}{128}(\theta + \sin\theta\cos\theta)\Bigg|_{0}^{\pi/4} = \frac{1}{128}\left(\frac{\pi}{4} + \frac{1}{2}\right).$$

12. Let $x = 2\sin\theta$. Then $dx = 2\cos\theta\,d\theta$ and hence

$$\int_{0}^{1} \frac{dx}{(4-x^2)^{3/2}} = \int_{0}^{\pi/6} \frac{2\cos\theta\,d\theta}{8\cos^3\theta} = \frac{1}{4}\int_{0}^{\pi/6} \sec^2\theta\,d\theta = \frac{1}{4}\tan\theta\Bigg|_{0}^{\pi/6} = \frac{1}{4\sqrt{3}}.$$

13. Let $x = \sin\theta$. Then $dx = \cos\theta\, d\theta$ and hence

$$\int \frac{x^2}{(1-x^2)^{3/2}} = \int \frac{\sin^2\theta}{\cos^3\theta}\cos\theta\, d\theta = \int \tan^2\theta\, d\theta$$

$$= \int(\sec^2\theta - 1)\, d\theta = \tan\theta - \theta + C = \frac{x}{\sqrt{1-x^2}} - \text{Sin}^{-1}x + C.$$

14. Let $x = 2\sec\theta$. Then $dx = 2\sec\theta\tan\theta\, d\theta$ and hence

$$\int \frac{dx}{(x^2-4)^{3/2}} = \int \frac{2\sec\theta\tan\theta\, d\theta}{8\tan^3\theta} = \frac{1}{4}\int \cot\theta\csc\theta\, d\theta = -\frac{1}{4}\csc\theta + C = -\frac{x}{4\sqrt{x^2-4}} + C.$$

15. Let $u = 1 - x^2$. Then $du = -2x\,dx$ and hence

$$\int_0^1 x^3\sqrt{1-x^2}\, dx = -\frac{1}{2}\int_1^0 (1-u)\sqrt{u}\, du = -\frac{1}{2}\int_1^0 (u^{1/2} - u^{3/2})\, du$$

$$= -\frac{1}{2}\left(\frac{2}{3}u^{3/2} - \frac{2}{5}u^{5/2}\right)\bigg|_1^0 = \frac{2}{15}.$$

16. Let $x = 2\tan\theta$. Then $dx = 2\sec^2\theta\, d\theta$ and hence

$$\int_0^2 \frac{x^2}{\sqrt{4+x^2}}\, dx = \int_0^{\pi/4} \frac{4\tan^2\theta}{2\sec\theta}2\sec^2\theta\, d\theta = 4\int_0^{\pi/4} \tan^2\theta\sec\theta\, d\theta$$

$$= 2\big(\sec\theta\tan\theta - \ln|\sec\theta + \tan\theta|\big)\bigg|_0^{\pi/4} = 2[\sqrt{2} - \ln(\sqrt{2}+1)].$$

17. Let $x = \sqrt{3}\tan\theta$. Then $dx = \sqrt{3}\sec^2\theta\, d\theta$ and hence

$$\int \frac{x^2\, dx}{(x^2+3)^{3/2}} = \int \frac{3\tan^2\theta}{3\sqrt{3}\sec^3\theta}\sqrt{3}\sec^2\theta\, d\theta = \int \frac{\sin^2\theta}{\cos\theta}\, d\theta = \int \frac{1-\cos^2\theta}{\cos\theta}\, d\theta = \int(\sec\theta - \cos\theta)\, d\theta$$

$$= \ln|\sec\theta + \tan\theta| - \sin\theta + C = \ln\left|\frac{\sqrt{x^2+3}+x}{\sqrt{3}}\right| - \frac{x}{\sqrt{x^2+3}} + C = \ln(\sqrt{x^2+3}+x) - \frac{x}{\sqrt{x^2+3}} + C.$$

Note that the missing constant, $\ln\sqrt{3}$, is absorbed in the constant C.

18. Let $x = \sin\theta$. Then $dx = \cos\theta\, d\theta$ and hence

$$\int \frac{\sqrt{1-x^2}}{x^4}\, dx = \int \frac{\cos\theta}{\sin^4\theta}\cos\theta\, d\theta = \underbrace{\int \cot^2\theta\csc^2\theta\, d\theta}_{u=\cot\theta}$$

$$= -\int u^2\, du = -\frac{1}{3}u^3 + C = -\frac{1}{3}\cot^3\theta + C = -\frac{1}{3}\frac{(1-x^2)^{3/2}}{x^3} + C.$$

19. Let $x = a\tan\theta$. Then $dx = a\sec^2\theta\, d\theta$ and hence

$$\int \frac{x^2}{\sqrt{x^2+a^2}}\, dx = \int \frac{a^2\tan^2\theta}{a\sec\theta}a\sec^2\theta\, d\theta = a^2\int \tan^2\theta\sec\theta\, d\theta = \frac{a^2}{2}\big(\sec\theta\tan\theta - \ln|\sec\theta + \tan\theta|\big) + C$$

$$= \frac{a^2}{2}\left(\frac{\sqrt{x^2+a^2}}{a}\frac{x}{a} - \ln\left|\frac{\sqrt{x^2+a^2}}{a} + \frac{x}{a}\right|\right) + C = \frac{x\sqrt{x^2+a^2}}{2} - \frac{a^2}{2}\ln(\sqrt{x^2+a^2}+x) + C.$$

719

20. Let $x = a \sec \theta$. Then $dx = a \sec \theta \tan \theta \, d\theta$ and hence

$$\int \frac{\sqrt{x^2 - a^2}}{x} \, dx = \int \frac{a \tan \theta}{a \sec \theta} \, a \sec \theta \tan \theta \, d\theta = a \int \tan^2 \theta \, d\theta = a \int (\sec^2 \theta - 1) \, d\theta$$

$$= a(\tan \theta - \theta) + C = a \left(\frac{\sqrt{x^2 - a^2}}{a} - \operatorname{Sec}^{-1} \frac{x}{a} \right) + C = \sqrt{x^2 - a^2} - a \operatorname{Sec}^{-1} \frac{x}{a} + C.$$

21. Let $x = a \sec \theta$. Then $dx = a \sec \theta \tan \theta \, d\theta$ and hence

$$\int \frac{\sqrt{x^2 - a^2}}{x^2} \, dx = \int \frac{a \tan \theta}{a^2 \sec^2 \theta} \, a \sec \theta \tan \theta \, d\theta = \int \frac{\tan^2 \theta}{\sec \theta} \, d\theta = \int \frac{\sec^2 \theta - 1}{\sec \theta} \, d\theta = \int (\sec \theta - \cos \theta) \, d\theta$$

$$= \ln|\sec \theta + \tan \theta| - \sin \theta + C = \ln \left| \frac{x}{a} + \frac{\sqrt{x^2 - a^2}}{a} \right| - \frac{\sqrt{x^2 - a^2}}{x} + C = \ln \left| x + \sqrt{x^2 - a^2} \right| - \frac{\sqrt{x^2 - a^2}}{x} + C.$$

Note that the missing constant, $\ln a$, is absorbed into the constant C.

22. Let $x = a \sec \theta$. Then $dx = a \sec \theta \tan \theta \, d\theta$ and hence

$$\begin{aligned}
\int \frac{x^2}{\sqrt{x^2 - a^2}} \, dx &= \int \frac{a^2 \sec^2 \theta}{a \tan \theta} \, a \sec \theta \tan \theta \, d\theta \\
&= a^2 \int \sec^3 \theta \, d\theta \\
&= \frac{a^2}{2} \sec \theta \tan \theta + \frac{a^2}{2} \ln|\sec \theta + \tan \theta| + C \quad \text{(Example 7, Sec. 10.1)} \\
&= \frac{a^2}{2} \frac{x}{a} \frac{\sqrt{x^2 - a^2}}{a} + \frac{a^2}{2} \ln \left| \frac{x}{a} + \frac{x^2 - a^2}{a} \right| + C \\
&= \frac{x}{2} \sqrt{x^2 - a^2} + \frac{a^2}{2} \ln \left| x + \sqrt{x^2 - a^2} \right| + C.
\end{aligned}$$

23. Let $u = x^2 + a^2$. Then $du = 2x \, dx$ and hence

$$\int \frac{x \, dx}{\sqrt{(x^2 + a^2)^3}} = \frac{1}{2} \int \frac{du}{u^{3/2}} = -u^{-1/2} + C = -\frac{1}{\sqrt{x^2 + a^2}} + C.$$

24. Let $x = a \tan \theta$. Then $dx = a \sec^2 \theta \, d\theta$ and hence

$$\int \frac{\sqrt{x^2 + a^2}}{x} \, dx = \int \frac{a \sec \theta}{a \tan \theta} \, a \underbrace{\sec^2 \theta}_{1 + \tan^2 \theta} \, d\theta = a \int (\csc \theta + \sec \theta \tan \theta) \, d\theta$$

$$= a \ln|\csc \theta - \cot \theta| + a \sec \theta + C = a \ln \left| \frac{\sqrt{x^2 + a^2} - a}{x} \right| + \sqrt{x^2 + a^2} + C.$$

25. Let $x = a \sec \theta$. Then $dx = a \sec \theta \tan \theta \, d\theta$ and hence

$$\int \frac{dx}{x \sqrt{x^2 - a^2}} = \int \frac{a \sec \theta \tan \theta \, d\theta}{(a \sec \theta)(a \tan \theta)} = \frac{1}{a} \theta + C = \frac{1}{a} \operatorname{Sec}^{-1} \frac{x}{a} + C.$$

26. Let $x = a \tan \theta$. Then $dx = a \sec^2 \theta \, d\theta$ and hence

$$\int \frac{dx}{x\sqrt{x^2 + a^2}} = \int \frac{a \sec^2 \theta \, d\theta}{(a \tan \theta)(a \sec \theta)} = \frac{1}{a} \int \csc \theta \, d\theta$$

$$= \frac{1}{a} \ln |\csc \theta - \cot \theta| + C = \frac{1}{a} \ln \left| \frac{\sqrt{x^2 + a^2} - a}{x} \right| + C.$$

27. Let $x = a \tan \theta$. Then $dx = a \sec^2 \theta \, d\theta$ and hence

$$\int \frac{\sqrt{x^2 + a^2}}{x^2} \, dx = \int \frac{a \sec \theta}{a^2 \tan^2 \theta} a \sec^2 \theta \, d\theta = \int \frac{\sec \theta (1 + \tan^2 \theta)}{\tan^2 \theta} \, d\theta$$

$$= \int \cot \theta \csc \theta \, d\theta + \int \sec \theta \, d\theta = -\csc \theta + \ln |\sec \theta + \tan \theta| + C$$

$$= -\frac{\sqrt{x^2 + a^2}}{x} + \ln \left| \frac{\sqrt{x^2 + a^2}}{a} + \frac{x}{a} \right| + C = -\frac{\sqrt{x^2 + a^2}}{x} + \ln(\sqrt{x^2 + a^2} + x) + C.$$

Note that the missing constant, $\ln a$, is absorbed in the constant C.

28. Let $x = a \tan \theta$. Then $dx = a \sec^2 \theta \, d\theta$ and hence

$$\int \frac{x^2 \, dx}{(x^2 + a^2)^{3/2}} = \int \frac{(a^2 \tan^2 \theta)(a \sec^2 \theta \, d\theta)}{a^3 \sec^3 \theta} = \int \frac{\sec^2 \theta - 1}{\sec \theta} \, d\theta = \int (\sec \theta - \cos \theta) \, d\theta$$

$$= \ln |\sec \theta + \tan \theta| - \sin \theta + C = \ln \left| \frac{\sqrt{x^2 + a^2}}{a} + \frac{x}{a} \right| - \frac{x}{\sqrt{x^2 + a^2}} + C = \ln \left| \sqrt{x^2 + a^2} + x \right| - \frac{x}{\sqrt{x^2 + a^2}} + C.$$

Note that the missing constant, $\ln a$, is absorbed in the constant C.

29. Let $x = 3 \tan \theta$. Then $dx = 3 \sec^2 \theta \, d\theta$ and hence

$$\int \frac{x^2 + 3}{\sqrt{x^2 + 9}} \, dx = \int \frac{9 \tan^2 \theta + 3}{3 \sec \theta} 3 \sec^2 \theta \, d\theta = 9 \int (\sec \theta)(\tan^2 \theta) \, d\theta + 3 \int \sec \theta \, d\theta$$

$$= \frac{9}{2} (\sec \theta \tan \theta - \ln |\sec \theta + \tan \theta|) + 3 \ln |\sec \theta + \tan \theta| + C = \frac{9}{2} \sec \theta \tan \theta - \frac{3}{2} \ln |\sec \theta + \tan \theta| + C$$

$$= \frac{9}{2} \frac{\sqrt{x^2 + 9}}{3} \frac{x}{3} - \frac{3}{2} \ln \left| \frac{\sqrt{x^2 + 9}}{3} + \frac{x}{3} \right| + C = \frac{1}{2} x \sqrt{x^2 + 9} - \frac{3}{2} \ln(\sqrt{x^2 + 9} + x) + C.$$

30. Let $x = 2 \tan \theta$. Then $dx = 2 \sec^2 \theta \, d\theta$ and hence

$$\int \frac{x^2 - 3}{x\sqrt{x^2 + 4}} \, dx = \int \frac{4 \tan^2 \theta - 3}{(2 \tan \theta)(2 \sec \theta)} 2 \sec^2 \theta \, d\theta = 2 \int \tan \theta \sec \theta \, d\theta - \frac{3}{2} \int \csc \theta \, d\theta$$

$$= 2 \sec \theta - \frac{3}{2} \ln |\csc \theta - \cot \theta| + C = \sqrt{x^2 + 4} - \frac{3}{2} \ln \left| \frac{\sqrt{x^2 + 4} - 2}{x} \right| + C.$$

31. Let $x = 3 \sin \theta$. Then $dx = 3 \cos \theta \, d\theta$ and hence

$$\int \frac{x^2 - 4x + 5}{x\sqrt{9 - x^2}} \, dx = \int \frac{9 \sin^2 \theta - 12 \sin \theta + 5}{(3 \sin \theta)(3 \cos \theta)} (3 \cos \theta \, d\theta) = 3 \int \sin \theta \, d\theta - 4 \int d\theta + \frac{5}{3} \int \csc \theta \, d\theta$$

$$= -3 \cos \theta - 4\theta + \frac{5}{3} \ln |\csc \theta - \cot \theta| + C = -\sqrt{9 - x^2} - 4 \operatorname{Sin}^{-1} \frac{x}{3} + \frac{5}{3} \ln \left| \frac{3}{x} - \frac{\sqrt{9 - x^2}}{x} \right| + C.$$

721

32. Let $x = \tan\theta$. Then $dx = \sec^2\theta\, d\theta$ and hence

$$\int \frac{1}{1+x^2}\, dx = \int \frac{1}{\sec^2\theta}\, \sec^2\theta\, d\theta = \int d\theta = \theta + C = \mathrm{Tan}^{-1}x + C.$$

33. First observe that $y = 5 - \sqrt{x^2+9} = 0 \Longrightarrow x^2 + 9 = 25 \Longrightarrow x = \pm 4$. Hence the area is equal to

$$2\int_0^4 \left(5 - \sqrt{x^2+9}\right)\, dx = 10x\big|_0^4 - 2\int_0^4 \sqrt{x^2+9}\, dx = 40 - 2\int_0^4 \sqrt{x^2+9}\, dx.$$

To find the last integral, let $x = 3\tan\theta$. Then $dx = 3\sec^2\, d\theta$ and hence

$$\int \sqrt{x^2+9}\, dx = \int (3\sec\theta)(3\sec^2\theta)\, d\theta = 9\int \sec^3\theta\, d\theta$$

$$= \frac{9}{2}\sec\theta\tan\theta + \frac{9}{2}\ln|\sec\theta + \tan\theta| + C = \frac{1}{2}x\sqrt{x^2+9} + \frac{9}{2}\ln\left|\sqrt{x^2+9} + x\right| + C.$$

Therefore

$$\text{Area} = 40 - 2\left[\frac{1}{2}x\sqrt{x^2+9} + \frac{9}{2}\ln\left|\sqrt{x^2+9} + x\right|\right]\Bigg|_0^4 = 20 - 9\ln 3.$$

34. The curves intersect when

$$\frac{x^2}{5} = \frac{x^2}{\sqrt{x^2+16}} \quad\Longrightarrow\quad x^2\left(5 - \sqrt{x^2+16}\right) = 0 \quad\Longrightarrow\quad x = 0, x = \pm 3.$$

Hence, since $\dfrac{x^2}{\sqrt{x^2+16}} > \dfrac{x^2}{5}$ for $0 \le x \le 3$ and since the region is symmetric about the y-axis, the area is equal to

$$2\int_0^3 \left(\frac{x^2}{\sqrt{x^2+16}} - \frac{x^2}{5}\right)\, dx = 2\left[\frac{x\sqrt{x^2+16}}{2} - 8\ln\left|\sqrt{x^2+16} + x\right| - \frac{1}{15}x^3\right]\Bigg|_0^3 \quad \text{(see Exercise 19)}$$

$$= \frac{171}{15} - 16\ln 2.$$

35. If $0 \le x \le 3$, $\dfrac{1}{\sqrt{x^2+16}} \ge -\dfrac{x}{60} + \dfrac{1}{4}$, with equality at $x = 3$. Hence the area between the curves is equal to

$$\int_0^3 \left[\frac{1}{\sqrt{x^2+16}} - \left(-\frac{x}{60} + \frac{1}{4}\right)\right]\, dx.$$

Now, let $x = 4\tan\theta$. Then $dx = 4\sec^2\theta\, d\theta$ and hence

$$\int \frac{1}{\sqrt{x^2+16}}\, dx = \int \frac{1}{4\sec\theta}\, 4\sec^2\theta\, d\theta = \int \sec\theta\, d\theta$$

$$= \ln|\sec\theta + \tan\theta| + C = \ln\left|\frac{\sqrt{x^2+16}}{4} + \frac{x}{4}\right| + C = \ln\left|\sqrt{x^2+16} + x\right| + C,$$

the missing constant, $\ln 4$, being absorbed in the constant C. Therefore the area is equal to

$$\left(\ln\left|\sqrt{x^2+16} + x\right| + \frac{x^2}{120} - \frac{1}{4}x\right)\Bigg|_0^3 = \ln 2 - \frac{27}{40}.$$

36. Let dV be the differential volume of a shell of thickness dx revolved about the y-axis. Then $dV = 2\pi rh\,dx = 2\pi xy\,dx = 2\pi x^2\sqrt{9-x^2}\,dx$ and hence, since the region is symmetric about the y-axis, the total volume of revolution is

$$V = 2 \cdot 2\pi \int_0^3 x^2\sqrt{9-x^2}\,dx.$$

To evaluate this integral, let $x = 3\sin\theta$. Then $dx = 3\cos\theta\,d\theta$ and hence

$$V = 4\pi \int_0^{\pi/2} (9\sin^2\theta)(3\cos\theta)3\cos\theta\,d\theta = 324\pi \int_0^{\pi/2} \sin^2\theta\cos^2\theta\,d\theta = 81\pi \int_0^{\pi/2} \sin^2 2\theta\,d\theta$$

$$= \frac{81\pi}{2} \int_0^{\pi/2} (1 - \cos 4\theta)\,d\theta = \frac{81\pi}{2} \left(\theta - \frac{1}{4}\sin 4\theta\right)\Big|_0^{\pi/2} = \frac{81\pi^2}{4}.$$

37. Let dV be the differential volume of a cross-section disk of thickness dx. Then $dV = \pi r^2 dx = \pi y^2 dx = \pi x^2\sqrt{16-x^2}\,dx$ and hence, since the region is symmetric about the y-axis, the total volume of revolution is

$$V = 2\pi \int_0^4 x^2\sqrt{16-x^2}\,dx.$$

To evaluate this integral, let $x = 4\sin\theta$. Then $dx = 4\cos\theta\,d\theta$ and hence, as in Exercise 36,

$$V = 2\pi \int_0^{\pi/2} (16\sin^2\theta)(4\cos\theta)(4\cos\theta)\,d\theta$$

$$= 512\pi \int_0^{\pi/2} \sin^2\theta\cos^2\theta\,d\theta = 64\pi \left(\theta - \frac{1}{4}\sin 4\theta\right)\Big|_0^{\pi/2} = 32\pi^2.$$

38. Let dV be the differential volume of a shell of thickness dx revolved about the y-axis. Then $dV = 2\pi rh\,dx = 2\pi xy\,dx = 2\pi x/(x^2+9)^{3/2}\,dx$ and hence the volume of revolution is

$$V = 2\pi \int_0^4 \frac{x}{\underbrace{(x^2+9)^{3/2}}_{u=x^2+9}}\,dx = \pi \int_9^{25} u^{-3/2}\,dx = \pi(-2u^{-1/2})\Big|_9^{25} = \frac{4\pi}{15}.$$

39. Let dV be the differential volume of a cross-section washer of thickness dx. Then $dV = \pi(R^2 - r^2)dx = \pi[1/9 - 1/(x^2+9)]dx$ and hence the volume of revolution is

$$V = \pi \int_0^3 \left[\frac{1}{9} - \frac{1}{x^2+9}\right]dx = \frac{\pi}{9} x\Big|_0^3 - \frac{\pi}{9}\int_0^3 \frac{1}{\underbrace{(x/3)^2+1}_{u=x/3}}\,dx$$

$$= \frac{\pi}{3} - \frac{\pi}{3}\int_0^1 \frac{1}{1+u^2}\,du = \frac{\pi}{3} - \frac{\pi}{3} \operatorname{Tan}^{-1}u\Big|_0^1 = \frac{\pi}{3} - \frac{\pi^2}{12}.$$

40. $\dfrac{x^2}{4} + \dfrac{y^2}{3} = 1 \Longrightarrow y = \dfrac{\sqrt{3}}{2}\sqrt{4-x^2}$. Hence the area of the ellipse is equal to

$$4\int_0^2 \frac{\sqrt{3}}{2}\sqrt{4-x^2}\,dx = 2\sqrt{3}\int_0^2 \sqrt{4-x^2}\,dx.$$

Using the result in Exercise 1, it follows that

$$\text{Area} = \frac{2\sqrt{3}}{2}\left(x\sqrt{4-x^2} + 4\operatorname{Sin}^{-1}\frac{x}{2}\right)\Big|_0^2 = 2\pi\sqrt{3}.$$

41. Let L be the horizontal length of the tank. Then the total volume of the tank is $\pi(5)^2 L$ and hence the volume of water in the tank is $(\text{Area})L$, where Area stands for the area of the circular end when the water is at a depth of 6 meters. Therefore the percentage of the filled capacity is

$$\frac{\text{Vol(water)}}{\text{Vol(tank)}} = \frac{(\text{Area})\,L}{\pi(5)^2 L} = \frac{\text{Area}}{25\pi}.$$

Now,

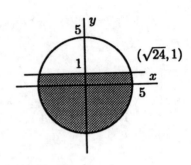

$$\begin{aligned}
\text{Area} &= \pi(5)^2 - 2\int_0^{\sqrt{24}} \left(\sqrt{25 - x^2} - 1\right)\,dx \\
&= 25\pi - 2\left[\frac{1}{2}\left(x\sqrt{25 - x^2} + 25\mathrm{Sin}^{-1}\frac{x}{5}\right) - x\right]\Bigg|_0^{\sqrt{24}} \\
&= 25\pi + \sqrt{24} - 25\mathrm{Sin}^{-1}\frac{\sqrt{24}}{5} \approx 49.20279.
\end{aligned}$$

Hence the percentage of filled capacity is $49.20279/25\pi \approx 0.6265$, or 62.65%.

42. $y = \ln x \Longrightarrow y' = 1/x \Longrightarrow$

$$\text{Length} = \int_1^3 \sqrt{1 + \left(\frac{1}{x}\right)^2}\,dx = \int_1^3 \frac{\sqrt{x^2 + 1}}{x}\,dx.$$

To evaluate this integral, let $x = \tan\theta$. Then $dx = \sec^2\theta\,d\theta$ and hence

$$\int \frac{\sqrt{x^2 + 1}}{x}\,dx = \int \frac{\sec\theta}{\tan\theta}\sec^2\theta\,d\theta = \int \frac{\sec\theta}{\tan\theta}(1 + \tan^2\theta)\,d\theta$$

$$= \int \csc\theta\,d\theta + \int \sec\theta\tan\theta\,d\theta = \ln|\csc\theta - \cot\theta| + \sec\theta + C = \ln\left|\frac{\sqrt{x^2+1}}{x} - \frac{1}{x}\right| + \sqrt{x^2 + 1} + C.$$

Hence

$$\text{Length} = \ln\left(\frac{\sqrt{10}}{3} - \frac{1}{3}\right) + \sqrt{10} - \ln|\sqrt{2} - 1| - \sqrt{2}.$$

43. $y = x^2 \Longrightarrow y' = 2x \Longrightarrow$

$$\text{Length} = \int_0^{1/2} \sqrt{1 + 4x^2}\,dx.$$

To evaluate this integral, let $x = (1/2)\tan\theta$. Then $dx = (1/2)\sec^2\theta\,d\theta$ and hence

$$\text{Length} = \int_0^{\pi/4} \sec\theta\left(\frac{1}{2}\sec^2\theta\right)\,d\theta = \frac{1}{2}\int \sec^3\theta\,d\theta$$

$$= \frac{1}{4}\left[\sec\theta\tan\theta + \ln|\sec\theta + \tan\theta|\right]\Big|_0^{\pi/4} = \frac{1}{4}\sqrt{2} + \frac{1}{4}\ln(\sqrt{2} + 1).$$

10.3 Integrals Involving Quadratic Expressions

1. $x^2 - 4x + 4 = (x-2)^2$. Let $u = x - 2$. Then $du = dx$ and hence

$$\int \frac{dx}{x^2 - 4x + 4} = \int \frac{du}{u^2} = -\frac{1}{u} + C = -\frac{1}{x-2} + C.$$

2. $x^2 - 6x + 12 = (x-3)^2 + 3$. Let $u = x - 3$. Then $du = dx$ and hence

$$\int \frac{dx}{x^2 - 6x + 12} = \int \frac{du}{u^2 + 3}.$$

Now, let $u = \sqrt{3}\tan\theta$. Then $du = \sqrt{3}\sec^2\theta\, d\theta$ and hence

$$\int \frac{du}{u^2 + 3} = \int \frac{\sqrt{3}\sec^2\theta\, d\theta}{3\sec^2\theta} = \frac{1}{\sqrt{3}}\theta + C = \frac{1}{\sqrt{3}}\operatorname{Tan}^{-1}\frac{u}{\sqrt{3}} + C$$

$$\implies \int \frac{dx}{x^2 - 6x + 12} = \frac{1}{\sqrt{3}}\operatorname{Tan}^{-1}\frac{x-3}{\sqrt{3}} + C.$$

3. $x^2 + 6x + 13 = (x+3)^2 + 4$. Let $u = x + 3$. Then $du = dx$ and hence

$$\int \frac{dx}{\sqrt{x^2 + 6x + 13}} = \int \frac{du}{\sqrt{u^2 + 4}}.$$

Now, let $u = 2\tan\theta$. Then $du = 2\sec^2\theta\, d\theta$ and hence

$$\int \frac{du}{\sqrt{u^2 + 4}} = \int \frac{2\sec^2\theta\, d\theta}{2\sec\theta} = \int \sec\theta\, d\theta = \ln|\sec\theta + \tan\theta| + C = \ln\left|\sqrt{u^2 + 4} + u\right| + C$$

$$\implies \int \frac{dx}{\sqrt{x^2 + 6x + 13}} = \ln(\sqrt{x^2 + 6x + 13} + x + 3) + C.$$

4. $5 + 4x - x^2 = 9 - (x-2)^2$. Let $u = x - 2$. Then $du = dx$ and hence

$$\int \frac{dx}{\sqrt{5 + 4x - x^2}} = \int \frac{du}{\sqrt{9 - u^2}}.$$

Now, let $u = 3\sin\theta$. Then $du = 3\cos\theta\, d\theta$ and hence

$$\int \frac{du}{\sqrt{9 - u^2}} = \int \frac{3\cos\theta\, d\theta}{3\cos\theta} = \theta + C = \operatorname{Sin}^{-1}\frac{u}{3} + C$$

$$\implies \int \frac{dx}{\sqrt{5 + 4x - x^2}} = \operatorname{Sin}^{-1}\frac{x-2}{3} + C.$$

5. $x^2 - 6x = (x-3)^2 - 9$. Let $u = x - 3$. Then $du = dx$ and hence

$$\int \frac{dx}{\sqrt{x^2 - 6x}} = \int \frac{du}{\sqrt{u^2 - 9}}.$$

Now, let $u = 3\sec\theta$. Then $du = 3\sec\theta\tan\theta\, d\theta$ and hence

$$\int \frac{du}{\sqrt{u^2 - 9}} = \int \frac{3\sec\theta\tan\theta\, d\theta}{3\tan\theta} = \int \sec\theta = \ln|\sec\theta + \tan\theta| + C = \ln\left|u + \sqrt{u^2 - 9}\right| + C$$

$$\implies \int \frac{dx}{\sqrt{x^2 - 6x}} = \ln\left|x - 3 + \sqrt{x^2 - 6x}\right| + C.$$

6. $x^2 + 6x + 15 = (x+3)^2 + 6$. Let $u = x + 3$. Then $du = dx$ and hence

$$\int \frac{x\,dx}{x^2 + 6x + 15} = \int \frac{u-3}{u^2+6}\,du = \int \frac{u}{u^2+6}\,du - 3\int \frac{1}{u^2+6}\,du = \frac{1}{2}\ln(u^2+6) - 3\int \frac{1}{u^2+6}\,du.$$

Now, let $u = \sqrt{6}\tan\theta$. Then $du = \sqrt{6}\sec^2\theta\,d\theta$ and hence

$$\int \frac{1}{u^2+6}\,du = \int \frac{\sqrt{6}\sec^2\theta\,d\theta}{6\sec^2\theta} = \frac{1}{\sqrt{6}}\theta + C = \frac{1}{\sqrt{6}}\operatorname{Tan}^{-1}\left(\frac{u}{\sqrt{6}}\right) + C$$

$$\implies \quad \int \frac{x\,dx}{x^2+6x+15} = \frac{1}{2}\ln(x^2+6x+15) - \frac{3}{\sqrt{6}}\operatorname{Tan}^{-1}\left(\frac{x+3}{\sqrt{6}}\right) + C.$$

7. $6x - x^2 = 9 - (x-3)^2$. Let $u = x - 3$. Then $du = dx$ and hence

$$\int \frac{x\,dx}{\sqrt{6x - x^2}} = \int \frac{u+3}{\sqrt{9 - u^2}}\,du = \int \frac{u}{\sqrt{9 - u^2}}\,du + 3\int \frac{1}{\sqrt{9 - u^2}}\,du$$

$$= -\sqrt{9 - u^2} + 3\operatorname{Sin}^{-1}\frac{u}{3} + C = -\sqrt{6x - x^2} + 3\operatorname{Sin}^{-1}\frac{x-3}{3} + C.$$

8. $x^2 - 6x + 13 = (x-3)^2 + 4$. Let $u = x - 3$. Then $du = dx$ and hence

$$\int \frac{1+x}{\sqrt{x^2 - 6x + 13}}\,dx = \int \frac{u+4}{\sqrt{u^2+4}}\,du = \int \frac{u}{\sqrt{u^2+4}}\,du + 4\int \frac{1}{\sqrt{u^2+4}}\,du$$

$$= \sqrt{u^2+4} + \ln\left|\sqrt{u^2+4} + u\right| + C = \sqrt{x^2 - 6x + 13} + 4\ln\left|\sqrt{x^2 - 6x + 13} + x - 3\right| + C.$$

9. $x^2 + 6x + 18 = (x+3)^2 + 9$. Let $u = x + 3$. Then $du = dx$ and hence

$$\int_0^1 \frac{2x+1}{\sqrt{18 + 6x + x^2}}\,dx = \int_3^4 \frac{2u-5}{\sqrt{u^2+9}}\,du = 2\int_3^4 \frac{u}{\sqrt{u^2+9}}\,du - 5\int \frac{1}{\sqrt{u^2+9}}\,du$$

$$= \left.\left(2\sqrt{u^2+9} - 5\ln\left|\sqrt{u^2+9} + u\right|\right)\right|_3^4 = 10 - 6\sqrt{2} - 5\ln 9 + 5\ln(3\sqrt{2} + 3).$$

10. $x^2 + 8x = (x+4)^2 - 16$. Let $u = x + 4$. Then $du = dx$ and hence

$$\int_0^1 \frac{3x-1}{\sqrt{x^2 + 8x}}\,dx = \int_4^5 \frac{3u-13}{\sqrt{u^2-16}}\,du = 3\int_4^5 \frac{u}{\sqrt{u^2-16}}\,du - 13\int_4^5 \frac{1}{\sqrt{u^2-16}}\,du$$

$$= \left.3\sqrt{u^2-16}\right|_4^5 - 13\int_4^5 \frac{1}{\sqrt{u^2-16}}\,du = 9 - 13\int_4^5 \frac{1}{\sqrt{u^2-16}}\,du.$$

Now, let $u = 4\sec\theta$. Then $du = 4\sec\theta\tan\theta\,d\theta$ and hence

$$\int \frac{1}{\sqrt{u^2-16}}\,du = \int \frac{4\sec\theta\tan\theta\,d\theta}{4\sec\theta} = \ln|\sec\theta + \tan\theta| + C = \ln\left|u + \sqrt{u^2-16}\right| + C$$

$$\implies \quad \int_4^5 \frac{1}{\sqrt{u^2-16}}\,du = \ln 8 - \ln 4 = \ln 2$$

$$\implies \quad \int_0^1 \frac{3x-1}{\sqrt{x^2+8x}}\,dx = 9 - 13\ln 2.$$

11. $x^2 - 8x - 7 = (x-4)^2 - 23$. Let $u = x - 4$. Then $du = dx$ and hence

$$\int_{-3}^{-2} \frac{x-4}{x^2 - 8x - 7} \, dx = \int_{-7}^{-6} \frac{u}{u^2 - 23} \, du = \frac{1}{2} \ln|u^2 - 23|\Big|_{-7}^{-6} = -\frac{1}{2} \ln 2 = -\ln\sqrt{2}.$$

12. $16 + 12x - 4x^2 = 25 - (2x-3)^2$. Let $u = 2x - 3$. Then $du = 2dx$ and hence

$$\int_0^{3/2} (2x-3)\sqrt{16 + 12x - 4x^2} \, dx = \frac{1}{2}\int_{-3}^0 u\sqrt{25 - u^2} \, du = -\frac{1}{6}(25 - u^2)^{3/2}\Big|_{-3}^0 = -\frac{61}{6}.$$

13. $x^2 + 2x + 2 = (x+1)^2 + 1$. Let $u = x + 1$. Then $du = dx$ and hence

$$\int_0^1 \frac{x}{(x^2 + 2x + 2)^2} \, dx = \int_1^2 \frac{u-1}{(u^2+1)^2} \, du = \int_1^2 \frac{u}{(u^2+1)^2} \, du - \int_1^2 \frac{1}{(u^2+1)^2} \, du$$

$$= \frac{1}{2}\frac{-1}{u^2+1}\Big|_1^2 - \int_1^2 \frac{1}{(u^2+1)^2} \, du = \frac{3}{20} - \int_1^2 \frac{1}{(u^2+1)^2} \, du.$$

Now, let $u = \tan\theta$. Then $du = \sec^2\theta$ and hence

$$\int \frac{1}{(u^2+1)^2} \, du = \int \frac{\sec^2\theta \, d\theta}{\sec^4\theta} = \int \cos^2\theta \, d\theta = \frac{1}{2}\int (1 + \cos 2\theta) \, d\theta = \frac{1}{2}\left(\theta + \frac{1}{2}\sin 2\theta\right) + C$$

$$= \frac{1}{2}(\theta + \sin\theta\cos\theta) + C = \frac{1}{2}\left(\text{Tan}^{-1}u + \frac{u}{\sqrt{u^2+1}}\frac{1}{\sqrt{u^2+1}}\right) + C = \frac{1}{2}\left(\text{Tan}^{-1}u + \frac{u}{u^2+1}\right) + C$$

$$\implies \int_1^2 \frac{1}{(u^2+1)^2} \, du = \frac{1}{2}\left(\text{Tan}^{-1}2 - \frac{1}{10} - \frac{\pi}{4}\right)$$

$$\implies \int_0^1 \frac{x}{(x^2+2x+2)^2} \, dx = \frac{3}{20} - \frac{1}{2}\left(\text{Tan}^{-1}2 - \frac{1}{10} - \frac{\pi}{4}\right) = \frac{1}{5} - \frac{1}{2}\text{Tan}^{-1}2 + \frac{\pi}{8}.$$

14. $6x^2 + 12x = 6[(x+1)^2 - 1]$. Let $u = x + 1$. Then $du = dx$ and hence

$$\int \frac{x^2 + 2x + 1}{6x^2 + 12x} \, dx = \int \frac{u^2}{6(u^2-1)} \, du = \frac{1}{6}\int \left(1 + \frac{1}{u^2-1}\right) du = \frac{1}{6}u + \frac{1}{6}\int \frac{1}{u^2-1} \, du.$$

Now, let $u = \sec\theta$. Then $du = \sec\theta\tan\theta \, d\theta$ and hence

$$\int \frac{1}{u^2-1} \, du = \int \frac{\sec\theta\tan\theta \, d\theta}{\tan^2\theta} = \int \csc\theta \, d\theta = \ln|\csc\theta - \cot\theta| + C$$

$$= \ln\left|\frac{u}{\sqrt{u^2-1}} - \frac{1}{\sqrt{u^2-1}}\right| + C = \ln\left|\frac{u-1}{\sqrt{(u-1)(u+1)}}\right| + C = \frac{1}{2}\ln\frac{u-1}{u+1} + C$$

$$\implies \int \frac{x^2 + 2x + 1}{6x^2 + 12x} \, dx = \frac{1}{6}(x+1) + \frac{1}{12}\ln\frac{x}{x+2} + C.$$

15. $x^2 - 18x + 88 = (x - 9)^2 + 7$. Let $u = x - 9$. Then $du = dx$ and hence

$$\int \frac{dx}{(9-x)(88-18x+x^2)^{3/2}} = -\int \frac{du}{u(u^2+7)^{3/2}}.$$

Now, let $u = \sqrt{7}\tan\theta$. Then $du = \sqrt{7}\sec^2\theta\,d\theta$ and hence

$$\int \frac{du}{u(u^2+7)^{3/2}} = \int \frac{\sqrt{7}\sec^2\theta\,d\theta}{(\sqrt{7}\tan\theta)7\sqrt{7}\sec^3\theta} = \frac{1}{7\sqrt{7}}\int \frac{\cos^2\theta}{\sin\theta}\,d\theta = \frac{1}{7\sqrt{7}}\int (\csc\theta - \sin\theta)\,d\theta$$

$$= \frac{1}{7\sqrt{7}}\left(\ln|\csc\theta - \cot\theta| + \cos\theta\right) + C = \frac{1}{7\sqrt{7}}\left(\ln\left|\frac{\sqrt{u^2+7}}{u} - \frac{\sqrt{7}}{u}\right| + \frac{\sqrt{7}}{\sqrt{u^2+7}}\right) + C$$

$$\implies \int \frac{dx}{(9-x)(88-18x+x^2)^{3/2}} = \frac{-1}{7\sqrt{7}}\left(\ln\left|\frac{\sqrt{x^2-18x+88} - \sqrt{7}}{x-9}\right| + \sqrt{\frac{7}{88-18x+x^2}}\right) + C.$$

16. $x^2 + 10x + 21 = (x+5)^2 - 4$. Let $u = x + 5$. Then $du = dx$ and hence

$$\int \frac{\sqrt{x^2+10x+21}}{x^3+15x^2+75x+125}\,dx = \int \frac{\sqrt{u^2-4}}{u^3}\,du.$$

Now, let $u = 2\sec\theta$. Then $du = 2\sec\theta\tan\theta\,d\theta$ and hence

$$\int \frac{\sqrt{u^2-4}}{u^3}\,du = \int \frac{2\tan\theta}{8\sec^3\theta}2\sec\theta\tan\theta\,d\theta = \frac{1}{2}\int \frac{\tan^2\theta}{\sec^2\theta}\,d\theta = \frac{1}{2}\int \sin^2\theta\,d\theta$$

$$= \frac{1}{4}\int (1-\cos 2\theta)\,d\theta = \frac{1}{4}\left(\theta - \frac{1}{2}\sin 2\theta\right) + C = \frac{1}{4}(\theta - \sin\theta\cos\theta) + C = \frac{1}{4}\operatorname{Sec}^{-1}\frac{u}{2} - \frac{\sqrt{u^2-4}}{2u^2} + C$$

$$\implies \int \frac{\sqrt{x^2+10x+21}}{x^3+15x^2+75x+125}\,dx = \frac{1}{4}\operatorname{Sec}^{-1}\frac{x+5}{2} - \frac{\sqrt{x^2+10x+21}}{2(x+5)^2} + C.$$

17. $4x^2 + 20x + 29 = (2x+5)^2 + 4$. Let $u = 2x + 5$. Then $du = 2dx$ and hence

$$\int \frac{4x^2+20x+25}{\sqrt{4x^2+20x+29}}\,dx = \frac{1}{2}\int \frac{u^2}{\sqrt{u^2+4}}\,du.$$

Now, let $u = 2\tan\theta$. Then $du = 2\sec^2\theta\,d\theta$ and hence

$$\int \frac{u^2}{\sqrt{u^2+4}}\,du = \int \frac{4\tan^2\theta}{2\sec\theta}2\sec^2\theta\,d\theta = 4\int \sec\theta\tan^2\theta\,d\theta$$

$$= 2(\sec\theta\tan\theta - \ln|\sec\theta + \tan\theta|) + C = 2\left(\frac{\sqrt{u^2+4}}{2}\frac{u}{2} - \ln\left|\sqrt{u^2+4} + u\right|\right) + C$$

$$\implies \int \frac{4x^2+20x+25}{\sqrt{4x^2+20x+29}}\,dx = \frac{(2x+5)\sqrt{4x^2+20x+29}}{4} - \ln\left|\sqrt{4x^2+20x+29} + 2x+5\right| + C.$$

18. $18 - 6x + x^2 = (x-3)^2 + 9$. Let $u = x - 3$. Then $du = dx$ and hence

$$\int \frac{dx}{(3-x)\sqrt{18-6x+x^2}} = -\int \frac{du}{u\sqrt{u^2+9}}.$$

Now, let $u = 3\tan\theta$. Then $du = 3\sec^2\theta\, d\theta$ and hence

$$\int \frac{du}{u\sqrt{u^2+9}} = \int \frac{3\sec^2\theta\, d\theta}{(3\tan\theta)(3\sec\theta)} = \frac{1}{3}\int \csc\theta\, d\theta = \frac{1}{3}\ln|\csc\theta - \cot\theta| + C = \frac{1}{3}\ln\left|\frac{\sqrt{u^2+9}}{u} - \frac{3}{u}\right| + C$$

$$\implies \int \frac{dx}{(3-x)\sqrt{18-6x+x^2}} = -\frac{1}{3}\ln\left|\frac{\sqrt{18-6x+x^2}-3}{x-3}\right| + C.$$

19. $x^2 - 4x + 8 = (x-2)^2 + 4$. Let $u = x - 2$. Then

$$\text{Area} = \int_2^4 \frac{1}{x^2-4x+8}\, dx = \int_0^2 \frac{1}{u^2+4}\, du = \frac{1}{2}\operatorname{Tan}^{-1}\frac{u}{2}\Big|_0^2 = \frac{\pi}{8}.$$

20. $2x^2 - 12x + 68 = 2(x-3)^2 + 50$. Let $u = x - 3$. Then

$$\text{Area} = \int_{-2}^3 \frac{5}{2x^2-12x+68}\, dx = 5\int_{-5}^0 \frac{1}{2u^2+50}\, du = \frac{1}{2}\operatorname{Tan}^{-1}\frac{u}{5}\Big|_{-5}^0 = \frac{\pi}{8}.$$

21. Let dV be the differential volume of a cross-section disk of thickness dx. Then $dV = \pi r^2 dx = \pi y^2 dx$ $= \pi(8x - x^2)^{-1/2}dx$ and hence the volume of revolution is

$$V = \pi\int_2^4 \frac{1}{\sqrt{8x-x^2}}\, dx = \pi\int_2^4 \frac{1}{\sqrt{16-(4-x)^2}}\, dx = -\pi\int_2^0 \frac{1}{\sqrt{16-u^2}}\, du.$$

Now, let $u = 4\sin\theta$. Then $du = 4\cos\theta\, d\theta$ and hence

$$V = -\pi\int_{\pi/6}^0 \frac{1}{4\cos\theta}\, 4\cos\theta\, d\theta = -\pi\theta\big|_{\pi/6}^0 = \frac{\pi^2}{6}.$$

22. Let dV be the differential volume of a shell of thickness dx revolved about the y-axis. Then $dV = 2\pi rh\, dx = 2\pi xy\, dx = 2\pi x(8x - x^2)^{-1/2}dx$ and hence the volume of revolution is

$$V = 2\pi\int_2^4 \frac{x}{\sqrt{8x-x^2}}\, dx = 2\pi\int_2^4 \frac{x}{\sqrt{16-(4-x)^2}}\, dx = -2\pi\int_2^0 \frac{4-u}{\sqrt{16-u^2}}\, dx.$$

Now, let $u = 4\sin\theta$. Then $du = 4\cos\theta\, d\theta$ and hence

$$V = -2\pi\int_{\pi/6}^0 \frac{4(1-\sin\theta)}{4\cos\theta}\, 4\cos\theta\, d\theta = -8\,\pi(\theta + \cos\theta)\big|_{\pi/6}^0 = -8\pi\left(1 - \frac{\pi}{6} - \frac{1}{2}\sqrt{3}\right).$$

10.4   The Method of Partial Fractions

1. $\dfrac{1}{x(x+1)} = \dfrac{A}{x} + \dfrac{B}{x+1}$. Then $x = 0 \implies 1 = A$, $x = -1 \implies -1 = B$, and hence

$$\int \frac{1}{x(x+1)}\, dx = \int \left(\frac{1}{x} - \frac{1}{x+1}\right) dx = \ln|x| - \ln|x+1| + C.$$

2. $\dfrac{3x+2}{x(x+1)} = \dfrac{A}{x} + \dfrac{B}{x+1}$. Then $x = 0 \Longrightarrow 2 = A$, $x = -1 \Longrightarrow 1 = B$, and hence

$$\int \frac{3x+2}{x(x+1)}\, dx = \int \left(\frac{2}{x} + \frac{1}{x+1}\right) dx = 2\ln|x| + \ln|x+1| + C.$$

3. $\dfrac{2x}{(x-3)(x-1)} = \dfrac{A}{x-3} + \dfrac{B}{x-1}$. Then $x = 3 \Longrightarrow 3 = A$, $x = 1 \Longrightarrow -1 = B$, and hence

$$\int_4^6 \frac{2x}{(x-3)(x-1)}\, dx = \int_4^6 \left(\frac{3}{x-3} + \frac{-1}{x-1}\right) dx = (3\ln|x-3| - \ln|x-1|)\,\big|_4^6 = 4\ln 3 - \ln 5.$$

4. $\dfrac{-x-8}{(x+1)(x-6)} = \dfrac{A}{x+1} + \dfrac{B}{x-6}$. Then $x = -1 \Longrightarrow 1 = A$, $x = 6 \Longrightarrow -2 = B$, and hence

$$\int_7^{11} \frac{-x-8}{(x+1)(x-6)}\, dx = \int_7^{11} \left(\frac{1}{x+1} - \frac{2}{x-6}\right) dx = (\ln|x+1| - 2\ln|x-6|)\,\big|_7^{11} = \ln 12 - 2\ln 5 - \ln 8.$$

5. $\dfrac{2x+4}{1-x^2} = \dfrac{-2x-4}{(x-1)(x+1)} = \dfrac{A}{x-1} + \dfrac{B}{x+1}$. Then $x = 1 \Longrightarrow -3 = A$, $x = -1 \Longrightarrow 1 = B$, and hence

$$\int \frac{2x+4}{1-x^2}\, dx = \int \left(\frac{-3}{x-1} + \frac{1}{x+1}\right) dx = -3\ln|x-1| + \ln|x+1| + C.$$

6. $\dfrac{7x+2}{x^2+x-2} = \dfrac{7x+2}{(x+2)(x-1)} = \dfrac{A}{x+2} + \dfrac{B}{x-1}$. Then $x = -2 \Longrightarrow 4 = A$, $x = 1 \Longrightarrow 3 = B$, and hence

$$\int \frac{7x+2}{x^2+x-2}\, dx = \int \left(\frac{4}{x+2} + \frac{3}{x-1}\right) dx = 4\ln|x+2| + 3\ln|x-1| + C.$$

7. $\dfrac{1}{1-x^2} = \dfrac{-1}{(x-1)(x+1)} = \dfrac{A}{x-1} + \dfrac{B}{x+1}$. Then $x = 1 \Longrightarrow -1/2 = A$, $x = -1 \Longrightarrow 1/2 = B$, and hence

$$\int \frac{1}{1-x^2}\, dx = \int \left(\frac{-1/2}{x-1} + \frac{1/2}{x+1}\right) dx = -\frac{1}{2}\ln|x-1| + \frac{1}{2}\ln|x+1| + C.$$

8. $\dfrac{x^2+1}{x^2-1} = 1 + \dfrac{2}{(x-1)(x+1)} = 1 + \dfrac{A}{x-1} + \dfrac{B}{x+1}$. Then $x = 1 \Longrightarrow 1 = A$, $x = -1 \Longrightarrow -1 = B$, and hence

$$\int \frac{x^2+1}{x^2-1}\, dx = \int \left(1 + \frac{1}{x-1} - \frac{1}{x+1}\right) dx = x + \ln|x-1| - \ln|x+1| + C.$$

9. $\dfrac{1}{x+x^3} = \dfrac{1}{x(1+x^2)} = \dfrac{A}{x} + \dfrac{Bx+C}{1+x^2} \Longrightarrow 1 = A(1+x^2) + x(Bx+C) = A + x(C) + x^2(A+B).$

Equating coefficients of x, we find that

$$\begin{aligned} A &= 1 \quad \text{coefficients of constant terms} \\ C &= 0 \quad \text{coefficients of } x \\ A+B &= 0 \quad \text{coefficients of } x^2 \end{aligned}$$

Therefore $A = 1$, $B = -1$, $C = 0$ and hence

$$\int_e^{e^2} \frac{1}{x + x^3}\, dx = \int_e^{e^2} \left(\frac{1}{x} + \frac{-x}{1+x^2} \right) dx = \left[\ln|x| - \frac{1}{2}\ln(1+x^2) \right]\Big|_e^{e^2} = 1 - \frac{1}{2}\ln(1+e^4) + \frac{1}{2}\ln(1+e^2).$$

10. $\dfrac{x^2 + 3x + 3}{x(1+x)} = 1 + \dfrac{2x+3}{x(1+x)} = 1 + \dfrac{A}{x} + \dfrac{B}{1+x}$. Thus

$$\begin{aligned} x = 0 &\implies 3 = A \\ x = -1 &\implies 1 = -B \implies B = -1 \end{aligned}$$

and therefore

$$\int_1^e \frac{x^2 + 3x + 3}{x(1+x)}\, dx = \int_1^e \left(1 + \frac{3}{x} - \frac{1}{1+x} \right) dx = (x + 3\ln|x| - \ln|1+x|)\big|_1^e = 2 + e - \ln(1+e) + \ln 2.$$

11. $\dfrac{1}{(x+1)(x^2+1)} = \dfrac{A}{x+1} + \dfrac{Bx+C}{x^2+1}$

$$\implies 1 = A(x^2+1) + (Bx+C)(x+1) = (A+C) + x(B+C) + x^2(A+B).$$

Equating coefficients of like powers of x, we find that

$$\begin{aligned} A + C &= 1 \quad \text{coefficients of constant terms} \\ B + C &= 0 \quad \text{coefficients of } x \\ A + B &= 0 \quad \text{coefficients of } x^2 \end{aligned}$$

It follows that $A = C = 1/2$, $B = -1/2$, and hence

$$\int \frac{1}{(x+1)(x^2+1)}\, dx = \int \left(\frac{\frac{1}{2}}{x+1} + \frac{-\frac{1}{2}x + \frac{1}{2}}{x^2+1} \right) dx$$

$$= \frac{1}{2}\int \frac{1}{x+1}\, dx - \frac{1}{2}\int \frac{x}{x^2+1}\, dx + \frac{1}{2}\int \frac{1}{x^2+1}\, dx = \frac{1}{2}\ln|x+1| - \frac{1}{4}\ln(x^2+1) + \frac{1}{2}\text{Tan}^{-1}x + C.$$

12. $\dfrac{x^2 + x + 1}{x + x^3} = \dfrac{x^2 + x + 1}{x(1+x^2)} = \dfrac{A}{x} + \dfrac{Bx+C}{1+x^2}$

$$\begin{aligned} &\implies x^2 + x + 1 = A(1+x^2) + x(Bx+C) = A + x(C) + x^2(A+B) \\ &\implies A = 1,\ C = 1,\ A + B = 1 \\ &\implies A = 1,\ B = 0,\ C = 1. \end{aligned}$$

Therefore

$$\int \frac{x^2 + x + 1}{x + x^3}\, dx = \int \left(\frac{1}{x} + \frac{1}{1+x^2} \right) dx = \ln|x| + \text{Tan}^{-1}x + C.$$

13. $\dfrac{3x^2 - 2x + 2}{(x-2)(x^2+1)} = \dfrac{A}{x-2} + \dfrac{Bx+C}{x^2+1}$

$$\begin{aligned} &\implies 3x^2 - 2x + 2 = A(x^2+1) + (x-2)(Bx+C) = (A-2C) + x(-2B+C) + x^2(A+B) \\ &\implies A - 2C = 2,\ -2B + C = -2,\ A + B = 3 \\ &\implies A = 2,\ B = 1,\ C = 0. \end{aligned}$$

Therefore

$$\int_3^4 \frac{3x^2 - 2x + 2}{(x-2)(x^2+1)} \, dx = \int_3^4 \left(\frac{2}{x-2} + \frac{x}{x^2+1} \right) dx$$

$$= \left(2\ln|x-2| + \frac{1}{2}\ln(x^2+1) \right)\Big|_3^4 = 2\ln 2 + \frac{1}{2}\ln 17 - \frac{1}{2}\ln 10.$$

14. $\dfrac{-2x^2 - x - 1}{(3x^2+1)(x-1)} = \dfrac{A}{x-1} + \dfrac{Bx+C}{3x^2+1}$

$\implies \ -2x^2 - x - 1 = A(3x^2+1) + (x-1)(Bx+C) = (A-C) + x(-B+C) + x^2(3A+B)$

$\implies \ A - C = -1, \ -B+C = -1, \ 3A+B = -2$

$\implies \ A = -1, \ B = 1, \ C = 0.$

Therefore

$$\int_2^5 \frac{-2x^2 - x - 1}{(3x^2+1)(x-1)} \, dx = \int_2^5 \left(\frac{-1}{x-1} + \frac{x}{3x^2+1} \right) dx$$

$$= \left[-\ln|x-1| + \frac{1}{6}\ln(3x^2+1) \right]\Big|_2^5 = -\ln 4 + \frac{1}{6}\ln 76 - \frac{1}{6}\ln 13.$$

15. $\dfrac{4x^2 + x + 1}{(x+2)(x^2+1)} = \dfrac{A}{x+2} + \dfrac{Bx+C}{x^2+1}$

$\implies \ 4x^2 + x + 1 = A(x^2+1) + (x+2)(Bx+C) = (A+2C) + x(2B+C) + x^2(A+B)$

$\implies \ A + 2C = 1, \ 2B+C = 1, \ A+B = 4$

$\implies \ A = 3, \ B = 1, \ C = -1.$

Therefore

$$\int \frac{4x^2 + x + 1}{(x+2)(x^2+1)} \, dx = \int \left(\frac{3}{x+2} + \frac{x-1}{x^2+1} \right) dx$$

$$= 3\int \frac{1}{x+2} \, dx + \int \frac{x}{x^2+1} \, dx - \int \frac{1}{x^2+1} \, dx = 3\ln|x+2| + \frac{1}{2}\ln(x^2+1) - \text{Tan}^{-1}x + C.$$

16. $\dfrac{1}{x(x+1)^2} = \dfrac{A}{x} + \dfrac{B}{x+1} + \dfrac{C}{(x+1)^2}$

$\implies \ 1 = A(x+1)^2 + Bx(x+1) + xC = A + x(2A+B+C) + x^2(A+B)$

$\implies \ A = 1, \ 2A+B+C = 0, \ A+B = 0$

$\implies \ A = 1, \ B = -1, \ C = -1.$

Therefore

$$\int \frac{1}{x(x+1)^2} \, dx = \int \left(\frac{1}{x} - \frac{1}{x+1} - \frac{1}{(x+1)^2} \right) dx = \ln|x| - \ln|x+1| + \frac{1}{x+1} + C.$$

17. $\dfrac{x^2 + 5x + 1}{x(x+1)^2} = \dfrac{A}{x} + \dfrac{B}{x+1} + \dfrac{C}{(x+1)^2}$

$\implies \ x^2 + 5x + 1 = A(x+1)^2 + Bx(x+1) + Cx = A + x(2A+B+C) + x^2(A+B)$

$\implies \ A = 1, \ 2A+B+C = 5, \ A+B = 1$

$\implies \ A = 1, \ B = 0, \ C = 3.$

Therefore

$$\int_e^{e^3} \frac{x^2+5x+1}{x(x+1)^2}\, dx = \int_e^{e^3} \left(\frac{1}{x} + \frac{3}{(x+1)^2}\right)\, dx = \left(\ln|x| - \frac{3}{x+1}\right)\Big|_e^{e^3} = 2 + \frac{3}{e+1} - \frac{3}{e^3+1}.$$

18. $\displaystyle \int_1^7 \frac{dx}{(x+1)^2} = -\frac{1}{x+1}\Big|_1^7 = \frac{3}{8}.$

19. $\displaystyle \frac{2x^2+3x+2}{x^3+2x^2+x} = \frac{2x^2+3x+2}{x(x+1)^2} = \frac{A}{x} + \frac{B}{x+1} + \frac{C}{(x+1)^2}$

$\implies 2x^2+3x+2 = A((x+1)^2 + Bx(x+1) + Cx = A + x(2A+B+C) + x^2(A+B)$

$\implies A = 2,\ 2A+B+C = 3,\ A+B = 2$

$\implies A = 2,\ B = 0,\ C = -1$

Therefore

$$\int \frac{2x^2+3x+2}{x^3+2x^2+x}\, dx = \int \left(\frac{2}{x} - \frac{1}{(x+1)^2}\right)\, dx = 2\ln|x| + \frac{1}{x+1} + C.$$

20. $\displaystyle \frac{x^3+x^2+1}{x^4+2x^2+1} = \frac{x^3+x^2+1}{(x^2+1)^2} = \frac{Ax+B}{x^2+1} + \frac{Cx+D}{(x^2+1)^2}$

$\implies x^3+x^2+1 = (A+B)(x^2+1) + Cx + D = (B+d) + x(A+C) + x^2(B) + x^3(A)$

$\implies B+D = 1,\ A+C = 0,\ B = 1,\ A = 1$

$\implies A = 1,\ B = 1,\ C = -1,\ D = 0.$

Therefore

$$\int \frac{x^3+x^2+1}{x^4+2x^2+1}\, dx = \int \left(\frac{x+1}{x^2+1} + \frac{-x}{(x^2+1)^2}\right)\, dx$$

$$= \int \frac{x}{x^2+1} + \int \frac{1}{x^2+1}\, dx - \int \frac{x}{(x^2+1)^2}\, dx = \frac{1}{2}\ln(x^2+1) + \mathrm{Tan}^{-1}x + \frac{1}{2}\frac{1}{x^2+1} + C.$$

21. $\displaystyle \frac{1}{x^3+3x^2+2x} = \frac{1}{x(x+1)(x+2)} = \frac{A}{x} + \frac{B}{x+1} + \frac{C}{x+2}.$ Then

$$\begin{array}{ccccc} x=0 & \implies & 1/2 & = & A \\ x=-1 & \implies & -1 & = & B \\ x=-2 & \implies & 1/2 & = & C. \end{array}$$

and therefore

$$\int_1^2 \frac{1}{x^3+3x^2+2x}\, dx = \int_1^2 \left(\frac{1/2}{x} - \frac{1}{x+1} + \frac{1/2}{x+2}\right)\, dx$$

$$= \left(\frac{1}{2}\ln|x| - \ln|x+1| + \frac{1}{2}\ln|x+2|\right)\Big|_1^2 = \frac{3}{2}\ln 2 - \frac{3}{2}\ln 3 + \ln 2.$$

22. $\displaystyle \frac{3x^2+6x+2}{x(x+1)(x+2)} = \frac{A}{x} + \frac{B}{x+1} + \frac{C}{x+2}.$ Then

$$\begin{array}{ccc} x=0 & \implies & 1=A \\ x=-1 & \implies & 1=B \\ x=-2 & \implies & 1=C \end{array}$$

and therefore

$$\int_6^9 \frac{3x^2 + 6x + 2}{x(x+1)(x+2)} \, dx = \int_6^9 \left(\frac{1}{x} + \frac{1}{x+1} + \frac{1}{x+2} \right) \, dx$$

$$= (\ln|x| + \ln|x+1| + \ln|x+2|) \big|_6^9 = \ln 9 + \ln 10 + \ln 11 - \ln 6 - \ln 7 - \ln 8.$$

23. $\dfrac{5x^3 - 9x^2 + 3x - 3}{x(x-3)(x^2+1)} = \dfrac{A}{x} + \dfrac{B}{x-3} + \dfrac{Cx+D}{x^2+1}$

$$\implies \quad 5x^3 - 9x^2 + 3x - 3 = A(x-3)(x^2+1) + Bx(x^2+1) + (Cx+D)x(x-3)$$

$$= (-3A) + x(A + B - 3D) + x^2(-3A - 3C + D) + x^3(A + B + C)$$

$$\implies \quad -3A = -3, \ A + B - 3D = 3, \ -3A - 3C + D = -9, \ A + B + C = 5$$

$$\implies \quad A = 1, \ B = 2, \ C = 2, \ D = 0.$$

Therefore

$$\int \frac{5x^3 - 9x^2 + 3x - 3}{x(x-3)(x^2+1)} \, dx = \int \left(\frac{1}{x} + \frac{2}{x-3} + \frac{2x}{x^2+1} \right) \, dx$$

$$= \ln|x| + 2\ln|x-3| + \ln(x^2+1) + C.$$

24. $\dfrac{6x^3 - 4x^2 + 3x - 3}{x(x-1)(1+x^2)} = \dfrac{A}{x} + \dfrac{B}{x-1} + \dfrac{Cx+D}{1+x^2}$

$$\implies \quad 6x^3 - 4x^2 + 3x - 3 = A(x-1)(1+x^2) + Bx(1+x^2) + (Cx+D)x(x-1)$$

$$= (-A) + x(A + B - D) + x^2(-A_C + D) + x^3(A + B + C)$$

$$\implies \quad -A = -3, \ A + B - D = 3, \ -A - C + D = -4, \ A + B + C = 6$$

$$\implies \quad A = 3, \ B = 1, \ C = 2, \ D = 1.$$

Therefore

$$\int \frac{6x^3 - 4x^2 + 3x - 3}{x(x-1)(1+x^2)} \, dx = \int \left(\frac{3}{x} + \frac{1}{x-1} + \frac{2x+1}{1+x^2} \right) \, dx$$

$$= 3\int \frac{1}{x} \, dx + \int \frac{1}{x-1} \, dx + \int \frac{2x}{1+x^2} \, dx + \int \frac{1}{1+x^2} \, dx = 3\ln|x| + \ln|x-1| + \ln|1+x^2| + \operatorname{Tan}^{-1}x + C.$$

25. $\dfrac{4}{x^3 + x} = \dfrac{4}{x(x^2+1)} = \dfrac{A}{x} + \dfrac{Bx+C}{x^2+1}$

$$\implies \quad 4 = A(x^2+1) + (Bx+C)x = A + x(C) + x^2(A+B)$$

$$\implies \quad A = 4, \ C = 0, \ A + B = 0$$

$$\implies \quad A = 4, \ B = -4, \ C = 0.$$

Therefore

$$\int \frac{4}{x^3 + x} \, dx = \int \left(\frac{4}{x} + \frac{-4x}{x^2+1} \right) \, dx = 4\ln|x| - 2\ln(x^2+1) + C.$$

26. $\dfrac{1}{x^3 - 6x^2 + 13x - 10} = \dfrac{1}{(x-2)(x^2-4x+5)} = \dfrac{A}{x-2} + \dfrac{Bx+C}{x^2-4x+5}$

$$\implies \quad 1 = A(x^2 - 4x + 5) + (Bx+C)(x-2) = (5A - 2C) + x(-4A - 2B + C) + x^2(A+B)$$

$$\implies \quad 5A - 2C = 1, \ -4A - 2B + C = 0, \ A + B = 0$$

$$\implies \quad A = 1, \ B = -1, \ C = 2.$$

Therefore

$$\int \frac{dx}{x^3 - 6x^2 + 13x - 10} = \int \left(\frac{1}{x-2} + \frac{-x+2}{x^2 - 4x + 5} \right) dx = \int \underbrace{\left(\frac{1}{x-2} - \frac{x-2}{(x-2)^2 + 1} \right)}_{u = x - 2} dx$$

$$= \int \left(\frac{1}{u} - \frac{u}{u^2 + 1} \right) du = \ln |u| - \frac{1}{2} \ln |u^2 + 1| + C = \ln |x - 2| - \frac{1}{2} \ln |x^2 - 4x + 5| + C.$$

27. First observe, by trial and error, that

$$\frac{1}{a^2 - x^2} = \frac{1}{2a} \left(\frac{1}{a-x} + \frac{1}{a+x} \right).$$

Therefore

$$\frac{1}{(a^2 - x^2)^2} = \quad = \quad \frac{1}{4a^2} \left[\frac{1}{(a-x)^2} + \frac{2}{(a-x)(a+x)} + \frac{1}{(a+x)^2} \right]$$

$$= \quad \frac{1}{4a^2} \left[\frac{1}{(a-x)^2} + \frac{1}{a} \left(\frac{1}{a-x} + \frac{1}{a+x} \right) + \frac{1}{(a+x)^2} \right]$$

$$= \quad \frac{1}{4a^3} \frac{1}{a-x} + \frac{1}{4a^2} \frac{1}{(a-x)^2} + \frac{1}{4a^3} \frac{1}{a+x} + \frac{1}{4a^2} \frac{1}{(a+x)^2}$$

Hence

$$\int \frac{1}{(a^2 - x^2)^2} \, dx = \frac{1}{4a^3} \int \frac{1}{a-x} \, dx + \frac{1}{4a^2} \int \frac{1}{(a-x)^2} \, dx + \frac{1}{4a^3} \int \frac{1}{a+x} \, dx + \frac{1}{4a^2} \int \frac{1}{(a+x)^2} \, dx$$

$$= -\frac{1}{4a^3} \ln |a - x| + \frac{1}{4a^2} \frac{1}{a-x} + \frac{1}{4a^3} \ln |a + x| - \frac{1}{4a^2} \frac{1}{a+x} + C.$$

28. $$\frac{ax^2 + bx + a}{x + x^3} = \frac{ax^2 + bx + a}{x(1 + x^2)} = \frac{A}{x} + \frac{Bx + C}{1 + x^2}$$

$$\implies \quad ax^2 + bx + a = A(1 + x^2) + x(Bx + C) = A + x(C) + x^2(A + B)$$

$$\implies \quad A = a, \ C = b, \ A + B = a$$

$$\implies \quad A = a, \ B = 0, \ C = b.$$

Therefore

$$\int \frac{ax^2 + bx + c}{x + x^3} \, dx = \int \left(\frac{a}{x} + \frac{b}{1 + x^2} \right) dx = a \ln |x| + b \text{Tan}^{-1} x + C.$$

29. $$\frac{2x^3 - 4x^2 - 7x - 18}{x^2 - 2x - 8} = 2x + \frac{9x - 18}{(x-4)(x+2)} = 2x + \frac{A}{x-4} + \frac{B}{x+2}. \text{ Thus}$$

$$x = 4 \implies 3 = A$$

$$x = -2 \implies 6 = B.$$

Therefore

$$\int \frac{2x^3 - 4x^2 - 7x - 18}{x^2 - 2x - 8} \, dx = \int \left(2x + \frac{3}{x-4} + \frac{6}{x+2} \right) dx = x^2 + 3 \ln |x - 4| + 6 \ln |x + 2| + C.$$

30. $\dfrac{2x^2 - 3x - 2}{x^3 + x^2 - 2x} = \dfrac{2x^2 - 3x - 2}{x(x+2)(x-1)} = \dfrac{A}{x} + \dfrac{B}{x+2} + \dfrac{C}{x-1}$. Then

$$
\begin{aligned}
x = 0 &\implies 1 = A \\
x = -2 &\implies 2 = B \\
x = 1 &\implies -1 = C.
\end{aligned}
$$

Therefore

$$
\int \frac{2x^2 - 3x - 2}{x^3 + x^2 - 2x}\, dx = \int \left(\frac{1}{x} + \frac{2}{x+2} - \frac{1}{x-1} \right) dx = \ln|x| + 2\ln|x+2| - \ln|x-1| + C.
$$

31. $\dfrac{5x^2 + 2x + 2}{x^3 - 1} = \dfrac{5x^2 + 2x + 2}{(x-1)(x^2 + x + 1)} = \dfrac{A}{x-1} + \dfrac{Bx + C}{x^2 + x + 1}$

$\implies 5x^2 + 2x + 2 = A(x^2 + x + 1) + (x-1)(Bx + C) = (A - C) + x(A - B + C) + x^2(A + B)$

$\implies A - C = 2,\ A - B + C = 2,\ A + B = 5$

$\implies A = 3,\ B = 2,\ C = 1.$

Therefore

$$
\int \frac{5x^2 + 2x + 2}{x^3 - 1}\, dx = \int \left(\frac{3}{x-1} + \frac{2x + 1}{x^2 + x + 1} \right) dx = 3\ln|x-1| + \ln(x^2 + x + 1) + C.
$$

32. $\dfrac{6x^2 - 21x - 9}{x^3 - 6x^2 + 3x + 10} = \dfrac{6x^2 - 21x - 9}{(x-2)(x-5)(x+1)} = \dfrac{A}{x-2} + \dfrac{B}{x-5} + \dfrac{C}{x+1}$. Then

$$
\begin{aligned}
x = 2 &\implies 3 = A \\
x = 5 &\implies 2 = B \\
x = 1 &\implies 1 = C.
\end{aligned}
$$

Therefore

$$
\int \frac{6x^2 - 21x - 9}{x^3 - 6x^2 + 3x + 10}\, dx = \int \left(\frac{3}{x-2} + \frac{2}{x-5} + \frac{1}{x+1} \right) dx = 3\ln|x-2| + 2\ln|x-5| + \ln|x+1| + C.
$$

33. $\dfrac{9x^2 + 26x - 16}{x^3 + 2x^2 - 8x} = \dfrac{9x^2 + 26x - 16}{x(x+4)(x-2)} = \dfrac{A}{x} + \dfrac{B}{x+4} + \dfrac{C}{x-2}$. Then

$$
\begin{aligned}
x = 0 &\implies 2 = A \\
x = -4 &\implies 1 = B \\
x = 2 &\implies 6 = C.
\end{aligned}
$$

Therefore

$$
\int \frac{9x^2 + 26x - 16}{x^3 + 2x^2 - 8x}\, dx = \int \left(\frac{2}{x} + \frac{1}{x+4} + \frac{6}{x-2} \right) dx = 2\ln|x| + \ln|x+4| + 6\ln|x-2| + C.
$$

34. $$\frac{x^4 - x^3 + 3x^2 - 10x + 8}{x^3 - x^2 - 4} = x + \frac{3x^2 - 6x + 8}{(x-2)(x^2+x+2)} = x + \frac{A}{x-2} + \frac{Bx+C}{x^2+x+2}$$

$$\implies 3x^2 - 6x + 8 = A(x^2 + x + 2) + (Bx + C)(x - 2)$$
$$= (2A - 2C) + x(A - 2B + C) + x^2(A + B)$$
$$\implies 2A - 2C = 8, \ A - 2B + C = -6, \ A + B = 3$$
$$\implies A = 1, \ B = 2, \ C = -3.$$

Therefore

$$\int \frac{x^4 - x^3 + 3x^2 - 10x + 8}{x^3 - x^2 - 4} \, dx = \int \left(x + \frac{1}{x-2} + \frac{2x-3}{x^2+x+2} \right) dx$$

$$= \frac{1}{2} x^2 + \ln|x - 2| + \int \underbrace{\frac{2x-3}{(x+1/2)^2 + 7/4}}_{u=x+1/2} \, dx$$

$$= \frac{1}{2} x^2 + \ln|x - 2| + \int \frac{2u - 4}{u^2 + 7/4} \, du$$

$$= \frac{1}{2} x^2 + \ln|x - 2| + \ln\left|u^2 + \frac{7}{4}\right| - \frac{8}{\sqrt{7}} \text{Tan}^{-1} \frac{2u}{\sqrt{7}} + C$$

$$= \frac{1}{2} x^2 + \ln|x - 2| + \ln|x^2 + x + 2| - \frac{8}{\sqrt{7}} \text{Tan}^{-1} \frac{2x+1}{\sqrt{7}} + C.$$

35. Area $= \displaystyle\int_1^2 \frac{7x+3}{x^2+x} \, dx = \int_1^2 \left(\frac{3}{x} + \frac{4}{x+1} \right) dx = [3\ln x + 4\ln(x+1)]\Big|_1^2 = 4\ln 3 - \ln 2.$

36. Let dV be the differential volume of a disk of thickness dx. Then

$$dV = \pi r^2 \, dx = \pi y^2 \, dx = \pi \frac{x^2}{x^2 + 1} \, dx$$

and hence the volume of revolution is

$$V = \pi \int_0^1 \underbrace{\frac{x^2}{(x^2+1)^2}}_{x=\tan\theta} \, dx = \pi \int_0^{\pi/4} \frac{\tan^2\theta}{\sec^4\theta} \sec^2\theta \, d\theta = \pi \int_0^{\pi/4} \sin^2\theta \, d\theta = \frac{\pi}{2} \int_0^{\pi/4} (1 - \cos 2\theta) \, d\theta$$

$$= \frac{\pi}{2} \left(\theta - \frac{1}{2} \sin 2\theta \right)\Big|_0^{\pi/4} = \frac{\pi}{2} (\theta - \sin\theta\cos\theta)\Big|_0^{\pi/4} = \frac{\pi}{2} \left(\frac{\pi}{4} - \frac{1}{2} \right).$$

37. Let dV be the differential volume of a shell of thickness dx revolved about the y-axis. Then

$$dV = 2\pi rh \, dx = 2\pi xy \, dx = \frac{2\pi x}{(x+1)(x+3)} \, dx$$

and hence the volume of revolution is

$$V = 2\pi \int_0^2 \frac{x}{(x+1)(x+3)} \, dx = 2\pi \int_0^2 \left(\frac{-1/2}{x+1} + \frac{3/2}{x+3} \right) dx$$

$$= 2\pi \left(-\frac{1}{2} \ln(x+1) + \frac{3}{2} \ln(x+3) \right)\Big|_0^2 = 2\pi \left(-2\ln 3 + \frac{3}{2} \ln 5 \right).$$

38. The average value is equal to

$$\int_0^1 \frac{4x^2}{(1+x^2)^2}\, dx = 4\int_0^1 \frac{x^2}{(1+x^2)^2} = \frac{\pi}{2} - 1$$

by Exercise 36.

39. Let dV be the differential volume of a cross-section disk of thickness dx. Then

$$dV = \pi r^2 dx = \pi y^2 dx = \pi\, \frac{4x^2+1}{x^3+x}\, dx$$

and hence the volume of revolution is

$$V = \pi \int_1^5 \frac{4x^2+1}{x^3+x}\, dx = \pi \int_1^5 \left(\frac{1}{x} + \frac{3x}{x^2+1}\right)\, dx$$

$$= \pi \left(\ln x + \frac{3}{2}\ln(x^2+1)\right)\Bigg|_1^5 = \pi\left(\ln 5 + \frac{3}{2}\ln 13\right).$$

40. Let dV be the differential volume of a shell of thickness dx revolved about the y-axis. Then

$$dV = 2\pi r h\, dx = 2\pi x y\, dx = \frac{2\pi x}{(x-1)^2}\, dx$$

and hence the volume of revolution is

$$V = 2\pi \int_3^7 \underbrace{\frac{x}{(x-1)^2}}_{u=x-1}\, dx = 2\pi \int_2^6 \frac{u+1}{u^2}\, du$$

$$= 2\pi \int_2^6 \left(\frac{1}{u} + \frac{1}{u^2}\right)\, du = 2\pi\left(\ln u - \frac{1}{u}\right)\Bigg|_2^6 = 2\pi\left(\ln 3 + \frac{1}{3}\right).$$

41. $\dfrac{A}{x-a} + \dfrac{B}{(x-a)^2} = \dfrac{A(x-a)+B}{(x-a)^2} = \dfrac{Ax+B-aA}{(x-a)^2}$, which has the required form with $C = A$ and $D = B - aA$.

42. $p(0) = q(0) \Longrightarrow a_0 = b_0$. Differentiating, we find that

$$p'(x) = na_n x^{n-1} + \cdots + a_1$$
$$q'(x) = nb_n^{n-1} + \cdots + b_1.$$

Since $p'(x) = q'(x)$, it follows that $p'(0) = a_1 = q'(0) = b_1$. Continuing in this manner, we find that $a_i = b_i$ for $i = 0, 1, \ldots, n$.

43. $\dfrac{dP}{dt} = 2P\left(\dfrac{100-P}{100}\right)$

$$\Longrightarrow \frac{dP}{P(100-P)} = \frac{1}{50}\, dt$$

$$\Longrightarrow \frac{1}{50}\int dt = \int \frac{dP}{P(100-P)} = \frac{1}{100}\int \left(\frac{1}{P} + \frac{1}{100-P}\right)\, dt$$

$$\implies \frac{1}{50}t = \frac{1}{100}(\ln|P| - \ln|100-P|) + C = \frac{1}{100}\ln\frac{P}{100-P} + C$$

$$\implies ce^{2t} = \frac{P}{100-P} = \frac{1}{\frac{100}{P}-1}$$

$$\implies P = \frac{100ce^{2t}}{1+ce^{2t}} = \frac{100c}{c+e^{-2t}}.$$

It follows that the carrying capacity is

$$K = \lim_{t\to\infty} P(t) = \lim_{t\to\infty}\frac{100c}{c+e^{-2t}} = 100.$$

44. $\int \frac{1}{x}\,dx = \ln|x| + C.$

$$\int \underbrace{\frac{2x+2}{x^2+2x+2}}_{u=x^2+2x+2}\,dx = \int \frac{du}{u} = \ln|u| + C = \ln|x^2+2x+2| + C.$$

$$\int \frac{x}{(x^2+2x+2)^2}\,dx = \int \frac{x}{\underbrace{[(x+1)^2+1]^2}_{u=x+1}}\,dx = \int \frac{u-1}{(u^2+1)^2}\,du.$$

Now, let $u = \tan\theta$. Then $du = \sec^2\theta\,d\theta$ and hence

$$\int \frac{u-1}{(u^2+1)^2}\,du = \int \frac{(\tan\theta-1)}{\sec^4\theta}\sec^2\theta\,d\theta = \int \frac{\tan\theta-1}{\sec^2\theta}\,d\theta$$

$$= \int \sin\theta\cos\theta\,d\theta - \int \cos^2\theta\,d\theta = \frac{1}{2}\int \sin 2\theta\,d\theta - \frac{1}{2}\int (1+\cos 2\theta)\,d\theta$$

$$= -\frac{1}{4}\cos 2\theta - \frac{1}{2}\left(\theta + \frac{1}{2}\sin 2\theta\right) + C = -\frac{1}{4}(1-2\sin^2\theta) - \frac{1}{2}(\theta + \sin\theta\cos\theta) + C$$

$$= -\frac{1}{4}\left(1 - 2\frac{u^2}{u^2+1}\right) - \frac{1}{2}\left(\text{Tan}^{-1}u - \frac{u}{\sqrt{u^2+1}}\frac{1}{\sqrt{u^2+1}}\right) + C = \frac{x^2+4x+2}{4(x^2+2x+2)} - \frac{1}{2}\text{Tan}^{-1}(x+1) + C.$$

45. $\dfrac{dP}{dt} = P(1-P) \implies \dfrac{dP}{P(1-P)} = dt \implies \int \left(\dfrac{1}{P} + \dfrac{1}{1-P}\right)dP = \int dt$

$$\implies \ln|P| - \ln|1-P| = t + C \implies \ln\frac{P}{1-P} = t + C \implies \frac{P}{1-P} = ce^t$$

$$\implies P = \frac{c}{c+e^{-t}} \implies \lim_{t\to\infty} P = 1.$$

10.5 Miscellaneous Substitutions

1. Let $x = u^2$. Then $dx = 2u\,du$ and hence

$$\int \frac{\sqrt{x}}{1+\sqrt{x}}\,dx = \int \frac{u}{1+u}2u\,du = 2\int\left(u - 1 + \frac{1}{1+u}\right)du$$

$$= 2\left(\frac{1}{2}u^2 - u + \ln|1+u|\right) + C = x - 2\sqrt{x} + 2\ln(1+\sqrt{x}) + C.$$

2. Let $x + 1 = u^2$. Then $dx = 2u\,du$ and hence

$$\int \frac{\sqrt{x+1}}{x}\,dx = \int \frac{u}{u^2-1}\,2u\,du = 2\int \left(1 + \frac{1/2}{u-1} - \frac{1/2}{u+1}\right)\,du$$

$$= 2\left(u + \frac{1}{2}\ln|u-1| - \frac{1}{2}\ln|u+1|\right) + C = 2\sqrt{x+1} + \ln\left|\sqrt{x+1}-1\right| - \ln\left|\sqrt{x+1}+1\right| + C.$$

3. Let $x = u^6$. Then $dx = 6u^5\,du$ and hence

$$\int \frac{dx}{\sqrt{x}+2\sqrt[3]{x}} = \int \frac{6u^5}{u^3+2u^2}\,du = 6\int \left(u^2 - 2u + 4 - \frac{8}{u+2}\right)\,du$$

$$= 6\left(\frac{1}{3}u^3 - u^2 + 4u - 8\ln|u+2|\right) + C = 2\sqrt{x} - 6\sqrt[3]{x} + 24\sqrt[6]{x} - 48\ln\left|\sqrt[6]{x}+2\right| + C.$$

4. Let $x = u^3$. Then $dx = 3u^2\,du$ and hence

$$\int \frac{\sqrt[3]{x}+1}{\sqrt[3]{x}-1}\,dx = \int \frac{u+1}{u-1}\,3u^2\,du = 3\int \left(u^2 + 2u + 2 + \frac{2}{u-1}\right)\,du$$

$$= 3\left(\frac{1}{3}u^3 + u^2 + 2u + 2\ln|u-1|\right) + C = x + 3\sqrt[3]{x^2} + 6\sqrt[3]{x} + 6\ln\left|\sqrt[3]{x}-1\right| + C.$$

5. Let $u = x + 1$. Then $du = dx$ and hence

$$\int_0^7 \frac{x}{\sqrt[3]{x+1}}\,dx = \int_1^8 \frac{u-1}{\sqrt[3]{u}}\,du = \int_1^8 (u^{2/3} - u^{-1/3})\,du = \left(\frac{3}{5}u^{5/3} - \frac{3}{2}u^{2/3}\right)\Big|_1^8 = \frac{141}{10}.$$

6. Let $u = x - 2$. Then $du = dx$ and hence

$$\int_3^6 \frac{x+2}{\sqrt{x-2}}\,dx = \int_1^4 \frac{u+4}{\sqrt{u}}\,du = \int_1^4 (u^{1/2} + 4u^{-1/2})\,du = \left(\frac{2}{3}u^{3/2} + 8u^{1/2}\right)\Big|_1^4 = \frac{38}{3}.$$

7. Let $u = \tan(x/2)$. Then

$$du = \frac{1}{2}\sec^2\frac{x}{2}\,dx = \frac{1}{2}\left(1+u^2\right)\,dx, \ \sin x = \frac{2u}{1+u^2},$$

and hence

$$\int \frac{dx}{2+\sin x} = \int \frac{1}{2+\frac{2u}{1+u^2}}\,\frac{2\,du}{1+u^2} = \int \frac{1}{(u+1/2)^2+3/4}\,du$$

$$= \frac{2}{\sqrt{3}}\mathrm{Tan}^{-1}\frac{u+1/2}{\sqrt{3}/2} + C = \frac{2}{\sqrt{3}}\mathrm{Tan}^{-1}\left(\frac{2\tan(x/2)+1}{\sqrt{3}}\right) + C.$$

8. Using the same substitution as in Exercise 7,

$$\int \frac{dx}{1-\sin x} = \int \frac{1}{1-\frac{2u}{1+u^2}}\,\frac{2\,du}{1+u^2} = 2\int \frac{1}{(u-1)^2}\,du = -\frac{2}{u-1} + C = -\frac{2}{\tan(x/2)-1} + C.$$

9. Let $u = \tan(x/2)$. Then

$$du = \frac{1}{2}\left(1 + u^2\right) dx, \ \cos x = \frac{1 - u^2}{1 + u^2},$$

and hence

$$\int \frac{dx}{1 - \cos x} = \int \frac{1}{1 - \frac{1-u^2}{1+u^2}} \frac{2}{1 + u^2} \, du = \int \frac{1}{u^2} \, du = -\frac{1}{u} + C = -\cot\frac{x}{2} + C.$$

10. Let $u^2 = 1 + x^2$. Then $2u\,du = 2x\,dx$ and hence

$$\int \frac{x^3}{\sqrt{1 + x^2}} \, dx = \int \frac{(u^2 - 1)}{u} u \, du = \frac{1}{3} u^3 - u + C = \frac{1}{3}\left(1 + x^2\right)^{3/2} - \left(1 + x^2\right)^{1/2} + C.$$

11. Let $u^3 = 1 + x$. Then $3u^2 du = dx$ and hence

$$\int_0^2 \frac{x^2}{\sqrt[3]{1 + x}} \, dx = \int_1^{\sqrt[3]{3}} \frac{(u^3 - 1)^2}{u} 3u^2 \, du = 3 \int_1^{\sqrt[3]{3}} (u^7 - 2u^4 + u) 3u^2 \, du$$

$$= 3\left(\frac{1}{8} u^8 - \frac{2}{5} u^5 + \frac{1}{2} u^2\right)\Bigg|_1^{\sqrt[3]{3}} = \frac{51}{40} \sqrt[3]{9} - \frac{27}{40}.$$

12. Let $x = u^2$. Then $dx = 2u\,du$ and hence

$$\int_4^{12} \frac{1 + \sqrt{x}}{1 - \sqrt{x}} \, dx = \int_2^{2\sqrt{3}} \frac{1 + u}{1 - u} 2u \, du = 2 \int_2^{2\sqrt{3}} \left(-u - 2 + \frac{2}{1 - u}\right) du$$

$$= \left[2\left(-\frac{1}{2} u^2 - 2u\right) - 4\ln|1 - u|\right]\Bigg|_2^{2\sqrt{3}} = -8\sqrt{3} - 4\ln(2\sqrt{3} - 1).$$

13. Let $u = \tan(x/2)$. Then

$$du = \frac{1}{2}\left(1 + u^2\right) dx, \ \sin x = \frac{2u}{1 + u^2}, \ \cos x = \frac{1 - u^2}{1 + u^2},$$

and hence

$$\int \frac{1}{\sin x + \cos x} \, dx = \int \frac{1}{\frac{2u}{1+u^2} + \frac{1-u^2}{1+u^2}} \frac{2}{1 + u^2} \, du = -2 \int \frac{1}{(u - 1)^2 - 2} \, du.$$

Now, let $u - 1 = \sqrt{2}\sec\theta$. Then $du = \sqrt{2}\sec\theta\tan\theta\,d\theta$ and hence

$$\int \frac{1}{(u - 1)^2 - 2} \, du = \int \frac{1}{2\sec^2\theta - 2} \sqrt{2}\sec\theta\tan\theta \, d\theta = \frac{1}{\sqrt{2}} \int \csc\theta \, d\theta = \frac{\sqrt{2}}{2} \ln|\csc\theta - \cot\theta| + C$$

$$= \frac{\sqrt{2}}{2} \ln\left|\frac{u - 1}{\sqrt{(u - 1)^2 - 2}} - \frac{\sqrt{2}}{\sqrt{(u - 1)^2 - 2}}\right| + C = \frac{\sqrt{2}}{2} \ln\left|\frac{u - 1 - \sqrt{2}}{\sqrt{(u - 1 - \sqrt{2})(u - 1 + \sqrt{2})}}\right| + C$$

$$= \frac{\sqrt{2}}{2} \ln\left|\sqrt{\frac{u - 1 - \sqrt{2}}{u - 1 + \sqrt{2}}}\right| + C = \frac{\sqrt{2}}{4} \ln\left|\frac{u - 1 - \sqrt{2}}{u - 1 + \sqrt{2}}\right| + C.$$

Hence

$$\int \frac{1}{\sin x + \cos x} \, dx = -2\frac{\sqrt{2}}{4} \ln\left|\frac{\tan(x/2) - 1 - \sqrt{2}}{\tan(x/2) - 1 + \sqrt{2}}\right| + C = \frac{\sqrt{2}}{2} \ln\left|\frac{\tan(x/2) - 1 + \sqrt{2}}{\tan(x/2) - 1 - \sqrt{2}}\right| + C.$$

14. Let $u = \tan(x/2)$. Then

$$du = \frac{1}{2}\left(1 + u^2\right) dx, \; \sin x = \frac{2u}{1+u^2}, \; \cos x = \frac{1-u^2}{1+u^2},$$

and hence

$$\int \frac{1 - \sin x}{1 + \cos x}\, dx = \int \frac{1 - \frac{2u}{1+u^2}}{1 + \frac{1-u^2}{1+u^2}}\, \frac{2}{1+u^2}\, du = \int \left(1 - \frac{2u}{1+u^2}\right) du$$

$$= u - \ln|u^2 + 1| + C = \tan(x/2) - \ln|\tan^2(x/2) + 1| + C.$$

15. Let $u^2 = 1 - 2x$. Then $2u\,du = -2\,dx$ and hence

$$\int \frac{x^3}{\sqrt{1-2x}}\, dx = \int \frac{\left(\frac{1-u^2}{2}\right)^3}{u}\, (-u\,du)$$

$$= -\frac{1}{8}\int \left(1 - 3u^2 + 3u^4 - u^6\right)\, du = -\frac{1}{8}\left(u - u^3 + \frac{3}{5}\,u^5 - \frac{1}{7}\,u^7\right) + C$$

$$= -\frac{1}{8}\left[(1 - 2x)^{1/2} - (1 - 2x)^{3/2} + \frac{3}{5}\,(1 - 2x)^{5/2} - \frac{1}{7}\,(1 - 2x)^{7/2}\right] + C.$$

16. Let $x = \tan\theta$. Then $dx = \sec^2\theta\,d\theta$ and hence

$$\int x^2(x^2 + 1)^{3/2}\, dx = \int (\tan^2\theta)(\sec^3\theta)(\sec^2\theta\,d\theta) = \int (\sec^7\theta - \sec^5\theta)\, d\theta$$

$$= \frac{1}{6}\,\tan\theta\sec^5\theta - \frac{1}{6}\int \sec^5\theta\,d\theta = \frac{1}{6}\,\tan\theta\sec^5\theta - \frac{1}{24}\,\tan\theta\sec^3\theta - \frac{1}{8}\int \sec^3\theta\,d\theta$$

$$= \frac{1}{6}\,\tan\theta\sec^5\theta - \frac{1}{24}\,\tan\theta\sec^3\theta - \frac{1}{16}\,\tan\theta\sec\theta - \frac{1}{16}\int \sec\theta\,d\theta$$

$$= \frac{1}{6}\,\tan\theta\sec^5\theta - \frac{1}{24}\,\tan\theta\sec^3\theta - \frac{1}{16}\,\tan\theta\sec\theta - \frac{1}{16}\ln|\sec\theta + \tan\theta| + C$$

$$= \frac{1}{6}\,x(1 + x^2)^{5/2} - \frac{1}{24}\,x(1 + x^2)^{3/2} - \frac{1}{16}\,x(1 + x^2)^{1/2} - \frac{1}{16}\ln\left|\sqrt{1 + x^2} + x\right| + C.$$

17. $\displaystyle\int \sqrt{\frac{1+x}{1-x}}\, dx = \int \sqrt{\frac{1+x}{1-x}\,\frac{1+x}{1+x}}\, dx = \int \frac{1+x}{\sqrt{1-x^2}}\, dx = \operatorname{Sin}^{-1}x - \sqrt{1 - x^2} + C.$

18. Let $u = 1 + e^x$. Then $du = e^x\,dx$ and hence

$$\int \frac{dx}{1+e^x} = \int \frac{1}{u}\,\frac{du}{u-1} = \int \left(-\frac{1}{u} + \frac{1}{u-1}\right) du$$

$$= -\ln|u| + \ln|u - 1| + C = x - \ln(1 + e^x) + C.$$

19. Let $u^2 = a + bx$. Then $2u\,du = b\,dx$ and hence

$$\int \frac{dx}{x\sqrt{a+bx}} = \int \frac{\frac{2u}{b}\,du}{\frac{u^2-a}{b}\,u} = 2\int \frac{1}{u^2 - a}\, du.$$

Now, if $a > 0$, then

$$\int \frac{dx}{x\sqrt{a+bx}} = 2\int \frac{1}{u^2-a}\,du = \frac{1}{\sqrt{a}}\int \left(\frac{1}{u-\sqrt{a}} - \frac{1}{u+\sqrt{a}}\right)du$$

$$= \frac{1}{\sqrt{a}}\left(\ln|u-\sqrt{a}| - \ln|u+\sqrt{a}|\right) + C = \frac{1}{\sqrt{a}}\ln\left|\frac{\sqrt{a+bx}-\sqrt{a}}{\sqrt{a+bx}+\sqrt{a}}\right| + C.$$

On the other hand, if $a < 0$, let $u = \sqrt{-a}\tan\theta$. Then $du = \sqrt{-a}\sec^2\theta\,d\theta$ and hence

$$\int \frac{dx}{x\sqrt{a+bx}} = 2\int \frac{1}{u^2-a}\,du = 2\int \frac{1}{-a\tan^2\theta - a}\,\sqrt{-a}\sec^2\theta\,d\theta$$

$$= \frac{2}{-a}\sqrt{-a}\int d\theta = \frac{2}{\sqrt{-a}}\theta + C = \frac{2}{\sqrt{-a}}\operatorname{Tan}^{-1}\frac{u}{\sqrt{-a}} = \frac{2}{\sqrt{-a}}\operatorname{Tan}^{-1}\sqrt{\frac{a+bx}{-a}} + C.$$

20. Let $u = 1 + x$. Then $du = dx$ and hence

$$\int_0^3 x\sqrt{1+x}\,dx = \int_1^4 (u-1)\sqrt{u}\,du = \int_1^4 (u^{3/2} - u^{1/2})\,du = \left(\frac{2}{5}u^{5/2} - \frac{2}{3}u^{3/2}\right)\Big|_1^4 = \frac{116}{15}.$$

21. Let $u^2 = 2 + x$. Then $2u\,du = dx$ and hence

$$\int_0^2 \frac{x}{\sqrt{2+x}}\,dx = \int_{\sqrt{2}}^2 \frac{u^2-2}{u}\,2u\,du = 2\left(\frac{1}{3}u^3 - 2u\right)\Big|_{\sqrt{2}}^2 = -\frac{8}{3} + \frac{8}{3}\sqrt{2}.$$

10.6 The Use of Integral Tables

1. Formula 102: Let $u = 5x$. Then $du = 5dx$ and hence

$$\int \frac{1}{1+e^{5x}}\,dx = \frac{1}{5}\int \frac{1}{1+e^u}\,du = \frac{1}{5}\ln\left(\frac{e^u}{1+e^u}\right) + C = \frac{1}{5}\ln\left(\frac{e^{5x}}{1+e^{5x}}\right) + C.$$

2. Formula 29: $\displaystyle\int \frac{dx}{x(3+x)} = -\frac{1}{3}\ln\left|\frac{3+x}{x}\right| + C.$

3. Formula 32:

$$\int_1^2 \frac{dx}{x^2(9+2x)^2} = \left(-\frac{9+4x}{81x(9+2x)} + \frac{4}{729}\ln\left|\frac{9+2x}{x}\right|\right)\Big|_1^2 = \frac{151}{23166} + \frac{4}{729}(\ln 13 - \ln 2 - \ln 11).$$

4. Formula 55: Let $u = x - 7$. Then $du = dx$ and hence

$$\int_7^{21/2} \frac{5\,dx}{\sqrt{14x-x^2}} = 5\int_7^{21/2} \frac{dx}{\underbrace{\sqrt{49-(x-7)^2}}_{u=x-7}} = 5\int_0^{7/2} \frac{du}{\sqrt{49-u^2}} = 5\operatorname{Sin}^{-1}\frac{u}{7}\Big|_0^{7/2} = \frac{5\pi}{6}.$$

5. Formula 42:

$$\int \frac{x+3}{2+5x^2}\,dx = \int \frac{x}{2+5x^2}\,dx + 3\int \frac{1}{2+5x^2}\,dx = \frac{1}{10}\ln(2+5x^2) + \frac{3}{\sqrt{10}}\operatorname{Tan}^{-1}\left(x\sqrt{\frac{5}{2}}\right) + C.$$

6. Formula 40: Let $u = x + 3$. Then $du = dx$ and hence

$$\int \frac{3}{\sqrt{6x + x^2}} = 3 \int \frac{dx}{\sqrt{(x+3)^2 - 9}} = 3 \int \frac{du}{\sqrt{u^2 - 9}}$$

$$= 3 \ln|u + \sqrt{u^2 - 9}| + C = 3 \ln|x + 3 + \sqrt{6x + x^2}| + C.$$

7. Formula 80: Let $u = \pi x$. Then $du = \pi dx$ and hence

$$\int_0^{1/4} \tan^4 \pi x \, dx = \frac{1}{\pi} \int_0^{\pi/4} \tan^4 u \, du = \frac{1}{\pi} \left(\frac{1}{3} \tan^3 u - \tan u + u \right)\bigg|_0^{\pi/4} = \frac{1}{4} - \frac{2}{3\pi}.$$

8. Formula 95: $\displaystyle\int_e^{4e} x^2 \ln x \, dx = \left(\frac{x^3}{3} \ln x - \frac{x^3}{9} \right)\bigg|_e^{4e} = e^3 \left(\frac{64}{3} \ln 4 + 14 \right).$

9. Formula 77: $\displaystyle\int \sin 6x \sin 3x \, dx = \frac{\sin 3x}{6} - \frac{\sin 9x}{18} + C.$

10. Formula 103: $\displaystyle\int \frac{dx}{8 + 3e^{5x}} = \frac{x}{8} - \frac{1}{40} \ln(8 + 3e^{5x}) + C.$

11. Formula 31: $\displaystyle\int \frac{dx}{x^2(2 - 3x)} = -\frac{1}{2x} - \frac{3}{4} \ln\left|\frac{2 - 3x}{x}\right| + C.$

12. Formula 54: Let $u = x - 3/2$. Then $du = dx$ and hence

$$\int \sqrt{12x - 4x^2} \, dx = 2 \int \sqrt{\frac{9}{4} - \left(x - \frac{3}{2}\right)^2} \, dx = 2 \int \sqrt{\frac{9}{4} - u^2} \, du$$

$$= u\sqrt{\frac{9}{4} - u^2} + \frac{9}{4} \operatorname{Sin}^{-1}\frac{2u}{3} + C = \left(x - \frac{3}{2}\right)\sqrt{3x - x^2} + \frac{9}{4} \operatorname{Sin}^{-1}\left(\frac{2x - 3}{3}\right) + C.$$

13. Formula 38: Let $u = 2x$. Then $du = 2dx$ and hence

$$\int \frac{9 \, dx}{25 - 4x^2} = \frac{9}{2} \int \frac{du}{25 - u^2} = \frac{9}{20} \ln\left|\frac{5 + u}{5 - u}\right| + C = \frac{9}{20} \ln\left|\frac{5 + 2x}{5 - 2x}\right| + C.$$

14. Formula 40: Let $u = x + 5$. Then $du = dx$ and hence

$$\int \frac{5 \, dx}{\sqrt{10x + x^2}} = 5 \int \frac{du}{\sqrt{u^2 - 25}} = 5 \ln|u + \sqrt{u^2 - 25}| + C = 5 \ln|x + 5 + \sqrt{10x + x^2}| + C.$$

15. Formula 79: Let $u = 5x$. Then $du = 5dx$ and hence

$$\int \tan^3 5x \, dx = \frac{1}{5} \int \tan^3 u \, du = \frac{1}{5} \left(\frac{1}{2} \tan^2 u + \ln|\cos u| \right) + C = \frac{1}{10} \tan^2 5x + \frac{1}{5} \ln|\cos 5x| + C.$$

16. Formula 104: $\displaystyle\int e^x \sin 3x \, dx = \frac{1}{10} e^x (\sin 3x - 3 \cos 3x) + C.$

17. Formula 37: $\displaystyle\int \frac{6x^2 \, dx}{\sqrt{6 + x}} = \frac{4}{5} (288 - 24x + 3x^2)\sqrt{6 + x} + C.$

18. Formula 61: $\displaystyle\int \frac{\sqrt{9-x^2}}{x^2}\,dx = -\frac{\sqrt{9-x^2}}{x} - \text{Sin}^{-1}\left(\frac{x}{3}\right) + C.$

19. Formula 57: $\displaystyle\int \frac{6\,dx}{(14-x^2)^{3/2}} = \frac{3x}{7\sqrt{14-x^2}} + C.$

20. Formula 92: $\displaystyle\int_0^{1/3} x\text{Cos}^{-1}3x\,dx = \frac{1}{36}\left[(18x^2-1)\text{Cos}^{-1}3x - 3x\sqrt{1-9x^2}\right]\Big|_0^{1/3} = \frac{\pi}{72}.$

21. Formula 73: Let $u = 2x$. Then $du = 2dx$ and hence

$$\int_0^{\pi/6} \frac{\pi}{4+4\sin 2x}\,dx = \frac{\pi}{8}\int_0^{\pi/3} \frac{1}{1+\sin u}\,du = -\frac{\pi}{8}\tan\left(\frac{\pi}{4}-\frac{u}{2}\right)\Big|_0^{\pi/3} = \frac{\pi}{8}\left(1 - \tan\frac{\pi}{12}\right).$$

22. Formula 41: $\displaystyle\int \frac{4\,dx}{x\sqrt{9-x^2}} = -\frac{4}{3}\ln\left|\frac{3+\sqrt{9-x^2}}{x}\right| + C.$

23. Formula 41: $\displaystyle\int \frac{3\,dx}{x\sqrt{x^2+16}} = -\frac{3}{4}\ln\left|\frac{4+\sqrt{x^2+16}}{x}\right| + C.$

24. Formula 99: Let $u = \pi x$. Then $du = \pi\,dx$ and hence

$$\int \sin\ln \pi x\,dx = \frac{1}{\pi}\int \sin\ln u\,du = \frac{x}{2}\left(\sin\ln \pi x - \cos\ln \pi x\right) + C.$$

25. Formula 20: Let $u = 3x + 2$. Then $du = 3dx$ and hence

$$\int_{-1}^{-2/3} \frac{dx}{9x^2+12x+5} = \frac{1}{3}\int_{-1}^0 \frac{du}{u^2+1} = \frac{1}{3}\text{Tan}^{-1}u\Big|_{-1}^0 = \frac{\pi}{12}.$$

26. Formula 22: Let $u = x + 3$. Then $du = dx$ and hence

$$\int_{-3}^0 \frac{dx}{\sqrt{7-6x-x^2}} = \int_0^3 \frac{du}{\sqrt{16-u^2}} = \text{Sin}^{-1}\frac{u}{4}\Big|_0^3 = \text{Sin}^{-1}\frac{3}{4}.$$

27. Formula 48: Let $u = x + 4$. Then $du = dx$ and hence

$$\int \frac{7\,dx}{(x+4)\sqrt{x^2+8x+20}} = 7\int \frac{du}{u\sqrt{u^2+4}}$$

$$= -\frac{7}{2}\ln\left(\frac{2+\sqrt{u^2+4}}{u}\right) + C = -\frac{7}{2}\ln\left(\frac{2+\sqrt{x^2+8x+20}}{x+4}\right) + C.$$

28. Formula 39: Let $u = 2x + 5$. Then $du = 2dx$ and hence

$$\int \frac{4\,dx}{4x^2+20x+16} = 2\int \frac{du}{u^2-9} = \frac{1}{3}\ln\left|\frac{u-3}{u+3}\right| + C = \frac{1}{3}\ln\left|\frac{x+1}{x+4}\right| + C.$$

29. Formulas 59 and 22:

$$\int \frac{(2x^2-7)\,dx}{\sqrt{16-x^2}} = 2\int \frac{x^2}{\sqrt{16-x^2}}\,dx - 7\int \frac{dx}{\sqrt{16-x^2}}$$

$$= 2\left(-\frac{x}{2}\sqrt{16-x^2} + 8\,\text{Sin}^{-1}\frac{x}{4}\right) - 7\,\text{Sin}^{-1}\frac{x}{4} + C = -x\sqrt{16-x^2} + 9\,\text{Sin}^{-1}\frac{x}{4} + C.$$

30. Formula 57: Let $u = x - 1$. Then $du = dx$ and hence

$$\int \frac{dx}{(2x-x^2)^{3/2}} = \int \frac{du}{(1-u^2)^{3/2}} = \frac{u}{\sqrt{1-u^2}} + C = \frac{x-1}{\sqrt{2x-x^2}} + C.$$

Review Exercises – Chapter 10

1. $\displaystyle\int x\sqrt{x^2+9}\,dx = \frac{1}{3}\,(x^2+9)^{3/2} + C.$

2. $\displaystyle\int \underbrace{xe^{3x}\,dx}_{u=x,\ dv=e^{3x}\,dx} = \frac{1}{3}\,xe^{3x} - \frac{1}{9}\,e^{3x} + C.$

3. Let $u = x^2$, $dv = \sin 2x\,dx$. Then $du = 2x\,dx$, $v = -\frac{1}{2}\cos 2x$ and hence

$$\int x^2 \sin 2x\,dx = -\frac{1}{2}\,x^2 \cos 2x + \int \underbrace{x \cos 2x\,dx}_{u=x,\,dv=\cos 2x\,dx}$$

$$= -\frac{1}{2}\,x^2 \cos 2x + \frac{1}{2}\,x \sin 2x - \frac{1}{2}\int \sin 2x\,dx = -\frac{1}{2}\,x^2 \cos 2x + \frac{1}{2}\,x \sin 2x + \frac{1}{4}\,\cos 2x + C.$$

Evaluating this expression, we find that

$$\int_0^\pi x^2 \sin 2x\,dx = -\frac{\pi^2}{2}.$$

4. Let $u = x^3$, $dv = \sin\dfrac{\pi x}{2}\,dx$. Then $du = 3x^2\,dx$, $v = -\dfrac{2}{\pi}\cos\dfrac{\pi x}{2}$ and hence

$$\int x^3 \sin\frac{\pi x}{2}\,dx = -\frac{2}{\pi}\,x^3 \cos\frac{\pi x}{2} + \frac{6}{\pi}\int \underbrace{x^2 \cos\frac{\pi x}{2}\,dx}_{u=x^2,\,dv=\cos(\pi x/2)}$$

$$= -\frac{2}{\pi}\,x^3 \cos\frac{\pi x}{2} + \frac{6}{\pi}\left[\frac{2}{\pi}\,x^2 \sin\frac{\pi x}{2} - \frac{4}{\pi}\int \underbrace{x \sin\frac{\pi x}{2}\,dx}_{u=x,\,dv=\sin(\pi x/2)}\right]$$

$$= -\frac{2}{\pi}\,x^3 \cos\frac{\pi x}{2} + \frac{12}{\pi^2}\,x^2 \sin\frac{\pi x}{2} - \frac{24}{\pi^2}\left[-\frac{2}{\pi}\,x \cos\frac{\pi x}{2} + \frac{2}{\pi}\int \cos\frac{\pi x}{2}\,dx\right]$$

$$= -\frac{2}{\pi}\,x^3 \cos\frac{\pi x}{2} + \frac{12}{\pi^2}\,x^2 \sin\frac{\pi x}{2} + \frac{48}{\pi^3}\,x \cos\frac{\pi x}{2} - \frac{96}{\pi^4}\,\sin\frac{\pi x}{2} + C.$$

Evaluating this expression, we find that

$$\int_{1/2}^1 x^3 \sin\frac{\pi x}{2}\,dx = \frac{24-3\sqrt{2}}{2\pi^2} + \frac{\sqrt{2}}{8\pi} - \frac{12\sqrt{2}}{\pi^3} + \frac{48\sqrt{2}-96}{\pi^4}.$$

5. Let $x = a\tan\theta$. Then $dx = a\sec^2\theta\,d\theta$ and hence

$$\int \sqrt{x^2+a^2}\,dx = \int (a\sec\theta)(a\sec^2\theta)\,d\theta = a^2\int \sec^3\theta\,d\theta$$

$$= \frac{a^2}{2}(\sec\theta\tan\theta + \ln|\sec\theta+\tan\theta|) + C = \frac{1}{2}\,x\sqrt{x^2+a^2} + \frac{a^2}{2}\,\ln\left|\sqrt{x^2+a^2}+x\right| + C.$$

Note that the missing constant, $\ln a$, has been absorbed in the constant C.

746

6. $\displaystyle\int \frac{dx}{x^2 - 6x + 9} = \int \frac{dx}{(x-3)^2} = -\frac{1}{x-3} + C.$

7. $\displaystyle\int_1^3 \frac{2x+3}{x(x+3)}\, dx = \int_1^3 \left(\frac{1}{x} + \frac{1}{x+3}\right)\, dx = \left(\ln|x| + \ln|x+3|\right)\big|_1^3 = \ln 3 + \ln 6 - \ln 4.$

8. Let $u^2 = x$. Then $2u\,du = dx$ and hence

$$\int_3^4 \frac{\sqrt{x+1}}{\sqrt{x}}\, dx = \int_{\sqrt{3}}^2 \frac{\sqrt{u^2+1}}{u}\, 2u\, du = \left(u\sqrt{u^2+1} + \ln\left|\sqrt{u^2+1} + u\right|\right)\Big|_{\sqrt{3}}^2 \quad \text{by Exercise 5}$$

$$= 2\sqrt{5} + \ln\left(2 + \sqrt{5}\right) - 2\sqrt{3} - \ln\left(\sqrt{3} + 2\right).$$

9. Let $u = x$, $dv = \cos \pi x\, dx$. Then $du = dx$, $v = \dfrac{1}{\pi} \sin \pi x$ and hence

$$\int_{-1/2}^{1/2} x \cos \pi x\, dx = \left(\frac{x}{\pi} \sin \pi x - \frac{1}{\pi} \int \sin \pi x\, dx\right)\Big|_{-1/2}^{1/2} = \left(\frac{x}{\pi} \sin \pi x + \frac{1}{\pi^2} \cos \pi x\right)\Big|_{-1/2}^{1/2} = 0.$$

10. Let $u = \text{Tan}^{-1}x$, $dv = x\,dx$. Then $du = \dfrac{1}{1+x^2}\, dx$, $v = \dfrac{1}{2}x^2$, and hence

$$\int_0^1 x\,\text{Tan}^{-1}x\, dx = \left(\frac{x^2}{2}\text{Tan}^{-1}x - \frac{1}{2}\int \frac{x^2}{1+x^2}\, dx\right)\Big|_0^1$$

$$= \frac{1}{2}\left[x^2\text{Tan}^{-1}x - \int\left(1 - \frac{1}{1+x^2}\right)\, dx\right]\Big|_0^1 = \frac{1}{2}\left(x^2\text{Tan}^{-1}x - x + \text{Tan}^{-1}x\right)\big|_0^1 = \frac{\pi}{4} - \frac{1}{2}.$$

11. Let $u = x$, $dv = \cos(2x+1)dx$. Then $du = dx$, $v = \dfrac{1}{2}\sin(2x+1)$, and hence

$$\int_{-1/2}^{(\pi-1)/2} x\cos(2x+1)\, dx = \left(\frac{1}{2}x\sin(2x+1) - \frac{1}{2}\int \sin(2x+1)\, dx\right)\Big|_{-1/2}^{(\pi-1)/2}$$

$$= \left(\frac{1}{2}x\sin(2x+1) + \frac{1}{4}\cos(2x+1)\right)\Big|_{-1/2}^{(\pi-1)/2} = -\frac{1}{2}.$$

12. Let $u = x^2$, $dv = e^{3x}dx$. Then $du = 2x\,dx$, $v = \dfrac{1}{3}e^{3x}$ and hence

$$\int x^2 e^{3x}\, dx = \frac{1}{3}x^2 e^{3x} - \frac{2}{3}\int \underbrace{xe^{3x}\, dx}_{u=x,\ dv=e^{3x}dx}$$

$$= \frac{1}{3}x^2 e^{3x} - \frac{2}{3}\left(\frac{1}{3}xe^{3x} - \frac{1}{9}e^{3x}\right) + C = \frac{1}{3}x^2 e^{3x} - \frac{2}{9}xe^{3x} + \frac{2}{27}e^{3x} + C.$$

Therefore

$$\int_0^{\ln 2} x^2 e^{3x}\, dx = \frac{8}{3}(\ln 2)^2 - \frac{16}{9}(\ln 2) + \frac{14}{27}.$$

13. Let $u = x^2$, $dv = e^{-5x}dx$. Then $du = 2xdx$, $v = -\frac{1}{5}e^{-5x}$ and hence

$$\int x^2 e^{-5x}\ dx = -\frac{1}{5}\ x^2 e^{-5x} + \frac{2}{5}\int \underbrace{xe^{-5x}\ dx}_{u=x,\ dv=e^{-5x}dx}$$

$$= -\frac{1}{5}\ x^2 e^{-5x} + \frac{2}{5}\left(-\frac{1}{5}\ xe^{-5x} + \frac{1}{5}\int e^{-5x}\ dx\right) = -\frac{1}{5}\ x^2 e^{-5x} - \frac{2}{25}\ xe^{-5x} - \frac{2}{125}\ e^{-5x} + C.$$

Therefore

$$\int_0^1 x^2 e^{-5x}\ dx = \frac{2}{125} - \frac{37}{125}\ e^{-5}.$$

14. Let $u = (\ln x)^3$, $dv = x^3 dx$. Then $du = 3(\ln x)^2\ \frac{1}{x}\ dx$, $v = \frac{1}{4}\ x^4$, and hence

$$\int (x\ln x)^3\ dx\ =\ \frac{1}{4}\ x^4(\ln x)^3 - \frac{3}{4}\int \underbrace{x^3(\ln x)^2\ dx}_{u=(\ln x)^2,\ dv=x^3 dx}$$

$$=\ \frac{1}{4}\ x^4(\ln x)^3 - \frac{3}{4}\left[\frac{1}{4}\ x^4(\ln x)^2 - \frac{1}{2}\int \underbrace{x^3(\ln x)\ dx}_{u=\ln x,\ dv=x^3 dx}\right]$$

$$=\ \frac{1}{4}\ x^4(\ln x)^3 - \frac{3}{16}\ x^4(\ln x)^2 + \frac{3}{8}\left[\frac{1}{4}\ x^4\ln x - \frac{1}{4}\int x^3\ dx\right]$$

$$=\ x^4\left[\frac{1}{4}\ (\ln x)^3 - \frac{3}{16}\ (\ln x)^2 + \frac{3}{32}\ \ln x - \frac{3}{128}\right] + C.$$

15. Let $u = (\ln x)^2$, $dv = xdx$. Then $du = 2(\ln x)\ \frac{1}{x}\ dx$, $v = \frac{1}{2}x^2$ and hence

$$\int x\ln^2 x\ dx = \frac{1}{2}\ x^2(\ln x)^2 - \int \underbrace{x\ln x\ dx}_{u=\ln x,\ dv=xdx}\ = \frac{1}{2}\ x^2(\ln x)^2 - \frac{1}{2}\ x^2(\ln x) + \frac{1}{4}\ x^2 + C.$$

16. Let $x = a\tan\theta$. Then $dx = a\sec^2\theta\ d\theta$ and hence

$$\int \frac{dx}{\sqrt{x^2 + a^2}} = \int \frac{a\sec^2\theta\ d\theta}{a\sec\theta} = \ln|\sec\theta + \tan\theta| + C$$

$$= \ln\left|\frac{\sqrt{x^2 + a^2}}{a} + \frac{x}{a}\right| + C = \ln(\sqrt{x^2 + a^2} + x) + C,$$

where the missing constant, $\ln a$, is absorbed in the constant C.

17. $\displaystyle\int \frac{dx}{x^2 - 4x + 9} = \int \frac{dx}{(x-2)^2 + 5} = \frac{1}{\sqrt{5}}\ \text{Tan}^{-1}\frac{x-2}{\sqrt{5}} + C.$

18. Let $u = x - 1/2$. Then $du = dx$ and hence

$$\int \frac{4x+5}{x^2 - x + 2}\ dx = \int \frac{4u+7}{u^2 + 7/4}\ du = 4\int \frac{u}{u^2 + 7/4}\ du + 7\int \frac{1}{u^2 + 7/4}\ du$$

$$= 2\ln|u^2 + 7/4| + 2\sqrt{7}\ \text{Tan}^{-1}\frac{2}{\sqrt{7}}\ u = 2\ln|x^2 - x + 2| + 2\sqrt{7}\ \text{Tan}^{-1}\frac{2x-1}{\sqrt{7}} + C.$$

19. Let $u^2 = x + 4$. Then $2u\,du = dx$ and hence

$$\int \frac{\sqrt{x+4}}{x}\,dx = \int \frac{u}{u^2 - 4}\,2u\,du = 2\int \left(1 + \frac{1}{u-2} - \frac{1}{u+2}\right)\,du$$

$$= 2(u + \ln|u - 2| - \ln|u + 2|) + C = 2\left(\sqrt{x+4} + \ln\left|\sqrt{x+4} - 2\right| - \ln\left|\sqrt{x+4} + 2\right|\right) + C.$$

20. Let $u = x + 1$. Then $du = dx$ and hence

$$\int \frac{1}{\sqrt[3]{x+1}}\,dx = \int u^{-1/3}\,du = \frac{3}{2}\,u^{2/3} + C = \frac{3}{2}\,(x+1)^{2/3} + C.$$

21. Let $u = x + 2$. Then $du = dx$ and hence

$$\int_0^2 x\sqrt{x+2}\,dx = \int_2^4 (u - 2)\sqrt{u}\,du = \int_2^4 (u^{3/2} - 2u^{1/2})\,du = \left(\frac{2}{5}\,u^{5/2} - \frac{4}{3}\,u^{3/2}\right)\Big|_2^4 = \frac{32}{15} + \frac{16}{15}\,\sqrt{2}.$$

22. Let $u = x + 3$. Then $du = dx$ and hence

$$\int_0^1 x(x+3)^6\,dx = \int_3^4 (u-3)u^6\,du = \int_3^4 (u^7 - 3u^6)\,du = \left(\frac{1}{8}\,u^8 - \frac{3}{7}\,u^7\right)\Big|_3^4 = \frac{72097}{56}.$$

23. Let $x = a\sin\theta$. Then $dx = a\cos\theta\,d\theta$ and hence

$$\int \frac{dx}{\sqrt{a^2 - x^2}} = \int \frac{a\cos\theta\,d\theta}{a\cos\theta} = \theta + C = \mathrm{Sin}^{-1}\frac{x}{a} + C.$$

24. Let $u = x + 4$. Then

$$\int \frac{dx}{\sqrt{x^2 + 8x + 25}} = \int \frac{du}{\underbrace{\sqrt{u^2 + 9}}_{u = 3\tan\theta}} = \int \frac{3\sec^2\theta\,d\theta}{3\sec\theta} = \int \sec\theta\,d\theta$$

$$= \ln|\sec\theta + \tan\theta| + C = \ln\left|\frac{\sqrt{u^2 + 9}}{3} + \frac{u}{3}\right| + C = \ln\left|\sqrt{x^2 + 8x + 25} + x + 4\right| + C.$$

25. $\displaystyle\int \frac{3x^2 + x + 1}{x^2(x+1)}\,dx = \int \left(\frac{1}{x^2} + \frac{3}{x+1}\right)\,dx = -\frac{1}{x} + 3\ln|x+1| + C.$

26. Let $u^6 = x$. Then $6u^5\,du = dx$ and hence

$$\int \frac{dx}{2\sqrt{x} + 3\sqrt[3]{x}} = \int \frac{6u^5\,du}{2u^3 + 3u^2} = 6\int \frac{u^3\,du}{2u + 3} = 6\int \left(\frac{1}{2}\,u^2 - \frac{3}{4}\,u + \frac{9}{8} - \frac{27}{8(2u+3)}\right)\,du$$

$$= 6\left(\frac{1}{6}\,u^3 - \frac{3}{8}\,u^2 + \frac{9}{8}\,u - \frac{27}{16}\,\ln|2u+3|\right) + C = \sqrt{x} - \frac{9}{4}\,\sqrt[3]{x} + \frac{27}{4}\,\sqrt[6]{x} - \frac{81}{8}\,\ln\left|2\sqrt[6]{x} + 3\right| + C.$$

27. $\displaystyle\int_0^{\pi/4} \tan^2 x\,dx = \int_0^{\pi/4} (\sec^2 x - 1)\,dx = (\tan x - x)\big|_0^{\pi/4} = 1 - \frac{\pi}{4}.$

28. Let $u = \ln x$, $dv = x^2 dx$. Then $du = \frac{1}{x} dx$, $v = \frac{1}{3} x^3$ and hence

$$\int_1^e x^2 \ln x \, dx = \left(\frac{1}{3} x^3 \ln x - \frac{1}{3} \int x^2 \, dx \right) \Big|_1^e = \left(\frac{1}{3} x^3 \ln x - \frac{1}{9} x^3 \right) \Big|_1^e = \frac{2}{9} e^3 + \frac{1}{9}.$$

29. Let $x = a \sin \theta$. Then $dx = a \cos \theta \, d\theta$ and hence

$$\int \sqrt{a^2 - x^2} \, dx = \int (a \cos \theta)(a \cos \theta \, d\theta) = \frac{a^2}{2} \int (1 + \cos 2\theta) \, d\theta$$

$$= \frac{a^2}{2} \left(\theta + \frac{1}{2} \sin 2\theta \right) + C = \frac{a^2}{2} (\theta + \sin \theta \cos \theta) + C = \frac{a^2}{2} \operatorname{Sin}^{-1} \frac{x}{a} + \frac{x \sqrt{a^2 - x^2}}{2} + C.$$

30. $\displaystyle \int \frac{dx}{\sqrt{1 + 4x - x^2}} = \int \frac{dx}{\sqrt{5 - (x - 2)^2}} = \operatorname{Sin}^{-1} \frac{x - 2}{\sqrt{5}} + C.$

31. $\displaystyle \int_0^1 \frac{4x^2 + x + 2}{(x - 2)(x^2 + 1)} \, dx = \int_0^1 \left(\frac{4}{x - 2} + \frac{1}{x^2 + 1} \right) \, dx = (4 \ln |x - 2| + \operatorname{Tan}^{-1} x) \Big|_0^1 = \frac{\pi}{4} - 4 \ln 2.$

32. Let $u^2 = x$. Then $2u \, du = dx$ and hence

$$\int_0^1 \frac{2 + \sqrt{x}}{2 - \sqrt{x}} \, dx = \int_0^1 \frac{2 + u}{2 - u} \, 2u \, du = 2 \int_0^1 \left(-u - 4 + \frac{8}{2 - u} \right) \, du$$

$$= 2 \left(-\frac{1}{2} u^2 - 4u - 8 \ln |2 - u| \right) \Big|_0^1 = 16 \ln 2 - 9.$$

33. $\displaystyle \int \frac{x + 1}{\sqrt[3]{x + 1}} \, dx = \int (x + 1)^{2/3} \, dx = \frac{3}{5} (x + 1)^{5/3} + C.$

34. Let $u = x$, $dv = \csc^2 \pi x \, dx$. Then $du = dx$, $v = -\frac{1}{\pi} \cot \pi x$ and hence

$$\int x \csc^2 \pi x \, dx = -\frac{x}{\pi} \cot \pi x + \frac{1}{\pi} \int \cot \pi x \, dx = -\frac{x}{\pi} \cot \pi x + \frac{1}{\pi^2} \ln |\sin \pi x| + C.$$

35. Let $x = a \sec \theta$. Then $dx = a \sec \theta \tan \theta \, d\theta$ and hence

$$\int \sqrt{x^2 - a^2} \, dx = \int (a \tan \theta)(a \sec \theta \tan \theta) \, d\theta = a^2 \int (\sec^3 \theta - \sec \theta) \, d\theta$$

$$= a^2 \left[\frac{1}{2} (\sec \theta \tan \theta + \ln |\sec \theta + \tan \theta|) - \ln |\sec \theta + \tan \theta| \right] + C$$

$$= \frac{1}{2} x \sqrt{x^2 - a^2} - \frac{a^2}{2} \ln |x + \sqrt{x^2 - a^2}| + C,$$

where the missing constant, $\ln a$, is absorbed in the constant C.

36. Let $u = x - 4$. Then $du = dx$ and hence

$$\int \frac{dx}{\sqrt{x^2 - 8x}} = \int \frac{dx}{\underbrace{\sqrt{u^2 - 16}}_{u = 4 \sec \theta}} = \int \frac{4 \sec \theta \tan \theta \, d\theta}{4 \tan \theta} = \ln |\sec \theta + \tan \theta| + C = \ln |x - 4 + \sqrt{x^2 - 8x}| + C,$$

where the missing constant, $\ln 4$, is absorbed in the constant C.

37. $\int \dfrac{x^2 + 2x + 3}{(x+2)(x^2+x+1)}\, dx = \int \left(\dfrac{1}{x+2} + \dfrac{1}{(x+1/2)^2 + 3/4} \right) dx = \ln|x+2| + \dfrac{2}{\sqrt{3}} \operatorname{Tan}^{-1} \dfrac{2x+1}{\sqrt{3}} + C.$

38. Let $u = x + 2$. Then $du = dx$ and hence

$$\int \frac{x-3}{\sqrt{x+2}}\, dx = \int \frac{u-5}{\sqrt{u}}\, du = \int (u^{1/2} - 5u^{-1/2})\, du$$

$$= \frac{2}{3} u^{3/2} - 10u^{1/2} + C = \frac{2}{3}(x+2)^{3/2} - 10(x+2)^{1/2} + C.$$

39. Let $x = 2\sec\theta$. Then $dx = 2\sec\theta\tan\theta\, d\theta$ and hence

$$
\begin{aligned}
\int (x^2-4)^{3/2}\, dx &= \int (8\tan^3\theta)(2\sec\theta\tan\theta\, d\theta) \\[2mm]
&= 16\int \tan^4\theta\sec\theta\, d\theta = 16\int (\sec^5\theta - 2\sec^3\theta + \sec\theta)\, d\theta \\[2mm]
&= 16\left[\frac{1}{4}\tan\theta\sec^3\theta + \frac{3}{4}\int \sec^3\theta\, d\theta \right] - 32\int \sec^3\theta\, d\theta + 16\int \sec\, d\theta \\[2mm]
&= 4\tan\theta\sec^3\theta - 20\int \sec^3\, d\theta + 16\int \sec\theta\, d\theta \\[2mm]
&= 4\tan\theta\sec^3\theta - 20\left[\frac{1}{2}\tan\theta\sec\theta + \frac{1}{2}\int \sec\theta\, d\theta \right] + 16\int \sec\theta\, d\theta \\[2mm]
&= 4\tan\theta\sec^3\theta - 10\tan\theta\sec\theta + 6\ln|\sec\theta + \tan\theta| + C \\[2mm]
&= \frac{1}{4}x^3\sqrt{x^2-4} - \frac{5}{2}x\sqrt{x^2-4} + 6\ln|x + \sqrt{x^2-4}| + C.
\end{aligned}
$$

40. Let $u = x + 2$. Then $du = dx$ and hence

$$\int \frac{x\, dx}{x^2+4x+8} = \int \frac{u-2}{u^2+4}\, du = \frac{1}{2}\ln|u^2+4| - \operatorname{Tan}^{-1}\frac{u}{2} + C = \frac{1}{2}\ln|x^2+4x+8| - \operatorname{Tan}^{-1}\frac{x+2}{2} + C.$$

41. Since $1 + x^3 = (1+x)(1-x+x^2)$, we find that

$$
\begin{aligned}
\int \frac{3x^3 + x + 2}{x(1+x^3)}\, dx &= \int \frac{2}{x}\, dx + \frac{2}{3}\int \frac{1}{1+x}\, dx + \frac{1}{3}\int \underbrace{\frac{x+1}{x^2-x+1}}_{u=x-1/2}\, dx \\[2mm]
&= 2\int \frac{1}{x}\, dx + \frac{2}{3}\int \frac{1}{1+x}\, dx + \frac{1}{3}\int \frac{u+3/2}{u^2+3/4}\, du \\[2mm]
&= 2\ln|x| + \frac{2}{3}\ln|x+1| + \frac{1}{6}\ln\left|u^2 + \frac{3}{4}\right| + \frac{1}{\sqrt{3}}\operatorname{Tan}^{-1}\frac{u}{\sqrt{3/4}} + C \\[2mm]
&= 2\ln|x| + \frac{2}{3}\ln|x+1| + \frac{1}{6}\ln(x^2-x+1) + \frac{1}{\sqrt{3}}\operatorname{Tan}^{-1}\frac{2x-1}{\sqrt{3}} + C.
\end{aligned}
$$

42. Let $u = \tan(x/2)$. Then

$$\sin x = \frac{2u}{1+u^2}, \quad du = \frac{1}{2}\sec^2\frac{x}{2}\, dx = \frac{1}{2}(1+u^2)\, dx,$$

and hence

$$\int \frac{dx}{1+2\sin x} = \int \frac{1}{1 + \frac{4u}{1+u^2}}\frac{2\, du}{1+u^2} = 2\int \frac{du}{u^2+4u+1} = 2\int \frac{du}{(u+2)^2 - 3}$$

$$= \frac{1}{\sqrt{3}} \int \left(\frac{1}{u+2-\sqrt{3}} - \frac{1}{u+2+\sqrt{3}} \right) du = \frac{1}{\sqrt{3}} \left(\ln|u+2-\sqrt{3}| - \ln|u+2+\sqrt{3}| \right) + C$$

$$= \frac{1}{\sqrt{3}} \left(\ln|\tan(x/2)+2-\sqrt{3}| - \ln|\tan(x/2)+2+\sqrt{3}| \right) + C.$$

43. Let $u = \csc x$, $dv = \csc^2 x \, dx$. Then $du = -\csc x \cot x \, dx$, $v = -\cot x$ and hence

$$\int \csc^3 x \, dx = \int \csc x \csc^2 x \, dx = -\csc x \cot x - \int \csc x \cot^2 x \, dx$$

$$= -\csc x \cot x - \int \csc x(\csc^2 x - 1) \, dx = -\csc x \cot x - \int \csc^3 x \, dx + \ln|\csc x - \cot x| + C$$

$$\implies \int \csc^3 x \, dx = -\frac{1}{2} \csc x \cot x + \frac{1}{2} \ln|\csc x - \cot x| + C.$$

44. Let $u = 1 - e^x$. Then $du = -e^x \, dx$ and hence

$$\int \frac{e^{2x}}{\sqrt{1-e^x}} \, dx = -\int \frac{(1-u) \, du}{\sqrt{u}} = -\int (u^{-1/2} - u^{1/2}) \, du$$

$$= -\left(2u^{1/2} - \frac{2}{3} u^{1/2} - \frac{2}{3} u^{3/2} \right) + C = -2(1-e^x)^{1/2} + \frac{2}{3} (1-e^x)^{3/2} + C.$$

45. Let $x = 3 \tan \theta$. Then $dx = 3 \sec^2 \theta \, d\theta$ and hence

$$\int \frac{dx}{x\sqrt{9+x^2}} = \int \frac{3 \sec^2 \theta \, d\theta}{3 \tan \theta 3 \sec \theta} = \frac{1}{3} \int \csc \theta \, d\theta = \frac{1}{3} \ln|\csc \theta - \cot \theta| + C = \frac{1}{3} \ln\left| \frac{\sqrt{9+x^2}-3}{x} \right| + C.$$

46. Let $u = x - 3$. Then $du = dx$ and hence

$$\int \frac{2+x}{\sqrt{x^2-6x+13}} \, dx = \int \frac{u+5}{\sqrt{u^2+4}} \, du = \int \frac{u}{\sqrt{u^2+4}} \, du + 5 \int \frac{1}{\sqrt{u^2+4}} \, du$$

$$= \sqrt{u^2+4} + 5 \ln|u + \sqrt{u^2+4}| + C = \sqrt{x^2-6x+13} + 5 \ln|x-3+\sqrt{x^2-6x+13}| + C.$$

47. $\int \frac{3x^2-x+10}{(x^2+2)(4-x)} \, dx = \int \left(\frac{3}{4-x} + \frac{1}{x^2+2} \right) dx = -3 \ln|4-x| + \frac{1}{\sqrt{2}} \text{Tan}^{-1} \frac{x}{\sqrt{2}} + C.$

48. Let $u = \tan(x/2)$. Then

$$\sin x = \frac{2u}{1+u^2}, \quad du = \frac{1}{2} \sec^2 \frac{x}{2} \, dx = \frac{1+u^2}{2} \, dx,$$

and hence

$$\int \frac{dx}{1-2\sin x} = \int \frac{1}{1-\frac{4u}{1+u^2}} \frac{2}{1+u^2} \, du = 2 \int \frac{1}{(u-2)^2-3} \, du$$

$$= \frac{1}{\sqrt{3}} \int \left(\frac{1}{u-2-\sqrt{3}} - \frac{1}{u-2+\sqrt{3}} \right) du = \frac{1}{\sqrt{3}} \left(\ln|\tan(x/2)-2-\sqrt{3}| - \ln|\tan(x/2)-2+\sqrt{3}| \right) + C.$$

49. Let $u = \tan(x/2)$. Then

$$\cos x = \frac{1-u^2}{1+u^2}, \quad du = \frac{1}{2} \sec^2 \frac{x}{2} \, dx = \frac{1+u^2}{2} \, dx$$

and hence

$$\int \frac{\cos x}{2 - \cos x} \, dx = \int \left(-1 + \frac{2}{2 - \cos x} \right) dx = -x + 2 \int \frac{1}{2 - \frac{1-u^2}{1+u^2}} \frac{2}{1+u^2} \, du$$

$$= -x + 4 \int \frac{1}{1 + 3u^2} \, du = -x + \frac{4}{\sqrt{3}} \mathrm{Tan}^{-1} u\sqrt{3} + C = -x + \frac{4}{\sqrt{3}} \mathrm{Tan}^{-1} \left(\sqrt{3} \, \tan \frac{x}{2} \right) + C.$$

50. Let $x = \tan \theta$. Then $dx = \sec^2 \theta \, d\theta$ and hence

$$\int \frac{x^2}{\sqrt{1+x^2}} \, dx = \int \frac{\tan^2 \theta}{\sec \theta} \sec^2 \theta \, d\theta = \int (\sec^3 \theta - \sec \theta) \, d\theta = \frac{1}{2} \tan \theta \sec \theta + \frac{1}{2} \int \sec \theta \, d\theta - \int \sec \theta \, d\theta$$

$$= \frac{1}{2} \tan \theta \sec \theta - \frac{1}{2} \ln|\sec \theta + \tan \theta| + C = \frac{1}{2} x\sqrt{1+x^2} - \frac{1}{2} \ln|\sqrt{1+x^2} + x| + C.$$

51. Let $u = e^x$. Then $du = e^x \, dx$ and hence

$$\int_{\ln 2}^{\ln 3} \frac{dx}{e^x - e^{-x}} = \int_{\ln 2}^{\ln 3} \frac{e^x}{(e^x - 1)(e^x + 1)} \, dx = \frac{1}{2} \int_2^3 \left(\frac{1}{u-1} - \frac{1}{u+1} \right) du$$

$$= \frac{1}{2} \left(\ln|u-1| - \ln|u+1| \right) \Big|_2^3 = \frac{1}{2} \left(\ln 2 - \ln 4 + \ln 3 \right) = \ln \sqrt{3/2}.$$

52. Let $u = \sec^3 x$, $dv = \sec^2 x \, dx$. Then $du = 3 \sec^3 x \tan x \, dx$, $v = \tan x$ and hence

$$\int \sec^5 x \, dx = \sec^3 x \tan x - 3 \int \sec^3 x \tan^2 x \, dx = \sec^3 x \tan x - 3 \int (\sec^5 x - \sec^3 x) \, dx$$

$$\implies \int \sec^5 = \frac{1}{4} \sec^3 x \tan x + \frac{3}{4} \int \sec^3 x \, dx$$

$$= \frac{1}{4} \sec^3 x \tan x + \frac{3}{4} \left(\frac{1}{2} \sec x \tan x + \frac{1}{2} \ln|\sec x + \tan x| \right) + C$$

$$= \frac{1}{4} \sec^3 x \tan x + \frac{3}{8} \sec x \tan x + \frac{3}{8} \ln|\sec x + \tan x| + C.$$

53. Let $u = a^2 + x^2$. Then $du = 2x\,dx$ and hence

$$\int \frac{x}{a^2 + x^2} \, dx = \frac{1}{2} \int \frac{du}{u} = \frac{1}{2} \ln|u| + C = \frac{1}{2} \ln(a^2 + x^2) + C.$$

54. Let $u = x + 4$. Then $du = dx$ and hence

$$\int \frac{2x - 1}{\sqrt{17 + 8x + x^2}} \, dx = \int \frac{2u - 9}{\sqrt{u^2 + 1}} \, du = \int \frac{2u}{\sqrt{u^2 + 1}} \, du - 9 \int \frac{1}{\sqrt{u^2 + 1}} \, du$$

$$= 2\sqrt{u^2 + 1} - 9 \ln|u + \sqrt{u^2 + 1}| + C = 2\sqrt{x^2 + 8x + 17} - 9 \ln|x + 4 + \sqrt{x^2 + 8x + 17}| + C.$$

55. $\int_2^3 \frac{5x^2 - 2x + 12}{x^3 - x^2 + 4x - 4} \, dx = \int_2^3 \left(\frac{3}{x-1} + \frac{2x}{x^2+4} \right) \, dx = \left(3\ln|x-1| + \ln(x^2+4) \right) \big|_2^3 = \ln 13.$

56. Let $u = x - 1/2$. Then $du = dx$ and hence

$$\int \frac{dx}{1+x^3} = \frac{1}{3} \int \left(\frac{1}{1+x} - \frac{x-2}{1-x+x^2} \right) \, dx = \frac{1}{3} \ln|1+x| - \frac{1}{3} \int \frac{u - 3/2}{u^2 + 3/4} \, du$$

$$= \frac{1}{3} \ln|1+x| - \frac{1}{6} \ln\left|u^2 + \frac{3}{4}\right| + \frac{1}{\sqrt{3}} \, \text{Tan}^{-1} \frac{u}{\sqrt{3/4}} + C$$

$$= \frac{1}{3} \ln|1+x| - \frac{1}{6} \ln|x^2 - x + 1| + \frac{1}{\sqrt{3}} \, \text{Tan}^{-1} \frac{2x-1}{\sqrt{3}} + C.$$

57. $\int \frac{1}{1+e^x} \, dx = \int \frac{e^{-x}}{e^{-x}+1} \, dx = -\ln(e^{-x}+1) + C.$

58. $\int_{-\sqrt{13}}^0 \frac{x}{\sqrt{36+x^2}} \, dx = \sqrt{36+x^2} \, \Big|_{-\sqrt{13}}^0 = 6 - 7 = -1.$

59. Let $u = x + 4$. Then $du = dx$ and hence

$$\int \frac{3x+1}{\sqrt{x^2+8x}} \, dx = \int \frac{3u-11}{\sqrt{u^2-16}} \, du = 3\sqrt{u^2-16} - 11\ln\left|u + \sqrt{u^2-16}\right| + C$$

$$= 3\sqrt{x^2+8x} - 11\ln\left|x + 4 + \sqrt{x^2+8x}\right| + C.$$

60. $\int \frac{6x^2 + 5x - 2}{x^3 + x^2 - 2x} \, dx = \int \left(\frac{1}{x} + \frac{2}{x+2} + \frac{3}{x-1} \right) \, dx = \ln|x| + 2\ln|x+2| + 3\ln|x-1| + C.$

61. Let $u = 1 + x^2$. Then $du = 2x\,dx$ and hence

$$\int \frac{x^5}{\sqrt{1+x^2}} \, dx = \frac{1}{2} \int \frac{(u-1)^2}{\sqrt{u}} \, du = \frac{1}{2} \int \left(u^{3/2} - 2u^{1/2} + u^{-1/2} \right) \, du$$

$$= \frac{1}{2} \left(\frac{2}{5} u^{5/2} - \frac{4}{3} u^{3/2} + 2u^{1/2} \right) + C = \frac{1}{5} (1+x^2)^{5/2} - \frac{2}{3} (1+x^2)^{3/2} + (1+x^2)^{1/2} + C.$$

62. Let $x = 5\sec\theta$. Then $dx = 5\sec\theta\tan\theta \, d\theta$ and hence

$$\int \frac{1}{x^2\sqrt{x^2-25}} \, dx = \int \frac{1}{(25\sec\theta)(5\tan\theta)} 5\sec\theta\tan\theta \, d\theta = \frac{1}{25} \int \cos\theta \, d\theta$$

$$= \frac{1}{25} \sin\theta + C = \frac{\sqrt{x^2-25}}{25x} + C.$$

63. Let $u = x^2 + 2$. Then $du = 2x\,dx$ and hence

$$\int \frac{x^5}{(x^2+2)^2} \, dx = \frac{1}{2} \int \frac{(u-2)^2}{u^2} \, du = \frac{1}{2} \int \left(1 - \frac{4}{u} + \frac{4}{u^2} \right) \, du$$

$$= \frac{1}{2} \left(u - 4\ln|u| - \frac{4}{u} \right) + C = \frac{1}{2} \left(x^2 - 4\ln(x^2+2) - \frac{4}{x^2+2} \right) + C.$$

64. $\displaystyle\int \sqrt{1-\cos x}\,dx = \int \frac{\sin x}{\sqrt{1+\cos x}}\,dx = -2\sqrt{1+\cos x}+C.$

65. Let $2x = 5\sec\theta$. Then $2dx = 5\sec\theta\tan\theta\,d\theta$ and hence

$$\int \frac{1}{\sqrt{4x^2-25}}\,dx = \int \frac{1}{5\tan\theta}\,\frac{5}{2}\,\sec\theta\tan\theta\,d\theta = \frac{1}{2}\int \sec\theta\,d\theta$$

$$= \frac{1}{2}\,\ln|\sec\theta+\tan\theta|+C = \frac{1}{2}\,\ln\left|2x+\sqrt{4x^2-25}\right|+C.$$

66. Let $u = x+\pi/2$. Then $du = dx$ and hence, by Exercise 64,

$$\int \sqrt{1+\sin x}\,dx = \int \sqrt{1-\cos u}\,du = -2\sqrt{1+\cos u}+C = -2\sqrt{1-\sin x}+C.$$

67. Let $u = (\ln x)^2$, $dv = dx$. Then $du = 2(\ln x)/x$, $v = x$ and hence

$$\int_1^e \ln^2 x\,dx = \left(x(\ln x)^2 - 2\int \ln x\,dx\right)\Big|_1^e = [x(\ln x)^2 - 2(x\ln x - x)]\big|_1^e = e-2.$$

68. $\displaystyle\int \frac{7\,dx}{13+x^2} = \frac{7}{\sqrt{13}}\,\mathrm{Tan}^{-1}\frac{x}{\sqrt{13}}+C$ (Formula 62)

69. $\displaystyle\int \frac{3\,dx}{x^2-5} = \frac{3}{2\sqrt{5}}\,\ln\left|\frac{x-\sqrt{5}}{x+\sqrt{5}}\right|+C = \frac{3}{2\sqrt{5}}\,\ln\left(\frac{x-\sqrt{5}}{x+\sqrt{5}}\right)+C$ since $x > \sqrt{5}$.   (Formula 39)

70. $\displaystyle\int \frac{\pi\,dx}{x\sqrt{36-x^2}} = -\frac{\pi}{6}\,\ln\left|\frac{6+\sqrt{36-x^2}}{x}\right|+C$ (Formula 41)

71. Let $u^2 = 9-\pi x$. Then $2u\,du = -\pi\,dx$ and hence

$$\int \frac{dx}{x\sqrt{9-\pi x}} = -\frac{1}{\pi}\int \frac{2u\,du}{\frac{9-u^2}{\pi}\,u} = -2\int \frac{1}{9-u^2}\,du = -\frac{1}{3}\int\left(\frac{1}{3-u}+\frac{1}{3+u}\right)du$$

$$= -\frac{1}{3}(-\ln|3-u|+\ln|3+u|)+C = -\frac{1}{3}\left(-\ln|3-\sqrt{9-\pi x}|+\ln|3+\sqrt{9-\pi x}|\right)+C.$$

72. Let $u = 7-3x$. Then $du = -3dx$ and hence

$$\int \frac{4dx}{(7-3x)^2} = -\frac{4}{3}\int \frac{du}{u^2} = -\frac{4}{3}\left(-\frac{1}{u}\right)+C = \frac{4}{3}\,\frac{1}{7-3x}+C.$$

73. $\displaystyle\int \frac{5}{x^2(7-2x)}\,dx = \frac{10}{49}\int \frac{1}{x}\,dx + \frac{5}{7}\int \frac{1}{x^2}\,dx + \frac{20}{49}\int \frac{1}{7-2x}\,dx$

$$= \frac{10}{49}\,\ln|x| - \frac{5}{7}\,\frac{1}{x} - \frac{10}{49}\,\ln|7-2x|+C$$

74. $\displaystyle\int \frac{1}{x(\pi+4x)^2}\,dx = \frac{1}{\pi^2}\int \frac{1}{x}\,dx - \frac{4}{\pi^2}\int \frac{1}{\pi+4x}\,dx - \frac{4}{\pi}\int \frac{1}{(\pi+4x)^2}\,dx$

$$= \frac{1}{\pi^2}\,\ln|x| - \frac{1}{\pi^2}\,\ln|\pi+4x| + \frac{1}{\pi}\,\frac{1}{\pi+4x}+C$$

755

75. Let $u = 9 + 2x$. Then $du = 2dx$ and hence

$$\int \frac{3x\,dx}{\sqrt{9+2x}} = \frac{3}{2}\int \frac{\frac{1}{2}(u-9)}{\sqrt{u}}\,du = \frac{3}{4}\int (u^{1/2} - 9u^{-1/2})\,du$$

$$= \frac{3}{4}\left(\frac{2}{3}u^{3/2} - 18u^{1/2}\right) + C = \frac{1}{2}(9+2x)^{3/2} - \frac{27}{2}(9+2x)^{1/2} + C.$$

76. $\displaystyle\int \frac{4x^2}{\sqrt{49-x^2}}\,dx = -2x\sqrt{49-x^2} + 98\,\text{Sin}^{-1}\frac{x}{7} + C$ (Formula 59)

77. $\displaystyle\int \frac{dx}{x^2\sqrt{8-x^2}}\,dx = -\frac{\sqrt{8-x^2}}{8x} + C$ (Formula 60)

78. Let $u = x - 5/2$. Then $du = dx$ and hence

$$\int \frac{dx}{\sqrt{5x-x^2}} = \int \frac{du}{\sqrt{\frac{25}{4}-u^2}} = \text{Sin}^{-1}\frac{2}{5}u + C = \text{Sin}^{-1}\frac{2x-5}{5} + C.$$

79. Let $u = \tan(x/2)$. Then

$$\cos x = \frac{1-u^2}{1+u^2}, \; du = \frac{1}{2}\sec^2\frac{x}{2}\,dx = \frac{1+u^2}{2}\,dx$$

and hence

$$\int \frac{dx}{9+4\cos x} = \int \frac{\frac{2}{1+u^2}\,du}{9+4\frac{1-u^2}{1+u^2}} = \int \frac{2\,du}{13+5u^2} = \frac{2}{\sqrt{65}}\text{Tan}^{-1}\left(\sqrt{\frac{5}{13}}\tan\frac{x}{2}\right) + C.$$

80. Let $u = \text{Sin}^{-1}2x$, $dv = xdx$. Then $du = \dfrac{2}{\sqrt{1-4x^2}}\,dx$, $v = \dfrac{1}{2}x^2$, and hence

$$\int x\text{Sin}^{-1}2x\,dx = \frac{1}{2}x^2\text{Sin}^{-1}2x - \int \underbrace{\frac{x^2}{\sqrt{1-4x^2}}}_{w=2x}\,dx = \frac{1}{2}x^2\text{Sin}^{-1}2x - \frac{1}{8}\int \frac{w^2}{\sqrt{1-w^2}}\,dw$$

$$= \frac{1}{2}x^2\text{Sin}^{-1}2x - \frac{1}{8}\left(-\frac{w}{2}\sqrt{1-w^2} + \frac{1}{2}\text{Sin}^{-1}w\right) + C = \frac{1}{2}x^2\text{Sin}^{-1}2x + \frac{1}{8}x\sqrt{1-4x^2} - \frac{1}{16}\text{Sin}^{-1}2x + C.$$

81. Let $u = \ln x$, $dv = x^3\,dx$. Then $du = \dfrac{1}{x}\,dx$, $v = \dfrac{1}{4}x^4$ and hence

$$\int x^3\ln x\,dx = \frac{1}{4}x^4\ln x - \int x^3\,dx = \frac{1}{4}x^4\ln x - \frac{1}{16}x^4 + C.$$

82. $\text{Area} = \displaystyle\int_{-1}^{0} \frac{1}{\sqrt{x^2-6x+5}}\,dx = \int_{-1}^{0} \frac{1}{\sqrt{(x-3)^2-4}} = \ln\left|x-3+\sqrt{x^2-6x+5}\right|\,\Big|_{-1}^{0}$

$= \ln\left|-3+\sqrt{5}\right| - \ln\left|-4+\sqrt{12}\right|.$

83. The graphs intersect when $xe^{-x} = x/e \Longrightarrow x(e^{-x+1} - 1) = 0 \Longrightarrow x = 0,\ x = 1$. Hence

$$\text{Area} \ = \ \int_0^1 \left(\underbrace{xe^{-x}}_{u=x,\,dx=e^{-x}dx} - \frac{x}{e} \right) dx = \left(-xe^{-x} - e^{-x} - \frac{x^2}{2e} \right)\Big|_0^1 = 1 - \frac{5}{2e}.$$

84. Let dV be the differential volume of a shell of thickness dx revolved about the y-axis. Then $dV = 2\pi r h\, dx = 2\pi xy\, dx = 2\pi x^2 \sin(\pi x/3)\, dx$ and hence the volume of revolution is

$$V = 2\pi \int_0^3 x^2 \sin \frac{\pi x}{3}\ dx.$$

Now, let $u = x^2$, $dv = \sin(\pi x/3)\, dx$. Then $du = 2x\, dx$, $v = (-3/\pi)\cos(\pi x/3)$ and hence

$$\int x^2 \sin \frac{\pi x}{3}\ dx \ = \ -\frac{3x^2}{\pi}\cos\frac{\pi x}{3} + \frac{6}{\pi}\int \underbrace{x\cos\frac{\pi x}{3}}_{u=x,\,dv=\cos(\pi x/3)dx}\ dx$$

$$= \ -\frac{3x^2}{\pi}\cos\frac{\pi x}{3} + \frac{6}{\pi}\left(\frac{3x}{\pi}\sin\frac{\pi x}{3} - \frac{3}{\pi}\int \sin\frac{\pi x}{3}\ dx\right)$$

$$= \ -\frac{3x^2}{\pi}\cos\frac{\pi x}{3} + \frac{18x}{\pi^2}\sin\frac{\pi x}{3} + \frac{54}{\pi^3}\cos\frac{\pi x}{3}.$$

Hence

$$V = 2\pi\left(\frac{27}{\pi} - \frac{54}{\pi^3} - \frac{54}{\pi^3}\right) = 54 - \frac{216}{\pi^2}.$$

85. Let dV be the differential volume of a cross-section disk of thickness dx. Then

$$dV = \pi r^2\ dx = \pi y^2\ dx = \pi\frac{x+5}{x^2(x-5)}\ dx$$

and hence the volume of revolution is

$$V = \pi \int_6^7 \frac{x+5}{x^2(x-5)}\ dx = \pi \int_6^7 \left(\frac{-2/5}{x} - \frac{1}{x^2} + \frac{2/5}{x-5}\right)\ dx$$

$$= \pi\left(-\frac{2}{5}\ln x + \frac{1}{x} + \frac{2}{5}\ln(x-5)\right)\Big|_6^7 = \pi\left(-\frac{2}{5}\ln 7 + \frac{2}{5}\ln 2 + \frac{2}{5}\ln 6 - \frac{1}{42}\right) = \pi\left(-\frac{1}{42} + \frac{2}{5}\ln\frac{12}{7}\right).$$

86. $y = x^2 \Longrightarrow y' = 2x \Longrightarrow$

$$\text{Arc length} \ = \ \int_0^1 \underbrace{\sqrt{1+4x^2}}_{u=2x}\ dx = \frac{1}{2}\int_0^2 \sqrt{1+u^2}\ du$$

$$= \frac{1}{4}\left[u\sqrt{1+u^2} + \ln\left(u + \sqrt{u^2+1}\right)\right]\Big|_0^2 = \frac{1}{4}\left[2\sqrt{5} + \ln(2+\sqrt{5})\right].$$

87. $y = -\sqrt{1-x^2} \Longrightarrow y' = x(1-x^2)^{-1/2} \Longrightarrow$

$$\text{Arc length} \ = \ \int_0^{1/2} \sqrt{1 + \frac{x^2}{1-x^2}}\ dx = \int_0^{1/2} \frac{1}{\sqrt{1-x^2}}\ dx = \text{Sin}^{-1}x\Big|_0^{1/2} = \frac{\pi}{6}.$$

88. The center of mass is at the point

$$\bar{x} = \frac{\int_0^2 \frac{1}{x^2+1}\, x\pi\, dx}{\int_0^2 \frac{1}{x^2+1}\, \pi\, dx} = \frac{(1/2)\ln(x^2+1)\big|_0^2}{\mathrm{Tan}^{-1}x\big|_0^2} = \frac{\ln 5}{2\,\mathrm{Tan}^{-1}2}.$$

89. The center of mass is at the point

$$\bar{x} = \frac{\int_0^4 \frac{4}{(x+1)(5-x)}\, x(3)\, dx}{\int_0^4 \frac{4}{(x+1)(5-x)}\, (3)\, dx} = \frac{\int_0^4 \left(\frac{-1/6}{x+1} + \frac{5/6}{5-x}\right)\, dx}{\int_0^4 \left(\frac{1/6}{x+1} + \frac{1/6}{5-x}\right)\, dx} = \frac{(-\ln(x+1) - 5\ln(5-x))\big|_0^4}{(\ln(x+1) - \ln(5-x))\big|_0^4} = 2.$$

Chapter 11

l'Hôpital's Rule and Improper Integrals

11.1 Indeterminate Forms: l'Hôpital's Rule

1. $\displaystyle\lim_{x\to 1}\frac{1-x}{e^x-e}=\lim_{x\to 1}\frac{-1}{e^x}=-\frac{1}{e}$

2. $\displaystyle\lim_{x\to 0}\frac{\sin 5x}{x}=\lim_{x\to 0}\frac{5\cos 5x}{1}=5$

3. $\displaystyle\lim_{x\to 0}\frac{\sin x^2}{x}=\lim_{x\to 0}\frac{2x\cos x^2}{1}=0$

4. $\displaystyle\lim_{x\to 2}\frac{x^3-x^2-x-2}{x-2}=\lim_{x\to 2}\frac{3x^2-2x-1}{1}=7$

5. $\displaystyle\lim_{x\to 0}\frac{1-\cos x}{x^2}=\lim_{x\to 0}\frac{\sin x}{2x}=\lim_{x\to 0}\frac{\cos x}{2}=1/2$

6. $\displaystyle\lim_{x\to\pi/2}\frac{1-\sin x}{\cos x}=\lim_{x\to\pi/2}\frac{-\cos x}{-\sin x}=0$

7. $\displaystyle\lim_{x\to 1^-}\frac{\sqrt{1-x^2}}{x-1}=-\lim_{x\to 1^-}\sqrt{\frac{1+x}{1-x}}=-\infty$

8. $\displaystyle\lim_{x\to 0^+}\frac{1-\cos x}{x^3}=\lim_{x\to 0^+}\frac{\sin x}{3x^2}=\lim_{x\to 0^+}\frac{\cos x}{6x}=+\infty$

9. $\displaystyle\lim_{\theta\to 0}\frac{\tan\theta-\theta}{\theta-\sin\theta}=\lim_{\theta\to 0}\frac{\sec^2\theta-1}{1-\cos\theta}=\lim_{\theta\to 0}\frac{1-\cos^2\theta}{\cos^2\theta(1-\cos\theta)}=\lim_{\theta\to 0}\frac{1+\cos\theta}{\cos^2\theta}=2$

10. $\displaystyle\lim_{x\to\infty}\frac{x^3+2x+1}{4x^3+1}=\lim_{x\to\infty}\frac{1+\frac{2}{x^2}+\frac{1}{x^3}}{4+\frac{1}{x^3}}=\frac{1}{4}$

11. $\displaystyle\lim_{x\to 1}\frac{\ln x}{x-1}=\lim_{x\to 1}\frac{1/x}{1}=1$

12. $\displaystyle\lim_{x\to 1}\frac{\sqrt{x}-\sqrt[4]{x}}{x-1}=\lim_{x\to 1}\frac{\frac{1}{2\sqrt{x}}-\frac{1}{4\sqrt[4]{x^3}}}{1}=\frac{1}{4}$

13. $\displaystyle\lim_{x\to 0^+}\frac{1+\cos\sqrt{x}}{\sin x}=+\infty$

14. $\displaystyle\lim_{x\to 0}\frac{\ln(1+x)}{1-e^x}=\lim_{x\to 0}\frac{\frac{1}{1+x}}{-e^x}=-1$

15. $\displaystyle\lim_{x\to 0^+}\frac{x^2}{x-\sin x}=\lim_{x\to 0^+}\frac{2x}{1-\cos x}=\lim_{x\to 0^+}\frac{2}{\sin x}=+\infty$

16. $\displaystyle\lim_{x\to\infty}\frac{\sqrt{1+x^2}}{x}=\lim_{x\to\infty}\sqrt{\frac{1}{x^2}+1}=1$

17. $\displaystyle\lim_{x\to\infty}\frac{\sqrt[3]{x^3+3}}{x^2}=\lim_{x\to\infty}\sqrt[3]{\frac{1}{x^3}+\frac{3}{x^6}}=0$

18. $\displaystyle\lim_{x\to 0}\frac{\sin x-x\cos x}{x}=\lim_{x\to 0}\frac{x\sin x}{1}=0$

19. $\displaystyle\lim_{x\to 1}\frac{e^{x^2}-e^x}{x^2-1}=\lim_{x\to 1}\frac{2xe^{x^2}-e^x}{2x}=\frac{e}{2}$

20. $\displaystyle\lim_{x\to 0^+}\frac{\cos x-x}{\sqrt{x}}=+\infty$

21. $\displaystyle\lim_{x\to 0}\frac{\sin x}{\sqrt[3]{x}}=\lim_{x\to 0}\frac{\cos x}{\frac{1}{3}x^{-2/3}}=3\lim_{x\to 0}x^{2/3}\cos x=0$

22. $\displaystyle\lim_{x\to\pi^+}\frac{\sin x}{\sqrt{x-\pi}}=\lim_{x\to\pi^+}\frac{\cos x}{1/(2\sqrt{x-\pi})}=\lim_{x\to\pi^+}(2\cos x)\sqrt{x-\pi}=0$

23. $\displaystyle\lim_{x\to 0}\frac{e^{x^2}-1}{x^2}=\lim_{x\to 0}\frac{2xe^{x^2}}{2x}=1$

24. $\displaystyle\lim_{x\to 0^+}\frac{\cos x-x}{x-\tan x}=-\infty$

25. $\displaystyle\lim_{x\to\infty}\frac{x-\sin x}{x\sin x}=\lim_{x\to\infty}\frac{1-\frac{\sin x}{x}}{\sin x}$ does not exist

26. $\displaystyle\lim_{x\to 1^-}\frac{x^{5/2}-1+\sqrt{1-x}}{\sqrt{1-x^2}}=\lim_{x\to 1^-}\frac{\frac{5}{2}x^{3/2}-\frac{1}{2\sqrt{1-x}}}{\frac{-x}{\sqrt{1-x^2}}}=\lim_{x\to 1^-}\frac{\frac{5}{2}x^{3/2}\sqrt{1-x^2}-\frac{1}{2}\sqrt{1+x}}{-x}=\sqrt{2}/2$

27. $\displaystyle\lim_{x\to\infty}\frac{9-x^3}{xe^{\pi x}}=\lim_{x\to\infty}\frac{-3x^2}{e^{\pi x}(\pi x+1)}=\lim_{x\to\infty}\frac{-6x}{e^{\pi x}(\pi^2 x+\pi+1)}=\lim_{x\to\infty}\frac{-6}{e^{\pi x}(\pi^2)+\pi e^{\pi x}(\pi^2 x+\pi+1)}=0$

28. $\displaystyle\lim_{x\to\infty}\frac{e^{2x}}{1+x^2}=\lim_{x\to\infty}\frac{2e^{2x}}{2x}=\lim_{x\to\infty}\frac{2e^{2x}}{1}=+\infty$

29. $\displaystyle\lim_{x\to 0}\frac{\operatorname{Sin}^{-1}x}{x}=\lim_{x\to 0}\frac{\frac{1}{\sqrt{1-x^2}}}{1}=1$

30. $\displaystyle\lim_{x\to\pi/2}\frac{\cos 3x}{\cos x}=\lim_{x\to\pi/2}\frac{-3\sin 3x}{-\sin x}=-3$

31. $\lim\limits_{x\to\pi/2} \dfrac{\tan(x/2)-1}{x-\pi/2} = \lim\limits_{x\to\pi/2} \dfrac{\frac{1}{2}\sec^2(x/2)}{1} = 1$

32. $\lim\limits_{x\to\infty} \dfrac{\tan(1/x)}{1/x} = \lim\limits_{x\to\infty} \dfrac{(-1/x^2)\sec^2(1/x)}{-1/x^2} = \lim\limits_{x\to\infty} \sec^2(1/x) = 1$

33. $\lim\limits_{x\to\infty} \dfrac{\sqrt{x}-3x^2}{x(6-x)} = \lim\limits_{x\to\infty} \dfrac{1/(x\sqrt{x})-3}{(6/x)-1} = 3$

34. $\lim\limits_{x\to\infty} \dfrac{(ax+b)^3}{(x+c)^3} = \lim\limits_{x\to\infty} \dfrac{(a+b/x)^3}{(1+c/x)^3} = a^3$

35. $\lim\limits_{x\to\infty} \dfrac{\sqrt{x}-\sqrt{a}}{\sqrt{x}+\sqrt{a}} = \lim\limits_{x\to\infty} \dfrac{1-\sqrt{a/x}}{1+\sqrt{a/x}} = 1$

36. $\lim\limits_{x\to\infty} \dfrac{x^3-7x^2+6x-5}{(x-3)(5-x^2)} = \lim\limits_{x\to\infty} \dfrac{1-2/x+6/x^2-5/x^3}{(1-3/x)(5/x^2-1)} = -1$

37. $\lim\limits_{x\to\infty} \dfrac{3x^2-4}{4x^2+3} = \lim\limits_{x\to\infty} \dfrac{3-4/x^2}{4+3/x^2} = 3/4$

38. $\lim\limits_{x\to\infty} \dfrac{x^2(x-1)(x+3)}{(x^3-6)(2x^2+x+1)} = \lim\limits_{x\to\infty} \dfrac{(1-1/x)(1/x+3/x^2)}{(1-6/x^3)(2+1/x+1/x^2)} = 0$

39. $\lim\limits_{x\to\infty} \dfrac{\sin x}{x^2+\pi} = 0$ since $0 \le \left|\dfrac{\sin x}{x^2+\pi}\right| \le \dfrac{1}{x^2}$ and $\lim\limits_{x\to\infty} \dfrac{1}{x^2} = 0$

40. $f(x) = \dfrac{x+4}{x-2} = 1 + \dfrac{6}{x-2} \implies f'(x) = -6/(x-2)^2$ and $f''(x) = 12/(x-2)^3$. It follows that the graph of $y = f(x)$ is concave down for $x < 2$, concave up for $x > 2$, and has no relative extrema. Since $\lim_{x\to 2^+} f(x) = +\infty$ and $\lim_{x\to 2^-} f(x) = -\infty$, the line $x = 2$ is a vertical asymptote. Since

$$\lim\limits_{x\to\infty} f(x) = \lim\limits_{x\to\infty} \dfrac{1+4/x}{1-2/x} = 1,$$

the line $y = 1$ is a horizontal asymptote.

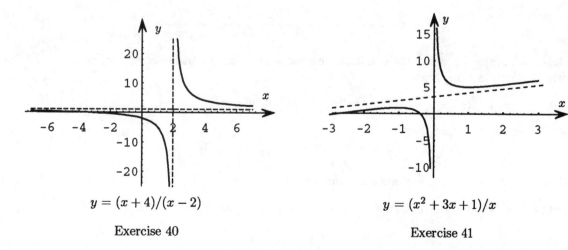

$y = (x+4)/(x-2)$

Exercise 40

$y = (x^2+3x+1)/x$

Exercise 41

41. $f(x) = (x^2 + 3x + 1)/x = x + 3 + 1/x \implies f'(x) = 1 - 1/x^2$ and $f''(x) = 2/x^3$. It follows that the graph of $y = f(x)$ has a relative minimum at $(1, 5)$, a relative maximum at $(-1, 1)$, and is concave down for $x < 0$ and concave up for $x > 0$. Since $\lim_{x \to 0^+} f(x) = +\infty$ and $\lim_{x \to 0^-} f(x) = -\infty$, the line $x = 0$ is a vertical asymptote. Since $\lim_{x \to \pm\infty} f(x) = \pm\infty$, the graph has no horizontal asymptote. However,

$$\lim_{x \to \pm\infty} (f(x) - x - 3) = \lim_{x \to \pm\infty} \frac{1}{x} = 0,$$

which means that the graph is asymptotic to the line $y = x + 3$.

42. $f(x) = \frac{x}{1 - e^x} \implies f'(x) = \frac{1 - e^x(1 - x)}{(1 - e^x)^2} = \frac{e^{-x} - (1 - x)}{e^{-x}(1 - e^x)^2}$. Now, $e^{-x} > 1 - x$ since $y = 1 - x$ is the tangent line to the graph of $y = e^{-x}$ at $x = 0$. In particular, $f' = 0 \iff x = 0$. But $x = 0$ is not in the domain of $f(x)$ and hence is not a critical number for the function $f(x)$. Since $\lim_{x \to 0} f(x) = \lim_{x \to 0} 1/(-e^x) = -1$, it follows that the graph of $y = f(x)$ has a singular point at $(0, -1)$. Moreover, $\lim_{x \to -\infty} x/(1 - e^x) = -\infty/1 = -\infty$ and the graph is increasing since $f'(x) > 0$ for all $x \neq 0$. Finally, since $\lim_{x \to \infty} x/(1 - e^x) = \lim_{x \to \infty} 1/(-e^x) = 0$, the line $y = 0$ is a horizontal asymptote for the graph of $y = f(x)$.

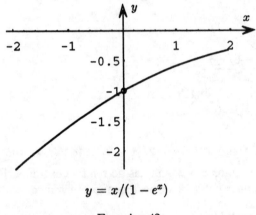

$y = x/(1 - e^x)$

Exercise 42

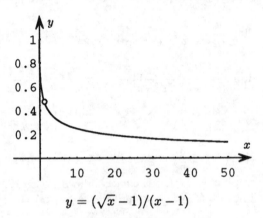

$y = (\sqrt{x} - 1)/(x - 1)$

Exercise 43

43. To obtain the graph of $y = f(x)$, first observe that if $x \neq 1$,

$$f(x) = \frac{\sqrt{x} - 1}{x - 1} = \frac{\sqrt{x} - 1}{(\sqrt{x} - 1)(\sqrt{x} + 1)} = \frac{1}{\sqrt{x} + 1}.$$

Hence the graph lies above the y-axis and is decreasing. For the behavior of $f(x)$ at $x = 1$ we find, using l'Hôpital's Rule, that

$$\lim_{x \to 1} \frac{\sqrt{x} - 1}{x - 1} = \lim_{x \to 1} \frac{\frac{1}{2}\sqrt{x}}{1} = \frac{1}{2},$$

while as $x \to \infty$ we find that

$$\lim_{x \to \infty} \frac{\sqrt{x} - 1}{x - 1} = \lim_{x \to \infty} \frac{\frac{1}{2}\sqrt{x}}{1} = 0.$$

Hence the line $y = 0$ is a horizontal asymptote.

44. Define $F(x)$ and $G(x)$ as in the text. Then, by Cauchy's Mean Value Theorem,

$$\frac{F(x)}{G(x)} = \frac{F'(c_x)}{G'(c_x)}.$$

for some c_x such that $x < c_x < a$. It follows, as in the text, that

$$\lim_{x \to a^-} \frac{f(x)}{g(x)} = \lim_{x \to a^-} \frac{F'(x)}{G'(x)} = \lim_{x \to a^-} \frac{f'(x)}{g'(x)}.$$

45. If the hypotheses of l'Hôpital's Rule are satisfied, then

$$\lim_{x \to a^-} \frac{f(x)}{g(x)} = \lim_{x \to a^-} \frac{f'(x)}{g'(x)} = \lim_{x \to a^+} \frac{f'(x)}{g'(x)} = \lim_{x \to a^+} \frac{f(x)}{g(x)}.$$

Therefore

$$\lim_{x \to a} \frac{f(x)}{g(x)} = \lim_{x \to a} \frac{f'(x)}{g'(x)}.$$

46. (a) This is clear since $x \to \infty \iff t \to 0^+$.

   (b) Since $\lim_{x \to \infty} f(x) = \lim_{x \to \infty} g(x) = 0$, $\lim_{t \to 0^+} \frac{f(1/t)}{g(1/t)}$ is indeterminate of the form $0/0$. Hence, applying l'Hôpital's Rule, it follows that

$$\lim_{t \to 0^+} \frac{f(1/t)}{g(1/t)} = \lim_{t \to 0^+} \frac{(-1/t^2)f'(1/t)}{(-1/t^2)g'(1/t)} = \lim_{t \to 0^+} \frac{f'(1/t)}{g'(1/t)} = \lim_{x \to \infty} \frac{f'(x)}{g'(x)}.$$

   (c) $\lim_{x \to \infty} \frac{f(x)}{g(x)} = \lim_{t \to 0^+} \frac{f(1/t)}{g(1/t)} = \lim_{x \to \infty} \frac{f'(x)}{g'(x)}.$

47. (a) Graph $(e^x - 1)/(x^3 - 2x)$ using the range $[-1, 1] \times [-2, 0]$.

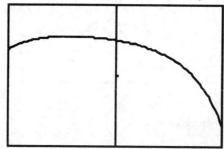

By magnifying the graph at the y-intercept we find that $\lim_{x \to 0}(f(x)/g(x)) = -1/2$. After magnifying the graph six times with factors of 10 it becomes oscillatory.

   (b) Graph $e^x/(3x^2 - 2)$ using the range $[-1, 1] \times [-1, 2]$.

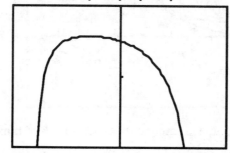

By magnifying the graph at the y-intercept we find that $\lim_{x \to 0}(f'(x)/g'(x)) = -1/2$.

   (c) The method in part (b) is more accurate since the graph doesn't become oscillatory as we magnify it.

48. Graph $f(x) = 2x(x-1)$ and $g(x) = 2x^2(x-1)^{2/3}$ using the range $[-0.5, 1.5] \times [-1, 2]$.

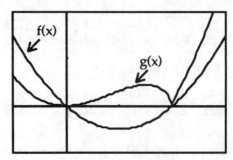

At $x = 0$, $f'(0) \neq 0$ and $g'(0) = 0$. Hence, by L'Hôpital's Rule, $\lim\limits_{x \to 0} \big(f(x)/g(x)\big)$ does not exist. At $x = 1$, $f'(1) \neq 0$, $\lim\limits_{x \to 1-} g'(x) = -\infty$ and $\lim\limits_{x \to 1+} g'(x) = +\infty$. Therefore $\lim\limits_{x \to 1-} \big(f'(x)/g'(x)\big) = \lim\limits_{x \to 1+} \big(f'(x)/g'(x)\big) = 0$. Thus $\lim\limits_{x \to 1-} \big(f(x)/g(x)\big) = 0$ and therefore $f(x)/g(x)$ has removable discontinuity at $x = 1$.

49. Graph $f(x) = x \sin \pi x$ and $g(x) = 2x(x-1)$ using the range $[0, 2] \times [-2, 1]$.

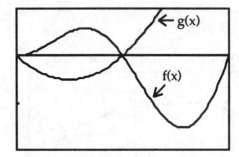

From the graph we see that $f'(1) \neq 0$ and $g'(1) \neq 0$. L'Hôpital's Rule implies therefore that both $f(x)/g(x)$ and $g(x)/f(x)$ are continuous at $x = 1$.

50. Graph $f(x) = \tan(\pi x/2)$ and $g(x) = \sec(\pi x/2)$ using the range $[0, 2] \times [-5, 5]$.

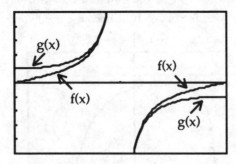

From the graph it is apparent that $\lim\limits_{x \to 1-} \tan(\pi x/2) = +\infty$ and $\lim\limits_{x \to 1+} \tan(\pi x/2) = -\infty$. The same holds for $g(x)$ as well. But $f(x)/g(x) = \sin(\pi x/2)$ so $f(x)/g(x)$ is continuous at $x = 1$, i.e., $f(1)/g(1) = 1$. The same holds for $g(x)/f(x)$ also.

51. Graph $f(x) = \ln(2x + 1)$ and $g(x) = x \arcsin x$ using the range $[-1, 1] \times [-2, 2]$.

764

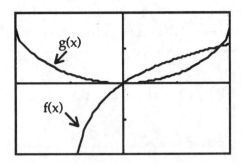

From the graph we see that $f'(0) \neq 0$, but $g'(0) = 0$ so L'Hôpital's Rule implies that $f(x)/g(x)$ is discontinuous at $x = 0$ while $g(x)/f(x)$ is continuous at $x = 0$.

52. Graph $f(x) = -3 + 6/x$ and $g(x) = -x^2/4 + 2x - 3$ using the range $[1, 3] \times [-2, 2]$.

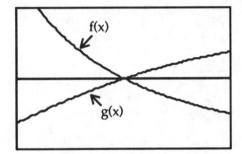

From the graph we see that $f'(2) \neq 0$ and $g'(2) \neq 0$. Hence L'Hôpital's Rule implies that both $f(x)/g(x)$ and $g(x)/f(x)$ are continuous at $x = 2$.

53. (a) If $\lim\limits_{x \to +\infty} (f'(x)/g'(x)) = 1$ then as $x \to +\infty$ the tangent lines to $f(x)$ and $g(x)$ become parallel.

(b) Let $f(x) = x^2$ and $g(x) = x^2 - 100x$. Graph these using the range $[500, 1050] \times [25000, 1100000]$.

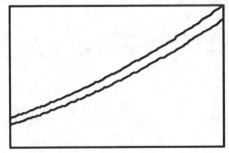

Note that the graphs appear to be nearly parallel.

(c) If $f(x) > g(x)$ and $\lim\limits_{x \to +\infty} (f(x) - g(x)) = +\infty$ then the distance between the graphs must increase as $x \to +\infty$, *i.e.*, $\lim\limits_{x \to +\infty} |f(x) - g(x)| = +\infty$. The functions in part (b) also have this property since $f(x) - g(x) = 100x$ and $\lim\limits_{x \to +\infty} 100x = +\infty$.

11.2 Other Indeterminate Forms

1. $\lim\limits_{x \to 0^+} \left(\dfrac{1}{x} - \dfrac{1}{x^2} \right) = \lim\limits_{x \to 0^+} \dfrac{x(x-1)}{x^2} = \lim\limits_{x \to 0^+} \left(1 - \dfrac{1}{x} \right) = -\infty$

2. $\displaystyle\lim_{x\to 0^+}(\csc x - \cot x) = \lim_{x\to 0^+}\frac{1-\cos x}{\sin x} = \lim_{x\to 0^+}\frac{\sin x}{\cos x} = 0$

3. $\displaystyle\lim_{x\to 0} x\cot x = \lim_{x\to 0}\frac{x\cos x}{\sin x} = \lim_{x\to 0}\frac{x(-\sin x)+\cos x}{\cos x} = 1$

4. $\displaystyle\lim_{x\to 0} x\sin(1/x) = 0$ since $0 \le |x\sin(1/x)| \le |x|$ and $\displaystyle\lim_{x\to 0}|x| = 0$

5. $\displaystyle\lim_{x\to 0^+} x\ln x = \lim_{x\to 0^+}\frac{\ln x}{1/x} = \lim_{x\to 0^+}\frac{1/x}{-1/x^2} = \lim_{x\to 0^+}(-x) = 0$

6. $\displaystyle\lim_{x\to 0}\left(\frac{1}{x}-\csc x\right) = \lim_{x\to 0}\frac{\sin x - x}{x\sin x} = \lim_{x\to 0}\frac{\cos x - 1}{x\cos x + \sin x} = \lim_{x\to 0}\frac{-\sin x}{-x\sin x + 2\cos x} = 0$

7. $\displaystyle\lim_{x\to 1^+}\left(\frac{1}{x-1}-\frac{x}{\sqrt{x-1}}\right) = \lim_{x\to 1^+}\left(\frac{1-x\sqrt{x-1}}{x-1}\right) = +\infty$

8. $\displaystyle\lim_{x\to\infty} e^{-x}\ln x = \lim_{x\to\infty}\frac{\ln x}{e^x} = \lim_{x\to\infty}\frac{1/x}{e^x} = 0$

9. $\displaystyle\lim_{x\to 0^+}(1/x)\mathrm{Tan}^{-1}x = \lim_{x\to 0^+}\frac{1/(1+x^2)}{1} = 1$

10. $\displaystyle\lim_{x\to\infty}\left(\frac{1}{x}\right)\ln\frac{1}{x} = -\lim_{x\to\infty}\frac{\ln x}{x} = -\lim_{x\to\infty}\frac{1/x}{1} = 0$

11. $\displaystyle\lim_{x\to 0^+} x\ln x = 0$ by Exercise 5 $\Longrightarrow \displaystyle\lim_{x\to 0^+} x^x = \lim_{x\to 0^+} e^{x\ln x} = e^0 = 1$

12. $\displaystyle\lim_{x\to 0^+}(\cos x)\ln x = -\infty \Longrightarrow \displaystyle\lim_{x\to 0^+} x^{\cos x} = \lim_{x\to 0^+} e^{(\cos x)\ln x} = 0$

13. $\displaystyle\lim_{x\to\infty}(1+x^2)e^{-2x} = \lim_{x\to\infty}\frac{1+x^2}{e^{2x}} = \lim_{x\to\infty}\frac{2x}{2e^{2x}} = \lim_{x\to\infty}\frac{1}{2e^{2x}} = 0$

14. $\displaystyle\lim_{x\to 0^+}\sqrt{x^x} = \sqrt{\lim_{x\to 0^+} x^x} = 1$ by Exercise 11

15. $\displaystyle\lim_{x\to 0^+}(\tan x)\ln(\tan x) = \lim_{x\to 0^+}\frac{\ln(\tan x)}{\cot x} = \lim_{x\to 0^+}\frac{\sec^2 x/\tan x}{-\csc^2 x} = \lim_{x\to 0^+}\frac{\sin x}{\cos x} = 0$

16. $\displaystyle\lim_{x\to 0^+}(1+2x)^{\cot x} = e^2$ since

$$\lim_{x\to 0^+}(\cot x)\ln(1+2x) = \lim_{x\to 0^+}\frac{\ln(1+2x)}{\tan x} = \lim_{x\to 0^+}\frac{2/(1+2x)}{\sec^2 x} = 2.$$

17. $\displaystyle\lim_{x\to\infty}\left(1+\frac{2}{x}\right)^{3x} = e^6$ since

$$\lim_{x\to\infty} 3x\ln\left(1+\frac{2}{x}\right) = 3\lim_{x\to\infty}\frac{\ln(1+2/x)}{1/x} = 3\lim_{x\to\infty}\frac{\frac{1}{1+2/x}(-2/x^2)}{-1/x^2} = 3\lim_{x\to\infty}\frac{2}{1+2/x} = 6.$$

18. $\displaystyle\lim_{x\to\infty}\left(1-\frac{2}{x}\right)^{5x} = e^{-10}$ since

$$\lim_{x\to\infty} 5x\ln(1-2/x) = 5\lim_{x\to\infty}\frac{\ln(1-2/x)}{1/x} = 5\lim_{x\to\infty}\frac{\frac{1}{1-2/x}(2/x^2)}{-1/x^2} = -10.$$

19. $\displaystyle\lim_{x\to 0^+}\cot x\,\mathrm{Tan}^{-1}x = \lim_{x\to 0^+}\frac{\mathrm{Tan}^{-1}x}{\tan x} = \lim_{x\to 0^+}\frac{1/(1+x^2)}{\sec^2 x} = 1$

20. $\displaystyle\lim_{x\to 0^+}\sqrt{x}\ln x = \lim_{x\to 0^+}\frac{\ln x}{x^{-1/2}} = \lim_{x\to 0^+}\frac{1/x}{(-1/2)x^{-3/2}} = -2\lim_{x\to 0^+}\sqrt{x} = 0$

21. $\displaystyle\lim_{x\to 0}(2-e^x)^{1/x} = e^{-1}$ since

$$\lim_{x\to 0}\frac{1}{x}\ln(2-e^x) = \lim_{x\to 0}\frac{\frac{1}{2-e^x}(-e^x)}{1} = -1.$$

22. $\displaystyle\lim_{x\to\infty}[\ln(x^2+3)-\ln x] = \lim_{x\to\infty}\ln\frac{x^2+3}{x} = \ln\lim_{x\to\infty}\left(x+\frac{3}{x}\right) = \infty$

23. $\displaystyle\lim_{x\to\infty}\left[\ln\sqrt{4x+2}-\ln\sqrt{x+3}\,\right] = \lim_{x\to\infty}\ln\sqrt{\frac{4+2/x}{1+3/x}} = \ln 2$

24. $\displaystyle\lim_{x\to 0}(e^x+x)^{1/x} = e^2$ since

$$\lim_{x\to 0}\frac{1}{x}\ln(e^x+x) = \lim_{x\to 0}\frac{\frac{1}{e^x+x}(e^x+1)}{1} = 2.$$

25. $\displaystyle\lim_{x\to 0}(e^x+x^2)^{1/x^2} = \infty$ since

$$\lim_{x\to 0}\frac{1}{x^2}\ln(e^x+x^2) = \lim_{x\to 0}\frac{\frac{1}{e^x+x^2}(e^x+2x)}{2x} = \infty.$$

26. $\displaystyle\lim_{x\to\infty}\left(\frac{x+1}{x}\right)^{-2x} = e^{-2}$ since

$$\lim_{x\to\infty}(-2x)\ln\frac{x+1}{x} = -2\lim_{x\to\infty}\frac{\ln(1+1/x)}{1/x} = -2\lim_{x\to\infty}\frac{\frac{1}{1+1/x}(-1/x^2)}{-1/x^2} = -2\lim_{x\to\infty}\frac{1}{1+1/x} = -2.$$

27. $\displaystyle\lim_{x\to\infty}\left[\ln\sqrt{x^6+3x^2}-\ln(2x^3)\right] = \ln\lim_{x\to\infty}\sqrt{\frac{1+3/x^4}{4}} = -\ln 2$

28. $\displaystyle\lim_{x\to 0^+}x^2\cot x^2 = \lim_{x\to 0^+}\frac{x^2}{\tan x^2} = \lim_{x\to 0^+}\frac{2x}{2x\sec^2 x^2} = \lim_{x\to 0^+}\cos^2 x^2 = 1$

29. $\displaystyle\lim_{x\to\infty}\left(\frac{x}{x+1}\right)^{x+1} = \lim_{x\to\infty}(1+1/x)^{-(x+1)} = e^{-1}$ since

$$-\lim_{x\to\infty}\frac{\ln(1+1/x)}{(x+1)^{-1}} = -\lim_{x\to\infty}\frac{\frac{1}{1+1/x}(-1/x^2)}{-(x+1)^{-2}} = -\lim_{x\to\infty}(1+1/x) = -1.$$

30. $\displaystyle\lim_{x\to 0^+}x^{\sqrt{x}} = e^0 = 1$ since

$$\lim_{x\to 0^+}\sqrt{x}\ln x = \lim_{x\to 0^+}\frac{\ln x}{x^{-1/2}} = \lim_{x\to 0^+}\frac{1/x}{(-1/2)x^{-3/2}} = \lim_{x\to 0^+}(-2\sqrt{x}) = 0.$$

31. $f(x) = x^2 e^x \implies f'(x) = xe^x(x+2)$, $f''(x) = e^x(x^2 + 4x + 2)$. Since $f' = 0 \iff x = 0, -2$, and $f''(0) > 0$ while $f''(-2) < 0$, the graph of $y = f(x)$ has a relative minimum at $(0,0)$ and a relative maximum at $(-2, 4/e^2)$. Moreover,

$$\lim_{x \to -\infty} x^2 e^x = \lim_{x \to -\infty} \frac{x^2}{e^{-x}} = \lim_{x \to -\infty} \frac{2x}{-e^{-x}} = \lim_{x \to -\infty} \frac{2}{e^{-x}} = \lim_{x \to -\infty} 2e^x = 0$$

and $\lim_{x \to \infty} x^2 e^x = \infty$. Hence the graph is asymptotic to the x-axis as $x \to -\infty$, but increases without bound as $x \to \infty$.

32. $f(x) = xe^{-2x} \implies f'(x) = e^{-2x}(1 - 2x)$, $f''(x) = -4e^{-2x}(1 - x)$. Since $f' = 0 \iff x = 1/2$ and $f''(1/2) < 0$, the graph of $y = f(x)$ has a relative maximum at $(1/2, 1/2e)$. Moreover,

$$\lim_{x \to -\infty} xe^{-2x} = -\infty$$

and

$$\lim_{x \to \infty} xe^{-2x} = \lim_{x \to \infty} \frac{x}{e^{2x}} = \lim_{x \to \infty} \frac{1}{2e^{2x}} = 0.$$

Hence the graph is asymptotic to the x-axis as x increases without bound.

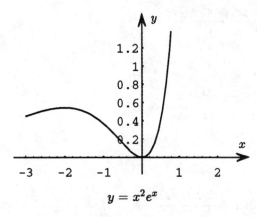

$$y = x^2 e^x$$

Exercise 31

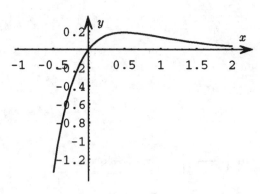

$$y = xe^{-2x}$$

Exercise 32

33. $f(x) = \frac{1}{x} \ln \frac{1}{x} = -\frac{\ln x}{x} \implies f'(x) = -\frac{1 - \ln x}{x^2}$ and $f''(x) = \frac{3 - 2\ln x}{x^3}$. Since $f' = 0 \iff 1 = \ln x \iff x = e$ and $f''(e) > 0$, the graph of $y = f(x)$ has a relative minimum at $(e, -1/e)$. Moreover,

$$\lim_{x \to 0^+} \left(-\frac{\ln x}{x} \right) = +\infty \quad \text{and} \quad \lim_{x \to \infty} \left(-\frac{\ln x}{x} \right) = -\lim_{x \to \infty} \frac{1/x}{1} = 0$$

and hence the graph, on the next page, is asymptotic to the x-axis as x increases without bound.

34. $f(x) = x \ln^2 x \implies f'(x) = (\ln x)(2 + \ln x)$ and $f''(x) = (2/x)(\ln x + 1)$. Since $f' = 0 \iff x = 1$, e^{-2} amd $f''(1) > 0$ and $f''(e^{-2}) < 0$, the graph of $y = f(x)$ has a relative minimum at $(1, 0)$ and a relative maximum at $(1/e^2, 4/e^2)$. Moreover, since

$$\lim_{x \to 0^+} x(\ln^2 x) = \lim_{x \to 0^+} \frac{(\ln x)^2}{1/x} = \lim_{x \to 0^+} \frac{2(\ln x)(1/x)}{-1/x^2} = 2 \lim_{x \to 0^+} \frac{\ln x}{1/x} = 2 \lim_{x \to 0^+} \frac{1/x}{-1/x^2} = -2 \lim_{x \to 0^+} x = 0$$

and $\lim_{x \to \infty} x(\ln x)^2 = +\infty$, the graph, shown on the next page, has a singularity at $x = 0$ and increases without bound as x increases without bound.

768

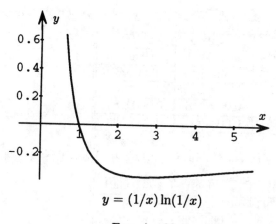

$$y = (1/x)\ln(1/x)$$

Exercise 33

$$y = x\ln^2 x$$

Exercise 34

35. $\displaystyle\lim_{x\to\infty} x^r \ln x = \lim_{x\to\infty} \frac{\ln x}{x^{-r}} = \lim_{x\to\infty} \frac{1/x}{-rx^{-r-1}} = -\frac{1}{r}\lim_{x\to\infty} x^r = 0$ if $r < 0$.

36. If n is a positive integer, then, after n applications of l'Hôpital's rule, we find that

$$\lim_{x\to\infty} x^n e^{-x} = \cdots = \lim_{x\to\infty} \frac{n!}{e^x} = 0,$$

while if n is negative, say $n = -m$, $m > 0$, then

$$\lim_{x\to\infty} x^n e^{-x} = \lim_{x\to\infty} \frac{1}{x^m e^x} = 0.$$

37. (a) Let $y = \left(1 + \dfrac{1}{t}\right)^{tx}$. Then

$$\lim_{t\to+\infty} \ln y = \lim_{t\to+\infty} \ln\left(1 + \frac{1}{t}\right)^{tx} = \lim_{t\to+\infty}\left[tx\ln\left(1 + \frac{1}{t}\right)\right] = \lim_{t\to+\infty} \frac{x\ln(1 + 1/t)}{1/t}.$$

Now, from Example 5 in the text, with $r = 1$, we know that

$$\lim_{t\to+\infty} \frac{x\ln(1 + 1/t)}{1/t} = x\lim_{t\to+\infty} \frac{\ln(1 + 1/t)}{1/t} = x(1) = x.$$

Hence $\lim_{t\to+\infty} \ln y = x$ and therefore

$$\lim_{t\to+\infty}\left(1 + \frac{1}{t}\right)^{tx} = e^x.$$

In particular, if we let $s = 1/t$, then $s \to 0^+$ as $t \to +\infty$ and therefore

$$\lim_{s\to 0^+}(1 + s)^{x/s} = \lim_{t\to+\infty}\left(1 + \frac{1}{t}\right)^{tx} = e^x.$$

(b) The following table gives the values of $h(x) = 1 - (1 + 1/t)^{tx}$ for $x = -1, -1/2, 0, 1/2, 1$.

x	$t = 10$	$t = 100$	$t = 1000$	$t = 10000$
-1	-0.480153	-0.0049793	-0.0004998	-0.0000500
$-1/2$	-0.0237262	-0.0024865	-0.0002499	-0.0000250
0	0	0	0	0
$1/2$	0.0237263	0.0024804	0.0002498	0.0000250
1	0.0458155	0.004956	0.0004995	0.0000500

(c) The following table gives the values of $g(x) = 1 - (1 + s)^{x/s}$ for $x = 0.1, 0.01, 0.001, 0.0001$.

x	$s = 0.1$	$s = 0.01$	$s = 0.001$	$s = 0.0001$
-1	-0.480153	-0.0049793	-0.0004998	-0.0000500
$-1/2$	-0.0237262	-0.0024865	-0.0002499	-0.0000250
0	0	0	0	0
$1/2$	0.0237263	0.0024804	0.0002498	0.0000250
1	0.0458155	0.004956	0.0004995	0.0000500

38. From Example 5 in the text, we know that $\lim\limits_{m \to +\infty} (1 + r/m)^m = e$. Hence

$$\lim_{m \to \infty} \left(1 + \frac{r}{m}\right)^{mt} = \left[\lim_{m \to \infty} \left(1 + \frac{r}{m}\right)^m\right]^t = e^{rt},$$

where m is the number of times per year the interest is compounded. So a m increases the frequency of compounding increases. In the limit as $m \to +\infty$ the interest is compounded continuously.

(a) Graph $A(t) = 1000e^{0.075t}$ and $A_m(t) = 1000(1 + 0.075/365)^{365t}$ with range $[0, 10] \times [1000, 2100]$.

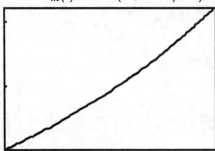

There is no visible difference between the graphs.

(b) For continuous compounding $A(10) = 2117.00$ and $A(25) = 6520.82$, while for daily compounding $A_m(10) = 2116.84$ and $A_m(25) = 6519.56$. For $t = 10$ the difference is 0.16 (16 cents), and for $t = 25$ the difference is 1.26 (1 dollar and 26 cents).

11.3 Improper Integrals

1. $\displaystyle\int_3^\infty \frac{1}{x}\,dx = \lim_{t \to \infty} \ln x \Big|_3^t = \lim_{t \to \infty} (\ln t - \ln 3)$ diverges

2. $\displaystyle\int_2^\infty \frac{1}{(x+1)^2}\,dx = \lim_{t \to \infty} \frac{-1}{x+1}\Big|_2^t = \lim_{t \to \infty}\left(\frac{-1}{t+1} + \frac{1}{3}\right) = \frac{1}{3}$

3. $\displaystyle\int_1^\infty \frac{x}{\sqrt{1+x^2}}\,dx = \lim_{t\to\infty}\sqrt{1+x^2}\Big|_1^t = \lim_{t\to\infty}\left(\sqrt{1+t^2}-\sqrt{2}\right)$ diverges

4. $\displaystyle\int_1^\infty \frac{dx}{4+x^2} = \lim_{t\to\infty}\frac{1}{2}\,\mathrm{Tan}^{-1}\frac{x}{2}\Big|_1^t = \lim_{t\to\infty}\left(\frac{1}{2}\,\mathrm{Tan}^{-1}\frac{t}{2}-\frac{1}{2}\,\mathrm{Tan}^{-1}\frac{1}{2}\right) = \frac{\pi}{4}-\frac{1}{2}\,\mathrm{Tan}^{-1}\frac{1}{2}$

5. $\displaystyle\int_0^\infty e^{-x}\,dx = \lim_{t\to\infty}(-e^{-x})\Big|_0^t = \lim_{t\to\infty}\left(\frac{-1}{e^t}+1\right) = 1$

6. $\displaystyle\int_2^\infty \frac{x}{(4+x^2)^{3/2}}\,dx = \lim_{t\to\infty}\frac{-1}{\sqrt{4+x^2}}\Big|_2^t = \lim_{t\to\infty}\left(\frac{-1}{\sqrt{4+t^2}}+\frac{1}{\sqrt{8}}\right) = \frac{1}{\sqrt{8}}$

7. $\displaystyle\int_0^\infty e^{-x}\sin x\,dx = \lim_{t\to\infty}\frac{1}{2}\,e^{-x}(-\sin x-\cos x)\Big|_0^t = \lim_{t\to\infty}\left[\frac{1}{2}\,e^{-t}(-\sin t-\cos t)-\frac{1}{2}\right] = \frac{1}{2}$

8. $\displaystyle\int_0^\infty e^x\cos x\,dx = \lim_{t\to\infty}\frac{e^x}{2}\,(\cos x+\sin x)\Big|_0^t = \lim_{t\to\infty}\left[\frac{e^t}{2}\,(\cos t+\sin t)-\frac{1}{2}\right]$ diverges

9. $\displaystyle\int_{-\infty}^2 e^{2x}\,dx = \lim_{t\to-\infty}\frac{1}{2}\,e^{2x}\Big|_t^2 = \lim_{t\to-\infty}\frac{1}{2}\,(e^4-e^{2t}) = \frac{1}{2}\,e^4$

10. $\displaystyle\int_{-\infty}^0 \frac{1}{1+x^2}\,dx = \lim_{t\to-\infty}\mathrm{Tan}^{-1}x\Big|_t^0 = \lim_{t\to-\infty}(-\mathrm{Tan}^{-1}t) = \frac{\pi}{2}$

11. $\displaystyle\int_0^\infty xe^{-x}\,dx = \lim_{t\to\infty}e^{-x}(-x-1)\Big|_0^t = \lim_{t\to\infty}\left(\frac{-t}{e^t}-\frac{1}{e^t}+1\right) = \lim_{t\to\infty}\frac{-1}{-e^{-t}}-\lim_{t\to\infty}e^{-t}+1 = 1$

12. $\displaystyle\int_{-\infty}^\infty \frac{dx}{4+x^2} = \lim_{t\to-\infty}\frac{1}{2}\,\mathrm{Tan}^{-1}\frac{x}{2}\Big|_t^0 + \lim_{t\to\infty}\frac{1}{2}\,\mathrm{Tan}^{-1}\frac{x}{2}\Big|_0^t = \lim_{t\to-\infty}\left(-\frac{1}{2}\,\mathrm{Tan}^{-1}\frac{t}{2}\right) + \lim_{t\to\infty}\frac{1}{2}\,\mathrm{Tan}^{-1}\frac{t}{2} = \frac{\pi}{2}$

13. $\displaystyle\int_0^1 \frac{1}{x}\,dx = \lim_{t\to0^+}\ln x\Big|_t^1 = \lim_{t\to0^+}(-\ln t)$ diverges

14. $\displaystyle\int_0^1 \frac{1}{x^2}\,dx = \lim_{t\to0^+}\left(-\frac{1}{x}\right)\Big|_t^1 = \lim_{t\to0^+}\left(-1+\frac{1}{t}\right)$ diverges

15. $\displaystyle\int_0^3 \frac{1}{\sqrt{9-x^2}}\,dx = \lim_{t\to3^-}\mathrm{Sin}^{-1}\frac{x}{3}\Big|_0^t = \lim_{t\to3^-}\left(\mathrm{Sin}^{-1}\frac{t}{3}\right) = \frac{\pi}{2}$

16. $\displaystyle\int_e^\infty \frac{1}{x\ln x}\,dx = \lim_{t\to\infty}(\ln|\ln x|)\Big|_e^t = \lim_{t\to\infty}(\ln|\ln t|)$ diverges

17. $\displaystyle\int_0^1 x\ln x\,dx = \lim_{t\to0^+}\frac{x^2}{4}\,(2\ln x-1)\Big|_t^1 = \lim_{t\to0^+}\left[-\frac{1}{4}-\frac{t^2}{4}\,(2\ln t-1)\right] = -\frac{1}{4}-\frac{1}{2}\lim_{t\to0^+}\frac{\ln t}{t^{-2}}$

$\displaystyle = -\frac{1}{4}-\frac{1}{2}\lim_{t\to0^+}\frac{1/t}{-2t^{-3}} = -\frac{1}{4}-\frac{1}{4}\lim_{t\to0^+}t^2 = -\frac{1}{4}$

18. $\displaystyle\int_0^1 \frac{dx}{\sqrt{1-x}} = \lim_{t\to1^-}-2\sqrt{1-x}\Big|_0^t = \lim_{t\to1^-}(-2\sqrt{1-t}+2) = 2$

19. $\int_0^e x^2 \ln x \, dx = \lim_{t \to 0+} \frac{x^3}{9} (3 \ln x - 1) \Big|_t^e = \lim_{t \to 0+} \left[\frac{2e^3}{9} - \frac{t^3}{3} \ln t + \frac{t^3}{9} \right]$

$= \frac{2e^3}{9} - \frac{1}{3} \lim_{t \to 0+} \frac{\ln t}{t^{-3}} = \frac{2e^3}{9} - \frac{1}{3} \lim_{t \to 0+} -\frac{1}{3} \lim \frac{1/t}{-3t^{-4}} = \frac{2e^3}{9} + \frac{1}{9} \lim_{t \to 0+} t^3 = \frac{2e^3}{9}$

20. $\int_0^1 x^{-2/3} \, dx = \lim_{t \to 0+} 3x^{1/3} \Big|_t^1 = \lim_{t \to 0+} (3 - 3t^{1/3}) = 3$

21. $\int_{-1}^1 \frac{1}{x} \, dx = \lim_{t \to 0-} \ln|x| \Big|_{-1}^t + \lim_{t \to 0+} \ln|x| \Big|_t^1 = \lim_{t \to 0-} \ln|t| + \lim_{t \to 0+} (-\ln t) \quad \text{diverges}$

22. $\int_0^3 \frac{dx}{x^2 + 2x - 3} = \frac{1}{4} \int_0^3 \left(\frac{1}{x-1} - \frac{1}{x+3} \right) dx =$

$\frac{1}{4} \lim_{t \to 1-} \ln \left| \frac{x-1}{x+3} \right| \Big|_0^t + \frac{1}{4} \lim_{t \to 1+} \ln \left| \frac{x-1}{x+3} \right| \Big|_t^3 = \frac{1}{4} \lim_{t \to 1-} \left(\ln \left| \frac{t-1}{t+3} \right| - \ln \frac{1}{3} \right) + \frac{1}{4} \lim_{t \to 1+} \left(\ln \frac{1}{3} - \ln \left| \frac{t-1}{t+3} \right| \right)$

diverges

23. $\int_0^\infty e^{-2x} \cos 2x \, dx = \lim_{t \to \infty} \frac{e^{-2x}}{8} (-2 \cos 2x + 2 \sin 2x) \Big|_0^t = \lim_{t \to \infty} \left(\frac{-2 \cos 2t + 2 \sin 2t}{8e^{2t}} + \frac{1}{4} \right) = \frac{1}{4}$

24. $\int_0^\infty e^{-3x} \sin 2x \, dx = \lim_{t \to \infty} \left[\frac{e^{-3x}}{13} (-3 \sin 2x - 2 \cos 2x) \right] \Big|_0^t = \lim_{t \to \infty} \left(\frac{-3 \sin 2t - 2 \cos 2t}{13e^{3t}} + \frac{2}{13} \right) = \frac{2}{13}$

25. $\int_0^\infty e^{-2\sqrt{x}} \sqrt{x} \, dx = \lim_{t \to \infty} \left(-e^{-2\sqrt{x}} \right) \Big|_0^t = \lim_{t \to \infty} \left(-e^{-2\sqrt{t}} + 1 \right) = 1$

26. $\int_0^\infty \sqrt{x} e^x \, dx$ diverges since $\sqrt{x} \, e^x$ increases without bound as $x \to \infty$.

27. This is a proper integral:

$$\int_0^1 \underbrace{x \ln(1+x) \, dx}_{u=1+x} = \int_1^2 (u-1) \ln u \, du = \left[\frac{u^2}{4} (2 \ln u - 1) - u(\ln u - 1) \right] \Big|_1^2 = \frac{1}{4}.$$

28. $\int_0^1 \ln^2 x \, dx = \lim_{t \to 0+} \left[x(\ln x)^2 - 2x \ln x + 2x \right] \Big|_t^1 = \lim_{t \to 0+} \left[2 - t(\ln t)^2 + 2t \ln t - 2t \right] = 2$

29. $\int_0^\infty \sin x \, dx = \lim_{t \to \infty} (-\cos x) \Big|_0^t = \lim_{t \to \infty} (1 - \cos t)$, which does not exist. Hence the improper integral diverges.

30. The integral $\int_1^{+\infty} (1/x^2) \, dx$ represents the area under the curve $y = 1/x^2$ from $x = 1$ to infinity. The function $g(x) = 1/\sqrt{x}$, on the other hand, is the inverse of $f(x) = 1/x^2$ and hence its graph, shown on the following page, is the reflection of the graph of $y = 1/x^2$ about the line $y = x$. Therefore

$$\int_1^{+\infty} \frac{1}{x^2} \, dx = \left(\text{Area under } y = \tfrac{1}{\sqrt{x}} \text{ from } x = 0 \text{ to } x = 1 \right) - \left(\text{Area of unit square} \right) = \int_0^1 \frac{1}{\sqrt{x}} \, dx - 1.$$

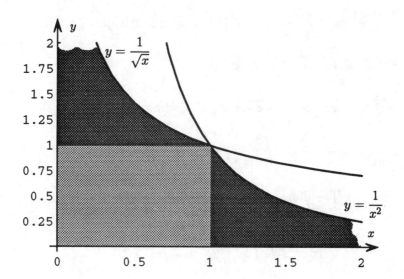

31. $\int_0^1 x^p\,dx$ converges if and only if $p > -1$:

$$p > -1 \implies \int_0^1 x^p\,dx = \lim_{t\to 0}\frac{x^{1+p}}{1+p}\Big|_t^1 = \lim_{t\to 0}\left(\frac{1-t^{1+p}}{1+p}\right) = 1/(1+p) \quad \text{since } 1+p > 0$$

$$p = -1 \implies \int_0^1 \frac{1}{x}\,dx = \lim_{t\to 0}\ln|x|\,\Big|_t^1 = \lim_{t\to 0}(-\ln|t|) \quad \text{diverges}$$

$$p < -1 \implies \int_0^1 x^p\,dx = \lim_{t\to 0}\frac{x^{1+p}}{1+p}\Big|_t^1 = \lim_{t\to 0}\left(\frac{1}{1+p}-\frac{t^{1+p}}{1+p}\right) \quad \text{diverges since } 1+p < 0$$

32. None; the integral diverges for all p since

$$\int_0^\infty x^p\,dx = \underbrace{\int_0^1 x^p\,dx}_{\text{diverges for } p\le -1} + \underbrace{\int_1^\infty x^p\,dx}_{\text{diverges for } p > -1}$$

33. Suppose $\int_a^\infty g(x)\,dx$ converges to some number A. Since $f(x) \le g(x)$, the area under the graph of $y = f(x)$ from $x = a$ to $x = t$ is bounded above by the area under the graph of $y = g(x)$. Hence

$$\int_a^\infty f(x)\,dx = \lim_{t\to\infty}\int_a^t f(x)\,dx \le \lim_{t\to\infty}\int_a^t g(x)\,dx = \int_a^\infty g(x)\,dx = A$$

and therefore $\int_a^\infty f(x)\,dx$ converges. It follows that if $\int_a^\infty f(x)\,dx$ diverges, so does $\int_a^\infty g(x)\,dx$.

34. converges; $\int_1^\infty \frac{1}{1+x^3}\,dx < \int_1^\infty \frac{1}{x^3}\,dx$, which converges ($p$-test, $p = 3 > 1$)

35. diverges; $\int_1^\infty \frac{e^x}{\sqrt{1+x^2}}\,dx > \int_1^\infty \frac{x}{\sqrt{1+x^2}}\,dx = \lim_{t\to\infty}\left(\sqrt{1+t^2}-\sqrt{2}\right)$, which diverges.

36. converges; $\int_1^\infty \frac{1}{\sqrt{1+x^5}}\,dx < \int_1^\infty \frac{1}{x^{5/2}}\,dx$, which converges ($p$-test, $p = 5/2 > 1$)

773

37. converges; $\displaystyle\int_1^\infty \frac{|\sin x|}{x^2}\, dx < \int_1^\infty \frac{1}{x^2}\, dx$, which converges ($p$-test, $p = 2 > 1$)

38. diverges; $\displaystyle\int_1^\infty e^{\sqrt{x}}\, dx > \int_1^\infty dx$, which diverges

39. diverges; $\displaystyle\int_1^\infty \sqrt{e^x + \sin x}\, dx > \int_1^\infty \sqrt{e - 1}\, dx$, which diverges

40. converges; $\displaystyle\int_1^\infty \frac{dx}{\sqrt{e^{x^2} + x + \cos x}} < \int_1^\infty \frac{1}{e^{x/2}}\, dx$, which converges

41. $\displaystyle\int_{-\infty}^\infty x^3\, dx = \lim_{t\to -\infty} \frac{1}{4}\, x^4 \Big|_t^0 + \lim_{t\to\infty} \frac{1}{4}\, x^4 \Big|_0^t$, which diverges, while

$$\lim_{t\to\infty} \int_{-t}^t x^3\, dx = \lim_{t\to\infty} \frac{1}{4}\, x^4 \Big|_{-t}^t = 0.$$

42. By definition,

$$
\begin{aligned}
\int_{-\infty}^\infty f(x)\, dx &= \lim_{t\to -\infty} \int_t^0 f(x)\, dx + \lim_{t\to\infty} \int_0^t f(x)\, dx \\
&= \lim_{t\to -\infty} \int_t^a f(x)\, dx + \int_a^0 f(x)\, dx + \int_0^a f(x)\, dx + \lim_{t\to\infty} \int_a^t f(x)\, dx \\
&= \lim_{t\to -\infty} \int_t^a f(x)\, dx + \lim_{t\to\infty} \int_a^t f(x)\, dx \\
&= \int_{-\infty}^a f(x)\, dx + \int_a^\infty f(x)\, dx
\end{aligned}
$$

43. (a) $y = 1/x \Longrightarrow y' = -1/x^2 \Longrightarrow$

$$\text{Surface Area} = \int_1^\infty 2\pi \frac{1}{x}\sqrt{1 + \frac{1}{x^4}}\, dx > 2\pi \int_1^\infty \frac{1}{x}\, dx,$$

which diverges. Hence the surface area is infinite.

(b) The volume of revolution is finite.

44. Let dV be the differential volume of a cross-section disk of thickness dx. Then $dV = \pi r^2 dx = \pi e^{-2x} dx$ and hence the volume of revolution is

$$V = \pi \int_0^\infty e^{-2x}\, dx = -\frac{\pi}{2} \lim_{t\to\infty} e^{-2x} \Big|_0^t = -\frac{\pi}{2} \lim_{t\to\infty} \left(e^{-2t} - 1\right) = \frac{\pi}{2}.$$

For the surface area, $y' = -e^{-x}$ and hence

$$\text{Surface Area} = 2\pi \int_0^\infty \underbrace{e^{-x}\sqrt{1 + e^{-2x}}\, dx}_{u = e^{-x}} = -2\pi \int_1^0 \sqrt{1 + u^2}\, du$$

$$= -\pi \left[u\sqrt{1 + u^2} + \ln(u + \sqrt{1 + u^2})\right]\Big|_1^0 = \pi \left[\sqrt{2} + \ln(1 + \sqrt{2})\right].$$

45. Area $= \displaystyle\int_1^\infty \frac{1}{x^4}\, dx = \lim_{t\to\infty} \left(-\frac{1}{3}\frac{1}{x^3}\right)\Big|_1^t = \lim_{t\to\infty}\left(-\frac{1}{3t^3}+\frac{1}{3}\right)=\frac{1}{3}$

46. Area $= \displaystyle\int_0^\infty xe^{-4x}\, dx = \lim_{t\to\infty} \frac{e^{-4x}}{16}(-4x-1)\Big|_0^t = \lim_{t\to\infty}\left(\frac{-4t-1}{16e^{4t}}+\frac{1}{16}\right)=\frac{1}{16}$

47. Area $= -\displaystyle\int_0^1 \ln x\, dx = -\lim_{t\to 0^+}(x\ln x - x)\Big|_t^1 = -\lim_{t\to)^+}(-1 - t\ln t + t)=1+\lim_{t\to 0^+}\frac{\ln t}{1/t}$

    $= 1 + \displaystyle\lim_{t\to 0^+}\frac{1/t}{-1/t^2}=1$

48. Let dV be the differential volume of a shell of thickness dx revolved about the y-axis. Then $dV = 2\pi r h\, dx = 2\pi xy\, dx = 2\pi x^{3/4}\, dx$ and hence the volume of revolution is

$$V = 2\pi \int_0^1 x^{3/4}\, dx = 2\pi\, \frac{4}{7}\, x^{7/4}\Big|_0^1 = \frac{8\pi}{7}.$$

49. Let dV be the differential volume of a shell of thickness dx revolved about the y-axis. Then $dV = 2\pi r h dx = 2\pi x(e^{-x}+\frac{1}{x^3})\, dx$ and hence the volume of revolution is

$$V = 2\pi \int_1^\infty \left(xe^{-x}+\frac{1}{x^2}\right)\, dx = 2\pi \lim_{t\to\infty}\left[e^{-x}(-x-1)-\frac{1}{x}\right]\Big|_1^t$$

$$= 2\pi \lim_{t\to\infty}\left(\frac{-t-1}{e^t}-\frac{1}{t}+\frac{2}{e}+1\right)=2\pi\left(\frac{2}{e}+1\right).$$

50. The centroid is at the point $(\bar{x},\bar{y})$, where

$$\bar{x}=\frac{\displaystyle\int_0^\infty xe^{-x}\, dx}{\displaystyle\int_0^\infty e^{-x}\, dx}=\frac{\displaystyle\lim_{t\to\infty}e^{-x}(-x-1)\big|_0^t}{\displaystyle\lim_{t\to\infty}(-e^{-x})\big|_0^t}=\frac{\displaystyle\lim_{t\to\infty}\left(\frac{-t-1}{e^t}+1\right)}{\displaystyle\lim_{t\to\infty}\left(\frac{-1}{e^t}+1\right)}=1$$

and

$$\bar{y}=\frac{\displaystyle\int_0^\infty e^{-2x}\, dx}{2(1)}=\frac{\displaystyle\lim_{t\to\infty}\left(-\frac{1}{2}e^{-2x}\right)\Big|_0^t}{2}=\frac{\displaystyle\lim_{t\to\infty}\left(\frac{-1}{2e^{2t}}+\frac{1}{2}\right)}{2}=\frac{1}{4}.$$

51. The centroid is at the point $(\bar{x},\bar{y})$, where

$$\bar{x}=\frac{\displaystyle\int_1^\infty x\left(\frac{1}{x^3}\right)\, dx}{\displaystyle\int_1^\infty x^3\, dx}=\frac{\displaystyle\lim_{t\to\infty}\left(-\frac{1}{x}\right)\Big|_1^t}{\displaystyle\lim_{t\to\infty}\left(-\frac{1}{2x^2}\right)\Big|_1^t}=\frac{\displaystyle\lim_{t\to\infty}\left(-\frac{1}{t}+1\right)}{\displaystyle\lim_{t\to\infty}\left(-\frac{1}{2t^2}+\frac{1}{2}\right)}=2$$

and

$$\bar{y}=\frac{\displaystyle\int_1^\infty \frac{1}{x^6}\, dx}{2(1/2)}=\lim_{t\to\infty}\left(-\frac{x^{-5}}{5}\right)\Big|_1^t=\lim_{t\to\infty}\left(-\frac{1}{5t^5}+\frac{1}{5}\right)=\frac{1}{5}.$$

52. (a) $\lim\limits_{x \to \infty} [f(x) - g(x)] = \lim\limits_{x \to \infty} \left(\dfrac{x^3 - 1}{x^2} - x \right) = \lim\limits_{x \to \infty} \left(-\dfrac{1}{x^2} \right) = 0$

 (b) Area $= \displaystyle\int_1^\infty \left(x - \dfrac{x^3 - 1}{x^2} \right) dx = \int_1^\infty \dfrac{1}{x^2} \, dx = \lim\limits_{t \to \infty} \left(-\dfrac{1}{x} \right)\Big|_1^t = \lim\limits_{t \to \infty} \left(-\dfrac{1}{t} + 1 \right) = 1$

53. (a) $\lim\limits_{x \to \infty} [f(x) - g(x)] = \lim\limits_{x \to \infty} \left(\dfrac{x^2 - 1}{x} - x \right) = \lim\limits_{x \to \infty} \left(-\dfrac{1}{x} \right) = 0$

 (b) Area $= \displaystyle\int_1^\infty \left(x - \dfrac{x^2 - 1}{x} \right) dx = \int_1^\infty \dfrac{1}{x} \, dx$, which diverges.

54. The integral converges when $n > -1$:

$$n \neq -1 \implies \int_0^1 x^n \ln x \, dx = \lim_{t \to 0^+} x^{n+1} \left[\frac{\ln x}{n+1} - \frac{1}{(n+1)^2} \right]\Big|_t^1$$

$$= -\frac{1}{(n+1)^2} - \frac{1}{n+1} \lim_{t \to 0^+} \frac{\ln t}{t^{-n-1}} = -\frac{1}{(n+1)^2} + \frac{1}{(n+1)^2} \lim_{t \to 0^+} t^{n+1} = -\frac{1}{(n+1)^2}$$

$$n = -1 \implies \int_0^1 x^{-1} \ln x \, dx = \lim_{t \to 0^+} \frac{1}{2} (\ln x)^2 \Big|_t^1 = \lim_{t \to 0^+} \left[-\frac{1}{2} (\ln t)^2 \right] = -\infty$$

55. (a) $\displaystyle\int_0^5 e^{-0.08t} \, dt = -\frac{25}{2} e^{-0.08t} \Big|_0^5 \approx \$4.12$ million

 (b) $\displaystyle\int_0^\infty e^{-0.08t} \, dt = -\frac{25}{2} \lim_{t \to \infty} \left(e^{-0.08t} - 1 \right) \approx \$12.5$ million

56. (a) $\displaystyle\int_0^5 (1000 + 200t)e^{-0.1t} \, dt = \left[-1000e^{-0.1t} + 20000e^{-0.1t}(-0.1t - 1) \right]\Big|_0^5 \approx \$5{,}738.77$

 (b) $\displaystyle\int_0^\infty (1000 + 200t)e^{-0.1t} \, dt = \lim_{t \to \infty} \left[-10000e^{-0.1t} + 20000e^{-0.1t}(-0.1t - 1) + 30000 \right] \approx \$30{,}000$

57. If $n = 1$, $\int_0^1 \ln x \, dx = \lim_{t \to 0^+} (x \ln x - x)\big|_t^1 = \lim_{t \to 0^+} (-1 - t \ln t - t) = -1$. Now, if $n > 1$, assume that $\int_0^1 (\ln x)^{n-1} \, dx = (-1)^{n-1}(n-1)!$. Then

$$\int_0^1 \underbrace{(\ln x)^n \, dx}_{u = (\ln x)^n, \, dv = dx} = x(\ln x)^n\big|_0^1 - n \int_0^1 (\ln x)^{n-1} \, dx = -n(-1)^{n-1}(n-1)! = (-1)^n n!.$$

58. $\displaystyle\int_1^\infty \frac{dx}{1 + e^x} = \lim_{t \to \infty} \int_1^t \frac{e^{-x}}{e^{-x} + 1} \, dx = -\lim_{t \to \infty} \ln(e^{-x} + 1)\big|_1^t = -\lim_{t \to \infty} \left[\ln(e^{-t} + 1) - \ln(e^{-1} + 1) \right]$

$= \ln \dfrac{1 + e}{e} = \ln(1 + e) - 1.$

59. If $y = \sqrt{r^2 - x^2}$, then the surface area obtained by revolving the semi-circle about the x-axis, which is the surface area of the sphere, is equal to

$$2 \int_0^r 2\pi y \sqrt{1 + (y')^2} \, dx = 4\pi \lim_{t \to r^-} \int_0^t \sqrt{r^2 - x^2} \sqrt{1 + \frac{x^2}{r^2 - x^2}} \, dx = 4\pi \lim_{t \to r^-} \int_0^t r \, dx = 4\pi r^2.$$

60. (a) If $r = 0.085$ and $P(t) = 215000$, then

$$P_n = 215000 \int_0^n e^{-0.085t} \, dt = 215000 \left(-\frac{1}{0.085} e^{-0.085t} \right) \Big|_0^n = 2529411.76(1 - e^{-0.085n})$$

and

$$P_{n,1} = 215000 \int_0^n \left(1 - \frac{0.085}{1} \right)^t \, dt = 215000 \int_0^n (0.915)^t \, dt$$

$$= \frac{215000}{\ln(0.915)} (0.915)^t \Big|_0^n = 2420321.41(1 - 0.915^n).$$

Graph these two functions using the range $[0, 20] \times [0, 2011000]$.

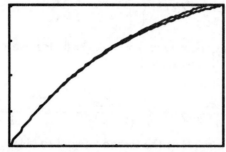

The difference is only apparent for n near 20. Note that $P_{15,1} = 17811778.70$, $P_{20,1} = 2010781.91$, $P_{15} = 1822615.78$, and $P_{20} = 2067329.91$. Differences range from $P_{15} - P_{15,1} = 40837.08$ to $P_{20} - P_{20,1} = 56548.00$.

(b) Now $\lim\limits_{n \to +\infty} P_n = 2529411.76$ and $\lim\limits_{n \to +\infty} P_{n,1} = 2422032.42$. This is a difference of 109091.35. If the interest is compounded monthly, then $m = 12$ and hence

$$P_{n,12} = 215000 \int_0^n (0.9929167)^{12t} \, dt$$

$$= \frac{215000}{12 \ln(0.9929167)} (0.9929167)^{12t} \Big|_0^n = 2520454.72(1 - 0.9929167^{12n}).$$

Therefore $\lim\limits_{n \to +\infty} P_{n,12} = 2520454.72$. This value differs from the value for continuous compounding by 8957.04.

For Exercises 61–67, S_{20} denotes the value of Simpson's Rule using 20 points.

61. Let $t = 1/x$. Then

$$\int_2^{+\infty} \frac{dx}{4 + x^2} = \int_{1/2}^0 \frac{-1/t^2}{4 + 1/t^2} \, dt = \int_0^{1/2} \frac{1/t^2}{4 + 1/t^2} \, dt.$$

Since

$$\lim_{t \to 0+} \frac{1/t^2}{4 + 1/t^2} = \lim_{t \to 0+} \frac{1}{4t^2 + 1} = 1,$$

we apply Simpson's Rule with $n = 20$ to the integral $\int_0^{1/2} f(t) dt$, where

$$f(t) = \begin{cases} \frac{1/t^2}{4 + 1/t^2}, & t > 0 \\ 1, & t = 0, \end{cases}$$

to obtain $S_{20} = 0.3926991$ and $|S_{20} - \pi/8| = 7.9 \times 10^{-11}$.

62. Let $t = 1/x$. Then

$$\int_1^{+\infty} e^{-\sqrt{x}}\,dx = \int_1^0 e^{-1/\sqrt{t}}\left(-\frac{1}{t^2}\,dt\right) = \int_0^1 \frac{e^{-1/\sqrt{t}}}{t^2}\,dt.$$

After four applications of l'Hôpital's Rule, we find that

$$\lim_{t\to 0^+} \frac{e^{-1/\sqrt{t}}}{t^2} = \cdots = 24\lim_{t\to 0^+}\frac{1}{e^{1/\sqrt{t}}} = 0.$$

Thus, applying Simpson's Rule with $n = 20$ to the integral $\int_0^1 f(t)\,dt$, where

$$f(t) = \begin{cases} \frac{e^{-1/\sqrt{t}}}{t^2}, & t > 0 \\ 0, & t = 0, \end{cases}$$

we obtain 1.49412. Alternatively, we may take $a = 0.000001$ instead of 0, in which case we find that $S_{20} = 1.4941130$ and $|S_{20} - 4e^{-1}| = 0.0225953$.

63. We first write

$$\int_0^{+\infty} \frac{x\,dx}{(1+x^2)^{3/2}} = \int_0^1 \frac{x\,dx}{(1+x^2)^{3/2}} + \int_1^{+\infty} \frac{x\,dx}{(1+x^2)^{3/2}}.$$

Now apply the transformation $x = 1/t$ to the second integral to obtain

$$\int_1^{+\infty} \frac{x\,dx}{(1+x^2)^{3/2}} = \int_1^0 \frac{(1/t)(-dt/t^2)}{(1+1/t^2)^{3/2}} = \int_0^1 \frac{dt}{(1+t^2)^{3/2}}.$$

Therefore

$$\int_0^{+\infty} \frac{x\,dx}{(1+x^2)^{3/2}} = \int_0^1 \frac{x\,dx}{(1+x^2)^{3/2}} + \int_0^1 \frac{dt}{(1+t^2)^{3/2}}.$$

Applying Simpson's Rule to both integrals using $n = 20$ gives 0.2928936 for the first integral and 0.7071068 for the second integral. Hence $S_{20} = 0.2928936 + 0.7071068 = 1.0000004$.

64. Let $x = 1/t$. Then $dx = -dt/t^2$ and hence

$$\int_1^{+\infty} xe^{-2x}\sin\pi x\,dx = \int_1^0 (1/t)e^{-2/t}\sin(\pi/t)(-dt/t^2) = \int_0^1 \frac{e^{-2/t}\sin(\pi/t)}{t^3}\,dt.$$

Now

$$\lim_{t\to 0^+} \frac{e^{-2/t}\sin(\pi/t)}{t^3} = \lim_{x\to +\infty} x^3 e^{-2x}\sin\pi x.$$

But $|x^3 e^{-2x}\sin\pi x| \le x^3 e^{-2x}$ for $x > 0$. Applying L'Hôpital's Rule shows that $\lim\limits_{x\to +\infty}(x^3/e^{2x}) = 0$. Therefore

$$\lim_{x\to +\infty} x^3 e^{-2x}\sin\pi x = 0.$$

So apply Simpson's Rule to the integral $\int_0^1 f(t)\,dt$ where

$$f(t) = \begin{cases} \frac{e^{-2/t}}{t^3}\sin(\pi/t), & t > 0 \\ 0, & t = 0, \end{cases}$$

or to the integral $\int_a^1 f(t)\,dt$ where $a = 0.000001$. With the latter technique $S_{20} = -0.0390781$, and if we denote the exact value $(-\pi(\pi^2+8)e^{-2})/(4+\pi^2)^2$ by EV, then $|S_{20} - EV| = 0.0004174$.

65. Let $x = t^2$. Then

$$\int_0^{1/2} \frac{\cos \pi x}{\sqrt{x}} \, dx = \int_0^{\sqrt{1/2}} \frac{\cos \pi t^2}{t} \, 2t\,dt = 2\int_0^{\sqrt{1/2}} \cos \pi t^2 dt.$$

Applying Simpson's Rule to the last integral we get $S_{20} = 1.1029373$ and $|S_{20} - EV| = 0.00000152$.

66. Let $x = t^{3/2}$. Then

$$\int_0^4 \frac{e^x}{x^{1/3}} \, dx = \int_0^{2\sqrt[3]{2}} \frac{e^{t^{3/2}}}{t^{1/2}} \frac{3}{2} \, t^{1/2} \, dt = \frac{3}{2}\int_0^{2\sqrt[3]{2}} e^{t^{3/2}} dt.$$

Clearly, $\lim_{t \to 0} e^{t^{3/2}} = 1$. Applying Simpson's Rule to the last integral we get $S_{20} = 38.253820207$ and $|S_{20} - EV| = 0.002006216$.

67. Let $x = t^2$. Then

$$\int_0^1 \frac{e^{-x^2/2} \sin 3x}{\sqrt{x}} \, dx = \int_0^1 \frac{e^{-t^4/2} \sin 3t^2}{t} \, 2t \, dt = 2\int_0^1 e^{-t^4/2} \sin 3t^2 dt.$$

Clearly,

$$\lim_{t \to 0} e^{-t^4/2} \sin 3t^2 = (1)(0) = 0.$$

Applying Simpson's Rule to the last integral we get $S_{20} = 0.910287634$ and $|S_{20} - EV| = 0.000012066$.

Review Exercises – Chapter 11

1. $\displaystyle \lim_{x \to 0} \frac{\sin 3x}{\sin 4x} = \lim_{x \to 0} \frac{3\cos 3x}{4\cos 4x} = \frac{3}{4}$

2. $\displaystyle \lim_{x \to 0} \frac{\sin x}{1 - e^x} = \lim_{x \to 0} \frac{\cos x}{-e^x} = -1$

3. $\displaystyle \lim_{x \to 0^+} \frac{\tan x}{x^2} = \lim_{x \to 0^+} \frac{\sec^2 x}{2x} = +\infty$

4. $\displaystyle \lim_{x \to 2} \frac{x - 2}{x^2 + x - 6} = \lim_{x \to 2} \frac{1}{2x + 1} = \frac{1}{5}$

5. $\displaystyle \lim_{x \to \infty} \frac{\sqrt{x^2 + 2x + 1}}{x} = \lim_{x \to \infty} \frac{x + 1}{x} = \lim_{x \to \infty} \left(1 + \frac{1}{x}\right) = 1$

6. $\displaystyle \lim_{x \to \infty} \frac{(x^4 + 1)^{3/2}}{x^7} = \lim_{x \to \infty} \frac{x^4 + 1}{x^4} \sqrt{\frac{x^4 + 1}{x^6}} = 0$

7. $\displaystyle \lim_{x \to 0} \frac{x - \sin x}{\tan x} = \lim_{x \to 0} \frac{1 - \cos x}{\sec^2 x} = 0$

8. $\displaystyle \lim_{x \to 0} \frac{\pi - \csc x}{\pi + \cot x} = \lim_{x \to 0} \frac{\pi \sin x - 1}{\pi \sin x + \cos x} = -1$

9. $\displaystyle \lim_{x \to 0^+} \frac{\sin \sqrt{x}}{\sqrt{x}} = \lim_{x \to 0^+} \frac{(\cos \sqrt{x}) \frac{1}{2\sqrt{x}}}{\frac{1}{2\sqrt{x}}} = \lim_{x \to 0^+} \cos \sqrt{x} = 1$

10. $\displaystyle\lim_{x\to\infty} \frac{\pi/2 - \mathrm{Tan}^{-1}x}{xe^{-x}} = \lim_{x\to\infty} \frac{-\frac{1}{1+x^2}}{-xe^{-x}+e^{-x}} = \lim_{x\to\infty} \frac{e^x}{(1+x^2)(1-x)} = \infty$

11. $\displaystyle\lim_{x\to\infty} \frac{\ln\sqrt{x}}{\sqrt{x}} = \lim_{x\to\infty} \frac{1/(2x)}{1/(2\sqrt{x})} = \lim_{x\to\infty} \frac{1}{\sqrt{x}} = 0$

12. $\displaystyle\lim_{x\to-\infty} \frac{e^{-x}}{x^2} = \lim_{x\to-\infty} \frac{-e^{-x}}{2x} = \lim_{x\to-\infty} \frac{e^{-x}}{2} = \infty$

13. $\displaystyle\lim_{x\to\infty} \frac{x^3+e^x}{xe^{2x}} = $ (four applications of l'Hôpital) $= \lim_{x\to\infty} \frac{1}{16e^x(x+2)} = 0$

14. $\displaystyle\lim_{x\to\infty} x^2e^{-x} = \lim_{x\to\infty} \frac{2x}{e^x} = \lim_{x\to\infty} \frac{2}{e^x} = 0$

15. $\displaystyle\lim_{x\to\infty} x^{1/x} = e^0 = 1$ since $\displaystyle\lim_{x\to\infty} \frac{\ln x}{x} = \lim_{x\to\infty} \frac{1}{x} = 0$

16. $\displaystyle\lim_{x\to0^+} (\tan x)(\ln x) = \lim_{x\to0^+} \frac{\ln x}{\cot x} = \lim_{x\to0^+} \frac{1/x}{-\csc^2 x} = -\lim_{x\to0^+} \frac{\sin^2 x}{x} = -\lim_{x\to0^+} \frac{2\sin x\cos x}{1} = 0$

17. $\displaystyle\lim_{x\to0^+} x^{\sin 2x} = e^0 = 1$ since

$$\lim_{x\to0^+} (\sin 2x)\ln x = \lim_{x\to0^+} \frac{\ln x}{\csc 2x} = \lim_{x\to0^+} \frac{1/x}{-\csc 2x\cot 2x}$$

$$= -\lim_{x\to0^+} \frac{\sin^2 2x}{x\cos 2x} = \lim_{x\to0^+} \frac{2\sin 2x\cos 2x}{-2x\sin 2x + \cos 2x} = 0.$$

18. $\displaystyle\lim_{x\to\infty} (\ln x)^{e^{-x}} = e^0 = 1$ since

$$\lim_{x\to\infty} e^{-x}\ln(\ln x) = \lim_{x\to\infty} \frac{\ln(\ln x)}{e^x} = \lim_{x\to\infty} \frac{1/(x\ln x)}{e^x} = 0.$$

19. $\displaystyle\int_1^\infty xe^{-x}\,dx = \lim_{t\to\infty} e^{-x}(-x-1)\Big|_1^t = \lim_{t\to\infty} \left(\frac{-t-1}{e^t} + \frac{2}{e}\right) = \frac{2}{e}$

20. $\displaystyle\int_{-\infty}^0 \frac{6x}{1+x^2}\,dx = \lim_{t\to-\infty} 3\ln(1+x^2)\Big|_t^0 = \lim_{t\to-\infty} \left[-3\ln(1+t^2)\right]$, which diverges.

21. $\displaystyle\int_0^2 \frac{2x+1}{x^2+x-6}\,dx = \lim_{t\to2^-} \ln|x^2+x-6|\Big|_0^t = \lim_{t\to2^-} \left[\ln|t^2+t-6| - \ln 6\right]$, which diverges.

22. $\displaystyle\int_0^4 \frac{\ln\sqrt{x}}{\sqrt{x}}\,dx = \lim_{t\to0^+} 2(\sqrt{x}\ln\sqrt{x} - \sqrt{x})\Big|_t^4 = 2\lim_{t\to0^+} \left[(2\ln 2 - 2) - (\sqrt{t}\ln t - \sqrt{t})\right]$

$$= 4\ln 2 - 4 - 2\lim_{t\to0^+} \frac{(1/2)\ln t}{t^{-1/2}} = 4\ln 2 - 4$$

23. $\displaystyle\int_{-\infty}^\infty xe^{-x^2}\,dx = \lim_{t\to-\infty} \left(-\frac{1}{2}e^{-x^2}\right)\Big|_t^0 + \lim_{t\to\infty} \left(-\frac{1}{2}e^{-x^2}\right)\Big|_0^t$

$$= \lim_{t\to-\infty} \left(-\frac{1}{2} + \frac{1}{2}e^{-t^2}\right) + \lim_{t\to\infty} \left(-\frac{1}{2}e^{-t^2} + \frac{1}{2}\right) = 0$$

24. $\int_1^\infty \frac{1}{x\sqrt{x^2-1}}\, dx = \lim_{t\to\infty} \left. \text{Sec}^{-1}x \right|_1^t = \lim_{t\to\infty} \text{Sec}^{-1}t = \frac{\pi}{2}$

25. $\int_0^{\pi/2} \tan x\, dx = \lim_{t\to\pi/2-} \left. \ln|\sec x| \right|_0^t = \lim_{t\to\pi/2-} \ln|\sec t|$, which diverges.

26. $\int_e^\infty \frac{1}{x\ln x}\, dx = \lim_{t\to\infty} \left. \ln(\ln x) \right|_e^t = \lim_{t\to\infty} \ln(\ln t)$, which diverges.

27. $\int_0^4 \frac{1}{\sqrt{16-x^2}}\, dx = \lim_{t\to4-} \left. \text{Sin}^{-1}\frac{x}{4} \right|_0^t = \lim_{t\to4-} \text{Sin}^{-1}\frac{t}{4} = \frac{\pi}{2}$

28. $\int_{-\infty}^\infty \text{Tan}^{-1}x\, dx$ diverges since $\lim_{x\to\infty} \text{Tan}^{-1}x = \pi/2$

29. $y = \frac{3x+1}{x-2} \implies y' = \frac{-7}{(x-2)^2} < 0$. Hence the graph of $y = \frac{3x+1}{x-2}$ is decreasing and has no relative extrema. Since $\lim_{x\to2-} y = -\infty$ and $\lim_{x\to2+} y = \infty$, the graph has the line $x = 2$ as a vertical asymptote. Since $\lim_{x\to\pm\infty} y = 3$, the graph has $y = 3$ as a horizontal asymptote.

30. $y = \frac{2x^2}{x^2-1} \implies y' = \frac{-4x}{(x^2-1)^2}$. It follows that the point $(0,0)$ is a relative maximum for the graph. Moreover, the graph is symmetric about the y-axis and is increasing for $x < 0$ and decreasing for $x > 0$. Since $\lim_{x\to-1-} y = \lim_{x\to1+} y = \infty$ and $\lim_{x\to-1+} y = \lim_{x\to1-} y = -\infty$, the lines $x = -1$ and $x = 1$ are vertical asymptotes. Finally, $\lim_{x\to\pm\infty} y = 2$ and hence the line $y = 2$ is a horizontal asymptote.

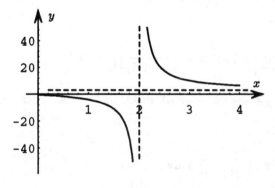

$y = (3x+1)/(x-2)$

Exercise 29

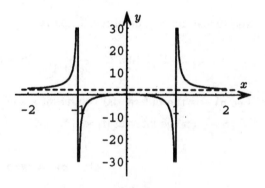

$y = 2x^2/(x^2-1)$

Exercise 30

31. $y = \frac{\ln x}{x} \implies y' = \frac{1-\ln x}{x^2} = 0 \iff x = e$. It follows that the point $(e, 1/e)$ is a relative maximum. Since $\lim_{x\to0+} y = -\infty$, the graph has the line $x = 0$ as a vertical asymptote, and since $\lim_{x\to\infty} y = \lim_{x\to\infty}(1/x) = 0$, the line $y = 0$ is a horizontal asymptote.

32. $y = \frac{\ln x}{\sqrt{x}} \implies y' = \frac{2-\ln x}{2x\sqrt{x}}$. It follows that the point $(e^2, 2/e^2)$ is a relative maximum for the graph. The line $x = 0$ is a vertical asymptote since $\lim_{x\to0+} y = -\infty$, and the line $y = 0$ is a horizontal asymptote since $\lim_{x\to\pm\infty} y = 0$.

33. Area $= \int_1^\infty \left(\frac{x^5+3x+1}{x^3} - x^2 \right) dx = \int_1^\infty \left(\frac{3}{x^2} + \frac{1}{x^3} \right) dx = \lim_{t\to\infty} \left. \left(-\frac{3}{x} - \frac{1}{2x^2} \right) \right|_1^t = \frac{7}{2}$

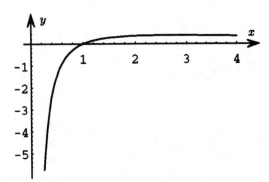

$$y = (\ln x)/x$$

Exercise 31

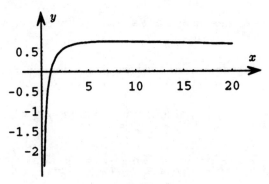

$$y = (\ln x)/\sqrt{x}$$

Exercise 32

34. Area $= \displaystyle\int_0^\infty x^2 e^{-x}\, dx = \lim_{t \to \infty} \left[-x^2 e^{-x} + 2\int xe^{-x}\, dx \right]\bigg|_0^t = \lim_{t \to \infty} \left[-e^{-x}(x^2 + 2x + 2) \right]\bigg|_0^t = 2$

35. Let dV be the differential volume of a cross-section disk of thickness dx. Then

$$dV = \pi r^2 dx = \pi y^2 dx = \frac{16\pi}{(x^2+1)^2}\, dx$$

and hence the volume of revolution is

$$V = 16\pi \int_0^\infty \frac{1}{(x^2+1)^2}\, dx = 16\pi \lim_{t \to \infty} \left[\frac{1}{2}\, \mathrm{Tan}^{-1}x + \frac{x}{2(x^2+1)} \right]\bigg|_0^t = 4\pi^2.$$

36. Let dV be the differential volume of a shell of thickness dx revolved about the y-axis. Then, since $\dfrac{1}{x^3} > \dfrac{1}{x^4}$ on the interval $[1, \infty)$,

$$dV = 2\pi rh\, dx = 2\pi x \left(\frac{1}{x^3} - \frac{1}{x^4} \right) dx$$

and hence the volume of revolution is

$$V = 2\pi \int_1^\infty \left(\frac{1}{x^2} - \frac{1}{x^3} \right) dx = 2\pi \lim_{t \to \infty} \left(-\frac{1}{x} + \frac{1}{2x^2} \right)\bigg|_1^t = \pi.$$